Solar System Data

Body	Mass (kg)	Mean Radius (m)	Period (s)	Distance from the Sun (m)
Mercury	3.18×10^{23}	2.43×10^6	7.60×10^6	5.79×10^{10}
Venus	4.88×10^{24}	6.06×10^6	1.94×10^7	1.08×10^{11}
Earth	5.98×10^{24}	6.37×10^6	3.156×10^7	1.496×10^{11}
Mars	6.42×10^{23}	3.37×10^6	5.94×10^7	2.28×10^{11}
Jupiter	1.90×10^{27}	6.99×10^7	3.74×10^8	7.78×10^{11}
Saturn	5.68×10^{26}	5.85×10^7	9.35×10^8	1.43×10^{12}
Uranus	8.68×10^{25}	2.33×10^7	2.64×10^9	2.87×10^{12}
Neptune	1.03×10^{26}	2.21×10^7	5.22×10^9	4.50×10^{12}
Pluto	$\approx 1.4 \times 10^{22}$	$\approx 1.5 \times 10^6$	7.82×10^9	5.91×10^{12}
Moon	7.36×10^{22}	1.74×10^6	—	—
Sun	1.991×10^{30}	6.96×10^8	—	—

Physical Data Often Used[a]

Average Earth–Moon distance	3.84×10^8 m
Average Earth–Sun distance	1.496×10^{11} m
Average radius of the Earth	6.37×10^6 m
Density of air (20°C and 1 atm)	1.20 kg/m^3
Density of water (20°C and 1 atm)	1.00×10^3 kg/m^3
Free-fall acceleration	9.80 m/s^2
Mass of the Earth	5.98×10^{24} kg
Mass of the Moon	7.36×10^{22} kg
Mass of the Sun	1.99×10^{30} kg
Standard atmospheric pressure	1.013×10^5 Pa

[a] These are the values of the constants as used in the text.

Some Prefixes for Powers of Ten

Power	Prefix	Abbreviation	Power	Prefix	Abbreviation
10^{-24}	yocto	y	10^1	deka	da
10^{-21}	zepto	z	10^2	hecto	h
10^{-18}	atto	a	10^3	kilo	k
10^{-15}	femto	f	10^6	mega	M
10^{-12}	pico	p	10^9	giga	G
10^{-9}	nano	n	10^{12}	tera	T
10^{-6}	micro	μ	10^{15}	peta	P
10^{-3}	milli	m	10^{18}	exa	E
10^{-2}	centi	c	10^{21}	zetta	Z
10^{-1}	deci	d	10^{24}	yotta	Y

Pedagogical Color Chart

Part 1 (Chapters 1–15) : Mechanics

Displacement and
position vectors

Linear ($\mathbf{p}$) and
angular ($\mathbf{L}$)
momentum vectors

Linear ($\mathbf{v}$) and angular ($\boldsymbol{\omega}$)
velocity vectors

Torque vectors ($\boldsymbol{\tau}$)

Velocity component vectors

Linear or rotational
motion directions

Force vectors ($\mathbf{F}$)

Force component vectors

Springs

Acceleration vectors ($\mathbf{a}$)

Pulleys

Acceleration component vectors

Part 4 (Chapters 23–34) : Electricity and Magnetism

Electric fields

Capacitors

Magnetic fields

Inductors (coils)

Positive charges

Voltmeters

Negative charges

Ammeters

Resistors

AC Generators

Batteries and other
DC power supplies

Ground symbol

Switches

Part 5 (Chapters 35–38) : Light and Optics

Light rays

Objects

Lenses and prisms

Images

Mirrors

Open the door to the fascinating world of
physics

Physics, the most fundamental of all natural sciences, will reveal to you the basic principles of the Universe. And while physics can seem challenging, its true beauty lies in the sheer simplicity of fundamental physical theories—theories and concepts that can alter and expand your view of the world around you. Other courses that follow will use the same principles, so it is important that you understand and are able to apply the various concepts and theories discussed in the text. **Physics for Scientists and Engineers, Sixth Edition** is your guide to this fascinating science.

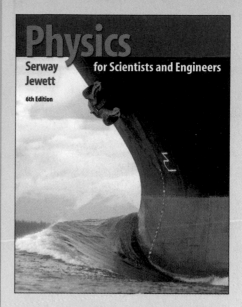

Your quick start for studying smart

Achieve success in your physics course by making the most of what **Physics for Scientists and Engineers, Sixth Edition** has to offer you. From a host of in-text features to a range of Web resources, you'll have everything you need to understand the natural forces and principles of physics:

▶ **Dynamic built-in study aids.**
Throughout every chapter the authors have built in a wide range of examples, exercises, and illustrations that will help you understand and appreciate the laws of physics. *See pages 2 and 3 for more information.*

▶ **A powerful Web-based learning system**.
The text is fully integrated with **PhysicsNow**, an interactive learning system that tailors itself to your needs in the course. It's like having a personal tutor available whenever you need it! *See pages 4–7 to explore* **PhysicsNow**.

Your *Quick Start for Studying Smart* begins with this special tour through the book. On the following pages you'll discover how **Physics for Scientists and Engineers, Sixth Edition** and **PhysicsNow** not only enhance your experience in this course, but help you to succeed!

Everything you need to succeed in your course is available to you in **Physics for Scientists and Engineers, Sixth Edition.** Authors Serway and Jewett have filled their text with learning tools and study aids that will clarify concepts and help you build a solid base of knowledge. The end result: confidence in the classroom, in your study sessions, and in your exams.

Quick Start for Studying Smart!

THE RIGHT APPROACH

Start out right! Early on in the text the authors outline a general problem-solving strategy that will enable you to increase your accuracy in solving problems, enhance your understanding of physical concepts, eliminate initial worry or lack of direction in approaching a problem, and organize your work. The problem-solving strategy is integrated into the *Coached Problems* found on **PhysicsNow** to reinforce this key skill. (See pages 4–7 for more information about the **PhysicsNow** Web-based and student-centered learning system.)

PROBLEM-SOLVING HINTS

Problem-Solving Hints help you approach homework assignments with greater confidence. General strategies and suggestions are included for solving the types of problems featured in the worked examples, end-of-chapter problems, and **PhysicsNow**. This feature helps you identify the essential steps in solving problems and increases your skills as a problem solver.

WORKED EXAMPLES

Reinforce your understanding of essential problem-solving techniques using a large number of realistic *Worked Examples*. In many cases, these examples serve as models for solving the end-of-chapter problems. Numerous *Worked Examples* include specific references to the general problem-solving strategy to illustrate the underlying concepts and methodology used in arriving at a correct solution. This will help you understand the logic behind the solution and the advantage of using a particular approach to solve the problem. **PhysicsNow** also features a number of worked examples to further enhance your understanding of problem solving and to give you even more practice solving problems.

GENERAL PROBLEM-SOLVING STRATEGY

Conceptualize

- The first thing to do when approaching a problem is to *think about* and *understand* the situation. Study carefully any diagrams, graphs, tables, or photographs that accompany the problem. Imagine a movie, running in your mind, of what happens in the problem.
- If a diagram is not provided, you should almost always make a quick drawing of the situation. Indicate any known values, perhaps in a table or directly on your sketch.
- Now focus on what algebraic or numerical information is given in the problem. Carefully read the problem statement, looking for key phrases such as "starts from at rest" ($v_i = 0$), "stops" ($v_f = 0$), or "freely falls" ($a_y = -g = -9.80 \text{ m/s}^2$).
- Now focus on the expected result of solving the problem. Exactly what is the question asking? Will the final result be numerical or algebraic? Do you know what units to expect?
- Don't forget to incorporate information from your own experiences and common sense. What should a reasonable answer look like? You wouldn't expect to calculate the speed of an automobile to be

Analyze

- Now you must analyze the problem and strive for a mathematical solution. Because you have already categorized the problem, it should not be too difficult to select relevant equations that apply to the type of situation in the problem. For example, if the problem involves a particle moving under constant acceleration, Equations 2.9 to 2.13 are relevant.
- Use algebra (and calculus, if necessary) to solve symbolically for the unknown variable in terms of what is given. Substitute in the appropriate numbers, calculate the result, and round it to the proper number of significant figures.

Finalize

- This is the most important part. Examine your numerical answer. Does it have the correct units? Does it meet your expectations from your conceptualization of the problem? What about the algebraic form of the result — before you substituted numerical values? Does it make sense? Examine the variables in the problem to see whether the answer would change in a physically meaningful way if they were drastically increased or decreased or even became zero. Looking at limiting cases to see whether they yield expected values is a very useful way to make sure that you are obtaining reasonable results.

  Think about how this problem compares with others you have done. How was it similar? In what critical ways did it differ? Why was this problem assigned? You should have learned something by doing it. Can you figure out what? If it is a new category of problem, be sure you understand it so that you can use it as a model for solving future problems in the same category.

  When solving complex problems, you may need to identify a series of sub-problems and apply the problem-solv-

---

PROBLEM-SOLVING HINTS

Applying Newton's Laws

The following procedure is recommended when dealing with problems involving Newton's laws:

- Draw a simple, neat diagram of the system to help *conceptualize* the problem.
- *Categorize* the problem: if any acceleration component is zero, the particle is in equilibrium in this direction and $\Sigma F = 0$. If not, the particle is undergoing an acceleration, the problem is one of nonequilibrium in this direction, and $\Sigma F = ma$.
- *Analyze* the problem by isolating the object whose motion is being analyzed. Draw a free-body diagram for this object. For systems containing more than one object, draw *separate* free-body diagrams for each object. *Do not* include in the free-body diagram forces exerted by the object on its surroundings.
- Establish convenient coordinate axes for each object and find the components of the forces along these axes. Apply Newton's second law, $\Sigma F = ma$, in component form. Check your dimensions to make sure that all terms have units of force.
- Solve the comp[...] have as many in[...] complete soluti[...]
- *Finalize* by maki[...] Also check the [...] variables. By do[...]

Example 4.3 The Long Jump

A long-jumper (Fig. 4.12) leaves the ground at an angle of 20.0° above the horizontal and at a speed of 11.0 m/s.

(A) How far does he jump in the horizontal direction? (Assume his motion is equivalent to that of a particle.)

Solution We *conceptualize* the motion of the long-jumper as equivalent to that of a simple projectile such as the ball in Example 4.2, and *categorize* this problem as a projectile motion problem. Because the initial speed and launch angle are given, and because the final height is the same as the initial height, we further categorize this problem as satisfying the conditions for which Equations 4.13 and 4.14 can be used. This is the most direct way to *analyze* this problem, although, in general, we will *analyze* by describing the motion in the *x* and *y* directions and using the gen[...]

Figure 4.12 (Example 4.3) Mike Powell, current holder of the world long jump record of 8.95 m.

This is the time at which the long-jumper is at the *top* of the jump. Because of the symmetry of the vertical motion,

provides a graphical representation of the flight of the long-jumper. As before, we set our origin of coordinates at the takeoff point and label the peak as Ⓐ and the landing point as Ⓑ. The horizontal motion is described by Equation 4.11:

$$x_f = x_B = (v_i \cos \theta_i) t_B = (11.0 \text{ m/s})(\cos 20.0°) t_B$$

The value of x_B can be found if the time of landing t_B is known. We can find t_B by remembering that $a_y = -g$ and by using the y part of Equation 4.8a. We also note that at the top of the jump the vertical component of velocity v_{yA} is zero:

$$v_{yf} = v_{yA} = v_i \sin \theta_i - gt_A$$

another 0.384 s passes before the jumper returns to the ground. Therefore, the time at which the jumper lands is $t_B = 2t_A = 0.768$ s. Substituting this value into the above expression for x_f gives

$$x_f = x_B = (11.0 \text{ m/s})(\cos 20.0°)(0.768 \text{ s}) = \boxed{7.94 \text{ m}}$$

This is a reasonable distance for a world-class athlete.

(B) What is the maximum height reached?

Solution We find the maximum height reached by using Equation 4.12:

$$y_{max} = y_A = (v_i \sin \theta_i) t_A - \tfrac{1}{2} gt_A^2$$
$$= (11.0 \text{ m/s})(\sin 20.0°)(0.384 \text{ s})$$
$$- \tfrac{1}{2}(9.80 \text{ m/s}^2)(0.384 \text{ s})^2 = \boxed{0.722 \text{ m}}$$

To *finalize* this problem, find the answers to parts (a) and (b) using Equations 4.13 and 4.14. The results should agree. Treating the long-jumper as a particle is an oversimplification. Nevertheless, the values obtained are consistent with experience in sports. We learn that we can model a complicated system such as a long-jumper as a particle and still obtain results that are reasonable.

where k is a dimensionless constant of proportionality. Knowing the dimensions of a, r, and v, we see that the dimensional equation must be

$$\frac{L}{T^2} = L^n\left(\frac{L}{T}\right)^m = \frac{L^{n+m}}{T^m}$$

...pression as

$$a = kr^{-1}v^2 = k\frac{v^2}{r}$$

When we discuss uniform circular motion later, we shall see that $k = 1$ if a consistent set of units is used. The constant k would not equal 1 if, for example, v were in km/h and you wanted a in m/s^2.

1.5 Conversion of Units

Sometimes it is necessary to convert units from one measurement system to another, or to convert within a system, for example, from kilometers to meters. Equalities between SI and U.S. customary units of length are as follows:

1 mile = 1 609 m = 1.609 km	1 ft = 0.304 8 m = 30.48 cm
1 m = 39.37 in. = 3.281 ft	1 in. = 0.025 4 m = 2.54 cm (exactly)

A more complete list of conversion factors can be found in Appendix A.

Units can be treated as algebraic quantities that can cancel each other. For example, suppose we wish to convert 15.0 in. to centimeters. Because 1 in. is defined as exactly 2.54 cm, we find that

Example 4.5 That's Quite an Arm!

A stone is thrown from the top of a building upward at an angle of 30.0° to the horizontal with an initial speed of 20.0 m/s, as shown in Figure 4.14. If the height of the building is 45.0 m,

(A) how long before the stone hits the ground?

Solution We *conceptualize* the problem by studying Figure 4.14, in which we have indicated the various parameters. By now, it should be natural to *categorize* this as a projectile motion problem.

To *analyze* the problem, let us once again separate motion into two components. The initial x and y components of the stone's velocity are

$$v_{xi} = v_i \cos\theta_i = (20.0 \text{ m/s})\cos 30.0° = 17.3 \text{ m/s}$$

$$v_{yi} = v_i \sin\theta_i = (20.0 \text{ m/s})\sin 30.0° = 10.0 \text{ m/s}$$

To find t, we can use $y_f = y_i + v_{yi}t + \frac{1}{2}a_yt^2$ (Eq. 4.9a) with $y_i = 0$, $y_f = -45.0$ m, $a_y = -g$, and $v_{yi} = 10.0$ m/s (there is a negative sign on the numerical value of y_f because we have chosen the top of the building as the origin):

$$-45.0 \text{ m} = (10.0 \text{ m/s})t - \frac{1}{2}(9.80 \text{ m/s}^2)t^2$$

Solving the quadratic equation for t gives, for the positive root, $t = 4.22$ s . To *finalize* this part, think: Does the negative root have any physical meaning?

(B) What is the speed of the stone just before it strikes the ground?

Solution We can use Equation 4.8a, $v_{yf} = v_{yi} + a_yt$, with $t = 4.22$ s to obtain the y component of the velocity just before the stone strikes the ground:

$$v_{yf} = 10.0 \text{ m/s} - (9.80 \text{ m/s}^2)(4.22 \text{ s}) = -31.4 \text{ m/s}$$

Because $v_{xf} = v_{xi} = 17.3$ m/s, the required speed is

$$v_f = \sqrt{v_{xf}^2 + v_{yf}^2} = \sqrt{(17.3)^2 + (-31.4)^2} \text{ m/s} = 35.9 \text{ m/s}$$

What If? What if a horizontal wind is blowing in the same direction as the ball is thrown and it causes the ball to have a horizontal acceleration component $a_x = 0.500$ m/s^2. Which part of this example, (a) or (b), will have a different answer?

Answer Recall that the motions in the x and y directions are independent. Thus, the horizontal wind cannot affect the vertical motion. The vertical motion determines the time of the projectile in the air, so the answer to (a) does not change. The wind will cause the horizontal velocity component to increase with time, so that the final speed will change in part (b).

We can find the new final horizontal velocity component by using Equation 4.8a:

$$v_{xf} = v_{xi} + a_xt = 17.3 \text{ m/s} + (0.500 \text{ m/s}^2)(4.22 \text{ s})$$
$$= 19.4 \text{ m/s}$$

and the new final speed:

$$v_f = \sqrt{v_{xf}^2 + v_{yf}^2} = \sqrt{(19.4)^2 + (-31.4)^2} \text{ m/s} = 36.9 \text{ m/s}$$

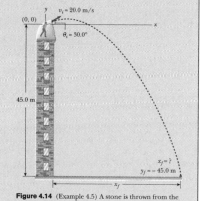

Figure 4.14 (Example 4.5) A stone is thrown from the top of a building.

You can study the projectile trajectory for this example link at http://www.pse6.com.

Quick Quiz 5.2 An object experiences no acceleration. Which of the following *cannot* be true for the object? (a) A single force acts on the object. (b) No forces act on the object. (c) Forces act on the object, but the forces cancel.

Quick Quiz 5.3 An object experiences a net force and exhibits an acceleration in response. Which of the following statements is *always* true? (a) The object moves in the direction of the force. (b) The acceleration is in the same direction as the velocity. (c) The acceleration is in the same [...] the object increases.

Quick Quiz 5.4 You push an object [...] with a constant force for a time interval Δ[...] object. You repeat the experiment, but with [...] terval is now required to reach the same fina[...] (e) Δt/4.

Answers to Quick Quizzes

5.1 (d). Choice (a) is true. Newton's first law tells us that motion requires no force: an object in motion continues to move at constant velocity in the absence of external forces. Choice (b) is also true. A stationary object can have several forces acting on it, but if the vector sum of all these external forces is zero, there is no net force and the object remains stationary.

5.2 (a). If a single force acts, this force constitutes the net force and there is an acceleration according to Newton's second law.

5.3 (c). Newton's second law relates only the force and the acceleration. Direction of motion is part of an object's *velocity*, and force determines the direction of acceleration, not that of velocity.

5.4 (d). With twice the force, the object will experience twice the acceleration. Because the force is constant, the acceler-

"You do not know anything until you have practiced."

R. P. Feynman, Nobel Laureate in Physics

Quick Start for Studying Smart!

Quick Start for Studying Smart!

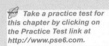

Take a practice test for this chapter by clicking on the Practice Test link at http://www.pse6.com.

SUMMARY

Scalar quantities are those that have only magnitude and no associated direction. **Vector quantities** have both magnitude and direction and obey the laws of vector addition. The magnitude of a vector is *always* a positive number.

When two or more vectors are added together, all of them must have the same units and all of them must be the same type of quantity. We can add two vectors **A** and **B** graphically. In this method (Fig. 3.6), the resultant vector $\mathbf{R} = \mathbf{A} + \mathbf{B}$ runs from the tail of **A** to the tip of **B**.

A second method of adding vectors involves **components** of the vectors. The x component A_x of the vector **A** is equal to the projection of **A** along the x axis of a coordinate system, as shown in Figure 3.13, where $A_x = A\cos\theta$. The y component A_y of **A** is the projection of **A** along the y axis, where $A_y = A\sin\theta$. Be sure you can determine which trigonometric functions you should use in all situations, especially when θ is defined as something other than the counterclockwise angle from the positive x axis.

If a vector **A** has an x component A_x and a y component A_y, the vector can be expressed in unit–vector form as $\mathbf{A} = A_x\mathbf{i} + A_y\mathbf{j}$. In this notation, $\mathbf{i}$ is a unit vector pointing in the positive x direction, and $\mathbf{j}$ is a unit vector pointing in the positive y direction. Because $\mathbf{i}$ and $\mathbf{j}$ are unit vectors, $|\mathbf{i}| = |\mathbf{j}| = 1$.

We can find the resultant of two or more vectors by resolving all vectors into their x and y components, adding their resultant x and y components, and then using the Pythagorean theorem to find the magnitude of the resultant vector. We can find the angle that the resultant vector makes with respect to the x axis by using a suitable trigonometric function.

QUESTIONS

1. Two vectors have unequal magnitudes. Can their sum be zero? Explain.

2. Can the magnitude of a particle's displacement be greater

4. Which of the following are vectors and which are not: force, temperature, the volume of water in a can, the ratings of a TV show, the height of a building, the velocity of ... he Universe?

... y plane. For what orientations of **A** ... ents be negative? For what orienta- ... s have opposite signs?

Example 4.5   That's Quite an Arm!

A stone is thrown from the top of a building upward at an angle of 30.0° to the horizontal with an initial speed of 20.0 m/s, as shown in Figure 4.14. If the height of the building is 45.0 m,

(A) how long before the stone hits the ground?

Solution We *conceptualize* the problem by studying Figure 4.14, in which we have indicated the various parameters. By now, it should be natural to *categorize* this as a projectile motion problem.

To *analyze* the problem, let us once again separate motion into two components. The initial x and y components of the stone's velocity are

$$v_{xi} = v_i\cos\theta_i = (20.0 \text{ m/s})\cos 30.0° = 17.3 \text{ m/s}$$

$$v_{yi} = v_i\sin\theta_i = (20.0 \text{ m/s})\sin 30.0° = 10.0 \text{ m/s}$$

To find t, we can use $y_f = y_i + v_{yi}t + \frac{1}{2}a_yt^2$ (Eq. 4.9a) with $y_i = 0$, $y_f = -45.0$ m, $a_y = -g$, and $v_{yi} = 10.0$ m/s (there is a negative sign on the numerical value of y_f because we have chosen the top of the building as the origin):

$$-45.0 \text{ m} = (10.0 \text{ m/s})t - \tfrac{1}{2}(9.80 \text{ m/s}^2)t^2$$

Solving the quadratic equation for t gives, for the positive root, $t = 4.22$ s. To *finalize* this part, think: Does the negative root have any physical meaning?

(B) What is the speed of the stone just before it strikes the ground?

Solution We can use Equation 4.8a, $v_{yf} = v_{yi} + a_yt$, with $t = 4.22$ s to obtain the y component of the velocity just before the stone strikes the ground:

$$v_{yf} = 10.0 \text{ m/s} - (9.80 \text{ m/s}^2)(4.22 \text{ s}) = -31.4 \text{ m/s}$$

Because $v_{xf} = v_{xi} = 17.3$ m/s, the required speed is

$$v_f = \sqrt{v_{xf}^2 + v_{yf}^2} = \sqrt{(17.3)^2 + (-31.4)^2} \text{ m/s} = 35.9 \text{ m/s}$$

To *finalize* this part, is it reasonable that the y component of the final velocity is negative? Is it reasonable that the final speed is larger than the initial speed of 20.0 m/s?

What If? What if a horizontal wind is blowing in the same direction as the ball is thrown and it causes the ball to have a horizontal acceleration component $a_x = 0.500$ m/s^2. Which part of this example, (a) or (b), will have a different answer?

Answer Recall that the motions in the x and y directions are independent. Thus, the horizontal wind cannot affect the vertical motion. The vertical motion determines the time of the projectile in the air, so the answer to (a) does not change. The wind will cause the horizontal velocity component to increase with time, so that the final speed will change in part (b).

We can find the new final horizontal velocity component by using Equation 4.8a:

$$v_{xf} = v_{xi} + a_xt = 17.3 \text{ m/s} + (0.500 \text{ m/s}^2)(4.22 \text{ s})$$
$$= 19.4 \text{ m/s}$$

and the new final speed:

$$v_f = \sqrt{v_{xf}^2 + v_{yf}^2} = \sqrt{(19.4)^2 + (-31.4)^2} \text{ m/s} = 36.9 \text{ m/s}$$

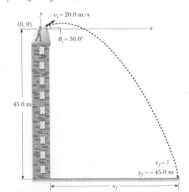

Figure 4.14 (Example 4.5) A stone is thrown from the top of a building.

Investigate this situation at the Interactive Worked Example link at http://www.pse6.com.

GO ONLINE AT www.pse6.com

Log on to **PhysicsNow** at **www.pse6.com** by using the free pincode packaged with this text.* You'll immediately notice the system's easy-to-use, browser-based format. Getting to where you need to go is as easy as a click of the mouse. The **PhysicsNow** system is made up of three interrelated parts:

- ► **How Much Do I Know?**
- ► **What Do I Need to Learn?**
- ► **What Have I Learned?**

These three interrelated elements work together, but are distinct enough to allow you the freedom to explore only those assets that meet your personal needs. You can use **PhysicsNow** like a traditional Web site, accessing all assets of a particular chapter and exploring on your own. The best way to maximize the system and *your* time is to start by taking the *Pre-Test*.

* Free PIN codes are only available with new copies of
Physics for Scientists and Engineers, Sixth Edition.

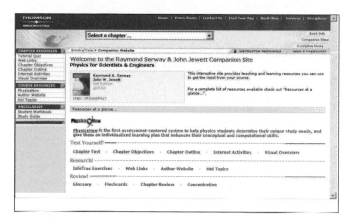

HOW MUCH DO I KNOW?

The Pre-Test is the first step in creating your *Personalized Learning Plan*. Each *Pre-Test* is based on the end-of-chapter homework problems and includes approximately 15 questions.

Once you've completed the *Pre-Test* you'll be presented with a detailed *Learning Plan* that outlines the elements you need to review to master the chapter's most essential concepts.

At each stage, the text is referenced to reinforce its value as a learning tool.

Turn the page to view problems from a sample *Personalized Learning Plan*.

Turn the page to view problems from a sample *Personalized Learning Plan*.

Quick Start for Studying Smart!

WHAT DO I NEED TO LEARN?

Once you've completed the *Pre-Test* you're ready to work the problems in your *Personalized Learning Plan*—problems that will help you master concepts essential to your success in this course.

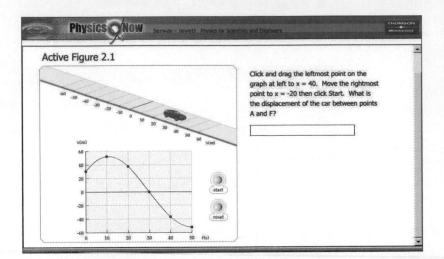

More than 200 *Active Figures* are taken from the text and animated to help you visualize physics in action. Each figure is paired with a question to help you focus on physics at work, and a brief quiz ensures that you understand the concept played out in the animations.

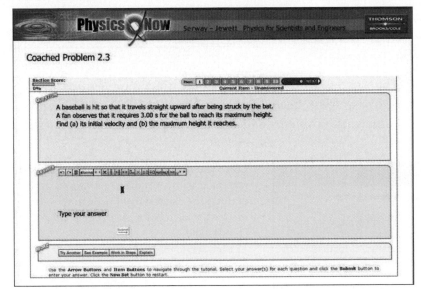

You'll continue to master the concepts though *Coached Problems.* These engaging problems reinforce the lessons in the text by taking a step-by-step approach to problem-solving methodology. Each *Coached Problem* gives you the option of working a question and receiving feedback, or seeing a solution worked for you. You'll find approximately five *Coached Problems* per chapter.

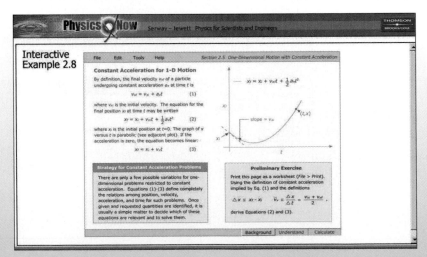

You'll strengthen your problem-solving and visualization skills by working through the *Interactive Examples.* Each step in the examples uses the authors' problem-solving methodology that is introduced in the text (see page 2 of this Visual Preface). You'll find *Interactive Examples* for each chapter of the text.

WHAT HAVE I LEARNED?

After working through the problems highlighted in your personal *Learning Plan* you'll move on to a *Chapter Quiz*. These multiple-choice quizzes present you with questions that are similar to those you might find in an exam. You can even e-mail your quiz results to your instructor.

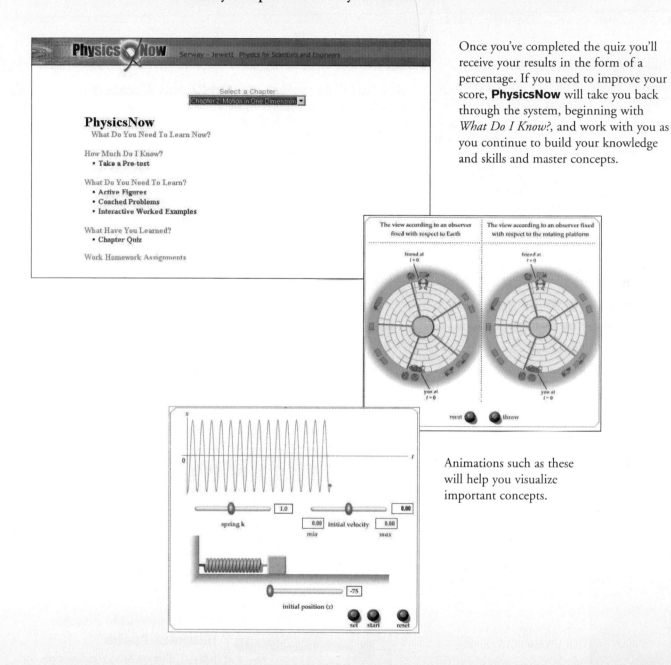

Once you've completed the quiz you'll receive your results in the form of a percentage. If you need to improve your score, **PhysicsNow** will take you back through the system, beginning with *What Do I Know?*, and work with you as you continue to build your knowledge and skills and master concepts.

Animations such as these will help you visualize important concepts.

Chart your own course for success . . .

Log on to **www.pse6.com** to take advantage of **PhysicsNow!**

Make the most of the course and your time with these exclusive study tools

Enrich your experience outside of the classroom with a host of resources designed to help you excel in the course. To purchase any of these supplements, contact your campus bookstore or visit our online BookStore at www.brookscole.com.

Quick Start for Studying Smart!

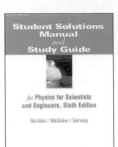

Student Solutions Manual with Study Guide

Volume I ISBN: 0-534-40855-9
Volume II ISBN: 0-534-40856-7

by John R. Gordon, Ralph McGrew, and Raymond Serway This two-volume manual features detailed solutions to 20% of the end-of-chapter problems from the text. The manual also features a list of important equations, concepts, and answers to selected end-of-chapter questions.

Essential Study Skills for Science Students

ISBN: 0-534-37595-2

by Daniel D. Chiras Written specifically for science students, this book discusses how to develop good study habits, sharpen memory, learn more quickly, get the most out of lectures, prepare for tests, produce excellent term papers, and improve critical-thinking skills.

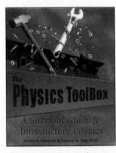

The Physics Toolbox: A Survival Guide for Introductory Physics

ISBN: 0-03-034652-5

by Kirsten A. Hubbard and Debora M. Katz This "paperback mentor" gives you the material critical for success in physics, including an introduction to the nature of physics and science, a look at what to expect and how to succeed, a verbal overview of the concepts you'll encounter, and an extensive review of the math you'll need to solve the problems.

WebTUTOR Advantage

WebTutor™ Advantage on WebCT and Blackboard

WebCT ISBN: 0-534-40859-1
Blackboard ISBN: 0-534-40950-4

WebTutor Advantage offers real-time access to a full array of study tools, including chapter outlines, summaries, learning objectives, glossary flashcards (with audio), practice quizzes, **InfoTrac® College Edition** exercises, and Web links. **WebTutor Advantage** also provides robust communication tools, such as a course calendar, asynchronous discussion, real-time chat, a whiteboard, and an integrated e-mail system. Also new to **WebTutor Advantage** is access to *NewsEdge,* an online news service that brings the latest news to the **WebTutor Advantage** site daily.

Contact your instructor for more information.

Additional Web resources . . .

available FREE to you with each new copy of this text:

InfoTrac® College Edition

When you purchase a new copy of **Physics for Scientists and Engineers, Sixth Edition**, you automatically receive a FREE four-month subscription to **InfoTrac College Edition!** Newly improved, this extensive online library opens the door to the full text (not just abstracts) of countless articles from thousands of publications including *American Scientist, Physical Review, Science, Science Weekly,* and more! Use the passcode included with the new copy of this text and log on to **www.infotrac-college.com** to explore the wealth of resources available to you—24 hours a day and seven days a week!

Available only to college and university students. Journals subject to change.

The Brooks/Cole Physics Resource Center

http://physics.brookscole.com
Here you'll find even more opportunities to hone your skills and expand your knowledge. **The Brooks/Cole Physics Resource Center** is filled with helpful content that will engage you while you master the material. You'll find additional online quizzes, Web links, animations, and *NewEdge*—an online news service that brings the latest news to this site daily.

*We dedicate this book to the courageous astronauts
who died on the space shuttle* Columbia *on February 1, 2003.
The women and men of the international team lost their lives
not in a contest between countries or a struggle for
necessities but in advancing one of humankind's noblest
creations—science.*

6th Edition • Volume 2

PHYSICS

for Scientists and Engineers

with Modern Physics

Raymond A. Serway

John W. Jewett, Jr.

California State Polytechnic University–Pomona

THOMSON

BROOKS/COLE

Australia • Canada • Mexico • Singapore • Spain
United Kingdom • United States

THOMSON

BROOKS/COLE

Editor-in-Chief: Michelle Julet
Publisher: David Harris
Physics Editor: Chris Hall
Development Editor: Susan Dust Pashos
Assistant Editor: Rebecca Heider, Alyssa White
Editorial Assistant: Seth Dobrin, Jessica Howard
Technology Project Manager: Sam Subity
Marketing Manager: Kelley McAllister
Marketing Assistant: Sandra Perin
Advertising Project Manager: Stacey Purviance
Project Manager, Editorial Production: Teri Hyde
Print/Media Buyer: Barbara Britton
Permissions Editor: Joohee Lee

Production Service: Sparkpoint Communications,
a division of J. B. Woolsey Associates
Text Designer: Lisa Devenish
Photo Researcher: Terri Wright
Copy Editor: Andrew Potter
Illustrator: Rolin Graphics
Cover Designer: Lisa Devenish
Cover Image: Water Displaced by Oil Tanker, © Stuart
Westmorland/CORBIS
Compositor: Progressive Information Technologies
Cover Printer: The Lehigh Press, Inc.
Printer: Quebecor World, Versailles

Printed in the United States
3 4 5 6 7 07 06 05 04

For more information about our products, contact us at:
Thomson Learning Academic Resource Center
1-800-423-0563

For permission to use material from this text, contact us by:
Phone: 1-800-730-2214
Fax: 1-800-730-2215
Web: http://www.thomsonrights.com

Library of Congress Control Number: 2003100126

Student Edition:
ISBN 0-534-40846-X

International Student Edition:
ISBN 0-534-42398-1
(Not for sale in the United States)

Instructor's Edition: ISBN 0-534-40843-5

Brooks/Cole—Thomson Learning
10 Davis Drive
Belmont, CA 94002
USA

Asia
Thomson Learning
5 Shenton Way #01-01
UIC Building
Singapore 068808

Australia
Nelson Thomson Learning
102 Dodds Street
South Melbourne, Victoria 3205
Australia

Canada
Nelson Thomson Learning
1120 Birchmount Road
Toronto, Ontario M1K 5G4
Canada

Europe/Middle East/Africa
Thomson Learning
High Holborn House
50/51 Bedford Row
London WC1R 4LR
United Kingdom

Latin America
Thomson Learning
Seneca, 53
Colonia Polanco
11560 Mexico D.F.
Mexico

Spain
Paraninfo Thomson Learning
Calle/Magallanes, 25
28015 Madrid, Spain

Contents Overview

PART 4 **Electricity and Magnetism** **705**

23 **Electric Fields** *706*
24 **Gauss's Law** *739*
25 **Electric Potential** *762*
26 **Capacitance and Dielectrics** *795*
27 **Current and Resistance** *831*
28 **Direct Current Circuits** *858*
29 **Magnetic Fields** *894*
30 **Sources of the Magnetic Field** *926*
31 **Faraday's Law** *967*
32 **Inductance** *1003*
33 **Alternating Current Circuits** *1033*
34 **Electromagnetic Waves** *1066*

PART 5 **Light and Optics** **1093**

35 **The Nature of Light and the Laws of Geometric Optics** *1094*
36 **Image Formation** *1126*
37 **Interference of Light Waves** *1176*
38 **Diffraction Patterns and Polarization** *1205*

PART 6 **Modern Physics** **1243**

39 **Relativity** *1244*
40 **Introduction to Quantum Physics** *1284*
41 **Quantum Mechanics** *1321*
42 **Atomic Physics** *1352*
43 **Molecules and Solids** *1398*
44 **Nuclear Structure** *1440*
45 **Applications of Nuclear Physics** *1479*
46 **Particle Physics and Cosmology** *1511*

Appendices *A.1*
Answers to Odd-Numbered Problems *A.37*
Index *I.1*

Steve Niedorf/Getty Images

Table of Contents

PART 4 Electricity and Magnetism 705

Chapter 23 Electric Fields *706*

23.1 Properties of Electric Charges *707*
23.2 Charging Objects by Induction *709*
23.3 Coulomb's Law *711*
23.4 The Electric Field *715*
23.5 Electric Field of a Continuous Charge Distribution *719*
23.6 Electric Field Lines *723*
23.7 Motion of Charged Particles in a Uniform Electric Field *725*

Chapter 24 Gauss's Law *739*

24.1 Electric Flux *740*
24.2 Gauss's Law *743*
24.3 Application of Gauss's Law to Various Charge Distributions *746*
24.4 Conductors in Electrostatic Equilibrium *750*
24.5 Formal Derivation of Gauss's Law *752*

Chapter 25 Electric Potential *762*

25.1 Potential Difference and Electric Potential *763*
25.2 Potential Differences in a Uniform Electric Field *765*

25.3 Electric Potential and Potential Energy Due to Point Charges *768*
25.4 Obtaining the Value of the Electric Field from the Electric Potential *772*
25.5 Electric Potential Due to Continuous Charge Distributions *774*
25.6 Electric Potential Due to a Charged Conductor *778*
25.7 The Millikan Oil-Drop Experiment *781*
25.8 Applications of Electrostatics *782*

Chapter 26 Capacitance and Dielectrics *795*

26.1 Definition of Capacitance *796*
26.2 Calculating Capacitance *797*
26.3 Combinations of Capacitors *802*
26.4 Energy Stored in a Charged Capacitor *807*
26.5 Capacitors with Dielectrics *810*
26.6 Electric Dipole in an Electric Field *815*
26.7 An Atomic Description of Dielectrics *817*

Chapter 27 Current and Resistance *831*

27.1 Electric Current *832*
27.2 Resistance *835*
27.3 A Model for Electrical Conduction *841*
27.4 Resistance and Temperature *843*
27.5 Superconductors *844*
27.6 Electrical Power *845*

Chapter 28 Direct Current Circuits *858*

28.1 Electromotive Force *859*
28.2 Resistors in Series and Parallel *862*
28.3 Kirchhoff's Rules *869*
28.4 *RC* Circuits *873*
28.5 Electrical Meters *879*
28.6 Household Wiring and Electrical Safety *880*

Chapter 29 Magnetic Fields *894*

29.1 Magnetic Fields and Forces *896*
29.2 Magnetic Force Acting on a Current-Carrying Conductor *900*
29.3 Torque on a Current Loop in a Uniform Magnetic Field *904*
29.4 Motion of a Charged Particle in a Uniform Magnetic Field *907*
29.5 Applications Involving Charged Particles Moving in a Magnetic Field *910*
29.6 The Hall Effect *914*

Chapter 30 Sources of the Magnetic Field *926*

30.1 The Biot–Savart Law *927*
30.2 The Magnetic Force Between Two Parallel Conductors *932*
30.3 Ampère's Law *933*

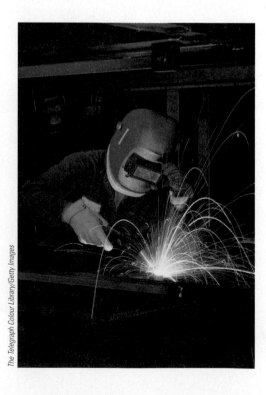

The Telegraph Colour Library/Getty Images

30.4 The Magnetic Field of a Solenoid *938*
30.5 Magnetic Flux *940*
30.6 Gauss's Law in Magnetism *941*
30.7 Displacement Current and the General Form of Ampère's Law *942*
30.8 Magnetism in Matter *944*
30.9 The Magnetic Field of the Earth *953*

Chapter 31 Faraday's Law 967

31.1 Faraday's Law of Induction *968*
31.2 Motional emf *973*
31.3 Lenz's Law *977*
31.4 Induced emf and Electric Fields *981*
31.5 Generators and Motors *982*
31.6 Eddy Currents *986*
31.7 Maxwell's Equations *988*

Chapter 32 Inductance 1003

32.1 Self-Inductance *1004*
32.2 RL Circuits *1006*
32.3 Energy in a Magnetic Field *1011*
32.4 Mutual Inductance *1013*
32.5 Oscillations in an LC Circuit *1015*
32.6 The RLC Circuit *1020*

Chapter 33 Alternating Current Circuits 1033

33.1 AC Sources *1033*
33.2 Resistors in an AC Circuit *1034*
33.3 Inductors in an AC Circuit *1038*
33.4 Capacitors in an AC Circuit *1041*
33.5 The RLC Series Circuit *1043*
33.6 Power in an AC Circuit *1047*
33.7 Resonance in a Series RLC Circuit *1049*
33.8 The Transformer and Power Transmission *1052*
33.9 Rectifiers and Filters *1054*

Chapter 34 Electromagnetic Waves 1066

34.1 Maxwell's Equations and Hertz's Discoveries *1067*
34.2 Plane Electromagnetic Waves *1069*
34.3 Energy Carried by Electromagnetic Waves *1074*
34.4 Momentum and Radiation Pressure *1076*
34.5 Production of Electromagnetic Waves by an Antenna *1079*
34.6 The Spectrum of Electromagnetic Waves *1080*

PART 5   Light and Optics   1093

Chapter 35 The Nature of Light and the Laws of Geometric Optics 1094

35.1 The Nature of Light *1095*
35.2 Measurements of the Speed of Light *1096*
35.3 The Ray Approximation in Geometric Optics *1097*
35.4 Reflection *1098*
35.5 Refraction *1102*
35.6 Huygens's Principle *1107*

35.7 Dispersion and Prisms *1109*
35.8 Total Internal Reflection *1111*
35.9 Fermat's Principle *1114*

Chapter 36 Image Formation 1126

36.1 Images Formed by Flat Mirrors *1127*
36.2 Images Formed by Spherical Mirrors *1131*
36.3 Images Formed by Refraction *1138*
36.4 Thin Lenses *1141*
36.5 Lens Aberrations *1152*
36.6 The Camera *1153*
36.7 The Eye *1155*
36.8 The Simple Magnifier *1159*
36.9 The Compound Microscope *1160*
36.10 The Telescope *1162*

Chapter 37 Interference of Light Waves 1176

37.1 Conditions for Interference *1177*
37.2 Young's Double-Slit Experiment *1177*
37.3 Intensity Distribution of the Double-Slit Interference Pattern *1182*
37.4 Phasor Addition of Waves *1184*
37.5 Change of Phase Due to Reflection *1188*
37.6 Interference in Thin Films *1189*
37.7 The Michelson Interferometer *1194*

Chapter 38 Diffraction Patterns and Polarization 1205

38.1 Introduction to Diffraction Patterns *1206*
38.2 Diffraction Patterns from Narrow Slits *1207*
38.3 Resolution of Single Slit and Circular Apertures *1214*
38.4 The Diffraction Grating *1217*
38.5 Diffraction of X-Rays by Crystals *1224*
38.6 Polarization of Light Waves *1225*

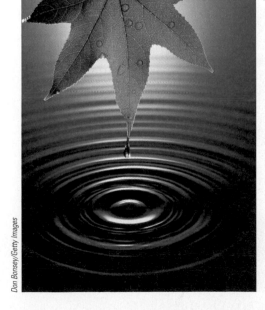

Don Bonsey/Getty Images

PART 6   Modern Physics   1243

Chapter 39   Relativity   1244

39.1   The Principle of Galilean Relativity   1246
39.2   The Michelson–Morley Experiment   1248
39.3   Einstein's Principle of Relativity   1250
39.4   Consequences of the Special Theory of Relativity   1251
39.5   The Lorentz Transformation Equations   1262
39.6   The Lorentz Velocity Transformation Equations   1264
39.7   Relativistic Linear Momentum and the Relativistic Form of Newton's Laws   1267
39.8   Relativistic Energy   1268
39.9   Mass and Energy   1272
39.10 The General Theory of Relativity   1273

Chapter 40   Introduction to Quantum Physics   1284

40.1   Blackbody Radiation and Planck's Hypothesis   1285
40.2   The Photoelectric Effect   1291
40.3   The Compton Effect   1297
40.4   Photons and Electromagnetic Waves   1300
40.5   The Wave Properties of Particles   1301
40.6   The Quantum Particle   1304
40.7   The Double-Slit Experiment Revisited   1307
40.8   The Uncertainty Principle   1309

Chapter 41   Quantum Mechanics   1321

41.1   An Interpretation of Quantum Mechanics   1322
41.2   A Particle in a Box   1326
41.3   The Particle Under Boundary Conditions   1330
41.4   The Schrödinger Equation   1331
41.5   A Particle in a Well of Finite Height   1334
41.6   Tunneling Through a Potential Energy Barrier   1336
41.7   The Scanning Tunneling Microscope   1340
41.8   The Simple Harmonic Oscillator   1341

© 1973 Kim Vandiver & Harold E. Edgerton/Courtesy of Palm Press, Inc.

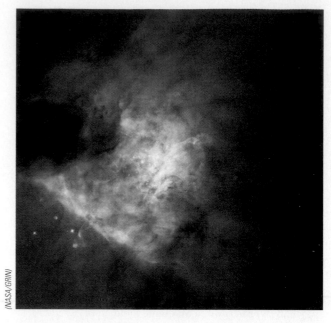

(NASA/GRIN)

Chapter 42   Atomic Physics   1351

42.1   Atomic Spectra of Gases   1352
42.2   Early Models of the Atom   1355
42.3   Bohr's Model of the Hydrogen Atom   1356
42.4   The Quantum Model of the Hydrogen Atom   1361
42.5   The Wave Functions for Hydrogen   1364
42.6   Physical Interpretation of the Quantum Numbers   1367
42.7   The Exclusion Principle and the Periodic Table   1374
42.8   More on Atomic Spectra: Visible and X-Ray   1380
42.9   Spontaneous and Stimulated Transitions   1383
42.10 Lasers   1385

Chapter 43   Molecules and Solids   1398

43.1   Molecular Bonds   1399
43.2   Energy States and Spectra of Molecules   1403
43.3   Bonding in Solids   1411
43.4   Free-Electron Theory of Metals   1415
43.5   Band Theory of Solids   1418
43.6   Electrical Conduction in Metals, Insulators, and Semiconductors   1420
43.7   Semiconductor Devices   1424
43.8   Superconductivity   1430

Chapter 44   Nuclear Structure   1440

44.1   Some Properties of Nuclei   1441
44.2   Nuclear Binding Energy   1447
44.3   Nuclear Models   1448
44.4   Radioactivity   1452
44.5   The Decay Process   1456
44.6   Natural Radioactivity   1465
44.7   Nuclear Reactions   1465
44.8   Nuclear Magnetic Resonance and Magnetic Resonance Imaging   1467

Chapter 45 **Applications of Nuclear Physics** *1479*

45.1 Interactions Involving Neutrons *1480*
45.2 Nuclear Fission *1481*
45.3 Nuclear Reactors *1483*
45.4 Nuclear Fusion *1487*
45.5 Radiation Damage *1495*
45.6 Radiation Detectors *1497*
45.7 Uses of Radiation *1500*

Chapter 46 **Particle Physics and Cosmology** *1511*

46.1 The Fundamental Forces in Nature *1512*
46.2 Positrons and Other Antiparticles *1513*
46.3 Mesons and the Beginning of Particle Physics *1516*
46.4 Classification of Particles *1518*
46.5 Conservation Laws *1520*
46.6 Strange Particles and Strangeness *1523*
46.7 Making Particles and Measuring Their Properties *1524*
46.8 Finding Patterns in the Particles *1527*
46.9 Quarks *1529*
46.10 Multicolored Quarks *1532*
46.11 The Standard Model *1534*
46.12 The Cosmic Connection *1536*
46.13 Problems and Perspectives *1542*

Appendix A **Tables** *A.1*
Table A.1 Conversion Factors *A.1*
Table A.2 Symbols, Dimensions, and Units of Physical Quantities *A.2*
Table A.3 Table of Atomic Masses *A.4*

Appendix B **Mathematics Review** *A.14*
B.1 Scientific Notation *A.14*
B.2 Algebra *A.15*
B.3 Geometry *A.20*
B.4 Trigonometry *A.21*
B.5 Series Expansions *A.23*
B.6 Differential Calculus *A.23*
B.7 Integral Calculus *A.25*
B.8 Propagation of Uncertainty *A.28*

Appendix C **Periodic Table of the Elements** *A.30*

Appendix D **SI Units** *A.32*

Appendix E **Nobel Prizes** *A.33*

Answers to Odd-Numbered Problems *A.37*

Index *I.1*

Austin McCrae

About the Authors

Raymond A. Serway received his doctorate at Illinois Institute of Technology and is Professor Emeritus at James Madison University. Dr. Serway began his teaching career at Clarkson University, where he conducted research and taught from 1967 to 1980. His second academic appointment was at James Madison University as Professor of Physics and Head of the Physics Department from 1980 to 1986. He remained at James Madison University until his retirement in 1997. He was the recipient of the Madison Scholar Award at James Madison University in 1990, the Distinguished Teaching Award at Clarkson University in 1977, and the Alumni Achievement Award from Utica College in 1985. As Guest Scientist at the IBM Research Laboratory in Zurich, Switzerland, he worked with K. Alex Müller, 1987 Nobel Prize recipient. Dr. Serway also held research appointments at Rome Air Development Center from 1961 to 1963, at IIT Research Institute from 1963 to 1967, and as a visiting scientist at Argonne National Laboratory, where he collaborated with his mentor and friend, Sam Marshall. In addition to earlier editions of this textbook, Dr. Serway is the co-author of the high-school textbook *Physics* with Jerry Faughn, published by Holt, Rinehart, & Winston and co-author of the third edition of *Principles of Physics* with John Jewett, the sixth edition of *College Physics* with Jerry Faughn, and the second edition of *Modern Physics* with Clem Moses and Curt Moyer. In addition, Dr. Serway has published more than 40 research papers in the field of condensed matter physics and has given more than 70 presentations at professional meetings. Dr. Serway and his wife Elizabeth enjoy traveling, golfing, gardening, and spending quality time with their four children and five grandchildren.

John W. Jewett, Jr. earned his doctorate at Ohio State University, specializing in optical and magnetic properties of condensed matter. Dr. Jewett began his academic career at Richard Stockton College of New Jersey, where he taught from 1974 to 1984. He is currently Professor of Physics at California State Polytechnic University, Pomona. Throughout his teaching career, Dr. Jewett has been active in promoting science education. In addition to receiving four National Science Foundation grants, he helped found and direct the Southern California Area Modern Physics Institute (SCAMPI). He also directed Science IMPACT (Institute for Modern Pedagogy and Creative Teaching), which works with teachers and schools to develop effective science curricula. Dr. Jewett's honors include the Stockton Merit Award at Richard Stockton College, the Outstanding Professor Award at California State Polytechnic University for 1991–1992, and the Excellence in Undergraduate Physics Teaching Award from the American Association of Physics Teachers (AAPT) in 1998. He has given over 80 presentations at professional meetings, including presentations at international conferences in China and Japan. In addition to his work on this textbook, he is co-author of the third edition of *Principles of Physics* with Ray Serway and author of *The World of Physics . . . Mysteries, Magic, and Myth*. Dr. Jewett enjoys playing piano, traveling, and collecting antiques that can be used as demonstration apparatus in physics lectures, as well as spending time with his wife Lisa and their children and grandchildren.

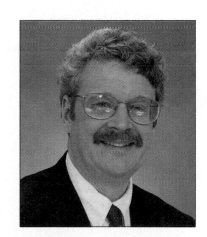

Preface

Mark Cooper/Corbis Stock Market

In writing this sixth edition of *Physics for Scientists and Engineers with Modern Physics,* we continue our ongoing efforts to improve the clarity of presentation and we again include new pedagogical features that help support the learning and teaching processes. Drawing on positive feedback from users of the fifth edition and reviewers' suggestions, we have refined the text in order to better meet the needs of students and teachers. We have for the first time integrated a powerful collection of media resources into many of the illustrations, examples, and end-of-chapter problems in the text. These resources compose the Web-based learning system *PhysicsNow* and are flagged by the media icon. Further details are described below.

This textbook is intended for a course in introductory physics for students majoring in science or engineering. The entire contents of the text in its extended version could be covered in a three-semester course, but it is possible to use the material in shorter sequences with the omission of selected chapters and sections. The mathematical background of the student taking this course should ideally include one semester of calculus. If that is not possible, the student should be enrolled in a concurrent course in introductory calculus.

Objectives

This introductory physics textbook has two main objectives: to provide the student with a clear and logical presentation of the basic concepts and principles of physics, and to strengthen an understanding of the concepts and principles through a broad range of interesting applications to the real world. To meet these objectives, we have placed emphasis on sound physical arguments and problem-solving methodology. At the same time, we have attempted to motivate the student through practical examples that demonstrate the role of physics in other disciplines, including engineering, chemistry, and medicine.

Changes in the Sixth Edition

A large number of changes and improvements have been made in preparing the sixth edition of this text. Some of the new features are based on our experiences and on current trends in science education. Other changes have been incorporated in response to comments and suggestions offered by users of the fifth edition and by reviewers of the manuscript. The following represent the major changes in the sixth edition:

Active Figures Many diagrams from the text have been animated to form **Active Figures,** part of the *PhysicsNow* integrated Web-based learning system. By visualizing phenomena and processes that cannot be fully represented on a static page, students greatly increase their conceptual understanding. **Active Figures** are identified with the media icon. An addition to the figure caption in blue type describes briefly the nature and contents of the animation.

Interactive Worked Examples Approximately 90 of the worked examples in the text have been identified as interactive, labeled with the media icon. As part of the *PhysicsNow* Web-based learning system, students can engage in an extension of the problem solved in the example. This often includes elements of both visualization and calculation, and may also involve prediction and intuition building. Often the interactivity is inspired by the **"What If?"** question we posed in the example text.

What If? Approximately one-third of the worked examples in the text contain this new feature. At the completion of the example solution, a **What If?** question offers a variation on the situation posed in the text of the example. For instance, this feature might explore the effects of changing the conditions of the situation, determine what happens when a quantity is taken to a particular limiting value, or question whether additional information can be determined about the problem situation. The answer to the question generally includes both a conceptual response and a mathematical response. This feature encourages students to think about the results of the example and assists in conceptual understanding of the principles. It also prepares students to encounter novel problems featured on exams. Some of the end-of-chapter problems also carry the **"What If?"** feature.

Quick Quizzes   The number of Quick Quiz questions in each chapter has been increased. Quick Quizzes provide students with opportunities to test their understanding of the physical concepts presented. The questions require students to make decisions on the basis of sound reasoning, and some of them have been written to help students overcome common misconceptions. Quick Quizzes have been cast in an objective format, including multiple choice, true–false, and ranking. Answers to all Quick Quiz questions are found at the end of each chapter. Additional Quick Quizzes that can be used in classroom teaching are available on the instructor's companion Web site. Many instructors choose to use such questions in a "peer instruction" teaching style, but they can be used in standard quiz format as well.

Pitfall Preventions   These new features are placed in the margins of the text and address common student misconceptions and situations in which students often follow unproductive paths. Over 200 Pitfall Preventions are provided to help students avoid common mistakes and misunderstandings.

General Problem-Solving Strategy   A general strategy to be followed by the student is outlined at the end of Chapter 2 and provides students with a structured process for solving problems. In Chapters 3 through 5, the strategy is employed explicitly in every example so that students learn how it is applied. In the remaining chapters, the strategy appears explicitly in one example per chapter so that students are encouraged throughout the course to follow the procedure.

Line-by-Line Revision   The entire text has been carefully edited to improve clarity of presentation and precision of language. We hope that the result is a book that is both accurate and enjoyable to read.

Problems   A substantial revision of the end-of-chapter problems was made in an effort to improve their variety and interest, while maintaining their clarity and quality. Approximately 17% of the problems (about 550) are new. All problems have been carefully edited. Solutions to approximately 20% of the end-of-chapter problems are included in the *Student Solutions Manual and Study Guide.* These problems are identified by boxes around their numbers. A smaller subset of solutions, identified by the media icon 🌀, are available on the World Wide Web (**http://www.pse6.com**) as coached solutions with hints. Targeted feedback is provided for students whose instructors adopt *Physics for Scientists and Engineers with Modern Physics,* sixth edition. See the next section for a complete description of other features of the problem set.

Content Changes   The content and organization of the textbook is essentially the same as that of the fifth edition. An exception is that Chapter 13 (Oscillatory Motion) in the fifth edition has been moved to the Chapter 15 position in the sixth edition, in order to form a cohesive four-chapter Part 2 on oscillations and waves. Many sections in various chapters have been streamlined, deleted, or combined with other sections to allow for a more balanced presentation. The chapters on Modern Physics, Chapters 39–46, have been extensively rewritten to provide more up-to-date material as well as modern applications. A more detailed list of content changes can be found on the instructor's companion Web site.

Content

The material in this book covers fundamental topics in classical physics and provides an introduction to modern physics. The book is divided into six parts. Part 1 (Chapters 1 to 14) deals with the fundamentals of Newtonian mechanics and the physics of fluids, Part 2 (Chapters 15 to 18) covers oscillations, mechanical waves, and sound, Part 3 (Chapters 19 to 22) addresses heat and thermodynamics, Part 4 (Chapters 23 to 34) treats electricity and magnetism, Part 5 (Chapters 35 to 38) covers light and optics, and Part 6 (Chapters 39 to 46) deals with relativity and modern physics. Each part opener includes an overview of the subject matter covered in that part, as well as some historical perspectives.

Text Features

Most instructors would agree that the textbook selected for a course should be the student's primary guide for understanding and learning the subject matter. Furthermore, the textbook should be easily accessible and should be styled and written to facilitate instruction and learning. With these points in mind, we have included many pedagogical features in the textbook that are intended to enhance its usefulness to both students and instructors. These features are as follows:

Style   To facilitate rapid comprehension, we have attempted to write the book in a style that is clear, logical, and engaging. We have chosen a writing style that is somewhat informal and relaxed so that students will find the text appealing and enjoyable to read. New terms are carefully defined, and we have avoided the use of jargon.

Previews   All chapters begin with a brief preview that includes a discussion of the chapter's objectives and content.

Important Statements and Equations   Most important statements and definitions are set in **boldface** type or are highlighted with a background screen for added emphasis and ease of review. Similarly, important equations are highlighted with a background screen to facilitate location.

Bruce Ayers/Getty Images

Problem-Solving Hints   In several chapters, we have included general strategies for solving the types of problems featured both in the examples and in the end-of-chapter problems. This feature helps students to identify necessary steps in problem solving and to eliminate any uncertainty they might have. Problem-solving strategies are highlighted with a light red background screen for emphasis and ease of location.

Marginal Notes   Comments and notes appearing in blue type in the margin can be used to locate important statements, equations, and concepts in the text.

Pedagogical Use of Color   Readers should consult the **pedagogical color chart** (second page inside the front cover) for a listing of the color-coded symbols used in the text diagrams, Web-based **Active Figures,** and diagrams within **Interactive Worked Examples.** This system is followed consistently whenever possible, with slight variations made necessary by the complexity of physical situations depicted in Part 4.

Mathematical Level   We have introduced calculus gradually, keeping in mind that students often take introductory courses in calculus and physics concurrently. Most steps are shown when basic equations are developed, and reference is often made to mathematical appendices at the end of the textbook. Vector products are introduced later in the text, where they are needed in physical applications. The dot product is introduced in Chapter 7, which addresses energy and energy transfer; the cross product is introduced in Chapter 11, which deals with angular momentum.

Worked Examples   A large number of worked examples of varying difficulty are presented to promote students' understanding of concepts. In many cases, the examples serve as models for solving the end-of-chapter problems. Because of the increased emphasis on understanding physical concepts, many examples are conceptual in nature and are labeled as such. The examples are set off in boxes, and the answers to examples with numerical solutions are highlighted with a background screen. We have already mentioned that a number of examples are designated as interactive and are part of the *PhysicsNow* Web-based learning system.

Questions   Questions of a conceptual nature requiring verbal or written responses are provided at the end of each chapter. Over 1 000 questions are included in this edition. Some questions provide the student with a means of self-testing the concepts presented in the chapter. Others could serve as a basis for initiating classroom discussions. Answers to selected questions are included in the *Student Solutions Manual and Study Guide,* and answers to all questions are found in the *Instructor's Solutions Manual.*

Significant Figures   Significant figures in both worked examples and end-of-chapter problems have been handled with care. Most numerical examples are worked out to either two or three significant figures, depending on the precision of the data provided. End-of-chapter problems regularly state data and answers to three-digit precision.

Problems   An extensive set of problems is included at the end of each chapter; in all, over 3 000 problems are given throughout the text. Answers to odd-numbered problems are provided at the end of the book in a section whose pages have colored edges for ease of location. For the convenience of both the student and the instructor, about two thirds of the problems are keyed to specific sections of the chapter. The remaining problems, labeled "Additional Problems," are not keyed to specific sections.

Usually, the problems within a given section are presented so that the straightforward problems (those with black problem numbers) appear first. For ease of identification, the numbers of intermediate-level problems are printed in blue, and those of challenging problems are printed in magenta.

- **Review Problems**   Many chapters include review problems requiring the student to combine concepts covered in the chapter with those discussed in previous chapters. These problems reflect the cohesive nature of the principles in the text and verify that physics is not a scattered set of ideas. When facing real-world issues such as global warming or nuclear weapons, it may be necessary to call on ideas in physics from several parts of a textbook such as this one.

- **Paired Problems**   To allow focused practice in solving problems stated in symbolic terms, some end-of-chapter numerical problems are paired with the same problems in symbolic form. Paired problems are identified by a common light red background screen.

- **Computer- and Calculator-Based Problems**   Many chapters include one or more problems whose solution requires the use of a computer or graphing calculator. Computer modeling of physical phenomena enables students to obtain graphical representations of variables and to perform numerical analyses.

- **Coached Problems with Hints**   These have been described above as part of the *PhysicsNow* Web-based learning system. These problems are identified by the media icon  and targeted feedback is provided to students of instructors adopting the sixth edition.

Units   The international system of units (SI) is used throughout the text. The U.S. customary system of units is used only to a limited extent in the chapters on mechanics, heat, and thermodynamics.

Summaries   Each chapter contains a summary that reviews the important concepts and equations discussed in that chapter. A marginal note in blue type next to each chapter summary directs students to a practice test (Post-Test) for the chapter.

Appendices and Endpapers   Several appendices are provided at the end of the textbook. Most of the appendix material represents a review of mathematical concepts and techniques used in the text, including scientific notation, algebra, geometry, trigonometry, differential calculus, and integral calculus. Reference to these appendices is made

Courtesy NASA

throughout the text. Most mathematical review sections in the appendices include worked examples and exercises with answers. In addition to the mathematical reviews, the appendices contain tables of physical data, conversion factors, atomic masses, and the SI units of physical quantities, as well as a periodic table of the elements. Other useful information, including fundamental constants and physical data, planetary data, a list of standard prefixes, mathematical symbols, the Greek alphabet, and standard abbreviations of units of measure, appears on the endpapers.

Student Ancillaries

Student Solutions Manual and Study Guide by John R. Gordon, Ralph McGrew, and Raymond Serway. This two-volume manual features detailed solutions to 20% of the end-of-chapter problems from the text. The manual also features a list of important equations, concepts, and notes from key sections of the text, in addition to answers to selected end-of-chapter questions. Volume 1 contains Chapters 1 through 22 and Volume 2 contains Chapters 23 through 46.

WebTutor™ on WebCT and Blackboard **WebTutor** offers students real-time access to a full array of study tools, including chapter outlines, summaries, learning objectives, glossary flashcards (with audio), practice quizzes, **InfoTrac® College Edition** exercises, and Web links.

InfoTrac® College Edition Adopters and their students automatically receive a four-month subscription to **InfoTrac® College Edition** with every new copy of this book. Newly improved, this extensive online library opens the door to the full text (not just abstracts) of countless articles from thousands of publications including *American Scientist, Physical Review, Science, Science Weekly,* and more! Available only to college and university students. Journals subject to change.

The Brooks/Cole Physics Resource Center You will find additional online quizzes, Web links and animations at **http://physics.brookscole.com.**

Ancillaries for Instructors

The first four ancillaries below are available to qualified adopters. Please consult your local sales representative for details.

Instructor's Solutions Manual by Ralph McGrew and James A. Currie. This two-volume manual contains complete worked solutions to all of the end-of-chapter problems in the textbook as well as answers to even-numbered problems. The solutions to problems new to the sixth edition are marked for easy identification by the instructor. New to this edition are complete answers to the conceptual questions in the main text. Volume 1 contains Chapters 1 through 22 and Volume 2 contains Chapters 23 through 46.

Printed Test Bank by Edward Adelson. This two-volume test bank contains approximately 2 300 multiple-choice questions. These questions are also available in electronic format with complete answers and solutions in the Brooks/Cole Assessment test program. Volume 1 contains Chapters 1 through 22 and Volume 2 contains Chapters 23 through 46.

Multimedia Manager This easy-to-use multimedia lecture tool allows you to quickly assemble art and database files with notes to create fluid lectures. The CD-ROM set (Volume 1, Chapters 1–22; Volume 2, Chapters 23–46) includes a database of animations, video clips, and digital art from the text as well as electronic files of the *Instructor's Solutions Manual and Test Bank.* The simple interface makes it easy for you to incorporate graphics, digital video, animations, and audio clips into your lectures.

Transparency Acetates Each volume contains approximately 100 acetates featuring art from the text. Volume 1 contains Chapters 1 through 22 and Volume 2 contains Chapters 23 through 46.

Brooks/Cole Assessment With a balance of efficiency, high performance, simplicity and versatility, **Brooks/Cole Assessment (BCA)** gives you the power to transform the learning and teaching experience. **BCA** is fully integrated testing, tutorial, and course management software accessible by instructors and students anytime, anywhere. Delivered for FREE in a browser-based format without the need for any proprietary software or plug-ins, **BCA** uses correct scientific notation to provide the drill of basic skills that students need, enabling the instructor to focus more time in higher-level learning activities (i.e., concepts and applications). Students can have unlimited practice in questions and problems, building their own confidence and skills. Results flow automatically to a grade book for tracking so that instructors will be better able to assess student understanding of the material, even prior to class or an actual test.

George Sample

WebTutor™ on WebCT and Blackboard With **WebTutor's** text-specific, preformatted content and total flexibility, instructors can easily create and manage their own personal Web site. **WebTutor's** course management tool gives instructors the ability to provide virtual office hours, post syllabi, set up threaded discussions, track student progress with the quizzing material, and much more. **WebTutor** also provides robust communication tools, such as a course calendar, asynchronous discussion, real-time chat, a whiteboard, and an integrated e-mail system.

Additional Options for Online Homework For detailed information and demonstrations, contact your Thomson•Brooks/Cole representative or visit the following:

- WebAssign: A Web-based Homework System
  http://www.webassign.net or contact WebAssign at *webassign@ncsu.edu*
- Homework Service
  http://hw.ph.utexas.edu/hw.html or contact *moore@physics.utexas.edu*
- CAPA: A Computer-Assisted Personalized Approach
  http://capa4.lite.msu.edu/homepage/

Instructor's Companion Web Site Consult the instructor's site at *http://www.pse6.com* for additional Quick Quiz questions, a detailed list of content changes since the fifth edition, a problem correlation guide, images from the text, and sample PowerPoint lectures. Instructors adopting the sixth edition of *Physics for Scientists and Engineers with Modern Physics* may download these materials after securing the appropriate password from their local Thomson•Brooks/Cole sales representative.

Teaching Options

The topics in this textbook are presented in the following sequence: classical mechanics, oscillations and mechanical waves, and heat and thermodynamics followed by electricity and magnetism, electromagnetic waves, optics, relativity, and modern physics. This presentation represents a traditional sequence, with the subject of mechanical waves being presented before electricity and magnetism. Some instructors may prefer to cover this material after completing electricity and magnetism (i.e., after Chapter 34). The chapter on relativity is placed near the end of the text because this topic often is treated as an introduction to the era of "modern physics." If time permits, instructors may choose to cover Chapter 39 after completing Chapter 13, as it concludes the material on Newtonian mechanics.

For those instructors teaching a two-semester sequence, some sections and chapters could be deleted without any loss of continuity. The following sections can be considered optional for this purpose:

2.7	Kinematic Equations Derived from Calculus	6.4	Motion in the Presence of Resistive Forces
4.6	Relative Velocity and Relative Acceleration	6.5	Numerical Modeling in Particle Dynamics
6.3	Motion in Accelerated Frames	7.9	Energy and the Automobile

8.6	Energy Diagrams and Equilibrium of a System
9.7	Rocket Propulsion
11.5	The Motion of Gyroscopes and Tops
11.6	Angular Momentum as a Fundamental Quantity
14.7	Other Applications of Fluid Dynamics
15.6	Damped Oscillations
15.7	Forced Oscillations
17.5	Digital Sound Recording
17.6	Motion Picture Sound
18.6	Standing Waves in Rods and Membranes
18.8	Nonsinusoidal Wave Patterns
21.7	Mean Free Path
22.8	Entropy on a Microscopic Scale
24.5	Formal Derivation of Gauss's Law
25.7	The Millikan Oil-Drop Experiment
25.8	Applications of Electrostatics
26.7	An Atomic Description of Dielectrics
27.5	Superconductors
28.5	Electrical Meters
28.6	Household Wiring and Electrical Safety
29.5	Applications Involving Charged Particles Moving in a Magnetic Field
29.6	The Hall Effect
30.8	Magnetism in Matter
30.9	The Magnetic Field of the Earth
31.6	Eddy Currents
33.9	Rectifiers and Filters
34.5	Production of Electromagnetic Waves by an Antenna
35.9	Fermat's Principle
36.5	Lens Aberrations
36.6	The Camera
36.7	The Eye
36.8	The Simple Magnifier
36.9	The Compound Microscope
36.10	The Telescope
38.5	Diffraction of X-Rays by Crystals
39.10	The General Theory of Relativity

Topham Picturepoint/The Image Works

Again we stress that Chapters 40–46 on modern physics have been thoroughly revised and updated. For those instructors covering modern physics and finding themselves pressed for time, the following sections could be deleted without loss of continuity:

41.7	The Scanning Tunneling Microscope
42.10	Lasers
43.7	Semiconductor Devices
43.8	Superconductivity
45.5	Radiation Damage
45.6	Radiation Detectors
45.7	Uses of Radiation

Acknowledgments

The sixth edition of this textbook was prepared with the guidance and assistance of many professors who reviewed selections of the manuscript, the pre-revision text, or both. We wish to acknowledge the following scholars and express our sincere appreciation for their suggestions, criticisms, and encouragement:

Edward Adelson, *Ohio State University*

Michael R. Cohen, *Shippensburg University*

Jerry D. Cook, *Eastern Kentucky University*

J. William Dawicke, *Milwaukee School of Engineering*

N. John DiNardo, *Drexel University*

Andrew Duffy, *Boston University*

Robert J. Endorf, *University of Cincinnati*

F. Paul Esposito, *University of Cincinnati*

Joe L. Ferguson, *Mississippi State University*

Perry Ganas, *California State University, Los Angeles*

John C. Hardy, *Texas A&M University*

Michael Hayes, *University of Pretoria (South Africa)*

John T. Ho, *The State University of New York, Buffalo*

Joseph W. Howard, *Salisbury University*

Robert Hunt, *Johnson County Community College*

Walter S. Jaronski, *Radford University*

Sangyong Jeon, *McGill University, Quebec*

Stan Jones, *University of Alabama*

L. R. Jordan, *Palm Beach Community College*

Teruki Kamon, *Texas A & M University*

Louis E. Keiner, *Coastal Carolina University*

Mario Klarič, *Midlands Technical College*

Laird Kramer, *Florida International University*

Edwin H. Lo, *American University*

James G. McLean, *The State University of New York, Geneseo*

Richard E. Miers, *Indiana University–Purdue University, Fort Wayne*

Oscar Romulo Ochoa, *The College of New Jersey*

Paul S. Ormsby, *Moraine Valley Community College*

Didarul I. Qadir, *Central Michigan University*

Judith D. Redling, *New Jersey Institute of Technology*

Richard W. Robinett, *Pennsylvania State University*

Om P. Rustgi, *SUNY College at Buffalo*

Mesgun Sebhatu, *Winthrop University*

Natalia Semushkina, *Shippensburg University*

Daniel Stump, *Michigan State University*

Uwe C. Täuber, *Virginia Polytechnic Institute*

Perry A. Tompkins, *Samford University*

Doug Welch, *McMaster University, Ontario*

Augden Windelborn, *Northern Illinois University*

Jerzy M. Wrobel, *University of Missouri, Kansas City*

Jianshi Wu, *Fayetteville State University*

Michael Zincani, *University of Dallas*

Frank Oberle/Getty Images

This title was carefully checked for accuracy by Michael Kotlarchyk *(Rochester Institute of Technology)*, Chris Vuille *(Embry-Riddle Aeronautical University)*, Laurencin Dunbar *(St. Louis Community College)*, William Dawicke *(Milwaukee School of Engineering)*, Ioan Kosztin *(University of Missouri)*, Tom Barrett *(Ohio State University)*, Z. M. Stadnik *(University of Ottawa)*, Ronald E. Jodoin *(Rochester Institute of Technology)*, Brian A. Raue *(Florida International University)*, Peter Moeck *(Portland State University)*, and Grant Hart *(Brigham Young University)*. We thank them for their diligent efforts under schedule pressure!

We are grateful to Ralph McGrew for organizing the end-of-chapter problems, writing many new problems, and his excellent suggestions for improving the content of the textbook. Problems new to this edition were written by Edward Adelson, Ronald Bieniek, Michael Browne, Andrew Duffy, Robert Forsythe, Perry Ganas, Michael Hones, John Jewett, Boris Korsunsky, Edwin Lo, Ralph McGrew, Raymond Serway, and Jerzy Wrobel, with the help of Bennett Simpson and JoAnne Maniago. Students Alexander Coto, Karl Payne, and Eric Peterman made corrections to problems taken from previous editions, as did teachers David Aspnes, Robert Beichner, Joseph Biegen, Tom Devlin, Vasili Haralambous, Frank Hayes, Erika Hermon, Ken Menningen, Henry Nebel, and Charles Teague. We are grateful to authors John R. Gordon and Ralph McGrew and compositor Michael Rudmin for preparing the *Student Solutions Manual and Study Guide*. Authors Ralph McGrew and James Currie and compositor Mary Toscano have prepared an excellent *Instructor's Solutions Manual*, and we thank them. Edward Adelson has carefully edited and improved the Test Bank for the sixth edition. Kurt Vandervoort prepared extra Quick Quiz questions for the instructor's companion Web site.

Special thanks and recognition go to the professional staff at the Brooks/Cole Publishing Company—in particular Susan Pashos, Rebecca Heider and Alyssa White (who managed the ancillary program and so much more), Jessica Howard, Seth Dobrin, Peter McGahey, Teri Hyde, Michelle Julet, David Harris, and Chris Hall—for their fine work during the development and production of this textbook. We are most appreciative of Sam Subity's masterful management of the **PhysicsNow** media program. Kelley McAllister is our energetic Marketing Manager, and Stacey Purviance coordinates our marketing communications. We recognize the skilled production service provided by the staff at Sparkpoint Communications, the excellent artwork produced by Rolin Graphics, and the dedicated photo research efforts of Terri Wright.

Finally, we are deeply indebted to our wives and children for their love, support, and long-term sacrifices.

Raymond A. Serway
Leesburg, Virginia

John W. Jewett, Jr.
Pomona, California

To the Student

It is appropriate to offer some words of advice that should be of benefit to you, the student. Before doing so, we assume that you have read the Preface, which describes the various features of the text that will help you through the course.

How to Study

Very often instructors are asked, "How should I study physics and prepare for examinations?" There is no simple answer to this question, but we would like to offer some suggestions that are based on our own experiences in learning and teaching over the years.

First and foremost, maintain a positive attitude toward the subject matter, keeping in mind that physics is the most fundamental of all natural sciences. Other science courses that follow will use the same physical principles, so it is important that you understand and are able to apply the various concepts and theories discussed in the text.

Concepts and Principles

It is essential that you understand the basic concepts and principles before attempting to solve assigned problems. You can best accomplish this goal by carefully reading the textbook before you attend your lecture on the covered material. When reading the text, you should jot down those points that are not clear to you. We've purposely left wide margins in the text to give you space for making notes. Also be sure to make a diligent attempt at answering the questions in the Quick Quizzes as you come to them in your reading. We have worked hard to prepare questions that help you judge for yourself how well you understand the material. Study carefully the **What If?** features that appear with many of the worked examples. These will help you to extend your understanding beyond the simple act of arriving at a numerical result. The Pitfall Preventions will also help guide you away from common misunderstandings about physics. During class, take careful notes and ask questions about those ideas that are unclear to you. Keep in mind that few people are able to absorb the full meaning of scientific material after only one reading. Several readings of the text and your notes may be necessary. Your lectures and laboratory work supplement reading of the textbook and should clarify some of the more difficult material. You should minimize your memorization of material. Successful memorization of passages from the text, equations, and derivations does not necessarily indicate that you understand the material. Your understanding of the material will be enhanced through a combination of efficient study habits, discussions with other students and with instructors, and your ability to solve the problems presented in the textbook. Ask questions whenever you feel clarification of a concept is necessary.

Study Schedule

It is important that you set up a regular study schedule, preferably a daily one. Make sure that you read the syllabus for the course and adhere to the schedule set by your instructor. The lectures will make much more sense if you read the corresponding text material before attending them. As a general rule, you should devote about two hours of study time for every hour you are in class. If you are having trouble with the course, seek the advice of the instructor or other students who have taken the course. You may find it necessary to seek further instruction from experienced students. Very often, instructors offer review sessions in addition to regular class periods. It is important that you avoid the practice of delaying study until a day or two before an exam. More often than not, this approach has disastrous results. Rather than undertake an all-night study session, briefly review the basic concepts and equations, and get a good night's rest. If you feel you need additional help in understanding the concepts, in preparing for exams, or in problem solving, we suggest that you acquire a copy of the *Student Solutions Manual and Study Guide* that accompanies this textbook; this manual should be available at your college bookstore.

Use the Features

You should make full use of the various features of the text discussed in the Preface. For example, marginal notes are useful for locating and describing important equations and concepts, and **boldfaced** type indicates important statements and definitions. Many useful tables are contained in the Appendices, but most are incorporated in the text where they are most often referenced. Appendix B is a convenient review of mathematical techniques.

Answers to odd-numbered problems are given at the end of the textbook, answers to Quick Quizzes are located at the end of each chapter, and answers to selected end-of-chapter questions are provided in the *Student Solutions Manual and Study Guide*. Problem-Solving Strategies and Hints are included in selected chapters throughout the text and give you additional information about how you should solve problems. The Table of Contents provides an overview of the entire text, while the Index enables you to locate specific material quickly. Footnotes sometimes are used to supplement the text or to cite other references on the subject discussed.

After reading a chapter, you should be able to define any new quantities introduced in that chapter and to discuss the principles and assumptions that were used to arrive at certain key relations. The chapter summaries and the review sections of the *Student Solutions Manual and Study Guide* should help you in this regard. In some cases, it may be necessary for you to refer to the index of the text to locate certain topics. You should be able to associate with each physical quantity the correct symbol used to represent that quantity and the unit in which the quantity is specified. Furthermore, you should be able to express each important equation in a concise and accurate prose statement.

Problem Solving

R. P. Feynman, Nobel laureate in physics, once said, "You do not know anything until you have practiced." In keeping with this statement, we strongly advise that you develop the skills necessary to solve a wide range of problems. Your ability to solve problems will be one of the main tests of your knowledge of physics, and therefore you should try to solve as many problems as possible. It is essential that you understand basic concepts and principles before attempting to solve problems. It is good practice to try to find alternate solutions to the same problem. For example, you can solve problems in mechanics using Newton's laws, but very often an alternative method that draws on energy considerations is more direct. You should not deceive yourself into thinking that you understand a problem merely because you have seen it solved in class. You must be able to solve the problem and similar problems on your own.

The approach to solving problems should be carefully planned. A systematic plan is especially important when a problem involves several concepts. First, read the problem several times until you are confident you understand what is being asked. Look for any key words that will help you interpret the problem and perhaps allow you to make certain assumptions. Your ability to interpret a question properly is an integral part of problem solving. Second, you should acquire the habit of writing down the information given in a problem and those quantities that need to be found; for example, you might construct a table listing both the quantities given and the quantities to be found. This procedure is sometimes used in the worked examples of the textbook. Finally, after you have decided on the method you feel is appropriate for a given problem, proceed with your solution. Specific problem-solving strategies (Hints) of this type are included in the text and are highlighted with a light red screen. We have also developed a General Problem-Solving Strategy to help guide you through complex problems. If you follow the steps of this procedure (*Conceptualize, Categorize, Analyze, Finalize*), you will not only find it easier to come up with a solution, but you will also gain more from your efforts. This Strategy is located at the end of Chapter 2 (page 47) and is used in all worked examples in Chapters 3 through 5 so that you can learn how to apply it. In the remaining chapters, the Strategy is used in one example per chapter as a reminder of its usefulness.

Often, students fail to recognize the limitations of certain equations or physical laws in a particular situation. It is very important that you understand and remember the assumptions that underlie a particular theory or formalism. For example, certain equations in kinematics apply only to a particle moving with constant acceleration. These equations are not valid for describing motion whose acceleration is not constant, such as the motion of an object connected to a spring or the motion of an object through a fluid.

Experiments

© Phil Degginger/Stone/Getty

Physics is a science based on experimental observations. In view of this fact, we recommend that you try to supplement the text by performing various types of "hands-on" experiments, either at home or in the laboratory. These can be used to test ideas and models discussed in class or in the textbook. For example, the common Slinky™ toy is excellent for studying traveling waves; a ball swinging on the end of a long string can be used to investigate pendulum motion; various masses attached to the end of a vertical spring or rubber band can be used to determine their elastic nature; an old pair of Polaroid sunglasses and some discarded lenses and a magnifying glass are the components of various experiments in optics; and an approximate measure of the free-fall acceleration can be determined simply by measuring with a stopwatch the time it takes for a ball to drop from a known height. The list of such experiments is endless. When physical models are not available, be imaginative and try to develop models of your own.

New Media

We strongly encourage you to use the *PhysicsNow* Web-based learning system that accompanies this textbook. It is far easier to understand physics if you see it in action, and these new materials will enable you to become a part of that action. *PhysicsNow* media described in the Preface are accessed at the URL *http://www.pse6.com,* and feature a three-step learning process consisting of a Pre-Test, a personalized learning plan, and a Post-Test.

In addition to other elements, *PhysicsNow* includes the following Active Figures and Interactive Worked Examples:

Chapter 2
Active Figures 2.1, 2.3, 2.9, 2.10, 2.11, and 2.13
Examples 2.8 and 2.12

Chapter 3
Active Figures 3.2, 3.3, 3.6, and 3.16
Example 3.5

Chapter 4
Active Figures 4.5, 4.7, and 4.11
Examples 4.4, 4.5, and 4.18

Chapter 5
Active Figure 5.16
Examples 5.9, 5.10, 5.12, and 5.14

Chapter 6
Active Figures 6.2, 6.8, 6.12, and 6.15
Examples 6.4, 6.5, and 6.7

Chapter 7
Active Figure 7.10
Examples 7.9 and 7.11

Chapter 8
Active Figures 8.3, 8.4, and 8.16
Examples 8.2 and 8.4

Chapter 9
Active Figures 9.8, 9.9, 9.13, 9.16, and 9.17
Examples 9.1, 9.5, and 9.8

Chapter 10
Active Figures 10.4, 10.14, and 10.30
Examples 10.12, 10.13, and 10.14

Chapter 11
Active Figures 11.1, 11.3, and 11.4
Examples 11.6 and 11.10

Chapter 12
Active Figures 12.14, 12.16, 12.17, and 12.18
Examples 12.3 and 12.4

Chapter 13
Active Figures 13.1, 13.5, and 13.7
Example 13.1

Chapter 14
Active Figures 14.9 and 14.10
Examples 14.2 and 14.10

Chapter 15
Active Figures 15.1, 15.2, 15.7, 15.9, 15.10, 15.11, 15.14, 15.17, and 15.22

Chapter 16
Active Figures 16.4, 16.7, 16.8, 16.10, 16.14, 16.15, and 16.17
Example 16.5

Chapter 17
Active Figures 17.2, 17.8, and 17.9
Examples 17.1 and 17.6

Chapter 18
Active Figures 18.1, 18.2, 18.4, 18.9, 18.10, 18.22, and 18.25
Examples 18.4 and 18.5

Chapter 19
Active Figures 19.8 and 19.12
Example 19.7

Chapter 20
Active Figures 20.4, 20.5, and 20.7
Example 20.10

Chapter 21
Active Figures 21.2, 21.4, 21.11, and 21.12
Example 21.4

Chapter 22
Active Figures 22.2, 22.5, 22.10, 22.11, 22.12, 22.13, and 22.19
Example 22.10

Chapter 23
Active Figures 23.7, 23.13, 23.24, and 23.26
Examples 23.3 and 23.11

Chapter 24
Active Figures 24.4 and 24.9
Examples 24.5 and 24.10

Chapter 25
Active Figures 25.10 and 25.27
Examples 25.2 and 25.3

Chapter 26
Active Figures 26.4, 26.9, and 26.10
Examples 26.4 and 26.5

Chapter 27
Active Figures 27.9 and 27.13
Examples 27.3 and 27.8

Chapter 28
Active Figures 28.2, 28.4, 28.6, 28.19, 28.21, 28.27, and 28.29
Examples 28.1, 28.5, and 28.9

Chapter 29
Active Figures 29.1, 29.14, 29.18, 29.19, 29.23, and 29.24
Example 29.7

Chapter 30
Active Figures 30.8, 30.9, and 30.21
Examples 30.1, 30.3, and 30.8

Chapter 31
Active Figures 31.1, 31.2, 31.10, 31.21, 31.23, and 31.26
Examples 31.4 and 31.5

Chapter 32
Active Figures 32.3, 32.4, 32.6, 32.7, 32.17, 32.18, 32.21, and 32.23
Examples 32.3 and 32.7

Chapter 33
Active Figures 33.2, 33.3, 33.6, 33.7, 33.9, 33.10, 33.13, 33.15, 33.19, 33.25, and 33.26
Examples 33.5 and 33.7

Chapter 34
Active Figure 34.3
Examples 34.1 and 34.4

Chapter 35
Active Figures 35.4, 35.6, 35.10, 35.11, 35.23, and 35.26
Examples 35.2 and 35.6

Chapter 36
Active Figures 36.2, 36.15, 36.20, 36.28, 36.44, and 36.45
Examples 36.4, 36.5, 36.9, 36.10, and 36.12

Chapter 37
Active Figures 37.2, 37.11, 37.13, 37.22
Examples 37.1 and 37.4

Chapter 38
Active Figures 38.4, 38.11, 38.17, 38.18, 38.30, and 38.31
Examples 38.1 and 38.7

Chapter 39
Active Figures 39.4, 39.6, and 39.11
Examples 39.4 and 39.9

Chapter 40
Active Figures 40.3, 40.7, 40.9, 40.10, 40.11, 40.20, 40.21, and 40.23
Examples 40.3 and 40.4

Chapter 41
Active Figures 41.4, 41.5, 41.8, and 41.10
Examples 41.2 and 41.6

Chapter 42
Active Figures 42.8, 42.9, 42.13, 42.24, 42.25, and 42.26
Examples 42.2 and 42.7

Chapter 43
Active Figures 43.3, 43.5, 43.6, 43.7, 43.8, and 43.15
Examples 43.1 and 43.2

Chapter 44
Active Figures 44.1, 44.9, 44.10, 44.11, 44.14, and 44.15
Examples 44.5 and 44.10

Chapter 45
Active Figures 45.3 and 45.14
Example 45.3

Chapter 46
Active Figure 46.12
Example 46.2

An Invitation to Physics

It is our sincere hope that you too will find physics an exciting and enjoyable experience and that you will profit from this experience, regardless of your chosen profession. Welcome to the exciting world of physics!

The scientist does not study nature because it is useful; he studies it because he delights in it, and he delights in it because it is beautiful. If nature were not beautiful, it would not be worth knowing, and if nature were not worth knowing, life would not be worth living.

—Henri Poincaré

Electricity and Magnetism

We now study the branch of physics concerned with electric and magnetic phenomena. The laws of electricity and magnetism have a central role in the operation of such devices as radios, televisions, electric motors, computers, high-energy accelerators, and other electronic devices. More fundamentally, the interatomic and intermolecular forces responsible for the formation of solids and liquids are electric in origin. Furthermore, such forces as the pushes and pulls between objects and the elastic force in a spring arise from electric forces at the atomic level.

Evidence in Chinese documents suggests that magnetism was observed as early as 2000 B.C. The ancient Greeks observed electric and magnetic phenomena possibly as early as 700 B.C. They found that a piece of amber, when rubbed, becomes electrified and attracts pieces of straw or feathers. The Greeks knew about magnetic forces from observations that the naturally occurring stone *magnetite* (Fe_3O_4) is attracted to iron. (The word *electric* comes from *elecktron*, the Greek word for "amber." The word *magnetic* comes from *Magnesia*, the name of the district of Greece where magnetite was first found.) In 1600, the Englishman William Gilbert discovered that electrification is not limited to amber but rather is a general phenomenon. In the years following this discovery, scientists electrified a variety of objects. Experiments by Charles Coulomb in 1785 confirmed the inverse-square law for electric forces.

It was not until the early part of the nineteenth century that scientists established that electricity and magnetism are related phenomena. In 1819, Hans Oersted discovered that a compass needle is deflected when placed near a circuit carrying an electric current. In 1831, Michael Faraday and, almost simultaneously, Joseph Henry showed that when a wire is moved near a magnet (or, equivalently, when a magnet is moved near a wire), an electric current is established in the wire. In 1873, James Clerk Maxwell used these observations and other experimental facts as a basis for formulating the laws of electromagnetism as we know them today. (*Electromagnetism* is a name given to the combined study of electricity and magnetism.) Shortly thereafter (around 1888), Heinrich Hertz verified Maxwell's predictions by producing electromagnetic waves in the laboratory. This achievement led to such practical developments as radio and television.

Maxwell's contributions to the field of electromagnetism were especially significant because the laws he formulated are basic to *all* forms of electromagnetic phenomena. His work is as important as Newton's work on the laws of motion and the theory of gravitation.

◀ *Lightning is a dramatic example of electrical phenomena occurring in nature. While we are most familiar with lightning originating from thunderclouds, it can occur in other situations, such as in a volcanic eruption (here, the Sakurajima volcano, Japan). (M. Zhilin/ M. Newman/Photo Researchers, Inc.)*

Chapter 23

Electric Fields

CHAPTER OUTLINE

23.1 Properties of Electric Charges

23.2 Charging Objects By Induction

23.3 Coulomb's Law

23.4 The Electric Field

23.5 Electric Field of a Continuous Charge Distribution

23.6 Electric Field Lines

23.7 Motion of Charged Particles in a Uniform Electric Field

▲ *Mother and daughter are both enjoying the effects of electrically charging their bodies. Each individual hair on their heads becomes charged and exerts a repulsive force on the other hairs, resulting in the "stand-up" hairdos that you see here. (Courtesy of Resonance Research Corporation)*

The electromagnetic force between charged particles is one of the fundamental forces of nature. We begin this chapter by describing some of the basic properties of one manifestation of the electromagnetic force, the electric force. We then discuss Coulomb's law, which is the fundamental law governing the electric force between any two charged particles. Next, we introduce the concept of an electric field associated with a charge distribution and describe its effect on other charged particles. We then show how to use Coulomb's law to calculate the electric field for a given charge distribution. We conclude the chapter with a discussion of the motion of a charged particle in a uniform electric field.

23.1   Properties of Electric Charges

A number of simple experiments demonstrate the existence of electric forces and charges. For example, after running a comb through your hair on a dry day, you will find that the comb attracts bits of paper. The attractive force is often strong enough to suspend the paper. The same effect occurs when certain materials are rubbed together, such as glass rubbed with silk or rubber with fur.

Another simple experiment is to rub an inflated balloon with wool. The balloon then adheres to a wall, often for hours. When materials behave in this way, they are said to be *electrified*, or to have become **electrically charged.** You can easily electrify your body by vigorously rubbing your shoes on a wool rug. Evidence of the electric charge on your body can be detected by lightly touching (and startling) a friend. Under the right conditions, you will see a spark when you touch, and both of you will feel a slight tingle. (Experiments such as these work best on a dry day because an excessive amount of moisture in the air can cause any charge you build up to "leak" from your body to the Earth.)

In a series of simple experiments, it was found that there are two kinds of electric charges, which were given the names **positive** and **negative** by Benjamin Franklin (1706–1790). We identify negative charge as that type possessed by electrons and positive charge as that possessed by protons. To verify that there are two types of charge, suppose a hard rubber rod that has been rubbed with fur is suspended by a sewing thread, as shown in Figure 23.1. When a glass rod that has been rubbed with silk is brought near the rubber rod, the two attract each other (Fig. 23.1a). On the other hand, if two charged rubber rods (or two charged glass rods) are brought near each other, as shown in Figure 23.1b, the two repel each other. This observation shows that the rubber and glass have two different types of charge on them. On the basis of these observations, we conclude that **charges of the same sign repel one another and charges with opposite signs attract one another.**

Using the convention suggested by Franklin, the electric charge on the glass rod is called positive and that on the rubber rod is called negative. Therefore, any charged object attracted to a charged rubber rod (or repelled by a charged glass rod) must

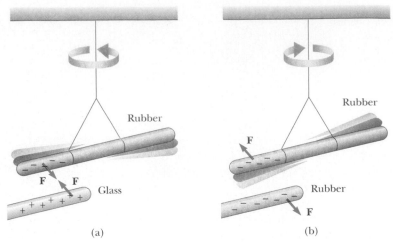

(a)   (b)

Figure 23.1 (a) A negatively charged rubber rod suspended by a thread is attracted to a positively charged glass rod. (b) A negatively charged rubber rod is repelled by another negatively charged rubber rod.

have a positive charge, and any charged object repelled by a charged rubber rod (or attracted to a charged glass rod) must have a negative charge.

Attractive electric forces are responsible for the behavior of a wide variety of commercial products. For example, the plastic in many contact lenses, *etafilcon*, is made up of molecules that electrically attract the protein molecules in human tears. These protein molecules are absorbed and held by the plastic so that the lens ends up being primarily composed of the wearer's tears. Because of this, the lens does not behave as a foreign object to the wearer's eye, and it can be worn comfortably. Many cosmetics also take advantage of electric forces by incorporating materials that are electrically attracted to skin or hair, causing the pigments or other chemicals to stay put once they are applied.

Another important aspect of electricity that arises from experimental observations is that **electric charge is always conserved** in an isolated system. That is, when one object is rubbed against another, charge is not created in the process. The electrified state is due to a *transfer* of charge from one object to the other. One object gains some amount of negative charge while the other gains an equal amount of positive charge. For example, when a glass rod is rubbed with silk, as in Figure 23.2, the silk obtains a negative charge that is equal in magnitude to the positive charge on the glass rod. We now know from our understanding of atomic structure that electrons are transferred from the glass to the silk in the rubbing process. Similarly, when rubber is rubbed with fur, electrons are transferred from the fur to the rubber, giving the rubber a net negative charge and the fur a net positive charge. This process is consistent with the fact that neutral, uncharged matter contains as many positive charges (protons within atomic nuclei) as negative charges (electrons).

In 1909, Robert Millikan (1868–1953) discovered that electric charge always occurs as some integral multiple of a fundamental amount of charge e (see Section 25.7). In modern terms, the electric charge q is said to be **quantized,** where q is the standard symbol used for charge as a variable. That is, electric charge exists as discrete "packets," and we can write $q = Ne$, where N is some integer. Other experiments in the same period showed that the electron has a charge $-e$ and the proton has a charge of equal magnitude but opposite sign $+e$. Some particles, such as the neutron, have no charge.

From our discussion thus far, we conclude that electric charge has the following important properties:

Electric charge is conserved

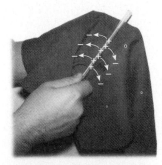

Figure 23.2 When a glass rod is rubbed with silk, electrons are transferred from the glass to the silk. Because of conservation of charge, each electron adds negative charge to the silk, and an equal positive charge is left behind on the rod. Also, because the charges are transferred in discrete bundles, the charges on the two objects are $\pm e$, or $\pm 2e$, or $\pm 3e$, and so on.

- There are two kinds of charges in nature; charges of opposite sign attract one another and charges of the same sign repel one another.
- Total charge in an isolated system is conserved.
- Charge is quantized.

Quick Quiz 23.1 If you rub an inflated balloon against your hair, the two materials attract each other, as shown in Figure 23.3. Is the amount of charge present in the system of the balloon and your hair after rubbing (a) less than, (b) the same as, or (c) more than the amount of charge present before rubbing?

Quick Quiz 23.2 Three objects are brought close to each other, two at a time. When objects A and B are brought together, they repel. When objects B and C are brought together, they also repel. Which of the following are true? (a) Objects A and C possess charges of the same sign. (b) Objects A and C possess charges of opposite sign. (c) All three of the objects possess charges of the same sign. (d) One of the objects is neutral. (e) We would need to perform additional experiments to determine the signs of the charges.

Charles D. Winters

Figure 23.3 (Quick Quiz 23.1) Rubbing a balloon against your hair on a dry day causes the balloon and your hair to become charged.

23.2 Charging Objects By Induction

It is convenient to classify materials in terms of the ability of electrons to move through the material:

Electrical **conductors** are materials in which some of the electrons are free electrons[1] that are not bound to atoms and can move relatively freely through the material; electrical **insulators** are materials in which all electrons are bound to atoms and cannot move freely through the material.

Materials such as glass, rubber, and wood fall into the category of electrical insulators. When such materials are charged by rubbing, only the area rubbed becomes charged, and the charged particles are unable to move to other regions of the material.

In contrast, materials such as copper, aluminum, and silver are good electrical conductors. When such materials are charged in some small region, the charge readily distributes itself over the entire surface of the material. If you hold a copper rod in your hand and rub it with wool or fur, it will not attract a small piece of paper. This might suggest that a metal cannot be charged. However, if you attach a wooden handle to the rod and then hold it by that handle as you rub the rod, the rod will remain charged and attract the piece of paper. The explanation for this is as follows: without the insulating wood, the electric charges produced by rubbing readily move from the copper through your body, which is also a conductor, and into the Earth. The insulating wooden handle prevents the flow of charge into your hand.

Semiconductors are a third class of materials, and their electrical properties are somewhere between those of insulators and those of conductors. Silicon and

---

[1] A metal atom contains one or more outer electrons, which are weakly bound to the nucleus. When many atoms combine to form a metal, the so-called _free electrons_ are these outer electrons, which are not bound to any one atom. These electrons move about the metal in a manner similar to that of gas molecules moving in a container.

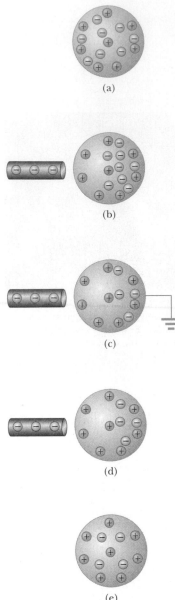

germanium are well-known examples of semiconductors commonly used in the fabrication of a variety of electronic chips used in computers, cellular telephones, and stereo systems. The electrical properties of semiconductors can be changed over many orders of magnitude by the addition of controlled amounts of certain atoms to the materials.

To understand how to charge a conductor by a process known as **induction,** consider a neutral (uncharged) conducting sphere insulated from the ground, as shown in Figure 23.4a. There are an equal number of electrons and protons in the sphere if the charge on the sphere is exactly zero. When a negatively charged rubber rod is brought near the sphere, electrons in the region nearest the rod experience a repulsive force and migrate to the opposite side of the sphere. This leaves the side of the sphere near the rod with an effective positive charge because of the diminished number of electrons, as in Figure 23.4b. (The left side of the sphere in Figure 23.4b is positively charged *as if* positive charges moved into this region, but remember that it is only electrons that are free to move.) This occurs even if the rod never actually touches the sphere. If the same experiment is performed with a conducting wire connected from the sphere to the Earth (Fig. 23.4c), some of the electrons in the conductor are so strongly repelled by the presence of the negative charge in the rod that they move out of the sphere through the wire and into the Earth. The symbol ⏚ at the end of the wire in Figure 23.4c indicates that the wire is connected to **ground,** which means a reservoir, such as the Earth, that can accept or provide electrons freely with negligible effect on its electrical characteristics. If the wire to ground is then removed (Fig. 23.4d), the conducting sphere contains an excess of *induced* positive charge because it has fewer electrons than it needs to cancel out the positive charge of the protons. When the rubber rod is removed from the vicinity of the sphere (Fig. 23.4e), this induced positive charge remains on the ungrounded sphere. Note that the rubber rod loses none of its negative charge during this process.

Charging an object by induction requires no contact with the object inducing the charge. This is in contrast to charging an object by rubbing (that is, by *conduction*), which does require contact between the two objects.

A process similar to induction in conductors takes place in insulators. In most neutral molecules, the center of positive charge coincides with the center of negative charge. However, in the presence of a charged object, these centers inside each molecule in an insulator may shift slightly, resulting in more positive charge on one side of the molecule than on the other. This realignment of charge within individual molecules produces a layer of charge on the surface of the insulator, as shown in Figure 23.5a. Knowing about induction in insulators, you should be able to explain why a comb that has been rubbed through hair attracts bits of electrically neutral paper and why a balloon that has been rubbed against your clothing is able to stick to an electrically neutral wall.

Figure 23.4 Charging a metallic object by *induction* (that is, the two objects never touch each other). (a) A neutral metallic sphere, with equal numbers of positive and negative charges. (b) The electrons on the neutral sphere are redistributed when a charged rubber rod is placed near the sphere. (c) When the sphere is grounded, some of its electrons leave through the ground wire. (d) When the ground connection is removed, the sphere has excess positive charge that is nonuniformly distributed. (e) When the rod is removed, the remaining electrons redistribute uniformly and there is a net uniform distribution of positive charge on the sphere.

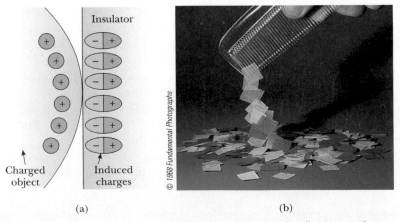

Figure 23.5 (a) The charged object on the left induces a charge distribution on the surface of an insulator due to realignment of charges in the molecules. (b) A charged comb attracts bits of paper because charges in molecules in the paper are realigned.

Quick Quiz 23.3 Three objects are brought close to each other, two at a time. When objects A and B are brought together, they attract. When objects B and C are brought together, they repel. From this, we conclude that (a) objects A and C possess charges of the same sign. (b) objects A and C possess charges of opposite sign. (c) all three of the objects possess charges of the same sign. (d) one of the objects is neutral. (e) we need to perform additional experiments to determine information about the charges on the objects.

23.3 Coulomb's Law

Charles Coulomb (1736–1806) measured the magnitudes of the electric forces between charged objects using the torsion balance, which he invented (Fig. 23.6). Coulomb confirmed that the electric force between two small charged spheres is proportional to the inverse square of their separation distance r—that is, $F_e \propto 1/r^2$. The operating principle of the torsion balance is the same as that of the apparatus used by Cavendish to measure the gravitational constant (see Section 13.2), with the electrically neutral spheres replaced by charged ones. The electric force between charged spheres A and B in Figure 23.6 causes the spheres to either attract or repel each other, and the resulting motion causes the suspended fiber to twist. Because the restoring torque of the twisted fiber is proportional to the angle through which the fiber rotates, a measurement of this angle provides a quantitative measure of the electric force of attraction or repulsion. Once the spheres are charged by rubbing, the electric force between them is very large compared with the gravitational attraction, and so the gravitational force can be neglected.

From Coulomb's experiments, we can generalize the following properties of the **electric force** between two stationary charged particles. The electric force

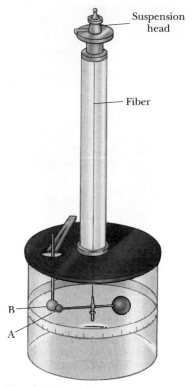

Figure 23.6 Coulomb's torsion balance, used to establish the inverse-square law for the electric force between two charges.

- is inversely proportional to the square of the separation r between the particles and directed along the line joining them;
- is proportional to the product of the charges q_1 and q_2 on the two particles;
- is attractive if the charges are of opposite sign and repulsive if the charges have the same sign;
- is a conservative force.

We will use the term **point charge** to mean a particle of zero size that carries an electric charge. The electrical behavior of electrons and protons is very well described by modeling them as point charges. From experimental observations on the electric force, we can express **Coulomb's law** as an equation giving the magnitude of the electric force (sometimes called the *Coulomb force*) between two point charges:

$$F_e = k_e \frac{|q_1||q_2|}{r^2}$$
(23.1)   **Coulomb's law**

where k_e is a constant called the **Coulomb constant.** In his experiments, Coulomb was able to show that the value of the exponent of r was 2 to within an uncertainty of a few percent. Modern experiments have shown that the exponent is 2 to within an uncertainty of a few parts in 10^{16}.

The value of the Coulomb constant depends on the choice of units. The SI unit of charge is the **coulomb** (C). The Coulomb constant k_e in SI units has the value

$$k_e = 8.987\,5 \times 10^9 \ \text{N} \cdot \text{m}^2/\text{C}^2$$
(23.2)   **Coulomb constant**

This constant is also written in the form

$$k_e = \frac{1}{4\pi\epsilon_0}$$
(23.3)

Table 23.1

Charge and Mass of the Electron, Proton, and Neutron		
Particle	Charge (C)	Mass (kg)
Electron (e)	$-1.602\ 191\ 7 \times 10^{-19}$	$9.109\ 5 \times 10^{-31}$
Proton (p)	$+1.602\ 191\ 7 \times 10^{-19}$	$1.672\ 61 \times 10^{-27}$
Neutron (n)	0	$1.674\ 92 \times 10^{-27}$

Charles Coulomb

French physicist (1736–1806)

Coulomb's major contributions to science were in the areas of electrostatics and magnetism. During his lifetime, he also investigated the strengths of materials and determined the forces that affect objects on beams, thereby contributing to the field of structural mechanics. In the field of ergonomics, his research provided a fundamental understanding of the ways in which people and animals can best do work. *(Photo courtesy of AIP Niels Bohr Library/E. Scott Barr Collection)*

where the constant ϵ_0 (lowercase Greek epsilon) is known as the **permittivity of free space** and has the value

$$\epsilon_0 = 8.854\ 2 \times 10^{-12}\ \text{C}^2/\text{N} \cdot \text{m}^2 \tag{23.4}$$

The smallest unit of charge e known in nature[2] is the charge on an electron $(-e)$ or a proton $(+e)$ and has a magnitude

$$e = 1.602\ 19 \times 10^{-19}\ \text{C} \tag{23.5}$$

Therefore, 1 C of charge is approximately equal to the charge of 6.24×10^{18} electrons or protons. This number is very small when compared with the number of free electrons in 1 cm^3 of copper, which is on the order of 10^{23}. Still, 1 C is a substantial amount of charge. In typical experiments in which a rubber or glass rod is charged by friction, a net charge on the order of 10^{-6} C is obtained. In other words, only a very small fraction of the total available charge is transferred between the rod and the rubbing material.

The charges and masses of the electron, proton, and neutron are given in Table 23.1.

Quick Quiz 23.4 Object A has a charge of $+2\ \mu\text{C}$, and object B has a charge of $+6\ \mu\text{C}$. Which statement is true about the electric forces on the objects? (a) $F_{AB} = -3F_{BA}$ (b) $F_{AB} = -F_{BA}$ (c) $3F_{AB} = -F_{BA}$ (d) $F_{AB} = 3F_{BA}$ (e) $F_{AB} = F_{BA}$ (f) $3F_{AB} = F_{BA}$

Example 23.1   The Hydrogen Atom

The electron and proton of a hydrogen atom are separated (on the average) by a distance of approximately 5.3×10^{-11} m. Find the magnitudes of the electric force and the gravitational force between the two particles.

Solution From Coulomb's law, we find that the magnitude of the electric force is

$$F_e = k_e \frac{|e||-e|}{r^2} = (8.99 \times 10^9\ \text{N} \cdot \text{m}^2/\text{C}^2) \frac{(1.60 \times 10^{-19}\ \text{C})^2}{(5.3 \times 10^{-11}\ \text{m})^2}$$

$$= 8.2 \times 10^{-8}\ \text{N}$$

Using Newton's law of universal gravitation and Table 23.1 for the particle masses, we find that the magnitude of the

gravitational force is

$$F_g = G \frac{m_e m_p}{r^2}$$

$$= (6.67 \times 10^{-11}\ \text{N} \cdot \text{m}^2/\text{kg}^2)$$

$$\times \frac{(9.11 \times 10^{-31}\ \text{kg})(1.67 \times 10^{-27}\ \text{kg})}{(5.3 \times 10^{-11}\ \text{m})^2}$$

$$= 3.6 \times 10^{-47}\ \text{N}$$

The ratio $F_e/F_g \approx 2 \times 10^{39}$. Thus, the gravitational force between charged atomic particles is negligible when compared with the electric force. Note the similarity of form of Newton's law of universal gravitation and Coulomb's law of electric forces. Other than magnitude, what is a fundamental difference between the two forces?

---

[2]   No unit of charge smaller than e has been detected on a free particle; however, current theories propose the existence of particles called *quarks* having charges $-e/3$ and $2e/3$. Although there is considerable experimental evidence for such particles inside nuclear matter, *free* quarks have never been detected. We discuss other properties of quarks in Chapter 46 of the extended version of this text.

When dealing with Coulomb's law, you must remember that force is a vector quantity and must be treated accordingly. The law expressed in vector form for the electric force exerted by a charge q_1 on a second charge q_2, written $\mathbf{F}_{12}$, is

$$\mathbf{F}_{12} = k_e \frac{q_1 q_2}{r^2} \hat{\mathbf{r}} \qquad (23.6)$$

where $\hat{\mathbf{r}}$ is a unit vector directed from q_1 toward q_2, as shown in Figure 23.7a. Because the electric force obeys Newton's third law, the electric force exerted by q_2 on q_1 is equal in magnitude to the force exerted by q_1 on q_2 and in the opposite direction; that is, $\mathbf{F}_{21} = -\mathbf{F}_{12}$. Finally, from Equation 23.6, we see that if q_1 and q_2 have the same sign, as in Figure 23.7a, the product $q_1 q_2$ is positive. If q_1 and q_2 are of opposite sign, as shown in Figure 23.7b, the product $q_1 q_2$ is negative. These signs describe the *relative* direction of the force but not the *absolute* direction. A negative product indicates an attractive force, so that the charges each experience a force toward the other—thus, the force on one charge is in a direction *relative* to the other. A positive product indicates a repulsive force such that each charge experiences a force away from the other. The *absolute* direction of the force in space is not determined solely by the sign of $q_1 q_2$—whether the force on an individual charge is in the positive or negative direction on a coordinate axis depends on the location of the other charge. For example, if an x axis lies along the two charges in Figure 23.7a, the product $q_1 q_2$ is positive, but $\mathbf{F}_{12}$ points in the $+x$ direction and $\mathbf{F}_{21}$ points in the $-x$ direction.

Vector form of Coulomb's law

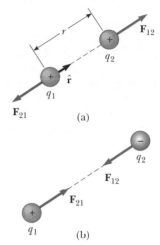

Active Figure 23.7 Two point charges separated by a distance r exert a force on each other that is given by Coulomb's law. The force $\mathbf{F}_{21}$ exerted by q_2 on q_1 is equal in magnitude and opposite in direction to the force $\mathbf{F}_{12}$ exerted by q_1 on q_2. (a) When the charges are of the same sign, the force is repulsive. (b) When the charges are of opposite signs, the force is attractive.

At the Active Figures link at http://www.pse6.com, you can move the charges to any position in two-dimensional space and observe the electric forces on them.

Quick Quiz 23.5 Object A has a charge of $+2\ \mu C$, and object B has a charge of $+6\ \mu C$. Which statement is true about the electric forces on the objects? (a) $\mathbf{F}_{AB} = -3\mathbf{F}_{BA}$ (b) $\mathbf{F}_{AB} = -\mathbf{F}_{BA}$ (c) $3\mathbf{F}_{AB} = -\mathbf{F}_{BA}$ (d) $\mathbf{F}_{AB} = 3\mathbf{F}_{BA}$ (e) $\mathbf{F}_{AB} = \mathbf{F}_{BA}$ (f) $3\mathbf{F}_{AB} = \mathbf{F}_{BA}$

When more than two charges are present, the force between any pair of them is given by Equation 23.6. Therefore, the resultant force on any one of them equals the vector sum of the forces exerted by the various individual charges. For example, if four charges are present, then the resultant force exerted by particles 2, 3, and 4 on particle 1 is

$$\mathbf{F}_1 = \mathbf{F}_{21} + \mathbf{F}_{31} + \mathbf{F}_{41}$$

Example 23.2 Find the Resultant Force

Consider three point charges located at the corners of a right triangle as shown in Figure 23.8, where $q_1 = q_3 = 5.0\ \mu C$, $q_2 = -2.0\ \mu C$, and $a = 0.10$ m. Find the resultant force exerted on q_3.

Solution First, note the direction of the individual forces exerted by q_1 and q_2 on q_3. The force $\mathbf{F}_{23}$ exerted by q_2 on q_3 is attractive because q_2 and q_3 have opposite signs. The force $\mathbf{F}_{13}$ exerted by q_1 on q_3 is repulsive because both charges are positive.

The magnitude of $\mathbf{F}_{23}$ is

$$F_{23} = k_e \frac{|q_2||q_3|}{a^2}$$

$$= (8.99 \times 10^9\ \text{N} \cdot \text{m}^2/\text{C}^2) \frac{(2.0 \times 10^{-6}\ \text{C})(5.0 \times 10^{-6}\ \text{C})}{(0.10\ \text{m})^2}$$

$$= 9.0\ \text{N}$$

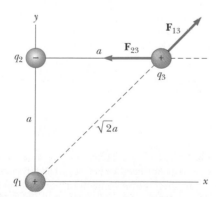

Figure 23.8 (Example 23.2) The force exerted by q_1 on q_3 is $\mathbf{F}_{13}$. The force exerted by q_2 on q_3 is $\mathbf{F}_{23}$. The resultant force $\mathbf{F}_3$ exerted on q_3 is the vector sum $\mathbf{F}_{13} + \mathbf{F}_{23}$.

In the coordinate system shown in Figure 23.8, the attractive force $\mathbf{F}_{23}$ is to the left (in the negative x direction).

The magnitude of the force $\mathbf{F}_{13}$ exerted by q_1 on q_3 is

$$F_{13} = k_e \frac{|q_1||q_3|}{(\sqrt{2}a)^2}$$

$$= (8.99 \times 10^9 \text{ N} \cdot \text{m}^2/\text{C}^2) \frac{(5.0 \times 10^{-6} \text{ C})(5.0 \times 10^{-6} \text{ C})}{2(0.10 \text{ m})^2}$$

$$= 11 \text{ N}$$

The repulsive force $\mathbf{F}_{13}$ makes an angle of $45°$ with the x axis. Therefore, the x and y components of $\mathbf{F}_{13}$ are equal, with magnitude given by $F_{13} \cos 45° = 7.9$ N.

Combining $\mathbf{F}_{13}$ with $\mathbf{F}_{23}$ by the rules of vector addition, we arrive at the x and y components of the resultant force acting on q_3:

$$F_{3x} = F_{13x} + F_{23x} = 7.9 \text{ N} + (-9.0 \text{ N}) = -1.1 \text{ N}$$

$$F_{3y} = F_{13y} + F_{23y} = 7.9 \text{ N} + 0 = 7.9 \text{ N}$$

We can also express the resultant force acting on q_3 in unit-vector form as

$$\mathbf{F}_3 = \boxed{(-1.1\hat{\mathbf{i}} + 7.9\hat{\mathbf{j}}) \text{ N}}$$

What If? What if the signs of all three charges were changed to the opposite signs? How would this affect the result for $\mathbf{F}_3$?

Answer The charge q_3 would still be attracted toward q_2 and repelled from q_1 with forces of the same magnitude. Thus, the final result for $\mathbf{F}_3$ would be exactly the same.

Example 23.3   Where Is the Resultant Force Zero?   Interactive

Three point charges lie along the x axis as shown in Figure 23.9. The positive charge $q_1 = 15.0$ μC is at $x = 2.00$ m, the positive charge $q_2 = 6.00$ μC is at the origin, and the resultant force acting on q_3 is zero. What is the x coordinate of q_3?

Solution Because q_3 is negative and q_1 and q_2 are positive, the forces $\mathbf{F}_{13}$ and $\mathbf{F}_{23}$ are both attractive, as indicated in Figure 23.9. From Coulomb's law, $\mathbf{F}_{13}$ and $\mathbf{F}_{23}$ have magnitudes

$$F_{13} = k_e \frac{|q_1||q_3|}{(2.00 - x)^2} \qquad F_{23} = k_e \frac{|q_2||q_3|}{x^2}$$

For the resultant force on q_3 to be zero, $\mathbf{F}_{23}$ must be equal in magnitude and opposite in direction to $\mathbf{F}_{13}$. Setting the magnitudes of the two forces equal, we have

$$k_e \frac{|q_2||q_3|}{x^2} = k_e \frac{|q_1||q_3|}{(2.00 - x)^2}$$

Noting that k_e and $|q_3|$ are common to both sides and so can be dropped, we solve for x and find that

$$(2.00 - x)^2|q_2| = x^2|q_1|$$

$$(4.00 - 4.00x + x^2)(6.00 \times 10^{-6} \text{ C}) = x^2(15.0 \times 10^{-6} \text{ C})$$

This can be reduced to the following quadratic equation:

$$3.00x^2 + 8.00x - 8.00 = 0$$

Solving this quadratic equation for x, we find that the positive root is $x = \boxed{0.775 \text{ m}}$. There is also a second root, $x = -3.44$ m. This is another location at which the magnitudes

of the forces on q_3 are equal, but both forces are in the same direction at this location.

What If? Suppose charge q_3 is constrained to move only along the x axis. From its initial position at $x = 0.775$ m, it is pulled a very small distance along the x axis. When released, will it return to equilibrium or be pulled further from equilibrium? That is, is the equilibrium stable or unstable?

Answer If the charge is moved to the right, $\mathbf{F}_{13}$ becomes larger and $\mathbf{F}_{23}$ becomes smaller. This results in a net force to the right, in the same direction as the displacement. Thus, the equilibrium is *unstable*.

Note that if the charge is constrained to stay at a *fixed* x coordinate but allowed to move up and down in Figure 23.9, the equilibrium is stable. In this case, if the charge is pulled upward (or downward) and released, it will move back toward the equilibrium position and undergo oscillation.

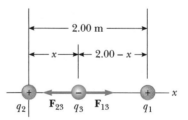

Figure 23.9 (Example 23.3) Three point charges are placed along the x axis. If the resultant force acting on q_3 is zero, then the force $\mathbf{F}_{13}$ exerted by q_1 on q_3 must be equal in magnitude and opposite in direction to the force $\mathbf{F}_{23}$ exerted by q_2 on q_3.

 At the Interactive Worked Example link at **http://www.pse6.com,** *you can predict where on the x axis the electric force is zero for random values of q₁ and q₂.*

Example 23.4   Find the Charge on the Spheres

Two identical small charged spheres, each having a mass of 3.0×10^{-2} kg, hang in equilibrium as shown in Figure 23.10a. The length of each string is 0.15 m, and the angle θ is $5.0°$. Find the magnitude of the charge on each sphere.

Solution Figure 23.10a helps us conceptualize this problem—the two spheres exert repulsive forces on each other. If they are held close to each other and released, they will move outward from the center and settle into the configuration in Figure 23.10a after the damped oscillations

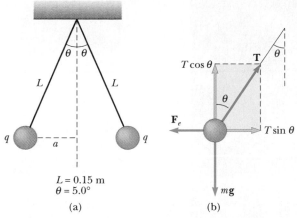

Figure 23.10 (Example 23.4) (a) Two identical spheres, each carrying the same charge q, suspended in equilibrium. (b) The free-body diagram for the sphere on the left.

due to air resistance have vanished. The key phrase "in equilibrium" helps us categorize this as an equilibrium problem, which we approach as we did equilibrium problems in Chapter 5 with the added feature that one of the forces on a sphere is an electric force. We analyze this problem by drawing the free-body diagram for the left-hand sphere in Figure 23.10b. The sphere is in equilibrium under the application of the forces **T** from the string, the electric force $\mathbf{F}_e$ from the other sphere, and the gravitational force $m\mathbf{g}$.

Because the sphere is in equilibrium, the forces in the horizontal and vertical directions must separately add up to zero:

$$(1) \qquad \sum F_x = T \sin\theta - F_e = 0$$

$$(2) \qquad \sum F_y = T \cos\theta - mg = 0$$

From Equation (2), we see that $T - mg/\cos\theta$; thus, T can be eliminated from Equation (1) if we make this substitution. This gives a value for the magnitude of the electric force F_e:

$$F_e = mg \tan\theta = (3.0 \times 10^{-2}\,\text{kg})(9.80\,\text{m/s}^2)\tan(5.0°)$$

$$= 2.6 \times 10^{-2}\,\text{N}$$

Considering the geometry of the right triangle in Figure 23.10a, we see that $\sin\theta = a/L$. Therefore,

$$a = L \sin\theta = (0.15\,\text{m})\sin(5.0°) = 0.013\,\text{m}$$

The separation of the spheres is $2a = 0.026$ m.

From Coulomb's law (Eq. 23.1), the magnitude of the electric force is

$$F_e = k_e \frac{|q|^2}{r^2}$$

where $r = 2a = 0.026$ m and $|q|$ is the magnitude of the charge on each sphere. (Note that the term $|q|^2$ arises here because the charge is the same on both spheres.) This equation can be solved for $|q|^2$ to give

$$|q|^2 = \frac{F_e r^2}{k_e} = \frac{(2.6 \times 10^{-2}\,\text{N})(0.026\,\text{m})^2}{8.99 \times 10^9\,\text{N}\cdot\text{m}^2/\text{C}^2} = 1.96 \times 10^{-15}\,\text{C}^2$$

$$|q| = \quad 4.4 \times 10^{-8}\,\text{C}$$

To finalize the problem, note that we found only the magnitude of the charge $|q|$ on the spheres. There is no way we could find the sign of the charge from the information given. In fact, the sign of the charge is not important. The situation will be exactly the same whether both spheres are positively charged or negatively charged.

What If? Suppose your roommate proposes solving this problem without the assumption that the charges are of equal magnitude. She claims that the symmetry of the problem is destroyed if the charges are not equal, so that the strings would make two different angles with the vertical, and the problem would be much more complicated. How would you respond?

Answer You should argue that the symmetry is not destroyed and the angles remain the same. Newton's third law requires that the electric forces on the two charges be the same, regardless of the equality or nonequality of the charges. The solution to the example remains the same through the calculation of $|q|^2$. In this situation, the value of 1.96×10^{-15} C^2 corresponds to the product $q_1 q_2$, where q_1 and q_2 are the values of the charges on the two spheres. The symmetry of the problem would be destroyed if the *masses* of the spheres were not the same. In this case, the strings would make different angles with the vertical and the problem would be more complicated.

23.4 The Electric Field

Two field forces have been introduced into our discussions so far—the gravitational force in Chapter 13 and the electric force here. As pointed out earlier, field forces can act through space, producing an effect even when no physical contact occurs between interacting objects. The gravitational field **g** at a point in space was defined in Section 13.5 to be equal to the gravitational force $\mathbf{F}_g$ acting on a test particle of mass m divided by that mass: $\mathbf{g} \equiv \mathbf{F}_g/m$. The concept of a field was developed by Michael Faraday (1791–1867) in the context of electric forces and is of such practical value that we shall devote much attention to it in the next several chapters. In this approach, an **electric field** is said to exist in the region of space around a charged object—the **source charge.** When another charged object—the **test charge**—enters this electric field, an

This dramatic photograph captures a lightning bolt striking a tree near some rural homes. Lightning is associated with very strong electric fields in the atmosphere.

©Johnny Autery

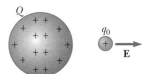

Figure 23.11 A small positive test charge q_0 placed near an object carrying a much larger positive charge Q experiences an electric field **E** directed as shown.

electric force acts on it. As an example, consider Figure 23.11, which shows a small positive test charge q_0 placed near a second object carrying a much greater positive charge Q. We define the electric field due to the source charge at the location of the test charge to be the electric force on the test charge *per unit charge*, or to be more specific

Definition of electric field

> **the electric field vector E** at a point in space is defined as the electric force $\mathbf{F}_e$ acting on a positive test charge q_0 placed at that point divided by the test charge:
>
> $$\mathbf{E} \equiv \frac{\mathbf{F}_e}{q_0} \tag{23.7}$$

▲ **PITFALL PREVENTION**

23.1 Particles Only

Equation 23.8 is only valid for a charged *particle*—an object of zero size. For a charged object of finite size in an electric field, the field may vary in magnitude and direction over the size of the object, so the corresponding force equation may be more complicated.

Note that **E** is the field produced by some charge or charge distribution *separate from* the test charge—it is not the field produced by the test charge itself. Also, note that the existence of an electric field is a property of its source—the presence of the test charge is not necessary for the field to exist. The test charge serves as a *detector* of the electric field.

Equation 23.7 can be rearranged as

$$\mathbf{F}_e = q\mathbf{E} \tag{23.8}$$

where we have used the general symbol q for a charge. This equation gives us the force on a charged particle placed in an electric field. If q is positive, the force is in the same

Table 23.2

Typical Electric Field Values	
Source	**E (N/C)**
Fluorescent lighting tube	10
Atmosphere (fair weather)	100
Balloon rubbed on hair	1 000
Atmosphere (under thundercloud)	10 000
Photocopier	100 000
Spark in air	>3 000 000
Near electron in hydrogen atom	5×10^{11}

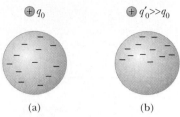

Figure 23.12 (a) For a small enough test charge q_0, the charge distribution on the sphere is undisturbed. (b) When the test charge q_0' is greater, the charge distribution on the sphere is disturbed as the result of the proximity of q_0'.

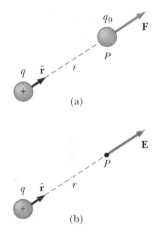

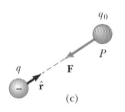

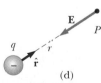

Active Figure 23.13 A test charge q_0 at point P is a distance r from a point charge q. (a) If q is positive, then the force on the test charge is directed away from q. (b) For the positive source charge, the electric field at P points radially outward from q. (c) If q is negative, then the force on the test charge is directed toward q. (d) For the negative source charge, the electric field at P points radially inward toward q.

At the Active Figures link at http://www.pse6.com, you can move point P to any position in two-dimensional space and observe the electric field due to q.

direction as the field. If q is negative, the force and the field are in opposite directions. Notice the similarity between Equation 23.8 and the corresponding equation for a particle with mass placed in a gravitational field, $\mathbf{F}_g = m\mathbf{g}$ (Eq. 5.6).

The vector $\mathbf{E}$ has the SI units of newtons per coulomb (N/C). The direction of $\mathbf{E}$, as shown in Figure 23.11, is the direction of the force a positive test charge experiences when placed in the field. We say that **an electric field exists at a point if a test charge at that point experiences an electric force.** Once the magnitude and direction of the electric field are known at some point, the electric force exerted on *any* charged particle placed at that point can be calculated from Equation 23.8. The electric field magnitudes for various field sources are given in Table 23.2.

When using Equation 23.7, we must assume that the test charge q_0 is small enough that it does not disturb the charge distribution responsible for the electric field. If a vanishingly small test charge q_0 is placed near a uniformly charged metallic sphere, as in Figure 23.12a, the charge on the metallic sphere, which produces the electric field, remains uniformly distributed. If the test charge is great enough ($q_0' \gg q_0$), as in Figure 23.12b, the charge on the metallic sphere is redistributed and the ratio of the force to the test charge is different: ($F_e'/q_0' \neq F_e/q_0$). That is, because of this redistribution of charge on the metallic sphere, the electric field it sets up is different from the field it sets up in the presence of the much smaller test charge q_0.

To determine the direction of an electric field, consider a point charge q as a source charge. This charge creates an electric field at all points in space surrounding it. A test charge q_0 is placed at point P, a distance r from the source charge, as in Figure 23.13a. We imagine using the test charge to determine the direction of the electric force and therefore that of the electric field. However, the electric field does not depend on the existence of the test charge—it is established solely by the source charge. According to Coulomb's law, the force exerted by q on the test charge is

$$\mathbf{F}_e = k_e \frac{qq_0}{r^2} \hat{\mathbf{r}}$$

where $\hat{\mathbf{r}}$ is a unit vector directed from q toward q_0. This force in Figure 23.13a is directed away from the source charge q. Because the electric field at P, the position of the test charge, is defined by $\mathbf{E} = \mathbf{F}_e/q_0$, we find that at P, the electric field created by q is

$$\mathbf{E} = k_e \frac{q}{r^2} \hat{\mathbf{r}} \tag{23.9}$$

If the source charge q is positive, Figure 23.13b shows the situation with the test charge removed—the source charge sets up an electric field at point P, directed away from q. If q is negative, as in Figure 23.13c, the force on the test charge is toward the source charge, so the electric field at P is directed toward the source charge, as in Figure 23.13d.

To calculate the electric field at a point P due to a group of point charges, we first calculate the electric field vectors at P individually using Equation 23.9 and then add them vectorially. In other words,

> at any point P, the total electric field due to a group of source charges equals the vector sum of the electric fields of all the charges.

This superposition principle applied to fields follows directly from the superposition property of electric forces, which, in turn, follows from the fact that we know that forces add as vectors from Chapter 5. Thus, the electric field at point P due to a group of source charges can be expressed as the vector sum

Electric field due to a finite number of point charges

$$\mathbf{E} = k_e \sum_i \frac{q_i}{r_i^2} \hat{\mathbf{r}}_i \qquad (23.10)$$

where r_i is the distance from the ith source charge q_i to the point P and $\hat{\mathbf{r}}_i$ is a unit vector directed from q_i toward P.

Quick Quiz 23.6   A test charge of $+3\ \mu C$ is at a point P where an external electric field is directed to the right and has a magnitude of 4×10^6 N/C. If the test charge is replaced with another test charge of $-3\ \mu C$, the external electric field at P (a) is unaffected (b) reverses direction (c) changes in a way that cannot be determined

---

Example 23.5   Electric Field Due to Two Charges

A charge $q_1 = 7.0\ \mu C$ is located at the origin, and a second charge $q_2 = -5.0\ \mu C$ is located on the x axis, 0.30 m from the origin (Fig. 23.14). Find the electric field at the point P, which has coordinates $(0, 0.40)$ m.

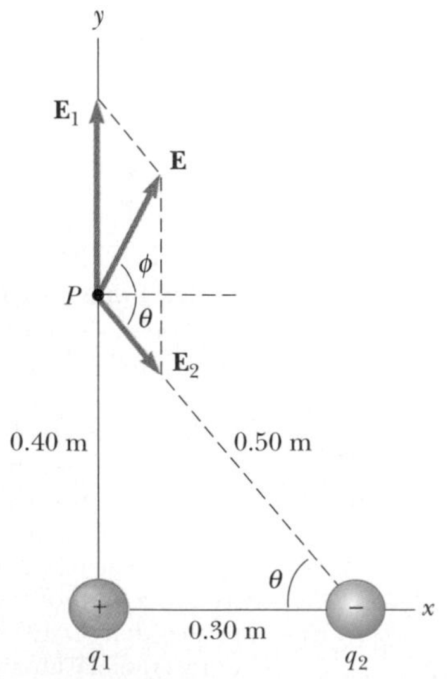

Figure 23.14 (Example 23.5) The total electric field $\mathbf{E}$ at P equals the vector sum $\mathbf{E}_1 + \mathbf{E}_2$, where $\mathbf{E}_1$ is the field due to the positive charge q_1 and $\mathbf{E}_2$ is the field due to the negative charge q_2.

Solution   First, let us find the magnitude of the electric field at P due to each charge. The fields $\mathbf{E}_1$ due to the 7.0-μC charge and $\mathbf{E}_2$ due to the -5.0-μC charge are shown in Figure 23.14. Their magnitudes are

$$E_1 = k_e \frac{|q_1|}{r_1^2} = (8.99 \times 10^9\ \text{N·m}^2/\text{C}^2) \frac{(7.0 \times 10^{-6}\ \text{C})}{(0.40\ \text{m})^2}$$
$$= 3.9 \times 10^5\ \text{N/C}$$

$$E_2 = k_e \frac{|q_2|}{r_2^2} = (8.99 \times 10^9\ \text{N·m}^2/\text{C}^2) \frac{(5.0 \times 10^{-6}\ \text{C})}{(0.50\ \text{m})^2}$$
$$= 1.8 \times 10^5\ \text{N/C}$$

The vector $\mathbf{E}_1$ has only a y component. The vector $\mathbf{E}_2$ has an x component given by $E_2 \cos\theta = \frac{3}{5}E_2$ and a negative y component given by $-E_2 \sin\theta = -\frac{4}{5}E_2$. Hence, we can express the vectors as

$$\mathbf{E}_1 = 3.9 \times 10^5 \hat{\mathbf{j}}\ \text{N/C}$$

$$\mathbf{E}_2 = (1.1 \times 10^5 \hat{\mathbf{i}} - 1.4 \times 10^5 \hat{\mathbf{j}})\ \text{N/C}$$

The resultant field $\mathbf{E}$ at P is the superposition of $\mathbf{E}_1$ and $\mathbf{E}_2$:

$$\mathbf{E} = \mathbf{E}_1 + \mathbf{E}_2 = \boxed{(1.1 \times 10^5 \hat{\mathbf{i}} + 2.5 \times 10^5 \hat{\mathbf{j}})\ \text{N/C}}$$

From this result, we find that $\mathbf{E}$ makes an angle ϕ of 66° with the positive x axis and has a magnitude of 2.7×10^5 N/C.

Example 23.6   Electric Field of a Dipole

An **electric dipole** is defined as a positive charge q and a negative charge $-q$ separated by a distance $2a$. For the dipole shown in Figure 23.15, find the electric field $\mathbf{E}$ at P due to the dipole, where P is a distance $y \gg a$ from the origin.

Solution At P, the fields $\mathbf{E}_1$ and $\mathbf{E}_2$ due to the two charges are equal in magnitude because P is equidistant from the charges. The total field is $\mathbf{E} = \mathbf{E}_1 + \mathbf{E}_2$, where

$$E_1 = E_2 = k_e \frac{q}{r^2} = k_e \frac{q}{y^2 + a^2}$$

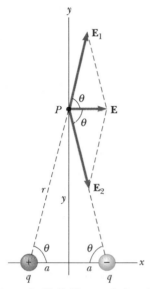

Figure 23.15 (Example 23.6) The total electric field $\mathbf{E}$ at P due to two charges of equal magnitude and opposite sign (an electric dipole) equals the vector sum $\mathbf{E}_1 + \mathbf{E}_2$. The field $\mathbf{E}_1$ is due to the positive charge q, and $\mathbf{E}_2$ is the field due to the negative charge $-q$.

The y components of $\mathbf{E}_1$ and $\mathbf{E}_2$ cancel each other, and the x components are both in the positive x direction and have the same magnitude. Therefore, $\mathbf{E}$ is parallel to the x axis and has a magnitude equal to $2E_1 \cos \theta$. From Figure 23.15 we see that $\cos \theta = a/r = a/(y^2 + a^2)^{1/2}$. Therefore,

$$E = 2E_1 \cos \theta = 2k_e \frac{q}{(y^2 + a^2)} \frac{a}{(y^2 + a^2)^{1/2}}$$

$$= k_e \frac{2qa}{(y^2 + a^2)^{3/2}}$$

Because $y \gg a$, we can neglect a^2 compared to y^2 and write

$$E \approx k_e \frac{2qa}{y^3}$$

Thus, we see that, at distances far from a dipole but along the perpendicular bisector of the line joining the two charges, the magnitude of the electric field created by the dipole varies as $1/r^3$, whereas the more slowly varying field of a point charge varies as $1/r^2$ (see Eq. 23.9). This is because at distant points, the fields of the two charges of equal magnitude and opposite sign almost cancel each other. The $1/r^3$ variation in E for the dipole also is obtained for a distant point along the x axis (see Problem 22) and for any general distant point.

The electric dipole is a good model of many molecules, such as hydrochloric acid (HCl). Neutral atoms and molecules behave as dipoles when placed in an external electric field. Furthermore, many molecules, such as HCl, are permanent dipoles. The effect of such dipoles on the behavior of materials subjected to electric fields is discussed in Chapter 26.

23.5 Electric Field of a Continuous Charge Distribution

Very often the distances between charges in a group of charges are much smaller than the distance from the group to some point of interest (for example, a point where the electric field is to be calculated). In such situations, the system of charges can be modeled as continuous. That is, the system of closely spaced charges is equivalent to a total charge that is continuously distributed along some line, over some surface, or throughout some volume.

To evaluate the electric field created by a continuous charge distribution, we use the following procedure: first, we divide the charge distribution into small elements, each of which contains a small charge Δq, as shown in Figure 23.16. Next, we use Equation 23.9 to calculate the electric field due to one of these elements at a point P. Finally, we evaluate the total electric field at P due to the charge distribution by summing the contributions of all the charge elements (that is, by applying the superposition principle).

The electric field at P due to one charge element carrying charge Δq is

$$\Delta \mathbf{E} = k_e \frac{\Delta q}{r^2} \hat{\mathbf{r}}$$

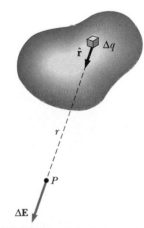

Figure 23.16 The electric field at P due to a continuous charge distribution is the vector sum of the fields $\Delta \mathbf{E}$ due to all the elements Δq of the charge distribution.

where r is the distance from the charge element to point P and $\hat{\mathbf{r}}$ is a unit vector directed from the element toward P. The total electric field at P due to all elements in the charge distribution is approximately

$$\mathbf{E} \approx k_e \sum_i \frac{\Delta q_i}{r_i^2} \hat{\mathbf{r}}_i$$

where the index i refers to the ith element in the distribution. Because the charge distribution is modeled as continuous, the total field at P in the limit $\Delta q_i \to 0$ is

Electric field due to a continuous charge distribution

$$\mathbf{E} = k_e \lim_{\Delta q_i \to 0} \sum_i \frac{\Delta q_i}{r_i^2} \hat{\mathbf{r}}_i = k_e \int \frac{dq}{r^2} \hat{\mathbf{r}} \qquad (23.11)$$

where the integration is over the entire charge distribution. This is a vector operation and must be treated appropriately.

We illustrate this type of calculation with several examples, in which we assume the charge is uniformly distributed on a line, on a surface, or throughout a volume. When performing such calculations, it is convenient to use the concept of a *charge density* along with the following notations:

- If a charge Q is uniformly distributed throughout a volume V, the **volume charge density** ρ is defined by

Volume charge density

$$\rho \equiv \frac{Q}{V}$$

where ρ has units of coulombs per cubic meter (C/m^3).

- If a charge Q is uniformly distributed on a surface of area A, the **surface charge density** σ (lowercase Greek sigma) is defined by

Surface charge density

$$\sigma \equiv \frac{Q}{A}$$

where σ has units of coulombs per square meter (C/m^2).

- If a charge Q is uniformly distributed along a line of length ℓ, the **linear charge density** λ is defined by

Linear charge density

$$\lambda \equiv \frac{Q}{\ell}$$

where λ has units of coulombs per meter (C/m).

- If the charge is nonuniformly distributed over a volume, surface, or line, the amounts of charge dq in a small volume, surface, or length element are

$$dq = \rho \, dV \qquad dq = \sigma \, dA \qquad dq = \lambda \, d\ell$$

PROBLEM-SOLVING HINTS

Finding the Electric Field

- **Units:** in calculations using the Coulomb constant k_e $(= 1/4\pi\epsilon_0)$, charges must be expressed in coulombs and distances in meters.
- **Calculating the electric field of point charges:** to find the total electric field at a given point, first calculate the electric field at the point due to each individual charge. The resultant field at the point is the vector sum of the fields due to the individual charges.
- **Continuous charge distributions:** when you are confronted with problems that involve a continuous distribution of charge, the vector sums for evaluating the

total electric field at some point must be replaced by vector integrals. Divide the charge distribution into infinitesimal pieces, and calculate the vector sum by integrating over the entire charge distribution. Examples 23.7 through 23.9 demonstrate this technique.

- **Symmetry:** with both distributions of point charges and continuous charge distributions, take advantage of any symmetry in the system to simplify your calculations.

Example 23.7   The Electric Field Due to a Charged Rod

A rod of length ℓ has a uniform positive charge per unit length λ and a total charge Q. Calculate the electric field at a point P that is located along the long axis of the rod and a distance a from one end (Fig. 23.17).

Solution Let us assume that the rod is lying along the x axis, that dx is the length of one small segment, and that dq is the charge on that segment. Because the rod has a charge per unit length λ, the charge dq on the small segment is $dq = \lambda\, dx$.

The field $d\mathbf{E}$ at P due to this segment is in the negative x direction (because the source of the field carries a positive charge), and its magnitude is

$$dE = k_e \frac{dq}{x^2} = k_e \frac{\lambda\, dx}{x^2}$$

Because every other element also produces a field in the negative x direction, the problem of summing their contributions

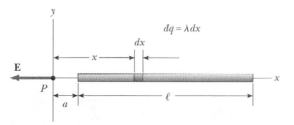

Figure 23.17 (Example 23.7) The electric field at P due to a uniformly charged rod lying along the x axis. The magnitude of the field at P due to the segment of charge dq is $k_e dq/x^2$. The total field at P is the vector sum over all segments of the rod.

is particularly simple in this case. The total field at P due to all segments of the rod, which are at different distances from P, is given by Equation 23.11, which in this case becomes[3]

$$E = \int_a^{\ell + a} k_e \lambda \frac{dx}{x^2}$$

where the limits on the integral extend from one end of the rod ($x = a$) to the other ($x = \ell + a$). The constants k_e and λ can be removed from the integral to yield

$$E = k_e \lambda \int_a^{\ell + a} \frac{dx}{x^2} = k_e \lambda \left[-\frac{1}{x} \right]_a^{\ell + a}$$

$$= k_e \lambda \left(\frac{1}{a} - \frac{1}{\ell + a} \right) = \boxed{\frac{k_e Q}{a(\ell + a)}}$$

where we have used the fact that the total charge $Q = \lambda \ell$.

What If? Suppose we move to a point P very far away from the rod. What is the nature of the electric field at such a point?

Answer If P is far from the rod ($a \gg \ell$), then ℓ in the denominator of the final expression for E can be neglected, and $E \approx k_e Q/a^2$. This is just the form you would expect for a point charge. Therefore, at large values of a/ℓ, the charge distribution appears to be a point charge of magnitude Q—we are so far away from the rod that we cannot distinguish that it has a size. The use of the limiting technique ($a/\ell \rightarrow \infty$) often is a good method for checking a mathematical expression.

Example 23.8   The Electric Field of a Uniform Ring of Charge

A ring of radius a carries a uniformly distributed positive total charge Q. Calculate the electric field due to the ring at a point P lying a distance x from its center along the central axis perpendicular to the plane of the ring (Fig. 23.18a).

Solution The magnitude of the electric field at P due to the segment of charge dq is

$$dE = k_e \frac{dq}{r^2}$$

This field has an x component $dE_x = dE \cos \theta$ along the x axis and a component $dE_\perp$ perpendicular to the x axis. As we see in Figure 23.18b, however, the resultant field at P must lie along the x axis because the perpendicular com-

---

[3]   It is important that you understand how to carry out integrations such as this. First, express the charge element dq in terms of the other variables in the integral. (In this example, there is one variable, x, and so we made the change $dq = \lambda\, dx$.) The integral must be over scalar quantities; therefore, you must express the electric field in terms of components, if necessary. (In this example the field has only an x component, so we do not bother with this detail.) Then, reduce your expression to an integral over a single variable (or to multiple integrals, each over a single variable). In examples that have spherical or cylindrical symmetry, the single variable will be a radial coordinate.

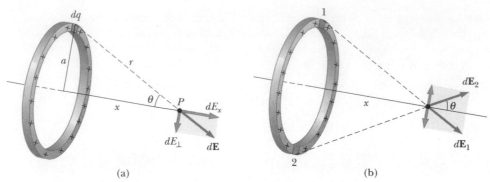

Figure 23.18 (Example 23.8) A uniformly charged ring of radius a. (a) The field at P on the x axis due to an element of charge dq. (b) The total electric field at P is along the x axis. The perpendicular component of the field at P due to segment 1 is canceled by the perpendicular component due to segment 2.

ponents of all the various charge segments sum to zero. That is, the perpendicular component of the field created by any charge element is canceled by the perpendicular component created by an element on the opposite side of the ring. Because $r = (x^2 + a^2)^{1/2}$ and $\cos \theta = x/r$, we find that

$$dE_x = dE \cos \theta = \left(k_e \frac{dq}{r^2} \right) \frac{x}{r} = \frac{k_e x}{(x^2 + a^2)^{3/2}} \, dq$$

All segments of the ring make the same contribution to the field at P because they are all equidistant from this point. Thus, we can integrate to obtain the total field at P:

$$E_x = \int \frac{k_e x}{(x^2 + a^2)^{3/2}} \, dq = \frac{k_e x}{(x^2 + a^2)^{3/2}} \int dq$$

$$= \frac{k_e x}{(x^2 + a^2)^{3/2}} \, Q$$

This result shows that the field is zero at $x = 0$. Does this finding surprise you?

What If? Suppose a negative charge is placed at the center of the ring in Figure 23.18 and displaced slightly by a distance $x \ll a$ along the x axis. When released, what type of motion does it exhibit?

Answer In the expression for the field due to a ring of charge, we let $x \ll a$, which results in

$$E_x = \frac{k_e Q}{a^3} x$$

Thus, from Equation 23.8, the force on a charge $-q$ placed near the center of the ring is

$$F_x = -\frac{k_e q Q}{a^3} x$$

Because this force has the form of Hooke's law (Eq. 15.1), the motion will be *simple harmonic*!

Example 23.9   The Electric Field of a Uniformly Charged Disk

A disk of radius R has a uniform surface charge density σ. Calculate the electric field at a point P that lies along the central perpendicular axis of the disk and a distance x from the center of the disk (Fig. 23.19).

Solution If we consider the disk as a set of concentric rings, we can use our result from Example 23.8—which gives the field created by a ring of radius a—and sum the contributions of all rings making up the disk. By symmetry, the field at an axial point must be along the central axis.

The ring of radius r and width dr shown in Figure 23.19 has a surface area equal to $2\pi r \, dr$. The charge dq on this ring is equal to the area of the ring multiplied by the surface charge density: $dq = 2\pi\sigma r \, dr$. Using this result in the equation given for E_x in Example 23.8 (with a replaced by r), we have for the field due to the ring

$$dE_x = \frac{k_e x}{(x^2 + r^2)^{3/2}} \, (2\pi\sigma r \, dr)$$

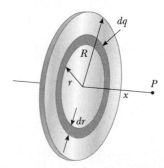

Figure 23.19 (Example 23.9) A uniformly charged disk of radius R. The electric field at an axial point P is directed along the central axis, perpendicular to the plane of the disk.

To obtain the total field at P, we integrate this expression over the limits $r = 0$ to $r = R$, noting that x is a constant.

This gives

$$E_x = k_e x \pi \sigma \int_0^R \frac{2r\,dr}{(x^2 + r^2)^{3/2}}$$

$$= k_e x \pi \sigma \int_0^R (x^2 + r^2)^{-3/2}\,d(r^2)$$

$$= k_e x \pi \sigma \left[\frac{(x^2 + r^2)^{-1/2}}{-1/2} \right]_0^R$$

$$= 2\pi k_e \sigma \left(1 - \frac{x}{(x^2 + R^2)^{1/2}} \right)$$

This result is valid for all values of $x > 0$. We can calculate the field close to the disk along the axis by assuming that $R \gg x$; thus, the expression in parentheses reduces to unity to give us the near-field approximation:

$$E_x = 2\pi k_e \sigma = \frac{\sigma}{2\epsilon_0}$$

where ϵ_0 is the permittivity of free space. In the next chapter we shall obtain the same result for the field created by a uniformly charged infinite sheet.

23.6 Electric Field Lines

We have defined the electric field mathematically through Equation 23.7. We now explore a means of representing the electric field pictorially. A convenient way of visualizing electric field patterns is to draw curved lines that are parallel to the electric field vector at any point in space. These lines, called *electric field lines* and first introduced by Faraday, are related to the electric field in a region of space in the following manner:

- The electric field vector **E** is tangent to the electric field line at each point. The line has a direction, indicated by an arrowhead, that is the same as that of the electric field vector.

- The number of lines per unit area through a surface perpendicular to the lines is proportional to the magnitude of the electric field in that region. Thus, the field lines are close together where the electric field is strong and far apart where the field is weak.

These properties are illustrated in Figure 23.20. The density of lines through surface A is greater than the density of lines through surface B. Therefore, the magnitude of the electric field is larger on surface A than on surface B. Furthermore, the fact that the lines at different locations point in different directions indicates that the field is nonuniform.

Is this relationship between strength of the electric field and the density of field lines consistent with Equation 23.9, the expression we obtained for E using Coulomb's law? To answer this question, consider an imaginary spherical surface of radius r concentric with a point charge. From symmetry, we see that the magnitude of the electric field is the same everywhere on the surface of the sphere. The number of lines N that emerge from the charge is equal to the number that penetrate the spherical surface. Hence, the number of lines per unit area on the sphere is $N/4\pi r^2$ (where the surface area of the sphere is $4\pi r^2$). Because E is proportional to the number of lines per unit area, we see that E varies as $1/r^2$; this finding is consistent with Equation 23.9.

Representative electric field lines for the field due to a single positive point charge are shown in Figure 23.21a. This two-dimensional drawing shows only the field lines that lie in the plane containing the point charge. The lines are actually directed radially outward from the charge in all directions; thus, instead of the flat "wheel" of lines shown, you should picture an entire spherical distribution of lines. Because a positive test charge placed in this field would be repelled by the positive source charge, the lines are directed radially away from the source charge. The electric field lines representing the field due to a single negative point charge are directed toward the charge (Fig. 23.21b). In either case, the lines are along the radial direction and extend all the way to infinity. Note that the lines become closer together as they approach the charge; this indicates that the strength of the field increases as we move toward the source charge.

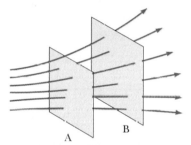

Figure 23.20 Electric field lines penetrating two surfaces. The magnitude of the field is greater on surface A than on surface B.

▲ PITFALL PREVENTION

23.2 Electric Field Lines are not Paths of Particles!

Electric field lines represent the field at various locations. Except in very special cases, they *do not* represent the path of a charged particle moving in an electric field.

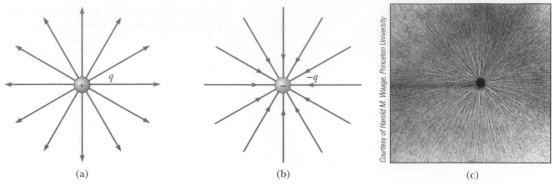

(a)   (b)   (c)

Figure 23.21 The electric field lines for a point charge. (a) For a positive point charge, the lines are directed radially outward. (b) For a negative point charge, the lines are directed radially inward. Note that the figures show only those field lines that lie in the plane of the page. (c) The dark areas are small pieces of thread suspended in oil, which align with the electric field produced by a small charged conductor at the center.

The rules for drawing electric field lines are as follows:

- The lines must begin on a positive charge and terminate on a negative charge. In the case of an excess of one type of charge, some lines will begin or end infinitely far away.
- The number of lines drawn leaving a positive charge or approaching a negative charge is proportional to the magnitude of the charge.
- No two field lines can cross.

We choose the number of field lines starting from any positively charged object to be Cq and the number of lines ending on any negatively charged object to be $C|q|$, where C is an arbitrary proportionality constant. Once C is chosen, the number of lines is fixed. For example, if object 1 has charge Q_1 and object 2 has charge Q_2, then the ratio of number of lines is $N_2/N_1 = Q_2/Q_1$. The electric field lines for two point charges of equal magnitude but opposite signs (an electric dipole) are shown in Figure 23.22. Because the charges are of equal magnitude, the number of lines that begin at the positive charge must equal the number that terminate at the negative charge. At points very near the charges, the lines are nearly radial. The high density of lines between the charges indicates a region of strong electric field.

▲ **PITFALL PREVENTION**

23.3 Electric Field Lines are not Real

Electric field lines are not material objects. They are used only as a pictorial representation to provide a qualitative description of the electric field. Only a finite number of lines from each charge can be drawn, which makes it appear as if the field were quantized and exists only in certain parts of space. The field, in fact, is continuous—existing at every point. You should avoid obtaining the wrong impression from a two-dimensional drawing of field lines used to describe a three-dimensional situation.

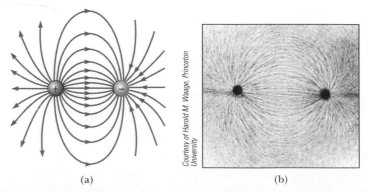

(a)   (b)

Figure 23.22 (a) The electric field lines for two point charges of equal magnitude and opposite sign (an electric dipole). The number of lines leaving the positive charge equals the number terminating at the negative charge. (b) The dark lines are small pieces of thread suspended in oil, which align with the electric field of a dipole.

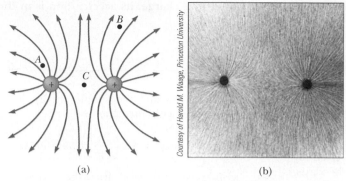

Figure 23.23 (a) The electric field lines for two positive point charges. (The locations *A*, *B*, and *C* are discussed in Quick Quiz 23.7.) (b) Pieces of thread suspended in oil, which align with the electric field created by two equal-magnitude positive charges.

Figure 23.23 shows the electric field lines in the vicinity of two equal positive point charges. Again, the lines are nearly radial at points close to either charge, and the same number of lines emerge from each charge because the charges are equal in magnitude. At great distances from the charges, the field is approximately equal to that of a single point charge of magnitude $2q$.

Finally, in Figure 23.24 we sketch the electric field lines associated with a positive charge $+2q$ and a negative charge $-q$. In this case, the number of lines leaving $+2q$ is twice the number terminating at $-q$. Hence, only half of the lines that leave the positive charge reach the negative charge. The remaining half terminate on a negative charge we assume to be at infinity. At distances that are much greater than the charge separation, the electric field lines are equivalent to those of a single charge $+q$.

Quick Quiz 23.7 Rank the magnitudes of the electric field at points *A*, *B*, and *C* shown in Figure 23.23a (greatest magnitude first).

Quick Quiz 23.8 Which of the following statements about electric field lines associated with electric charges is false? (a) Electric field lines can be either straight or curved. (b) Electric field lines can form closed loops. (c) Electric field lines begin on positive charges and end on negative charges. (d) Electric field lines can never intersect with one another.

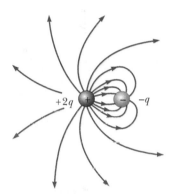

Active Figure 23.24 The electric field lines for a point charge $+2q$ and a second point charge $-q$. Note that two lines leave $+2q$ for every one that terminates on $-q$.

At the Active Figures link at http://www.pse6.com, *you can choose the values and signs for the two charges and observe the electric field lines for the configuration that you have chosen.*

23.7 Motion of Charged Particles in a Uniform Electric Field

When a particle of charge q and mass m is placed in an electric field **E**, the electric force exerted on the charge is $q\mathbf{E}$ according to Equation 23.8. If this is the only force exerted on the particle, it must be the net force and causes the particle to accelerate according to Newton's second law. Thus,

$$\mathbf{F}_e = q\mathbf{E} = m\mathbf{a}$$

The acceleration of the particle is therefore

$$\mathbf{a} = \frac{q\mathbf{E}}{m} \tag{23.12}$$

If **E** is uniform (that is, constant in magnitude and direction), then the acceleration is constant. If the particle has a positive charge, its acceleration is in the direction of the electric field. If the particle has a negative charge, its acceleration is in the direction opposite the electric field.

Example 23.10 An Accelerating Positive Charge

A positive point charge q of mass m is released from rest in a uniform electric field $\mathbf{E}$ directed along the x axis, as shown in Figure 23.25. Describe its motion.

Solution The acceleration is constant and is given by $q\mathbf{E}/m$. The motion is simple linear motion along the x axis. Therefore, we can apply the equations of kinematics in one dimension (see Chapter 2):

$$x_f = x_i + v_i t + \tfrac{1}{2}at^2$$

$$v_f = v_i + at$$

$$v_f{}^2 = v_i{}^2 + 2a(x_f - x_i)$$

Choosing the initial position of the charge as $x_i = 0$ and assigning $v_i = 0$ because the particle starts from rest, the position of the particle as a function of time is

$$x_f = \tfrac{1}{2}at^2 = \frac{qE}{2m}\,t^2$$

The speed of the particle is given by

$$v_f = at = \frac{qE}{m}\,t$$

The third kinematic equation gives us

$$v_f{}^2 = 2ax_f = \left(\frac{2qE}{m}\right)x_f$$

from which we can find the kinetic energy of the charge after it has moved a distance $\Delta x = x_f - x_i$:

$$K = \tfrac{1}{2}mv_f{}^2 = \tfrac{1}{2}m\left(\frac{2qE}{m}\right)\Delta x = qE\,\Delta x$$

We can also obtain this result from the work–kinetic energy theorem because the work done by the electric force is $F_e \Delta x = qE\,\Delta x$ and $W = \Delta K$.

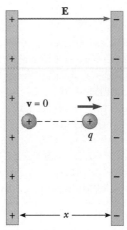

Figure 23.25 (Example 23.10) A positive point charge q in a uniform electric field $\mathbf{E}$ undergoes constant acceleration in the direction of the field.

The electric field in the region between two oppositely charged flat metallic plates is approximately uniform (Fig. 23.26). Suppose an electron of charge $-e$ is projected horizontally into this field from the origin with an initial velocity $v_i\hat{\mathbf{i}}$ at time $t = 0$. Because the electric field $\mathbf{E}$ in Figure 23.26 is in the positive y direction, the acceleration of the electron is in the negative y direction. That is,

$$\mathbf{a} = -\frac{eE}{m_e}\,\hat{\mathbf{j}} \tag{23.13}$$

Because the acceleration is constant, we can apply the equations of kinematics in two dimensions (see Chapter 4) with $v_{xi} = v_i$ and $v_{yi} = 0$. After the electron has been in the

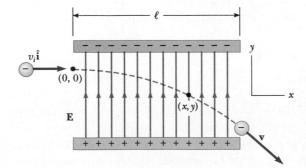

At the Active Figures link at http://www.pse6.com, you can choose the strength of the electric field and the mass and charge of the projected particle.

Active Figure 23.26 An electron is projected horizontally into a uniform electric field produced by two charged plates. The electron undergoes a downward acceleration (opposite $\mathbf{E}$), and its motion is parabolic while it is between the plates.

electric field for a time interval, the components of its velocity at time t are

$$v_x = v_i = \text{constant} \tag{23.14}$$

$$v_y = a_y t = -\frac{eE}{m_e} t \tag{23.15}$$

Its position coordinates at time t are

$$x_f = v_i t \tag{23.16}$$

$$y_f = \tfrac{1}{2} a_y t^2 = -\tfrac{1}{2} \frac{eE}{m_e} t^2 \tag{23.17}$$

Substituting the value $t = x_f/v_i$ from Equation 23.16 into Equation 23.17, we see that y_f is proportional to $x_f{}^2$. Hence, the trajectory is a parabola. This should not be a surprise—consider the analogous situation of throwing a ball horizontally in a uniform gravitational field (Chapter 4). After the electron leaves the field, the electric force vanishes and the electron continues to move in a straight line in the direction of **v** in Figure 23.26 with a speed $v > v_i$.

Note that we have neglected the gravitational force acting on the electron. This is a good approximation when we are dealing with atomic particles. For an electric field of 10^4 N/C, the ratio of the magnitude of the electric force eE to the magnitude of the gravitational force mg is on the order of 10^{14} for an electron and on the order of 10^{11} for a proton.

 **PITFALL PREVENTION**

23.4 Just Another Force

Electric forces and fields may seem abstract to you. However, once $\mathbf{F}_e$ is evaluated, it causes a particle to move according to our well-established understanding of forces and motion from Chapters 5 and 6. Keeping this link with the past in mind will help you solve problems in this chapter.

Example 23.11   An Accelerated Electron   `Interactive`

An electron enters the region of a uniform electric field as shown in Figure 23.26, with $v_i = 3.00 \times 10^6$ m/s and $E - 200$ N/C. The horizontal length of the plates is $\ell = 0.100$ m.

(A) Find the acceleration of the electron while it is in the electric field.

Solution The charge on the electron has an absolute value of 1.60×10^{-19} C, and $m_e = 9.11 \times 10^{-31}$ kg. Therefore, Equation 23.13 gives

$$\mathbf{a} = -\frac{eE}{m_e}\,\hat{\mathbf{j}} = -\frac{(1.60 \times 10^{-19}\,\text{C})(200\,\text{N/C})}{9.11 \times 10^{-31}\,\text{kg}}\,\hat{\mathbf{j}}$$

$$= \boxed{-3.51 \times 10^{13}\,\hat{\mathbf{j}}\,\text{m/s}^2}$$

(B) If the electron enters the field at time $t = 0$, find the time at which it leaves the field.

Solution The horizontal distance across the field is $\ell = 0.100$ m. Using Equation 23.16 with $x_f = \ell$, we find that the time at which the electron exits the electric field is

$$t = \frac{\ell}{v_i} = \frac{0.100\,\text{m}}{3.00 \times 10^6\,\text{m/s}} = \boxed{3.33 \times 10^{-8}\,\text{s}}$$

(C) If the vertical position of the electron as it enters the field is $y_i = 0$, what is its vertical position when it leaves the field?

Solution Using Equation 23.17 and the results from parts (A) and (B), we find that

$$y_f = \tfrac{1}{2} a_y t^2 = -\tfrac{1}{2}(3.51 \times 10^{13}\,\text{m/s}^2)(3.33 \times 10^{-8}\,\text{s})^2$$

$$= -0.019\,5\,\text{m} = \boxed{-1.95\,\text{cm}}$$

If the electron enters just below the negative plate in Figure 23.26 and the separation between the plates is less than the value we have just calculated, the electron will strike the positive plate.

 At the Interactive Worked Example link at http://www.pse6.com, you can predict the required initial velocity for the exiting electron to just miss the right edge of the lower plate, for random values of the electric field.

The Cathode Ray Tube

The example we just worked describes a portion of a cathode ray tube (CRT). This tube, illustrated in Figure 23.27, is commonly used to obtain a visual display of electronic information in oscilloscopes, radar systems, television receivers, and computer monitors. The CRT is a vacuum tube in which a beam of electrons is accelerated and deflected under the influence of electric or magnetic fields. The electron beam is

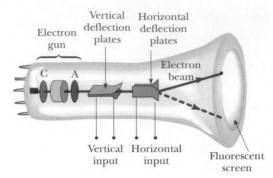

Figure 23.27 Schematic diagram of a cathode ray tube. Electrons leaving the cathode C are accelerated to the anode A. In addition to accelerating electrons, the electron gun is also used to focus the beam of electrons, and the plates deflect the beam.

produced by an assembly called an *electron gun* located in the neck of the tube. These electrons, if left undisturbed, travel in a straight-line path until they strike the front of the CRT, the "screen," which is coated with a material that emits visible light when bombarded with electrons.

In an oscilloscope, the electrons are deflected in various directions by two sets of plates placed at right angles to each other in the neck of the tube. (A television CRT steers the beam with a magnetic field, as discussed in Chapter 29.) An external electric circuit is used to control the amount of charge present on the plates. The placing of positive charge on one horizontal plate and negative charge on the other creates an electric field between the plates and allows the beam to be steered from side to side. The vertical deflection plates act in the same way, except that changing the charge on them deflects the beam vertically.

SUMMARY

Take a practice test for this chapter by clicking on the Practice Test link at http://www.pse6.com.

Electric charges have the following important properties:

- Charges of opposite sign attract one another and charges of the same sign repel one another.

- Total charge in an isolated system is conserved.

- Charge is quantized.

Conductors are materials in which electrons move freely. **Insulators** are materials in which electrons do not move freely.

Coulomb's law states that the electric force exerted by a charge q_1 on a second charge q_2 is

$$\mathbf{F}_{12} = k_e \frac{q_1 q_2}{r^2} \,\hat{\mathbf{r}} \tag{23.6}$$

where r is the distance between the two charges and $\hat{\mathbf{r}}$ is a unit vector directed from q_1 toward q_2. The constant k_e, which is called the Coulomb constant, has the value $k_e = 8.99 \times 10^9 \, \text{N} \cdot \text{m}^2/\text{C}^2$.

The smallest unit of free charge e known to exist in nature is the charge on an electron $(-e)$ or proton $(+e)$, where $e = 1.602\ 19 \times 10^{-19} \, \text{C}$.

The electric field **E** at some point in space is defined as the electric force $\mathbf{F}_e$ that acts on a small positive test charge placed at that point divided by the magnitude q_0 of the test charge:

$$\mathbf{E} \equiv \frac{\mathbf{F}_e}{q_0} \tag{23.7}$$

Thus, the electric force on a charge q placed in an electric field **E** is given by

$$\mathbf{F}_e = q\mathbf{E} \tag{23.8}$$

At a distance r from a point charge q, the electric field due to the charge is given by

$$\mathbf{E} = k_e \frac{q}{r^2} \hat{\mathbf{r}} \qquad (23.9)$$

where $\hat{\mathbf{r}}$ is a unit vector directed from the charge toward the point in question. The electric field is directed radially outward from a positive charge and radially inward toward a negative charge.

The electric field due to a group of point charges can be obtained by using the superposition principle. That is, the total electric field at some point equals the vector sum of the electric fields of all the charges:

$$\mathbf{E} = k_e \sum_i \frac{q_i}{r_i^2} \hat{\mathbf{r}}_i \qquad (23.10)$$

The electric field at some point due to a continuous charge distribution is

$$\mathbf{E} = k_e \int \frac{dq}{r^2} \hat{\mathbf{r}} \qquad (23.11)$$

where dq is the charge on one element of the charge distribution and r is the distance from the element to the point in question.

Electric field lines describe an electric field in any region of space. The number of lines per unit area through a surface perpendicular to the lines is proportional to the magnitude of $\mathbf{E}$ in that region.

A charged particle of mass m and charge q moving in an electric field $\mathbf{E}$ has an acceleration

$$\mathbf{a} = \frac{q\mathbf{E}}{m} \qquad (23.12)$$

QUESTIONS

1. Explain what is meant by the term "a neutral atom." Explain what "a negatively charged atom" means.

2. A charged comb often attracts small bits of dry paper that then fly away when they touch the comb. Explain.

3. Sparks are often seen or heard on a dry day when fabrics are removed from a clothes dryer in dim light. Explain.

4. Hospital personnel must wear special conducting shoes while working around oxygen in an operating room. Why? Contrast with what might happen if people wore rubber-soled shoes.

5. Explain from an atomic viewpoint why charge is usually transferred by electrons.

6. A light, uncharged metallic sphere suspended from a thread is attracted to a charged rubber rod. After it touches the rod, the sphere is repelled by the rod. Explain.

7. A foreign student who grew up in a tropical country but is studying in the United States may have had no experience with static electricity sparks or shocks until he or she first experiences an American winter. Explain.

8. Explain the similarities and differences between Newton's law of universal gravitation and Coulomb's law.

9. A balloon is negatively charged by rubbing and then clings to a wall. Does this mean that the wall is positively charged? Why does the balloon eventually fall?

10. A light strip of aluminum foil is draped over a horizontal wooden pencil. When a rod carrying a positive charge is brought close to the foil, the two parts of the foil stand apart. Why? What kind of charge is on the foil?

11. When defining the electric field, why is it necessary to specify that the magnitude of the test charge be very small?

12. How could you experimentally distinguish an electric field from a gravitational field?

13. A large metallic sphere insulated from ground is charged with an electrostatic generator while a student standing on an insulating stool holds the sphere. Why is it safe to do this? Why would it not be safe for another person to touch the sphere after it had been charged?

14. Is it possible for an electric field to exist in empty space? Explain. Consider point A in Figure 23.23(a). Does charge exist at this point? Does a force exist at this point? Does a field exist at this point?

15. When is it valid to approximate a charge distribution by a point charge?

16. Explain why electric field lines never cross. *Suggestion:* Begin by explaining why the electric field at a particular point must have only one direction.

17. Figures 23.14 and 23.15 show three electric field vectors at the same point. With a little extrapolation, Figure

23.21 would show many electric field lines at the same point. Is it really true that "no two field lines can cross"? Are the diagrams drawn correctly? Explain your answers.

18. A free electron and a free proton are released in identical electric fields. Compare the electric forces on the two particles. Compare their accelerations.

19. Explain what happens to the magnitude of the electric field created by a point charge as r approaches zero.

20. An object with negative charge is placed in a region of space where the electric field is directed vertically upward. What is the direction of the electric force exerted on this charge?

21. A charge $4q$ is at a distance r from a charge $-q$. Compare the number of electric field lines leaving the charge $4q$ with the number entering the charge $-q$. Where do the extra lines beginning on $4q$ end?

22. Consider two equal point charges separated by some distance d. At what point (other than ∞) would a third test charge experience no net force?

23. Explain the differences between linear, surface, and volume charge densities, and give examples of when each would be used.

24. If the electron in Figure 23.26 is projected into the electric field with an arbitrary velocity $\mathbf{v}_i$ (at an arbitrary angle to $\mathbf{E}$), will its trajectory still be parabolic? Explain.

25. Would life be different if the electron were positively charged and the proton were negatively charged? Does the choice of signs have any bearing on physical and chemical interactions? Explain.

26. Why should a ground wire be connected to the metal support rod for a television antenna?

27. Suppose someone proposes the idea that people are bound to the Earth by electric forces rather than by gravity. How could you prove this idea is wrong?

28. Consider two electric dipoles in empty space. Each dipole has zero net charge. Does an electric force exist between the dipoles—that is, can two objects with zero net charge exert electric forces on each other? If so, is the force one of attraction or of repulsion?

PROBLEMS

1, 2, 3 = straightforward, intermediate, challenging ☐ = full solution available in the *Student Solutions Manual and Study Guide*

⚛ = coached solution with hints available at http://www.pse6.com 💻 = computer useful in solving problem

▨ = paired numerical and symbolic problems

Section 23.1 Properties of Electric Charges

1. (a) Find to three significant digits the charge and the mass of an ionized hydrogen atom, represented as H^+. *Suggestion:* Begin by looking up the mass of a neutral atom on the periodic table of the elements. (b) Find the charge and the mass of Na^+, a singly ionized sodium atom. (c) Find the charge and the average mass of a chloride ion Cl^- that joins with the Na^+ to make one molecule of table salt. (d) Find the charge and the mass of $Ca^{++} = Ca^{2+}$, a doubly ionized calcium atom. (e) You can model the center of an ammonia molecule as an N^{3-} ion. Find its charge and mass. (f) The plasma in a hot star contains quadruply ionized nitrogen atoms, N^{4+}. Find their charge and mass. (g) Find the charge and the mass of the nucleus of a nitrogen atom. (h) Find the charge and the mass of the molecular ion H_2O^-.

2. (a) Calculate the number of electrons in a small, electrically neutral silver pin that has a mass of 10.0 g. Silver has 47 electrons per atom, and its molar mass is 107.87 g/mol. (b) Electrons are added to the pin until the net negative charge is 1.00 mC. How many electrons are added for every 10^9 electrons already present?

Section 23.2 Charging Objects by Induction
Section 23.3 Coulomb's Law

3. ⚛ The Nobel laureate Richard Feynman once said that if two persons stood at arm's length from each other and each person had 1% more electrons than protons, the force of repulsion between them would be enough to lift a "weight" equal to that of the entire Earth. Carry out an order-of-magnitude calculation to substantiate this assertion.

4. Two protons in an atomic nucleus are typically separated by a distance of 2×10^{-15} m. The electric repulsion force between the protons is huge, but the attractive nuclear force is even stronger and keeps the nucleus from bursting apart. What is the magnitude of the electric force between two protons separated by 2.00×10^{-15} m?

5. (a) Two protons in a molecule are separated by 3.80×10^{-10} m. Find the electric force exerted by one proton on the other. (b) How does the magnitude of this force compare to the magnitude of the gravitational force between the two protons? (c) **What If?** What must be the charge-to-mass ratio of a particle if the magnitude of the gravitational force between two of these particles equals the magnitude of electric force between them?

6. Two small silver spheres, each with a mass of 10.0 g, are separated by 1.00 m. Calculate the fraction of the electrons in one sphere that must be transferred to the other in order to produce an attractive force of 1.00×10^4 N (about 1 ton) between the spheres. (The number of electrons per atom of silver is 47, and the number of atoms per gram is Avogadro's number divided by the molar mass of silver, 107.87 g/mol.)

7. Three point charges are located at the corners of an equilateral triangle as shown in Figure P23.7. Calculate the resultant electric force on the 7.00-μC charge.

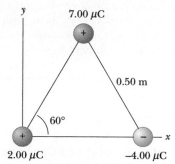

Figure P23.7 Problems 7 and 18.

8. Suppose that 1.00 g of hydrogen is separated into electrons and protons. Suppose also that the protons are placed at the Earth's north pole and the electrons are placed at the south pole. What is the resulting compressional force on the Earth?

9. Two identical conducting small spheres are placed with their centers 0.300 m apart. One is given a charge of 12.0 nC and the other a charge of − 18.0 nC. (a) Find the electric force exerted by one sphere on the other. (b) **What If?** The spheres are connected by a conducting wire. Find the electric force between the two after they have come to equilibrium.

10. Two small beads having positive charges $3q$ and q are fixed at the opposite ends of a horizontal, insulating rod, extending from the origin to the point $x = d$. As shown in Figure P23.10, a third small charged bead is free to slide on the rod. At what position is the third bead in equilibrium? Can it be in stable equilibrium?

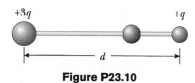

Figure P23.10

11. **Review problem.** In the Bohr theory of the hydrogen atom, an electron moves in a circular orbit about a proton, where the radius of the orbit is 0.529×10^{-10} m. (a) Find the electric force between the two. (b) If this force causes the centripetal acceleration of the electron, what is the speed of the electron?

12. **Review problem.** Two identical particles, each having charge $+ q$, are fixed in space and separated by a distance d. A third point charge $- Q$ is free to move and lies initially at rest on the perpendicular bisector of the two fixed charges a distance x from the midpoint between the two fixed charges (Fig. P23.12). (a) Show that if x is small compared with d, the motion of $- Q$ will be simple harmonic along the perpendicular bisector. Determine the period of that motion. (b) How fast will the charge $- Q$ be moving when it is at the midpoint between the two fixed charges, if initially it is released at a distance $a \ll d$ from the midpoint?

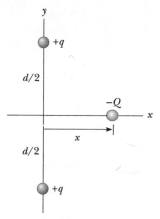

Figure P23.12

Section 23.4   The Electric Field

13. What are the magnitude and direction of the electric field that will balance the weight of (a) an electron and (b) a proton? (Use the data in Table 23.1.)

14. An object having a net charge of 24.0 μC is placed in a uniform electric field of 610 N/C directed vertically. What is the mass of this object if it "floats" in the field?

15. In Figure P23.15, determine the point (other than infinity) at which the electric field is zero.

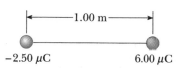

Figure P23.15

16. An airplane is flying through a thundercloud at a height of 2 000 m. (This is a very dangerous thing to do because of updrafts, turbulence, and the possibility of electric discharge.) If a charge concentration of + 40.0 C is above the plane at a height of 3 000 m within the cloud and a charge concentration of − 40.0 C is at height 1 000 m, what is the electric field at the aircraft?

17. Two point charges are located on the x axis. The first is a charge $+ Q$ at $x = - a$. The second is an unknown charge located at $x = + 3a$. The net electric field these charges produce at the origin has a magnitude of $2k_eQ/a^2$. What are the two possible values of the unknown charge?

18. Three charges are at the corners of an equilateral triangle as shown in Figure P23.7. (a) Calculate the electric field at the position of the 2.00-μC charge due to the 7.00-μC and − 4.00-μC charges. (b) Use your answer to part (a) to determine the force on the 2.00-μC charge.

19. Three point charges are arranged as shown in Figure P23.19. (a) Find the vector electric field that the 6.00-nC and − 3.00-nC charges together create at the origin. (b) Find the vector force on the 5.00-nC charge.

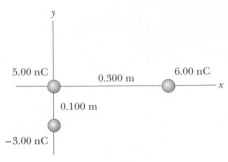

Figure P23.19

20. Two 2.00-μC point charges are located on the x axis. One is at $x = 1.00$ m, and the other is at $x = -1.00$ m. (a) Determine the electric field on the y axis at $y = 0.500$ m. (b) Calculate the electric force on a -3.00-μC charge placed on the y axis at $y = 0.500$ m.

21. Four point charges are at the corners of a square of side a as shown in Figure P23.21. (a) Determine the magnitude and direction of the electric field at the location of charge q. (b) What is the resultant force on q?

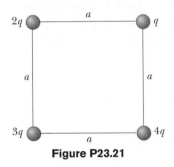

Figure P23.21

22. Consider the electric dipole shown in Figure P23.22. Show that the electric field at a *distant* point on the $+x$ axis is $E_x \approx 4k_e qa/x^3$.

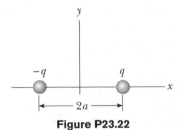

Figure P23.22

23. Consider n equal positive point charges each of magnitude Q/n placed symmetrically around a circle of radius R. (a) Calculate the magnitude of the electric field at a point a distance x on the line passing through the center of the circle and perpendicular to the plane of the circle. (b) Explain why this result is identical to that of the calculation done in Example 23.8.

24. Consider an infinite number of identical charges (each of charge q) placed along the x axis at distances a, $2a$, $3a$, $4a$, . . . , from the origin. What is the electric field at the origin due to this distribution? *Suggestion:* Use the fact that

$$1 + \frac{1}{2^2} + \frac{1}{3^2} + \frac{1}{4^2} + \cdots = \frac{\pi^2}{6}$$

Section 23.5   Electric Field of a Continuous Charge Distribution

25. A rod 14.0 cm long is uniformly charged and has a total charge of -22.0 μC. Determine the magnitude and direction of the electric field along the axis of the rod at a point 36.0 cm from its center.

26. A continuous line of charge lies along the x axis, extending from $x = +x_0$ to positive infinity. The line carries charge with a uniform linear charge density λ_0. What are the magnitude and direction of the electric field at the origin?

27. A uniformly charged ring of radius 10.0 cm has a total charge of 75.0 μC. Find the electric field on the axis of the ring at (a) 1.00 cm, (b) 5.00 cm, (c) 30.0 cm, and (d) 100 cm from the center of the ring.

28. A line of charge starts at $x = +x_0$ and extends to positive infinity. The linear charge density is $\lambda = \lambda_0 x_0/x$. Determine the electric field at the origin.

29. Show that the maximum magnitude E_{max} of the electric field along the axis of a uniformly charged ring occurs at $x = a/\sqrt{2}$ (see Fig. 23.18) and has the value $Q/(6\sqrt{3}\pi\epsilon_0 a^2)$.

30. A uniformly charged disk of radius 35.0 cm carries charge with a density of 7.90×10^{-3} C/m^2. Calculate the electric field on the axis of the disk at (a) 5.00 cm, (b) 10.0 cm, (c) 50.0 cm, and (d) 200 cm from the center of the disk.

31. Example 23.9 derives the exact expression for the electric field at a point on the axis of a uniformly charged disk. Consider a disk, of radius $R = 3.00$ cm, having a uniformly distributed charge of $+5.20$ μC. (a) Using the result of Example 23.9, compute the electric field at a point on the axis and 3.00 mm from the center. **What If?** Compare this answer with the field computed from the near-field approximation $E = \sigma/2\epsilon_0$. (b) Using the result of Example 23.9, compute the electric field at a point on the axis and 30.0 cm from the center of the disk. **What If?** Compare this with the electric field obtained by treating the disk as a $+5.20$-μC point charge at a distance of 30.0 cm.

32. The electric field along the axis of a uniformly charged disk of radius R and total charge Q was calculated in Example 23.9. Show that the electric field at distances x that are large compared with R approaches that of a point charge $Q = \sigma\pi R^2$. (*Suggestion:* First show that $x/(x^2 + R^2)^{1/2} = (1 + R^2/x^2)^{-1/2}$ and use the binomial expansion $(1 + \delta)^n \approx 1 + n\delta$ when $\delta \ll 1$.)

33. A uniformly charged insulating rod of length 14.0 cm is bent into the shape of a semicircle as shown in Figure P23.33. The rod has a total charge of -7.50 μC. Find the magnitude and direction of the electric field at O, the center of the semicircle.

Figure P23.33

34. (a) Consider a uniformly charged thin-walled right circular cylindrical shell having total charge Q, radius R, and height h. Determine the electric field at a point a distance d from the right side of the cylinder as shown in Figure P23.34. (*Suggestion:* Use the result of Example 23.8 and treat the cylinder as a collection of ring charges.) (b) **What If?** Consider now a solid cylinder with the same dimensions and carrying the same charge, uniformly distributed through its volume. Use the result of Example 23.9 to find the field it creates at the same point.

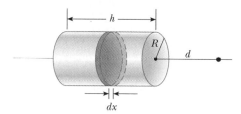

Figure P23.34

35. A thin rod of length ℓ and uniform charge per unit length λ lies along the x axis, as shown in Figure P23.35. (a) Show that the electric field at P, a distance y from the rod along its perpendicular bisector, has no x component and is given by $E = 2k_e\lambda \sin\theta_0/y$. (b) **What If?** Using your result to part (a), show that the field of a rod of infinite length is $E = 2k_e\lambda/y$. (*Suggestion:* First calculate the field at P due to an element of length dx, which has a charge $\lambda\ dx$. Then change variables from x to θ, using the relationships $x = y\tan\theta$ and $dx = y\sec^2\theta\ d\theta$, and integrate over θ.)

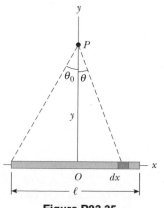

Figure P23.35

36. Three solid plastic cylinders all have radius 2.50 cm and length 6.00 cm. One (a) carries charge with uniform density 15.0 nC/m² everywhere on its surface. Another (b) carries charge with the same uniform density on its curved lateral surface only. The third (c) carries charge with uniform density 500 nC/m³ throughout the plastic. Find the charge of each cylinder.

37. Eight solid plastic cubes, each 3.00 cm on each edge, are glued together to form each one of the objects (i, ii, iii, and iv) shown in Figure P23.37. (a) Assuming each object carries charge with uniform density 400 nC/m³ throughout its volume, find the charge of each object. (b) Assuming each object carries charge with uniform density 15.0 nC/m² everywhere on its exposed surface, find the charge on each object. (c) Assuming charge is placed only on the edges where perpendicular surfaces meet, with uniform density 80.0 pC/m, find the charge of each object.

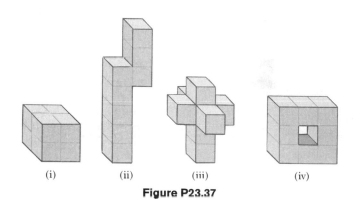

(i) (ii) (iii) (iv)

Figure P23.37

Section 23.6 Electric Field Lines

38. A positively charged disk has a uniform charge per unit area as described in Example 23.9. Sketch the electric field lines in a plane perpendicular to the plane of the disk passing through its center.

39. A negatively charged rod of finite length carries charge with a uniform charge per unit length. Sketch the electric field lines in a plane containing the rod.

40. Figure P23.40 shows the electric field lines for two point charges separated by a small distance. (a) Determine the ratio q_1/q_2. (b) What are the signs of q_1 and q_2?

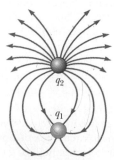

Figure P23.40

41. 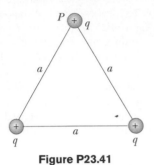 Three equal positive charges q are at the corners of an equilateral triangle of side a as shown in Figure P23.41. (a) Assume that the three charges together create an electric field. Sketch the field lines in the plane of the charges. Find the location of a point (other than ∞) where the electric field is zero. (b) What are the magnitude and direction of the electric field at P due to the two charges at the base?

Figure P23.41

Section 23.7   Motion of Charged Particles in a Uniform Electric Field

42. An electron and a proton are each placed at rest in an electric field of 520 N/C. Calculate the speed of each particle 48.0 ns after being released.

43. A proton accelerates from rest in a uniform electric field of 640 N/C. At some later time, its speed is 1.20×10^6 m/s (nonrelativistic, because v is much less than the speed of light). (a) Find the acceleration of the proton. (b) How long does it take the proton to reach this speed? (c) How far has it moved in this time? (d) What is its kinetic energy at this time?

44. A proton is projected in the positive x direction into a region of a uniform electric field $\mathbf{E} = -6.00 \times 10^5 \hat{\mathbf{i}}$ N/C at $t = 0$. The proton travels 7.00 cm before coming to rest. Determine (a) the acceleration of the proton, (b) its initial speed, and (c) the time at which the proton comes to rest.

45. 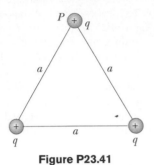 The electrons in a particle beam each have a kinetic energy K. What are the magnitude and direction of the electric field that will stop these electrons in a distance d?

46. A positively charged bead having a mass of 1.00 g falls from rest in a vacuum from a height of 5.00 m in a uniform vertical electric field with a magnitude of 1.00×10^4 N/C. The bead hits the ground at a speed of 21.0 m/s. Determine (a) the direction of the electric field (up or down), and (b) the charge on the bead.

47. A proton moves at 4.50×10^5 m/s in the horizontal direction. It enters a uniform vertical electric field with a magnitude of 9.60×10^3 N/C. Ignoring any gravitational effects, find (a) the time interval required for the proton to travel 5.00 cm horizontally, (b) its vertical displacement during the time interval in which it travels 5.00 cm horizontally, and (c) the horizontal and vertical components of its velocity after it has traveled 5.00 cm horizontally.

48. Two horizontal metal plates, each 100 mm square, are aligned 10.0 mm apart, with one above the other. They are given equal-magnitude charges of opposite sign so that a uniform downward electric field of 2 000 N/C exists in the region between them. A particle of mass 2.00×10^{-16} kg and with a positive charge of 1.00×10^{-6} C leaves the center of the bottom negative plate with an initial speed of 1.00×10^5 m/s at an angle of $37.0°$ above the horizontal. Describe the trajectory of the particle. Which plate does it strike? Where does it strike, relative to its starting point?

49. Protons are projected with an initial speed $v_i = 9.55 \times 10^3$ m/s into a region where a uniform electric field $\mathbf{E} = -720 \hat{\mathbf{j}}$ N/C is present, as shown in Figure P23.49. The protons are to hit a target that lies at a horizontal distance of 1.27 mm from the point where the protons cross the plane and enter the electric field in Figure P23.49. Find (a) the two projection angles θ that will result in a hit and (b) the total time of flight (the time interval during which the proton is above the plane in Figure P23.49) for each trajectory.

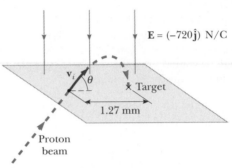

Figure P23.49

Additional Problems

50. Two known charges, $-12.0 \ \mu C$ and $45.0 \ \mu C$, and an unknown charge are located on the x axis. The charge $-12.0 \ \mu C$ is at the origin, and the charge $45.0 \ \mu C$ is at $x = 15.0$ cm. The unknown charge is to be placed so that each charge is in equilibrium under the action of the electric forces exerted by the other two charges. Is this situation possible? Is it possible in more than one way? Find the required location, magnitude, and sign of the unknown charge.

51. A uniform electric field of magnitude 640 N/C exists between two parallel plates that are 4.00 cm apart. A proton is released from the positive plate at the same instant that an electron is released from the negative plate. (a) Determine the distance from the positive plate at which the two pass each other. (Ignore the electrical attraction between the proton and electron.) (b) **What If?** Repeat part (a) for a sodium ion (Na^+) and a chloride ion (Cl^-).

52. Three point charges are aligned along the x axis as shown in Figure P23.52. Find the electric field at (a) the position (2.00, 0) and (b) the position (0, 2.00).

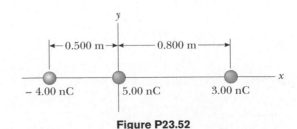

Figure P23.52

53. A researcher studying the properties of ions in the upper atmosphere wishes to construct an apparatus with the following characteristics: Using an electric field, a beam of ions, each having charge q, mass m, and initial velocity $v\hat{\mathbf{i}}$, is turned through an angle of 90° as each ion undergoes displacement $R\hat{\mathbf{i}} + R\hat{\mathbf{j}}$. The ions enter a chamber as shown in Figure P23.53, and leave through the exit port with the same speed they had when they entered the chamber. The electric field acting on the ions is to have constant magnitude. (a) Suppose the electric field is produced by two concentric cylindrical electrodes not shown in the diagram, and hence is radial. What magnitude should the field have? **What If?** (b) If the field is produced by two flat plates and is uniform in direction, what value should the field have in this case?

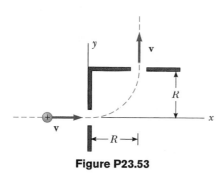

Figure P23.53

54. A small, 2.00-g plastic ball is suspended by a 20.0-cm-long string in a uniform electric field as shown in Figure P23.54. If the ball is in equilibrium when the string makes a 15.0° angle with the vertical, what is the net charge on the ball?

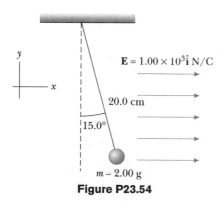

$\mathbf{E} = 1.00 \times 10^3 \hat{\mathbf{i}}$ N/C

20.0 cm

15.0°

$m = 2.00$ g

Figure P23.54

55. A charged cork ball of mass 1.00 g is suspended on a light string in the presence of a uniform electric field as shown in Figure P23.55. When $\mathbf{E} = (3.00\hat{\mathbf{i}} + 5.00\hat{\mathbf{j}}) \times 10^5$ N/C, the ball is in equilibrium at $\theta = 37.0°$. Find (a) the charge on the ball and (b) the tension in the string.

56. A charged cork ball of mass m is suspended on a light string in the presence of a uniform electric field as shown in Figure P23.55. When $\mathbf{E} = (A\hat{\mathbf{i}} + B\hat{\mathbf{j}})$ N/C, where A and B are positive numbers, the ball is in equilibrium at

the angle θ. Find (a) the charge on the ball and (b) the tension in the string.

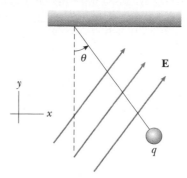

Figure P23.55 Problems 55 and 56.

57. Four identical point charges ($q = +10.0 \ \mu$C) are located on the corners of a rectangle as shown in Figure P23.57. The dimensions of the rectangle are $L = 60.0$ cm and $W = 15.0$ cm. Calculate the magnitude and direction of the resultant electric force exerted on the charge at the lower left corner by the other three charges.

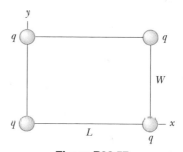

Figure P23.57

58. Inez is putting up decorations for her sister's quinceañera (fifteenth birthday party). She ties three light silk ribbons together to the top of a gateway and hangs a rubber balloon from each ribbon (Fig. P23.58). To include the

Figure P23.58

effects of the gravitational and buoyant forces on it, each balloon can be modeled as a particle of mass 2.00 g, with its center 50.0 cm from the point of support. To show off the colors of the balloons, Inez rubs the whole surface of each balloon with her woolen scarf, to make them hang separately with gaps between them. The centers of the hanging balloons form a horizontal equilateral triangle with sides 30.0 cm long. What is the common charge each balloon carries?

59. **Review problem.** Two identical metallic blocks resting on a frictionless horizontal surface are connected by a light metallic spring having a spring constant k as shown in Figure P23.59a and an unstretched length L_i. A total charge Q is slowly placed on the system, causing the spring to stretch to an equilibrium length L, as shown in Figure P23.59b. Determine the value of Q, assuming that all the charge resides on the blocks and modeling the blocks as point charges.

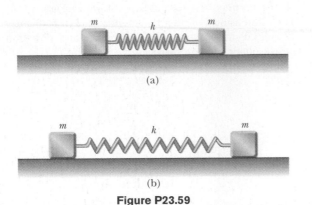

(a)

(b)

Figure P23.59

60. Consider a regular polygon with 29 sides. The distance from the center to each vertex is a. Identical charges q are placed at 28 vertices of the polygon. A single charge Q is placed at the center of the polygon. What is the magnitude and direction of the force experienced by the charge Q? (*Suggestion:* You may use the result of Problem 63 in Chapter 3.)

61. Identical thin rods of length $2a$ carry equal charges $+Q$ uniformly distributed along their lengths. The rods lie along the x axis with their centers separated by a distance $b > 2a$ (Fig. P23.61). Show that the magnitude of the force exerted by the left rod on the right one is given by

$$F = \left(\frac{k_e Q^2}{4a^2} \right) \ln \left(\frac{b^2}{b^2 - 4a^2} \right)$$

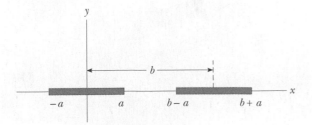

Figure P23.61

62. Two small spheres, each of mass 2.00 g, are suspended by light strings 10.0 cm in length (Fig. P23.62). A uniform electric field is applied in the x direction. The spheres have charges equal to -5.00×10^{-8} C and $+5.00 \times 10^{-8}$ C. Determine the electric field that enables the spheres to be in equilibrium at an angle $\theta = 10.0°$.

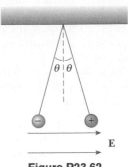

Figure P23.62

63. A line of positive charge is formed into a semicircle of radius $R = 60.0$ cm as shown in Figure P23.63. The charge per unit length along the semicircle is described by the expression $\lambda = \lambda_0 \cos \theta$. The total charge on the semicircle is 12.0 μC. Calculate the total force on a charge of 3.00 μC placed at the center of curvature.

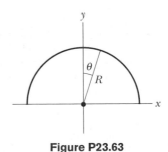

Figure P23.63

64. Three charges of equal magnitude q are fixed in position at the vertices of an equilateral triangle (Fig. P23.64). A fourth charge Q is free to move along the positive x axis

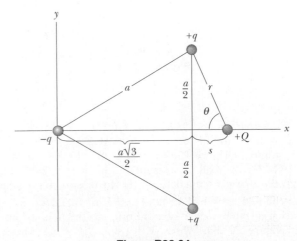

Figure P23.64

under the influence of the forces exerted by the three fixed charges. Find a value for s for which Q is in equilibrium. You will need to solve a transcendental equation.

65. Two small spheres of mass m are suspended from strings of length ℓ that are connected at a common point. One sphere has charge Q; the other has charge $2Q$. The strings make angles θ_1 and θ_2 with the vertical. (a) How are θ_1 and θ_2 related? (b) Assume θ_1 and θ_2 are small. Show that the distance r between the spheres is given by

$$r \approx \left(\frac{4k_eQ^2\ell}{mg} \right)^{1/3}$$

66. Review problem. Four identical particles, each having charge $+q$, are fixed at the corners of a square of side L. A fifth point charge $-Q$ lies a distance z along the line perpendicular to the plane of the square and passing through the center of the square (Fig. P23.66). (a) Show that the force exerted by the other four charges on $-Q$ is

$$\mathbf{F} = -\frac{4k_e q Qz}{[z^2 + (L^2/2)]^{3/2}}\,\hat{\mathbf{k}}$$

Note that this force is directed toward the center of the square whether z is positive ($-Q$ above the square) or negative ($-Q$ below the square). (b) If z is small compared with L, the above expression reduces to $\mathbf{F} \approx -(\text{constant})z\hat{\mathbf{k}}$. Why does this imply that the motion of the charge $-Q$ is simple harmonic, and what is the period of this motion if the mass of $-Q$ is m?

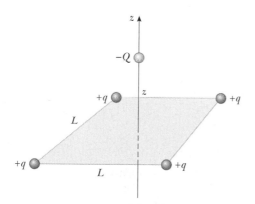

Figure P23.66

67. Review problem. A 1.00-g cork ball with charge 2.00 μC is suspended vertically on a 0.500-m-long light string in the presence of a uniform, downward-directed electric field of magnitude $E = 1.00 \times 10^5$ N/C. If the ball is displaced slightly from the vertical, it oscillates like a simple pendulum. (a) Determine the period of this oscillation. (b) Should gravity be included in the calculation for part (a)? Explain.

68. Two identical beads each have a mass m and charge q. When placed in a hemispherical bowl of radius R with frictionless, nonconducting walls, the beads move, and at equilibrium they are a distance R apart (Fig. P23.68). Determine the charge on each bead.

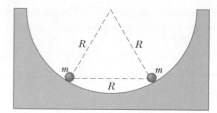

Figure P23.68

69. Eight point charges, each of magnitude q, are located on the corners of a cube of edge s, as shown in Figure P23.69. (a) Determine the x, y, and z components of the resultant force exerted by the other charges on the charge located at point A. (b) What are the magnitude and direction of this resultant force?

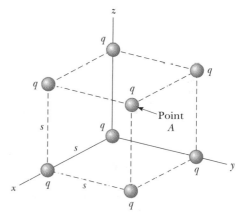

Figure P23.69 Problems 69 and 70.

70. Consider the charge distribution shown in Figure P23.69. (a) Show that the magnitude of the electric field at the center of any face of the cube has a value of $2.18k_eq/s^2$. (b) What is the direction of the electric field at the center of the top face of the cube?

71. Review problem. A negatively charged particle $-q$ is placed at the center of a uniformly charged ring, where the ring has a total positive charge Q as shown in Example 23.8. The particle, confined to move along the x axis, is displaced a small distance x along the axis (where $x \ll a$) and released. Show that the particle oscillates in simple harmonic motion with a frequency given by

$$f = \frac{1}{2\pi}\left(\frac{k_eqQ}{ma^3}\right)^{1/2}$$

72. A line of charge with uniform density 35.0 nC/m lies along the line $y = -15.0$ cm, between the points with coordinates $x = 0$ and $x = 40.0$ cm. Find the electric field it creates at the origin.

73. Review problem. An electric dipole in a uniform electric field is displaced slightly from its equilibrium position, as shown in Figure P23.73, where θ is small. The separation of the charges is $2a$, and the moment of inertia of the dipole is I. Assuming the dipole is released from this position, show that its angular orientation exhibits simple harmonic motion with a frequency

$$f = \frac{1}{2\pi} \sqrt{\frac{2qaE}{I}}$$

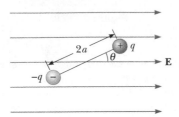

Figure P23.73

Answers to Quick Quizzes

23.1 (b). The amount of charge present in the isolated system after rubbing is the same as that before because charge is conserved; it is just distributed differently.

23.2 (a), (c), and (e). The experiment shows that A and B have charges of the same sign, as do objects B and C. Thus, all three objects have charges of the same sign. We cannot determine from this information, however, whether the charges are positive or negative.

23.3 (e). In the first experiment, objects A and B may have charges with opposite signs, or one of the objects may be neutral. The second experiment shows that B and C have charges with the same signs, so that B must be charged. But we still do not know if A is charged or neutral.

23.4 (e). From Newton's third law, the electric force exerted by object B on object A is equal in magnitude to the force exerted by object A on object B.

23.5 (b). From Newton's third law, the electric force exerted by object B on object A is equal in magnitude to the force exerted by object A on object B and in the opposite direction.

23.6 (a). There is no effect on the electric field if we assume that the source charge producing the field is not disturbed by our actions. Remember that the electric field is created by source charge(s) (unseen in this case), not the test charge(s).

23.7 *A*, *B*, *C*. The field is greatest at point *A* because this is where the field lines are closest together. The absence of lines near point *C* indicates that the electric field there is zero.

23.8 (b). Electric field lines begin and end on charges and cannot close on themselves to form loops.

Chapter 24

Gauss's Law

CHAPTER OUTLINE

24.1 Electric Flux

24.2 Gauss's Law

24.3 Application of Gauss's Law to Various Charge Distributions

24.4 Conductors in Electrostatic Equilibrium

24.5 Formal Derivation of Gauss's Law

▲ *In a table-top plasma ball, the colorful lines emanating from the sphere give evidence of strong electric fields. Using Gauss's law, we show in this chapter that the electric field surrounding a charged sphere is identical to that of a point charge. (Getty Images)*

In the preceding chapter we showed how to calculate the electric field generated by a given charge distribution. In this chapter, we describe *Gauss's law* and an alternative procedure for calculating electric fields. The law is based on the fact that the fundamental electrostatic force between point charges exhibits an inverse-square behavior. Although a consequence of Coulomb's law, Gauss's law is more convenient for calculating the electric fields of highly symmetric charge distributions and makes possible useful qualitative reasoning when dealing with complicated problems.

24.1 Electric Flux

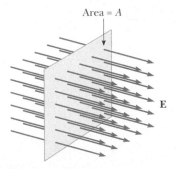

Area = A

E

Figure 24.1 Field lines representing a uniform electric field penetrating a plane of area A perpendicular to the field. The electric flux Φ_E through this area is equal to EA.

The concept of electric field lines was described qualitatively in Chapter 23. We now treat electric field lines in a more quantitative way.

Consider an electric field that is uniform in both magnitude and direction, as shown in Figure 24.1. The field lines penetrate a rectangular surface of area A, whose plane is oriented perpendicular to the field. Recall from Section 23.6 that the number of lines per unit area (in other words, the *line density*) is proportional to the magnitude of the electric field. Therefore, the total number of lines penetrating the surface is proportional to the product EA. This product of the magnitude of the electric field E and surface area A perpendicular to the field is called the **electric flux** Φ_E (uppercase Greek phi):

$$\Phi_E = EA \tag{24.1}$$

From the SI units of E and A, we see that Φ_E has units of newton-meters squared per coulomb ($N \cdot m^2/C$.) **Electric flux is proportional to the number of electric field lines penetrating some surface.**

Example 24.1 Electric Flux Through a Sphere

What is the electric flux through a sphere that has a radius of 1.00 m and carries a charge of $+1.00 \ \mu C$ at its center?

Solution The magnitude of the electric field 1.00 m from this charge is found using Equation 23.9:

$$E = k_e \frac{q}{r^2} = (8.99 \times 10^9 \ N \cdot m^2/C^2) \frac{1.00 \times 10^{-6} \ C}{(1.00 \ m)^2}$$

$$= 8.99 \times 10^3 \ N/C$$

The field points radially outward and is therefore everywhere perpendicular to the surface of the sphere. The flux through the sphere (whose surface area $A = 4\pi r^2 = 12.6 \ m^2$) is thus

$$\Phi_E = EA = (8.99 \times 10^3 \ N/C)(12.6 \ m^2)$$

$$= 1.13 \times 10^5 \ N \cdot m^2/C$$

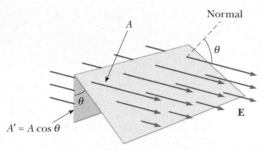

Figure 24.2 Field lines representing a uniform electric field penetrating an area A that is at an angle θ to the field. Because the number of lines that go through the area A' is the same as the number that go through A, the flux through A' is equal to the flux through A and is given by $\Phi_E = EA\cos\theta$.

If the surface under consideration is not perpendicular to the field, the flux through it must be less than that given by Equation 24.1. We can understand this by considering Figure 24.2, where the normal to the surface of area A is at an angle θ to the uniform electric field. Note that the number of lines that cross this area A is equal to the number that cross the area A', which is a projection of area A onto a plane oriented perpendicular to the field. From Figure 24.2 we see that the two areas are related by $A' = A\cos\theta$. Because the flux through A equals the flux through A', we conclude that the flux through A is

$$\Phi_E = EA' = EA\cos\theta \tag{24.2}$$

From this result, we see that the flux through a surface of fixed area A has a maximum value EA when the surface is perpendicular to the field (when the normal to the surface is parallel to the field, that is, $\theta = 0°$ in Figure 24.2); the flux is zero when the surface is parallel to the field (when the normal to the surface is perpendicular to the field, that is, $\theta = 90°$).

We assumed a uniform electric field in the preceding discussion. In more general situations, the electric field may vary over a surface. Therefore, our definition of flux given by Equation 24.2 has meaning only over a small element of area. Consider a general surface divided up into a large number of small elements, each of area ΔA. The variation in the electric field over one element can be neglected if the element is sufficiently small. It is convenient to define a vector $\Delta\mathbf{A}_i$ whose magnitude represents the area of the ith element of the surface and whose direction is defined to be *perpendicular* to the surface element, as shown in Figure 24.3. The electric field $\mathbf{E}_i$ at the location of this element makes an angle θ_i with the vector $\Delta\mathbf{A}_i$. The electric flux $\Delta\Phi_E$ through this element is

$$\Delta\Phi_E = E_i \,\Delta A_i \cos\theta_i = \mathbf{E}_i \cdot \Delta\mathbf{A}_i$$

where we have used the definition of the scalar product (or dot product; see Chapter 7) of two vectors ($\mathbf{A} \cdot \mathbf{B} = AB\cos\theta$). By summing the contributions of all elements, we obtain the total flux through the surface. If we let the area of each element approach zero, then the number of elements approaches infinity and the sum is replaced by an integral. Therefore, the general definition of electric flux is[1]

$$\Phi_E = \lim_{\Delta A_i \to 0} \sum \mathbf{E}_i \cdot \Delta\mathbf{A}_i = \int_{\text{surface}} \mathbf{E} \cdot d\mathbf{A} \tag{24.3}$$

Equation 24.3 is a *surface integral*, which means it must be evaluated over the surface in question. In general, the value of Φ_E depends both on the field pattern and on the surface.

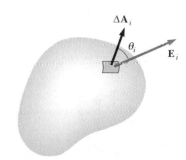

Figure 24.3 A small element of surface area ΔA_i. The electric field makes an angle θ_i with the vector $\Delta\mathbf{A}_i$, defined as being normal to the surface element, and the flux through the element is equal to $E_i\,\Delta A_i\cos\theta_i$.

Definition of electric flux

---

[1]   Drawings with field lines have their inaccuracies because a limited number of field lines are typically drawn in a diagram. Consequently, a small area element drawn on a diagram (depending on its location) may happen to have too few field lines penetrating it to represent the flux accurately. We stress that the basic definition of electric flux is Equation 24.3. The use of lines is only an aid for visualizing the concept.

At the Active Figures link at http://www.pse6.com, *you can select any segment on the surface and see the relationship between the electric field vector* **E** *and the area vector* $\Delta\mathbf{A}_i$.

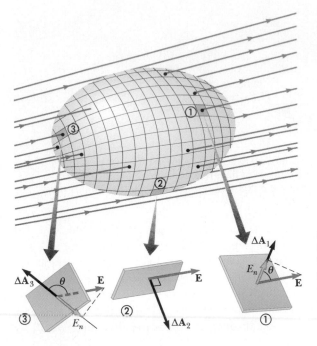

Active Figure 24.4 A closed surface in an electric field. The area vectors $\Delta\mathbf{A}_i$ are, by convention, normal to the surface and point outward. The flux through an area element can be positive (element ①), zero (element ②), or negative (element ③).

Karl Friedrich Gauss

German mathematician and astronomer (1777–1855)

Gauss received a doctoral degree in mathematics from the University of Helmstedt in 1799. In addition to his work in electromagnetism, he made contributions to mathematics and science in number theory, statistics, non-Euclidean geometry, and cometary orbital mechanics. He was a founder of the German Magnetic Union, which studies the Earth's magnetic field on a continual basis.

We are often interested in evaluating the flux through a *closed surface*, which is defined as one that divides space into an inside and an outside region, so that one cannot move from one region to the other without crossing the surface. The surface of a sphere, for example, is a closed surface.

Consider the closed surface in Figure 24.4. The vectors $\Delta\mathbf{A}_i$ point in different directions for the various surface elements, but at each point they are normal to the surface and, by convention, always point outward. At the element labeled ①, the field lines are crossing the surface from the inside to the outside and $\theta < 90°$; hence, the flux $\Delta\Phi_E = \mathbf{E}\cdot\Delta\mathbf{A}_1$ through this element is positive. For element ②, the field lines graze the surface (perpendicular to the vector $\Delta\mathbf{A}_2$); thus, $\theta = 90°$ and the flux is zero. For elements such as ③, where the field lines are crossing the surface from outside to inside, $180° > \theta > 90°$ and the flux is negative because $\cos\theta$ is negative. The *net* flux through the surface is proportional to the net number of lines leaving the surface, where the net number means *the number leaving the surface minus the number entering the surface*. If more lines are leaving than entering, the net flux is positive. If more lines are entering than leaving, the net flux is negative. Using the symbol $\oint$ to represent an integral over a closed surface, we can write the net flux Φ_E through a closed surface as

$$\Phi_E = \oint \mathbf{E}\cdot d\mathbf{A} = \oint E_n\, dA \qquad (24.4)$$

where E_n represents the component of the electric field normal to the surface. If the field is normal to the surface at each point and constant in magnitude, the calculation is straightforward, as it was in Example 24.1. Example 24.2 also illustrates this point.

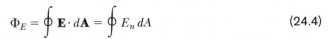

Quick Quiz 24.1 Suppose the radius of the sphere in Example 24.1 is changed to 0.500 m. What happens to the flux through the sphere and the magnitude of the electric field at the surface of the sphere? (a) The flux and field both increase. (b) The flux and field both decrease. (c) The flux increases and the field decreases. (d) The flux decreases and the field increases. (e) The flux remains the same and the field increases. (f) The flux decreases and the field remains the same.

Example 24.2   Flux Through a Cube

Consider a uniform electric field $\mathbf{E}$ oriented in the x direction. Find the net electric flux through the surface of a cube of edge length ℓ, oriented as shown in Figure 24.5.

Solution The net flux is the sum of the fluxes through all faces of the cube. First, note that the flux through four of

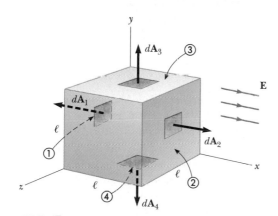

Figure 24.5 (Example 24.2) A closed surface in the shape of a cube in a uniform electric field oriented parallel to the x axis. Side ④ is the bottom of the cube, and side ① is opposite side ②.

the faces (③, ④, and the unnumbered ones) is zero because $\mathbf{E}$ is perpendicular to $d\mathbf{A}$ on these faces.

The net flux through faces ① and ② is

$$\Phi_E = \int_1 \mathbf{E} \cdot d\mathbf{A} + \int_2 \mathbf{E} \cdot d\mathbf{A}$$

For face ①, $\mathbf{E}$ is constant and directed inward but $d\mathbf{A}_1$ is directed outward ($\theta = 180°$); thus, the flux through this face is

$$\int_1 \mathbf{E} \cdot d\mathbf{A} = \int_1 E(\cos 180°) \, dA = -E \int_1 dA = -EA = -E\ell^2$$

because the area of each face is $A = \ell^2$.

For face ②, $\mathbf{E}$ is constant and outward and in the same direction as $d\mathbf{A}_2$ ($\theta = 0°$); hence, the flux through this face is

$$\int_2 \mathbf{E} \cdot d\mathbf{A} = \int_2 E(\cos 0°) \, dA = E \int_2 dA = +EA = E\ell^2$$

Therefore, the net flux over all six faces is

$$\Phi_E = -E\ell^2 + E\ell^2 + 0 + 0 + 0 + 0 = \boxed{0}$$

24.2   Gauss's Law

In this section we describe a general relationship between the net electric flux through a closed surface (often called a *gaussian surface*) and the charge enclosed by the surface. This relationship, known as *Gauss's law*, is of fundamental importance in the study of electric fields.

Let us again consider a positive point charge q located at the center of a sphere of radius r, as shown in Figure 24.6. From Equation 23.9 we know that the magnitude of the electric field everywhere on the surface of the sphere is $E = k_e q/r^2$. As noted in Example 24.1, the field lines are directed radially outward and hence are perpendicular to the surface at every point on the surface. That is, at each surface point, $\mathbf{E}$ is parallel to the vector $\Delta \mathbf{A}_i$ representing a local element of area ΔA_i surrounding the surface point. Therefore,

$$\mathbf{E} \cdot \Delta \mathbf{A}_i = E \, \Delta A_i$$

and from Equation 24.4 we find that the net flux through the gaussian surface is

$$\Phi_E = \oint \mathbf{E} \cdot d\mathbf{A} = \oint E \, dA = E \oint dA$$

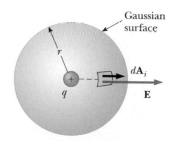

Figure 24.6 A spherical gaussian surface of radius r surrounding a point charge q. When the charge is at the center of the sphere, the electric field is everywhere normal to the surface and constant in magnitude.

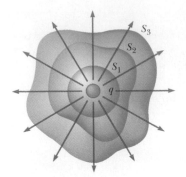

Figure 24.7 Closed surfaces of various shapes surrounding a charge q. The net electric flux is the same through all surfaces.

where we have moved E outside of the integral because, by symmetry, E is constant over the surface and given by $E = k_e q / r^2$. Furthermore, because the surface is spherical, $\oint dA = A = 4\pi r^2$. Hence, the net flux through the gaussian surface is

$$\Phi_E = \frac{k_e q}{r^2} (4\pi r^2) = 4\pi k_e q$$

Recalling from Section 23.3 that $k_e = 1/4\pi\epsilon_0$, we can write this equation in the form

$$\Phi_E = \frac{q}{\epsilon_0} \tag{24.5}$$

We can verify that this expression for the net flux gives the same result as Example 24.1: $\Phi_E = (1.00 \times 10^{-6}\ \text{C})/(8.85 \times 10^{-12}\ \text{C}^2/\text{N} \cdot \text{m}^2) = 1.13 \times 10^5\ \text{N} \cdot \text{m}^2/\text{C}$.

Note from Equation 24.5 that the net flux through the spherical surface is proportional to the charge inside. The flux is independent of the radius r because the area of the spherical surface is proportional to r^2, whereas the electric field is proportional to $1/r^2$. Thus, in the product of area and electric field, the dependence on r cancels.

Now consider several closed surfaces surrounding a charge q, as shown in Figure 24.7. Surface S_1 is spherical, but surfaces S_2 and S_3 are not. From Equation 24.5, the flux that passes through S_1 has the value q/ϵ_0. As we discussed in the preceding section, flux is proportional to the number of electric field lines passing through a surface. The construction shown in Figure 24.7 shows that the number of lines through S_1 is equal to the number of lines through the nonspherical surfaces S_2 and S_3. Therefore, we conclude that **the net flux through *any* closed surface surrounding a point charge q is given by q/ϵ_0 and is independent of the shape of that surface.**

Now consider a point charge located *outside* a closed surface of arbitrary shape, as shown in Figure 24.8. As you can see from this construction, any electric field line that enters the surface leaves the surface at another point. The number of electric field lines entering the surface equals the number leaving the surface. Therefore, we conclude that **the net electric flux through a closed surface that surrounds no charge is zero.** If we apply this result to Example 24.2, we can easily see that the net flux through the cube is zero because there is no charge inside the cube.

Let us extend these arguments to two generalized cases: (1) that of many point charges and (2) that of a continuous distribution of charge. We once again use the superposition principle, which states that **the electric field due to many charges is the vector sum of the electric fields produced by the individual charges.** Therefore, we can express the flux through any closed surface as

$$\oint \mathbf{E} \cdot d\mathbf{A} = \oint (\mathbf{E}_1 + \mathbf{E}_2 + \cdots) \cdot d\mathbf{A}$$

where $\mathbf{E}$ is the total electric field at any point on the surface produced by the vector addition of the electric fields at that point due to the individual charges. Consider the system of charges shown in Figure 24.9. The surface S surrounds only one charge, q_1; hence, the net flux through S is q_1/ϵ_0. The flux through S due to charges q_2, q_3, and q_4 outside it is zero because each electric field line that enters S at one point leaves it at

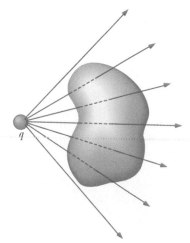

Figure 24.8 A point charge located *outside* a closed surface. The number of lines entering the surface equals the number leaving the surface.

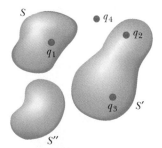

Active Figure 24.9 The net electric flux through any closed surface depends only on the charge *inside* that surface. The net flux through surface S is q_1/ϵ_0, the net flux through surface S' is $(q_2 + q_3)/\epsilon_0$, and the net flux through surface S'' is zero. Charge q_4 does not contribute to the flux through any surface because it is outside all surfaces.

At the Active Figures link at http://www.pse6.com, you can change the size and shape of a closed surface and see the effect on the electric flux of surrounding combinations of charge with that surface.

another. The surface S' surrounds charges q_2 and q_3; hence, the net flux through it is $(q_2 + q_3)/\epsilon_0$. Finally, the net flux through surface S'' is zero because there is no charge inside this surface. That is, *all* the electric field lines that enter S'' at one point leave at another. Notice that charge q_4 does not contribute to the net flux through any of the surfaces because it is outside all of the surfaces.

Gauss's law, which is a generalization of what we have just described, states that the net flux through *any* closed surface is

$$\Phi_E = \oint \mathbf{E} \cdot d\mathbf{A} = \frac{q_{in}}{\epsilon_0} \qquad (24.6)$$

Gauss's law

where q_{in} represents the net charge inside the surface and $\mathbf{E}$ represents the electric field at any point on the surface.

A formal proof of Gauss's law is presented in Section 24.5. When using Equation 24.6, you should note that although the charge q_{in} is the net charge inside the gaussian surface, $\mathbf{E}$ represents the *total electric field,* which includes contributions from charges both inside and outside the surface.

In principle, Gauss's law can be solved for $\mathbf{E}$ to determine the electric field due to a system of charges or a continuous distribution of charge. In practice, however, this type of solution is applicable only in a limited number of highly symmetric situations. In the next section we use Gauss's law to evaluate the electric field for charge distributions that have spherical, cylindrical, or planar symmetry. If one chooses the gaussian surface surrounding the charge distribution carefully, the integral in Equation 24.6 can be simplified.

 PITFALL PREVENTION

24.1 Zero Flux is not Zero Field

We see two situations in which there is zero flux through a closed surface—either there are no charged particles enclosed by the surface or there are charged particles enclosed, but the net charge inside the surface is zero. For either situation, it is *incorrect* to conclude that the electric field on the surface is zero. Gauss's law states that the electric *flux* is proportional to the enclosed charge, not the electric *field.*

Quick Quiz 24.3 If the net flux through a gaussian surface is *zero*, the following four statements *could be true*. Which of the statements *must be true*? (a) There are no charges inside the surface. (b) The net charge inside the surface is zero. (c) The electric field is zero everywhere on the surface. (d) The number of electric field lines entering the surface equals the number leaving the surface.

Quick Quiz 24.4 Consider the charge distribution shown in Figure 24.9. The charges contributing to the total electric *flux* through surface S' are (a) q_1 only (b) q_4 only (c) q_2 and q_3 (d) all four charges (e) none of the charges.

Quick Quiz 24.5 Again consider the charge distribution shown in Figure 24.9. The charges contributing to the total electric *field* at a chosen point on the surface S' are (a) q_1 only (b) q_4 only (c) q_2 and q_3 (d) all four charges (e) none of the charges.

Conceptual Example 24.3   Flux Due to a Point Charge

A spherical gaussian surface surrounds a point charge q. Describe what happens to the total flux through the surface if

(A) the charge is tripled,

(B) the radius of the sphere is doubled,

(C) the surface is changed to a cube, and

(D) the charge is moved to another location inside the surface.

Solution

(A) The flux through the surface is tripled because flux is proportional to the amount of charge inside the surface.

(B) The flux does not change because all electric field lines from the charge pass through the sphere, regardless of its radius.

(C) The flux does not change when the shape of the gaussian surface changes because all electric field lines from the charge pass through the surface, regardless of its shape.

(D) The flux does not change when the charge is moved to another location inside that surface because Gauss's law refers to the total charge enclosed, regardless of where the charge is located inside the surface.

24.3   Application of Gauss's Law to Various Charge Distributions

As mentioned earlier, Gauss's law is useful in determining electric fields when the charge distribution is characterized by a high degree of symmetry. The following examples demonstrate ways of choosing the gaussian surface over which the surface integral given by Equation 24.6 can be simplified and the electric field determined. In choosing the surface, we should always take advantage of the symmetry of the charge distribution so that we can remove E from the integral and solve for it. The goal in this type of calculation is to determine a surface that satisfies one or more of the following conditions:

1. The value of the electric field can be argued by symmetry to be constant over the surface.

2. The dot product in Equation 24.6 can be expressed as a simple algebraic product $E \, dA$ because $\mathbf{E}$ and $d\mathbf{A}$ are parallel.

3. The dot product in Equation 24.6 is zero because $\mathbf{E}$ and $d\mathbf{A}$ are perpendicular.

4. The field can be argued to be zero over the surface.

All four of these conditions are used in examples throughout the remainder of this chapter.

 **PITFALL PREVENTION**

24.2 Gaussian Surfaces are not Real

A gaussian surface is an imaginary surface that you choose to satisfy the conditions listed here. It does not have to coincide with a physical surface in the situation.

Example 24.4   The Electric Field Due to a Point Charge

Starting with Gauss's law, calculate the electric field due to an isolated point charge q.

Solution A single charge represents the simplest possible charge distribution, and we use this familiar case to show how to solve for the electric field with Gauss's law. Figure 24.10 and our discussion of the electric field due to a point charge in Chapter 23 help us to conceptualize the physical situation. Because the space around the single charge has spherical symmetry, we categorize this problem as one in which there is enough symmetry to apply Gauss's law. To analyze any Gauss's law problem, we consider the details of the electric field and choose a gaussian surface that satisfies some or all of the conditions that we have listed above. We choose a spherical gaussian surface of radius r centered on the point charge, as shown in Figure 24.10. The electric field due to a positive point charge is directed radially outward by

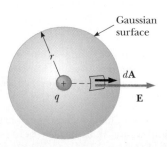

Figure 24.10 (Example 24.4) The point charge q is at the center of the spherical gaussian surface, and $\mathbf{E}$ is parallel to $d\mathbf{A}$ at every point on the surface.

symmetry and is therefore normal to the surface at every point. Thus, as in condition (2), $\mathbf{E}$ is parallel to $d\mathbf{A}$ at each point. Therefore, $\mathbf{E} \cdot d\mathbf{A} = E \, dA$ and Gauss's law gives

$$\Phi_E = \oint \mathbf{E} \cdot d\mathbf{A} = \oint E \, dA = \frac{q}{\epsilon_0}$$

By symmetry, E is constant everywhere on the surface, which satisfies condition (1), so it can be removed from the integral. Therefore,

$$\oint E \, dA = E \oint dA = E(4\pi r^2) = \frac{q}{\epsilon_0}$$

where we have used the fact that the surface area of a sphere is $4\pi r^2$. Now, we solve for the electric field:

$$E = \frac{q}{4\pi\epsilon_0 r^2} = k_e \frac{q}{r^2}$$

To finalize this problem, note that this is the familiar electric field due to a point charge that we developed from Coulomb's law in Chapter 23.

What If? What if the charge in Figure 24.10 were not at the center of the spherical gaussian surface?

Answer In this case, while Gauss's law would still be valid, the situation would not possess enough symmetry to evaluate the electric field. Because the charge is not at the center, the magnitude of $\mathbf{E}$ would vary over the surface of the sphere and the vector $\mathbf{E}$ would not be everywhere perpendicular to the surface.

Example 24.5   A Spherically Symmetric Charge Distribution

An insulating solid sphere of radius a has a uniform volume charge density ρ and carries a total positive charge Q (Fig. 24.11).

(A) Calculate the magnitude of the electric field at a point outside the sphere.

Solution Because the charge distribution is spherically symmetric, we again select a spherical gaussian surface of radius r, concentric with the sphere, as shown in Figure 24.11a. For this choice, conditions (1) and (2) are satisfied, as they were for the point charge in Example 24.4. Following the line of reasoning given in Example 24.4, we find that

$$(1) \qquad E = k_e \frac{Q}{r^2} \qquad \text{(for } r > a)$$

Note that this result is identical to the one we obtained for a point charge. Therefore, we conclude that, **for a uniformly charged sphere, the field in the region external to the sphere is** *equivalent* **to that of a point charge located at the center of the sphere.**

(B) Find the magnitude of the electric field at a point inside the sphere.

Solution In this case we select a spherical gaussian surface having radius $r < a$, concentric with the insulating sphere (Fig. 24.11b). Let us denote the volume of this smaller sphere by V'. To apply Gauss's law in this situation, it is important to recognize that the charge q_{in} within the gaussian surface of volume V' is less than Q. To calculate q_{in}, we use the fact that $q_{\text{in}} = \rho V'$:

$$q_{\text{in}} = \rho V' = \rho (\tfrac{4}{3} \pi r^3)$$

By symmetry, the magnitude of the electric field is constant everywhere on the spherical gaussian surface and is normal to the surface at each point—both conditions

(1) and (2) are satisfied. Therefore, Gauss's law in the region $r < a$ gives

$$\oint E \, dA = E \oint dA = E \, (4\pi r^2) = \frac{q_{\text{in}}}{\epsilon_0}$$

Solving for E gives

$$E = \frac{q_{\text{in}}}{4\pi\epsilon_0 r^2} = \frac{\rho(\tfrac{4}{3}\pi r^3)}{4\pi\epsilon_0 r^2} = \frac{\rho}{3\epsilon_0} \, r$$

Because $\rho = Q/\tfrac{4}{3}\pi a^3$ by definition and because $k_e = 1/4\pi\epsilon_0$, this expression for E can be written as

$$(2) \qquad E = \frac{Qr}{4\pi\epsilon_0 a^3} = k_e \frac{Q}{a^3} \, r \qquad \text{(for } r < a)$$

Note that this result for E differs from the one we obtained in part (A). It shows that $E \rightarrow 0$ as $r \rightarrow 0$. Therefore, the result eliminates the problem that would exist at $r = 0$ if E varied as $1/r^2$ inside the sphere as it does outside the sphere. That is, if $E \propto 1/r^2$ for $r < a$, the field would be infinite at $r = 0$, which is physically impossible.

What If? Suppose we approach the radial position $r = a$ from inside the sphere and from outside. Do we measure the same value of the electric field from both directions?

Answer From Equation (1), we see that the field approaches a value from the outside given by

$$E = \lim_{r \rightarrow a} \left(k_e \frac{Q}{r^2} \right) = k_e \frac{Q}{a^2}$$

From the inside, Equation (2) gives us

$$E = \lim_{r \rightarrow a} \left(k_e \frac{Q}{a^3} \, r \right) = k_e \frac{Q}{a^3} \, a = k_e \frac{Q}{a^2}$$

Thus, the value of the field is the same as we approach the surface from both directions. A plot of E versus r is shown in Figure 24.12. Note that the magnitude of the field is continuous, but the derivative of the field magnitude is not.

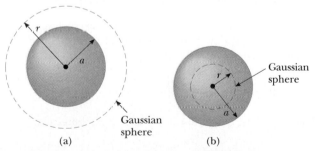

(a)   (b)

Figure 24.11 (Example 24.5) A uniformly charged insulating sphere of radius a and total charge Q. (a) For points outside the sphere, a large spherical gaussian surface is drawn concentric with the sphere. In diagrams such as this, the dotted line represents the intersection of the gaussian surface with the plane of the page. (b) For points inside the sphere, a spherical gaussian surface smaller than the sphere is drawn.

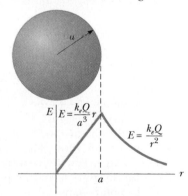

Figure 24.12 (Example 24.5) A plot of E versus r for a uniformly charged insulating sphere. The electric field inside the sphere ($r < a$) varies linearly with r. The field outside the sphere ($r > a$) is the same as that of a point charge Q located at $r = 0$.

 At the Interactive Worked Example link at **http://www.pse6.com,** *you can investigate the electric field inside and outside the sphere.*

Example 24.6   The Electric Field Due to a Thin Spherical Shell

A thin spherical shell of radius a has a total charge Q distributed uniformly over its surface (Fig. 24.13a). Find the electric field at points

(A) outside and

(B) inside the shell.

Solution

(A) The calculation for the field outside the shell is identical to that for the solid sphere shown in Example 24.5a. If we construct a spherical gaussian surface of radius $r > a$ concentric with the shell (Fig. 24.13b), the charge inside this surface is Q. Therefore, the field at a point outside the shell is equivalent to that due to a point charge Q located at the center:

$$E = k_e \frac{Q}{r^2} \qquad \text{(for } r > a)$$

(B) The electric field inside the spherical shell is zero. This follows from Gauss's law applied to a spherical surface of radius $r < a$ concentric with the shell (Fig. 24.13c). Because of the spherical symmetry of the charge distribution and because the net charge inside the surface is zero—satisfaction of conditions (1) and (2) again—application of Gauss's law shows that $E = 0$ in the region $r < a$. We obtain the same results using Equation 23.11 and integrating over the charge distribution. This calculation is rather complicated. Gauss's law allows us to determine these results in a much simpler way.

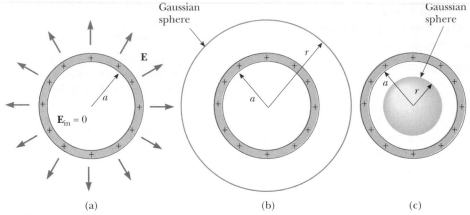

(a)   (b)   (c)

Figure 24.13 (Example 24.6) (a) The electric field inside a uniformly charged spherical shell is zero. The field outside is the same as that due to a point charge Q located at the center of the shell. (b) Gaussian surface for $r > a$. (c) Gaussian surface for $r < a$.

Example 24.7   A Cylindrically Symmetric Charge Distribution

Find the electric field a distance r from a line of positive charge of infinite length and constant charge per unit length λ (Fig. 24.14a).

Solution The symmetry of the charge distribution requires that $\mathbf{E}$ be perpendicular to the line charge and directed outward, as shown in Figure 24.14a and b. To reflect the symmetry of the charge distribution, we select a cylindrical gaussian surface of radius r and length ℓ that is coaxial with the line charge. For the curved part of this surface, $\mathbf{E}$ is constant in magnitude and perpendicular to the surface at each point—satisfaction of conditions (1) and (2). Furthermore, the flux through the ends of the gaussian cylinder is zero because $\mathbf{E}$ is parallel to these surfaces—the first application we have seen of condition (3).

We take the surface integral in Gauss's law over the entire gaussian surface. Because of the zero value of $\mathbf{E} \cdot d\mathbf{A}$ for the ends of the cylinder, however, we can restrict our attention to only the curved surface of the cylinder.

The total charge inside our gaussian surface is $\lambda\ell$. Applying Gauss's law and conditions (1) and (2), we find that for the curved surface

$$\Phi_E = \oint \mathbf{E} \cdot d\mathbf{A} = E \oint dA = EA = \frac{q_{\text{in}}}{\epsilon_0} = \frac{\lambda\ell}{\epsilon_0}$$

The area of the curved surface is $A = 2\pi r\ell$; therefore,

$$E(2\pi r\ell) = \frac{\lambda\ell}{\epsilon_0}$$

$$E = \frac{\lambda}{2\pi\epsilon_0 r} = 2k_e \frac{\lambda}{r} \qquad (24.7)$$

Thus, we see that the electric field due to a cylindrically symmetric charge distribution varies as $1/r$, whereas the field external to a spherically symmetric charge distribution varies as $1/r^2$. Equation 24.7 was also derived by integration of the field of a point charge. (See Problem 35 in Chapter 23.)

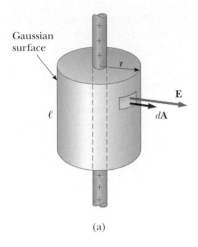

Gaussian surface

r

ℓ

E

$d\mathbf{A}$

(a)

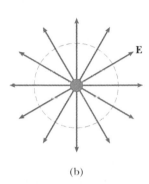

E

(b)

What If? What if the line segment in this example were not infinitely long?

Answer If the line charge in this example were of finite length, the result for E would not be that given by Equation 24.7. A finite line charge does not possess sufficient symmetry for us to make use of Gauss's law. This is because the magnitude of the electric field is no longer constant over the surface of the gaussian cylinder—the field near the ends of the line would be different from that far from the ends. Thus, condition (1) would not be satisfied in this situation. Furthermore, **E** is not perpendicular to the cylindrical surface at all points—the field vectors near the ends would have a component parallel to the line. Thus, condition (2) would not be satisfied. For points close to a finite line charge and far from the ends, Equation 24.7 gives a good approximation of the value of the field.

It is left for you to show (see Problem 29) that the electric field inside a uniformly charged rod of finite radius and infinite length is proportional to r.

Figure 24.14 (Example 24.7) (a) An infinite line of charge surrounded by a cylindrical gaussian surface concentric with the line. (b) An end view shows that the electric field at the cylindrical surface is constant in magnitude and perpendicular to the surface.

Example 24.8   A Plane of Charge

Find the electric field due to an infinite plane of positive charge with uniform surface charge density σ.

Solution By symmetry, **E** must be perpendicular to the plane and must have the same magnitude at all points equidistant from the plane. The fact that the direction of **E** is away from positive charges indicates that the direction of **E** on one side of the plane must be opposite its direction on the other side, as shown in Figure 24.15. A gaussian surface that reflects the symmetry is a small cylinder whose axis is perpendicular to the plane and whose ends

each have an area A and are equidistant from the plane. Because **E** is parallel to the curved surface—and, therefore, perpendicular to $d\mathbf{A}$ everywhere on the surface—condition (3) is satisfied and there is no contribution to the surface integral from this surface. For the flat ends of the cylinder, conditions (1) and (2) are satisfied. The flux through each end of the cylinder is EA; hence, the total flux through the entire gaussian surface is just that through the ends, $\Phi_E = 2EA$.

Noting that the total charge inside the surface is $q_{in} = \sigma A$, we use Gauss's law and find that the total flux through the gaussian surface is

$$\Phi_E = 2EA = \frac{q_{in}}{\epsilon_0} = \frac{\sigma A}{\epsilon_0}$$

leading to

$$E = \frac{\sigma}{2\epsilon_0} \qquad (24.8)$$

Because the distance from each flat end of the cylinder to the plane does not appear in Equation 24.8, we conclude that $E = \sigma/2\epsilon_0$ at *any* distance from the plane. That is, the field is uniform everywhere.

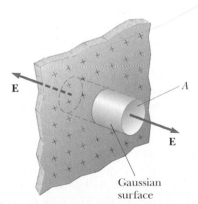

E

A

E

Gaussian surface

Figure 24.15 (Example 24.8) A cylindrical gaussian surface penetrating an infinite plane of charge. The flux is EA through each end of the gaussian surface and zero through its curved surface.

What If? Suppose we place two infinite planes of charge parallel to each other, one positively charged and the other negatively charged. Both planes have the same surface charge density. What does the electric field look like now?

Answer In this situation, the electric fields due to the two planes add in the region between the planes, resulting in a uniform field of magnitude σ/ϵ_0, and cancel elsewhere to give a field of zero. This is a practical way to achieve uniform electric fields, such as those needed in the CRT tube discussed in Section 23.7.

Conceptual Example 24.9   Don't Use Gauss's Law Here!

Explain why Gauss's law cannot be used to calculate the electric field near an electric dipole, a charged disk, or a triangle with a point charge at each corner.

Solution The charge distributions of all these configurations do not have sufficient symmetry to make the use of Gauss's law practical. We cannot find a closed surface surrounding any of these distributions that satisfies one or more of conditions (1) through (4) listed at the beginning of this section.

24.4 Conductors in Electrostatic Equilibrium

As we learned in Section 23.2, a good electrical conductor contains charges (electrons) that are not bound to any atom and therefore are free to move about within the material. When there is no net motion of charge within a conductor, the conductor is in **electrostatic equilibrium.** A conductor in electrostatic equilibrium has the following properties:

Properties of a conductor in electrostatic equilibrium

1. The electric field is zero everywhere inside the conductor.

2. If an isolated conductor carries a charge, the charge resides on its surface.

3. The electric field just outside a charged conductor is perpendicular to the surface of the conductor and has a magnitude σ/ϵ_0, where σ is the surface charge density at that point.

4. On an irregularly shaped conductor, the surface charge density is greatest at locations where the radius of curvature of the surface is smallest.

We verify the first three properties in the discussion that follows. The fourth property is presented here so that we have a complete list of properties for conductors in electrostatic equilibrium, but cannot be verified until Chapter 25.

We can understand the first property by considering a conducting slab placed in an external field **E** (Fig. 24.16). The electric field inside the conductor *must* be zero under the assumption that we have electrostatic equilibrium. If the field were not zero, free electrons in the conductor would experience an electric force ($\mathbf{F} = q\mathbf{E}$) and would accelerate due to this force. This motion of electrons, however, would mean that the conductor is not in electrostatic equilibrium. Thus, the existence of electrostatic equilibrium is consistent only with a zero field in the conductor.

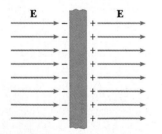

Figure 24.16 A conducting slab in an external electric field **E**. The charges induced on the two surfaces of the slab produce an electric field that opposes the external field, giving a resultant field of zero inside the slab.

Let us investigate how this zero field is accomplished. Before the external field is applied, free electrons are uniformly distributed throughout the conductor. When the external field is applied, the free electrons accelerate to the left in Figure 24.16, causing a plane of negative charge to be present on the left surface. The movement of electrons to the left results in a plane of positive charge on the right surface. These planes of charge create an additional electric field inside the conductor that opposes the external field. As the electrons move, the surface charge densities on the left and right surfaces increase until the magnitude of the internal field equals that of the external field, resulting in a net field of zero inside the conductor. The time it takes a good conductor to reach equilibrium is on the order of 10^{-16} s, which for most purposes can be considered instantaneous.

We can use Gauss's law to verify the second property of a conductor in electrostatic equilibrium. Figure 24.17 shows an arbitrarily shaped conductor. A gaussian surface is drawn inside the conductor and can be as close to the conductor's surface as we wish. As we have just shown, the electric field everywhere inside the conductor is zero when it is in electrostatic equilibrium. Therefore, the electric field must be zero at every point on the gaussian surface, in accordance with condition (4) in Section 24.3. Thus, the net flux through this gaussian surface is zero. From this result and Gauss's law, we conclude that the net charge inside the gaussian surface is zero. Because there can be no net charge inside the gaussian surface (which is arbitrarily close to the conductor's surface), **any net charge on the conductor must reside on its surface.** Gauss's law does not indicate how this excess charge is distributed on the conductor's surface, only that it resides exclusively on the surface.

We can also use Gauss's law to verify the third property. First, note that if the field vector **E** had a component parallel to the conductor's surface, free electrons would experience an electric force and move along the surface; in such a case, the conductor would not be in equilibrium. Thus, the field vector must be perpendicular to the surface. To determine the magnitude of the electric field, we draw a gaussian surface in the shape of a small cylinder whose end faces are parallel to the surface of the conductor (Fig. 24.18). Part of the cylinder is just outside the conductor, and part is inside. The field is perpendicular to the conductor's surface from the condition of electrostatic equilibrium. Thus, we satisfy condition (3) in Section 24.3 for the curved part of the cylindrical gaussian surface—there is no flux through this part of the gaussian surface because **E** is parallel to the surface. There is no flux through the flat face of the cylinder inside the conductor because here **E** = 0; this satisfies condition (4). Hence, the net flux through the gaussian surface is that through only the flat face outside the conductor, where the field is perpendicular to the gaussian surface. Using conditions (1) and (2) for this face, the flux is EA, where E is the electric field just outside the conductor and A is the area of the cylinder's face. Applying Gauss's law to this surface, we obtain

$$\Phi_E = \oint E \, dA = EA = \frac{q_{in}}{\epsilon_0} = \frac{\sigma A}{\epsilon_0}$$

where we have used the fact that $q_{in} = \sigma A$. Solving for E gives for the electric field just outside a charged conductor

$$E = \frac{\sigma}{\epsilon_0} \tag{24.9}$$

Figure 24.19 shows electric field lines made visible by pieces of thread floating in oil. Notice that the field lines are perpendicular to both the cylindrical conducting surface and the straight conducting surface.

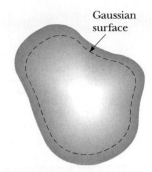

Figure 24.17 A conductor of arbitrary shape. The broken line represents a gaussian surface that can be as close to the surface of the conductor as we wish.

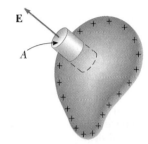

Figure 24.18 A gaussian surface in the shape of a small cylinder is used to calculate the electric field just outside a charged conductor. The flux through the gaussian surface is EA. Remember that **E** is zero inside the conductor.

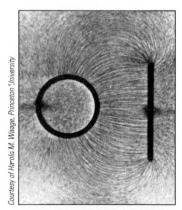

Figure 24.19 Electric field pattern surrounding a charged conducting plate placed near an oppositely charged conducting cylinder. Small pieces of thread suspended in oil align with the electric field lines. Note that (1) the field lines are perpendicular to both conductors and (2) there are no lines inside the cylinder ($E = 0$).

Quick Quiz 24.6 Your little brother likes to rub his feet on the carpet and then touch you to give you a shock. While you are trying to escape the shock treatment, you discover a hollow metal cylinder in your basement, large enough to climb inside. In which of the following cases will you *not* be shocked? (a) You climb inside the cylinder, making contact with the inner surface, and your charged brother touches the outer metal surface. (b) Your charged brother is inside touching the inner metal surface and you are outside, touching the outer metal surface. (c) Both of you are outside the cylinder, touching its outer metal surface but not touching each other directly.

Example 24.10 **A Sphere Inside a Spherical Shell**

A solid conducting sphere of radius a carries a net positive charge $2Q$. A conducting spherical shell of inner radius b and outer radius c is concentric with the solid sphere and carries a net charge $-Q$. Using Gauss's law, find the electric field in the regions labeled ①, ②, ③, and ④ in Figure 24.20 and the charge distribution on the shell when the entire system is in electrostatic equilibrium.

Solution First note that the charge distributions on both the sphere and the shell are characterized by spherical symmetry around their common center. To determine the electric field at various distances r from this center, we construct a spherical gaussian surface for each of the four regions of interest. Such a surface for region ② is shown in Figure 24.20.

To find E inside the solid sphere (region ①), consider a gaussian surface of radius $r < a$. Because there can be no charge inside a conductor in electrostatic equilibrium, we see that $q_{in} = 0$; thus, on the basis of Gauss's law and symmetry, $E_1 = 0$ for $r < a$.

In region ②—between the surface of the solid sphere and the inner surface of the shell—we construct a spherical gaussian surface of radius r where $a < r < b$ and note that the charge inside this surface is $+2Q$ (the charge on the solid sphere). Because of the spherical symmetry, the electric field lines must be directed radially outward and be

constant in magnitude on the gaussian surface. Following Example 24.4 and using Gauss's law, we find that

$$E_2 A = E_2(4\pi r^2) = \frac{q_{in}}{\epsilon_0} = \frac{2Q}{\epsilon_0}$$

$$E_2 = \frac{2Q}{4\pi\epsilon_0 r^2} = \boxed{\frac{2k_eQ}{r^2}} \qquad \text{(for } a < r < b)$$

In region ④, where $r > c$, the spherical gaussian surface we construct surrounds a total charge of $q_{in} = 2Q + (-Q) = Q$. Therefore, application of Gauss's law to this surface gives

$$E_4 = \boxed{\frac{k_eQ}{r^2}} \qquad \text{(for } r > c)$$

In region ③, the electric field must be zero because the spherical shell is also a conductor in equilibrium. Figure 24.21 shows a graphical representation of the variation of electric field with r.

If we construct a gaussian surface of radius r where $b < r < c$, we see that q_{in} must be zero because $E_3 = 0$. From this argument, we conclude that the charge on the inner surface of the spherical shell must be $-2Q$ to cancel the charge $+2Q$ on the solid sphere. Because the net charge on the shell is $-Q$, we conclude that its outer surface must carry a charge $+Q$.

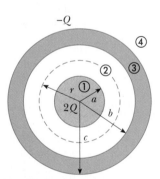

Figure 24.20 (Example 24.10) A solid conducting sphere of radius a and carrying a charge $2Q$ surrounded by a conducting spherical shell carrying a charge $-Q$.

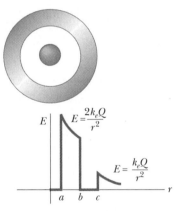

Figure 24.21 (Example 24.10) A plot of E versus r for the two-conductor system shown in Figure 24.20.

 Explore the electric field of the system in Figure 24.20 at the Interactive Worked Example link at **http://www.pse6.com.**

24.5 Formal Derivation of Gauss's Law

One way of deriving Gauss's law involves *solid angles*. Consider a spherical surface of radius r containing an area element ΔA. The solid angle $\Delta\Omega$ (Ω: uppercase Greek omega) subtended at the center of the sphere by this element is defined to be

$$\Delta\Omega \equiv \frac{\Delta A}{r^2}$$

From this equation, we see that $\Delta\Omega$ has no dimensions because ΔA and r^2 both have dimensions L^2. The dimensionless unit of a solid angle is the **steradian.** (You may want to compare this equation to Equation 10.1b, the definition of the radian.) Because the

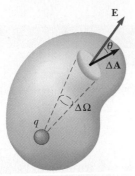

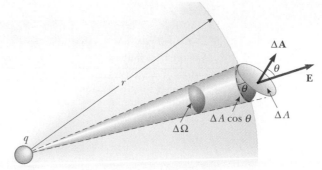

Figure 24.22 A closed surface of arbitrary shape surrounds a point charge q. The net electric flux through the surface is independent of the shape of the surface.

Figure 24.23 The area element ΔA subtends a solid angle $\Delta\Omega = (\Delta A \cos\theta)/r^2$ at the charge q.

surface area of a sphere is $4\pi r^2$, the total solid angle subtended by the sphere is

$$\Omega = \frac{4\pi r^2}{r^2} = 4\pi \text{ steradians}$$

Now consider a point charge q surrounded by a closed surface of arbitrary shape (Fig. 24.22). The total electric flux through this surface can be obtained by evaluating $\mathbf{E} \cdot \Delta\mathbf{A}$ for each small area element ΔA and summing over all elements. The flux through each element is

$$\Delta\Phi_E = \mathbf{E} \cdot \Delta\mathbf{A} = (E \cos\theta)\Delta A = k_e q \frac{\Delta A \cos\theta}{r^2}$$

where r is the distance from the charge to the area element, θ is the angle between the electric field $\mathbf{E}$ and $\Delta\mathbf{A}$ for the element, and $E = k_e q/r^2$ for a point charge. In Figure 24.23, we see that the projection of the area element perpendicular to the radius vector is $\Delta A \cos\theta$. Thus, the quantity $(\Delta A \cos\theta)/r^2$ is equal to the solid angle $\Delta\Omega$ that the surface element ΔA subtends at the charge q. We also see that $\Delta\Omega$ is equal to the solid angle subtended by the area element of a spherical surface of radius r. Because the total solid angle at a point is 4π steradians, the total flux through the closed surface is

$$\Phi_E = k_e q \oint \frac{dA \cos\theta}{r^2} = k_e q \oint d\Omega = 4\pi k_e q = \frac{q}{\epsilon_0}$$

Thus we have derived Gauss's law, Equation 24.6. Note that this result is independent of the shape of the closed surface and independent of the position of the charge within the surface.

SUMMARY

Electric flux is proportional to the number of electric field lines that penetrate a surface. If the electric field is uniform and makes an angle θ with the normal to a surface of area A, the electric flux through the surface is

$$\Phi_E = EA \cos\theta \tag{24.2}$$

In general, the electric flux through a surface is

$$\Phi_E = \int_{\text{surface}} \mathbf{E} \cdot d\mathbf{A} \tag{24.3}$$

Take a practice test for this chapter by clicking on the Practice Test link at http://www.pse6.com.

Table 24.1

Typical Electric Field Calculations Using Gauss's Law		
Charge Distribution	**Electric Field**	**Location**
Insulating sphere of radius R, uniform charge density, and total charge Q	$\begin{cases} k_e \dfrac{Q}{r^2} \\[2mm] k_e \dfrac{Q}{R^2} r \end{cases}$	$r > R$ $r < R$
Thin spherical shell of radius R and total charge Q	$\begin{cases} k_e \dfrac{Q}{r^2} \\[2mm] 0 \end{cases}$	$r > R$ $r < R$
Line charge of infinite length and charge per unit length λ	$2k_e \dfrac{\lambda}{r}$	Outside the line
Infinite charged plane having surface charge density σ	$\dfrac{\sigma}{2\epsilon_0}$	Everywhere outside the plane
Conductor having surface charge density σ	$\begin{cases} \dfrac{\sigma}{\epsilon_0} \\[2mm] 0 \end{cases}$	Just outside the conductor Inside the conductor

You should be able to apply Equations 24.2 and 24.3 in a variety of situations, particularly those in which symmetry simplifies the calculation.

Gauss's law says that the net electric flux Φ_E through any closed gaussian surface is equal to the *net* charge q_{in} inside the surface divided by ϵ_0:

$$\Phi_E = \oint \mathbf{E} \cdot d\mathbf{A} = \frac{q_{in}}{\epsilon_0} \qquad (24.6)$$

Using Gauss's law, you can calculate the electric field due to various symmetric charge distributions. Table 24.1 lists some typical results.

A conductor in **electrostatic equilibrium** has the following properties:

1. The electric field is zero everywhere inside the conductor.
2. Any net charge on the conductor resides entirely on its surface.
3. The electric field just outside the conductor is perpendicular to its surface and has a magnitude σ/ϵ_0, where σ is the surface charge density at that point.
4. On an irregularly shaped conductor, the surface charge density is greatest where the radius of curvature of the surface is the smallest.

QUESTIONS

1. The Sun is lower in the sky during the winter months than it is in the summer. How does this change the flux of sunlight hitting a given area on the surface of the Earth? How does this affect the weather?

2. If the electric field in a region of space is zero, can you conclude that no electric charges are in that region? Explain.

3. If more electric field lines leave a gaussian surface than enter it, what can you conclude about the net charge enclosed by that surface?

4. A uniform electric field exists in a region of space in which there are no charges. What can you conclude about the net electric flux through a gaussian surface placed in this region of space?

5. If the total charge inside a closed surface is known but the distribution of the charge is unspecified, can you use Gauss's law to find the electric field? Explain.

6. Explain why the electric flux through a closed surface with a given enclosed charge is independent of the size or shape of the surface.

7. Consider the electric field due to a nonconducting infinite plane having a uniform charge density. Explain why the electric field does not depend on the distance from the plane, in terms of the spacing of the electric field lines.

8. Use Gauss's law to explain why electric field lines must begin or end on electric charges. (*Suggestion:* Change the size of the gaussian surface.)

9. On the basis of the repulsive nature of the force between like charges and the freedom of motion of charge within a conductor, explain why excess charge on an isolated conductor must reside on its surface.

10. A person is placed in a large hollow metallic sphere that is insulated from ground. If a large charge is placed on the sphere, will the person be harmed upon touching the inside of the sphere? Explain what will happen if the person also has an initial charge whose sign is opposite that of the charge on the sphere.

11. Two solid spheres, both of radius R, carry identical total charges, Q. One sphere is a good conductor while the other is an insulator. If the charge on the insulating sphere is uniformly distributed throughout its interior volume, how do the electric fields outside these two spheres compare? Are the fields identical inside the two spheres?

12. A common demonstration involves charging a rubber balloon, which is an insulator, by rubbing it on your hair, and touching the balloon to a ceiling or wall, which is also an insulator. The electrical attraction between the charged balloon and the neutral wall results in the balloon sticking to the wall. Imagine now that we have two infinitely large flat sheets of insulating material. One is charged and the other is neutral. If these are brought into contact, will an attractive force exist between them, as there was for the balloon and the wall?

13. You may have heard that one of the safer places to be during a lightning storm is inside a car. Why would this be the case?

PROBLEMS

1, 2, 3 = straightforward, intermediate, challenging   ☐ = full solution available in the *Student Solutions Manual and Study Guide*

🕸 = coached solution with hints available at http://www.pse6.com   🖥 = computer useful in solving problem

▒ = paired numerical and symbolic problems

Section 24.1   Electric Flux

1. An electric field with a magnitude of 3.50 kN/C is applied along the x axis. Calculate the electric flux through a rectangular plane 0.350 m wide and 0.700 m long assuming that (a) the plane is parallel to the yz plane; (b) the plane is parallel to the xy plane; (c) the plane contains the y axis, and its normal makes an angle of 40.0° with the x axis.

2. A vertical electric field of magnitude 2.00×10^4 N/C exists above the Earth's surface on a day when a thunderstorm is brewing. A car with a rectangular size of 6.00 m by 3.00 m is traveling along a roadway sloping downward at 10.0°. Determine the electric flux through the bottom of the car.

3. A 40.0-cm-diameter loop is rotated in a uniform electric field until the position of maximum electric flux is found. The flux in this position is measured to be 5.20×10^5 N·m²/C. What is the magnitude of the electric field?

4. Consider a closed triangular box resting within a horizontal electric field of magnitude $E = 7.80 \times 10^4$ N/C as shown in Figure P24.4. Calculate the electric flux through (a) the vertical rectangular surface, (b) the slanted surface, and (c) the entire surface of the box.

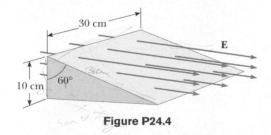

Figure P24.4

5. A uniform electric field $a\hat{\mathbf{i}} + b\hat{\mathbf{j}}$ intersects a surface of area A. What is the flux through this area if the surface lies (a) in the yz plane? (b) in the xz plane? (c) in the xy plane?

6. A point charge q is located at the center of a uniform ring having linear charge density λ and radius a, as shown in Figure P24.6. Determine the total electric flux through a sphere centered at the point charge and having radius R, where $R < a$.

Figure P24.6

7. A pyramid with horizontal square base, 6.00 m on each side, and a height of 4.00 m is placed in a vertical electric field of 52.0 N/C. Calculate the total electric flux through the pyramid's four slanted surfaces.

8. A cone with base radius R and height h is located on a horizontal table. A horizontal uniform field E penetrates the cone, as shown in Figure P24.8. Determine the electric flux that enters the left-hand side of the cone.

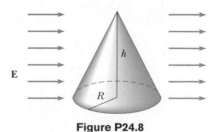

Figure P24.8

Section 24.2   Gauss's Law

9. The following charges are located inside a submarine: 5.00 μC, − 9.00 μC, 27.0 μC, and − 84.0 μC. (a) Calculate

the net electric flux through the hull of the submarine. (b) Is the number of electric field lines leaving the submarine greater than, equal to, or less than the number entering it?

10. The electric field everywhere on the surface of a thin spherical shell of radius 0.750 m is measured to be 890 N/C and points radially toward the center of the sphere. (a) What is the net charge within the sphere's surface? (b) What can you conclude about the nature and distribution of the charge inside the spherical shell?

11. Four closed surfaces, S_1 through S_4, together with the charges $-2Q$, Q, and $-Q$ are sketched in Figure P24.11. (The colored lines are the intersections of the surfaces with the page.) Find the electric flux through each surface.

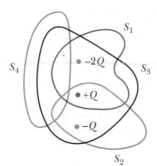

Figure P24.11

12. (a) A point charge q is located a distance d from an infinite plane. Determine the electric flux through the plane due to the point charge. (b) **What If?** A point charge q is located a *very small* distance from the center of a *very large* square on the line perpendicular to the square and going through its center. Determine the approximate electric flux through the square due to the point charge. (c) Explain why the answers to parts (a) and (b) are identical.

13. Calculate the total electric flux through the paraboloidal surface due to a uniform electric field of magnitude E_0 in the direction shown in Figure P24.13.

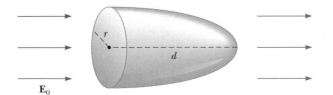

Figure P24.13

14. A point charge of 12.0 μC is placed at the center of a spherical shell of radius 22.0 cm. What is the total electric flux through (a) the surface of the shell and (b) any hemispherical surface of the shell? (c) Do the results depend on the radius? Explain.

15. [www] A point charge Q is located just above the center of the flat face of a hemisphere of radius R as shown in Figure P24.15. What is the electric flux (a) through the curved surface and (b) through the flat face?

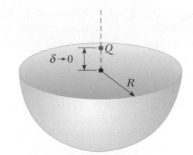

Figure P24.15

16. In the air over a particular region at an altitude of 500 m above the ground the electric field is 120 N/C directed downward. At 600 m above the ground the electric field is 100 N/C downward. What is the average volume charge density in the layer of air between these two elevations? Is it positive or negative?

17. A point charge $Q = 5.00\ \mu$C is located at the center of a cube of edge $L = 0.100$ m. In addition, six other identical point charges having $q = -1.00\ \mu$C are positioned symmetrically around Q as shown in Figure P24.17. Determine the electric flux through one face of the cube.

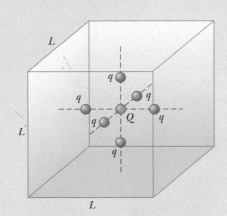

Figure P24.17 Problems 17 and 18.

18. A positive point charge Q is located at the center of a cube of edge L. In addition, six other identical negative point charges q are positioned symmetrically around Q as shown in Figure P24.17. Determine the electric flux through one face of the cube.

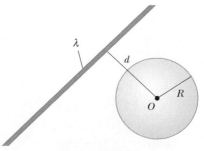

Figure P24.19

19. An infinitely long line charge having a uniform charge per unit length λ lies a distance d from point O as shown in Figure P24.19. Determine the total electric flux through the surface of a sphere of radius R centered at O resulting from this line charge. Consider both cases, where $R < d$ and $R > d$.

20. An uncharged nonconducting hollow sphere of radius 10.0 cm surrounds a 10.0-μC charge located at the origin of a cartesian coordinate system. A drill with a radius of 1.00 mm is aligned along the z axis, and a hole is drilled in the sphere. Calculate the electric flux through the hole.

21. A charge of 170 μC is at the center of a cube of edge 80.0 cm. (a) Find the total flux through each face of the cube. (b) Find the flux through the whole surface of the cube. (c) **What If?** Would your answers to parts (a) or (b) change if the charge were not at the center? Explain.

22. The line ag in Figure P24.22 is a diagonal of a cube. A point charge q is located on the extension of line ag, very close to vertex a of the cube. Determine the electric flux through each of the sides of the cube which meet at the point a.

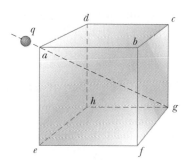

Figure P24.22

Section 24.3   Application of Gauss's Law to Various Charge Distributions

23. Determine the magnitude of the electric field at the surface of a lead-208 nucleus, which contains 82 protons and 126 neutrons. Assume the lead nucleus has a volume 208 times that of one proton, and consider a proton to be a sphere of radius 1.20×10^{-15} m.

24. A solid sphere of radius 40.0 cm has a total positive charge of 26.0 μC uniformly distributed throughout its volume. Calculate the magnitude of the electric field (a) 0 cm, (b) 10.0 cm, (c) 40.0 cm, and (d) 60.0 cm from the center of the sphere.

25. A 10.0-g piece of Styrofoam carries a net charge of -0.700 μC and floats above the center of a large horizontal sheet of plastic that has a uniform charge density on its surface. What is the charge per unit area on the plastic sheet?

26. A cylindrical shell of radius 7.00 cm and length 240 cm has its charge uniformly distributed on its curved surface. The magnitude of the electric field at a point 19.0 cm radially outward from its axis (measured from the midpoint of the shell) is 36.0 kN/C. Find (a) the net charge on the shell and (b) the electric field at a point 4.00 cm from the axis, measured radially outward from the midpoint of the shell.

27. A particle with a charge of -60.0 nC is placed at the center of a nonconducting spherical shell of inner radius 20.0 cm and outer radius 25.0 cm. The spherical shell carries charge with a uniform density of -1.33 μC/m^3. A proton moves in a circular orbit just outside the spherical shell. Calculate the speed of the proton.

28. A nonconducting wall carries a uniform charge density of 8.60 μC/cm^2. What is the electric field 7.00 cm in front of the wall? Does your result change as the distance from the wall is varied?

29. Consider a long cylindrical charge distribution of radius R with a uniform charge density ρ. Find the electric field at distance r from the axis where $r < R$.

30. A solid plastic sphere of radius 10.0 cm has charge with uniform density throughout its volume. The electric field 5.00 cm from the center is 86.0 kN/C radially inward. Find the magnitude of the electric field 15.0 cm from the center.

31. Consider a thin spherical shell of radius 14.0 cm with a total charge of 32.0 μC distributed uniformly on its surface. Find the electric field (a) 10.0 cm and (b) 20.0 cm from the center of the charge distribution.

32. In nuclear fission, a nucleus of uranium-238, which contains 92 protons, can divide into two smaller spheres, each having 46 protons and a radius of 5.90×10^{-15} m. What is the magnitude of the repulsive electric force pushing the two spheres apart?

33. Fill two rubber balloons with air. Suspend both of them from the same point and let them hang down on strings of equal length. Rub each with wool or on your hair, so that they hang apart with a noticeable separation from each other. Make order-of-magnitude estimates of (a) the force on each, (b) the charge on each, (c) the field each creates at the center of the other, and (d) the total flux of electric field created by each balloon. In your solution state the quantities you take as data and the values you measure or estimate for them.

34. An insulating solid sphere of radius a has a uniform volume charge density and carries a total positive charge Q. A spherical gaussian surface of radius r, which shares a common center with the insulating sphere, is inflated starting from $r = 0$. (a) Find an expression for the electric flux passing through the surface of the gaussian sphere as a function of r for $r < a$. (b) Find an expression for the electric flux for $r > a$. (c) Plot the flux versus r.

35. A uniformly charged, straight filament 7.00 m in length has a total positive charge of 2.00 μC. An uncharged cardboard cylinder 2.00 cm in length and 10.0 cm in radius surrounds the filament at its center, with the filament as the axis of the cylinder. Using reasonable approximations, find (a) the electric field at the surface of the cylinder and (b) the total electric flux through the cylinder.

36. An insulating sphere is 8.00 cm in diameter and carries a 5.70-μC charge uniformly distributed throughout its interior volume. Calculate the charge enclosed by a concentric spherical surface with radius (a) $r = 2.00$ cm and (b) $r = 6.00$ cm.

37. A large flat horizontal sheet of charge has a charge per unit area of 9.00 μC/m^2. Find the electric field just above the middle of the sheet.

38. The charge per unit length on a long, straight filament is $-90.0 \ \mu C/m$. Find the electric field (a) 10.0 cm, (b) 20.0 cm, and (c) 100 cm from the filament, where distances are measured perpendicular to the length of the filament.

Section 24.4   Conductors in Electrostatic Equilibrium

39. A long, straight metal rod has a radius of 5.00 cm and a charge per unit length of 30.0 nC/m. Find the electric field (a) 3.00 cm, (b) 10.0 cm, and (c) 100 cm from the axis of the rod, where distances are measured perpendicular to the rod.

40. On a clear, sunny day, a vertical electric field of about 130 N/C points down over flat ground. What is the surface charge density on the ground for these conditions?

41. A very large, thin, flat plate of aluminum of area A has a total charge Q uniformly distributed over its surfaces. Assuming the same charge is spread uniformly over the *upper* surface of an otherwise identical glass plate, compare the electric fields just above the center of the upper surface of each plate.

42. A solid copper sphere of radius 15.0 cm carries a charge of 40.0 nC. Find the electric field (a) 12.0 cm, (b) 17.0 cm, and (c) 75.0 cm from the center of the sphere. (d) **What If?** How would your answers change if the sphere were hollow?

43. A square plate of copper with 50.0-cm sides has no net charge and is placed in a region of uniform electric field of 80.0 kN/C directed perpendicularly to the plate. Find (a) the charge density of each face of the plate and (b) the total charge on each face.

44. A solid conducting sphere of radius 2.00 cm has a charge of 8.00 μC. A conducting spherical shell of inner radius 4.00 cm and outer radius 5.00 cm is concentric with the solid sphere and has a total charge of $-4.00 \ \mu C$. Find the electric field at (a) $r = 1.00$ cm, (b) $r = 3.00$ cm, (c) $r = 4.50$ cm, and (d) $r = 7.00$ cm from the center of this charge configuration.

45. Two identical conducting spheres each having a radius of 0.500 cm are connected by a light 2.00-m-long conducting wire. A charge of 60.0 μC is placed on one of the conductors. Assume that the surface distribution of charge on each sphere is uniform. Determine the tension in the wire.

46. The electric field on the surface of an irregularly shaped conductor varies from 56.0 kN/C to 28.0 kN/C. Calculate the local surface charge density at the point on the surface where the radius of curvature of the surface is (a) greatest and (b) smallest.

47. A long, straight wire is surrounded by a hollow metal cylinder whose axis coincides with that of the wire. The wire has a charge per unit length of λ, and the cylinder has a net charge per unit length of 2λ. From this information, use Gauss's law to find (a) the charge per unit length on the inner and outer surfaces of the cylinder and (b) the electric field outside the cylinder, a distance r from the axis.

48. A conducting spherical shell of radius 15.0 cm carries a net charge of $-6.40 \ \mu C$ uniformly distributed on its surface. Find the electric field at points (a) just outside the shell and (b) inside the shell.

49. A thin square conducting plate 50.0 cm on a side lies in the xy plane. A total charge of 4.00×10^{-8} C is placed on the plate. Find (a) the charge density on the plate, (b) the electric field just above the plate, and (c) the electric field just below the plate. You may assume that the charge density is uniform.

50. A conducting spherical shell of inner radius a and outer radius b carries a net charge Q. A point charge q is placed at the center of this shell. Determine the surface charge density on (a) the inner surface of the shell and (b) the outer surface of the shell.

51. A hollow conducting sphere is surrounded by a larger concentric spherical conducting shell. The inner sphere has charge $-Q$, and the outer shell has net charge $+3Q$. The charges are in electrostatic equilibrium. Using Gauss's law, find the charges and the electric fields everywhere.

52. A positive point charge is at a distance $R/2$ from the center of an uncharged thin conducting spherical shell of radius R. Sketch the electric field lines set up by this arrangement both inside and outside the shell.

Section 24.5   Formal Derivation of Gauss's Law

53. A sphere of radius R surrounds a point charge Q, located at its center. (a) Show that the electric flux through a circular cap of half-angle θ (Fig. P24.53) is

$$\Phi_E = \frac{Q}{2\epsilon_0} \, (1 - \cos \theta)$$

What is the flux for (b) $\theta = 90°$ and (c) $\theta = 180°$?

Figure P24.53

Additional Problems

54. A nonuniform electric field is given by the expression $\mathbf{E} = ay\hat{\mathbf{i}} + bz\hat{\mathbf{j}} + cx\hat{\mathbf{k}}$, where a, b, and c are constants. Determine the electric flux through a rectangular surface in the xy plane, extending from $x = 0$ to $x = w$ and from $y = 0$ to $y = h$.

55. A solid insulating sphere of radius a carries a net positive charge $3Q$, uniformly distributed throughout its volume. Concentric with this sphere is a conducting spherical shell with inner radius b and outer radius c, and having a net charge $-Q$, as shown in Figure P24.55. (a) Construct a spherical gaussian surface of radius $r > c$ and find the net charge enclosed by this surface. (b) What is the direction of the electric field at $r > c$? (c) Find the electric field at $r > c$. (d) Find the electric field in the region with radius r where $c > r > b$. (e) Construct a spherical gaussian surface of radius r, where $c > r > b$, and find the net charge enclosed by this surface. (f) Construct a spherical gaussian surface of radius r, where $b > r > a$, and find the net charge enclosed by this surface. (g) Find the electric field in the region $b > r > a$. (h) Construct a spherical gaussian surface of radius $r < a$, and find an expression for the net charge enclosed by this surface, as a function of r. Note that the charge inside this surface is less than $3Q$. (i) Find the electric field in the region $r < a$. (j) Determine the charge on the inner surface of the conducting shell. (k) Determine the charge on the outer surface of the conducting shell. (l) Make a plot of the magnitude of the electric field versus r.

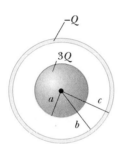

Figure P24.55

56. Consider two identical conducting spheres whose surfaces are separated by a small distance. One sphere is given a large net positive charge while the other is given a small net positive charge. It is found that the force between them is attractive even though both spheres have net charges of the same sign. Explain how this is possible.

57. A solid, insulating sphere of radius a has a uniform charge density ρ and a total charge Q. Concentric with this sphere is an uncharged, conducting hollow sphere whose inner and outer radii are b and c, as shown in Figure P24.57. (a) Find the magnitude of the electric field in the regions $r < a$, $a < r < b$, $b < r < c$, and $r > c$. (b) Determine the induced charge per unit area on the inner and outer surfaces of the hollow sphere.

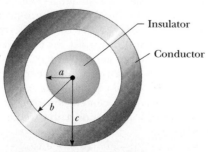

Figure P24.57 Problems 57 and 58.

58. For the configuration shown in Figure P24.57, suppose that $a = 5.00$ cm, $b = 20.0$ cm, and $c = 25.0$ cm. Furthermore, suppose that the electric field at a point 10.0 cm from the center is measured to be 3.60×10^3 N/C radially inward while the electric field at a point 50.0 cm from the center is 2.00×10^2 N/C radially outward. From this information, find (a) the charge on the insulating sphere, (b) the net charge on the hollow conducting sphere, and (c) the charges on the inner and outer surfaces of the hollow conducting sphere.

59. A particle of mass m and charge q moves at high speed along the x axis. It is initially near $x = -\infty$, and it ends up near $x = +\infty$. A second charge Q is fixed at the point $x = 0$, $y = -d$. As the moving charge passes the stationary charge, its x component of velocity does not change appreciably, but it acquires a small velocity in the y direction. Determine the angle through which the moving charge is deflected. *Suggestion:* The integral you encounter in determining v_y can be evaluated by applying Gauss's law to a long cylinder of radius d, centered on the stationary charge.

60. **Review problem.** An early (incorrect) model of the hydrogen atom, suggested by J. J. Thomson, proposed that a positive cloud of charge $+e$ was uniformly distributed throughout the volume of a sphere of radius R, with the electron an equal-magnitude negative point charge $-e$ at the center. (a) Using Gauss's law, show that the electron would be in equilibrium at the center and, if displaced from the center a distance $r < R$, would experience a restoring force of the form $F = -Kr$, where K is a constant. (b) Show that $K = k_e e^2/R^3$. (c) Find an expression for the frequency f of simple harmonic oscillations that an electron of mass m_e would undergo if displaced a small distance ($< R$) from the center and released. (d) Calculate a numerical value for R that would result in a frequency of 2.47×10^{15} Hz, the frequency of the light radiated in the most intense line in the hydrogen spectrum.

61. An infinitely long cylindrical insulating shell of inner radius a and outer radius b has a uniform volume charge density ρ. A line of uniform linear charge density λ is placed along the axis of the shell. Determine the electric field everywhere.

62. Two infinite, nonconducting sheets of charge are parallel to each other, as shown in Figure P24.62. The sheet on the left has a uniform surface charge density σ, and the one

Figure P24.62

on the right has a uniform charge density $-\sigma$. Calculate the electric field at points (a) to the left of, (b) in between, and (c) to the right of the two sheets.

63. *What If?* Repeat the calculations for Problem 62 when both sheets have *positive* uniform surface charge densities of value σ.

64. A sphere of radius $2a$ is made of a nonconducting material that has a uniform volume charge density ρ. (Assume that the material does not affect the electric field.) A spherical cavity of radius a is now removed from the sphere, as shown in Figure P24.64. Show that the electric field within the cavity is uniform and is given by $E_x = 0$ and $E_y = \rho a/3\epsilon_0$. (*Suggestion:* The field within the cavity is the superposition of the field due to the original uncut sphere, plus the field due to a sphere the size of the cavity with a uniform negative charge density $-\rho$.)

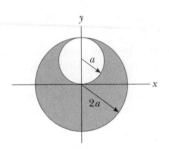

Figure P24.64

65. A uniformly charged spherical shell with surface charge density σ contains a circular hole in its surface. The radius of the hole is small compared with the radius of the sphere. What is the electric field at the center of the hole? (*Suggestion:* This problem, like Problem 64, can be solved by using the idea of superposition.)

66. A closed surface with dimensions $a = b = 0.400$ m and $c = 0.600$ m is located as in Figure P24.66. The left edge of the closed surface is located at position $x = a$. The electric field throughout the region is nonuniform and given by $\mathbf{E} = (3.0 + 2.0x^2)\hat{\mathbf{i}}$ N/C, where x is in meters. Calculate the net electric flux leaving the closed surface. What net charge is enclosed by the surface?

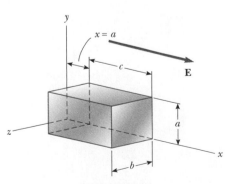

Figure P24.66

67. A solid insulating sphere of radius R has a nonuniform charge density that varies with r according to the expression $\rho = Ar^2$, where A is a constant and $r < R$ is measured from the center of the sphere. (a) Show that the magnitude of the electric field outside $(r > R)$ the sphere is $E = AR^5/5\epsilon_0 r^2$. (b) Show that the magnitude of the electric field inside $(r < R)$ the sphere is $E = Ar^3/5\epsilon_0$. (*Suggestion:* The total charge Q on the sphere is equal to the integral of $\rho\, dV$, where r extends from 0 to R; also, the charge q within a radius $r < R$ is less than Q. To evaluate the integrals, note that the volume element dV for a spherical shell of radius r and thickness dr is equal to $4\pi r^2 dr$.)

68. A point charge Q is located on the axis of a disk of radius R at a distance b from the plane of the disk (Fig. P24.68). Show that if one fourth of the electric flux from the charge passes through the disk, then $R = \sqrt{3}b$.

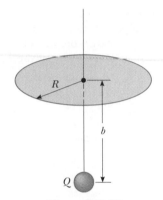

Figure P24.68

69. A spherically symmetric charge distribution has a charge density given by $\rho = a/r$, where a is constant. Find the electric field as a function of r. (*Suggestion:* The charge within a sphere of radius R is equal to the integral of $\rho\, dV$, where r extends from 0 to R. To evaluate the integral, note that the volume element dV for a spherical shell of radius r and thickness dr is equal to $4\pi r^2 dr$.)

70. An infinitely long insulating cylinder of radius R has a volume charge density that varies with the radius as

$$\rho = \rho_0 \left(a - \frac{r}{b}\right)$$

where ρ_0, a, and b are positive constants and r is the distance from the axis of the cylinder. Use Gauss's law to determine the magnitude of the electric field at radial distances (a) $r < R$ and (b) $r > R$.

71. **Review problem.** A slab of insulating material (infinite in two of its three dimensions) has a uniform positive charge density ρ. An edge view of the slab is shown in Figure P24.71. (a) Show that the magnitude of the electric field a distance x from its center and inside the slab is $E = \rho x/\epsilon_0$. (b) **What If?** Suppose an electron of charge $-e$ and mass m_e can move freely within the slab. It is released from rest at a distance x from the center. Show that the electron exhibits simple harmonic motion with a frequency

$$f = \frac{1}{2\pi}\sqrt{\frac{\rho e}{m_e \epsilon_0}}$$

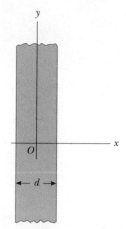

Figure P24.71 Problems 71 and 72.

72. A slab of insulating material has a nonuniform positive charge density $\rho = Cx^2$, where x is measured from the center of the slab as shown in Figure P24.71, and C is a constant. The slab is infinite in the y and z directions. Derive expressions for the electric field in (a) the exterior regions and (b) the interior region of the slab ($-d/2 < x < d/2$).

73. (a) Using the mathematical similarity between Coulomb's law and Newton's law of universal gravitation, show that Gauss's law for gravitation can be written as

$$\oint \mathbf{g} \cdot d\mathbf{A} = -4\pi G m_{in}$$

where m_{in} is the net mass inside the gaussian surface and $\mathbf{g} = \mathbf{F}_g/m$ represents the gravitational field at any point on the gaussian surface. (b) Determine the gravitational field at a distance r from the center of the Earth where $r < R_E$, assuming that the Earth's mass density is uniform.

Answers to Quick Quizzes

24.1 (e). The same number of field lines pass through a sphere of any size. Because points on the surface of the sphere are closer to the charge, the field is stronger.

24.2 (d). All field lines that enter the container also leave the container so that the total flux is zero, regardless of the nature of the field or the container.

24.3 (b) and (d). Statement (a) is not necessarily true because an equal number of positive and negative charges could be present inside the surface. Statement (c) is not necessarily true, as can be seen from Figure 24.8: a nonzero electric field exists everywhere on the surface, but the charge is not enclosed within the surface; thus, the net flux is zero.

24.4 (c). The charges q_1 and q_4 are outside the surface and contribute zero net flux through S'.

24.5 (d). We don't need the surfaces to realize that any given point in space will experience an electric field due to all local source charges.

24.6 (a). Charges added to the metal cylinder by your brother will reside on the outer surface of the conducting cylinder. If you are on the inside, these charges cannot transfer to you from the inner surface. For this same reason, you are safe in a metal automobile during a lightning storm.

Chapter 25

Electric Potential

CHAPTER OUTLINE

25.1 Potential Difference and Electric Potential

25.2 Potential Differences in a Uniform Electric Field

25.3 Electric Potential and Potential Energy Due to Point Charges

25.4 Obtaining the Value of the Electric Field from the Electric Potential

25.5 Electric Potential Due to Continuous Charge Distributions

25.6 Electric Potential Due to a Charged Conductor

25.7 The Millikan Oil-Drop Experiment

25.8 Applications of Electrostatics

▲ *Processes occurring during thunderstorms cause large differences in electric potential between a thundercloud and the ground. The result of this potential difference is an electrical discharge that we call lightning, such as this display over Tucson, Arizona. (© Keith Kent/ Photo Researchers, Inc.)*

$\top$he concept of potential energy was introduced in Chapter 8 in connection with such conservative forces as the gravitational force and the elastic force exerted by a spring. By using the law of conservation of energy, we were able to avoid working directly with forces when solving various problems in mechanics. The concept of potential energy is also of great value in the study of electricity. Because the electrostatic force is conservative, electrostatic phenomena can be conveniently described in terms of an electric potential energy. This idea enables us to define a scalar quantity known as *electric potential*. Because the electric potential at any point in an electric field is a scalar quantity, we can use it to describe electrostatic phenomena more simply than if we were to rely only on the electric field and electric forces. The concept of electric potential is of great practical value in the operation of electric circuits and devices we will study in later chapters.

25.1 Potential Difference and Electric Potential

When a test charge q_0 is placed in an electric field $\mathbf{E}$ created by some source charge distribution, the electric force acting on the test charge is $q_0\mathbf{E}$. The force $q_0\mathbf{E}$ is conservative because the force between charges described by Coulomb's law is conservative. When the test charge is moved in the field by some external agent, the work done by the field on the charge is equal to the negative of the work done by the external agent causing the displacement. This is analogous to the situation of lifting an object with mass in a gravitational field—the work done by the external agent is mgh and the work done by the gravitational force is $-mgh$.

When analyzing electric and magnetic fields, it is common practice to use the notation $d\mathbf{s}$ to represent an infinitesimal displacement vector that is oriented tangent to a path through space. This path may be straight or curved, and an integral performed along this path is called either a *path integral* or a *line integral* (the two terms are synonymous).

For an infinitesimal displacement $d\mathbf{s}$ of a charge, the work done by the electric field on the charge is $\mathbf{F} \cdot d\mathbf{s} = q_0\mathbf{E} \cdot d\mathbf{s}$. As this amount of work is done by the field, the potential energy of the charge–field system is changed by an amount $dU = -q_0\mathbf{E} \cdot d\mathbf{s}$. For a finite displacement of the charge from point A to point B, the change in potential energy of the system $\Delta U = U_B - U_A$ is

$$\Delta U = -q_0 \int_A^B \mathbf{E} \cdot d\mathbf{s} \qquad (25.1)$$

Change in electric potential energy of a system

The integration is performed along the path that q_0 follows as it moves from A to B. Because the force $q_0\mathbf{E}$ is conservative, **this line integral does not depend on the path taken from A to B.**

For a given position of the test charge in the field, the charge–field system has a potential energy U relative to the configuration of the system that is defined as $U = 0$. Dividing the potential energy by the test charge gives a physical quantity that depends only on the source charge distribution. The potential energy per unit charge U/q_0 is

▲ **PITFALL PREVENTION**

25.1 Potential and Potential Energy

The *potential is characteristic of the field only*, independent of a charged test particle that may be placed in the field. *Potential energy is characteristic of the charge–field system* due to an interaction between the field and a charged particle placed in the field.

Potential difference between two points

independent of the value of q_0 and has a value at every point in an electric field. This quantity U/q_0 is called the **electric potential** (or simply the **potential**) V. Thus, the electric potential at any point in an electric field is

$$V = \frac{U}{q_0} \tag{25.2}$$

The fact that potential energy is a scalar quantity means that electric potential also is a scalar quantity.

As described by Equation 25.1, if the test charge is moved between two positions A and B in an electric field, the charge–field system experiences a change in potential energy. The **potential difference** $\Delta V = V_B - V_A$ between two points A and B in an electric field is defined as the change in potential energy of the system when a test charge is moved between the points divided by the test charge q_0:

$$\Delta V \equiv \frac{\Delta U}{q_0} = -\int_A^B \mathbf{E} \cdot d\mathbf{s} \tag{25.3}$$

Just as with potential energy, only *differences* in electric potential are meaningful. To avoid having to work with potential differences, however, we often take the value of the electric potential to be zero at some convenient point in an electric field.

Potential difference should not be confused with difference in potential energy. The potential difference between A and B depends only on the source charge distribution (consider points A and B *without* the presence of the test charge), while the difference in potential energy exists only if a test charge is moved between the points. **Electric potential is a scalar characteristic of an electric field, independent of any charges that may be placed in the field.**

If an external agent moves a test charge from A to B without changing the kinetic energy of the test charge, the agent performs work which changes the potential energy of the system: $W = \Delta U$. The test charge q_0 is used as a mental device to define the electric potential. Imagine an arbitrary charge q located in an electric field. From Equation 25.3, the work done by an external agent in moving a charge q through an electric field at constant velocity is

$$W = q \, \Delta V \tag{25.4}$$

▲ **PITFALL PREVENTION**

25.2 Voltage

A variety of phrases are used to describe the potential difference between two points, the most common being **voltage,** arising from the unit for potential. A voltage *applied* to a device, such as a television, or *across* a device is the same as the potential difference across the device. If we say that the voltage applied to a lightbulb is 120 volts, we mean that the potential difference between the two electrical contacts on the lightbulb is 120 volts.

Because electric potential is a measure of potential energy per unit charge, the SI unit of both electric potential and potential difference is joules per coulomb, which is defined as a **volt** (V):

$$1\,\text{V} \equiv 1\,\frac{\text{J}}{\text{C}}$$

That is, 1 J of work must be done to move a 1-C charge through a potential difference of 1 V.

Equation 25.3 shows that potential difference also has units of electric field times distance. From this, it follows that the SI unit of electric field (N/C) can also be expressed in volts per meter:

$$1\,\frac{\text{N}}{\text{C}} = 1\,\frac{\text{V}}{\text{m}}$$

Therefore, **we can interpret the electric field as a measure of the rate of change with position of the electric potential.**

A unit of energy commonly used in atomic and nuclear physics is the **electron volt** (eV), which is defined as **the energy a charge–field system gains or loses when a charge of magnitude e (that is, an electron or a proton) is moved through a potential difference of 1 V.** Because $1\,\text{V} = 1\,\text{J/C}$ and because the fundamental charge is 1.60×10^{-19} C, the electron volt is related to the joule as follows:

The electron volt

$$1\,\text{eV} = 1.60 \times 10^{-19}\,\text{C} \cdot \text{V} = 1.60 \times 10^{-19}\,\text{J} \tag{25.5}$$

For instance, an electron in the beam of a typical television picture tube may have a speed of 3.0×10^7 m/s. This corresponds to a kinetic energy of 4.1×10^{-16} J, which is equivalent to 2.6×10^3 eV. Such an electron has to be accelerated from rest through a potential difference of 2.6 kV to reach this speed.

> **Quick Quiz 25.1** In Figure 25.1, two points A and B are located within a region in which there is an electric field. The potential difference $\Delta V = V_B - V_A$ is (a) positive (b) negative (c) zero.
>
> **Quick Quiz 25.2** In Figure 25.1, a negative charge is placed at A and then moved to B. The change in potential energy of the charge–field system for this process is (a) positive (b) negative (c) zero.

25.2 Potential Differences in a Uniform Electric Field

Equations 25.1 and 25.3 hold in all electric fields, whether uniform or varying, but they can be simplified for a uniform field. First, consider a uniform electric field directed along the negative y axis, as shown in Figure 25.2a. Let us calculate the potential difference between two points A and B separated by a distance $|\mathbf{s}| = d$, where $\mathbf{s}$ is parallel to the field lines. Equation 25.3 gives

$$V_B - V_A = \Delta V = -\int_A^B \mathbf{E} \cdot d\mathbf{s} = -\int_A^B (E \cos 0°)\, ds = -\int_A^B E\, ds$$

Because E is constant, we can remove it from the integral sign; this gives

$$\Delta V = -E \int_A^B ds = -Ed \qquad (25.6)$$

The negative sign indicates that the electric potential at point B is lower than at point A; that is, $V_B < V_A$. **Electric field lines always point in the direction of decreasing electric potential,** as shown in Figure 25.2a.

PITFALL PREVENTION

25.3 The Electron Volt

The electron volt is a unit of *energy*, NOT of potential. The energy of any system may be expressed in eV, but this unit is most convenient for describing the emission and absorption of visible light from atoms. Energies of nuclear processes are often expressed in MeV.

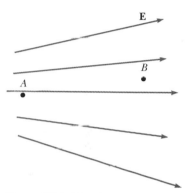

Figure 25.1 (Quick Quiz 25.1) Two points in an electric field.

Potential difference between two points in a uniform electric field

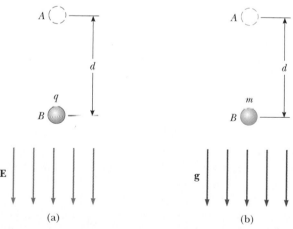

Figure 25.2 (a) When the electric field $\mathbf{E}$ is directed downward, point B is at a lower electric potential than point A. When a positive test charge moves from point A to point B, the charge–field system loses electric potential energy. (b) When an object of mass m moves downward in the direction of the gravitational field $\mathbf{g}$, the object–field system loses gravitational potential energy.

Now suppose that a test charge q_0 moves from A to B. We can calculate the change in the potential energy of the charge–field system from Equations 25.3 and 25.6:

$$\Delta U = q_0 \, \Delta V = - q_0 E d \qquad (25.7)$$

From this result, we see that if q_0 is positive, then ΔU is negative. We conclude that **a system consisting of a positive charge and an electric field loses electric potential energy when the charge moves in the direction of the field.** This means that an electric field does work on a positive charge when the charge moves in the direction of the electric field. (This is analogous to the work done by the gravitational field on a falling object, as shown in Figure 25.2b.) If a positive test charge is released from rest in this electric field, it experiences an electric force $q_0\mathbf{E}$ in the direction of $\mathbf{E}$ (downward in Fig. 25.2a). Therefore, it accelerates downward, gaining kinetic energy. **As the charged particle gains kinetic energy, the charge–field system loses an equal amount of potential energy.** This should not be surprising—it is simply conservation of energy in an isolated system as introduced in Chapter 8.

If q_0 is negative, then ΔU in Equation 25.7 is positive and the situation is reversed: **A system consisting of a negative charge and an electric field gains electric potential energy when the charge moves in the direction of the field.** If a negative charge is released from rest in an electric field, it accelerates in a direction opposite the direction of the field. In order for the negative charge to move in the direction of the field, an external agent must apply a force and do positive work on the charge.

Now consider the more general case of a charged particle that moves between A and B in a uniform electric field such that the vector $\mathbf{s}$ is not parallel to the field lines, as shown in Figure 25.3. In this case, Equation 25.3 gives

$$\Delta V = - \int_A^B \mathbf{E} \cdot d\mathbf{s} = - \mathbf{E} \cdot \int_A^B d\mathbf{s} = - \mathbf{E} \cdot \mathbf{s} \qquad (25.8)$$

where again we are able to remove $\mathbf{E}$ from the integral because it is constant. The change in potential energy of the charge–field system is

$$\Delta U = q_0 \, \Delta V = - q_0 \mathbf{E} \cdot \mathbf{s} \qquad (25.9)$$

Finally, we conclude from Equation 25.8 that all points in a plane perpendicular to a uniform electric field are at the same electric potential. We can see this in Figure 25.3, where the potential difference $V_B - V_A$ is equal to the potential difference $V_C - V_A$. (Prove this to yourself by working out the dot product $\mathbf{E} \cdot \mathbf{s}$ for $\mathbf{s}_{A \to B}$, where the angle θ between $\mathbf{E}$ and $\mathbf{s}$ is arbitrary as shown in Figure 25.3, and the dot product for $\mathbf{s}_{A \to C}$, where $\theta = 0$.) Therefore, $V_B = V_C$. **The name equipotential surface is given to any surface consisting of a continuous distribution of points having the same electric potential.**

The equipotential surfaces of a uniform electric field consist of a family of parallel planes that are all perpendicular to the field. Equipotential surfaces for fields with other symmetries are described in later sections.

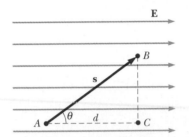

Figure 25.3 A uniform electric field directed along the positive x axis. Point B is at a lower electric potential than point A. Points B and C are at the *same* electric potential.

Change in potential energy when a charged particle is moved in a uniform electric field

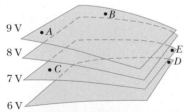

Figure 25.4 (Quick Quiz 25.3) Four equipotential surfaces.

Quick Quiz 25.3 The labeled points in Figure 25.4 are on a series of equipotential surfaces associated with an electric field. Rank (from greatest to least) the work done by the electric field on a positively charged particle that moves from A to B; from B to C; from C to D; from D to E.

Quick Quiz 25.4 For the equipotential surfaces in Figure 25.4, what is the *approximate* direction of the electric field? (a) Out of the page (b) Into the page (c) Toward the right edge of the page (d) Toward the left edge of the page (e) Toward the top of the page (f) Toward the bottom of the page.

Example 25.1   The Electric Field Between Two Parallel Plates of Opposite Charge

A battery produces a specified potential difference ΔV between conductors attached to the battery terminals. A 12-V battery is connected between two parallel plates, as shown in Figure 25.5. The separation between the plates is $d = 0.30$ cm, and we assume the electric field between the plates to be

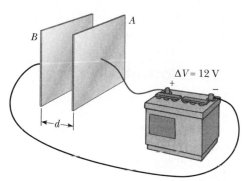

Figure 25.5 (Example 25.1) A 12-V battery connected to two parallel plates. The electric field between the plates has a magnitude given by the potential difference ΔV divided by the plate separation d.

uniform. (This assumption is reasonable if the plate separation is small relative to the plate dimensions and if we do not consider locations near the plate edges.) Find the magnitude of the electric field between the plates.

Solution The electric field is directed from the positive plate (A) to the negative one (B), and the positive plate is at a higher electric potential than the negative plate is. The potential difference between the plates must equal the potential difference between the battery terminals. We can understand this by noting that all points on a conductor in equilibrium are at the same electric potential[1]; no potential difference exists between a terminal and any portion of the plate to which it is connected. Therefore, the magnitude of the electric field between the plates is, from Equation 25.6,

$$E = \frac{|V_B - V_A|}{d} = \frac{12 \text{ V}}{0.30 \times 10^{-2} \text{ m}} = \boxed{4.0 \times 10^3 \text{ V/m}}$$

The configuration of plates in Figure 25.5 is called a *parallel-plate capacitor*, and is examined in greater detail in Chapter 26.

Example 25.2   Motion of a Proton in a Uniform Electric Field   `Interactive`

A proton is released from rest in a uniform electric field that has a magnitude of 8.0×10^4 V/m (Fig. 25.6). The proton undergoes a displacement of 0.50 m in the direction of **E**.

(A) Find the change in electric potential between points A and B.

Solution Because the positively charged proton moves in the direction of the field, we expect it to move to a position of lower electric potential. From Equation 25.6, we have

$$\Delta V = -Ed = -(8.0 \times 10^4 \text{ V/m})(0.50 \text{ m}) = \boxed{-4.0 \times 10^4 \text{ V}}$$

(B) Find the change in potential energy of the proton–field system for this displacement.

Solution Using Equation 25.3,

$$\Delta U = q_0 \, \Delta V = e \, \Delta V$$
$$= (1.6 \times 10^{-19} \text{ C})(-4.0 \times 10^4 \text{ V})$$
$$= \boxed{-6.4 \times 10^{-15} \text{ J}}$$

The negative sign means the potential energy of the system decreases as the proton moves in the direction of the electric field. As the proton accelerates in the direction of the field, it gains kinetic energy and at the same time the system loses electric potential energy.

(C) Find the speed of the proton after completing the 0.50 m displacement in the electric field.

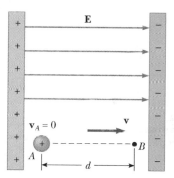

Figure 25.6 (Example 25.2) A proton accelerates from A to B in the direction of the electric field.

Solution The charge–field system is isolated, so the mechanical energy of the system is conserved:

$$\Delta K + \Delta U = 0$$
$$\left(\tfrac{1}{2}mv^2 - 0\right) + e \, \Delta V = 0$$
$$v = \sqrt{\frac{-(2e \, \Delta V)}{m}}$$
$$= \sqrt{\frac{-2(1.6 \times 10^{-19} \text{ C})(-4.0 \times 10^4 \text{ V})}{1.67 \times 10^{-27} \text{ kg}}}$$
$$= \boxed{2.8 \times 10^6 \text{ m/s}}$$

What If? What if the situation is exactly the same as that shown in Figure 25.6, but no proton is present? Could both parts (A) and (B) of this example still be answered?

---

[1] The electric field vanishes within a conductor in electrostatic equilibrium; thus, the path integral between any two points in the conductor must be zero. A more complete discussion of this point is given in Section 25.6.

Answer Part (A) of the example would remain exactly the same because the potential difference between points A and B is established by the source charges in the parallel plates. The potential difference does not depend on the presence of the proton, which plays the role of a test charge. Part (B) of the example would be meaningless if the proton is not present. A change in potential energy is related to a change in the charge–field system. In the absence of the proton, the system of the electric field alone does not change.

 At the Interactive Worked Example link at **http://www.pse6.com,** you can predict and observe the speed of the proton as it arrives at the negative plate for random values of the electric field.

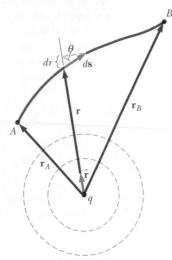

Figure 25.7 The potential difference between points A and B due to a point charge q depends *only* on the initial and final radial coordinates r_A and r_B. The two dashed circles represent intersections of spherical equipotential surfaces with the page.

⚠ **PITFALL PREVENTION**

25.4 Similar Equation Warning

Do not confuse Equation 25.11 for the electric potential of a point charge with Equation 23.9 for the electric field of a point charge. Potential is proportional to $1/r$, while the field is proportional to $1/r^2$. The effect of a charge on the space surrounding it can be described in two ways. The charge sets up a vector electric field $\mathbf{E}$, which is related to the force experienced by a test charge placed in the field. It also sets up a scalar potential V, which is related to the potential energy of the two-charge system when a test charge is placed in the field.

25.3 Electric Potential and Potential Energy Due to Point Charges

In Section 23.4 we discussed the fact that an isolated positive point charge q produces an electric field that is directed radially outward from the charge. To find the electric potential at a point located a distance r from the charge, we begin with the general expression for potential difference:

$$V_B - V_A = -\int_A^B \mathbf{E} \cdot d\mathbf{s}$$

where A and B are the two arbitrary points shown in Figure 25.7. At any point in space, the electric field due to the point charge is $\mathbf{E} = k_e q\hat{\mathbf{r}}/r^2$ (Eq. 23.9), where $\hat{\mathbf{r}}$ is a unit vector directed from the charge toward the point. The quantity $\mathbf{E} \cdot d\mathbf{s}$ can be expressed as

$$\mathbf{E} \cdot d\mathbf{s} = k_e \frac{q}{r^2} \hat{\mathbf{r}} \cdot d\mathbf{s}$$

Because the magnitude of $\hat{\mathbf{r}}$ is 1, the dot product $\hat{\mathbf{r}} \cdot d\mathbf{s} = ds \cos\theta$, where θ is the angle between $\hat{\mathbf{r}}$ and $d\mathbf{s}$. Furthermore, $ds \cos\theta$ is the projection of $d\mathbf{s}$ onto $\mathbf{r}$; thus, $ds \cos\theta = dr$. That is, any displacement $d\mathbf{s}$ along the path from point A to point B produces a change dr in the magnitude of $\mathbf{r}$, the position vector of the point relative to the charge creating the field. Making these substitutions, we find that $\mathbf{E} \cdot d\mathbf{s} = (k_e q/r^2)\,dr$; hence, the expression for the potential difference becomes

$$V_B - V_A = -k_e q \int_{r_A}^{r_B} \frac{dr}{r^2} = \left. \frac{k_e q}{r} \right]_{r_A}^{r_B}$$

$$V_B - V_A = k_e q \left[\frac{1}{r_B} - \frac{1}{r_A} \right] \tag{25.10}$$

This equation shows us that the integral of $\mathbf{E} \cdot d\mathbf{s}$ is *independent* of the path between points A and B. Multiplying by a charge q_0 that moves between points A and B, we see that the integral of $q_0\mathbf{E} \cdot d\mathbf{s}$ is also independent of path. This latter integral is the work done by the electric force, which tells us that the electric force is conservative (see Section 8.3). We define a field that is related to a conservative force as a **conservative field.** Thus, Equation 25.10 tells us that the electric field of a fixed point charge is conservative. Furthermore, Equation 25.10 expresses the important result that the potential difference between any two points A and B in a field created by a point charge depends only on the radial coordinates r_A and r_B. It is customary to choose the reference of electric potential for a point charge to be $V = 0$ at $r_A = \infty$. With this reference choice, the electric potential created by a point charge at any distance r from the charge is

$$V = k_e \frac{q}{r} \tag{25.11}$$

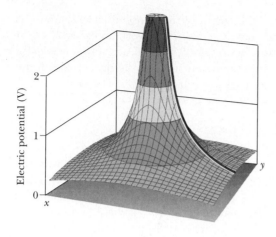

Figure 25.8 The electric potential in the plane around a single positive charge is plotted on the vertical axis. (The electric potential function for a negative charge would look like a hole instead of a hill.) The red line shows the $1/r$ nature of the electric potential, as given by Equation 25.11.

Figure 25.8 shows a plot of the electric potential on the vertical axis for a positive charge located in the xy plane. Consider the following analogy to gravitational potential: imagine trying to roll a marble toward the top of a hill shaped like the surface in Figure 25.8. Pushing the marble up the hill is analogous to pushing one positively charged object toward another positively charged object. Similarly, the electric potential graph of the region surrounding a negative charge is analogous to a "hole" with respect to any approaching positively charged objects. A charged object must be infinitely distant from another charge before the surface in Figure 25.8 is "flat" and has an electric potential of zero.

We obtain the electric potential resulting from two or more point charges by applying the superposition principle. That is, the total electric potential at some point P due to several point charges is the sum of the potentials due to the individual charges. For a group of point charges, we can write the total electric potential at P in the form

$$V = k_e \sum_i \frac{q_i}{r_i} \tag{25.12}$$

Electric potential due to several point charges

where the potential is again taken to be zero at infinity and r_i is the distance from the point P to the charge q_i. Note that the sum in Equation 25.12 is an algebraic sum of scalars rather than a vector sum (which we use to calculate the electric field of a group of charges). Thus, it is often much easier to evaluate V than to evaluate $\mathbf{E}$. The electric potential around a dipole is illustrated in Figure 25.9. Notice the steep slope of the potential between the charges, representing a region of strong electric field.

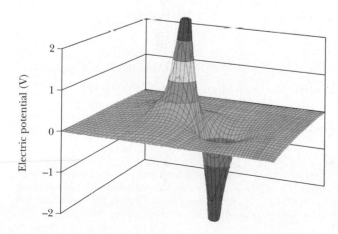

Figure 25.9 The electric potential in the plane containing a dipole.

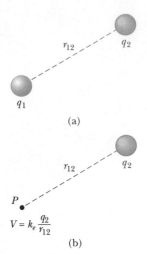

Active Figure 25.10 (a) If two point charges are separated by a distance r_{12}, the potential energy of the pair of charges is given by $k_e q_1 q_2 / r_{12}$. (b) If charge q_1 is removed, a potential $k_e q_2 / r_{12}$ exists at point P due to charge q_2.

At the Active Figures link at http://www.pse6.com, you can move charge q_1 or point P and see the result on the electric potential energy of the system for part (a) and the electric potential due to charge q_2 for part (b).

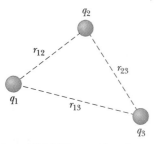

Figure 25.11 Three point charges are fixed at the positions shown. The potential energy of this system of charges is given by Equation 25.14.

▲ **PITFALL PREVENTION**

25.5 Which Work?

There is a difference between work done *by one member of a system on another member* and work done *on a system by an external agent*. In the present discussion, we are considering the group of charges to be the system and an external agent is doing work on the system to move the charges from an infinite separation to a small separation.

We now consider the potential energy of a system of two charged particles. If V_2 is the electric potential at a point P due to charge q_2, then the work an external agent must do to bring a second charge q_1 from infinity to P without acceleration is $q_1 V_2$. This work represents a transfer of energy into the system and the energy appears in the system as potential energy U when the particles are separated by a distance r_{12} (Fig. 25.10a). Therefore, we can express the potential energy of the system as[2]

$$U = k_e \frac{q_1 q_2}{r_{12}} \qquad (25.13)$$

Note that if the charges are of the same sign, U is positive. This is consistent with the fact that positive work must be done by an external agent on the system to bring the two charges near one another (because charges of the same sign repel). If the charges are of opposite sign, U is negative; this means that negative work is done by an external agent against the attractive force between the charges of opposite sign as they are brought near each other—a force must be applied opposite to the displacement to prevent q_1 from accelerating toward q_2.

In Figure 25.10b, we have removed the charge q_1. At the position that this charge previously occupied, point P, we can use Equations 25.2 and 25.13 to define a potential due to charge q_2 as $V = U/q_1 = k_e q_2 / r_{12}$. This expression is consistent with Equation 25.11.

If the system consists of more than two charged particles, we can obtain the total potential energy by calculating U for every pair of charges and summing the terms algebraically. As an example, the total potential energy of the system of three charges shown in Figure 25.11 is

$$U = k_e \left(\frac{q_1 q_2}{r_{12}} + \frac{q_1 q_3}{r_{13}} + \frac{q_2 q_3}{r_{23}} \right) \qquad (25.14)$$

Physically, we can interpret this as follows: imagine that q_1 is fixed at the position shown in Figure 25.11 but that q_2 and q_3 are at infinity. The work an external agent must do to bring q_2 from infinity to its position near q_1 is $k_e q_1 q_2 / r_{12}$, which is the first term in Equation 25.14. The last two terms represent the work required to bring q_3 from infinity to its position near q_1 and q_2. (The result is independent of the order in which the charges are transported.)

Quick Quiz 25.5 A spherical balloon contains a positively charged object at its center. As the balloon is inflated to a greater volume while the charged object remains at the center, does the electric potential at the surface of the balloon (a) increase, (b) decrease, or (c) remain the same? Does the electric flux through the surface of the balloon (d) increase, (e) decrease, or (f) remain the same?

Quick Quiz 25.6 In Figure 25.10a, take q_1 to be a negative source charge and q_2 to be the test charge. If q_2 is initially positive and is changed to a charge of the same magnitude but negative, the potential at the position of q_2 due to q_1 (a) increases (b) decreases (c) remains the same.

Quick Quiz 25.7 Consider the situation in Quick Quiz 25.6 again. When q_2 is changed from positive to negative, the potential energy of the two-charge system (a) increases (b) decreases (c) remains the same.

---

[2]   The expression for the electric potential energy of a system made up of two point charges, Equation 25.13, is of the *same* form as the equation for the gravitational potential energy of a system made up of two point masses, $-G m_1 m_2 / r$ (see Chapter 13). The similarity is not surprising in view of the fact that both expressions are derived from an inverse-square force law.

Example 25.3   The Electric Potential Due to Two Point Charges   Interactive

A charge $q_1 = 2.00\ \mu C$ is located at the origin, and a charge $q_2 = -6.00\ \mu C$ is located at $(0, 3.00)$ m, as shown in Figure 25.12a.

(A) Find the total electric potential due to these charges at the point P, whose coordinates are $(4.00, 0)$ m.

Solution For two charges, the sum in Equation 25.12 gives

$$V_P = k_e\left(\frac{q_1}{r_1} + \frac{q_2}{r_2}\right)$$

$$V_P = (8.99 \times 10^9\ \text{N} \cdot \text{m}^2/\text{C}^2)$$

$$\times\left(\frac{2.00 \times 10^{-6}\ \text{C}}{4.00\ \text{m}} - \frac{6.00 \times 10^{-6}\ \text{C}}{5.00\ \text{m}}\right)$$

$$= \boxed{-6.29 \times 10^3\ \text{V}}$$

(B) Find the change in potential energy of the system of two charges plus a charge $q_3 = 3.00\ \mu C$ as the latter charge moves from infinity to point P (Fig. 25.12b).

Solution When the charge q_3 is at infinity, let us define $U_i = 0$ for the system, and when the charge is at P, $U_f = q_3 V_P$; therefore,

$$\Delta U = q_3 V_P - 0 = (3.00 \times 10^{-6}\ \text{C})(-6.29 \times 10^3\ \text{V})$$

$$= \boxed{-1.89 \times 10^{-2}\ \text{J}}$$

Therefore, because the potential energy of the system has decreased, positive work would have to be done by an external agent to remove the charge from point P back to infinity.

What If? You are working through this example with a classmate and she says, "Wait a minute! In part (B), we ignored the potential energy associated with the pair of charges q_1 and q_2!" How would you respond?

Answer Given the statement of the problem, it is not necessary to include this potential energy, because part (B) asks for the *change* in potential energy of the system as q_3 is brought in from infinity. Because the configuration of charges q_1 and q_2 does not change in the process, there is no ΔU associated with these charges. However, if part (B) had asked to find the change in potential energy when *all three* charges start out infinitely far apart and are then brought to the positions in Figure 25.12b, we would need to calculate the change as follows, using Equation 25.14:

$$U = k_e\left(\frac{q_1 q_2}{r_{12}} + \frac{q_1 q_3}{r_{13}} + \frac{q_2 q_3}{r_{23}}\right)$$

$$= (8.99 \times 10^9\ \text{N} \cdot \text{m}^2/\text{C}^2)$$

$$\times\left(\frac{(2.00 \times 10^{-6}\ \text{C})(-6.00 \times 10^{-6}\ \text{C})}{3.00\ \text{m}}\right.$$

$$+ \frac{(2.00 \times 10^{-6}\ \text{C})(3.00 \times 10^{-6}\ \text{C})}{4.00\ \text{m}}$$

$$\left. + \frac{(3.00 \times 10^{-6}\ \text{C})(-6.00 \times 10^{-6}\ \text{C})}{5.00\ \text{m}}\right)$$

$$= -5.48 \times 10^{-2}\ \text{J}$$

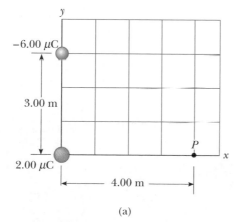

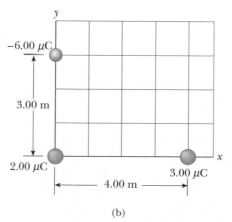

(a)   (b)

Figure 25.12 (Example 25.3) (a) The electric potential at P due to the two charges q_1 and q_2 is the algebraic sum of the potentials due to the individual charges. (b) A third charge $q_3 = 3.00\ \mu C$ is brought from infinity to a position near the other charges.

Explore the value of the electric potential at point P and the electric potential energy of the system in Figure 25.12b at the Interactive Worked Example link at **http://www.pse6.com.**

25.4   Obtaining the Value of the Electric Field from the Electric Potential

The electric field **E** and the electric potential V are related as shown in Equation 25.3. We now show how to calculate the value of the electric field if the electric potential is known in a certain region.

From Equation 25.3 we can express the potential difference dV between two points a distance ds apart as

$$dV = -\mathbf{E} \cdot d\mathbf{s} \tag{25.15}$$

If the electric field has only one component E_x, then $\mathbf{E} \cdot d\mathbf{s} = E_x\, dx$. Therefore, Equation 25.15 becomes $dV = -E_x\, dx$, or

$$E_x = -\frac{dV}{dx} \tag{25.16}$$

That is, the x component of the electric field is equal to the negative of the derivative of the electric potential with respect to x. Similar statements can be made about the y and z components. Equation 25.16 is the mathematical statement of the fact that the electric field is a measure of the rate of change with position of the electric potential, as mentioned in Section 25.1.

Experimentally, electric potential and position can be measured easily with a voltmeter (see Section 28.5) and a meter stick. Consequently, an electric field can be determined by measuring the electric potential at several positions in the field and making a graph of the results. According to Equation 25.16, the slope of a graph of V versus x at a given point provides the magnitude of the electric field at that point.

When a test charge undergoes a displacement $d\mathbf{s}$ along an equipotential surface, then $dV = 0$ because the potential is constant along an equipotential surface. From Equation 25.15, we see that $dV = -\mathbf{E} \cdot d\mathbf{s} = 0$; thus, **E** must be perpendicular to the displacement along the equipotential surface. This shows that the **equipotential surfaces must always be perpendicular to the electric field lines passing through them.**

As mentioned at the end of Section 25.2, the equipotential surfaces for a uniform electric field consist of a family of planes perpendicular to the field lines. Figure 25.13a shows some representative equipotential surfaces for this situation.

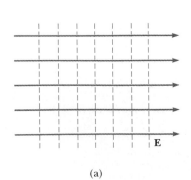

(a)

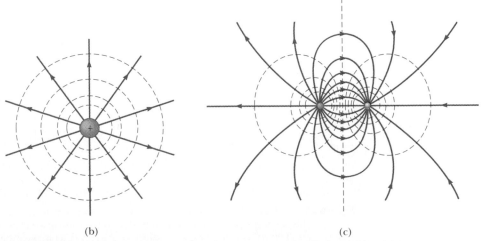

(b)   (c)

Figure 25.13 Equipotential surfaces (the dashed blue lines are intersections of these surfaces with the page) and electric field lines (red-brown lines) for (a) a uniform electric field produced by an infinite sheet of charge, (b) a point charge, and (c) an electric dipole. In all cases, the equipotential surfaces are *perpendicular* to the electric field lines at every point.

If the charge distribution creating an electric field has spherical symmetry such that the volume charge density depends only on the radial distance r, then the electric field is radial. In this case, $\mathbf{E} \cdot d\mathbf{s} = E_r \, dr$, and we can express dV in the form $dV = -E_r \, dr$. Therefore,

$$E_r = -\frac{dV}{dr} \qquad\qquad (25.17)$$

For example, the electric potential of a point charge is $V = k_e q/r$. Because V is a function of r only, the potential function has spherical symmetry. Applying Equation 25.17, we find that the electric field due to the point charge is $E_r = k_e q/r^2$, a familiar result. Note that the potential changes only in the radial direction, not in any direction perpendicular to r. Thus, V (like E_r) is a function only of r. Again, this is consistent with the idea that **equipotential surfaces are perpendicular to field lines.** In this case the equipotential surfaces are a family of spheres concentric with the spherically symmetric charge distribution (Fig. 25.13b).

The equipotential surfaces for an electric dipole are sketched in Figure 25.13c.

In general, the electric potential is a function of all three spatial coordinates. If $V(r)$ is given in terms of the Cartesian coordinates, the electric field components E_x, E_y, and E_z can readily be found from $V(x, y, z)$ as the partial derivatives[3]

$$E_x = -\frac{\partial V}{\partial x} \qquad E_y = -\frac{\partial V}{\partial y} \qquad E_z = -\frac{\partial V}{\partial z} \qquad (25.18)$$

Finding the electric field from the potential

For example, if $V = 3x^2 y + y^2 + yz$, then

$$\frac{\partial V}{\partial x} = \frac{\partial}{\partial x}(3x^2 y + y^2 + yz) = \frac{\partial}{\partial x}(3x^2 y) = 3y \frac{d}{dx}(x^2) = 6xy$$

Quick Quiz 25.8 In a certain region of space, the electric potential is zero everywhere along the x axis. From this we can conclude that the x component of the electric field in this region is (a) zero (b) in the $+x$ direction (c) in the $-x$ direction.

Quick Quiz 25.9 In a certain region of space, the electric field is zero. From this we can conclude that the electric potential in this region is (a) zero (b) constant (c) positive (d) negative.

Example 25.4   The Electric Potential Due to a Dipole

An electric dipole consists of two charges of equal magnitude and opposite sign separated by a distance $2a$, as shown in Figure 25.14. The dipole is along the x axis and is centered at the origin.

(A) Calculate the electric potential at point P.

Solution For point P in Figure 25.14,

$$V = k_e \sum \frac{q_i}{r_i} = k_e \left(\frac{q}{x - a} - \frac{q}{x + a} \right) = \frac{2k_e qa}{x^2 - a^2}$$

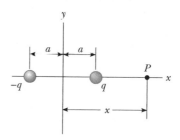

Figure 25.14 (Example 25.4) An electric dipole located on the x axis.

[3]   In vector notation, $\mathbf{E}$ is often written in Cartesian coordinate systems as

$$\mathbf{E} = -\nabla V = -\left(\hat{\mathbf{i}} \frac{\partial}{\partial x} + \hat{\mathbf{j}} \frac{\partial}{\partial y} + \hat{\mathbf{k}} \frac{\partial}{\partial z} \right) V$$

where ∇ is called the *gradient operator*.

(B) Calculate V and E_x at a point far from the dipole.

Solution If point P is far from the dipole, such that $x \gg a$, then a^2 can be neglected in the term $x^2 - a^2$ and V becomes

$$V \approx \frac{2k_e q a}{x^2} \qquad (x \gg a)$$

Using Equation 25.16 and this result, we can calculate the magnitude of the electric field at a point far from the dipole:

$$E_x = -\frac{dV}{dx} = \frac{4k_e q a}{x^3} \qquad (x \gg a)$$

(C) Calculate V and E_x if point P is located anywhere between the two charges.

Solution Using Equation 25.12,

$$V = k_e \sum \frac{q_i}{r_i} = k_e \left(\frac{q}{a-x} - \frac{q}{a+x} \right) = \frac{2k_e q x}{a^2 - x^2}$$

and using Equation 25.16,

$$E_x = -\frac{dV}{dx} = -\frac{d}{dx}\left(\frac{2k_e q x}{a^2 - x^2} \right) = -2k_e q \left(\frac{a^2 + x^2}{(a^2 - x^2)^2} \right)$$

We can check these results by considering the situation at the center of the dipole, where $x = 0$, $V = 0$, and $E_x = -2k_e q / a^2$.

What If? What if point P in Figure 25.14 happens to be located to the left of the negative charge? Would the answer to part (A) be the same?

Answer The potential should be negative because a point to the left of the dipole is closer to the negative charge than to the positive charge. If we redo the calculation in part (A) with P on the left side of $-q$, we have

$$V = k_e \sum \frac{q_i}{r_i} = k_e \left(\frac{q}{x+a} - \frac{q}{x-a} \right) = -\frac{2k_e q a}{x^2 - a^2}$$

Thus, the potential has the same value but is negative for points on the left of the dipole.

25.5 Electric Potential Due to Continuous Charge Distributions

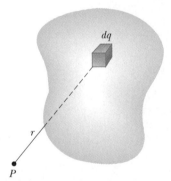

Figure 25.15 The electric potential at the point P due to a continuous charge distribution can be calculated by dividing the charge distribution into elements of charge dq and summing the electric potential contributions over all elements.

Electric potential due to a continuous charge distribution

We can calculate the electric potential due to a continuous charge distribution in two ways. If the charge distribution is known, we can start with Equation 25.11 for the electric potential of a point charge. We then consider the potential due to a small charge element dq, treating this element as a point charge (Fig. 25.15). The electric potential dV at some point P due to the charge element dq is

$$dV = k_e \frac{dq}{r} \qquad (25.19)$$

where r is the distance from the charge element to point P. To obtain the total potential at point P, we integrate Equation 25.19 to include contributions from all elements of the charge distribution. Because each element is, in general, a different distance from point P and because k_e is constant, we can express V as

$$V = k_e \int \frac{dq}{r} \qquad (25.20)$$

In effect, we have replaced the sum in Equation 25.12 with an integral. Note that this expression for V uses a particular reference: the electric potential is taken to be zero when point P is infinitely far from the charge distribution.

If the electric field is already known from other considerations, such as Gauss's law, we can calculate the electric potential due to a continuous charge distribution using Equation 25.3. If the charge distribution has sufficient symmetry, we first evaluate $\mathbf{E}$ at any point using Gauss's law and then substitute the value obtained into Equation 25.3 to determine the potential difference ΔV between any two points. We then choose the electric potential V to be zero at some convenient point.

Calculating Electric Potential

- Remember that electric potential is a scalar quantity, so vector components do not exist. Therefore, when using the superposition principle to evaluate the electric potential at a point due to a system of point charges, simply take the algebraic sum of the potentials due to the various charges. However, you must keep track of signs. The potential is positive for positive charges and negative for negative charges.

- Just as with gravitational potential energy in mechanics, only *changes* in electric potential are significant; hence, the point where you choose the potential to be zero is arbitrary. When dealing with point charges or a charge distribution of finite size, we usually define $V = 0$ to be at a point infinitely far from the charges.

- You can evaluate the electric potential at some point P due to a continuous distribution of charge by dividing the charge distribution into infinitesimal elements of charge dq located at a distance r from P. Then, treat one charge element as a point charge, such that the potential at P due to the element is $dV = k_e dq/r$. Obtain the total potential at P by integrating dV over the entire charge distribution. In performing the integration for most problems, you must express dq and r in terms of a single variable. To simplify the integration, consider the geometry involved in the problem carefully. Study Examples 25.5 through 25.7 below for guidance.

- Another method that you can use to obtain the electric potential due to a finite continuous charge distribution is to start with the definition of potential difference given by Equation 25.3. If you know or can easily obtain $\mathbf{E}$ (from Gauss's law), then you can evaluate the line integral of $\mathbf{E} \cdot d\mathbf{s}$. This method is demonstrated in Example 25.8.

Example 25.5   Electric Potential Due to a Uniformly Charged Ring

(A) Find an expression for the electric potential at a point P located on the perpendicular central axis of a uniformly charged ring of radius a and total charge Q.

Solution Figure 25.16, in which the ring is oriented so that its plane is perpendicular to the x axis and its center is at the origin, helps us to conceptualize this problem. Because the ring consists of a continuous distribution of charge rather

than a set of discrete charges, we categorize this problem as one in which we need to use the integration technique represented by Equation 25.20. To analyze the problem, we take point P to be at a distance x from the center of the ring, as shown in Figure 25.16. The charge element dq is at a distance $\sqrt{x^2 + a^2}$ from point P. Hence, we can express V as

$$V = k_e \int \frac{dq}{r} = k_e \int \frac{dq}{\sqrt{x^2 + a^2}}$$

Because each element dq is at the same distance from point P, we can bring $\sqrt{x^2 + a^2}$ in front of the integral sign, and V reduces to

$$V = \frac{k_e}{\sqrt{x^2 + a^2}} \int dq = \frac{k_e Q}{\sqrt{x^2 + a^2}} \qquad (25.21)$$

The only variable in this expression for V is x. This is not surprising because our calculation is valid only for points along the x axis, where y and z are both zero.

(B) Find an expression for the magnitude of the electric field at point P.

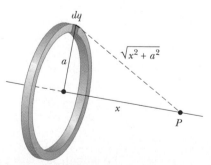

Figure 25.16 (Example 25.5) A uniformly charged ring of radius a lies in a plane perpendicular to the x axis. All elements dq of the ring are the same distance from a point P lying on the x axis.

Solution From symmetry, we see that along the x axis **E** can have only an x component. Therefore, we can use Equation 25.16:

$$E_x = -\frac{dV}{dx} = -k_e Q \frac{d}{dx}(x^2 + a^2)^{-1/2}$$

$$= -k_e Q(-\tfrac{1}{2})(x^2 + a^2)^{-3/2}(2x)$$

$$E_x = \frac{k_e Qx}{(x^2 + a^2)^{3/2}} \qquad (25.22)$$

To finalize this problem, we see that this result for the electric field agrees with that obtained by direct integration (see Example 23.8). Note that $E_x = 0$ at $x = 0$ (the center of the ring). Could you have guessed this?

Example 25.6 Electric Potential Due to a Uniformly Charged Disk

A uniformly charged disk has radius a and surface charge density σ. Find

(A) the electric potential and

(B) the magnitude of the electric field along the perpendicular central axis of the disk.

Solution (A) Again, we choose the point P to be at a distance x from the center of the disk and take the plane of the disk to be perpendicular to the x axis. We can simplify the problem by dividing the disk into a series of charged rings of infinitesimal width dr. The electric potential due to each ring is given by Equation 25.21. Consider one such ring of radius r and width dr, as indicated in Figure 25.17. The surface area of the ring is $dA = 2\pi r\, dr$. From the definition of surface charge density (see Section 23.5), we know that the charge on the ring is $dq = \sigma\, dA = \sigma 2\pi r\, dr$. Hence, the potential at the point P due to this ring is

$$dV = \frac{k_e\, dq}{\sqrt{r^2 + x^2}} = \frac{k_e \sigma 2\pi r\, dr}{\sqrt{r^2 + x^2}}$$

where x is a constant and r is a variable. To find the *total* electric potential at P, we sum over all rings making up the disk. That is, we integrate dV from $r = 0$ to $r = a$:

$$V = \pi k_e \sigma \int_0^a \frac{2r\, dr}{\sqrt{r^2 + x^2}} = \pi k_e \sigma \int_0^a (r^2 + x^2)^{-1/2} 2r\, dr$$

This integral is of the common form $\int u^n\, du$ and has the value $u^{n+1}/(n+1)$, where $n = -\tfrac{1}{2}$ and $u = r^2 + x^2$. This gives

$$V = 2\pi k_e \sigma\, [(x^2 + a^2)^{1/2} - x] \qquad (25.23)$$

(B) As in Example 25.5, we can find the electric field at any axial point using Equation 25.16:

$$E_x = -\frac{dV}{dx} = 2\pi k_e \sigma \left(1 - \frac{x}{\sqrt{x^2 + a^2}}\right) \qquad (25.24)$$

The calculation of V and **E** for an arbitrary point off the axis is more difficult to perform, and we do not treat this situation in this text.

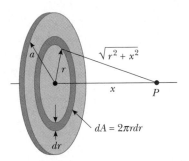

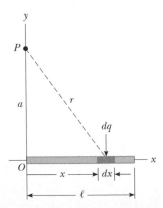

Figure 25.17 (Example 25.6) A uniformly charged disk of radius a lies in a plane perpendicular to the x axis. The calculation of the electric potential at any point P on the x axis is simplified by dividing the disk into many rings of radius r and width dr, with area $2\pi r\, dr$.

Example 25.7 Electric Potential Due to a Finite Line of Charge

A rod of length ℓ located along the x axis has a total charge Q and a uniform linear charge density $\lambda = Q/\ell$. Find the electric potential at a point P located on the y axis a distance a from the origin (Fig. 25.18).

Solution The length element dx has a charge $dq = \lambda\, dx$. Because this element is a distance $r = \sqrt{x^2 + a^2}$ from point P, we can express the potential at point P due to this element as

$$dV = k_e \frac{dq}{r} = k_e \frac{\lambda\, dx}{\sqrt{x^2 + a^2}}$$

To obtain the total potential at P, we integrate this expression over the limits $x = 0$ to $x = \ell$. Noting that k_e and λ are

Figure 25.18 (Example 25.7) A uniform line charge of length ℓ located along the x axis. To calculate the electric potential at P, the line charge is divided into segments each of length dx and each carrying a charge $dq = \lambda\, dx$.

constants, we find that

$$V = k_e\lambda \int_0^\ell \frac{dx}{\sqrt{x^2 + a^2}} = k_e \frac{Q}{\ell} \int_0^\ell \frac{dx}{\sqrt{x^2 + a^2}}$$

This integral has the following value (see Appendix B):

$$\int \frac{dx}{\sqrt{x^2 + a^2}} = \ln\left(x + \sqrt{x^2 + a^2}\right)$$

Evaluating V, we find

$$V = \frac{k_e Q}{\ell} \ln\left(\frac{\ell + \sqrt{\ell^2 + a^2}}{a}\right) \qquad (25.25)$$

What If? What if we were asked to find the electric field at point P? Would this be a simple calculation?

Answer Calculating the electric field by means of Equation 23.11 would be a little messy. There is no symmetry to appeal to, and the integration over the line of charge would represent a vector addition of electric fields at point P. Using Equation 25.18, we could find E_y by replacing a with y in Equation 25.25 and performing the differentiation with respect to y. Because the charged rod in Figure 25.18 lies entirely to the right of $x = 0$, the electric field at point P would have an x component to the left if the rod is charged positively. We cannot use Equation 25.18 to find the x component of the field, however, because we evaluated the potential due to the rod at a specific value of x ($x = 0$) rather than a general value of x. We would need to find the potential as a function of both x and y to be able to find the x and y components of the electric field using Equation 25.25.

Example 25.8   Electric Potential Due to a Uniformly Charged Sphere

An insulating solid sphere of radius R has a uniform positive volume charge density and total charge Q.

(A) Find the electric potential at a point outside the sphere, that is, for $r > R$. Take the potential to be zero at $r = \infty$.

Solution In Example 24.5, we found that the magnitude of the electric field outside a uniformly charged sphere of radius R is

$$E_r = k_e \frac{Q}{r^2} \qquad (\text{for } r > R)$$

where the field is directed radially outward when Q is positive. This is the same as the field due to a point charge, which we studied in Section 23.4. In this case, to obtain the electric potential at an exterior point, such as B in Figure 25.19, we use Equation 25.10, choosing point A as $r = \infty$:

$$V_B - V_A = k_e Q \left[\frac{1}{r_B} - \frac{1}{r_A}\right]$$

$$V_B - 0 = k_e Q \left[\frac{1}{r_B} - 0\right]$$

$$V_B = k_e \frac{Q}{r} \qquad (\text{for } r > R)$$

Because the potential must be continuous at $r = R$, we can use this expression to obtain the potential at the surface of the sphere. That is, the potential at a point such as C shown in Figure 25.19 is

$$V_C = k_e \frac{Q}{R} \qquad (\text{for } r = R)$$

(B) Find the potential at a point inside the sphere, that is, for $r < R$.

Solution In Example 24.5 we found that the electric field inside an insulating uniformly charged sphere is

$$E_r = \frac{k_e Q}{R^3} r \qquad (\text{for } r < R)$$

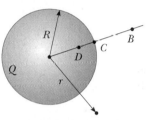

Figure 25.19 (Example 25.8) A uniformly charged insulating sphere of radius R and total charge Q. The electric potentials at points B and C are equivalent to those produced by a point charge Q located at the center of the sphere, but this is not true for point D.

We can use this result and Equation 25.3 to evaluate the potential difference $V_D - V_C$ at some interior point D:

$$V_D - V_C = -\int_R^r E_r\, dr = -\frac{k_e Q}{R^3} \int_R^r r\, dr = \frac{k_e Q}{2R^3}\left(R^2 - r^2\right)$$

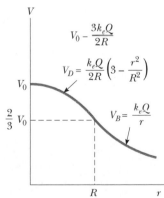

Figure 25.20 (Example 25.8) A plot of electric potential V versus distance r from the center of a uniformly charged insulating sphere of radius R. The curve for V_D inside the sphere is parabolic and joins smoothly with the curve for V_B outside the sphere, which is a hyperbola. The potential has a maximum value V_0 at the center of the sphere. We could make this graph three dimensional (similar to Figures 25.8 and 25.9) by revolving it around the vertical axis.

Substituting $V_C = k_e Q/R$ into this expression and solving for V_D, we obtain

$$V_D = \frac{k_e Q}{2R}\left(3 - \frac{r^2}{R^2}\right) \quad \text{(for } r < R) \quad (25.26)$$

At $r = R$, this expression gives a result that agrees with that for the potential at the surface, that is, V_C. A plot of V versus r for this charge distribution is given in Figure 25.20.

25.6   Electric Potential Due to a Charged Conductor

In Section 24.4 we found that when a solid conductor in equilibrium carries a net charge, the charge resides on the outer surface of the conductor. Furthermore, we showed that the electric field just outside the conductor is perpendicular to the surface and that the field inside is zero.

We now show that **every point on the surface of a charged conductor in equilibrium is at the same electric potential.** Consider two points A and B on the surface of a charged conductor, as shown in Figure 25.21. Along a surface path connecting these points, **E** is always perpendicular to the displacement $d\mathbf{s}$; therefore $\mathbf{E} \cdot d\mathbf{s} = 0$. Using this result and Equation 25.3, we conclude that the potential difference between A and B is necessarily zero:

$$V_B - V_A = -\int_A^B \mathbf{E} \cdot d\mathbf{s} = 0$$

This result applies to any two points on the surface. Therefore, V is constant everywhere on the surface of a charged conductor in equilibrium. That is,

the surface of any charged conductor in electrostatic equilibrium is an equipotential surface. Furthermore, because the electric field is zero inside the conductor, we conclude that the electric potential is constant everywhere inside the conductor and equal to its value at the surface.

Because this is true, no work is required to move a test charge from the interior of a charged conductor to its surface.

Consider a solid metal conducting sphere of radius R and total positive charge Q, as shown in Figure 25.22a. The electric field outside the sphere is $k_e Q/r^2$ and points radially outward. From Example 25.8, we know that the electric potential at the interior and surface of the sphere must be $k_e Q/R$ relative to infinity. The potential outside the sphere is $k_e Q/r$. Figure 25.22b is a plot of the electric potential as a function of r, and Figure 25.22c shows how the electric field varies with r.

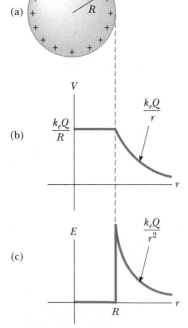

Figure 25.22 (a) The excess charge on a conducting sphere of radius R is uniformly distributed on its surface. (b) Electric potential versus distance r from the center of the charged conducting sphere. (c) Electric field magnitude versus distance r from the center of the charged conducting sphere.

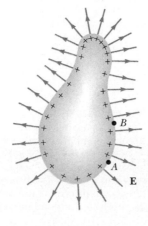

Figure 25.21 An arbitrarily shaped conductor carrying a positive charge. When the conductor is in electrostatic equilibrium, all of the charge resides at the surface, **E** = 0 inside the conductor, and the direction of **E** just outside the conductor is perpendicular to the surface. The electric potential is constant inside the conductor and is equal to the potential at the surface. Note from the spacing of the positive signs that the surface charge density is nonuniform.

When a net charge is placed on a spherical conductor, the surface charge density is uniform, as indicated in Figure 25.22a. However, if the conductor is nonspherical, as in Figure 25.21, the surface charge density is high where the radius of curvature is small (as noted in Section 24.4), and it is low where the radius of curvature is large. Because the electric field just outside the conductor is proportional to the surface charge density, we see that **the electric field is large near convex points having small radii of curvature and reaches very high values at sharp points.** This is demonstrated in Figure 25.23, in which small pieces of thread suspended in oil show the electric field lines. Notice that the density of field lines is highest at the sharp tip of the left-hand conductor and at the highly curved ends of the right-hand conductor. In Example 25.9, the relationship between electric field and radius of curvature is explored mathematically.

Figure 25.24 shows the electric field lines around two spherical conductors: one carrying a net charge Q, and a larger one carrying zero net charge. In this case, the surface charge density is not uniform on either conductor. The sphere having zero net charge has negative charges induced on its side that faces the charged sphere and positive charges induced on its side opposite the charged sphere. The broken blue curves in the figure represent the cross sections of the equipotential surfaces for this charge configuration. As usual, the field lines are perpendicular to the conducting surfaces at all points, and the equipotential surfaces are perpendicular to the field lines everywhere.

▲ **PITFALL PREVENTION**

25.6 Potential May Not Be Zero

The electric potential inside the conductor is not necessarily zero in Figure 25.22, even though the electric field is zero. From Equation 25.15, we see that a zero value of the field results in no *change* in the potential from one point to another inside the conductor. Thus, the potential everywhere inside the conductor, including the surface, has the same value, which may or may not be zero, depending on where the zero of potential is defined.

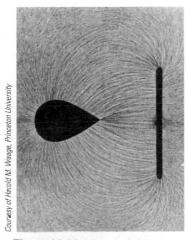

Courtesy of Harold M. Waage, Princeton University

Figure 25.23 Electric field pattern of a charged conducting plate placed near an oppositely charged pointed conductor. Small pieces of thread suspended in oil align with the electric field lines. The field surrounding the pointed conductor is most intense near the pointed end and at other places where the radius of curvature is small.

Quick Quiz 25.10 Consider starting at the center of the left-hand sphere (sphere 1, of radius a) in Figure 25.24 and moving to the far right of the diagram, passing through the center of the right-hand sphere (sphere 2, of radius c) along the way. The centers of the spheres are a distance b apart. Draw a graph of the electric potential as a function of position relative to the center of the left-hand sphere.

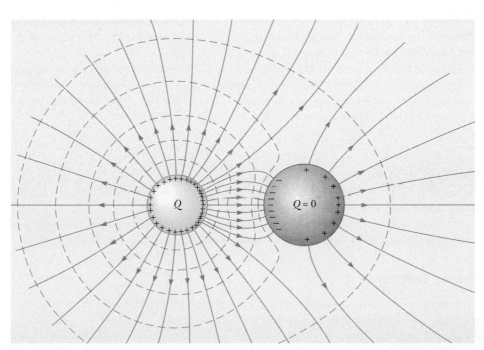

Figure 25.24 The electric field lines (in red-brown) around two spherical conductors. The smaller sphere has a net charge Q, and the larger one has zero net charge. The broken blue curves are intersections of equipotential surfaces with the page.

Example 25.9 Two Connected Charged Spheres

Two spherical conductors of radii r_1 and r_2 are separated by a distance much greater than the radius of either sphere. The spheres are connected by a conducting wire, as shown in Figure 25.25. The charges on the spheres in equilibrium are q_1 and q_2, respectively, and they are uniformly charged. Find the ratio of the magnitudes of the electric fields at the surfaces of the spheres.

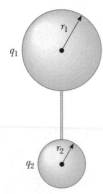

q_1

q_2

Figure 25.25 (Example 25.9) Two charged spherical conductors connected by a conducting wire. The spheres are at the *same* electric potential V.

Solution Because the spheres are connected by a conducting wire, they must both be at the same electric potential:

$$V = k_e \frac{q_1}{r_1} = k_e \frac{q_2}{r_2}$$

Therefore, the ratio of charges is

$$(1) \qquad \frac{q_1}{q_2} = \frac{r_1}{r_2}$$

Because the spheres are very far apart and their surfaces uniformly charged, we can express the magnitude of the electric fields at their surfaces as

$$E_1 = k_e \frac{q_1}{r_1^2} \qquad \text{and} \qquad E_2 = k_e \frac{q_2}{r_2^2}$$

Taking the ratio of these two fields and making use of Equation (1), we find that

$$(2) \qquad \frac{E_1}{E_2} = \frac{r_2}{r_1}$$

Hence, the field is more intense in the vicinity of the smaller sphere even though the electric potentials of both spheres are the same.

A Cavity Within a Conductor

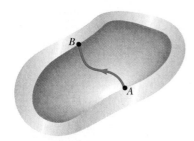

Figure 25.26 A conductor in electrostatic equilibrium containing a cavity. The electric field in the cavity is zero, regardless of the charge on the conductor.

Now suppose a conductor of arbitrary shape contains a cavity as shown in Figure 25.26. Let us assume that no charges are inside the cavity. **In this case, the electric field inside the cavity must be zero** regardless of the charge distribution on the outside surface of the conductor. Furthermore, the field in the cavity is zero even if an electric field exists outside the conductor.

To prove this point, we use the fact that every point on the conductor is at the same electric potential, and therefore any two points A and B on the surface of the cavity must be at the same potential. Now imagine that a field **E** exists in the cavity and evaluate the potential difference $V_B - V_A$ defined by Equation 25.3:

$$V_B - V_A = -\int_A^B \mathbf{E} \cdot d\mathbf{s}$$

Because $V_B - V_A = 0$, the integral of $\mathbf{E} \cdot d\mathbf{s}$ must be zero for all paths between any two points A and B on the conductor. The only way that this can be true for *all* paths is if **E** is zero *everywhere* in the cavity. Thus, we conclude that **a cavity surrounded by conducting walls is a field-free region as long as no charges are inside the cavity.**

Corona Discharge

A phenomenon known as **corona discharge** is often observed near a conductor such as a high-voltage power line. When the electric field in the vicinity of the conductor is sufficiently strong, electrons resulting from random ionizations of air molecules near the conductor accelerate away from their parent molecules. These rapidly moving electrons can ionize additional molecules near the conductor, creating more free electrons. The observed glow (or corona discharge) results from the recombination of these free electrons with the ionized air molecules. If a conductor has an irregular shape, the electric field can be very high near sharp points or edges of the conductor; consequently, the ionization process and corona discharge are most likely to occur around such points.

Corona discharge is used in the electrical transmission industry to locate broken or faulty components. For example, a broken insulator on a transmission tower has sharp edges where corona discharge is likely to occur. Similarly, corona discharge will occur at the sharp end of a broken conductor strand. Observation of these discharges is difficult because the visible radiation emitted is weak and most of the radiation is in the ultraviolet. (We will discuss ultraviolet radiation and other portions of the electromagnetic spectrum in Section 34.6.) Even use of traditional ultraviolet cameras is of little help because the radiation from the corona discharge is overwhelmed by ultraviolet radiation from the Sun. Newly developed dual-spectrum devices combine a narrowband ultraviolet camera with a visible light camera to show a daylight view of the corona discharge in the actual location on the transmission tower or cable. The ultraviolet part of the camera is designed to operate in a wavelength range in which radiation from the Sun is very weak.

25.7 The Millikan Oil-Drop Experiment

During the period from 1909 to 1913, Robert Millikan performed a brilliant set of experiments in which he measured e, the magnitude of the elementary charge on an electron, and demonstrated the quantized nature of this charge. His apparatus, diagrammed in Figure 25.27, contains two parallel metallic plates. Oil droplets from an atomizer are allowed to pass through a small hole in the upper plate. Millikan used x-rays to ionize the air in the chamber, so that freed electrons would adhere to the oil drops, giving them a negative charge. A horizontally directed light beam is used to illuminate the oil droplets, which are viewed through a telescope whose long axis is perpendicular to the light beam. When the droplets are viewed in this manner, they appear as shining stars against a dark background, and the rate at which individual drops fall can be determined.

Let us assume that a single drop having a mass m and carrying a charge q is being viewed and that its charge is negative. If no electric field is present between the plates,

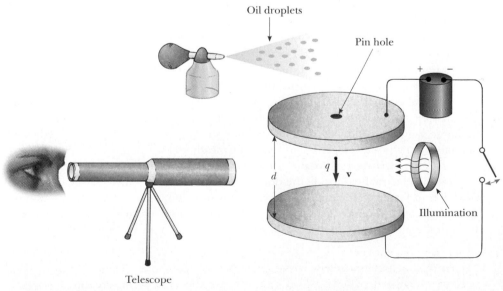

Oil droplets

Pin hole

d q $\downarrow$ $\mathbf{v}$

Illumination

Telescope

Active Figure 25.27 Schematic drawing of the Millikan oil-drop apparatus.

At the Active Figures link at http://www.pse6.com, you can do a simplified version of the experiment for yourself. You will be able to take data on a number of oil drops and determine the elementary charge from your data.

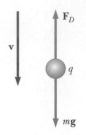

(a) Field off

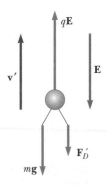

(b) Field on

Figure 25.28 The forces acting on a negatively charged oil droplet in the Millikan experiment.

the two forces acting on the charge are the gravitational force $m\mathbf{g}$ acting downward[4] and a viscous drag force $\mathbf{F}_D$ acting upward as indicated in Figure 25.28a. The drag force is proportional to the drop's speed. When the drop reaches its terminal speed v, the two forces balance each other ($mg = F_D$).

Now suppose that a battery connected to the plates sets up an electric field between the plates such that the upper plate is at the higher electric potential. In this case, a third force $q\mathbf{E}$ acts on the charged drop. Because q is negative and $\mathbf{E}$ is directed downward, this electric force is directed upward, as shown in Figure 25.28b. If this force is sufficiently great, the drop moves upward and the drag force $\mathbf{F}_D'$ acts downward. When the upward electric force $q\mathbf{E}$ balances the sum of the gravitational force and the downward drag force $\mathbf{F}_D'$, the drop reaches a new terminal speed v' in the upward direction.

With the field turned on, a drop moves slowly upward, typically at rates of hundredths of a centimeter per second. The rate of fall in the absence of a field is comparable. Hence, one can follow a single droplet for hours, alternately rising and falling, by simply turning the electric field on and off.

After recording measurements on thousands of droplets, Millikan and his co-workers found that all droplets, to within about 1% precision, had a charge equal to some integer multiple of the elementary charge e:

$$q = ne \qquad n = 0, -1, -2, -3, \ldots$$

where $e = 1.60 \times 10^{-19}$ C. Millikan's experiment yields conclusive evidence that charge is quantized. For this work, he was awarded the Nobel Prize in Physics in 1923.

25.8 Applications of Electrostatics

The practical application of electrostatics is represented by such devices as lightning rods and electrostatic precipitators and by such processes as xerography and the painting of automobiles. Scientific devices based on the principles of electrostatics include electrostatic generators, the field-ion microscope, and ion-drive rocket engines.

The Van de Graaff Generator

Experimental results show that when a charged conductor is placed in contact with the inside of a hollow conductor, all of the charge on the charged conductor is transferred to the hollow conductor. In principle, the charge on the hollow conductor and its electric potential can be increased without limit by repetition of the process.

In 1929 Robert J. Van de Graaff (1901–1967) used this principle to design and build an electrostatic generator. This type of generator is used extensively in nuclear physics research. A schematic representation of the generator is given in Figure 25.29. Charge is delivered continuously to a high-potential electrode by means of a moving belt of insulating material. The high-voltage electrode is a hollow metal dome mounted on an insulating column. The belt is charged at point A by means of a corona discharge between comb-like metallic needles and a grounded grid. The needles are maintained at a positive electric potential of typically 10^4 V. The positive charge on the moving belt is transferred to the dome by a second comb of needles at point B. Because the electric field inside the dome is negligible, the positive charge on the belt is easily transferred to the conductor regardless of its potential. In practice, it is possible to increase the electric potential of the dome until electrical discharge occurs through the air. Because the "breakdown" electric field in air is about 3×10^6 V/m, a

---

[4]   There is also a buoyant force on the oil drop due to the surrounding air. This force can be incorporated as a correction in the gravitational force $m\mathbf{g}$ on the drop, so we will not consider it in our analysis.

sphere 1 m in radius can be raised to a maximum potential of 3×10^6 V. The potential can be increased further by increasing the radius of the dome and by placing the entire system in a container filled with high-pressure gas.

Van de Graaff generators can produce potential differences as large as 20 million volts. Protons accelerated through such large potential differences receive enough energy to initiate nuclear reactions between themselves and various target nuclei. Smaller generators are often seen in science classrooms and museums. If a person insulated from the ground touches the sphere of a Van de Graaff generator, his or her body can be brought to a high electric potential. The hair acquires a net positive charge, and each strand is repelled by all the others, as in the opening photograph of Chapter 23.

The Electrostatic Precipitator

One important application of electrical discharge in gases is the *electrostatic precipitator*. This device removes particulate matter from combustion gases, thereby reducing air pollution. Precipitators are especially useful in coal-burning power plants and in industrial operations that generate large quantities of smoke. Current systems are able to eliminate more than 99% of the ash from smoke.

Figure 25.30a shows a schematic diagram of an electrostatic precipitator. A high potential difference (typically 40 to 100 kV) is maintained between a wire running down the center of a duct and the walls of the duct, which are grounded. The wire is maintained at a negative electric potential with respect to the walls, so the electric field is directed toward the wire. The values of the field near the wire become high enough to cause a corona discharge around the wire; the air near the wire contains positive ions, electrons, and such negative ions as O_2^-. The air to be cleaned enters the duct and moves near the wire. As the electrons and negative ions created by the discharge are accelerated toward the outer wall by the electric field, the dirt particles in the air become charged by collisions and ion capture. Because most of the charged dirt particles are negative, they too are drawn to the duct walls by the electric field. When the duct is periodically shaken, the particles break loose and are collected at the bottom.

In addition to reducing the level of particulate matter in the atmosphere (compare Figs. 25.30b and c), the electrostatic precipitator recovers valuable materials in the form of metal oxides.

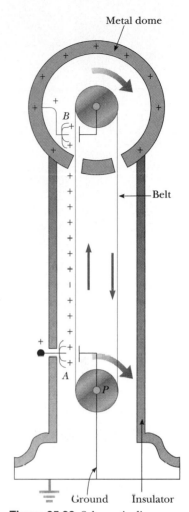

Figure 25.29 Schematic diagram of a Van de Graaff generator. Charge is transferred to the metal dome at the top by means of a moving belt. The charge is deposited on the belt at point A and transferred to the hollow conductor at point B.

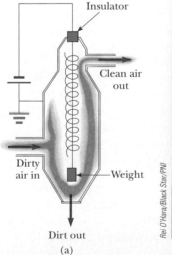

(a)

(b)

(c)

Figure 25.30 (a) Schematic diagram of an electrostatic precipitator. The high negative electric potential maintained on the central coiled wire creates a corona discharge in the vicinity of the wire. Compare the air pollution when the electrostatic precipitator is (b) operating and (c) turned off.

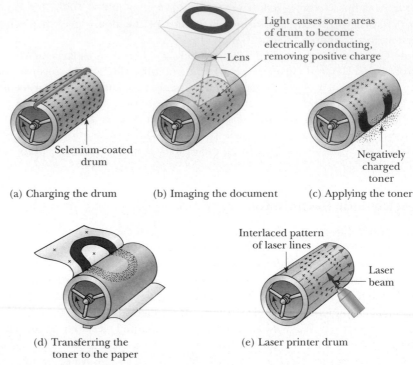

Figure 25.31 The xerographic process: (a) The photoconductive surface of the drum is positively charged. (b) Through the use of a light source and lens, an image is formed on the surface in the form of positive charges. (c) The surface containing the image is covered with a negatively charged powder, which adheres only to the image area. (d) A piece of paper is placed over the surface and given a positive charge. This transfers the image to the paper as the negatively charged powder particles migrate to the paper. The paper is then heat-treated to "fix" the powder. (e) A laser printer operates similarly except the image is produced by turning a laser beam on and off as it sweeps across the selenium-coated drum.

Xerography and Laser Printers

The basic idea of xerography[5] was developed by Chester Carlson, who was granted a patent for the xerographic process in 1940. The unique feature of this process is the use of a photoconductive material to form an image. (A *photoconductor* is a material that is a poor electrical conductor in the dark but becomes a good electrical conductor when exposed to light.)

The xerographic process is illustrated in Figure 25.31a to d. First, the surface of a plate or drum that has been coated with a thin film of photoconductive material (usually selenium or some compound of selenium) is given a positive electrostatic charge in the dark. An image of the page to be copied is then focused by a lens onto the charged surface. The photoconducting surface becomes conducting only in areas where light strikes it. In these areas, the light produces charge carriers in the photoconductor that move the positive charge off the drum. However, positive charges remain on those areas of the photoconductor not exposed to light, leaving a latent image of the object in the form of a positive surface charge distribution.

Next, a negatively charged powder called a *toner* is dusted onto the photoconducting surface. The charged powder adheres only to those areas of the surface that contain the positively charged image. At this point, the image becomes visible. The toner (and hence the image) is then transferred to the surface of a sheet of positively charged paper.

Finally, the toner is "fixed" to the surface of the paper as the toner melts while passing through high-temperature rollers. This results in a permanent copy of the original.

A laser printer (Fig. 25.31e) operates by the same principle, with the exception that a computer-directed laser beam is used to illuminate the photoconductor instead of a lens.

---

[5] The prefix *xero-* is from the Greek word meaning "dry." Note that liquid ink is not used in xerography.

SUMMARY

🌐 *Take a practice test for this chapter by clicking on the Practice Test link at http://www.pse6.com.*

When a positive test charge q_0 is moved between points A and B in an electric field $\mathbf{E}$, the **change in the potential energy of the charge–field system** is

$$\Delta U = -q_0 \int_A^B \mathbf{E} \cdot d\mathbf{s} \tag{25.1}$$

The **electric potential** $V = U/q_0$ is a scalar quantity and has the units of J/C, where $1 \text{ J/C} \equiv 1 \text{ V}$.

The **potential difference** ΔV between points A and B in an electric field $\mathbf{E}$ is defined as

$$\Delta V = \frac{\Delta U}{q_0} = -\int_A^B \mathbf{E} \cdot d\mathbf{s} \tag{25.3}$$

The potential difference between two points A and B in a uniform electric field $\mathbf{E}$, where $\mathbf{s}$ is a vector that points from A to B and is parallel to $\mathbf{E}$ is

$$\Delta V = -Ed \tag{25.6}$$

where $d = |\mathbf{s}|$.

An **equipotential surface** is one on which all points are at the same electric potential. Equipotential surfaces are perpendicular to electric field lines.

If we define $V = 0$ at $r_A = \infty$, the electric potential due to a point charge at any distance r from the charge is

$$V = k_e \frac{q}{r} \tag{25.11}$$

We can obtain the electric potential associated with a group of point charges by summing the potentials due to the individual charges.

The **potential energy associated with a pair of point charges** separated by a distance r_{12} is

$$U = k_e \frac{q_1 q_2}{r_{12}} \tag{25.13}$$

This energy represents the work done by an external agent when the charges are brought from an infinite separation to the separation r_{12}. We obtain the potential energy of a distribution of point charges by summing terms like Equation 25.13 over all pairs of particles.

If we know the electric potential as a function of coordinates x, y, z, we can obtain the components of the electric field by taking the negative derivative of the electric potential with respect to the coordinates. For example, the x component of the electric field is

$$E_x = -\frac{dV}{dx} \tag{25.16}$$

The **electric potential due to a continuous charge distribution** is

$$V = k_e \int \frac{dq}{r} \tag{25.20}$$

Every point on the surface of a charged conductor in electrostatic equilibrium is at the same electric potential. The potential is constant everywhere inside the conductor and equal to its value at the surface.

Table 25.1 lists electric potentials due to several charge distributions.

Table 25.1

Electric Potential Due to Various Charge Distributions

Charge Distribution	Electric Potential	Location
Uniformly charged ring of radius a	$V = k_e \dfrac{Q}{\sqrt{x^2 + a^2}}$	Along perpendicular central axis of ring, distance x from ring center
Uniformly charged disk of radius a	$V = 2\pi k_e \sigma \left[(x^2 + a^2)^{1/2} - x \right]$	Along perpendicular central axis of disk, distance x from disk center
Uniformly charged, *insulating* solid sphere of radius R and total charge Q	$V = k_e \dfrac{Q}{r}$ $\quad$ $V = \dfrac{k_e Q}{2R} \left(3 - \dfrac{r^2}{R^2} \right)$	$r \geq R$ $\quad$ $r < R$
Isolated *conducting* sphere of radius R and total charge Q	$V = k_e \dfrac{Q}{r}$ $\quad$ $V = k_e \dfrac{Q}{R}$	$r > R$ $\quad$ $r \leq R$

QUESTIONS

1. Distinguish between electric potential and electric potential energy.

2. A negative charge moves in the direction of a uniform electric field. Does the potential energy of the charge–field system increase or decrease? Does the charge move to a position of higher or lower potential?

3. Give a physical explanation of the fact that the potential energy of a pair of charges with the same sign is positive whereas the potential energy of a pair of charges with opposite signs is negative.

4. A uniform electric field is parallel to the x axis. In what direction can a charge be displaced in this field without any external work being done on the charge?

5. Explain why equipotential surfaces are always perpendicular to electric field lines.

6. Describe the equipotential surfaces for (a) an infinite line of charge and (b) a uniformly charged sphere.

7. Explain why, under static conditions, all points in a conductor must be at the same electric potential.

8. The electric field inside a hollow, uniformly charged sphere is zero. Does this imply that the potential is zero inside the sphere? Explain.

9. The potential of a point charge is defined to be zero at an infinite distance. Why can we not define the potential of an infinite line of charge to be zero at $r = \infty$?

10. Two charged conducting spheres of different radii are connected by a conducting wire as shown in Figure 25.25. Which sphere has the greater charge density?

11. What determines the maximum potential to which the dome of a Van de Graaff generator can be raised?

12. Explain the origin of the glow sometimes observed around the cables of a high-voltage power line.

13. Why is it important to avoid sharp edges or points on conductors used in high-voltage equipment?

14. How would you shield an electronic circuit or laboratory from stray electric fields? Why does this work?

15. Two concentric spherical conducting shells of radii $a = 0.400$ m and $b = 0.500$ m are connected by a thin wire as shown in Figure Q25.15. If a total charge $Q = 10.0$ μC is placed on the system, how much charge settles on each sphere?

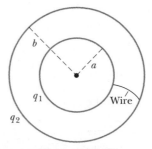

Figure Q25.15

16. Study Figure 23.4 and the accompanying text discussion of charging by induction. You may also compare to Figure

25.24. When the grounding wire is touched to the rightmost point on the sphere in Figure 23.4c, electrons are drained away from the sphere to leave the sphere positively charged. Suppose instead that the grounding wire is touched to the leftmost point on the sphere. Will electrons still drain away, moving closer to the negatively charged rod as they do so? What kind of charge, if any, will remain on the sphere?

PROBLEMS

1, 2, 3 = straightforward, intermediate, challenging   ☐ = full solution available in the *Student Solutions Manual and Study Guide*

🌐 = coached solution with hints available at http://www.pse6.com   💻 = computer useful in solving problem

▬ = paired numerical and symbolic problems

Section 25.1   Potential Difference and Electric Potential

1. How much work is done (by a battery, generator, or some other source of potential difference) in moving Avogadro's number of electrons from an initial point where the electric potential is 9.00 V to a point where the potential is −5.00 V? (The potential in each case is measured relative to a common reference point.)

2. An ion accelerated through a potential difference of 115 V experiences an increase in kinetic energy of 7.37×10^{-17} J. Calculate the charge on the ion. $q = \dfrac{\Delta u}{\Delta v} = \quad = C$

3. (a) Calculate the speed of a proton that is accelerated from rest through a potential difference of 120 V. (b) Calculate the speed of an electron that is accelerated through the same potential difference.

4. What potential difference is needed to stop an electron having an initial speed of 4.20×10^5 m/s?

Section 25.2   Potential Differences in a Uniform Electric Field

5. A uniform electric field of magnitude 250 V/m is directed in the positive x direction. A +12.0-μC charge moves from the origin to the point $(x, y) = (20.0 \text{ cm}, 50.0 \text{ cm})$. (a) What is the change in the potential energy of the charge–field system? (b) Through what potential difference does the charge move?

6. The difference in potential between the accelerating plates in the electron gun of a TV picture tube is about 25 000 V. If the distance between these plates is 1.50 cm, what is the magnitude of the uniform electric field in this region?

7. 🌐 An electron moving parallel to the x axis has an initial speed of 3.70×10^6 m/s at the origin. Its speed is reduced to 1.40×10^5 m/s at the point $x = 2.00$ cm. Calculate the potential difference between the origin and that point. Which point is at the higher potential? $V_y = 3.70 \times 10^6 \text{ m/s}$ $V_f = 1.40 \times 10^5 \text{ m/s}$ $x = 2.00 \text{ cm}$

38.9 V

8. Suppose an electron is released from rest in a uniform electric field whose magnitude is 5.90×10^3 V/m. (a) Through what potential difference will it have passed after moving 1.00 cm? (b) How fast will the electron be moving after it has traveled 1.00 cm?

9. A uniform electric field of magnitude 325 V/m is directed in the negative y direction in Figure P25.9. The coordinates of point A are $(-0.200, -0.300)$ m, and those of point B are $(0.400, 0.500)$ m. Calculate the potential difference $V_B - V_A$, using the blue path.

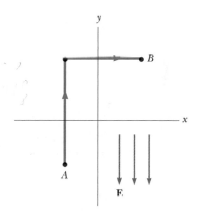

Figure P25.9

10. Starting with the definition of work, prove that at every point on an equipotential surface the surface must be perpendicular to the electric field there.

11. **Review problem.** A block having mass m and charge $+Q$ is connected to a spring having constant k. The block lies on a frictionless horizontal track, and the system is immersed in a uniform electric field of magnitude E, directed as shown in Figure P25.11. If the block is released from rest when the spring is unstretched (at $x = 0$), (a) by what maximum amount does the spring expand? (b) What is the equilibrium position of the block? (c) Show that the block's motion is simple harmonic, and determine its period. (d) **What If?** Repeat part (a) if the coefficient of kinetic friction between block and surface is μ_k.

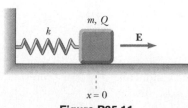

Figure P25.11

12. On planet Tehar, the free-fall acceleration is the same as that on Earth but there is also a strong downward electric field that is uniform close to the planet's surface. A 2.00-kg ball having a charge of 5.00 μC is thrown upward at a speed of 20.1 m/s, and it hits the ground after an interval of 4.10 s. What is the potential difference between the starting point and the top point of the trajectory?

13. An insulating rod having linear charge density $\lambda = 40.0 \ \mu$C/m and linear mass density $\mu = 0.100$ kg/m is released from rest in a uniform electric field $E = 100$ V/m directed perpendicular to the rod (Fig. P25.13). (a) Determine the speed of the rod after it has traveled 2.00 m. (b) **What If?** How does your answer to part (a) change if the electric field is not perpendicular to the rod? Explain.

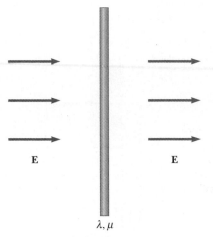

λ, μ

Figure P25.13

14. A particle having charge $q = +2.00 \ \mu$C and mass $m = 0.010 \ 0$ kg is connected to a string that is $L = 1.50$ m long and is tied to the pivot point P in Figure P25.14. The particle, string and pivot point all lie on a frictionless horizontal table. The particle is released from rest when the string makes an angle $\theta = 60.0°$ with a uniform electric field of magnitude $E = 300$ V/m. Determine the speed of the particle when the string is parallel to the electric field (point a in Fig. P25.14).

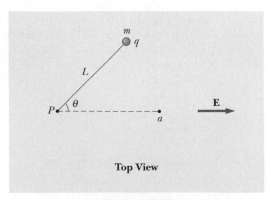

Top View

Figure P25.14

Section 25.3 Electric Potential and Potential Energy Due to Point Charges

Note: Unless stated otherwise, assume the reference level of potential is $V = 0$ at $r = \infty$.

15. (a) Find the potential at a distance of 1.00 cm from a proton. (b) What is the potential difference between two points that are 1.00 cm and 2.00 cm from a proton? (c) **What If?** Repeat parts (a) and (b) for an electron.

16. Given two 2.00-μC charges, as shown in Figure P25.16, and a positive test charge $q = 1.28 \times 10^{-18}$ C at the origin, (a) what is the net force exerted by the two 2.00-μC charges on the test charge q? (b) What is the electric field at the origin due to the two 2.00-μC charges? (c) What is the electric potential at the origin due to the two 2.00-μC charges?

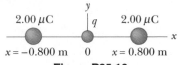

Figure P25.16

17. At a certain distance from a point charge, the magnitude of the electric field is 500 V/m and the electric potential is -3.00 kV. (a) What is the distance to the charge? (b) What is the magnitude of the charge?

18. A charge $+q$ is at the origin. A charge $-2q$ is at $x = 2.00$ m on the x axis. For what finite value(s) of x is (a) the electric field zero? (b) the electric potential zero?

19. The three charges in Figure P25.19 are at the vertices of an isosceles triangle. Calculate the electric potential at the midpoint of the base, taking $q = 7.00 \ \mu$C.

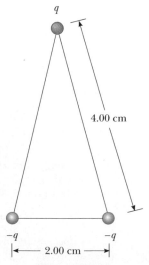

Figure P25.19

20. Two point charges, $Q_1 = +5.00$ nC and $Q_2 = -3.00$ nC, are separated by 35.0 cm. (a) What is the potential energy of the pair? What is the significance of the algebraic sign of your answer? (b) What is the electric potential at a point midway between the charges?

21. *Compare this problem with Problem 57 in Chapter 23.* Four identical point charges ($q = +10.0$ μC) are located on the corners of a rectangle as shown in Figure P23.57. The dimensions of the rectangle are $L = 60.0$ cm and $W = 15.0$ cm. Calculate the change in electric potential energy of the system as the charge at the lower left corner in Figure P23.57 is brought to this position from infinitely far away. Assume that the other three charges in Figure P23.57 remain fixed in position.

22. *Compare this problem with Problem 20 in Chapter 23.* Two point charges each of magnitude 2.00 μC are located on the x axis. One is at $x = 1.00$ m, and the other is at $x = -1.00$ m. (a) Determine the electric potential on the y axis at $y = 0.500$ m. (b) Calculate the change in electric potential energy of the system as a third charge of -3.00 μC is brought from infinitely far away to a position on the y axis at $y = 0.500$ m.

23. Show that the amount of work required to assemble four identical point charges of magnitude Q at the corners of a square of side s is $5.41 k_e Q^2/s$.

24. *Compare this problem with Problem 23 in Chapter 23.* Five equal negative point charges $-q$ are placed symmetrically around a circle of radius R. Calculate the electric potential at the center of the circle.

25. *Compare this problem with Problem 41 in Chapter 23.* Three equal positive charges q are at the corners of an equilateral triangle of side a as shown in Figure P23.41. (a) At what point, if any, in the plane of the charges is the electric potential zero? (b) What is the electric potential at the point P due to the two charges at the base of the triangle?

26. Review problem. Two insulating spheres have radii 0.300 cm and 0.500 cm, masses 0.100 kg and 0.700 kg, and uniformly distributed charges of -2.00 μC and 3.00 μC. They are released from rest when their centers are separated by 1.00 m. (a) How fast will each be moving when they collide? (*Suggestion:* consider conservation of energy and of linear momentum.) (b) **What If?** If the spheres were conductors, would the speeds be greater or less than those calculated in part (a)? Explain.

27. Review problem. Two insulating spheres have radii r_1 and r_2, masses m_1 and m_2, and uniformly distributed charges $-q_1$ and q_2. They are released from rest when their centers are separated by a distance d. (a) How fast is each moving when they collide? (*Suggestion:* consider conservation of energy and conservation of linear momentum.) (b) **What If?** If the spheres were conductors, would their speeds be greater or less than those calculated in part (a)? Explain.

28. Two particles, with charges of 20.0 nC and -20.0 nC, are placed at the points with coordinates (0, 4.00 cm) and (0, -4.00 cm), as shown in Figure P25.28. A particle with charge 10.0 nC is located at the origin. (a) Find the electric potential energy of the configuration of the three fixed charges. (b) A fourth particle, with a mass of 2.00×10^{-13} kg and a charge of 40.0 nC, is released from rest at the point (3.00 cm, 0). Find its speed after it has moved freely to a very large distance away.

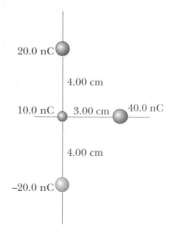

20.0 nC

4.00 cm

10.0 nC   3.00 cm   40.0 nC

4.00 cm

-20.0 nC

Figure P25.28

29. Review problem. A light unstressed spring has length d. Two identical particles, each with charge q, are connected to the opposite ends of the spring. The particles are held stationary a distance d apart and then released at the same time. The system then oscillates on a horizontal frictionless table. The spring has a bit of internal kinetic friction, so the oscillation is damped. The particles eventually stop vibrating when the distance between them is $3d$. Find the increase in internal energy that appears in the spring during the oscillations. Assume that the system of the spring and two charges is isolated.

30. Two point charges of equal magnitude are located along the y axis equal distances above and below the x axis, as shown in Figure P25.30. (a) Plot a graph of the potential at points along the x axis over the interval $-3a < x < 3a$. You should plot the potential in units of $k_e Q/a$. (b) Let the charge located at $-a$ be negative and plot the potential along the y axis over the interval $-4a < y < 4a$.

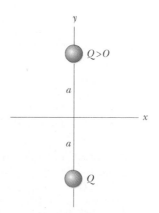

y

$Q > 0$

a

x

a

Q

Figure P25.30

31. A small spherical object carries a charge of 8.00 nC. At what distance from the center of the object is the potential equal to 100 V? 50.0 V? 25.0 V? Is the spacing of the equipotentials proportional to the change in potential?

32. In 1911 Ernest Rutherford and his assistants Geiger and Marsden conducted an experiment in which they scattered alpha particles from thin sheets of gold. An alpha particle, having charge $+2e$ and mass 6.64×10^{-27} kg, is a product of certain radioactive decays. The results of the experiment led Rutherford to the idea that most of the mass of an atom is in a very small nucleus, with electrons in orbit around it—his planetary model of the atom. Assume an alpha particle, initially very far from a gold nucleus, is fired with a velocity of 2.00×10^7 m/s directly toward the nucleus (charge $+79e$). How close does the alpha particle get to the nucleus before turning around? Assume the gold nucleus remains stationary.

33. An electron starts from rest 3.00 cm from the center of a uniformly charged insulating sphere of radius 2.00 cm and total charge 1.00 nC. What is the speed of the electron when it reaches the surface of the sphere?

34. Calculate the energy required to assemble the array of charges shown in Figure P25.34, where $a = 0.200$ m, $b = 0.400$ m, and $q = 6.00$ μC.

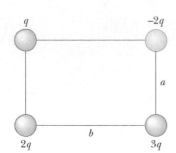

Figure P25.34

35. Four identical particles each have charge q and mass m. They are released from rest at the vertices of a square of side L. How fast is each charge moving when their distance from the center of the square doubles?

36. How much work is required to assemble eight identical point charges, each of magnitude q, at the corners of a cube of side s?

Section 25.4 Obtaining the Value of the Electric Field from the Electric Potential

37. The potential in a region between $x = 0$ and $x = 6.00$ m is $V = a + bx$, where $a = 10.0$ V and $b = -7.00$ V/m. Determine (a) the potential at $x = 0$, 3.00 m, and 6.00 m, and (b) the magnitude and direction of the electric field at $x = 0$, 3.00 m, and 6.00 m.

38. The electric potential inside a charged spherical conductor of radius R is given by $V = k_e Q/R$, and the potential outside is given by $V = k_e Q/r$. Using $E_r = -dV/dr$, derive the electric field (a) inside and (b) outside this charge distribution.

39. Over a certain region of space, the electric potential is $V = 5x - 3x^2 y + 2yz^2$. Find the expressions for the x, y, and z components of the electric field over this region. What is the magnitude of the field at the point P that has coordinates $(1, 0, -2)$ m?

40. Figure P25.40 shows several equipotential lines each labeled by its potential in volts. The distance between the lines of the square grid represents 1.00 cm. (a) Is the magnitude of the field larger at A or at B? Why? (b) What is **E** at B? (c) Represent what the field looks like by drawing at least eight field lines.

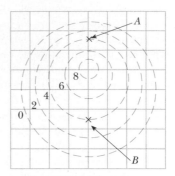

Figure P25.40

41. It is shown in Example 25.7 that the potential at a point P a distance a above one end of a uniformly charged rod of length ℓ lying along the x axis is

$$V = \frac{k_e Q}{\ell} \ln \left(\frac{\ell + \sqrt{\ell^2 + a^2}}{a} \right)$$

Use this result to derive an expression for the y component of the electric field at P. (*Suggestion:* Replace a with y.)

Section 25.5 Electric Potential Due to Continuous Charge Distributions

42. Consider a ring of radius R with the total charge Q spread uniformly over its perimeter. What is the potential difference between the point at the center of the ring and a point on its axis a distance $2R$ from the center?

43. A rod of length L (Fig. P25.43) lies along the x axis with its left end at the origin. It has a nonuniform charge density $\lambda = \alpha x$, where α is a positive constant. (a) What are the units of α? (b) Calculate the electric potential at A.

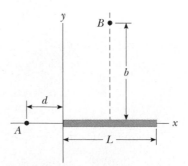

Figure P25.43 Problems 43 and 44.

44. For the arrangement described in the previous problem, calculate the electric potential at point B, which lies on the perpendicular bisector of the rod a distance b above the x axis.

45. *Compare this problem with Problem 33 in Chapter 23.* A uniformly charged insulating rod of length 14.0 cm is bent into the shape of a semicircle as shown in Figure P23.33. The rod has a total charge of $-7.50 \ \mu C$. Find the electric potential at O, the center of the semicircle.

46. Calculate the electric potential at point P on the axis of the annulus shown in Figure P25.46, which has a uniform charge density σ.

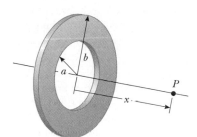

Figure P25.46

47. A wire having a uniform linear charge density λ is bent into the shape shown in Figure P25.47. Find the electric potential at point O.

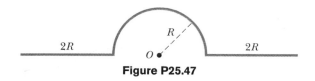

Figure P25.47

Section 25.6   Electric Potential Due to a Charged Conductor

48. How many electrons should be removed from an initially uncharged spherical conductor of radius 0.300 m to produce a potential of 7.50 kV at the surface?

49. A spherical conductor has a radius of 14.0 cm and charge of 26.0 μC. Calculate the electric field and the electric potential (a) $r = 10.0$ cm, (b) $r = 20.0$ cm, and (c) $r = 14.0$ cm from the center.

50. Electric charge can accumulate on an airplane in flight. You may have observed needle-shaped metal extensions on the wing tips and tail of an airplane. Their purpose is to allow charge to leak off before much of it accumulates. The electric field around the needle is much larger than the field around the body of the airplane, and can become large enough to produce dielectric breakdown of the air, discharging the airplane. To model this process, assume that two charged spherical conductors are connected by a long conducting wire, and a charge of 1.20 μC is placed on the combination. One sphere, representing the body of the airplane, has a radius of 6.00 cm, and the other, representing the tip of the needle, has a radius of 2.00 cm. (a) What is the electric potential of each sphere? (b) What is the electric field at the surface of each sphere?

Section 25.8   Applications of Electrostatics

51. Lightning can be studied with a Van de Graaff generator, essentially consisting of a spherical dome on which charge is continuously deposited by a moving belt. Charge can be added until the electric field at the surface of the dome becomes equal to the dielectric strength of air. Any more charge leaks off in sparks, as shown in Figure P25.51. Assume the dome has a diameter of 30.0 cm and is surrounded by dry air with dielectric strength 3.00×10^6 V/m. (a) What is the maximum potential of the dome? (b) What is the maximum charge on the dome?

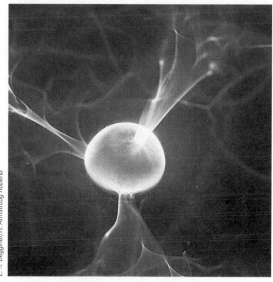

Figure P25.51 Problems 51 and 52.

52. The spherical dome of a Van de Graaff generator can be raised to a maximum potential of 600 kV; then additional charge leaks off in sparks, by producing dielectric breakdown of the surrounding dry air, as shown in Figure P25.51. Determine (a) the charge on the dome and (b) the radius of the dome.

Additional Problems

53. The liquid-drop model of the atomic nucleus suggests that high-energy oscillations of certain nuclei can split the nucleus into two unequal fragments plus a few neutrons. The fission products acquire kinetic energy from their mutual Coulomb repulsion. Calculate the electric potential energy (in electron volts) of two spherical fragments from a uranium nucleus having the following charges and radii: $38e$ and 5.50×10^{-15} m; $54e$ and 6.20×10^{-15} m. Assume that the charge is distributed uniformly throughout the volume of each spherical fragment and that just before separating they are at rest with their surfaces in contact. The electrons surrounding the nucleus can be ignored.

54. On a dry winter day you scuff your leather-soled shoes across a carpet and get a shock when you extend the tip of one finger toward a metal doorknob. In a dark room you see a spark perhaps 5 mm long. Make order-of-magnitude estimates of (a) your electric potential and (b) the charge on your body before you touch the doorknob. Explain your reasoning.

55. The Bohr model of the hydrogen atom states that the single electron can exist only in certain allowed orbits around the proton. The radius of each Bohr orbit is $r = n^2(0.0529 \text{ nm})$ where $n = 1, 2, 3, \ldots$. Calculate the electric potential energy of a hydrogen atom when the electron (a) is in the first allowed orbit, with $n = 1$, (b) is in the second allowed orbit, $n = 2$, and (c) has escaped from the atom, with $r = \infty$. Express your answers in electron volts.

56. An electron is released from rest on the axis of a uniform positively charged ring, 0.100 m from the ring's center. If the linear charge density of the ring is $+ 0.100 \, \mu\text{C/m}$ and the radius of the ring is 0.200 m, how fast will the electron be moving when it reaches the center of the ring?

57. As shown in Figure P25.57, two large parallel vertical conducting plates separated by distance d are charged so that their potentials are $+ V_0$ and $- V_0$. A small conducting ball of mass m and radius R (where $R \ll d$) is hung midway between the plates. The thread of length L supporting the ball is a conducting wire connected to ground, so the potential of the ball is fixed at $V = 0$. The ball hangs straight down in stable equilibrium when V_0 is sufficiently small. Show that the equilibrium of the ball is unstable if V_0 exceeds the critical value $k_e d^2 mg/(4RL)$. (*Suggestion:* consider the forces on the ball when it is displaced a distance $x \ll L$.)

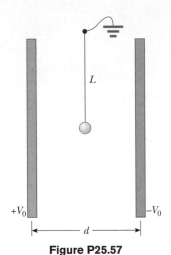

Figure P25.57

58. *Compare this problem with Problem 34 in Chapter 23.* (a) A uniformly charged cylindrical shell has total charge Q, radius R, and height h. Determine the electric potential at a point a distance d from the right end of the cylinder, as shown in Figure P25.58. (*Suggestion:* use the result of Example 25.5 by treating the cylinder as a collection of ring charges.) (b) **What If?** Use the result of Example 25.6 to solve the same problem for a solid cylinder.

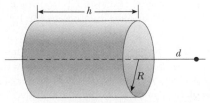

Figure P25.58

59. Calculate the work that must be done to charge a spherical shell of radius R to a total charge Q.

60. Two parallel plates having charges of equal magnitude but opposite sign are separated by 12.0 cm. Each plate has a surface charge density of 36.0 nC/m². A proton is released from rest at the positive plate. Determine (a) the potential difference between the plates, (b) the kinetic energy of the proton when it reaches the negative plate, (c) the speed of the proton just before it strikes the negative plate, (d) the acceleration of the proton, and (e) the force on the proton. (f) From the force, find the magnitude of the electric field and show that it is equal to the electric field found from the charge densities on the plates.

61. A Geiger tube is a radiation detector that essentially consists of a closed, hollow metal cylinder (the cathode) of inner radius r_a and a coaxial cylindrical wire (the anode) of radius r_b (Fig. P25.61). The charge per unit length on the anode is λ, while the charge per unit length on the cathode is $- \lambda$. A gas fills the space between the electrodes. When a high-energy elementary particle passes through this space, it can ionize an atom of the gas. The strong electric field makes the resulting ion and electron accelerate in opposite directions. They strike other molecules of the gas to ionize them, producing an avalanche of electrical discharge. The pulse of electric current between the wire and the cylinder is counted by an external circuit. (a) Show that the magnitude of the potential difference between the wire and the cylinder is

$$\Delta V = 2k_e \lambda \, \ln \left(\frac{r_a}{r_b} \right)$$

(b) Show that the magnitude of the electric field in the space between cathode and anode is given by

$$E = \frac{\Delta V}{\ln (r_a/r_b)} \left(\frac{1}{r} \right)$$

where r is the distance from the axis of the anode to the point where the field is to be calculated.

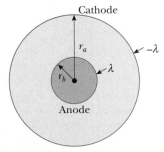

Figure P25.61 Problems 61 and 62.

62. The results of Problem 61 apply also to an electrostatic precipitator (Figures 25.30 and P25.61). An applied voltage $\Delta V = V_a - V_b = 50.0$ kV is to produce an electric field of magnitude 5.50 MV/m at the surface of the central wire. Assume the outer cylindrical wall has uniform radius $r_a = 0.850$ m. (a) What should be the radius r_b of the central wire? You will need to solve a transcendental equation. (b) What is the magnitude of the electric field at the outer wall?

63. From Gauss's law, the electric field set up by a uniform line of charge is

$$\mathbf{E} = \left(\frac{\lambda}{2\pi\epsilon_0 r}\right)\hat{\mathbf{r}}$$

where $\hat{\mathbf{r}}$ is a unit vector pointing radially away from the line and λ is the linear charge density along the line. Derive an expression for the potential difference between $r = r_1$ and $r = r_2$.

64. Four balls, each with mass m, are connected by four nonconducting strings to form a square with side a, as shown in Figure P25.64. The assembly is placed on a horizontal nonconducting frictionless surface. Balls 1 and 2 each have charge q, and balls 3 and 4 are uncharged. Find the maximum speed of balls 3 and 4 after the string connecting balls 1 and 2 is cut.

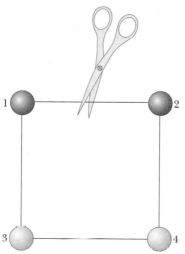

Figure P25.64

65. A point charge q is located at $x = -R$, and a point charge $-2q$ is located at the origin. Prove that the equipotential surface that has zero potential is a sphere centered at $(-4R/3, 0, 0)$ and having a radius $r = 2R/3$.

66. Consider two thin, conducting, spherical shells as shown in Figure P25.66. The inner shell has a radius $r_1 = 15.0$ cm and a charge of 10.0 nC. The outer shell has a radius $r_2 = 30.0$ cm and a charge of -15.0 nC. Find (a) the electric field $\mathbf{E}$ and (b) the electric potential V in regions A, B, and C, with $V = 0$ at $r = \infty$.

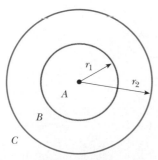

Figure P25.66

67. The x axis is the symmetry axis of a stationary uniformly charged ring of radius R and charge Q (Fig. P25.67).

A point charge Q of mass M is located initially at the center of the ring. When it is displaced slightly, the point charge accelerates along the x axis to infinity. Show that the ultimate speed of the point charge is

$$v = \left(\frac{2k_e Q^2}{MR}\right)^{1/2}$$

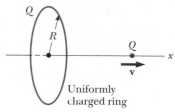

Figure P25.67

68. The thin, uniformly charged rod shown in Figure P25.68 has a linear charge density λ. Find an expression for the electric potential at P.

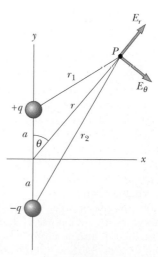

Figure P25.68

69. An electric dipole is located along the y axis as shown in Figure P25.69. The magnitude of its electric dipole moment is defined as $p = 2qa$. (a) At a point P, which is far from the dipole ($r \gg a$), show that the electric potential is

$$V = \frac{k_e p \cos\theta}{r^2}$$

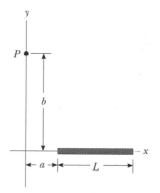

Figure P25.69

(b) Calculate the radial component E_r and the perpendicular component E_θ of the associated electric field. Note that $E_\theta = -(1/r)(\partial V/\partial \theta)$. Do these results seem reasonable for $\theta = 90°$ and $0°$? for $r = 0$? (c) For the dipole arrangement shown, express V in terms of Cartesian coordinates using $r = (x^2 + y^2)^{1/2}$ and

$$\cos \theta = \frac{y}{(x^2 + y^2)^{1/2}}$$

Using these results and again taking $r \gg a$, calculate the field components E_x and E_y.

70. When an uncharged conducting sphere of radius a is placed at the origin of an xyz coordinate system that lies in an initially uniform electric field $\mathbf{E} = E_0\hat{\mathbf{k}}$, the resulting electric potential is $V(x, y, z) = V_0$ for points inside the sphere and

$$V(x, y, z) = V_0 - E_0 z + \frac{E_0 a^3 z}{(x^2 + y^2 + z^2)^{3/2}}$$

for points outside the sphere, where V_0 is the (constant) electric potential on the conductor. Use this equation to determine the x, y, and z components of the resulting electric field.

71. A disk of radius R (Fig. P25.71) has a nonuniform surface charge density $\sigma = Cr$, where C is a constant and r is measured from the center of the disk. Find (by direct integration) the potential at P.

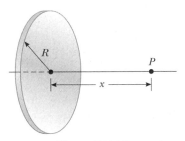

Figure P25.71

72. A solid sphere of radius R has a uniform charge density ρ and total charge Q. Derive an expression for its total electric potential energy. (*Suggestion:* imagine that the sphere is constructed by adding successive layers of concentric shells of charge $dq = (4\pi r^2\, dr)\rho$ and use $dU = V\, dq$.)

73. 🖳 Charge is uniformly distributed with a density of $100.0\ \mu C/m^3$ throughout the volume of a cube 10.00 cm on each edge. (a) Find the electric potential at a distance of 5.000 cm from the center of one face of the cube, measured along a perpendicular to the face. Determine the potential to four significant digits. Use a numerical method that divides the cube into a sufficient number of smaller cubes, treated as point charges. Symmetry considerations will reduce the number of actual calculations. (b) **What If?** If the charge on the cube is redistributed into a uniform sphere of charge with the same center, by how much does the potential change?

Answers to Quick Quizzes

25.1 (b). When moving straight from A to B, $\mathbf{E}$ and $d\mathbf{s}$ both point toward the right. Thus, the dot product $\mathbf{E} \cdot d\mathbf{s}$ in Equation 25.3 is positive and ΔV is negative.

25.2 (a). From Equation 25.3, $\Delta U = q_0\, \Delta V$, so if a negative test charge is moved through a negative potential difference, the potential energy is positive. Work must be done to move the charge in the direction opposite to the electric force on it.

25.3 $B \to C$, $C \to D$, $A \to B$, $D \to E$. Moving from B to C decreases the electric potential by 2 V, so the electric field performs 2 J of work on each coulomb of positive charge that moves. Moving from C to D decreases the electric potential by 1 V, so 1 J of work is done by the field. It takes no work to move the charge from A to B because the electric potential does not change. Moving from D to E increases the electric potential by 1 V, and thus the field does -1 J of work per unit of positive charge that moves.

25.4 (f). The electric field points in the direction of decreasing electric potential.

25.5 (b) and (f). The electric potential is inversely proportion to the radius (see Eq. 25.11). Because the same number of field lines passes through a closed surface of any shape or size, the electric flux through the surface remains constant.

25.6 (c). The potential is established only by the source charge and is independent of the test charge.

25.7 (a). The potential energy of the two-charge system is initially negative, due to the products of charges of opposite sign in Equation 25.13. When the sign of q_2 is changed, both charges are negative, and the potential energy of the system is positive.

25.8 (a). If the potential is constant (zero in this case), its derivative along this direction is zero.

25.9 (b). If the electric field is zero, there is no change in the electric potential and it must be constant. This constant value *could be* zero but does not *have to be* zero.

25.10 The graph would look like the sketch below. Notice the flat plateaus at each conductor, representing the constant electric potential inside a conductor.

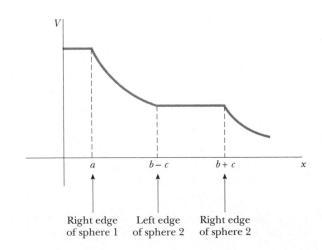

Chapter 26

Capacitance and Dielectrics

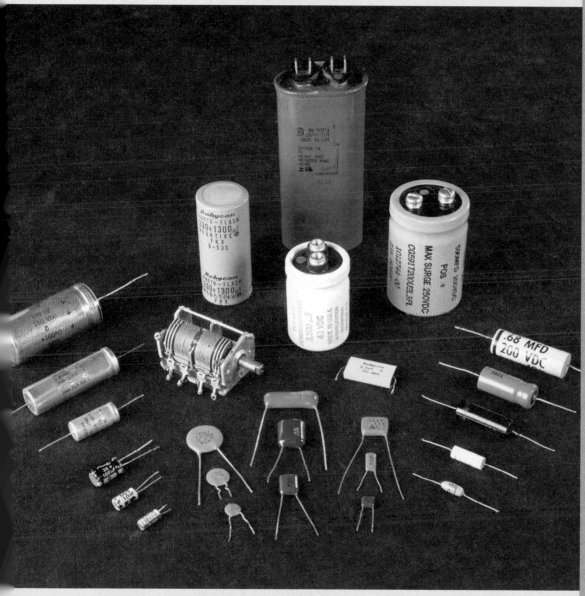

CHAPTER OUTLINE

26.1 Definition of Capacitance

26.2 Calculating Capacitance

26.3 Combinations of Capacitors

26.4 Energy Stored in a Charged Capacitor

26.5 Capacitors with Dielectrics

26.6 Electric Dipole in an Electric Field

26.7 An Atomic Description of Dielectrics

▲ *All of these devices are capacitors, which store electric charge and energy. A capacitor is one type of circuit element that we can combine with others to make electric circuits. (Paul Silverman/Fundamental Photographs)*

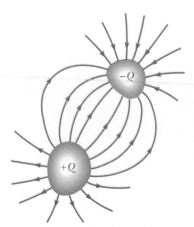

Figure 26.1 A capacitor consists of two conductors. When the capacitor is charged, the conductors carry charges of equal magnitude and opposite sign.

▲ **PITFALL PREVENTION**

26.1 Capacitance Is a Capacity

To understand capacitance, think of similar notions that use a similar word. The *capacity* of a milk carton is the volume of milk that it can store. The *heat capacity* of an object is the amount of energy an object can store per unit of temperature difference. The *capacitance* of a capacitor is the amount of charge the capacitor can store per unit of potential difference.

Definition of capacitance

In this chapter, we will introduce the first of three simple *circuit elements* that can be connected with wires to form an electric circuit. Electric circuits are the basis for the vast majority of the devices that we use in current society. We shall discuss *capacitors*—devices that store electric charge. This discussion will be followed by the study of *resistors* in Chapter 27 and *inductors* in Chapter 32. In later chapters, we will study more sophisticated circuit elements such as *diodes* and *transistors*.

Capacitors are commonly used in a variety of electric circuits. For instance, they are used to tune the frequency of radio receivers, as filters in power supplies, to eliminate sparking in automobile ignition systems, and as energy-storing devices in electronic flash units.

A capacitor consists of two conductors separated by an insulator. The capacitance of a given capacitor depends on its geometry and on the material—called a *dielectric*—that separates the conductors.

26.1 Definition of Capacitance

Consider two conductors carrying charges of equal magnitude and opposite sign, as shown in Figure 26.1. Such a combination of two conductors is called a **capacitor.** The conductors are called *plates*. A potential difference ΔV exists between the conductors due to the presence of the charges.

What determines how much charge is on the plates of a capacitor for a given voltage? Experiments show that the quantity of charge Q on a capacitor[1] is linearly proportional to the potential difference between the conductors; that is, $Q \propto \Delta V$. The proportionality constant depends on the shape and separation of the conductors.[2] We can write this relationship as $Q = C\,\Delta V$ if we define capacitance as follows:

The **capacitance** C of a capacitor is defined as the ratio of the magnitude of the charge on either conductor to the magnitude of the potential difference between the conductors:

$$C \equiv \frac{Q}{\Delta V} \qquad (26.1)$$

---

[1]   Although the total charge on the capacitor is zero (because there is as much excess positive charge on one conductor as there is excess negative charge on the other), it is common practice to refer to the magnitude of the charge on either conductor as "the charge on the capacitor."

[2]   The proportionality between ΔV and Q can be proved from Coulomb's law or by experiment.

Note that by definition *capacitance is always a positive quantity*. Furthermore, the charge Q and the potential difference ΔV are always expressed in Equation 26.1 as positive quantities. Because the potential difference increases linearly with the stored charge, the ratio $Q/\Delta V$ is constant for a given capacitor. Therefore, capacitance is a measure of a capacitor's ability to store charge. Because positive and negative charges are separated in the system of two conductors in a capacitor, there is electric potential energy stored in the system.

From Equation 26.1, we see that capacitance has SI units of coulombs per volt. The SI unit of capacitance is the **farad** (F), which was named in honor of Michael Faraday:

$$1 \text{ F} = 1 \text{ C/V}$$

The farad is a very large unit of capacitance. In practice, typical devices have capacitances ranging from microfarads (10^{-6} F) to picofarads (10^{-12} F). We shall use the symbol μF to represent microfarads. To avoid the use of Greek letters, in practice, physical capacitors often are labeled "mF" for microfarads and "mmF" for micromicrofarads or, equivalently, "pF" for picofarads.

Let us consider a capacitor formed from a pair of parallel plates, as shown in Figure 26.2. Each plate is connected to one terminal of a battery, which acts as a source of potential difference. If the capacitor is initially uncharged, the battery establishes an electric field in the connecting wires when the connections are made. Let us focus on the plate connected to the negative terminal of the battery. The electric field applies a force on electrons in the wire just outside this plate; this force causes the electrons to move onto the plate. This movement continues until the plate, the wire, and the terminal are all at the same electric potential. Once this equilibrium point is attained, a potential difference no longer exists between the terminal and the plate, and as a result no electric field is present in the wire, and the movement of electrons stops. The plate now carries a negative charge. A similar process occurs at the other capacitor plate, with electrons moving from the plate to the wire, leaving the plate positively charged. In this final configuration, the potential difference across the capacitor plates is the same as that between the terminals of the battery.

Suppose that we have a capacitor rated at 4 pF. This rating means that the capacitor can store 4 pC of charge for each volt of potential difference between the two conductors. If a 9-V battery is connected across this capacitor, one of the conductors ends up with a net charge of -36 pC and the other ends up with a net charge of $+36$ pC.

> **Quick Quiz 26.1** A capacitor stores charge Q at a potential difference ΔV. If the voltage applied by a battery to the capacitor is doubled to $2\Delta V$, (a) the capacitance falls to half its initial value and the charge remains the same (b) the capacitance and the charge both fall to half their initial values (c) the capacitance and the charge both double (d) the capacitance remains the same and the charge doubles.

26.2 Calculating Capacitance

We can derive an expression for the capacitance of a pair of oppositely charged conductors in the following manner: assume a charge of magnitude Q, and calculate the potential difference using the techniques described in the preceding chapter. We then use the expression $C = Q/\Delta V$ to evaluate the capacitance. As we might expect, we can perform this calculation relatively easily if the geometry of the capacitor is simple.

▲ **PITFALL PREVENTION**

26.2 Potential Difference is ΔV, not V

We use the symbol ΔV for the potential difference across a circuit element or a device because this is consistent with our definition of potential difference and with the meaning of the delta sign. It is a common, but confusing, practice to use the symbol V without the delta sign for a potential difference. Keep this in mind if you consult other texts.

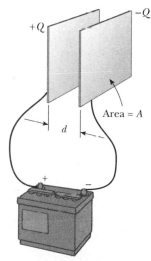

Figure 26.2 A parallel-plate capacitor consists of two parallel conducting plates, each of area A, separated by a distance d. When the capacitor is charged by connecting the plates to the terminals of a battery, the plates carry equal amounts of charge. One plate carries positive charge, and the other carries negative charge.

▲ **PITFALL PREVENTION**

26.3 Too Many C's

Do not confuse italic C for capacitance with non-italic C for the unit coulomb.

While the most common situation is that of two conductors, a single conductor also has a capacitance. For example, imagine a spherical charged conductor. The electric field lines around this conductor are exactly the same as if there were a conducting shell of infinite radius, concentric with the sphere and carrying a charge of the same magnitude but opposite sign. Thus, we can identify the imaginary shell as the second conductor of a two-conductor capacitor. We now calculate the capacitance for this situation. The electric potential of the sphere of radius R is simply $k_e Q/R$, and setting $V = 0$ for the infinitely large shell, we have

Capacitance of an isolated charged sphere

$$C = \frac{Q}{\Delta V} = \frac{Q}{k_e Q/R} = \frac{R}{k_e} = 4\pi\epsilon_0 R \qquad (26.2)$$

This expression shows that the capacitance of an isolated charged sphere is proportional to its radius and is independent of both the charge on the sphere and the potential difference.

The capacitance of a pair of conductors depends on the geometry of the conductors. Let us illustrate this with three familiar geometries, namely, parallel plates, concentric cylinders, and concentric spheres. In these examples, we assume that the charged conductors are separated by a vacuum. The effect of a dielectric material placed between the conductors is treated in Section 26.5.

Parallel-Plate Capacitors

Two parallel metallic plates of equal area A are separated by a distance d, as shown in Figure 26.2. One plate carries a charge Q, and the other carries a charge $-Q$. Let us consider how the geometry of these conductors influences the capacity of the combination to store charge. Recall that charges of the same sign repel one another. As a capacitor is being charged by a battery, electrons flow into the negative plate and out of the positive plate. If the capacitor plates are large, the accumulated charges are able to distribute themselves over a substantial area, and the amount of charge that can be stored on a plate for a given potential difference increases as the plate area is increased. Thus, we expect the capacitance to be proportional to the plate area A.

Now let us consider the region that separates the plates. If the battery has a constant potential difference between its terminals, then the electric field between the plates must increase as d is decreased. Let us imagine that we move the plates closer together and consider the situation before any charges have had a chance to move in response to this change. Because no charges have moved, the electric field between the plates has the same value but extends over a shorter distance. Thus, the magnitude of the potential difference between the plates $\Delta V = Ed$ (Eq. 25.6) is now smaller. The difference between this new capacitor voltage and the terminal voltage of the battery now exists as a potential difference across the wires connecting the battery to the capacitor. This potential difference results in the electric field in the wires that drives more charge onto the plates, increasing the potential difference between the plates. When the potential difference between the plates again matches that of the battery, the potential difference across the wires falls back to zero, and the flow of charge stops. Thus, moving the plates closer together causes the charge on the capacitor to increase. If d is increased, the charge decreases. As a result, we expect the capacitance of the pair of plates to be inversely proportional to d.

We can verify these physical arguments with the following derivation. The surface charge density on either plate is $\sigma = Q/A$. If the plates are very close together (in comparison with their length and width), we can assume that the electric field is uniform between the plates and is zero elsewhere. According to the **What If?** feature in Example 24.8, the value of the electric field between the plates is

$$E = \frac{\sigma}{\epsilon_0} = \frac{Q}{\epsilon_0 A}$$

Because the field between the plates is uniform, the magnitude of the potential difference between the plates equals Ed (see Eq. 25.6); therefore,

$$\Delta V = Ed = \frac{Qd}{\epsilon_0 A}$$

Substituting this result into Equation 26.1, we find that the capacitance is

$$C = \frac{Q}{\Delta V} = \frac{Q}{Qd/\epsilon_0 A}$$

$$C = \frac{\epsilon_0 A}{d} \qquad\qquad (26.3) \qquad \textbf{Capacitance of parallel plates}$$

That is, **the capacitance of a parallel-plate capacitor is proportional to the area of its plates and inversely proportional to the plate separation,** just as we expected from our conceptual argument.

A careful inspection of the electric field lines for a parallel-plate capacitor reveals that the field is uniform in the central region between the plates, as shown in Figure 26.3a. However, the field is nonuniform at the edges of the plates. Figure 26.3b is a photograph of the electric field pattern of a parallel-plate capacitor. Note the nonuniform nature of the electric field at the ends of the plates. Such end effects can be neglected if the plate separation is small compared with the length of the plates.

Figure 26.4 shows a battery connected to a single parallel-plate capacitor with a switch in the circuit. Let us identify the circuit as a system. When the switch is closed, the battery establishes an electric field in the wires and charges flow between the wires and the capacitor. As this occurs, there is a transformation of energy within the system. Before the switch is closed, energy is stored as chemical energy in the battery. This energy is transformed during the chemical reaction that occurs within the battery when it is operating in an electric circuit. When the switch is closed, some of the chemical energy in the battery is converted to electric potential energy related to the separation of positive and negative charges on the plates. As a result, we can describe a capacitor as a device that stores energy as well as charge. We will explore this energy storage in more detail in Section 26.4.

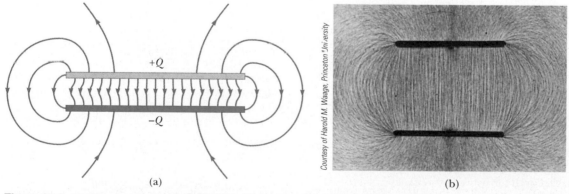

(a)   (b)

Courtesy of Harold M. Waage, Princeton University

Figure 26.3 (a) The electric field between the plates of a parallel-plate capacitor is uniform near the center but nonuniform near the edges. (b) Electric field pattern of two oppositely charged conducting parallel plates. Small pieces of thread on an oil surface align with the electric field.

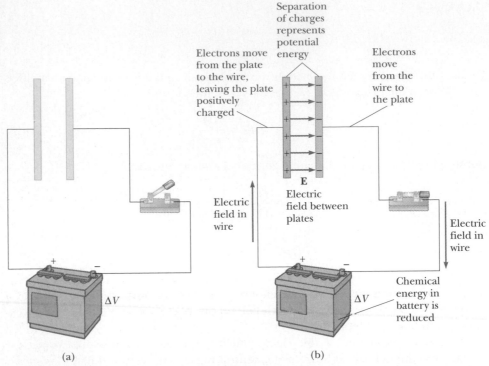

(a)   (b)

Active Figure 26.4 (a) A circuit consisting of a capacitor, a battery, and a switch. (b) When the switch is closed, the battery establishes an electric field in the wire that causes electrons to move from the left plate into the wire and into the right plate from the wire. As a result, a separation of charge exists on the plates, which represents an increase in electric potential energy of the system of the circuit. This energy in the system has been transformed from chemical energy in the battery.

Quick Quiz 26.2 Many computer keyboard buttons are constructed of capacitors, as shown in Figure 26.5. When a key is pushed down, the soft insulator between the movable plate and the fixed plate is compressed. When the key is pressed, the capacitance (a) increases, (b) decreases, or (c) changes in a way that we cannot determine because the complicated electric circuit connected to the keyboard button may cause a change in ΔV.

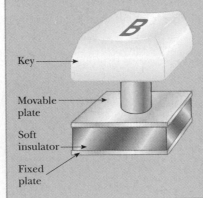

Key

Movable plate

Soft insulator

Fixed plate

Figure 26.5 (Quick Quiz 26.2) One type of computer keyboard button.

Example 26.1 Parallel-Plate Capacitor

A parallel-plate capacitor with air between the plates has an area $A = 2.00 \times 10^{-4}$ m² and a plate separation $d = 1.00$ mm. Find its capacitance.

Solution From Equation 26.3, we find that

$$C = \frac{\epsilon_0 A}{d} = \frac{(8.85 \times 10^{-12} \text{ C}^2/\text{N} \cdot \text{m}^2)(2.00 \times 10^{-4} \text{ m}^2)}{1.00 \times 10^{-3} \text{ m}}$$

$$= 1.77 \times 10^{-12} \text{ F} = \boxed{1.77 \text{ pF}}$$

Cylindrical and Spherical Capacitors

From the definition of capacitance, we can, in principle, find the capacitance of any geometric arrangement of conductors. The following examples demonstrate the use of this definition to calculate the capacitance of the other familiar geometries that we mentioned: cylinders and spheres.

Example 26.2   The Cylindrical Capacitor

A solid cylindrical conductor of radius a and charge Q is coaxial with a cylindrical shell of negligible thickness, radius $b > a$, and charge $-Q$ (Fig. 26.6a). Find the capacitance of this cylindrical capacitor if its length is ℓ.

Solution It is difficult to apply physical arguments to this configuration, although we can reasonably expect the capacitance to be proportional to the cylinder length ℓ for the same reason that parallel-plate capacitance is proportional to plate area: stored charges have more room in which to be distributed. If we assume that ℓ is much greater than a and b, we can neglect end effects. In this case, the electric field is perpendicular to the long axis of the cylinders and is confined to the region between them (Fig. 26.6b). We must first calculate the potential difference between the two cylinders, which is given in general by

$$V_b - V_a = -\int_a^b \mathbf{E} \cdot d\mathbf{s}$$

where $\mathbf{E}$ is the electric field in the region between the cylinders. In Chapter 24, we showed using Gauss's law that the magnitude of the electric field of a cylindrical charge distribution having linear charge density λ is $E = 2k_e\lambda/r$ (Eq. 24.7). The same result applies here because, according to Gauss's law, the charge on the outer cylinder does not contribute to the electric field inside it. Using this result and noting from Figure 26.6b that $\mathbf{E}$ is along r, we find that

$$V_b - V_a = -\int_a^b E_r\, dr = -2k_e\lambda \int_a^b \frac{dr}{r} = -2k_e\lambda \ln\left(\frac{b}{a}\right)$$

Substituting this result into Equation 26.1 and using the fact that $\lambda = Q/\ell$, we obtain

$$C = \frac{Q}{\Delta V} = \frac{Q}{(2k_eQ/\ell)\ln(b/a)} = \boxed{\frac{\ell}{2k_e \ln(b/a)}} \qquad (26.4)$$

where ΔV is the magnitude of the potential difference between the cylinders, given by $\Delta V = |V_a - V_b| = 2k_e\lambda\ln(b/a)$, a positive quantity. As predicted, the capacitance is proportional to the length of the cylinders. As we might expect, the capacitance also depends on the radii of the two cylindrical conductors. From Equation 26.4, we see that the capacitance per unit length of a combination of concentric cylindrical conductors is

$$\frac{C}{\ell} = \frac{1}{2k_e \ln(b/a)} \qquad (26.5)$$

An example of this type of geometric arrangement is a *coaxial cable*, which consists of two concentric cylindrical conductors separated by an insulator. You are likely to have a coaxial cable attached to your television set or VCR if you are a subscriber to cable television. The cable carries electrical signals in the inner and outer conductors. Such a geometry is especially useful for shielding the signals from any possible external influences.

What If? Suppose $b = 2.00a$ for the cylindrical capacitor. We would like to increase the capacitance, and we can do so by choosing to increase ℓ by 10% or by increasing a by 10%. Which choice is more effective at increasing the capacitance?

Answer According to Equation 26.4, C is proportional to ℓ, so increasing ℓ by 10% results in a 10% increase in C. For the result of the change in a, let us first evaluate C for $b = 2.00a$:

$$C = \frac{\ell}{2k_e\ln(b/a)} = \frac{\ell}{2k_e\ln(2.00)} = \frac{\ell}{2k_e(0.693)}$$

$$= 0.721\,\frac{\ell}{k_e}$$

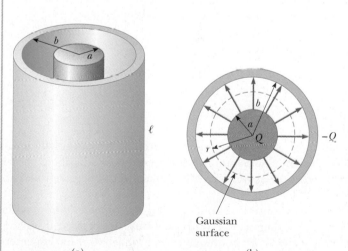

Figure 26.6 (Example 26.2) (a) A cylindrical capacitor consists of a solid cylindrical conductor of radius a and length ℓ surrounded by a coaxial cylindrical shell of radius b. (b) End view. The electric field lines are radial. The dashed line represents the end of the cylindrical gaussian surface of radius r and length ℓ.

Now, for a 10% increase in a, the new value is $a' = 1.10a$, so

$$C' = \frac{\ell}{2k_e \ln(b/a')} = \frac{\ell}{2k_e \ln(2.00a/1.10a)}$$

$$= \frac{\ell}{2k_e \ln(2.00/1.10)} = \frac{\ell}{2k_e(0.598)} = 0.836 \frac{\ell}{k_e}$$

The ratio of the new and old capacitances is

$$\frac{C'}{C} = \frac{0.836 \, \ell/k_e}{0.721 \, \ell/k_e} = 1.16$$

corresponding to a 16% increase in capacitance. Thus, it is more effective to increase a than to increase ℓ.

Note two more extensions of this problem. First, the advantage goes to increasing a only for a range of relationships between a and b. It is a valuable exercise to show that if $b > 2.85a$, increasing ℓ by 10% is more effective than increasing a (Problem 77). Second, if we increase b, we *reduce* the capacitance, so we would need to decrease b to increase the capacitance. Increasing a and decreasing b both have the effect of bringing the plates closer together, which increases the capacitance.

Example 26.3 The Spherical Capacitor

A spherical capacitor consists of a spherical conducting shell of radius b and charge $-Q$ concentric with a smaller conducting sphere of radius a and charge Q (Fig. 26.7). Find the capacitance of this device.

Solution As we showed in Chapter 24, the field outside a spherically symmetric charge distribution is radial and given by the expression $k_e Q/r^2$. In this case, this result applies to the field *between* the spheres ($a < r < b$). From Gauss's law we see that only the inner sphere contributes to this field. Thus, the potential difference between the spheres is

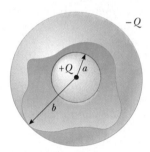

Figure 26.7 (Example 26.3) A spherical capacitor consists of an inner sphere of radius a surrounded by a concentric spherical shell of radius b. The electric field between the spheres is directed radially outward when the inner sphere is positively charged.

$$V_b - V_a = -\int_a^b E_r \, dr = -k_e Q \int_a^b \frac{dr}{r^2} = k_e Q \left[\frac{1}{r} \right]_a^b$$

$$= k_e Q \left(\frac{1}{b} - \frac{1}{a} \right)$$

The magnitude of the potential difference is

$$\Delta V = |V_b - V_a| = k_e Q \frac{(b-a)}{ab}$$

Substituting this value for ΔV into Equation 26.1, we obtain

$$C = \frac{Q}{\Delta V} = \frac{ab}{k_e(b-a)} \qquad (26.6)$$

What If? What if the radius b of the outer sphere approaches infinity? What does the capacitance become?

Answer In Equation 26.6, we let $b \to \infty$:

$$C = \lim_{b \to \infty} \frac{ab}{k_e(b-a)} = \frac{ab}{k_e(b)} = \frac{a}{k_e} = 4\pi\epsilon_0 a$$

Note that this is the same expression as Equation 26.2, the capacitance of an isolated spherical conductor.

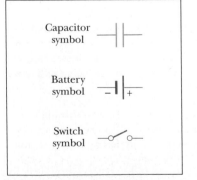

Figure 26.8 Circuit symbols for capacitors, batteries, and switches. Note that capacitors are in blue and batteries and switches are in red.

26.3 Combinations of Capacitors

Two or more capacitors often are combined in electric circuits. We can calculate the equivalent capacitance of certain combinations using methods described in this section. Throughout this section, we assume that the capacitors to be combined are initially uncharged.

In studying electric circuits, we use a simplified pictorial representation called a **circuit diagram.** Such a diagram uses **circuit symbols** to represent various circuit elements. The circuit symbols are connected by straight lines that represent the wires between the circuit elements. The circuit symbols for capacitors and batteries, as well as the color codes used for them in this text, are given in Figure 26.8. The symbol for the capacitor reflects the geometry of the most common model for a capacitor—a pair of parallel plates. The positive terminal of the battery is at the higher potential and is represented in the circuit symbol by the longer line.

Parallel Combination

Two capacitors connected as shown in Figure 26.9a are known as a *parallel combination* of capacitors. Figure 26.9b shows a circuit diagram for this combination of capacitors. The left plates of the capacitors are connected by a conducting wire to the positive terminal of the battery and are therefore both at the same electric potential as the positive terminal. Likewise, the right plates are connected to the negative terminal and are therefore both at the same potential as the negative terminal. Thus, **the individual potential differences across capacitors connected in parallel are the same and are equal to the potential difference applied across the combination.**

In a circuit such as that shown in Figure 26.9, the voltage applied across the combination is the terminal voltage of the battery. Situations can occur in which the parallel combination is in a circuit with other circuit elements; in such situations, we must determine the potential difference across the combination by analyzing the entire circuit.

When the capacitors are first connected in the circuit shown in Figure 26.9, electrons are transferred between the wires and the plates; this transfer leaves the left plates positively charged and the right plates negatively charged. The flow of charge ceases when the voltage across the capacitors is equal to that across the battery terminals. The capacitors reach their maximum charge when the flow of charge ceases. Let us call the maximum charges on the two capacitors Q_1 and Q_2. The *total charge Q* stored by the two capacitors is

$$Q = Q_1 + Q_2 \tag{26.7}$$

That is, **the total charge on capacitors connected in parallel is the sum of the charges on the individual capacitors.** Because the voltages across the capacitors are the same, the charges that they carry are

$$Q_1 = C_1 \, \Delta V \qquad Q_2 = C_2 \, \Delta V$$

Suppose that we wish to replace these two capacitors by one *equivalent capacitor* having a capacitance C_{eq}, as in Figure 26.9c. The effect this equivalent capacitor has on the circuit must be exactly the same as the effect of the combination of the two

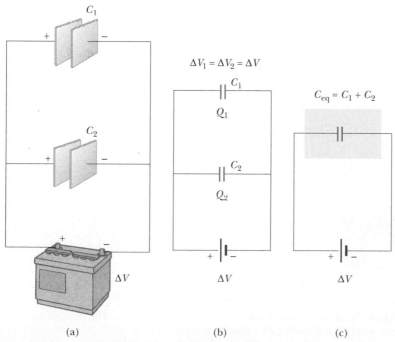

(a)   (b)   (c)

Active Figure 26.9 (a) A parallel combination of two capacitors in an electric circuit in which the potential difference across the battery terminals is ΔV. (b) The circuit diagram for the parallel combination. (c) The equivalent capacitance is $C_{eq} = C_1 + C_2$.

At the Active Figures link at http://www.pse6.com, you can adjust the battery voltage and the individual capacitances to see the resulting charges and voltages on the capacitors. You can combine up to four capacitors in parallel.

individual capacitors. That is, the equivalent capacitor must store Q units of charge when connected to the battery. We can see from Figure 26.9c that the voltage across the equivalent capacitor also is ΔV because the equivalent capacitor is connected directly across the battery terminals. Thus, for the equivalent capacitor,

$$Q = C_{eq}\, \Delta V$$

Substituting these three relationships for charge into Equation 26.7, we have

$$C_{eq}\, \Delta V = C_1\, \Delta V + C_2\, \Delta V$$

$$C_{eq} = C_1 + C_2 \qquad \text{(parallel combination)}$$

If we extend this treatment to three or more capacitors connected in parallel, we find the equivalent capacitance to be

Capacitors in parallel

$$C_{eq} = C_1 + C_2 + C_3 + \cdots \qquad \text{(parallel combination)} \qquad (26.8)$$

Thus, **the equivalent capacitance of a parallel combination of capacitors is the algebraic sum of the individual capacitances and is greater than any of the individual capacitances.** This makes sense because we are essentially combining the areas of all the capacitor plates when we connect them with conducting wire, and capacitance of parallel plates is proportional to area (Eq. 26.3).

Series Combination

Two capacitors connected as shown in Figure 26.10a and the equivalent circuit diagram in Figure 26.10b are known as a *series combination* of capacitors. The left plate of capacitor 1 and the right plate of capacitor 2 are connected to the terminals of a battery. The other two plates are connected to each other and to nothing else; hence, they form an isolated conductor that is initially uncharged and must continue to have zero net charge. To analyze this combination, let us begin by considering the uncharged capacitors and follow what happens just after a battery is connected to the circuit. When the battery is connected, electrons are transferred out of the left plate of C_1 and into the right plate of C_2. As this negative charge accumulates on the right plate of C_2, an equivalent amount of negative charge is forced off the left plate of C_2, and this left plate therefore has an excess positive charge. The negative charge leaving

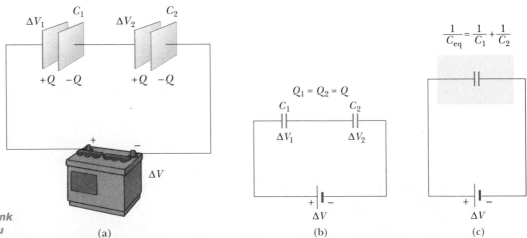

At the Active Figures link at http://www.pse6.com, you can adjust the battery voltage and the individual capacitances to see the resulting charges and voltages on the capacitors. You can combine up to four capacitors in series.

Active Figure 26.10 (a) A series combination of two capacitors. The charges on the two capacitors are the same. (b) The circuit diagram for the series combination. (c) The equivalent capacitance can be calculated from the relationship

$$\frac{1}{C_{eq}} = \frac{1}{C_1} + \frac{1}{C_2}.$$

the left plate of C_2 causes negative charges to accumulate on the right plate of C_1. As a result, all the right plates end up with a charge $-Q$, and all the left plates end up with a charge $+Q$. Thus, **the charges on capacitors connected in series are the same.**

From Figure 26.10a, we see that the voltage ΔV across the battery terminals is split between the two capacitors:

$$\Delta V = \Delta V_1 + \Delta V_2 \tag{26.9}$$

where ΔV_1 and ΔV_2 are the potential differences across capacitors C_1 and C_2, respectively. In general, **the total potential difference across any number of capacitors connected in series is the sum of the potential differences across the individual capacitors.**

Suppose that the equivalent single capacitor in Figure 26.10c has the same effect on the circuit as the series combination when it is connected to the battery. After it is fully charged, the equivalent capacitor must have a charge of $-Q$ on its right plate and a charge of $+Q$ on its left plate. Applying the definition of capacitance to the circuit in Figure 26.10c, we have

$$\Delta V = \frac{Q}{C_{eq}}$$

Because we can apply the expression $Q = C\,\Delta V$ to each capacitor shown in Figure 26.10b, the potential differences across them are

$$\Delta V_1 = \frac{Q}{C_1} \qquad \Delta V_2 = \frac{Q}{C_2}$$

Substituting these expressions into Equation 26.9, we have

$$\frac{Q}{C_{eq}} = \frac{Q}{C_1} + \frac{Q}{C_2}$$

Canceling Q, we arrive at the relationship

$$\frac{1}{C_{eq}} = \frac{1}{C_1} + \frac{1}{C_2} \qquad \text{(series combination)}$$

When this analysis is applied to three or more capacitors connected in series, the relationship for the equivalent capacitance is

$$\frac{1}{C_{eq}} = \frac{1}{C_1} + \frac{1}{C_2} + \frac{1}{C_3} + \cdots \qquad \text{(series combination)} \tag{26.10}$$

Capacitors in series

This shows that **the inverse of the equivalent capacitance is the algebraic sum of the inverses of the individual capacitances and the equivalent capacitance of a series combination is always less than any individual capacitance in the combination.**

Quick Quiz 26.3 Two capacitors are identical. They can be connected in series or in parallel. If you want the *smallest* equivalent capacitance for the combination, do you connect them in (a) series, in (b) parallel, or (c) do the combinations have the same capacitance?

Quick Quiz 26.4 Consider the two capacitors in Quick Quiz 26.3 again. Each capacitor is charged to a voltage of 10 V. If you want the largest combined potential difference across the combination, do you connect them in (a) series, in (b) parallel, or (c) do the combinations have the same potential difference?

PROBLEM-SOLVING HINTS

Capacitors

- Be careful with units. When you calculate capacitance in farads, make sure that distances are expressed in meters. When checking consistency of units, remember that the unit for electric fields can be either N/C or V/m.

- When two or more capacitors are connected in parallel, the potential difference across each is the same. The charge on each capacitor is proportional to its capacitance; hence, the capacitances can be added directly to give the equivalent capacitance of the parallel combination. The equivalent capacitance is always larger than the individual capacitances.

- When two or more capacitors are connected in series, they carry the same charge, and the sum of the potential differences equals the total potential difference applied to the combination. The sum of the reciprocals of the capacitances equals the reciprocal of the equivalent capacitance, which is always less than the capacitance of the smallest individual capacitor.

Example 26.4   Equivalent Capacitance

Interactive

Find the equivalent capacitance between a and b for the combination of capacitors shown in Figure 26.11a. All capacitances are in microfarads.

Solution Using Equations 26.8 and 26.10, we reduce the combination step by step as indicated in the figure. The 1.0-μF and 3.0-μF capacitors are in parallel and combine according to the expression $C_{eq} = C_1 + C_2 = 4.0~\mu$F. The 2.0-$\mu$F and 6.0-$\mu$F capacitors also are in parallel and have an equivalent capacitance of 8.0 μF. Thus, the upper branch in Figure 26.11b consists of two 4.0-μF capacitors in series, which combine as follows:

$$\frac{1}{C_{eq}} = \frac{1}{C_1} + \frac{1}{C_2} = \frac{1}{4.0~\mu F} + \frac{1}{4.0~\mu F} = \frac{1}{2.0~\mu F}$$

$$C_{eq} = 2.0~\mu F$$

The lower branch in Figure 26.11b consists of two 8.0-μF capacitors in series, which combine to yield an equivalent capacitance of 4.0 μF. Finally, the 2.0-μF and 4.0-μF capacitors in Figure 26.11c are in parallel and thus have an equivalent capacitance of 6.0 μF.

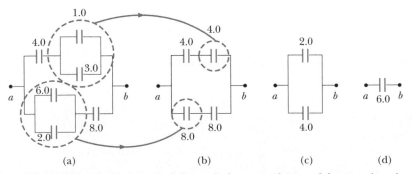

(a)   (b)   (c)   (d)

Figure 26.11 (Example 26.4) To find the equivalent capacitance of the capacitors in part (a), we reduce the various combinations in steps as indicated in parts (b), (c), and (d), using the series and parallel rules described in the text.

 Practice reducing a combination of capacitors to a single equivalent capacitance at the Interactive Worked Example link at **http://www.pse6.com.**

26.4 Energy Stored in a Charged Capacitor

Almost everyone who works with electronic equipment has at some time verified that a capacitor can store energy. If the plates of a charged capacitor are connected by a conductor, such as a wire, charge moves between each plate and its connecting wire until the capacitor is uncharged. The discharge can often be observed as a visible spark. If you should accidentally touch the opposite plates of a charged capacitor, your fingers act as a pathway for discharge, and the result is an electric shock. The degree of shock you receive depends on the capacitance and on the voltage applied to the capacitor. Such a shock could be fatal if high voltages are present, such as in the power supply of a television set. Because the charges can be stored in a capacitor even when the set is turned off, unplugging the television does not make it safe to open the case and touch the components inside.

To calculate the energy stored in the capacitor, we shall assume a charging process that is different from the actual process described in Section 26.1 but which gives the same final result. We can make this assumption because the energy in the final configuration does not depend on the actual charge-transfer process. We imagine that the charge is transferred mechanically through the space between the plates. We reach in and grab a small amount of positive charge on the plate connected to the negative terminal and apply a force that causes this positive charge to move over to the plate connected to the positive terminal. Thus, we do work on the charge as we transfer it from one plate to the other. At first, no work is required to transfer a small amount of charge dq from one plate to the other.[3] However, once this charge has been transferred, a small potential difference exists between the plates. Therefore, work must be done to move additional charge through this potential difference. As more and more charge is transferred from one plate to the other, the potential difference increases in proportion, and more work is required.

Suppose that q is the charge on the capacitor at some instant during the charging process. At the same instant, the potential difference across the capacitor is $\Delta V = q/C$. From Section 25.2, we know that the work necessary to transfer an increment of charge dq from the plate carrying charge $-q$ to the plate carrying charge q (which is at the higher electric potential) is

$$dW = \Delta V \, dq = \frac{q}{C} \, dq$$

This is illustrated in Figure 26.12. The total work required to charge the capacitor from $q = 0$ to some final charge $q = Q$ is

$$W = \int_0^Q \frac{q}{C} \, dq = \frac{1}{C} \int_0^Q q \, dq = \frac{Q^2}{2C}$$

The work done in charging the capacitor appears as electric potential energy U stored in the capacitor. Using Equation 26.1, we can express the potential energy stored in a charged capacitor in the following forms:

$$U = \frac{Q^2}{2C} = \tfrac{1}{2} Q \, \Delta V = \tfrac{1}{2} C (\Delta V)^2 \qquad (26.11)$$

This result applies to any capacitor, regardless of its geometry. We see that for a given capacitance, the stored energy increases as the charge increases and as the potential difference increases. In practice, there is a limit to the maximum energy (or charge) that can be stored because, at a sufficiently great value of ΔV, discharge ultimately occurs between the plates. For this reason, capacitors are usually labeled with a maximum operating voltage.

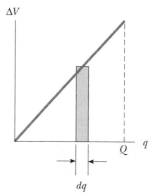

Figure 26.12 A plot of potential difference versus charge for a capacitor is a straight line having a slope $1/C$. The work required to move charge dq through the potential difference ΔV existing at the time across the capacitor plates is given approximately by the area of the shaded rectangle. The total work required to charge the capacitor to a final charge Q is the triangular area under the straight line, $W = \tfrac{1}{2} Q \Delta V$. (Don't forget that $1 \text{ V} = \text{J}/\text{C}$; hence, the unit for the triangular area is the joule.)

Energy stored in a charged capacitor

---

[3] We shall use lowercase q for the time-varying charge on the capacitor while it is charging, to distinguish it from uppercase Q, which is the total charge on the capacitor after it is completely charged.

We can consider the energy stored in a capacitor as being stored in the electric field created between the plates as the capacitor is charged. This description is reasonable because the electric field is proportional to the charge on the capacitor. For a parallel-plate capacitor, the potential difference is related to the electric field through the relationship $\Delta V = Ed$. Furthermore, its capacitance is $C = \epsilon_0 A/d$ (Eq. 26.3). Substituting these expressions into Equation 26.11, we obtain

$$U = \tfrac{1}{2}\frac{\epsilon_0 A}{d}\,(E^2 d^2) = \tfrac{1}{2}(\epsilon_0 Ad)E^2 \tag{26.12}$$

Because the volume occupied by the electric field is Ad, the *energy per unit volume* $u_E = U/Ad$, known as the *energy density*, is

$$u_E = \tfrac{1}{2}\epsilon_0 E^2 \tag{26.13}$$

Energy density in an electric field

Although Equation 26.13 was derived for a parallel-plate capacitor, the expression is generally valid, regardless of the source of the electric field. That is, **the energy density in any electric field is proportional to the square of the magnitude of the electric field at a given point.**

 PITFALL PREVENTION

26.4   Not a New Kind of Energy

The energy given by Equation 26.13 is not a new kind of energy. It is familiar electric potential energy associated with a system of separated source charges. Equation 26.13 provides a new *interpretation*, or a new way of *modeling* the energy, as energy associated with the electric field, regardless of the source of the field.

Quick Quiz 26.5   You have three capacitors and a battery. In which of the following combinations of the three capacitors will the maximum possible energy be stored when the combination is attached to the battery? (a) series (b) parallel (c) Both combinations will store the same amount of energy.

Quick Quiz 26.6   You charge a parallel-plate capacitor, remove it from the battery, and prevent the wires connected to the plates from touching each other. When you pull the plates apart to a larger separation, do the following quantities increase, decrease, or stay the same? (a) C; (b) Q; (c) E between the plates; (d) ΔV; (e) energy stored in the capacitor.

Quick Quiz 26.7   Repeat Quick Quiz 26.6, but this time answer the questions for the situation in which the battery remains connected to the capacitor while you pull the plates apart.

Example 26.5   Rewiring Two Charged Capacitors   `Interactive`

Two capacitors C_1 and C_2 (where $C_1 > C_2$) are charged to the same initial potential difference ΔV_i. The charged capacitors are removed from the battery, and their plates are connected with opposite polarity as in Figure 26.13a. The switches S_1 and S_2 are then closed, as in Figure 26.13b.

(A) Find the final potential difference ΔV_f between a and b after the switches are closed.

Solution Figure 26.13 helps us conceptualize the initial and final configurations of the system. In Figure 26.13b, it might appear as if the capacitors are connected in parallel, but there is no battery in this circuit that is applying a voltage across the combination. Thus, we *cannot* categorize this as a problem in which capacitors are connected in parallel. We *can* categorize this as a problem involving an isolated system for electric charge—the left-hand plates of the capac-

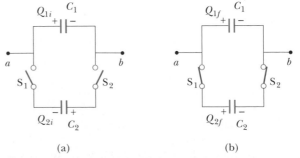

Figure 26.13 (Example 26.5) (a) Two capacitors are charged to the same initial potential difference and connected together with plates of opposite sign to be in contact when the switches are closed. (b) When the switches are closed, the charges redistribute.

itors form an isolated system because they are not connected to the right-hand plates by conductors. To analyze the problem, note that the charges on the left-hand plates before the switches are closed are

$$Q_{1i} = C_1 \Delta V_i \quad \text{and} \quad Q_{2i} = -C_2 \Delta V_i$$

The negative sign for Q_{2i} is necessary because the charge on the left plate of capacitor C_2 is negative. The total charge Q in the system is

$$(1) \qquad Q = Q_{1i} + Q_{2i} = (C_1 - C_2)\Delta V_i$$

After the switches are closed, the total charge Q in the system remains the same but the charges on the individual capacitors change to new values Q_{1f} and Q_{2f}. Because the system is isolated,

$$(2) \qquad Q = Q_{1f} + Q_{2f}$$

The charges redistribute until the potential difference is the same across both capacitors, ΔV_f. To satisfy this requirement, the charges on the capacitors after the switches are closed are

$$Q_{1f} = C_1 \Delta V_f \quad \text{and} \quad Q_{2f} = C_2 \Delta V_f$$

Dividing the first equation by the second, we have

$$(3) \qquad Q_{1f} = \frac{C_1}{C_2} Q_{2f}$$

Combining Equations (2) and (3), we obtain

$$Q = Q_{1f} + Q_{2f} = \frac{C_1}{C_2} Q_{2f} + Q_{2f} = Q_{2f}\left(1 + \frac{C_1}{C_2}\right)$$

$$(4) \qquad Q_{2f} = Q\left(\frac{C_2}{C_1 + C_2}\right)$$

Using Equations (3) and (4) to find Q_{1f} in terms of Q, we have

$$(5) \qquad Q_{1f} = \frac{C_1}{C_2} Q_{2f} = \frac{C_1}{C_2} Q\left(\frac{C_2}{C_1 + C_2}\right)$$

$$= Q\left(\frac{C_1}{C_1 + C_2}\right)$$

Finally, using Equation 26.1 to find the voltage across each capacitor, we find that

$$(6) \qquad \Delta V_{1f} = \frac{Q_{1f}}{C_1} = \frac{Q[C_1/(C_1 + C_2)]}{C_1} = \frac{Q}{C_1 + C_2}$$

$$(7) \qquad \Delta V_{2f} = \frac{Q_{2f}}{C_2} = \frac{Q[C_2/(C_1 + C_2)]}{C_2} = \frac{Q}{C_1 + C_2}$$

As noted earlier, $\Delta V_{1f} = \Delta V_{2f} = \Delta V_f$.

To express ΔV_f in terms of the given quantities C_1, C_2, and ΔV_i, we substitute the value of Q from Equation (1) into either Equation (6) or (7) to obtain

$$\Delta V_f = \left(\frac{C_1 - C_2}{C_1 + C_2}\right)\Delta V_i$$

(B) Find the total energy stored in the capacitors before and after the switches are closed and the ratio of the final energy to the initial energy.

Solution Before the switches are closed, the total energy stored in the capacitors is

$$U_i = \tfrac{1}{2}C_1(\Delta V_i)^2 + \tfrac{1}{2}C_2(\Delta V_i)^2 = \boxed{\tfrac{1}{2}(C_1 + C_2)(\Delta V_i)^2}$$

After the switches are closed, the total energy stored in the capacitors is

$$U_f = \tfrac{1}{2}C_1(\Delta V_f)^2 + \tfrac{1}{2}C_2(\Delta V_f)^2 = \tfrac{1}{2}(C_1 + C_2)(\Delta V_f)^2$$

Using the results of part (A), we can express this as

$$U_f = \boxed{\tfrac{1}{2}\frac{(C_1 - C_2)^2(\Delta V_i)^2}{(C_1 + C_2)}}$$

Therefore, the ratio of the final energy stored to the initial energy stored is

$$(8) \qquad \frac{U_f}{U_i} = \frac{\tfrac{1}{2}(C_1 - C_2)^2(\Delta V_i)^2/(C_1 + C_2)}{\tfrac{1}{2}(C_1 + C_2)(\Delta V_i)^2}$$

$$= \boxed{\left(\frac{C_1 - C_2}{C_1 + C_2}\right)^2}$$

To finalize this problem, note that this ratio is *less* than unity, indicating that the final energy is *less* than the initial energy. At first, you might think that the law of energy conservation has been violated, but this is not the case. The "missing" energy is transferred out of the system of the capacitors by the mechanism of electromagnetic waves, as we shall see in Chapter 34.

What If? What if the two capacitors have the same capacitance? What would we expect to happen when the switches are closed?

Answer The equal-magnitude charges on the two capacitors should simply cancel each other and the capacitors will be uncharged afterward.

Let us test our results to see if this is the case mathematically. In Equation (1), because the charges are of equal magnitude and opposite sign, we see that $Q = 0$. Thus, Equations (4) and (5) show us that $Q_{1f} = Q_{2f} = 0$, consistent with our prediction. Furthermore, Equations (6) and (7) show us that $\Delta V_{1f} = \Delta V_{2f} = 0$, which is consistent with uncharged capacitors. Finally, if $C_1 = C_2$, Equation (8) shows us that $U_f = 0$, which is also consistent with uncharged capacitors.

 At the Interactive Worked Example link at **http://www.pse6.com,** *explore this situation for various initial values of the voltage and the capacitances.*

Adam Hart-Davis/SPL/Custom Medical Stock

Figure 26.14 In a hospital or at an emergency scene, you might see a patient being revived with a defibrillator. The defibrillator's paddles are applied to the patient's chest, and an electric shock is sent through the chest cavity. The aim of this technique is to restore the heart's normal rhythm pattern.

One device in which capacitors have an important role is the *defibrillator* (Fig. 26.14). Up to 360 J is stored in the electric field of a large capacitor in a defibrillator when it is fully charged. The defibrillator can deliver all this energy to a patient in about 2 ms. (This is roughly equivalent to 3 000 times the power delivered to a 60-W lightbulb!) Under the proper conditions, the defibrillator can be used to stop cardiac fibrillation (random contractions) in heart attack victims. When fibrillation occurs, the heart produces a rapid, irregular pattern of beats. A fast discharge of energy through the heart can return the organ to its normal beat pattern. Emergency medical teams use portable defibrillators that contain batteries capable of charging a capacitor to a high voltage. (The circuitry actually permits the capacitor to be charged to a much higher voltage than that of the battery.) The stored energy is released through the heart by conducting electrodes, called paddles, that are placed on both sides of the victim's chest. The paramedics must wait between applications of the energy due to the time necessary for the capacitors to become fully charged. In this case and others (e.g., camera flash units and lasers used for fusion experiments), capacitors serve as energy reservoirs which can be slowly charged and then discharged quickly to provide large amounts of energy in a short pulse.

A camera's flash unit also uses a capacitor, although the total amount of energy stored is much less than that stored in a defibrillator. After the flash unit's capacitor is charged, tripping the camera's shutter causes the stored energy to be sent through a special lightbulb that briefly illuminates the subject being photographed.

26.5 Capacitors with Dielectrics

A **dielectric** is a nonconducting material, such as rubber, glass, or waxed paper. When a dielectric is inserted between the plates of a capacitor, the capacitance increases. If the dielectric completely fills the space between the plates, the capacitance increases by a dimensionless factor κ, which is called the **dielectric constant** of the material. The dielectric constant varies from one material to another. In this section, we analyze this change in capacitance in terms of electrical parameters such as electric charge, electric field, and potential difference; in Section 26.7, we shall discuss the microscopic origin of these changes.

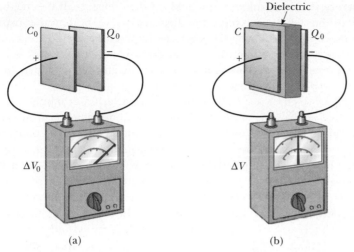

Figure 26.15 A charged capacitor (a) before and (b) after insertion of a dielectric between the plates. The charge on the plates remains unchanged, but the potential difference decreases from ΔV_0 to $\Delta V = \Delta V_0/\kappa$. Thus, the capacitance increases from C_0 to κC_0.

We can perform the following experiment to illustrate the effect of a dielectric in a capacitor. Consider a parallel-plate capacitor that without a dielectric has a charge Q_0 and a capacitance C_0. The potential difference across the capacitor is $\Delta V_0 = Q_0/C_0$. Figure 26.15a illustrates this situation. The potential difference is measured by a *voltmeter*, which we shall study in greater detail in Chapter 28. Note that no battery is shown in the figure; also, we must assume that no charge can flow through an ideal voltmeter. Hence, there is no path by which charge can flow and alter the charge on the capacitor. If a dielectric is now inserted between the plates, as in Figure 26.15b, the voltmeter indicates that the voltage between the plates decreases to a value ΔV. The voltages with and without the dielectric are related by the factor κ as follows:

$$\Delta V = \frac{\Delta V_0}{\kappa}$$

Because $\Delta V < \Delta V_0$, we see that $\kappa > 1$.

Because the charge Q_0 on the capacitor does not change, we conclude that the capacitance must change to the value

$$C = \frac{Q_0}{\Delta V} = \frac{Q_0}{\Delta V_0/\kappa} = \kappa \frac{Q_0}{\Delta V_0}$$

$$\boxed{C = \kappa C_0} \tag{26.14}$$

That is, the capacitance *increases* by the factor κ when the dielectric completely fills the region between the plates.[4] For a parallel-plate capacitor, where $C_0 = \epsilon_0 A/d$ (Eq. 26.3), we can express the capacitance when the capacitor is filled with a dielectric as

$$C = \kappa \frac{\epsilon_0 A}{d} \tag{26.15}$$

From Equations 26.3 and 26.15, it would appear that we could make the capacitance very large by decreasing d, the distance between the plates. In practice, the lowest value of d is limited by the electric discharge that could occur through the dielectric medium separating the plates. For any given separation d, the maximum voltage that can be applied to a capacitor without causing a discharge depends on the

▲ **PITFALL PREVENTION**

26.5   Is the Capacitor Connected to a Battery?

In problems in which you are modifying a capacitor (by insertion of a dielectric, for example), you must note whether modifications to the capacitor are being made while the capacitor is connected to a battery or after it is disconnected. If the capacitor remains connected to the battery, the voltage across the capacitor necessarily remains the same. If you disconnect the capacitor from the battery before making any modifications to the capacitor, the capacitor is an isolated system and its charge remains the same.

◀ **Capacitance of a capacitor filled with a material of dielectric constant κ**

---

[4]   If the dielectric is introduced while the potential difference is held constant by a battery, the charge increases to a value $Q = \kappa Q_0$. The additional charge comes from the wires attached to the capacitor, and the capacitance again increases by the factor κ.

Figure 26.16 Dielectric breakdown in air. Sparks are produced when the high voltage between the wires causes the electric field to exceed the dielectric strength of air.

dielectric strength (maximum electric field) of the dielectric. If the magnitude of the electric field in the dielectric exceeds the dielectric strength, then the insulating properties break down and the dielectric begins to conduct. Figure 26.16 shows the effect of exceeding the dielectric strength of air. Sparks appear between the two wires, due to ionization of atoms and recombination with electrons in the air, similar to the process that produced corona discharge in Section 25.6.

Physical capacitors have a specification called by a variety of names, including *working voltage*, *breakdown voltage*, and *rated voltage*. This parameter represents the largest voltage that can be applied to the capacitor without exceeding the dielectric strength of the dielectric material in the capacitor. Consequently, when selecting a capacitor for a given application, you must consider the capacitance of the device along with the expected voltage across the capacitor in the circuit, making sure that the expected voltage will be smaller than the rated voltage of the capacitor. You can see the rated voltage on several of the capacitors in the opening photograph for this chapter.

Insulating materials have values of κ greater than unity and dielectric strengths greater than that of air, as Table 26.1 indicates. Thus, we see that a dielectric provides the following advantages:

- Increase in capacitance

- Increase in maximum operating voltage

- Possible mechanical support between the plates, which allows the plates to be close together without touching, thereby decreasing d and increasing C.

Types of Capacitors

Commercial capacitors are often made from metallic foil interlaced with thin sheets of either paraffin-impregnated paper or Mylar as the dielectric material. These alternate layers of metallic foil and dielectric are rolled into a cylinder to form a small package (Fig. 26.17a). High-voltage capacitors commonly consist of a number of interwoven

Table 26.1

Approximate Dielectric Constants and Dielectric Strengths of Various Materials at Room Temperature		
Material	**Dielectric Constant κ**	**Dielectric Strength[a] (10^6 V/m)**
Air (dry)	1.000 59	3
Bakelite	4.9	24
Fused quartz	3.78	8
Mylar	3.2	7
Neoprene rubber	6.7	12
Nylon	3.4	14
Paper	3.7	16
Paraffin-impregnated paper	3.5	11
Polystyrene	2.56	24
Polyvinyl chloride	3.4	40
Porcelain	6	12
Pyrex glass	5.6	14
Silicone oil	2.5	15
Strontium titanate	233	8
Teflon	2.1	60
Vacuum	1.000 00	—
Water	80	—

[a] The dielectric strength equals the maximum electric field that can exist in a dielectric without electrical breakdown. Note that these values depend strongly on the presence of impurities and flaws in the materials.

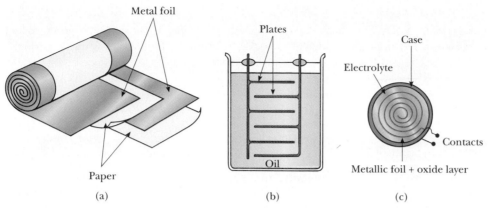

Figure 26.17 Three commercial capacitor designs. (a) A tubular capacitor, whose plates are separated by paper and then rolled into a cylinder. (b) A high-voltage capacitor consisting of many parallel plates separated by insulating oil. (c) An electrolytic capacitor.

metallic plates immersed in silicone oil (Fig. 26.17b). Small capacitors are often constructed from ceramic materials.

Often, an *electrolytic capacitor* is used to store large amounts of charge at relatively low voltages. This device, shown in Figure 26.17c, consists of a metallic foil in contact with an *electrolyte*—a solution that conducts electricity by virtue of the motion of ions contained in the solution. When a voltage is applied between the foil and the electrolyte, a thin layer of metal oxide (an insulator) is formed on the foil, and this layer serves as the dielectric. Very large values of capacitance can be obtained in an electrolytic capacitor because the dielectric layer is very thin, and thus the plate separation is very small.

Electrolytic capacitors are not reversible as are many other capacitors—they have a polarity, which is indicated by positive and negative signs marked on the device. When electrolytic capacitors are used in circuits, the polarity must be aligned properly. If the polarity of the applied voltage is opposite that which is intended, the oxide layer is removed and the capacitor conducts electricity instead of storing charge.

Variable capacitors (typically 10 to 500 pF) usually consist of two interwoven sets of metallic plates, one fixed and the other movable, and contain air as the dielectric (Fig. 26.18). These types of capacitors are often used in radio tuning circuits.

Figure 26.18 A variable capacitor. When one set of metal plates is rotated so as to lie between a fixed set of plates, the capacitance of the device changes.

Quick Quiz 26.8 If you have ever tried to hang a picture or a mirror, you know it can be difficult to locate a wooden stud in which to anchor your nail or screw. A carpenter's stud-finder is basically a capacitor with its plates arranged side by side instead of facing one another, as shown in Figure 26.19. When the device is moved over a stud, does the capacitance increase or decrease?

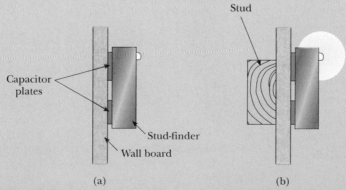

Figure 26.19 (Quick Quiz 26.8) A stud-finder. (a) The materials between the plates of the capacitor are the wallboard and air. (b) When the capacitor moves across a stud in the wall, the materials between the plates are the wallboard and the wood. The change in the dielectric constant causes a signal light to illuminate.

Quick Quiz 26.9 A fully charged parallel-plate capacitor remains connected to a battery while you slide a dielectric between the plates. Do the following quantities increase, decrease, or stay the same? (a) C; (b) Q; (c) E between the plates; (d) ΔV.

Example 26.6 A Paper-Filled Capacitor

A parallel-plate capacitor has plates of dimensions 2.0 cm by 3.0 cm separated by a 1.0-mm thickness of paper.

(A) Find its capacitance.

Solution Because $\kappa = 3.7$ for paper (see Table 26.1), we have

$$C = \kappa \frac{\epsilon_0 A}{d}$$

$$= 3.7 \left(\frac{(8.85 \times 10^{-12}\,\text{C}^2/\text{N·m}^2)(6.0 \times 10^{-4}\,\text{m}^2)}{1.0 \times 10^{-3}\,\text{m}} \right)$$

$$= 20 \times 10^{-12}\,\text{F} = \boxed{20\ \text{pF}}$$

(B) What is the maximum charge that can be placed on the capacitor?

Solution From Table 26.1 we see that the dielectric strength of paper is 16×10^6 V/m. Because the thickness of the paper is 1.0 mm, the maximum voltage that can be applied before breakdown is

$$\Delta V_{max} = E_{max} d = (16 \times 10^6\,\text{V/m})(1.0 \times 10^{-3}\,\text{m})$$

$$= 16 \times 10^3\,\text{V}$$

Hence, the maximum charge is

$$Q_{max} = C\,\Delta V_{max} = (20 \times 10^{-12}\,\text{F})(16 \times 10^3\,\text{V})$$

$$= \boxed{0.32\ \mu\text{C}}$$

Example 26.7 Energy Stored Before and After

A parallel-plate capacitor is charged with a battery to a charge Q_0, as shown in Figure 26.20a. The battery is then removed, and a slab of material that has a dielectric constant κ is inserted between the plates, as shown in Figure 26.20b. Find the energy stored in the capacitor before and after the dielectric is inserted.

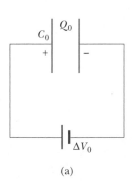

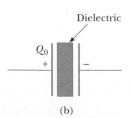

Figure 26.20 (Example 26.7) (a) A battery charges up a parallel-plate capacitor. (b) The battery is removed and a slab of dielectric material is inserted between the plates.

Solution From Equation 26.11, we see that the energy stored in the absence of the dielectric is

$$U_0 = \frac{Q_0^2}{2C_0}$$

After the battery is removed and the dielectric inserted, the *charge on the capacitor remains the same.* Hence, the energy stored in the presence of the dielectric is

$$U = \frac{Q_0^2}{2C}$$

But the capacitance in the presence of the dielectric is $C = \kappa C_0$, so U becomes

$$U = \frac{Q_0^2}{2\kappa C_0} = \boxed{\frac{U_0}{\kappa}}$$

Because $\kappa > 1$, the final energy is less than the initial energy. We can account for the "missing" energy by noting that the dielectric, when inserted, is pulled into the device· (see Section 26.7). An external agent must do negative work to keep the dielectric from accelerating. This work is simply the difference $U - U_0$. (Alternatively, the positive work done by the system on the external agent is $U_0 - U$.)

26.6   Electric Dipole in an Electric Field

We have discussed the effect on the capacitance of placing a dielectric between the plates of a capacitor. In Section 26.7, we shall describe the microscopic origin of this effect. Before we can do so, however, we need to expand upon the discussion of the electric dipole that we began in Section 23.4 (see Example 23.6). The electric dipole consists of two charges of equal magnitude and opposite sign separated by a distance $2a$, as shown in Figure 26.21. The **electric dipole moment** of this configuration is defined as the vector **p** directed from $-q$ toward $+q$ along the line joining the charges and having magnitude $2aq$:

$$p \equiv 2aq \tag{26.16}$$

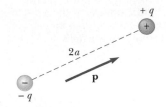

Figure 26.21 An electric dipole consists of two charges of equal magnitude and opposite sign separated by a distance of $2a$. The electric dipole moment **p** is directed from $-q$ toward $+q$.

Now suppose that an electric dipole is placed in a uniform electric field **E**, as shown in Figure 26.22. We identify **E** as the field *external* to the dipole, distinguishing it from the field *due to* the dipole, which we discussed in Section 23.4. The field **E** is established by some other charge distribution, and we place the dipole into this field. Let us imagine that the dipole moment makes an angle θ with the field.

The electric forces acting on the two charges are equal in magnitude ($F = qE$) and opposite in direction as shown in Figure 26.22. Thus, the net force on the dipole is zero. However, the two forces produce a net torque on the dipole; as a result, the dipole rotates in the direction that brings the dipole moment vector into greater alignment with the field. The torque due to the force on the positive charge about an axis through O in Figure 26.22 has magnitude $Fa \sin \theta$, where $a \sin \theta$ is the moment arm of F about O. This force tends to produce a clockwise rotation. The torque about O on the negative charge is also of magnitude $Fa \sin \theta$; here again, the force tends to produce a clockwise rotation. Thus, the magnitude of the net torque about O is

$$\tau = 2Fa \sin \theta$$

Because $F = qE$ and $p = 2aq$, we can express τ as

$$\tau = 2aqE \sin \theta = pE \sin \theta \tag{26.17}$$

It is convenient to express the torque in vector form as the cross product of the vectors **p** and **E**:

$$\boxed{\tau = \mathbf{p} \times \mathbf{E}} \tag{26.18}$$

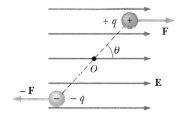

Figure 26.22 An electric dipole in a uniform external electric field. The dipole moment **p** is at an angle θ to the field, causing the dipole to experience a torque.

Torque on an electric dipole in an external electric field

We can determine the potential energy of the system—an electric dipole in an external electric field—as a function of the orientation of the dipole with respect to the field. To do this, we recognize that work must be done by an external agent to rotate the dipole through an angle so as to cause the dipole moment vector to become less aligned with the field. The work done is then stored as potential energy in the system. The work dW required to rotate the dipole through an angle $d\theta$ is $dW = \tau\, d\theta$ (Eq. 10.22). Because $\tau = pE \sin \theta$ and because the work results in an increase in the potential energy U, we find that for a rotation from θ_i to θ_f the change in potential energy of the system is

$$U_f - U_i = \int_{\theta_i}^{\theta_f} \tau\, d\theta = \int_{\theta_i}^{\theta_f} pE \sin \theta\, d\theta = pE \int_{\theta_i}^{\theta_f} \sin \theta\, d\theta$$

$$= pE[-\cos \theta]_{\theta_i}^{\theta_f} = pE(\cos \theta_i - \cos \theta_f)$$

The term that contains $\cos \theta_i$ is a constant that depends on the initial orientation of the dipole. It is convenient for us to choose a reference angle of $\theta_i = 90°$, so that $\cos \theta_i = \cos 90° = 0$. Furthermore, let us choose $U_i = 0$ at $\theta_i = 90°$ as our reference of potential energy. Hence, we can express a general value of $U = U_f$ as

$$U = -pE \cos \theta \tag{26.19}$$

We can write this expression for the potential energy of a dipole in an electric field as the dot product of the vectors **p** and **E**:

Potential energy of the system of an electric dipole in an external electric field

$$U = -\mathbf{p} \cdot \mathbf{E} \qquad (26.20)$$

To develop a conceptual understanding of Equation 26.19, compare this expression with the expression for the potential energy of the system of an object in the gravitational field of the Earth, $U = mgh$ (see Chapter 8). The gravitational expression includes a parameter associated with the object we place in the field—its mass m. Likewise, Equation 26.19 includes a parameter of the object in the electric field—its dipole moment p. The gravitational expression includes the magnitude of the gravitational field g. Similarly, Equation 26.19 includes the magnitude of the electric field E. So far, these two contributions to the potential energy expressions appear analogous. However, the final contribution is somewhat different in the two cases. In the gravitational expression, the potential energy depends on how high we lift the object, measured by h. In Equation 26.19, the potential energy depends on the angle θ through which we rotate the dipole. In both cases, we are making a change in the configuration of the system. In the gravitational case, the change involves moving an object in a *translational* sense, whereas in the electrical case, the change involves moving an object in a *rotational* sense. In both cases, however, once the change is made, the system tends to return to the original configuration when the object is released: the object of mass m falls back to the ground, and the dipole begins to rotate back toward the configuration in which it is aligned with the field. Thus, apart from the type of motion, the expressions for potential energy in these two cases are similar.

Molecules are said to be *polarized* when a separation exists between the average position of the negative charges and the average position of the positive charges in the molecule. In some molecules, such as water, this condition is always present—such molecules are called **polar molecules.** Molecules that do not possess a permanent polarization are called **nonpolar molecules.**

We can understand the permanent polarization of water by inspecting the geometry of the water molecule. In the water molecule, the oxygen atom is bonded to the hydrogen atoms such that an angle of 105° is formed between the two bonds (Fig. 26.23). The center of the negative charge distribution is near the oxygen atom, and the center of the positive charge distribution lies at a point midway along the line joining the hydrogen atoms (the point labeled × in Fig. 26.23). We can model the water molecule and other polar molecules as dipoles because the average positions of the positive and negative charges act as point charges. As a result, we can apply our discussion of dipoles to the behavior of polar molecules.

Microwave ovens take advantage of the polar nature of the water molecule. When in operation, microwave ovens generate a rapidly changing electric field that causes the polar molecules to swing back and forth, absorbing energy from the field in the process. Because the jostling molecules collide with each other, the energy they absorb from the field is converted to internal energy, which corresponds to an increase in temperature of the food.

Another household scenario in which the dipole structure of water is exploited is washing with soap and water. Grease and oil are made up of nonpolar molecules, which are generally not attracted to water. Plain water is not very useful for removing this type of grime. Soap contains long molecules called *surfactants*. In a long molecule, the polarity characteristics of one end of the molecule can be different from those at the other end. In a surfactant molecule, one end acts like a nonpolar molecule and the other acts like a polar molecule. The nonpolar end can attach to a grease or oil molecule, and the polar end can attach to a water molecule. Thus, the soap serves as a chain, linking the dirt and water molecules together. When the water is rinsed away, the grease and oil go with it.

A symmetric molecule (Fig. 26.24a) has no permanent polarization, but polarization can be induced by placing the molecule in an electric field. A field directed to the left, as shown in Figure 26.24b, would cause the center of the positive charge distribution to shift to the left from its initial position and the center of the negative charge distribution to shift to the right. This *induced polarization* is the effect that predominates in most materials used as dielectrics in capacitors.

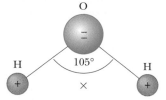

Figure 26.23 The water molecule, H_2O, has a permanent polarization resulting from its nonlinear geometry. The center of the positive charge distribution is at the point ×.

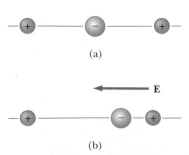

Figure 26.24 (a) A linear symmetric molecule has no permanent polarization. (b) An external electric field induces a polarization in the molecule.

Example 26.8 The H₂O Molecule

The water (H$_2$O) molecule has an electric dipole moment of 6.3×10^{-30} C·m. A sample contains 10^{21} water molecules, with the dipole moments all oriented in the direction of an electric field of magnitude 2.5×10^5 N/C. How much work is required to rotate the dipoles from this orientation ($\theta = 0°$) to one in which all the moments are perpendicular to the field ($\theta = 90°$)?

Solution The work required to rotate one molecule 90° is equal to the difference in potential energy between the 90° orientation and the 0° orientation. Using Equation 26.19,

we obtain

$$W = U_{90°} - U_{0°} = (-pE \cos 90°) - (-pE \cos 0°)$$
$$= pE = (6.3 \times 10^{-30} \text{ C·m})(2.5 \times 10^5 \text{ N/C})$$
$$= 1.6 \times 10^{-24} \text{ J}$$

Because there are 10^{21} molecules in the sample, the *total* work required is

$$W_{\text{total}} = (10^{21})(1.6 \times 10^{-24} \text{ J}) = \boxed{1.6 \times 10^{-3} \text{ J}}$$

26.7 An Atomic Description of Dielectrics

In Section 26.5 we found that the potential difference ΔV_0 between the plates of a capacitor is reduced to $\Delta V_0/\kappa$ when a dielectric is introduced. The potential difference is reduced because the magnitude of the electric field decreases between the plates. In particular, if $\mathbf{E}_0$ is the electric field without the dielectric, the field in the presence of a dielectric is

$$\mathbf{E} = \frac{\mathbf{E}_0}{\kappa} \tag{26.21}$$

Let us first consider a dielectric made up of polar molecules placed in the electric field between the plates of a capacitor. The dipoles (that is, the polar molecules making up the dielectric) are randomly oriented in the absence of an electric field, as shown in Figure 26.25a. When an external field $\mathbf{E}_0$ due to charges on the capacitor plates is applied, a torque is exerted on the dipoles, causing them to partially align with the field, as shown in Figure 26.25b. We can now describe the dielectric as being polarized. The degree of alignment of the molecules with the electric field depends on temperature and on the magnitude of the field. In general, the alignment increases with decreasing temperature and with increasing electric field.

If the molecules of the dielectric are nonpolar, then the electric field due to the plates produces some charge separation and an *induced dipole moment*. These induced dipole moments tend to align with the external field, and the dielectric is polarized. Thus, we can polarize a dielectric with an external field regardless of whether the molecules are polar or nonpolar.

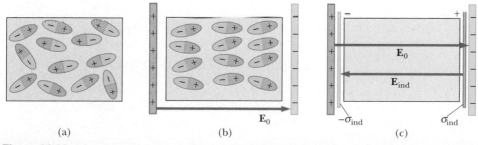

(a) (b) (c)

Figure 26.25 (a) Polar molecules are randomly oriented in the absence of an external electric field. (b) When an external electric field is applied, the molecules partially align with the field. (c) The charged edges of the dielectric can be modeled as an additional pair of parallel plates establishing an electric field $\mathbf{E}_{\text{ind}}$ in the direction opposite to that of $\mathbf{E}_0$.

With these ideas in mind, consider a slab of dielectric material placed between the plates of a capacitor so that it is in a uniform electric field $\mathbf{E}_0$, as shown in Figure 26.25b. The electric field due to the plates is directed to the right and polarizes the dielectric. The net effect on the dielectric is the formation of an *induced* positive surface charge density σ_{ind} on the right face and an equal-magnitude negative surface charge density $-\sigma_{ind}$ on the left face, as shown in Figure 26.25c. Because we can model these surface charge distributions as being due to parallel plates, the induced surface charges on the dielectric give rise to an induced electric field $\mathbf{E}_{ind}$ in the direction opposite the external field $\mathbf{E}_0$. Therefore, the net electric field $\mathbf{E}$ in the dielectric has a magnitude

$$E = E_0 - E_{ind} \tag{26.22}$$

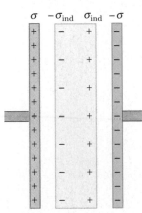

In the parallel-plate capacitor shown in Figure 26.26, the external field E_0 is related to the charge density σ on the plates through the relationship $E_0 = \sigma/\epsilon_0$. The induced electric field in the dielectric is related to the induced charge density σ_{ind} through the relationship $E_{ind} = \sigma_{ind}/\epsilon_0$. Because $E = E_0/\kappa = \sigma/\kappa\epsilon_0$, substitution into Equation 26.22 gives

$$\frac{\sigma}{\kappa\epsilon_0} = \frac{\sigma}{\epsilon_0} - \frac{\sigma_{ind}}{\epsilon_0}$$

$$\sigma_{ind} = \left(\frac{\kappa - 1}{\kappa}\right)\sigma \tag{26.23}$$

Figure 26.26 Induced charge on a dielectric placed between the plates of a charged capacitor. Note that the induced charge density on the dielectric is *less* than the charge density on the plates.

Because $\kappa > 1$, this expression shows that the charge density σ_{ind} induced on the dielectric is less than the charge density σ on the plates. For instance, if $\kappa = 3$ we see that the induced charge density is two-thirds the charge density on the plates. If no dielectric is present, then $\kappa = 1$ and $\sigma_{ind} = 0$ as expected. However, if the dielectric is replaced by an electrical conductor, for which $E = 0$, then Equation 26.22 indicates that $E_0 = E_{ind}$; this corresponds to $\sigma_{ind} = \sigma$. That is, the surface charge induced on the conductor is equal in magnitude but opposite in sign to that on the plates, resulting in a net electric field of zero in the conductor (see Fig. 24.16).

We can use the existence of the induced surface charge distributions on the dielectric to explain the result of Example 26.7. As we saw there, the energy of a capacitor not connected to a battery is lowered when a dielectric is inserted between the plates; this means that negative work is done on the dielectric by the external agent inserting the dielectric into the capacitor. This, in turn, implies that a force must be acting on the dielectric that draws it into the capacitor. This force originates from the nonuniform nature of the electric field of the capacitor near its edges, as indicated in Figure 26.27.

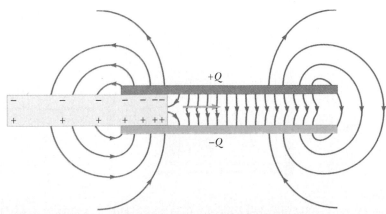

Figure 26.27 The nonuniform electric field near the edges of a parallel-plate capacitor causes a dielectric to be pulled into the capacitor. Note that the field acts on the induced surface charges on the dielectric, which are nonuniformly distributed.

The horizontal component of this *fringe field* acts on the induced charges on the surface of the dielectric, producing a net horizontal force directed into the space between the capacitor plates.

Example 26.9   Effect of a Metallic Slab

A parallel-plate capacitor has a plate separation d and plate area A. An uncharged metallic slab of thickness a is inserted midway between the plates.

(A) Find the capacitance of the device.

Solution We can solve this problem by noting that any charge that appears on one plate of the capacitor must induce a charge of equal magnitude and opposite sign on the near side of the slab, as shown in Figure 26.28a. Consequently, the net charge on the slab remains zero, and the electric field inside the slab is zero. Hence, the capacitor is equivalent to two capacitors in series, each having a plate separation $(d - a)/2$, as shown in Figure 26.28b.

Using Eq. 26.3 and the rule for adding two capacitors in series (Eq. 26.10), we obtain

$$\frac{1}{C} = \frac{1}{C_1} + \frac{1}{C_2} = \frac{1}{\left[\dfrac{\epsilon_0 A}{(d-a)/2}\right]} + \frac{1}{\left[\dfrac{\epsilon_0 A}{(d-a)/2}\right]}$$

$$C = \frac{\epsilon_0 A}{d - a}$$

Note that C approaches infinity as a approaches d. Why?

(B) Show that the capacitance of the original capacitor is unaffected by the insertion of the metallic slab if the slab is infinitesimally thin.

Solution In the result for part (A), we let $a \rightarrow 0$:

$$C = \lim_{a \rightarrow 0} \frac{\epsilon_0 A}{d - a} = \frac{\epsilon_0 A}{d}$$

which is the original capacitance.

What If? What if the metallic slab in part (A) is not midway between the plates? How does this affect the capacitance?

Answer Let us imagine that the slab in Figure 26.27a is moved upward so that the distance between the upper edge of the slab and the upper plate is b. Then, the distance between the lower edge of the slab and the lower plate is $d - b - a$. As in part (A), we find the total capacitance of the series combination:

$$\frac{1}{C} = \frac{1}{C_1} + \frac{1}{C_2} = \frac{1}{(\epsilon_0 A/b)} + \frac{1}{\epsilon_0 A/(d - b - a)}$$

$$= \frac{b}{\epsilon_0 A} + \frac{d - b - a}{\epsilon_0 A} = \frac{d - a}{\epsilon_0 A}$$

$$C = \frac{\epsilon_0 A}{d - a}$$

This is the same result as in part (A). It is independent of the value of b, so it does not matter where the slab is located. In Figure 26.28b, when the central structure is moved up or down, the decrease in plate separation of one capacitor is compensated by the increase in plate separation for the other.

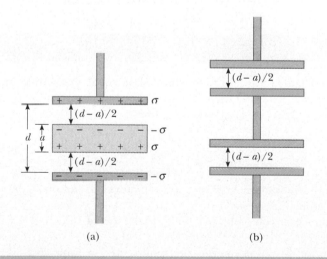

(a)   (b)

Figure 26.28 (Example 26.9) (a) A parallel-plate capacitor of plate separation d partially filled with a metallic slab of thickness a. (b) The equivalent circuit of the device in part (a) consists of two capacitors in series, each having a plate separation $(d - a)/2$.

Example 26.10   A Partially Filled Capacitor

A parallel-plate capacitor with a plate separation d has a capacitance C_0 in the absence of a dielectric. What is the capacitance when a slab of dielectric material of dielectric constant κ and thickness $\frac{1}{3} d$ is inserted between the plates (Fig. 26.29a)?

Solution In Example 26.9, we found that we could insert a metallic slab between the plates of a capacitor and consider the combination as two capacitors in series. The resulting capacitance was independent of the location of the slab. Furthermore, if the thickness of the slab approaches zero,

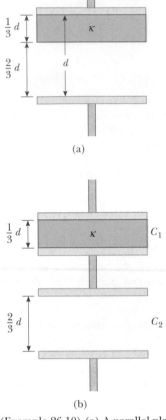

(a)

(b)

Figure 26.29 (Example 26.10) (a) A parallel-plate capacitor of plate separation d partially filled with a dielectric of thickness $d/3$. (b) The equivalent circuit of the capacitor consists of two capacitors connected in series.

then the capacitance of the system approaches the capacitance when the slab is absent. From this, we conclude that we can insert an infinitesimally thin metallic slab anywhere between the plates of a capacitor without affecting the capacitance. Thus, let us imagine sliding an infinitesimally thin metallic slab along the bottom face of the dielectric shown in Figure 26.29a. We can then consider this system to be the series combination of the two capacitors shown in Figure 26.29b: one having a plate separation $d/3$ and filled with a dielectric, and the other having a plate separation $2d/3$ and air between its plates.

From Equations 26.15 and 26.3, the two capacitances are

$$C_1 = \frac{\kappa \epsilon_0 A}{d/3} \quad \text{and} \quad C_2 = \frac{\epsilon_0 A}{2d/3}$$

Using Equation 26.10 for two capacitors combined in series, we have

$$\frac{1}{C} = \frac{1}{C_1} + \frac{1}{C_2} = \frac{d/3}{\kappa \epsilon_0 A} + \frac{2d/3}{\epsilon_0 A}$$

$$\frac{1}{C} = \frac{d}{3\epsilon_0 A}\left(\frac{1}{\kappa} + 2\right) = \frac{d}{3\epsilon_0 A}\left(\frac{1 + 2\kappa}{\kappa}\right)$$

$$C = \left(\frac{3\kappa}{2\kappa + 1}\right)\frac{\epsilon_0 A}{d}$$

Because the capacitance without the dielectric is $C_0 = \epsilon_0 A/d$, we see that

$$C = \left(\frac{3\kappa}{2\kappa + 1}\right)C_0$$

SUMMARY

Take a practice test for this chapter by clicking on the Practice Test link at http://www.pse6.com.

A **capacitor** consists of two conductors carrying charges of equal magnitude and opposite sign. The **capacitance** C of any capacitor is the ratio of the charge Q on either conductor to the potential difference ΔV between them:

$$C \equiv \frac{Q}{\Delta V} \tag{26.1}$$

The capacitance depends only on the geometry of the conductors and not on an external source of charge or potential difference.

The SI unit of capacitance is coulombs per volt, or the **farad** (F), and 1 F = 1 C/V.

Capacitance expressions for various geometries are summarized in Table 26.2.

If two or more capacitors are connected in parallel, then the potential difference is the same across all of them. The equivalent capacitance of a parallel combination of capacitors is

$$C_{eq} = C_1 + C_2 + C_3 + \cdots \tag{26.8}$$

If two or more capacitors are connected in series, the charge is the same on all of them, and the equivalent capacitance of the series combination is given by

Table 26.2

Capacitance and Geometry		
Geometry	**Capacitance**	**Equation**
Isolated sphere of radius R (second spherical conductor assumed to have infinite radius)	$C = 4\pi\epsilon_0 R$	26.2
Parallel-plate capacitor of plate area A and plate separation d	$C = \epsilon_0 \dfrac{A}{d}$	26.3
Cylindrical capacitor of length ℓ and inner and outer radii a and b, respectively	$C = \dfrac{\ell}{2k_e \ln(b/a)}$	26.4
Spherical capacitor with inner and outer radii a and b, respectively	$C = \dfrac{ab}{k_e(b-a)}$	26.6

$$\frac{1}{C_{eq}} = \frac{1}{C_1} + \frac{1}{C_2} + \frac{1}{C_3} + \cdots \qquad (26.10)$$

These two equations enable you to simplify many electric circuits by replacing multiple capacitors with a single equivalent capacitance.

Energy is stored in a capacitor because the charging process is equivalent to the transfer of charges from one conductor at a lower electric potential to another conductor at a higher potential. The energy stored in a capacitor with charge Q is

$$U = \frac{Q^2}{2C} = \tfrac{1}{2}Q\,\Delta V = \tfrac{1}{2}C(\Delta V)^2 \qquad (26.11)$$

When a dielectric material is inserted between the plates of a capacitor, the capacitance increases by a dimensionless factor κ, called the **dielectric constant:**

$$C - \kappa C_0 \qquad (26.14)$$

where C_0 is the capacitance in the absence of the dielectric. The increase in capacitance is due to a decrease in the magnitude of the electric field in the presence of the dielectric. The decrease in the magnitude of **E** arises from an internal electric field produced by aligned dipoles in the dielectric.

The **electric dipole moment p** of an electric dipole has a magnitude

$$p \equiv 2aq \qquad (26.16)$$

The direction of the electric dipole moment vector is from the negative charge toward the positive charge.

The torque acting on an electric dipole in a uniform electric field **E** is

$$\boldsymbol{\tau} = \mathbf{p} \times \mathbf{E} \qquad (26.18)$$

The potential energy of the system of an electric dipole in a uniform external electric field **E** is

$$U = -\mathbf{p} \cdot \mathbf{E} \qquad (26.20)$$

QUESTIONS

1. The plates of a capacitor are connected to a battery. What happens to the charge on the plates if the connecting wires are removed from the battery? What happens to the charge if the wires are removed from the battery and connected to each other?

2. A farad is a very large unit of capacitance. Calculate the length of one side of a square, air-filled capacitor that has a capacitance of 1 F and a plate separation of 1 m.

3. A pair of capacitors are connected in parallel while an identical pair are connected in series. Which pair would be more dangerous to handle after being connected to the same battery? Explain.

4. If you are given three different capacitors C_1, C_2, C_3, how many different combinations of capacitance can you produce?

5. What advantage might there be in using two identical capacitors in parallel connected in series with another identical parallel pair, rather than using a single capacitor?

6. Is it always possible to reduce a combination of capacitors to one equivalent capacitor with the rules we have developed? Explain.

7. The sum of the charges on both plates of a capacitor is zero. What does a capacitor store?

8. Because the charges on the plates of a parallel-plate capacitor are opposite in sign, they attract each other. Hence, it would take positive work to increase the plate separation. What type of energy in the system changes due to the external work done in this process?

9. Why is it dangerous to touch the terminals of a high-voltage capacitor even after the applied potential difference has been turned off? What can be done to make the capacitor safe to handle after the voltage source has been removed?

10. Explain why the work needed to move a charge Q through a potential difference ΔV is $W = Q \Delta V$ whereas the energy stored in a charged capacitor is $U = \frac{1}{2} Q \Delta V$. Where does the $\frac{1}{2}$ factor come from?

11. If the potential difference across a capacitor is doubled, by what factor does the energy stored change?

12. It is possible to obtain large potential differences by first charging a group of capacitors connected in parallel and then activating a switch arrangement that in effect dis-connects the capacitors from the charging source and from each other and reconnects them in a series arrangement. The group of charged capacitors is then discharged in series. What is the maximum potential difference that can be obtained in this manner by using ten capacitors each of 500 μF and a charging source of 800 V?

13. Assume you want to increase the maximum operating voltage of a parallel-plate capacitor. Describe how you can do this for a fixed plate separation.

14. An air-filled capacitor is charged, then disconnected from the power supply, and finally connected to a voltmeter. Explain how and why the potential difference changes when a dielectric is inserted between the plates of the capacitor.

15. Using the polar molecule description of a dielectric, explain how a dielectric affects the electric field inside a capacitor.

16. Explain why a dielectric increases the maximum operating voltage of a capacitor although the physical size of the capacitor does not change.

17. What is the difference between dielectric strength and the dielectric constant?

18. Explain why a water molecule is permanently polarized. What type of molecule has no permanent polarization?

19. If a dielectric-filled capacitor is heated, how will its capacitance change? (Ignore thermal expansion and assume that the dipole orientations are temperature-dependent.)

20. If you were asked to design a capacitor where small size and large capacitance were required, what factors would be important in your design?

PROBLEMS

1, 2, 3 = straightforward, intermediate, challenging ☐ = full solution available in the *Student Solutions Manual and Study Guide*

= coached solution with hints available at http://www.pse6.com ▨ = computer useful in solving problem

= paired numerical and symbolic problems

Section 26.1 Definition of Capacitance

1. (a) How much charge is on each plate of a 4.00-μF capacitor when it is connected to a 12.0-V battery? (b) If this same capacitor is connected to a 1.50-V battery, what charge is stored?

2. Two conductors having net charges of $+10.0$ μC and -10.0 μC have a potential difference of 10.0 V between them. (a) Determine the capacitance of the system. (b) What is the potential difference between the two conductors if the charges on each are increased to $+100$ μC and -100 μC?

Section 26.2 Calculating Capacitance

3. An isolated charged conducting sphere of radius 12.0 cm creates an electric field of 4.90×10^4 N/C at a distance 21.0 cm from its center. (a) What is its surface charge density? (b) What is its capacitance?

4. (a) If a drop of liquid has capacitance 1.00 pF, what is its radius? (b) If another drop has radius 2.00 mm, what is its capacitance? (c) What is the charge on the smaller drop if its potential is 100 V?

5. Two conducting spheres with diameters of 0.400 m and 1.00 m are separated by a distance that is large compared with the diameters. The spheres are connected by a thin wire and are charged to 7.00 μC. (a) How is this total charge shared between the spheres? (Ignore any charge on the wire.) (b) What is the potential of the system of spheres when the reference potential is taken to be $V = 0$ at $r = \infty$?

6. Regarding the Earth and a cloud layer 800 m above the Earth as the "plates" of a capacitor, calculate the

capacitance. Assume the cloud layer has an area of 1.00 km² and that the air between the cloud and the ground is pure and dry. Assume charge builds up on the cloud and on the ground until a uniform electric field of 3.00×10^6 N/C throughout the space between them makes the air break down and conduct electricity as a lightning bolt. What is the maximum charge the cloud can hold?

7. 🌐 An air-filled capacitor consists of two parallel plates, each with an area of 7.60 cm², separated by a distance of 1.80 mm. A 20.0-V potential difference is applied to these plates. Calculate (a) the electric field between the plates, (b) the surface charge density, (c) the capacitance, and (d) the charge on each plate.

8. A 1-megabit computer memory chip contains many 60.0-fF capacitors. Each capacitor has a plate area of 21.0×10^{-12} m². Determine the plate separation of such a capacitor (assume a parallel-plate configuration). The order of magnitude of the diameter of an atom is 10^{-10} m = 0.1 nm. Express the plate separation in nanometers.

9. When a potential difference of 150 V is applied to the plates of a parallel-plate capacitor, the plates carry a surface charge density of 30.0 nC/cm². What is the spacing between the plates?

10. A variable air capacitor used in a radio tuning circuit is made of N semicircular plates each of radius R and positioned a distance d from its neighbors, to which it is electrically connected. As shown in Figure P26.10, a second identical set of plates is enmeshed with its plates halfway between those of the first set. The second set can rotate as a unit. Determine the capacitance as a function of the angle of rotation θ, where $\theta = 0$ corresponds to the maximum capacitance.

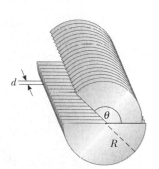

Figure P26.10

11. 🌐 A 50.0-m length of coaxial cable has an inner conductor that has a diameter of 2.58 mm and carries a charge of 8.10 μC. The surrounding conductor has an inner diameter of 7.27 mm and a charge of -8.10 μC. (a) What is the capacitance of this cable? (b) What is the potential difference between the two conductors? Assume the region between the conductors is air.

12. A 20.0-μF spherical capacitor is composed of two concentric metal spheres, one having a radius twice as large as the

other. The region between the spheres is a vacuum. Determine the volume of this region.

13. An air-filled spherical capacitor is constructed with inner and outer shell radii of 7.00 and 14.0 cm, respectively. (a) Calculate the capacitance of the device. (b) What potential difference between the spheres results in a charge of 4.00 μC on the capacitor?

14. A small object of mass m carries a charge q and is suspended by a thread between the vertical plates of a parallel-plate capacitor. The plate separation is d. If the thread makes an angle θ with the vertical, what is the potential difference between the plates?

15. Find the capacitance of the Earth. (*Suggestion:* The outer conductor of the "spherical capacitor" may be considered as a conducting sphere at infinity where V approaches zero.)

Section 26.3   Combinations of Capacitors

16. Two capacitors, $C_1 = 5.00$ μF and $C_2 = 12.0$ μF, are connected in parallel, and the resulting combination is connected to a 9.00-V battery. (a) What is the equivalent capacitance of the combination? What are (b) the potential difference across each capacitor and (c) the charge stored on each capacitor?

17. **What If?** The two capacitors of Problem 16 are now connected in series and to a 9.00-V battery. Find (a) the equivalent capacitance of the combination, (b) the potential difference across each capacitor, and (c) the charge on each capacitor.

18. Evaluate the equivalent capacitance of the configuration shown in Figure P26.18. All the capacitors are identical, and each has capacitance C.

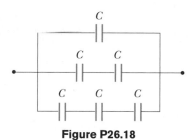

Figure P26.18

19. Two capacitors when connected in parallel give an equivalent capacitance of 9.00 pF and give an equivalent capacitance of 2.00 pF when connected in series. What is the capacitance of each capacitor?

20. Two capacitors when connected in parallel give an equivalent capacitance of C_p and an equivalent capacitance of C_s when connected in series. What is the capacitance of each capacitor?

21. 🌐 Four capacitors are connected as shown in Figure P26.21. (a) Find the equivalent capacitance between points a and b. (b) Calculate the charge on each capacitor if $\Delta V_{ab} = 15.0$ V.

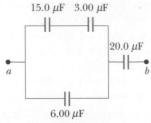

Figure P26.21

22. Three capacitors are connected to a battery as shown in Figure P26.22. Their capacitances are $C_1 = 3C$, $C_2 = C$, and $C_3 = 5C$. (a) What is the equivalent capacitance of this set of capacitors? (b) State the ranking of the capacitors according to the charge they store, from largest to smallest. (c) Rank the capacitors according to the potential differences across them, from largest to smallest. (d) **What If?** If C_3 is increased, what happens to the charge stored by each of the capacitors?

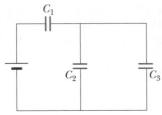

Figure P26.22

23. Consider the circuit shown in Figure P26.23, where $C_1 = 6.00 \ \mu F$, $C_2 = 3.00 \ \mu F$, and $\Delta V = 20.0$ V. Capacitor C_1 is first charged by the closing of switch S_1. Switch S_1 is then opened, and the charged capacitor is connected to the uncharged capacitor by the closing of S_2. Calculate the initial charge acquired by C_1 and the final charge on each capacitor.

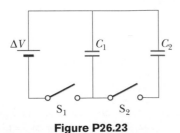

Figure P26.23

24. According to its design specification, the timer circuit delaying the closing of an elevator door is to have a capacitance of 32.0 μF between two points A and B. (a) When one circuit is being constructed, the inexpensive but durable capacitor installed between these two points is found to have capacitance 34.8 μF. To meet the specification, one additional capacitor can be placed between the two points. Should it be in series or in parallel with the 34.8-μF capacitor? What should be its capacitance? (b) **What If?** The next circuit comes down the assembly

line with capacitance 29.8 μF between A and B. What additional capacitor should be installed in series or in parallel in that circuit, to meet the specification?

25. A group of identical capacitors is connected first in series and then in parallel. The combined capacitance in parallel is 100 times larger than for the series connection. How many capacitors are in the group?

26. Consider three capacitors C_1, C_2, C_3, and a battery. If C_1 is connected to the battery, the charge on C_1 is 30.8 μC. Now C_1 is disconnected, discharged, and connected in series with C_2. When the series combination of C_2 and C_1 is connected across the battery, the charge on C_1 is 23.1 μC. The circuit is disconnected and the capacitors discharged. Capacitor C_3, capacitor C_1, and the battery are connected in series, resulting in a charge on C_1 of 25.2 μC. If, after being disconnected and discharged, C_1, C_2, and C_3 are connected in series with one another and with the battery, what is the charge on C_1?

27. Find the equivalent capacitance between points a and b for the group of capacitors connected as shown in Figure P26.27. Take $C_1 = 5.00 \ \mu F$, $C_2 = 10.0 \ \mu F$, and $C_3 = 2.00 \ \mu F$.

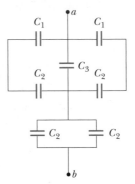

Figure P26.27 Problems 27 and 28.

28. For the network described in the previous problem, if the potential difference between points a and b is 60.0 V, what charge is stored on C_3?

29. Find the equivalent capacitance between points a and b in the combination of capacitors shown in Figure P26.29.

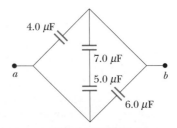

Figure P26.29

30. Some physical systems possessing capacitance continuously distributed over space can be modeled as an infinite array of discrete circuit elements. Examples are a microwave waveguide and the axon of a nerve cell. To practice analysis of an infinite array, determine the equivalent capacitance C between terminals X and Y of the infinite set of capacitors represented in Figure P26.30. Each capacitor has capacitance C_0. *Suggestion:* Imagine that the ladder is cut at the line AB, and note that the equivalent capacitance of the infinite section to the right of AB is also C.

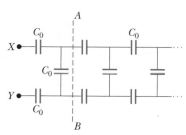

Figure P26.30

Section 26.4 Energy Stored in a Charged Capacitor

31. (a) A 3.00-μF capacitor is connected to a 12.0-V battery. How much energy is stored in the capacitor? (b) If the capacitor had been connected to a 6.00-V battery, how much energy would have been stored?

32. The immediate cause of many deaths is ventricular fibrillation, uncoordinated quivering of the heart as opposed to proper beating. An electric shock to the chest can cause momentary paralysis of the heart muscle, after which the heart will sometimes start organized beating again. A *defibrillator* (Fig. 26.14) is a device that applies a strong electric shock to the chest over a time interval of a few milliseconds. The device contains a capacitor of several microfarads, charged to several thousand volts. Electrodes called paddles, about 8 cm across and coated with conducting paste, are held against the chest on both sides of the heart. Their handles are insulated to prevent injury to the operator, who calls, "Clear!" and pushes a button on one paddle to discharge the capacitor through the patient's chest. Assume that an energy of 300 J is to be delivered from a 30.0-μF capacitor. To what potential difference must it be charged?

33. Two capacitors, $C_1 = 25.0$ μF and $C_2 = 5.00$ μF, are connected in parallel and charged with a 100-V power supply. (a) Draw a circuit diagram and calculate the total energy stored in the two capacitors. (b) **What If?** What potential difference would be required across the same two capacitors connected in series in order that the combination stores the same amount of energy as in (a)? Draw a circuit diagram of this circuit.

34. A parallel-plate capacitor is charged and then disconnected from a battery. By what fraction does the stored energy change (increase or decrease) when the plate separation is doubled?

35. As a person moves about in a dry environment, electric charge accumulates on his body. Once it is at high voltage, either positive or negative, the body can discharge via sometimes noticeable sparks and shocks. Consider a human body well separated from ground, with the typical capacitance 150 pF. (a) What charge on the body will produce a potential of 10.0 kV? (b) Sensitive electronic devices can be destroyed by electrostatic discharge from a person. A particular device can be destroyed by a discharge releasing an energy of 250 μJ. To what voltage on the body does this correspond?

36. A uniform electric field $E = 3\ 000$ V/m exists within a certain region. What volume of space contains an energy equal to 1.00×10^{-7} J? Express your answer in cubic meters and in liters.

37. A parallel-plate capacitor has a charge Q and plates of area A. What force acts on one plate to attract it toward the other plate? Because the electric field between the plates is $E = Q/A\epsilon_0$, you might think that the force is $F = QE = Q^2/A\epsilon_0$. This is wrong, because the field E includes contributions from both plates, and the field created by the positive plate cannot exert any force on the positive plate. Show that the force exerted on each plate is actually $F = Q^2/2\epsilon_0 A$. (*Suggestion:* Let $C = \epsilon_0 A/x$ for an arbitrary plate separation x; then require that the work done in separating the two charged plates be $W = \int F\, dx$.) The force exerted by one charged plate on another is sometimes used in a machine shop to hold a workpiece stationary.

38. The circuit in Figure P26.38 consists of two identical parallel metal plates connected by identical metal springs to a 100-V battery. With the switch open, the plates are uncharged, are separated by a distance $d = 8.00$ mm, and have a capacitance $C = 2.00$ μF. When the switch is closed, the distance between the plates decreases by a factor of 0.500. (a) How much charge collects on each plate and (b) what is the spring constant for each spring? (*Suggestion:* Use the result of Problem 37.)

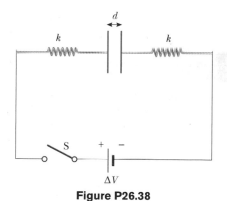

Figure P26.38

39. **Review problem.** A certain storm cloud has a potential of 1.00×10^8 V relative to a tree. If, during a lightning storm, 50.0 C of charge is transferred through this potential difference and 1.00% of the energy is absorbed by the tree, how much sap in the tree can be boiled away? Model the sap as water initially at 30.0°C. Water has a specific heat of 4 186 J/kg°C, a boiling point of 100°C, and a latent heat of vaporization of 2.26×10^6 J/kg.

40. Two identical parallel-plate capacitors, each with capacitance C, are charged to potential difference ΔV and connected in parallel. Then the plate separation in one of the capacitors is doubled. (a) Find the total energy of the system of two capacitors *before* the plate separation is doubled. (b) Find the potential difference across each capacitor *after* the plate separation is doubled. (c) Find the

total energy of the system *after* the plate separation is doubled. (d) Reconcile the difference in the answers to parts (a) and (c) with the law of conservation of energy.

41. Show that the energy associated with a conducting sphere of radius R and charge Q surrounded by a vacuum is $U = k_e Q^2/2R$.

42. Consider two conducting spheres with radii R_1 and R_2. They are separated by a distance much greater than either radius. A total charge Q is shared between the spheres, subject to the condition that the electric potential energy of the system has the smallest possible value. The total charge Q is equal to $q_1 + q_2$, where q_1 represents the charge on the first sphere and q_2 the charge on the second. Because the spheres are very far apart, you can assume that the charge of each is uniformly distributed over its surface. You may use the result of Problem 41. (a) Determine the values of q_1 and q_2 in terms of Q, R_1, and R_2. (b) Show that the potential difference between the spheres is zero. (We saw in Chapter 25 that two conductors joined by a conducting wire will be at the same potential in a static situation. This problem illustrates the general principle that static charge on a conductor will distribute itself so that the electric potential energy of the system is a minimum.)

Section 26.5 Capacitors with Dielectrics

43. Determine (a) the capacitance and (b) the maximum potential difference that can be applied to a Teflon-filled parallel-plate capacitor having a plate area of 1.75 cm^2 and plate separation of $0.040\,0$ mm.

44. (a) How much charge can be placed on a capacitor with air between the plates before it breaks down, if the area of each of the plates is 5.00 cm^2? (b) **What If?** Find the maximum charge if polystyrene is used between the plates instead of air.

45. A commercial capacitor is to be constructed as shown in Figure 26.17a. This particular capacitor is made from two strips of aluminum separated by a strip of paraffin-coated paper. Each strip of foil and paper is 7.00 cm wide. The foil is $0.004\,00$ mm thick, and the paper is $0.025\,0$ mm thick and has a dielectric constant of 3.70. What length should the strips have, if a capacitance of 9.50×10^{-8} F is desired before the capacitor is rolled up? (Adding a second strip of paper and rolling the capacitor effectively doubles its capacitance, by allowing charge storage on both sides of each strip of foil.)

46. The supermarket sells rolls of aluminum foil, of plastic wrap, and of waxed paper. Describe a capacitor made from supermarket materials. Compute order-of-magnitude estimates for its capacitance and its breakdown voltage.

47. A parallel-plate capacitor in air has a plate separation of 1.50 cm and a plate area of 25.0 cm^2. The plates are charged to a potential difference of 250 V and disconnected from the source. The capacitor is then immersed in distilled water. Determine (a) the charge on the plates before and after immersion, (b) the capacitance and potential difference after immersion, and (c) the change in energy of the capacitor. Assume the liquid is an insulator.

48. A wafer of titanium dioxide ($\kappa = 173$) of area 1.00 cm^2 has a thickness of 0.100 mm. Aluminum is evaporated on the parallel faces to form a parallel-plate capacitor. (a) Calculate the capacitance. (b) When the capacitor is charged with a 12.0-V battery, what is the magnitude of charge delivered to each plate? (c) For the situation in part (b), what are the free and induced surface charge densities? (d) What is the magnitude of the electric field?

49. Each capacitor in the combination shown in Figure P26.49 has a breakdown voltage of 15.0 V. What is the breakdown voltage of the combination?

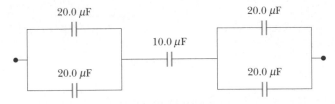

Figure P26.49

Section 26.6 Electric Dipole in an Electric Field

50. A small rigid object carries positive and negative 3.50-nC charges. It is oriented so that the positive charge has coordinates $(-1.20 \text{ mm}, 1.10 \text{ mm})$ and the negative charge is at the point $(1.40 \text{ mm}, -1.30 \text{ mm})$. (a) Find the electric dipole moment of the object. The object is placed in an electric field $\mathbf{E} = (7\,800\hat{\mathbf{i}} - 4\,900\hat{\mathbf{j}}) \text{ N/C}$. (b) Find the torque acting on the object. (c) Find the potential energy of the object–field system when the object is in this orientation. (d) If the orientation of the object can change, find the difference between the maximum and minimum potential energies of the system.

51. A small object with electric dipole moment $\mathbf{p}$ is placed in a nonuniform electric field $\mathbf{E} = E(x)\hat{\mathbf{i}}$. That is, the field is in the x direction and its magnitude depends on the coordinate x. Let θ represent the angle between the dipole moment and the x direction. (a) Prove that the dipole feels a net force

$$F = p\left(\frac{dE}{dx}\right)\cos\theta$$

in the direction toward which the field increases. (b) Consider a spherical balloon centered at the origin, with radius 15.0 cm and carrying charge 2.00 μC. Evaluate dE/dx at the point (16 cm, 0, 0). Assume a water droplet at this point has an induced dipole moment of $6.30\hat{\mathbf{i}} \text{ nC} \cdot \text{m}$. Find the force on it.

Section 26.7 An Atomic Description of Dielectrics

52. A detector of radiation called a Geiger tube consists of a closed, hollow, conducting cylinder with a fine wire along its axis. Suppose that the internal diameter of the cylinder is 2.50 cm and that the wire along the axis has a diameter of 0.200 mm. The dielectric strength of the gas between the central wire and the cylinder is 1.20×10^6 V/m. Calculate the maximum potential difference that can be applied between the wire and the cylinder before breakdown occurs in the gas.

53. The general form of Gauss's law describes how a charge creates an electric field in a material, as well as in vacuum. It is

$$\oint \mathbf{E} \cdot d\mathbf{A} = \frac{q}{\epsilon}$$

where $\epsilon = \kappa\epsilon_0$ is the permittivity of the material. (a) A sheet with charge Q uniformly distributed over its area A is surrounded by a dielectric. Show that the sheet creates a uniform electric field at nearby points, with magnitude $E = Q/2A\epsilon$. (b) Two large sheets of area A, carrying opposite charges of equal magnitude Q, are a small distance d apart. Show that they create uniform electric field in the space between them, with magnitude $E = Q/A\epsilon$. (c) Assume that the negative plate is at zero potential. Show that the positive plate is at potential $Qd/A\epsilon$. (d) Show that the capacitance of the pair of plates is $A\epsilon/d = \kappa A\epsilon_0/d$.

Additional Problems

54. For the system of capacitors shown in Figure P26.54, find (a) the equivalent capacitance of the system, (b) the potential across each capacitor, (c) the charge on each capacitor, and (d) the total energy stored by the group.

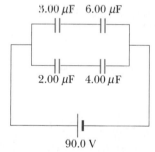

3.00 μF   6.00 μF

2.00 μF   4.00 μF

90.0 V

Figure P26.54

55. Four parallel metal plates P_1, P_2, P_3, and P_4, each of area 7.50 cm², are separated successively by a distance $d = 1.19$ mm, as shown in Figure P26.55. P_1 is connected to the negative terminal of a battery, and P_2 to the positive terminal. The battery maintains a potential difference of 12.0 V. (a) If P_3 is connected to the negative terminal, what is the

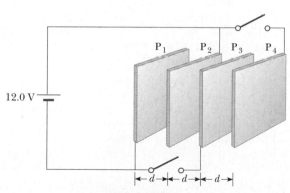

P_1 P_2 P_3 P_4

12.0 V

$\leftarrow d \rightarrow \leftarrow d \rightarrow \leftarrow d \rightarrow$

Figure P26.55

capacitance of the three-plate system $P_1P_2P_3$? (b) What is the charge on P_2? (c) If P_4 is now connected to the positive terminal of the battery, what is the capacitance of the four-plate system $P_1P_2P_3P_4$? (d) What is the charge on P_4?

56. One conductor of an overhead electric transmission line is a long aluminum wire 2.40 cm in radius. Suppose that at a particular moment it carries charge per length 1.40 μC/m and is at potential 345 kV. Find the potential 12.0 m below the wire. Ignore the other conductors of the transmission line and assume the electric field is everywhere purely radial.

57. Two large parallel metal plates are oriented horizontally and separated by a distance $3d$. A grounded conducting wire joins them, and initially each plate carries no charge. Now a third identical plate carrying charge Q is inserted between the two plates, parallel to them and located a distance d from the upper plate, as in Figure P26.57. (a) What induced charge appears on each of the two original plates? (b) What potential difference appears between the middle plate and each of the other plates? Each plate has area A.

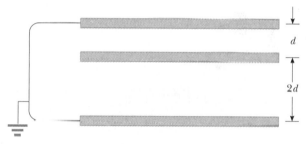

d

$2d$

Figure P26.67

58. A 2.00-nF parallel-plate capacitor is charged to an initial potential difference $\Delta V_i = 100$ V and then isolated. The dielectric material between the plates is mica, with a dielectric constant of 5.00. (a) How much work is required to withdraw the mica sheet? (b) What is the potential difference of the capacitor after the mica is withdrawn?

59. A parallel-plate capacitor is constructed using a dielectric material whose dielectric constant is 3.00 and whose dielectric strength is 2.00×10^8 V/m. The desired capacitance is 0.250 μF, and the capacitor must withstand a maximum potential difference of 4 000 V. Find the minimum area of the capacitor plates.

60. A 10.0-μF capacitor has plates with vacuum between them. Each plate carries a charge of magnitude 1 000 μC. A particle with charge -3.00 μC and mass 2.00×10^{-16} kg is fired from the positive plate toward the negative plate with an initial speed of 2.00×10^6 m/s. Does it reach the negative plate? If so, find its impact speed. If not, what fraction of the way across the capacitor does it travel?

61. A parallel-plate capacitor is constructed by filling the space between two square plates with blocks of three dielectric materials, as in Figure P26.61. You may assume that $\ell \gg d$. (a) Find an expression for the capacitance of the device in terms of the plate area A and d, κ_1, κ_2, and κ_3. (b) Calculate the capacitance using the values $A = 1.00$ cm², $d = 2.00$ mm, $\kappa_1 = 4.90$, $\kappa_2 = 5.60$, and $\kappa_3 = 2.10$.

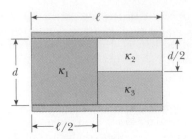

Figure P26.61

62. A 10.0-μF capacitor is charged to 15.0 V. It is next connected in series with an uncharged 5.00-μF capacitor. The series combination is finally connected across a 50.0-V battery, as diagrammed in Figure P26.62. Find the new potential differences across the 5-μF and 10-μF capacitors.

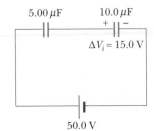

Figure P26.62

63. (a) Two spheres have radii a and b and their centers are a distance d apart. Show that the capacitance of this system is

$$C = \frac{4\pi\epsilon_0}{\dfrac{1}{a} + \dfrac{1}{b} - \dfrac{2}{d}}$$

provided that d is large compared with a and b. (*Suggestion:* Because the spheres are far apart, assume that the potential of each equals the sum of the potentials due to each sphere, and when calculating those potentials assume that $V = k_e Q/r$ applies.) (b) Show that as d approaches infinity the above result reduces to that of two spherical capacitors in series.

64. A capacitor is constructed from two square plates of sides ℓ and separation d. A material of dielectric constant κ is inserted a distance x into the capacitor, as shown in Figure P26.64. Assume that d is much smaller than x. (a) Find the equivalent capacitance of the device. (b) Calculate the energy stored in the capacitor, letting ΔV repre-

sent the potential difference. (c) Find the direction and magnitude of the force exerted on the dielectric, assuming a constant potential difference ΔV. Ignore friction. (d) Obtain a numerical value for the force assuming that $\ell = 5.00$ cm, $\Delta V = 2\,000$ V, $d = 2.00$ mm, and the dielectric is glass ($\kappa = 4.50$). (*Suggestion:* The system can be considered as two capacitors connected in parallel.)

65. A capacitor is constructed from two square plates of sides ℓ and separation d, as suggested in Figure P26.64. You may assume that d is much less than ℓ. The plates carry charges $+ Q_0$ and $- Q_0$. A block of metal has a width ℓ, a length ℓ, and a thickness slightly less than d. It is inserted a distance x into the capacitor. The charges on the plates are not disturbed as the block slides in. In a static situation, a metal prevents an electric field from penetrating inside it. The metal can be thought of as a perfect dielectric, with $\kappa \to \infty$. (a) Calculate the stored energy as a function of x. (b) Find the direction and magnitude of the force that acts on the metallic block. (c) The area of the advancing front face of the block is essentially equal to ℓd. Considering the force on the block as acting on this face, find the stress (force per area) on it. (d) For comparison, express the energy density in the electric field between the capacitor plates in terms of Q_0, ℓ, d, and ϵ_0.

66. When considering the energy supply for an automobile, the energy per unit mass of the energy source is an important parameter. Using the following data, compare the energy per unit mass (J/kg) for gasoline, lead–acid batteries, and capacitors. (The ampere A will be introduced in the next chapter as the SI unit of electric current. 1 A = 1 C/s.)
Gasoline: 126 000 Btu/gal; density = 670 kg/m^3.
Lead–acid battery: 12.0 V; 100 A·h; mass = 16.0 kg.
Capacitor: potential difference at full charge = 12.0 V; capacitance = 0.100 F; mass = 0.100 kg.

67. An isolated capacitor of unknown capacitance has been charged to a potential difference of 100 V. When the charged capacitor is then connected in parallel to an uncharged 10.0-μF capacitor, the potential difference across the combination is 30.0 V. Calculate the unknown capacitance.

68. To repair a power supply for a stereo amplifier, an electronics technician needs a 100-μF capacitor capable of withstanding a potential difference of 90 V between the plates. The only available supply is a box of five 100-μF capacitors, each having a maximum voltage capability of 50 V. Can the technician substitute a combination of these capacitors that has the proper electrical characteristics? If so, what will be the maximum voltage across any of the capacitors used? (*Suggestion:* The technician may not have to use all the capacitors in the box.)

69. A parallel-plate capacitor of plate separation d is charged to a potential difference ΔV_0. A dielectric slab of thickness d and dielectric constant κ is introduced between the plates while the battery remains connected to the plates. (a) Show that the ratio of energy stored after the dielectric is introduced to the energy stored in the empty capacitor is $U/U_0 = \kappa$. Give a physical explanation for this increase in stored energy. (b) What happens to the charge on the capacitor? (Note that this situation is not the same as in

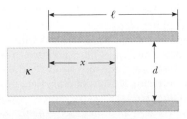

Figure P26.64 Problems 64 and 65.

Example 26.7, in which the battery was removed from the circuit before the dielectric was introduced.)

70. A vertical parallel-plate capacitor is half filled with a dielectric for which the dielectric constant is 2.00 (Fig. P26.70a). When this capacitor is positioned horizontally, what fraction of it should be filled with the same dielectric (Fig. P26.70b) in order for the two capacitors to have equal capacitance?

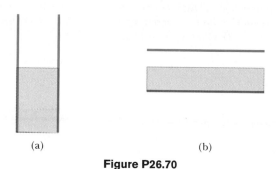

(a)   (b)

Figure P26.70

71. Capacitors $C_1 = 6.00\ \mu F$ and $C_2 = 2.00\ \mu F$ are charged as a parallel combination across a 250-V battery. The capacitors are disconnected from the battery and from each other. They are then connected positive plate to negative plate and negative plate to positive plate. Calculate the resulting charge on each capacitor.

72. Calculate the equivalent capacitance between the points a and b in Figure P26.72. Note that this is not a simple series or parallel combination (*Suggestion:* Assume a potential difference ΔV between points a and b. Write expressions for ΔV_{ab} in terms of the charges and capacitances for the various possible pathways from a to b, and require conservation of charge for those capacitor plates that are connected to each other.)

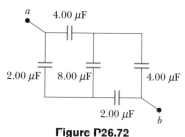

Figure P26.72

73. The inner conductor of a coaxial cable has a radius of 0.800 mm, and the outer conductor's inside radius is 3.00 mm. The space between the conductors is filled with polyethylene, which has a dielectric constant of 2.30 and a dielectric strength of 18.0×10^6 V/m. What is the maximum potential difference that this cable can withstand?

74. You are optimizing coaxial cable design for a major manufacturer. Show that for a given outer conductor radius b, maximum potential difference capability is attained when the radius of the inner conductor is $a = b/e$ where e is the base of natural logarithms.

75. Determine the equivalent capacitance of the combination shown in Figure P26.75. (*Suggestion:* Consider the symmetry involved.)

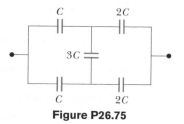

Figure P26.75

76. Consider two long, parallel, and oppositely charged wires of radius d with their centers separated by a distance D. Assuming the charge is distributed uniformly on the surface of each wire, show that the capacitance per unit length of this pair of wires is

$$\frac{C}{\ell} = \frac{\pi\epsilon_0}{\ln[(D-d)/d]}$$

77. Example 26.2 explored a cylindrical capacitor of length ℓ and radii a and b of the two conductors. In the **What If?** section, it was claimed that increasing ℓ by 10% is more effective in terms of increasing the capacitance than increasing a by 10% if $b > 2.85a$. Verify this claim mathematically.

Answers to Quick Quizzes

26.1   (d). The capacitance is a property of the physical system and does not vary with applied voltage. According to Equation 26.1, if the voltage is doubled, the charge is doubled.

26.2   (a). When the key is pressed, the plate separation is decreased and the capacitance increases. Capacitance depends only on how a capacitor is constructed and not on the external circuit.

26.3   (a). When connecting capacitors in series, the inverses of the capacitances add, resulting in a smaller overall equivalent capacitance.

26.4   (a). When capacitors are connected in series, the voltages add, for a total of 20 V in this case. If they are combined in parallel, the voltage across the combination is still 10 V.

26.5   (b). For a given voltage, the energy stored in a capacitor is proportional to C: $U = C(\Delta V)^2/2$. Thus, you want to maximize the equivalent capacitance. You do this by connecting the three capacitors in parallel, so that the capacitances add.

26.6   (a) C decreases (Eq. 26.3). (b) Q stays the same because there is no place for the charge to flow. (c) E remains constant (see Eq. 24.8 and the paragraph following it). (d) ΔV increases because $\Delta V = Q/C$, Q is constant (part b), and C decreases (part a). (e) The energy stored in the capacitor is proportional to both Q and ΔV (Eq.

26.11) and thus increases. The additional energy comes from the work you do in pulling the two plates apart.

26.7 (a) C decreases (Eq. 26.3). (b) Q decreases. The battery supplies a constant potential difference ΔV; thus, charge must flow out of the capacitor if $C = Q/\Delta V$ is to decrease. (c) E decreases because the charge density on the plates decreases. (d) ΔV remains constant because of the presence of the battery. (e) The energy stored in the capacitor decreases (Eq. 26.11).

26.8 Increase. The dielectric constant of wood (and of all other insulating materials, for that matter) is greater than 1; therefore, the capacitance increases (Eq. 26.14). This increase is sensed by the stud-finder's special circuitry, which causes an indicator on the device to light up.

26.9 (a) C increases (Eq. 26.14). (b) Q increases. Because the battery maintains a constant ΔV, Q must increase if C increases. (c) E between the plates remains constant because $\Delta V = Ed$ and neither ΔV nor d changes. The electric field due to the charges on the plates increases because more charge has flowed onto the plates. The induced surface charges on the dielectric create a field that opposes the increase in the field caused by the greater number of charges on the plates (see Section 26.7). (d) The battery maintains a constant ΔV.

Current and Resistance

CHAPTER OUTLINE

27.1 Electric Current

27.2 Resistance

27.3 A Model for Electrical Conduction

27.4 Resistance and Temperature

27.5 Superconductors

27.6 Electrical Power

▲ These power lines transfer energy from the power company to homes and businesses. The energy is transferred at a very high voltage, possibly hundreds of thousands of volts in some cases. Despite the fact that this makes power lines very dangerous, the high voltage results in less loss of power due to resistance in the wires. (Telegraph Colour Library/FPG)

Thus far our treatment of electrical phenomena has been confined to the study of charges in equilibrium situations, or *electrostatics*. We now consider situations involving electric charges that are *not* in equilibrium. We use the term *electric current*, or simply *current*, to describe the rate of flow of charge through some region of space. Most practical applications of electricity deal with electric currents. For example, the battery in a flashlight produces a current in the filament of the bulb when the switch is turned on. A variety of home appliances operate on alternating current. In these common situations, current exists in a conductor, such as a copper wire. It also is possible for currents to exist outside a conductor. For instance, a beam of electrons in a television picture tube constitutes a current.

This chapter begins with the definition of current. A microscopic description of current is given, and some of the factors that contribute to the opposition to the flow of charge in conductors are discussed. A classical model is used to describe electrical conduction in metals, and some of the limitations of this model are cited. We also define electrical resistance and introduce a new circuit element, the resistor. We conclude by discussing the rate at which energy is transferred to a device in an electric circuit.

27.1 Electric Current

In this section, we study the flow of electric charges through a piece of material. The amount of flow depends on the material through which the charges are passing and the potential difference across the material. Whenever there is a net flow of charge through some region, an electric **current** is said to exist.

It is instructive to draw an analogy between water flow and current. In many localities it is common practice to install low-flow showerheads in homes as a water-conservation measure. We quantify the flow of water from these and similar devices by specifying the amount of water that emerges during a given time interval, which is often measured in liters per minute. On a grander scale, we can characterize a river current by describing the rate at which the water flows past a particular location. For example, the flow over the brink at Niagara Falls is maintained at rates between $1\ 400\ \text{m}^3/\text{s}$ and $2\ 800\ \text{m}^3/\text{s}$.

There is also an analogy between thermal conduction and current. In Section 20.7, we discussed the flow of energy by heat through a sample of material. The rate of energy flow is determined by the material as well as the temperature difference across the material, as described by Equation 20.14.

To define current more precisely, suppose that charges are moving perpendicular to a surface of area A, as shown in Figure 27.1. (This area could be the cross-sectional area of a wire, for example.) **The current is the rate at which charge flows through this surface.** If ΔQ is the amount of charge that passes through this area in a time interval Δt, the **average current** I_{av} is equal to the charge that passes through A per unit time:

$$I_{av} = \frac{\Delta Q}{\Delta t} \tag{27.1}$$

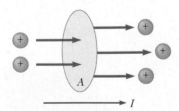

Figure 27.1 Charges in motion through an area A. The time rate at which charge flows through the area is defined as the current I. The direction of the current is the direction in which positive charges flow when free to do so.

If the rate at which charge flows varies in time, then the current varies in time; we define the **instantaneous current** I as the differential limit of average current:

$$I \equiv \frac{dQ}{dt}$$ (27.2)

Electric current

The SI unit of current is the **ampere** (A):

$$1 \text{ A} = \frac{1 \text{ C}}{1 \text{ s}}$$ (27.3)

That is, 1 A of current is equivalent to 1 C of charge passing through the surface area in 1 s.

The charges passing through the surface in Figure 27.1 can be positive or negative, or both. **It is conventional to assign to the current the same direction as the flow of positive charge.** In electrical conductors, such as copper or aluminum, the current is due to the motion of negatively charged electrons. Therefore, when we speak of current in an ordinary conductor, **the direction of the current is opposite the direction of flow of electrons.** However, if we are considering a beam of positively charged protons in an accelerator, the current is in the direction of motion of the protons. In some cases—such as those involving gases and electrolytes, for instance—the current is the result of the flow of both positive and negative charges.

If the ends of a conducting wire are connected to form a loop, all points on the loop are at the same electric potential, and hence the electric field is zero within and at the surface of the conductor. Because the electric field is zero, there is no net transport of charge through the wire, and therefore there is no current. However, if the ends of the conducting wire are connected to a battery, all points on the loop are not at the same potential. The battery sets up a potential difference between the ends of the loop, creating an electric field within the wire. The electric field exerts forces on the conduction electrons in the wire, causing them to move in the wire, thus creating a current.

It is common to refer to a moving charge (positive or negative) as a mobile **charge carrier.** For example, the mobile charge carriers in a metal are electrons.

Microscopic Model of Current

We can relate current to the motion of the charge carriers by describing a microscopic model of conduction in a metal. Consider the current in a conductor of cross-sectional area A (Fig. 27.2). The volume of a section of the conductor of length Δx (the gray region shown in Fig. 27.2) is $A \Delta x$. If n represents the number of mobile charge carriers per unit volume (in other words, the charge carrier density), the number of carriers in the gray section is $nA \Delta x$. Therefore, the total charge ΔQ in this section is

$$\Delta Q = \text{number of carriers in section} \times \text{charge per carrier} = (nA \Delta x)q$$

where q is the charge on each carrier. If the carriers move with a speed v_d, the displacement they experience in the x direction in a time interval Δt is $\Delta x = v_d \Delta t$. Let us choose Δt to be the time interval required for the charges in the cylinder to move through a displacement whose magnitude is equal to the length of the cylinder. This time interval is also that required for all of the charges in the cylinder to pass through the circular area at one end. With this choice, we can write ΔQ in the form

$$\Delta Q = (nAv_d \Delta t)q$$

If we divide both sides of this equation by Δt, we see that the average current in the conductor is

$$I_{av} = \frac{\Delta Q}{\Delta t} = nqv_d A$$ (27.4)

PITFALL PREVENTION

27.1 "Current Flow" Is Redundant

The phrase *current flow* is commonly used, although it is strictly incorrect, because current *is* a flow (of charge). This is similar to the phrase *heat transfer*, which is also redundant because heat *is* a transfer (of energy). We will avoid this phrase and speak of *flow of charge* or *charge flow.*

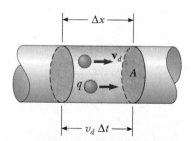

Figure 27.2 A section of a uniform conductor of cross-sectional area A. The mobile charge carriers move with a speed v_d, and the displacement they experience in the x direction in a time interval Δt is $\Delta x = v_d \Delta t$. If we choose Δt to be the time interval during which the charges are displaced, on the average, by the length of the cylinder, the number of carriers in the section of length Δx is $nAv_d \Delta t$, where n is the number of carriers per unit volume.

Current in a conductor in terms of microscopic quantities

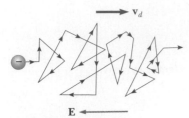

Figure 27.3 A schematic representation of the zigzag motion of an electron in a conductor. The changes in direction are the result of collisions between the electron and atoms in the conductor. Note that the net motion of the electron is opposite the direction of the electric field. Because of the acceleration of the charge carriers due to the electric force, the paths are actually parabolic. However, the drift speed is much smaller than the average speed, so the parabolic shape is not visible on this scale.

The speed of the charge carriers v_d is an average speed called the **drift speed.** To understand the meaning of drift speed, consider a conductor in which the charge carriers are free electrons. If the conductor is isolated—that is, the potential difference across it is zero—then these electrons undergo random motion that is analogous to the motion of gas molecules. As we discussed earlier, when a potential difference is applied across the conductor (for example, by means of a battery), an electric field is set up in the conductor; this field exerts an electric force on the electrons, producing a current. However, the electrons do not move in straight lines along the conductor. Instead, they collide repeatedly with the metal atoms, and their resultant motion is complicated and zigzag (Fig. 27.3). Despite the collisions, the electrons move slowly along the conductor (in a direction opposite that of **E**) at the drift velocity $\mathbf{v}_d$.

We can think of the atom–electron collisions in a conductor as an effective internal friction (or drag force) similar to that experienced by the molecules of a liquid flowing through a pipe stuffed with steel wool. The energy transferred from the electrons to the metal atoms during collisions causes an increase in the vibrational energy of the atoms and a corresponding increase in the temperature of the conductor.

Quick Quiz 27.1 Consider positive and negative charges moving horizontally through the four regions shown in Figure 27.4. Rank the current in these four regions, from lowest to highest.

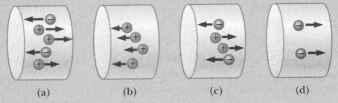

Figure 27.4 (Quick Quiz 27.1) Charges move through four regions.

Quick Quiz 27.2 Electric charge is conserved. As a consequence, when current arrives at a junction of wires, the charges can take either of two paths out of the junction and the numerical sum of the currents in the two paths equals the current that entered the junction. Thus, current is (a) a vector (b) a scalar (c) neither a vector nor a scalar.

Example 27.1 Drift Speed in a Copper Wire

The 12-gauge copper wire in a typical residential building has a cross-sectional area of 3.31×10^{-6} m². If it carries a current of 10.0 A, what is the drift speed of the electrons? Assume that each copper atom contributes one free electron to the current. The density of copper is 8.95 g/cm³.

Solution From the periodic table of the elements in Appendix C, we find that the molar mass of copper is 63.5 g/mol. Recall that 1 mol of any substance contains Avogadro's number of atoms (6.02×10^{23}). Knowing the density of copper, we can calculate the volume occupied by 63.5 g ($= 1$ mol) of copper:

$$V = \frac{m}{\rho} = \frac{63.5 \text{ g}}{8.95 \text{ g/cm}^3} = 7.09 \text{ cm}^3$$

Because each copper atom contributes one free electron to the current, we have

$$n = \frac{6.02 \times 10^{23} \text{ electrons}}{7.09 \text{ cm}^3} \left(\frac{1.00 \times 10^6 \text{ cm}^3}{1 \text{ m}^3} \right)$$

$$= 8.49 \times 10^{28} \text{ electrons/m}^3$$

From Equation 27.4, we find that the drift speed is

$$v_d = \frac{I}{nqA}$$

where q is the absolute value of the charge on each electron. Thus,

$$v_d = \frac{I}{nqA}$$

$$= \frac{10.0 \text{ C/s}}{(8.49 \times 10^{28} \text{ m}^{-3})(1.60 \times 10^{-19} \text{ C})(3.31 \times 10^{-6} \text{ m}^2)}$$

$$= 2.22 \times 10^{-4} \text{ m/s}$$

Example 27.1 shows that typical drift speeds are very low. For instance, electrons traveling with a speed of 2.22×10^{-4} m/s would take about 75 min to travel 1 m! In view of this, you might wonder why a light turns on almost instantaneously when a switch is thrown. In a conductor, changes in the electric field that drives the free electrons travel through the conductor with a speed close to that of light. Thus, when you flip on a light switch, electrons already in the filament of the lightbulb experience electric forces and begin moving after a time interval on the order of nanoseconds.

27.2 Resistance

In Chapter 24 we found that the electric field inside a conductor is zero. However, this statement is true *only* if the conductor is in static equilibrium. The purpose of this section is to describe what happens when the charges in the conductor are not in equilibrium, in which case there is an electric field in the conductor.

Consider a conductor of cross-sectional area A carrying a current I. The **current density** J in the conductor is defined as the current per unit area. Because the current $I = nqv_dA$, the current density is

$$J \equiv \frac{I}{A} = nqv_d \qquad (27.5)$$

where J has SI units of A/m^2. This expression is valid only if the current density is uniform and only if the surface of cross-sectional area A is perpendicular to the direction of the current. In general, current density is a vector quantity:

$$\mathbf{J} = nq\mathbf{v}_d \qquad (27.6)$$

From this equation, we see that current density is in the direction of charge motion for positive charge carriers and opposite the direction of motion for negative charge carriers.

A current density J and an electric field E are established in a conductor whenever a potential difference is maintained across the conductor. In some materials, the current density is proportional to the electric field:

$$\mathbf{J} = \sigma\mathbf{E} \qquad (27.7)$$

where the constant of proportionality σ is called the **conductivity** of the conductor.[1] Materials that obey Equation 27.7 are said to follow **Ohm's law,** named after Georg Simon Ohm (1789–1854). More specifically, Ohm's law states that

> for many materials (including most metals), the ratio of the current density to the electric field is a constant σ that is independent of the electric field producing the current.

Materials that obey Ohm's law and hence demonstrate this simple relationship between **E** and **J** are said to be *ohmic*. Experimentally, however, it is found that not all materials have this property. Materials and devices that do not obey Ohm's law are said to be *nonohmic*. Ohm's law is not a fundamental law of nature but rather an empirical relationship valid only for certain materials.

We can obtain an equation useful in practical applications by considering a segment of straight wire of uniform cross-sectional area A and length ℓ, as shown in

[1]   Do not confuse conductivity σ with surface charge density, for which the same symbol is used.

PITFALL PREVENTION

27.3 We've Seen Something Like Equation 27.8 Before

In Chapter 5, we introduced Newton's second law, $\Sigma F = ma$, for a net force on an object of mass m. This can be written as

$$m = \frac{\Sigma F}{a}$$

In that chapter, we defined mass as *resistance to a change in motion in response to an external force*. Mass as resistance to changes in motion is analogous to electrical resistance to charge flow, and Equation 27.8 is analogous to the form of Newton's second law shown here.

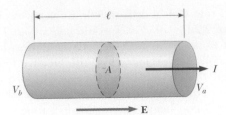

Figure 27.5 A uniform conductor of length ℓ and cross-sectional area A. A potential difference $\Delta V = V_b - V_a$ maintained across the conductor sets up an electric field **E**, and this field produces a current I that is proportional to the potential difference.

Figure 27.5. A potential difference $\Delta V = V_b - V_a$ is maintained across the wire, creating in the wire an electric field and a current. If the field is assumed to be uniform, the potential difference is related to the field through the relationship[2]

$$\Delta V = E\ell$$

Therefore, we can express the magnitude of the current density in the wire as

$$J = \sigma E = \sigma \frac{\Delta V}{\ell}$$

Because $J = I/A$, we can write the potential difference as

$$\Delta V = \frac{\ell}{\sigma} J = \left(\frac{\ell}{\sigma A}\right) I = RI$$

 PITFALL PREVENTION

27.4 Equation 27.8 Is Not Ohm's Law

Many individuals call Equation 27.8 Ohm's law, but this is incorrect. This equation is simply the definition of resistance, and provides an important relationship between voltage, current, and resistance. Ohm's law is related to a linear relationship between **J** and **E** (Eq. 27.7) or, equivalently, between I and ΔV, which, from Equation 27.8, indicates that the resistance is constant, independent of the applied voltage.

The quantity $R = \ell/\sigma A$ is called the **resistance** of the conductor. We can define the resistance as the ratio of the potential difference across a conductor to the current in the conductor:

$$R \equiv \frac{\Delta V}{I} \tag{27.8}$$

We will use this equation over and over again when studying electric circuits. From this result we see that resistance has SI units of volts per ampere. One volt per ampere is defined to be one **ohm** (Ω):

$$1 \, \Omega \equiv \frac{1 \, V}{1 \, A} \tag{27.9}$$

This expression shows that if a potential difference of 1 V across a conductor causes a current of 1 A, the resistance of the conductor is 1 Ω. For example, if an electrical appliance connected to a 120-V source of potential difference carries a current of 6 A, its resistance is 20 Ω.

The inverse of conductivity is **resistivity**[3] ρ:

Resistivity is the inverse of conductivity

$$\rho = \frac{1}{\sigma} \tag{27.10}$$

where ρ has the units ohm-meters ($\Omega \cdot m$). Because $R = \ell/\sigma A$, we can express the resistance of a uniform block of material along the length ℓ as

Resistance of a uniform material along the length ℓ

$$R = \rho \frac{\ell}{A} \tag{27.11}$$

---

[2] This result follows from the definition of potential difference:

$$V_b - V_a = -\int_a^b \mathbf{E} \cdot d\mathbf{s} = E \int_0^\ell dx = E\ell$$

[3] Do not confuse resistivity ρ with mass density or charge density, for which the same symbol is used.

Table 27.1

Resistivities and Temperature Coefficients of Resistivity for Various Materials		
Material	**Resistivity[a]($\Omega \cdot m$)**	**Temperature Coefficient[b] $\alpha[(°C)^{-1}]$**
Silver	1.59×10^{-8}	3.8×10^{-3}
Copper	1.7×10^{-8}	3.9×10^{-3}
Gold	2.44×10^{-8}	3.4×10^{-3}
Aluminum	2.82×10^{-8}	3.9×10^{-3}
Tungsten	5.6×10^{-8}	4.5×10^{-3}
Iron	10×10^{-8}	5.0×10^{-3}
Platinum	11×10^{-8}	3.92×10^{-3}
Lead	22×10^{-8}	3.9×10^{-3}
Nichrome[c]	1.50×10^{-6}	0.4×10^{-3}
Carbon	3.5×10^{-5}	-0.5×10^{-3}
Germanium	0.46	-48×10^{-3}
Silicon	640	-75×10^{-3}
Glass	10^{10} to 10^{14}	
Hard rubber	$\sim 10^{13}$	
Sulfur	10^{15}	
Quartz (fused)	75×10^{16}	

[a] All values at 20°C.

[b] See Section 27.4.

[c] A nickel–chromium alloy commonly used in heating elements.

Every ohmic material has a characteristic resistivity that depends on the properties of the material and on temperature. Additionally, as you can see from Equation 27.11, the resistance of a sample depends on geometry as well as on resistivity. Table 27.1 gives the resistivities of a variety of materials at 20°C. Note the enormous range, from very low values for good conductors such as copper and silver, to very high values for good insulators such as glass and rubber. An ideal conductor would have zero resistivity, and an ideal insulator would have infinite resistivity.

Equation 27.11 shows that the resistance of a given cylindrical conductor such as a wire is proportional to its length and inversely proportional to its cross-sectional area. If the length of a wire is doubled, then its resistance doubles. If its cross-sectional area is doubled, then its resistance decreases by one half. The situation is analogous to the flow of a liquid through a pipe. As the pipe's length is increased, the resistance to flow increases. As the pipe's cross-sectional area is increased, more liquid crosses a given cross section of the pipe per unit time interval. Thus, more liquid flows for the same pressure differential applied to the pipe, and the resistance to flow decreases.

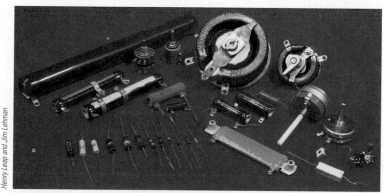

An assortment of resistors used in electrical circuits.

 PITFALL PREVENTION

27.5 Resistance and Resistivity

Resistivity is property of a *substance*, while resistance is a property of an *object*. We have seen similar pairs of variables before. For example, density is a property of a substance, while mass is a property of an object. Equation 27.11 relates resistance to resistivity, and we have seen a previous equation (Equation 1.1) which relates mass to density.

SuperStock

Figure 27.6 The colored bands on a resistor represent a code for determining resistance. The first two colors give the first two digits in the resistance value. The third color represents the power of ten for the multiplier of the resistance value. The last color is the tolerance of the resistance value. As an example, the four colors on the circled resistors are red ($=2$), black ($=0$), orange ($=10^3$), and gold ($=5\%$), and so the resistance value is $20 \times 10^3 \ \Omega = 20 \ k\Omega$ with a tolerance value of $5\% = 1 \ k\Omega$. (The values for the colors are from Table 27.2.)

Table 27.2

Color Coding for Resistors			
Color	**Number**	**Multiplier**	**Tolerance**
Black	0	1	
Brown	1	10^1	
Red	2	10^2	
Orange	3	10^3	
Yellow	4	10^4	
Green	5	10^5	
Blue	6	10^6	
Violet	7	10^7	
Gray	8	10^8	
White	9	10^9	
Gold		10^{-1}	5%
Silver		10^{-2}	10%
Colorless			20%

Most electric circuits use circuit elements called **resistors** to control the current level in the various parts of the circuit. Two common types of resistors are the *composition resistor*, which contains carbon, and the *wire-wound resistor*, which consists of a coil of wire. Values of resistors in ohms are normally indicated by color-coding, as shown in Figure 27.6 and Table 27.2.

Ohmic materials and devices have a linear current–potential difference relationship over a broad range of applied potential differences (Fig. 27.7a). The slope of the I-versus-ΔV curve in the linear region yields a value for $1/R$. Nonohmic materials have a nonlinear current–potential difference relationship. One common semiconducting device that has nonlinear I-versus-ΔV characteristics is the *junction diode* (Fig. 27.7b). The resistance of this device is low for currents in one direction (positive ΔV) and high for currents in the reverse direction (negative ΔV). In fact, most modern electronic devices, such as transistors, have nonlinear current–potential difference relationships; their proper operation depends on the particular way in which they violate Ohm's law.

Quick Quiz 27.3 Suppose that a current-carrying ohmic metal wire has a cross-sectional area that gradually becomes smaller from one end of the wire to the other. The current must have the same value in each section of the wire so that charge does not accumulate at any one point. How do the drift velocity and the resistance per

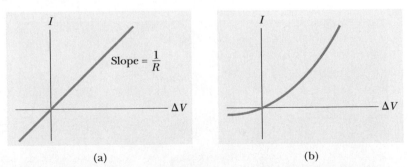

(a)   (b)

Figure 27.7 (a) The current–potential difference curve for an ohmic material. The curve is linear, and the slope is equal to the inverse of the resistance of the conductor. (b) A nonlinear current–potential difference curve for a junction diode. This device does not obey Ohm's law.

unit length vary along the wire as the area becomes smaller? (a) The drift velocity and resistance both increase. (b) The drift velocity and resistance both decrease. (c) The drift velocity increases and the resistance decreases. (d) The drift velocity decreases and the resistance increases.

Quick Quiz 27.4 A cylindrical wire has a radius r and length ℓ. If both r and ℓ are doubled, the resistance of the wire (a) increases (b) decreases (c) remains the same.

Quick Quiz 27.5 In Figure 27.7b, as the applied voltage increases, the resistance of the diode (a) increases (b) decreases (c) remains the same.

Example 27.2 The Resistance of a Conductor

Calculate the resistance of an aluminum cylinder that has a length of 10.0 cm and a cross-sectional area of $2.00 \times 10^{-4}\ \mathrm{m^2}$. Repeat the calculation for a cylinder of the same dimensions and made of glass having a resistivity of $3.0 \times 10^{10}\ \Omega \cdot \mathrm{m}$.

Solution From Equation 27.11 and Table 27.1, we can calculate the resistance of the aluminum cylinder as follows:

$$R = \rho \frac{\ell}{A} = (2.82 \times 10^{-8}\ \Omega \cdot \mathrm{m}) \left(\frac{0.100\ \mathrm{m}}{2.00 \times 10^{-4}\ \mathrm{m^2}} \right)$$

$$= \boxed{1.41 \times 10^{-5}\ \Omega}$$

Similarly, for glass we find that

$$R = \rho \frac{\ell}{A} = (3.0 \times 10^{10}\ \Omega \cdot \mathrm{m}) \left(\frac{0.100\ \mathrm{m}}{2.00 \times 10^{-4}\ \mathrm{m^2}} \right)$$

$$= \boxed{1.5 \times 10^{13}\ \Omega}$$

As you might guess from the large difference in resistivities, the resistances of identically shaped cylinders of aluminum and glass differ widely. The resistance of the glass cylinder is 18 orders of magnitude greater than that of the aluminum cylinder.

Example 27.3 The Resistance of Nichrome Wire
`Interactive`

(A) Calculate the resistance per unit length of a 22-gauge Nichrome wire, which has a radius of 0.321 mm.

Solution The cross-sectional area of this wire is

$$A = \pi r^2 = \pi (0.321 \times 10^{-3}\ \mathrm{m})^2 = 3.24 \times 10^{-7}\ \mathrm{m^2}$$

The resistivity of Nichrome is $1.5 \times 10^{-6}\ \Omega \cdot \mathrm{m}$ (see Table 27.1). Thus, we can use Equation 27.11 to find the resistance per unit length:

$$\frac{R}{\ell} = \frac{\rho}{A} = \frac{1.5 \times 10^{-6}\ \Omega \cdot \mathrm{m}}{3.24 \times 10^{-7}\ \mathrm{m^2}} = \boxed{4.6\ \Omega/\mathrm{m}}$$

(B) If a potential difference of 10 V is maintained across a 1.0-m length of the Nichrome wire, what is the current in the wire?

Solution Because a 1.0-m length of this wire has a resistance of 4.6 Ω, Equation 27.8 gives

$$I = \frac{\Delta V}{R} = \frac{10\ \mathrm{V}}{4.6\ \Omega} = \boxed{2.2\ \mathrm{A}}$$

Note from Table 27.1 that the resistivity of Nichrome wire is about 100 times that of copper. A copper wire of the same radius would have a resistance per unit length of only 0.052 Ω/m. A 1.0-m length of copper wire of the same radius would carry the same current (2.2 A) with an applied potential difference of only 0.11 V.

Because of its high resistivity and its resistance to oxidation, Nichrome is often used for heating elements in toasters, irons, and electric heaters.

 Explore the resistance of different materials at the Interactive Worked Example link at **http://www.pse6.com.**

Example 27.4 The Radial Resistance of a Coaxial Cable

Coaxial cables are used extensively for cable television and other electronic applications. A coaxial cable consists of two concentric cylindrical conductors. The region between the conductors is completely filled with silicon, as shown in Figure 27.8a, and current leakage through the silicon, in the *radial* direction, is unwanted. (The cable is designed to conduct current along its length—this is *not* the current we are considering here.) The radius of the inner conductor is $a = 0.500$ cm, the radius of the outer one is $b = 1.75$ cm, and the length is $L = 15.0$ cm.

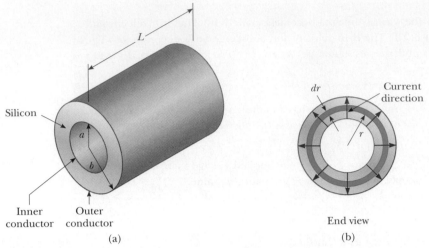

Figure 27.8 (Example 27.4) A coaxial cable. (a) Silicon fills the gap between the two conductors. (b) End view, showing current leakage.

Calculate the resistance of the silicon between the two conductors.

Solution Conceptualize by imagining two currents, as suggested in the text of the problem. The desired current is along the cable, carried within the conductors. The undesired current corresponds to charge leakage through the silicon and its direction is radial. Because we know the resistivity and the geometry of the silicon, we categorize this as a problem in which we find the resistance of the silicon from these parameters, using Equation 27.11. Because the area through which the charges pass depends on the radial position, we must use integral calculus to determine the answer.

To analyze the problem, we divide the silicon into concentric elements of infinitesimal thickness dr (Fig. 27.8b). We start by using the differential form of Equation 27.11, replacing ℓ with r for the distance variable: $dR = \rho\, dr/A$, where dR is the resistance of an element of silicon of thickness dr and surface area A. In this example, we take as our representative concentric element a hollow silicon cylinder of radius r, thickness dr, and length L, as in Figure 27.8. Any charge that passes from the inner conductor to the outer one must pass radially through this concentric element, and the area through which this charge passes is $A = 2\pi r L$. (This is the curved surface area—circumference multiplied by length—of our hollow silicon cylinder of thickness dr.) Hence, we can write the resistance of our hollow cylinder of silicon as

$$dR = \frac{\rho}{2\pi r L}\, dr$$

Because we wish to know the total resistance across the entire thickness of the silicon, we must integrate this expression from $r = a$ to $r = b$:

$$(1) \qquad R = \int_a^b dR = \frac{\rho}{2\pi L} \int_a^b \frac{dr}{r} = \frac{\rho}{2\pi L} \ln\left(\frac{b}{a}\right)$$

Substituting in the values given, and using $\rho = 640\ \Omega \cdot \text{m}$ for silicon, we obtain

$$R = \frac{640\ \Omega \cdot \text{m}}{2\pi(0.150\ \text{m})} \ln\left(\frac{1.75\ \text{cm}}{0.500\ \text{cm}}\right) = \boxed{851\ \Omega}$$

To finalize this problem, let us compare this resistance to that of the inner conductor of the cable along the 15.0-cm length. Assuming that the conductor is made of copper, we have

$$R = \rho\,\frac{\ell}{A} = (1.7 \times 10^{-8}\ \Omega \cdot \text{m})\left(\frac{0.150\ \text{m}}{\pi(5.00 \times 10^{-3}\ \text{m})^2}\right)$$
$$= 3.2 \times 10^{-5}\ \Omega$$

This resistance is much smaller than the radial resistance. As a consequence, almost all of the current corresponds to charge moving along the length of the cable, with a very small fraction leaking in the radial direction.

What If? Suppose the coaxial cable is enlarged to twice the overall diameter with two possibilities: (1) the ratio b/a is held fixed, or (2) the difference $b - a$ is held fixed. For which possibility does the leakage current between the inner and outer conductors increase when the voltage is applied between the two conductors?

Answer In order for the current to increase, the resistance must decrease. For possibility (1), in which b/a is held fixed, Equation (1) tells us that the resistance is unaffected. For possibility (2), we do not have an equation involving the difference $b - a$ to inspect. Looking at Figure 27.8b, however, we see that increasing b and a while holding the voltage constant results in charge flowing through the same thickness of silicon but through a larger overall area perpendicular to the flow. This larger area will result in lower resistance and a higher current.

27.3   A Model for Electrical Conduction

In this section we describe a classical model of electrical conduction in metals that was first proposed by Paul Drude (1863–1906) in 1900. This model leads to Ohm's law and shows that resistivity can be related to the motion of electrons in metals. Although the Drude model described here does have limitations, it nevertheless introduces concepts that are still applied in more elaborate treatments.

Consider a conductor as a regular array of atoms plus a collection of free electrons, which are sometimes called *conduction* electrons. The conduction electrons, although bound to their respective atoms when the atoms are not part of a solid, gain mobility when the free atoms condense into a solid. In the absence of an electric field, the conduction electrons move in random directions through the conductor with average speeds on the order of 10^6 m/s. The situation is similar to the motion of gas molecules confined in a vessel. In fact, some scientists refer to conduction electrons in a metal as an *electron gas*. There is no current in the conductor in the absence of an electric field because the drift velocity of the free electrons is zero. That is, on the average, just as many electrons move in one direction as in the opposite direction, and so there is no net flow of charge.

This situation changes when an electric field is applied. Now, in addition to undergoing the random motion just described, the free electrons drift slowly in a direction opposite that of the electric field, with an average drift speed v_d that is much smaller (typically 10^{-4} m/s) than their average speed between collisions (typically 10^6 m/s).

Figure 27.9 provides a crude description of the motion of free electrons in a conductor. In the absence of an electric field, there is no net displacement after many collisions (Fig. 27.9a). An electric field **E** modifies the random motion and causes the electrons to drift in a direction opposite that of **E** (Fig. 27.9b).

In our model, we assume that the motion of an electron after a collision is independent of its motion before the collision. We also assume that the excess energy acquired by the electrons in the electric field is lost to the atoms of the conductor when the electrons and atoms collide. The energy given up to the atoms increases their vibrational energy, and this causes the temperature of the conductor to increase. The temperature increase of a conductor due to resistance is utilized in electric toasters and other familiar appliances.

We are now in a position to derive an expression for the drift velocity. When a free electron of mass m_e and charge $q(= -e)$ is subjected to an electric field **E**, it experiences a force $\mathbf{F} = q\mathbf{E}$. Because this force is related to the acceleration of the electron through Newton's second law, $\mathbf{F} = m_e\mathbf{a}$, we conclude that the acceleration of the electron is

$$\mathbf{a} = \frac{q\mathbf{E}}{m_e} \tag{27.12}$$

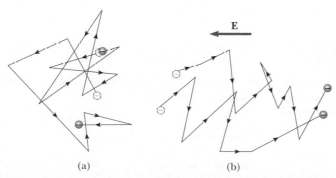

(a)   (b)

Active Figure 27.9   (a) A schematic diagram of the random motion of two charge carriers in a conductor in the absence of an electric field. The drift velocity is zero. (b) The motion of the charge carriers in a conductor in the presence of an electric field. Note that the random motion is modified by the field, and the charge carriers have a drift velocity.

At the Active Figures link at http://www.pse6.com, you can adjust the electric field to see the resulting effect on the motion of an electron.

This acceleration, which occurs for only a short time interval between collisions, enables the electron to acquire a small drift velocity. If $\mathbf{v}_i$ is the electron's initial velocity the instant after a collision (which occurs at a time that we define as $t = 0$), then the velocity of the electron at time t (at which the next collision occurs) is

$$\mathbf{v}_f = \mathbf{v}_i + \mathbf{a}t = \mathbf{v}_i + \frac{q\mathbf{E}}{m_e} t \qquad (27.13)$$

We now take the average value of $\mathbf{v}_f$ over all possible collision times t and all possible values of $\mathbf{v}_i$. If we assume that the initial velocities are randomly distributed over all possible values, we see that the average value of $\mathbf{v}_i$ is zero. The term $(q\mathbf{E}/m_e)t$ is the velocity change of the electron due to the electric field during one trip between atoms. The average value of the second term of Equation 27.13 is $(q\mathbf{E}/m_e)\tau$, where τ is the *average time interval between successive collisions*. Because the average value of $\mathbf{v}_f$ is equal to the drift velocity, we have

Drift velocity in terms of microscopic quantities

$$\overline{\mathbf{v}_f} = \mathbf{v}_d = \frac{q\mathbf{E}}{m_e} \tau \qquad (27.14)$$

We can relate this expression for drift velocity to the current in the conductor. Substituting Equation 27.14 into Equation 27.6, we find that the magnitude of the current density is

Current density in terms of microscopic quantities

$$J = nqv_d = \frac{nq^2E}{m_e} \tau \qquad (27.15)$$

where n is the number of charge carriers per unit volume. Comparing this expression with Ohm's law, $J = \sigma E$, we obtain the following relationships for conductivity and resistivity of a conductor:

Conductivity in terms of microscopic quantities

$$\sigma = \frac{nq^2\tau}{m_e} \qquad (27.16)$$

Resistivity in terms of microscopic quantities

$$\rho = \frac{1}{\sigma} = \frac{m_e}{nq^2\tau} \qquad (27.17)$$

According to this classical model, conductivity and resistivity do not depend on the strength of the electric field. This feature is characteristic of a conductor obeying Ohm's law.

The average time interval τ between collisions is related to the average distance between collisions ℓ (that is, the *mean free path*; see Section 21.7) and the average speed $\overline{v}$ through the expression

$$\tau = \frac{\ell}{\overline{v}} \qquad (27.18)$$

Example 27.5   Electron Collisions in a Wire

(A) Using the data and results from Example 27.1 and the classical model of electron conduction, estimate the average time interval between collisions for electrons in household copper wiring.

Solution From Equation 27.17, we see that

$$\tau = \frac{m_e}{nq^2\rho}$$

where $\rho = 1.7 \times 10^{-8}\ \Omega \cdot$ m for copper and the carrier density is $n = 8.49 \times 10^{28}$ electrons/m^3 for the wire described

in Example 27.1. Substitution of these values into the expression above gives

$$\tau = \frac{9.11 \times 10^{-31}\ \text{kg}}{(8.49 \times 10^{28}\ \text{m}^{-3})(1.6 \times 10^{-19}\ \text{C})^2\,(1.7 \times 10^{-8}\ \Omega \cdot \text{m})}$$

$$= \boxed{2.5 \times 10^{-14}\ \text{s}}$$

(B) Assuming that the average speed for free electrons in copper is 1.6×10^6 m/s and using the result from part (A), calculate the mean free path for electrons in copper.

Solution From Equation 27.18,

$$\ell = \bar{v}\tau = (1.6 \times 10^6 \text{ m/s})(2.5 \times 10^{-14} \text{ s})$$

$$= \boxed{4.0 \times 10^{-8} \text{ m}}$$

which is equivalent to 40 nm (compared with atomic spacings of about 0.2 nm). Thus, although the time interval between collisions is very short, an electron in the wire travels about 200 atomic spacings between collisions.

27.4 Resistance and Temperature

Over a limited temperature range, the resistivity of a conductor varies approximately linearly with temperature according to the expression

$$\rho = \rho_0[1 + \alpha(T - T_0)] \qquad (27.19)$$

Variation of ρ with temperature

where ρ is the resistivity at some temperature T (in degrees Celsius), ρ_0 is the resistivity at some reference temperature T_0 (usually taken to be 20°C), and α is the **temperature coefficient of resistivity.** From Equation 27.19, we see that the temperature coefficient of resistivity can be expressed as

$$\alpha - \frac{1}{\rho_0}\frac{\Delta\rho}{\Delta T} \qquad (27.20)$$

Temperature coefficient of resistivity

where $\Delta\rho = \rho - \rho_0$ is the change in resistivity in the temperature interval $\Delta T = T - T_0$.

The temperature coefficients of resistivity for various materials are given in Table 27.1. Note that the unit for α is degrees Celsius^{-1} [(°C)$^{-1}$]. Because resistance is proportional to resistivity (Eq. 27.11), we can write the variation of resistance as

$$R = R_0[1 + \alpha(T - T_0)] \qquad (27.21)$$

Use of this property enables us to make precise temperature measurements, as shown in Example 27.6.

Quick Quiz 27.6 When does a lightbulb carry more current: (a) just after it is turned on and the glow of the metal filament is increasing, or (b) after it has been on for a few milliseconds and the glow is steady?

Example 27.6 A Platinum Resistance Thermometer

A resistance thermometer, which measures temperature by measuring the change in resistance of a conductor, is made from platinum and has a resistance of 50.0 Ω at 20.0°C. When immersed in a vessel containing melting indium, its resistance increases to 76.8 Ω. Calculate the melting point of the indium.

Solution Solving Equation 27.21 for ΔT and using the α value for platinum given in Table 27.1, we obtain

$$\Delta T = \frac{R - R_0}{\alpha R_0} = \frac{76.8 \text{ }\Omega - 50.0 \text{ }\Omega}{[3.92 \times 10^{-3}(°C)^{-1}](50.0 \text{ }\Omega)}$$

$$= 137°C$$

Because $T_0 = 20.0°C$, we find that T, the temperature of the melting indium sample, is $\boxed{157°C}$.

For metals like copper, resistivity is nearly proportional to temperature, as shown in Figure 27.10. However, a nonlinear region always exists at very low temperatures, and the resistivity usually reaches some finite value as the temperature approaches absolute zero. This residual resistivity near absolute zero is caused primarily by the

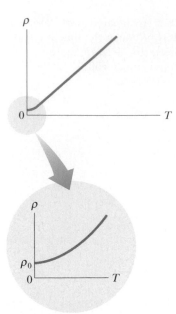

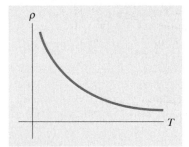

Figure 27.10 Resistivity versus temperature for a metal such as copper. The curve is linear over a wide range of temperatures, and ρ increases with increasing temperature. As T approaches absolute zero (inset), the resistivity approaches a finite value ρ_0.

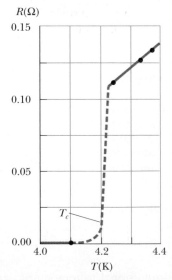

Figure 27.11 Resistivity versus temperature for a pure semiconductor, such as silicon or germanium.

collision of electrons with impurities and imperfections in the metal. In contrast, high-temperature resistivity (the linear region) is predominantly characterized by collisions between electrons and metal atoms.

Notice that three of the α values in Table 27.1 are negative; this indicates that the resistivity of these materials decreases with increasing temperature (Fig. 27.11), which is indicative of a class of materials called *semiconductors*. This behavior is due to an increase in the density of charge carriers at higher temperatures.

Because the charge carriers in a semiconductor are often associated with impurity atoms, the resistivity of these materials is very sensitive to the type and concentration of such impurities. We shall return to the study of semiconductors in Chapter 43.

27.5 Superconductors

There is a class of metals and compounds whose resistance decreases to zero when they are below a certain temperature T_c, known as the **critical temperature.** These materials are known as **superconductors.** The resistance–temperature graph for a superconductor follows that of a normal metal at temperatures above T_c (Fig. 27.12). When the temperature is at or below T_c, the resistivity drops suddenly to zero. This phenomenon was discovered in 1911 by the Dutch physicist Heike Kamerlingh-Onnes (1853–1926) as he worked with mercury, which is a superconductor below 4.2 K. Recent measurements have shown that the resistivities of superconductors below their T_c values are less than 4×10^{-25} $\Omega \cdot$m—around 10^{17} times smaller than the resistivity of copper and in practice considered to be zero.

Today thousands of superconductors are known, and as Table 27.3 illustrates, the critical temperatures of recently discovered superconductors are substantially higher than initially thought possible. Two kinds of superconductors are recognized. The more recently identified ones are essentially ceramics with high critical temperatures, whereas superconducting materials such as those observed by Kamerlingh-Onnes are metals. If a room-temperature superconductor is ever identified, its impact on technology could be tremendous.

The value of T_c is sensitive to chemical composition, pressure, and molecular structure. It is interesting to note that copper, silver, and gold, which are excellent conductors, do not exhibit superconductivity.

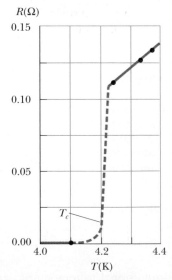

Figure 27.12 Resistance versus temperature for a sample of mercury (Hg). The graph follows that of a normal metal above the critical temperature T_c. The resistance drops to zero at T_c, which is 4.2 K for mercury.

Table 27.3

Critical Temperatures for Various Superconductors	
Material	**T_c(K)**
$HgBa_2Ca_2Cu_3O_8$	134
Tl–Ba–Ca–Cu–O	125
Bi–Sr–Ca–Cu–O	105
$YBa_2Cu_3O_7$	92
Nb_3Ge	23.2
Nb_3Sn	18.05
Nb	9.46
Pb	7.18
Hg	4.15
Sn	3.72
Al	1.19
Zn	0.88

One of the truly remarkable features of superconductors is that once a current is set up in them, it persists *without any applied potential difference* (because $R = 0$). Steady currents have been observed to persist in superconducting loops for several years with no apparent decay!

An important and useful application of superconductivity is in the development of superconducting magnets, in which the magnitudes of the magnetic field are about ten times greater than those produced by the best normal electromagnets. Such superconducting magnets are being considered as a means of storing energy. Superconducting magnets are currently used in medical magnetic resonance imaging (MRI) units, which produce high-quality images of internal organs without the need for excessive exposure of patients to x-rays or other harmful radiation.

For further information on superconductivity, see Section 43.8.

27.6 Electrical Power

If a battery is used to establish an electric current in a conductor, there is a continuous transformation of chemical energy in the battery to kinetic energy of the electrons to internal energy in the conductor, resulting in an increase in the temperature of the conductor.

In typical electric circuits, energy is transferred from a source such as a battery, to some device, such as a lightbulb or a radio receiver. Let us determine an expression that will allow us to calculate the rate of this energy transfer. First, consider the simple circuit in Figure 27.13, where we imagine energy is being delivered to a resistor. (Resistors are designated by the circuit symbol ———⋎⋎⋎———.) Because the connecting wires also have resistance, some energy is delivered to the wires and some energy to the resistor. Unless noted otherwise, we shall assume that the resistance of the wires is so small compared to the resistance of the circuit element that we ignore the energy delivered to the wires.

Imagine following a positive quantity of charge Q that is moving clockwise around the circuit in Figure 27.13 from point *a* through the battery and resistor back to point *a*. We identify the entire circuit as our system. As the charge moves from *a* to *b* through the battery, the electric potential energy of the system *increases* by an amount $Q\,\Delta V$ while the chemical potential energy in the battery *decreases* by the same amount. (Recall from Eq. 25.9 that $\Delta U = q\,\Delta V$.) However, as the charge moves from *c* to *d* through the resistor, the system *loses* this electric potential energy during collisions of electrons with atoms in the resistor. In this process, the energy is transformed to internal energy corresponding to increased vibrational motion of the atoms in the resistor. Because we have neglected the resistance of the interconnecting wires, no energy transformation occurs for paths *bc* and *da*. When the charge returns to point *a*, the net result is that some of the chemical energy in the battery has been delivered to the resistor and resides in the resistor as internal energy associated with molecular vibration.

The resistor is normally in contact with air, so its increased temperature will result in a transfer of energy by heat into the air. In addition, the resistor emits thermal

A small permanent magnet levitated above a disk of the superconductor $YBa_2Cu_3O_7$, which is at 77 K.

▲ **PITFALL PREVENTION**

27.6 Misconceptions About Current

There are several common misconceptions associated with current in a circuit like that in Figure 27.13. One is that current comes out of one terminal of the battery and is then "used up" as it passes through the resistor, leaving current in only one part of the circuit. The truth is that the current is the same *everywhere* in the circuit. A related misconception has the current coming out of the resistor being smaller than that going in, because some of the current is "used up." Another misconception has current coming out of both terminals of the battery, in opposite directions, and then "clashing" in the resistor, delivering the energy in this manner. This is not the case—the charges flow in the same rotational sense at *all* points in the circuit.

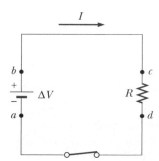

Active Figure 27.13 A circuit consisting of a resistor of resistance R and a battery having a potential difference ΔV across its terminals. Positive charge flows in the clockwise direction.

At the Active Figures link at http://www.pse6.com, you can adjust the battery voltage and the resistance to see the resulting current in the circuit and power delivered to the resistor.

▲ **PITFALL PREVENTION**

27.7 Charges Do Not Move All the Way Around a Circuit in a Short Time

Due to the very small magnitude of the drift velocity, it might take *hours* for a single electron to make one complete trip around the circuit. In terms of understanding the energy transfer in a circuit, however, it is useful to *imagine* a charge moving all the way around the circuit.

radiation, representing another means of escape for the energy. After some time interval has passed, the resistor reaches a constant temperature, at which time the input of energy from the battery is balanced by the output of energy by heat and radiation. Some electrical devices include *heat sinks*[4] connected to parts of the circuit to prevent these parts from reaching dangerously high temperatures. These are pieces of metal with many fins. The high thermal conductivity of the metal provides a rapid transfer of energy by heat away from the hot component, while the large number of fins provides a large surface area in contact with the air, so that energy can transfer by radiation and into the air by heat at a high rate.

Let us consider now the rate at which the system loses electric potential energy as the charge Q passes through the resistor:

$$\frac{dU}{dt} = \frac{d}{dt}(Q\Delta V) = \frac{dQ}{dt}\Delta V = I\Delta V$$

where I is the current in the circuit. The system regains this potential energy when the charge passes through the battery, at the expense of chemical energy in the battery. The rate at which the system loses potential energy as the charge passes through the resistor is equal to the rate at which the system gains internal energy in the resistor. Thus, the power $\mathcal{P}$, representing the rate at which energy is delivered to the resistor, is

Power delivered to a device

$$\mathcal{P} = I\Delta V \qquad (27.22)$$

We derived this result by considering a battery delivering energy to a resistor. However, Equation 27.22 can be used to calculate the power delivered by a voltage source to *any* device carrying a current I and having a potential difference ΔV between its terminals.

Using Equation 27.22 and the fact that $\Delta V = IR$ for a resistor, we can express the power delivered to the resistor in the alternative forms

Power delivered to a resistor

$$\mathcal{P} = I^2 R = \frac{(\Delta V)^2}{R} \qquad (27.23)$$

When I is expressed in amperes, ΔV in volts, and R in ohms, the SI unit of power is the watt, as it was in Chapter 7 in our discussion of mechanical power. The process by which power is lost as internal energy in a conductor of resistance R is often called *joule heating*[5]; this transformation is also often referred to as an I^2R loss.

▲ **PITFALL PREVENTION**

27.8 Energy Is Not "Dissipated"

In some books, you may see Equation 27.23 described as the power "dissipated in" a resistor, suggesting that energy disappears. Instead we say energy is "delivered to" a resistor. The notion of *dissipation* arises because a warm resistor will expel energy by radiation and heat, so that energy delivered by the battery leaves the circuit. (It does not disappear!)

When transporting energy by electricity through power lines, such as those shown in the opening photograph for this chapter, we cannot make the simplifying assumption that the lines have zero resistance. Real power lines do indeed have resistance, and power is delivered to the resistance of these wires. Utility companies seek to minimize the power transformed to internal energy in the lines and maximize the energy delivered to the consumer. Because $\mathcal{P} = I\Delta V$, the same amount of power can be transported either at high currents and low potential differences or at low currents and high potential differences. Utility companies choose to transport energy at low currents and high potential differences primarily for economic reasons. Copper wire is very expensive, and so it is cheaper to use high-resistance wire (that is, wire having a small cross-sectional area; see Eq. 27.11). Thus, in the expression for the power delivered to a resistor, $\mathcal{P} = I^2 R$, the resistance of the wire is fixed at a relatively high value for economic considerations. The I^2R loss can be reduced by keeping the current I as low as possible, which means transferring the energy at a high voltage. In some instances, power is transported at potential differences as great as 765 kV. Once the electricity reaches your city, the potential difference is usually reduced to 4 kV by a device called a *transformer*. Another

---

[4]   This is another misuse of the word *heat* that is ingrained in our common language.

[5]   It is commonly called *joule heating* even though the process of heat does not occur. This is another example of incorrect usage of the word *heat* that has become entrenched in our language.

transformer drops the potential difference to 240 V before the electricity finally reaches your home. Of course, each time the potential difference decreases, the current increases by the same factor, and the power remains the same. We shall discuss transformers in greater detail in Chapter 33.

Demands on our dwindling energy supplies have made it necessary for us to be aware of the energy requirements of our electrical devices. Every electrical appliance carries a label that contains the information you need to calculate the appliance's power requirements. In many cases, the power consumption in watts is stated directly, as it is on a lightbulb. In other cases, the amount of current used by the device and the potential difference at which it operates are given. This information and Equation 27.22 are sufficient for calculating the power requirement of any electrical device.

Quick Quiz 27.7 The same potential difference is applied to the two lightbulbs shown in Figure 27.14. Which one of the following statements is true? (a) The 30-W bulb carries the greater current and has the higher resistance. (b) The 30-W bulb carries the greater current, but the 60-W bulb has the higher resistance. (c) The 30-W bulb has the higher resistance, but the 60-W bulb carries the greater current. (d) The 60-W bulb carries the greater current and has the higher resistance.

Figure 27.14 (Quick Quiz 27.7) These lightbulbs operate at their rated power only when they are connected to a 120-V source.

Quick Quiz 27.8 For the two lightbulbs shown in Figure 27.15, rank the current values at points *a* through *f*, from greatest to least.

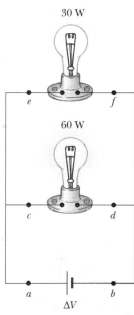

Figure 27.15 (Quick Quiz 27.8) Two lightbulbs connected across the same potential difference.

Example 27.7 Power in an Electric Heater

An electric heater is constructed by applying a potential difference of 120 V to a Nichrome wire that has a total resistance of 8.00 Ω. Find the current carried by the wire and the power rating of the heater.

Solution Because $\Delta V = IR$, we have

$$I = \frac{\Delta V}{R} = \frac{120 \text{ V}}{8.00 \text{ }\Omega} = \boxed{15.0 \text{ A}}$$

We can find the power rating using the expression $\mathcal{P} = I^2 R$:

$$\mathcal{P} = I^2 R = (15.0 \text{ A})^2 (8.00 \text{ }\Omega) = 1.80 \times 10^3 \text{ W}$$

$$\mathcal{P} = \boxed{1.80 \text{ kW}}$$

What If? What if the heater were accidentally connected to a 240-V supply? (This is difficult to do because the shape and orientation of the metal contacts in 240-V plugs are different from those in 120-V plugs.) How would this affect the current carried by the heater and the power rating of the heater?

Answer If we doubled the applied potential difference, Equation 27.8 tells us that the current would double. According to Equation 27.23, $\mathcal{P} = (\Delta V)^2 / R$, the power would be four times larger.

Example 27.8 Linking Electricity and Thermodynamics

(A) What is the required resistance of an immersion heater that will increase the temperature of 1.50 kg of water from 10.0°C to 50.0°C in 10.0 min while operating at 110 V?

(B) Estimate the cost of heating the water.

Solution This example allows us to link our new understanding of power in electricity with our experience with specific heat in thermodynamics (Chapter 20). An immersion heater is a resistor that is inserted into a container of water. As energy is delivered to the immersion heater, raising its temperature, energy leaves the surface of the resistor by heat, going into the water. When the immersion heater reaches a constant temperature, the rate of energy delivered to the resistance by electrical transmission is equal to the rate of energy delivered by heat to the water.

(A) To simplify the analysis, we ignore the initial period during which the temperature of the resistor increases, and also ignore any variation of resistance with temperature. Thus, we imagine a constant rate of energy transfer for the entire 10.0 min. Setting the rate of energy delivered to the resistor equal to the rate of energy entering the water by heat, we have

$$\mathcal{P} = \frac{(\Delta V)^2}{R} = \frac{Q}{\Delta t}$$

where Q represents an amount of energy transfer by heat into the water and we have used Equation 27.23 to express

the electrical power. The amount of energy transfer by heat necessary to raise the temperature of the water is given by Equation 20.4, $Q = mc\,\Delta T$. Thus,

$$\frac{(\Delta V)^2}{R} = \frac{mc\,\Delta T}{\Delta t} \quad \longrightarrow \quad R = \frac{(\Delta V)^2\,\Delta t}{mc\,\Delta T}$$

Substituting the values given in the statement of the problem, we have

$$R = \frac{(110 \text{ V})^2(600 \text{ s})}{(1.50 \text{ kg})(4186 \text{ J/kg}\cdot°\text{C})(50.0°\text{C} - 10.0°\text{C})}$$

$$= \boxed{28.9 \ \Omega}$$

(B) Because the energy transferred equals power multiplied by time interval, the amount of energy transferred is

$$\mathcal{P}\,\Delta t = \frac{(\Delta V)^2}{R}\,\Delta t = \frac{(110 \text{ V})^2}{28.9 \ \Omega}(10.0 \text{ min})\left(\frac{1 \text{ h}}{60.0 \text{ min}}\right)$$

$$= 69.8 \text{ Wh} = 0.069\ 8 \text{ kWh}$$

If the energy is purchased at an estimated price of 10.0¢ per kilowatt-hour, the cost is

$$\text{Cost} = (0.069\ 8 \text{ kWh})(\$0.100/\text{kWh}) = \$0.006\ 98$$

$$\approx \boxed{0.7 \ ¢}$$

 At the Interactive Worked Example link at http://www.pse6.com, *you can explore the heating of the water.*

Example 27.9 Current in an Electron Beam

In a certain particle accelerator, electrons emerge with an energy of 40.0 MeV (1 MeV = 1.60×10^{-13} J). The electrons emerge not in a steady stream but rather in pulses at the rate of 250 pulses/s. This corresponds to a time interval between pulses of 4.00 ms (Fig. 27.16). Each pulse has a duration of 200 ns, and the electrons in the pulse constitute a current of 250 mA. The current is zero between pulses.

(A) How many electrons are delivered by the accelerator per pulse?

Solution We use Equation 27.2 in the form $dQ = I\,dt$ and integrate to find the charge per pulse. While the pulse is on, the current is constant; thus,

$$Q_{\text{pulse}} = I\int dt = I\,\Delta t = (250 \times 10^{-3} \text{ A})(200 \times 10^{-9} \text{ s})$$

$$= 5.00 \times 10^{-8} \text{ C}$$

Dividing this quantity of charge per pulse by the electronic charge gives the number of electrons per pulse:

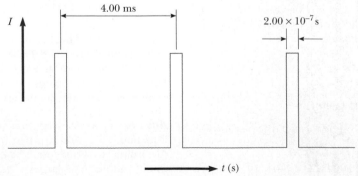

Figure 27.16 (Example 27.9) Current versus time for a pulsed beam of electrons.

$$\text{Electrons per pulse} = \frac{5.00 \times 10^{-8}\,\text{C/pulse}}{1.60 \times 10^{-19}\,\text{C/electron}}$$

$$= \boxed{3.13 \times 10^{11}\,\text{electrons/pulse}}$$

(B) What is the average current per pulse delivered by the accelerator?

Solution Average current is given by Equation 27.1, $I_{av} = \Delta Q / \Delta t$. Because the time interval between pulses is 4.00 ms, and because we know the charge per pulse from part (A), we obtain

$$I_{av} = \frac{Q_{pulse}}{\Delta t} = \frac{5.00 \times 10^{-8}\,\text{C}}{4.00 \times 10^{-3}\,\text{s}} = \boxed{12.5\,\mu\text{A}}$$

This represents only 0.005% of the peak current, which is 250 mA.

(C) What is the peak power delivered by the electron beam?

Solution By definition, power is energy delivered per unit time interval. Thus, the peak power is equal to the energy delivered by a pulse divided by the pulse duration:

$$(1)\qquad \mathcal{P}_{peak} = \frac{\text{pulse energy}}{\text{pulse duration}}$$

$$= \frac{(3.13 \times 10^{11}\,\text{electrons/pulse})(40.0\,\text{MeV/electron})}{2.00 \times 10^{-7}\,\text{s/pulse}}$$

$$\times \left(\frac{1.60 \times 10^{-13}\,\text{J}}{1\,\text{MeV}}\right)$$

$$= 1.00 \times 10^{7}\,\text{W} - \boxed{10.0\,\text{MW}}$$

We could also compute this power directly. We assume that each electron has zero energy before being accelerated. Thus, by definition, each electron must go through a potential difference of 40.0 MV to acquire a final energy of 40.0 MeV. Hence, we have

$$(2)\qquad \mathcal{P}_{peak} = I_{peak}\,\Delta V$$

$$= (250 \times 10^{-3}\,\text{A})(40.0 \times 10^{6}\,\text{V})$$

$$= \boxed{10.0\,\text{MW}}$$

What If? What if the requested quantity in part (C) were the *average* power rather than the *peak* power?

Answer Instead of Equation (1), we would use the time interval between pulses rather than the duration of a pulse:

$$\mathcal{P}_{av} = \frac{\text{pulse energy}}{\text{time interval between pulses}}$$

$$= \frac{(3.13 \times 10^{11}\,\text{electrons/pulse})(40.0\,\text{MeV/electron})}{4.00 \times 10^{-3}\,\text{s/pulse}}$$

$$\times \left(\frac{1.60 \times 10^{-13}\,\text{J}}{1\,\text{MeV}}\right)$$

$$= 500\,\text{W}$$

Instead of Equation (2), we would use the average current found in part (B):

$$\mathcal{P}_{av} = I_{av}\,\Delta V = (12.5 \times 10^{-6}\,\text{A})(40.0 \times 10^{6}\,\text{V})$$

$$= 500\,\text{W}$$

Notice that these two calculations agree with each other and that the average power is much lower than the peak power.

SUMMARY

The **electric current** I in a conductor is defined as

$$I \equiv \frac{dQ}{dt} \qquad (27.2)$$

where dQ is the charge that passes through a cross section of the conductor in a time interval dt. The SI unit of current is the **ampere** (A), where 1 A = 1 C/s.

The average current in a conductor is related to the motion of the charge carriers through the relationship

$$I_{av} = nqv_d A \qquad (27.4)$$

where n is the density of charge carriers, q is the charge on each carrier, v_d is the drift speed, and A is the cross-sectional area of the conductor.

The magnitude of the **current density** J in a conductor is the current per unit area:

$$J \equiv \frac{I}{A} = nqv_d \qquad (27.5)$$

Take a practice test for this chapter by clicking on the Practice Test link at http://www.pse6.com.

The current density in an ohmic conductor is proportional to the electric field according to the expression

$$\mathbf{J} = \sigma\mathbf{E} \tag{27.7}$$

The proportionality constant σ is called the **conductivity** of the material of which the conductor is made. The inverse of σ is known as **resistivity** ρ (that is, $\rho = 1/\sigma$). Equation 27.7 is known as **Ohm's law,** and a material is said to obey this law if the ratio of its current density $\mathbf{J}$ to its applied electric field $\mathbf{E}$ is a constant that is independent of the applied field.

The **resistance** R of a conductor is defined as

$$R \equiv \frac{\Delta V}{I} \tag{27.8}$$

where ΔV is the potential difference across it, and I is the current it carries.

The SI unit of resistance is volts per ampere, which is defined to be 1 **ohm** (Ω); that is, $1\ \Omega = 1\ \text{V/A}$. If the resistance is independent of the applied potential difference, the conductor obeys Ohm's law.

For a uniform block of material of cross sectional area A and length ℓ, the resistance over the length ℓ is

$$R = \rho\frac{\ell}{A} \tag{27.11}$$

where ρ is the resistivity of the material.

In a classical model of electrical conduction in metals, the electrons are treated as molecules of a gas. In the absence of an electric field, the average velocity of the electrons is zero. When an electric field is applied, the electrons move (on the average) with a **drift velocity** $\mathbf{v}_d$ that is opposite the electric field and given by the expression

$$\mathbf{v}_d = \frac{q\mathbf{E}}{m_e}\tau \tag{27.14}$$

where τ is the average time interval between electron–atom collisions, m_e is the mass of the electron, and q is its charge. According to this model, the resistivity of the metal is

$$\rho = \frac{m_e}{nq^2\tau} \tag{27.17}$$

where n is the number of free electrons per unit volume.

The resistivity of a conductor varies approximately linearly with temperature according to the expression

$$\rho = \rho_0[1 + \alpha(T - T_0)] \tag{27.19}$$

where α is the **temperature coefficient of resistivity** and ρ_0 is the resistivity at some reference temperature T_0.

If a potential difference ΔV is maintained across a circuit element, the **power,** or rate at which energy is supplied to the element, is

$$\mathcal{P} = I\,\Delta V \tag{27.22}$$

Because the potential difference across a resistor is given by $\Delta V = IR$, we can express the power delivered to a resistor in the form

$$\mathcal{P} = I^2R = \frac{(\Delta V)^2}{R} \tag{27.23}$$

The energy delivered to a resistor by electrical transmission appears in the form of internal energy in the resistor.

QUESTIONS

1. In an analogy between electric current and automobile traffic flow, what would correspond to charge? What would correspond to current?

2. Newspaper articles often contain a statement such as "10 000 volts of electricity surged through the victim's body." What is wrong with this statement?

3. What factors affect the resistance of a conductor?

4. What is the difference between resistance and resistivity?

5. Two wires A and B of circular cross section are made of the same metal and have equal lengths, but the resistance of wire A is three times greater than that of wire B. What is the ratio of their cross-sectional areas? How do their radii compare?

6. Do all conductors obey Ohm's law? Give examples to justify your answer.

7. We have seen that an electric field must exist inside a conductor that carries a current. How is it possible in view of the fact that in electrostatics we concluded that the electric field must be zero inside a conductor?

8. A very large potential difference is not necessarily required to produce long sparks in air. With a device called *Jacob's ladder*, a potential difference of about 10 kV produces an electric arc a few millimeters long between the bottom ends of two curved rods that project upward from the power supply. (The device is seen in classic mad-scientist horror movies and in Figure Q27.8.) The arc rises, climbing the rods and getting longer and longer. It disappears when it reaches the top; then a new spark immediately forms at the bottom and the process repeats. Explain these phenomena. Why does the arc rise? Why does a new arc appear only after the previous one is gone?

9. When the voltage across a certain conductor is doubled, the current is observed to increase by a factor of three. What can you conclude about the conductor?

10. In the water analogy of an electric circuit, what corresponds to the power supply, resistor, charge, and potential difference?

11. Use the atomic theory of matter to explain why the resistance of a material should increase as its temperature increases.

12. Why might a "good" electrical conductor also be a "good" thermal conductor?

13. How does the resistance for copper and for silicon change with temperature? Why are the behaviors of these two materials different?

14. Explain how a current can persist in a superconductor without any applied voltage.

15. What single experimental requirement makes superconducting devices expensive to operate? In principle, can this limitation be overcome?

16. What would happen to the drift velocity of the electrons in a wire and to the current in the wire if the electrons could move freely without resistance through the wire?

17. If charges flow very slowly through a metal, why does it not require several hours for a light to come on when you throw a switch?

18. In a conductor, changes in the electric field that drives the electrons through the conductor propagate with a speed close to the speed of light, although the drift velocity of the electrons is very small. Explain how these statements can both be true. Does one particular electron move from one end of the conductor to the other?

19. Two conductors of the same length and radius are connected across the same potential difference. One conductor has twice the resistance of the other. To which conductor is more power delivered?

20. Two lightbulbs both operate from 120 V. One has a power of 25 W and the other 100 W. Which bulb has higher resistance? Which bulb carries more current?

21. Car batteries are often rated in ampere-hours. Does this designate the amount of current, power, energy, or charge that can be drawn from the battery?

22. If you were to design an electric heater using Nichrome wire as the heating element, what parameters of the wire could you vary to meet a specific power output, such as 1 000 W?

Figure Q27.8

PROBLEMS

1, 2, 3 = straightforward, intermediate, challenging   ☐ = full solution available in the *Student Solutions Manual and Study Guide*

 = coached solution with hints available at http://www.pse6.com   💻 = computer useful in solving problem

 = paired numerical and symbolic problems

Section 27.1   Electric Current

1. In a particular cathode ray tube, the measured beam current is 30.0 μA. How many electrons strike the tube screen every 40.0 s?

2. A teapot with a surface area of 700 cm² is to be silver plated. It is attached to the negative electrode of an electrolytic cell containing silver nitrate ($Ag^+ NO_3^-$). If the cell is powered by a 12.0-V battery and has a resistance of 1.80 Ω, how long does it take for a 0.133-mm layer of silver to build up on the teapot? (The density of silver is 10.5×10^3 kg/m³.)

3. ☐ Suppose that the current through a conductor decreases exponentially with time according to the equation $I(t) = I_0 e^{-t/\tau}$ where I_0 is the initial current (at $t = 0$), and τ is a constant having dimensions of time. Consider a fixed observation point within the conductor. (a) How much charge passes this point between $t = 0$ and $t = \tau$? (b) How much charge passes this point between $t = 0$ and $t = 10\tau$? (c) **What If?** How much charge passes this point between $t = 0$ and $t = \infty$?

4. In the Bohr model of the hydrogen atom, an electron in the lowest energy state follows a circular path 5.29×10^{-11} m from the proton. (a) Show that the speed of the electron is 2.19×10^6 m/s. (b) What is the effective current associated with this orbiting electron?

5. A small sphere that carries a charge q is whirled in a circle at the end of an insulating string. The angular frequency of rotation is ω. What average current does this rotating charge represent?

6. The quantity of charge q (in coulombs) that has passed through a surface of area 2.00 cm² varies with time according to the equation $q = 4t^3 + 5t + 6$, where t is in seconds. (a) What is the instantaneous current through the surface at $t = 1.00$ s? (b) What is the value of the current density?

7. An electric current is given by the expression $I(t) = 100 \sin(120\pi t)$, where I is in amperes and t is in seconds. What is the total charge carried by the current from $t = 0$ to $t = (1/240)$ s?

8. Figure P27.8 represents a section of a circular conductor of nonuniform diameter carrying a current of 5.00 A. The

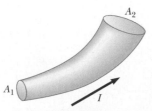

Figure P27.8

radius of cross section A_1 is 0.400 cm. (a) What is the magnitude of the current density across A_1? (b) If the current density across A_2 is one-fourth the value across A_1, what is the radius of the conductor at A_2?

9. The electron beam emerging from a certain high-energy electron accelerator has a circular cross section of radius 1.00 mm. (a) The beam current is 8.00 μA. Find the current density in the beam, assuming that it is uniform throughout. (b) The speed of the electrons is so close to the speed of light that their speed can be taken as $c = 3.00 \times 10^8$ m/s with negligible error. Find the electron density in the beam. (c) How long does it take for Avogadro's number of electrons to emerge from the accelerator?

10. A Van de Graaff generator produces a beam of 2.00-MeV deuterons, which are heavy hydrogen nuclei containing a proton and a neutron. (a) If the beam current is 10.0 μA, how far apart are the deuterons? (b) Is the electric force of repulsion among them a significant factor in beam stability? Explain.

11. An aluminum wire having a cross-sectional area of 4.00×10^{-6} m² carries a current of 5.00 A. Find the drift speed of the electrons in the wire. The density of aluminum is 2.70 g/cm³. Assume that one conduction electron is supplied by each atom.

Section 27.2   Resistance

12. Calculate the current density in a gold wire at 20°C, if an electric field of 0.740 V/m exists in the wire.

13. A lightbulb has a resistance of 240 Ω when operating with a potential difference of 120 V across it. What is the current in the lightbulb?

14. A resistor is constructed of a carbon rod that has a uniform cross-sectional area of 5.00 mm². When a potential difference of 15.0 V is applied across the ends of the rod, the rod carries a current of 4.00×10^{-3} A. Find (a) the resistance of the rod and (b) the rod's length.

15. ☐ A 0.900-V potential difference is maintained across a 1.50-m length of tungsten wire that has a cross-sectional area of 0.600 mm². What is the current in the wire?

16. A conductor of uniform radius 1.20 cm carries a current of 3.00 A produced by an electric field of 120 V/m. What is the resistivity of the material?

17. Suppose that you wish to fabricate a uniform wire out of 1.00 g of copper. If the wire is to have a resistance of $R = 0.500$ Ω, and if all of the copper is to be used, what will be (a) the length and (b) the diameter of this wire?

18. Gold is the most ductile of all metals. For example, one gram of gold can be drawn into a wire 2.40 km long. What is the resistance of such a wire at 20°C? You can find the necessary reference information in this textbook.

19. (a) Make an order-of-magnitude estimate of the resistance between the ends of a rubber band. (b) Make an order-of-magnitude estimate of the resistance between the 'heads' and 'tails' sides of a penny. In each case state what quantities you take as data and the values you measure or estimate for them. (c) WARNING! Do not try this at home! What is the order of magnitude of the current that each would carry if it were connected across a 120-V power supply?

20. A solid cube of silver (density = 10.5 g/cm^3) has a mass of 90.0 g. (a) What is the resistance between opposite faces of the cube? (b) Assume each silver atom contributes one conduction electron. Find the average drift speed of electrons when a potential difference of 1.00×10^{-5} V is applied to opposite faces. The atomic number of silver is 47, and its molar mass is 107.87 g/mol.

21. A metal wire of resistance R is cut into three equal pieces that are then connected side by side to form a new wire the length of which is equal to one-third the original length. What is the resistance of this new wire?

22. Aluminum and copper wires of equal length are found to have the same resistance. What is the ratio of their radii?

23. A current density of 6.00×10^{-13} A/m^2 exists in the atmosphere at a location where the electric field is 100 V/m. Calculate the electrical conductivity of the Earth's atmosphere in this region.

24. The rod in Figure P27.24 is made of two materials. The figure is not drawn to scale. Each conductor has a square cross section 3.00 mm on a side. The first material has a resistivity of 4.00×10^{-3} $\Omega \cdot$m and is 25.0 cm long, while the second material has a resistivity of 6.00×10^{-3} $\Omega \cdot$m and is 40.0 cm long. What is the resistance between the ends of the rod?

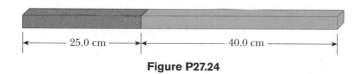

|← 25.0 cm →|← 40.0 cm →|

Figure P27.24

Section 27.3 A Model for Electrical Conduction

25. If the magnitude of the drift velocity of free electrons in a copper wire is 7.84×10^{-4} m/s, what is the electric field in the conductor?

26. If the current carried by a conductor is doubled, what happens to the (a) charge carrier density? (b) current density? (c) electron drift velocity? (d) average time interval between collisions?

27. Use data from Example 27.1 to calculate the collision mean free path of electrons in copper. Assume the average thermal speed of conduction electrons is 8.60×10^5 m/s.

Section 27.4 Resistance and Temperature

28. While taking photographs in Death Valley on a day when the temperature is 58.0°C, Bill Hiker finds that a certain voltage applied to a copper wire produces a current of 1.000 A. Bill then travels to Antarctica and applies the same voltage to the same wire. What current does he register there if the temperature is −88.0°C? Assume that no change occurs in the wire's shape and size.

29. A certain lightbulb has a tungsten filament with a resistance of 19.0 Ω when cold and 140 Ω when hot. Assume that the resistivity of tungsten varies linearly with temperature even over the large temperature range involved here, and find the temperature of the hot filament. Assume the initial temperature is 20.0°C.

30. A carbon wire and a Nichrome wire are connected in series, so that the same current exists in both wires. If the combination has a resistance of 10.0 kΩ at 0°C, what is the resistance of each wire at 0°C so that the resistance of the combination does not change with temperature? The total or equivalent resistance of resistors in series is the sum of their individual resistances.

31. An aluminum wire with a diameter of 0.100 mm has a uniform electric field of 0.200 V/m imposed along its entire length. The temperature of the wire is 50.0°C. Assume one free electron per atom. (a) Use the information in Table 27.1 and determine the resistivity. (b) What is the current density in the wire? (c) What is the total current in the wire? (d) What is the drift speed of the conduction electrons? (e) What potential difference must exist between the ends of a 2.00-m length of the wire to produce the stated electric field?

32. **Review problem.** An aluminum rod has a resistance of 1.234 Ω at 20.0°C. Calculate the resistance of the rod at 120°C by accounting for the changes in both the resistivity and the dimensions of the rod.

33. What is the fractional change in the resistance of an iron filament when its temperature changes from 25.0°C to 50.0°C?

34. The resistance of a platinum wire is to be calibrated for low-temperature measurements. A platinum wire with resistance 1.00 Ω at 20.0°C is immersed in liquid nitrogen at 77 K (−196°C). If the temperature response of the platinum wire is linear, what is the expected resistance of the platinum wire at −196°C? ($\alpha_{platinum} = 3.92 \times 10^{-3}$/°C)

35. The temperature of a sample of tungsten is raised while a sample of copper is maintained at 20.0°C. At what temperature will the resistivity of the tungsten be four times that of the copper?

Section 27.6 Electrical Power

36. A toaster is rated at 600 W when connected to a 120-V source. What current does the toaster carry, and what is its resistance?

37. A Van de Graaff generator (see Figure 25.29) is operating so that the potential difference between the high-voltage electrode B and the charging needles at A is 15.0 kV. Calculate the power required to drive the belt against electrical forces at an instant when the effective current delivered to the high-voltage electrode is 500 μA.

38. In a hydroelectric installation, a turbine delivers 1 500 hp to a generator, which in turn transfers 80.0% of the mechanical energy out by electrical transmission. Under these conditions, what current does the generator deliver at a terminal potential difference of 2 000 V?

39. [🌐] What is the required resistance of an immersion heater that increases the temperature of 1.50 kg of water from 10.0°C to 50.0°C in 10.0 min while operating at 110 V?

40. One rechargeable battery of mass 15.0 g delivers to a CD player an average current of 18.0 mA at 1.60 V for 2.40 h before the battery needs to be recharged. The recharger maintains a potential difference of 2.30 V across the battery and delivers a charging current of 13.5 mA for 4.20 h. (a) What is the efficiency of the battery as an energy storage device? (b) How much internal energy is produced in the battery during one charge–discharge cycle? (b) If the battery is surrounded by ideal thermal insulation and has an overall effective specific heat of 975 J/kg°C, by how much will its temperature increase during the cycle?

41. Suppose that a voltage surge produces 140 V for a moment. By what percentage does the power output of a 120-V, 100-W lightbulb increase? Assume that its resistance does not change.

42. A 500-W heating coil designed to operate from 110 V is made of Nichrome wire 0.500 mm in diameter. (a) Assuming that the resistivity of the Nichrome remains constant at its 20.0°C value, find the length of wire used. (b) **What If?** Now consider the variation of resistivity with temperature. What power will the coil of part (a) actually deliver when it is heated to 1 200°C?

43. A coil of Nichrome wire is 25.0 m long. The wire has a diameter of 0.400 mm and is at 20.0°C. If it carries a current of 0.500 A, what are (a) the magnitude of the electric field in the wire, and (b) the power delivered to it? (c) **What If?** If the temperature is increased to 340°C and the voltage across the wire remains constant, what is the power delivered?

44. Batteries are rated in terms of ampere-hours (A·h). For example, a battery that can produce a current of 2.00 A for 3.00 h is rated at 6.00 A·h. (a) What is the total energy, in kilowatt-hours, stored in a 12.0-V battery rated at 55.0 A·h? (b) At $0.060 0 per kilowatt-hour, what is the value of the electricity produced by this battery?

45. A 10.0-V battery is connected to a 120-Ω resistor. Ignoring the internal resistance of the battery, calculate the power delivered to the resistor.

46. Residential building codes typically require the use of 12-gauge copper wire (diameter 0.205 3 cm) for wiring receptacles. Such circuits carry currents as large as 20 A. A wire of smaller diameter (with a higher gauge number) could carry this much current, but the wire could rise to a high temperature and cause a fire. (a) Calculate the rate at which internal energy is produced in 1.00 m of 12-gauge copper wire carrying a current of 20.0 A. (b) **What If?** Repeat the calculation for an aluminum wire. Would a 12-gauge aluminum wire be as safe as a copper wire?

47. An 11.0-W energy-efficient fluorescent lamp is designed to produce the same illumination as a conventional 40.0-W incandescent lightbulb. How much money does the user of the energy-efficient lamp save during 100 hours of use? Assume a cost of $0.080 0/kWh for energy from the power company.

48. We estimate that 270 million plug-in electric clocks are in the United States, approximately one clock for each person. The clocks convert energy at the average rate 2.50 W. To supply this energy, how many metric tons of coal are burned per hour in coal-fired electric generating plants that are, on average, 25.0% efficient? The heat of combustion for coal is 33.0 MJ/kg.

49. Compute the cost per day of operating a lamp that draws a current of 1.70 A from a 110-V line. Assume the cost of energy from the power company is $0.060 0/kWh.

50. **Review problem.** The heating element of a coffee maker operates at 120 V and carries a current of 2.00 A. Assuming that the water absorbs all of the energy delivered to the resistor, calculate how long it takes to raise the temperature of 0.500 kg of water from room temperature (23.0°C) to the boiling point.

51. A certain toaster has a heating element made of Nichrome wire. When the toaster is first connected to a 120-V source (and the wire is at a temperature of 20.0°C), the initial current is 1.80 A. However, the current begins to decrease as the heating element warms up. When the toaster reaches its final operating temperature, the current drops to 1.53 A. (a) Find the power delivered to the toaster when it is at its operating temperature. (b) What is the final temperature of the heating element?

52. The cost of electricity varies widely through the United States; $0.120/kWh is one typical value. At this unit price, calculate the cost of (a) leaving a 40.0-W porch light on for two weeks while you are on vacation, (b) making a piece of dark toast in 3.00 min with a 970-W toaster, and (c) drying a load of clothes in 40.0 min in a 5 200-W dryer.

53. Make an order-of-magnitude estimate of the cost of one person's routine use of a hair dryer for 1 yr. If you do not use a blow dryer yourself, observe or interview someone who does. State the quantities you estimate and their values.

Additional Problems

54. One lightbulb is marked '25 W 120 V,' and another '100 W 120 V'; this means that each bulb has its respective power delivered to it when plugged into a constant 120-V potential difference. (a) Find the resistance of each bulb. (b) How long does it take for 1.00 C to pass through the dim bulb? Is the charge different in any way upon its exit from the bulb versus its entry? (c) How long does it take for 1.00 J to pass through the dim bulb? By what mechanisms does this energy enter and exit the bulb? (d) Find how much it costs to run the dim bulb continuously for 30.0 days if the electric company sells its product at $0.070 0 per kWh. What product *does* the electric company sell? What is its price for one SI unit of this quantity?

55. A charge Q is placed on a capacitor of capacitance C. The capacitor is connected into the circuit shown in Figure P27.55, with an open switch, a resistor, and an initially uncharged capacitor of capacitance $3C$. The switch is then closed and the circuit comes to equilibrium. In terms of Q and C, find (a) the final potential difference between the plates of each capacitor, (b) the charge on each capacitor,

and (c) the final energy stored in each capacitor. (d) Find the internal energy appearing in the resistor.

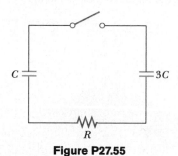

Figure P27.55

56. A high-voltage transmission line with a diameter of 2.00 cm and a length of 200 km carries a steady current of 1 000 A. If the conductor is copper wire with a free charge density of 8.49×10^{28} electrons/m^3, how long does it take one electron to travel the full length of the line?

57. A more general definition of the temperature coefficient of resistivity is

$$\alpha = \frac{1}{\rho} \frac{d\rho}{dT}$$

where ρ is the resistivity at temperature T. (a) Assuming that α is constant, show that

$$\rho = \rho_0 e^{\alpha(T - T_0)}$$

where ρ_0 is the resistivity at temperature T_0. (b) Using the series expansion $e^x \approx 1 + x$ for $x \ll 1$, show that the resistivity is given approximately by the expression $\rho = \rho_0[1 + \alpha(T - T_0)]$ for $\alpha(T - T_0) \ll 1$.

58. A high-voltage transmission line carries 1 000 A starting at 700 kV for a distance of 100 mi. If the resistance in the wire is 0.500 Ω/mi, what is the power loss due to resistive losses?

59. An experiment is conducted to measure the electrical resistivity of Nichrome in the form of wires with different lengths and cross-sectional areas. For one set of measurements, a student uses 30-gauge wire, which has a cross-sectional area of 7.30×10^{-8} m^2. The student measures the potential difference across the wire and the current in the wire with a voltmeter and an ammeter, respectively. For each of the measurements given in the table taken on wires of three different lengths, calculate the resistance of the wires and the corresponding values of the resistivity. What is the average value of the resistivity, and how does this value compare with the value given in Table 27.1?

L (m)	ΔV (V)	I (A)	R (Ω)	ρ ($\Omega \cdot$ m)
0.540	5.22	0.500		
1.028	5.82	0.276		
1.543	5.94	0.187		

60. An electric utility company supplies a customer's house from the main power lines (120 V) with two copper wires, each of which is 50.0 m long and has a resistance of 0.108 Ω per 300 m. (a) Find the voltage at the customer's house for a load current of 110 A. For this load current, find (b) the power the customer is receiving and (c) the electric power lost in the copper wires.

61. A straight cylindrical wire lying along the x axis has a length of 0.500 m and a diameter of 0.200 mm. It is made of a material that obeys Ohm's law with a resistivity of $\rho = 4.00 \times 10^{-8} \; \Omega \cdot$ m. Assume that a potential of 4.00 V is maintained at $x = 0$, and that $V = 0$ at $x = 0.500$ m. Find (a) the electric field **E** in the wire, (b) the resistance of the wire, (c) the electric current in the wire, and (d) the current density **J** in the wire. Express vectors in vector notation. (e) Show that $\mathbf{E} = \rho \mathbf{J}$.

62. A straight cylindrical wire lying along the x axis has a length L and a diameter d. It is made of a material that obeys Ohm's law with a resistivity ρ. Assume that potential V is maintained at $x = 0$, and that the potential is zero at $x = L$. In terms of L, d, V, ρ, and physical constants, derive expressions for (a) the electric field in the wire, (b) the resistance of the wire, (c) the electric current in the wire, and (d) the current density in the wire. Express vectors in vector notation. (e) Prove that $\mathbf{E} = \rho \mathbf{J}$.

63. The potential difference across the filament of a lamp is maintained at a constant level while equilibrium temperature is being reached. It is observed that the steady-state current in the lamp is only one tenth of the current drawn by the lamp when it is first turned on. If the temperature coefficient of resistivity for the lamp at 20.0°C is 0.004 50 (°C)$^{-1}$, and if the resistance increases linearly with increasing temperature, what is the final operating temperature of the filament?

64. The current in a resistor decreases by 3.00 A when the voltage applied across the resistor decreases from 12.0 V to 6.00 V. Find the resistance of the resistor.

65. An electric car is designed to run off a bank of 12.0-V batteries with total energy storage of 2.00×10^7 J. (a) If the electric motor draws 8.00 kW, what is the current delivered to the motor? (b) If the electric motor draws 8.00 kW as the car moves at a steady speed of 20.0 m/s, how far will the car travel before it is "out of juice"?

66. **Review problem.** When a straight wire is heated, its resistance is given by $R = R_0[1 + \alpha(T - T_0)]$ according to Equation 27.21, where α is the temperature coefficient of resistivity. (a) Show that a more precise result, one that includes the fact that the length and area of the wire change when heated, is

$$R = \frac{R_0[1 + \alpha(T - T_0)][1 + \alpha'(T - T_0)]}{[1 + 2\alpha'(T - T_0)]}$$

where α' is the coefficient of linear expansion (see Chapter 19). (b) Compare these two results for a 2.00-m-long copper wire of radius 0.100 mm, first at 20.0°C and then heated to 100.0°C.

67. The temperature coefficients of resistivity in Table 27.1 were determined at a temperature of 20°C. What would they be at 0°C? Note that the temperature coefficient of resistivity at 20°C satisfies $\rho = \rho_0[1 + \alpha(T - T_0)]$, where ρ_0 is the resistivity of the material at $T_0 = 20$°C. The temperature

coefficient of resistivity α' at 0°C must satisfy the expression $\rho = \rho_0'[1 + \alpha'T]$, where ρ_0' is the resistivity of the material at 0°C.

68. An oceanographer is studying how the ion concentration in sea water depends on depth. She does this by lowering into the water a pair of concentric metallic cylinders (Fig. P27.68) at the end of a cable and taking data to determine the resistance between these electrodes as a function of depth. The water between the two cylinders forms a cylindrical shell of inner radius r_a, outer radius r_b, and length L much larger than r_b. The scientist applies a potential difference ΔV between the inner and outer surfaces, producing an outward radial current I. Let ρ represent the resistivity of the water. (a) Find the resistance of the water between the cylinders in terms of L, ρ, r_a, and r_b. (b) Express the resistivity of the water in terms of the measured quantities L, r_a, r_b, ΔV, and I.

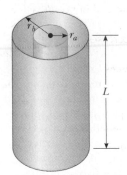

Figure P27.68

69. In a certain stereo system, each speaker has a resistance of 4.00 Ω. The system is rated at 60.0 W in each channel, and each speaker circuit includes a fuse rated 4.00 A. Is this system adequately protected against overload? Explain your reasoning.

70. A close analogy exists between the flow of energy by heat because of a temperature difference (see Section 20.7) and the flow of electric charge because of a potential difference. The energy dQ and the electric charge dq can both be transported by free electrons in the conducting material. Consequently, a good electrical conductor is usually a good thermal conductor as well. Consider a thin conducting slab of thickness dx, area A, and electrical conductivity σ, with a potential difference dV between opposite faces. Show that the current $I = dq/dt$ is given by the equation on the left below:

Charge Conduction	Thermal Conduction (Eq. 20.14)
$\dfrac{dq}{dt} = \sigma A \left| \dfrac{dV}{dx} \right|$	$\dfrac{dQ}{dt} = kA \left| \dfrac{dT}{dx} \right|$

In the analogous thermal conduction equation on the right, the rate of energy flow dQ/dt (in SI units of joules per second) is due to a temperature gradient dT/dx, in a material of thermal conductivity k. State analogous rules relating the direction of the electric current to the change in potential, and relating the direction of energy flow to the change in temperature.

71. Material with uniform resistivity ρ is formed into a wedge as shown in Figure P27.71. Show that the resistance between face A and face B of this wedge is

$$R = \rho \frac{L}{w(y_2 - y_1)} \ln\left(\frac{y_2}{y_1}\right)$$

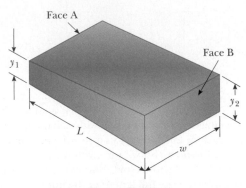

Figure P27.71

72. A material of resistivity ρ is formed into the shape of a truncated cone of altitude h as shown in Figure P27.72. The bottom end has radius b, and the top end has radius a. Assume that the current is distributed uniformly over any circular cross section of the cone, so that the current density does not depend on radial position. (The current density does vary with position along the axis of the cone.) Show that the resistance between the two ends is described by the expression

$$R = \frac{\rho}{\pi}\left(\frac{h}{ab}\right)$$

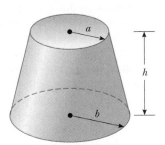

Figure P27.72

73. The dielectric material between the plates of a parallel-plate capacitor always has some nonzero conductivity σ. Let A represent the area of each plate and d the distance between them. Let κ represent the dielectric constant of the material. (a) Show that the resistance R and the capacitance C of the capacitor are related by

$$RC = \frac{\kappa \epsilon_0}{\sigma}$$

(b) Find the resistance between the plates of a 14.0-nF capacitor with a fused quartz dielectric.

74. 💻 The current–voltage characteristic curve for a semiconductor diode as a function of temperature T is given by the equation

$$I = I_0(e^{e\,\Delta V/k_B T} - 1)$$

Here the first symbol e represents Euler's number, the base of natural logarithms. The second e is the charge on the electron. The k_B stands for Boltzmann's constant, and T is the absolute temperature. Set up a spreadsheet to calculate I and $R = \Delta V/I$ for $\Delta V = 0.400$ V to 0.600 V in increments of 0.005 V. Assume $I_0 = 1.00$ nA. Plot R versus ΔV for $T = 280$ K, 300 K, and 320 K.

75. **Review problem.** A parallel-plate capacitor consists of square plates of edge length ℓ that are separated by a distance d, where $d \ll \ell$. A potential difference ΔV is maintained between the plates. A material of dielectric constant κ fills half of the space between the plates. The dielectric slab is now withdrawn from the capacitor, as shown in Figure P27.75. (a) Find the capacitance when the left edge of the dielectric is at a distance x from the center of the capacitor. (b) If the dielectric is removed at a constant speed v, what is the current in the circuit as the dielectric is being withdrawn?

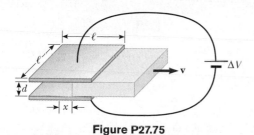

Figure P27.75

Answers to Quick Quizzes

27.1 d, b = c, a. The current in part (d) is equivalent to two positive charges moving to the left. Parts (b) and (c) each represent four positive charges moving in the same direction because negative charges moving to the left are equivalent to positive charges moving to the right. The current in part (a) is equivalent to five positive charges moving to the right.

27.2 (b). The currents in the two paths add numerically to equal the current coming into the junction, without regard for the directions of the two wires coming out of the junction. This is indicative of scalar addition. Even though we can assign a direction to a current, it is not a vector. This suggests a deeper meaning for vectors besides that of a quantity with magnitude and direction.

27.3 (a). The current in each section of the wire is the same even though the wire constricts. As the cross-sectional area A decreases, the drift velocity must increase in order for the constant current to be maintained, in accordance with Equation 27.4. As A decreases, Equation 27.11 tells us that R increases.

27.4 (b). The doubling of the radius causes the area A to be four times as large, so Equation 27.11 tells us that the resistance decreases.

27.5 (b). The slope of the tangent to the graph line at a point is the reciprocal of the resistance at that point. Because the slope is increasing, the resistance is decreasing.

27.6 (a). When the filament is at room temperature, its resistance is low, and hence the current is relatively large. As the filament warms up, its resistance increases, and the current decreases. Older lightbulbs often fail just as they are turned on because this large initial current "spike" produces rapid temperature increase and mechanical stress on the filament, causing it to break.

27.7 (c). Because the potential difference ΔV is the same across the two bulbs and because the power delivered to a conductor is $\mathcal{P} = I\Delta V$, the 60-W bulb, with its higher power rating, must carry the greater current. The 30-W bulb has the higher resistance because it draws less current at the same potential difference.

27.8 $I_a = I_b > I_c = I_d > I_e = I_f$. The current I_a leaves the positive terminal of the battery and then splits to flow through the two bulbs; thus, $I_a = I_c + I_e$. From Quick Quiz 27.7, we know that the current in the 60-W bulb is greater than that in the 30-W bulb. Because charge does not build up in the bulbs, we know that the same amount of charge flowing into a bulb from the left must flow out on the right; consequently, $I_c = I_d$ and $I_e = I_f$. The two currents leaving the bulbs recombine to form the current back into the battery, $I_f + I_d = I_b$.

Chapter 28

Direct Current Circuits

CHAPTER OUTLINE

28.1 Electromotive Force

28.2 Resistors in Series and Parallel

28.3 Kirchhoff's Rules

28.4 *RC* Circuits

28.5 Electrical Meters

28.6 Household Wiring and Electrical Safety

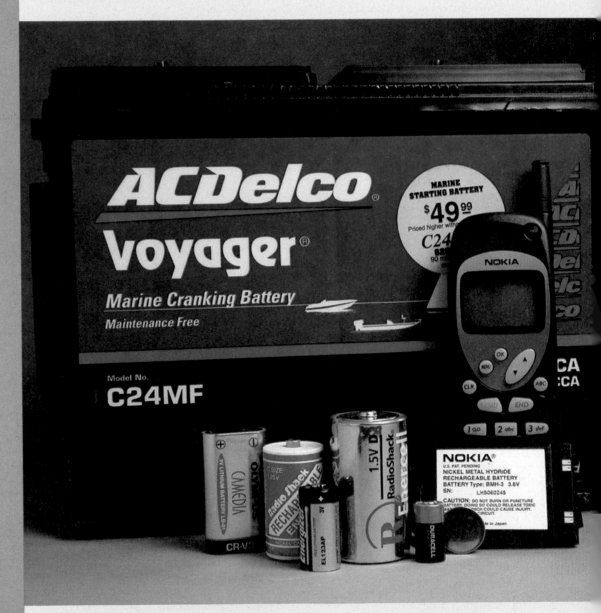

▲ An assortment of batteries that can be used to provide energy for various devices. Batteries provide a voltage with a fixed polarity, resulting in a direct current in a circuit, that is, a current for which the drift velocity of the charges is always in the same direction. (George Semple)

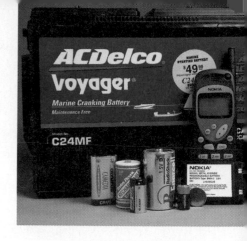

This chapter is concerned with the analysis of simple electric circuits that contain batteries, resistors, and capacitors in various combinations. We will see some circuits in which resistors can be combined using simple rules. The analysis of more complicated circuits is simplified using two rules known as *Kirchhoff's rules*, which follow from the laws of conservation of energy and conservation of electric charge for isolated systems. Most of the circuits analyzed are assumed to be in *steady state*, which means that currents in the circuit are constant in magnitude and direction. A current that is constant in direction is called a *direct current* (DC). We will study *alternating current* (AC), in which the current changes direction periodically, in Chapter 33. Finally, we describe electrical meters for measuring current and potential difference, and discuss electrical circuits in the home.

28.1 Electromotive Force

In Section 27.6 we discussed a closed circuit in which a battery produces a potential difference and causes charges to move. We will generally use a battery in our discussion and in our circuit diagrams as a source of energy for the circuit. Because the potential difference at the battery terminals is constant in a particular circuit, the current in the circuit is constant in magnitude and direction and is called **direct current.** A battery is called either a *source of electromotive force* or, more commonly, a *source of emf.* (The phrase *electromotive force* is an unfortunate historical term, describing not a force but rather a potential difference in volts.) **The emf $\mathcal{E}$ of a battery is the maximum possible voltage that the battery can provide between its terminals.** You can think of a source of emf as a "charge pump." When an electric potential difference exists between two points, the source moves charges "uphill" from the lower potential to the higher.

Consider the circuit shown in Figure 28.1, consisting of a battery connected to a resistor. We shall generally assume that the connecting wires have no resistance.

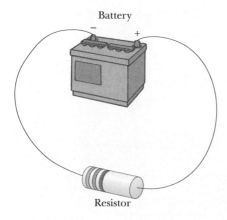

Figure 28.1 A circuit consisting of a resistor connected to the terminals of a battery.

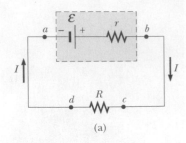

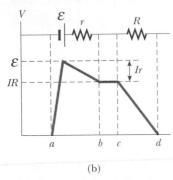

Active Figure 28.2 (a) Circuit diagram of a source of emf $\mathcal{E}$ (in this case, a battery), of internal resistance r, connected to an external resistor of resistance R. (b) Graphical representation showing how the electric potential changes as the circuit in part (a) is traversed clockwise.

🔗 **At the Active Figures link at http://www.pse6.com, you can adjust the emf and resistances r and R to see the effect on the current and on the graph in part (b).**

⚠ **PITFALL PREVENTION**

28.1 What Is Constant in a Battery?

It is a common misconception that a battery is a source of constant current. Equation 28.3 clearly shows that this is not true. The current in the circuit depends on the resistance connected to the battery. It is also not true that a battery is a source of constant terminal voltage, as shown by Equation 28.1. **A battery is a source of constant emf.**

The positive terminal of the battery is at a higher potential than the negative terminal. Because a real battery is made of matter, there is resistance to the flow of charge within the battery. This resistance is called **internal resistance** r. For an idealized battery with zero internal resistance, the potential difference across the battery (called its *terminal voltage*) equals its emf. However, for a real battery, the terminal voltage is *not* equal to the emf for a battery in a circuit in which there is a current. To understand why this is so, consider the circuit diagram in Figure 28.2a, where the battery of Figure 28.1 is represented by the dashed rectangle containing an ideal, resistance-free emf $\mathcal{E}$ in series with an internal resistance r. Now imagine moving through the battery from a to b and measuring the electric potential at various locations. As we pass from the negative terminal to the positive terminal, the potential *increases* by an amount $\mathcal{E}$. However, as we move through the resistance r, the potential *decreases* by an amount Ir, where I is the current in the circuit. Thus, the terminal voltage of the battery $\Delta V = V_b - V_a$ is[1]

$$\Delta V = \mathcal{E} - Ir \qquad (28.1)$$

From this expression, note that $\mathcal{E}$ is equivalent to the **open-circuit voltage**—that is, the *terminal voltage when the current is zero*. The emf is the voltage labeled on a battery—for example, the emf of a D cell is 1.5 V. The actual potential difference between the terminals of the battery depends on the current in the battery, as described by Equation 28.1.

Figure 28.2b is a graphical representation of the changes in electric potential as the circuit is traversed in the clockwise direction. By inspecting Figure 28.2a, we see that the terminal voltage ΔV must equal the potential difference across the external resistance R, often called the **load resistance.** The load resistor might be a simple resistive circuit element, as in Figure 28.1, or it could be the resistance of some electrical device (such as a toaster, an electric heater, or a lightbulb) connected to the battery (or, in the case of household devices, to the wall outlet). The resistor represents a *load* on the battery because the battery must supply energy to operate the device. The potential difference across the load resistance is $\Delta V = IR$. Combining this expression with Equation 28.1, we see that

$$\mathcal{E} = IR + Ir \qquad (28.2)$$

Solving for the current gives

$$I = \frac{\mathcal{E}}{R + r} \qquad (28.3)$$

This equation shows that the current in this simple circuit depends on both the load resistance R external to the battery and the internal resistance r. If R is much greater than r, as it is in many real-world circuits, we can neglect r.

If we multiply Equation 28.2 by the current I, we obtain

$$I\mathcal{E} = I^2 R + I^2 r \qquad (28.4)$$

This equation indicates that, because power $\mathcal{P} = I\,\Delta V$ (see Eq. 27.22), the total power output $I\mathcal{E}$ of the battery is delivered to the external load resistance in the amount $I^2 R$ and to the internal resistance in the amount $I^2 r$.

> **Quick Quiz 28.1** In order to maximize the percentage of the power that is delivered from a battery to a device, the internal resistance of the battery should be (a) as low as possible (b) as high as possible (c) The percentage does not depend on the internal resistance.

---

[1]   The terminal voltage in this case is less than the emf by an amount Ir. In some situations, the terminal voltage may *exceed* the emf by an amount Ir. This happens when the direction of the current is *opposite* that of the emf, as in the case of charging a battery with another source of emf.

Example 28.1   Terminal Voltage of a Battery

Interactive

A battery has an emf of 12.0 V and an internal resistance of 0.05 Ω. Its terminals are connected to a load resistance of 3.00 Ω.

(A) Find the current in the circuit and the terminal voltage of the battery.

Solution Equation 28.3 gives us the current:

$$I = \frac{\mathcal{E}}{R + r} = \frac{12.0\,\text{V}}{3.05\,\Omega} = \boxed{3.93\,\text{A}}$$

and from Equation 28.1, we find the terminal voltage:

$$\Delta V = \mathcal{E} - Ir = 12.0\,\text{V} - (3.93\,\text{A})(0.05\,\Omega) = \boxed{11.8\,\text{V}}$$

To check this result, we can calculate the voltage across the load resistance R:

$$\Delta V = IR = (3.93\,\text{A})(3.00\,\Omega) = 11.8\,\text{V}$$

(B) Calculate the power delivered to the load resistor, the power delivered to the internal resistance of the battery, and the power delivered by the battery.

Solution The power delivered to the load resistor is

$$\mathcal{P}_R = I^2R = (3.93\,\text{A})^2(3.00\,\Omega) = \boxed{46.3\,\text{W}}$$

The power delivered to the internal resistance is

$$\mathcal{P}_r = I^2r = (3.93\,\text{A})^2(0.05\,\Omega) = \boxed{0.772\,\text{W}}$$

Hence, the power delivered by the battery is the sum of these quantities, or 47.1 W. You should check this result, using the expression $\mathcal{P} = I\mathcal{E}$.

What If? As a battery ages, its internal resistance increases. Suppose the internal resistance of this battery rises to 2.00 Ω toward the end of its useful life. How does this alter the ability of the battery to deliver energy?

Answer Let us connect the same 3.00-Ω load resistor to the battery. The current in the battery now is

$$I = \frac{\mathcal{E}}{R + r} = \frac{12.0\,\text{V}}{(3.00\,\Omega + 2.00\,\Omega)} = 2.40\,\text{A}$$

and the terminal voltage is

$$\Delta V = \mathcal{E} - Ir = 12.0\,\text{V} - (2.40\,\text{A})(2.00\,\Omega) = 7.2\,\text{V}$$

Notice that the terminal voltage is only 60% of the emf. The powers delivered to the load resistor and internal resistance are

$$\mathcal{P}_R = I^2R = (2.40\,\text{A})^2(3.00\,\Omega) = \boxed{17.3\,\text{W}}$$

$$\mathcal{P}_r = I^2r = (2.40\,\text{A})^2(2.00\,\Omega) = 11.5\,\text{W}$$

Notice that 40% of the power from the battery is delivered to the internal resistance. In part (B), this percentage is 1.6%. Consequently, even though the emf remains fixed, the increasing internal resistance significantly reduces the ability of the battery to deliver energy.

 At the Interactive Worked Example link at **http://www.pse6.com,** *you can vary the load resistance and internal resistance, observing the power delivered to each.*

Example 28.2   Matching the Load

Show that the maximum power delivered to the load resistance R in Figure 28.2a occurs when the load resistance matches the internal resistance—that is, when $R = r$.

Solution The power delivered to the load resistance is equal to I^2R, where I is given by Equation 28.3:

$$\mathcal{P} = I^2R = \frac{\mathcal{E}^2R}{(R + r)^2}$$

When $\mathcal{P}$ is plotted versus R as in Figure 28.3, we find that $\mathcal{P}$ reaches a maximum value of $\mathcal{E}^2/4r$ at $R = r$. When R is large, there is very little current, so that the power I^2R delivered to the load resistor is small. When R is small, the current is large and there is significant loss of power I^2r as energy is delivered to the internal resistance. When $R = r$, these effects balance to give a maximum transfer of power.

We can also prove that the power maximizes at $R = r$ by differentiating $\mathcal{P}$ with respect to R, setting the result equal

to zero, and solving for R. The details are left as a problem for you to solve (Problem 57).

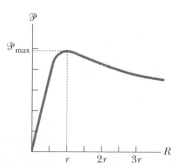

Figure 28.3 (Example 28.2) Graph of the power $\mathcal{P}$ delivered by a battery to a load resistor of resistance R as a function of R. The power delivered to the resistor is a maximum when the load resistance equals the internal resistance of the battery.

28.2 Resistors in Series and Parallel

Suppose that you and your friends are at a crowded basketball game in a sports arena and decide to leave early. You have two choices: (1) your group can exit through a single door and push your way down a long hallway containing several concession stands, each surrounded by a large crowd of people waiting to buy food or souvenirs; or (2) each member of your group can exit through a separate door in the main hall of the arena, where each will have to push his or her way through a single group of people standing by the door. In which scenario will less time be required for your group to leave the arena?

It should be clear that your group will be able to leave faster through the separate doors than down the hallway where each of you has to push through several groups of people. We could describe the groups of people in the hallway as being in *series*, because each of you must push your way through all of the groups. The groups of people around the doors in the arena can be described as being in *parallel*. Each member of your group must push through only one group of people, and each member pushes through a *different* group of people. This simple analogy will help us understand the behavior of currents in electric circuits containing more than one resistor.

When two or more resistors are connected together as are the lightbulbs in Figure 28.4a, they are said to be in *series*. Figure 28.4b is the circuit diagram for the lightbulbs, which are shown as resistors, and the battery. In a series connection, if an amount of charge Q exits resistor R_1, charge Q must also enter the second resistor R_2. (This is analogous to all members of your group pushing through each crowd in the single hallway of the sports arena.) Otherwise, charge will accumulate on the wire between the resistors. Thus, the same amount of charge passes through both resistors in a given time interval. Hence,

> for a series combination of two resistors, the currents are the same in both resistors because the amount of charge that passes through R_1 must also pass through R_2 in the same time interval.

The potential difference applied across the series combination of resistors will divide between the resistors. In Figure 28.4b, because the voltage drop[2] from a to b equals IR_1 and the voltage drop from b to c equals IR_2, the voltage drop from a to c is

$$\Delta V = IR_1 + IR_2 = I(R_1 + R_2)$$

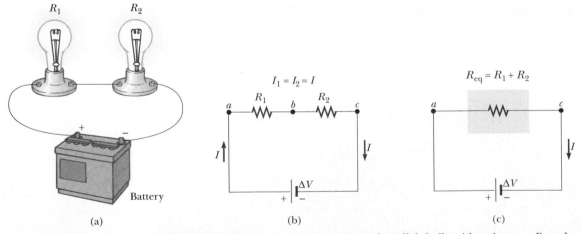

(a)   (b)   (c)

Active Figure 28.4 (a) A series connection of two lightbulbs with resistances R_1 and R_2. (b) Circuit diagram for the two-resistor circuit. The current in R_1 is the same as that in R_2. (c) The resistors replaced with a single resistor having an equivalent resistance $R_{eq} = R_1 + R_2$.

At the Active Figures link at http://www.pse6.com, you can adjust the battery voltage and resistances R_1 and R_2 to see the effect on the currents and voltages in the individual resistors.

[2] The term *voltage drop* is synonymous with a decrease in electric potential across a resistor and is used often by individuals working with electric circuits.

The potential difference across the battery is also applied to the **equivalent resistance** R_{eq} in Figure 28.4c:

$$\Delta V = IR_{eq}$$

where we have indicated that the equivalent resistance has the same effect on the circuit because it results in the same current in the battery as the combination of resistors. Combining these equations, we see that we can replace the two resistors in series with a single equivalent resistance whose value is the *sum* of the individual resistances:

$$\Delta V = IR_{eq} = I(R_1 + R_2) \quad \longrightarrow \quad R_{eq} = R_1 + R_2 \qquad (28.5)$$

The resistance R_{eq} is equivalent to the series combination $R_1 + R_2$ in the sense that the circuit current is unchanged when R_{eq} replaces $R_1 + R_2$.

The equivalent resistance of three or more resistors connected in series is

$$R_{eq} = R_1 + R_2 + R_3 + \cdots \qquad (28.6)$$

The equivalent resistance of several resistors in series

This relationship indicates that **the equivalent resistance of a series connection of resistors is the numerical sum of the individual resistances and is always greater than any individual resistance.**

Looking back at Equation 28.3, the denominator is the simple algebraic sum of the external and internal resistances. This is consistent with the fact that internal and external resistances are in series in Figure 28.2a.

Note that if the filament of one lightbulb in Figure 28.4 were to fail, the circuit would no longer be complete (resulting in an open-circuit condition) and the second bulb would also go out. This is a general feature of a series circuit—if one device in the series creates an open circuit, all devices are inoperative.

▲ **PITFALL PREVENTION**

28.2 Lightbulbs Don't Burn

We will describe the end of the life of a lightbulb by saying that *the filament fails*, rather than by saying that the lightbulb "burns out." The word *burn* suggests a combustion process, which is not what occurs in a lightbulb.

Quick Quiz 28.2 In Figure 28.4, imagine positive charges pass first through R_1 and then through R_2. Compared to the current in R_1, the current in R_2 is (a) smaller, (b) larger, or (c) the same.

Quick Quiz 28.3 If a piece of wire is used to connect points b and c in Figure 28.4b, does the brightness of bulb R_1 (a) increase, (b) decrease, or (c) remain the same?

Quick Quiz 28.4 With the switch in the circuit of Figure 28.5 closed (left), there is no current in R_2, because the current has an alternate zero-resistance path through the switch. There is current in R_1 and this current is measured with the ammeter (a device for measuring current) at the right side of the circuit. If the switch is opened (Fig. 28.5, right), there is current in R_2. What happens to the reading on the ammeter when the switch is opened? (a) the reading goes up; (b) the reading goes down; (c) the reading does not change.

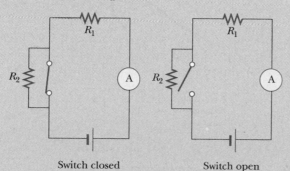

Switch closed   Switch open

Figure 28.5 (Quick Quiz 28.4) What happens when the switch is opened?

*At the Active Figures link
at http://www.pse6.com, you
can adjust the battery voltage
and resistances R_1 and R_2 to
see the effect on the currents
and voltages in the individual
resistors.*

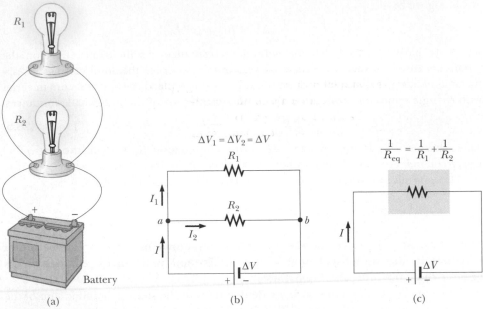

(a)   (b)   (c)

Active Figure 28.6 (a) A parallel connection of two lightbulbs with resistances R_1 and
R_2. (b) Circuit diagram for the two-resistor circuit. The potential difference across R_1 is
the same as that across R_2. (c) The resistors replaced with a single resistor having an
equivalent resistance given by Equation 28.7.

▲ PITFALL PREVENTION

28.3 Local and Global Changes

**A local change in one part of a
circuit may result in a global
change throughout the circuit.**
For example, if a single resistance
is changed in a circuit containing
several resistors and batteries, the
currents in all resistors and batteries,
the terminal voltages of all batteries,
and the voltages across all
resistors may change as a result.

▲ PITFALL PREVENTION

28.4 Current Does Not Take the Path of Least Resistance

You may have heard a phrase like
"current takes the path of least
resistance" in reference to a parallel
combination of current
paths, such that there are two or
more paths for the current to
take. The phrase is incorrect.
The current takes *all* paths.
Those paths with lower resistance
will have large currents, but even
very high-resistance paths will
carry *some* of the current.

Now consider two resistors connected in *parallel,* as shown in Figure 28.6. When
charges reach point *a* in Figure 28.6b, called a *junction,* they split into two parts, with some
going through R_1 and the rest going through R_2. A **junction** is any point in a circuit
where a current can split (just as your group might split up and leave the sports arena
through several doors, as described earlier.) This split results in less current in each individual
resistor than the current leaving the battery. Because electric charge is conserved,
the current I that enters point *a* must equal the total current leaving that point:

$$I = I_1 + I_2$$

where I_1 is the current in R_1 and I_2 is the current in R_2.

As can be seen from Figure 28.6, both resistors are connected directly across the
terminals of the battery. Therefore,

when resistors are connected in parallel, the potential differences across the resistors
is the same.

Because the potential differences across the resistors are the same, the expression
$\Delta V = IR$ gives

$$I = I_1 + I_2 = \frac{\Delta V}{R_1} + \frac{\Delta V}{R_2} = \Delta V \left(\frac{1}{R_1} + \frac{1}{R_2} \right) = \frac{\Delta V}{R_{eq}}$$

where R_{eq} is an equivalent single resistance which will have the same effect on the
circuit as the two resistors in parallel; that is, it will draw the same current from the
battery (Fig. 28.6c). From this result, we see that the equivalent resistance of two resistors
in parallel is given by

$$\frac{1}{R_{eq}} = \frac{1}{R_1} + \frac{1}{R_2} \qquad (28.7)$$

or

$$R_{eq} = \frac{1}{\dfrac{1}{R_1} + \dfrac{1}{R_2}} = \frac{R_1 R_2}{R_1 + R_2}$$

An extension of this analysis to three or more resistors in parallel gives

$$\frac{1}{R_{eq}} = \frac{1}{R_1} + \frac{1}{R_2} + \frac{1}{R_3} + \cdots$$

(28.8)   **The equivalent resistance of several resistors in parallel**

We can see from this expression that **the inverse of the equivalent resistance of two or more resistors connected in parallel is equal to the sum of the inverses of the individual resistances. Furthermore, the equivalent resistance is always less than the smallest resistance in the group.**

Household circuits are always wired such that the appliances are connected in parallel. Each device operates independently of the others so that if one is switched off, the others remain on. In addition, in this type of connection, all of the devices operate on the same voltage.

Quick Quiz 28.5
In Figure 28.4, imagine that we add a third resistor in series with the first two. Does the current in the battery (a) increase, (b) decrease, or (c) remain the same? Does the terminal voltage of the battery (d) increase, (e) decrease, or (f) remain the same?

Quick Quiz 28.6
In Figure 28.6, imagine that we add a third resistor in parallel with the first two. Does the current in the battery (a) increase, (b) decrease, or (c) remain the same? Does the terminal voltage of the battery (d) increase, (e) decrease, or (f) remain the same?

Quick Quiz 28.7
With the switch in the circuit of Figure 28.7 open (left), there is no current in R_2. There is current in R_1 and this current is measured with the ammeter at the right side of the circuit. If the switch is closed (Fig. 28.7, right), there is current in R_2. What happens to the reading on the ammeter when the switch is closed? (a) the reading goes up; (b) the reading goes down; (c) the reading does not change.

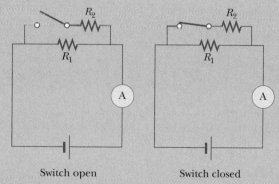

Switch open   Switch closed

Figure 28.7 (Quick Quiz 28.7) What happens when the switch is closed?

Conceptual Example 28.3   Landscape Lights

A homeowner wishes to install 12-volt landscape lighting in his back yard. To save money, he purchases inexpensive 18-gauge cable, which has a relatively high resistance per unit length. This cable consists of two side-by-side wires separated by insulation, like the cord on an appliance. He runs a 200-foot length of this cable from the power supply to the farthest point at which he plans to position a light fixture. He attaches light fixtures across the two wires on the cable at 10-foot intervals, so the light fixtures are in parallel. Because of the cable's resistance, the brightness of the bulbs in the light fixtures is not as desired. Which problem does the homeowner have? (a) All of the bulbs glow equally less brightly than they would if lower-resistance cable had been used. (b) The brightness of the bulbs decreases as you move farther from the power supply.

Solution A circuit diagram for the system appears in Figure 28.8. The horizontal resistors (such as R_A and R_B) represent the resistance of the wires in the cable between the light fixtures while the vertical resistors (such as R_C) represent the resistance of the light fixtures themselves. Part of the terminal voltage of the power supply is dropped across resistors R_A and R_B. Thus, the voltage across light

fixture R_C is less than the terminal voltage. There is a further voltage drop across resistors R_D and R_E. Consequently, the voltage across light fixture R_F is smaller than that across R_C. This continues on down the line of light fixtures, so the correct choice is (b). Each successive light fixture has a smaller voltage across it and glows less brightly than the one before.

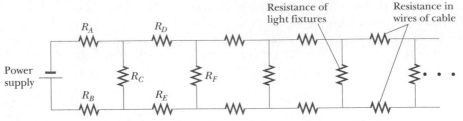

Figure 28.8 (Conceptual Example 28.3) The circuit diagram for a set of landscape light fixtures connected in parallel across the two wires of a two-wire cable. The horizontal resistors represent resistance in the wires of the cable. The vertical resistors represent the light fixtures.

Example 28.4 Find the Equivalent Resistance

Four resistors are connected as shown in Figure 28.9a.

(A) Find the equivalent resistance between points a and c.

Solution The combination of resistors can be reduced in steps, as shown in Figure 28.9. The 8.0-Ω and 4.0-Ω resistors are in series; thus, the equivalent resistance between a and b is 12.0 Ω (see Eq. 28.5). The 6.0-Ω and 3.0-Ω resistors are in parallel, so from Equation 28.7 we find that the equivalent resistance from b to c is 2.0 Ω. Hence, the equivalent resistance from a to c is 14.0 Ω.

(B) What is the current in each resistor if a potential difference of 42 V is maintained between a and c?

Solution The currents in the 8.0-Ω and 4.0-Ω resistors are the same because they are in series. In addition, this is the same as the current that would exist in the 14.0-Ω equivalent resistor subject to the 42-V potential difference. Therefore, using Equation 27.8 ($R = \Delta V/I$) and the result from part (A), we obtain

$$I = \frac{\Delta V_{ac}}{R_{eq}} = \frac{42 \text{ V}}{14.0 \ \Omega} = 3.0 \text{ A}$$

This is the current in the 8.0-Ω and 4.0-Ω resistors. When this 3.0-A current enters the junction at b, however, it splits, with part passing through the 6.0-Ω resistor (I_1) and part through the 3.0-Ω resistor (I_2). Because the potential difference is ΔV_{bc} across each of these parallel resistors, we see that $(6.0 \ \Omega)I_1 = (3.0 \ \Omega)I_2$, or $I_2 = 2I_1$. Using this result and the fact that $I_1 + I_2 = 3.0$ A, we find that $I_1 = 1.0$ A and

$I_2 = 2.0$ A. We could have guessed this at the start by noting that the current in the 3.0-Ω resistor has to be twice that in the 6.0-Ω resistor, in view of their relative resistances and the fact that the same voltage is applied to each of them.

As a final check of our results, note that $\Delta V_{bc} = (6.0 \ \Omega)I_1 = (3.0 \ \Omega)I_2 = 6.0$ V and $\Delta V_{ab} = (12.0 \ \Omega)I = 36$ V; therefore, $\Delta V_{ac} = \Delta V_{ab} + \Delta V_{bc} = 42$ V, as it must.

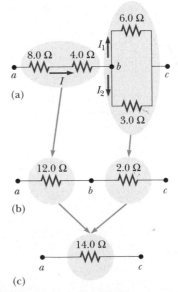

Figure 28.9 (Example 28.4) The original network of resistors is reduced to a single equivalent resistance.

Example 28.5 Finding R_{eq} by Symmetry Arguments

Consider five resistors connected as shown in Figure 28.10a. Find the equivalent resistance between points a and b.

Solution If we inspect this system of resistors, we realize that we cannot reduce it by using our rules for series and parallel

connections. We can, however, assume a current entering junction a and then apply symmetry arguments. Because of the symmetry in the circuit (all 1-Ω resistors in the outside loop), the currents in branches ac and ad must be equal; hence, the electric potentials at points c and d must be equal.

This means that $\Delta V_{cd} = 0$ and there is no current between points c and d. As a result, points c and d may be connected together without affecting the circuit, as in Figure 28.10b. Thus, the 5-Ω resistor may be removed from the circuit and the remaining circuit then reduced as in Figures 28.10c and d. From this reduction we see that the equivalent resistance of the combination is 1 Ω. Note that the result is 1 Ω regardless of the value of the resistor connected between c and d.

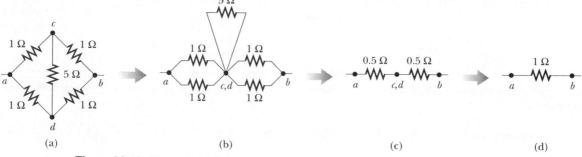

(a)   (b)   (c)   (d)

Figure 28.10 (Example 28.5) Because of the symmetry in this circuit, the 5-Ω resistor does not contribute to the resistance between points a and b and therefore can be disregarded when we calculate the equivalent resistance.

Example 28.6   Three Resistors in Parallel

Interactive

Three resistors are connected in parallel as shown in Figure 28.11a. A potential difference of 18.0 V is maintained between points a and b.

(A) Find the current in each resistor.

Solution The resistors are in parallel, and so the potential difference across each must be 18.0 V. Applying the relationship $\Delta V = IR$ to each resistor gives

$$I_1 = \frac{\Delta V}{R_1} = \frac{18.0\ \text{V}}{3.00\ \Omega} = \boxed{6.00\ \text{A}}$$

$$I_2 = \frac{\Delta V}{R_2} = \frac{18.0\ \text{V}}{6.00\ \Omega} = \boxed{3.00\ \text{A}}$$

$$I_3 = \frac{\Delta V}{R_3} = \frac{18.0\ \text{V}}{9.00\ \Omega} = \boxed{2.00\ \text{A}}$$

(B) Calculate the power delivered to each resistor and the total power delivered to the combination of resistors.

Solution We apply the relationship $\mathcal{P} = I^2 R$ to each resistor and obtain

3.00-Ω: $\mathcal{P}_1 = I_1{}^2 R_1 = (6.00\ \text{A})^2 (3.00\ \Omega) = \boxed{108\ \text{W}}$

6.00-Ω: $\mathcal{P}_2 = I_2{}^2 R_2 = (3.00\ \text{A})^2 (6.00\ \Omega) = \boxed{54.0\ \text{W}}$

9.00 Ω: $\mathcal{P}_3 = I_3{}^2 R_3 = (2.00\ \text{A})^2 (9.00\ \Omega) = \boxed{36.0\ \text{W}}$

This shows that the smallest resistor receives the most power. Summing the three quantities gives a total power of 198 W.

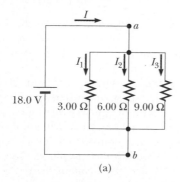

(a)

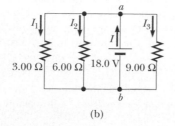

(b)

Figure 28.11 (Example 28.6) (a) Three resistors connected in parallel. The voltage across each resistor is 18.0 V. (b) Another circuit with three resistors and a battery. Is this equivalent to the circuit in part (a) of the figure?

(C) Calculate the equivalent resistance of the circuit.

Solution We can use Equation 28.8 to find R_{eq}:

$$\frac{1}{R_{eq}} = \frac{1}{3.00\ \Omega} + \frac{1}{6.00\ \Omega} + \frac{1}{9.00\ \Omega}$$

$$R_{eq} = \frac{18.0\ \Omega}{11.0} = \boxed{1.64\ \Omega}$$

What If? What if the circuit is as shown in Figure 28.11b instead of as in Figure 28.11a? How does this affect the calculation?

Answer There is no effect on the calculation. The physical placement of the battery is not important. In Figure 28.11b, the battery still applies a potential difference of 18.0 V between points a and b, so the two circuits in Figure 28.11 are electrically identical.

 At the Interactive Worked Example link at http://www.pse6.com, you can explore different configurations of the battery and resistors.

Conceptual Example 28.7 Operation of a Three-Way Lightbulb

Figure 28.12 illustrates how a three-way lightbulb is constructed to provide three levels of light intensity. The socket of the lamp is equipped with a three-way switch for selecting different light intensities. The bulb contains two filaments. When the lamp is connected to a 120-V source, one filament receives 100 W of power, and the other receives 75 W. Explain how the two filaments are used to provide three different light intensities.

Solution The three light intensities are made possible by applying the 120 V to one filament alone, to the other filament alone, or to the two filaments in parallel. When switch S_1 is closed and switch S_2 is opened, current exists only in the 75-W filament. When switch S_1 is open and switch S_2 is closed, current exists only in the 100-W filament. When both switches are closed, current exists in both filaments, and the total power is 175 W.

If the filaments were connected in series and one of them were to break, no charges could pass through the bulb, and the bulb would give no illumination, regardless of the switch position. However, with the filaments connected

in parallel, if one of them (for example, the 75-W filament) breaks, the bulb will still operate in two of the switch positions as current exists in the other (100-W) filament.

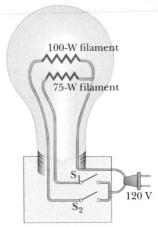

Figure 28.12 (Conceptual Example 28.7) A three-way lightbulb.

Application Strings of Lights

Strings of lights are used for many ornamental purposes, such as decorating Christmas trees.[3] Over the years, both parallel and series connections have been used for strings of lights powered by 120 V. Series-wired bulbs are safer than parallel-wired bulbs for indoor Christmas-tree use because series-wired bulbs operate with less energy per bulb and at a lower temperature. However, if the filament of a single bulb fails (or if the bulb is removed from its socket), all the lights on the string go out. The popularity of series-wired light strings diminished because troubleshooting a failed bulb was a tedious, time-consuming chore that involved trial-and-error substitution of a good bulb in each socket along the string until the defective bulb was found.

In a parallel-wired string, each bulb operates at 120 V. By design, the bulbs are brighter and hotter than those on

a series-wired string. As a result, these bulbs are inherently more dangerous (more likely to start a fire, for instance), but if one bulb in a parallel-wired string fails or is removed, the rest of the bulbs continue to glow. (A 25-bulb string of 4-W bulbs results in a power of 100 W; the total power becomes substantial when several strings are used.)

A new design was developed for so-called "miniature" lights wired in series, to prevent the failure of one bulb from causing the entire string to go out. This design creates a connection (called a jumper) across the filament after it fails. When the filament breaks in one of these miniature lightbulbs, the break in the filament represents the largest resistance in the series, much larger than that of the intact filaments. As a result, most of the applied 120 V appears across the bulb with the broken filament. Inside the

---

[3] These and other household devices, such as the three-way lightbulb in Conceptual Example 28.7 and the kitchen appliances discussed in Section 28.6, actually operate on alternating current (AC), to be introduced in Chapter 33.

lightbulb, a small jumper loop covered by an insulating material is wrapped around the filament leads. When the filament fails and 120 V appears across the bulb, an arc burns the insulation on the jumper and connects the filament leads. This connection now completes the circuit through the bulb even though its filament is no longer active (Fig. 28.13).

Suppose that all the bulbs in a 50-bulb miniature-light string are operating. A 2.40-V potential drop occurs across each bulb because the bulbs are in series. A typical power input to this style of bulb is 0.340 W. The filament resistance of each bulb at the operating temperature is $(2.40 \text{ V})^2/(0.340 \text{ W}) = 16.9 \, \Omega$. The current in each bulb is $2.40 \text{ V}/16.9 \, \Omega = 0.142 \text{ A}$. When a bulb fails, the resistance across its terminals is reduced to zero because of the alternate

jumper connection mentioned in the preceding paragraph. All the other bulbs not only stay on but glow more brightly because the total resistance of the string is reduced and consequently the current in each bulb increases.

Let us assume that the resistance of a bulb remains at $16.9 \, \Omega$ even though its temperature rises as a result of the increased current. If one bulb fails, the potential difference across each of the remaining bulbs increases to $120 \text{ V}/49 = 2.45 \text{ V}$, the current increases from 0.142 A to 0.145 A, and the power increases to 0.355 W. As more bulbs fail, the current keeps rising, the filament of each bulb operates at a higher temperature, and the lifetime of the bulb is reduced. For this reason, you should check for failed (nonglowing) bulbs in such a series-wired string and replace them as soon as possible, in order to maximize the lifetimes of all the bulbs.

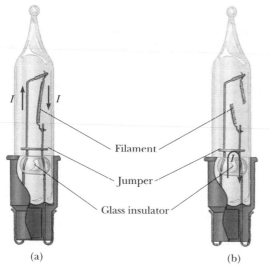

(a)

(b)   (c)

Figure 28.13 (a) Schematic diagram of a modern "miniature" holiday lightbulb, with a jumper connection to provide a current path if the filament breaks. When the filament is intact, charges flow in the filament. (b) A holiday lightbulb with a broken filament. In this case, charges flow in the jumper connection. (c) A Christmas tree lightbulb.

28.3 Kirchhoff's Rules

As we saw in the preceding section, simple circuits can be analyzed using the expression $\Delta V = IR$ and the rules for series and parallel combinations of resistors. Very often, however, it is not possible to reduce a circuit to a single loop. The procedure for analyzing more complex circuits is greatly simplified if we use two principles called **Kirchhoff's rules:**

1. **Junction rule.** The sum of the currents entering any junction in a circuit must equal the sum of the currents leaving that junction:

$$\sum I_{\text{in}} = \sum I_{\text{out}} \qquad (28.9)$$

2. **Loop rule.** The sum of the potential differences across all elements around any closed circuit loop must be zero:

$$\sum_{\substack{\text{closed} \\ \text{loop}}} \Delta V = 0 \qquad (28.10)$$

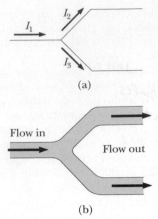

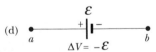

Figure 28.14 (a) Kirchhoff's junction rule. Conservation of charge requires that all charges entering a junction must leave that junction. Therefore, $I_1 = I_2 + I_3$. (b) A mechanical analog of the junction rule: the amount of water flowing out of the branches on the right must equal the amount flowing into the single branch on the left.

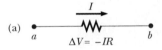

Figure 28.15 Rules for determining the potential differences across a resistor and a battery. (The battery is assumed to have no internal resistance.) Each circuit element is traversed from left to right.

Kirchhoff's first rule is a statement of conservation of electric charge. All charges that enter a given point in a circuit must leave that point because charge cannot build up at a point. If we apply this rule to the junction shown in Figure 28.14a, we obtain

$$I_1 = I_2 + I_3$$

Figure 28.14b represents a mechanical analog of this situation, in which water flows through a branched pipe having no leaks. Because water does not build up anywhere in the pipe, the flow rate into the pipe equals the total flow rate out of the two branches on the right.

Kirchhoff's second rule follows from the law of conservation of energy. Let us imagine moving a charge around a closed loop of a circuit. When the charge returns to the starting point, the charge–circuit system must have the same total energy as it had before the charge was moved. The sum of the increases in energy as the charge passes through some circuit elements must equal the sum of the decreases in energy as it passes through other elements. The potential energy decreases whenever the charge moves through a potential drop $-IR$ across a resistor or whenever it moves in the reverse direction through a source of emf. The potential energy increases whenever the charge passes through a battery from the negative terminal to the positive terminal.

When applying Kirchhoff's second rule in practice, we imagine *traveling* around the loop and consider changes in *electric potential*, rather than the changes in *potential energy* described in the preceding paragraph. You should note the following sign conventions when using the second rule:

- Because charges move from the high-potential end of a resistor toward the low-potential end, if a resistor is traversed in the direction of the current, the potential difference ΔV across the resistor is $-IR$ (Fig. 28.15a).

- If a resistor is traversed in the direction *opposite* the current, the potential difference ΔV across the resistor is $+IR$ (Fig. 28.15b).

- If a source of emf (assumed to have zero internal resistance) is traversed in the direction of the emf (from $-$ to $+$), the potential difference ΔV is $+\mathcal{E}$ (Fig. 28.15c). The emf of the battery increases the electric potential as we move through it in this direction.

- If a source of emf (assumed to have zero internal resistance) is traversed in the direction opposite the emf (from $+$ to $-$), the potential difference ΔV is $-\mathcal{E}$ (Fig. 28.15d). In this case the emf of the battery reduces the electric potential as we move through it.

Limitations exist on the numbers of times you can usefully apply Kirchhoff's rules in analyzing a circuit. You can use the junction rule as often as you need, so long as each time you write an equation you include in it a current that has not been used in a preceding junction-rule equation. In general, the number of times you can use the junction rule is one fewer than the number of junction points in the circuit. You can apply the loop rule as often as needed, as long as a new circuit element (resistor or battery) or a new current appears in each new equation. In general, **in order to solve a particular circuit problem, the number of independent equations you need to obtain from the two rules equals the number of unknown currents.**

Complex networks containing many loops and junctions generate great numbers of independent linear equations and a correspondingly great number of unknowns. Such situations can be handled formally through the use of matrix algebra. Computer software can also be used to solve for the unknowns.

The following examples illustrate how to use Kirchhoff's rules. In all cases, it is assumed that the circuits have reached steady-state conditions—that is, the currents in the various branches are constant. **Any capacitor acts as an open branch in a circuit;** that is, the current in the branch containing the capacitor is zero under steady-state conditions.

PROBLEM-SOLVING HINTS

Kirchhoff's Rules

- Draw a circuit diagram, and label all the known and unknown quantities. You must assign a *direction* to the current in each branch of the circuit. Although the assignment of current directions is arbitrary, you must adhere rigorously to the assigned directions when applying Kirchhoff's rules.

- Apply the junction rule to any junctions in the circuit that provide new relationships among the various currents.

- Apply the loop rule to as many loops in the circuit as are needed to solve for the unknowns. To apply this rule, you must correctly identify the potential difference as you imagine crossing each element while traversing the closed loop (either clockwise or counterclockwise). Watch out for errors in sign!

- Solve the equations simultaneously for the unknown quantities. Do not be alarmed if a current turns out to be negative; *its magnitude will be correct and the direction is opposite to that which you assigned.*

Gustav Kirchhoff
German Physicist (1824–1887)

Kirchhoff, a professor at Heidelberg, and Robert Bunsen invented the spectroscope and founded the science of spectroscopy, which we shall study in Chapter 42. They discovered the elements cesium and rubidium and invented astronomical spectroscopy. *(AIP ESVA/W.F. Meggers Collection)*

Quick Quiz 28.8 In using Kirchhoff's rules, you generally assign a separate unknown current to (a) each resistor in the circuit (b) each loop in the circuit (c) each branch in the circuit (d) each battery in the circuit.

Example 28.8   A Single-Loop Circuit

A single-loop circuit contains two resistors and two batteries, as shown in Figure 28.16. (Neglect the internal resistances of the batteries.)

(A) Find the current in the circuit.

Solution We do not need Kirchhoff's rules to analyze this simple circuit, but let us use them anyway just to see how they are applied. There are no junctions in this single-loop circuit; thus, the current is the same in all elements. Let us assume that the current is clockwise, as shown in Figure 28.16. Traversing the circuit in the clockwise direction, starting at a, we see that $a \rightarrow b$ represents a potential difference of $+\mathcal{E}_1$, $b \rightarrow c$ represents a potential difference of $-IR_1$, $c \rightarrow d$ represents a potential difference of $-\mathcal{E}_2$, and

$d \rightarrow a$ represents a potential difference of $-IR_2$. Applying Kirchhoff's loop rule gives

$$\sum \Delta V = 0$$

$$\mathcal{E}_1 - IR_1 - \mathcal{E}_2 - IR_2 = 0$$

Solving for I and using the values given in Figure 28.16, we obtain

$$(1) \qquad I = \frac{\mathcal{E}_1 - \mathcal{E}_2}{R_1 + R_2} = \frac{6.0 \text{ V} - 12 \text{ V}}{8.0 \ \Omega + 10 \ \Omega} = \boxed{-0.33 \text{ A}}$$

The negative sign for I indicates that the direction of the current is opposite the assumed direction. Notice that the emfs in the numerator subtract because the batteries have opposite polarities in Figure 28.16. In the denominator, the resistances add because the two resistors are in series.

(B) What power is delivered to each resistor? What power is delivered by the 12-V battery?

Solution Using Equation 27.23,

$$\mathcal{P}_1 = I^2 R_1 = (0.33 \text{ A})^2 (8.0 \ \Omega) = \boxed{0.87 \text{ W}}$$

$$\mathcal{P}_2 = I^2 R_2 = (0.33 \text{ A})^2 (10 \ \Omega) = \boxed{1.1 \text{ W}}$$

Hence, the total power delivered to the resistors is $\mathcal{P}_1 + \mathcal{P}_2 = 2.0$ W.

The 12-V battery delivers power $I\mathcal{E}_2 = 4.0$ W. Half of this power is delivered to the two resistors, as we just calculated. The other half is delivered to the 6-V battery, which is being

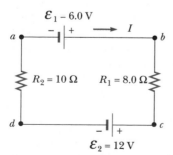

Figure 28.16 (Example 28.8) A series circuit containing two batteries and two resistors, where the polarities of the batteries are in opposition.

charged by the 12-V battery. If we had included the internal resistances of the batteries in our analysis, some of the power would appear as internal energy in the batteries; as a result, we would have found that less power was being delivered to the 6-V battery.

What If? What if the polarity of the 12.0-V battery were reversed? How would this affect the circuit?

Answer While we could repeat the Kirchhoff's rules calculation, let us examine Equation (1) and modify it accordingly. Because the polarities of the two batteries are now in the same direction, the signs of $\mathcal{E}_1$ and $\mathcal{E}_2$ are the same and Equation (1) becomes

$$I = \frac{\mathcal{E}_1 + \mathcal{E}_2}{R_1 + R_2} = \frac{6.0\,\text{V} + 12\,\text{V}}{8.0\,\Omega + 10\,\Omega} = 1.0\,\text{A}$$

The new powers delivered to the resistors are

$$\mathcal{P}_1 = I^2 R_1 = (1.0\,\text{A})^2 (8.0\,\Omega) = 8.0\,\text{W}$$

$$\mathcal{P}_2 = I^2 R_2 = (1.0\,\text{A})^2 (10\,\Omega) = 10\,\text{W}$$

This totals 18 W, nine times as much as in the original circuit, in which the batteries were opposing each other.

Example 28.9 Applying Kirchhoff's Rules
`Interactive`

Find the currents I_1, I_2, and I_3 in the circuit shown in Figure 28.17.

Solution Conceptualize by noting that we cannot simplify the circuit by the rules of adding resistances in series and in parallel. (If the 10.0-V battery were taken away, we could reduce the remaining circuit with series and parallel combinations.) Thus, we categorize this problem as one in which we must use Kirchhoff's rules. To analyze the circuit, we arbitrarily choose the directions of the currents as labeled in Figure 28.17. Applying Kirchhoff's junction rule to junction c gives

$$(1) \qquad I_1 + I_2 = I_3$$

We now have one equation with three unknowns—I_1, I_2, and I_3. There are three loops in the circuit—*abcda*, *befcb*, and *aefda*. We therefore need only two loop equations to determine the unknown currents. (The third loop equation would give no new information.) Applying Kirchhoff's loop rule to loops *abcda* and *befcb* and traversing these loops clockwise, we obtain the expressions

$$(2) \qquad abcda \quad 10.0\,\text{V} - (6.0\,\Omega)I_1 - (2.0\,\Omega)I_3 = 0$$

$$(3) \quad befcb \quad -14.0\,\text{V} + (6.0\,\Omega)I_1 - 10.0\,\text{V} - (4.0\,\Omega)I_2 = 0$$

Note that in loop *befcb* we obtain a positive value when traversing the 6.0-Ω resistor because our direction of travel is opposite the assumed direction of I_1. Expressions (1), (2), and (3) represent three independent equations with three unknowns. Substituting Equation (1) into Equation (2) gives

$$10.0\,\text{V} - (6.0\,\Omega)I_1 - (2.0\,\Omega)(I_1 + I_2) = 0$$

$$(4) \qquad 10.0\,\text{V} = (8.0\,\Omega)I_1 + (2.0\,\Omega)I_2$$

Dividing each term in Equation (3) by 2 and rearranging gives

$$(5) \qquad -12.0\,\text{V} = -(3.0\,\Omega)I_1 + (2.0\,\Omega)I_2$$

Subtracting Equation (5) from Equation (4) eliminates I_2, giving

$$22.0\,\text{V} = (11.0\,\Omega)I_1$$

$$I_1 = \boxed{2.0\,\text{A}}$$

Using this value of I_1 in Equation (5) gives a value for I_2:

$$(2.0\,\Omega)I_2 = (3.0\,\Omega)I_1 - 12.0\,\text{V}$$

$$= (3.0\,\Omega)(2.0\,\text{A}) - 12.0\,\text{V} = -6.0\,\text{V}$$

$$I_2 = \boxed{-3.0\,\text{A}}$$

Finally,

$$I_3 = I_1 + I_2 = \boxed{-1.0\,\text{A}}$$

To finalize the problem, note that I_2 and I_3 are both negative. This indicates only that the currents are opposite the direction we chose for them. However, the numerical values are correct. What would have happened had we left the current directions as labeled in Figure 28.17 but traversed the loops in the opposite direction?

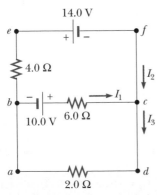

Figure 28.17 (Example 28.9) A circuit containing different branches.

Practice applying Kirchhoff's rules at the Interactive Worked Example link at http://www.pse6.com.

Example 28.10   A Multiloop Circuit

(A) Under steady-state conditions, find the unknown currents I_1, I_2, and I_3 in the multiloop circuit shown in Figure 28.18.

Solution First note that because the capacitor represents an open circuit, there is no current between g and b along path $ghab$ under steady-state conditions. Therefore, when the charges associated with I_1 reach point g, they all go toward point b through the 8.00-V battery; hence, $I_{gb} = I_1$. Labeling the currents as shown in Figure 28.18 and applying Equation 28.9 to junction c, we obtain

$$(1) \qquad I_1 + I_2 = I_3$$

Equation 28.10 applied to loops $defcd$ and $cfgbc$, traversed clockwise, gives

$$(2) \qquad defcd \quad 4.00\text{ V} - (3.00\ \Omega)I_2 - (5.00\ \Omega)I_3 = 0$$

$$(3) \qquad cfgbc \quad (3.00\ \Omega)I_2 - (5.00\ \Omega)I_1 + 8.00\text{ V} = 0$$

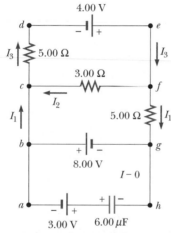

Figure 28.18 (Example 28.10) A multiloop circuit. Kirchhoff's loop rule can be applied to *any* closed loop, including the one containing the capacitor.

From Equation (1) we see that $I_1 = I_3 - I_2$, which, when substituted into Equation (3), gives

$$(4) \qquad (8.00\ \Omega)I_2 - (5.00\ \Omega)I_3 + 8.00\text{ V} = 0$$

Subtracting Equation (4) from Equation (2), we eliminate I_3 and find that

$$I_2 = -\frac{4.00\text{ V}}{11.0\ \Omega} = \boxed{-0.364\text{ A}}$$

Because our value for I_2 is negative, we conclude that the direction of I_2 is from c to f in the 3.00-Ω resistor. Despite this interpretation of the direction, however, we must continue to use this negative value for I_2 in subsequent calculations because our equations were established with our original choice of direction.

Using $I_2 = -0.364$ A in Equations (3) and (1) gives

$$I_1 = \boxed{1.38\text{ A}} \qquad I_3 = \boxed{1.02\text{ A}}$$

(B) What is the charge on the capacitor?

Solution We can apply Kirchhoff's loop rule to loop $bghab$ (or any other loop that contains the capacitor) to find the potential difference ΔV_{cap} across the capacitor. We use this potential difference in the loop equation without reference to a sign convention because the charge on the capacitor depends only on the magnitude of the potential difference. Moving clockwise around this loop, we obtain

$$-8.00\text{ V} + \Delta V_{cap} - 3.00\text{ V} = 0$$

$$\Delta V_{cap} = 11.0\text{ V}$$

Because $Q = C\,\Delta V_{cap}$ (see Eq. 26.1), the charge on the capacitor is

$$Q = (6.00\ \mu\text{F})(11.0\text{ V}) = \boxed{66.0\ \mu\text{C}}$$

Why is the left side of the capacitor positively charged?

28.4 *RC* Circuits

So far we have analyzed direct current circuits in which the current is constant. In DC circuits containing capacitors, the current is always in the same direction but may vary in time. A circuit containing a series combination of a resistor and a capacitor is called an **RC circuit.**

Charging a Capacitor

Figure 28.19 shows a simple series *RC* circuit. Let us assume that the capacitor in this circuit is initially uncharged. There is no current while switch S is open (Fig. 28.19b). If the switch is closed at $t = 0$, however, charge begins to flow, setting up a current in the circuit, and the capacitor begins to charge.[4] Note that during charging, charges do

---

[4]   In previous discussions of capacitors, we assumed a steady-state situation, in which no current was present in any branch of the circuit containing a capacitor. Now we are considering the case *before* the steady-state condition is realized; in this situation, charges are moving and a current exists in the wires connected to the capacitor.

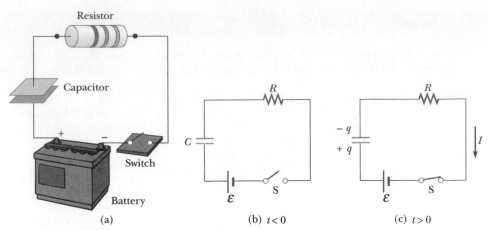

Active Figure 28.19 (a) A capacitor in series with a resistor, switch, and battery. (b) Circuit diagram representing this system at time $t < 0$, before the switch is closed. (c) Circuit diagram at time $t > 0$, after the switch has been closed.

not jump across the capacitor plates because the gap between the plates represents an open circuit. Instead, charge is transferred between each plate and its connecting wires due to the electric field established in the wires by the battery, until the capacitor is fully charged. As the plates are being charged, the potential difference across the capacitor increases. The value of the maximum charge on the plates depends on the voltage of the battery. Once the maximum charge is reached, the current in the circuit is zero because the potential difference across the capacitor matches that supplied by the battery.

To analyze this circuit quantitatively, let us apply Kirchhoff's loop rule to the circuit after the switch is closed. Traversing the loop in Fig. 28.19c clockwise gives

$$\mathcal{E} - \frac{q}{C} - IR = 0 \qquad (28.11)$$

where q/C is the potential difference across the capacitor and IR is the potential difference across the resistor. We have used the sign conventions discussed earlier for the signs on $\mathcal{E}$ and IR. For the capacitor, notice that we are traveling in the direction from the positive plate to the negative plate; this represents a decrease in potential. Thus, we use a negative sign for this potential difference in Equation 28.11. Note that q and I are *instantaneous* values that depend on time (as opposed to steady-state values) as the capacitor is being charged.

We can use Equation 28.11 to find the initial current in the circuit and the maximum charge on the capacitor. At the instant the switch is closed ($t = 0$), the charge on the capacitor is zero, and from Equation 28.11 we find that the initial current I_0 in the circuit is a maximum and is equal to

$$I_0 = \frac{\mathcal{E}}{R} \qquad \text{(current at } t = 0) \qquad (28.12)$$

At this time, the potential difference from the battery terminals appears entirely across the resistor. Later, when the capacitor is charged to its maximum value Q, charges cease to flow, the current in the circuit is zero, and the potential difference from the battery terminals appears entirely across the capacitor. Substituting $I = 0$ into Equation 28.11 gives the charge on the capacitor at this time:

$$Q = C\mathcal{E} \qquad \text{(maximum charge)} \qquad (28.13)$$

To determine analytical expressions for the time dependence of the charge and current, we must solve Equation 28.11—a single equation containing two variables, q and I. The current in all parts of the series circuit must be the same. Thus, the current in the resistance R must be the same as the current between the capacitor plates and the wires. This current is equal to the time rate of change of the charge on the capacitor plates. Thus, we substitute $I = dq/dt$ into Equation 28.11 and rearrange the equation:

$$\frac{dq}{dt} = \frac{\mathcal{E}}{R} - \frac{q}{RC}$$

To find an expression for q, we solve this separable differential equation. We first combine the terms on the right-hand side:

$$\frac{dq}{dt} = \frac{C\mathcal{E}}{RC} - \frac{q}{RC} = -\frac{q - C\mathcal{E}}{RC}$$

Now we multiply by dt and divide by $q - C\mathcal{E}$ to obtain

$$\frac{dq}{q - C\mathcal{E}} = -\frac{1}{RC}\,dt$$

Integrating this expression, using the fact that $q = 0$ at $t = 0$, we obtain

$$\int_0^q \frac{dq}{(q - C\mathcal{E})} = -\frac{1}{RC} \int_0^t dt$$

$$\ln\left(\frac{q - C\mathcal{E}}{-C\mathcal{E}}\right) = -\frac{t}{RC}$$

From the definition of the natural logarithm, we can write this expression as

$$q(t) = C\mathcal{E}(1 - e^{-t/RC}) = Q(1 - e^{-t/RC}) \qquad (28.14)$$

Charge as a function of time for a capacitor being charged

where e is the base of the natural logarithm and we have made the substitution from Equation 28.13.

We can find an expression for the charging current by differentiating Equation 28.14 with respect to time. Using $I = dq/dt$, we find that

$$I(t) = \frac{\mathcal{E}}{R}\,e^{-t/RC} \qquad (28.15)$$

Current as a function of time for a capacitor being charged

Plots of capacitor charge and circuit current versus time are shown in Figure 28.20. Note that the charge is zero at $t = 0$ and approaches the maximum value $C\mathcal{E}$ as $t \rightarrow \infty$. The current has its maximum value $I_0 = \mathcal{E}/R$ at $t = 0$ and decays exponentially to zero as $t \rightarrow \infty$. The quantity RC, which appears in the exponents of Equations 28.14 and 28.15, is called the **time constant** τ of the circuit. It represents the time interval during which the current decreases to $1/e$ of its initial value; that is, in a time interval τ, $I = e^{-1}I_0 = 0.368I_0$. In a time interval 2τ, $I = e^{-2}I_0 = 0.135I_0$, and so forth. Likewise, in a time interval τ, the charge increases from zero to $C\mathcal{E}[1 - e^{-1}] = 0.632C\mathcal{E}$.

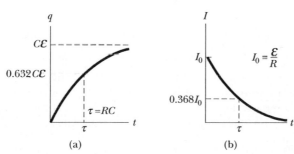

(a)　　　　　　　　(b)

Figure 28.20 (a) Plot of capacitor charge versus time for the circuit shown in Figure 28.19. After a time interval equal to one time constant τ has passed, the charge is 63.2% of the maximum value $C\mathcal{E}$. The charge approaches its maximum value as t approaches infinity. (b) Plot of current versus time for the circuit shown in Figure 28.19. The current has its maximum value $I_0 = \mathcal{E}/R$ at $t = 0$ and decays to zero exponentially as t approaches infinity. After a time interval equal to one time constant τ has passed, the current is 36.8% of its initial value.

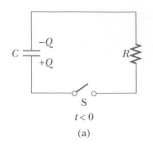

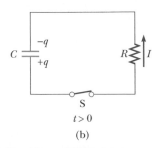

Active Figure 28.21 (a) A charged capacitor connected to a resistor and a switch, which is open for $t < 0$. (b) After the switch is closed at $t = 0$, a current that decreases in magnitude with time is set up in the direction shown, and the charge on the capacitor decreases exponentially with time.

At the Active Figures link at http://www.pse6.com, you can adjust the values of R and C to see the effect on the discharging of the capacitor.

The following dimensional analysis shows that τ has the units of time:

$$[\tau] = [RC] = \left[\frac{\Delta V}{I} \times \frac{Q}{\Delta V} \right] = \left[\frac{Q}{Q/\Delta t} \right] = [\Delta t] = \text{T}$$

Because $\tau = RC$ has units of time, the combination τ/RC is dimensionless, as it must be in order to be an exponent of e in Equations 28.14 and 28.15.

The energy output of the battery as the capacitor is fully charged is $Q\mathcal{E} = C\mathcal{E}^2$. After the capacitor is fully charged, the energy stored in the capacitor is $\frac{1}{2}Q\mathcal{E} = \frac{1}{2}C\mathcal{E}^2$, which is just half the energy output of the battery. It is left as a problem (Problem 64) to show that the remaining half of the energy supplied by the battery appears as internal energy in the resistor.

Discharging a Capacitor

Now consider the circuit shown in Figure 28.21, which consists of a capacitor carrying an initial charge Q, a resistor, and a switch. When the switch is open, a potential difference Q/C exists across the capacitor and there is zero potential difference across the resistor because $I = 0$. If the switch is closed at $t = 0$, the capacitor begins to discharge through the resistor. At some time t during the discharge, the current in the circuit is I and the charge on the capacitor is q (Fig. 28.21b). The circuit in Figure 28.21 is the same as the circuit in Figure 28.19 except for the absence of the battery. Thus, we eliminate the emf $\mathcal{E}$ from Equation 28.11 to obtain the appropriate loop equation for the circuit in Figure 28.21:

$$-\frac{q}{C} - IR = 0 \tag{28.16}$$

When we substitute $I = dq/dt$ into this expression, it becomes

$$-R\frac{dq}{dt} = \frac{q}{C}$$

$$\frac{dq}{q} = -\frac{1}{RC}\,dt$$

Integrating this expression, using the fact that $q = Q$ at $t = 0$ gives

$$\int_{Q}^{q} \frac{dq}{q} = -\frac{1}{RC}\int_{0}^{t} dt$$

$$\ln\left(\frac{q}{Q}\right) = -\frac{t}{RC}$$

Charge as a function of time for a discharging capacitor

$$q(t) = Qe^{-t/RC} \tag{28.17}$$

Differentiating this expression with respect to time gives the instantaneous current as a function of time:

Current as a function of time for a discharging capacitor

$$I(t) = \frac{dq}{dt} = \frac{d}{dt}\left(Qe^{-t/RC}\right) = -\frac{Q}{RC}e^{-t/RC} \tag{28.18}$$

where $Q/RC = I_0$ is the initial current. The negative sign indicates that as the capacitor discharges, the current direction is opposite its direction when the capacitor was being charged. (Compare the current directions in Figs. 28.19c and 28.21b.) We see that both the charge on the capacitor and the current decay exponentially at a rate characterized by the time constant $\tau = RC$.

Quick Quiz 28.9 Consider the circuit in Figure 28.19 and assume that the battery has no internal resistance. Just after the switch is closed, the potential difference across which of the following is equal to the emf of the battery? (a) C (b) R (c) neither C nor R. After a very long time, the potential difference across which of the following is equal to the emf of the battery? (d) C (e) R (f) neither C nor R.

Quick Quiz 28.10 Consider the circuit in Figure 28.22 and assume that the battery has no internal resistance. Just after the switch is closed, the current in the battery is (a) zero (b) $\mathcal{E}/2R$ (c) $2\mathcal{E}/R$ (d) $\mathcal{E}/R$ (e) impossible to determine. After a very long time, the current in the battery is (f) zero (g) $\mathcal{E}/2R$ (h) $2\mathcal{E}/R$ (i) $\mathcal{E}/R$ (j) impossible to determine.

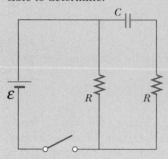

Figure 28.22 (Quick Quiz 28.10) How does the current vary after the switch is closed?

Conceptual Example 28.11   Intermittent Windshield Wipers

Many automobiles are equipped with windshield wipers that can operate intermittently during a light rainfall. How does the operation of such wipers depend on the charging and discharging of a capacitor?

Solution The wipers are part of an RC circuit whose time constant can be varied by selecting different values of R through a multiposition switch. As it increases with time, the voltage across the capacitor reaches a point at which it triggers the wipers and discharges, ready to begin another charging cycle. The time interval between the individual sweeps of the wipers is determined by the value of the time constant.

Example 28.12   Charging a Capacitor in an RC Circuit   `Interactive`

An uncharged capacitor and a resistor are connected in series to a battery, as shown in Figure 28.23. If $\mathcal{E} = 12.0$ V, $C = 5.00\ \mu\text{F}$, and $R = 8.00 \times 10^5\ \Omega$, find the time constant of the circuit, the maximum charge on the capacitor, the maximum current in the circuit, and the charge and current as functions of time.

Solution The time constant of the circuit is $\tau = RC = (8.00 \times 10^5\ \Omega)(5.00 \times 10^{-6}\ \text{F}) = 4.00$ s. The maximum charge on the capacitor is $Q = C\mathcal{E} = (5.00\ \mu\text{F})(12.0\ \text{V}) = 60.0\ \mu\text{C}$. The maximum current in the circuit is $I_0 = \mathcal{E}/R = (12.0\ \text{V})/(8.00 \times 10^5\ \Omega) = 15.0\ \mu\text{A}$. Using these values and Equations 28.14 and 28.15, we find that

$$q(t) = (60.0\ \mu\text{C})(1 - e^{-t/4.00\ \text{s}})$$

$$I(t) = (15.0\ \mu\text{A})e^{-t/4.00\ \text{s}}$$

Graphs of these functions are provided in Figure 28.24.

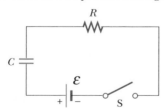

Figure 28.23 (Example 28.12) The switch in this series RC circuit, open for times $t < 0$, is closed at $t = 0$.

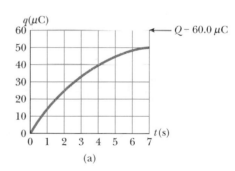

(a)

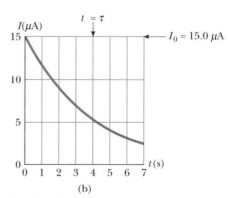

(b)

Figure 28.24 (Example 28.12) Plots of (a) charge versus time and (b) current versus time for the RC circuit shown in Figure 28.23, with $\mathcal{E} = 12.0$ V, $R = 8.00 \times 10^5\ \Omega$, and $C = 5.00\ \mu\text{F}$.

 At the Interactive Worked Example link at* http://www.pse6.com, *you can vary R, C, and $\mathcal{E}$ and observe the charge and current as functions of time while charging or discharging the capacitor.

Example 28.13 Discharging a Capacitor in an *RC* Circuit

Consider a capacitor of capacitance C that is being discharged through a resistor of resistance R, as shown in Figure 28.21.

(A) After how many time constants is the charge on the capacitor one-fourth its initial value?

Solution The charge on the capacitor varies with time according to Equation 28.17, $q(t) = Qe^{-t/RC}$. To find the time interval during which q drops to one-fourth its initial value, we substitute $q(t) = Q/4$ into this expression and solve for t:

$$\frac{Q}{4} = Qe^{-t/RC}$$

$$\frac{1}{4} = e^{-t/RC}$$

Taking logarithms of both sides, we find

$$-\ln 4 = -\frac{t}{RC}$$

$$t = RC\,(\ln 4) = 1.39RC = \boxed{1.39\tau}$$

(B) The energy stored in the capacitor decreases with time as the capacitor discharges. After how many time constants is this stored energy one-fourth its initial value?

Solution Using Equations 26.11 ($U = Q^2/2C$) and 28.17, we can express the energy stored in the capacitor at any time t as

$$U = \frac{q^2}{2C} = \frac{Q^2}{2C}\,e^{-2t/RC} = U_0 e^{-2t/RC}$$

where $U_0 = Q^2/2C$ is the initial energy stored in the capacitor. As in part (A), we now set $U = U_0/4$ and solve for t:

$$\frac{U_0}{4} = U_0 e^{-2t/RC}$$

$$\frac{1}{4} = e^{-2t/RC}$$

Again, taking logarithms of both sides and solving for t gives

$$t = \tfrac{1}{2}RC \ln 4 = 0.693RC = \boxed{0.693\tau}$$

What If? What if we wanted to describe the circuit in terms of the time interval required for the charge to fall to one-half its original value, rather than by the time constant τ? This would give a parameter for the circuit called its *half-life* $t_{1/2}$. How is the half-life related to the time constant?

Answer After one half-life, the charge has fallen from Q to $Q/2$. Thus, from Equation 28.17,

$$\frac{Q}{2} = Qe^{-t_{1/2}/RC}$$

$$\tfrac{1}{2} = e^{-t_{1/2}/RC}$$

leading to

$$t_{1/2} = 0.693\tau$$

The concept of half-life will be important to us when we study nuclear decay in Chapter 44. The radioactive decay of an unstable sample behaves in a mathematically similar manner to a discharging capacitor in an *RC* circuit.

Example 28.14 Energy Delivered to a Resistor

A 5.00-μF capacitor is charged to a potential difference of 800 V and then discharged through a 25.0-kΩ resistor. How much energy is delivered to the resistor in the time interval required to fully discharge the capacitor?

Solution We shall solve this problem in two ways. The first way is to note that the initial energy in the circuit equals the energy stored in the capacitor, $C\mathcal{E}^2/2$ (see Eq. 26.11). Once the capacitor is fully discharged, the energy stored in it is zero. Because energy in an isolated system is conserved, the initial energy stored in the capacitor is transformed into internal energy in the resistor. Using the given values of C and $\mathcal{E}$, we find

$$\text{Energy} = \tfrac{1}{2}C\mathcal{E}^2 = \tfrac{1}{2}(5.00 \times 10^{-6}\,\text{F})(800\,\text{V})^2 = \boxed{1.60\,\text{J}}$$

The second way, which is more difficult but perhaps more instructive, is to note that as the capacitor discharges through the resistor, the rate at which energy is delivered to the resistor is given by I^2R, where I is the instantaneous current given by Equation 28.18. Because power is defined as the rate at which energy is transferred, we conclude that the energy delivered to the resistor must equal the time integral of $I^2R\,dt$:

$$\text{Energy} = \int_0^\infty I^2R\,dt = \int_0^\infty (-I_0 e^{-t/RC})^2\,R\,dt$$

To evaluate this integral, we note that the initial current I_0 is equal to $\mathcal{E}/R$ and that all parameters except t are constant. Thus, we find

$$(1) \qquad \text{Energy} = \frac{\mathcal{E}^2}{R}\int_0^\infty e^{-2t/RC}\,dt$$

This integral has a value of $RC/2$ (see Problem 35); hence, we find

$$\text{Energy} = \tfrac{1}{2}C\mathcal{E}^2$$

which agrees with the result we obtained using the simpler approach, as it must. Note that we can use this second approach to find the total energy delivered to the resistor at *any* time after the switch is closed by simply replacing the upper limit in the integral with that specific value of t.

28.5 Electrical Meters

The Galvanometer

The **galvanometer** is the main component in analog meters for measuring current and voltage. (Many analog meters are still in use although digital meters, which operate on a different principle, are currently in wide use.) Figure 28.25 illustrates the essential features of a common type called the *D'Arsonval galvanometer.* It consists of a coil of wire mounted so that it is free to rotate on a pivot in a magnetic field provided by a permanent magnet. The basic operation of the galvanometer uses the fact that a torque acts on a current loop in the presence of a magnetic field (Chapter 29). The torque experienced by the coil is proportional to the current in it: the larger the current, the greater the torque and the more the coil rotates before the spring tightens enough to stop the rotation. Hence, the deflection of a needle attached to the coil is proportional to the current. Once the instrument is properly calibrated, it can be used in conjunction with other circuit elements to measure either currents or potential differences.

The Ammeter

A device that measures current is called an **ammeter.** The charges constituting the current to be measured must pass directly through the ammeter, so the ammeter must be connected in series with other elements in the circuit, as shown in Figure 28.26. When using an ammeter to measure direct currents, you must connect it so that charges enter the instrument at the positive terminal and exit at the negative terminal.

Ideally, an ammeter should have zero resistance so that the current being measured is not altered. In the circuit shown in Figure 28.26, this condition requires that the resistance of the ammeter be much less than $R_1 + R_2$. Because any ammeter always has some internal resistance, the presence of the ammeter in the circuit slightly reduces the current from the value it would have in the meter's absence.

A typical off-the-shelf galvanometer is often not suitable for use as an ammeter, primarily because it has a resistance of about 60 Ω. An ammeter resistance this great considerably alters the current in a circuit. You can understand this by considering the following example. The current in a simple series circuit containing a 3-V battery and a 3-Ω resistor is 1 A. If you insert a 60-Ω galvanometer in this circuit to measure the current, the total resistance becomes 63 Ω and the current is reduced to 0.048 A!

A second factor that limits the use of a galvanometer as an ammeter is the fact that a typical galvanometer gives a full-scale deflection for currents on the order of 1 mA or less. Consequently, such a galvanometer cannot be used directly to measure currents greater than this value. However, it can be converted to a useful ammeter by placing a shunt resistor R_p in parallel with the galvanometer, as shown in Figure 28.27. The value of R_p must be much less than the galvanometer resistance so that most of the current to be measured is directed to the shunt resistor.

The Voltmeter

A device that measures potential difference is called a **voltmeter.** The potential difference between any two points in a circuit can be measured by attaching the terminals of the voltmeter between these points without breaking the circuit, as shown in Figure 28.28. The potential difference across resistor R_2 is measured by connecting the voltmeter in parallel with R_2. Again, it is necessary to observe the polarity of the instrument. The positive terminal of the voltmeter must be connected to the end of the resistor that is at the higher potential, and the negative terminal to the end of the resistor at the lower potential.

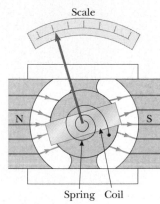

Scale

Spring   Coil

Figure 28.25 The principal components of a D'Arsonval galvanometer. When the coil situated in a magnetic field carries a current, the magnetic torque causes the coil to twist. The angle through which the coil rotates is proportional to the current in the coil because of the counteracting torque of the spring.

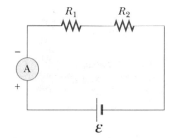

Figure 28.26 Current can be measured with an ammeter connected in series with the elements in which the measurement of a current is desired. An ideal ammeter has zero resistance.

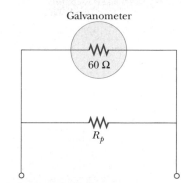

Active Figure 28.27 A galvanometer is represented here by its internal resistance of 60 Ω. When a galvanometer is to be used as an ammeter, a shunt resistor R_p is connected in parallel with the galvanometer.

At the Active Figures link at http://www.pse6.com, you can predict the value of R_p needed to cause full-scale deflection in the circuit of Figure 28.26, and test your result.

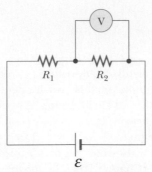

Figure 28.28 The potential difference across a resistor can be measured with a voltmeter connected in parallel with the resistor. An ideal voltmeter has infinite resistance.

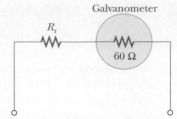

Active Figure 28.29 When the galvanometer is used as a voltmeter, a resistor R_s is connected in series with the galvanometer.

At the Active Figures link at http://www.pse6.com, you can predict the value of R_s needed to cause full-scale deflection in the circuit of Figure 28.28, and test your result.

An ideal voltmeter has infinite resistance so that no current exists in it. In Figure 28.28, this condition requires that the voltmeter have a resistance much greater than R_2. In practice, if this condition is not met, corrections should be made for the known resistance of the voltmeter.

A galvanometer can also be used as a voltmeter by adding an external resistor R_s in series with it, as shown in Figure 28.29. In this case, the external resistor must have a value much greater than the resistance of the galvanometer to ensure that the galvanometer does not significantly alter the voltage being measured.

28.6 Household Wiring and Electrical Safety

Household circuits represent a practical application of some of the ideas presented in this chapter. In our world of electrical appliances, it is useful to understand the power requirements and limitations of conventional electrical systems and the safety measures that prevent accidents.

In a conventional installation, the utility company distributes electric power to individual homes by means of a pair of wires, with each home connected in parallel to these wires. One wire is called the *live wire*,[5] as illustrated in Figure 28.30, and the other is called the *neutral wire*. The neutral wire is grounded; that is, its electric potential is taken to be zero. The potential difference between the live and neutral wires is about 120 V. This voltage alternates in time, and the potential of the live wire oscillates relative to ground. Much of what we have learned so far for the constant-emf situation (direct current) can also be applied to the alternating current that power companies supply to businesses and households. (Alternating voltage and current are discussed in Chapter 33.)

A meter is connected in series with the live wire entering the house to record the household's energy consumption. After the meter, the wire splits so that there are several separate circuits in parallel distributed throughout the house. Each circuit contains a circuit breaker (or, in older installations, a fuse). The wire and circuit breaker for each circuit are carefully selected to meet the current demands for that circuit. If a circuit is to carry currents as large as 30 A, a heavy wire and an appropriate circuit breaker must be selected to handle this current. A circuit used to power only lamps and small appliances often requires only 20 A. Each circuit has its own circuit breaker to provide protection for that part of the entire electrical system of the house.

Figure 28.30 Wiring diagram for a household circuit. The resistances represent appliances or other electrical devices that operate with an applied voltage of 120 V.

---

[5] *Live wire* is a common expression for a conductor whose electric potential is above or below ground potential.

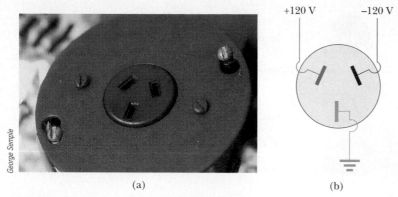

Figure 28.31 (a) An outlet for connection to a 240-V supply. (b) The connections for each of the openings in a 240-V outlet.

As an example, consider a circuit in which a toaster oven, a microwave oven, and a coffee maker are connected (corresponding to R_1, R_2, and R_3 in Fig. 28.30). We can calculate the current in each appliance by using the expression $\mathcal{P} = I \, \Delta V$. The toaster oven, rated at 1 000 W, draws a current of 1 000 W/120 V = 8.33 A. The microwave oven, rated at 1 300 W, draws 10.8 A, and the coffee maker, rated at 800 W, draws 6.67 A. If the three appliances are operated simultaneously, they draw a total current of 25.8 A. Therefore, the circuit should be wired to handle at least this much current. If the rating of the circuit breaker protecting the circuit is too small—say, 20 A—the breaker will be tripped when the third appliance is turned on, preventing all three appliances from operating. To avoid this situation, the toaster oven and coffee maker can be operated on one 20-A circuit and the microwave oven on a separate 20-A circuit.

Many heavy-duty appliances, such as electric ranges and clothes dryers, require 240 V for their operation. The power company supplies this voltage by providing a third wire that is 120 V below ground potential (Fig. 28.31). The potential difference between this live wire and the other live wire (which is 120 V above ground potential) is 240 V. An appliance that operates from a 240-V line requires half as much current compared to operating it at 120 V; therefore, smaller wires can be used in the higher-voltage circuit without overheating.

Electrical Safety

When the live wire of an electrical outlet is connected directly to ground, the circuit is completed and a short-circuit condition exists. A *short circuit* occurs when almost zero resistance exists between two points at different potentials; this results in a very large current. When this happens accidentally, a properly operating circuit breaker opens the circuit and no damage is done. However, a person in contact with ground can be electrocuted by touching the live wire of a frayed cord or other exposed conductor. An exceptionally effective (and dangerous!) ground contact is made when the person either touches a water pipe (normally at ground potential) or stands on the ground with wet feet. The latter situation represents effective ground contact because normal, nondistilled water is a conductor due to the large number of ions associated with impurities. This situation should be avoided at all cost.

Electric shock can result in fatal burns, or it can cause the muscles of vital organs, such as the heart, to malfunction. The degree of damage to the body depends on the magnitude of the current, the length of time it acts, the part of the body touched by the live wire, and the part of the body in which the current exists. Currents of 5 mA or less cause a sensation of shock but ordinarily do little or no damage. If the current is larger than about 10 mA, the muscles contract and the person may be unable to release the live wire. If a current of about 100 mA passes through the body for only a few seconds, the result can be fatal. Such a large current paralyzes the respiratory

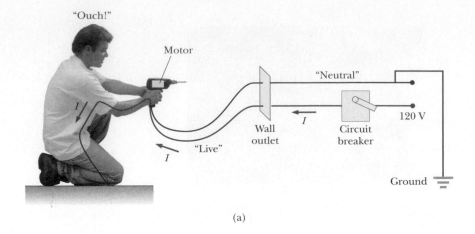

(a)

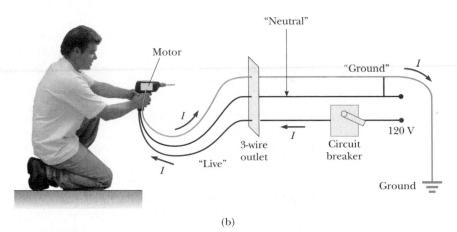

(b)

Figure 28.32 (a) A diagram of the circuit for an electric drill with only two connecting wires. The normal current path is from the live wire through the motor connections and back to ground through the neutral wire. In the situation shown, the live wire has come into contact with the drill case. As a result, the person holding the drill acts as a current path to ground and receives an electric shock. (b) This shock can be avoided by connecting the drill case to ground through a third ground wire. In this situation, the drill case remains at ground potential and no current exists in the person.

muscles and prevents breathing. In some cases, currents of about 1 A can produce serious (and sometimes fatal) burns. In practice, no contact with live wires is regarded as safe whenever the voltage is greater than 24 V.

Many 120-V outlets are designed to accept a three-pronged power cord. (This feature is required in all new electrical installations.) One of these prongs is the live wire at a nominal potential of 120 V. The second is the neutral wire, nominally at 0 V, and carries current to ground. The third, round prong is a safety ground wire that normally carries no current but is both grounded and connected directly to the casing of the appliance (see Figure 28.32). If the live wire is accidentally shorted to the casing (which can occur if the wire insulation wears off), most of the current takes the low-resistance path through the appliance to ground. In contrast, if the casing of the appliance is not properly grounded and a short occurs, anyone in contact with the appliance experiences an electric shock because the body provides a low-resistance path to ground.

Special power outlets called *ground-fault interrupters* (GFIs) are now used in kitchens, bathrooms, basements, exterior outlets, and other hazardous areas of new homes. These devices are designed to protect persons from electric shock by sensing small currents (≈ 5 mA) leaking to ground. (The principle of their operation is described in Chapter 31.) When an excessive leakage current is detected, the current is shut off in less than 1 ms.

SUMMARY

The **emf** of a battery is equal to the voltage across its terminals when the current is zero. That is, the emf is equivalent to the **open-circuit voltage** of the battery.

The **equivalent resistance** of a set of resistors connected in **series** is

$$R_{eq} = R_1 + R_2 + R_3 + \cdots \tag{28.6}$$

The **equivalent resistance** of a set of resistors connected in **parallel** is found from the relationship

$$\frac{1}{R_{eq}} = \frac{1}{R_1} + \frac{1}{R_2} + \frac{1}{R_3} + \cdots \tag{28.8}$$

If it is possible to combine resistors into series or parallel equivalents, the preceding two equations make it easy to determine how the resistors influence the rest of the circuit.

Circuits involving more than one loop are conveniently analyzed with the use of **Kirchhoff's rules:**

1. **Junction rule.** The sum of the currents entering any junction in an electric circuit must equal the sum of the currents leaving that junction:

$$\sum I_{in} = \sum I_{out} \tag{28.9}$$

2. **Loop rule.** The sum of the potential differences across all elements around any circuit loop must be zero:

$$\sum_{\substack{closed \\ loop}} \Delta V = 0 \tag{28.10}$$

The first rule is a statement of conservation of charge; the second is equivalent to a statement of conservation of energy.

When a resistor is traversed in the direction of the current, the potential difference ΔV across the resistor is $-IR$. When a resistor is traversed in the direction opposite the current, $\Delta V = +IR$. When a source of emf is traversed in the direction of the emf (negative terminal to positive terminal), the potential difference is $+\mathcal{E}$. When a source of emf is traversed opposite the emf (positive to negative), the potential difference is $-\mathcal{E}$. The use of these rules together with Equations 28.9 and 28.10 allows you to analyze electric circuits.

If a capacitor is charged with a battery through a resistor of resistance R, the charge on the capacitor and the current in the circuit vary in time according to the expressions

$$q(t) = Q(1 - e^{-t/RC}) \tag{28.14}$$

$$I(t) = \frac{\mathcal{E}}{R} e^{-t/RC} \tag{28.15}$$

where $Q = C\mathcal{E}$ is the maximum charge on the capacitor. The product RC is called the **time constant** τ of the circuit. If a charged capacitor is discharged through a resistor of resistance R, the charge and current decrease exponentially in time according to the expressions

$$q(t) = Qe^{-t/RC} \tag{28.17}$$

$$I(t) = -I_0 e^{-t/RC} \tag{28.18}$$

where Q is the initial charge on the capacitor and $I_0 = Q/RC$ is the initial current in the circuit.

Take a practice test for this chapter by clicking on the Practice Test link at http://www.pse6.com.

QUESTIONS

1. Explain the difference between load resistance in a circuit and internal resistance in a battery.

2. Under what condition does the potential difference across the terminals of a battery equal its emf? Can the terminal voltage ever exceed the emf? Explain.

3. Is the direction of current through a battery always from the negative terminal to the positive terminal? Explain.

4. How would you connect resistors so that the equivalent resistance is larger than the greatest individual resistance? Give an example involving three resistors.

5. How would you connect resistors so that the equivalent resistance is smaller than the least individual resistance? Give an example involving three resistors.

6. Given three lightbulbs and a battery, sketch as many different electric circuits as you can.

7. When resistors are connected in series, which of the following would be the same for each resistor: potential difference, current, power?

8. When resistors are connected in parallel, which of the following would be the same for each resistor: potential difference, current, power?

9. What advantage might there be in using two identical resistors in parallel connected in series with another identical parallel pair, rather than just using a single resistor?

10. An incandescent lamp connected to a 120-V source with a short extension cord provides more illumination than the same lamp connected to the same source with a very long extension cord. Explain.

11. Why is it possible for a bird to sit on a high-voltage wire without being electrocuted?

12. When can the potential difference across a resistor be positive?

13. Referring to Figure Q28.13, describe what happens to the lightbulb after the switch is closed. Assume that the capacitor has a large capacitance and is initially uncharged, and assume that the light illuminates when connected directly across the battery terminals.

14. What is the internal resistance of an ideal ammeter? Of an ideal voltmeter? Do real meters ever attain these ideals?

15. A "short circuit" is a path of very low resistance in a circuit in parallel with some other part of the circuit. Discuss the effect of the short circuit on the portion of the circuit it parallels. Use a lamp with a frayed cord as an example.

16. If electric power is transmitted over long distances, the resistance of the wires becomes significant. Why? Which method of transmission would result in less energy wasted—high current and low voltage or low current and high voltage? Explain your answer.

17. Are the two headlights of a car wired in series or in parallel? How can you tell?

18. Embodied in Kirchhoff's rules are two conservation laws. What are they?

19. Figure Q28.19 shows a series combination of three lightbulbs, all rated at 120 V with power ratings of 60 W, 75 W, and 200 W. Why is the 60-W lamp the brightest and the 200-W lamp the dimmest? Which bulb has the greatest resistance? How would their intensities differ if they were connected in parallel?

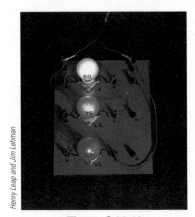

Figure Q28.19

Henry Leap and Jim Lehman

20. A student claims that the second lightbulb in series is less bright than the first, because the first bulb uses up some of the current. How would you respond to this statement?

21. Is a circuit breaker wired in series or in parallel with the device it is protecting?

22. So that your grandmother can listen to *A Prairie Home Companion*, you take her bedside radio to the hospital where she is staying. You are required to have a maintenance worker test it for electrical safety. Finding that it develops 120 V on one of its knobs, he does not let you take it up to your grandmother's room. She complains that she has had the radio for many years and nobody has ever gotten a shock from it. You end up having to buy a new plastic radio. Is this fair? Will the old radio be safe back in her bedroom?

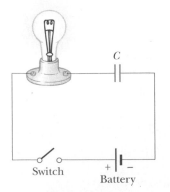

Figure Q28.13

23. Suppose you fall from a building and on the way down grab a high-voltage wire. If the wire supports you as you hang from it, will you be electrocuted? If the wire then breaks, should you continue to hold onto an end of the wire as you fall?

24. What advantage does 120-V operation offer over 240 V? What disadvantages?

25. When electricians work with potentially live wires, they often use the backs of their hands or fingers to move wires. Why do you suppose they use this technique?

26. What procedure would you use to try to save a person who is "frozen" to a live high-voltage wire without endangering your own life?

27. If it is the current through the body that determines how serious a shock will be, why do we see warnings of *high voltage* rather than *high current* near electrical equipment?

28. Suppose you are flying a kite when it strikes a high-voltage wire. What factors determine how great a shock you receive?

29. A series circuit consists of three identical lamps connected to a battery as shown in Figure Q28.29. When the switch S is closed, what happens (a) to the intensities of lamps A and B; (b) to the intensity of lamp C; (c) to the current in the circuit; and (d) to the voltage across the three lamps? (e) Does the power delivered to the circuit increase, decrease, or remain the same?

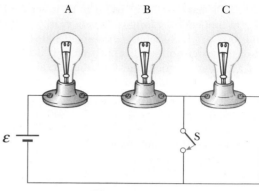

Figure Q28.29

30. If your car's headlights are on when you start the ignition, why do they dim while the car is starting?

31. A ski resort consists of a few chair lifts and several interconnected downhill runs on the side of a mountain, with a lodge at the bottom. The lifts are analogous to batteries, and the runs are analogous to resistors. Describe how two runs can be in series. Describe how three runs can be in parallel. Sketch a junction of one lift and two runs. State Kirchhoff's junction rule for ski resorts. One of the skiers happens to be carrying a skydiver's altimeter. She never takes the same set of lifts and runs twice, but keeps passing you at the fixed location where you are working. State Kirchhoff's loop rule for ski resorts.

PROBLEMS

1, 2, 3 = straightforward, intermediate, challenging   ☐ = full solution available in the *Student Solutions Manual and Study Guide*

🌐 = coached solution with hints available at http://www.pse.com   🖳 = computer useful in solving problem

▨ = paired numerical and symbolic problems

Section 28.1 Electromotive Force

1. 🌐 A battery has an emf of 15.0 V. The terminal voltage of the battery is 11.6 V when it is delivering 20.0 W of power to an external load resistor R. (a) What is the value of R? (b) What is the internal resistance of the battery?

2. (a) What is the current in a 5.60-Ω resistor connected to a battery that has a 0.200-Ω internal resistance if the terminal voltage of the battery is 10.0 V? (b) What is the emf of the battery?

3. Two 1.50-V batteries—with their positive terminals in the same direction—are inserted in series into the barrel of a flashlight. One battery has an internal resistance of 0.255 Ω, the other an internal resistance of 0.153 Ω. When the switch is closed, a current of 600 mA occurs in the lamp. (a) What is the lamp's resistance? (b) What fraction of the chemical energy transformed appears as internal energy in the batteries?

4. An automobile battery has an emf of 12.6 V and an internal resistance of 0.080 0 Ω. The headlights together present equivalent resistance 5.00 Ω (assumed constant). What is the potential difference across the headlight bulbs (a) when they are the only load on the battery and (b) when the starter motor is operated, taking an additional 35.0 A from the battery?

Section 28.2 Resistors in Series and Parallel

5. The current in a loop circuit that has a resistance of R_1 is 2.00 A. The current is reduced to 1.60 A when an additional resistor $R_2 = 3.00 \ \Omega$ is added in series with R_1. What is the value of R_1?

6. (a) Find the equivalent resistance between points a and b in Figure P28.6. (b) A potential difference of 34.0 V is applied between points a and b. Calculate the current in each resistor.

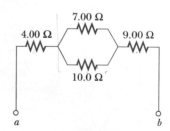

Figure P28.6

7. A lightbulb marked "75 W [at] 120 V" is screwed into a socket at one end of a long extension cord, in which each of the two conductors has resistance 0.800 Ω. The other end of the extension cord is plugged into a 120-V outlet. Draw a circuit diagram and find the actual power delivered to the bulb in this circuit.

8. Four copper wires of equal length are connected in series. Their cross-sectional areas are 1.00 cm², 2.00 cm², 3.00 cm², and 5.00 cm². A potential difference of 120 V is applied across the combination. Determine the voltage across the 2.00-cm² wire.

9. 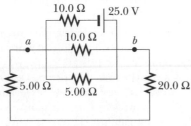 Consider the circuit shown in Figure P28.9. Find (a) the current in the 20.0-Ω resistor and (b) the potential difference between points *a* and *b*.

Figure P28.9

10. For the purpose of measuring the electric resistance of shoes through the body of the wearer to a metal ground plate, the American National Standards Institute (ANSI) specifies the circuit shown in Figure P28.10. The potential difference ΔV across the 1.00-MΩ resistor is measured with a high-resistance voltmeter. (a) Show that the resistance of the footwear is given by

$$R_{shoes} = 1.00 \text{ M}\Omega \left(\frac{50.0 \text{ V} - \Delta V}{\Delta V} \right)$$

(b) In a medical test, a current through the human body should not exceed 150 μA. Can the current delivered by the ANSI-specified circuit exceed 150 μA? To decide, consider a person standing barefoot on the ground plate.

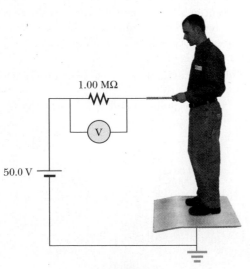

Figure P28.10

11. Three 100-Ω resistors are connected as shown in Figure P28.11. The maximum power that can safely be delivered to any one resistor is 25.0 W. (a) What is the maximum voltage that can be applied to the terminals *a* and *b*? For the voltage determined in part (a), what is the power delivered to each resistor? What is the total power delivered?

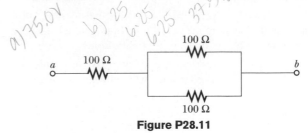

Figure P28.11

12. Using only three resistors—2.00 Ω, 3.00 Ω, and 4.00 Ω—find 17 resistance values that may be obtained by various combinations of one or more resistors. Tabulate the combinations in order of increasing resistance.

13. The current in a circuit is tripled by connecting a 500-Ω resistor in parallel with the resistance of the circuit. Determine the resistance of the circuit in the absence of the 500-Ω resistor.

14. A 6.00-V battery supplies current to the circuit shown in Figure P28.14. When the double-throw switch S is open, as shown in the figure, the current in the battery is 1.00 mA. When the switch is closed in position 1, the current in the battery is 1.20 mA. When the switch is closed in position 2, the current in the battery is 2.00 mA. Find the resistances R_1, R_2, and R_3.

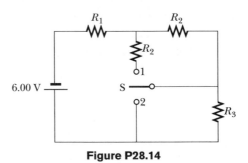

Figure P28.14

15. Calculate the power delivered to each resistor in the circuit shown in Figure P28.15.

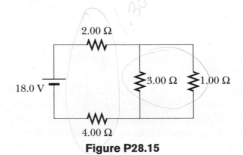

Figure P28.15

16. Two resistors connected in series have an equivalent resistance of 690 Ω. When they are connected in parallel, their equivalent resistance is 150 Ω. Find the resistance of each resistor.

17. An electric teakettle has a multiposition switch and two heating coils. When only one of the coils is switched on, the well-insulated kettle brings a full pot of water to a boil over the time interval Δt. When only the other coil is switched on, it requires a time interval of $2\Delta t$ to boil the same amount of water. Find the time interval required to boil the same amount of water if both coils are switched on (a) in a parallel connection and (b) in a series connection.

18. In Figures 28.4 and 28.6, let $R_1 = 11.0$ Ω, $R_2 = 22.0$ Ω, and let the battery have a terminal voltage of 33.0 V. (a) In the parallel circuit shown in Figure 28.6, to which resistor is more power delivered? (b) Verify that the sum of the power (I^2R) delivered to each resistor equals the power supplied by the battery ($\mathcal{P} = I\Delta V$). (c) In the series circuit, which resistor uses more power? (d) Verify that the sum of the power (I^2R) used by each resistor equals the power supplied by the battery ($\mathcal{P} = I\Delta V$). (e) Which circuit configuration uses more power?

19. Four resistors are connected to a battery as shown in Figure P28.19. The current in the battery is I, the battery emf is $\mathcal{E}$, and the resistor values are $R_1 = R$, $R_2 = 2R$, $R_3 = 4R$, $R_4 = 3R$. (a) Rank the resistors according to the potential difference across them, from largest to smallest. Note any cases of equal potential differences. (b) Determine the potential difference across each resistor in terms of $\mathcal{E}$. (c) Rank the resistors according to the current in them, from largest to smallest. Note any cases of equal currents. (d) Determine the current in each resistor in terms of I. (e) **What If?** If R_3 is increased, what happens to the current in each of the resistors? (f) In the limit that $R_3 \to \infty$, what are the new values of the current in each resistor in terms of I, the original current in the battery?

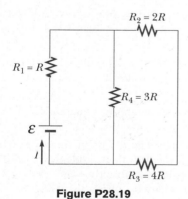

Figure P28.19

Section 28.3   Kirchhoff's Rules

Note: The currents are not necessarily in the direction shown for some circuits.

20. The ammeter shown in Figure P28.20 reads 2.00 A. Find I_1, I_2, and $\mathcal{E}$.

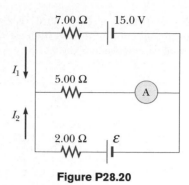

Figure P28.20

21. Determine the current in each branch of the circuit shown in Figure P28.21.

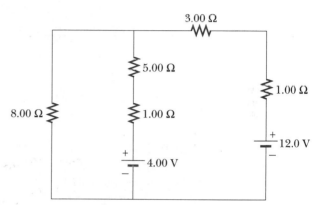

Figure P28.21   Problems 21, 22, and 23.

22. In Figure P28.21, show how to add just enough ammeters to measure every different current. Show how to add just enough voltmeters to measure the potential difference across each resistor and across each battery.

23. The circuit considered in Problem 21 and shown in Figure P28.21 is connected for 2.00 min. (a) Find the energy delivered by each battery. (b) Find the energy delivered to each resistor. (c) Identify the net energy transformation that occurs in the operation of the circuit and the total amount of energy transformed.

24. Using Kirchhoff's rules, (a) find the current in each resistor in Figure P28.24. (b) Find the potential difference between points c and f. Which point is at the higher potential?

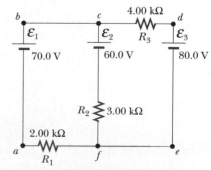

Figure P28.24

25. Taking $R = 1.00 \text{ k}\Omega$ and $\mathcal{E} = 250 \text{ V}$ in Figure P28.25, determine the direction and magnitude of the current in the horizontal wire between a and e.

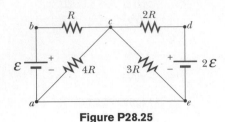

Figure P28.25

26. In the circuit of Figure P28.26, determine the current in each resistor and the voltage across the 200-Ω resistor.

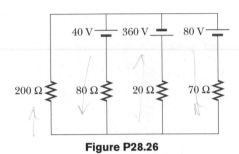

Figure P28.26

27. A dead battery is charged by connecting it to the live battery of another car with jumper cables (Fig. P28.27). Determine the current in the starter and in the dead battery.

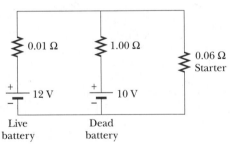

Figure P28.27

28. For the network shown in Figure P28.28, show that the resistance $R_{ab} = (27/17) \ \Omega$.

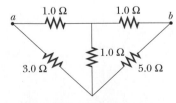

Figure P28.28

29. For the circuit shown in Figure P28.29, calculate (a) the current in the 2.00-Ω resistor and (b) the potential difference between points a and b.

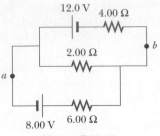

Figure P28.29

30. Calculate the power delivered to each resistor shown in Figure P28.30.

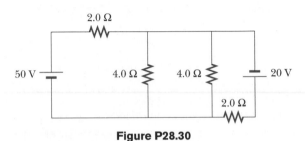

Figure P28.30

Section 28.4 *RC* Circuits

31. Consider a series *RC* circuit (see Fig. 28.19) for which $R = 1.00 \text{ M}\Omega$, $C = 5.00 \ \mu\text{F}$, and $\mathcal{E} = 30.0 \text{ V}$. Find (a) the time constant of the circuit and (b) the maximum charge on the capacitor after the switch is closed. (c) Find the current in the resistor 10.0 s after the switch is closed.

32. A 2.00-nF capacitor with an initial charge of 5.10 μC is discharged through a 1.30-kΩ resistor. (a) Calculate the current in the resistor 9.00 μs after the resistor is connected across the terminals of the capacitor. (b) What charge remains on the capacitor after 8.00 μs? (c) What is the maximum current in the resistor?

33. A fully charged capacitor stores energy U_0. How much energy remains when its charge has decreased to half its original value?

34. A capacitor in an *RC* circuit is charged to 60.0% of its maximum value in 0.900 s. What is the time constant of the circuit?

35. Show that the integral in Equation (1) of Example 28.14 has the value $RC/2$.

36. In the circuit of Figure P28.36, the switch S has been open for a long time. It is then suddenly closed. Determine the time constant (a) before the switch is closed and (b) after the switch is closed. (c) Let the switch be closed at $t = 0$. Determine the current in the switch as a function of time.

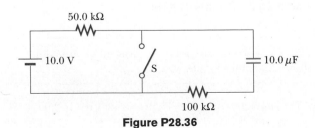

Figure P28.36

37. The circuit in Figure P28.37 has been connected for a long time. (a) What is the voltage across the capacitor? (b) If the battery is disconnected, how long does it take the capacitor to discharge to one tenth of its initial voltage?

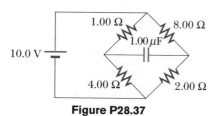

Figure P28.37

38. In places such as a hospital operating room and a factory for electronic circuit boards, electric sparks must be avoided. A person standing on a grounded floor and touching nothing else can typically have a body capacitance of 150 pF, in parallel with a foot capacitance of 80.0 pF produced by the dielectric soles of his or her shoes. The person acquires static electric charge from interactions with furniture, clothing, equipment, packaging materials, and essentially everything else. The static charge is conducted to ground through the equivalent resistance of the two shoe soles in parallel with each other. A pair of rubber-soled street shoes can present an equivalent resistance of 5 000 MΩ. A pair of shoes with special static-dissipative soles can have an equivalent resistance of 1.00 MΩ. Consider the person's body and shoes as forming an *RC* circuit with the ground. (a) How long does it take the rubber-soled shoes to reduce a 3 000-V static charge to 100 V? (b) How long does it take the static-dissipative shoes to do the same thing?

39. A 4.00-MΩ resistor and a 3.00-μF capacitor are connected in series with a 12.0-V power supply. (a) What is the time constant for the circuit? (b) Express the current in the circuit and the charge on the capacitor as functions of time.

40. Dielectric materials used in the manufacture of capacitors are characterized by conductivities that are small but not zero. Therefore, a charged capacitor slowly loses its charge by "leaking" across the dielectric. If a capacitor having capacitance *C* leaks charge such that the potential difference has decreased to half its initial ($t = 0$) value at a time *t*, what is the equivalent resistance of the dielectric?

Section 28.5   Electrical Meters

41. Assume that a galvanometer has an internal resistance of 60.0 Ω and requires a current of 0.500 mA to produce full-scale deflection. What resistance must be connected in parallel with the galvanometer if the combination is to serve as an ammeter that has a full-scale deflection for a current of 0.100 A?

42. A typical galvanometer, which requires a current of 1.50 mA for full-scale deflection and has a resistance of 75.0 Ω, may be used to measure currents of much greater values. To enable an operator to measure large currents without damage to the galvanometer, a relatively small shunt resistor is wired in parallel with the galvanometer, as suggested in Figure 28.27. Most of the current then goes through the shunt resistor. Calculate the value of the shunt resistor that allows the galvanometer to be used to measure a current of 1.00 A at full scale deflection. (*Suggestion:* use Kirchhoff's rules.)

43. The same galvanometer described in the previous problem may be used to measure voltages. In this case a large resistor is wired in series with the galvanometer, as suggested in Figure 28.29. The effect is to limit the current in the galvanometer when large voltages are applied. Most of the potential drop occurs across the resistor placed in series. Calculate the value of the resistor that allows the galvanometer to measure an applied voltage of 25.0 V at full-scale deflection.

44. *Meter loading.* Work this problem to five-digit precision. Refer to Figure P28.44. (a) When a 180.00-Ω resistor is connected across a battery of emf 6.000 0 V and internal resistance 20.000 Ω, what is the current in the resistor? What is the potential difference across it? (b) Suppose now an ammeter of resistance 0.500 00 Ω and a voltmeter of resistance 20 000 Ω are added to the circuit as shown in Figure P28.44b. Find the reading of each. (c) **What If?** Now one terminal of one wire is moved, as shown in Figure P28.44c. Find the new meter readings.

45. Design a multirange ammeter capable of full-scale deflection for 25.0 mA, 50.0 mA, and 100 mA. Assume the meter movement is a galvanometer that has a resistance of 25.0 Ω and gives a full-scale deflection for 1.00 mA.

46. Design a multirange voltmeter capable of full-scale deflection for 20.0 V, 50.0 V, and 100 V. Assume the meter movement is a galvanometer that has a resistance of 60.0 Ω and gives a full-scale deflection for a current of 1.00 mA.

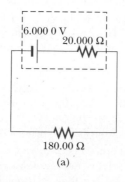

(a)

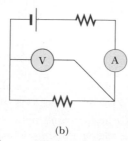

(b)

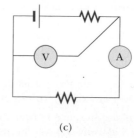

(c)

Figure P28.44

47. A particular galvanometer serves as a 2.00-V full-scale voltmeter when a 2 500-Ω resistor is connected in series with it. It serves as a 0.500-A full-scale ammeter when a 0.220-Ω resistor is connected in parallel with it. Determine the internal resistance of the galvanometer and the current required to produce full-scale deflection.

Section 28.6 Household Wiring and Electrical Safety

48. An 8.00-ft extension cord has two 18-gauge copper wires, each having a diameter of 1.024 mm. At what rate is energy delivered to the resistance in the cord when it is carrying a current of (a) 1.00 A and (b) 10.0 A?

49. An electric heater is rated at 1 500 W, a toaster at 750 W, and an electric grill at 1 000 W. The three appliances are connected to a common 120-V household circuit. (a) How much current does each draw? (b) Is a circuit with a 25.0-A circuit breaker sufficient in this situation? Explain your answer.

50. Aluminum wiring has sometimes been used instead of copper for economy. According to the National Electrical Code, the maximum allowable current for 12-gauge copper wire with rubber insulation is 20 A. What should be the maximum allowable current in a 12-gauge aluminum wire if the power per unit length delivered to the resistance in the aluminum wire is the same as that delivered in the copper wire?

51. Turn on your desk lamp. Pick up the cord, with your thumb and index finger spanning the width of the cord. (a) Compute an order-of-magnitude estimate for the current in your hand. You may assume that at a typical instant the conductor inside the lamp cord next to your thumb is at potential $\sim 10^2$ V and that the conductor next to your index finger is at ground potential (0 V). The resistance of your hand depends strongly on the thickness and the moisture content of the outer layers of your skin. Assume that the resistance of your hand between fingertip and thumb tip is $\sim 10^4$ Ω. You may model the cord as having rubber insulation. State the other quantities you measure or estimate and their values. Explain your reasoning. (b) Suppose that your body is isolated from any other charges or currents. In order-of-magnitude terms describe the potential of your thumb where it contacts the cord, and the potential of your finger where it touches the cord.

Additional Problems

52. Four 1.50-V AA batteries in series are used to power a transistor radio. If the batteries can move a charge of 240 C, how long will they last if the radio has a resistance of 200 Ω?

53. A battery has an emf of 9.20 V and an internal resistance of 1.20 Ω. (a) What resistance across the battery will extract from it a power of 12.8 W? (b) a power of 21.2 W?

54. Calculate the potential difference between points a and b in Figure P28.54 and identify which point is at the higher potential.

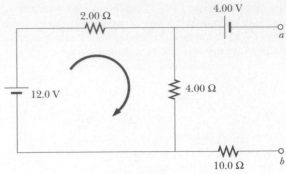

Figure P28.54

55. Assume you have a battery of emf $\mathcal{E}$ and three identical lightbulbs, each having constant resistance R. What is the total power delivered by the battery if the bulbs are connected (a) in series? (b) in parallel? (c) For which connection will the bulbs shine the brightest?

56. A group of students on spring break manages to reach a deserted island in their wrecked sailboat. They splash ashore with fuel, a European gasoline-powered 240-V generator, a box of North American 100-W 120-V lightbulbs, a 500-W 120-V hot pot, lamp sockets, and some insulated wire. While waiting to be rescued, they decide to use the generator to operate some lightbulbs. (a) Draw a diagram of a circuit they can use, containing the minimum number of lightbulbs with 120 V across each bulb, and no higher voltage. Find the current in the generator and its power output. (b) One student catches a fish and wants to cook it in the hot pot. Draw a diagram of a circuit containing the hot pot and the minimum number of lightbulbs with 120 V across each device, and not more. Find the current in the generator and its power output.

57. A battery has an emf $\mathcal{E}$ and internal resistance r. A variable load resistor R is connected across the terminals of the battery. (a) Determine the value of R such that the potential difference across the terminals is a maximum. (b) Determine the value of R so that the current in the circuit is a maximum. (c) Determine the value of R so that the power delivered to the load resistor is a maximum. Choosing the load resistance for maximum power transfer is a case of what is called *impedance matching* in general. Impedance matching is important in shifting gears on a bicycle, in connecting a loudspeaker to an audio amplifier, in connecting a battery charger to a bank of solar photoelectric cells, and in many other applications.

58. A 10.0-μF capacitor is charged by a 10.0-V battery through a resistance R. The capacitor reaches a potential difference of 4.00 V in a time 3.00 s after charging begins. Find R.

59. When two unknown resistors are connected in series with a battery, the battery delivers 225 W and carries a total current of 5.00 A. For the same total current, 50.0 W is delivered when the resistors are connected in parallel. Determine the values of the two resistors.

60. When two unknown resistors are connected in series with a battery, the battery delivers total power $\mathcal{P}_s$ and carries a total current of I. For the same total current, a total power

$\mathcal{P}_p$ is delivered when the resistors are connected in parallel. Determine the values of the two resistors.

61. A power supply has an open-circuit voltage of 40.0 V and an internal resistance of 2.00 Ω. It is used to charge two storage batteries connected in series, each having an emf of 6.00 V and internal resistance of 0.300 Ω. If the charging current is to be 4.00 A, (a) what additional resistance should be added in series? (b) At what rate does the internal energy increase in the supply, in the batteries, and in the added series resistance? (c) At what rate does the chemical energy increase in the batteries?

62. Two resistors R_1 and R_2 are in parallel with each other. Together they carry total current I. (a) Determine the current in each resistor. (b) Prove that this division of the total current I between the two resistors results in less power delivered to the combination than any other division. It is a general principle that *current in a direct current circuit distributes itself so that the total power delivered to the circuit is a minimum.*

63. The value of a resistor R is to be determined using the ammeter–voltmeter setup shown in Figure P28.63. The ammeter has a resistance of 0.500 Ω, and the voltmeter has a resistance of 20 000 Ω. Within what range of actual values of R will the measured values be correct to within 5.00% if the measurement is made using the circuit shown in (a) Figure P28.63a and (b) Figure P28.63b?

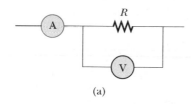

(a)

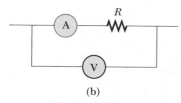

(b)

Figure P28.63

64. A battery is used to charge a capacitor through a resistor, as shown in Figure 28.19. Show that half the energy supplied by the battery appears as internal energy in the resistor and that half is stored in the capacitor.

65. The values of the components in a simple series RC circuit containing a switch (Fig. 28.19) are $C = 1.00$ μF, $R = 2.00 \times 10^6$ Ω, and $\mathcal{E} = 10.0$ V. At the instant 10.0 s after the switch is closed, calculate (a) the charge on the capacitor, (b) the current in the resistor, (c) the rate at which energy is being stored in the capacitor, and (d) the rate at which energy is being delivered by the battery.

66. The switch in Figure P28.66a closes when $\Delta V_c > 2\,\Delta V/3$ and opens when $\Delta V_c < \Delta V/3$. The voltmeter reads a voltage as plotted in Figure P28.66b. What is the period T of the waveform in terms of R_1, R_2, and C?

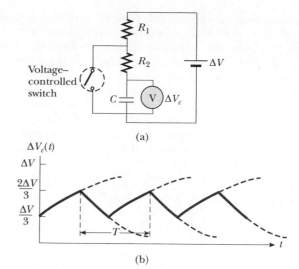

(a)

(b)

Figure P28.66

67. Three 60.0-W, 120-V lightbulbs are connected across a 120-V power source, as shown in Figure P28.67. Find (a) the total power delivered to the three bulbs and (b) the voltage across each. Assume that the resistance of each bulb is constant (even though in reality the resistance might increase markedly with current).

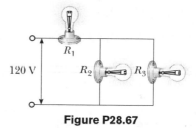

Figure P28.67

68. Switch S has been closed for a long time, and the electric circuit shown in Figure P28.68 carries a constant current. Take $C_1 = 3.00$ μF, $C_2 = 6.00$ μF, $R_1 = 4.00$ kΩ, and $R_2 = 7.00$ kΩ. The power delivered to R_2 is 2.40 W. (a) Find the charge on C_1. (b) Now the switch is opened. After many milliseconds, by how much has the charge on C_2 changed?

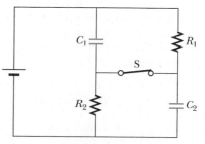

Figure P28.68

69. Four resistors are connected in parallel across battery. They carry currents of 150 14.00 mA, and 4.00 mA. (a) If the resi resistance is replaced with one havin tance, what is the ratio of the new curre to the original current? (b) **What If?** If instead the resistor with the smallest resistance is replaced with one having twice the resistance, what is the ratio of the new total current to the original current? (c) On a February night, energy leaves a house by several heat leaks, including the following: 1 500 W by conduction through the ceiling; 450 W by infiltration (air flow) around the windows; 140 W by conduction through the basement wall above the foundation sill; and 40.0 W by conduction through the plywood door to the attic. To produce the biggest saving in heating bills, which one of these energy transfers should be reduced first?

70. Figure P28.70 shows a circuit model for the transmission of an electrical signal, such as cable TV, to a large number of subscribers. Each subscriber connects a load resistance R_L between the transmission line and the ground. The ground is assumed to be at zero potential and able to carry any current between any ground connections with negligible resistance. The resistance of the transmission line itself between the connection points of different subscribers is modeled as the constant resistance R_T. Show that the equivalent resistance across the signal source is

$$R_{eq} = \frac{1}{2}[(4R_TR_L + R_T^2)^{1/2} + R_T]$$

Suggestion: Because the number of subscribers is large, the equivalent resistance would not change noticeably if the first subscriber cancelled his service. Consequently, the equivalent resistance of the section of the circuit to the right of the first load resistor is nearly equal to R_{eq}.

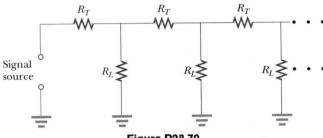

Figure P28.70

71. In Figure P28.71, suppose the switch has been closed for a time sufficiently long for the capacitor to become fully charged. Find (a) the steady-state current in each resistor and (b) the charge Q on the capacitor. (c) The switch is now opened at $t = 0$. Write an equation for the current I_{R_2} through R_2 as a function of time and (d) find the time interval required for the charge on the capacitor to fall to one-fifth its initial value.

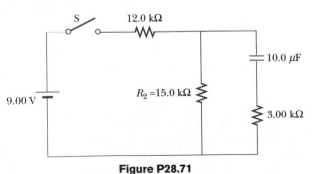

Figure P28.71

72. A regular tetrahedron is a pyramid with a triangular base. Six 10.0-Ω resistors are placed along its six edges, with junctions at its four vertices. A 12.0-V battery is connected to any two of the vertices. Find (a) the equivalent resistance of the tetrahedron between these vertices and (b) the current in the battery.

73. The circuit shown in Figure P28.73 is set up in the laboratory to measure an unknown capacitance C with the use of a voltmeter of resistance $R = 10.0$ MΩ and a battery whose emf is 6.19 V. The data given in the table are the measured voltages across the capacitor as a function of time, where $t = 0$ represents the instant at which the switch is opened. (a) Construct a graph of $\ln(\mathcal{E}/\Delta V)$ versus t, and perform a linear least-squares fit to the data. (b) From the slope of your graph, obtain a value for the time constant of the circuit and a value for the capacitance.

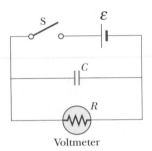

Figure P28.73

ΔV (V)	t (s)	$\ln(\mathcal{E}/\Delta V)$
6.19	0	
5.55	4.87	
4.93	11.1	
4.34	19.4	
3.72	30.8	
3.09	46.6	
2.47	67.3	
1.83	102.2	

74. The student engineer of a campus radio station wishes to verify the effectiveness of the lightning rod on the antenna mast (Fig. P28.74). The unknown resistance R_x is between points C and E. Point E is a true ground but is inaccessible for direct measurement since this stratum is several meters below the Earth's surface. Two identical rods are driven into the ground at A and B, introducing an unknown resistance R_y. The procedure is as follows. Measure resistance R_1 between points A and B, then connect A and B with a heavy conducting wire and measure resistance R_2 between

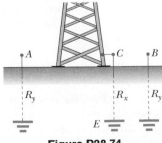

Figure P28.74

points A and C. (a) Derive an equation for R_x in terms of the observable resistances, R_1 and R_2. (b) A satisfactory ground resistance would be $R_x < 2.00\ \Omega$. Is the grounding of the station adequate if measurements give $R_1 = 13.0\ \Omega$ and $R_2 = 6.00\ \Omega$?

75. The circuit in Figure P28.75 contains two resistors, $R_1 = 2.00\ \text{k}\Omega$ and $R_2 = 3.00\ \text{k}\Omega$, and two capacitors, $C_1 = 2.00\ \mu\text{F}$ and $C_2 = 3.00\ \mu\text{F}$, connected to a battery with emf $\mathcal{E} = 120\ \text{V}$. No charge is on either capacitor before switch S is closed. Determine the charges q_1 and q_2 on capacitors C_1 and C_2, respectively, after the switch is closed. (*Suggestion:* First reconstruct the circuit so that it becomes a simple RC circuit containing a single resistor and single capacitor in series, connected to the battery, and then determine the total charge q stored in the equivalent circuit.)

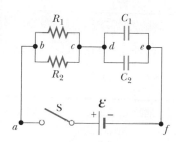

Figure P28.75

76. This problem[6] illustrates how a digital voltmeter affects the voltage across a capacitor in an RC circuit. A digital voltmeter of internal resistance r is used to measure the voltage across a capacitor after the switch in Figure P28.76 is closed. Because the meter has finite resistance, part of the current supplied by the battery passes through the meter. (a) Apply Kirchhoff's rules to this circuit, and use the fact that $i_C = dq/dt$ to show that this leads to the differential equation

$$R_{\text{eq}}\frac{dq}{dt} + \frac{q}{C} = \frac{r}{r + R}\mathcal{E}$$

where $R_{\text{eq}} = rR/(r + R)$. (b) Show that the solution to this differential equation is

$$q = \frac{r}{r + R}C\mathcal{E}\ (1 - e^{-t/R_{\text{eq}}C})$$

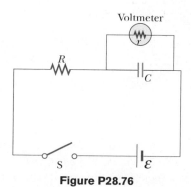

Figure P28.76

---

[6] After Joseph Priest, "Meter Resistance: Don't Forget It!" *The Physics Teacher,* January 2003, p. 40.

and that the voltage across the capacitor as a function of time is

$$V_C = \frac{r}{r + R}\ \mathcal{E}(1 - e^{-t/R_{\text{eq}}C})$$

(c) **What If?** If the capacitor is fully charged, and the switch is then opened, how does the voltage across the capacitor behave in this case?

Answers to Quick Quizzes

28.1 (a). Power is delivered to the internal resistance of a battery, so decreasing the internal resistance will decrease this "lost" power and increase the percentage of the power delivered to the device.

28.2 (c). In a series circuit, the current is the same in all resistors in series. Current is not "used up" as charges pass through a resistor.

28.3 (a). Connecting b to c "shorts out" bulb R_2 and changes the total resistance of the circuit from $R_1 + R_2$ to just R_1. Because the resistance of the circuit has decreased (and the emf supplied by the battery does not change), the current in the circuit increases.

28.4 (b). When the switch is opened, resistors R_1 and R_2 are in series, so that the total circuit resistance is larger than when the switch was closed. As a result, the current decreases.

28.5 (b), (d). Adding another series resistor increases the total resistance of the circuit and thus reduces the current in the circuit. The potential difference across the battery terminals increases because the reduced current results in a smaller voltage decrease across the internal resistance.

28.6 (a), (e). If the second resistor were connected in parallel, the total resistance of the circuit would decrease, and the current in the battery would increase. The potential difference across the terminals would decrease because the increased current results in a greater voltage drop across the internal resistance.

28.7 (a). When the switch is closed, resistors R_1 and R_2 are in parallel, so that the total circuit resistance is smaller than when the switch was open. As a result, the current increases.

28.8 (c). A current is assigned to a given branch of a circuit. There may be multiple resistors and batteries in a given branch.

28.9 (b), (d). Just after the switch is closed, there is no charge on the capacitor, so there is no voltage across it. Charges begin to flow in the circuit to charge up the capacitor, so that all of the voltage $\Delta V = IR$ appears across the resistor. After a long time, the capacitor is fully charged and the current drops to zero. Thus, the battery voltage is now entirely across the capacitor.

28.10 (c), (i). Just after the switch is closed, there is no charge on the capacitor. Current exists in both branches of the circuit as the capacitor begins to charge, so the right half of the circuit is equivalent to two resistances R in parallel for an equivalent resistance of $\frac{1}{2}R$. After a long time, the capacitor is fully charged and the current in the right-hand branch drops to zero. Now, current exists only in a resistance R across the battery.

Chapter 29

Magnetic Fields

CHAPTER OUTLINE

29.1 Magnetic Fields and Forces

29.2 Magnetic Force Acting on a Current-Carrying Conductor

29.3 Torque on a Current Loop in a Uniform Magnetic Field

29.4 Motion of a Charged Particle in a Uniform Magnetic Field

29.5 Applications Involving Charged Particles Moving in a Magnetic Field

29.6 The Hall Effect

▲ *Magnetic fingerprinting allows fingerprints to be seen on surfaces that otherwise would not allow prints to be lifted. The powder spread on the surface is coated with an organic material that adheres to the greasy residue in a fingerprint. A magnetic "brush" removes the excess powder and makes the fingerprint visible. (James King-Holmes/Photo Researchers, Inc.)*

Many historians of science believe that the compass, which uses a magnetic needle, was used in China as early as the 13th century B.C., its invention being of Arabic or Indian origin. The early Greeks knew about magnetism as early as 800 B.C. They discovered that the stone magnetite (Fe_3O_4) attracts pieces of iron. Legend ascribes the name *magnetite* to the shepherd Magnes, the nails of whose shoes and the tip of whose staff stuck fast to chunks of magnetite while he pastured his flocks.

In 1269 a Frenchman named Pierre de Maricourt found that the directions of a needle near a spherical natural magnet formed lines that encircled the sphere and passed through two points diametrically opposite each other, which he called the *poles* of the magnet. Subsequent experiments showed that every magnet, regardless of its shape, has two poles, called *north* (N) and *south* (S) poles, that exert forces on other magnetic poles similar to the way that electric charges exert forces on one another. That is, like poles (N–N or S–S) repel each other, and opposite poles (N–S) attract each other.

The poles received their names because of the way a magnet, such as that in a compass, behaves in the presence of the Earth's magnetic field. If a bar magnet is suspended from its midpoint and can swing freely in a horizontal plane, it will rotate until its north pole points to the Earth's geographic North Pole and its south pole points to the Earth's geographic South Pole.[1]

In 1600 William Gilbert (1540–1603) extended de Maricourt's experiments to a variety of materials. Using the fact that a compass needle orients in preferred directions, he suggested that the Earth itself is a large permanent magnet. In 1750 experimenters used a torsion balance to show that magnetic poles exert attractive or repulsive forces on each other and that these forces vary as the inverse square of the distance between interacting poles. Although the force between two magnetic poles is otherwise similar to the force between two electric charges, electric charges can be isolated (witness the electron and proton) whereas **a single magnetic pole has never been isolated.** That is, **magnetic poles are always found in pairs.** All attempts thus far to detect an isolated magnetic pole have been unsuccessful. No matter how many times a permanent magnet is cut in two, each piece always has a north and a south pole.[2]

The relationship between magnetism and electricity was discovered in 1819 when, during a lecture demonstration, the Danish scientist Hans Christian Oersted found that an electric current in a wire deflected a nearby compass needle.[3] In the 1820s,

Hans Christian Oersted

Danish Physicist and Chemist (1777–1851)

Oersted is best known for observing that a compass needle deflects when placed near a wire carrying a current. This important discovery was the first evidence of the connection between electric and magnetic phenomena. Oersted was also the first to prepare pure aluminum. *(North Wind Picture Archives)*

---

[1] Note that the Earth's geographic North Pole is magnetically a south pole, whereas its geographic South Pole is magnetically a north pole. Because *opposite* magnetic poles attract each other, the pole on a magnet that is attracted to the Earth's geographic North Pole is the magnet's *north* pole and the pole attracted to the Earth's geographic South Pole is the magnet's *south* pole.

[2] There is some theoretical basis for speculating that magnetic *monopoles*—isolated north or south poles—may exist in nature, and attempts to detect them are an active experimental field of investigation.

[3] The same discovery was reported in 1802 by an Italian jurist, Gian Dominico Romognosi, but was overlooked, probably because it was published in an obscure journal.

further connections between electricity and magnetism were demonstrated independently by Faraday and Joseph Henry (1797–1878). They showed that an electric current can be produced in a circuit either by moving a magnet near the circuit or by changing the current in a nearby circuit. These observations demonstrate that a changing magnetic field creates an electric field. Years later, theoretical work by Maxwell showed that the reverse is also true: a changing electric field creates a magnetic field.

This chapter examines the forces that act on moving charges and on current-carrying wires in the presence of a magnetic field. The source of the magnetic field is described in Chapter 30.

29.1 Magnetic Fields and Forces

In our study of electricity, we described the interactions between charged objects in terms of electric fields. Recall that an electric field surrounds any electric charge. In addition to containing an electric field, the region of space surrounding any *moving* electric charge also contains a magnetic field. A magnetic field also surrounds a magnetic substance making up a permanent magnet.

Historically, the symbol **B** has been used to represent a magnetic field, and this is the notation we use in this text. The direction of the magnetic field **B** at any location is the direction in which a compass needle points at that location. As with the electric field, we can represent the magnetic field by means of drawings with *magnetic field lines.*

Figure 29.1 shows how the magnetic field lines of a bar magnet can be traced with the aid of a compass. Note that the magnetic field lines outside the magnet point away from north poles and toward south poles. One can display magnetic field patterns of a bar magnet using small iron filings, as shown in Figure 29.2.

We can define a magnetic field **B** at some point in space in terms of the magnetic force F_B that the field exerts on a charged particle moving with a velocity **v**, which we call the test object. For the time being, let us assume that no electric or gravitational fields are present at the location of the test object. Experiments on various charged particles moving in a magnetic field give the following results:

Properties of the magnetic force on a charge moving in a magnetic field B

- The magnitude F_B of the magnetic force exerted on the particle is proportional to the charge q and to the speed v of the particle.

- The magnitude and direction of **F**$_B$ depend on the velocity of the particle and on the magnitude and direction of the magnetic field **B**.

- When a charged particle moves parallel to the magnetic field vector, the magnetic force acting on the particle is zero.

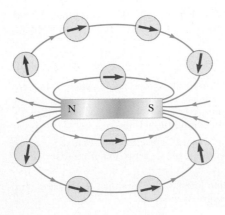

At the Active Figures link at http://www.pse6.com, you can move the compass around and trace the magnetic field lines for yourself.

Active Figure 29.1 Compass needles can be used to trace the magnetic field lines in the region outside a bar magnet.

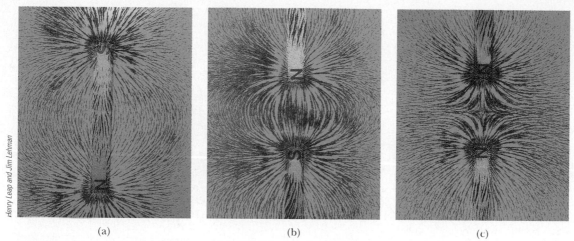

Figure 29.2 (a) Magnetic field pattern surrounding a bar magnet as displayed with iron filings. (b) Magnetic field pattern between *opposite* poles (N–S) of two bar magnets. (c) Magnetic field pattern between *like* poles (N–N) of two bar magnets.

- When the particle's velocity vector makes any angle $\theta \neq 0$ with the magnetic field, the magnetic force acts in a direction perpendicular to both **v** and **B**; that is, $\mathbf{F}_B$ is perpendicular to the plane formed by **v** and **B** (Fig. 29.3a).

- The magnetic force exerted on a positive charge is in the direction opposite the direction of the magnetic force exerted on a negative charge moving in the same direction (Fig. 29.3b).

- The magnitude of the magnetic force exerted on the moving particle is proportional to $\sin \theta$, where θ is the angle the particle's velocity vector makes with the direction of **B**.

We can summarize these observations by writing the magnetic force in the form

$$\mathbf{F}_B = q\mathbf{v} \times \mathbf{B}$$

(29.1)   **Vector expression for the magnetic force on a charged particle moving in a magnetic field**

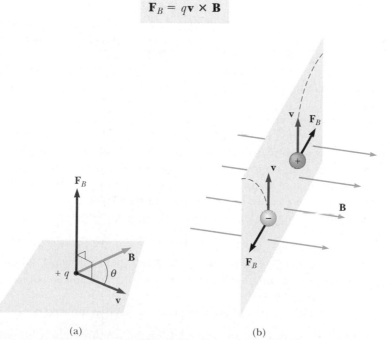

Figure 29.3 The direction of the magnetic force $\mathbf{F}_B$ acting on a charged particle moving with a velocity **v** in the presence of a magnetic field **B**. (a) The magnetic force is perpendicular to both **v** and **B**. (b) Oppositely directed magnetic forces $\mathbf{F}_B$ are exerted on two oppositely charged particles moving at the same velocity in a magnetic field. The dashed lines show the paths of the particles, which we will investigate in Section 29.4.

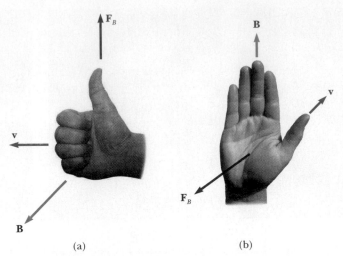

(a)   (b)

Figure 29.4 Two right-hand rules for determining the direction of the magnetic force $\mathbf{F}_B = q\mathbf{v} \times \mathbf{B}$ acting on a particle with charge q moving with a velocity $\mathbf{v}$ in a magnetic field $\mathbf{B}$. (a) In this rule, the fingers point in the direction of $\mathbf{v}$, with $\mathbf{B}$ coming out of your palm, so that you can curl your fingers in the direction of $\mathbf{B}$. The direction of $\mathbf{v} \times \mathbf{B}$, and the force on a positive charge, is the direction in which the thumb points. (b) In this rule, the vector $\mathbf{v}$ is in the direction of your thumb and $\mathbf{B}$ in the direction of your fingers. The force $\mathbf{F}_B$ on a positive charge is in the direction of your palm, as if you are pushing the particle with your hand.

which by definition of the cross product (see Section 11.1) is perpendicular to both $\mathbf{v}$ and $\mathbf{B}$. We can regard this equation as an operational definition of the magnetic field at some point in space. That is, the magnetic field is defined in terms of the force acting on a moving charged particle.

Figure 29.4 reviews two right-hand rules for determining the direction of the cross product $\mathbf{v} \times \mathbf{B}$ and determining the direction of $\mathbf{F}_B$. The rule in Figure 29.4a depends on our right-hand rule for the cross product in Figure 11.2. Point the four fingers of your right hand along the direction of $\mathbf{v}$ with the palm facing $\mathbf{B}$ and curl them toward $\mathbf{B}$. The extended thumb, which is at a right angle to the fingers, points in the direction of $\mathbf{v} \times \mathbf{B}$. Because $\mathbf{F}_B = q\mathbf{v} \times \mathbf{B}$, $\mathbf{F}_B$ is in the direction of your thumb if q is positive and opposite the direction of your thumb if q is negative. (If you need more help understanding the cross product, you should review pages 337 to 339, including Fig. 11.2.)

An alternative rule is shown in Figure 29.4b. Here the thumb points in the direction of $\mathbf{v}$ and the extended fingers in the direction of $\mathbf{B}$. Now, the force $\mathbf{F}_B$ on a positive charge extends outward from your palm. The advantage of this rule is that the force on the charge is in the direction that you would push on something with your hand—outward from your palm. The force on a negative charge is in the opposite direction. Feel free to use either of these two right-hand rules.

The magnitude of the magnetic force on a charged particle is

$$F_B = |q|vB \sin \theta \tag{29.2}$$

Magnitude of the magnetic force on a charged particle moving in a magnetic field

where θ is the smaller angle between $\mathbf{v}$ and $\mathbf{B}$. From this expression, we see that F_B is zero when $\mathbf{v}$ is parallel or antiparallel to $\mathbf{B}$ ($\theta = 0$ or $180°$) and maximum when $\mathbf{v}$ is perpendicular to $\mathbf{B}$ ($\theta = 90°$).

There are several important differences between electric and magnetic forces:

- The electric force acts along the direction of the electric field, whereas the magnetic force acts perpendicular to the magnetic field.

- The electric force acts on a charged particle regardless of whether the particle is moving, whereas the magnetic force acts on a charged particle only when the particle is in motion.

- The electric force does work in displacing a charged particle, whereas the magnetic force associated with a steady magnetic field does no work when a particle is displaced because the force is perpendicular to the displacement.

From the last statement and on the basis of the work–kinetic energy theorem, we conclude that the kinetic energy of a charged particle moving through a magnetic field cannot be altered by the magnetic field alone. In other words, when a charged particle moves with a velocity **v** through a magnetic field, the field can alter the direction of the velocity vector but cannot change the speed or kinetic energy of the particle.

From Equation 29.2, we see that the SI unit of magnetic field is the newton per coulomb-meter per second, which is called the **tesla** (T):

$$1 \text{ T} = 1 \frac{\text{N}}{\text{C} \cdot \text{m/s}}$$

The tesla

Because a coulomb per second is defined to be an ampere, we see that

$$1 \text{ T} = 1 \frac{\text{N}}{\text{A} \cdot \text{m}}$$

A non-SI magnetic-field unit in common use, called the *gauss* (G), is related to the tesla through the conversion $1 \text{ T} = 10^4 \text{ G}$. Table 29.1 shows some typical values of magnetic fields.

Quick Quiz 29.1 The north-pole end of a bar magnet is held near a positively charged piece of plastic. Is the plastic (a) attracted, (b) repelled, or (c) unaffected by the magnet?

Quick Quiz 29.2 A charged particle moves with velocity **v** in a magnetic field **B**. The magnetic force on the particle is a maximum when **v** is (a) parallel to **B**, (b) perpendicular to **B**, (c) zero.

Quick Quiz 29.3 An electron moves in the plane of this paper toward the top of the page. A magnetic field is also in the plane of the page and directed toward the right. The direction of the magnetic force on the electron is (a) toward the top of the page, (b) toward the bottom of the page, (c) toward the left edge of the page, (d) toward the right edge of the page, (e) upward out of the page, (f) downward into the page.

Table 29.1

Some Approximate Magnetic Field Magnitudes	
Source of Field	**Field Magnitude (T)**
Strong superconducting laboratory magnet	30
Strong conventional laboratory magnet	2
Medical MRI unit	1.5
Bar magnet	10^{-2}
Surface of the Sun	10^{-2}
Surface of the Earth	0.5×10^{-4}
Inside human brain (due to nerve impulses)	10^{-13}

Example 29.1   **An Electron Moving in a Magnetic Field**

An electron in a television picture tube moves toward the front of the tube with a speed of 8.0×10^6 m/s along the x axis (Fig. 29.5). Surrounding the neck of the tube are coils of wire that create a magnetic field of magnitude 0.025 T, directed at an angle of 60° to the x axis and lying in the xy plane.

(A) Calculate the magnetic force on the electron using Equation 29.2.

Solution Using Equation 29.2, we find the magnitude of the magnetic force:

$$
\begin{aligned}
F_B &= |q| v B \sin \theta \\
&= (1.6 \times 10^{-19}\, \text{C})(8.0 \times 10^6\, \text{m/s})(0.025\, \text{T})(\sin 60°) \\
&= \boxed{2.8 \times 10^{-14}\, \text{N}}
\end{aligned}
$$

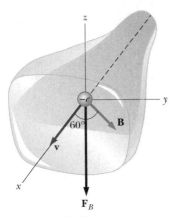

Figure 29.5 (Example 29.1) The magnetic force $\mathbf{F}_B$ acting on the electron is in the negative z direction when $\mathbf{v}$ and $\mathbf{B}$ lie in the xy plane.

Because $\mathbf{v} \times \mathbf{B}$ is in the positive z direction (from the right-hand rule) and the charge is negative, $\mathbf{F}_B$ is in the negative z direction.

(B) Find a vector expression for the magnetic force on the electron using Equation 29.1.

Solution We begin by writing a vector expression for the velocity of the electron:

$$\mathbf{v} = (8.0 \times 10^6\, \hat{\mathbf{i}})\, \text{m/s}$$

and one for the magnetic field:

$$
\begin{aligned}
\mathbf{B} &= (0.025 \cos 60° \hat{\mathbf{i}} + 0.025 \sin 60° \hat{\mathbf{j}})\text{T} \\
&= (0.013\, \hat{\mathbf{i}} + 0.022\, \hat{\mathbf{j}})\text{T}
\end{aligned}
$$

The force on the electron, using Equation 29.1, is

$$
\begin{aligned}
\mathbf{F}_B &= q\mathbf{v} \times \mathbf{B} \\
&= (-e)[(8.0 \times 10^6\, \hat{\mathbf{i}})\, \text{m/s}] \times [(0.013\, \hat{\mathbf{i}} + 0.022\, \hat{\mathbf{j}})\text{T}] \\
&= (-e)[(8.0 \times 10^6\, \hat{\mathbf{i}})\, \text{m/s}] \times [(0.013\, \hat{\mathbf{i}})\text{T}] \\
&\quad + (-e)[(8.0 \times 10^6\, \hat{\mathbf{i}})\, \text{m/s}] \times [(0.022\, \hat{\mathbf{j}})\text{T}] \\
&= (-e)(8.0 \times 10^6\, \text{m/s})(0.013\, \text{T})(\hat{\mathbf{i}} \times \hat{\mathbf{i}}) \\
&\quad + (-e)(8.0 \times 10^6\, \text{m/s})(0.022\, \text{T})(\hat{\mathbf{i}} \times \hat{\mathbf{j}}) \\
&= (-1.6 \times 10^{-19}\, \text{C})(8.0 \times 10^6\, \text{m/s})(0.022\, \text{T})\, \hat{\mathbf{k}}
\end{aligned}
$$

where we have used Equations 11.7a and 11.7b to evaluate $\hat{\mathbf{i}} \times \hat{\mathbf{i}}$ and $\hat{\mathbf{i}} \times \hat{\mathbf{j}}$. Carrying out the multiplication, we find,

$$\mathbf{F}_B = \boxed{(-2.8 \times 10^{-14}\, \text{N})\hat{\mathbf{k}}}$$

This expression agrees with the result in part (A). The magnitude is the same as we found there, and the force vector is in the negative z direction.

29.2   Magnetic Force Acting on a Current-Carrying Conductor

If a magnetic force is exerted on a single charged particle when the particle moves through a magnetic field, it should not surprise you that a current-carrying wire also experiences a force when placed in a magnetic field. This follows from the fact that the current is a collection of many charged particles in motion; hence, the resultant force exerted by the field on the wire is the vector sum of the individual forces exerted on all the charged particles making up the current. The force exerted on the particles is transmitted to the wire when the particles collide with the atoms making up the wire.

Before we continue our discussion, some explanation of the notation used in this book is in order. To indicate the direction of $\mathbf{B}$ in illustrations, we sometimes present perspective views, such as those in Figure 29.5. If $\mathbf{B}$ lies in the plane of the page or is present in a perspective drawing, we use blue vectors or blue field lines with arrowheads. In non-perspective illustrations, we depict a magnetic field perpendicular to and directed out of the page with a series of blue dots, which represent the tips of arrows coming toward you (see Fig. 29.6a). In this case, we label the field $\mathbf{B}_{out}$. If $\mathbf{B}$ is directed perpendicularly into the page, we use blue crosses, which represent the feathered tails of arrows fired away from you, as in Figure 29.6b. In this case, we label the field $\mathbf{B}_{in}$, where the subscript "in" indicates "into the page." The same notation with crosses and dots is also used for other quantities that might be perpendicular to the page, such as forces and current directions.

One can demonstrate the magnetic force acting on a current-carrying conductor by hanging a wire between the poles of a magnet, as shown in Figure 29.7a. For ease in visualization, part of the horseshoe magnet in part (a) is removed to show the end face of the south pole in parts (b), (c), and (d) of Figure 29.7. The magnetic field is directed into the page and covers the region within the shaded squares. When the current in the wire is zero, the wire remains vertical, as shown in Figure 29.7b. However, when the wire carries a current directed upward, as shown in Figure 29.7c, the wire deflects to the left. If we reverse the current, as shown in Figure 29.7d, the wire deflects to the right.

Let us quantify this discussion by considering a straight segment of wire of length L and cross-sectional area A, carrying a current I in a uniform magnetic field $\mathbf{B}$, as shown in Figure 29.8. The magnetic force exerted on a charge q moving with a drift velocity $\mathbf{v}_d$ is $q\mathbf{v}_d \times \mathbf{B}$. To find the total force acting on the wire, we multiply the force $q\mathbf{v}_d \times \mathbf{B}$ exerted on one charge by the number of charges in the segment. Because the volume of the segment is AL, the number of charges in the segment is nAL, where n is the number of charges per unit volume. Hence, the total magnetic force on the wire of length L is

$$\mathbf{F}_B = (q\mathbf{v}_d \times \mathbf{B})\,nAL$$

We can write this expression in a more convenient form by noting that, from Equation 27.4, the current in the wire is $I = nqv_dA$. Therefore,

$$\mathbf{F}_B = I\mathbf{L} \times \mathbf{B} \tag{29.3}$$

where $\mathbf{L}$ is a vector that points in the direction of the current I and has a magnitude equal to the length L of the segment. Note that this expression applies only to a straight segment of wire in a uniform magnetic field.

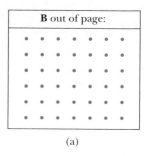

(a)

(b)

Figure 29.6 (a) Magnetic field lines coming out of the paper are indicated by dots, representing the tips of arrows coming outward. (b) Magnetic field lines going into the paper are indicated by crosses, representing the feathers of arrows going inward.

Force on a segment of current-carrying wire in a uniform magnetic field

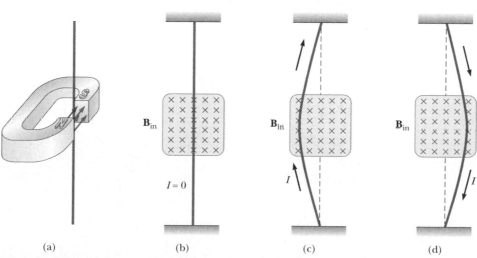

(a)   (b)   (c)   (d)

Figure 29.7 (a) A wire suspended vertically between the poles of a magnet. (b) The setup shown in part (a) as seen looking at the south pole of the magnet, so that the magnetic field (blue crosses) is directed into the page. When there is no current in the wire, it remains vertical. (c) When the current is upward, the wire deflects to the left. (d) When the current is downward, the wire deflects to the right.

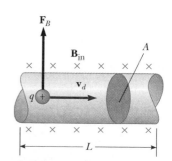

Figure 29.8 A segment of a current-carrying wire in a magnetic field $\mathbf{B}$. The magnetic force exerted on each charge making up the current is $q\mathbf{v}_d \times \mathbf{B}$ and the net force on the segment of length L is $I\mathbf{L} \times \mathbf{B}$.

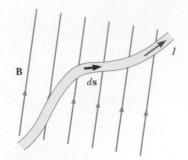

Figure 29.9 A wire segment of arbitrary shape carrying a current I in a magnetic field **B** experiences a magnetic force. The magnetic force on any segment $d\mathbf{s}$ is $I\,d\mathbf{s} \times \mathbf{B}$ and is directed out of the page. You should use the right-hand rule to confirm this force direction.

Now consider an arbitrarily shaped wire segment of uniform cross section in a magnetic field, as shown in Figure 29.9. It follows from Equation 29.3 that the magnetic force exerted on a small segment of vector length $d\mathbf{s}$ in the presence of a field **B** is

$$d\mathbf{F}_B = I\,d\mathbf{s} \times \mathbf{B} \qquad (29.4)$$

where $d\mathbf{F}_B$ is directed out of the page for the directions of **B** and $d\mathbf{s}$ in Figure 29.9. We can consider Equation 29.4 as an alternative definition of **B**. That is, we can define the magnetic field **B** in terms of a measurable force exerted on a current element, where the force is a maximum when **B** is perpendicular to the element and zero when **B** is parallel to the element.

To calculate the total force $\mathbf{F}_B$ acting on the wire shown in Figure 29.9, we integrate Equation 29.4 over the length of the wire:

$$\mathbf{F}_B = I \int_a^b d\mathbf{s} \times \mathbf{B} \qquad (29.5)$$

where a and b represent the end points of the wire. When this integration is carried out, the magnitude of the magnetic field and the direction the field makes with the vector $d\mathbf{s}$ may differ at different points.

We now treat two interesting special cases involving Equation 29.5. In both cases, the magnetic field is assumed to be uniform in magnitude and direction.

Case 1. A curved wire carries a current I and is located in a uniform magnetic field **B**, as shown in Figure 29.10a. Because the field is uniform, we can take **B** outside the integral in Equation 29.5, and we obtain

$$\mathbf{F}_B = I \left(\int_a^b d\mathbf{s} \right) \times \mathbf{B} \qquad (29.6)$$

But the quantity $\int_a^b d\mathbf{s}$ represents the *vector sum* of all the length elements from a to b. From the law of vector addition, the sum equals the vector $\mathbf{L}'$, directed from a to b. Therefore, Equation 29.6 reduces to

$$\mathbf{F}_B = I\mathbf{L}' \times \mathbf{B} \qquad (29.7)$$

From this we conclude that **the magnetic force on a curved current-carrying wire in a uniform magnetic field is equal to that on a straight wire connecting the end points and carrying the same current.**

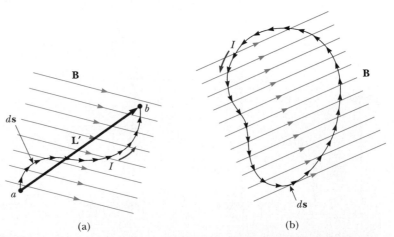

(a) (b)

Figure 29.10 (a) A curved wire carrying a current I in a uniform magnetic field. The total magnetic force acting on the wire is equivalent to the force on a straight wire of length L' running between the ends of the curved wire. (b) A current-carrying loop of arbitrary shape in a uniform magnetic field. The net magnetic force on the loop is zero.

Case 2. An arbitrarily shaped closed loop carrying a current I is placed in a uniform magnetic field, as shown in Figure 29.10b. We can again express the magnetic force acting on the loop in the form of Equation 29.6, but this time we must take the vector sum of the length elements $d\mathbf{s}$ over the entire loop:

$$\mathbf{F}_B = I\left(\oint d\mathbf{s}\right) \times \mathbf{B}$$

Because the set of length elements forms a closed polygon, the vector sum must be zero. This follows from the procedure for adding vectors by the graphical method. Because $\oint d\mathbf{s} = 0$, we conclude that $\mathbf{F}_B = 0$; that is, **the net magnetic force acting on any closed current loop in a uniform magnetic field is zero.**

Quick Quiz 29.4 The four wires shown in Figure 29.11 all carry the same current from point A to point B through the same magnetic field. In all four parts of the figure, the points A and B are 10 cm apart. Rank the wires according to the magnitude of the magnetic force exerted on them, from greatest to least.

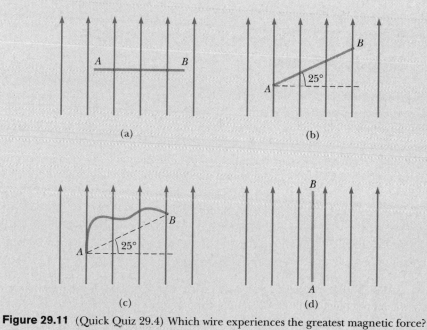

Figure 29.11 (Quick Quiz 29.4) Which wire experiences the greatest magnetic force?

Quick Quiz 29.5 A wire carries current in the plane of this paper toward the top of the page. The wire experiences a magnetic force toward the right edge of the page. The direction of the magnetic field causing this force is (a) in the plane of the page and toward the left edge, (b) in the plane of the page and toward the bottom edge, (c) upward out of the page, (d) downward into the page.

Example 29.2 Force on a Semicircular Conductor

A wire bent into a semicircle of radius R forms a closed circuit and carries a current I. The wire lies in the xy plane, and a uniform magnetic field is directed along the positive y axis, as shown in Figure 29.12. Find the magnitude and direction of the magnetic force acting

on the straight portion of the wire and on the curved portion.

Solution The magnetic force $\mathbf{F}_1$ acting on the straight portion has a magnitude $F_1 = ILB = 2IRB$ because $L = 2R$ and

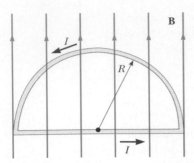

Figure 29.12 (Example 29.2) The net magnetic force acting on a closed current loop in a uniform magnetic field is zero. In the setup shown here, the magnetic force on the straight portion of the loop is $2IRB$ and directed out of the page, and the magnetic force on the curved portion is $2IRB$ directed into the page.

the wire is oriented perpendicular to **B**. The direction of $\mathbf{F}_1$ is out of the page based on the right-hand rule for the cross product $\mathbf{L} \times \mathbf{B}$.

To find the magnetic force $\mathbf{F}_2$ acting on the curved part, we use the results of Case 1. The magnetic force on the curved portion is the same as that on a straight wire of length $2R$ carrying current I to the left. Thus, $F_2 = ILB = 2IRB$. The direction of $\mathbf{F}_2$ is into the page based on the right-hand rule for the cross product $\mathbf{L} \times \mathbf{B}$.

Because the wire lies in the xy plane, the two forces on the loop can be expressed as

$$\mathbf{F}_1 = \boxed{2IRB\,\hat{\mathbf{k}}}$$

$$\mathbf{F}_2 = \boxed{-2IRB\,\hat{\mathbf{k}}}$$

The net magnetic force on the loop is

$$\sum \mathbf{F} = \mathbf{F}_1 + \mathbf{F}_2 = 2IRB\,\hat{\mathbf{k}} - 2IRB\,\hat{\mathbf{k}} = 0$$

Note that this is consistent with Case 2, because the wire forms a closed loop in a uniform magnetic field.

29.3 Torque on a Current Loop in a Uniform Magnetic Field

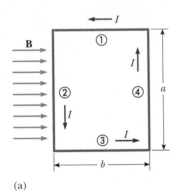

(a)

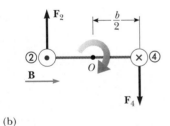

(b)

Figure 29.13 (a) Overhead view of a rectangular current loop in a uniform magnetic field. No magnetic forces are acting on sides ① and ③ because these sides are parallel to **B**. Forces are acting on sides ② and ④, however. (b) Edge view of the loop sighting down sides ② and ④ shows that the magnetic forces $\mathbf{F}_2$ and $\mathbf{F}_4$ exerted on these sides create a torque that tends to twist the loop clockwise. The purple dot in the left circle represents current in wire ② coming toward you; the purple cross in the right circle represents current in wire ④ moving away from you.

In the preceding section, we showed how a magnetic force is exerted on a current-carrying conductor placed in a magnetic field. With this as a starting point, we now show that a torque is exerted on a current loop placed in a magnetic field. The results of this analysis will be of great value when we discuss motors in Chapter 31.

Consider a rectangular loop carrying a current I in the presence of a uniform magnetic field directed parallel to the plane of the loop, as shown in Figure 29.13a. No magnetic forces act on sides ① and ③ because these wires are parallel to the field; hence, $\mathbf{L} \times \mathbf{B} = 0$ for these sides. However, magnetic forces do act on sides ② and ④ because these sides are oriented perpendicular to the field. The magnitude of these forces is, from Equation 29.3,

$$F_2 = F_4 = IaB$$

The direction of $\mathbf{F}_2$, the magnetic force exerted on wire ②, is out of the page in the view shown in Figure 29.13a, and that of $\mathbf{F}_4$, the magnetic force exerted on wire ④, is into the page in the same view. If we view the loop from side ③ and sight along sides ② and ④, we see the view shown in Figure 29.13b, and the two magnetic forces $\mathbf{F}_2$ and $\mathbf{F}_4$ are directed as shown. Note that the two forces point in opposite directions but are *not* directed along the same line of action. If the loop is pivoted so that it can rotate about point O, these two forces produce about O a torque that rotates the loop clockwise. The magnitude of this torque τ_{max} is

$$\tau_{max} = F_2 \frac{b}{2} + F_4 \frac{b}{2} = (IaB)\frac{b}{2} + (IaB)\frac{b}{2} = IabB$$

where the moment arm about O is $b/2$ for each force. Because the area enclosed by the loop is $A = ab$, we can express the maximum torque as

$$\tau_{max} = IAB \qquad (29.8)$$

This maximum-torque result is valid only when the magnetic field is parallel to the plane of the loop. The sense of the rotation is clockwise when viewed from side ③, as indicated in Figure 29.13b. If the current direction were reversed, the force directions would also reverse, and the rotational tendency would be counterclockwise.

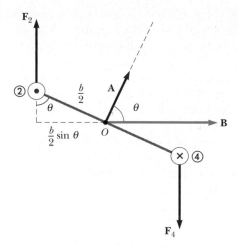

Active Figure 29.14 An end view of the loop in Figure 29.13b rotated through an angle with respect to the magnetic field. If **B** is at an angle θ with respect to vector **A**, which is perpendicular to the plane of the loop, the torque is $IAB \sin \theta$ where the magnitude of **A** is A, the area of the loop.

At the Active Figures link at http://www.pse6.com, you can choose the current in the loop, the magnetic field, and the initial orientation of the loop and observe the subsequent motion.

Now suppose that the uniform magnetic field makes an angle $\theta < 90°$ with a line perpendicular to the plane of the loop, as in Figure 29.14. For convenience, we assume that **B** is perpendicular to sides ② and ④. In this case, the magnetic forces $\mathbf{F}_1$ and $\mathbf{F}_3$ exerted on sides ① and ③ cancel each other and produce no torque because they pass through a common origin. However, the magnetic forces $\mathbf{F}_2$ and $\mathbf{F}_4$ acting on sides ② and ④ produce a torque about *any point*. Referring to the end view shown in Figure 29.14, we note that the moment arm of $\mathbf{F}_2$ about the point O is equal to $(b/2) \sin \theta$. Likewise, the moment arm of $\mathbf{F}_4$ about O is also $(b/2) \sin \theta$. Because $F_2 = F_4 = IaB$, the magnitude of the net torque about O is

$$\tau = F_2 \frac{b}{2} \sin \theta + F_4 \frac{b}{2} \sin \theta$$

$$= IaB \left(\frac{b}{2} \sin \theta \right) + IaB \left(\frac{b}{2} \sin \theta \right) = IabB \sin \theta$$

$$= IAB \sin \theta$$

where $A = ab$ is the area of the loop. This result shows that the torque has its maximum value IAB when the field is perpendicular to the normal to the plane of the loop ($\theta = 90°$), as we saw when discussing Figure 29.13, and is zero when the field is parallel to the normal to the plane of the loop ($\theta = 0$).

A convenient expression for the torque exerted on a loop placed in a uniform magnetic field **B** is

$$\boldsymbol{\tau} = I\mathbf{A} \times \mathbf{B} \tag{29.9}$$

where **A**, the vector shown in Figure 29.14, is perpendicular to the plane of the loop and has a magnitude equal to the area of the loop. We determine the direction of **A** using the right-hand rule described in Figure 29.15. When you curl the fingers of your right hand in the direction of the current in the loop, your thumb points in the direction of **A**. As we see in Figure 29.14, the loop tends to rotate in the direction of decreasing values of θ (that is, such that the area vector **A** rotates toward the direction of the magnetic field).

The product $I\mathbf{A}$ is defined to be the **magnetic dipole moment** $\boldsymbol{\mu}$ (often simply called the "magnetic moment") of the loop:

$$\boldsymbol{\mu} = I\mathbf{A} \tag{29.10}$$

The SI unit of magnetic dipole moment is ampere-meter^2 ($\text{A} \cdot \text{m}^2$). Using this definition, we can express the torque exerted on a current-carrying loop in a magnetic field **B** as

$$\boldsymbol{\tau} = \boldsymbol{\mu} \times \mathbf{B} \tag{29.11}$$

Note that this result is analogous to Equation 26.18, $\boldsymbol{\tau} = \mathbf{p} \times \mathbf{E}$, for the torque exerted on an electric dipole in the presence of an electric field **E**, where **p** is the electric dipole moment.

Figure 29.15 Right-hand rule for determining the direction of the vector **A**. The direction of the magnetic moment $\boldsymbol{\mu}$ is the same as the direction of **A**.

Torque on a current loop in a magnetic field

Magnetic dipole moment of a current loop

Torque on a magnetic moment in a magnetic field

Although we obtained the torque for a particular orientation of **B** with respect to the loop, the equation $\boldsymbol{\tau} = \boldsymbol{\mu} \times \mathbf{B}$ is valid for any orientation. Furthermore, although we derived the torque expression for a rectangular loop, the result is valid for a loop of any shape.

If a coil consists of N turns of wire, each carrying the same current and enclosing the same area, the total magnetic dipole moment of the coil is N times the magnetic dipole moment for one turn. The torque on an N-turn coil is N times that on a one-turn coil. Thus, we write $\boldsymbol{\tau} = N\boldsymbol{\mu}_{\text{loop}} \times \mathbf{B} = \boldsymbol{\mu}_{\text{coil}} \times \mathbf{B}$.

In Section 26.6, we found that the potential energy of a system of an electric dipole in an electric field is given by $U = -\mathbf{p} \cdot \mathbf{E}$. This energy depends on the orientation of the dipole in the electric field. Likewise, the potential energy of a system of a magnetic dipole in a magnetic field depends on the orientation of the dipole in the magnetic field and is given by

$$U = -\boldsymbol{\mu} \cdot \mathbf{B} \tag{29.12}$$

Potential energy of a system of a magnetic moment in a magnetic field

From this expression, we see that the system has its lowest energy $U_{\text{min}} = -\mu B$ when $\boldsymbol{\mu}$ points in the same direction as **B**. The system has its highest energy $U_{\text{max}} = +\mu B$ when $\boldsymbol{\mu}$ points in the direction opposite **B**.

Quick Quiz 29.6 Rank the magnitudes of the torques acting on the rectangular loops shown edge-on in Figure 29.16, from highest to lowest. All loops are identical and carry the same current.

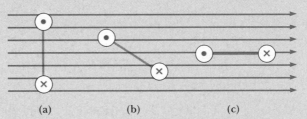

(a)   (b)   (c)

Figure 29.16 (Quick Quiz 29.6) Which current loop (seen edge-on) experiences the greatest torque? (Quick Quiz 29.7) Which current loop (seen edge-on) experiences the greatest net force?

Quick Quiz 29.7 Rank the magnitudes of the net forces acting on the rectangular loops shown in Figure 29.16, from highest to lowest. All loops are identical and carry the same current.

Example 29.3   The Magnetic Dipole Moment of a Coil

A rectangular coil of dimensions 5.40 cm × 8.50 cm consists of 25 turns of wire and carries a current of 15.0 mA. A 0.350-T magnetic field is applied parallel to the plane of the loop.

(A) Calculate the magnitude of its magnetic dipole moment.

Solution Because the coil has 25 turns, we modify Equation 29.10 to obtain

$\mu_{\text{coil}} = NIA = (25)(15.0 \times 10^{-3}\,\text{A})(0.054\,0\,\text{m})(0.085\,0\,\text{m})$

$= 1.72 \times 10^{-3}\,\text{A} \cdot \text{m}^2$

(B) What is the magnitude of the torque acting on the loop?

Solution Because **B** is perpendicular to $\boldsymbol{\mu}_{\text{coil}}$, Equation 29.11 gives

$\tau = \mu_{\text{coil}} B = (1.72 \times 10^{-3}\,\text{A} \cdot \text{m}^2)(0.350\,\text{T})$

$= 6.02 \times 10^{-4}\,\text{N} \cdot \text{m}$

Example 29.4   Satellite Attitude Control

Many satellites use coils called *torquers* to adjust their orientation. These devices interact with the Earth's magnetic field to create a torque on the spacecraft in the x, y, or z direction. The major advantage of this type of attitude-control system is that it uses solar-generated electricity and so does not consume any thruster fuel.

If a typical device has a magnetic dipole moment of $250 \, \text{A} \cdot \text{m}^2$, what is the maximum torque applied to a satellite when its torquer is turned on at an altitude where the magnitude of the Earth's magnetic field is $3.0 \times 10^{-5} \, \text{T}$?

Solution We once again apply Equation 29.11, recognizing that the maximum torque is obtained when the magnetic dipole moment of the torquer is perpendicular to the Earth's magnetic field:

$$\tau_{\text{max}} = \mu B = (250 \, \text{A} \cdot \text{m}^2)(3.0 \times 10^{-5} \, \text{T})$$

$$= 7.5 \times 10^{-3} \, \text{N} \cdot \text{m}$$

Example 29.5   The D'Arsonval Galvanometer

An end view of a D'Arsonval galvanometer (see Section 28.5) is shown in Figure 29.17. When the turns of wire making up the coil carry a current, the magnetic field created by the magnet exerts on the coil a torque that turns it (along with its attached pointer) against the spring. Show that the angle of deflection of the pointer is directly proportional to the current in the coil.

Solution We can use Equation 29.11 to find the torque τ_m that the magnetic field exerts on the coil. If we assume that the magnetic field through the coil is perpendicular to the normal to the plane of the coil, Equation 29.11 becomes

$$\tau_m = \mu B$$

(This is a reasonable assumption because the circular cross section of the magnet ensures radial magnetic field lines.) This magnetic torque is opposed by the torque due to the spring, which is given by the rotational version of Hooke's law, $\tau_s = -\kappa\phi$, where κ is the torsional spring constant and ϕ is the angle through which the spring turns. Because the coil does not have an angular acceleration when the pointer is at rest, the sum of these torques must be zero:

$$(1) \qquad \tau_m + \tau_s = \mu B - \kappa\phi = 0$$

Equation 29.10 allows us to relate the magnetic moment of the N turns of wire to the current through them:

$$\mu = NIA$$

We can substitute this expression for μ in Equation (1) to obtain

$$(NIA)B - \kappa\phi = 0$$

$$\phi = \frac{NAB}{\kappa} I$$

Thus, the angle of deflection of the pointer is directly proportional to the current in the loop. The factor NAB/κ tells us that deflection also depends on the design of the meter.

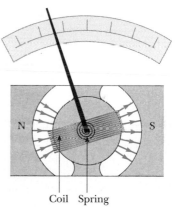

Figure 29.17 (Example 29.5) Structure of a moving-coil galvanometer.

29.4   Motion of a Charged Particle in a Uniform Magnetic Field

In Section 29.1 we found that the magnetic force acting on a charged particle moving in a magnetic field is perpendicular to the velocity of the particle and that consequently the work done by the magnetic force on the particle is zero. Now consider the special case of a positively charged particle moving in a uniform magnetic field with the initial velocity vector of the particle perpendicular to the field. Let us assume that the direction of the magnetic field is into the page, as in Figure 29.18. As the particle changes the direction of its velocity in response to the magnetic force, the magnetic force remains perpendicular to the velocity. As we found in Section 6.1, if the force is always perpendicular to the velocity, the path of

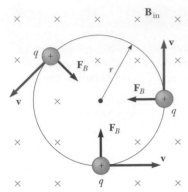

Active Figure 29.18 When the velocity of a charged particle is perpendicular to a uniform magnetic field, the particle moves in a circular path in a plane perpendicular to **B**. The magnetic force $\mathbf{F}_B$ acting on the charge is always directed toward the center of the circle.

At the Active Figures link at http://www.pse6.com, you can adjust the mass, speed, and charge of the particle and the magnitude of the magnetic field to observe the resulting circular motion.

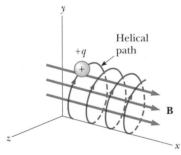

Active Figure 29.19 A charged particle having a velocity vector that has a component parallel to a uniform magnetic field moves in a helical path.

At the Active Figures link at http://www.pse6.com, you can adjust the x component of the velocity of the particle and observe the resulting helical motion.

the particle is a circle! Figure 29.18 shows the particle moving in a circle in a plane perpendicular to the magnetic field.

The particle moves in a circle because the magnetic force $\mathbf{F}_B$ is perpendicular to **v** and **B** and has a constant magnitude qvB. As Figure 29.18 illustrates, the rotation is counterclockwise for a positive charge. If q were negative, the rotation would be clockwise. We can use Equation 6.1 to equate this magnetic force to the product of the particle mass and the centripetal acceleration:

$$\sum F = ma_c$$

$$F_B = qvB = \frac{mv^2}{r}$$

$$r = \frac{mv}{qB} \qquad (29.13)$$

That is, the radius of the path is proportional to the linear momentum mv of the particle and inversely proportional to the magnitude of the charge on the particle and to the magnitude of the magnetic field. The angular speed of the particle (from Eq. 10.10) is

$$\omega = \frac{v}{r} = \frac{qB}{m} \qquad (29.14)$$

The period of the motion (the time interval the particle requires to complete one revolution) is equal to the circumference of the circle divided by the linear speed of the particle:

$$T = \frac{2\pi r}{v} = \frac{2\pi}{\omega} = \frac{2\pi m}{qB} \qquad (29.15)$$

These results show that the angular speed of the particle and the period of the circular motion do not depend on the linear speed of the particle or on the radius of the orbit. The angular speed ω is often referred to as the **cyclotron frequency** because charged particles circulate at this angular frequency in the type of accelerator called a *cyclotron*, which is discussed in Section 29.5.

If a charged particle moves in a uniform magnetic field with its velocity at some arbitrary angle with respect to **B**, its path is a helix. For example, if the field is directed in the x direction, as shown in Figure 29.19, there is no component of force in the x direction. As a result, $a_x = 0$, and the x component of velocity remains constant. However, the magnetic force $q\mathbf{v} \times \mathbf{B}$ causes the components v_y and v_z to change in time, and the resulting motion is a helix whose axis is parallel to the magnetic field. The projection of the path onto the yz plane (viewed along the x axis) is a circle. (The projections of the path onto the xy and xz planes are sinusoids!) Equations 29.13 to 29.15 still apply provided that v is replaced by $v_\perp = \sqrt{v_y{}^2 + v_z{}^2}$.

Quick Quiz 29.8 A charged particle is moving perpendicular to a magnetic field in a circle with a radius r. An identical particle enters the field, with **v** perpendicular to **B**, but with a higher speed v than the first particle. Compared to the radius of the circle for the first particle, the radius of the circle for the second particle is (a) smaller (b) larger (c) equal in size.

Quick Quiz 29.9 A charged particle is moving perpendicular to a magnetic field in a circle with a radius r. The magnitude of the magnetic field is increased. Compared to the initial radius of the circular path, the radius of the new path is (a) smaller (b) larger (c) equal in size.

Example 29.6   A Proton Moving Perpendicular to a Uniform Magnetic Field

A proton is moving in a circular orbit of radius 14 cm in a uniform 0.35-T magnetic field perpendicular to the velocity of the proton. Find the linear speed of the proton.

Solution From Equation 29.13, we have

$$v = \frac{qBr}{m_p} = \frac{(1.60 \times 10^{-19}\,\text{C})(0.35\,\text{T})(0.14\,\text{m})}{1.67 \times 10^{-27}\,\text{kg}}$$

$$= \boxed{4.7 \times 10^6\,\text{m/s}}$$

What If? What if an electron, rather than a proton, moves in a direction perpendicular to the same magnetic field with this same linear speed? Will the radius of its orbit be different?

Answer An electron has a much smaller mass than a proton, so the magnetic force should be able to change its velocity much easier than for the proton. Thus, we should expect the radius to be smaller. Looking at Equation 29.13, we see that r is proportional to m with q, B, and v the same for the electron as for the proton. Consequently, the radius will be smaller by the same factor as the ratio of masses m_e/m_p.

Example 29.7   Bending an Electron Beam   **Interactive**

In an experiment designed to measure the magnitude of a uniform magnetic field, electrons are accelerated from rest through a potential difference of 350 V. The electrons travel along a curved path because of the magnetic force exerted on them, and the radius of the path is measured to be 7.5 cm. (Fig. 29.20 shows such a curved beam of electrons.) If the magnetic field is perpendicular to the beam,

(A) what is the magnitude of the field?

Solution Conceptualize the circular motion of the electrons with the help of Figures 29.18 and 29.20. We categorize this problem as one involving both uniform circular motion and a magnetic force. Looking at Equation 29.13, we see that we need the speed v of the electron if we are to find the magnetic field magnitude, and v is not given. Consequently, we must find the speed of the electron based on the potential difference through which it is accelerated. Therefore, we also categorize this as a problem in conservation of mechanical energy for an isolated system. To begin analyzing the problem, we find the electron speed. For the isolated electron–electric field system, the loss of potential energy as the electron moves through the 350-V potential difference appears as an increase in the kinetic energy of the electron. Because $K_i = 0$ and $K_f = \frac{1}{2}m_e v^2$, we have

$$\Delta K + \Delta U = 0 \quad \longrightarrow \quad \tfrac{1}{2}m_e v^2 + (-e)\,\Delta V = 0$$

$$v = \sqrt{\frac{2e\,\Delta V}{m_e}} = \sqrt{\frac{2(1.60 \times 10^{-19}\,\text{C})(350\,\text{V})}{9.11 \times 10^{-31}\,\text{kg}}}$$

$$= 1.11 \times 10^7\,\text{m/s}$$

Now, using Equation 29.13, we find

$$B = \frac{m_e v}{er} = \frac{(9.11 \times 10^{-31}\,\text{kg})(1.11 \times 10^7\,\text{m/s})}{(1.60 \times 10^{-19}\,\text{C})(0.075\,\text{m})}$$

$$= \boxed{8.4 \times 10^{-4}\,\text{T}}$$

(B) What is the angular speed of the electrons?

Solution Using Equation 29.14, we find that

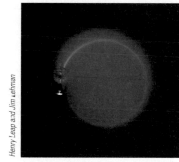

Figure 29.20 (Example 29.7) The bending of an electron beam in a magnetic field.

$$\omega = \frac{v}{r} = \frac{1.11 \times 10^7\,\text{m/s}}{0.075\,\text{m}} = \boxed{1.5 \times 10^8\,\text{rad/s}}$$

To finalize this problem, note that the angular speed can be represented as $\omega = (1.5 \times 10^8\,\text{rad/s})(1\,\text{rev}/2\pi\,\text{rad}) = 2.4 \times 10^7\,\text{rev/s}$. The electrons travel around the circle 24 million times per second! This is consistent with the very high speed that we found in part (A).

What If? What if a sudden voltage surge causes the accelerating voltage to increase to 400 V? How does this affect the angular speed of the electrons, assuming that the magnetic field remains constant?

Answer The increase in accelerating voltage ΔV will cause the electrons to enter the magnetic field with a higher speed v. This will cause them to travel in a circle with a larger radius r. The angular speed is the ratio of v to r. Both v and r increase by the same factor, so that the effects cancel and the angular speed remains the same. Equation 29.14 is an expression for the cyclotron frequency, which is the same as the angular speed of the electrons. The cyclotron frequency depends only on the charge q, the magnetic field B, and the mass m_e, none of which have changed. Thus, the voltage surge has no effect on the angular speed. (However, in reality, the voltage surge may also increase the magnetic field if the magnetic field is powered by the same source as the accelerating voltage. In this case, the angular speed will increase according to Equation 29.14.)

 At the Interactive Worked Example link at http://www.pse6.com, you can investigate the relationship between the radius of the circular path of the electrons and the magnetic field.

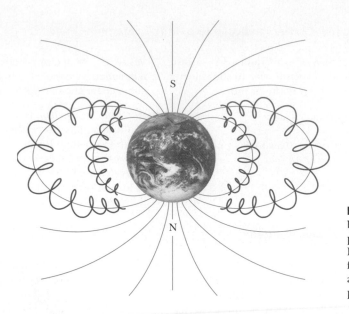

Figure 29.22 The Van Allen belts are made up of charged particles trapped by the Earth's nonuniform magnetic field. The magnetic field lines are in blue and the particle paths in red.

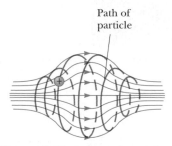

Figure 29.21 A charged particle moving in a nonuniform magnetic field (a magnetic bottle) spirals about the field and oscillates between the end points. The magnetic force exerted on the particle near either end of the bottle has a component that causes the particle to spiral back toward the center.

When charged particles move in a nonuniform magnetic field, the motion is complex. For example, in a magnetic field that is strong at the ends and weak in the middle, such as that shown in Figure 29.21, the particles can oscillate back and forth between two positions. A charged particle starting at one end spirals along the field lines until it reaches the other end, where it reverses its path and spirals back. This configuration is known as a *magnetic bottle* because charged particles can be trapped within it. The magnetic bottle has been used to confine a *plasma*, a gas consisting of ions and electrons. Such a plasma-confinement scheme could fulfill a crucial role in the control of nuclear fusion, a process that could supply us with an almost endless source of energy. Unfortunately, the magnetic bottle has its problems. If a large number of particles are trapped, collisions between them cause the particles to eventually leak from the system.

The Van Allen radiation belts consist of charged particles (mostly electrons and protons) surrounding the Earth in doughnut-shaped regions (Fig. 29.22). The particles, trapped by the Earth's nonuniform magnetic field, spiral around the field lines from pole to pole, covering the distance in just a few seconds. These particles originate mainly from the Sun, but some come from stars and other heavenly objects. For this reason, the particles are called *cosmic rays*. Most cosmic rays are deflected by the Earth's magnetic field and never reach the atmosphere. However, some of the particles become trapped; it is these particles that make up the Van Allen belts. When the particles are located over the poles, they sometimes collide with atoms in the atmosphere, causing the atoms to emit visible light. Such collisions are the origin of the beautiful Aurora Borealis, or Northern Lights, in the northern hemisphere and the Aurora Australis in the southern hemisphere. Auroras are usually confined to the polar regions because the Van Allen belts are nearest the Earth's surface there. Occasionally, though, solar activity causes larger numbers of charged particles to enter the belts and significantly distort the normal magnetic field lines associated with the Earth. In these situations an aurora can sometimes be seen at lower latitudes.

29.5 Applications Involving Charged Particles Moving in a Magnetic Field

A charge moving with a velocity **v** in the presence of both an electric field **E** and a magnetic field **B** experiences both an electric force $q\mathbf{E}$ and a magnetic force $q\mathbf{v} \times \mathbf{B}$. The total force (called the Lorentz force) acting on the charge is

Lorentz force

$$\mathbf{F} = q\mathbf{E} + q\mathbf{v} \times \mathbf{B} \qquad (29.16)$$

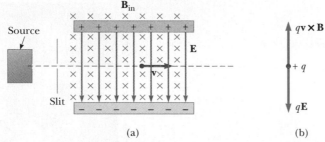

(a)   (b)

Active Figure 29.23 (a) A velocity selector. When a positively charged particle is moving with velocity **v** in the presence of a magnetic field directed into the page and an electric field directed downward, it experiences a downward electric force $q\mathbf{E}$ and an upward magnetic force $q\mathbf{v} \times \mathbf{B}$. (b) When these forces balance, the particle moves in a horizontal line through the fields.

At the Active Figures link at http://www.pse6.com, you can adjust the electric and magnetic fields to try to achieve straight line motion for the charge.

Velocity Selector

In many experiments involving moving charged particles, it is important that the particles all move with essentially the same velocity. This can be achieved by applying a combination of an electric field and a magnetic field oriented as shown in Figure 29.23. A uniform electric field is directed vertically downward (in the plane of the page in Fig. 29.23a), and a uniform magnetic field is applied in the direction perpendicular to the electric field (into the page in Fig. 29.23a). If q is positive and the velocity **v** is to the right, the magnetic force $q\mathbf{v} \times \mathbf{B}$ is upward and the electric force $q\mathbf{E}$ is downward. When the magnitudes of the two fields are chosen so that $qE = qvB$, the particle moves in a straight horizontal line through the region of the fields. From the expression $qE = qvB$, we find that

$$v = \frac{E}{B} \qquad (29.17)$$

Only those particles having speed v pass undeflected through the mutually perpendicular electric and magnetic fields. The magnetic force exerted on particles moving at speeds greater than this is stronger than the electric force, and the particles are deflected upward. Those moving at speeds less than this are deflected downward.

The Mass Spectrometer

A **mass spectrometer** separates ions according to their mass-to-charge ratio. In one version of this device, known as the *Bainbridge mass spectrometer,* a beam of ions first passes through a velocity selector and then enters a second uniform magnetic field $\mathbf{B}_0$ that has the same direction as the magnetic field in the selector (Fig. 29.24). Upon entering the second magnetic field, the ions move in a semicircle of radius r before striking a detector array at P. If the ions are positively charged, the beam deflects upward, as Figure 29.24 shows. If the ions are negatively charged, the beam deflects

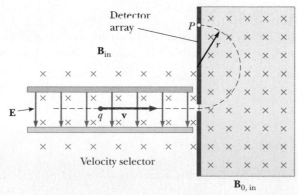

Active Figure 29.24 A mass spectrometer. Positively charged particles are sent first through a velocity selector and then into a region where the magnetic field $\mathbf{B}_0$ causes the particles to move in a semicircular path and strike a detector array at P.

At the Active Figures link at http://www.pse6.com, you can predict where particles will strike the detector array.

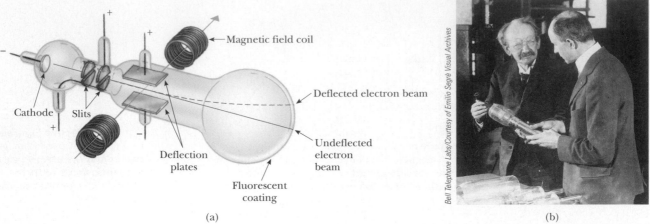

(a)

(b)

Figure 29.25 (a) Thomson's apparatus for measuring e/m_e. Electrons are accelerated from the cathode, pass through two slits, and are deflected by both an electric field and a magnetic field (directed perpendicular to the electric field). The beam of electrons then strikes a fluorescent screen. (b) J. J. Thomson *(left)* in the Cavendish Laboratory, University of Cambridge. The man on the right, Frank Baldwin Jewett, is a distant relative of John W. Jewett, Jr., co-author of this text.

downward. From Equation 29.13, we can express the ratio m/q as

$$\frac{m}{q} = \frac{rB_0}{v}$$

Using Equation 29.17, we find that

$$\frac{m}{q} = \frac{rB_0B}{E}$$ (29.18)

Therefore, we can determine m/q by measuring the radius of curvature and knowing the field magnitudes B, B_0, and E. In practice, one usually measures the masses of various isotopes of a given ion, with the ions all carrying the same charge q. In this way, the mass ratios can be determined even if q is unknown.

A variation of this technique was used by J. J. Thomson (1856–1940) in 1897 to measure the ratio e/m_e for electrons. Figure 29.25a shows the basic apparatus he used. Electrons are accelerated from the cathode and pass through two slits. They then drift into a region of perpendicular electric and magnetic fields. The magnitudes of the two fields are first adjusted to produce an undeflected beam. When the magnetic field is turned off, the electric field produces a measurable beam deflection that is recorded on the fluorescent screen. From the size of the deflection and the measured values of E and B, the charge-to-mass ratio can be determined. The results of this crucial experiment represent the discovery of the electron as a fundamental particle of nature.

Quick Quiz 29.10 Three types of particles enter a mass spectrometer like the one shown in Figure 29.24. Figure 29.26 shows where the particles strike the detector array. Rank the particles that arrive at *a*, *b*, and *c* by speed and m/q ratio.

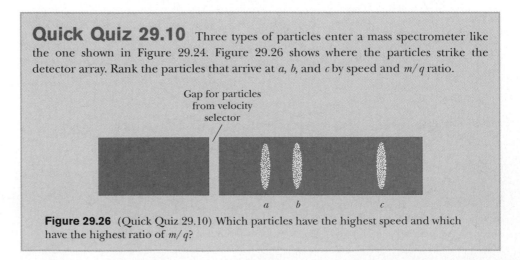

Figure 29.26 (Quick Quiz 29.10) Which particles have the highest speed and which have the highest ratio of m/q?

The Cyclotron

A **cyclotron** is a device that can accelerate charged particles to very high speeds. The energetic particles produced are used to bombard atomic nuclei and thereby produce nuclear reactions of interest to researchers. A number of hospitals use cyclotron facilities to produce radioactive substances for diagnosis and treatment.

Both electric and magnetic forces have a key role in the operation of a cyclotron. A schematic drawing of a cyclotron is shown in Figure 29.27a. The charges move inside two semicircular containers D_1 and D_2, referred to as *dees*, because of their shape like the letter D. A high-frequency alternating potential difference is applied to the dees, and a uniform magnetic field is directed perpendicular to them. A positive ion released at P near the center of the magnet in one dee moves in a semicircular path (indicated by the dashed red line in the drawing) and arrives back at the gap in a time interval $T/2$, where T is the time interval needed to make one complete trip around the two dees, given by Equation 29.15. The frequency of the applied potential difference is adjusted so that the polarity of the dees is reversed in the same time interval during which the ion travels around one dee. If the applied potential difference is adjusted such that D_2 is at a lower electric potential than D_1 by an amount ΔV, the ion accelerates across the gap to D_2 and its kinetic energy increases by an amount $q \Delta V$. It then moves around D_2 in a semicircular path of greater radius (because its speed has increased). After a time interval $T/2$, it again arrives at the gap between the dees. By this time, the polarity across the dees has again been reversed, and the ion is given another "kick" across the gap. The motion continues so that for each half-circle trip around one dee, the ion gains additional kinetic energy equal to $q \Delta V$. When the radius of its path is nearly that of the dees, the energetic ion leaves the system through the exit slit. Note that the operation of the cyclotron is based on the fact that T is independent of the speed of the ion and of the radius of the circular path (Eq. 29.15).

We can obtain an expression for the kinetic energy of the ion when it exits the cyclotron in terms of the radius R of the dees. From Equation 29.13 we know that $v = qBR/m$. Hence, the kinetic energy is

$$K = \tfrac{1}{2}mv^2 = \frac{q^2 B^2 R^2}{2m} \tag{29.19}$$

When the energy of the ions in a cyclotron exceeds about 20 MeV, relativistic effects come into play. (Such effects are discussed in Chapter 39.) We observe that T increases and that the moving ions do not remain in phase with the applied potential

▲ **PITFALL PREVENTION**

29.1 The Cyclotron Is Not State-of-the-Art Technology

The cyclotron is important historically because it was the first particle accelerator to achieve very high particle speeds. Cyclotrons are still in use in medical applications, but most accelerators currently in research use are not cyclotrons. Research accelerators work on a different principle and are generally called *synchrotrons*.

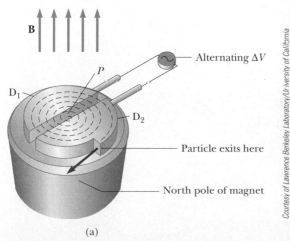

(a)

(b)

Courtesy of Lawrence Berkeley Laboratory/University of California

Figure 29.27 (a) A cyclotron consists of an ion source at P, two dees D_1 and D_2 across which an alternating potential difference is applied, and a uniform magnetic field. (The south pole of the magnet is not shown.) The red dashed curved lines represent the path of the particles. (b) The first cyclotron, invented by E. O. Lawrence and M. S. Livingston in 1934.

difference. Some accelerators overcome this problem by modifying the period of the applied potential difference so that it remains in phase with the moving ions.

29.6 The Hall Effect

When a current-carrying conductor is placed in a magnetic field, a potential difference is generated in a direction perpendicular to both the current and the magnetic field. This phenomenon, first observed by Edwin Hall (1855–1938) in 1879, is known as the *Hall effect*. It arises from the deflection of charge carriers to one side of the conductor as a result of the magnetic force they experience. The Hall effect gives information regarding the sign of the charge carriers and their density; it can also be used to measure the magnitude of magnetic fields.

The arrangement for observing the Hall effect consists of a flat conductor carrying a current I in the x direction, as shown in Figure 29.28. A uniform magnetic field **B** is applied in the y direction. If the charge carriers are electrons moving in the negative x direction with a drift velocity $\mathbf{v}_d$, they experience an upward magnetic force $\mathbf{F}_B = q\mathbf{v}_d \times \mathbf{B}$, are deflected upward, and accumulate at the upper edge of the flat conductor, leaving an excess of positive charge at the lower edge (Fig. 29.29a). This accumulation of charge at the edges establishes an electric field in the conductor and increases until the electric force on carriers remaining in the bulk of the conductor balances the magnetic force acting on the carriers. When this equilibrium condition is reached, the electrons are no longer deflected upward. A sensitive voltmeter or potentiometer connected across the sample, as shown in Figure 29.29, can measure the potential difference—known as the **Hall voltage** ΔV_H—generated across the conductor.

If the charge carriers are positive and hence move in the positive x direction (for rightward current), as shown in Figures 29.28 and 29.29b, they also experience an upward magnetic force $q\mathbf{v}_d \times \mathbf{B}$. This produces a buildup of positive charge on the upper edge and leaves an excess of negative charge on the lower edge. Hence, the sign of the Hall voltage generated in the sample is opposite the sign of the Hall voltage resulting from the deflection of electrons. The sign of the charge carriers can therefore be determined from a measurement of the polarity of the Hall voltage.

In deriving an expression for the Hall voltage, we first note that the magnetic force exerted on the carriers has magnitude qv_dB. In equilibrium, this force is balanced by the electric force qE_H, where E_H is the magnitude of the electric field due to the charge separation (sometimes referred to as the *Hall field*). Therefore,

$$qv_dB = qE_H$$
$$E_H = v_dB$$

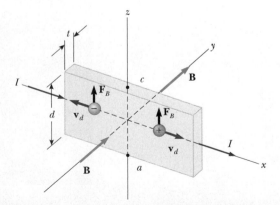

Figure 29.28 To observe the Hall effect, a magnetic field is applied to a current-carrying conductor. When I is in the x direction and **B** in the y direction, both positive and negative charge carriers are deflected upward in the magnetic field. The Hall voltage is measured between points a and c.

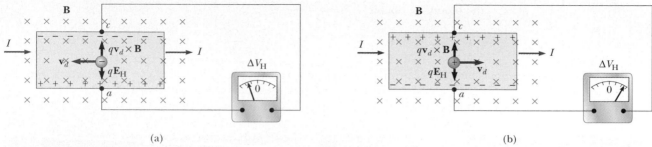

(a)   (b)

Figure 29.29 (a) When the charge carriers in a Hall-effect apparatus are negative, the upper edge of the conductor becomes negatively charged, and c is at a lower electric potential than a. (b) When the charge carriers are positive, the upper edge becomes positively charged, and c is at a higher potential than a. In either case, the charge carriers are no longer deflected when the edges become sufficiently charged that there is a balance on the charge carriers between the electrostatic force qE_H and the magnetic deflection force qvB.

If d is the width of the conductor, the Hall voltage is

$$\Delta V_H = E_H d = v_d B d \qquad (29.20)$$

Thus, the measured Hall voltage gives a value for the drift speed of the charge carriers if d and B are known.

We can obtain the charge carrier density n by measuring the current in the sample. From Equation 27.4, we can express the drift speed as

$$v_d = \frac{I}{nqA} \qquad (29.21)$$

where A is the cross-sectional area of the conductor. Substituting Equation 29.21 into Equation 29.20, we obtain

$$\Delta V_H = \frac{IBd}{nqA} \qquad (29.22)$$

Because $A = td$, where t is the thickness of the conductor, we can also express Equation 29.22 as

$$\Delta V_H = \frac{IB}{nqt} = \frac{R_H IB}{t} \qquad (29.23)$$

The Hall voltage

where $R_H = 1/nq$ is the **Hall coefficient.** This relationship shows that a properly calibrated conductor can be used to measure the magnitude of an unknown magnetic field.

Because all quantities in Equation 29.23 other than nq can be measured, a value for the Hall coefficient is readily obtainable. The sign and magnitude of R_H give the sign of the charge carriers and their number density. In most metals, the charge carriers are electrons, and the charge-carrier density determined from Hall-effect measurements is in good agreement with calculated values for such metals as lithium (Li), sodium (Na), copper (Cu), and silver (Ag), whose atoms each give up one electron to act as a current carrier. In this case, n is approximately equal to the number of conducting electrons per unit volume. However, this classical model is not valid for metals such as iron (Fe), bismuth (Bi), and cadmium (Cd) or for semiconductors. These discrepancies can be explained only by using a model based on the quantum nature of solids.

An interesting medical application related to the Hall effect is the electromagnetic blood flowmeter, first developed in the 1950s and continually improved since then. Imagine that we replace the conductor in Figure 29.29 with an artery carrying blood. The blood contains charged ions that experience electric and magnetic forces like the charge carriers in the conductor. The speed of flow of these ions can be related to the volume rate of flow of blood. Solving Equation 29.20 for the speed v_d of the ions in the blood, we obtain

$$v_d = \frac{\Delta V_H}{Bd}$$

Thus, by measuring the voltage across the artery, the diameter of the artery, and the applied magnetic field, the speed of the blood can be calculated.

Example 29.8   The Hall Effect for Copper

A rectangular copper strip 1.5 cm wide and 0.10 cm thick carries a current of 5.0 A. Find the Hall voltage for a 1.2-T magnetic field applied in a direction perpendicular to the strip.

Solution   If we assume that one electron per atom is available for conduction, we can take the charge carrier density to be 8.49×10^{28} electrons/m³ (see Example 27.1). Substituting this value and the given data into Equation 29.23 gives

$$\Delta V_H = \frac{IB}{nqt}$$

$$= \frac{(5.0 \text{ A})(1.2 \text{ T})}{(8.49 \times 10^{28} \text{ m}^{-3})(1.6 \times 10^{-19} \text{ C})(0.001\ 0 \text{ m})}$$

$$\Delta V_H = \boxed{0.44\ \mu\text{V}}$$

Such an extremely small Hall voltage is expected in good conductors. (Note that the width of the conductor is not needed in this calculation.)

What If?   What if the strip has the same dimensions but is made of a semiconductor? Will the Hall voltage be smaller or larger?

Answer   In semiconductors, n is much smaller than it is in metals that contribute one electron per atom to the current; hence, the Hall voltage is usually larger because it varies as the inverse of n. Currents on the order of 0.1 mA are generally used for such materials. Consider a piece of silicon that has the same dimensions as the copper strip in this example and whose value for n is 1.0×10^{20} electrons/m³. Taking $B = 1.2$ T and $I = 0.10$ mA, we find that $\Delta V_H = 7.5$ mV. A potential difference of this magnitude is readily measured.

SUMMARY

Take a practice test for this chapter by clicking on the Practice Test link at http://www.pse6.com.

The magnetic force that acts on a charge q moving with a velocity $\mathbf{v}$ in a magnetic field $\mathbf{B}$ is

$$\mathbf{F}_B = q\mathbf{v} \times \mathbf{B} \tag{29.1}$$

The direction of this magnetic force is perpendicular both to the velocity of the particle and to the magnetic field. The magnitude of this force is

$$F_B = |q|vB \sin \theta \tag{29.2}$$

where θ is the smaller angle between $\mathbf{v}$ and $\mathbf{B}$. The SI unit of $\mathbf{B}$ is the **tesla** (T), where $1 \text{ T} = 1 \text{ N/A} \cdot \text{m}$.

When a charged particle moves in a magnetic field, the work done by the magnetic force on the particle is zero because the displacement is always perpendicular to the direction of the force. The magnetic field can alter the direction of the particle's velocity vector, but it cannot change its speed.

If a straight conductor of length L carries a current I, the force exerted on that conductor when it is placed in a uniform magnetic field $\mathbf{B}$ is

$$\mathbf{F}_B = I\mathbf{L} \times \mathbf{B} \tag{29.3}$$

where the direction of $\mathbf{L}$ is in the direction of the current and $|\mathbf{L}| = L$.

If an arbitrarily shaped wire carrying a current I is placed in a magnetic field, the magnetic force exerted on a very small segment $d\mathbf{s}$ is

$$d\mathbf{F}_B = I\, d\mathbf{s} \times \mathbf{B} \tag{29.4}$$

To determine the total magnetic force on the wire, one must integrate Equation 29.4, keeping in mind that both $\mathbf{B}$ and $d\mathbf{s}$ may vary at each point. Integration gives for the force exerted on a current-carrying conductor of arbitrary shape in a uniform magnetic field

$$\mathbf{F}_B = I\mathbf{L}' \times \mathbf{B} \qquad (29.7)$$

where $\mathbf{L}'$ is a vector directed from one end of the conductor to the opposite end. Because integration of Equation 29.4 for a closed loop yields a zero result, the net magnetic force on any closed loop carrying a current in a uniform magnetic field is zero.

The **magnetic dipole moment** $\boldsymbol{\mu}$ of a loop carrying a current I is

$$\boldsymbol{\mu} = I\mathbf{A} \qquad (29.10)$$

where the area vector $\mathbf{A}$ is perpendicular to the plane of the loop and $|\mathbf{A}|$ is equal to the area of the loop. The SI unit of $\boldsymbol{\mu}$ is $A \cdot m^2$.

The torque $\boldsymbol{\tau}$ on a current loop placed in a uniform magnetic field $\mathbf{B}$ is

$$\boldsymbol{\tau} = \boldsymbol{\mu} \times \mathbf{B} \qquad (29.11)$$

The potential energy of the system of a magnetic dipole in a magnetic field is

$$U = -\boldsymbol{\mu} \cdot \mathbf{B} \qquad (29.12)$$

If a charged particle moves in a uniform magnetic field so that its initial velocity is perpendicular to the field, the particle moves in a circle, the plane of which is perpendicular to the magnetic field. The radius of the circular path is

$$r = \frac{mv}{qB} \qquad (29.13)$$

where m is the mass of the particle and q is its charge. The angular speed of the charged particle is

$$\omega = \frac{qB}{m} \qquad (29.14)$$

QUESTIONS

1. At a given instant, a proton moves in the positive x direction through a magnetic field in the negative z direction. What is the direction of the magnetic force? Does the proton continue to move in the positive x direction? Explain.

2. Two charged particles are projected into a magnetic field perpendicular to their velocities. If the charges are deflected in opposite directions, what can you say about them?

3. If a charged particle moves in a straight line through some region of space, can you say that the magnetic field in that region is zero?

4. Suppose an electron is chasing a proton up this page when they suddenly enter a magnetic field perpendicular to the page. What happens to the particles?

5. How can the motion of a moving charged particle be used to distinguish between a magnetic field and an electric field? Give a specific example to justify your argument.

6. List several similarities and differences between electric and magnetic forces.

7. Justify the following statement: "It is impossible for a constant (in other words, a time-independent) magnetic field to alter the speed of a charged particle."

8. In view of your answer to Question 7, what is the role of a magnetic field in a cyclotron?

9. The electron beam in Figure Q29.9 is projected to the right. The beam deflects downward in the presence of a magnetic field produced by a pair of current-carrying coils. (a) What is the direction of the magnetic field? (b) What would happen to the beam if the magnetic field were reversed in direction?

Figure Q29.9

Courtesy of Central Scientific Company

10. A current-carrying conductor experiences no magnetic force when placed in a certain manner in a uniform magnetic field. Explain.

11. Is it possible to orient a current loop in a uniform magnetic field such that the loop does not tend to rotate? Explain.

12. Explain why it is not possible to determine the charge and the mass of a charged particle separately by measuring accelerations produced by electric and magnetic forces on the particle.

13. How can a current loop be used to determine the presence of a magnetic field in a given region of space?

14. Charged particles from outer space, called cosmic rays, strike the Earth more frequently near the poles than near the equator. Why?

15. What is the net force on a compass needle in a uniform magnetic field?

16. What type of magnetic field is required to exert a resultant force on a magnetic dipole? What is the direction of the resultant force?

17. A proton moving horizontally enters a uniform magnetic field perpendicular to the proton's velocity, as shown in Figure Q29.17. Describe the subsequent motion of the proton. How would an electron behave under the same circumstances?

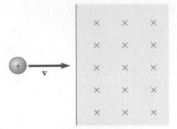

Figure Q29.17

18. In the cyclotron, why do particles having different speeds take the same amount of time to complete a one-half circle trip around one dee?

19. The *bubble chamber* is a device used for observing tracks of particles that pass through the chamber, which is immersed in a magnetic field. If some of the tracks are spirals and others are straight lines, what can you say about the particles?

20. Can a constant magnetic field set into motion an electron initially at rest? Explain your answer.

21. You are designing a magnetic probe that uses the Hall effect to measure magnetic fields. Assume that you are restricted to using a given material and that you have already made the probe as thin as possible. What, if anything, can be done to increase the Hall voltage produced for a given magnitude of magnetic field?

PROBLEMS

1, 2, 3 = straightforward, intermediate, challenging   ☐ = full solution available in the *Student Solutions Manual and Study Guide*

🌐 = coached solution with hints available at http://www.pse6.com   💻 = computer useful in solving problem

▨ = paired numerical and symbolic problems

Section 29.1 Magnetic Fields and Forces

1. 🌐 Determine the initial direction of the deflection of charged particles as they enter the magnetic fields as shown in Figure P29.1.

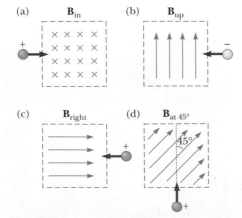

Figure P29.1

2. Consider an electron near the Earth's equator. In which direction does it tend to deflect if its velocity is directed

(a) downward, (b) northward, (c) westward, or (d) southeastward?

3. An electron moving along the positive x axis perpendicular to a magnetic field experiences a magnetic deflection in the negative y direction. What is the direction of the magnetic field?

4. A proton travels with a speed of 3.00×10^6 m/s at an angle of 37.0° with the direction of a magnetic field of 0.300 T in the $+y$ direction. What are (a) the magnitude of the magnetic force on the proton and (b) its acceleration?

5. A proton moves perpendicular to a uniform magnetic field **B** at 1.00×10^7 m/s and experiences an acceleration of 2.00×10^{13} m/s² in the $+x$ direction when its velocity is in the $+z$ direction. Determine the magnitude and direction of the field.

6. An electron is accelerated through 2 400 V from rest and then enters a uniform 1.70-T magnetic field. What are (a) the maximum and (b) the minimum values of the magnetic force this charge can experience?

7. A proton moving at 4.00×10^6 m/s through a magnetic field of 1.70 T experiences a magnetic force of magnitude 8.20×10^{-13} N. What is the angle between the proton's velocity and the field?

8. At the equator, near the surface of the Earth, the magnetic field is approximately 50.0 μT northward, and the electric field is about 100 N/C downward in fair weather. Find the gravitational, electric, and magnetic forces on an electron in this environment, assuming the electron has an instantaneous velocity of 6.00×10^6 m/s directed to the east.

9. A proton moves with a velocity of $\mathbf{v} = (2\hat{\mathbf{i}} - 4\hat{\mathbf{j}} + \hat{\mathbf{k}})$ m/s in a region in which the magnetic field is $\mathbf{B} = (\hat{\mathbf{i}} + 2\hat{\mathbf{j}} - 3\hat{\mathbf{k}})$ T. What is the magnitude of the magnetic force this charge experiences?

10. An electron has a velocity of 1.20×10^4 m/s (in the positive x direction), and an acceleration of 2.00×10^{12} m/s^2 (in the positive z direction) in a uniform electric and magnetic field. If the electric field has a magnitude of 20.0 N/C (in the positive z direction), what can you determine about the magnetic field in the region? What can you not determine?

Section 29.2   Magnetic Force Acting on a Current-Carrying Conductor

11. A wire having a mass per unit length of 0.500 g/cm carries a 2.00-A current horizontally to the south. What are the direction and magnitude of the minimum magnetic field needed to lift this wire vertically upward?

12. A wire carries a steady current of 2.40 A. A straight section of the wire is 0.750 m long and lies along the x axis within a uniform magnetic field, $\mathbf{B} = 1.60\hat{\mathbf{k}}$ T. If the current is in the $+x$ direction, what is the magnetic force on the section of wire?

13. A wire 2.80 m in length carries a current of 5.00 A in a region where a uniform magnetic field has a magnitude of 0.390 T. Calculate the magnitude of the magnetic force on the wire assuming the angle between the magnetic field and the current is (a) 60.0°, (b) 90.0°, (c) 120°.

14. A conductor suspended by two flexible wires as shown in Figure P29.14 has a mass per unit length of 0.040 0 kg/m. What current must exist in the conductor in order for the tension in the supporting wires to be zero when the magnetic field is 3.60 T into the page? What is the required direction for the current?

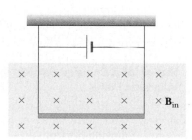

Figure P29.14

1.09 m/s

15. **Review Problem.** A rod of mass 0.720 kg and radius 6.00 cm rests on two parallel rails (Fig. P29.15) that are $d = 12.0$ cm apart and $L = 45.0$ cm long. The rod carries a current of $I = 48.0$ A (in the direction shown) and rolls along the rails without slipping. A uniform magnetic field

of magnitude 0.240 T is directed perpendicular to the rod and the rails. If it starts from rest, what is the speed of the rod as it leaves the rails?

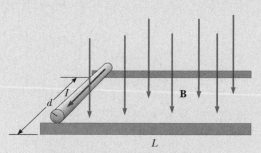

Figure P29.15 Problems 15 and 16.

16. **Review Problem.** A rod of mass m and radius R rests on two parallel rails (Fig. P29.15) that are a distance d apart and have a length L. The rod carries a current I (in the direction shown) and rolls along the rails without slipping. A uniform magnetic field B is directed perpendicular to the rod and the rails. If it starts from rest, what is the speed of the rod as it leaves the rails?

17. *A nonuniform magnetic field exerts a net force on a magnetic dipole.* A strong magnet is placed under a horizontal conducting ring of radius r that carries current I as shown in Figure P29.17. If the magnetic field $\mathbf{B}$ makes an angle θ with the vertical at the ring's location, what are the magnitude and direction of the resultant force on the ring?

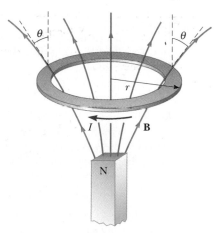

Figure P29.17

18. In Figure P29.18, the cube is 40.0 cm on each edge. Four straight segments of wire—ab, bc, cd, and da—form a closed loop that carries a current $I = 5.00$ A, in the direction shown. A uniform magnetic field of magnitude $B = 0.020\,0$ T is in the positive y direction. Determine the magnitude and direction of the magnetic force on each segment.

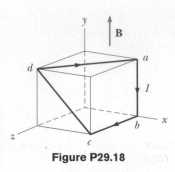

Figure P29.18

19. Assume that in Atlanta, Georgia, the Earth's magnetic field is 52.0 μT northward at 60.0° below the horizontal. A tube in a neon sign carries current 35.0 mA, between two diagonally opposite corners of a shop window, which lies in a north–south vertical plane. The current enters the tube at the bottom south corner of the window. It exits at the opposite corner, which is 1.40 m farther north and 0.850 m higher up. Between these two points, the glowing tube spells out DONUTS. Use the theorem proved as Case 1 in the text to determine the total vector magnetic force on the tube.

Section 29.3 Torque on a Current Loop in a Uniform Magnetic Field

20. A current of 17.0 mA is maintained in a single circular loop of 2.00 m circumference. A magnetic field of 0.800 T is directed parallel to the plane of the loop. (a) Calculate the magnetic moment of the loop. (b) What is the magnitude of the torque exerted by the magnetic field on the loop?

21. A small bar magnet is suspended in a uniform 0.250-T magnetic field. The maximum torque experienced by the bar magnet is 4.60×10^{-3} N·m. Calculate the magnetic moment of the bar magnet.

22. A long piece of wire with a mass of 0.100 kg and a total length of 4.00 m is used to make a square coil with a side of 0.100 m. The coil is hinged along a horizontal side, carries a 3.40-A current, and is placed in a vertical magnetic field with a magnitude of 0.010 0 T. (a) Determine the angle that the plane of the coil makes with the vertical when the coil is in equilibrium. (b) Find the torque acting on the coil due to the magnetic force at equilibrium.

23. A rectangular coil consists of $N = 100$ closely wrapped turns and has dimensions $a = 0.400$ m and $b = 0.300$ m. The coil is hinged along the y axis, and its plane makes an angle $\theta = 30.0°$ with the x axis (Fig. P29.23). What is the

magnitude of the torque exerted on the coil by a uniform magnetic field $B = 0.800$ T directed along the x axis when the current is $I = 1.20$ A in the direction shown? What is the expected direction of rotation of the coil?

24. A 40.0-cm length of wire carries a current of 20.0 A. It is bent into a loop and placed with its normal perpendicular to a magnetic field with a magnitude of 0.520 T. What is the torque on the loop if it is bent into (a) an equilateral triangle? **What If?** What is the torque if the loop is (b) a square or (c) a circle? (d) Which torque is greatest?

25. A current loop with magnetic dipole moment $\boldsymbol{\mu}$ is placed in a uniform magnetic field **B**, with its moment making angle θ with the field. With the arbitrary choice of $U = 0$ for $\theta = 90°$, prove that the potential energy of the dipole–field system is $U = -\boldsymbol{\mu} \cdot \mathbf{B}$. You may imitate the discussion in Chapter 26 of the potential energy of an electric dipole in an electric field.

26. The needle of a magnetic compass has magnetic moment 9.70 mA·m^2. At its location, the Earth's magnetic field is 55.0 μT north at 48.0° below the horizontal. (a) Identify the orientations of the compass needle that represent minimum potential energy and maximum potential energy of the needle–field system. (b) How much work must be done on the needle to move it from the former to the latter orientation?

27. A wire is formed into a circle having a diameter of 10.0 cm and placed in a uniform magnetic field of 3.00 mT. The wire carries a current of 5.00 A. Find (a) the maximum torque on the wire and (b) the range of potential energies of the wire–field system for different orientations of the circle.

28. The rotor in a certain electric motor is a flat rectangular coil with 80 turns of wire and dimensions 2.50 cm by 4.00 cm. The rotor rotates in a uniform magnetic field of 0.800 T. When the plane of the rotor is perpendicular to the direction of the magnetic field, it carries a current of 10.0 mA. In this orientation, the magnetic moment of the rotor is directed opposite the magnetic field. The rotor then turns through one-half revolution. This process is repeated to cause the rotor to turn steadily at 3 600 rev/min. (a) Find the maximum torque acting on the rotor. (b) Find the peak power output of the motor. (c) Determine the amount of work performed by the magnetic field on the rotor in every full revolution. (d) What is the average power of the motor?

Section 29.4 Motion of a Charged Particle in a Uniform Magnetic Field

29. The magnetic field of the Earth at a certain location is directed vertically downward and has a magnitude of 50.0 μT. A proton is moving horizontally toward the west in this field with a speed of 6.20×10^6 m/s. (a) What are the direction and magnitude of the magnetic force the field exerts on this charge? (b) What is the radius of the circular arc followed by this proton?

30. A singly charged positive ion has a mass of 3.20×10^{-26} kg. After being accelerated from rest through a potential difference of 833 V, the ion enters a magnetic field of 0.920 T along a direction perpendicular to the direction

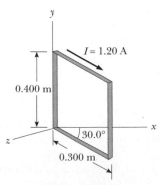

y

$I = 1.20$ A

0.400 m

30.0°

x

z

0.300 m

Figure P29.23

of the field. Calculate the radius of the path of the ion in the field.

31. **Review Problem.** One electron collides elastically with a second electron initially at rest. After the collision, the radii of their trajectories are 1.00 cm and 2.40 cm. The trajectories are perpendicular to a uniform magnetic field of magnitude 0.044 0 T. Determine the energy (in keV) of the incident electron.

32. A proton moving in a circular path perpendicular to a constant magnetic field takes 1.00 μs to complete one revolution. Determine the magnitude of the magnetic field.

33. A proton (charge $+e$, mass m_p), a deuteron (charge $+e$, mass $2m_p$), and an alpha particle (charge $+2e$, mass $4m_p$) are accelerated through a common potential difference ΔV. Each of the particles enters a uniform magnetic field **B**, with its velocity in a direction perpendicular to **B**. The proton moves in a circular path of radius r_p. Determine the radii of the circular orbits for the deuteron, r_d, and the alpha particle, r_α, in terms of r_p.

34. **Review Problem.** An electron moves in a circular path perpendicular to a constant magnetic field of magnitude 1.00 mT. The angular momentum of the electron about the center of the circle is 4.00×10^{-25} J·s. Determine (a) the radius of the circular path and (b) the speed of the electron.

35. Calculate the cyclotron frequency of a proton in a magnetic field of magnitude 5.20 T.

36. A singly charged ion of mass m is accelerated from rest by a potential difference ΔV. It is then deflected by a uniform magnetic field (perpendicular to the ion's velocity) into a semicircle of radius R. Now a doubly charged ion of mass m' is accelerated through the same potential difference and deflected by the same magnetic field into a semicircle of radius $R' = 2R$. What is the ratio of the masses of the ions?

37. A cosmic-ray proton in interstellar space has an energy of 10.0 MeV and executes a circular orbit having a radius equal to that of Mercury's orbit around the Sun (5.80×10^{10} m). What is the magnetic field in that region of space?

38. Figure 29.21 shows a charged particle traveling in a nonuniform magnetic field forming a magnetic bottle. (a) Explain why the positively charged particle in the figure must be moving clockwise. The particle travels along a helix whose radius decreases and whose pitch decreases as the particle moves into a stronger magnetic field. If the particle is moving to the right along the x axis, its velocity in this direction will be reduced to zero and it will be reflected from the right-hand side of the bottle, acting as a "magnetic mirror." The particle ends up bouncing back and forth between the ends of the bottle. (b) Explain qualitatively why the axial velocity is reduced to zero as the particle moves into the region of strong magnetic field at the end of the bottle. (c) Explain why the tangential velocity increases as the particle approaches the end of the bottle. (d) Explain why the orbiting particle has a magnetic dipole moment. (e) Sketch the magnetic moment and use the result of Problem 17 to explain again

how the nonuniform magnetic field exerts a force on the orbiting particle along the x axis.

39. A singly charged positive ion moving at 4.60×10^5 m/s leaves a circular track of radius 7.94 mm along a direction perpendicular to the 1.80-T magnetic field of a bubble chamber. Compute the mass (in atomic mass units) of this ion, and, from that value, identify it.

Section 29.5   Applications Involving Charged Particles Moving in a Magnetic Field

40. A velocity selector consists of electric and magnetic fields described by the expressions $\mathbf{E} = E\hat{\mathbf{k}}$ and $\mathbf{B} = B\hat{\mathbf{j}}$, with $B = 15.0$ mT. Find the value of E such that a 750-eV electron moving along the positive x axis is undeflected.

41. Singly charged uranium-238 ions are accelerated through a potential difference of 2.00 kV and enter a uniform magnetic field of 1.20 T directed perpendicular to their velocities. (a) Determine the radius of their circular path. (b) Repeat for uranium-235 ions. **What If?** How does the ratio of these path radii depend on the accelerating voltage and on the magnitude of the magnetic field?

42. Consider the mass spectrometer shown schematically in Figure 29.24. The magnitude of the electric field between the plates of the velocity selector is 2 500 V/m, and the magnetic field in both the velocity selector and the deflection chamber has a magnitude of 0.035 0 T. Calculate the radius of the path for a singly charged ion having a mass $m = 2.18 \times 10^{-26}$ kg.

43. A cyclotron designed to accelerate protons has a magnetic field of magnitude 0.450 T over a region of radius 1.20 m. What are (a) the cyclotron frequency and (b) the maximum speed acquired by the protons?

44. What is the required radius of a cyclotron designed to accelerate protons to energies of 34.0 MeV using a magnetic field of 5.20 T?

45. A cyclotron designed to accelerate protons has an outer radius of 0.350 m. The protons are emitted nearly at rest from a source at the center and are accelerated through 600 V each time they cross the gap between the dees. The dees are between the poles of an electromagnet where the field is 0.800 T. (a) Find the cyclotron frequency. (b) Find the speed at which protons exit the cyclotron and (c) their maximum kinetic energy. (d) How many revolutions does a proton make in the cyclotron? (e) For what time interval does one proton accelerate?

46. At the Fermilab accelerator in Batavia, Illinois, protons having momentum 4.80×10^{-16} kg·m/s are held in a circular orbit of radius 1.00 km by an upward magnetic field. What is the magnitude of this field?

47. The picture tube in a television uses magnetic deflection coils rather than electric deflection plates. Suppose an electron beam is accelerated through a 50.0-kV potential difference and then through a region of uniform magnetic field 1.00 cm wide. The screen is located 10.0 cm from the center of the coils and is 50.0 cm wide. When the field is turned off, the electron beam hits the center of the screen. What field magnitude is necessary to deflect the beam to the side of the screen? Ignore relativistic corrections.

Section 29.6 The Hall Effect

48. A flat ribbon of silver having a thickness $t = 0.200$ mm is used in a Hall-effect measurement of a uniform magnetic field perpendicular to the ribbon, as shown in Figure P29.48. The Hall coefficient for silver is $R_H = 0.840 \times 10^{-10}$ m^3/C. (a) What is the density of charge carriers in silver? (b) If a current $I = 20.0$ A produces a Hall voltage $\Delta V_H = 15.0$ μV, what is the magnitude of the applied magnetic field?

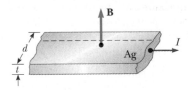

Figure P29.48

49. A flat copper ribbon 0.330 mm thick carries a steady current of 50.0 A and is located in a uniform 1.30-T magnetic field directed perpendicular to the plane of the ribbon. If a Hall voltage of 9.60 μV is measured across the ribbon, what is the charge density of the free electrons? What effective number of free electrons per atom does this result indicate?

50. A Hall-effect probe operates with a 120-mA current. When the probe is placed in a uniform magnetic field of magnitude 0.080 0 T, it produces a Hall voltage of 0.700 μV. (a) When it is measuring an unknown magnetic field, the Hall voltage is 0.330 μV. What is the magnitude of the unknown field? (b) The thickness of the probe in the direction of **B** is 2.00 mm. Find the density of the charge carriers, each of which has charge of magnitude e.

51. In an experiment that is designed to measure the Earth's magnetic field using the Hall effect, a copper bar 0.500 cm thick is positioned along an east–west direction. If a current of 8.00 A in the conductor results in a Hall voltage of 5.10×10^{-12} V, what is the magnitude of the Earth's magnetic field? (Assume that $n = 8.49 \times 10^{28}$ electrons/m^3 and that the plane of the bar is rotated to be perpendicular to the direction of **B**.)

Additional Problems

52. Assume that the region to the right of a certain vertical plane contains a vertical magnetic field of magnitude 1.00 mT, and the field is zero in the region to the left of the plane. An electron, originally traveling perpendicular to the boundary plane, passes into the region of the field. (a) Determine the time interval required for the electron to leave the "field-filled" region, noting that its path is a semicircle. (b) Find the kinetic energy of the electron if the maximum depth of penetration into the field is 2.00 cm.

53. Sodium melts at 99°C. Liquid sodium, an excellent thermal conductor, is used in some nuclear reactors to cool the reactor core. The liquid sodium is moved through pipes by pumps that exploit the force on a moving charge in a magnetic field. The principle is as follows. Assume the liquid metal to be in an electrically insulating pipe having a rectangular cross section of width w and height h. A uniform magnetic field perpendicular to the pipe affects a section of length L (Fig. P29.53). An electric current directed perpendicular to the pipe and to the magnetic field produces a current density J in the liquid sodium. (a) Explain why this arrangement produces on the liquid a force that is directed along the length of the pipe. (b) Show that the section of liquid in the magnetic field experiences a pressure increase JLB.

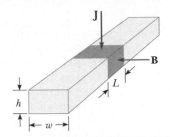

Figure P29.53

54. A 0.200-kg metal rod carrying a current of 10.0 A glides on two horizontal rails 0.500 m apart. What vertical magnetic field is required to keep the rod moving at a constant speed if the coefficient of kinetic friction between the rod and rails is 0.100?

55. Protons having a kinetic energy of 5.00 MeV are moving in the positive x direction and enter a magnetic field $\mathbf{B} = 0.050\,0\hat{\mathbf{k}}$ T directed out of the plane of the page and extending from $x = 0$ to $x = 1.00$ m, as shown in Figure P29.55. (a) Calculate the y component of the protons' momentum as they leave the magnetic field. (b) Find the angle α between the initial velocity vector of the proton beam and the velocity vector after the beam emerges from the field. Ignore relativistic effects and note that 1 eV = 1.60×10^{-19} J.

Figure P29.55

56. (a) A proton moving in the $+x$ direction with velocity $\mathbf{v} = v_i\hat{\mathbf{i}}$ experiences a magnetic force $\mathbf{F} = F_i\hat{\mathbf{j}}$ in the $+y$ direction. Explain what you can and cannot infer about **B** from this information. (b) **What If?** In terms of F_i, what would be the force on a proton in the same field moving with velocity $\mathbf{v} = -v_i\hat{\mathbf{i}}$? (c) What would be the force on an electron in the same field moving with velocity $\mathbf{v} = v_i\hat{\mathbf{i}}$?

57. A positive charge $q = 3.20 \times 10^{-19}$ C moves with a velocity $\mathbf{v} = (2\hat{\mathbf{i}} + 3\hat{\mathbf{j}} - \hat{\mathbf{k}})$ m/s through a region where both a uniform magnetic field and a uniform electric field exist. (a) Calculate the total force on the moving charge (in unit-vector notation), taking $\mathbf{B} = (2\hat{\mathbf{i}} + 4\hat{\mathbf{j}} + \hat{\mathbf{k}})$ T and $\mathbf{E} = (4\hat{\mathbf{i}} - \hat{\mathbf{j}} - 2\hat{\mathbf{k}})$ V/m. (b) What angle does the force vector make with the positive x axis?

58. **Review Problem.** A wire having a linear mass density of 1.00 g/cm is placed on a horizontal surface that has a coefficient of kinetic friction of 0.200. The wire carries a current of 1.50 A toward the east and slides horizontally to the north. What are the magnitude and direction of the smallest magnetic field that enables the wire to move in this fashion?

59. Electrons in a beam are accelerated from rest through a potential difference ΔV. The beam enters an experimental chamber through a small hole. As shown in Figure P29.59, the electron velocity vectors lie within a narrow cone of half angle ϕ oriented along the beam axis. We wish to use a uniform magnetic field directed parallel to the axis to focus the beam, so that all of the electrons can pass through a small exit port on the opposite side of the chamber after they travel the length d of the chamber. What is the required magnitude of the magnetic field? *Hint:* Because every electron passes through the same potential difference and the angle ϕ is small, they all require the same time interval to travel the axial distance d.

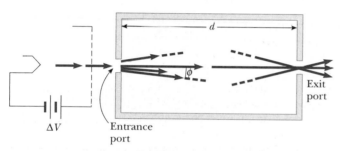

Figure P29.59

60. **Review Problem.** A proton is at rest at the plane vertical boundary of a region containing a uniform vertical magnetic field B. An alpha particle moving horizontally makes a head-on elastic collision with the proton. Immediately after the collision, both particles enter the magnetic field, moving perpendicular to the direction of the field. The radius of the proton's trajectory is R. Find the radius of the alpha particle's trajectory. The mass of the alpha particle is four times that of the proton, and its charge is twice that of the proton.

61. The circuit in Figure P29.61 consists of wires at the top and bottom and identical metal springs in the left and right sides. The upper portion of the circuit is fixed. The wire at the bottom has a mass of 10.0 g and is 5.00 cm

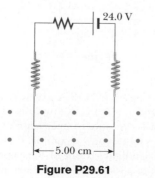

Figure P29.61

long. The springs stretch 0.500 cm under the weight of the wire and the circuit has a total resistance of 12.0 Ω. When a magnetic field is turned on, directed out of the page, the springs stretch an additional 0.300 cm. What is the magnitude of the magnetic field?

62. A hand-held electric mixer contains an electric motor. Model the motor as a single flat compact circular coil carrying electric current in a region where a magnetic field is produced by an external permanent magnet. You need consider only one instant in the operation of the motor. (We will consider motors again in Chapter 31.) The coil moves because the magnetic field exerts torque on the coil, as described in Section 29.3. Make order-of-magnitude estimates of the magnetic field, the torque on the coil, the current in it, its area, and the number of turns in the coil, so that they are related according to Equation 29.11. Note that the input power to the motor is electric, given by $\mathcal{P} = I\,\Delta V$, and the useful output power is mechanical, $\mathcal{P} = \tau\omega$.

63. A nonconducting sphere has mass 80.0 g and radius 20.0 cm. A flat compact coil of wire with 5 turns is wrapped tightly around it, with each turn concentric with the sphere. As shown in Figure P29.63, the sphere is placed on an inclined plane that slopes downward to the left, making an angle θ with the horizontal, so that the coil is parallel to the inclined plane. A uniform magnetic field of 0.350 T vertically upward exists in the region of the sphere. What current in the coil will enable the sphere to rest in equilibrium on the inclined plane? Show that the result does not depend on the value of θ.

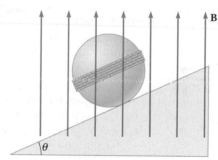

Figure P29.63

64. A metal rod having a mass per unit length λ carries a current I. The rod hangs from two vertical wires in a uniform vertical magnetic field as shown in Figure P29.64. The wires make an angle θ with the vertical when in equilibrium. Determine the magnitude of the magnetic field.

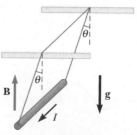

Figure P29.64

65. A cyclotron is sometimes used for carbon dating, as described in Chapter 44. Carbon-14 and carbon-12 ions are obtained from a sample of the material to be dated, and accelerated in the cyclotron. If the cyclotron has a magnetic field of magnitude 2.40 T, what is the difference in cyclotron frequencies for the two ions?

66. A uniform magnetic field of magnitude 0.150 T is directed along the positive x axis. A positron moving at 5.00×10^6 m/s enters the field along a direction that makes an angle of $85.0°$ with the x axis (Fig. P29.66). The motion of the particle is expected to be a helix, as described in Section 29.4. Calculate (a) the pitch p and (b) the radius r of the trajectory.

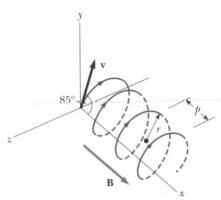

Figure P29.66

67. Consider an electron orbiting a proton and maintained in a fixed circular path of radius $R = 5.29 \times 10^{-11}$ m by the Coulomb force. Treating the orbiting charge as a current loop, calculate the resulting torque when the system is in a magnetic field of 0.400 T directed perpendicular to the magnetic moment of the electron.

68. A singly charged ion completes five revolutions in a uniform magnetic field of magnitude 5.00×10^{-2} T in 1.50 ms. Calculate the mass of the ion in kilograms.

69. A proton moving in the plane of the page has a kinetic energy of 6.00 MeV. A magnetic field of magnitude $B = 1.00$ T is directed into the page. The proton enters the magnetic field with its velocity vector at an angle $\theta = 45.0°$ to the linear boundary of the field as shown in Figure P29.69. (a) Find x, the distance from the point of entry to where the proton will leave the field. (b) Determine θ', the angle between the boundary and the proton's velocity vector as it leaves the field.

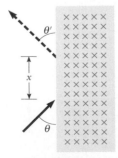

Figure P29.69

70. Table P29.70 shows measurements of a Hall voltage and corresponding magnetic field for a probe used to measure magnetic fields. (a) Plot these data, and deduce a relationship between the two variables. (b) If the measurements were taken with a current of 0.200 A and the sample is made from a material having a charge-carrier density of $1.00 \times 10^{26}/\text{m}^3$, what is the thickness of the sample?

Table P29.70

ΔV_H (μV)	B (T)
0	0.00
11	0.10
19	0.20
28	0.30
42	0.40
50	0.50
61	0.60
68	0.70
79	0.80
90	0.90
102	1.00

71. A heart surgeon monitors the flow rate of blood through an artery using an electromagnetic flowmeter (Fig. P29.71). Electrodes A and B make contact with the outer surface of the blood vessel, which has interior diameter 3.00 mm. (a) For a magnetic field magnitude of 0.040 0 T, an emf of 160 μV appears between the electrodes. Calculate the speed of the blood. (b) Verify that electrode A is positive, as shown. Does the sign of the emf depend on whether the mobile ions in the blood are predominantly positively or negatively charged? Explain.

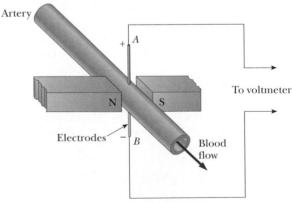

Figure P29.71

72. As shown in Figure P29.72, a particle of mass m having positive charge q is initially traveling with velocity $v\hat{\mathbf{j}}$. At the origin of coordinates it enters a region between $y = 0$ and $y = h$ containing a uniform magnetic field $B\hat{\mathbf{k}}$ directed perpendicularly out of the page. (a) What is the critical value of v such that the particle just reaches $y = h$?

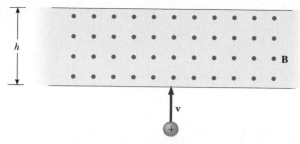

Figure P29.72

Describe the path of the particle under this condition, and predict its final velocity. (b) Specify the path the particle takes and its final velocity, if v is less than the critical value. (c) **What If?** Specify the path the particle takes and its final velocity if v is greater than the critical value.

Answers to Quick Quizzes

29.1 (c). The magnetic force exerted by a magnetic field on a charge is proportional to the charge's velocity relative to the field. If the charge is stationary, as in this situation, there is no magnetic force.

29.2 (b). The maximum value of $\sin\theta$ occurs for $\theta = 90°$.

29.3 (e). The right-hand rule gives the direction. Be sure to account for the negative charge on the electron.

29.4 (a), (b) = (c), (d). The magnitude of the force depends on the value of $\sin\theta$. The maximum force occurs when the wire is perpendicular to the field (a), and there is

zero force when the wire is parallel (d). Choices (b) and (c) represent the same force because Case 1 tells us that a straight wire between A and B will have the same force on it as the curved wire.

29.5 (c). Use the right-hand rule to determine the direction of the magnetic field.

29.6 (c), (b), (a). Because all loops enclose the same area and carry the same current, the magnitude of μ is the same for all. For (c), μ points upward and is perpendicular to the magnetic field and $\tau = \mu B$, the maximum torque possible. For the loop in (a), μ points along the direction of **B** and the torque is zero. For (b), the torque is intermediate between zero and the maximum value.

29.7 (a) = (b) = (c). Because the magnetic field is uniform, there is zero net force on all three loops.

29.8 (b). The magnetic force on the particle increases in proportion to v, but the centripetal acceleration increases according to the square of v. The result is a larger radius, as we can see from Equation 29.13.

29.9 (a). The magnetic force on the particle increases in proportion to B. The result is a smaller radius, as we can see from Equation 29.13.

29.10 Speed: (a) = (b) = (c). m/q ratio, from greatest to least: (c), (b), (a). The velocity selector ensures that all three types of particles have the same speed. We cannot determine individual masses or charges, but we can rank the particles by m/q ratio. Equation 29.18 indicates that those particles traveling through the circle of greatest radius have the greatest m/q ratio.

Chapter 30

Sources of the Magnetic Field

CHAPTER OUTLINE

30.1 The Biot–Savart Law

30.2 The Magnetic Force Between Two Parallel Conductors

30.3 Ampère's Law

30.4 The Magnetic Field of a Solenoid

30.5 Magnetic Flux

30.6 Gauss's Law in Magnetism

30.7 Displacement Current and the General Form of Ampère's Law

30.8 Magnetism in Matter

30.9 The Magnetic Field of the Earth

▲ *A proposed method for launching future payloads into space is the use of rail guns, in which projectiles are accelerated by means of magnetic forces. This photo shows the firing of a projectile at a speed of over 3 km/s from an experimental rail gun at Sandia National Research Laboratories, Albuquerque, New Mexico. (Defense Threat Reduction Agency [DTRA])*

In the preceding chapter, we discussed the magnetic force exerted on a charged particle moving in a magnetic field. To complete the description of the magnetic interaction, this chapter explores the origin of the magnetic field—moving charges. We begin by showing how to use the law of Biot and Savart to calculate the magnetic field produced at some point in space by a small current element. Using this formalism and the principle of superposition, we then calculate the total magnetic field due to various current distributions. Next, we show how to determine the force between two current-carrying conductors, which leads to the definition of the ampere. We also introduce Ampère's law, which is useful in calculating the magnetic field of a highly symmetric configuration carrying a steady current.

This chapter is also concerned with the complex processes that occur in magnetic materials. All magnetic effects in matter can be explained on the basis of atomic magnetic moments, which arise both from the orbital motion of electrons and from an intrinsic property of electrons known as spin.

30.1 The Biot–Savart Law

Shortly after Oersted's discovery in 1819 that a compass needle is deflected by a current-carrying conductor, Jean-Baptiste Biot (1774–1862) and Félix Savart (1791–1841) performed quantitative experiments on the force exerted by an electric current on a nearby magnet. From their experimental results, Biot and Savart arrived at a mathematical expression that gives the magnetic field at some point in space in terms of the current that produces the field. That expression is based on the following experimental observations for the magnetic field $d\mathbf{B}$ at a point P associated with a length element $d\mathbf{s}$ of a wire carrying a steady current I (Fig. 30.1):

- The vector $d\mathbf{B}$ is perpendicular both to $d\mathbf{s}$ (which points in the direction of the current) and to the unit vector $\hat{\mathbf{r}}$ directed from $d\mathbf{s}$ toward P.
- The magnitude of $d\mathbf{B}$ is inversely proportional to r^2, where r is the distance from $d\mathbf{s}$ to P.
- The magnitude of $d\mathbf{B}$ is proportional to the current and to the magnitude ds of the length element $d\mathbf{s}$.
- The magnitude of $d\mathbf{B}$ is proportional to $\sin\theta$, where θ is the angle between the vectors $d\mathbf{s}$ and $\hat{\mathbf{r}}$.

These observations are summarized in the mathematical expression known today as the **Biot–Savart law:**

$$d\mathbf{B} = \frac{\mu_0}{4\pi}\frac{I\,d\mathbf{s}\times\hat{\mathbf{r}}}{r^2}$$

(30.1)

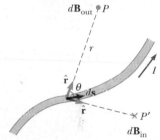

Figure 30.1 The magnetic field $d\mathbf{B}$ at a point due to the current I through a length element $d\mathbf{s}$ is given by the Biot–Savart law. The direction of the field is out of the page at P and into the page at P'.

> ▲ **PITFALL PREVENTION**
>
> **30.1 The Biot–Savart Law**
>
> The magnetic field described by the Biot–Savart law is the field *due to* a given current-carrying conductor. Do not confuse this field with any *external* field that may be applied to the conductor from some other source.

Biot–Savart law

where μ_0 is a constant called the **permeability of free space:**

Permeability of free space

$$\mu_0 = 4\pi \times 10^{-7} \text{ T}\cdot\text{m/A} \qquad (30.2)$$

Note that the field $d\mathbf{B}$ in Equation 30.1 is the field created by the current in only a small length element $d\mathbf{s}$ of the conductor. To find the *total* magnetic field $\mathbf{B}$ created at some point by a current of finite size, we must sum up contributions from all current elements $I\,d\mathbf{s}$ that make up the current. That is, we must evaluate $\mathbf{B}$ by integrating Equation 30.1:

$$\mathbf{B} = \frac{\mu_0 I}{4\pi} \int \frac{d\mathbf{s} \times \hat{\mathbf{r}}}{r^2} \qquad (30.3)$$

where the integral is taken over the entire current distribution. This expression must be handled with special care because the integrand is a cross product and therefore a vector quantity. We shall see one case of such an integration in Example 30.1.

Although we developed the Biot–Savart law for a current-carrying wire, it is also valid for a current consisting of charges flowing through space, such as the electron beam in a television set. In that case, $d\mathbf{s}$ represents the length of a small segment of space in which the charges flow.

Interesting similarities exist between Equation 30.1 for the magnetic field due to a current element and Equation 23.9 for the electric field due to a point charge. The magnitude of the magnetic field varies as the inverse square of the distance from the source, as does the electric field due to a point charge. However, the directions of the two fields are quite different. The electric field created by a point charge is radial, but the magnetic field created by a current element is perpendicular to both the length element $d\mathbf{s}$ and the unit vector $\hat{\mathbf{r}}$, as described by the cross product in Equation 30.1. Hence, if the conductor lies in the plane of the page, as shown in Figure 30.1, $d\mathbf{B}$ points out of the page at P and into the page at P'.

Another difference between electric and magnetic fields is related to the source of the field. An electric field is established by an isolated electric charge. The Biot–Savart law gives the magnetic field of an isolated current element at some point, but such an isolated current element cannot exist the way an isolated electric charge can. A current element *must* be part of an extended current distribution because we must have a complete circuit in order for charges to flow. Thus, the Biot–Savart law (Eq. 30.1) is only the first step in a calculation of a magnetic field; it must be followed by an integration over the current distribution, as in Equation 30.3.

Quick Quiz 30.1 Consider the current in the length of wire shown in Figure 30.2. Rank the points A, B, and C, in terms of magnitude of the magnetic field due to the current in the length element shown, from greatest to least.

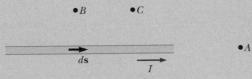

Figure 30.2 (Quick Quiz 30.1) Where is the magnetic field the greatest?

Example 30.1   **Magnetic Field Surrounding a Thin, Straight Conductor** | Interactive |

Consider a thin, straight wire carrying a constant current I and placed along the x axis as shown in Figure 30.3. Determine the magnitude and direction of the magnetic field at point P due to this current.

Solution From the Biot–Savart law, we expect that the magnitude of the field is proportional to the current in the wire and decreases as the distance a from the wire to point P increases. We start by considering a length element ds located a distance r from P. The direction of the magnetic field at point P due to the current in this element is out of the page because $ds \times \hat{r}$ is out of the page. In fact, because *all* of the current elements $I \, ds$ lie in the plane of the page, they all produce a magnetic field directed out of the page at point P. Thus, we have the direction of the magnetic field at point P, and we need only find the magnitude. Taking the origin at O and letting point P be along the positive y axis, with $\hat{k}$ being a unit vector pointing out of the page, we see that

$$ds \times \hat{r} = |ds \times \hat{r}|\hat{k} = (dx \sin \theta)\hat{k}$$

where $|ds \times \hat{r}|$ represents the magnitude of $ds \times \hat{r}$. Because $\hat{r}$ is a unit vector, the magnitude of the cross

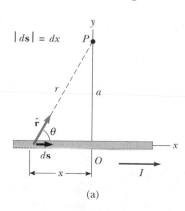

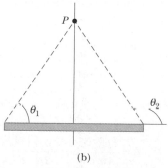

Figure 30.3 (Example 30.1) (a) A thin, straight wire carrying a current I. The magnetic field at point P due to the current in each element ds of the wire is out of the page, so the net field at point P is also out of the page. (b) The angles θ_1 and θ_2 used for determining the net field. When the wire is infinitely long, $\theta_1 = 0$ and $\theta_2 = 180°$.

product is simply the magnitude of ds, which is the length dx. Substitution into Equation 30.1 gives

$$d\mathbf{B} = (dB)\hat{k} = \frac{\mu_0 I}{4\pi} \frac{dx \sin \theta}{r^2} \hat{k}$$

Because all current elements produce a magnetic field in the $\hat{k}$ direction, let us restrict our attention to the magnitude of the field due to one current element, which is

$$(1) \qquad dB = \frac{\mu_0 I}{4\pi} \frac{dx \sin \theta}{r^2}$$

To integrate this expression, we must relate the variables θ, x, and r. One approach is to express x and r in terms of θ. From the geometry in Figure 30.3a, we have

$$(2) \qquad r = \frac{a}{\sin \theta} = a \csc \theta$$

Because $\tan \theta = a/(-x)$ from the right triangle in Figure 30.3a (the negative sign is necessary because ds is located at a negative value of x), we have

$$x = -a \cot \theta$$

Taking the derivative of this expression gives

$$(3) \qquad dx = a \csc^2 \theta \, d\theta$$

Substitution of Equations (2) and (3) into Equation (1) gives

$$(4) \qquad dB = \frac{\mu_0 I}{4\pi} \frac{a \csc^2 \theta \sin \theta \, d\theta}{a^2 \csc^2 \theta} = \frac{\mu_0 I}{4\pi a} \sin \theta \, d\theta$$

an expression in which the only variable is θ. We now obtain the magnitude of the magnetic field at point P by integrating Equation (4) over all elements, where the subtending angles range from θ_1 to θ_2 as defined in Figure 30.3b:

$$B = \frac{\mu_0 I}{4\pi a} \int_{\theta_1}^{\theta_2} \sin \theta \, d\theta = \frac{\mu_0 I}{4\pi a} (\cos \theta_1 - \cos \theta_2) \qquad (30.4)$$

We can use this result to find the magnetic field of *any* straight current-carrying wire if we know the geometry and hence the angles θ_1 and θ_2. Consider the special case of an infinitely long, straight wire. If we let the wire in Figure 30.3b become infinitely long, we see that $\theta_1 = 0$ and $\theta_2 = \pi$ for length elements ranging between positions $x = -\infty$ and $x = +\infty$. Because $(\cos \theta_1 - \cos \theta_2) = (\cos 0 - \cos \pi) = 2$, Equation 30.4 becomes

$$B = \frac{\mu_0 I}{2\pi a} \qquad (30.5)$$

Equations 30.4 and 30.5 both show that the magnitude of the magnetic field is proportional to the current and decreases with increasing distance from the wire, as we expected. Notice that Equation 30.5 has the same mathematical form as the expression for the magnitude of the electric field due to a long charged wire (see Eq. 24.7).

🌐 **At the Interactive Worked Example link at http://www.pse6.com,** *you can explore the field for different lengths of wire.*

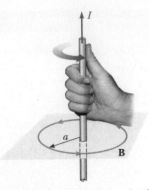

Figure 30.4 The right-hand rule for determining the direction of the magnetic field surrounding a long, straight wire carrying a current. Note that the magnetic field lines form circles around the wire.

The result of Example 30.1 is important because a current in the form of a long, straight wire occurs often. Figure 30.4 is a perspective view of the magnetic field surrounding a long, straight current-carrying wire. Because of the symmetry of the wire, the magnetic field lines are circles concentric with the wire and lie in planes perpendicular to the wire. The magnitude of **B** is constant on any circle of radius a and is given by Equation 30.5. A convenient rule for determining the direction of **B** is to grasp the wire with the right hand, positioning the thumb along the direction of the current. The four fingers wrap in the direction of the magnetic field.

Another observation we can make in Figure 30.4 is that the magnetic field line shown has no beginning and no end. It forms a closed loop. This is a major difference between magnetic field lines and electric field lines, which begin on positive charges and end on negative charges. We will explore this feature of magnetic field lines further in Section 30.6.

Example 30.2   Magnetic Field Due to a Curved Wire Segment

Calculate the magnetic field at point O for the current-carrying wire segment shown in Figure 30.5. The wire consists of two straight portions and a circular arc of radius R, which subtends an angle θ. The arrowheads on the wire indicate the direction of the current.

Solution The magnetic field at O due to the current in the straight segments AA' and CC' is zero because $d\mathbf{s}$ is parallel to $\hat{\mathbf{r}}$ along these paths; this means that $d\mathbf{s} \times \hat{\mathbf{r}} = 0$. Each length element $d\mathbf{s}$ along path AC is at the same distance R from O, and the current in each contributes a field element $d\mathbf{B}$ directed into the page at O. Furthermore, at every point on AC, $d\mathbf{s}$ is perpendicular to $\hat{\mathbf{r}}$; hence, $|d\mathbf{s} \times \hat{\mathbf{r}}| = ds$. Using this information and Equation 30.1, we can find the magnitude of the field at O due to the current in an ele-

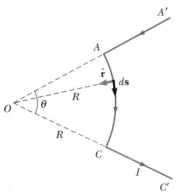

Figure 30.5 (Example 30.2) The magnetic field at O due to the current in the curved segment AC is into the page. The contribution to the field at O due to the current in the two straight segments is zero.

ment of length ds:

$$dB = \frac{\mu_0 I}{4\pi} \frac{ds}{R^2}$$

Because I and R are constants in this situation, we can easily integrate this expression over the curved path AC:

$$B = \frac{\mu_0 I}{4\pi R^2} \int ds = \frac{\mu_0 I}{4\pi R^2} s = \boxed{\frac{\mu_0 I}{4\pi R} \theta} \qquad (30.6)$$

where we have used the fact that $s = R\theta$ with θ measured in radians. The direction of **B** is into the page at O because $d\mathbf{s} \times \hat{\mathbf{r}}$ is into the page for every length element.

What If? What if you were asked to find the magnetic field at the center of a circular wire loop of radius R that carries a current I? Can we answer this question at this point in our understanding of the source of magnetic fields?

Answer Yes, we can. We argued that the straight wires in Figure 30.5 do not contribute to the magnetic field. The only contribution is from the curved segment. If we imagine increasing the angle θ, the curved segment will become a full circle when $\theta = 2\pi$. Thus, we can find the magnetic field at the center of a wire loop by letting $\theta = 2\pi$ in Equation 30.6:

$$B = \frac{\mu_0 I}{4\pi R} 2\pi = \frac{\mu_0 I}{2R}$$

We will confirm this result as a limiting case of a more general result in Example 30.3.

Example 30.3   Magnetic Field on the Axis of a Circular Current Loop   Interactive

Consider a circular wire loop of radius R located in the yz plane and carrying a steady current I, as in Figure 30.6. Calculate the magnetic field at an axial point P a distance x from the center of the loop.

Solution In this situation, every length element $d\mathbf{s}$ is perpendicular to the vector $\hat{\mathbf{r}}$ at the location of the element. Thus, for any element, $|d\mathbf{s} \times \hat{\mathbf{r}}| = (ds)(1)\sin 90° = ds$. Furthermore, all length elements around the loop are at the

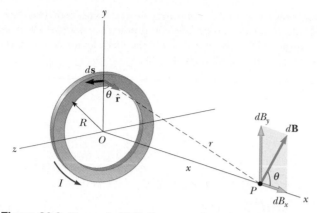

Figure 30.6 (Example 30.3) Geometry for calculating the magnetic field at a point P lying on the axis of a current loop. By symmetry, the total field **B** is along this axis.

same distance r from P, where $r^2 = x^2 + R^2$. Hence, the magnitude of $d\mathbf{B}$ due to the current in any length element $d\mathbf{s}$ is

$$dB = \frac{\mu_0 I}{4\pi} \frac{|d\mathbf{s} \times \hat{\mathbf{r}}|}{r^2} = \frac{\mu_0 I}{4\pi} \frac{ds}{(x^2 + R^2)}$$

The direction of $d\mathbf{B}$ is perpendicular to the plane formed by $\hat{\mathbf{r}}$ and $d\mathbf{s}$, as shown in Figure 30.6. We can resolve this vector into a component dB_x along the x axis and a component dB_y perpendicular to the x axis. When the components dB_y are summed over all elements around the loop, the resultant component is zero. That is, by symmetry the current in any element on one side of the loop sets up a perpendicular component of $d\mathbf{B}$ that cancels the perpendicular component set up by the current through the element diametrically opposite it. Therefore, *the resultant field at P must be along the x axis* and we can find it by integrating the components $dB_x = dB \cos \theta$. That is,

$$\mathbf{B} = B_x\hat{\mathbf{i}} \text{ where}$$

$$B_x = \oint dB \cos \theta = \frac{\mu_0 I}{4\pi} \oint \frac{ds \cos \theta}{x^2 + R^2}$$

and we must take the integral over the entire loop. Because θ, x, and R are constants for all elements of the loop and because $\cos \theta = R/(x^2 + R^2)^{1/2}$, we obtain

$$B_x = \frac{\mu_0 IR}{4\pi(x^2 + R^2)^{3/2}} \oint ds = \frac{\mu_0 IR^2}{2(x^2 + R^2)^{3/2}} \quad (30.7)$$

where we have used the fact that $\oint ds = 2\pi R$ (the circumference of the loop).

To find the magnetic field at the center of the loop, we set $x = 0$ in Equation 30.7. At this special point, therefore,

$$B = \frac{\mu_0 I}{2R} \quad (\text{at } x = 0) \quad (30.8)$$

which is consistent with the result of the **What If?** feature in Example 30.2.

The pattern of magnetic field lines for a circular current loop is shown in Figure 30.7a. For clarity, the lines are drawn for only one plane—one that contains the axis of the loop. Note that the field-line pattern is axially symmetric and looks like the pattern around a bar magnet, shown in Figure 30.7c.

What If? What if we consider points on the *x* axis very far from the loop? How does the magnetic field behave at these distant points?

Answer In this case, in which $x \gg R$, we can neglect the term R^2 in the denominator of Equation 30.7 and obtain

$$B \approx \frac{\mu_0 IR^2}{2x^3} \quad (\text{for } x \gg R) \quad (30.9)$$

Because the magnitude of the magnetic moment μ of the loop is defined as the product of current and loop area (see

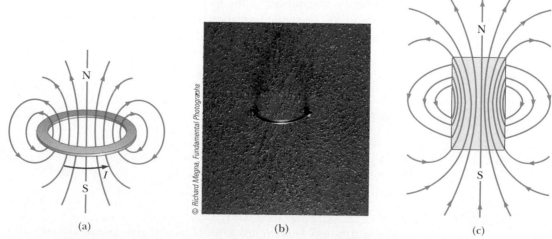

Figure 30.7 (Example 30.3) (a) Magnetic field lines surrounding a current loop. (b) Magnetic field lines surrounding a current loop, displayed with iron filings. (c) Magnetic field lines surrounding a bar magnet. Note the similarity between this line pattern and that of a current loop.

Eq. 29.10), $\mu = I(\pi R^2)$ for our circular loop. We can express Equation 30.9 as

$$B \approx \frac{\mu_0}{2\pi} \frac{\mu}{x^3} \qquad (30.10)$$

This result is similar in form to the expression for the electric field due to an electric dipole, $E = k_e(2qa/y^3)$ (see Example 23.6), where $2qa = p$ is the electric dipole moment as defined in Equation 26.16.

 At the Interactive Worked Example link at **http://www.pse6.com,** *you can explore the field for different loop radii.*

30.2 The Magnetic Force Between Two Parallel Conductors

In Chapter 29 we described the magnetic force that acts on a current-carrying conductor placed in an external magnetic field. Because a current in a conductor sets up its own magnetic field, it is easy to understand that two current-carrying conductors exert magnetic forces on each other. Such forces can be used as the basis for defining the ampere and the coulomb.

Consider two long, straight, parallel wires separated by a distance a and carrying currents I_1 and I_2 in the same direction, as in Figure 30.8. We can determine the force exerted on one wire due to the magnetic field set up by the other wire. Wire 2, which carries a current I_2 and is identified arbitrarily as the source wire, creates a magnetic field $\mathbf{B}_2$ at the location of wire 1, the test wire. The direction of $\mathbf{B}_2$ is perpendicular to wire 1, as shown in Figure 30.8. According to Equation 29.3, the magnetic force on a length ℓ of wire 1 is $\mathbf{F}_1 = I_1\boldsymbol{\ell} \times \mathbf{B}_2$. Because $\boldsymbol{\ell}$ is perpendicular to $\mathbf{B}_2$ in this situation, the magnitude of $\mathbf{F}_1$ is $F_1 = I_1\ell B_2$. Because the magnitude of $\mathbf{B}_2$ is given by Equation 30.5, we see that

$$F_1 = I_1\ell B_2 = I_1\ell \left(\frac{\mu_0 I_2}{2\pi a}\right) = \frac{\mu_0 I_1 I_2}{2\pi a}\ \ell \qquad (30.11)$$

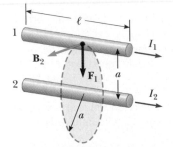

Active Figure 30.8 Two parallel wires that each carry a steady current exert a magnetic force on each other. The field $\mathbf{B}_2$ due to the current in wire 2 exerts a magnetic force of magnitude $F_1 = I_1\ell B_2$ on wire 1. The force is attractive if the currents are parallel (as shown) and repulsive if the currents are antiparallel.

 At the Active Figures link at **http://www.pse6.com,** *you can adjust the currents in the wires and the distance between them to see the effect on the force.*

The direction of $\mathbf{F}_1$ is toward wire 2 because $\boldsymbol{\ell} \times \mathbf{B}_2$ is in that direction. If the field set up at wire 2 by wire 1 is calculated, the force $\mathbf{F}_2$ acting on wire 2 is found to be equal in magnitude and opposite in direction to $\mathbf{F}_1$. This is what we expect because Newton's third law must be obeyed.[1] When the currents are in opposite directions (that is, when one of the currents is reversed in Fig. 30.8), the forces are reversed and the wires repel each other. Hence, **parallel conductors carrying currents in the same direction attract each other, and parallel conductors carrying currents in opposite directions repel each other.**

Because the magnitudes of the forces are the same on both wires, we denote the magnitude of the magnetic force between the wires as simply F_B. We can rewrite this magnitude in terms of the force per unit length:

$$\frac{F_B}{\ell} = \frac{\mu_0 I_1 I_2}{2\pi a} \qquad (30.12)$$

The force between two parallel wires is used to define the **ampere** as follows:

Definition of the ampere

When the magnitude of the force per unit length between two long parallel wires that carry identical currents and are separated by 1 m is 2×10^{-7} N/m, the current in each wire is defined to be 1 A.

---

[1] Although the total force exerted on wire 1 is equal in magnitude and opposite in direction to the total force exerted on wire 2, Newton's third law does not apply when one considers two small elements of the wires that are not exactly opposite each other. This apparent violation of Newton's third law and of the law of conservation of momentum is described in more advanced treatments on electricity and magnetism.

The value 2×10^{-7} N/m is obtained from Equation 30.12 with $I_1 = I_2 = 1$ A and $a = 1$ m. Because this definition is based on a force, a mechanical measurement can be used to standardize the ampere. For instance, the National Institute of Standards and Technology uses an instrument called a *current balance* for primary current measurements. The results are then used to standardize other, more conventional instruments, such as ammeters.

The SI unit of charge, the **coulomb,** is defined in terms of the ampere:

> When a conductor carries a steady current of 1 A, the quantity of charge that flows through a cross section of the conductor in 1 s is 1 C.

In deriving Equations 30.11 and 30.12, we assumed that both wires are long compared with their separation distance. In fact, only one wire needs to be long. The equations accurately describe the forces exerted on each other by a long wire and a straight parallel wire of limited length ℓ.

Quick Quiz 30.2 For $I_1 = 2$ A and $I_2 = 6$ A in Figure 30.8, which is true: (a) $F_1 = 3F_2$, (b) $F_1 = F_2/3$, (c) $F_1 = F_2$?

Quick Quiz 30.3 A loose spiral spring carrying no current is hung from the ceiling. When a switch is thrown so that a current exists in the spring, do the coils move (a) closer together, (b) farther apart, or (c) do they not move at all?

30.3 Ampère's Law

Oersted's 1819 discovery about deflected compass needles demonstrates that a current-carrying conductor produces a magnetic field. Figure 30.9a shows how this effect can be demonstrated in the classroom. Several compass needles are placed in a horizontal plane near a long vertical wire. When no current is

Andre-Marie Ampère
French Physicist (1775–1836)

Ampère is credited with the discovery of electromagnetism—the relationship between electric currents and magnetic fields. Ampère's genius, particularly in mathematics, became evident by the time he was 12 years old; his personal life, however, was filled with tragedy. His father, a wealthy city official, was guillotined during the French Revolution, and his wife died young, in 1803. Ampère died at the age of 61 of pneumonia. His judgment of his life is clear from the epitaph he chose for his gravestone: *Tandem Felix* (Happy at Last). *(Leonard de Selva/CORBIS)*

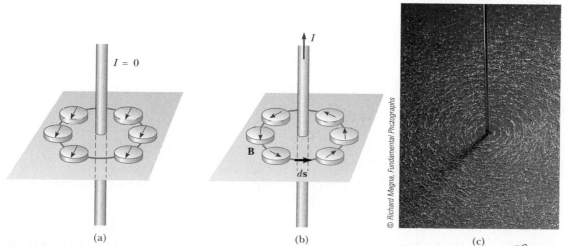

© Richard Megna, Fundamental Photographs

(a)   (b)   (c)

Active Figure 30.9 (a) When no current is present in the wire, all compass needles point in the same direction (toward the Earth's north pole). (b) When the wire carries a strong current, the compass needles deflect in a direction tangent to the circle, which is the direction of the magnetic field created by the current. (c) Circular magnetic field lines surrounding a current-carrying conductor, displayed with iron filings.

At the Active Figures link at http://www.pse6.com, *you can change the value of the current to see the effect on the compasses.*

present in the wire, all the needles point in the same direction (that of the Earth's magnetic field), as expected. When the wire carries a strong, steady current, the needles all deflect in a direction tangent to the circle, as in Figure 30.9b. These observations demonstrate that the direction of the magnetic field produced by the current in the wire is consistent with the right-hand rule described in Figure 30.4. When the current is reversed, the needles in Figure 30.9b also reverse.

Because the compass needles point in the direction of **B**, we conclude that the lines of **B** form circles around the wire, as discussed in the preceding section. By symmetry, the magnitude of **B** is the same everywhere on a circular path centered on the wire and lying in a plane perpendicular to the wire. By varying the current and distance a from the wire, we find that B is proportional to the current and inversely proportional to the distance from the wire, as Equation 30.5 describes.

Now let us evaluate the product $\mathbf{B} \cdot d\mathbf{s}$ for a small length element $d\mathbf{s}$ on the circular path defined by the compass needles, and sum the products for all elements over the closed circular path.[2] Along this path, the vectors $d\mathbf{s}$ and **B** are parallel at each point (see Fig. 30.9b), so $\mathbf{B} \cdot d\mathbf{s} = B\,ds$. Furthermore, the magnitude of **B** is constant on this circle and is given by Equation 30.5. Therefore, the sum of the products $B\,ds$ over the closed path, which is equivalent to the line integral of $\mathbf{B} \cdot d\mathbf{s}$, is

$$\oint \mathbf{B} \cdot d\mathbf{s} = B \oint ds = \frac{\mu_0 I}{2\pi r}(2\pi r) = \mu_0 I$$

where $\oint ds = 2\pi r$ is the circumference of the circular path. Although this result was calculated for the special case of a circular path surrounding a wire, it holds for a closed path of *any* shape (an *amperian loop*) surrounding a current that exists in an unbroken circuit. The general case, known as **Ampère's law,** can be stated as follows:

> The line integral of $\mathbf{B} \cdot d\mathbf{s}$ around any closed path equals $\mu_0 I$, where I is the total steady current passing through any surface bounded by the closed path.
>
> $$\oint \mathbf{B} \cdot d\mathbf{s} = \mu_0 I \qquad (30.13)$$

Ampère's law describes the creation of magnetic fields by all continuous current configurations, but at our mathematical level it is useful only for calculating the magnetic field of current configurations having a high degree of symmetry. Its use is similar to that of Gauss's law in calculating electric fields for highly symmetric charge distributions.

> **Quick Quiz 30.4** Rank the magnitudes of $\oint \mathbf{B} \cdot d\mathbf{s}$ for the closed paths in Figure 30.10, from least to greatest.

---

[2] You may wonder why we would choose to do this. The origin of Ampère's law is in nineteenth century science, in which a "magnetic charge" (the supposed analog to an isolated electric charge) was imagined to be moved around a circular field line. The work done on the charge was related to $\mathbf{B} \cdot d\mathbf{s}$, just as the work done moving an electric charge in an electric field is related to $\mathbf{E} \cdot d\mathbf{s}$. Thus, Ampère's law, a valid and useful principle, arose from an erroneous and abandoned work calculation!

▲ **PITFALL PREVENTION**

30.2 Avoiding Problems with Signs

When using Ampère's law, apply the following right-hand rule. Point your thumb in the direction of the current through the amperian loop. Your curled fingers then point in the direction that you should integrate around the loop in order to avoid having to define the current as negative.

Ampère's law

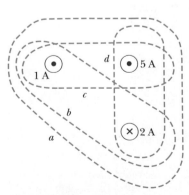

Figure 30.10 (Quick Quiz 30.4) Four closed paths around three current-carrying wires.

Quick Quiz 30.5 Rank the magnitudes of $\oint \mathbf{B} \cdot d\mathbf{s}$ for the closed paths in Figure 30.11, from least to greatest.

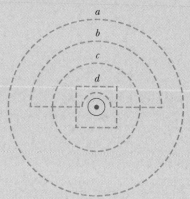

Figure 30.11 (Quick Quiz 30.5) Several closed paths near a single current-carrying wire.

Example 30.4   The Magnetic Field Created by a Long Current-Carrying Wire

A long, straight wire of radius R carries a steady current I that is uniformly distributed through the cross section of the wire (Fig. 30.12). Calculate the magnetic field a distance r from the center of the wire in the regions $r \geq R$ and $r < R$.

Solution Figure 30.12 helps us to conceptualize the wire and the current. Because the wire has a high degree of symmetry, we categorize this as an Ampère's law problem. For the $r \geq R$ case, we should arrive at the same result we obtained in Example 30.1, in which we applied the Biot–Savart law to the same situation. To analyze the problem, let us choose for our path of integration circle 1 in Figure 30.12. From symmetry, $\mathbf{B}$ must be constant in magnitude and parallel to $d\mathbf{s}$ at every point on this circle. Because the total current passing through the plane of the circle is I, Ampère's law gives

$$\oint \mathbf{B} \cdot d\mathbf{s} = B \oint ds = B(2\pi r) = \mu_0 I$$

$$B = \frac{\mu_0 I}{2\pi r} \qquad \text{(for } r \geq R) \qquad (30.14)$$

which is identical in form to Equation 30.5. Note how much easier it is to use Ampère's law than to use the Biot–Savart law. This is often the case in highly symmetric situations.

Now consider the interior of the wire, where $r < R$. Here the current I' passing through the plane of circle 2 is less than the total current I. Because the current is uniform over the cross section of the wire, the fraction of the current enclosed by circle 2 must equal the ratio of the area πr^2 enclosed by circle 2 to the cross-sectional area πR^2 of the wire:[3]

$$\frac{I'}{I} = \frac{\pi r^2}{\pi R^2}$$

$$I' = \frac{r^2}{R^2} I$$

Following the same procedure as for circle 1, we apply Ampère's law to circle 2:

$$\oint \mathbf{B} \cdot d\mathbf{s} = B(2\pi r) = \mu_0 I' = \mu_0 \left(\frac{r^2}{R^2} I \right)$$

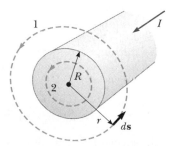

Figure 30.12 (Example 30.4) A long, straight wire of radius R carrying a steady current I uniformly distributed across the cross section of the wire. The magnetic field at any point can be calculated from Ampère's law using a circular path of radius r, concentric with the wire.

---

[3]   Another way to look at this problem is to realize that the current enclosed by circle 2 must equal the product of the current density $J = I/\pi R^2$ and the area πr^2 of this circle.

$$B = \left(\frac{\mu_0 I}{2\pi R^2}\right) r \qquad \text{(for } r < R) \qquad (30.15)$$

To finalize this problem, note that this result is similar in form to the expression for the electric field inside a uniformly charged sphere (see Example 24.5). The magnitude of the magnetic field versus r for this configuration is plotted in Figure 30.13. Note that inside the wire, $B \to 0$ as $r \to 0$. Furthermore, we see that Equations 30.14 and 30.15 give the same value of the magnetic field at $r = R$, demonstrating that the magnetic field is continuous at the surface of the wire.

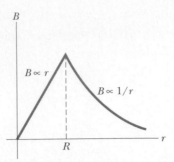

Figure 30.13 (Example 30.4) Magnitude of the magnetic field versus r for the wire shown in Figure 30.12. The field is proportional to r inside the wire and varies as $1/r$ outside the wire.

Example 30.5   The Magnetic Field Created by a Toroid

A device called a *toroid* (Fig. 30.14) is often used to create an almost uniform magnetic field in some enclosed area. The device consists of a conducting wire wrapped around a ring (a *torus*) made of a nonconducting material. For a toroid having N closely spaced turns of wire, calculate the magnetic field in the region occupied by the torus, a distance r from the center.

Solution To calculate this field, we must evaluate $\oint \mathbf{B} \cdot d\mathbf{s}$ over the circular amperian loop of radius r in the plane of Figure 30.14. By symmetry, we see that the magnitude of the field is constant on this circle and tangent to it, so $\mathbf{B} \cdot d\mathbf{s} = B \, ds$. Furthermore, the wire passes through the loop N times,

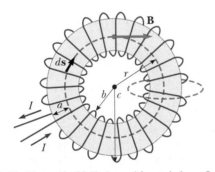

Figure 30.14 (Example 30.5) A toroid consisting of many turns of wire. If the turns are closely spaced, the magnetic field in the interior of the torus (the gold-shaded region) is tangent to the dashed circle and varies as $1/r$. The dimension a is the cross-sectional radius of the torus. The field outside the toroid is very small and can be described by using the amperian loop at the right side, perpendicular to the page.

so that the total current through the loop is NI. Therefore, the right side of Equation 30.13 is $\mu_0 NI$ in this case.

Ampère's law applied to the circle gives

$$\oint \mathbf{B} \cdot d\mathbf{s} = B \oint ds = B(2\pi r) = \mu_0 NI$$

$$B = \frac{\mu_0 NI}{2\pi r} \qquad (30.16)$$

This result shows that B varies as $1/r$ and hence is *nonuniform* in the region occupied by the torus. However, if r is very large compared with the cross-sectional radius a of the torus, then the field is approximately uniform inside the torus.

For an ideal toroid, in which the turns are closely spaced, the external magnetic field is close to zero. It is not exactly zero, however. In Figure 30.14, imagine the radius r of the amperian loop to be either smaller than b or larger than c. In either case, the loop encloses zero net current, so $\oint \mathbf{B} \cdot d\mathbf{s} = 0$. We might be tempted to claim that this proves that $\mathbf{B} = 0$, but it does not. Consider the amperian loop on the right side of the toroid in Figure 30.14. The plane of this loop is perpendicular to the page, and the toroid passes through the loop. As charges enter the toroid as indicated by the current directions in Figure 30.14, they work their way counterclockwise around the toroid. Thus, a current passes through the perpendicular amperian loop! This current is small, but it is not zero. As a result, the toroid acts as a current loop and produces a weak external field of the form shown in Figure 30.7. The reason that $\oint \mathbf{B} \cdot d\mathbf{s} = 0$ for the amperian loops of radius $r < b$ and $r > c$ in the plane of the page is that the field lines are perpendicular to $d\mathbf{s}$, *not* because $\mathbf{B} = 0$.

Example 30.6   Magnetic Field Created by an Infinite Current Sheet

So far we have imagined currents carried by wires of small cross section. Let us now consider an example in which a current exists in an extended object. A thin, infinitely large sheet lying in the yz plane carries a current of linear current density $\mathbf{J}_s$. The current is in the y direction, and J_s represents the current per unit length measured along the z axis. Find the magnetic field near the sheet.

Solution This situation is similar to those involving Gauss's law (see Example 24.8). You may recall that the electric field

due to an infinite sheet of charge does not depend on distance from the sheet. Thus, we might expect a similar result here for the magnetic field.

To evaluate the line integral in Ampère's law, we construct a rectangular path through the sheet, as in Figure 30.15. The rectangle has dimensions ℓ and w, with the sides of length ℓ parallel to the sheet surface. The net current in the plane of the rectangle is $J_s\ell$. We apply Ampère's law over the rectangle and note that the two sides of length w do not contribute to the line integral because the component of $\mathbf{B}$

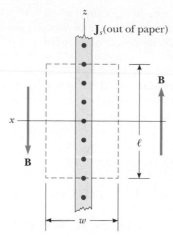

Figure 30.15 (Example 30.6) End view of an infinite current sheet lying in the yz plane, where the current is in the y direction (out of the page). This view shows the direction of **B** on both sides of the sheet.

along the direction of these paths is zero. By symmetry, the magnetic field is constant over the sides of length ℓ because every point on the infinitely large sheet is equivalent, and

hence the field should not vary from point to point. The only choices of field direction that are reasonable in this situation are either perpendicular or parallel to the sheet. However, a perpendicular field would pass *through* the current, which is inconsistent with the Biot–Savart law. Assuming a field that is constant in magnitude and parallel to the plane of the sheet, we obtain

$$\oint \mathbf{B} \cdot d\mathbf{s} = \mu_0 I = \mu_0 J_s \ell$$

$$2B\ell = \mu_0 J_s \ell$$

$$B = \boxed{\mu_0 \frac{J_s}{2}}$$

This result shows that the magnetic field is independent of distance from the current sheet, as we suspected. The expression for the magnitude of the magnetic field is similar in form to that for the magnitude of the electric field due to an infinite sheet of charge (Example 24.8):

$$E = \frac{\sigma}{2\epsilon_0}$$

Example 30.7 The Magnetic Force on a Current Segment

Wire 1 in Figure 30.16 is oriented along the y axis and carries a steady current I_1. A rectangular loop located to the right of the wire and in the xy plane carries a current I_2. Find the magnetic force exerted by wire 1 on the top wire of length b in the loop, labeled "Wire 2" in the figure.

Solution You may be tempted to use Equation 30.12 to obtain the force exerted on a small segment of length dx of wire 2. However, this equation applies only to two *parallel* wires and cannot be used here. The correct approach is to consider the force exerted by wire 1 on a small segment $d\mathbf{s}$ of wire 2 by using Equation 29.4. This force is given by $d\mathbf{F}_B = I \, d\mathbf{s} \times \mathbf{B}$, where $I = I_2$ and **B** is the magnetic field created by the current in wire 1 at the position of $d\mathbf{s}$. From Ampère's law, the field at a distance x

from wire 1 (see Eq. 30.14) is

$$\mathbf{B} = \frac{\mu_0 I_1}{2\pi x} (-\hat{\mathbf{k}})$$

where the unit vector $-\hat{\mathbf{k}}$ is used to indicate that the field due to the current in wire 1 at the position of $d\mathbf{s}$ points into the page. Because wire 2 is along the x axis, $d\mathbf{s} = dx\hat{\mathbf{i}}$, and we find that

$$d\mathbf{F}_B = \frac{\mu_0 I_1 I_2}{2\pi x} [\hat{\mathbf{i}} \times (-\hat{\mathbf{k}})] \, dx = \frac{\mu_0 I_1 I_2}{2\pi} \frac{dx}{x} \hat{\mathbf{j}}$$

Integrating over the limits $x = a$ to $x = a + b$ gives

$$\mathbf{F}_B = \frac{\mu_0 I_1 I_2}{2\pi} \ln x \Big]_a^{a+b} \hat{\mathbf{j}}$$

$$(1) \qquad \mathbf{F}_B = \boxed{\frac{\mu_0 I_1 I_2}{2\pi} \ln\left(1 + \frac{b}{a}\right)\hat{\mathbf{j}}}$$

The force on wire 2 points in the positive y direction, as indicated by the unit vector $\hat{\mathbf{j}}$ and as shown in Figure 30.16.

What If? What if the wire loop is moved to the left in Figure 30.16 until $a = 0$? What happens to the magnitude of the force on the wire?

Answer The force should become stronger because the loop is moving into a region of stronger magnetic field. Equation (1) shows that the force not only becomes stronger but the magnitude of the force becomes *infinite* as $a \rightarrow 0$! Thus, as the loop is moved to the left in Figure 30.16, the loop should be torn apart by the infinite upward force on the top side and the corresponding downward force on the bottom side! Furthermore, the force on the left side is

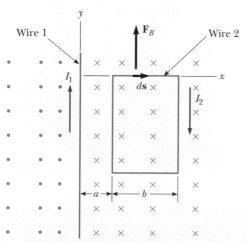

Figure 30.16 (Example 30.7) A wire on one side of a rectangular loop lying near a current-carrying wire experiences a force.

toward the left and should also become infinite. This is larger than the force toward the right on the right side because this side is still far from the wire, so the loop should be pulled into the wire with infinite force!

Does this really happen? In reality, it is impossible for $a \rightarrow 0$ because both wire 1 and wire 2 have finite sizes, so that the separation of the centers of the two wires is at least the sum of their radii.

A similar situation occurs when we re-examine the magnetic field due to a long straight wire, given by Equation 30.5. If we could move our observation point infinitesimally close to the wire, the magnetic field would become infinite! But in reality, the wire has a radius, and as soon as we enter the wire, the magnetic field starts to fall off as described by Equation 30.15—approaching zero as we approach the center of the wire.

30.4   The Magnetic Field of a Solenoid

A **solenoid** is a long wire wound in the form of a helix. With this configuration, a reasonably uniform magnetic field can be produced in the space surrounded by the turns of wire—which we shall call the *interior* of the solenoid—when the solenoid carries a current. When the turns are closely spaced, each can be approximated as a circular loop, and the net magnetic field is the vector sum of the fields resulting from all the turns.

Figure 30.17 shows the magnetic field lines surrounding a loosely wound solenoid. Note that the field lines in the interior are nearly parallel to one another, are uniformly distributed, and are close together, indicating that the field in this space is strong and almost uniform.

If the turns are closely spaced and the solenoid is of finite length, the magnetic field lines are as shown in Figure 30.18a. This field line distribution is similar to that surrounding a bar magnet (see Fig. 30.18b). Hence, one end of the solenoid behaves like the north pole of a magnet, and the opposite end behaves like the south pole. As the length of the solenoid increases, the interior field becomes more uniform and the exterior field becomes weaker. An *ideal solenoid* is approached when the turns are closely spaced and the length is much greater than the radius of the turns. Figure 30.19 shows a longitudinal cross section of part of such a solenoid carrying a current I. In this case, the external field is close to zero, and the interior field is uniform over a great volume.

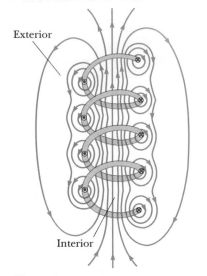

Figure 30.17 The magnetic field lines for a loosely wound solenoid.

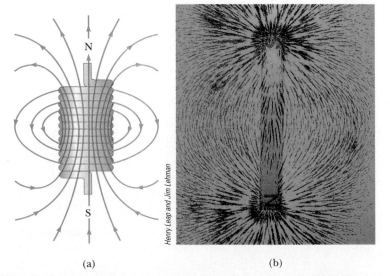

(a)   (b)

Henry Leap and Jim Lehman

Figure 30.18 (a) Magnetic field lines for a tightly wound solenoid of finite length, carrying a steady current. The field in the interior space is strong and nearly uniform. Note that the field lines resemble those of a bar magnet, meaning that the solenoid effectively has north and south poles. (b) The magnetic field pattern of a bar magnet, displayed with small iron filings on a sheet of paper.

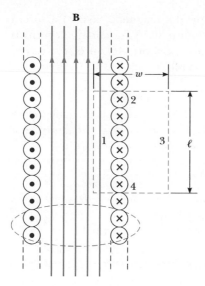

Figure 30.19 Cross-sectional view of an ideal solenoid, where the interior magnetic field is uniform and the exterior field is close to zero. Ampère's law applied to the circular path near the bottom whose plane is perpendicular to the page can be used to show that there is a weak field outside the solenoid. Ampère's law applied to the rectangular dashed path in the plane of the page can be used to calculate the magnitude of the interior field.

If we consider the amperian loop perpendicular to the page in Figure 30.19, surrounding the ideal solenoid, we see that it encloses a small current as the charges in the wire move coil by coil along the length of the solenoid. Thus, there is a nonzero magnetic field outside the solenoid. It is a weak field, with circular field lines, like those due to a line of current as in Figure 30.4. For an ideal solenoid, this is the only field external to the solenoid. We can eliminate this field in Figure 30.19 by adding a second layer of turns of wire outside the first layer, with the current carried along the axis of the solenoid in the opposite direction compared to the first layer. Then the net current along the axis is zero.

We can use Ampère's law to obtain a quantitative expression for the interior magnetic field in an ideal solenoid. Because the solenoid is ideal, **B** in the interior space is uniform and parallel to the axis, and the magnetic field lines in the exterior space form circles around the solenoid. The planes of these circles are perpendicular to the page. Consider the rectangular path of length ℓ and width w shown in Figure 30.19. We can apply Ampère's law to this path by evaluating the integral of $\mathbf{B} \cdot d\mathbf{s}$ over each side of the rectangle. The contribution along side 3 is zero because the magnetic field lines are perpendicular to the path in this region. The contributions from sides 2 and 4 are both zero, again because **B** is perpendicular to $d\mathbf{s}$ along these paths, both inside and outside the solenoid. Side 1 gives a contribution to the integral because along this path **B** is uniform and parallel to $d\mathbf{s}$. The integral over the closed rectangular path is therefore

$$\oint \mathbf{B} \cdot d\mathbf{s} = \int_{\text{path 1}} \mathbf{B} \cdot d\mathbf{s} = B \int_{\text{path 1}} ds = B\ell$$

The right side of Ampère's law involves the total current I through the area bounded by the path of integration. In this case, the total current through the rectangular path equals the current through each turn multiplied by the number of turns. If N is the number of turns in the length ℓ, the total current through the rectangle is NI. Therefore, Ampère's law applied to this path gives

$$\oint \mathbf{B} \cdot d\mathbf{s} = B\ell = \mu_0 NI$$

$$B = \mu_0 \frac{N}{\ell} I = \mu_0 nI \qquad (30.17) \qquad \textbf{Magnetic field inside a solenoid}$$

where $n = N/\ell$ is the number of turns per unit length.

We also could obtain this result by reconsidering the magnetic field of a toroid (see Example 30.5). If the radius r of the torus in Figure 30.14 containing N turns is much greater than the toroid's cross-sectional radius a, a short section of the toroid approximates a solenoid for which $n = N/2\pi r$. In this limit, Equation 30.16 agrees with Equation 30.17.

Equation 30.17 is valid only for points near the center (that is, far from the ends) of a very long solenoid. As you might expect, the field near each end is smaller than the value given by Equation 30.17. At the very end of a long solenoid, the magnitude of the field is half the magnitude at the center (see Problem 32).

Quick Quiz 30.6 Consider a solenoid that is very long compared to the radius. Of the following choices, the most effective way to increase the magnetic field in the interior of the solenoid is to (a) double its length, keeping the number of turns per unit length constant, (b) reduce its radius by half, keeping the number of turns per unit length constant, (c) overwrapping the entire solenoid with an additional layer of current-carrying wire.

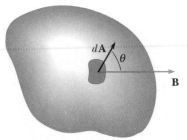

Figure 30.20 The magnetic flux through an area element dA is $\mathbf{B} \cdot d\mathbf{A} = B\,dA\cos\theta$, where $d\mathbf{A}$ is a vector perpendicular to the surface.

Definition of magnetic flux

30.5 Magnetic Flux

The flux associated with a magnetic field is defined in a manner similar to that used to define electric flux (see Eq. 24.3). Consider an element of area dA on an arbitrarily shaped surface, as shown in Figure 30.20. If the magnetic field at this element is $\mathbf{B}$, the magnetic flux through the element is $\mathbf{B} \cdot d\mathbf{A}$, where $d\mathbf{A}$ is a vector that is perpendicular to the surface and has a magnitude equal to the area dA. Therefore, the total magnetic flux Φ_B through the surface is

$$\Phi_B = \int \mathbf{B} \cdot d\mathbf{A} \tag{30.18}$$

Consider the special case of a plane of area A in a uniform field $\mathbf{B}$ that makes an angle θ with $d\mathbf{A}$. The magnetic flux through the plane in this case is

$$\Phi_B = BA \cos \theta \tag{30.19}$$

If the magnetic field is parallel to the plane, as in Figure 30.21a, then $\theta = 90°$ and the flux through the plane is zero. If the field is perpendicular to the plane, as in Figure 30.21b, then $\theta = 0$ and the flux through the plane is BA (the maximum value).

The unit of magnetic flux is $T \cdot m^2$, which is defined as a *weber* (Wb); 1 Wb $= 1\ T \cdot m^2$.

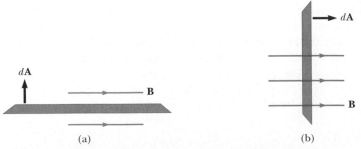

(a)   (b)

At the Active Figures link at http://www.pse6.com, you can rotate the plane and change the value of the field to see the effect on the flux.

Active Figure 30.21 Magnetic flux through a plane lying in a magnetic field. (a) The flux through the plane is zero when the magnetic field is parallel to the plane surface. (b) The flux through the plane is a maximum when the magnetic field is perpendicular to the plane.

Example 30.8   Magnetic Flux Through a Rectangular Loop

A rectangular loop of width a and length b is located near a long wire carrying a current I (Fig. 30.22). The distance between the wire and the closest side of the loop is c. The wire is parallel to the long side of the loop. Find the total magnetic flux through the loop due to the current in the wire.

Solution From Equation 30.14, we know that the magnitude of the magnetic field created by the wire at a distance r from the wire is

$$B = \frac{\mu_0 I}{2\pi r}$$

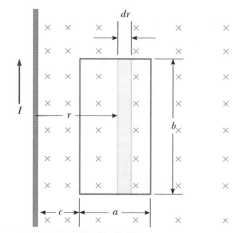

Figure 30.22 (Example 30.8) The magnetic field due to the wire carrying a current I is not uniform over the rectangular loop.

The factor $1/r$ indicates that the field varies over the loop, and Figure 30.22 shows that the field is directed into the page at the location of the loop. Because **B** is parallel to $d\mathbf{A}$ at any point within the loop, the magnetic flux through an area element dA is

$$\Phi_B = \int B \, dA = \int \frac{\mu_0 I}{2\pi r} \, dA$$

To integrate, we first express the area element (the tan region in Fig. 30.22) as $dA = b \, dr$. Because r is now the only variable in the integral, we have

$$\Phi_B = \frac{\mu_0 I b}{2\pi} \int_c^{a+c} \frac{dr}{r} = \frac{\mu_0 I b}{2\pi} \ln r \Big]_c^{a+c}$$

$$(1) \qquad = \frac{\mu_0 I b}{2\pi} \ln\left(\frac{a+c}{c}\right) = \frac{\mu_0 I b}{2\pi} \ln\left(1 + \frac{a}{c}\right)$$

What If? Suppose we move the loop in Figure 30.22 very far away from the wire. What happens to the magnetic flux?

Answer The flux should become smaller as the loop moves into weaker and weaker fields.

As the loop moves far away, the value of c is much larger than that of a, so that $a/c \to 0$. Thus, the natural logarithm in Equation (1) approaches the limit

$$\ln\left(1 + \frac{a}{c}\right) \longrightarrow \ln(1 + 0) = \ln(1) = 0$$

and we find that $\Phi_B \to 0$ as we expected.

 At the Interactive Worked Example link at http://www.pse6.com, you can investigate the flux as the loop parameters change.

30.6   Gauss's Law in Magnetism

In Chapter 24 we found that the electric flux through a closed surface surrounding a net charge is proportional to that charge (Gauss's law). In other words, the number of electric field lines leaving the surface depends only on the net charge within it. This property is based on the fact that electric field lines originate and terminate on electric charges.

The situation is quite different for magnetic fields, which are continuous and form closed loops. In other words, magnetic field lines do not begin or end at any point—as illustrated in Figures 30.4 and 30.23. Figure 30.23 shows the magnetic field lines of a bar magnet. Note that for any closed surface, such as the one outlined by the dashed line in Figure 30.23, the number of lines entering the surface equals the number leaving the surface; thus, the net magnetic flux is zero. In contrast, for a closed surface surrounding one charge of an electric dipole (Fig. 30.24), the net electric flux is not zero.

Gauss's law in magnetism states that

the net magnetic flux through any closed surface is always zero:

$$\oint \mathbf{B} \cdot d\mathbf{A} = 0 \qquad\qquad (30.20)$$ **Gauss's law in magnetism**

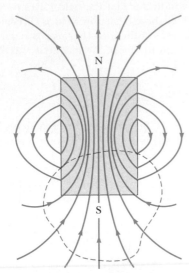

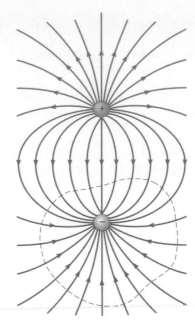

Figure 30.23 The magnetic field lines of a bar magnet form closed loops. Note that the net magnetic flux through a closed surface surrounding one of the poles (or any other closed surface) is zero. (The dashed line represents the intersection of the surface with the page.)

Figure 30.24 The electric field lines surrounding an electric dipole begin on the positive charge and terminate on the negative charge. The electric flux through a closed surface surrounding one of the charges is not zero.

This statement is based on the experimental fact, mentioned in the opening of Chapter 29, that isolated magnetic poles (monopoles) have never been detected and perhaps do not exist. Nonetheless, scientists continue the search because certain theories that are otherwise successful in explaining fundamental physical behavior suggest the possible existence of monopoles.

30.7 Displacement Current and the General Form of Ampère's Law

We have seen that charges in motion produce magnetic fields. When a current-carrying conductor has high symmetry, we can use Ampère's law to calculate the magnetic field it creates. In Equation 30.13, $\oint \mathbf{B} \cdot d\mathbf{s} = \mu_0 I$, the line integral is over any closed path through which the conduction current passes, where the conduction current is defined by the expression $I = dq/dt$. (In this section we use the term *conduction current* to refer to the current carried by the wire, to distinguish it from a new type of current that we shall introduce shortly.) We now show that **Ampère's law in this form is valid only if any electric fields present are constant in time.** Maxwell recognized this limitation and modified Ampère's law to include time-varying electric fields.

We can understand the problem by considering a capacitor that is being charged as illustrated in Figure 30.25. When a conduction current is present, the charge on the positive plate changes but *no conduction current exists in the gap between the plates.* Now consider the two surfaces S_1 and S_2 in Figure 30.25, bounded by the same path P. Ampère's law states that $\oint \mathbf{B} \cdot d\mathbf{s}$ around this path must equal $\mu_0 I$, where I is the total current through *any* surface bounded by the path P.

When the path P is considered as bounding S_1, $\oint \mathbf{B} \cdot d\mathbf{s} = \mu_0 I$ because the conduction current passes through S_1. When the path is considered as bounding S_2, however, $\oint \mathbf{B} \cdot d\mathbf{s} = 0$ because no conduction current passes through S_2. Thus, we have a contradictory situation that arises from the discontinuity of the current! Maxwell solved this problem by postulating an additional term on the right side

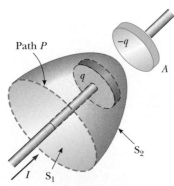

Figure 30.25 Two surfaces S_1 and S_2 near the plate of a capacitor are bounded by the same path P. The conduction current in the wire passes only through S_1. This leads to a contradiction in Ampère's law that is resolved only if one postulates a displacement current through S_2.

of Equation 30.13, which includes a factor called the **displacement current** I_d, defined as[4]

$$I_d \equiv \epsilon_0 \frac{d\Phi_E}{dt} \tag{30.21}$$

Displacement current

where ϵ_0 is the permittivity of free space (see Section 23.3) and $\Phi_E = \int \mathbf{E} \cdot d\mathbf{A}$ is the electric flux (see Eq. 24.3).

As the capacitor is being charged (or discharged), the changing electric field between the plates may be considered equivalent to a current that acts as a continuation of the conduction current in the wire. When the expression for the displacement current given by Equation 30.21 is added to the conduction current on the right side of Ampère's law, the difficulty represented in Figure 30.25 is resolved. No matter which surface bounded by the path P is chosen, either a conduction current or a displacement current passes through it. With this new term I_d, we can express the general form of Ampère's law (sometimes called the **Ampère–Maxwell law**) as[5]

$$\oint \mathbf{B} \cdot d\mathbf{s} = \mu_0(I + I_d) = \mu_0 I + \mu_0 \epsilon_0 \frac{d\Phi_E}{dt} \tag{30.22}$$

Ampère–Maxwell law

We can understand the meaning of this expression by referring to Figure 30.26. The electric flux through surface S_2 is $\Phi_E = \int \mathbf{E} \cdot d\mathbf{A} = EA$, where A is the area of the capacitor plates and E is the magnitude of the uniform electric field between the plates. If q is the charge on the plates at any instant, then $E = q/(\epsilon_0 A)$. (See Section 26.2.) Therefore, the electric flux through S_2 is simply

$$\Phi_E = EA = \frac{q}{\epsilon_0}$$

Hence, the displacement current through S_2 is

$$I_d = \epsilon_0 \frac{d\Phi_E}{dt} = \frac{dq}{dt} \tag{30.23}$$

That is, the displacement current I_d through S_2 is precisely equal to the conduction current I through S_1!

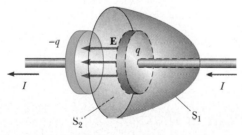

Figure 30.26 Because it exists only in the wires attached to the capacitor plates, the conduction current $I = dq/dt$ passes through S_1 but not through S_2. Only the displacement current $I_d = \epsilon_0 d\Phi_E/dt$ passes through S_2. The two currents must be equal for continuity.

---

[4] *Displacement* in this context does not have the meaning it does in Chapter 2. Despite the inaccurate implications, the word is historically entrenched in the language of physics, so we continue to use it.

[5] Strictly speaking, this expression is valid only in a vacuum. If a magnetic material is present, one must change μ_0 and ϵ_0 on the right-hand side of Equation 30.22 to the permeability μ_m (see Section 30.8) and permittivity ϵ characteristic of the material. Alternatively, one may include a magnetizing current I_m on the right hand side of Equation 30.22 to make Ampère's law fully general. On a microscopic scale, I_m is as real as I.

By considering surface S_2, we can identify the displacement current as the source of the magnetic field on the surface boundary. The displacement current has its physical origin in the time-varying electric field. The central point of this formalism is that

> magnetic fields are produced both by conduction currents and by time-varying electric fields.

This result was a remarkable example of theoretical work by Maxwell, and it contributed to major advances in the understanding of electromagnetism.

Quick Quiz 30.7 In an RC circuit, the capacitor begins to discharge. During the discharge, in the region of space between the plates of the capacitor, there is (a) conduction current but no displacement current, (b) displacement current but no conduction current, (c) both conduction and displacement current, (d) no current of any type.

Quick Quiz 30.8 The capacitor in an RC circuit begins to discharge. During the discharge, in the region of space between the plates of the capacitor, there is (a) an electric field but no magnetic field, (b) a magnetic field but no electric field, (c) both electric and magnetic fields, (d) no fields of any type.

Example 30.9 Displacement Current in a Capacitor

A sinusoidally varying voltage is applied across an 8.00-μF capacitor. The frequency of the voltage is 3.00 kHz, and the voltage amplitude is 30.0 V. Find the displacement current in the capacitor.

Solution The angular frequency of the source, from Equation 15.12, is given by $\omega = 2\pi f = 2\pi(3.00 \times 10^3 \text{ Hz}) = 1.88 \times 10^4 \text{ s}^{-1}$. Hence, the voltage across the capacitor in terms of t is

$$\Delta V = \Delta V_{\text{max}} \sin \omega t = (30.0 \text{ V}) \sin(1.88 \times 10^4\ t)$$

We can use Equation 30.23 and the fact that the charge on the capacitor is $q = C\,\Delta V$ to find the displacement current:

$$I_d = \frac{dq}{dt} = \frac{d}{dt}(C\,\Delta V) = C\frac{d}{dt}(\Delta V)$$

$$= (8.00 \times 10^{-6} \text{ F})\frac{d}{dt}[(30.0 \text{ V}) \sin(1.88 \times 10^4\ t)]$$

$$= \boxed{(4.52 \text{ A}) \cos(1.88 \times 10^4\ t)}$$

The displacement current varies sinusoidally with time and has a maximum value of 4.52 A.

30.8 Magnetism in Matter

The magnetic field produced by a current in a coil of wire gives us a hint as to what causes certain materials to exhibit strong magnetic properties. Earlier we found that a coil like the one shown in Figure 30.18 has a north pole and a south pole. In general, *any* current loop has a magnetic field and thus has a magnetic dipole moment, including the atomic-level current loops described in some models of the atom.

The Magnetic Moments of Atoms

We begin our discussion with a classical model of the atom in which electrons move in circular orbits around the much more massive nucleus. In this model, an orbiting electron constitutes a tiny current loop (because it is a moving charge),

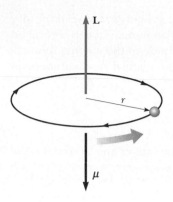

Figure 30.27 An electron moving in the direction of the gray arrow in a circular orbit of radius r has an angular momentum **L** in one direction and a magnetic moment $\boldsymbol{\mu}$ in the opposite direction. Because the electron carries a negative charge, the direction of the current due to its motion about the nucleus is opposite the direction of that motion.

and the magnetic moment of the electron is associated with this orbital motion. Although this model has many deficiencies, some of its predictions are in good agreement with the correct theory, which is expressed in terms of quantum physics.

In our classical model, we assume that an electron moves with constant speed v in a circular orbit of radius r about the nucleus, as in Figure 30.27. Because the electron travels a distance of $2\pi r$ (the circumference of the circle) in a time interval T, its orbital speed is $v = 2\pi r/T$. The current I associated with this orbiting electron is its charge e divided by T. Using $T = 2\pi/\omega$ and $\omega = v/r$, we have

$$I = \frac{e}{T} = \frac{e\omega}{2\pi} = \frac{ev}{2\pi r}$$

The magnitude of the magnetic moment associated with this current loop is $\mu = IA$, where $A = \pi r^2$ is the area enclosed by the orbit. Therefore,

$$\mu = IA = \left(\frac{ev}{2\pi r}\right)\pi r^2 = \tfrac{1}{2}evr \tag{30.24}$$

Because the magnitude of the orbital angular momentum of the electron is $L = m_e vr$ (Eq. 11.12 with $\phi = 90°$), the magnetic moment can be written as

$$\mu = \left(\frac{e}{2m_e}\right)L \tag{30.25}$$

Orbital magnetic moment

This result demonstrates that **the magnetic moment of the electron is proportional to its orbital angular momentum.** Because the electron is negatively charged, the vectors $\boldsymbol{\mu}$ and **L** point in *opposite* directions. Both vectors are perpendicular to the plane of the orbit, as indicated in Figure 30.27.

A fundamental outcome of quantum physics is that orbital angular momentum is quantized and is equal to multiples of $\hbar = h/2\pi = 1.05 \times 10^{-34}\,\text{J·s}$, where h is Planck's constant (introduced in Section 11.6). The smallest nonzero value of the electron's magnetic moment resulting from its orbital motion is

$$\mu = \sqrt{2}\,\frac{e}{2m_e}\,\hbar \tag{30.26}$$

We shall see in Chapter 42 how expressions such as Equation 30.26 arise.

Because all substances contain electrons, you may wonder why most substances are not magnetic. The main reason is that in most substances, the magnetic moment of one electron in an atom is canceled by that of another electron orbiting in the opposite direction. The net result is that, for most materials, **the magnetic effect produced by the orbital motion of the electrons is either zero or very small.**

In addition to its orbital magnetic moment, an electron (as well as protons, neutrons, and other particles) has an intrinsic property called **spin** that also contributes to its magnetic moment. Classically, the electron might be viewed as

 PITFALL PREVENTION

30.3 The Electron Does Not Spin

Do not be misled; the electron is *not* physically spinning. It has an intrinsic angular momentum *as if it were spinning*, but the notion of rotation for a point particle is meaningless. Rotation applies only to a *rigid object*, with an extent in space, as in Chapter 10. Spin angular momentum is actually a relativistic effect.

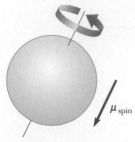

Figure 30.28 Classical model of a spinning electron. We can adopt this model to remind ourselves that electrons have an intrinsic angular momentum. The model should not be pushed too far, however—it gives an incorrect magnitude for the magnetic moment, incorrect quantum numbers, and too many degrees of freedom.

spinning about its axis as shown in Figure 30.28, but you should be very careful with the classical interpretation. The magnitude of the angular momentum **S** associated with spin is on the same order of magnitude as the magnitude of the angular momentum **L** due to the orbital motion. The magnitude of the spin angular momentum of an electron predicted by quantum theory is

$$S = \frac{\sqrt{3}}{2}\,\hbar$$

The magnetic moment characteristically associated with the spin of an electron has the value

$$\mu_{\text{spin}} = \frac{e\hbar}{2m_e} \tag{30.27}$$

This combination of constants is called the **Bohr magneton μ_{B}:**

$$\mu_{\text{B}} = \frac{e\hbar}{2m_e} = 9.27 \times 10^{-24} \, \text{J/T} \tag{30.28}$$

Thus, atomic magnetic moments can be expressed as multiples of the Bohr magneton. (Note that $1 \, \text{J/T} = 1 \, \text{A} \cdot \text{m}^2$.)

In atoms containing many electrons, the electrons usually pair up with their spins opposite each other; thus, the spin magnetic moments cancel. However, atoms containing an odd number of electrons must have at least one unpaired electron and therefore some spin magnetic moment. The total magnetic moment of an atom is the vector sum of the orbital and spin magnetic moments, and a few examples are given in Table 30.1. Note that helium and neon have zero moments because their individual spin and orbital moments cancel.

The nucleus of an atom also has a magnetic moment associated with its constituent protons and neutrons. However, the magnetic moment of a proton or neutron is much smaller than that of an electron and can usually be neglected. We can understand this by inspecting Equation 30.28 and replacing the mass of the electron with the mass of a proton or a neutron. Because the masses of the proton and neutron are much greater than that of the electron, their magnetic moments are on the order of 10^3 times smaller than that of the electron.

Table 30.1

Magnetic Moments of Some Atoms and Ions	
Atom or Ion	**Magnetic Moment (10^{-24} J/T)**
H	9.27
He	0
Ne	0
Ce^{3+}	19.8
Yb^{3+}	37.1

Magnetization vector M

Magnetization Vector and Magnetic Field Strength

The magnetic state of a substance is described by a quantity called the **magnetization vector M. The magnitude of this vector is defined as the magnetic moment per unit volume of the substance.** As you might expect, the total magnetic field **B** at a point within a substance depends on both the applied (external) field **B**$_0$ and the magnetization of the substance.

Consider a region in which a magnetic field **B**$_0$ is produced by a current-carrying conductor. If we now fill that region with a magnetic substance, the total magnetic field **B** in the region is **B** = **B**$_0$ + **B**$_m$, where **B**$_m$ is the field produced by the magnetic substance.

Let us determine the relationship between **B**$_m$ and **M**. Imagine that the field **B**$_m$ is created by a solenoid rather than by the magnetic material. Then, $B_m = \mu_0 n I$, where I is the current in the imaginary solenoid and n is the number of turns per unit length. Let us manipulate this expression as follows:

$$B_m = \mu_0 n I = \mu_0 \frac{N}{\ell} I = \mu_0 \frac{NIA}{\ell A}$$

where N is the number of turns in length ℓ, and we have multiplied the numerator and denominator by A, the cross sectional area of the solenoid in the last step. We recognize the numerator NIA as the total magnetic moment of all the loops in length ℓ and the denominator ℓA as the volume of the solenoid associated with this length:

$$B_m = \mu_0 \frac{\mu}{V}$$

The ratio of total magnetic moment to volume is what we have defined as magnetization in the case where the field is due to a material rather than a solenoid. Thus, we can express the contribution $\mathbf{B}_m$ to the total field in terms of the magnetization vector of the substance as $\mathbf{B}_m = \mu_0\mathbf{M}$. When a substance is placed in a magnetic field, the total magnetic field in the region is expressed as

$$\mathbf{B} = \mathbf{B}_0 + \mu_0\mathbf{M} \tag{30.29}$$

When analyzing magnetic fields that arise from magnetization, it is convenient to introduce a field quantity called the **magnetic field strength H** within the substance. The magnetic field strength is related to the magnetic field due to the conduction currents in wires. To emphasize the distinction between the field strength $\mathbf{H}$ and the field $\mathbf{B}$, the latter is often called the *magnetic flux density* or the *magnetic induction*. The magnetic field strength is the magnetic moment per unit volume due to currents; thus, it is similar to the vector $\mathbf{M}$ and has the same units.

Magnetic field strength H

Recognizing the similarity between $\mathbf{M}$ and $\mathbf{H}$, we can define $\mathbf{H}$ as $\mathbf{H} \equiv \mathbf{B}_0/\mu_0$. Thus, Equation 30.29 can be written

$$\mathbf{B} = \mu_0(\mathbf{H} + \mathbf{M}) \tag{30.30}$$

The quantities $\mathbf{H}$ and $\mathbf{M}$ have the same units. Because $\mathbf{M}$ is magnetic moment per unit volume, its SI units are (ampere)(meter)^2/(meter)^3, or amperes per meter.

To better understand these expressions, consider the torus region of a toroid that carries a current I. If this region is a vacuum, $\mathbf{M} = 0$ (because no magnetic material is present), the total magnetic field is that arising from the current alone, and $\mathbf{B} = \mathbf{B}_0 = \mu_0\mathbf{H}$. Because $B_0 = \mu_0 nI$ in the torus region, where n is the number of turns per unit length of the toroid, $H = B_0/\mu_0 = \mu_0 nI/\mu_0$ or

$$H = nI \tag{30.31}$$

In this case, the magnetic field in the torus region is due only to the current in the windings of the toroid.

If the torus is now made of some substance and the current I is kept constant, $\mathbf{H}$ in the torus region remains unchanged (because it depends on the current only) and has magnitude nI. The total field $\mathbf{B}$, however, is different from that when the torus region was a vacuum. From Equation 30.30, we see that part of $\mathbf{B}$ arises from the term $\mu_0\mathbf{H}$ associated with the current in the toroid, and part arises from the term $\mu_0\mathbf{M}$ due to the magnetization of the substance of which the torus is made.

Classification of Magnetic Substances

Substances can be classified as belonging to one of three categories, depending on their magnetic properties. **Paramagnetic** and **ferromagnetic** materials are those made of atoms that have permanent magnetic moments. **Diamagnetic** materials are those made of atoms that do not have permanent magnetic moments.

For paramagnetic and diamagnetic substances, the magnetization vector $\mathbf{M}$ is proportional to the magnetic field strength $\mathbf{H}$. For these substances placed in an external magnetic field, we can write

$$\mathbf{M} = \chi\mathbf{H} \tag{30.32}$$

where χ (Greek letter chi) is a dimensionless factor called the **magnetic susceptibility.** It can be considered a measure of how *susceptible* a material is to being magnetized. For paramagnetic substances, χ is positive and $\mathbf{M}$ is in the same direction as $\mathbf{H}$. For

Magnetic susceptibility χ

Table 30.2

Magnetic Susceptibilities of Some Paramagnetic and Diamagnetic Substances at 300 K			
Paramagnetic Substance	χ	Diamagnetic Substance	χ
Aluminum	2.3×10^{-5}	Bismuth	-1.66×10^{-5}
Calcium	1.9×10^{-5}	Copper	-9.8×10^{-6}
Chromium	2.7×10^{-4}	Diamond	-2.2×10^{-5}
Lithium	2.1×10^{-5}	Gold	-3.6×10^{-5}
Magnesium	1.2×10^{-5}	Lead	-1.7×10^{-5}
Niobium	2.6×10^{-4}	Mercury	-2.9×10^{-5}
Oxygen	2.1×10^{-6}	Nitrogen	-5.0×10^{-9}
Platinum	2.9×10^{-4}	Silver	-2.6×10^{-5}
Tungsten	6.8×10^{-5}	Silicon	-4.2×10^{-6}

diamagnetic substances, χ is negative and **M** is opposite **H**. The susceptibilities of some substances are given in Table 30.2.

Substituting Equation 30.32 for **M** into Equation 30.30 gives

$$\mathbf{B} = \mu_0(\mathbf{H} + \mathbf{M}) = \mu_0(\mathbf{H} + \chi\mathbf{H}) = \mu_0(1 + \chi)\mathbf{H}$$

or

$$\mathbf{B} = \mu_m\mathbf{H} \tag{30.33}$$

Magnetic permeability

where the constant μ_m is called the **magnetic permeability** of the substance and is related to the susceptibility by

$$\mu_m = \mu_0(1 + \chi) \tag{30.34}$$

Substances may be classified in terms of how their magnetic permeability μ_m compares with μ_0 (the permeability of free space), as follows:

$$\text{Paramagnetic} \qquad \mu_m > \mu_0$$

$$\text{Diamagnetic} \qquad \mu_m < \mu_0$$

Because χ is very small for paramagnetic and diamagnetic substances (see Table 30.2), μ_m is nearly equal to μ_0 for these substances. For ferromagnetic substances, however, μ_m is typically several thousand times greater than μ_0 (meaning that χ is very large for ferromagnetic substances).

Although Equation 30.33 provides a simple relationship between **B** and **H**, we must interpret it with care when dealing with ferromagnetic substances. We find that **M** is not a linear function of **H** for ferromagnetic substances. This is because the value of μ_m is not only a characteristic of the ferromagnetic substance but also depends on the previous state of the substance and on the process it underwent as it moved from its previous state to its present one. We shall investigate this more deeply after the following example.

Example 30.10 An Iron-Filled Toroid

A toroid wound with 60.0 turns/m of wire carries a current of 5.00 A. The torus is iron, which has a magnetic permeability of $\mu_m = 5\,000\mu_0$ under the given conditions. Find H and B inside the iron.

Solution Using Equations 30.31 and 30.33, we obtain

$$H = nI = (60.0 \text{ turns/m})(5.00 \text{ A}) = \boxed{300 \text{ A} \cdot \text{turns/m}}$$

$$B = \mu_m H = 5\,000\mu_0 H$$
$$= 5\,000(4\pi \times 10^{-7} \text{ T} \cdot \text{m/A})(300 \text{ A} \cdot \text{turns/m})$$
$$= \boxed{1.88 \text{ T}}$$

This value of B is 5 000 times the value in the absence of iron!

Ferromagnetism

A small number of crystalline substances exhibit strong magnetic effects called **ferromagnetism.** Some examples of ferromagnetic substances are iron, cobalt, nickel, gadolinium, and dysprosium. These substances contain permanent atomic magnetic moments that tend to align parallel to each other even in a weak external magnetic field. Once the moments are aligned, the substance remains magnetized after the external field is removed. This permanent alignment is due to a strong coupling between neighboring moments, a coupling that can be understood only in quantum-mechanical terms.

All ferromagnetic materials are made up of microscopic regions called **domains,** regions within which all magnetic moments are aligned. These domains have volumes of about 10^{-12} to 10^{-8} m^3 and contain 10^{17} to 10^{21} atoms. The boundaries between the various domains having different orientations are called **domain walls.** In an unmagnetized sample, the magnetic moments in the domains are randomly oriented so that the net magnetic moment is zero, as in Figure 30.29a. When the sample is placed in an external magnetic field $\mathbf{B}_0$, the size of those domains with magnetic moments aligned with the field grows, which results in a magnetized sample, as in Figure 30.29b. As the external field becomes very strong, as in Figure 30.29c, the domains in which the magnetic moments are not aligned with the field become very small. When the external field is removed, the sample may retain a net magnetization in the direction of the original field. At ordinary temperatures, thermal agitation is not sufficient to disrupt this preferred orientation of magnetic moments.

A typical experimental arrangement that is used to measure the magnetic properties of a ferromagnetic material consists of a torus made of the material wound with N turns of wire, as shown in Figure 30.30, where the windings are represented in black and are referred to as the *primary coil.* This apparatus is sometimes referred to as a **Rowland ring.** A *secondary coil* (the red wires in Fig. 30.30) connected to a galvanometer is used to measure the total magnetic flux through the torus. The magnetic field $\mathbf{B}$ in the torus is measured by increasing the current in the toroid from zero to I. As the current changes, the magnetic flux through the secondary coil changes by an amount BA, where A is the cross-sectional area of the toroid. As shown in Chapter 31, because of this changing flux, an emf that is proportional to the rate of change in magnetic flux is induced in the secondary coil. If the galvanometer is properly calibrated, a value for $\mathbf{B}$ corresponding to any value of the current in the primary coil can be obtained. The magnetic field $\mathbf{B}$ is measured first in the absence of the torus and then with the torus in place. The magnetic properties of the torus material are then obtained from a comparison of the two measurements.

Now consider a torus made of unmagnetized iron. If the current in the primary coil is increased from zero to some value I, the magnitude of the magnetic field strength H increases linearly with I according to the expression $H = nI$. Furthermore,

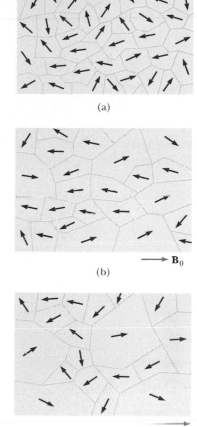

(a)

(b)

$\longrightarrow \mathbf{B}_0$

(c)

$\mathbf{B}_0$

Figure 30.29 (a) Random orientation of atomic magnetic dipoles in the domains of an unmagnetized substance. (b) When an external field $\mathbf{B}_0$ is applied, the domains with components of magnetic moment in the same direction as $\mathbf{B}_0$ grow larger, giving the sample a net magnetization. (c) As the field is made even stronger, the domains with magnetic moment vectors not aligned with the external field become very small.

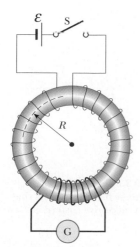

Figure 30.30 A toroidal winding arrangement used to measure the magnetic properties of a material. The torus is made of the material under study, and the circuit containing the galvanometer measures the magnetic flux.

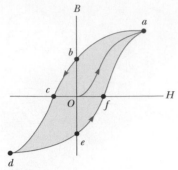

Figure 30.31 Magnetization curve for a ferromagnetic material.

the magnitude of the total field B also increases with increasing current, as shown by the curve from point O to point a in Figure 30.31. At point O, the domains in the iron are randomly oriented, corresponding to $B_m = 0$. As the increasing current in the primary coil causes the external field $\mathbf{B}_0$ to increase, the aligned domains grow in size until nearly all magnetic moments are aligned at point a. At this point the iron core is approaching *saturation*, which is the condition in which all magnetic moments in the iron are aligned.

Next, suppose that the current is reduced to zero, and the external field is consequently eliminated. The B-versus-H curve, called a **magnetization curve,** now follows the path ab in Figure 30.31. Note that at point b, $\mathbf{B}$ is not zero even though the external field $\mathbf{B}_0$ is zero. The reason is that the iron is now magnetized due to the alignment of a large number of its magnetic moments (that is, $\mathbf{B} = \mathbf{B}_m$). At this point, the iron is said to have a *remanent* magnetization.

If the current in the primary coil is reversed so that the direction of the external magnetic field is reversed, the magnetic moments reorient until the sample is again unmagnetized at point c, where $B = 0$. An increase in the reverse current causes the iron to be magnetized in the opposite direction, approaching saturation at point d in Figure 30.31. A similar sequence of events occurs as the current is reduced to zero and then increased in the original (positive) direction. In this case the magnetization curve follows the path def. If the current is increased sufficiently, the magnetization curve returns to point a, where the sample again has its maximum magnetization.

The effect just described, called **magnetic hysteresis,** shows that the magnetization of a ferromagnetic substance depends on the history of the substance as well as on the magnitude of the applied field. (The word *hysteresis* means "lagging behind.") It is often said that a ferromagnetic substance has a "memory" because it remains magnetized after the external field is removed. The closed loop in Figure 30.31 is referred to as a hysteresis loop. Its shape and size depend on the properties of the ferromagnetic substance and on the strength of the maximum applied field. The hysteresis loop for "hard" ferromagnetic materials is characteristically wide like the one shown in Figure 30.32a, corresponding to a large remanent magnetization. Such materials cannot be easily demagnetized by an external field. "Soft" ferromagnetic materials, such as iron, have a very narrow hysteresis loop and a small remanent magnetization (Fig. 30.32b.) Such materials are easily magnetized and demagnetized. An ideal soft ferromagnet would exhibit no hysteresis and hence would have no remanent magnetization. A ferromagnetic substance can be demagnetized by carrying it through successive hysteresis loops, due to a decreasing applied magnetic field, as shown in Figure 30.33.

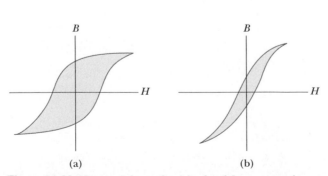

Figure 30.32 Hysteresis loops for (a) a hard ferromagnetic material and (b) a soft ferromagnetic material.

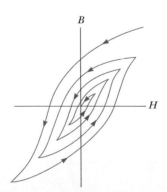

Figure 30.33 Demagnetizing a ferromagnetic material by carrying it through successive hysteresis loops.

The magnetization curve is useful for another reason: **the area enclosed by the magnetization curve represents the energy input required to take the material through the hysteresis cycle.** The energy acquired by the material in the magnetization process originates from the source of the external field—that is, the emf in the circuit of the toroidal coil. When the magnetization cycle is repeated, dissipative processes within the material due to realignment of the magnetic moments result in an increase in internal energy, made evident by an increase in the temperature of the substance. For this reason, devices subjected to alternating fields (such as AC adapters for cell phones, power tools, and so on) use cores made of soft ferromagnetic substances, which have narrow hysteresis loops and correspondingly little energy loss per cycle.

Magnetic computer disks store information by alternating the direction of **B** for portions of a thin layer of ferromagnetic material. Floppy disks have the layer on a circular sheet of plastic. Hard disks have several rigid platters with magnetic coatings on each side. Audio tapes and videotapes work the same way as floppy disks except that the ferromagnetic material is on a very long strip of plastic. Tiny coils of wire in a recording head are placed close to the magnetic material (which is moving rapidly past the head). Varying the current in the coils creates a magnetic field that magnetizes the recording material. To retrieve the information, the magnetized material is moved past a playback coil. The changing magnetism of the material induces a current in the coil, as shown in Chapter 31. This current is then amplified by audio or video equipment, or it is processed by computer circuitry.

When the temperature of a ferromagnetic substance reaches or exceeds a critical temperature called the **Curie temperature,** the substance loses its residual magnetization and becomes paramagnetic (Fig. 30.34). Below the Curie temperature, the magnetic moments are aligned and the substance is ferromagnetic. Above the Curie temperature, the thermal agitation is great enough to cause a random orientation of the moments, and the substance becomes paramagnetic. Curie temperatures for several ferromagnetic substances are given in Table 30.3.

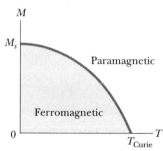

Figure 30.34 Magnetization versus absolute temperature for a ferromagnetic substance. The magnetic moments are aligned below the Curie temperature T_{Curie}, where the substance is ferromagnetic. The substance becomes paramagnetic (magnetic moments unaligned) above T_{Curie}.

Paramagnetism

Paramagnetic substances have a small but positive magnetic susceptibility $(0 < \chi \ll 1)$ resulting from the presence of atoms (or ions) that have permanent magnetic moments. These moments interact only weakly with each other and are randomly oriented in the absence of an external magnetic field. When a paramagnetic substance is placed in an external magnetic field, its atomic moments tend to line up with the field. However, this alignment process must compete with thermal motion, which tends to randomize the magnetic moment orientations.

Pierre Curie (1859–1906) and others since him have found experimentally that, under a wide range of conditions, the magnetization of a paramagnetic substance is proportional to the applied magnetic field and inversely proportional to the absolute temperature:

$$M = C \, \frac{B_0}{T} \tag{30.35}$$

Table 30.3

Curie Temperatures for Several Ferromagnetic Substance	
Substance	$T_{Curie}(\text{K})$
Iron	1 043
Cobalt	1 394
Nickel	631
Gadolinium	317
Fe_2O_3	893

This relationship is known as **Curie's law** after its discoverer, and the constant C is called **Curie's constant.** The law shows that when $B_0 = 0$, the magnetization is zero, corresponding to a random orientation of magnetic moments. As the ratio of magnetic field to temperature becomes great, the magnetization approaches its saturation value, corresponding to a complete alignment of its moments, and Equation 30.35 is no longer valid.

Diamagnetism

When an external magnetic field is applied to a diamagnetic substance, a weak magnetic moment is induced in the direction opposite the applied field. This causes diamagnetic substances to be weakly repelled by a magnet. Although diamagnetism is present in all matter, its effects are much smaller than those of paramagnetism or ferromagnetism, and are evident only when those other effects do not exist.

We can attain some understanding of diamagnetism by considering a classical model of two atomic electrons orbiting the nucleus in opposite directions but with the same speed. The electrons remain in their circular orbits because of the attractive electrostatic force exerted by the positively charged nucleus. Because the magnetic moments of the two electrons are equal in magnitude and opposite in direction, they cancel each other, and the magnetic moment of the atom is zero. When an external magnetic field is applied, the electrons experience an additional magnetic force $q\mathbf{v} \times \mathbf{B}$. This added magnetic force combines with the electrostatic force to increase the orbital speed of the electron whose magnetic moment is antiparallel to the field and to decrease the speed of the electron whose magnetic moment is parallel to the field. As a result, the two magnetic moments of the electrons no longer cancel, and the substance acquires a net magnetic moment that is opposite the applied field.

As you recall from Chapter 27, a superconductor is a substance in which the electrical resistance is zero below some critical temperature. Certain types of superconductors also exhibit perfect diamagnetism in the superconducting state. As a result, an applied magnetic field is expelled by the superconductor so that the field is zero in its interior. This phenomenon is known as the **Meissner effect.** If a permanent magnet is brought near a superconductor, the two objects repel each other. This is illustrated in Figure 30.35, which shows a small permanent magnet levitated above a superconductor maintained at 77 K.

Photo courtesy Argonne National Laboratory

Figure 30.35 An illustration of the Meissner effect, shown by this magnet suspended above a cooled ceramic superconductor disk, has become our most visual image of high-temperature superconductivity. Superconductivity is the loss of all resistance to electrical current, and is a key to more efficient energy use. In the Meissner effect, the magnet induces superconducting current in the disk, which is cooled to $-321°F$ (77 K). The currents create a magnetic force that repels and levitates the disk.

Leon Lewandowski

High Field Magnet Laboratory, University of Nijmegen, The Netherlands.

(*Left*) Paramagnetism: liquid oxygen, a paramagnetic material, is attracted to the poles of a magnet. (*Right*) Diamagnetism: a frog is levitated in a 16-T magnetic field at the Nijmegen High Field Magnet Laboratory, Netherlands. The levitation force is exerted on the diamagnetic water molecules in the frog's body. The frog suffered no ill effects from the levitation experience.

Example 30.11   Saturation Magnetization

Estimate the saturation magnetization in a long cylinder of iron, assuming one unpaired electron spin per atom.

Solution The saturation magnetization is obtained when all the magnetic moments in the sample are aligned. If the sample contains n atoms per unit volume, then the saturation magnetization M_s has the value

$$M_s = n\mu$$

where μ is the magnetic moment per atom. Because the molar mass of iron is 55 g/mol and its density is 7.9 g/cm^3,

the value of n for iron is 8.6×10^{28} atoms/m^3. Assuming that each atom contributes one Bohr magneton (due to one unpaired spin) to the magnetic moment, we obtain

$$M_s = (8.6 \times 10^{28} \text{ atoms/m}^3)(9.27 \times 10^{-24} \text{ A} \cdot \text{m}^2/\text{atom})$$

$$= 8.0 \times 10^5 \text{ A/m}$$

This is about half the experimentally determined saturation magnetization for iron, which indicates that actually two unpaired electron spins are present per atom.

30.9   The Magnetic Field of the Earth

When we speak of a compass magnet having a north pole and a south pole, we should say more properly that it has a "north-seeking" pole and a "south-seeking" pole. By this we mean that one pole of the magnet seeks, or points to, the north geographic pole of the Earth. Because the north pole of a magnet is attracted toward the north geographic pole of the Earth, we conclude that **the Earth's south magnetic pole is located near the north geographic pole, and the Earth's north magnetic pole is located near the south geographic pole.** In fact, the configuration of the Earth's magnetic field, pictured in Figure 30.36, is very much like the one that would be achieved by burying a gigantic bar magnet deep in the interior of the Earth.

If a compass needle is suspended in bearings that allow it to rotate in the vertical plane as well as in the horizontal plane, the needle is horizontal with respect to the Earth's surface only near the equator. As the compass is moved northward, the needle rotates so that it points more and more toward the surface of the Earth. Finally, at a point near Hudson Bay in Canada, the north pole of the needle points directly downward. This site, first found in 1832, is considered to be the location of the south magnetic pole of the Earth. It is approximately 1 300 mi from the Earth's geographic North Pole, and its exact position varies slowly with time. Similarly, the north magnetic pole of the Earth is about 1 200 mi away from the Earth's geographic South Pole.

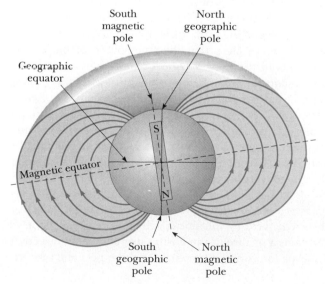

Figure 30.36 The Earth's magnetic field lines. Note that a south magnetic pole is near the north geographic pole, and a north magnetic pole is near the south geographic pole.

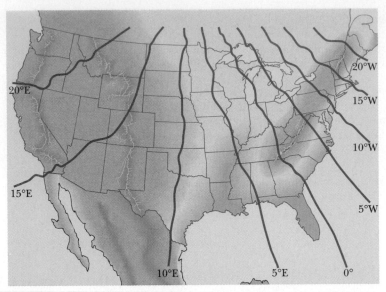

Figure 30.37 A map of the continental United States showing several lines of constant magnetic declination.

Because of this distance between the north geographic and south magnetic poles, it is only approximately correct to say that a compass needle points north. The difference between true north, defined as the geographic North Pole, and north indicated by a compass varies from point to point on the Earth, and the difference is referred to as *magnetic declination*. For example, along a line through Florida and the Great Lakes, a compass indicates true north, whereas in the state of Washington, it aligns 25° east of true north. Figure 30.37 shows some representative values of the magnetic declination for the continental United States.

Quick Quiz 30.10 If we wanted to cancel the Earth's magnetic field by running an enormous current loop around the equator, which way would the current have to to be directed: (a) east to west or (b) west to east?

Although the magnetic field pattern of the Earth is similar to the one that would be set up by a bar magnet deep within the Earth, it is easy to understand why the source of the Earth's magnetic field cannot be large masses of permanently magnetized material. The Earth does have large deposits of iron ore deep beneath its surface, but the high temperatures in the Earth's core prevent the iron from retaining any permanent magnetization. Scientists consider it more likely that the true source of the Earth's magnetic field is convection currents in the Earth's core. Charged ions or electrons circulating in the liquid interior could produce a magnetic field just as a current loop does. There is also strong evidence that the magnitude of a planet's magnetic field is related to the planet's rate of rotation. For example, Jupiter rotates faster than the Earth, and space probes indicate that Jupiter's magnetic field is stronger than ours. Venus, on the other hand, rotates more slowly than the Earth, and its magnetic field is found to be weaker. Investigation into the cause of the Earth's magnetism is ongoing.

There is an interesting sidelight concerning the Earth's magnetic field. It has been found that the direction of the field has been reversed several times during the last million years. Evidence for this is provided by basalt, a type of rock that contains iron and that forms from material spewed forth by volcanic activity on the ocean floor. As the lava cools, it solidifies and retains a picture of the Earth's magnetic field direction. The rocks are dated by other means to provide a timeline for these periodic reversals of the magnetic field.

SUMMARY

Take a practice test for this chapter by clicking on the Practice Test link at http://www.pse6.com.

The **Biot–Savart law** says that the magnetic field $d\mathbf{B}$ at a point P due to a length element $d\mathbf{s}$ that carries a steady current I is

$$d\mathbf{B} = \frac{\mu_0}{4\pi} \frac{I\, d\mathbf{s} \times \hat{\mathbf{r}}}{r^2} \tag{30.1}$$

where μ_0 is the **permeability of free space,** r is the distance from the element to the point P, and $\hat{\mathbf{r}}$ is a unit vector pointing from $d\mathbf{s}$ toward point P. We find the total field at P by integrating this expression over the entire current distribution.

The magnetic force per unit length between two parallel wires separated by a distance a and carrying currents I_1 and I_2 has a magnitude

$$\frac{F_B}{\ell} = \frac{\mu_0 I_1 I_2}{2\pi a} \tag{30.12}$$

The force is attractive if the currents are in the same direction and repulsive if they are in opposite directions.

Ampère's law says that the line integral of $\mathbf{B} \cdot d\mathbf{s}$ around any closed path equals $\mu_0 I$, where I is the total steady current through any surface bounded by the closed path:

$$\oint \mathbf{B} \cdot d\mathbf{s} = \mu_0 I \tag{30.13}$$

Using Ampère's law, one finds that the magnitude of the magnetic field at a distance r from a long, straight wire carrying an electric current I is

$$B = \frac{\mu_0 I}{2\pi r} \tag{30.14}$$

The field lines are circles concentric with the wire.

The magnitudes of the fields inside a toroid and solenoid are

$$B = \frac{\mu_0 N I}{2\pi r} \quad \text{(toroid)} \tag{30.16}$$

$$B = \mu_0 \frac{N}{\ell} I = \mu_0 n I \quad \text{(solenoid)} \tag{30.17}$$

where N is the total number of turns.

The **magnetic flux** Φ_B through a surface is defined by the surface integral

$$\Phi_B = \int \mathbf{B} \cdot d\mathbf{A} \tag{30.18}$$

Gauss's law of magnetism states that the net magnetic flux through any closed surface is zero.

The general form of Ampère's law, which is also called the **Ampère–Maxwell law,** is

$$\oint \mathbf{B} \cdot d\mathbf{s} = \mu_0 I + \mu_0 \epsilon_0 \frac{d\Phi_E}{dt} \tag{30.22}$$

This law describes the fact that magnetic fields are produced both by conduction currents and by changing electric fields.

When a substance is placed in an external magnetic field $\mathbf{B}_0$, the total magnetic field $\mathbf{B}$ is a combination of the external field and a magnetic field due to magnetic moments of atoms and electrons within the substance:

$$\mathbf{B} = \mathbf{B}_0 + \mu_0 \mathbf{M} \tag{30.29}$$

where **M** is the **magnetization vector.** The magnetization vector is the magnetic moment per unit volume in the substance.

The effect of external currents on the magnetic field in a substance is described by the **magnetic field strength** $\mathbf{H} = \mathbf{B}_0/\mu_0$. The magnetization vector is related to the magnetic field strength as follows:

$$\mathbf{M} = \chi\mathbf{H} \qquad (30.32)$$

where χ is the **magnetic susceptibility.**

Substances can be classified into one of three categories that describe their magnetic behavior. **Diamagnetic** substances are those in which the magnetization is weak and opposite the field $\mathbf{B}_0$, so that the susceptibility is negative. **Paramagnetic** substances are those in which the magnetization is weak and in the same direction as the field $\mathbf{B}_0$, so that the susceptibility is positive. In **ferromagnetic** substances, interactions between atoms cause magnetic moments to align and create a strong magnetization that remains after the external field is removed.

QUESTIONS

1. Is the magnetic field created by a current loop uniform? Explain.

2. A current in a conductor produces a magnetic field that can be calculated using the Biot–Savart law. Because current is defined as the rate of flow of charge, what can you conclude about the magnetic field produced by stationary charges? What about that produced by moving charges?

3. Explain why two parallel wires carrying currents in opposite directions repel each other.

4. Parallel current-carrying wires exert magnetic forces on each other. What about perpendicular wires? Imagine two such wires oriented perpendicular to each other, and almost touching. Does a magnetic force exist between the wires?

5. Is Ampère's law valid for all closed paths surrounding a conductor? Why is it not useful for calculating **B** for all such paths?

6. Compare Ampère's law with the Biot–Savart law. Which is more generally useful for calculating **B** for a current-carrying conductor?

7. Is the magnetic field inside a toroid uniform? Explain.

8. Describe the similarities between Ampère's law in magnetism and Gauss's law in electrostatics.

9. A hollow copper tube carries a current along its length. Why is **B** = 0 inside the tube? Is **B** nonzero outside the tube?

10. Describe the change in the magnetic field in the space enclosed by a solenoid carrying a steady current I if (a) the length of the solenoid is doubled but the number of turns remains the same and (b) the number of turns is doubled but the length remains the same.

11. A flat conducting loop is located in a uniform magnetic field directed along the *x* axis. For what orientation of the loop is the flux through it a maximum? A minimum?

12. What new concept did Maxwell's generalized form of Ampère's law include?

13. Many loops of wire are wrapped around a nail and the ends of the wire are connected to a battery. Identify the source of **M**, of **H**, and of **B**.

14. A magnet attracts a piece of iron. The iron can then attract another piece of iron. On the basis of domain alignment, explain what happens in each piece of iron.

15. Why does hitting a magnet with a hammer cause the magnetism to be reduced?

16. A Hindu ruler once suggested that he be entombed in a magnetic coffin with the polarity arranged so that he would be forever suspended between heaven and Earth. Is such magnetic levitation possible? Discuss.

17. Why is **M** = 0 in a vacuum? What is the relationship between **B** and **H** in a vacuum?

18. Explain why some atoms have permanent magnetic dipole moments and others do not.

19. What factors contribute to the total magnetic dipole moment of an atom?

20. Why is the susceptibility of a diamagnetic substance negative?

21. Why can the effect of diamagnetism be neglected in a paramagnetic substance?

22. Explain the significance of the Curie temperature for a ferromagnetic substance.

23. Discuss the difference among ferromagnetic, paramagnetic, and diamagnetic substances.

24. A current in a solenoid having air in the interior creates a magnetic field $\mathbf{B} = \mu_0\mathbf{H}$. Describe qualitatively what happens to the magnitude of **B** as (a) aluminum, (b) copper, and (c) iron are placed in the interior.

25. What is the difference between hard and soft ferromagnetic materials?

26. Should the surface of a computer disk be made from a hard or a soft ferromagnetic substance?

27. Explain why it is desirable to use hard ferromagnetic materials to make permanent magnets.

28. Would you expect the tape from a tape recorder to be attracted to a magnet? (Try it, but not with a recording you wish to save.)

29. Given only a strong magnet and a screwdriver, how would you first magnetize and then demagnetize the screwdriver?

30. Which way would a compass point if you were at the north magnetic pole of the Earth?

31. Figure Q30.31 shows two permanent magnets, each having a hole through its center. Note that the upper magnet is levitated above the lower one. (a) How does this occur? (b) What purpose does the pencil serve? (c) What can you say about the poles of the magnets from this observation? (d) If the upper magnet were inverted, what do you suppose would happen?

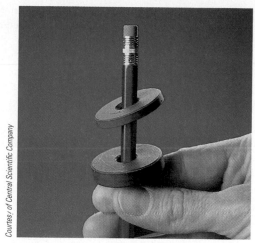

Courtesy of Central Scientific Company

Figure Q30.31

PROBLEMS

1, 2, 3 = straightforward, intermediate, challenging ☐ = full solution available in the *Student Solutions Manual and Study Guide*

🌐 = coached solution with hints available at http://www.pse6.com 🖥 = computer useful in solving problem

▨ = paired numerical and symbolic problems

Section 30.1 The Biot–Savart Law

1. In Niels Bohr's 1913 model of the hydrogen atom, an electron circles the proton at a distance of 5.29×10^{-11} m with a speed of 2.19×10^6 m/s. Compute the magnitude of the magnetic field that this motion produces at the location of the proton.

2. A lightning bolt may carry a current of 1.00×10^4 A for a short period of time. What is the resulting magnetic field 100 m from the bolt? Suppose that the bolt extends far above and below the point of observation.

3. (a) A conductor in the shape of a square loop of edge length $\ell = 0.400$ m carries a current $I = 10.0$ A as in Fig. P30.3. Calculate the magnitude and direction of the magnetic field at the center of the square. (b) **What If?** If this conductor is formed into a single circular turn and carries the same current, what is the value of the magnetic field at the center?

4. Calculate the magnitude of the magnetic field at a point 100 cm from a long, thin conductor carrying a current of 1.00 A.

5. 🌐 Determine the magnetic field at a point P located a distance x from the corner of an infinitely long wire bent at a right angle, as shown in Figure P30.5. The wire carries a steady current I.

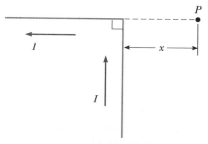

Figure P30.5

6. A conductor consists of a circular loop of radius R and two straight, long sections, as shown in Figure P30.6. The wire

Figure P30.6

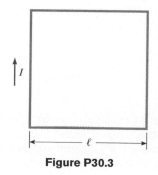

Figure P30.3

lies in the plane of the paper and carries a current I. Find an expression for the vector magnetic field at the center of the loop.

7. The segment of wire in Figure P30.7 carries a current of $I = 5.00$ A, where the radius of the circular arc is $R = 3.00$ cm. Determine the magnitude and direction of the magnetic field at the origin.

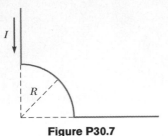

Figure P30.7

8. 🖥 Consider a flat circular current loop of radius R carrying current I. Choose the x axis to be along the axis of the loop, with the origin at the center of the loop. Plot a graph of the ratio of the magnitude of the magnetic field at coordinate x to that at the origin, for $x = 0$ to $x = 5R$. It may be useful to use a programmable calculator or a computer to solve this problem.

9. Two very long, straight, parallel wires carry currents that are directed perpendicular to the page, as in Figure P30.9. Wire 1 carries a current I_1 into the page (in the $-z$ direction) and passes through the x axis at $x = +a$. Wire 2 passes through the x axis at $x = -2a$ and carries an unknown current I_2. The total magnetic field at the origin due to the current-carrying wires has the magnitude $2\mu_0 I_1/(2\pi a)$. The current I_2 can have either of two possible values. (a) Find the value of I_2 with the smaller magnitude, stating it in terms of I_1 and giving its direction. (b) Find the other possible value of I_2.

Figure P30.9

10. A very long straight wire carries current I. In the middle of the wire a right-angle bend is made. The bend forms

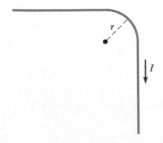

Figure P30.10

an arc of a circle of radius r, as shown in Figure P30.10. Determine the magnetic field at the center of the arc.

11. One very long wire carries current 30.0 A to the left along the x axis. A second very long wire carries current 50.0 A to the right along the line ($y = 0.280$ m, $z = 0$). (a) Where in the plane of the two wires is the total magnetic field equal to zero? (b) A particle with a charge of -2.00 μC is moving with a velocity of $150\hat{\mathbf{i}}$ Mm/s along the line ($y = 0.100$ m, $z = 0$). Calculate the vector magnetic force acting on the particle. (c) **What If?** A uniform electric field is applied to allow this particle to pass through this region undeflected. Calculate the required vector electric field.

12. Consider the current-carrying loop shown in Figure P30.12, formed of radial lines and segments of circles whose centers are at point P. Find the magnitude and direction of **B** at P.

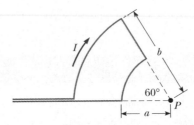

Figure P30.12

13. A wire carrying a current I is bent into the shape of an equilateral triangle of side L. (a) Find the magnitude of the magnetic field at the center of the triangle. (b) At a point halfway between the center and any vertex, is the field stronger or weaker than at the center?

14. Determine the magnetic field (in terms of I, a, and d) at the origin due to the current loop in Figure P30.14.

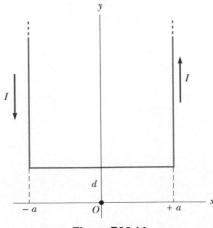

Figure P30.14

15. Two long, parallel conductors carry currents $I_1 = 3.00$ A and $I_2 = 3.00$ A, both directed into the page in Figure P30.15. Determine the magnitude and direction of the resultant magnetic field at P.

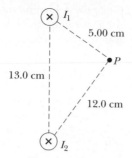

Figure P30.15

Section 30.2   The Magnetic Force Between Two Parallel Conductors

16. Two long, parallel conductors, separated by 10.0 cm, carry currents in the same direction. The first wire carries current $I_1 = 5.00$ A and the second carries $I_2 = 8.00$ A. (a) What is the magnitude of the magnetic field created by I_1 at the location of I_2? (b) What is the force per unit length exerted by I_1 on I_2? (c) What is the magnitude of the magnetic field created by I_2 at the location of I_1? (d) What is the force per length exerted by I_2 on I_1?

17. In Figure P30.17, the current in the long, straight wire is $I_1 = 5.00$ A and the wire lies in the plane of the rectangular loop, which carries the current $I_2 = 10.0$ A. The dimensions are $c = 0.100$ m, $a = 0.150$ m, and $\ell = 0.450$ m. Find the magnitude and direction of the net force exerted on the loop by the magnetic field created by the wire.

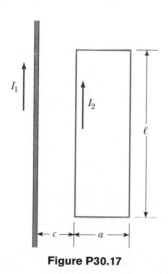

Figure P30.17

18. Two long, parallel wires are attracted to each other by a force per unit length of 320 μN/m when they are separated by a vertical distance of 0.500 m. The current in the upper wire is 20.0 A to the right. Determine the location of the line in the plane of the two wires along which the total magnetic field is zero.

19. Three long wires (wire 1, wire 2, and wire 3) hang vertically. The distance between wire 1 and wire 2 is 20.0 cm. On the left, wire 1 carries an upward current of 1.50 A. To the right, wire 2 carries a downward current of 4.00 A.

Wire 3 is located such that when it carries a certain current, each wire experiences no net force. Find (a) the position of wire 3, and (b) the magnitude and direction of the current in wire 3.

20. The unit of magnetic flux is named for Wilhelm Weber. The practical-size unit of magnetic field is named for Johann Karl Friedrich Gauss. Both were scientists at Göttingen, Germany. Along with their individual accomplishments, together they built a telegraph in 1833. It consisted of a battery and switch, at one end of a transmission line 3 km long, operating an electromagnet at the other end. (André Ampère suggested electrical signaling in 1821; Samuel Morse built a telegraph line between Baltimore and Washington in 1844.) Suppose that Weber and Gauss's transmission line was as diagrammed in Figure P30.20. Two long, parallel wires, each having a mass per unit length of 40.0 g/m, are supported in a horizontal plane by strings 6.00 cm long. When both wires carry the same current I, the wires repel each other so that the angle θ between the supporting strings is 16.0°. (a) Are the currents in the same direction or in opposite directions? (b) Find the magnitude of the current.

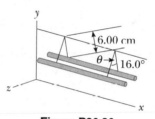

Figure P30.20

Section 30.3   Ampère's Law

21. Four long, parallel conductors carry equal currents of $I = 5.00$ A. Figure P30.21 is an end view of the conductors. The current direction is into the page at points A and B (indicated by the crosses) and out of the page at C and D (indicated by the dots). Calculate the magnitude and direction of the magnetic field at point P, located at the center of the square of edge length 0.200 m.

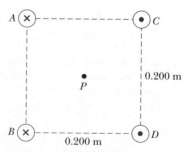

Figure P30.21

22. A long straight wire lies on a horizontal table and carries a current of 1.20 μA. In a vacuum, a proton moves parallel to the wire (opposite the current) with a constant speed of 2.30×10^4 m/s at a distance d above the wire. Determine the value of d. You may ignore the magnetic field due to the Earth.

23. Figure P30.23 is a cross-sectional view of a coaxial cable. The center conductor is surrounded by a rubber layer, which is surrounded by an outer conductor, which is surrounded by another rubber layer. In a particular application, the current in the inner conductor is 1.00 A out of the page and the current in the outer conductor is 3.00 A into the page. Determine the magnitude and direction of the magnetic field at points *a* and *b*.

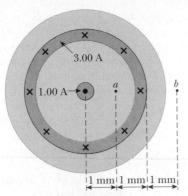

Figure P30.23

24. The magnetic field 40.0 cm away from a long straight wire carrying current 2.00 A is 1.00 μT. (a) At what distance is it 0.100 μT? (b) **What If?** At one instant, the two conductors in a long household extension cord carry equal 2.00-A currents in opposite directions. The two wires are 3.00 mm apart. Find the magnetic field 40.0 cm away from the middle of the straight cord, in the plane of the two wires. (c) At what distance is it one tenth as large? (d) The center wire in a coaxial cable carries current 2.00 A in one direction and the sheath around it carries current 2.00 A in the opposite direction. What magnetic field does the cable create at points outside?

25. A packed bundle of 100 long, straight, insulated wires forms a cylinder of radius $R = 0.500$ cm. (a) If each wire carries 2.00 A, what are the magnitude and direction of the magnetic force per unit length acting on a wire located 0.200 cm from the center of the bundle? (b) **What If?** Would a wire on the outer edge of the bundle experience a force greater or smaller than the value calculated in part (a)?

26. The magnetic coils of a tokamak fusion reactor are in the shape of a toroid having an inner radius of 0.700 m and an outer radius of 1.30 m. The toroid has 900 turns of large-diameter wire, each of which carries a current of 14.0 kA. Find the magnitude of the magnetic field inside the toroid along (a) the inner radius and (b) the outer radius.

27. Consider a column of electric current passing through plasma (ionized gas). Filaments of current within the column are magnetically attracted to one another. They can crowd together to yield a very great current density and a very strong magnetic field in a small region. Sometimes the current can be cut off momentarily by this *pinch effect*. (In a metallic wire a pinch effect is not important, because the current-carrying electrons repel one another with electric forces.) The pinch effect can be demonstrated by making an empty aluminum can carry a large current parallel to its axis. Let *R* represent the radius

of the can and *I* the upward current, uniformly distributed over its curved wall. Determine the magnetic field (a) just inside the wall and (b) just outside. (c) Determine the pressure on the wall.

28. Niobium metal becomes a superconductor when cooled below 9 K. Its superconductivity is destroyed when the surface magnetic field exceeds 0.100 T. Determine the maximum current a 2.00-mm-diameter niobium wire can carry and remain superconducting, in the absence of any external magnetic field.

29. A long cylindrical conductor of radius *R* carries a current *I* as shown in Figure P30.29. The current density *J*, however, is not uniform over the cross section of the conductor but is a function of the radius according to $J = br$, where *b* is a constant. Find an expression for the magnetic field *B* (a) at a distance $r_1 < R$ and (b) at a distance $r_2 > R$, measured from the axis.

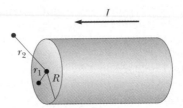

Figure P30.29

30. In Figure P30.30, both currents in the infinitely long wires are in the negative *x* direction. (a) Sketch the magnetic field pattern in the *yz* plane. (b) At what distance *d* along the *z* axis is the magnetic field a maximum?

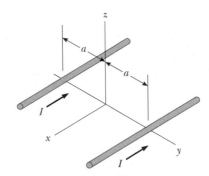

Figure P30.30

Section 30.4   The Magnetic Field of a Solenoid

31. What current is required in the windings of a long solenoid that has 1 000 turns uniformly distributed over a length of 0.400 m, to produce at the center of the solenoid a magnetic field of magnitude 1.00×10^{-4} T?

32. Consider a solenoid of length ℓ and radius *R*, containing *N* closely spaced turns and carrying a steady current *I*. (a) In terms of these parameters, find the magnetic field at a point along the axis as a function of distance *a* from the end of the solenoid. (b) Show that as ℓ becomes very long, *B* approaches $\mu_0 NI/2\ell$ at each end of the solenoid.

33. A single-turn square loop of wire, 2.00 cm on each edge, carries a clockwise current of 0.200 A. The loop is inside a solenoid, with the plane of the loop perpendicular to the magnetic field of the solenoid. The solenoid has 30 turns/cm and carries a clockwise current of 15.0 A. Find the force on each side of the loop and the torque acting on the loop.

Section 30.5 Magnetic Flux

34. Consider the hemispherical closed surface in Figure P30.34. The hemisphere is in a uniform magnetic field that makes an angle θ with the vertical. Calculate the magnetic flux through (a) the flat surface S_1 and (b) the hemispherical surface S_2.

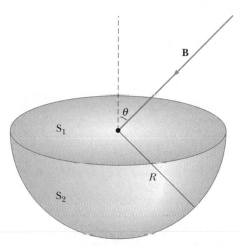

Figure P30.34

35. A cube of edge length $\ell = 2.50$ cm is positioned as shown in Figure P30.35. A uniform magnetic field given by $\mathbf{B} = (5\hat{\mathbf{i}} + 4\hat{\mathbf{j}} + 3\hat{\mathbf{k}})$ T exists throughout the region. (a) Calculate the flux through the shaded face. (b) What is the total flux through the six faces?

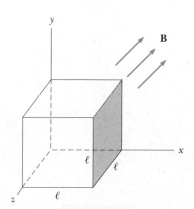

Figure P30.35

36. A solenoid 2.50 cm in diameter and 30.0 cm long has 300 turns and carries 12.0 A. (a) Calculate the flux through the surface of a disk of radius 5.00 cm that is positioned perpendicular to and centered on the axis of

the solenoid, as shown in Figure P30.36a. (b) Figure P30.36b shows an enlarged end view of the same solenoid. Calculate the flux through the blue area, which is defined by an annulus that has an inner radius of 0.400 cm and outer radius of 0.800 cm.

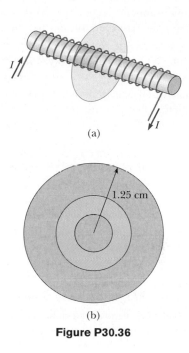

Figure P30.36

Section 30.7 Displacement Current and the General Form of Ampère's Law

37. A 0.100-A current is charging a capacitor that has square plates 5.00 cm on each side. The plate separation is 4.00 mm. Find (a) the time rate of change of electric flux between the plates and (b) the displacement current between the plates.

38. A 0.200-A current is charging a capacitor that has circular plates 10.0 cm in radius. If the plate separation is 4.00 mm, (a) what is the time rate of increase of electric field between the plates? (b) What is the magnetic field between the plates 5.00 cm from the center?

Section 30.8 Magnetism in Matter

39. In Bohr's 1913 model of the hydrogen atom, the electron is in a circular orbit of radius 5.29×10^{-11} m and its speed is 2.19×10^6 m/s. (a) What is the magnitude of the magnetic moment due to the electron's motion? (b) If the electron moves in a horizontal circle, counterclockwise as seen from above, what is the direction of this magnetic moment vector?

40. A magnetic field of 1.30 T is to be set up in an iron-core toroid. The toroid has a mean radius of 10.0 cm, and magnetic permeability of 5 000 μ_0. What current is required if the winding has 470 turns of wire? The thickness of the iron ring is small compared to 10 cm, so the field in the material is nearly uniform.

41. A toroid with a mean radius of 20.0 cm and 630 turns (see Fig. 30.30) is filled with powdered steel whose magnetic

susceptibility χ is 100. The current in the windings is 3.00 A. Find B (assumed uniform) inside the toroid.

42. A particular paramagnetic substance achieves 10.0% of its saturation magnetization when placed in a magnetic field of 5.00 T at a temperature of 4.00 K. The density of magnetic atoms in the sample is 8.00×10^{27} atoms/m^3, and the magnetic moment per atom is 5.00 Bohr magnetons. Calculate the Curie constant for this substance.

43. Calculate the magnetic field strength H of a magnetized substance in which the magnetization is 0.880×10^6 A/m and the magnetic field has magnitude 4.40 T.

44. At *saturation*, when nearly all of the atoms have their magnetic moments aligned, the magnetic field in a sample of iron can be 2.00 T. If each electron contributes a magnetic moment of 9.27×10^{-24} A·m^2 (one Bohr magneton), how many electrons per atom contribute to the saturated field of iron? Iron contains approximately 8.50×10^{28} atoms/m^3.

45. (a) Show that Curie's law can be stated in the following way: The magnetic susceptibility of a paramagnetic substance is inversely proportional to the absolute temperature, according to $\chi = C\mu_0/T$, where C is Curie's constant. (b) Evaluate Curie's constant for chromium.

Section 30.9 The Magnetic Field of the Earth

46. A circular coil of 5 turns and a diameter of 30.0 cm is oriented in a vertical plane with its axis perpendicular to the horizontal component of the Earth's magnetic field. A horizontal compass placed at the center of the coil is made to deflect 45.0° from magnetic north by a current of 0.600 A in the coil. (a) What is the horizontal component of the Earth's magnetic field? (b) The current in the coil is switched off. A "dip needle" is a magnetic compass mounted so that it can rotate in a vertical north–south plane. At this location a dip needle makes an angle of 13.0° from the vertical. What is the total magnitude of the Earth's magnetic field at this location?

47. The magnetic moment of the Earth is approximately 8.00×10^{22} A·m^2. (a) If this were caused by the complete magnetization of a huge iron deposit, how many unpaired electrons would this correspond to? (b) At two unpaired electrons per iron atom, how many kilograms of iron would this correspond to? (Iron has a density of 7 900 kg/m^3, and approximately 8.50×10^{28} iron atoms/m^3.)

Additional Problems

48. The magnitude of the Earth's magnetic field at either pole is approximately 7.00×10^{-5} T. Suppose that the field fades away, before its next reversal. Scouts, sailors, and conservative politicians around the world join together in a program to replace the field. One plan is to use a current loop around the equator, without relying on magnetization of any materials inside the Earth. Determine the current that would generate such a field if this plan were carried out. (Take the radius of the Earth as $R_E = 6.37 \times 10^6$ m.)

49. A very long, thin strip of metal of width w carries a current I along its length as shown in Figure P30.49. Find the magnetic field at the point P in the diagram. The point P is in the plane of the strip at distance b away from it.

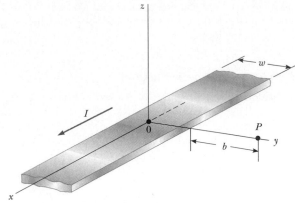

Figure P30.49

50. Suppose you install a compass on the center of the dashboard of a car. Compute an order-of-magnitude estimate for the magnetic field at this location produced by the current when you switch on the headlights. How does it compare with the Earth's magnetic field? You may suppose the dashboard is made mostly of plastic.

51. For a research project, a student needs a solenoid that produces an interior magnetic field of 0.030 0 T. She decides to use a current of 1.00 A and a wire 0.500 mm in diameter. She winds the solenoid in layers on an insulating form 1.00 cm in diameter and 10.0 cm long. Determine the number of layers of wire needed and the total length of the wire.

52. A thin copper bar of length $\ell = 10.0$ cm is supported horizontally by two (nonmagnetic) contacts. The bar carries current $I_1 = 100$ A in the $-x$ direction, as shown in Figure P30.52. At a distance $h = 0.500$ cm below one end of the bar, a long straight wire carries a current $I_2 = 200$ A in the z direction. Determine the magnetic force exerted on the bar.

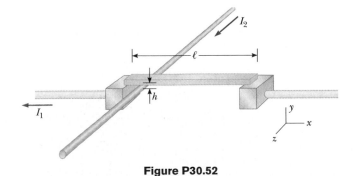

Figure P30.52

53. A nonconducting ring of radius 10.0 cm is uniformly charged with a total positive charge 10.0 μC. The ring rotates at a constant angular speed 20.0 rad/s about an

axis through its center, perpendicular to the plane of the ring. What is the magnitude of the magnetic field on the axis of the ring 5.00 cm from its center?

54. A nonconducting ring of radius R is uniformly charged with a total positive charge q. The ring rotates at a constant angular speed ω about an axis through its center, perpendicular to the plane of the ring. What is the magnitude of the magnetic field on the axis of the ring a distance $R/2$ from its center?

55. Two circular coils of radius R, each with N turns, are perpendicular to a common axis. The coil centers are a distance R apart. Each coil carries a steady current I in the same direction, as shown in Figure P30.55. (a) Show that the magnetic field on the axis at a distance x from the center of one coil is

$$B = \frac{N\mu_0 I R^2}{2}\left[\frac{1}{(R^2 + x^2)^{3/2}} + \frac{1}{(2R^2 + x^2 - 2Rx)^{3/2}}\right]$$

(b) Show that dB/dx and d^2B/dx^2 are both zero at the point midway between the coils. This means the magnetic field in the region midway between the coils is uniform. Coils in this configuration are called *Helmholtz coils*.

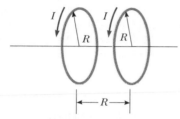

Figure P30.55 Problems 55 and 56.

56. Two identical, flat, circular coils of wire each have 100 turns and a radius of 0.500 m. The coils are arranged as a set of Helmholtz coils (see Fig. P30.55), parallel and with separation 0.500 m. Each coil carries a current of 10.0 A. Determine the magnitude of the magnetic field at a point on the common axis of the coils and halfway between them.

57. We have seen that a long solenoid produces a uniform magnetic field directed along the axis of a cylindrical region. However, to produce a uniform magnetic field directed parallel to a *diameter* of a cylindrical region, one can use the *saddle coils* illustrated in Figure P30.57. The loops are wrapped over a somewhat flattened tube. Assume the straight sections of wire are very long. The end view of the tube shows how the windings are applied. The overall current distribution is the superposition of two overlapping circular cylinders of uniformly distributed current, one toward you and one away from you. The current density J is the same for each cylinder. The position of the axis of one cylinder is described by a position vector **a** relative to the other cylinder. Prove that the magnetic field inside the hollow tube is $\mu_0 Ja/2$ downward. *Suggestion:* The use of vector methods simplifies the calculation.

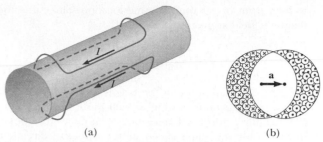

(a)   (b)

Figure P30.57 (a) General view of one turn of each saddle coil. (b) End view of the coils carrying current into the paper on the left and out of the paper on the right.

58. A very large parallel-plate capacitor carries charge with uniform charge per unit area $+\sigma$ on the upper plate and $-\sigma$ on the lower plate. The plates are horizontal and both move horizontally with speed v to the right. (a) What is the magnetic field between the plates? (b) What is the magnetic field close to the plates but outside of the capacitor? (c) What is the magnitude and direction of the magnetic force per unit area on the upper plate? (d) At what extrapolated speed v will the magnetic force on a plate balance the electric force on the plate? Calculate this speed numerically.

59. Two circular loops are parallel, coaxial, and almost in contact, 1.00 mm apart (Fig. P30.59). Each loop is 10.0 cm in radius. The top loop carries a clockwise current of 140 A. The bottom loop carries a counterclockwise current of 140 A. (a) Calculate the magnetic force exerted by the bottom loop on the top loop. (b) The upper loop has a mass of 0.021 0 kg. Calculate its acceleration, assuming that the only forces acting on it are the force in part (a) and the gravitational force. *Suggestion:* Think about how one loop looks to a bug perched on the other loop.

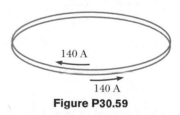

Figure P30.59

60. What objects experience a force in an electric field? Chapter 23 gives the answer: any electric charge, stationary or moving, other than the charge that created the field. What creates an electric field? Any electric charge, stationary or moving, as you studied in Chapter 23. What objects experience a force in a magnetic field? An electric current or a moving electric charge, other than the current or charge that created the field, as discussed in Chapter 29. What creates a magnetic field? An electric current, as you studied in Section 30.1, or a moving electric charge, as shown in this problem. (a) To display how a moving charge creates a magnetic field, consider a charge q moving with velocity **v**. Define the vector $\mathbf{r} = r\hat{\mathbf{r}}$

to lead from the charge to some location. Show that the magnetic field at that location is

$$\mathbf{B} = \frac{\mu_0}{4\pi}\, \frac{q\mathbf{v} \times \hat{\mathbf{r}}}{r^2}$$

(b) Find the magnitude of the magnetic field 1.00 mm to the side of a proton moving at 2.00×10^7 m/s. (c) Find the magnetic force on a second proton at this point, moving with the same speed in the opposite direction. (d) Find the electric force on the second proton.

61. *Rail guns* have been suggested for launching projectiles into space without chemical rockets, and for ground-to-air antimissile weapons of war. A tabletop model rail gun (Fig. P30.61) consists of two long parallel horizontal rails 3.50 cm apart, bridged by a bar *BD* of mass 3.00 g. The bar is originally at rest at the midpoint of the rails and is free to slide without friction. When the switch is closed, electric current is quickly established in the circuit *ABCDEA*. The rails and bar have low electric resistance, and the current is limited to a constant 24.0 A by the power supply. (a) Find the magnitude of the magnetic field 1.75 cm from a single very long straight wire carrying current 24.0 A. (b) Find the magnitude and direction of the magnetic field at point *C* in the diagram, the midpoint of the bar, immediately after the switch is closed. *Suggestion:* Consider what conclusions you can draw from the Biot–Savart law. (c) At other points along the bar *BD*, the field is in the same direction as at point *C*, but larger in magnitude. Assume that the average effective magnetic field along *BD* is five times larger than the field at *C*. With this assumption, find the magnitude and direction of the force on the bar. (d) Find the acceleration of the bar when it is in motion. (e) Does the bar move with constant acceleration? (f) Find the velocity of the bar after it has traveled 130 cm to the end of the rails.

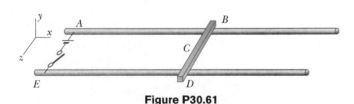

Figure P30.61

62. 🖥 Fifty turns of insulated wire 0.100 cm in diameter are tightly wound to form a flat spiral. The spiral fills a disk surrounding a circle of radius 5.00 cm and extending to a radius 10.00 cm at the outer edge. Assume the wire carries current *I* at the center of its cross section. Approximate each turn of wire as a circle. Then a loop of current exists at radius 5.05 cm, another at 5.15 cm, and so on. Numerically calculate the magnetic field at the center of the coil.

63. Two long, parallel conductors carry currents in the same direction as shown in Figure P30.63. Conductor A carries a current of 150 A and is held firmly in position. Conductor B carries a current I_B and is allowed to slide freely up and down (parallel to A) between a set of nonconducting guides. If the mass per unit length of conductor B

is 0.100 g/cm, what value of current I_B will result in equilibrium when the distance between the two conductors is 2.50 cm?

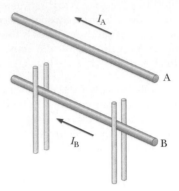

Figure P30.63

64. Charge is sprayed onto a large nonconducting belt above the left-hand roller in Figure P30.64. The belt carries the charge with a uniform surface charge density σ as it moves with a speed v between the rollers as shown. The charge is removed by a wiper at the right-hand roller. Consider a point just above the surface of the moving belt. (a) Find an expression for the magnitude of the magnetic field **B** at this point. (b) If the belt is positively charged, what is the direction of **B**? (Note that the belt may be considered as an infinite sheet.)

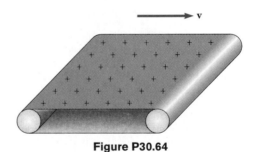

Figure P30.64

65. An infinitely long straight wire carrying a current I_1 is partially surrounded by a loop as shown in Figure P30.65.

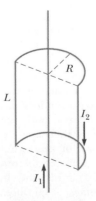

Figure P30.65

The loop has a length L, radius R, and carries a current I_2. The axis of the loop coincides with the wire. Calculate the force exerted on the loop.

66. Measurements of the magnetic field of a large tornado were made at the Geophysical Observatory in Tulsa, Oklahoma, in 1962. The tornado's field was measured to be $B = 1.50 \times 10^{-8}$ T pointing north when the tornado was 9.00 km east of the observatory. What current was carried up or down the funnel of the tornado, modeled as a long straight wire?

67. A wire is formed into the shape of a square of edge length L (Fig. P30.67). Show that when the current in the loop is I, the magnetic field at point P, a distance x from the center of the square along its axis is

$$B = \frac{\mu_0 I L^2}{2\pi(x^2 + L^2/4)\sqrt{x^2 + L^2/2}}$$

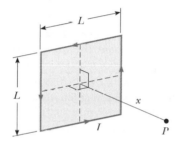

Figure P30.67

68. The force on a magnetic dipole $\boldsymbol{\mu}$ aligned with a nonuniform magnetic field in the x direction is given by $F_x = |\boldsymbol{\mu}| dB/dx$. Suppose that two flat loops of wire each have radius R and carry current I. (a) The loops are arranged coaxially and separated by a variable distance x, large compared to R. Show that the magnetic force between them varies as $1/x^4$. (b) Evaluate the magnitude of this force if $I = 10.0$ A, $R = 0.500$ cm, and $x = 5.00$ cm.

69. A wire carrying a current I is bent into the shape of an exponential spiral, $r = e^\theta$, from $\theta = 0$ to $\theta = 2\pi$ as suggested in Figure P30.69. To complete a loop, the ends of the spiral are connected by a straight wire along the x axis. Find the magnitude and direction of **B** at the origin. *Suggestions:* Use the Biot–Savart law. The angle β between a radial line and its tangent line at any point on the curve $r = f(\theta)$ is related to the function in the following way:

$$\tan \beta = \frac{r}{dr/d\theta}$$

Thus in this case $r = e^\theta$, $\tan \beta = 1$ and $\beta = \pi/4$. Therefore, the angle between $d\mathbf{s}$ and $\hat{\mathbf{r}}$ is $\pi - \beta = 3\pi/4$. Also

$$ds = \frac{dr}{\sin(\pi/4)} = \sqrt{2}\, dr$$

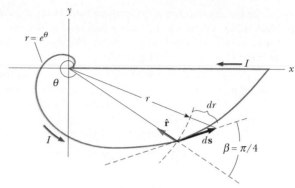

Figure P30.69

70. Table P30.70 contains data taken for a ferromagnetic material. (a) Construct a magnetization curve from the data. Remember that $\mathbf{B} = \mathbf{B}_0 + \mu_0\mathbf{M}$. (b) Determine the ratio B/B_0 for each pair of values of B and B_0, and construct a graph of B/B_0 versus B_0. (The fraction B/B_0 is called the relative permeability, and it is a measure of the induced magnetic field.)

Table P30.70

B (T)	B_0 (T)
0.2	4.8×10^{-5}
0.4	7.0×10^{-5}
0.6	8.8×10^{-5}
0.8	1.2×10^{-4}
1.0	1.8×10^{-4}
1.2	3.1×10^{-4}
1.4	8.7×10^{-4}
1.6	3.4×10^{-3}
1.8	1.2×10^{-1}

71. A sphere of radius R has a uniform volume charge density ρ. Determine the magnetic field at the center of the sphere when it rotates as a rigid object with angular speed ω about an axis through its center (Fig. P30.71).

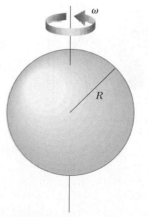

Figure P30.71 Problems 71 and 72.

72. A sphere of radius R has a uniform volume charge density ρ. Determine the magnetic dipole moment of the sphere when it rotates as a rigid body with angular speed ω about an axis through its center (Fig. P30.71).

73. A long cylindrical conductor of radius a has two cylindrical cavities of diameter a through its entire length, as shown in Figure P30.73. A current I is directed out of the page and is uniform through a cross section of the conductor. Find the magnitude and direction of the magnetic field in terms of μ_0, I, r, and a at (a) point P_1 and (b) point P_2.

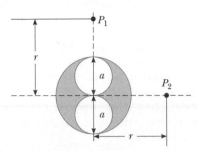

Figure P30.73

Answers to Quick Quizzes

30.1 *B, C, A.* Point *B* is closest to the current element. Point *C* is farther away and the field is further reduced by the $\sin \theta$ factor in the cross product $d\mathbf{s} \times \hat{\mathbf{r}}$. The field at *A* is zero because $\theta = 0$.

30.2 (c). $F_1 = F_2$ as required by Newton's third law. Another way to arrive at this answer is to realize that Equation 30.11 gives the same result whether the multiplication of currents is (2 A)(6 A) or (6 A)(2 A).

30.3 (a). The coils act like wires carrying parallel currents in the same direction and hence attract one another.

30.4 *b, d, a, c.* Equation 30.13 indicates that the value of the line integral depends only on the net current through each closed path. Path *b* encloses 1 A, path *d* encloses 3 A, path *a* encloses 4 A, and path *c* encloses 6 A.

30.5 *b,* then *a = c = d.* Paths *a, c,* and *d* all give the same nonzero value $\mu_0 I$ because the size and shape of the paths do not matter. Path *b* does not enclose the current, and hence its line integral is zero.

30.6 (c). The magnetic field in a very long solenoid is independent of its length or radius. Overwrapping with an additional layer of wire increases the number of turns per unit length.

30.7 (b). There can be no conduction current because there is no conductor between the plates. There is a time-varying electric field because of the decreasing charge on the plates, and the time-varying electric flux represents a displacement current.

30.8 (c). There is a time-varying electric field because of the decreasing charge on the plates. This time-varying electric field produces a magnetic field.

30.9 (a). The loop that looks like Figure 30.32a is better because the remanent magnetization at the point corresponding to point *b* in Figure 30.31 is greater.

30.10 (b). The lines of the Earth's magnetic field enter the planet in Hudson Bay and emerge from Antarctica; thus, the field lines resulting from the current would have to go in the opposite direction. Compare Figure 30.7a with Figure 30.36.

Calvin and Hobbes by Bill Watterson

Chapter 31

Faraday's Law

CHAPTER OUTLINE

31.1 Faraday's Law of Induction

31.2 Motional emf

31.3 Lenz's Law

31.4 Induced emf and Electric Fields

31.5 Generators and Motors

31.6 Eddy Currents

31.7 Maxwell's Equations

▲ In a commercial electric power plant, large generators produce energy that is transferred out of the plant by electrical transmission. These generators use magnetic induction to generate a potential difference when coils of wire in the generator are rotated in a magnetic field. The source of energy to rotate the coils might be falling water, burning fossil fuels, or a nuclear reaction. (Michael Melford/Getty Images)

The focus of our studies in electricity and magnetism so far has been the electric fields produced by stationary charges and the magnetic fields produced by moving charges. This chapter explores the effects produced by magnetic fields that vary in time.

Experiments conducted by Michael Faraday in England in 1831 and independently by Joseph Henry in the United States that same year showed that an emf can be induced in a circuit by a changing magnetic field. The results of these experiments led to a very basic and important law of electromagnetism known as *Faraday's law of induction*. An emf (and therefore a current as well) can be induced in various processes that involve a change in a magnetic flux.

With the treatment of Faraday's law, we complete our introduction to the fundamental laws of electromagnetism. These laws can be summarized in a set of four equations called *Maxwell's equations*. Together with the *Lorentz force law*, they represent a complete theory for describing the interaction of charged objects.

31.1 Faraday's Law of Induction

To see how an emf can be induced by a changing magnetic field, consider a loop of wire connected to a sensitive ammeter, as illustrated in Figure 31.1. When a magnet is moved toward the loop, the galvanometer needle deflects in one direction, arbitrarily shown to the right in Figure 31.1a. When the magnet is brought to rest and held stationary relative to the loop (Fig. 31.1b), no deflection is observed. When the magnet is moved away from the loop, the needle deflects in the opposite direction, as shown in Figure 31.1c. Finally, if the magnet is held stationary and the loop is moved either toward or away from it, the needle deflects. From these observations, we conclude that the loop detects that the magnet is moving relative to it and we relate this detection to a change in magnetic field. Thus, it seems that a relationship exists between current and changing magnetic field.

These results are quite remarkable in view of the fact that **a current is set up even though no batteries are present in the circuit!** We call such a current an *induced current* and say that it is produced by an *induced emf*.

Now let us describe an experiment conducted by Faraday and illustrated in Figure 31.2. A primary coil is connected to a switch and a battery. The coil is wrapped around an iron ring, and a current in the coil produces a magnetic field when the switch is closed. A secondary coil also is wrapped around the ring and is connected to a sensitive ammeter. No battery is present in the secondary circuit, and the secondary coil is not electrically connected to the primary coil. Any current detected in the secondary circuit must be induced by some external agent.

Initially, you might guess that no current is ever detected in the secondary circuit. However, something quite amazing happens when the switch in the primary circuit is either opened or thrown closed. At the instant the switch is closed, the galvanometer needle deflects in one direction and then returns to zero. At the instant the switch is opened, the needle deflects in the opposite direction and again returns to zero.

Michael Faraday

British Physicist and Chemist (1791–1867)

Faraday is often regarded as the greatest experimental scientist of the 1800s. His many contributions to the study of electricity include the invention of the electric motor, electric generator, and transformer, as well as the discovery of electromagnetic induction and the laws of electrolysis. Greatly influenced by religion, he refused to work on the development of poison gas for the British military. *(By kind permission of the President and Council of the Royal Society)*

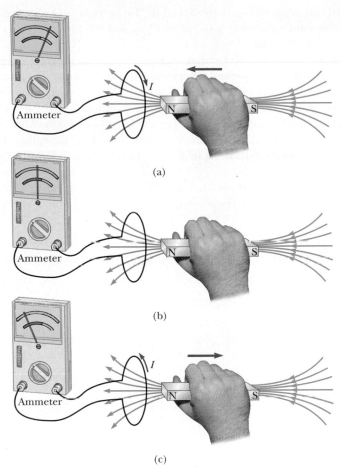

(a)

(b)

(c)

Active Figure 31.1 (a) When a magnet is moved toward a loop of wire connected to a sensitive ammeter, the ammeter deflects as shown, indicating that a current is induced in the loop. (b) When the magnet is held stationary, there is no induced current in the loop, even when the magnet is inside the loop. (c) When the magnet is moved away from the loop, the ammeter deflects in the opposite direction, indicating that the induced current is opposite that shown in part (a). Changing the direction of the magnet's motion changes the direction of the current induced by that motion.

At the Active Figures link at http://www.pse6.com, you can move the magnet and observe the current in the ammeter.

Finally, the galvanometer reads zero when there is either a steady current or no current in the primary circuit. The key to understanding what happens in this experiment is to note first that when the switch is closed, the current in the primary circuit produces a magnetic field that penetrates the secondary circuit. Furthermore, when

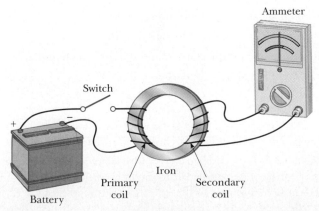

Active Figure 31.2 Faraday's experiment. When the switch in the primary circuit is closed, the ammeter in the secondary circuit deflects momentarily. The emf induced in the secondary circuit is caused by the changing magnetic field through the secondary coil.

At the Active Figures link at http://www.pse6.com, you can open and close the switch and observe the current in the ammeter.

the switch is closed, the magnetic field produced by the current in the primary circuit changes from zero to some value over some finite time, and this changing field induces a current in the secondary circuit.

As a result of these observations, Faraday concluded that **an electric current can be induced in a circuit (the secondary circuit in our setup) by a changing magnetic field.** The induced current exists for only a short time while the magnetic field through the secondary coil is changing. Once the magnetic field reaches a steady value, the current in the secondary coil disappears. In effect, the secondary circuit behaves as though a source of emf were connected to it for a short time. It is customary to say that **an induced emf is produced in the secondary circuit by the changing magnetic field.**

The experiments shown in Figures 31.1 and 31.2 have one thing in common: in each case, an emf is induced in the circuit when the magnetic flux through the circuit changes with time. In general,

> The emf induced in a circuit is directly proportional to the time rate of change of the magnetic flux through the circuit.

This statement, known as **Faraday's law of induction,** can be written

$$\mathcal{E} = -\frac{d\Phi_B}{dt} \tag{31.1}$$

where $\Phi_B = \int \mathbf{B} \cdot d\mathbf{A}$ is the magnetic flux through the circuit. (See Section 30.5.)

If the circuit is a coil consisting of N loops all of the same area and if Φ_B is the magnetic flux through one loop, an emf is induced in every loop. The loops are in series, so their emfs add; thus, the total induced emf in the coil is given by the expression

$$\mathcal{E} = -N\frac{d\Phi_B}{dt} \tag{31.2}$$

The negative sign in Equations 31.1 and 31.2 is of important physical significance, as discussed in Section 31.3.

Suppose that a loop enclosing an area A lies in a uniform magnetic field $\mathbf{B}$, as in Figure 31.3. The magnetic flux through the loop is equal to $BA\cos\theta$; hence, the induced emf can be expressed as

$$\mathcal{E} = -\frac{d}{dt}(BA\cos\theta) \tag{31.3}$$

From this expression, we see that an emf can be induced in the circuit in several ways:

- The magnitude of $\mathbf{B}$ can change with time.
- The area enclosed by the loop can change with time.
- The angle θ between $\mathbf{B}$ and the normal to the loop can change with time.
- Any combination of the above can occur.

Faraday's law

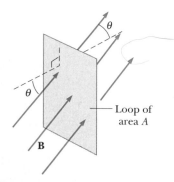

Figure 31.3 A conducting loop that encloses an area A in the presence of a uniform magnetic field $\mathbf{B}$. The angle between $\mathbf{B}$ and the normal to the loop is θ.

Quick Quiz 31.1 A circular loop of wire is held in a uniform magnetic field, with the plane of the loop perpendicular to the field lines. Which of the following will *not* cause a current to be induced in the loop? (a) crushing the loop; (b) rotating the loop about an axis perpendicular to the field lines; (c) keeping the orientation of the loop fixed and moving it along the field lines; (d) pulling the loop out of the field.

Quick Quiz 31.2 Figure 31.4 shows a graphical representation of the field magnitude versus time for a magnetic field that passes through a fixed loop and is oriented perpendicular to the plane of the loop. The magnitude of the magnetic field at any time is uniform over the area of the loop. Rank the magnitudes of the emf generated in the loop at the five instants indicated, from largest to smallest.

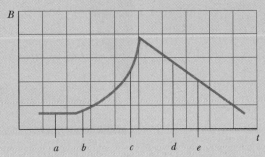

Figure 31.4 (Quick Quiz 31.2) The time behavior of a magnetic field through a loop.

Quick Quiz 31.3 Suppose you would like to steal power for your home from the electric company by placing a loop of wire near a transmission cable, so as to induce an emf in the loop (an illegal procedure). Should you (a) place your loop so that the transmission cable passes through your loop, or (b) simply place your loop near the transmission cable?

Some Applications of Faraday's Law

The ground fault interrupter (GFI) is an interesting safety device that protects users of electrical appliances against electric shock. Its operation makes use of Faraday's law. In the GFI shown in Figure 31.5, wire 1 leads from the wall outlet to the appliance to be protected, and wire 2 leads from the appliance back to the wall outlet. An iron ring surrounds the two wires, and a sensing coil is wrapped around part of the ring. Because the currents in the wires are in opposite directions, the net magnetic flux through the sensing coil due to the currents is zero. However, if the return current in wire 2 changes, the net magnetic flux through the sensing coil is no longer zero. (This can happen, for example, if the appliance becomes wet, enabling current to leak to ground.) Because household current is alternating (meaning that its direction keeps reversing), the magnetic flux through the sensing coil changes with time, inducing an emf in the coil. This induced emf is used to trigger a circuit breaker, which stops the current before it is able to reach a harmful level.

Another interesting application of Faraday's law is the production of sound in an electric guitar (Fig. 31.6). The coil in this case, called the *pickup coil*, is placed near the vibrating guitar string, which is made of a metal that can be magnetized. A permanent

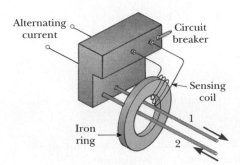

Figure 31.5 Essential components of a ground fault interrupter.

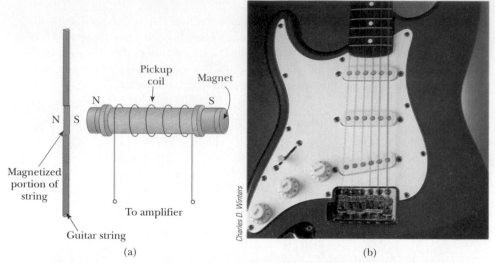

Figure 31.6 (a) In an electric guitar, a vibrating magnetized string induces an emf in a pickup coil. (b) The pickups (the circles beneath the metallic strings) of this electric guitar detect the vibrations of the strings and send this information through an amplifier and into speakers. (A switch on the guitar allows the musician to select which set of six pickups is used.)

magnet inside the coil magnetizes the portion of the string nearest the coil. When the string vibrates at some frequency, its magnetized segment produces a changing magnetic flux through the coil. The changing flux induces an emf in the coil that is fed to an amplifier. The output of the amplifier is sent to the loudspeakers, which produce the sound waves we hear.

Example 31.1   One Way to Induce an emf in a Coil

A coil consists of 200 turns of wire. Each turn is a square of side 18 cm, and a uniform magnetic field directed perpendicular to the plane of the coil is turned on. If the field changes linearly from 0 to 0.50 T in 0.80 s, what is the magnitude of the induced emf in the coil while the field is changing?

Solution The area of one turn of the coil is $(0.18 \text{ m})^2 = 0.032\,4 \text{ m}^2$. The magnetic flux through the coil at $t = 0$ is zero because $B = 0$ at that time. At $t = 0.80$ s, the magnetic flux through one turn is $\Phi_B = BA = (0.50 \text{ T})(0.032\,4 \text{ m}^2) = 0.016\,2 \text{ T} \cdot \text{m}^2$. Therefore, the magnitude of the induced emf is, from Equation 31.2,

$$|\mathcal{E}| = N\frac{\Delta\Phi_B}{\Delta t} = 200 \frac{(0.016\,2 \text{ T} \cdot \text{m}^2 - 0)}{0.80 \text{ s}}$$

$$= 4.1 \text{ T} \cdot \text{m}^2/\text{s} = \boxed{4.1 \text{ V}}$$

You should be able to show that $1 \text{ T} \cdot \text{m}^2/\text{s} = 1 \text{ V}$.

What If? What if you were asked to find the magnitude of the induced current in the coil while the field is changing? Can you answer this question?

Answer If the ends of the coil are not connected to a circuit, the answer to this question is easy—the current is zero! (Charges will move within the wire of the coil, but they cannot move into or out of the ends of the coil.) In order for a steady current to exist, the ends of the coil must be connected to an external circuit. Let us assume that the coil is connected to a circuit and that the total resistance of the coil and the circuit is 2.0 Ω. Then, the current in the coil is

$$I = \frac{\mathcal{E}}{R} = \frac{4.1 \text{ V}}{2.0 \text{ Ω}} = 2.0 \text{ A}$$

Example 31.2   An Exponentially Decaying B Field

A loop of wire enclosing an area A is placed in a region where the magnetic field is perpendicular to the plane of the loop. The magnitude of **B** varies in time according to the expression $B = B_{max}e^{-at}$, where a is some constant. That is, at $t = 0$ the field is B_{max}, and for $t > 0$, the field decreases

exponentially (Fig. 31.7). Find the induced emf in the loop as a function of time.

Solution Because **B** is perpendicular to the plane of the loop, the magnetic flux through the loop at time $t > 0$ is

$$\Phi_B = BA \cos 0 = AB_{max}e^{-at}$$

Because AB_{max} and a are constants, the induced emf calculated from Equation 31.1 is

$$\mathcal{E} = -\frac{d\Phi_B}{dt} = -AB_{max}\frac{d}{dt}e^{-at}$$

$$= aAB_{max}e^{-at}$$

This expression indicates that the induced emf decays exponentially in time. Note that the maximum emf occurs at $t = 0$, where $\mathcal{E}_{max} = aAB_{max}$. The plot of $\mathcal{E}$ versus t is similar to the B-versus-t curve shown in Figure 31.7.

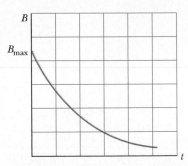

Figure 31.7 (Example 31.2) Exponential decrease in the magnitude of the magnetic field with time. The induced emf and induced current vary with time in the same way.

Conceptual Example 31.3   Which Bulb Is Shorted Out?

Two bulbs are connected to opposite sides of a circular loop of wire, as shown in Figure 31.8a. A changing magnetic field (confined to the smaller circular area shown in the figure) induces an emf in the loop that causes the two bulbs to light. When the switch is closed, the resistance-free wires connected to the switch short out bulb 2 and it goes out. What happens if the wires containing the closed switch remain connected at points a and b, but the switch and the wires are lifted up and moved to the other side of the field, as in Figure 3.18b? The wire is still connected to bulb 2 as it was before, so does it continue to stay dark?

Solution When the wire is moved to the other side, even though the connections have not changed, bulb 1 goes out and bulb 2 glows. The bulb that is shorted depends on which side of the changing field the switch is positioned! In Figure 31.8a, because the branch containing bulb 2 is infinitely more resistant than the branch containing the resistance-free switch, we can imagine removing the branch with the bulb without altering the circuit. Then we have a simple loop containing only bulb 1, which glows.

When the wire is moved, as in Figure 31.8b, there are two possible paths for current below points a and b. We can imagine removing the branch with bulb 1, leaving only a single loop with bulb 2.

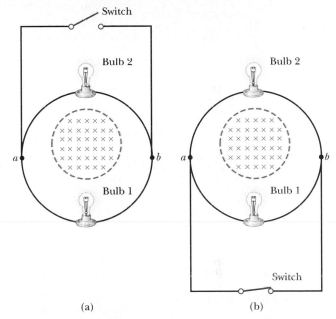

Figure 31.8 (Conceptual Example 31.3) (a) When the wire with the switch is located as shown, bulb 2 goes out when the switch is closed. (b) What happens when the switch and the wires are moved to the other side of the magnetic field?

31.2 Motional emf

In Examples 31.1 and 31.2, we considered cases in which an emf is induced in a stationary circuit placed in a magnetic field when the field changes with time. In this section we describe what is called **motional emf,** which is the emf induced in a conductor moving through a constant magnetic field.

The straight conductor of length ℓ shown in Figure 31.9 is moving through a uniform magnetic field directed into the page. For simplicity, we assume that the conductor is moving in a direction perpendicular to the field with constant velocity under the influence of some external agent. The electrons in the conductor experience a force $\mathbf{F}_B = q\mathbf{v} \times \mathbf{B}$ that is directed along the length ℓ, perpendicular to both $\mathbf{v}$ and $\mathbf{B}$ (Eq. 29.1). Under the influence of this force, the electrons move to the lower end of the conductor and accumulate there, leaving a net positive charge at the upper end. As a result of this charge separation, an electric field $\mathbf{E}$ is produced

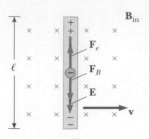

Figure 31.9 A straight electrical conductor of length ℓ moving with a velocity **v** through a uniform magnetic field **B** directed perpendicular to **v**. Due to the magnetic force on electrons, the ends of the conductor become oppositely charged. This establishes an electric field in the conductor. In steady state, the electric and magnetic forces on an electron in the wire are balanced.

inside the conductor. The charges accumulate at both ends until the downward magnetic force qvB on charges remaining in the conductor is balanced by the upward electric force qE. At this point, electrons move only with random thermal motion. The condition for equilibrium requires that

$$qE = qvB \quad \text{or} \quad E = vB$$

The electric field produced in the conductor is related to the potential difference across the ends of the conductor according to the relationship $\Delta V = E\ell$ (Eq. 25.6). Thus, for the equilibrium condition,

$$\Delta V = E\ell = B\ell v \tag{31.4}$$

where the upper end of the conductor in Figure 31.9 is at a higher electric potential than the lower end. Thus, **a potential difference is maintained between the ends of the conductor as long as the conductor continues to move through the uniform magnetic field.** If the direction of the motion is reversed, the polarity of the potential difference is also reversed.

A more interesting situation occurs when the moving conductor is part of a closed conducting path. This situation is particularly useful for illustrating how a changing magnetic flux causes an induced current in a closed circuit. Consider a circuit consisting of a conducting bar of length ℓ sliding along two fixed parallel conducting rails, as shown in Figure 31.10a.

For simplicity, we assume that the bar has zero resistance and that the stationary part of the circuit has a resistance R. A uniform and constant magnetic field **B** is applied perpendicular to the plane of the circuit. As the bar is pulled to the right with a velocity **v** under the influence of an applied force $\mathbf{F}_{app}$, free charges in the bar experience a magnetic force directed along the length of the bar. This force sets up an induced current because the charges are free to move in the closed conducting path. In this case, the rate of change of magnetic flux through the loop and the corresponding induced motional emf across the moving bar are proportional to the change in area of the loop. If the bar is pulled to the right with a constant velocity, the work done by the applied force appears as internal energy in the resistor R. (See Section 27.6.)

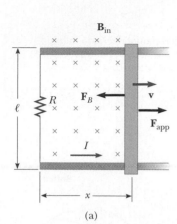

(a)

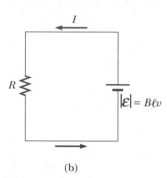

(b)

Active Figure 31.10 (a) A conducting bar sliding with a velocity **v** along two conducting rails under the action of an applied force $\mathbf{F}_{app}$. The magnetic force $\mathbf{F}_B$ opposes the motion, and a counterclockwise current I is induced in the loop. (b) The equivalent circuit diagram for the setup shown in part (a).

At the Active Figures link at http://www.pse6.com, you can adjust the applied force, the magnetic field, and the resistance to see the effects on the motion of the bar.

Because the area enclosed by the circuit at any instant is ℓx, where x is the position of the bar, the magnetic flux through that area is

$$\Phi_B = B\ell x$$

Using Faraday's law, and noting that x changes with time at a rate $dx/dt = v$, we find that the induced motional emf is

$$\mathcal{E} = -\frac{d\Phi_B}{dt} = -\frac{d}{dt}(B\ell x) = -B\ell\frac{dx}{dt}$$

Motional emf

$$\mathcal{E} = -B\ell v \tag{31.5}$$

Because the resistance of the circuit is R, the magnitude of the induced current is

$$I = \frac{|\mathcal{E}|}{R} = \frac{B\ell v}{R} \tag{31.6}$$

The equivalent circuit diagram for this example is shown in Figure 31.10b.

Let us examine the system using energy considerations. Because no battery is in the circuit, we might wonder about the origin of the induced current and the energy delivered to the resistor. We can understand the source of this current and energy by noting that the applied force does work on the conducting bar, thereby moving charges through a magnetic field. Their movement through the field causes the charges to move along the bar with some average drift velocity, and hence a current is established. The change in energy in the system during some time interval must be equal to the transfer of energy into the system by work, consistent with the general principle of conservation of energy described by Equation 7.17.

Let us verify this mathematically. As the bar moves through the uniform magnetic field **B**, it experiences a magnetic force $\mathbf{F}_B$ of magnitude $I\ell B$ (see Section 29.2). The direction of this force is opposite the motion of the bar, to the left in Figure 31.10a. Because the bar moves with constant velocity, the applied force must be equal in magnitude and opposite in direction to the magnetic force, or to the right in Figure 31.10a. (If $\mathbf{F}_B$ acted in the direction of motion, it would cause the bar to accelerate, violating the principle of conservation of energy.) Using Equation 31.6 and the fact that $F_{app} = I\ell B$, we find that the power delivered by the applied force is

$$\mathcal{P} = F_{app}v = (I\ell B)v = \frac{B^2 \ell^2 v^2}{R} = \frac{\mathcal{E}^2}{R} \tag{31.7}$$

From Equation 27.23, we see that this power input is equal to the rate at which energy is delivered to the resistor, so that Equation 7.17 is confirmed in this situation.

Quick Quiz 31.4 As an airplane flies from Los Angeles to Seattle, it passes through the Earth's magnetic field. As a result, a motional emf is developed between the wingtips. Which wingtip is positively charged? (a) the left wing (b) the right wing.

Quick Quiz 31.5 In Figure 31.10, a given applied force of magnitude F_{app} results in a constant speed v and a power input $\mathcal{P}$. Imagine that the force is increased so that the constant speed of the bar is doubled to $2v$. Under these conditions, the new force and the new power input are (a) $2F$ and $2\mathcal{P}$ (b) $4F$ and $2\mathcal{P}$ (c) $2F$ and $4\mathcal{P}$ (d) $4F$ and $4\mathcal{P}$.

Quick Quiz 31.6 You wish to move a rectangular loop of wire into a region of uniform magnetic field at a given speed so as to induce an emf in the loop. The plane of the loop remains perpendicular to the magnetic field lines. In which orientation should you hold the loop while you move it into the region of magnetic field in order to generate the largest emf? (a) with the long dimension of the loop parallel to the velocity vector (b) with the short dimension of the loop parallel to the velocity vector (c) either way—the emf is the same regardless of orientation.

Example 31.4 Motional emf Induced in a Rotating Bar   **Interactive**

A conducting bar of length ℓ rotates with a constant angular speed ω about a pivot at one end. A uniform magnetic field **B** is directed perpendicular to the plane of rotation, as shown in Figure 31.11. Find the motional emf induced between the ends of the bar.

Solution Consider a segment of the bar of length dr having a velocity **v**. According to Equation 31.5, the magnitude of

the emf induced in this segment is

$$d\mathcal{E} = Bv \, dr$$

Because every segment of the bar is moving perpendicular to **B**, an emf $d\mathcal{E}$ of the same form is generated across each segment. Summing the emfs induced across all segments, which are in series, gives the total emf between the ends

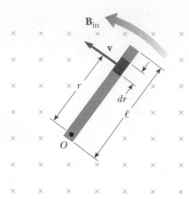

Figure 31.11 (Example 31.4) A conducting bar rotating around a pivot at one end in a uniform magnetic field that is perpendicular to the plane of rotation. A motional emf is induced across the ends of the bar.

of the bar:

$$\mathcal{E} = \int Bv \, dr$$

To integrate this expression, note that the linear speed v of an element is related to the angular speed ω through the relationship $v = r\omega$ (Eq. 10.10). Therefore, because B and ω are constants, we find that

$$\mathcal{E} = B \int v \, dr = B\omega \int_0^\ell r \, dr = \boxed{\tfrac{1}{2} B\omega\ell^2}$$

What If? Suppose, after reading through this example, you come up with a brilliant idea. A Ferris wheel has radial metallic spokes between the hub and the circular rim. These spokes move in the magnetic field of the Earth, so each spoke acts like the bar in Figure 31.11. You plan to use the emf generated by the rotation of the Ferris wheel to power the lightbulbs on the wheel! Will this idea work?

Answer The fact that this is not done in practice suggests that others may have thought of this idea and rejected it. Let us estimate the emf that is generated in this situation. We know the magnitude of the magnetic field of the Earth from Table 29.1, $B = 0.5 \times 10^{-4}$ T. A typical spoke on a Ferris wheel might have a length on the order of 10 m. Suppose the period of rotation is on the order of 10 s. This gives an angular speed of

$$\omega = \frac{2\pi}{T} = \frac{2\pi}{10 \text{ s}} = 0.63 \text{ s}^{-1} \sim 1 \text{ s}^{-1}$$

Assuming that the magnetic field lines of the Earth are horizontal at the location of the Ferris wheel and perpendicular to the spokes, the emf generated is

$$\mathcal{E} = \tfrac{1}{2} B\omega\ell^2 = \tfrac{1}{2} (0.5 \times 10^{-4} \text{ T})(1 \text{ s}^{-1})(10 \text{ m})^2$$

$$= 2.5 \times 10^{-3} \text{ V} \sim 1 \text{ mV}$$

This is a tiny emf, far smaller than that required to operate lightbulbs.

An additional difficulty is related to energy. Assuming you could find lightbulbs that operate using a potential difference on the order of millivolts, a spoke must be part of a circuit in order to provide a voltage to the bulbs. Consequently, the spoke must carry a current. Because this current-carrying spoke is in a magnetic field, a magnetic force is exerted on the spoke and the direction of the force is opposite to its direction of motion. As a result, the motor of the Ferris wheel must supply more energy to perform work against this magnetic drag force. The motor must ultimately provide the energy that is operating the lightbulbs and you have not gained anything for free!

 At the Interactive Worked Example link at http://www.pse6.com, *you can explore the induced emf for different angular speeds and field magnitudes.*

Example 31.5   Magnetic Force Acting on a Sliding Bar   `Interactive`

The conducting bar illustrated in Figure 31.12 moves on two frictionless parallel rails in the presence of a uniform magnetic field directed into the page. The bar has mass m and its length is ℓ. The bar is given an initial velocity $\mathbf{v}_i$ to the right and is released at $t = 0$.

(A) Using Newton's laws, find the velocity of the bar as a function of time.

(B) Show that the same result is reached by using an energy approach.

Solution (A) Conceptualize this situation as follows. As the bar slides to the right in Figure 31.12, a counterclockwise current is established in the circuit consisting of the bar, the rails, and the resistor. The upward current in the bar results in a magnetic force to the left on the bar as shown in the figure. As a result, the bar will slow down, so our mathematical solution should demonstrate this. The text of part (A) already categorizes this as a problem in using Newton's laws. To analyze the problem, we determine from

Equation 29.3 that the magnetic force is $F_B = -I\ell B$, where the negative sign indicates that the retarding force is to the left. Because this is the *only* horizontal force acting on the bar, Newton's second law applied to motion in the

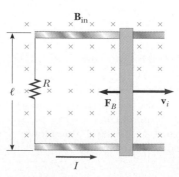

Figure 31.12 (Example 31.5) A conducting bar of length ℓ on two fixed conducting rails is given an initial velocity $\mathbf{v}_i$ to the right.

horizontal direction gives

$$F_x = ma = m\,\frac{dv}{dt} = -I\ell B$$

From Equation 31.6, we know that $I = B\ell v/R$, and so we can write this expression as

$$m\,\frac{dv}{dt} = -\frac{B^2\ell^2}{R}\,v$$

$$\frac{dv}{v} = -\left(\frac{B^2\ell^2}{mR}\right)dt$$

Integrating this equation using the initial condition that $v = v_i$ at $t = 0$, we find that

$$\int_{v_i}^{v}\frac{dv}{v} = \frac{-B^2\ell^2}{mR}\int_0^t dt$$

$$\ln\left(\frac{v}{v_i}\right) = -\left(\frac{B^2\ell^2}{mR}\right)t = -\frac{t}{\tau}$$

where the constant $\tau = mR/B^2\ell^2$. From this result, we see that the velocity can be expressed in the exponential form

$$(1)\qquad v = v_i e^{-t/\tau}$$

To finalize the problem, note that this expression for v indicates that the velocity of the bar decreases with time under the action of the magnetic retarding force, as we expect from our conceptualization of the problem.

(B) The text of part (B) immediately categorizes this as a problem in energy conservation. Consider the sliding bar as one system possessing kinetic energy, which decreases because energy is transferring *out* of the system by electrical transmission through the rails. The resistor is another system possessing internal energy, which rises because energy is transferring *into* this system. Because energy is not leaving the combination of two systems, the rate of energy transfer out of the bar equals the rate of energy transfer into the resistor. Thus,

$$\mathscr{P}_{\text{resistor}} = -\mathscr{P}_{\text{bar}}$$

where the negative sign is necessary because energy is leaving the bar and $\mathscr{P}_{\text{bar}}$ is a negative number. Substituting for the electrical power delivered to the resistor and the

time rate of change of kinetic energy for the bar, we have

$$I^2 R = -\frac{d}{dt}\left(\tfrac{1}{2}\,mv^2\right)$$

Using Equation 31.6 for the current and carrying out the derivative, we find

$$\frac{B^2\ell^2 v^2}{R} = -mv\,\frac{dv}{dt}$$

Rearranging terms gives

$$\frac{dv}{v} = -\left(\frac{B^2\ell^2}{mR}\right)dt$$

To finalize this part of the problem, note that this is the same expression that we obtained in part (A).

What If? Suppose you wished to increase the distance through which the bar moves between the time when it is initially projected and the time when it essentially comes to rest. You can do this by changing one of three variables: v_i, R, or B, by a factor of 2 or $\frac{1}{2}$. Which variable should you change in order to maximize the distance, and would you double it or halve it?

Answer Increasing v_i would make the bar move farther. Increasing R would decrease the current and, therefore, the magnetic force, making the bar move farther. Decreasing B would decrease the magnetic force and make the bar move farther. But which is most effective?

We use Equation (1) to find the distance that the bar moves by integration:

$$v = \frac{dx}{dt} = v_i e^{-t/\tau}$$

$$x = \int_0^{\infty} v_i e^{-t/\tau}\,dt = -v_i\tau e^{-t/\tau}\Big|_0^{\infty}$$

$$= -v_i\tau(0 - 1) = v_i\tau = v_i\left(\frac{mR}{B^2\ell^2}\right)$$

From this expression, we see that doubling v_i or R will double the distance. But changing B by a factor of $\frac{1}{2}$ causes the distance to be four times as great!

 At the Interactive Worked Example link at **http://www.pse6.com,** *you can study the motion of the bar after it is released.*

31.3 Lenz's Law

Faraday's law (Eq. 31.1) indicates that the induced emf and the change in flux have opposite algebraic signs. This has a very real physical interpretation that has come to be known as **Lenz's law**[1]:

> The induced current in a loop is in the direction that creates a magnetic field that opposes the change in magnetic flux through the area enclosed by the loop.

Lenz's law

---

[1]   Developed by the German physicist Heinrich Lenz (1804–1865).

That is, the induced current tends to keep the original magnetic flux through the circuit from changing. We shall show that this law is a consequence of the law of conservation of energy.

To understand Lenz's law, let us return to the example of a bar moving to the right on two parallel rails in the presence of a uniform magnetic field (the *external* magnetic field, Fig. 31.13a.) As the bar moves to the right, the magnetic flux through the area enclosed by the circuit increases with time because the area increases. Lenz's law states that the induced current must be directed so that the magnetic field it produces opposes the change in the external magnetic flux. Because the magnetic flux due to an external field directed into the page is increasing, the induced current, if it is to oppose this change, must produce a field directed out of the page. Hence, the induced current must be directed counterclockwise when the bar moves to the right. (Use the right-hand rule to verify this direction.) If the bar is moving to the left, as in Figure 31.13b, the external magnetic flux through the area enclosed by the loop decreases with time. Because the field is directed into the page, the direction of the induced current must be clockwise if it is to produce a field that also is directed into the page. In either case, the induced current tends to maintain the original flux through the area enclosed by the current loop.

Let us examine this situation using energy considerations. Suppose that the bar is given a slight push to the right. In the preceding analysis, we found that this motion sets up a counterclockwise current in the loop. What happens if we assume that the

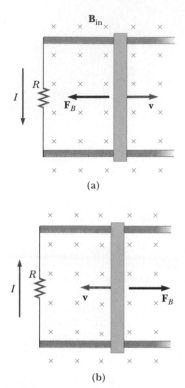

(a)

(b)

Figure 31.13 (a) As the conducting bar slides on the two fixed conducting rails, the magnetic flux due to the external magnetic field into the page through the area enclosed by the loop increases in time. By Lenz's law, the induced current must be counterclockwise so as to produce a counteracting magnetic field directed out of the page. (b) When the bar moves to the left, the induced current must be clockwise. Why?

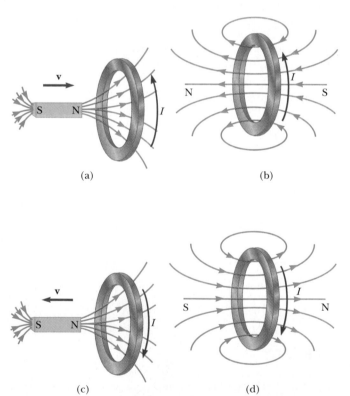

(a)   (b)

(c)   (d)

Figure 31.14 (a) When the magnet is moved toward the stationary conducting loop, a current is induced in the direction shown. The magnetic field lines shown are those due to the bar magnet. (b) This induced current produces its own magnetic field directed to the left that counteracts the increasing external flux. The magnetic field lines shown are those due to the induced current in the ring. (c) When the magnet is moved away from the stationary conducting loop, a current is induced in the direction shown. The magnetic field lines shown are those due to the bar magnet. (d) This induced current produces a magnetic field directed to the right and so counteracts the decreasing external flux. The magnetic field lines shown are those due to the induced current in the ring.

current is clockwise, such that the direction of the magnetic force exerted on the bar is to the right? This force would accelerate the rod and increase its velocity. This, in turn, would cause the area enclosed by the loop to increase more rapidly; this would result in an increase in the induced current, which would cause an increase in the force, which would produce an increase in the current, and so on. In effect, the system would acquire energy with no input of energy. This is clearly inconsistent with all experience and violates the law of conservation of energy. Thus, we are forced to conclude that the current must be counterclockwise.

Let us consider another situation, one in which a bar magnet moves toward a stationary metal loop, as in Figure 31.14a. As the magnet moves to the right toward the loop, the external magnetic flux through the loop increases with time. To counteract this increase in flux due to a field toward the right, the induced current produces its own magnetic field to the left, as illustrated in Figure 31.14b; hence, the induced current is in the direction shown. Knowing that like magnetic poles repel each other, we conclude that the left face of the current loop acts like a north pole and that the right face acts like a south pole.

If the magnet moves to the left, as in Figure 31.14c, its flux through the area enclosed by the loop decreases in time. Now the induced current in the loop is in the direction shown in Figure 31.14d because this current direction produces a magnetic field in the same direction as the external field. In this case, the left face of the loop is a south pole and the right face is a north pole.

Quick Quiz 31.7   Figure 31.15 shows a magnet being moved in the vicinity of a solenoid connected to a sensitive ammeter. The south pole of the magnet is the pole nearest the solenoid, and the ammeter indicates a clockwise (viewed from above) current in the solenoid. Is the person (a) inserting the magnet or (b) pulling it out?

Quick Quiz 31.8   Figure 31.16 shows a circular loop of wire being dropped toward a wire carrying a current to the left. The direction of the induced current in the loop of wire is (a) clockwise (b) counterclockwise (c) zero (d) impossible to determine.

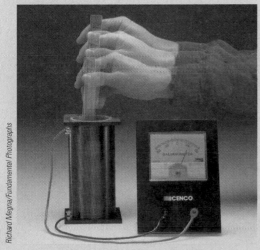

Richard Megna/Fundamental Photographs

Figure 31.15 (Quick Quiz 31.7)

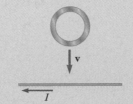

Figure 31.16 (Quick Quiz 31.8)

Conceptual Example 31.6   Application of Lenz's Law

A metal ring is placed near a solenoid, as shown in Figure 31.17a. Find the direction of the induced current in the ring

(A) at the instant the switch in the circuit containing the solenoid is thrown closed,

(B) after the switch has been closed for several seconds, and

(C) at the instant the switch is thrown open.

Solution (A) At the instant the switch is thrown closed, the situation changes from one in which no magnetic flux exists in the ring to one in which flux exists and the magnetic field is to the left as shown in Figure 31.17b. To counteract this change in the flux, the current induced in the ring must set up a magnetic field directed from left to right in Figure 31.17b. This requires a current directed as shown.

(B) After the switch has been closed for several seconds, no change in the magnetic flux through the loop occurs; hence, the induced current in the ring is zero.

(C) Opening the switch changes the situation from one in which magnetic flux exists in the ring to one in which there is no magnetic flux. The direction of the induced current is as shown in Figure 31.17c because current in this direction

produces a magnetic field that is directed right to left and so counteracts the decrease in the flux produced by the solenoid.

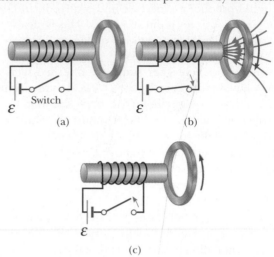

Figure 31.17 (Example 31.6) A current is induced in a metal ring near a solenoid when the switch is opened or thrown closed.

Conceptual Example 31.7   A Loop Moving Through a Magnetic Field

A rectangular metallic loop of dimensions ℓ and w and resistance R moves with constant speed v to the right, as in Figure 31.18a. The loop passes through a uniform magnetic field **B** directed into the page and extending a distance $3w$ along the x axis. Defining x as the position of the right side of the loop along the x axis, plot as functions of x

(A) the magnetic flux through the area enclosed by the loop,

(B) the induced motional emf, and

(C) the external applied force necessary to counter the magnetic force and keep v constant.

Solution (A) Figure 31.18b shows the flux through the area enclosed by the loop as a function of x. Before the loop enters the field, the flux is zero. As the loop enters the field, the flux increases linearly with position until the left edge of

the loop is just inside the field. Finally, the flux through the loop decreases linearly to zero as the loop leaves the field.

(B) Before the loop enters the field, no motional emf is induced in it because no field is present (Fig. 31.18c). As the right side of the loop enters the field, the magnetic flux directed into the page increases. Hence, according to Lenz's law, the induced current is counterclockwise because it must produce its own magnetic field directed out of the page. The motional emf $-B\ell v$ (from Eq. 31.5) arises from the magnetic force experienced by charges in the right side of the loop. When the loop is entirely in the field, the change in magnetic flux is zero, and hence the motional emf vanishes. This happens because, once the left side of the loop enters the field, the motional emf induced in it cancels the motional emf present in the right side of the loop. As the right side of the loop leaves the field, the flux begins to decrease, a clockwise

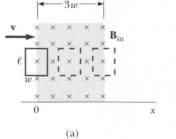

(a)

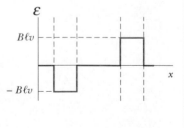

(c)

Figure 31.18 (Conceptual Example 31.7) (a) A conducting rectangular loop of width w and length ℓ moving with a velocity **v** through a uniform magnetic field extending a distance $3w$. (b) Magnetic flux through the area enclosed by the loop as a function of loop position. (c) Induced emf as a function of loop position. (d) Applied force required for constant velocity as a function of loop position.

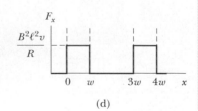

(b)   (d)

current is induced, and the induced emf is $B\ell v$. As soon as the left side leaves the field, the emf decreases to zero.

(C) The external force that must be applied to the loop to maintain this motion is plotted in Figure 31.18d. Before the loop enters the field, no magnetic force acts on it; hence, the applied force must be zero if v is constant. When the right side of the loop enters the field, the applied force necessary to maintain constant speed must be equal in magnitude and opposite in direction to the magnetic force exerted on that side. When the loop is entirely in the field, the flux through the loop is not changing with time. Hence, the net emf induced in the loop is zero, and the current also is zero. Therefore, no external force is needed to maintain the motion. Finally, as the right side leaves the field, the applied force must be equal in magnitude and opposite in direction to the magnetic force acting on the left side of the loop.

From this analysis, we conclude that power is supplied only when the loop is either entering or leaving the field. Furthermore, this example shows that the motional emf induced in the loop can be zero even when there is motion through the field! A motional emf is induced *only* when the magnetic flux through the loop *changes in time*.

31.4 Induced emf and Electric Fields

We have seen that a changing magnetic flux induces an emf and a current in a conducting loop. In our study of electricity, we related a current to an electric field that applies electric forces on charged particles. In the same way, we can relate an induced current in a conducting loop to an electric field by claiming that **an electric field is created in the conductor as a result of the changing magnetic flux.**

We also noted in our study of electricity that the existence of an electric field is independent of the presence of any test charges. This suggests that even in the absence of a conducting loop, a changing magnetic field would still generate an electric field in empty space.

This induced electric field is *nonconservative*, unlike the electrostatic field produced by stationary charges. We can illustrate this point by considering a conducting loop of radius r situated in a uniform magnetic field that is perpendicular to the plane of the loop, as in Figure 31.19. If the magnetic field changes with time, then, according to Faraday's law (Eq. 31.1), an emf $\mathcal{E} = -d\Phi_B/dt$ is induced in the loop. The induction of a current in the loop implies the presence of an induced electric field **E**, which must be tangent to the loop because this is the direction in which the charges in the wire move in response to the electric force. The work done by the electric field in moving a test charge q once around the loop is equal to $q\mathcal{E}$. Because the electric force acting on the charge is $q\mathbf{E}$, the work done by the electric field in moving the charge once around the loop is $qE(2\pi r)$, where $2\pi r$ is the circumference of the loop. These two expressions for the work done must be equal; therefore, we see that

$$q\mathcal{E} = qE(2\pi r)$$

$$E = \frac{\mathcal{E}}{2\pi r}$$

Using this result, along with Equation 31.1 and the fact that $\Phi_B = BA = \pi r^2 B$ for a circular loop, we find that the induced electric field can be expressed as

$$E = -\frac{1}{2\pi r}\frac{d\Phi_B}{dt} = -\frac{r}{2}\frac{dB}{dt} \tag{31.8}$$

If the time variation of the magnetic field is specified, we can easily calculate the induced electric field from Equation 31.8.

The emf for any closed path can be expressed as the line integral of $\mathbf{E} \cdot d\mathbf{s}$ over that path: $\mathcal{E} = \oint \mathbf{E} \cdot d\mathbf{s}$. In more general cases, E may not be constant, and the path may not be a circle. Hence, Faraday's law of induction, $\mathcal{E} = -d\Phi_B/dt$, can be written in the general form

$$\oint \mathbf{E} \cdot d\mathbf{s} = -\frac{d\Phi_B}{dt} \tag{31.9}$$

The induced electric field E in Equation 31.9 is a nonconservative field that is generated by a changing magnetic field. The field **E** that satisfies Equation 31.9

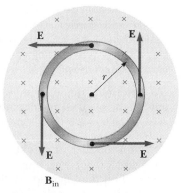

Figure 31.19 A conducting loop of radius r in a uniform magnetic field perpendicular to the plane of the loop. If **B** changes in time, an electric field is induced in a direction tangent to the circumference of the loop.

▲ **PITFALL PREVENTION**

31.2 Induced Electric Fields

The changing magnetic field does *not* need to be in existence at the location of the induced electric field. In Figure 31.19, even a loop outside the region of magnetic field will experience an induced electric field. For another example, consider Figure 31.8. The light bulbs glow (if the switch is open) even though the wires are outside the region of the magnetic field.

Faraday's law in general form

cannot possibly be an electrostatic field because if the field were electrostatic, and hence conservative, the line integral of $\mathbf{E} \cdot d\mathbf{s}$ over a closed loop would be zero (Section 25.1); this would be in contradiction to Equation 31.9.

Quick Quiz 31.9 In a region of space, the magnetic field increases at a constant rate. This changing magnetic field induces an electric field that (a) increases in time (b) is conservative (c) is in the direction of the magnetic field (d) has a constant magnitude.

Example 31.8   Electric Field Induced by a Changing Magnetic Field in a Solenoid

A long solenoid of radius R has n turns of wire per unit length and carries a time-varying current that varies sinusoidally as $I = I_{max} \cos \omega t$, where I_{max} is the maximum current and ω is the angular frequency of the alternating current source (Fig. 31.20).

(A) Determine the magnitude of the induced electric field outside the solenoid at a distance $r > R$ from its long central axis.

Solution First let us consider an external point and take the path for our line integral to be a circle of radius r centered on the solenoid, as illustrated in Figure 31.20. By symmetry we see that the magnitude of $\mathbf{E}$ is constant on this path and that $\mathbf{E}$ is tangent to it. The magnetic flux through the area enclosed by this path is $BA = B\pi R^2$; hence, Equation 31.9 gives

$$\oint \mathbf{E} \cdot d\mathbf{s} = -\frac{d}{dt}(B\pi R^2) = -\pi R^2 \frac{dB}{dt}$$

$$(1) \qquad \oint \mathbf{E} \cdot d\mathbf{s} = E(2\pi r) = -\pi R^2 \frac{dB}{dt}$$

The magnetic field inside a long solenoid is given by Equation 30.17, $B = \mu_0 nI$. When we substitute the expression $I = I_{max} \cos \omega t$ into this equation for B and then substitute the result into Equation (1), we find that

$$E(2\pi r) = -\pi R^2 \mu_0 nI_{max} \frac{d}{dt}(\cos \omega t)$$

$$= \pi R^2 \mu_0 nI_{max} \omega \sin \omega t$$

$$(2) \qquad E = \frac{\mu_0 nI_{max}\omega R^2}{2r} \sin \omega t \qquad \text{(for } r > R)$$

Hence, the amplitude of the electric field outside the solenoid falls off as $1/r$ and varies sinusoidally with time.

(B) What is the magnitude of the induced electric field inside the solenoid, a distance r from its axis?

Solution For an interior point $(r < R)$, the flux through an integration loop is given by $B\pi r^2$. Using the same procedure as in part (A), we find that

$$E(2\pi r) = -\pi r^2 \frac{dB}{dt} = \pi r^2 \mu_0 nI_{max} \omega \sin \omega t$$

$$(3) \qquad E = \frac{\mu_0 nI_{max}\omega}{2} r \sin \omega t \qquad \text{(for } r < R)$$

This shows that the amplitude of the electric field induced inside the solenoid by the changing magnetic flux through the solenoid increases linearly with r and varies sinusoidally with time.

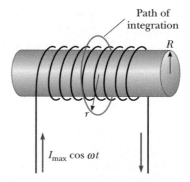

Figure 31.20 (Example 31.8) A long solenoid carrying a time-varying current given by $I = I_{max} \cos \omega t$. An electric field is induced both inside and outside the solenoid.

31.5 Generators and Motors

Electric generators take in energy by work and transfer it out by electrical transmission. To understand how they operate, let us consider the **alternating current** (AC) **generator.** In its simplest form, it consists of a loop of wire rotated by some external means in a magnetic field (Fig. 31.21a).

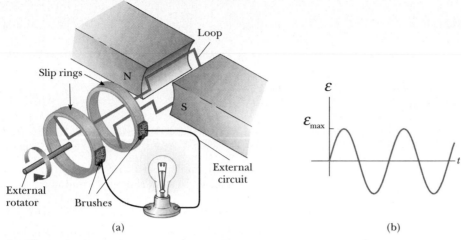

Active Figure 31.21 (a) Schematic diagram of an AC generator. An emf is induced in a loop that rotates in a magnetic field. (b) The alternating emf induced in the loop plotted as a function of time.

At the Active Figures link at http://www.pse6.com, you can adjust the speed of rotation and the strength of the field to see the effects on the emf generated.

In commercial power plants, the energy required to rotate the loop can be derived from a variety of sources. For example, in a hydroelectric plant, falling water directed against the blades of a turbine produces the rotary motion; in a coal-fired plant, the energy released by burning coal is used to convert water to steam, and this steam is directed against the turbine blades. As a loop rotates in a magnetic field, the magnetic flux through the area enclosed by the loop changes with time; this induces an emf and a current in the loop according to Faraday's law. The ends of the loop are connected to slip rings that rotate with the loop. Connections from these slip rings, which act as output terminals of the generator, to the external circuit are made by stationary brushes in contact with the slip rings.

Suppose that, instead of a single turn, the loop has N turns (a more practical situation), all of the same area A, and rotates in a magnetic field with a constant angular speed ω. If θ is the angle between the magnetic field and the normal to the plane of the loop, as in Figure 31.22, then the magnetic flux through the loop at any time t is

$$\Phi_B = BA \cos \theta = BA \cos \omega t$$

where we have used the relationship $\theta = \omega t$ between angular position and angular speed (see Eq. 10.3). (We have set the clock so that $t = 0$ when $\theta = 0$.) Hence, the induced emf in the coil is

$$\mathcal{E} = -N \frac{d\Phi_B}{dt} = -NAB \frac{d}{dt} (\cos \omega t) = NAB\omega \sin \omega t \qquad (31.10)$$

This result shows that the emf varies sinusoidally with time, as plotted in Figure 31.21b. From Equation 31.10 we see that the maximum emf has the value

$$\mathcal{E}_{max} = NAB\omega \qquad (31.11)$$

which occurs when $\omega t = 90°$ or $270°$. In other words, $\mathcal{E} = \mathcal{E}_{max}$ when the magnetic field is in the plane of the coil and the time rate of change of flux is a maximum. Furthermore, the emf is zero when $\omega t = 0$ or $180°$, that is, when **B** is perpendicular to the plane of the coil and the time rate of change of flux is zero.

The frequency for commercial generators in the United States and Canada is 60 Hz, whereas in some European countries it is 50 Hz. (Recall that $\omega = 2\pi f$, where f is the frequency in hertz.)

Figure 31.22 A loop enclosing an area A and containing N turns, rotating with constant angular speed ω in a magnetic field. The emf induced in the loop varies sinusoidally in time.

Quick Quiz 31.10 In an AC generator, a coil with N turns of wire spins in a magnetic field. Of the following choices, which will *not* cause an increase in the emf generated in the coil? (a) replacing the coil wire with one of lower resistance (b) spinning the coil faster (c) increasing the magnetic field (d) increasing the number of turns of wire on the coil.

Example 31.9 emf Induced in a Generator

An AC generator consists of 8 turns of wire, each of area $A = 0.090\ 0\ \text{m}^2$, and the total resistance of the wire is 12.0 Ω. The loop rotates in a 0.500-T magnetic field at a constant frequency of 60.0 Hz.

(A) Find the maximum induced emf.

Solution First, note that $\omega = 2\pi f = 2\pi(60.0\ \text{Hz}) = 377\ \text{s}^{-1}$. Thus, Equation 31.11 gives

$$\mathcal{E}_{max} = NAB\omega = 8(0.090\ 0\ \text{m}^2)(0.500\ \text{T})(377\ \text{s}^{-1})$$

$$= \boxed{136\ \text{V}}$$

(B) What is the maximum induced current when the output terminals are connected to a low-resistance conductor?

Solution From Equation 27.8 and the results to part (A), we have

$$I_{max} = \frac{\mathcal{E}_{max}}{R} = \frac{136\ \text{V}}{12.0\ \Omega} = \boxed{11.3\ \text{A}}$$

The **direct current** (DC) **generator** is illustrated in Figure 31.23a. Such generators are used, for instance, in older cars to charge the storage batteries. The components are essentially the same as those of the AC generator except that the contacts to the rotating loop are made using a split ring called a *commutator*.

In this configuration, the output voltage always has the same polarity and pulsates with time, as shown in Figure 31.23b. We can understand the reason for this by noting that the contacts to the split ring reverse their roles every half cycle. At the same time, the polarity of the induced emf reverses; hence, the polarity of the split ring (which is the same as the polarity of the output voltage) remains the same.

A pulsating DC current is not suitable for most applications. To obtain a more steady DC current, commercial DC generators use many coils and commutators distributed so that the sinusoidal pulses from the various coils are out of phase. When these pulses are superimposed, the DC output is almost free of fluctuations.

Motors are devices into which energy is transferred by electrical transmission while energy is transferred out by work. Essentially, a motor is a generator operating

At the Active Figures link at http://www.pse6.com, you can adjust the speed of rotation and the strength of the field to see the effects on the emf generated.

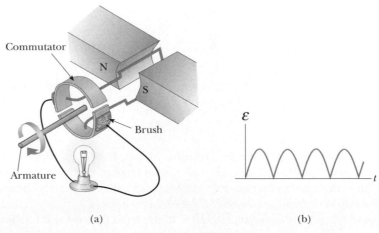

(a)   (b)

Active Figure 31.23 (a) Schematic diagram of a DC generator. (b) The magnitude of the emf varies in time but the polarity never changes.

in reverse. Instead of generating a current by rotating a coil, a current is supplied to the coil by a battery, and the torque acting on the current-carrying coil causes it to rotate.

Useful mechanical work can be done by attaching the rotating coil to some external device. However, as the coil rotates in a magnetic field, the changing magnetic flux induces an emf in the coil; this induced emf always acts to reduce the current in the coil. If this were not the case, Lenz's law would be violated. The back emf increases in magnitude as the rotational speed of the coil increases. (The phrase *back emf* is used to indicate an emf that tends to reduce the supplied current.) Because the voltage available to supply current equals the difference between the supply voltage and the back emf, the current in the rotating coil is limited by the back emf.

When a motor is turned on, there is initially no back emf; thus, the current is very large because it is limited only by the resistance of the coil. As the coil begins to rotate, the induced back emf opposes the applied voltage, and the current in the coil is reduced. If the mechanical load increases, the motor slows down; this causes the back emf to decrease. This reduction in the back emf increases the current in the coil and therefore also increases the power needed from the external voltage source. For this reason, the power requirements for starting a motor and for running it are greater for heavy loads than for light ones. If the motor is allowed to run under no mechanical load, the back emf reduces the current to a value just large enough to overcome energy losses due to internal energy and friction. If a very heavy load jams the motor so that it cannot rotate, the lack of a back emf can lead to dangerously high current in the motor's wire. This is a dangerous situation, and is explored in the **What If?** section of Example 31.10.

A current application of motors in automobiles is seen in the development of *hybrid drive systems*. In these automobiles, a gasoline engine and an electric motor are combined to increase the fuel economy of the vehicle and reduce its emissions. Figure 31.24 shows the engine compartment of the Toyota Prius, which is one of a small number of hybrids available in the United States. In this automobile, power to the wheels can come from either the gasoline engine or the electric motor. In normal driving, the electric motor accelerates the vehicle from rest until it is moving at a speed of about 15 mi/h (24 km/h). During this acceleration period, the engine is not running, so that gasoline is not used and there is no emission. When a hybrid vehicle brakes, the motor acts as a generator and returns some of the kinetic energy of the vehicle back to the battery as stored energy. In a normal vehicle, this kinetic energy is simply lost as it is transformed to internal energy in the brakes and roadway.

Figure 31.24 The engine compartment of the Toyota Prius, a hybrid vehicle.

Example 31.10 **The Induced Current in a Motor**

Assume that a motor in which the coil has a total resistance of 10 Ω is supplied by a voltage of 120 V. When the motor is running at its maximum speed, the back emf is 70 V. Find the current in the coil

(A) when the motor is turned on and

(B) when it has reached maximum speed.

Solution (A) When the motor is first turned on, the back emf is zero (because the coil is motionless). Thus, the current in the coil is a maximum and equal to

$$I = \frac{\mathcal{E}}{R} = \frac{120 \text{ V}}{10 \text{ }\Omega} = \boxed{12 \text{ A}}$$

(B) At the maximum speed, the back emf has its maximum value. Thus, the effective supply voltage is that of the external source minus the back emf. Hence, the current is reduced to

$$I = \frac{\mathcal{E} - \mathcal{E}_{\text{back}}}{R} = \frac{120 \text{ V} - 70 \text{ V}}{10 \text{ }\Omega} = \frac{50 \text{ V}}{10 \text{ }\Omega} = \boxed{5.0 \text{ A}}$$

What If? Suppose that this motor is in a circular saw. You are operating the saw and the blade becomes jammed in a piece of wood so that the motor cannot turn. By what percentage does the power input to the motor increase when it is jammed?

Answer You may have everyday experiences with motors becoming warm when they are prevented from turning. This is due to the increased power input to the motor. The higher rate of energy transfer results in an increase in the internal energy of the coil, an undesirable effect. When the motor is jammed, the current is that given in part (A). Let us set up the ratio of power input to the motor when jammed to that when it is not jammed:

$$\frac{\mathcal{P}_{\text{jammed}}}{\mathcal{P}_{\text{not jammed}}} = \frac{I_{(A)}^2 R}{I_{(B)}^2 R} = \frac{I_{(A)}^2}{I_{(B)}^2}$$

where the subscripts (A) and (B) refer to the currents in parts (A) and (B) of the example. Substituting these values,

$$\frac{\mathcal{P}_{\text{jammed}}}{\mathcal{P}_{\text{not jammed}}} = \frac{(12 \text{ A})^2}{(5.0 \text{ A})^2} = 5.76$$

This represents a 476% increase in the input power! Such a high power input can cause the coil to become so hot that it is damaged.

31.6 Eddy Currents

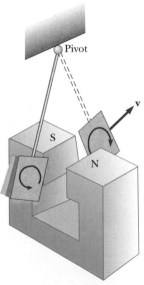

Figure 31.25 Formation of eddy currents in a conducting plate moving through a magnetic field. As the plate enters or leaves the field, the changing magnetic flux induces an emf, which causes eddy currents in the plate.

As we have seen, an emf and a current are induced in a circuit by a changing magnetic flux. In the same manner, circulating currents called **eddy currents** are induced in bulk pieces of metal moving through a magnetic field. This can easily be demonstrated by allowing a flat copper or aluminum plate attached at the end of a rigid bar to swing back and forth through a magnetic field (Fig. 31.25). As the plate enters the field, the changing magnetic flux induces an emf in the plate, which in turn causes the free electrons in the plate to move, producing the swirling eddy currents. According to Lenz's law, the direction of the eddy currents is such that they create magnetic fields that oppose the change that causes the currents. For this reason, the eddy currents must produce effective magnetic poles on the plate, which are repelled by the poles of the magnet; this gives rise to a repulsive force that opposes the motion of the plate. (If the opposite were true, the plate would accelerate and its energy would increase after each swing, in violation of the law of conservation of energy.)

As indicated in Figure 31.26a, with **B** directed into the page, the induced eddy current is counterclockwise as the swinging plate enters the field at position 1. This is because the flux due to the external magnetic field into the page through the plate is increasing, and hence by Lenz's law the induced current must provide its own magnetic field out of the page. The opposite is true as the plate leaves the field at position 2, where the current is clockwise. Because the induced eddy current always produces a magnetic retarding force $\mathbf{F}_B$ when the plate enters or leaves the field, the swinging plate eventually comes to rest.

If slots are cut in the plate, as shown in Figure 31.26b, the eddy currents and the corresponding retarding force are greatly reduced. We can understand this

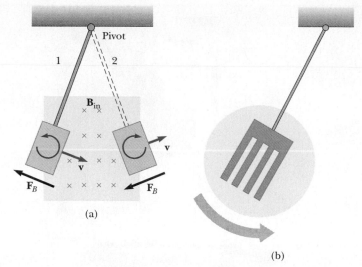

Active Figure 31.26 (a) As the conducting plate enters the field (position 1), the eddy currents are counterclockwise. As the plate leaves the field (position 2), the currents are clockwise. In either case, the force on the plate is opposite the velocity, and eventually the plate comes to rest. (b) When slots are cut in the conducting plate, the eddy currents are reduced and the plate swings more freely through the magnetic field.

Choose to let a solid or a slotted plate swing through the magnetic field and observe the effect at the Active Figures link at http://www.pse6.com.

by realizing that the cuts in the plate prevent the formation of any large current loops.

The braking systems on many subway and rapid-transit cars make use of electromagnetic induction and eddy currents. An electromagnet attached to the train is positioned near the steel rails. (An electromagnet is essentially a solenoid with an iron core.) The braking action occurs when a large current is passed through the electromagnet. The relative motion of the magnet and rails induces eddy currents in the rails, and the direction of these currents produces a drag force on the moving train. Because the eddy currents decrease steadily in magnitude as the train slows down, the braking effect is quite smooth. As a safety measure, some power tools use eddy currents to stop rapidly spinning blades once the device is turned off.

Eddy currents are often undesirable because they represent a transformation of mechanical energy to internal energy. To reduce this energy loss, conducting parts are often laminated—that is, they are built up in thin layers separated by a nonconducting material such as lacquer or a metal oxide. This layered structure increases the resistance of eddy current paths and effectively confines the currents to individual layers. Such a laminated structure is used in transformer cores (see Section 33.8) and motors to minimize eddy currents and thereby increase the efficiency of these devices.

Quick Quiz 31.11 In equal-arm balances from the early twentieth century (Fig. 31.27), it is sometimes observed that an aluminum sheet hangs from one of the arms and passes between the poles of a magnet. This causes the oscillations of the equal-arm balance to decay rapidly. In the absence of such magnetic braking, the oscillation might continue for a very long time, so that the experimenter would have to wait to take a reading. The oscillations decay because (a) the aluminum sheet is attracted to the magnet; (b) currents in the aluminum sheet set up a magnetic field that opposes the oscillations; (c) aluminum is paramagnetic.

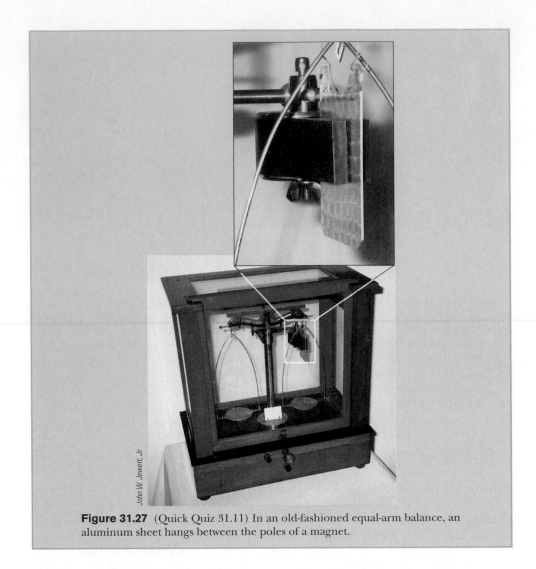

John W. Jewett, Jr.

Figure 31.27 (Quick Quiz 31.11) In an old-fashioned equal-arm balance, an aluminum sheet hangs between the poles of a magnet.

31.7 Maxwell's Equations

We conclude this chapter by presenting four equations that are regarded as the basis of all electrical and magnetic phenomena. These equations, developed by James Clerk Maxwell, are as fundamental to electromagnetic phenomena as Newton's laws are to mechanical phenomena. In fact, the theory that Maxwell developed was more far-reaching than even he imagined because it turned out to be in agreement with the special theory of relativity, as Einstein showed in 1905.

Maxwell's equations represent the laws of electricity and magnetism that we have already discussed, but they have additional important consequences. In Chapter 34 we shall show that these equations predict the existence of electromagnetic waves (traveling patterns of electric and magnetic fields), which travel with a speed $c = 1/\sqrt{\mu_0 \epsilon_0} = 3.00 \times 10^8$ m/s, the speed of light. Furthermore, the theory shows that such waves are radiated by accelerating charges.

For simplicity, we present **Maxwell's equations** as applied to free space, that is, in the absence of any dielectric or magnetic material. The four equations are

Gauss's law

$$\oint_S \mathbf{E} \cdot d\mathbf{A} = \frac{q}{\epsilon_0}$$

(31.12)

$$\oint_S \mathbf{B} \cdot d\mathbf{A} = 0 \qquad (31.13)$$ **Gauss's law in magnetism**

$$\oint \mathbf{E} \cdot d\mathbf{s} = -\frac{d\Phi_B}{dt} \qquad (31.14)$$ **Faraday's law**

$$\oint \mathbf{B} \cdot d\mathbf{s} = \mu_0 I + \epsilon_0 \mu_0 \frac{d\Phi_E}{dt} \qquad (31.15)$$ **Ampère–Maxwell law**

Equation 31.12 is Gauss's law: **the total electric flux through any closed surface equals the net charge inside that surface divided by ϵ_0.** This law relates an electric field to the charge distribution that creates it.

Equation 31.13, which can be considered Gauss's law in magnetism, states that **the net magnetic flux through a closed surface is zero.** That is, the number of magnetic field lines that enter a closed volume must equal the number that leave that volume. This implies that magnetic field lines cannot begin or end at any point. If they did, it would mean that isolated magnetic monopoles existed at those points. The fact that isolated magnetic monopoles have not been observed in nature can be taken as a confirmation of Equation 31.13.

Equation 31.14 is Faraday's law of induction, which describes the creation of an electric field by a changing magnetic flux. This law states that **the emf, which is the line integral of the electric field around any closed path, equals the rate of change of magnetic flux through any surface area bounded by that path.** One consequence of Faraday's law is the current induced in a conducting loop placed in a time-varying magnetic field.

Equation 31.15, usually called the Ampère–Maxwell law, is the generalized form of Ampère's law, and describes the creation of a magnetic field by an electric field and electric currents. **the line integral of the magnetic field around any closed path is the sum of μ_0 times the net current through that path and $\epsilon_0\mu_0$ times the rate of change of electric flux through any surface bounded by that path.**

Once the electric and magnetic fields are known at some point in space, the force acting on a particle of charge q can be calculated from the expression

$$\mathbf{F} = q\mathbf{E} + q\mathbf{v} \times \mathbf{B} \qquad (31.16)$$ **The Lorentz force law**

This relationship is called the **Lorentz force law.** (We saw this relationship earlier as Equation 29.16.) Maxwell's equations, together with this force law, completely describe all classical electromagnetic interactions.

It is interesting to note the symmetry of Maxwell's equations. Equations 31.12 and 31.13 are symmetric, apart from the absence of the term for magnetic monopoles in Equation 31.13. Furthermore, Equations 31.14 and 31.15 are symmetric in that the line integrals of **E** and **B** around a closed path are related to the rate of change of magnetic flux and electric flux, respectively. Maxwell's equations are of fundamental importance not only to electromagnetism but to all of science. Heinrich Hertz once wrote, "One cannot escape the feeling that these mathematical formulas have an independent existence and an intelligence of their own, that they are wiser than we are, wiser even than their discoverers, that we get more out of them than we put into them."

SUMMARY

Faraday's law of induction states that the emf induced in a circuit is directly proportional to the time rate of change of magnetic flux through the circuit:

$$\mathcal{E} = -\frac{d\Phi_B}{dt} \qquad (31.1)$$

where $\Phi_B = \oint \mathbf{B} \cdot d\mathbf{A}$ is the magnetic flux.

Take a practice test for this chapter by clicking on the Practice Test link at http://www.pse6.com.

When a conducting bar of length ℓ moves at a velocity $\mathbf{v}$ through a magnetic field $\mathbf{B}$, where $\mathbf{B}$ is perpendicular to the bar and to $\mathbf{v}$, the **motional emf** induced in the bar is

$$\mathcal{E} = -B\ell v \tag{31.5}$$

Lenz's law states that the induced current and induced emf in a conductor are in such a direction as to set up a magnetic field that opposes the change that produced them.

A general form of **Faraday's law of induction** is

$$\mathcal{E} = \oint \mathbf{E} \cdot d\mathbf{s} = -\frac{d\Phi_B}{dt} \tag{31.9}$$

where $\mathbf{E}$ is the nonconservative electric field that is produced by the changing magnetic flux.

When used with the Lorentz force law, $\mathbf{F} = q\mathbf{E} + q\mathbf{v} \times \mathbf{B}$, **Maxwell's equations** describe all electromagnetic phenomena:

$$\oint_S \mathbf{E} \cdot d\mathbf{A} = \frac{q}{\epsilon_0} \tag{31.12}$$

$$\oint_S \mathbf{B} \cdot d\mathbf{A} = 0 \tag{31.13}$$

$$\oint \mathbf{E} \cdot d\mathbf{s} = -\frac{d\Phi_B}{dt} \tag{31.14}$$

$$\oint \mathbf{B} \cdot d\mathbf{s} = \mu_0 I + \epsilon_0 \mu_0 \frac{d\Phi_E}{dt} \tag{31.15}$$

QUESTIONS

1. What is the difference between magnetic flux and magnetic field?

2. A loop of wire is placed in a uniform magnetic field. For what orientation of the loop is the magnetic flux a maximum? For what orientation is the flux zero?

3. As the bar in Figure Q31.3 moves to the right, an electric field is set up directed downward in the bar. Explain why the electric field would be upward if the bar were moving to the left.

4. As the bar in Figure Q31.3 moves perpendicular to the field, is an external force required to keep it moving with constant speed?

5. The bar in Figure Q31.5 moves on rails to the right with a velocity $\mathbf{v}$, and the uniform, constant magnetic field is directed out of the page. Why is the induced current clockwise? If the bar were moving to the left, what would be the direction of the induced current?

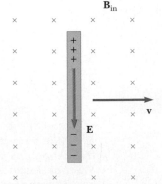

Figure Q31.3 Questions 3 and 4.

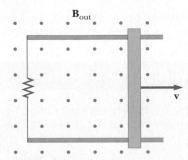

Figure Q31.5 Questions 5 and 6.

6. Explain why an applied force is necessary to keep the bar in Figure Q31.5 moving with a constant speed.

7. Wearing a metal bracelet in a region of strong magnetic field could be hazardous. Explain.

8. When a small magnet is moved toward a solenoid, an emf is induced in the coil. However, if the magnet is moved around inside a toroid, no measurable emf is induced. Explain.

9. How is energy produced in dams that is then transferred out by electrical transmission? (That is, how is the energy of motion of the water converted to energy that is transmitted by AC electricity?)

10. Will dropping a magnet down a long copper tube produce a current in the walls of the tube? Explain.

11. A piece of aluminum is dropped vertically downward between the poles of an electromagnet. Does the magnetic field affect the velocity of the aluminum?

12. What happens when the rotational speed of a generator coil is increased?

13. When the switch in Figure Q31.13a is closed, a current is set up in the coil and the metal ring springs upward (Fig. Q31.13b). Explain this behavior.

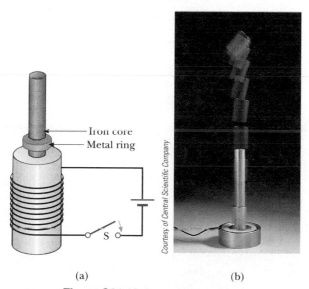

(a)   (b)

Figure Q31.13 Questions 13 and 14.

14. Assume that the battery in Figure Q31.13a is replaced by an AC source and the switch is held closed. If held down, the metal ring on top of the solenoid becomes hot. Why?

15. A bar magnet is held above a loop of wire in a horizontal plane, as shown in Figure Q31.15. The south end of the magnet is toward the loop of wire. The magnet is dropped toward the loop. Find the direction of the current through the resistor (a) while the magnet is falling toward the loop and (b) after the magnet has passed through the loop and moves away from it.

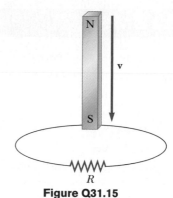

Figure Q31.15

16. Find the direction of the current in the resistor in Figure Q31.16 (a) at the instant the switch is closed, (b) after the switch has been closed for several minutes, and (c) at the instant the switch is opened.

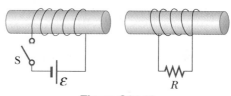

Figure Q31.16

17. Quick Quiz 31.4 describes the emf induced between the wingtips of an airplane by its motion in the Earth's magnetic field. Can this emf be used to power a light in the passenger compartment? Explain your answer.

18. Do Maxwell's equations allow for the existence of magnetic monopoles? Explain.

19. *Induction welding* has many important industrial applications. One example is the manufacture of airtight tubes, represented in Figure Q31.19. A sheet of metal is rolled into a cylinder and forced between compression rollers to bring its edges into contact. The tube then enters a coil carrying a time-varying current. The seam is welded when induced currents around the tube raise its temperature. Typically, a sinusoidal current with a frequency of 10 kHz is used. (a) What causes a current in the tube? (b) Why is a high frequency like 10 kHz chosen, rather than the 120 Hz commonly used for power transmission? (c) Why do the induced currents raise the temperature mainly of the seam, rather than all of the metal of the tube? (d) Why is it necessary to bring the edges of the sheet together with the compression rollers before the seam can be welded?

Figure Q31.19

PROBLEMS

1, 2, 3 = straightforward, intermediate, challenging   ☐ = full solution available in the *Student Solutions Manual and Study Guide*

🌐 = coached solution with hints available at http://www.pse.com   💻 = computer useful in solving problem

▨ = paired numerical and symbolic problems

Section 31.1   Faraday's Law of Induction
Section 31.3   Lenz's Law

1. A 50-turn rectangular coil of dimensions 5.00 cm × 10.0 cm is allowed to fall from a position where $B = 0$ to a new position where $B = 0.500$ T and the magnetic field is directed perpendicular to the plane of the coil. Calculate the magnitude of the average emf that is induced in the coil if the displacement occurs in 0.250 s.

2. A flat loop of wire consisting of a single turn of cross-sectional area 8.00 cm^2 is perpendicular to a magnetic field that increases uniformly in magnitude from 0.500 T to 2.50 T in 1.00 s. What is the resulting induced current if the loop has a resistance of 2.00 Ω?

3. A 25-turn circular coil of wire has diameter 1.00 m. It is placed with its axis along the direction of the Earth's magnetic field of 50.0 μT, and then in 0.200 s it is flipped 180°. An average emf of what magnitude is generated in the coil?

4. A rectangular loop of area A is placed in a region where the magnetic field is perpendicular to the plane of the loop. The magnitude of the field is allowed to vary in time according to $B = B_{max}e^{-t/\tau}$, where B_{max} and τ are constants. The field has the constant value B_{max} for $t < 0$. (a) Use Faraday's law to show that the emf induced in the loop is given by

$$\mathcal{E} = \frac{AB_{max}}{\tau}e^{-t/\tau}$$

(b) Obtain a numerical value for $\mathcal{E}$ at $t = 4.00$ s when $A = 0.160$ m^2, $B_{max} = 0.350$ T, and $\tau = 2.00$ s. (c) For the values of A, B_{max}, and τ given in (b), what is the maximum value of $\mathcal{E}$?

5. 🌐 A strong electromagnet produces a uniform magnetic field of 1.60 T over a cross-sectional area of 0.200 m^2. We place a coil having 200 turns and a total resistance of 20.0 Ω around the electromagnet. We then smoothly reduce the current in the electromagnet until it reaches zero in 20.0 ms. What is the current induced in the coil?

6. A magnetic field of 0.200 T exists within a solenoid of 500 turns and a diameter of 10.0 cm. How rapidly (that is, within what period of time) must the field be reduced to zero, if the average induced emf within the coil during this time interval is to be 10.0 kV?

7. 🌐 An aluminum ring of radius 5.00 cm and resistance 3.00×10^{-4} Ω is placed on top of a long air-core solenoid with 1 000 turns per meter and radius 3.00 cm, as shown in Figure P31.7. Over the area of the end of the solenoid, assume that the axial component of the field produced by the solenoid is half as strong as at the center of the solenoid. Assume the solenoid produces negligible field outside its cross-sectional area. The current in the solenoid is increasing at a rate of 270 A/s. (a) What is the induced current in the ring? At the center of the ring, what are (b) the magnitude and (c) the direction of the magnetic field produced by the induced current in the ring?

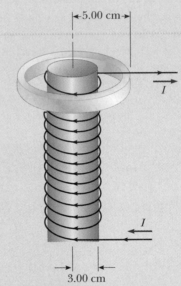

Figure P31.7 Problems 7 and 8.

8. An aluminum ring of radius r_1 and resistance R is placed around the top of a long air-core solenoid with n turns per meter and smaller radius r_2 as shown in Figure P31.7. Assume that the axial component of the field produced by the solenoid over the area of the end of the solenoid is half as strong as at the center of the solenoid. Assume that the solenoid produces negligible field outside its cross-sectional area. The current in the solenoid is increasing at a rate of $\Delta I/\Delta t$. (a) What is the induced current in the ring? (b) At the center of the ring, what is the magnetic field produced by the induced current in the ring? (c) What is the direction of this field?

9. (a) A loop of wire in the shape of a rectangle of width w and length L and a long, straight wire carrying a current I lie on a tabletop as shown in Figure P31.9. (a) Determine the magnetic flux through the loop due to the current I. (b) Suppose the current is changing with time according to $I = a + bt$, where a and b are constants. Determine the emf that is induced in the loop if $b = 10.0$ A/s, $h = 1.00$ cm, $w = 10.0$ cm, and $L = 100$ cm. What is the direction of the induced current in the rectangle?

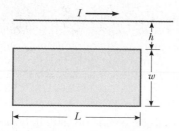

Figure P31.9 Problems 9 and 71.

10. A coil of 15 turns and radius 10.0 cm surrounds a long solenoid of radius 2.00 cm and 1.00×10^3 turns/meter (Fig. P31.10). The current in the solenoid changes as $I = (5.00 \text{ A}) \sin(120t)$. Find the induced emf in the 15-turn coil as a function of time.

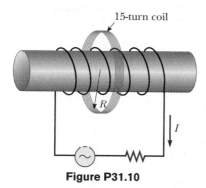

Figure P31.10

11. Find the current through section PQ of length $a = 65.0$ cm in Figure P31.11. The circuit is located in a magnetic field whose magnitude varies with time according to the expression $B = (1.00 \times 10^{-3} \text{ T/s})t$. Assume the resistance per length of the wire is 0.100 Ω/m.

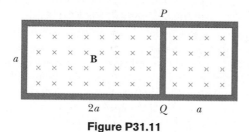

Figure P31.11

12. A 30-turn circular coil of radius 4.00 cm and resistance 1.00 Ω is placed in a magnetic field directed perpendicular to the plane of the coil. The magnitude of the magnetic field varies in time according to the expression $B = 0.010\,0t + 0.040\,0t^2$, where t is in seconds and B is in tesla. Calculate the induced emf in the coil at $t = 5.00$ s.

13. A long solenoid has $n = 400$ turns per meter and carries a current given by $I = (30.0 \text{ A})(1 - e^{-1.60t})$. Inside the solenoid and coaxial with it is a coil that has a radius of 6.00 cm and consists of a total of $N = 250$ turns of fine wire (Fig. P31.13). What emf is induced in the coil by the changing current?

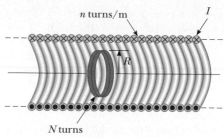

Figure P31.13

14. An instrument based on induced emf has been used to measure projectile speeds up to 6 km/s. A small magnet is imbedded in the projectile, as shown in Figure P31.14. The projectile passes through two coils separated by a distance d. As the projectile passes through each coil a pulse of emf is induced in the coil. The time interval between pulses can be measured accurately with an oscilloscope, and thus the speed can be determined. (a) Sketch a graph of ΔV versus t for the arrangement shown. Consider a current that flows counterclockwise as viewed from the starting point of the projectile as positive. On your graph, indicate which pulse is from coil 1 and which is from coil 2. (b) If the pulse separation is 2.40 ms and $d = 1.50$ m, what is the projectile speed?

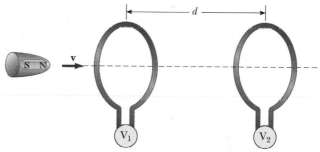

Figure P31.14

15. A coil formed by wrapping 50 turns of wire in the shape of a square is positioned in a magnetic field so that the normal to the plane of the coil makes an angle of 30.0° with the direction of the field. When the magnetic field is increased uniformly from 200 μT to 600 μT in 0.400 s, an emf of magnitude 80.0 mV is induced in the coil. What is the total length of the wire?

16. When a wire carries an AC current with a known frequency, you can use a *Rogowski coil* to determine the amplitude I_{max} of the current without disconnecting the wire to shunt the

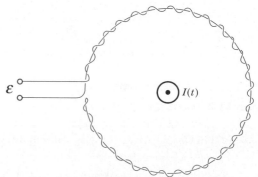

Figure P31.16

current in a meter. The Rogowski coil, shown in Figure P31.16, simply clips around the wire. It consists of a toroidal conductor wrapped around a circular return cord. The toroid has n turns per unit length and a cross-sectional area A. The current to be measured is given by $I(t) = I_{max} \sin \omega t$. (a) Show that the amplitude of the emf induced in the Rogowski coil is $\mathcal{E}_{max} = \mu_0 nA\omega I_{max}$. (b) Explain why the wire carrying the unknown current need not be at the center of the Rogowski coil, and why the coil will not respond to nearby currents that it does not enclose.

17. A toroid having a rectangular cross section ($a = 2.00$ cm by $b = 3.00$ cm) and inner radius $R = 4.00$ cm consists of 500 turns of wire that carries a sinusoidal current $I = I_{max} \sin \omega t$, with $I_{max} = 50.0$ A and a frequency $f = \omega/2\pi = 60.0$ Hz. A coil that consists of 20 turns of wire links with the toroid, as in Figure P31.17. Determine the emf induced in the coil as a function of time.

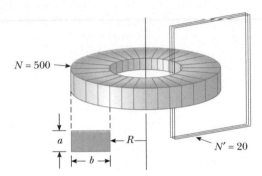

Figure P31.17

18. A piece of insulated wire is shaped into a figure 8, as in Figure P31.18. The radius of the upper circle is 5.00 cm and that of the lower circle is 9.00 cm. The wire has a uniform resistance per unit length of 3.00 Ω/m. A uniform magnetic field is applied perpendicular to the plane of the two circles, in the direction shown. The magnetic field is increasing at a constant rate of 2.00 T/s. Find the magnitude and direction of the induced current in the wire.

Figure P31.18

Section 31.2   Motional emf
Section 31.3   Lenz's Law

Problem 71 in Chapter 29 can be assigned with this section.

19. An automobile has a vertical radio antenna 1.20 m long. The automobile travels at 65.0 km/h on a horizontal road where the Earth's magnetic field is 50.0 μT directed toward the north and downward at an angle of 65.0° below the horizontal. (a) Specify the direction that the automobile should move in order to generate the maximum motional emf in the antenna, with the top of the antenna positive relative to the bottom. (b) Calculate the magnitude of this induced emf.

20. Consider the arrangement shown in Figure P31.20. Assume that $R = 6.00\ \Omega$, $\ell = 1.20$ m, and a uniform 2.50-T magnetic field is directed into the page. At what speed should the bar be moved to produce a current of 0.500 A in the resistor?

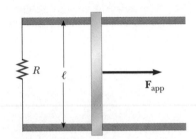

Figure P31.20 Problems 20, 21, and 22.

21. Figure P31.20 shows a top view of a bar that can slide without friction. The resistor is 6.00 Ω and a 2.50-T magnetic field is directed perpendicularly downward, into the paper. Let $\ell = 1.20$ m. (a) Calculate the applied force required to move the bar to the right at a constant speed of 2.00 m/s. (b) At what rate is energy delivered to the resistor?

22. A conducting rod of length ℓ moves on two horizontal, frictionless rails, as shown in Figure P31.20. If a constant force of 1.00 N moves the bar at 2.00 m/s through a magnetic field **B** that is directed into the page, (a) what is the current through the 8.00-Ω resistor R? (b) What is the rate at which energy is delivered to the resistor? (c) What is the mechanical power delivered by the force $\mathbf{F}_{app}$?

23. Very large magnetic fields can be produced using a procedure called *flux compression*. A metallic cylindrical tube of radius R is placed coaxially in a long solenoid of somewhat larger radius. The space between the tube and the solenoid is filled with a highly explosive material. When the explosive is set off, it collapses the tube to a cylinder of radius $r < R$. If the collapse happens very rapidly, induced current in the tube maintains the magnetic flux nearly constant inside the tube. If the initial magnetic field in the solenoid is 2.50 T, and $R/r = 12.0$, what maximum value of magnetic field can be achieved?

24. The *homopolar generator*, also called the *Faraday disk*, is a low-voltage, high-current electric generator. It consists of a rotating conducting disk with one stationary brush (a sliding electrical contact) at its axle and another at a point on its circumference, as shown in Figure P31.24. A magnetic field is applied perpendicular to the plane of the disk. Assume the field is 0.900 T, the angular speed is 3 200 rev/min, and the radius of the disk is 0.400 m. Find the emf generated between the brushes. When superconducting coils are used to produce a large magnetic field, a homopolar generator can have a power output of several

megawatts. Such a generator is useful, for example, in purifying metals by electrolysis. If a voltage is applied to the output terminals of the generator, it runs in reverse as a *homopolar motor* capable of providing great torque, useful in ship propulsion.

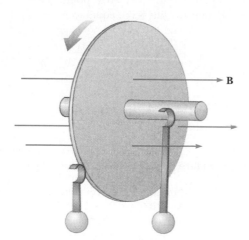

Figure P31.24

25. **Review problem.** A flexible metallic wire with linear density 3.00×10^{-3} kg/m is stretched between two fixed clamps 64.0 cm apart and held under tension 267 N. A magnet is placed near the wire as shown in Figure P31.25. Assume that the magnet produces a uniform field of 4.50 mT over a 2.00-cm length at the center of the wire, and a negligible field elsewhere. The wire is set vibrating at its fundamental (lowest) frequency. The section of the wire in the magnetic field moves with a uniform amplitude of 1.50 cm. Find (a) the frequency and (b) the amplitude of the electromotive force induced between the ends of the wire.

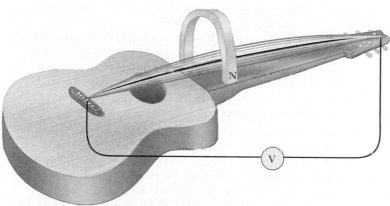

Figure P31.25

26. The square loop in Figure P31.26 is made of wires with total series resistance 10.0 Ω. It is placed in a uniform 0.100-T magnetic field directed perpendicularly into the plane of the paper. The loop, which is hinged at each corner, is pulled as shown until the separation between points A and B is 3.00 m. If this process takes 0.100 s, what is the average current generated in the loop? What is the direction of the current?

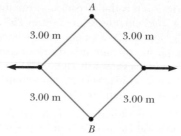

Figure P31.26

27. A helicopter (Figure P31.27) has blades of length 3.00 m, extending out from a central hub and rotating at 2.00 rev/s. If the vertical component of the Earth's magnetic field is 50.0 µT, what is the emf induced between the blade tip and the center hub?

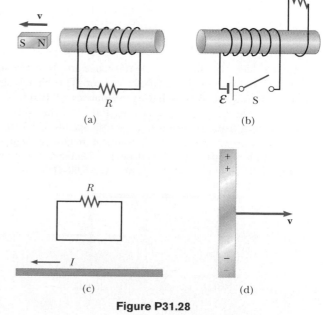

Figure P31.27

28. Use Lenz's law to answer the following questions concerning the direction of induced currents. (a) What is the direction of the induced current in resistor R in Figure P31.28a when the bar magnet is moved to the left? (b) What is the direction of the current induced in the resistor R immediately after the switch S in Figure P31.28b

Figure P31.28

is closed? (c) What is the direction of the induced current in R when the current I in Figure P31.28c decreases rapidly to zero? (d) A copper bar is moved to the right while its axis is maintained in a direction perpendicular to a magnetic field, as shown in Figure P31.28d. If the top of the bar becomes positive relative to the bottom, what is the direction of the magnetic field?

29. A rectangular coil with resistance R has N turns, each of length ℓ and width w as shown in Figure P31.29. The coil moves into a uniform magnetic field **B** with constant velocity **v**. What are the magnitude and direction of the total magnetic force on the coil (a) as it enters the magnetic field, (b) as it moves within the field, and (c) as it leaves the field?

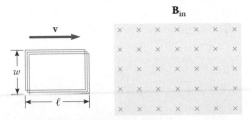

Figure P31.29

30. In Figure P31.30, the bar magnet is moved toward the loop. Is $V_a - V_b$ positive, negative, or zero? Explain.

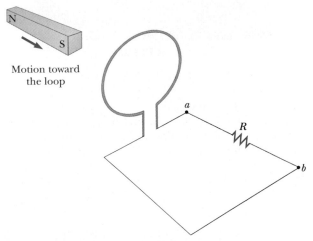

Figure P31.30

31. Two parallel rails with negligible resistance are 10.0 cm apart and are connected by a 5.00-Ω resistor. The circuit also contains two metal rods having resistances of 10.0 Ω and 15.0 Ω sliding along the rails (Fig. P31.31). The rods are pulled away from the resistor at constant speeds of 4.00 m/s and 2.00 m/s, respectively. A uniform magnetic field of magnitude 0.010 0 T is applied perpendicular to the plane of the rails. Determine the current in the 5.00-Ω resistor.

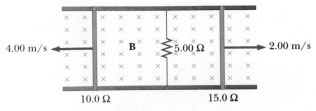

Figure P31.31

Section 31.4   Induced emf and Electric Fields

32. For the situation shown in Figure P31.32, the magnetic field changes with time according to the expression $B = (2.00t^3 - 4.00t^2 + 0.800)$T, and $r_2 = 2R = 5.00$ cm. (a) Calculate the magnitude and direction of the force exerted on an electron located at point P_2 when $t = 2.00$ s. (b) At what time is this force equal to zero?

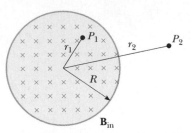

Figure P31.32 Problems 32 and 33.

33. A magnetic field directed into the page changes with time according to $B = (0.030\ 0t^2 + 1.40)$T, where t is in seconds. The field has a circular cross section of radius $R = 2.50$ cm (Fig. P31.32). What are the magnitude and direction of the electric field at point P_1 when $t = 3.00$ s and $r_1 = 0.020\ 0$ m?

34. A long solenoid with 1 000 turns per meter and radius 2.00 cm carries an oscillating current given by $I = (5.00$ A$) \sin(100\pi t)$. What is the electric field induced at a radius $r = 1.00$ cm from the axis of the solenoid? What is the direction of this electric field when the current is increasing counterclockwise in the coil?

Section 31.5   Generators and Motors

Problems 28 and 62 in Chapter 29 can be assigned with this section.

35. A coil of area 0.100 m^2 is rotating at 60.0 rev/s with the axis of rotation perpendicular to a 0.200-T magnetic field. (a) If the coil has 1 000 turns, what is the maximum emf generated in it? (b) What is the orientation of the coil with respect to the magnetic field when the maximum induced voltage occurs?

36. In a 250-turn automobile alternator, the magnetic flux in each turn is $\Phi_B = (2.50 \times 10^{-4}$ Wb$) \cos(\omega t)$, where ω is the angular speed of the alternator. The alternator is geared to rotate three times for each engine revolution. When the engine is running at an angular speed of 1 000 rev/min, determine (a) the induced emf in the alternator as a function of time and (b) the maximum emf in the alternator.

37. A long solenoid, with its axis along the x axis, consists of 200 turns per meter of wire that carries a steady current of 15.0 A. A coil is formed by wrapping 30 turns of thin wire around a circular frame that has a radius of 8.00 cm. The coil is placed inside the solenoid and mounted on an axis that is a diameter of the coil and coincides with the y axis. The coil is then rotated with an angular speed of 4.00π rad/s. (The plane of the coil is in the yz plane

at $t = 0$.) Determine the emf generated in the coil as a function of time.

38. A bar magnet is spun at constant angular speed ω around an axis as shown in Figure P31.38. A stationary flat rectangular conducting loop surrounds the magnet, and at $t = 0$, the magnet is oriented as shown. Make a qualitative graph of the induced current in the loop as a function of time, plotting counterclockwise currents as positive and clockwise currents as negative.

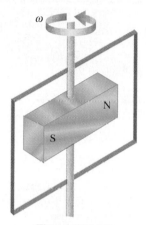

Figure P31.38

39. A motor in normal operation carries a direct current of 0.850 A when connected to a 120-V power supply. The resistance of the motor windings is 11.8 Ω. While in normal operation, (a) what is the back emf generated by the motor? (b) At what rate is internal energy produced in the windings? (c) **What If?** Suppose that a malfunction stops the motor shaft from rotating. At what rate will internal energy be produced in the windings in this case? (Most motors have a thermal switch that will turn off the motor to prevent overheating when this occurs.)

40. A semicircular conductor of radius $R = 0.250$ m is rotated about the axis AC at a constant rate of 120 rev/min (Fig. P31.40). A uniform magnetic field in all of the lower half of the figure is directed out of the plane of rotation and has a magnitude of 1.30 T. (a) Calculate the maximum value of the emf induced in the conductor. (b) What is the value of the average induced emf for each complete rotation? (c) **What If?** How would the answers to (a) and (b) change if **B** were allowed to extend a distance R above the axis of rotation? Sketch the emf versus time (d) when the field is as drawn in Figure P31.40 and (e) when the field is extended as described in (c).

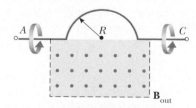

Figure P31.40

41. The rotating loop in an AC generator is a square 10.0 cm on a side. It is rotated at 60.0 Hz in a uniform field of 0.800 T. Calculate (a) the flux through the loop as a function of time, (b) the emf induced in the loop, (c) the

current induced in the loop for a loop resistance of 1.00 Ω, (d) the power delivered to the loop, and (e) the torque that must be exerted to rotate the loop.

Section 31.6   Eddy Currents

42. Figure P31.42 represents an electromagnetic brake that uses eddy currents. An electromagnet hangs from a railroad car near one rail. To stop the car, a large current is sent through the coils of the electromagnet. The moving electromagnet induces eddy currents in the rails, whose fields oppose the change in the field of the electromagnet. The magnetic fields of the eddy currents exert force on the current in the electromagnet, thereby slowing the car. The direction of the car's motion and the direction of the current in the electromagnet are shown correctly in the picture. Determine which of the eddy currents shown on the rails is correct. Explain your answer.

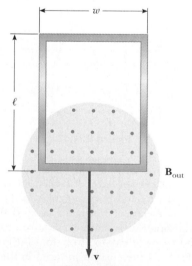

Figure P31.42

43. [www] A conducting rectangular loop of mass M, resistance R, and dimensions w by ℓ falls from rest into a magnetic field **B** as shown in Figure P31.43. During the time interval

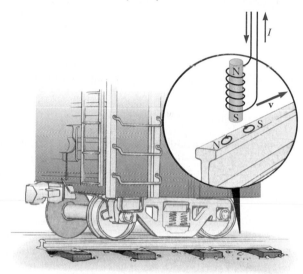

Figure P31.43

before the top edge of the loop reaches the field, the loop approaches a terminal speed v_T. (a) Show that

$$v_T = \frac{MgR}{B^2 w^2}$$

(b) Why is v_T proportional to R? (c) Why is it inversely proportional to B^2?

Section 31.7   Maxwell's Equations

44. An electron moves through a uniform electric field $\mathbf{E} = (2.50\hat{\mathbf{i}} + 5.00\hat{\mathbf{j}})$ V/m and a uniform magnetic field $\mathbf{B} = (0.400\hat{\mathbf{k}})$T. Determine the acceleration of the electron when it has a velocity $\mathbf{v} = 10.0\hat{\mathbf{i}}$ m/s.

45. A proton moves through a uniform electric field given by $\mathbf{E} = 50.0\hat{\mathbf{j}}$ V/m and a uniform magnetic field $\mathbf{B} = (0.200\hat{\mathbf{i}} + 0.300\hat{\mathbf{j}} + 0.400\hat{\mathbf{k}})$T. Determine the acceleration of the proton when it has a velocity $\mathbf{v} = 200\hat{\mathbf{i}}$ m/s.

Additional Problems

46. A steel guitar string vibrates (Figure 31.6). The component of magnetic field perpendicular to the area of a pickup coil nearby is given by

$$B = 50.0 \text{ mT} + (3.20 \text{ mT}) \sin(2\pi 523\, t/s)$$

The circular pickup coil has 30 turns and radius 2.70 mm. Find the emf induced in the coil as a function of time.

47. Figure P31.47 is a graph of the induced emf versus time for a coil of N turns rotating with angular speed ω in a uniform magnetic field directed perpendicular to the axis of rotation of the coil. **What If?** Copy this sketch (on a larger scale), and on the same set of axes show the graph of emf versus t (a) if the number of turns in the coil is doubled; (b) if instead the angular speed is doubled; and (c) if the angular speed is doubled while the number of turns in the coil is halved.

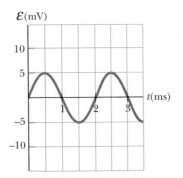

Figure P31.47

48. A technician wearing a brass bracelet enclosing area $0.005\,00$ m^2 places her hand in a solenoid whose magnetic field is 5.00 T directed perpendicular to the plane of the bracelet. The electrical resistance around the circumference of the bracelet is $0.020\,0$ Ω. An unexpected power failure causes the field to drop to 1.50 T in a time of 20.0 ms. Find (a) the current induced in the bracelet and (b) the power delivered to the bracelet. *Note:* As this

problem implies, you should not wear any metal objects when working in regions of strong magnetic fields.

49. Two infinitely long solenoids (seen in cross section) pass through a circuit as shown in Figure P31.49. The magnitude of **B** inside each is the same and is increasing at the rate of 100 T/s. What is the current in each resistor?

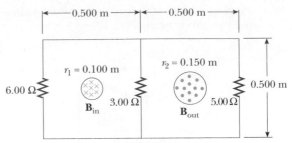

Figure P31.49

50. A conducting rod of length $\ell = 35.0$ cm is free to slide on two parallel conducting bars as shown in Figure P31.50. Two resistors $R_1 = 2.00$ Ω and $R_2 = 5.00$ Ω are connected across the ends of the bars to form a loop. A constant magnetic field $B = 2.50$ T is directed perpendicularly into the page. An external agent pulls the rod to the left with a constant speed of $v = 8.00$ m/s. Find (a) the currents in both resistors, (b) the total power delivered to the resistance of the circuit, and (c) the magnitude of the applied force that is needed to move the rod with this constant velocity.

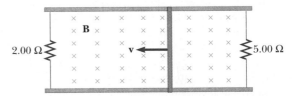

Figure P31.50

51. Suppose you wrap wire onto the core from a roll of cellophane tape to make a coil. Describe how you can use a bar magnet to produce an induced voltage in the coil. What is the order of magnitude of the emf you generate? State the quantities you take as data and their values.

52. A bar of mass m, length d, and resistance R slides without friction in a horizontal plane, moving on parallel rails as shown in Figure P31.52. A battery that maintains a constant emf $\mathcal{E}$ is connected between the rails, and a constant magnetic field **B** is directed perpendicularly to the plane of the page. Assuming the bar starts from rest, show that at time t it moves with a speed

$$v = \frac{\mathcal{E}}{Bd}\left(1 - e^{-B^2 d^2 t/mR}\right)$$

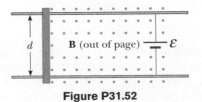

Figure P31.52

53. **Review problem**. A particle with a mass of 2.00×10^{-16} kg and a charge of 30.0 nC starts from rest, is accelerated by a strong electric field, and is fired from a small source inside a region of uniform constant magnetic field 0.600 T. The velocity of the particle is perpendicular to the field. The circular orbit of the particle encloses a magnetic flux of 15.0 μWb. (a) Calculate the speed of the particle. (b) Calculate the potential difference through which the particle accelerated inside the source.

54. An *induction furnace* uses electromagnetic induction to produce eddy currents in a conductor, thereby raising the conductor's temperature. Commercial units operate at frequencies ranging from 60 Hz to about 1 MHz and deliver powers from a few watts to several megawatts. Induction heating can be used for welding in a vacuum enclosure, to avoid oxidation and contamination of the metal. At high frequencies, induced currents occur only near the surface of the conductor—this is the "skin effect." By creating an induced current for a short time at an appropriately high frequency, one can heat a sample down to a controlled depth. For example, the surface of a farm tiller can be tempered to make it hard and brittle for effective cutting while keeping the interior metal soft and ductile to resist breakage.

   To explore induction heating, consider a flat conducting disk of radius R, thickness b, and resistivity ρ. A sinusoidal magnetic field $B_{max} \cos \omega t$ is applied perpendicular to the disk. Assume that the frequency is so low that the skin effect is not important. Assume the eddy currents occur in circles concentric with the disk. (a) Calculate the average power delivered to the disk. (b) **What If?** By what factor does the power change when the amplitude of the field doubles? (c) When the frequency doubles? (d) When the radius of the disk doubles?

55. The plane of a square loop of wire with edge length $a = 0.200$ m is perpendicular to the Earth's magnetic field at a point where $B = 15.0$ μT, as shown in Figure P31.55. The total resistance of the loop and the wires connecting it to a sensitive ammeter is 0.500 Ω. If the loop is suddenly collapsed by horizontal forces as shown, what total charge passes through the ammeter?

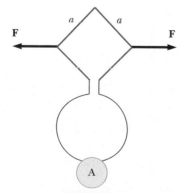

Figure P31.55

56. Magnetic field values are often determined by using a device known as a *search coil*. This technique depends on the measurement of the total charge passing through a coil in a time interval during which the magnetic flux linking the windings changes either because of the motion of the coil or because of a change in the value of B. (a) Show that as the flux through the coil changes from Φ_1 to Φ_2, the charge transferred through the coil will be given by $Q = N(\Phi_2 - \Phi_1)/R$, where R is the resistance of the coil and a sensitive ammeter connected across it and N is the number of turns. (b) As a specific example, calculate B when a 100-turn coil of resistance 200 Ω and cross-sectional area 40.0 cm^2 produces the following results. A total charge of 5.00×10^{-4} C passes through the coil when it is rotated in a uniform field from a position where the plane of the coil is perpendicular to the field to a position where the coil's plane is parallel to the field.

57. In Figure P31.57, the rolling axle, 1.50 m long, is pushed along horizontal rails at a constant speed $v = 3.00$ m/s. A resistor $R = 0.400$ Ω is connected to the rails at points a and b, directly opposite each other. (The wheels make good electrical contact with the rails, and so the axle, rails, and R form a closed-loop circuit. The only significant resistance in the circuit is R.) A uniform magnetic field $B = 0.080\ 0$ T is vertically downward. (a) Find the induced current I in the resistor. (b) What horizontal force F is required to keep the axle rolling at constant speed? (c) Which end of the resistor, a or b, is at the higher electric potential? (d) **What If?** After the axle rolls past the resistor, does the current in R reverse direction? Explain your answer.

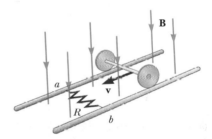

Figure P31.57

58. A conducting rod moves with a constant velocity **v** in a direction perpendicular to a long, straight wire carrying a current I as shown in Figure P31.58. Show that the magnitude of the emf generated between the ends of the rod is

$$|\mathcal{E}| = \frac{\mu_0 v I \ell}{2 \pi r}$$

In this case, note that the emf decreases with increasing r, as you might expect.

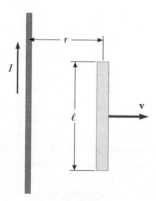

Figure P31.58

59. A circular loop of wire of radius r is in a uniform magnetic field, with the plane of the loop perpendicular to the direction of the field (Fig. P31.59). The magnetic field varies with time according to $B(t) = a + bt$, where a and b are constants. (a) Calculate the magnetic flux through the loop at $t = 0$. (b) Calculate the emf induced in the loop. (c) If the resistance of the loop is R, what is the induced current? (d) At what rate is energy being delivered to the resistance of the loop?

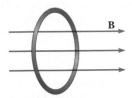

Figure P31.59

60. In Figure P31.60, a uniform magnetic field decreases at a constant rate $dB/dt = -K$, where K is a positive constant. A circular loop of wire of radius a containing a resistance R and a capacitance C is placed with its plane normal to the field. (a) Find the charge Q on the capacitor when it is fully charged. (b) Which plate is at the higher potential? (c) Discuss the force that causes the separation of charges.

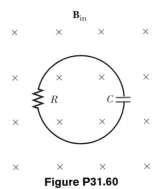

Figure P31.60

61. A rectangular coil of 60 turns, dimensions 0.100 m by 0.200 m and total resistance 10.0 Ω, rotates with angular speed 30.0 rad/s about the y axis in a region where a 1.00-T magnetic field is directed along the x axis. The rotation is initiated so that the plane of the coil is perpendicular to the direction of **B** at $t = 0$. Calculate (a) the maximum induced emf in the coil, (b) the maximum rate of change of magnetic flux through the coil, (c) the induced emf at $t = 0.050\ 0$ s, and (d) the torque exerted by the magnetic field on the coil at the instant when the emf is a maximum.

62. A small circular washer of radius 0.500 cm is held directly below a long, straight wire carrying a current of 10.0 A. The washer is located 0.500 m above the top of a table (Fig. P31.62). (a) If the washer is dropped from rest, what is the magnitude of the average induced emf in the washer from the time it is released to the moment it hits the table-top? Assume that the magnetic field is nearly constant over the area of the washer, and equal to the magnetic field at the center of the washer. (b) What is the direction of the induced current in the washer?

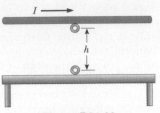

Figure P31.62

63. A conducting rod of length ℓ moves with velocity **v** parallel to a long wire carrying a steady current I. The axis of the rod is maintained perpendicular to the wire with the near end a distance r away, as shown in Figure P31.63. Show that the magnitude of the emf induced in the rod is

$$|\mathcal{E}| = \frac{\mu_0 I v}{2\pi} \ln\left(1 + \frac{\ell}{r}\right)$$

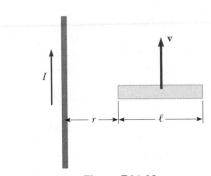

Figure P31.63

64. A rectangular loop of dimensions ℓ and w moves with a constant velocity **v** away from a long wire that carries a current I in the plane of the loop (Fig. P31.64). The total resistance of the loop is R. Derive an expression that gives the current in the loop at the instant the near side is a distance r from the wire.

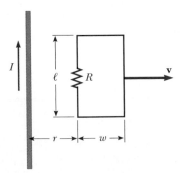

Figure P31.64

65. The magnetic flux through a metal ring varies with time t according to $\Phi_B = 3(at^3 - bt^2)$ T·m², with $a = 2.00$ s^{-3} and $b = 6.00$ s^{-2}. The resistance of the ring is 3.00 Ω. Determine the maximum current induced in the ring during the interval from $t = 0$ to $t = 2.00$ s.

66. Review problem. The bar of mass m in Figure P31.66 is pulled horizontally across parallel rails by a massless string that passes over an ideal pulley and is attached to a suspended object of mass M. The uniform magnetic field has a magnitude B, and the distance between the rails is ℓ. The rails are connected at one end by a load resistor R. Derive an expression that gives the horizontal speed of the bar as a function of time, assuming that the suspended object is released with the bar at rest at $t = 0$. Assume no friction between rails and bar.

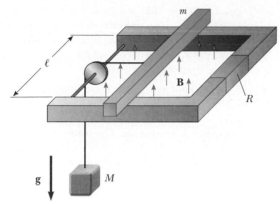

Figure P31.66

67. A solenoid wound with 2 000 turns/m is supplied with current that varies in time according to $I = (4A)\sin(120\pi t)$, where t is in seconds. A small coaxial circular coil of 40 turns and radius $r = 5.00$ cm is located inside the solenoid near its center. (a) Derive an expression that describes the manner in which the emf in the small coil varies in time. (b) At what average rate is energy delivered to the small coil if the windings have a total resistance of 8.00 Ω?

68. Figure P31.68 shows a stationary conductor whose shape is similar to the letter e. The radius of its circular portion is $a = 50.0$ cm. It is placed in a constant magnetic field of 0.500 T directed out of the page. A straight conducting rod, 50.0 cm long, is pivoted about point O and rotates with a constant angular speed of 2.00 rad/s. (a) Determine the induced emf in the loop POQ. Note that the area of the loop is $\theta a^2/2$. (b) If all of the conducting material has a resistance per length of 5.00 Ω/m, what is the induced current in the loop POQ at the instant 0.250 s after point P passes point Q?

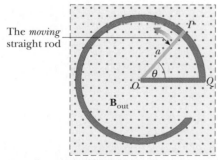

Figure P31.68

69. A *betatron* accelerates electrons to energies in the MeV range by means of electromagnetic induction. Electrons in a vacuum chamber are held in a circular orbit by a magnetic field perpendicular to the orbital plane. The magnetic field is gradually increased to induce an electric field around the orbit. (a) Show that the electric field is in the correct direction to make the electrons speed up. (b) Assume that the radius of the orbit remains constant. Show that the average magnetic field over the area enclosed by the orbit must be twice as large as the magnetic field at the circumference of the circle.

70. A wire 30.0 cm long is held parallel to and 80.0 cm above a long wire carrying 200 A and resting on the floor (Fig. P31.70). The 30.0-cm wire is released and falls, remaining parallel with the current-carrying wire as it falls. Assume that the falling wire accelerates at 9.80 m/s² and derive an equation for the emf induced in it. Express your result as a function of the time t after the wire is dropped. What is the induced emf 0.300 s after the wire is released?

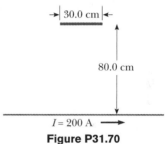

Figure P31.70

71. A long, straight wire carries a current that is given by $I = I_{max}\sin(\omega t + \phi)$ and lies in the plane of a rectangular coil of N turns of wire, as shown in Figure P31.9. The quantities I_{max}, ω, and ϕ are all constants. Determine the emf induced in the coil by the magnetic field created by the current in the straight wire. Assume $I_{max} = 50.0$ A, $\omega = 200\pi$ s^{-1}, $N = 100$, $h = w = 5.00$ cm, and $L = 20.0$ cm.

72. A dime is suspended from a thread and hung between the poles of a strong horseshoe magnet as shown in Figure P31.72. The dime rotates at constant angular speed ω

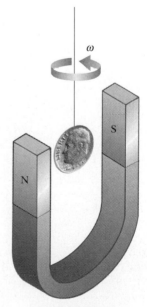

Figure P31.72

about a vertical axis. Letting θ represent the angle between the direction of **B** and the normal to the face of the dime, sketch a graph of the torque due to induced currents as a function of θ for $0 < \theta < 2\pi$.

Answers to Quick Quizzes

31.1 (c). In all cases except this one, there is a change in the magnetic flux through the loop.

31.2 c, $d = e$, b, a. The magnitude of the emf is proportional to the rate of change of the magnetic flux. For the situation described, the rate of change of magnetic flux is proportional to the rate of change of the magnetic field. This rate of change is the slope of the graph in Figure 31.4. The magnitude of the slope is largest at c. Points d and e are on a straight line, so the slope is the same at each point. Point b represents a point of relatively small slope, while a is at a point of zero slope because the curve is horizontal at that point.

31.3 (b). The magnetic field lines around the transmission cable will be circular, centered on the cable. If you place your loop around the cable, there are no field lines passing through the loop, so no emf is induced. The loop must be placed next to the cable, with the plane of the loop parallel to the cable to maximize the flux through its area.

31.4 (a). The Earth's magnetic field has a downward component in the northern hemisphere. As the plane flies north, the right-hand rule illustrated in Figure 29.4 indicates that positive charge experiences a force directed toward the west. Thus, the left wingtip becomes positively charged and the right wingtip negatively charged.

31.5 (c). The force on the wire is of magnitude $F_{app} = F_B = I\ell B$, with I given by Equation 31.6. Thus, the force is proportional to the speed and the force doubles. Because $\mathcal{P} = F_{app}v$, the doubling of the force *and* the speed results in the power being four times as large.

31.6 (b). According to Equation 31.5, because B and v are fixed, the emf depends only on the length of the wire moving in the magnetic field. Thus, you want the long dimension moving through the magnetic field lines so that it is perpendicular to the velocity vector. In this case, the short dimension is parallel to the velocity vector.

31.7 (a). Because the current induced in the solenoid is clockwise when viewed from above, the magnetic field lines produced by this current point downward in Figure 31.15. Thus, the upper end of the solenoid acts as a south pole. For this situation to be consistent with Lenz's law, the south pole of the bar magnet must be approaching the solenoid.

31.8 (b). At the position of the loop, the magnetic field lines due to the wire point into the page. The loop is entering a region of stronger magnetic field as it drops toward the wire, so the flux is increasing. The induced current must set up a magnetic field that opposes this increase. To do this, it creates a magnetic field directed out of the page. By the right-hand rule for current loops, this requires a counterclockwise current in the loop.

31.9 (d). The constant rate of change of B will result in a constant rate of change of the magnetic flux. According to Equation 31.9, if $d\Phi_B/dt$ is constant, **E** is constant in magnitude.

31.10 (a). While reducing the resistance may increase the current that the generator provides to a load, it does not alter the emf. Equation 31.11 shows that the emf depends on ω, B, and N, so all other choices increase the emf.

31.11 (b). When the aluminum sheet moves between the poles of the magnet, eddy currents are established in the aluminum. According to Lenz's law, these currents are in a direction so as to oppose the original change, which is the movement of the aluminum sheet in the magnetic field. The same principle is used in common laboratory triple-beam balances. See if you can find the magnet and the aluminum sheet the next time you use a triple-beam balance.

Chapter 32

Inductance

CHAPTER OUTLINE

32.1 Self-Inductance

32.2 *RL* Circuits

32.3 Energy in a Magnetic Field

32.4 Mutual Inductance

32.5 Oscillations in an *LC* Circuit

32.6 The *RLC* Circuit

▲ *An airport metal detector contains a large coil of wire around the frame. This coil has a property called inductance. When a passenger carries metal through the detector, the inductance of the coil changes, and the change in inductance signals an alarm to sound. (Jack Hollingsworth/Getty Images)*

In Chapter 31, we saw that an emf and a current are induced in a circuit when the magnetic flux through the area enclosed by the circuit changes with time. This phenomenon of electromagnetic induction has some practical consequences. In this chapter, we first describe an effect known as *self-induction,* in which a time-varying current in a circuit produces an induced emf opposing the emf that initially set up the time-varying current. Self-induction is the basis of the *inductor,* an electrical circuit element. We discuss the energy stored in the magnetic field of an inductor and the energy density associated with the magnetic field.

Next, we study how an emf is induced in a circuit as a result of a changing magnetic flux produced by a second circuit; this is the basic principle of *mutual induction.* Finally, we examine the characteristics of circuits that contain inductors, resistors, and capacitors in various combinations.

32.1   Self-Inductance

In this chapter, we need to distinguish carefully between emfs and currents that are caused by batteries or other sources and those that are induced by changing magnetic fields. When we use a term without an adjective (such as *emf* and *current*) we are describing the parameters associated with a physical source. We use the adjective *induced* to describe those emfs and currents caused by a changing magnetic field.

Consider a circuit consisting of a switch, a resistor, and a source of emf, as shown in Figure 32.1. When the switch is thrown to its closed position, the current does not immediately jump from zero to its maximum value $\mathcal{E}/R$. Faraday's law of electromagnetic induction (Eq. 31.1) can be used to describe this effect as follows: as the current increases with time, the magnetic flux through the circuit loop due to this current also increases with time. This increasing flux creates an induced emf in the circuit. The direction of the induced emf is such that it would cause an induced current in the loop (if the loop did not already carry a current), which would establish a magnetic field opposing the change in the original magnetic field. Thus, the direction of the induced emf is opposite the direction of the emf of the battery; this results in a gradual rather than instantaneous increase in the current to its final equilibrium value. Because of the direction of the induced emf, it is also called a *back emf,* similar to that in a motor, as discussed in Chapter 31. This effect is called **self-induction** because the changing flux through the circuit and the resultant induced emf arise from the circuit itself. The emf $\mathcal{E}_L$ set up in this case is called a **self-induced emf.**

As a second example of self-induction, consider Figure 32.2, which shows a coil wound on a cylindrical core. Assume that the current in the coil either increases or decreases with time. When the current is in the direction shown, a magnetic field directed from right to left is set up inside the coil, as seen in Figure 32.2a. As the current changes with time, the magnetic flux through the coil also changes and induces an emf in the coil. From Lenz's law, the polarity of this induced emf must be such that it opposes the change in the magnetic field from the current. If the current

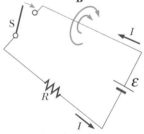

Figure 32.1 After the switch is closed, the current produces a magnetic flux through the area enclosed by the loop. As the current increases toward its equilibrium value, this magnetic flux changes in time and induces an emf in the loop.

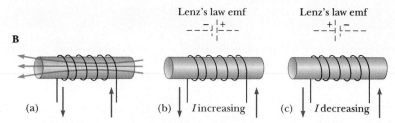

Figure 32.2 (a) A current in the coil produces a magnetic field directed to the left. (b) If the current increases, the increasing magnetic flux creates an induced emf in the coil having the polarity shown by the dashed battery. (c) The polarity of the induced emf reverses if the current decreases.

Joseph Henry
American Physicist (1797–1878)

Henry became the first director of the Smithsonian Institution and first president of the Academy of Natural Science. He improved the design of the electromagnet and constructed one of the first motors. He also discovered the phenomenon of self-induction but failed to publish his findings. The unit of inductance, the henry, is named in his honor. *(North Wind Picture Archives)*

is increasing, the polarity of the induced emf is as pictured in Figure 32.2b, and if the current is decreasing, the polarity of the induced emf is as shown in Figure 32.2c.

To obtain a quantitative description of self-induction, we recall from Faraday's law that the induced emf is equal to the negative of the time rate of change of the magnetic flux. The magnetic flux is proportional to the magnetic field due to the current, which in turn is proportional to the current in the circuit. Therefore, **a self-induced emf is always proportional to the time rate of change of the current.** For any coil, we find that

$$\mathcal{E}_L = -L \frac{dI}{dt} \qquad (32.1)$$

where L is a proportionality constant—called the **inductance** of the coil—that depends on the geometry of the coil and other physical characteristics. Combining this expression with Faraday's law, $\mathcal{E}_L = -N \, d\Phi_B/dt$, we see that the inductance of a closely spaced coil of N turns (a toroid or an ideal solenoid) carrying a current I and containing N turns is

$$L = \frac{N\Phi_B}{I} \qquad (32.2)$$

Inductance of an *N*-turn coil

where it is assumed that the same magnetic flux passes through each turn.

From Equation 32.1, we can also write the inductance as the ratio

$$L = -\frac{\mathcal{E}_L}{dI/dt} \qquad (32.3)$$

Inductance

Recall that resistance is a measure of the opposition to current ($R = \Delta V/I$); in comparison, inductance is a measure of the opposition to a *change* in current.

The SI unit of inductance is the **henry** (H), which, as we can see from Equation 32.3, is 1 volt-second per ampere:

$$1 \text{ H} = 1 \frac{\text{V} \cdot \text{s}}{\text{A}}$$

As shown in Examples 32.1 and 32.2, the inductance of a coil depends on its geometry. This is analogous to the capacitance of a capacitor depending on the geometry of its plates, as we found in Chapter 26. Inductance calculations can be quite difficult to perform for complicated geometries; however, the examples below involve simple situations for which inductances are easily evaluated.

Quick Quiz 32.1 A coil with zero resistance has its ends labeled a and b. The potential at a is higher than at b. Which of the following could be consistent with this situation? (a) The current is constant and is directed from a to b; (b) The current is constant and is directed from b to a; (c) The current is increasing and is directed from a to b; (d) The current is decreasing and is directed from a to b; (e) The current is increasing and is directed from b to a; (f) The current is decreasing and is directed from b to a.

Example 32.1 Inductance of a Solenoid

Find the inductance of a uniformly wound solenoid having N turns and length ℓ. Assume that ℓ is much longer than the radius of the windings and that the core of the solenoid is air.

Solution We can assume that the interior magnetic field due to the current is uniform and given by Equation 30.17:

$$B = \mu_0 nI = \mu_0 \frac{N}{\ell} I$$

where $n = N/\ell$ is the number of turns per unit length. The magnetic flux through each turn is

$$\Phi_B = BA = \mu_0 \frac{NA}{\ell} I$$

where A is the cross-sectional area of the solenoid. Using this expression and Equation 32.2, we find that

$$L = \frac{N\Phi_B}{I} = \frac{\mu_0 N^2 A}{\ell} \qquad (32.4)$$

This result shows that L depends on geometry and is proportional to the square of the number of turns. Because $N = n\ell$, we can also express the result in the form

$$L = \mu_0 \frac{(n\ell)^2}{\ell} A = \mu_0 n^2 A\ell = \mu_0 n^2 V \qquad (32.5)$$

where $V = A\ell$ is the interior volume of the solenoid.

What If? What would happen to the inductance if you inserted a ferromagnetic material inside the solenoid?

Answer The inductance would increase. For a given current, the magnetic flux in the solenoid is much greater because of the increase in the magnetic field originating from the magnetization of the ferromagnetic material. For example, if the material has a magnetic permeability of $500\mu_0$, the inductance increases by a factor of 500.

Example 32.2 Calculating Inductance and emf

(A) Calculate the inductance of an air-core solenoid containing 300 turns if the length of the solenoid is 25.0 cm and its cross-sectional area is 4.00 cm^2.

Solution Using Equation 32.4, we obtain

$$L = \frac{\mu_0 N^2 A}{\ell}$$

$$= \frac{(4\pi \times 10^{-7}\,\text{T} \cdot \text{m/A})(300)^2(4.00 \times 10^{-4}\,\text{m}^2)}{25.0 \times 10^{-2}\,\text{m}}$$

$$= 1.81 \times 10^{-4}\,\text{T} \cdot \text{m}^2/\text{A} = \boxed{0.181\ \text{mH}}$$

(B) Calculate the self-induced emf in the solenoid if the current it carries is decreasing at the rate of 50.0 A/s.

Solution Using Equation 32.1 and given that $dI/dt = -50.0$ A/s, we obtain

$$\mathcal{E}_L = -L\frac{dI}{dt} = -(1.81 \times 10^{-4}\,\text{H})(-50.0\,\text{A/s})$$

$$= \boxed{9.05\ \text{mV}}$$

32.2 *RL* Circuits

If a circuit contains a coil, such as a solenoid, the self-inductance of the coil prevents the current in the circuit from increasing or decreasing instantaneously. A circuit element that has a large self-inductance is called an **inductor** and has the circuit symbol ───⌁000⌁───. We always assume that the self-inductance of the remainder of a circuit is negligible compared with that of the inductor. Keep in mind, however, that even a circuit without a coil has some self-inductance that can affect the behavior of the circuit.

Because the inductance of the inductor results in a back emf, **an inductor in a circuit opposes changes in the current in that circuit.** The inductor attempts to keep the current the same as it was before the change occurred. If the battery voltage in the circuit is increased so that the current rises, the inductor opposes this change, and the rise is not instantaneous. If the battery voltage is decreased, the presence of the inductor results in a slow drop in the current rather than an immediate drop. Thus, the inductor causes the circuit to be "sluggish" as it reacts to changes in the voltage.

Consider the circuit shown in Figure 32.3, which contains a battery of negligible internal resistance. This is an **RL circuit** because the elements connected to the battery are a resistor and an inductor. Suppose that the switch S is open for $t < 0$ and then closed at $t = 0$. The current in the circuit begins to increase, and a back emf (Eq. 32.1) that opposes the increasing current is induced in the inductor. Because the current is increasing, dI/dt in Equation 32.1 is positive; thus, $\mathcal{E}_L$ is negative. This negative value reflects the decrease in electric potential that occurs in going from a to b across the inductor, as indicated by the positive and negative signs in Figure 32.3.

With this in mind, we can apply Kirchhoff's loop rule to this circuit, traversing the circuit in the clockwise direction:

$$\mathcal{E} - IR - L\frac{dI}{dt} = 0 \tag{32.6}$$

where IR is the voltage drop across the resistor. (We developed Kirchhoff's rules for circuits with steady currents, but they can also be applied to a circuit in which the current is changing if we imagine them to represent the circuit at one *instant* of time.) We must now look for a solution to this differential equation, which is similar to that for the *RC* circuit. (See Section 28.4.)

A mathematical solution of Equation 32.6 represents the current in the circuit as a function of time. To find this solution, we change variables for convenience, letting $x = (\mathcal{E}/R) - I$, so that $dx = -dI$. With these substitutions, we can write Equation 32.6 as

$$x + \frac{L}{R}\frac{dx}{dt} = 0$$

$$\frac{dx}{x} = -\frac{R}{L}dt$$

Integrating this last expression, we have

$$\int_{x_0}^{x}\frac{dx}{x} = -\frac{R}{L}\int_{0}^{t}dt$$

$$\ln\frac{x}{x_0} = -\frac{R}{L}t$$

where x_0 is the value of x at time $t = 0$. Taking the antilogarithm of this result, we obtain

$$x = x_0 e^{-Rt/L}$$

Because $I = 0$ at $t = 0$, we note from the definition of x that $x_0 = \mathcal{E}/R$. Hence, this last expression is equivalent to

$$\frac{\mathcal{E}}{R} - I = \frac{\mathcal{E}}{R}e^{-Rt/L}$$

$$I = \frac{\mathcal{E}}{R}(1 - e^{-Rt/L})$$

This expression shows how the inductor effects the current. The current does not increase instantly to its final equilibrium value when the switch is closed but instead increases according to an exponential function. If we remove the inductance in the circuit, which we can do by letting L approach zero, the exponential term becomes zero and we see that there is no time dependence of the current in this case—the current increases instantaneously to its final equilibrium value in the absence of the inductance.

We can also write this expression as

$$I = \frac{\mathcal{E}}{R}(1 - e^{-t/\tau}) \tag{32.7}$$

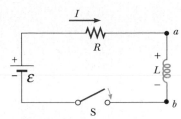

Active Figure 32.3 A series *RL* circuit. As the current increases toward its maximum value, an emf that opposes the increasing current is induced in the inductor.

 At the Active Figures link at **http://www.pse6.com**, *you can adjust the values of R and L to see the effect on the current. A graphical display as in Figure 32.4 is available.*

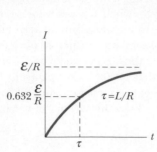

Active Figure 32.4 Plot of the current versus time for the *RL* circuit shown in Figure 32.3. The switch is open for $t < 0$ and then closed at $t = 0$, and the current increases toward its maximum value $\mathcal{E}/R$. The time constant τ is the time interval required for I to reach 63.2% of its maximum value.

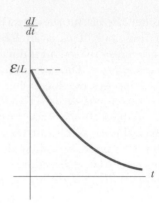

Figure 32.5 Plot of dI/dt versus time for the *RL* circuit shown in Figure 32.3. The time rate of change of current is a maximum at $t = 0$, which is the instant at which the switch is closed. The rate decreases exponentially with time as I increases toward its maximum value.

At the Active Figures link at http://www.pse6.com, you can observe this graph develop after the switch in Figure 32.3 is closed.

where the constant τ is the **time constant** of the *RL* circuit:

Time constant of an *RL* circuit

$$\tau = \frac{L}{R} \tag{32.8}$$

Physically, τ is the time interval required for the current in the circuit to reach $(1 - e^{-1}) = 0.632 = 63.2\%$ of its final value $\mathcal{E}/R$. The time constant is a useful parameter for comparing the time responses of various circuits.

Figure 32.4 shows a graph of the current versus time in the *RL* circuit. Note that the equilibrium value of the current, which occurs as t approaches infinity, is $\mathcal{E}/R$. We can see this by setting dI/dt equal to zero in Equation 32.6 and solving for the current I. (At equilibrium, the change in the current is zero.) Thus, we see that the current initially increases very rapidly and then gradually approaches the equilibrium value $\mathcal{E}/R$ as t approaches infinity.

Let us also investigate the time rate of change of the current. Taking the first time derivative of Equation 32.7, we have

$$\frac{dI}{dt} = \frac{\mathcal{E}}{L} e^{-t/\tau} \tag{32.9}$$

From this result, we see that the time rate of change of the current is a maximum (equal to $\mathcal{E}/L$) at $t = 0$ and falls off exponentially to zero as t approaches infinity (Fig. 32.5).

Now let us consider the *RL* circuit shown in Figure 32.6. The curved lines on the switch S represent a switch that is connected either to a or b at all times. (If the switch is connected to neither a nor b, the current in the circuit suddenly stops.) Suppose that the switch has been set at position a long enough to allow the current to reach its equilibrium value $\mathcal{E}/R$. In this situation, the circuit is described by the outer loop in Figure 32.6. If the switch is thrown from a to b, the circuit is now described by just the right-hand loop in Figure 32.6. Thus, we have a circuit with no battery ($\mathcal{E} = 0$). Applying Kirchhoff's loop rule to the right-hand loop at the instant the switch is thrown from a to b, we obtain

$$IR + L\frac{dI}{dt} = 0$$

It is left as a problem (Problem 16) to show that the solution of this differential equation is

$$I = \frac{\mathcal{E}}{R} e^{-t/\tau} = I_0 e^{-t/\tau} \tag{32.10}$$

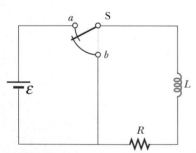

Active Figure 32.6 An *RL* circuit. When the switch S is in position a, the battery is in the circuit. When the switch is thrown to position b, the battery is no longer part of the circuit. The switch is designed so that it is never open, which would cause the current to stop.

At the Active Figures link at http://www.pse6.com, you can adjust the values of R and L to see the effect on the current. A graphical display as in Figure 32.7 is available.

where $\mathcal{E}$ is the emf of the battery and $I_0 = \mathcal{E}/R$ is the current at the instant at which the switch is thrown to b.

If the circuit did not contain an inductor, the current would immediately decrease to zero when the battery is removed. When the inductor is present, it opposes the decrease in the current and causes the current to decrease exponentially. A graph of the current in the circuit versus time (Fig. 32.7) shows that the current is continuously decreasing with time. Note that the slope dI/dt is always negative and has its maximum value at $t = 0$. The negative slope signifies that $\mathcal{E}_L = -L\,(dI/dt)$ is now positive.

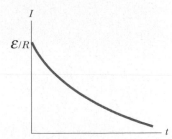

Active Figure 32.7 Current versus time for the right-hand loop of the circuit shown in Figure 32.6. For $t < 0$, the switch S is at position a. At $t = 0$, the switch is thrown to position b, and the current has its maximum value $\mathcal{E}/R$.

At the Active Figures link at **http://www.pse6.com**, *you can observe this graph develop after the switch in Figure 32.6 is thrown to position b.*

Quick Quiz 32.2 The circuit in Figure 32.8 consists of a resistor, an inductor, and an ideal battery with no internal resistance. At the instant just after the switch is closed, across which circuit element is the voltage equal to the emf of the battery? (a) the resistor (b) the inductor (c) both the inductor and resistor. After a very long time, across which circuit element is the voltage equal to the emf of the battery? (d) the resistor (e) the inductor (f) both the inductor and resistor.

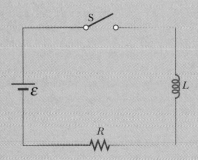

Figure 32.8 (Quick Quiz 32.2)

Quick Quiz 32.3 The circuit in Figure 32.9 includes a power source that provides a sinusoidal voltage. Thus, the magnetic field in the inductor is constantly changing. The inductor is a simple air-core solenoid. The switch in the circuit is closed and the lightbulb glows steadily. An iron rod is inserted into the interior of the solenoid, which increases the magnitude of the magnetic field in the solenoid. As this happens, the brightness of the lightbulb (a) increases, (b) decreases, (c) is unaffected.

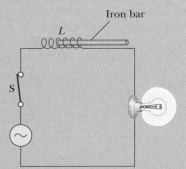

Figure 32.9 (Quick Quiz 32.3)

Quick Quiz 32.4 Two circuits like the one shown in Figure 32.6 are identical except for the value of L. In circuit A the inductance of the inductor is L_A, and in circuit B it is L_B. Switch S is thrown to position a at $t = 0$. At $t = 10$ s, the switch is thrown to position b. The resulting currents for the two circuits are as graphed in Figure 32.10.

If we assume that the time constant of each circuit is much less than 10 s, which of the following is true? (a) $L_A > L_B$; (b) $L_A < L_B$; (c) not enough information to tell.

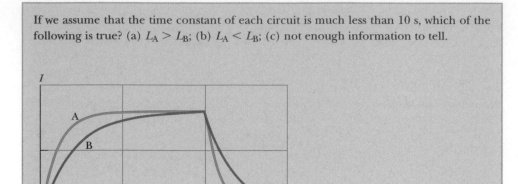

Figure 32.10 (Quick Quiz 32.4)

Example 32.3   Time Constant of an RL Circuit   `Interactive`

(A) Find the time constant of the circuit shown in Figure 32.11a.

Solution The time constant is given by Equation 32.8:

$$\tau = \frac{L}{R} = \frac{30.0 \times 10^{-3}\,\text{H}}{6.00\,\Omega} = \boxed{5.00\,\text{ms}}$$

(B) The switch in Figure 32.11a is closed at $t = 0$. Calculate the current in the circuit at $t = 2.00$ ms.

Solution Using Equation 32.7 for the current as a function of time (with t and τ in milliseconds), we find that at $t = 2.00$ ms,

$$I = \frac{\mathcal{E}}{R}(1 - e^{-t/\tau}) = \frac{12.0\,\text{V}}{6.00\,\Omega}(1 - e^{-0.400}) = \boxed{0.659\,\text{A}}$$

A plot of the current versus time for this circuit is given in Figure 32.11b.

(C) Compare the potential difference across the resistor with that across the inductor.

Solution At the instant the switch is closed, there is no current and thus no potential difference across the resistor. At this instant, the battery voltage appears entirely across the inductor in the form of a back emf of 12.0 V as the inductor tries to maintain the zero-current condition. (The left end of the inductor is at a higher electric potential than the right end.) As time passes, the emf across the inductor decreases and the current in the resistor (and hence the potential difference across it) increases, as shown in Figure 32.12. The sum of the two potential differences at all times is 12.0 V.

What If? In Figure 32.12, we see that the voltages across the resistor and inductor are equal at a time just before 4.00 ms. What if we wanted to delay the condition in which the voltages are equal to some later instant, such as $t = 10.0$ ms? Which parameter, L or R, would require the least adjustment, in terms of a percentage change, to achieve this?

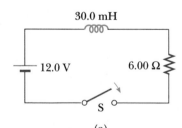

30.0 mH

12.0 V   6.00 Ω

S

(a)

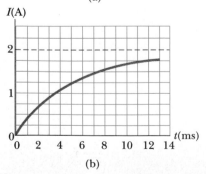

(b)

Figure 32.11 (Example 32.3) (a) The switch in this RL circuit is open for $t < 0$ and then closed at $t = 0$. (b) A graph of the current versus time for the circuit in part (a).

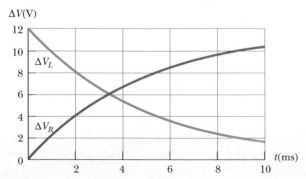

Figure 32.12 (Example 32.3) The time behavior of the potential differences across the resistor and inductor in Figure 32.11a.

Answer From Figure 32.12, we see that the voltages are equal when the voltage across the inductor has fallen to half of its original value. Thus, the time interval required for the voltages to become equal is the *half-life* $t_{1/2}$ of the decay. We introduced the half-life in the **What If?** section of Example 28.13, in describing the exponential decay in *RC* circuits, where we found that $t_{1/2} = 0.693\tau$. If we want the half-life of our *RL* circuit to be 10.0 ms, then the time constant must be

$$\tau = \frac{t_{1/2}}{0.693} = \frac{10.0 \text{ ms}}{0.693} = 14.4 \text{ ms}$$

We can achieve this time constant by holding *L* fixed and adjusting *R*; in this case,

$$\tau = \frac{L}{R} = \frac{30.0 \times 10^{-3} \text{ H}}{R} = 14.4 \text{ ms}$$

$$R = \frac{30.0 \times 10^{-3} \text{ H}}{14.4 \text{ ms}} = 2.08 \ \Omega$$

This corresponds to a 65% decrease compared to the initial resistance. Now hold *R* fixed and adjust *L*:

$$\tau = \frac{L}{R} = \frac{L}{6.00 \ \Omega} = 14.4 \text{ ms}$$

$$L = (6.00 \ \Omega)(14.4 \text{ ms}) = 86.4 \times 10^{-3} \text{ H}$$

This represents a 188% increase in inductance! Thus, a much smaller percentage adjustment in *R* can achieve the desired effect than in *L*.

 At the Interactive Worked Example link at http://www.pse6.com, you can explore the decay of current in an RL circuit.

32.3 Energy in a Magnetic Field

Because the emf induced in an inductor prevents a battery from establishing an instantaneous current, the battery must provide more energy than in a circuit without the inductor. Part of the energy supplied by the battery appears as internal energy in the resistor, while the remaining energy is stored in the magnetic field of the inductor. If we multiply each term in Equation 32.6 by *I* and rearrange the expression, we have

$$I\mathcal{E} = I^2 R + LI\frac{dI}{dt} \tag{32.11}$$

Recognizing $I\mathcal{E}$ as the rate at which energy is supplied by the battery and $I^2 R$ as the rate at which energy is delivered to the resistor, we see that $LI(dI/dt)$ must represent the rate at which energy is being stored in the inductor. If we let *U* denote the energy stored in the inductor at any time, then we can write the rate dU/dt at which energy is stored as

$$\frac{dU}{dt} = LI\frac{dI}{dt}$$

To find the total energy stored in the inductor, we can rewrite this expression as $dU = LI\, dI$ and integrate:

$$U = \int dU = \int_0^I LI\, dI = L\int_0^I I\, dI$$

$$U = \tfrac{1}{2}LI^2 \tag{32.12}$$

where *L* is constant and has been removed from the integral. This expression represents the energy stored in the magnetic field of the inductor when the current is *I*. Note that this equation is similar in form to Equation 26.11 for the energy stored in the electric field of a capacitor, $U = \tfrac{1}{2}C(\Delta V)^2$. In either case, we see that energy is required to establish a field.

We can also determine the energy density of a magnetic field. For simplicity, consider a solenoid whose inductance is given by Equation 32.5:

$$L = \mu_0 n^2 A\ell$$

The magnetic field of a solenoid is given by Equation 30.17:

$$B = \mu_0 nI$$

PITFALL PREVENTION

32.1 Compare Energy in a Capacitor, Resistor, and Inductor

Different energy-storage mechanisms are at work in capacitors, inductors, and resistors. A capacitor stores a given amount of energy for a fixed charge on its plates; as more charge is delivered, more energy is delivered. An inductor stores a given amount of energy for constant current; as the current increases, more energy is delivered. Energy delivered to a resistor is transformed to internal energy.

Energy stored in an inductor

Substituting the expression for L and $I = B/\mu_0 n$ into Equation 32.12 gives

$$U = \tfrac{1}{2} L I^2 = \tfrac{1}{2} \mu_0 n^2 A \ell \left(\frac{B}{\mu_0 n} \right)^2 = \frac{B^2}{2\mu_0} A \ell \qquad (32.13)$$

Because $A\ell$ is the volume of the solenoid, the magnetic energy density, or the energy stored per unit volume in the magnetic field of the inductor is

Magnetic energy density

$$u_B = \frac{U}{A\ell} = \frac{B^2}{2\mu_0} \qquad (32.14)$$

Although this expression was derived for the special case of a solenoid, it is valid for any region of space in which a magnetic field exists. Note that Equation 32.14 is similar in form to Equation 26.13 for the energy per unit volume stored in an electric field, $u_E = \tfrac{1}{2} \epsilon_0 E^2$. In both cases, the energy density is proportional to the square of the field magnitude.

Quick Quiz 32.5 You are performing an experiment that requires the highest possible energy density in the interior of a very long solenoid. Which of the following increases the energy density? (More than one choice may be correct.) (a) increasing the number of turns per unit length on the solenoid (b) increasing the cross-sectional area of the solenoid (c) increasing only the length of the solenoid while keeping the number of turns per unit length fixed (d) increasing the current in the solenoid.

Example 32.4   What Happens to the Energy in the Inductor?

Consider once again the RL circuit shown in Figure 32.6. Recall that the current in the right-hand loop decays exponentially with time according to the expression $I = I_0 e^{-t/\tau}$, where $I_0 = \mathcal{E}/R$ is the initial current in the circuit and $\tau = L/R$ is the time constant. Show that all the energy initially stored in the magnetic field of the inductor appears as internal energy in the resistor as the current decays to zero.

Solution The rate dU/dt at which energy is delivered to the resistor (which is the power) is equal to $I^2 R$, where I is the instantaneous current:

$$\frac{dU}{dt} = I^2 R = (I_0 e^{-Rt/L})^2 R = I_0^2 R e^{-2Rt/L}$$

To find the total energy delivered to the resistor, we solve for dU and integrate this expression over the limits $t = 0$ to $t \rightarrow \infty$. (The upper limit is infinity because it takes an infinite amount of time for the current to reach zero.)

$$(1) \qquad U = \int_0^\infty I_0^2 R e^{-2Rt/L} \, dt = I_0^2 R \int_0^\infty e^{-2Rt/L} \, dt$$

The value of the definite integral can be shown to be $L/2R$ (see Problem 34) and so U becomes

$$U = I_0^2 R \left(\frac{L}{2R} \right) = \tfrac{1}{2} L I_0^2$$

Note that this is equal to the initial energy stored in the magnetic field of the inductor, given by Equation 32.12, as we set out to prove.

Example 32.5   The Coaxial Cable

Coaxial cables are often used to connect electrical devices, such as your stereo system, and in receiving signals in TV cable systems. Model a long coaxial cable as consisting of two thin concentric cylindrical conducting shells of radii a and b and length ℓ, as in Figure 32.13. The conducting shells carry the same current I in opposite directions. Imagine that the inner conductor carries current to a device and that the outer one acts as a return path carrying the current back to the source.

(A) Calculate the self-inductance L of this cable.

Solution Conceptualize the situation with the help of Figure 32.13. While we do not have a visible coil in this geometry,

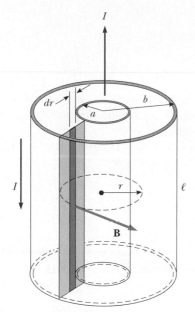

Figure 32.13 (Example 32.5) Section of a long coaxial cable. The inner and outer conductors carry equal currents in opposite directions.

imagine a thin radial slice of the coaxial cable, such as the light blue rectangle in Figure 32.13. If we consider the inner and outer conductors to be connected at the ends of the cable (above and below the figure), this slice represents one large conducting loop. The current in the loop sets up a magnetic field between the inner and outer conductors that passes through this loop. If the current changes, the magnetic field changes and the induced emf opposes the original change in the current in the conductors. We categorize this situation as one in which we can calculate an inductance, but we must return to the fundamental definition of inductance, Equation 32.2. To analyze the problem and obtain L, we must find the magnetic flux through the light blue rectangle in Figure 32.13. Ampère's law (see Section 30.3) tells us that the magnetic field in the region between the shells is due to the inner conductor and its magnitude is $B = \mu_0 I / 2\pi r$, where r is

measured from the common center of the shells. The magnetic field is zero outside the outer shell ($r > b$) because the net current passing through the area enclosed by a circular path surrounding the cable is zero, and hence from Ampère's law, $\oint \mathbf{B} \cdot d\mathbf{s} = 0$. The magnetic field is zero inside the inner shell because the shell is hollow and no current is present within a radius $r < a$.

The magnetic field is perpendicular to the light blue rectangle of length ℓ and width $b - a$, the cross section of interest. Because the magnetic field varies with radial position across this rectangle, we must use calculus to find the total magnetic flux. Dividing this rectangle into strips of width dr, such as the dark blue strip in Figure 32.13, we see that the area of each strip is $\ell \, dr$ and that the flux through each strip is $B \, dA = B\ell \, dr$. Hence, we find the total flux through the entire cross section by integrating:

$$\Phi_B = \int B \, dA = \int_a^b \frac{\mu_0 I}{2\pi r} \ell \, dr = \frac{\mu_0 I \ell}{2\pi} \int_a^b \frac{dr}{r} = \frac{\mu_0 I \ell}{2\pi} \ln\left(\frac{b}{a}\right)$$

Using this result, we find that the self-inductance of the cable is

$$L = \frac{\Phi_B}{I} = \frac{\mu_0 \ell}{2\pi} \ln\left(\frac{b}{a}\right)$$

(B) Calculate the total energy stored in the magnetic field of the cable.

Solution Using Equation 32.12 and the results to part (A) gives

$$U = \tfrac{1}{2} L I^2 = \frac{\mu_0 \ell I^2}{4\pi} \ln\left(\frac{b}{a}\right)$$

To finalize the problem, note that the inductance increases if ℓ increases, if b increases, or if a decreases. This is consistent with our conceptualization—any of these changes increases the size of the loop represented by our radial slice and through which the magnetic field passes; this increases the inductance.

32.4 Mutual Inductance

Very often, the magnetic flux through the area enclosed by a circuit varies with time because of time-varying currents in nearby circuits. This condition induces an emf through a process known as *mutual induction,* so called because it depends on the interaction of two circuits.

Consider the two closely wound coils of wire shown in cross-sectional view in Figure 32.14. The current I_1 in coil 1, which has N_1 turns, creates a magnetic field. Some of the magnetic field lines pass through coil 2, which has N_2 turns. The magnetic flux caused by the current in coil 1 and passing through coil 2 is represented by Φ_{12}. In analogy to Equation 32.2, we define the **mutual inductance** M_{12} of coil 2 with

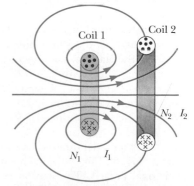

Figure 32.14 A cross-sectional view of two adjacent coils. A current in coil 1 sets up a magnetic field and some of the magnetic field lines pass through coil 2.

respect to coil 1:

Definition of mutual inductance

$$M_{12} \equiv \frac{N_2 \Phi_{12}}{I_1} \qquad (32.15)$$

Mutual inductance depends on the geometry of both circuits and on their orientation with respect to each other. As the circuit separation distance increases, the mutual inductance decreases because the flux linking the circuits decreases.

If the current I_1 varies with time, we see from Faraday's law and Equation 32.15 that the emf induced by coil 1 in coil 2 is

$$\mathcal{E}_2 = -N_2 \frac{d\Phi_{12}}{dt} = -N_2 \frac{d}{dt}\left(\frac{M_{12}I_1}{N_2}\right) = -M_{12}\frac{dI_1}{dt} \qquad (32.16)$$

In the preceding discussion, we assumed that the current is in coil 1. We can also imagine a current I_2 in coil 2. The preceding discussion can be repeated to show that there is a mutual inductance M_{21}. If the current I_2 varies with time, the emf induced by coil 2 in coil 1 is

$$\mathcal{E}_1 = -M_{21}\frac{dI_2}{dt} \qquad (32.17)$$

In mutual induction, the emf induced in one coil is always proportional to the rate at which the current in the other coil is changing. Although the proportionality constants M_{12} and M_{21} have been treated separately, it can be shown that they are equal. Thus, with $M_{12} = M_{21} = M$, Equations 32.16 and 32.17 become

$$\mathcal{E}_2 = -M\frac{dI_1}{dt} \qquad \text{and} \qquad \mathcal{E}_1 = -M\frac{dI_2}{dt}$$

These two equations are similar in form to Equation 32.1 for the self-induced emf $\mathcal{E} = -L(dI/dt)$. The unit of mutual inductance is the henry.

Quick Quiz 32.6 In Figure 32.14, coil 1 is moved closer to coil 2, with the orientation of both coils remaining fixed. Because of this movement, the mutual induction of the two coils (a) increases (b) decreases (c) is unaffected.

Example 32.6   "Wireless" Battery Charger

An electric toothbrush has a base designed to hold the toothbrush handle when not in use. As shown in Figure 32.15a, the handle has a cylindrical hole that fits loosely over a matching cylinder on the base. When the handle is placed on the base, a changing current in a solenoid inside the base cylinder induces a current in a coil inside the handle. This induced current charges the battery in the handle.

We can model the base as a solenoid of length ℓ with N_B turns (Fig. 32.15b), carrying a current I, and having a cross-sectional area A. The handle coil contains N_H turns and completely surrounds the base coil. Find the mutual inductance of the system.

Solution Because the base solenoid carries a current I, the magnetic field in its interior is

$$B = \frac{\mu_0 N_B I}{\ell}$$

Because the magnetic flux Φ_{BH} through the handle's coil caused by the magnetic field of the base coil is BA, the mutual inductance is

$$M = \frac{N_H \Phi_{BH}}{I} = \frac{N_H BA}{I} = \mu_0 \frac{N_B N_H A}{\ell}$$

Wireless charging is used in a number of other "cordless" devices. One significant example is the inductive charging that is used by some electric car manufacturers that avoids direct metal-to-metal contact between the car and the charging apparatus.

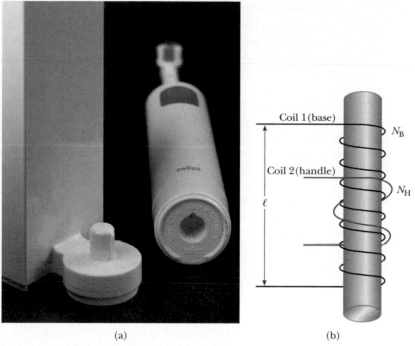

Figure 32.15 (Example 32.6) (a) This electric toothbrush uses the mutual induction of solenoids as part of its battery-charging system. (b) A coil of N_H turns wrapped around the center of a solenoid of N_B turns.

32.5 Oscillations in an *LC* Circuit

When a capacitor is connected to an inductor as illustrated in Figure 32.16, the combination is an **LC circuit.** If the capacitor is initially charged and the switch is then closed, we find that both the current in the circuit and the charge on the capacitor oscillate between maximum positive and negative values. If the resistance of the circuit is zero, no energy is transformed to internal energy. In the following analysis, we neglect the resistance in the circuit. We also assume an idealized situation in which energy is not radiated away from the circuit. We shall discuss this radiation in Chapter 34, but we neglect it for now. With these idealizations—zero resistance and no radiation—the oscillations in the circuit persist indefinitely.

Assume that the capacitor has an initial charge Q_{max} (the maximum charge) and that the switch is open for $t < 0$ and then closed at $t = 0$. Let us investigate what happens from an energy viewpoint.

When the capacitor is fully charged, the energy U in the circuit is stored in the electric field of the capacitor and is equal to $Q_{max}^2/2C$ (Eq. 26.11). At this time, the current in the circuit is zero, and therefore no energy is stored in the inductor. After the switch is closed, the rate at which charges leave or enter the capacitor plates (which is also the rate at which the charge on the capacitor changes) is equal to the current in the circuit. As the capacitor begins to discharge after the switch is closed, the energy stored in its electric field decreases. The discharge of the capacitor represents a current in the circuit, and hence some energy is now stored in the magnetic field of the inductor. Thus, energy is transferred from the electric field of the capacitor to the magnetic field of the inductor. When the capacitor is fully discharged, it stores no energy. At this time, the current reaches its maximum value, and all of the energy is stored in the inductor. The current continues in the same direction, decreasing in magnitude, with the capacitor eventually becoming fully charged again but with the polarity of its plates now opposite the initial polarity. This is followed by another discharge until the circuit returns to its original state of maximum charge Q_{max} and the plate polarity shown in Figure 32.16. The energy continues to oscillate between inductor and capacitor.

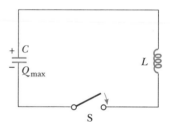

Figure 32.16 A simple *LC* circuit. The capacitor has an initial charge Q_{max}, and the switch is open for $t < 0$ and then closed at $t = 0$.

The oscillations of the LC circuit are an electromagnetic analog to the mechanical oscillations of a block–spring system, which we studied in Chapter 15. Much of what is discussed there is applicable to LC oscillations. For example, we investigated the effect of driving a mechanical oscillator with an external force, which leads to the phenomenon of *resonance*. The same phenomenon is observed in the LC circuit. For example, a radio tuner has an LC circuit with a natural frequency. When the circuit is driven by the electromagnetic oscillations of a radio signal detected by the antenna, the tuner circuit responds with a large amplitude of electrical oscillation only for the station frequency that matches the natural frequency. Therefore, only the signal from one radio station is passed on to the amplifier, even though signals from all stations are driving the circuit at the same time. When you turn the knob on the radio tuner to change the station, you are changing the natural frequency of the circuit so that it will exhibit a resonance response to a different driving frequency.

A graphical description of the fields in the inductor and the capacitor in an LC circuit is shown in Figure 32.17. The right side of the figure shows the analogous oscillating block–spring system studied in Chapter 15. In each case, the situation is shown at intervals of one-fourth the period of oscillation T. The potential energy $\frac{1}{2}kx^2$ stored in a stretched spring is analogous to the electric potential energy $\frac{1}{2}C(\Delta V_{max})^2$ stored in the capacitor. The kinetic energy $\frac{1}{2}mv^2$ of the moving block is analogous to the magnetic energy $\frac{1}{2}LI^2$ stored in the inductor, which requires the presence of moving charges. In Figure 32.17a, all of the energy is stored as electric potential energy in the capacitor at $t = 0$. In Figure 32.17b, which is one fourth of a period later, all of the energy is stored as magnetic energy $\frac{1}{2}LI_{max}^2$ in the inductor, where I_{max} is the maximum current in the circuit. In Figure 32.17c, the energy in the LC circuit is stored completely in the capacitor, with the polarity of the plates now opposite what it was in Figure 32.17a. In parts d and e, the system returns to the initial configuration over the second half of the cycle. At times other than those shown in the figure, part of the energy is stored in the electric field of the capacitor and part is stored in the magnetic field of the inductor. In the analogous mechanical oscillation, part of the energy is potential energy in the spring and part is kinetic energy of the block.

Let us consider some arbitrary time t after the switch is closed, so that the capacitor has a charge $Q < Q_{max}$ and the current is $I < I_{max}$. At this time, both circuit elements store energy, but the sum of the two energies must equal the total initial energy U stored in the fully charged capacitor at $t = 0$:

Total energy stored in an *LC* circuit

$$U = U_C + U_L = \frac{Q^2}{2C} + \tfrac{1}{2}LI^2 \tag{32.18}$$

Because we have assumed the circuit resistance to be zero and we ignore electromagnetic radiation, no energy is transformed to internal energy and none is transferred out of the system of the circuit. Therefore, *the total energy of the system must remain constant in time.* This means that $dU/dt = 0$. Therefore, by differentiating Equation 32.18 with respect to time while noting that Q and I vary with time, we obtain

$$\frac{dU}{dt} = \frac{d}{dt}\left(\frac{Q^2}{2C} + \tfrac{1}{2}LI^2\right) = \frac{Q}{C}\frac{dQ}{dt} + LI\frac{dI}{dt} = 0 \tag{32.19}$$

We can reduce this to a differential equation in one variable by remembering that the current in the circuit is equal to the rate at which the charge on the capacitor changes: $I = dQ/dt$. From this, it follows that $dI/dt = d^2Q/dt^2$. Substitution of these relationships into Equation 32.19 gives

$$\frac{Q}{C} + L\frac{d^2Q}{dt^2} = 0$$

$$\frac{d^2Q}{dt^2} = -\frac{1}{LC}Q \tag{32.20}$$

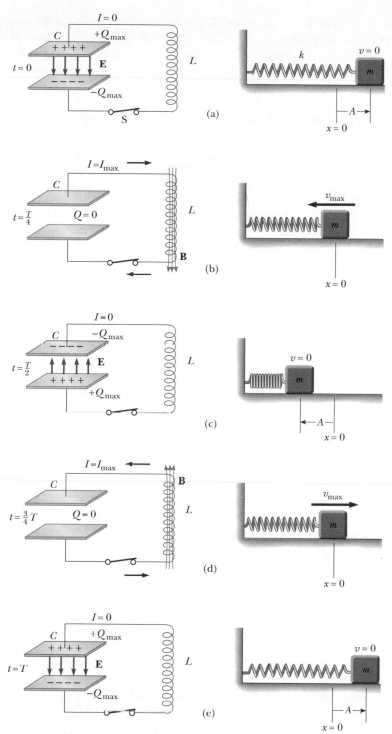

Active Figure 32.17 Energy transfer in a resistanceless, nonradiating *LC* circuit. The capacitor has a charge Q_{max} at $t = 0$, the instant at which the switch is closed. The mechanical analog of this circuit is a block–spring system.

At the Active Figures link at **http://www.pse6.com, you can adjust the values of C and L to see the effect on the oscillating current. The block on the spring oscillates in a mechanical analog of the electrical oscillations. A graphical display as in Figure 32.18 is available, as is an energy bar graph.**

We can solve for Q by noting that this expression is of the same form as the analogous Equations 15.3 and 15.5 for a block-spring system:

$$\frac{d^2x}{dt^2} = -\frac{k}{m}x = -\omega^2 x$$

where k is the spring constant, m is the mass of the block, and $\omega = \sqrt{k/m}$. The solution of this equation has the general form (Eq. 15.6),

$$x = A\cos(\omega t + \phi)$$

where ω is the angular frequency of the simple harmonic motion, A is the amplitude of motion (the maximum value of x), and ϕ is the phase constant; the values of A and ϕ depend on the initial conditions. Because Equation 32.20 is of the same form as the differential equation of the simple harmonic oscillator, we see that it has the solution

Charge as a function of time for an ideal *LC* circuit

$$Q = Q_{max} \cos(\omega t + \phi) \qquad (32.21)$$

where Q_{max} is the maximum charge of the capacitor and the angular frequency ω is

Angular frequency of oscillation in an *LC* circuit

$$\omega = \frac{1}{\sqrt{LC}} \qquad (32.22)$$

Note that the angular frequency of the oscillations depends solely on the inductance and capacitance of the circuit. This is the *natural frequency* of oscillation of the *LC* circuit.

Because Q varies sinusoidally with time, the current in the circuit also varies sinusoidally. We can easily show this by differentiating Equation 32.21 with respect to time:

Current as a function of time for an ideal *LC* current

$$I = \frac{dQ}{dt} = -\omega Q_{max} \sin(\omega t + \phi) \qquad (32.23)$$

To determine the value of the phase angle ϕ, we examine the initial conditions, which in our situation require that at $t = 0$, $I = 0$ and $Q = Q_{max}$. Setting $I = 0$ at $t = 0$ in Equation 32.23, we have

$$0 = -\omega Q_{max} \sin \phi$$

which shows that $\phi = 0$. This value for ϕ also is consistent with Equation 32.21 and with the condition that $Q = Q_{max}$ at $t = 0$. Therefore, in our case, the expressions for Q and I are

$$Q = Q_{max} \cos \omega t \qquad (32.24)$$

$$I = -\omega Q_{max} \sin \omega t = -I_{max} \sin \omega t \qquad (32.25)$$

Graphs of Q versus t and I versus t are shown in Figure 32.18. Note that the charge on the capacitor oscillates between the extreme values Q_{max} and $-Q_{max}$, and that the current oscillates between I_{max} and $-I_{max}$. Furthermore, the current is 90° out of phase with the charge. That is, when the charge is a maximum, the current is zero, and when the charge is zero, the current has its maximum value.

Let us return to the energy discussion of the *LC* circuit. Substituting Equations 32.24 and 32.25 in Equation 32.18, we find that the total energy is

$$U = U_C + U_L = \frac{Q_{max}^2}{2C} \cos^2 \omega t + \frac{1}{2} L I_{max}^2 \sin^2 \omega t \qquad (32.26)$$

This expression contains all of the features described qualitatively at the beginning of this section. It shows that the energy of the *LC* circuit continuously oscillates between energy stored in the electric field of the capacitor and energy stored in the magnetic

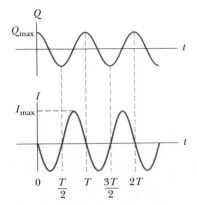

At the Active Figures link at http://www.pse6.com, you can observe this graph develop for the LC circuit in Figure 32.17.

Active Figure 32.18 Graphs of charge versus time and current versus time for a resistanceless, nonradiating *LC* circuit. Note that Q and I are 90° out of phase with each other.

field of the inductor. When the energy stored in the capacitor has its maximum value $Q_{max}^2/2C$, the energy stored in the inductor is zero. When the energy stored in the inductor has its maximum value $\frac{1}{2}LI_{max}^2$, the energy stored in the capacitor is zero.

Plots of the time variations of U_C and U_L are shown in Figure 32.19. The sum $U_C + U_L$ is a constant and equal to the total energy $Q_{max}^2/2C$, or $\frac{1}{2}LI_{max}^2$. Analytical verification of this is straightforward. The amplitudes of the two graphs in Figure 32.19 must be equal because the maximum energy stored in the capacitor (when $I = 0$) must equal the maximum energy stored in the inductor (when $Q = 0$). This is mathematically expressed as

$$\frac{Q_{max}^2}{2C} = \frac{LI_{max}^2}{2}$$

Using this expression in Equation 32.26 for the total energy gives

$$U = \frac{Q_{max}^2}{2C}(\cos^2 \omega t + \sin^2 \omega t) = \frac{Q_{max}^2}{2C} \qquad (32.27)$$

because $\cos^2 \omega t + \sin^2 \omega t = 1$.

In our idealized situation, the oscillations in the circuit persist indefinitely; however, remember that the total energy U of the circuit remains constant only if energy transfers and transformations are neglected. In actual circuits, there is always some resistance, and hence some energy is transformed to internal energy. We mentioned at the beginning of this section that we are also ignoring radiation from the circuit. In reality, radiation is inevitable in this type of circuit, and the total energy in the circuit continuously decreases as a result of this process.

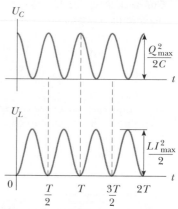

Figure 32.19 Plots of U_C versus t and U_L versus t for a resistanceless, nonradiating *LC* circuit. The sum of the two curves is a constant and equal to the total energy stored in the circuit.

Example 32.7   Oscillations in an *LC* Circuit   **Interactive**

In Figure 32.20, the capacitor is initially charged when switch S_1 is open and S_2 is closed. Switch S_2 is then opened, removing the battery from the circuit, and the capacitor remains charged. Switch S_1 is then closed, so that the capacitor is connected directly across the inductor.

(A) Find the frequency of oscillation of the circuit.

Solution Using Equation 32.22 gives for the frequency

$$f = \frac{\omega}{2\pi} = \frac{1}{2\pi\sqrt{LC}}$$

$$= \frac{1}{2\pi[(2.81 \times 10^{-3}\,\text{H})(9.00 \times 10^{-12}\,\text{F})]^{1/2}}$$

$$= \boxed{1.00 \times 10^6\,\text{Hz}}$$

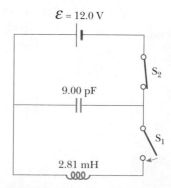

Figure 32.20 (Example 32.7) First the capacitor is fully charged with the switch S_1 open and S_2 closed. Then, S_2 is opened and S_1 is closed.

(B) What are the maximum values of charge on the capacitor and current in the circuit?

Solution The initial charge on the capacitor equals the maximum charge, and because $C = Q/\Delta V$, we have

$$Q_{max} = C\,\Delta V = (9.00 \times 10^{-12}\,\text{F})(12.0\,\text{V})$$

$$= 1.08 \times 10^{-10}\,\text{C}$$

From Equation 32.25, we can see how the maximum current is related to the maximum charge:

$$I_{max} = \omega Q_{max} = 2\pi f Q_{max}$$

$$= (2\pi \times 10^6\,\text{s}^{-1})(1.08 \times 10^{-10}\,\text{C})$$

$$I_{max} = 6.79 \times 10^{-4}\,\text{A}$$

(C) Determine the charge and current as functions of time.

Solution Equations 32.24 and 32.25 give the following expressions for the time variation of Q and I:

$$Q = Q_{max}\cos \omega t$$

$$= (1.08 \times 10^{-10}\,\text{C})\cos[(2\pi \times 10^6\,\text{s}^{-1})t]$$

$$I = -I_{max}\sin \omega t$$

$$= (-6.79 \times 10^{-4}\,\text{A})\sin[(2\pi \times 10^6\,\text{s}^{-1})t]$$

 At the Interactive Worked Example link at **http://www.pse6.com,** *you can study the oscillations in an LC circuit.*

32.6 The *RLC* Circuit

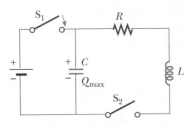

Active Figure 32.21 A series *RLC* circuit. Switch S_1 is closed and the capacitor is charged. S_1 is then opened and, at $t = 0$, switch S_2 is closed.

 **At the Active Figures link at http://www.pse6.com, you can adjust the values of R, L, and C to see the effect on the decaying charge on the capacitor. A graphical display as in Figure 32.23a is available, as is an energy bar graph.**

We now turn our attention to a more realistic circuit consisting of a resistor, an inductor, and a capacitor connected in series, as shown in Figure 32.21. Let us assume that the resistance of the resistor represents all of the resistance in the circuit. Now imagine that switch S_1 is closed and S_2 is open, so that the capacitor has an initial charge Q_{max}. Next, S_1 is opened and S_2 is closed. Once S_2 is closed and a current is established, the total energy stored in the capacitor and inductor at any time is given by Equation 32.18. However, this total energy is no longer constant, as it was in the *LC* circuit, because the resistor causes transformation to internal energy. (We continue to ignore electromagnetic radiation from the circuit in this discussion.) Because the rate of energy transformation to internal energy within a resistor is $I^2 R$, we have

$$\frac{dU}{dt} = -I^2 R$$

where the negative sign signifies that the energy U of the circuit is decreasing in time. Substituting this result into Equation 32.19 gives

$$LI\frac{dI}{dt} + \frac{Q}{C}\frac{dQ}{dt} = -I^2 R \tag{32.28}$$

To convert this equation into a form that allows us to compare the electrical oscillations with their mechanical analog, we first use the fact that $I = dQ/dt$ and move all terms to the left-hand side to obtain

$$LI\frac{d^2Q}{dt^2} + I^2 R + \frac{Q}{C}I = 0$$

Now we divide through by I:

$$L\frac{d^2Q}{dt^2} + IR + \frac{Q}{C} = 0$$

$$L\frac{d^2Q}{dt^2} + R\frac{dQ}{dt} + \frac{Q}{C} = 0 \tag{32.29}$$

The *RLC* circuit is analogous to the damped harmonic oscillator discussed in Section 15.6 and illustrated in Figure 32.22. The equation of motion for a damped

Figure 32.22 A block–spring system moving in a viscous medium with damped harmonic motion is analogous to an *RLC* circuit.

Table 32.1

Analogies Between Electrical and Mechanical Systems		
Electric Circuit		**One-Dimensional Mechanical System**
Charge	$Q \leftrightarrow x$	Position
Current	$I \leftrightarrow v_x$	Velocity
Potential difference	$\Delta V \leftrightarrow F_x$	Force
Resistance	$R \leftrightarrow b$	Viscous damping coefficient
Capacitance	$C \leftrightarrow 1/k$	(k = spring constant)
Inductance	$L \leftrightarrow m$	Mass
Current = time derivative of charge	$I = \dfrac{dQ}{dt} \leftrightarrow v_x = \dfrac{dx}{dt}$	Velocity = time derivative of position
Rate of change of current = second time derivative of charge	$\dfrac{dI}{dt} = \dfrac{d^2Q}{dt^2} \leftrightarrow a_x = \dfrac{dv_x}{dt} = \dfrac{d^2x}{dt^2}$	Acceleration = second time derivative of position
Energy in inductor	$U_L = \tfrac{1}{2}LI^2 \leftrightarrow K = \tfrac{1}{2}mv^2$	Kinetic energy of moving object
Energy in capacitor	$U_C = \tfrac{1}{2}\dfrac{Q^2}{C} \leftrightarrow U = \tfrac{1}{2}kx^2$	Potential energy stored in a spring
Rate of energy loss due to resistance	$I^2R \leftrightarrow bv^2$	Rate of energy loss due to friction
RLC circuit	$L\dfrac{d^2Q}{dt^2} + R\dfrac{dQ}{dt} + \dfrac{Q}{C} = 0 \leftrightarrow m\dfrac{d^2x}{dt^2} + b\dfrac{dx}{dt} + kx = 0$	Damped object on a spring

block–spring system is, from Equation 15.31,

$$m\frac{d^2x}{dt^2} + b\frac{dx}{dt} + kx = 0 \qquad (32.30)$$

Comparing Equations 32.29 and 32.30, we see that Q corresponds to the position x of the block at any instant, L to the mass m of the block, R to the damping coefficient b, and C to $1/k$, where k is the force constant of the spring. These and other relationships are listed in Table 32.1.

Because the analytical solution of Equation 32.29 is cumbersome, we give only a qualitative description of the circuit behavior. In the simplest case, when $R = 0$, Equation 32.29 reduces to that of a simple *LC* circuit, as expected, and the charge and the current oscillate sinusoidally in time. This is equivalent to removal of all damping in the mechanical oscillator.

When R is small, a situation analogous to light damping in the mechanical oscillator, the solution of Equation 32.29 is

$$Q = Q_{\max}\, e^{-Rt/2L} \cos \omega_d t \qquad (32.31)$$

where ω_d, the angular frequency at which the circuit oscillates, is given by

$$\omega_d = \left[\frac{1}{LC} - \left(\frac{R}{2L}\right)^2\right]^{1/2} \qquad (32.32)$$

That is, the value of the charge on the capacitor undergoes a damped harmonic oscillation in analogy with a block–spring system moving in a viscous medium. From Equation 32.32, we see that, when $R \ll \sqrt{4L/C}$ (so that the second term in the brackets is much smaller than the first), the frequency ω_d of the damped oscillator is close to that of the undamped oscillator, $1/\sqrt{LC}$. Because $I = dQ/dt$, it follows that the current also

At the Active Figures link at http://www.pse6.com, you can observe this graph develop for the damped RLC circuit in Figure 32.21.

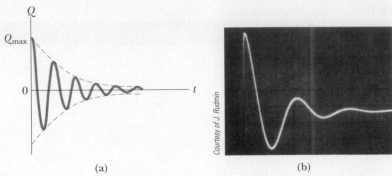

(a)

(b)

Courtesy of J. Rudmin

Active Figure 32.23 (a) Charge versus time for a damped *RLC* circuit. The charge decays in this way when $R > \sqrt{4L/C}$. The *Q*-versus-*t* curve represents a plot of Equation 32.31. (b) Oscilloscope pattern showing the decay in the oscillations of an *RLC* circuit.

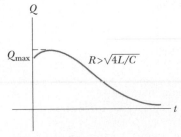

Figure 32.24 Plot of Q versus t for an overdamped *RLC* circuit, which occurs for values of $R > \sqrt{4L/C}$.

undergoes damped harmonic oscillation. A plot of the charge versus time for the damped oscillator is shown in Figure 32.23a. Note that the maximum value of Q decreases after each oscillation, just as the amplitude of a damped block–spring system decreases in time.

When we consider larger values of R, we find that the oscillations damp out more rapidly; in fact, there exists a critical resistance value $R_c = \sqrt{4L/C}$ above which no oscillations occur. A system with $R = R_c$ is said to be *critically damped*. When R exceeds R_c, the system is said to be *overdamped* (Fig. 32.24).

Take a practice test for this chapter by clicking on the Practice Test link at http://www.pse6.com.

SUMMARY

When the current in a coil changes with time, an emf is induced in the coil according to Faraday's law. The **self-induced emf** is

$$\mathcal{E}_L = -L\,\frac{dI}{dt} \tag{32.1}$$

where L is the **inductance** of the coil. Inductance is a measure of how much opposition a coil offers to a change in the current in the coil. Inductance has the SI unit of **henry** (H), where $1\ \text{H} = 1\ \text{V}\cdot\text{s}/\text{A}$.

The inductance of any coil is

$$L = \frac{N\Phi_B}{I} \tag{32.2}$$

where Φ_B is the magnetic flux through the coil and N is the total number of turns. The inductance of a device depends on its geometry. For example, the inductance of an air-core solenoid is

$$L = \frac{\mu_0 N^2 A}{\ell} \tag{32.4}$$

where A is the cross-sectional area, and ℓ is the length of the solenoid.

If a resistor and inductor are connected in series to a battery of emf $\mathcal{E}$, and if a switch in the circuit is open for $t < 0$ and then closed at $t = 0$, the current in the circuit varies in time according to the expression

$$I = \frac{\mathcal{E}}{R}\left(1 - e^{-t/\tau}\right) \tag{32.7}$$

where $\tau = L/R$ is the **time constant** of the RL circuit. That is, the current increases to an equilibrium value of $\mathcal{E}/R$ after a time interval that is long compared with τ. If the battery in the circuit is replaced by a resistanceless wire, the current decays exponentially with time according to the expression

$$I = \frac{\mathcal{E}}{R} e^{-t/\tau} \tag{32.10}$$

where $\mathcal{E}/R$ is the initial current in the circuit.

The energy stored in the magnetic field of an inductor carrying a current I is

$$U = \tfrac{1}{2} LI^2 \tag{32.12}$$

This energy is the magnetic counterpart to the energy stored in the electric field of a charged capacitor.

The energy density at a point where the magnetic field is B is

$$u_B = \frac{B^2}{2\mu_0} \tag{32.14}$$

The **mutual inductance** of a system of two coils is given by

$$M_{12} = \frac{N_2 \Phi_{12}}{I_1} = M_{21} = \frac{N_1 \Phi_{21}}{I_2} = M \tag{32.15}$$

This mutual inductance allows us to relate the induced emf in a coil to the changing source current in a nearby coil using the relationships

$$\mathcal{E}_2 = -M \frac{dI_1}{dt} \quad \text{and} \quad \mathcal{E}_1 = -M \frac{dI_2}{dt} \tag{32.16, 32.17}$$

In an LC circuit that has zero resistance and does not radiate electromagnetically (an idealization), the values of the charge on the capacitor and the current in the circuit vary in time according to the expressions

$$Q = Q_{\max} \cos(\omega t + \phi) \tag{32.21}$$

$$I = \frac{dQ}{dt} = -\omega Q_{\max} \sin(\omega t + \phi) \tag{32.23}$$

where $Q_{\max}$ is the maximum charge on the capacitor, ϕ is a phase constant, and ω is the angular frequency of oscillation:

$$\omega = \frac{1}{\sqrt{LC}} \tag{32.22}$$

The energy in an LC circuit continuously transfers between energy stored in the capacitor and energy stored in the inductor. The total energy of the LC circuit at any time t is

$$U = U_C + U_L = \frac{Q_{\max}^2}{2C} \cos^2 \omega t + \frac{LI_{\max}^2}{2} \sin^2 \omega t \tag{32.26}$$

At $t = 0$, all of the energy is stored in the electric field of the capacitor ($U = Q_{\max}^2/2C$). Eventually, all of this energy is transferred to the inductor ($U = LI_{\max}^2/2$). However, the total energy remains constant because energy transformations are neglected in the ideal LC circuit.

In an RLC circuit with small resistance, the charge on the capacitor varies with time according to

$$Q = Q_{\max} e^{-Rt/2L} \cos \omega_d t \tag{32.31}$$

where

$$\omega_d = \left[\frac{1}{LC} - \left(\frac{R}{2L} \right)^2 \right]^{1/2} \tag{32.32}$$

QUESTIONS

1. Why is the induced emf that appears in an inductor called a "counter" or "back" emf?

2. The current in a circuit containing a coil, resistor, and battery has reached a constant value. Does the coil have an inductance? Does the coil affect the value of the current?

3. What parameters affect the inductance of a coil? Does the inductance of a coil depend on the current in the coil?

4. How can a long piece of wire be wound on a spool so that the wire has a negligible self-inductance?

5. For the series *RL* circuit shown in Figure Q32.5, can the back emf ever be greater than the battery emf? Explain.

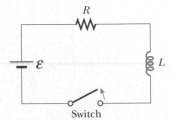

Figure Q32.5 Questions 5 and 6.

6. Suppose the switch in Figure Q32.5 has been closed for a long time and is suddenly opened. Does the current instantaneously drop to zero? Why does a spark appear at the switch contacts at the moment the switch is opened?

7. A switch controls the current in a circuit that has a large inductance. Is a spark (see Figure Q32.7) more likely to be produced at the switch when the switch is being closed or when it is being opened, or doesn't it matter? The electric arc can melt and oxidize the contact surfaces, resulting in high resistivity of the contacts and eventual destruction of the switch. Before electronic ignitions were invented, distributor contact points in automobiles had to be replaced regularly. Switches in power distribution networks and switches controlling large motors, generators, and electromagnets can suffer from arcing and can be very dangerous to operate.

8. Consider this thesis: "Joseph Henry, America's first professional physicist, caused the most recent basic change in the human view of the Universe when he discovered self-induction during a school vacation at the Albany Academy about 1830. Before that time, one could think of the Universe as composed of just one thing: matter. The energy that temporarily maintains the current after a battery is removed from a coil, on the other hand, is not energy that belongs to any chunk of matter. It is energy in the massless magnetic field surrounding the coil. With Henry's discovery, Nature forced us to admit that the Universe consists of fields as well as matter." Argue for or against the statement. What in your view comprises the Universe?

9. If the current in an inductor is doubled, by what factor does the stored energy change?

10. Discuss the similarities between the energy stored in the electric field of a charged capacitor and the energy stored in the magnetic field of a current-carrying coil.

11. What is the inductance of two inductors connected in series? Does it matter if they are solenoids or toroids?

12. The centers of two circular loops are separated by a fixed distance. For what relative orientation of the loops is their mutual inductance a maximum? a minimum? Explain.

13. Two solenoids are connected in series so that each carries the same current at any instant. Is mutual induction present? Explain.

14. In the *LC* circuit shown in Figure 32.16, the charge on the capacitor is sometimes zero, but at such instants the current in the circuit is not zero. How is this possible?

15. If the resistance of the wires in an *LC* circuit were not zero, would the oscillations persist? Explain.

16. How can you tell whether an *RLC* circuit is overdamped or underdamped?

17. What is the significance of critical damping in an *RLC* circuit?

18. Can an object exert a force on itself? When a coil induces an emf in itself, does it exert a force on itself?

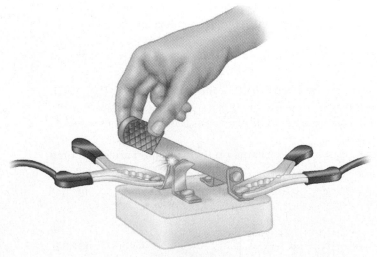

Figure Q32.7

PROBLEMS

1, 2, 3 = straightforward, intermediate, challenging ☐ = full solution available in the *Student Solutions Manual and Study Guide*

🌀 = coached solution with hints available at http://www.pse6.com 💻 = computer useful in solving problem

▨ = paired numerical and symbolic problems

Section 32.1 Self-Inductance

1. A coil has an inductance of 3.00 mH, and the current in it changes from 0.200 A to 1.50 A in a time of 0.200 s. Find the magnitude of the average induced emf in the coil during this time.

2. A coiled telephone cord forms a spiral with 70 turns, a diameter of 1.30 cm, and an unstretched length of 60.0 cm. Determine the self-inductance of one conductor in the unstretched cord.

3. A 2.00-H inductor carries a steady current of 0.500 A. When the switch in the circuit is opened, the current is effectively zero after 10.0 ms. What is the average induced emf in the inductor during this time?

4. Calculate the magnetic flux through the area enclosed by a 300-turn, 7.20-mH coil when the current in the coil is 10.0 mA.

5. 🌀 A 10.0-mH inductor carries a current $I = I_{max} \sin \omega t$, with $I_{max} = 5.00$ A and $\omega/2\pi = 60.0$ Hz. What is the back emf as a function of time?

6. An emf of 24.0 mV is induced in a 500-turn coil at an instant when the current is 4.00 A and is changing at the rate of 10.0 A/s. What is the magnetic flux through each turn of the coil?

7. An inductor in the form of a solenoid contains 420 turns, is 16.0 cm in length, and has a cross-sectional area of 3.00 cm². What uniform rate of decrease of current through the inductor induces an emf of 175 μV?

8. The current in a 90.0-mH inductor changes with time as $I = 1.00t^2 - 6.00t$ (in SI units). Find the magnitude of the induced emf at (a) $t = 1.00$ s and (b) $t = 4.00$ s. (c) At what time is the emf zero?

9. A 40.0-mA current is carried by a uniformly wound air-core solenoid with 450 turns, a 15.0-mm diameter, and 12.0-cm length. Compute (a) the magnetic field inside the solenoid, (b) the magnetic flux through each turn, and (c) the inductance of the solenoid. (d) **What If?** If the current were different, which of these quantities would change?

10. A solenoid has 120 turns uniformly wrapped around a wooden core, which has a diameter of 10.0 mm and a length of 9.00 cm. (a) Calculate the inductance of the solenoid. (b) **What If?** The wooden core is replaced with a soft iron rod that has the same dimensions, but a magnetic permeability $\mu_m = 800\mu_0$. What is the new inductance?

11. A piece of copper wire with thin insulation, 200 m long and 1.00 mm in diameter, is wound onto a plastic tube to form a long solenoid. This coil has a circular cross section and consists of tightly wound turns in one layer. If the current in the solenoid drops linearly from 1.80 A to zero in 0.120 seconds, an emf of 80.0 mV is induced in the coil. What is the length of the solenoid, measured along its axis?

12. A toroid has a major radius R and a minor radius r, and is tightly wound with N turns of wire, as shown in Figure P32.12. If $R \gg r$, the magnetic field in the region enclosed by the wire of the torus, of cross-sectional area $A = \pi r^2$, is essentially the same as the magnetic field of a solenoid that has been bent into a large circle of radius R. Modeling the field as the uniform field of a long solenoid, show that the self-inductance of such a toroid is approximately

$$ L \approx \frac{\mu_0 N^2 A}{2\pi R} $$

(An exact expression of the inductance of a toroid with a rectangular cross section is derived in Problem 64.)

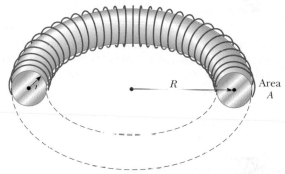

Figure P32.12

13. A self-induced emf in a solenoid of inductance L changes in time as $\mathcal{E} = \mathcal{E}_0 e^{-kt}$. Find the total charge that passes through the solenoid, assuming the charge is finite.

Section 32.2 RL Circuits

14. Calculate the resistance in an RL circuit in which $L = 2.50$ H and the current increases to 90.0% of its final value in 3.00 s.

15. A 12.0-V battery is connected into a series circuit containing a 10.0-Ω resistor and a 2.00-H inductor. How long will it take the current to reach (a) 50.0% and (b) 90.0% of its final value?

16. Show that $I = I_0 e^{-t/\tau}$ is a solution of the differential equation

$$ IR + L\frac{dI}{dt} = 0 $$

where $\tau = L/R$ and I_0 is the current at $t = 0$.

17. Consider the circuit in Figure P32.17, taking $\mathcal{E} = 6.00$ V, $L = 8.00$ mH, and $R = 4.00$ Ω. (a) What is the inductive time constant of the circuit? (b) Calculate the current in

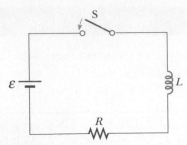

Figure P32.17 Problems 17, 18, 19, and 22.

the circuit 250 μs after the switch is closed. (c) What is the value of the final steady-state current? (d) How long does it take the current to reach 80.0% of its maximum value?

18. In the circuit shown in Figure P32.17, let $L = 7.00$ H, $R = 9.00\ \Omega$, and $\mathcal{E} = 120$ V. What is the self-induced emf 0.200 s after the switch is closed?

19. For the RL circuit shown in Figure P32.17, let the inductance be 3.00 H, the resistance 8.00 Ω, and the battery emf 36.0 V. (a) Calculate the ratio of the potential difference across the resistor to that across the inductor when the current is 2.00 A. (b) Calculate the voltage across the inductor when the current is 4.50 A.

20. A 12.0-V battery is connected in series with a resistor and an inductor. The circuit has a time constant of 500 μs, and the maximum current is 200 mA. What is the value of the inductance?

21. An inductor that has an inductance of 15.0 H and a resistance of 30.0 Ω is connected across a 100-V battery. What is the rate of increase of the current (a) at $t = 0$ and (b) at $t = 1.50$ s?

22. When the switch in Figure P32.17 is closed, the current takes 3.00 ms to reach 98.0% of its final value. If $R = 10.0\ \Omega$, what is the inductance?

23. The switch in Figure P32.23 is open for $t < 0$ and then closed at time $t = 0$. Find the current in the inductor and the current in the switch as functions of time thereafter.

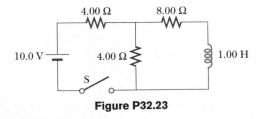

Figure P32.23

24. A series RL circuit with $L = 3.00$ H and a series RC circuit with $C = 3.00\ \mu$F have equal time constants. If the two circuits contain the same resistance R, (a) what is the value of R and (b) what is the time constant?

25. A current pulse is fed to the partial circuit shown in Figure P32.25. The current begins at zero, then becomes 10.0 A between $t = 0$ and $t = 200\ \mu$s, and then is zero once again. Determine the current in the inductor as a function of time.

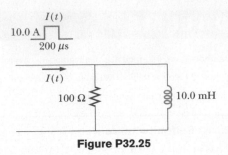

Figure P32.25

26. One application of an RL circuit is the generation of time-varying high voltage from a low-voltage source, as shown in Figure P32.26. (a) What is the current in the circuit a long time after the switch has been in position a? (b) Now the switch is thrown quickly from a to b. Compute the initial voltage across each resistor and across the inductor. (c) How much time elapses before the voltage across the inductor drops to 12.0 V?

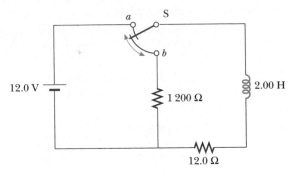

Figure P32.26

27. A 140-mH inductor and a 4.90-Ω resistor are connected with a switch to a 6.00-V battery as shown in Figure P32.27. (a) If the switch is thrown to the left (connecting the battery), how much time elapses before the current reaches 220 mA? (b) What is the current in the inductor 10.0 s after the switch is closed? (c) Now the switch is quickly thrown from a to b. How much time elapses before the current falls to 160 mA?

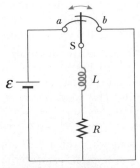

Figure P32.27

28. Consider two ideal inductors L_1 and L_2 that have *zero* internal resistance and are far apart, so that their magnetic fields do not influence each other. (a) Assuming these inductors

are connected in series, show that they are equivalent to a single ideal inductor having $L_{eq} = L_1 + L_2$. (b) Assuming these same two inductors are connected in parallel, show that they are equivalent to a single ideal inductor having $1/L_{eq} = 1/L_1 + 1/L_2$. (c) **What If?** Now consider two inductors L_1 and L_2 that have *nonzero* internal resistances R_1 and R_2, respectively. Assume they are still far apart so that their mutual inductance is zero. Assuming these inductors are connected in series, show that they are equivalent to a single inductor having $L_{eq} = L_1 + L_2$ and $R_{eq} = R_1 + R_2$. (d) If these same inductors are now connected in parallel, is it necessarily true that they are equivalent to a single ideal inductor having $1/L_{eq} = 1/L_1 + 1/L_2$ and $1/R_{eq} = 1/R_1 + 1/R_2$? Explain your answer.

Section 32.3 Energy in a Magnetic Field

29. Calculate the energy associated with the magnetic field of a 200-turn solenoid in which a current of 1.75 A produces a flux of 3.70×10^{-4} Wb in each turn.

30. The magnetic field inside a superconducting solenoid is 4.50 T. The solenoid has an inner diameter of 6.20 cm and a length of 26.0 cm. Determine (a) the magnetic energy density in the field and (b) the energy stored in the magnetic field within the solenoid.

31. An air-core solenoid with 68 turns is 8.00 cm long and has a diameter of 1.20 cm. How much energy is stored in its magnetic field when it carries a current of 0.770 A?

32. At $t = 0$, an emf of 500 V is applied to a coil that has an inductance of 0.800 H and a resistance of 30.0 Ω. (a) Find the energy stored in the magnetic field when the current reaches half its maximum value. (b) After the emf is connected, how long does it take the current to reach this value?

33. On a clear day at a certain location, a 100-V/m vertical electric field exists near the Earth's surface. At the same place, the Earth's magnetic field has a magnitude of 0.500×10^{-4} T. Compute the energy densities of the two fields.

34. Complete the calculation in Example 32.4 by proving that

$$\int_0^\infty e^{-2Rt/L}\, dt = \frac{L}{2R}$$

35. An RL circuit in which $L = 4.00$ H and $R = 5.00\ \Omega$ is connected to a 22.0-V battery at $t = 0$. (a) What energy is stored in the inductor when the current is 0.500 A? (b) At what rate is energy being stored in the inductor when $I = 1.00$ A? (c) What power is being delivered to the circuit by the battery when $I = 0.500$ A?

36. A 10.0-V battery, a 5.00-Ω resistor, and a 10.0-H inductor are connected in series. After the current in the circuit has reached its maximum value, calculate (a) the power being supplied by the battery, (b) the power being delivered to the resistor, (c) the power being delivered to the inductor, and (d) the energy stored in the magnetic field of the inductor.

37. A uniform electric field of magnitude 680 kV/m throughout a cylindrical volume results in a total energy of 3.40 μJ. What magnetic field over this same region stores the same total energy?

38. Assume that the magnitude of the magnetic field outside a sphere of radius R is $B = B_0(R/r)^2$, where B_0 is a constant. Determine the total energy stored in the magnetic field outside the sphere and evaluate your result for $B_0 = 5.00 \times 10^{-5}$ T and $R = 6.00 \times 10^6$ m, values appropriate for the Earth's magnetic field.

Section 32.4 Mutual Inductance

39. Two coils are close to each other. The first coil carries a time-varying current given by $I(t) = (5.00\ \text{A})\, e^{-0.025\,0t} \sin(377t)$. At $t = 0.800$ s, the emf measured across the second coil is -3.20 V. What is the mutual inductance of the coils?

40. Two coils, held in fixed positions, have a mutual inductance of 100 μH. What is the peak voltage in one when a sinusoidal current given by $I(t) = (10.0\ \text{A}) \sin(1\,000t)$ is in the other coil?

41. An emf of 96.0 mV is induced in the windings of a coil when the current in a nearby coil is increasing at the rate of 1.20 A/s. What is the mutual inductance of the two coils?

42. On a printed circuit board, a relatively long straight conductor and a conducting rectangular loop lie in the same plane, as shown in Figure P31.9. Taking $h = 0.400$ mm, $w = 1.30$ mm, and $L = 2.70$ mm, find their mutual inductance.

43. Two solenoids A and B, spaced close to each other and sharing the same cylindrical axis, have 400 and 700 turns, respectively. A current of 3.50 A in coil A produces an average flux of 300 μWb through each turn of A and a flux of 90.0 μWb through each turn of B. (a) Calculate the mutual inductance of the two solenoids. (b) What is the self-inductance of A? (c) What emf is induced in B when the current in A increases at the rate of 0.500 A/s?

44. A large coil of radius R_1 and having N_1 turns is coaxial with a small coil of radius R_2 and having N_2 turns. The centers of the coils are separated by a distance x that is much larger than R_1 and R_2. What is the mutual inductance of the coils? *Suggestion:* John von Neumann proved that the same answer must result from considering the flux through the first coil of the magnetic field produced by the second coil, or from considering the flux through the second coil of the magnetic field produced by the first coil. In this problem it is easy to calculate the flux through the small coil, but it is difficult to calculate the flux through the large coil, because to do so you would have to know the magnetic field away from the axis.

45. Two inductors having self-inductances L_1 and L_2 are connected in parallel as shown in Figure P32.45a. The mutual inductance between the two inductors is M. Determine the equivalent self-inductance L_{eq} for the system (Figure P32.45b).

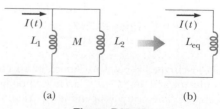

(a)   (b)

Figure P32.45

Section 32.5   Oscillations in an *LC* Circuit

46. A 1.00-μF capacitor is charged by a 40.0-V power supply. The fully charged capacitor is then discharged through a 10.0-mH inductor. Find the maximum current in the resulting oscillations.

47. An *LC* circuit consists of a 20.0-mH inductor and a 0.500-μF capacitor. If the maximum instantaneous current is 0.100 A, what is the greatest potential difference across the capacitor?

48. In the circuit of Figure P32.48, the battery emf is 50.0 V, the resistance is 250 Ω, and the capacitance is 0.500 μF. The switch S is closed for a long time and no voltage is measured across the capacitor. After the switch is opened, the potential difference across the capacitor reaches a maximum value of 150 V. What is the value of the inductance?

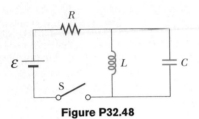

Figure P32.48

49. A fixed inductance $L = 1.05$ μH is used in series with a variable capacitor in the tuning section of a radiotelephone on a ship. What capacitance tunes the circuit to the signal from a transmitter broadcasting at 6.30 MHz?

50. Calculate the inductance of an *LC* circuit that oscillates at 120 Hz when the capacitance is 8.00 μF.

51. An *LC* circuit like the one in Figure 32.16 contains an 82.0-mH inductor and a 17.0-μF capacitor that initially carries a 180-μC charge. The switch is open for $t < 0$ and then closed at $t = 0$. (a) Find the frequency (in hertz) of the resulting oscillations. At $t = 1.00$ ms, find (b) the charge on the capacitor and (c) the current in the circuit.

52. The switch in Figure P32.52 is connected to point a for a long time. After the switch is thrown to point b, what are (a) the frequency of oscillation of the *LC* circuit, (b) the maximum charge that appears on the capacitor, (c) the maximum current in the inductor, and (d) the total energy the circuit possesses at $t = 3.00$ s?

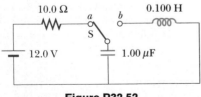

Figure P32.52

53. An *LC* circuit like that in Figure 32.16 consists of a 3.30-H inductor and an 840-pF capacitor, initially carrying a 105-μC charge. The switch is open for $t < 0$ and then closed at $t = 0$. Compute the following quantities at $t = 2.00$ ms:

(a) the energy stored in the capacitor; (b) the energy stored in the inductor; (c) the total energy in the circuit.

Section 32.6   The *RLC* Circuit

54. In Figure 32.21, let $R = 7.60$ Ω, $L = 2.20$ mH, and $C = 1.80$ μF. (a) Calculate the frequency of the damped oscillation of the circuit. (b) What is the critical resistance?

55. Consider an *LC* circuit in which $L = 500$ mH and $C = 0.100$ μF. (a) What is the resonance frequency ω_0? (b) If a resistance of 1.00 kΩ is introduced into this circuit, what is the frequency of the (damped) oscillations? (c) What is the percent difference between the two frequencies?

56. Show that Equation 32.28 in the text is Kirchhoff's loop rule as applied to the circuit in Figure 32.21.

57. The energy of an *RLC* circuit decreases by 1.00% during each oscillation when $R = 2.00$ Ω. If this resistance is removed, the resulting *LC* circuit oscillates at a frequency of 1.00 kHz. Find the values of the inductance and the capacitance.

58. Electrical oscillations are initiated in a series circuit containing a capacitance C, inductance L, and resistance R. (a) If $R \ll \sqrt{4L/C}$ (weak damping), how much time elapses before the amplitude of the current oscillation falls off to 50.0% of its initial value? (b) How long does it take the energy to decrease to 50.0% of its initial value?

Additional Problems

59. **Review problem**. This problem extends the reasoning of Section 26.4, Problem 26.37, Example 30.6, and Section 32.3. (a) Consider a capacitor with vacuum between its large, closely spaced, oppositely charged parallel plates. Show that the force on one plate can be accounted for by thinking of the electric field between the plates as exerting a "negative pressure" equal to the energy density of the electric field. (b) Consider two infinite plane sheets carrying electric currents in opposite directions with equal linear current densities J_s. Calculate the force per area acting on one sheet due to the magnetic field created by the other sheet. (c) Calculate the net magnetic field between the sheets and the field outside of the volume between them. (d) Calculate the energy density in the magnetic field between the sheets. (e) Show that the force on one sheet can be accounted for by thinking of the magnetic field between the sheets as exerting a positive pressure equal to its energy density. This result for magnetic pressure applies to all current configurations, not just to sheets of current.

60. Initially, the capacitor in a series *LC* circuit is charged. A switch is closed at $t = 0$, allowing the capacitor to discharge, and at time t the energy stored in the capacitor is one fourth of its initial value. Determine L, assuming C is known.

61. A 1.00-mH inductor and a 1.00-μF capacitor are connected in series. The current in the circuit is described by $I = 20.0t$, where t is in seconds and I is in amperes. The capacitor initially has no charge. Determine (a) the

voltage across the inductor as a function of time, (b) the voltage across the capacitor as a function of time, and (c) the time when the energy stored in the capacitor first exceeds that in the inductor.

62. An inductor having inductance L and a capacitor having capacitance C are connected in series. The current in the circuit increases linearly in time as described by $I = Kt$, where K is a constant. The capacitor is initially uncharged. Determine (a) the voltage across the inductor as a function of time, (b) the voltage across the capacitor as a function of time, and (c) the time when the energy stored in the capacitor first exceeds that in the inductor.

63. A capacitor in a series LC circuit has an initial charge Q and is being discharged. Find, in terms of L and C, the flux through each of the N turns in the coil, when the charge on the capacitor is $Q/2$.

64. The toroid in Figure P32.64 consists of N turns and has a rectangular cross section. Its inner and outer radii are a and b, respectively. (a) Show that the inductance of the toroid is

$$L = \frac{\mu_0 N^2 h}{2\pi} \ln \frac{b}{a}$$

(b) Using this result, compute the self-inductance of a 500-turn toroid for which $a = 10.0$ cm, $b = 12.0$ cm, and $h = 1.00$ cm. (c) **What If?** In Problem 12, an approximate expression for the inductance of a toroid with $R \gg r$ was derived. To get a feel for the accuracy of that result, use the expression in Problem 12 to compute the approximate inductance of the toroid described in part (b). Compare the result with the answer to part (b).

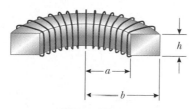

Figure P32.64

65. (a) A flat circular coil does not really produce a uniform magnetic field in the area it encloses, but estimate the self-inductance of a flat, compact circular coil, with radius R and N turns, by assuming that the field at its center is uniform over its area. (b) A circuit on a laboratory table consists of a 1.5-volt battery, a 270-Ω resistor, a switch, and three 30-cm-long patch cords connecting them. Suppose the circuit is arranged to be circular. Think of it as a flat coil with one turn. Compute the order of magnitude of its self-inductance and (c) of the time constant describing how fast the current increases when you close the switch.

66. A soft iron rod ($\mu_m = 800\mu_0$) is used as the core of a solenoid. The rod has a diameter of 24.0 mm and is 10.0 cm long. A 10.0-m piece of 22-gauge copper wire (diameter = 0.644 mm) is wrapped around the rod in a single uniform layer, except for a 10.0-cm length at each end, which is to

be used for connections. (a) How many turns of this wire can be wrapped around the rod? For an accurate answer you should add the diameter of the wire to the diameter of the rod in determining the circumference of each turn. Also note that the wire spirals diagonally along the surface of the rod. (b) What is the resistance of this inductor? (c) What is its inductance?

67. A wire of nonmagnetic material, with radius R, carries current uniformly distributed over its cross section. The total current carried by the wire is I. Show that the magnetic energy per unit length inside the wire is $\mu_0 I^2/16\pi$.

68. An 820-turn wire coil of resistance 24.0 Ω is placed around a 12 500-turn solenoid 7.00 cm long, as shown in Figure P32.68. Both coil and solenoid have cross-sectional areas of 1.00×10^{-4} m². (a) How long does it take the solenoid current to reach 63.2% of its maximum value? Determine (b) the average back emf caused by the self-inductance of the solenoid during this time interval, (c) the average rate of change in magnetic flux through the coil during this time interval, and (d) the magnitude of the average induced current in the coil.

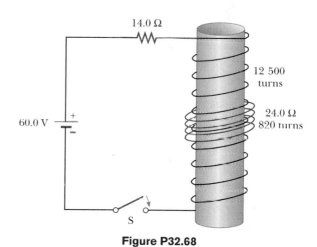

Figure P32.68

69. At $t = 0$, the open switch in Figure P32.69 is closed. By using Kirchhoff's rules for the instantaneous currents and voltages in this two-loop circuit, show that the current in the inductor at time $t > 0$ is

$$I(t) = \frac{\mathcal{E}}{R_1} [1 - e^{-(R'/L)t}]$$

where $R' = R_1 R_2/(R_1 + R_2)$.

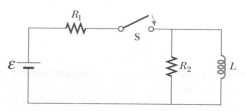

Figure P32.69 Problems 69 and 70.

70. In Figure P32.69 take $\mathcal{E} = 6.00$ V, $R_1 = 5.00\ \Omega$, and $R_2 = 1.00\ \Omega$. The inductor has negligible resistance. When the switch is opened after having been closed for a long time, the current in the inductor drops to 0.250 A in 0.150 s. What is the inductance of the inductor?

71. In Figure P32.71, the switch is closed for $t < 0$, and steady-state conditions are established. The switch is opened at $t = 0$. (a) Find the initial voltage $\mathcal{E}_0$ across L just after $t = 0$. Which end of the coil is at the higher potential: a or b? (b) Make freehand graphs of the currents in R_1 and in R_2 as a function of time, treating the steady-state directions as positive. Show values before and after $t = 0$. (c) How long after $t = 0$ does the current in R_2 have the value 2.00 mA?

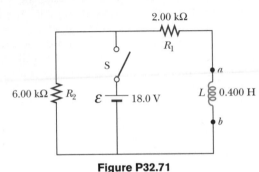

Figure P32.71

72. The open switch in Figure P32.72 is closed at $t = 0$. Before the switch is closed, the capacitor is uncharged, and all currents are zero. Determine the currents in L, C, and R and the potential differences across L, C, and R (a) at the instant after the switch is closed, and (b) long after it is closed.

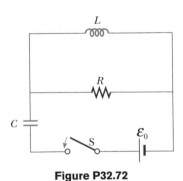

Figure P32.72

73. To prevent damage from arcing in an electric motor, a discharge resistor is sometimes placed in parallel with the armature. If the motor is suddenly unplugged while running, this resistor limits the voltage that appears across the armature coils. Consider a 12.0-V DC motor with an armature that has a resistance of 7.50 Ω and an inductance of 450 mH. Assume the back emf in the armature coils is 10.0 V when the motor is running at normal speed. (The equivalent circuit for the armature is shown in Figure P32.73.) Calculate the maximum resistance R that limits the voltage across the armature to 80.0 V when the motor is unplugged.

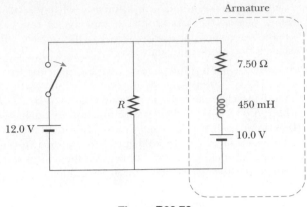

Figure P32.73

74. An air-core solenoid 0.500 m in length contains 1 000 turns and has a cross-sectional area of 1.00 cm². (a) Ignoring end effects, find the self-inductance. (b) A secondary winding wrapped around the center of the solenoid has 100 turns. What is the mutual inductance? (c) The secondary winding carries a constant current of 1.00 A, and the solenoid is connected to a load of 1.00 kΩ. The constant current is suddenly stopped. How much charge flows through the load resistor?

75. The lead-in wires from a television antenna are often constructed in the form of two parallel wires (Fig. P32.75). (a) Why does this configuration of conductors have an inductance? (b) What constitutes the flux loop for this configuration? (c) Ignoring any magnetic flux inside the wires, show that the inductance of a length x of this type of lead-in is

$$L = \frac{\mu_0 x}{\pi}\ \ln\left(\frac{w - a}{a}\right)$$

where a is the radius of the wires and w is their center-to-center separation.

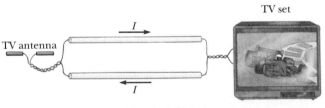

Figure P32.75

Review problems. Problems 76 through 79 apply ideas from this chapter and earlier chapters to some properties of superconductors, which were introduced in Section 27.5.

76. *The resistance of a superconductor.* In an experiment carried out by S. C. Collins between 1955 and 1958, a current was maintained in a superconducting lead ring for 2.50 yr with no observed loss. If the inductance of the ring was 3.14×10^{-8} H, and the sensitivity of the experiment was

1 part in 10^9, what was the maximum resistance of the ring? (*Suggestion:* Treat this as a decaying current in an *RL* circuit, and recall that $e^{-x} \approx 1 - x$ for small x.)

77. A novel method of storing energy has been proposed. A huge underground superconducting coil, 1.00 km in diameter, would be fabricated. It would carry a maximum current of 50.0 kA through each winding of a 150-turn Nb_3Sn solenoid. (a) If the inductance of this huge coil were 50.0 H, what would be the total energy stored? (b) What would be the compressive force per meter length acting between two adjacent windings 0.250 m apart?

78. *Superconducting power transmission.* The use of superconductors has been proposed for power transmission lines. A single coaxial cable (Fig. P32.78) could carry 1.00×10^3 MW (the output of a large power plant) at 200 kV, DC, over a distance of 1 000 km without loss. An inner wire of radius 2.00 cm, made from the superconductor Nb_3Sn, carries the current *I* in one direction. A surrounding superconducting cylinder, of radius 5.00 cm, would carry the return current *I*. In such a system, what is the magnetic field (a) at the surface of the inner conductor and (b) at the inner surface of the outer conductor? (c) How much energy would be stored in the space between the conductors in a 1 000-km superconducting line? (d) What is the pressure exerted on the outer conductor?

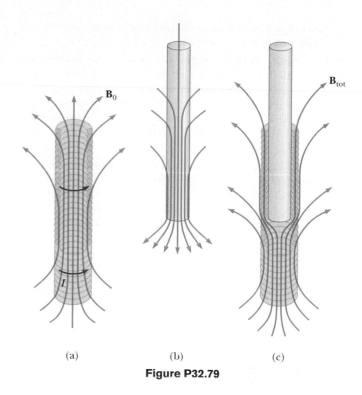

(a)   (b)   (c)

Figure P32.79

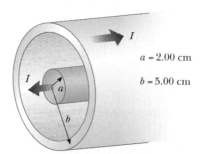

Figure P32.78

79. *The Meissner effect.* Compare this problem with Problem 65 in Chapter 26, on the force attracting a perfect dielectric into a strong electric field. A fundamental property of a Type I superconducting material is *perfect diamagnetism*, or demonstration of the *Meissner effect*, illustrated in Figure 30.35, and described as follows. The superconducting material has **B** = 0 everywhere inside it. If a sample of the material is placed into an externally produced magnetic field, or if it is cooled to become superconducting while it is in a magnetic field, electric currents appear on the surface of the sample. The currents have precisely the strength and orientation required to make the total magnetic field zero throughout the interior of the sample. The following problem will help you to understand the magnetic force that can then act on the superconducting sample.

A vertical solenoid with a length of 120 cm and a diameter of 2.50 cm consists of 1 400 turns of copper wire carrying a counterclockwise current of 2.00 A, as in Figure P32.79a. (a) Find the magnetic field in the vacuum inside the solenoid. (b) Find the energy density of the magnetic

field, and note that the units J/m^3 of energy density are the same as the units N/m^2 of pressure. (c) Now a superconducting bar 2.20 cm in diameter is inserted partway into the solenoid. Its upper end is far outside the solenoid, where the magnetic field is negligible. The lower end of the bar is deep inside the solenoid. Identify the direction required for the current on the curved surface of the bar, so that the total magnetic field is zero within the bar. The field created by the supercurrents is sketched in Figure P32.79b, and the total field is sketched in Figure P32.79c. (d) The field of the solenoid exerts a force on the current in the superconductor. Identify the direction of the force on the bar. (e) Calculate the magnitude of the force by multiplying the energy density of the solenoid field times the area of the bottom end of the superconducting bar.

Answers to Quick Quizzes

32.1 (c), (f). For the constant current in (a) and (b), there is no potential difference across the resistanceless inductor. In (c), if the current increases, the emf induced in the inductor is in the opposite direction, from *b* to *a*, making *a* higher in potential than *b*. Similarly, in (f), the decreasing current induces an emf in the same direction as the current, from *b* to *a*, again making the potential higher at *a* than *b*.

32.2 (b), (d). As the switch is closed, there is no current, so there is no voltage across the resistor. After a long time, the current has reached its final value, and the inductor has no further effect on the circuit.

32.3 (b). When the iron rod is inserted into the solenoid, the inductance of the coil increases. As a result, more potential difference appears across the coil than before.

Consequently, less potential difference appears across the bulb, so the bulb is dimmer.

32.4 (b). Figure 32.10 shows that circuit B has the greater time constant because in this circuit it takes longer for the current to reach its maximum value and then longer for this current to decrease to zero after the switch is thrown to position b. Equation 32.8 indicates that, for equal resistances R_A and R_B, the condition $\tau_B > \tau_A$ means that $L_A < L_B$.

32.5 (a), (d). Because the energy density depends on the magnitude of the magnetic field, to increase the energy density, we must increase the magnetic field. For a solenoid, $B = \mu_0 nI$, where n is the number of turns per unit length. In (a), we increase n to increase the magnetic field. In

(b), the change in cross-sectional area has no effect on the magnetic field. In (c), increasing the length but keeping n fixed has no effect on the magnetic field. Increasing the current in (d) increases the magnetic field in the solenoid.

32.6 (a). M_{12} increases because the magnetic flux through coil 2 increases.

32.7 (b). If the current is at its maximum value, the charge on the capacitor is zero.

32.8 (c). If the current is zero, this is the instant at which the capacitor is fully charged and the current is about to reverse direction.

Chapter 33

Alternating Current Circuits

CHAPTER OUTLINE

33.1 AC Sources

33.2 Resistors in an AC Circuit

33.3 Inductors in an AC Circuit

33.4 Capacitors in an AC Circuit

33.5 The *RLC* Series Circuit

33.6 Power in an AC Circuit

33.7 Resonance in a Series *RLC* Circuit

33.8 The Transformer and Power Transmission

33.9 Rectifiers and Filters

▲ *These large transformers are used to increase the voltage at a power plant for distribution of energy by electrical transmission to the power grid. Voltages can be changed relatively easily because power is distributed by alternating current rather than direct current. (Lester Lefkowitz/Getty Images)*

In this chapter we describe alternating current (AC) circuits. Every time we turn on a television set, a stereo, or any of a multitude of other electrical appliances in a home, we are calling on alternating currents to provide the power to operate them. We begin our study by investigating the characteristics of simple series circuits that contain resistors, inductors, and capacitors and that are driven by a sinusoidal voltage. We shall find that the maximum alternating current in each element is proportional to the maximum alternating voltage across the element. In addition, when the applied voltage is sinusoidal, the current in each element is also sinusoidal, but not necessarily in phase with the applied voltage. The primary aim of this chapter can be summarized as follows: if an AC source applies an alternating voltage to a series circuit containing resistors, inductors, and capacitors, we want to know the amplitude and time characteristics of the alternating current. We conclude the chapter with two sections concerning transformers, power transmission, and electrical filters.

33.1 AC Sources

An AC circuit consists of circuit elements and a power source that provides an alternating voltage Δv. This time-varying voltage is described by

$$\Delta v = \Delta V_{\max} \sin \omega t$$

where $\Delta V_{\max}$ is the maximum output voltage of the AC source, or the **voltage amplitude.** There are various possibilities for AC sources, including generators, as discussed in Section 31.5, and electrical oscillators. In a home, each electrical outlet serves as an AC source.

From Equation 15.12, the angular frequency of the AC voltage is

$$\omega = 2\pi f = \frac{2\pi}{T}$$

where f is the frequency of the source and T is the period. The source determines the frequency of the current in any circuit connected to it. Because the output voltage of an AC source varies sinusoidally with time, the voltage is positive during one half of the cycle and negative during the other half, as in Figure 33.1. Likewise, the current in any circuit driven by an AC source is an alternating current that also varies sinusoidally with time. Commercial electric-power plants in the United States use a frequency of 60 Hz, which corresponds to an angular frequency of 377 rad/s.

33.2 Resistors in an AC Circuit

Consider a simple AC circuit consisting of a resistor and an AC source ⎯⎯$(\sim)$⎯⎯ , as shown in Figure 33.2. At any instant, the algebraic sum of the voltages around a closed loop in a circuit must be zero (Kirchhoff's loop rule). Therefore, $\Delta v + \Delta v_R = 0$, so

Δv

$\Delta V_{\max}$

T

t

Figure 33.1 The voltage supplied by an AC source is sinusoidal with a period T.

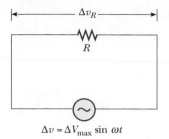

$\Delta v = \Delta V_{max} \sin \omega t$

Active Figure 33.2 A circuit consisting of a resistor of resistance R connected to an AC source, designated by the symbol ───⊗───.

At the Active Figures link at http://www.pse6.com, you can adjust the resistance, the frequency, and the maximum voltage. The results can be studied with the graph and phasor diagram in Figure 33.3.

that the magnitude of the source voltage equals the magnitude of the voltage across the resistor:

$$\Delta v = \Delta v_R = \Delta V_{max} \sin \omega t \qquad (33.1)$$

where Δv_R is the **instantaneous voltage across the resistor.** Therefore, from Equation 27.8, $R = \Delta V/I$, the instantaneous current in the resistor is

$$i_R = \frac{\Delta v_R}{R} = \frac{\Delta V_{max}}{R} \sin \omega t = I_{max} \sin \omega t \qquad (33.2)$$

where I_{max} is the maximum current:

$$I_{max} = \frac{\Delta V_{max}}{R}$$

Maximum current in a resistor

From Equations 33.1 and 33.2, we see that the instantaneous voltage across the resistor is

$$\Delta v_R = I_{max}R \sin \omega t \qquad (33.3)$$

Voltage across a resistor

A plot of voltage and current versus time for this circuit is shown in Figure 33.3a. At point a, the current has a maximum value in one direction, arbitrarily called the positive direction. Between points a and b, the current is decreasing in magnitude but is still in the positive direction. At b, the current is momentarily zero; it then begins to increase in the negative direction between points b and c. At c, the current has reached its maximum value in the negative direction.

The current and voltage are in step with each other because they vary identically with time. Because i_R and Δv_R both vary as $\sin \omega t$ and reach their maximum values at the same time, as shown in Figure 33.3a, they are said to be **in phase,** similar to the way that two waves can be in phase, as discussed in our study of wave motion in

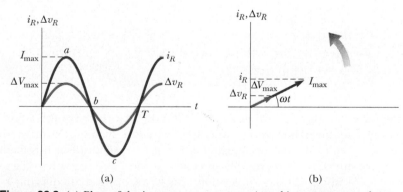

(a)   (b)

Active Figure 33.3 (a) Plots of the instantaneous current i_R and instantaneous voltage Δv_R across a resistor as functions of time. The current is in phase with the voltage, which means that the current is zero when the voltage is zero, maximum when the voltage is maximum, and minimum when the voltage is minimum. At time $t = T$, one cycle of the time-varying voltage and current has been completed. (b) Phasor diagram for the resistive circuit showing that the current is in phase with the voltage.

At the Active Figures link at http://www.pse6.com, you can adjust the resistance, the frequency, and the maximum voltage of the circuit in Figure 33.2. The results can be studied with the graph and phasor diagram in this figure.

Chapter 18. Thus, **for a sinusoidal applied voltage, the current in a resistor is always in phase with the voltage across the resistor.** For resistors in AC circuits, there are no new concepts to learn. Resistors behave essentially the same way in both DC and AC circuits. This will not be the case for capacitors and inductors.

To simplify our analysis of circuits containing two or more elements, we use graphical constructions called *phasor diagrams*. A **phasor** is a vector whose length is proportional to the maximum value of the variable it represents (ΔV_{max} for voltage and I_{max} for current in the present discussion) and which rotates counterclockwise at an angular speed equal to the angular frequency associated with the variable. The projection of the phasor onto the vertical axis represents the instantaneous value of the quantity it represents.

Figure 33.3b shows voltage and current phasors for the circuit of Figure 33.2 at some instant of time. The projections of the phasor arrows onto the vertical axis are determined by a sine function of the angle of the phasor with respect to the horizontal axis. For example, the projection of the current phasor in Figure 33.3b is $I_{max} \sin \omega t$. Notice that this is the same expression as Equation 33.2. Thus, we can use the projections of phasors to represent current values that vary sinusoidally in time. We can do the same with time-varying voltages. The advantage of this approach is that the phase relationships among currents and voltages can be represented as vector additions of phasors, using our vector addition techniques from Chapter 3.

In the case of the single-loop resistive circuit of Figure 33.2, the current and voltage phasors lie along the same line, as in Figure 33.3b, because i_R and Δv_R are in phase. The current and voltage in circuits containing capacitors and inductors have different phase relationships.

▲ PITFALL PREVENTION

33.2   We've Seen This Idea Before

To help with this discussion of phasors, review Section 15.4, in which we represented the simple harmonic motion of a real object to the projection of uniform circular motion of an imaginary object onto a coordinate axis. Phasors are a direct analog to this discussion.

Quick Quiz 33.1   Consider the voltage phasor in Figure 33.4, shown at three instants of time. Choose the part of the figure that represents the instant of time at which the instantaneous value of the voltage has the largest magnitude.

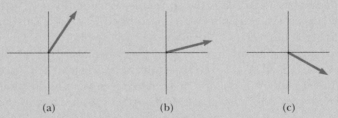

(a)   (b)   (c)

Figure 33.4 (Quick Quizzes 33.1 and 33.2) A voltage phasor is shown at three instants of time.

Quick Quiz 33.2   For the voltage phasor in Figure 33.4, choose the part of the figure that represents the instant of time at which the instantaneous value of the voltage has the smallest magnitude.

For the simple resistive circuit in Figure 33.2, note that **the average value of the current over one cycle is zero.** That is, the current is maintained in the positive direction for the same amount of time and at the same magnitude as it is maintained in the negative direction. However, the direction of the current has no effect on the behavior of the resistor. We can understand this by realizing that collisions between electrons and the fixed atoms of the resistor result in an increase in the resistor's temperature. Although this temperature increase depends on the magnitude of the current, it is independent of the direction of the current.

We can make this discussion quantitative by recalling that the rate at which energy is delivered to a resistor is the power $\mathcal{P} = i^2 R$, where i is the instantaneous current in

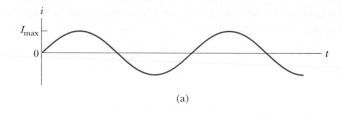

(a)

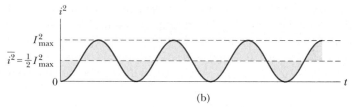

(b)

Figure 33.5 (a) Graph of the current in a resistor as a function of time. (b) Graph of the current squared in a resistor as a function of time. Notice that the gray shaded regions *under* the curve and *above* the dashed line for $I_{max}^2/2$ have the same area as the gray shaded regions *above* the curve and *below* the dashed line for $I_{max}^2/2$. Thus, the average value of i^2 is $I_{max}^2/2$.

the resistor. Because this rate is proportional to the square of the current, it makes no difference whether the current is direct or alternating—that is, whether the sign associated with the current is positive or negative. However, the temperature increase produced by an alternating current having a maximum value I_{max} is not the same as that produced by a direct current equal to I_{max}. This is because the alternating current is at this maximum value for only an instant during each cycle (Fig. 33.5a). What is of importance in an AC circuit is an average value of current, referred to as the **rms current.** As we learned in Section 21.1, the notation *rms* stands for *root-mean-square*, which in this case means the square root of the mean (average) value of the square of the current: $I_{rms} = \sqrt{\overline{i^2}}$. Because i^2 varies as $\sin^2 \omega t$ and because the average value of i^2 is $\frac{1}{2}I_{max}^2$ (see Fig. 33.5b), the rms current is[1]

$$I_{rms} = \frac{I_{max}}{\sqrt{2}} = 0.707 I_{max} \qquad (33.4)$$

rms current

This equation states that an alternating current whose maximum value is 2.00 A delivers to a resistor the same power as a direct current that has a value of $(0.707)(2.00\text{ A}) = 1.41$ A. Thus, the average power delivered to a resistor that carries an alternating current is

$$\mathcal{P}_{av} = I_{rms}^2 R$$

Average power delivered to a resistor

[1] That the square root of the average value of i^2 is equal to $I_{max}/\sqrt{2}$ can be shown as follows. The current in the circuit varies with time according to the expression $i = I_{max} \sin \omega t$, so $i^2 = I_{max}^2 \sin^2 \omega t$. Therefore, we can find the average value of i^2 by calculating the average value of $\sin^2 \omega t$. A graph of $\cos^2 \omega t$ versus time is identical to a graph of $\sin^2 \omega t$ versus time, except that the points are shifted on the time axis. Thus, the time average of $\sin^2 \omega t$ is equal to the time average of $\cos^2 \omega t$ when taken over one or more complete cycles. That is,

$$(\sin^2 \omega t)_{av} = (\cos^2 \omega t)_{av}$$

Using this fact and the trigonometric identity $\sin^2 \theta + \cos^2 \theta = 1$, we obtain

$$(\sin^2 \omega t)_{av} + (\cos^2 \omega t)_{av} = 2(\sin^2 \omega t)_{av} = 1$$

$$(\sin^2 \omega t)_{av} = \tfrac{1}{2}$$

When we substitute this result in the expression $i^2 = I_{max}^2 \sin^2 \omega t$, we obtain $(i^2)_{av} = \overline{i^2} = I_{rms}^2 = I_{max}^2/2$, or $I_{rms} = I_{max}/\sqrt{2}$. The factor $1/\sqrt{2}$ is valid only for sinusoidally varying currents. Other waveforms, such as sawtooth variations, have different factors.

Alternating voltage is also best discussed in terms of rms voltage, and the relationship is identical to that for current:

rms voltage

$$\Delta V_{rms} = \frac{\Delta V_{max}}{\sqrt{2}} = 0.707 \, \Delta V_{max} \qquad (33.5)$$

When we speak of measuring a 120-V alternating voltage from an electrical outlet, we are referring to an rms voltage of 120 V. A quick calculation using Equation 33.5 shows that such an alternating voltage has a maximum value of about 170 V. One reason we use rms values when discussing alternating currents and voltages in this chapter is that AC ammeters and voltmeters are designed to read rms values. Furthermore, with rms values, many of the equations we use have the same form as their direct current counterparts.

> **Quick Quiz 33.3** Which of the following statements might be true for a resistor connected to a sinusoidal AC source? (a) $\mathcal{P}_{av} = 0$ and $i_{av} = 0$ (b) $\mathcal{P}_{av} = 0$ and $i_{av} > 0$ (c) $\mathcal{P}_{av} > 0$ and $i_{av} = 0$ (d) $\mathcal{P}_{av} > 0$ and $i_{av} > 0$.

Example 33.1 What Is the rms Current?

The voltage output of an AC source is given by the expression $\Delta v = (200 \text{ V}) \sin \omega t$. Find the rms current in the circuit when this source is connected to a 100-Ω resistor.

Solution Comparing this expression for voltage output with the general form $\Delta v = \Delta V_{max} \sin \omega t$, we see that $\Delta V_{max} = 200$ V. Thus, the rms voltage is

$$\Delta V_{rms} = \frac{\Delta V_{max}}{\sqrt{2}} = \frac{200 \text{ V}}{\sqrt{2}} = 141 \text{ V}$$

Therefore,

$$I_{rms} = \frac{\Delta V_{rms}}{R} = \frac{141 \text{ V}}{100 \, \Omega} = \boxed{1.41 \text{ A}}$$

33.3 Inductors in an AC Circuit

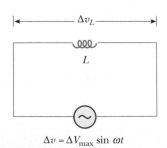

$$\Delta v = \Delta V_{max} \sin \omega t$$

Active Figure 33.6 A circuit consisting of an inductor of inductance L connected to an AC source.

At the Active Figures link at http://www.pse6.com, you can adjust the inductance, the frequency, and the maximum voltage. The results can be studied with the graph and phasor diagram in Figure 33.7.

Now consider an AC circuit consisting only of an inductor connected to the terminals of an AC source, as shown in Figure 33.6. If $\Delta v_L = \mathcal{E}_L = -L(di/dt)$ is the self-induced instantaneous voltage across the inductor (see Eq. 32.1), then Kirchhoff's loop rule applied to this circuit gives $\Delta v + \Delta v_L = 0$, or

$$\Delta v - L \frac{di}{dt} = 0$$

When we substitute $\Delta V_{max} \sin \omega t$ for Δv and rearrange, we obtain

$$\Delta v = L \frac{di}{dt} = \Delta V_{max} \sin \omega t \qquad (33.6)$$

Solving this equation for di, we find that

$$di = \frac{\Delta V_{max}}{L} \sin \omega t \, dt$$

Integrating this expression[2] gives the instantaneous current i_L in the inductor as a function of time:

$$i_L = \frac{\Delta V_{max}}{L} \int \sin \omega t \, dt = -\frac{\Delta V_{max}}{\omega L} \cos \omega t \qquad (33.7)$$

---

[2] We neglect the constant of integration here because it depends on the initial conditions, which are not important for this situation.

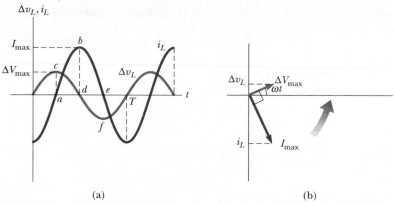

(a)   (b)

Active Figure 33.7 (a) Plots of the instantaneous current i_L and instantaneous voltage Δv_L across an inductor as functions of time. The current lags behind the voltage by 90°. (b) Phasor diagram for the inductive circuit, showing that the current lags behind the voltage by 90°.

When we use the trigonometric identity $\cos \omega t = -\sin(\omega t - \pi/2)$, we can express Equation 33.7 as

$$i_L = \frac{\Delta V_{max}}{\omega L} \sin\left(\omega t - \frac{\pi}{2}\right) \qquad (33.8)$$

Current in an inductor

Comparing this result with Equation 33.6, we see that the instantaneous current i_L in the inductor and the instantaneous voltage Δv_L across the inductor are *out* of phase by $(\pi/2)$ rad = 90°.

A plot of voltage and current versus time is provided in Figure 33.7a. In general, inductors in an AC circuit produce a current that is out of phase with the AC voltage. For example, when the current i_L in the inductor is a maximum (point b in Figure 33.7a), it is momentarily not changing, so the voltage across the inductor is zero (point d). At points like a and e, the current is zero and the rate of change of current is at a maximum. Thus, the voltage across the inductor is also at a maximum (points c and f). Note that the voltage reaches its maximum value one quarter of a period before the current reaches its maximum value. Thus, we see that

> for a sinusoidal applied voltage, the current in an inductor always lags behind the voltage across the inductor by 90° (one-quarter cycle in time).

As with the relationship between current and voltage for a resistor, we can represent this relationship for an inductor with a phasor diagram as in Figure 33.7b. Notice that the phasors are at 90° to one another, representing the 90° phase difference between current and voltage.

From Equation 33.7 we see that the current in an inductive circuit reaches its maximum value when $\cos \omega t = -1$:

$$I_{max} = \frac{\Delta V_{max}}{\omega L} \qquad (33.9)$$

Maximum current in an inductor

This looks similar to the relationship between current, voltage, and resistance in a DC circuit, $I = \Delta V/R$ (Eq. 27.8). In fact, because I_{max} has units of amperes and ΔV_{max} has units of volts, ωL must have units of ohms. Therefore, ωL has the same units as resistance and is related to current and voltage in the same way as resistance. It must behave in a manner similar to resistance, in the sense that it represents opposition to the flow of charge. Notice that because ωL depends on the applied frequency ω, the inductor *reacts* differently, in terms of offering resistance to current, for different frequencies. For this reason, we define ωL as the **inductive reactance:**

Inductive reactance

$$X_L \equiv \omega L \tag{33.10}$$

and we can write Equation 33.9 as

$$I_{max} = \frac{\Delta V_{max}}{X_L} \tag{33.11}$$

The expression for the rms current in an inductor is similar to Equation 33.9, with I_{max} replaced by I_{rms} and ΔV_{max} replaced by ΔV_{rms}.

Equation 33.10 indicates that, for a given applied voltage, the inductive reactance increases as the frequency increases. This is consistent with Faraday's law—the greater the rate of change of current in the inductor, the larger is the back emf. The larger back emf translates to an increase in the reactance and a decrease in the current.

Using Equations 33.6 and 33.11, we find that the instantaneous voltage across the inductor is

Voltage across an inductor

$$\Delta v_L = -L \frac{di}{dt} = -\Delta V_{max} \sin \omega t = -I_{max} X_L \sin \omega t \tag{33.12}$$

Quick Quiz 33.4 Consider the AC circuit in Figure 33.8. The frequency of the AC source is adjusted while its voltage amplitude is held constant. The lightbulb will glow the brightest at (a) high frequencies (b) low frequencies (c) The brightness will be the same at all frequencies.

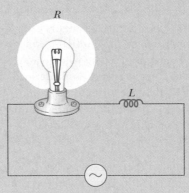

Figure 33.8 (Quick Quiz 33.4) At what frequencies will the bulb glow the brightest?

Example 33.2   A Purely Inductive AC Circuit

In a purely inductive AC circuit (see Fig. 33.6), $L = 25.0$ mH and the rms voltage is 150 V. Calculate the inductive reactance and rms current in the circuit if the frequency is 60.0 Hz.

Solution Equation 33.10 gives

$$X_L = \omega L = 2\pi f L = 2\pi(60.0 \text{ Hz})(25.0 \times 10^{-3} \text{ H})$$

$$= \boxed{9.42 \ \Omega}$$

From an rms version of Equation 33.11, the rms current is

$$I_{rms} = \frac{\Delta V_{L,rms}}{X_L} = \frac{150 \text{ V}}{9.42 \ \Omega} = \boxed{15.9 \text{ A}}$$

What If? What if the frequency increases to 6.00 kHz? What happens to the rms current in the circuit?

Answer If the frequency increases, the inductive reactance increases because the current is changing at a higher rate. The increase in inductive reactance results in a lower current.

Let us calculate the new inductive reactance:

$$X_L = 2\pi(6.00 \times 10^3 \text{ Hz})(25.0 \times 10^{-3} \text{ H}) = 942 \ \Omega$$

The new current is

$$I_{rms} = \frac{150 \text{ V}}{942 \ \Omega} = 0.159 \text{ A}$$

33.4   Capacitors in an AC Circuit

Figure 33.9 shows an AC circuit consisting of a capacitor connected across the terminals of an AC source. Kirchhoff's loop rule applied to this circuit gives $\Delta v + \Delta v_C = 0$, so that the magnitude of the source voltage is equal to the magnitude of the voltage across the capacitor:

$$\Delta v = \Delta v_C = \Delta V_{max} \sin \omega t \qquad (33.13)$$

where Δv_C is the instantaneous voltage across the capacitor. We know from the definition of capacitance that $C = q/\Delta v_C$; hence, Equation 33.13 gives

$$q = C \Delta V_{max} \sin \omega t \qquad (33.14)$$

where q is the instantaneous charge on the capacitor. Because $i = dq/dt$, differentiating Equation 33.14 with respect to time gives the instantaneous current in the circuit:

$$i_C = \frac{dq}{dt} = \omega C \Delta V_{max} \cos \omega t \qquad (33.15)$$

Using the trigonometric identity

$$\cos \omega t = \sin \left(\omega t + \frac{\pi}{2} \right)$$

we can express Equation 33.15 in the alternative form

$$i_C = \omega C \Delta V_{max} \sin \left(\omega t + \frac{\pi}{2} \right) \qquad (33.16)$$

Comparing this expression with Equation 33.13, we see that the current is $\pi/2$ rad $= 90°$ out of phase with the voltage across the capacitor. A plot of current and voltage versus time (Fig. 33.10a) shows that the current reaches its maximum value one quarter of a cycle sooner than the voltage reaches its maximum value.

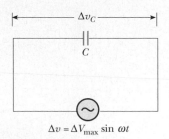

Active Figure 33.9 A circuit consisting of a capacitor of capacitance C connected to an AC source.

At the Active Figures link at http://www.pse6.com, you can adjust the capacitance, the frequency, and the maximum voltage. The results can be studied with the graph and phasor diagram in Figure 33.10.

Current in a capacitor

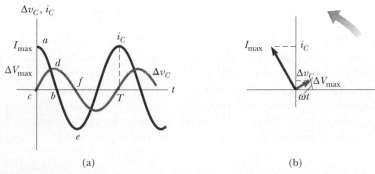

(a)   (b)

Active Figure 33.10 (a) Plots of the instantaneous current i_C and instantaneous voltage Δv_C across a capacitor as functions of time. The voltage lags behind the current by 90°. (b) Phasor diagram for the capacitive circuit, showing that the current leads the voltage by 90°.

At the Active Figures link at http://www.pse6.com, you can adjust the capacitance, the frequency, and the maximum voltage of the circuit in Figure 33.9. The results can be studied with the graph and phasor diagram in this figure.

Looking more closely, consider a point such as b where the current is zero. This occurs when the capacitor has just reached its maximum charge, so the voltage across the capacitor is a maximum (point d). At points such as a and e, the current is a maximum, which occurs at those instants at which the charge on the capacitor has just gone to zero and it begins to charge up with the opposite polarity. Because the charge is zero, the voltage across the capacitor is zero (points c and f). Thus, the current and voltage are out of phase.

As with inductors, we can represent the current and voltage for a capacitor on a phasor diagram. The phasor diagram in Figure 33.10b shows that

> for a sinusoidally applied voltage, the current always leads the voltage across a capacitor by 90°.

From Equation 33.15, we see that the current in the circuit reaches its maximum value when $\cos \omega t = 1$:

$$I_{max} = \omega C \Delta V_{max} = \frac{\Delta V_{max}}{(1/\omega C)} \tag{33.17}$$

As in the case with inductors, this looks like Equation 27.8, so that the denominator must play the role of resistance, with units of ohms. We give the combination $1/\omega C$ the symbol X_C, and because this function varies with frequency, we define it as the **capacitive reactance:**

Capacitive reactance

$$X_C \equiv \frac{1}{\omega C} \tag{33.18}$$

and we can write Equation 33.17 as

Maximum current in a capacitor

$$I_{max} = \frac{\Delta V_{max}}{X_C} \tag{33.19}$$

The rms current is given by an expression similar to Equation 33.19, with I_{max} replaced by I_{rms} and ΔV_{max} replaced by ΔV_{rms}.

Combining Equations 33.13 and 33.19, we can express the instantaneous voltage across the capacitor as

Voltage across a capacitor

$$\Delta v_C = \Delta V_{max} \sin \omega t = I_{max} X_C \sin \omega t \tag{33.20}$$

Equations 33.18 and 33.19 indicate that as the frequency of the voltage source increases, the capacitive reactance decreases and therefore the maximum current increases. Again, note that the frequency of the current is determined by the frequency of the voltage source driving the circuit. As the frequency approaches zero, the capacitive reactance approaches infinity, and hence the current approaches zero. This makes sense because the circuit approaches direct current conditions as ω approaches zero, and the capacitor represents an open circuit.

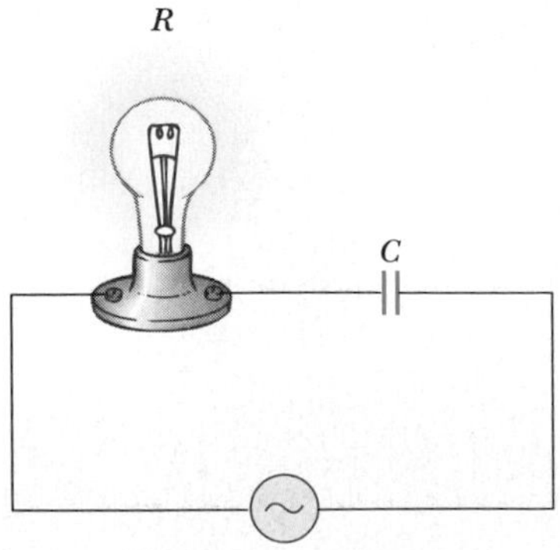

Figure 33.11 (Quick Quiz 33.5)

Quick Quiz 33.5 Consider the AC circuit in Figure 33.11. The frequency of the AC source is adjusted while its voltage amplitude is held constant. The lightbulb will glow the brightest at (a) high frequencies (b) low frequencies (c) The brightness will be same at all frequencies.

Quick Quiz 33.6 Consider the AC circuit in Figure 33.12. The frequency of the AC source is adjusted while its voltage amplitude is held constant. The lightbulb will glow the brightest at (a) high frequencies (b) low frequencies (c) The brightness will be same at all frequencies.

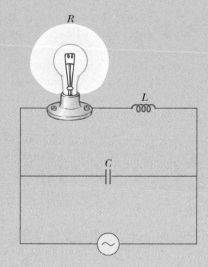

Figure 33.12 (Quick Quiz 33.6)

Example 33.3   A Purely Capacitive AC Circuit

An 8.00-μF capacitor is connected to the terminals of a 60.0-Hz AC source whose rms voltage is 150 V. Find the capacitive reactance and the rms current in the circuit.

Solution Using Equation 33.18 and the fact that $\omega = 2\pi f = 377 \text{ s}^{-1}$ gives

$$X_C = \frac{1}{\omega C} = \frac{1}{(377 \text{ s}^{-1})(8.00 \times 10^{-6} \text{ F})} = \boxed{332 \ \Omega}$$

Hence, from a modified Equation 33.19, the rms current is

$$I_{\text{rms}} = \frac{\Delta V_{\text{rms}}}{X_C} = \frac{150 \text{ V}}{332 \ \Omega} = \boxed{0.452 \text{ A}}$$

What If? What if the frequency is doubled? What happens to the rms current in the circuit?

Answer If the frequency increases, the capacitive reactance decreases—just the opposite as in the case of an inductor. The decrease in capacitive reactance results in an increase in the current.

Let us calculate the new capacitive reactance:

$$X_C = \frac{1}{\omega C} = \frac{1}{2(377 \text{ s}^{-1})(8.00 \times 10^{-6} \text{ F})} = 166 \ \Omega$$

The new current is

$$I_{\text{rms}} = \frac{150 \text{ V}}{166 \ \Omega} = 0.904 \text{ A}$$

33.5   The *RLC* Series Circuit

Figure 33.13a shows a circuit that contains a resistor, an inductor, and a capacitor connected in series across an alternating voltage source. As before, we assume that the applied voltage varies sinusoidally with time. It is convenient to assume that the instantaneous applied voltage is given by

$$\Delta v = \Delta V_{\text{max}} \sin \omega t$$

while the current varies as

$$i = I_{\text{max}} \sin(\omega t - \phi)$$

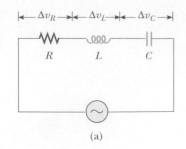

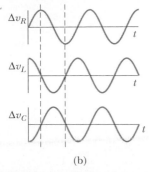

Active Figure 33.13 (a) A series circuit consisting of a resistor, an inductor, and a capacitor connected to an AC source. (b) Phase relationships for instantaneous voltages in the series *RLC* circuit.

At the Active Figures link at http://www.pse6.com, you can adjust the resistance, the inductance, and the capacitance. The results can be studied with the graph in this figure and the phasor diagram in Figure 33.15.

where ϕ is some **phase angle** between the current and the applied voltage. Based on our discussions of phase in Sections 33.3 and 33.4, we expect that the current will generally not be in phase with the voltage in an *RLC* circuit. Our aim is to determine ϕ and I_{max}. Figure 33.13b shows the voltage versus time across each element in the circuit and their phase relationships.

First, we note that because the elements are in series, the current everywhere in the circuit must be the same at any instant. That is, **the current at all points in a series AC circuit has the same amplitude and phase.** Based on the preceding sections, we know that the voltage across each element has a different amplitude and phase. In particular, the voltage across the resistor is in phase with the current, the voltage across the inductor leads the current by 90°, and the voltage across the capacitor lags behind the current by 90°. Using these phase relationships, we can express the instantaneous voltages across the three circuit elements as

$$\Delta v_R = I_{max} R \sin \omega t = \Delta V_R \sin \omega t \tag{33.21}$$

$$\Delta v_L = I_{max} X_L \sin \left(\omega t + \frac{\pi}{2} \right) = \Delta V_L \cos \omega t \tag{33.22}$$

$$\Delta v_C = I_{max} X_C \sin \left(\omega t - \frac{\pi}{2} \right) = -\Delta V_C \cos \omega t \tag{33.23}$$

where ΔV_R, ΔV_L, and ΔV_C are the maximum voltage values across the elements:

$$\Delta V_R = I_{max} R \qquad \Delta V_L = I_{max} X_L \qquad \Delta V_C = I_{max} X_C$$

At this point, we could proceed by noting that the instantaneous voltage Δv across the three elements equals the sum

$$\Delta v = \Delta v_R + \Delta v_L + \Delta v_C$$

Although this analytical approach is correct, it is simpler to obtain the sum by examining the phasor diagram, shown in Figure 33.14. Because the current at any instant is the same in all elements, we combine the three phasor pairs shown in Figure 33.14 to obtain Figure 33.15a, in which a single phasor I_{max} is used to represent the current in each element. Because phasors are rotating vectors, we can combine the three parts of Figure 33.14 by using vector addition. To obtain the vector sum of the three voltage phasors in Figure 33.15a, we redraw the phasor diagram as in Figure 33.15b. From this diagram, we see that the vector sum of the voltage amplitudes ΔV_R, ΔV_L, and ΔV_C equals a phasor whose length is the maximum applied voltage ΔV_{max}, and which makes an angle ϕ with the current phasor I_{max}. The voltage phasors ΔV_L and ΔV_C are in opposite directions along the same line, so we can construct the difference phasor $\Delta V_L - \Delta V_C$, which is perpendicular to the phasor ΔV_R. From either one of the right triangles

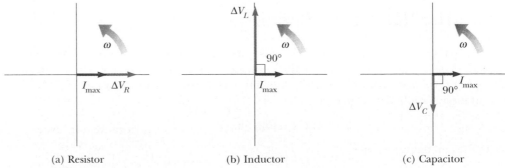

(a) Resistor   (b) Inductor   (c) Capacitor

Figure 33.14 Phase relationships between the voltage and current phasors for (a) a resistor, (b) an inductor, and (c) a capacitor connected in series.

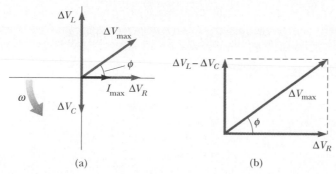

(a)   (b)

Active Figure 33.15 (a) Phasor diagram for the series *RLC* circuit shown in Figure 33.13a. The phasor ΔV_R is in phase with the current phasor I_{max}, the phasor ΔV_L leads I_{max} by 90°, and the phasor ΔV_C lags I_{max} by 90°. The total voltage ΔV_{max} makes an angle ϕ with I_{max}. (b) Simplified version of the phasor diagram shown in part (a).

At the Active Figures link at http://www.pse6.com, *you can adjust the resistance, the inductance, and the capacitance of the circuit in Figure 33.13a. The results can be studied with the graphs in Figure 33.13b and the phasor diagram in this figure.*

in Figure 33.15b, we see that

$$\Delta V_{max} = \sqrt{\Delta V_R{}^2 + (\Delta V_L - \Delta V_C)^2} = \sqrt{(I_{max}\,R)^2 + (I_{max}\,X_L - I_{max}\,X_C)^2}$$

$$\Delta V_{max} = I_{max}\sqrt{R^2 + (X_L - X_C)^2} \tag{33.24}$$

Therefore, we can express the maximum current as

$$I_{max} = \frac{\Delta V_{max}}{\sqrt{R^2 + (X_L - X_C)^2}}$$

Maximum current in an *RLC* circuit

Once again, this has the same mathematical form as Equation 27.8. The denominator of the fraction plays the role of resistance and is called the **impedance** *Z* of the circuit:

$$Z \equiv \sqrt{R^2 + (X_L - X_C)^2} \tag{33.25}$$

Impedance

where impedance also has units of ohms. Therefore, we can write Equation 33.24 in the form

$$\Delta V_{max} = I_{max}\,Z \tag{33.26}$$

We can regard Equation 33.26 as the AC equivalent of Equation 27.8. Note that the impedance and therefore the current in an AC circuit depend upon the resistance, the inductance, the capacitance, and the frequency (because the reactances are frequency-dependent).

By removing the common factor I_{max} from each phasor in Figure 33.15a, we can construct the *impedance triangle* shown in Figure 33.16. From this phasor diagram we find that the phase angle ϕ between the current and the voltage is

$$\phi = \tan^{-1}\left(\frac{X_L - X_C}{R}\right) \tag{33.27}$$

Phase angle

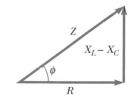

Figure 33.16 An impedance triangle for a series *RLC* circuit gives the relationship $Z = \sqrt{R^2 + (X_L - X_C)^2}$.

Also, from Figure 33.16, we see that $\cos\phi = R/Z$. When $X_L > X_C$ (which occurs at high frequencies), the phase angle is positive, signifying that the current lags behind the applied voltage, as in Figure 33.15a. We describe this situation by saying that the circuit is *more inductive than capacitive*. When $X_L < X_C$, the phase angle is negative, signifying that the current leads the applied voltage, and the circuit is *more capacitive than inductive*. When $X_L = X_C$, the phase angle is zero and the circuit is *purely resistive*.

Table 33.1 gives impedance values and phase angles for various series circuits containing different combinations of elements.

Table 33.1

Impedance Values and Phase Angles for Various Circuit-Element Combinations[a]

Circuit Elements	Impedance Z	Phase Angle ϕ
R	R	$0°$
C	X_C	$-90°$
L	X_L	$+90°$
R C	$\sqrt{R^2 + X_C^2}$	Negative, between $-90°$ and $0°$
R L	$\sqrt{R^2 + X_L^2}$	Positive, between $0°$ and $90°$
R L C	$\sqrt{R^2 + (X_L - X_C)^2}$	Negative if $X_C > X_L$ Positive if $X_C < X_L$

[a]   In each case, an AC voltage (not shown) is applied across the elements.

Quick Quiz 33.7 Label each part of Figure 33.17 as being $X_L > X_C$, $X_L = X_C$, or $X_L < X_C$.

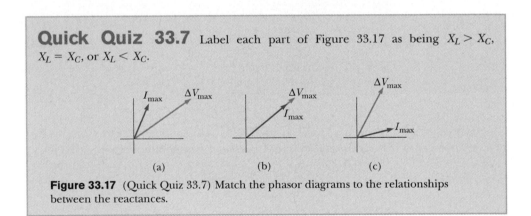

(a)   (b)   (c)

Figure 33.17 (Quick Quiz 33.7) Match the phasor diagrams to the relationships between the reactances.

Example 33.4   Finding *L* from a Phasor Diagram

In a series *RLC* circuit, the applied voltage has a maximum value of 120 V and oscillates at a frequency of 60.0 Hz. The circuit contains an inductor whose inductance can be varied, a 200-Ω resistor, and a 4.00-μF capacitor. What value of L should an engineer analyzing the circuit choose such that the voltage across the capacitor lags the applied voltage by 30.0°?

Solution The phase relationships for the voltages across the elements are shown in Figure 33.18. From the figure we see that the phase angle is $\phi = -60.0°$. (The phasors representing I_{max} and ΔV_R are in the same direction.) From Equation 33.27, we find that

$$X_L = X_C + R \tan \phi$$

Substituting Equations 33.10 and 33.18 (with $\omega = 2\pi f$) into this expression gives

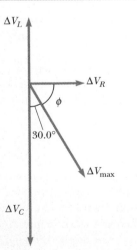

Figure 33.18 (Example 33.4) The phasor diagram for the given information.

$$2\pi f L = \frac{1}{2\pi f C} + R \tan \phi$$

$$L = \frac{1}{2\pi f}\left(\frac{1}{2\pi f C} + R \tan \phi\right)$$

Substituting the given values into the equation gives

$$L = \boxed{0.84 \text{ H}.}$$

Example 33.5   Analyzing a Series *RLC* Circuit

A series *RLC* AC circuit has $R = 425\ \Omega$, $L = 1.25$ H, $C = 3.50\ \mu$F, $\omega = 377$ s^{-1}, and $\Delta V_{max} = 150$ V.

(A) Determine the inductive reactance, the capacitive reactance, and the impedance of the circuit.

Solution The reactances are $X_L = \omega L = \boxed{471\ \Omega}$ and $X_C = 1/\omega C = \boxed{758\ \Omega}$.
The impedance is

$$Z = \sqrt{R^2 + (X_L - X_C)^2}$$
$$= \sqrt{(425\ \Omega)^2 + (471\ \Omega - 758\ \Omega)^2} = \boxed{513\ \Omega}$$

(B) Find the maximum current in the circuit.

Solution

$$I_{max} = \frac{\Delta V_{max}}{Z} = \frac{150 \text{ V}}{513\ \Omega} = \boxed{0.292 \text{ A}}$$

(C) Find the phase angle between the current and voltage.

Solution

$$\phi = \tan^{-1}\left(\frac{X_L - X_C}{R}\right) = \tan^{-1}\left(\frac{471\ \Omega - 758\ \Omega}{425\ \Omega}\right)$$
$$= \boxed{-34.0°}$$

Because the capacitive reactance is larger than the inductive reactance, the circuit is more capacitive than inductive. In this case, the phase angle ϕ is negative and the current leads the applied voltage.

(D) Find both the maximum voltage and the instantaneous voltage across each element.

Solution The maximum voltages are

$$\Delta V_R = I_{max} R = (0.292 \text{ A})(425\ \Omega) = \boxed{124 \text{ V}}$$

$$\Delta V_L = I_{max} X_L = (0.292 \text{ A})(471\ \Omega) = \boxed{138 \text{ V}}$$

$$\Delta V_C = I_{max} X_C = (0.292 \text{ A})(758\ \Omega) = \boxed{221 \text{ V}}$$

Using Equations 33.21, 33.22, and 33.23, we find that we can write the instantaneous voltages across the three elements as

$$\Delta v_R = \boxed{(124 \text{ V}) \sin 377t}$$

$$\Delta v_L = \boxed{(138 \text{ V}) \cos 377t}$$

$$\Delta v_C = \boxed{(-221 \text{ V}) \cos 377t}$$

What If? What if you added up the maximum voltages across the three circuit elements? Is this a physically meaningful quantity?

Answer The sum of the maximum voltages across the elements is $\Delta V_R + \Delta V_L + \Delta V_C = 484$ V. Note that this sum is much greater than the maximum voltage of the source, 150 V. The sum of the maximum voltages is a meaningless quantity because when sinusoidally varying quantities are added, *both their amplitudes and their phases* must be taken into account. We know that the maximum voltages across the various elements occur at different times. That is, the voltages must be added in a way that takes account of the different phases.

 At the Interactive Worked Example link at **http://www.pse6.com**, *you can investigate the RLC circuit for various values of the circuit elements.*

33.6   Power in an AC Circuit

Let us now take an energy approach to analyzing AC circuits, considering the transfer of energy from the AC source to the circuit. In Example 28.1 we found that the power delivered by a battery to a DC circuit is equal to the product of the current and the emf of the battery. Likewise, the instantaneous power delivered by an AC source to a circuit is the product of the source current and the applied voltage. For the *RLC* circuit shown in Figure 33.13a, we can express the instantaneous power $\mathcal{P}$ as

$$\mathcal{P} = i\,\Delta v = I_{max}\sin(\omega t - \phi)\,\Delta V_{max}\sin\omega t$$
$$= I_{max}\,\Delta V_{max}\sin\omega t\sin(\omega t - \phi) \qquad (33.28)$$

This result is a complicated function of time and therefore is not very useful from a practical viewpoint. What is generally of interest is the average power over one or more cycles. Such an average can be computed by first using the trigonometric identity $\sin(\omega t - \phi) = \sin\omega t\cos\phi - \cos\omega t\sin\phi$. Substituting this into Equation 33.28 gives

$$\mathcal{P} = I_{max}\,\Delta V_{max}\sin^2\omega t\cos\phi - I_{max}\,\Delta V_{max}\sin\omega t\cos\omega t\sin\phi \qquad (33.29)$$

We now take the time average of $\mathcal{P}$ over one or more cycles, noting that I_{max}, ΔV_{max}, ϕ, and ω are all constants. The time average of the first term on the right in Equation 33.29 involves the average value of $\sin^2\omega t$, which is $\frac{1}{2}$ (as shown in footnote 1). The time average of the second term on the right is identically zero because $\sin\omega t\cos\omega t = \frac{1}{2}\sin 2\omega t$, and the average value of $\sin 2\omega t$ is zero. Therefore, we can express the **average power** $\mathcal{P}_{av}$ as

$$\mathcal{P}_{av} = \tfrac{1}{2}I_{max}\,\Delta V_{max}\cos\phi \qquad (33.30)$$

It is convenient to express the average power in terms of the rms current and rms voltage defined by Equations 33.4 and 33.5:

$$\boxed{\mathcal{P}_{av} = I_{rms}\,\Delta V_{rms}\cos\phi} \qquad (33.31)$$

Average power delivered to an *RLC* circuit

where the quantity $\cos\phi$ is called the **power factor.** By inspecting Figure 33.15b, we see that the maximum voltage across the resistor is given by $\Delta V_R = \Delta V_{max}\cos\phi = I_{max}R$. Using Equation 33.5 and the fact that $\cos\phi = I_{max}R/\Delta V_{max}$, we find that we can express $\mathcal{P}_{av}$ as

$$\mathcal{P}_{av} = I_{rms}\,\Delta V_{rms}\cos\phi = I_{rms}\left(\frac{\Delta V_{max}}{\sqrt{2}}\right)\frac{I_{max}R}{\Delta V_{max}} = I_{rms}\frac{I_{max}R}{\sqrt{2}}$$

After making the substitution $I_{max} = \sqrt{2}\,I_{rms}$ from Equation 33.4, we have

$$\boxed{\mathcal{P}_{av} = I_{rms}^2\,R} \qquad (33.32)$$

In words, **the average power delivered by the source is converted to internal energy in the resistor,** just as in the case of a DC circuit. When the load is purely resistive, then $\phi = 0$, $\cos\phi = 1$, and from Equation 33.31 we see that

$$\mathcal{P}_{av} = I_{rms}\,\Delta V_{rms}$$

We find that **no power losses are associated with pure capacitors and pure inductors in an AC circuit.** To see why this is true, let us first analyze the power in an AC circuit containing only a source and a capacitor. When the current begins to increase in one direction in an AC circuit, charge begins to accumulate on the capacitor, and a voltage appears across it. When this voltage reaches its maximum value, the energy stored in the capacitor is $\frac{1}{2}C(\Delta V_{max})^2$. However, this energy storage is only momentary. The capacitor is charged and discharged twice during each cycle: charge is delivered to the capacitor during two quarters of the cycle and is returned to the voltage source during the remaining two quarters. Therefore, **the average power supplied by the source is zero.** In other words, **no power losses occur in a capacitor in an AC circuit.**

Let us now consider the case of an inductor. When the current reaches its maximum value, the energy stored in the inductor is a maximum and is given by $\frac{1}{2}LI_{max}^2$. When the current begins to decrease in the circuit, this stored energy is returned to the source as the inductor attempts to maintain the current in the circuit.

Equation 33.31 shows that the power delivered by an AC source to any circuit depends on the phase—a result that has many interesting applications. For example, a factory that uses large motors in machines, generators, or transformers has a large inductive load (because of all the windings). To deliver greater power to such devices in the factory without using excessively high voltages, technicians introduce capacitance in the circuits to shift the phase.

Quick Quiz 33.8 An AC source drives an *RLC* circuit with a fixed voltage amplitude. If the driving frequency is ω_1, the circuit is more capacitive than inductive and the phase angle is $-10°$. If the driving frequency is ω_2, the circuit is more inductive than capacitive and the phase angle is $+10°$. The largest amount of power is delivered to the circuit at (a) ω_1 (b) ω_2 (c) The same amount of power is delivered at both frequencies.

Example 33.6   Average Power in an *RLC* Series Circuit

Calculate the average power delivered to the series *RLC* circuit described in Example 33.5.

Solution First, let us calculate the rms voltage and rms current, using the values of ΔV_{max} and I_{max} from Example 33.5:

$$\Delta V_{rms} = \frac{\Delta V_{max}}{\sqrt{2}} = \frac{150 \text{ V}}{\sqrt{2}} = 106 \text{ V}$$

$$I_{rms} = \frac{I_{max}}{\sqrt{2}} = \frac{0.292 \text{ A}}{\sqrt{2}} = 0.206 \text{ A}$$

Because $\phi = -34.0°$, the power factor is $\cos(34.0°) = 0.829$; hence, the average power delivered is

$$\mathcal{P}_{av} = I_{rms}\, \Delta V_{rms} \cos \phi = (0.206 \text{ A})(106 \text{ V})(0.829)$$

$$= \boxed{18.1 \text{ W}}$$

We can obtain the same result using Equation 33.32.

33.7   Resonance in a Series *RLC* Circuit

A series *RLC* circuit is said to be **in resonance** when the current has its maximum value. In general, the rms current can be written

$$I_{rms} = \frac{\Delta V_{rms}}{Z} \tag{33.33}$$

where Z is the impedance. Substituting the expression for Z from Equation 33.25 into 33.33 gives

$$I_{rms} = \frac{\Delta V_{rms}}{\sqrt{R^2 + (X_L - X_C)^2}} \tag{33.34}$$

Because the impedance depends on the frequency of the source, the current in the *RLC* circuit also depends on the frequency. The frequency ω_0 at which $X_L - X_C = 0$ is called the **resonance frequency** of the circuit. To find ω_0, we use the condition $X_L = X_C$, from which we obtain $\omega_0 L = 1/\omega_0 C$, or

$$\omega_0 = \frac{1}{\sqrt{LC}} \tag{33.35}$$

Resonance frequency

This frequency also corresponds to the natural frequency of oscillation of an *LC* circuit (see Section 32.5). Therefore, the current in a series *RLC* circuit reaches its maximum value when the frequency of the applied voltage matches the natural oscillator frequency—which depends only on L and C. Furthermore, at this frequency the current is in phase with the applied voltage.

> **Quick Quiz 33.9** The impedance of a series RLC circuit at resonance is (a) larger than R (b) less than R (c) equal to R (d) impossible to determine.

A plot of rms current versus frequency for a series RLC circuit is shown in Figure 33.19a. The data assume a constant $\Delta V_{rms} = 5.0$ mV, that $L = 5.0$ μH, and that $C = 2.0$ nF. The three curves correspond to three values of R. In each case, the current reaches its maximum value at the resonance frequency ω_0. Furthermore, the curves become narrower and taller as the resistance decreases.

By inspecting Equation 33.34, we must conclude that, when $R = 0$, the current becomes infinite at resonance. However, real circuits always have some resistance, which limits the value of the current to some finite value.

It is also interesting to calculate the average power as a function of frequency for a series RLC circuit. Using Equations 33.32, 33.33, and 33.25, we find that

$$\mathcal{P}_{av} = I_{rms}^2\, R = \frac{(\Delta V_{rms})^2}{Z^2}\, R = \frac{(\Delta V_{rms})^2\, R}{R^2 + (X_L - X_C)^2} \tag{33.36}$$

Because $X_L = \omega L$, $X_C = 1/\omega C$, and $\omega_0^2 = 1/LC$, we can express the term $(X_L - X_C)^2$ as

$$(X_L - X_C)^2 = \left(\omega L - \frac{1}{\omega C}\right)^2 = \frac{L^2}{\omega^2}(\omega^2 - \omega_0^2)^2$$

Using this result in Equation 33.36 gives

Average power as a function of frequency in an RLC circuit

$$\mathcal{P}_{av} = \frac{(\Delta V_{rms})^2\, R\omega^2}{R^2\omega^2 + L^2(\omega^2 - \omega_0^2)^2} \tag{33.37}$$

This expression shows that **at resonance, when $\omega = \omega_0$, the average power is a maximum** and has the value $(\Delta V_{rms})^2/R$. Figure 33.19b is a plot of average power

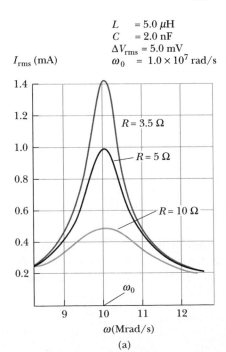

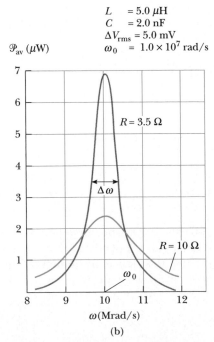

$$
\begin{array}{ll}
L & = 5.0\ \mu\text{H} \\
C & = 2.0\ \text{nF} \\
\Delta V_{rms} & = 5.0\ \text{mV} \\
\omega_0 & = 1.0 \times 10^7\ \text{rad/s}
\end{array}
$$

(a)

(b)

Active Figure 33.19 (a) The rms current versus frequency for a series RLC circuit, for three values of R. The current reaches its maximum value at the resonance frequency ω_0. (b) Average power delivered to the circuit versus frequency for the series RLC circuit, for two values of R.

versus frequency for two values of R in a series *RLC* circuit. As the resistance is made smaller, the curve becomes sharper in the vicinity of the resonance frequency. This curve sharpness is usually described by a dimensionless parameter known as the **quality factor,**[3] denoted by Q:

$$Q = \frac{\omega_0}{\Delta\omega}$$

where $\Delta\omega$ is the width of the curve measured between the two values of ω for which $\mathcal{P}_{av}$ has half its maximum value, called the *half-power points* (see Fig. 33.19b.) It is left as a problem (Problem 72) to show that the width at the half-power points has the value $\Delta\omega = R/L$, so

Quality factor

$$Q = \frac{\omega_0 L}{R} \tag{33.38}$$

The curves plotted in Figure 33.20 show that a high-Q circuit responds to only a very narrow range of frequencies, whereas a low-Q circuit can detect a much broader range of frequencies. Typical values of Q in electronic circuits range from 10 to 100.

The receiving circuit of a radio is an important application of a resonant circuit. One tunes the radio to a particular station (which transmits an electromagnetic wave or signal of a specific frequency) by varying a capacitor, which changes the resonance frequency of the receiving circuit. When the resonance frequency of the circuit matches that of the incoming electromagnetic wave, the current in the receiving circuit increases. This signal caused by the incoming wave is then amplified and fed to a speaker. Because many signals are often present over a range of frequencies, it is important to design a high-Q circuit to eliminate unwanted signals. In this manner, stations whose frequencies are near but not equal to the resonance frequency give signals at the receiver that are negligibly small relative to the signal that matches the resonance frequency.

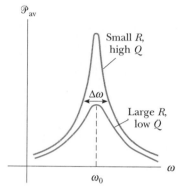

Figure 33.20 Average power versus frequency for a series *RLC* circuit. The width $\Delta\omega$ of each curve is measured between the two points where the power is half its maximum value. The power is a maximum at the resonance frequency ω_0.

Quick Quiz 33.10 An airport metal detector (see page 1003) is essentially a resonant circuit. The portal you step through is an inductor (a large loop of conducting wire) within the circuit. The frequency of the circuit is tuned to its resonance frequency when there is no metal in the inductor. Any metal on your body increases the effective inductance of the loop and changes the current in it. If you want the detector to detect a small metallic object, should the circuit have (a) a high quality factor or (b) a low quality factor?

Example 33.7   A Resonating Series *RLC* Circuit   `Interactive`

Consider a series *RLC* circuit for which $R = 150\ \Omega$, $L = 20.0$ mH, $\Delta V_{rms} = 20.0$ V, and $\omega = 5\,000\ \text{s}^{-1}$. Determine the value of the capacitance for which the current is a maximum.

Solution The current has its maximum value at the resonance frequency ω_0, which should be made to match the "driving" frequency of 5 000 s^{-1}:

$$\omega_0 = 5.00 \times 10^3\ \text{s}^{-1} = \frac{1}{\sqrt{LC}}$$

$$C = \frac{1}{\omega_0^2 L} = \frac{1}{(5.00 \times 10^3\ \text{s}^{-1})^2(20.0 \times 10^{-3}\ \text{H})}$$

$$= 2.00\ \mu\text{F}$$

 At the Interactive Worked Example link at **http://www.pse6.com,** *you can explore resonance in an RLC circuit.*

---

[3] The quality factor is also defined as the ratio $2\pi E/\Delta E$ where E is the energy stored in the oscillating system and ΔE is the energy decrease per cycle of oscillation due to the resistance.

33.8 The Transformer and Power Transmission

As discussed in Section 27.6, when electric power is transmitted over great distances, it is economical to use a high voltage and a low current to minimize the I^2R loss in the transmission lines. Consequently, 350-kV lines are common, and in many areas even higher-voltage (765-kV) lines are used. At the receiving end of such lines, the consumer requires power at a low voltage (for safety and for efficiency in design). Therefore, a device is required that can change the alternating voltage and current without causing appreciable changes in the power delivered. The AC transformer is that device.

In its simplest form, the **AC transformer** consists of two coils of wire wound around a core of iron, as illustrated in Figure 33.21. (Compare this to Faraday's experiment in Figure 31.2.) The coil on the left, which is connected to the input alternating voltage source and has N_1 turns, is called the *primary winding* (or the *primary*). The coil on the right, consisting of N_2 turns and connected to a load resistor R, is called the *secondary winding* (or the *secondary*). The purpose of the iron core is to increase the magnetic flux through the coil and to provide a medium in which nearly all the magnetic field lines through one coil pass through the other coil. Eddy-current losses are reduced by using a laminated core. Iron is used as the core material because it is a soft ferromagnetic substance and hence reduces hysteresis losses. Transformation of energy to internal energy in the finite resistance of the coil wires is usually quite small. Typical transformers have power efficiencies from 90% to 99%. In the discussion that follows, we assume an *ideal transformer*, one in which the energy losses in the windings and core are zero.

First, let us consider what happens in the primary circuit. If we assume that the resistance of the primary is negligible relative to its inductive reactance, then the primary circuit is equivalent to a simple circuit consisting of an inductor connected to an AC source. Because the current is 90° out of phase with the voltage, the power factor $\cos \phi$ is zero, and hence the average power delivered from the source to the primary circuit is zero. Faraday's law states that the voltage ΔV_1 across the primary is

$$\Delta V_1 = -N_1 \frac{d\Phi_B}{dt} \tag{33.39}$$

where Φ_B is the magnetic flux through each turn. If we assume that all magnetic field lines remain within the iron core, the flux through each turn of the primary equals the flux through each turn of the secondary. Hence, the voltage across the secondary is

$$\Delta V_2 = -N_2 \frac{d\Phi_B}{dt} \tag{33.40}$$

Solving Equation 33.39 for $d\Phi_B/dt$ and substituting the result into Equation 33.40, we find that

$$\Delta V_2 = \frac{N_2}{N_1} \Delta V_1 \tag{33.41}$$

When $N_2 > N_1$, the output voltage ΔV_2 exceeds the input voltage ΔV_1. This setup is referred to as a *step-up transformer*. When $N_2 < N_1$, the output voltage is less than the input voltage, and we have a *step-down transformer*.

When the switch in the secondary circuit is closed, a current I_2 is induced in the secondary. If the load in the secondary circuit is a pure resistance, the induced current is in phase with the induced voltage. The power supplied to the secondary circuit must be provided by the AC source connected to the primary circuit, as shown in Figure 33.22. In an ideal transformer, where there are no losses, the power

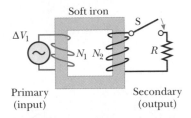

Soft iron

ΔV_1

N_1 N_2

R

S

Primary (input)

Secondary (output)

Figure 33.21 An ideal transformer consists of two coils wound on the same iron core. An alternating voltage ΔV_1 is applied to the primary coil, and the output voltage ΔV_2 is across the resistor of resistance R.

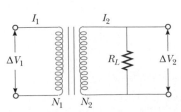

I_1 I_2

ΔV_1 R_L ΔV_2

N_1 N_2

Figure 33.22 Circuit diagram for a transformer.

$I_1 \, \Delta V_1$ supplied by the source is equal to the power $I_2 \, \Delta V_2$ in the secondary circuit. That is,

$$I_1 \, \Delta V_1 = I_2 \, \Delta V_2 \tag{33.42}$$

The value of the load resistance R_L determines the value of the secondary current because $I_2 = \Delta V_2 / R_L$. Furthermore, the current in the primary is $I_1 = \Delta V_1 / R_{\text{eq}}$, where

$$R_{\text{eq}} = \left(\frac{N_1}{N_2} \right)^2 R_L \tag{33.43}$$

is the equivalent resistance of the load resistance when viewed from the primary side. From this analysis we see that a transformer may be used to match resistances between the primary circuit and the load. In this manner, maximum power transfer can be achieved between a given power source and the load resistance. For example, a transformer connected between the 1-kΩ output of an audio amplifier and an 8-Ω speaker ensures that as much of the audio signal as possible is transferred into the speaker. In stereo terminology, this is called *impedance matching*.

We can now also understand why transformers are useful for transmitting power over long distances. Because the generator voltage is stepped up, the current in the transmission line is reduced, and hence $I^2 R$ losses are reduced. In practice, the voltage is stepped up to around 230 000 V at the generating station, stepped down to around 20 000 V at a distributing station, then to 4 000 V for delivery to residential areas, and finally to 120–240 V at the customer's site.

There is a practical upper limit to the voltages that can be used in transmission lines. Excessive voltages could ionize the air surrounding the transmission lines, which could result in a conducting path to ground or to other objects in the vicinity. This, of course, would present a serious hazard to any living creatures. For this reason, a long string of insulators is used to keep high voltage wires away from their supporting metal towers. Other insulators are used to maintain separation between wires.

Many common household electronic devices require low voltages to operate properly. A small transformer that plugs directly into the wall, like the one illustrated in Figure 33.23, can provide the proper voltage. The photograph shows the two windings wrapped around a common iron core that is found inside all these little "black boxes." This particular transformer converts the 120-V AC in the wall socket to 12.5-V AC. (Can you determine the ratio of the numbers of turns in the two coils?) Some black boxes also make use of diodes to convert the alternating current to direct current. (See Section 33.9.)

Nikola Tesla
American Physicist (1856–1943)

Tesla was born in Croatia but spent most of his professional life as an inventor in the United States. He was a key figure in the development of alternating current electricity, high-voltage transformers, and the transport of electrical power using AC transmission lines. Tesla's viewpoint was at odds with the ideas of Thomas Edison, who committed himself to the use of direct current in power transmission. Tesla's AC approach won out. *(UPI/CORBIS)*

Figure 33.23 The primary winding in this transformer is directly attached to the prongs of the plug. The secondary winding is connected to the power cord on the right, which runs to an electronic device. Many of these power-supply transformers also convert alternating current to direct current.

This transformer is smaller than the one in the opening photograph for this chapter. In addition, it is a step-down transformer. It drops the voltage from 4 000 V to 240 V for delivery to a group of residences.

Example 33.8 The Economics of AC Power

An electricity-generating station needs to deliver energy at a rate of 20 MW to a city 1.0 km away.

(A) If the resistance of the wires is 2.0 Ω and the energy costs about 10¢/kWh, estimate what it costs the utility company for the energy converted to internal energy in the wires during one day. A common voltage for commercial power generators is 22 kV, but a step-up transformer is used to boost the voltage to 230 kV before transmission.

Solution Conceptualize by noting that the resistance of the wires is in series with the resistance representing the load (homes and businesses). Thus, there will be a voltage drop in the wires, which means that some of the transmitted energy is converted to internal energy in the wires and never reaches the load. Because this is an estimate, let us categorize this as a problem in which the power factor is equal to 1. To analyze the problem, we begin by calculating I_{rms} from Equation 33.31:

$$I_{rms} = \frac{\mathcal{P}_{av}}{\Delta V_{rms}} = \frac{20 \times 10^6 \text{ W}}{230 \times 10^3 \text{ V}} = 87 \text{ A}$$

Now, we determine the rate at which energy is delivered to the resistance in the wires from Equation 33.32:

$$\mathcal{P}_{av} = I_{rms}^2 R = (87 \text{ A})^2(2.0 \ \Omega) = 15 \text{ kW}$$

Over the course of a day, the energy loss due to the resistance of the wires is $(15 \text{ kW})(24 \text{ h}) = 360$ kWh, at a cost of $36.

(B) Repeat the calculation for the situation in which the power plant delivers the energy at its original voltage of 22 kV.

Solution Again using Equation 33.31, we find

$$I_{rms} = \frac{\mathcal{P}_{av}}{\Delta V_{rms}} = \frac{20 \times 10^6 \text{ W}}{22 \times 10^3 \text{ V}} = 910 \text{ A}$$

and, from Equation 33.32,

$$\mathcal{P}_{av} = I_{rms}^2 R = (910 \text{ A})^2(2.0 \ \Omega) = 1.7 \times 10^3 \text{ kW}$$

$$\text{Cost per day} = (1.7 \times 10^3 \text{ kW})(24 \text{ h})(\$0.10/\text{kWh})$$

$$= \$4\,100$$

To finalize the example, note the tremendous savings that are possible through the use of transformers and high-voltage transmission lines. This, in combination with the efficiency of using alternating current to operate motors, led to the universal adoption of alternating current instead of direct current for commercial power grids.

33.9 Rectifiers and Filters

Portable electronic devices such as radios and compact disc (CD) players are often powered by direct current supplied by batteries. Many devices come with AC–DC converters such as that in Figure 33.23. Such a converter contains a transformer that steps the voltage down from 120 V to typically 9 V and a circuit that converts alternating current to direct current. The process of converting alternating current to direct current is called **rectification,** and the converting device is called a **rectifier.**

The most important element in a rectifier circuit is a **diode,** a circuit element that conducts current in one direction but not the other. Most diodes used in modern electronics are semiconductor devices. The circuit symbol for a diode is ——▶|—— , where the arrow indicates the direction of the current in the diode. A diode has low resistance to current in one direction (the direction of the arrow) and high resistance to current in the opposite direction. We can understand how a diode rectifies a current by considering Figure 33.24a, which shows a diode and a resistor connected to the secondary of a transformer. The transformer reduces the voltage from 120-V AC to the lower voltage that is needed for the device having a resistance R (the load resistance). Because the diode conducts current in only one direction, the alternating current in the load resistor is reduced to the form shown by the solid curve in Figure 33.24b. The diode conducts current only when the side of the symbol containing the arrowhead has a positive potential relative to the other side. In this situation, the diode acts as a *half-wave rectifier* because current is present in the circuit during only half of each cycle.

When a capacitor is added to the circuit, as shown by the dashed lines and the capacitor symbol in Figure 33.24a, the circuit is a simple DC power supply. The time variation in the current in the load resistor (the dashed curve in Fig. 33.24b) is close to

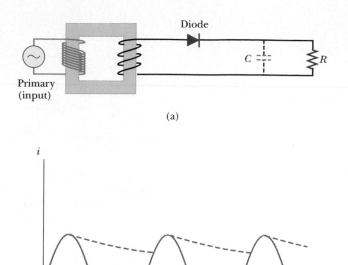

(a)

(b)

Figure 33.24 (a) A half-wave rectifier with an optional filter capacitor. (b) Current versus time in the resistor. The solid curve represents the current with no filter capacitor, and the dashed curve is the current when the circuit includes the capacitor.

being zero, as determined by the RC time constant of the circuit. As the current in the circuit begins to rise at $t = 0$ in Figure 33.24b, the capacitor charges up. When the current begins to fall, however, the capacitor discharges through the resistor, so that the current in the resistor does not fall as fast as the current from the transformer.

The RC circuit in Figure 33.24a is one example of a **filter circuit,** which is used to smooth out or eliminate a time-varying signal. For example, radios are usually powered by a 60-Hz alternating voltage. After rectification, the voltage still contains a small AC component at 60 Hz (sometimes called *ripple*), which must be filtered. By "filtered," we mean that the 60-Hz ripple must be reduced to a value much less than that of the audio signal to be amplified, because without filtering, the resulting audio signal includes an annoying hum at 60 Hz.

We can also design filters that will respond differently to different frequencies. Consider the simple series RC circuit shown in Figure 33.25a. The input voltage is across the series combination of the two elements. The output is the voltage across the resistor. A plot of the ratio of the output voltage to the input voltage as a function of the logarithm of angular frequency (see Fig. 33.25b) shows that at low frequencies ΔV_{out} is much smaller than ΔV_{in}, whereas at high frequencies the two voltages are equal. Because the circuit

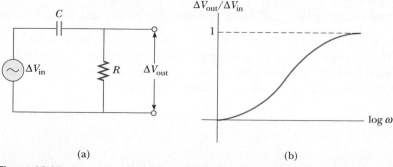

(a)

(b)

Active Figure 33.25 (a) A simple RC high-pass filter. (b) Ratio of output voltage to input voltage for an RC high-pass filter as a function of the angular frequency of the AC source.

At the Active Figures link at http://www.pse6.com, you can adjust the resistance and the capacitance of the circuit in part (a). You can then determine the output voltage for a given frequency or sweep through the frequencies to generate a curve like that in part (b).

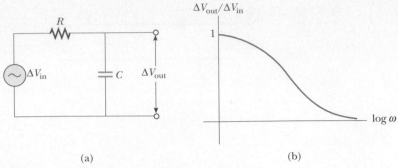

(a)   (b)

Active Figure 33.26 (a) A simple RC low-pass filter. (b) Ratio of output voltage to input voltage for an RC low-pass filter as a function of the angular frequency of the AC source.

preferentially passes signals of higher frequency while blocking low-frequency signals, the circuit is called an RC high-pass filter. (See Problem 51 for an analysis of this filter.)

Physically, a high-pass filter works because a capacitor "blocks out" direct current and AC current at low frequencies. At low frequencies, the capacitive reactance is large and much of the applied voltage appears across the capacitor rather than across the output resistor. As the frequency increases, the capacitive reactance drops and more of the applied voltage appears across the resistor.

Now consider the circuit shown in Figure 33.26a, where we have interchanged the resistor and capacitor and the output voltage is taken across the capacitor. At low frequencies, the reactance of the capacitor and the voltage across the capacitor is high. As the frequency increases, the voltage across the capacitor drops. Thus, this is an RC low-pass filter. The ratio of output voltage to input voltage (see Problem 52), plotted as a function of the logarithm of ω in Figure 33.26b, shows this behavior.

You may be familiar with crossover networks, which are an important part of the speaker systems for high-fidelity audio systems. These networks use low-pass filters to direct low frequencies to a special type of speaker, the "woofer," which is designed to reproduce the low notes accurately. The high frequencies are sent to the "tweeter" speaker.

Quick Quiz 33.11 Suppose you are designing a high-fidelity system containing both large loudspeakers (woofers) and small loudspeakers (tweeters). If you wish to deliver low-frequency signals to a woofer, what device would you place in series with it? (a) an inductor (b) a capacitor (c) a resistor. If you wish to deliver high-frequency signals to a tweeter, what device would you place in series with it? (d) an inductor (e) a capacitor (f) a resistor.

SUMMARY

If an AC circuit consists of a source and a resistor, the current is in phase with the voltage. That is, the current and voltage reach their maximum values at the same time.

The **rms current** and **rms voltage** in an AC circuit in which the voltages and current vary sinusoidally are given by the expressions

$$I_{rms} = \frac{I_{max}}{\sqrt{2}} = 0.707 I_{max} \tag{33.4}$$

$$\Delta V_{rms} = \frac{\Delta V_{max}}{\sqrt{2}} = 0.707 \, \Delta V_{max} \tag{33.5}$$

where I_{max} and ΔV_{max} are the maximum values.

If an AC circuit consists of a source and an inductor, the current lags behind the voltage by 90°. That is, the voltage reaches its maximum value one quarter of a period before the current reaches its maximum value.

If an AC circuit consists of a source and a capacitor, the current leads the voltage by 90°. That is, the current reaches its maximum value one quarter of a period before the voltage reaches its maximum value.

In AC circuits that contain inductors and capacitors, it is useful to define the **inductive reactance** X_L and the **capacitive reactance** X_C as

$$X_L \equiv \omega L \tag{33.10}$$

$$X_C \equiv \frac{1}{\omega C} \tag{33.18}$$

where ω is the angular frequency of the AC source. The SI unit of reactance is the ohm.

The **impedance** Z of an RLC series AC circuit is

$$Z \equiv \sqrt{R^2 + (X_L - X_C)^2} \tag{33.25}$$

This expression illustrates that we cannot simply add the resistance and reactances in a circuit. We must account for the fact that the applied voltage and current are out of phase, with the **phase angle** ϕ between the current and voltage being

$$\phi = \tan^{-1}\left(\frac{X_L - X_C}{R}\right) \tag{33.27}$$

The sign of ϕ can be positive or negative, depending on whether X_L is greater or less than X_C. The phase angle is zero when $X_L = X_C$.

The **average power** delivered by the source in an RLC circuit is

$$\mathcal{P}_{av} = I_{rms}\,\Delta V_{rms}\cos\phi \tag{33.31}$$

An equivalent expression for the average power is

$$\mathcal{P}_{av} = I_{rms}^2\,R \tag{33.32}$$

The average power delivered by the source results in increasing internal energy in the resistor. No power loss occurs in an ideal inductor or capacitor.

The rms current in a series RLC circuit is

$$I_{rms} = \frac{\Delta V_{rms}}{\sqrt{R^2 + (X_L - X_C)^2}} \tag{33.34}$$

A series RLC circuit is in resonance when the inductive reactance equals the capacitive reactance. When this condition is met, the current given by Equation 33.34 reaches its maximum value. The **resonance frequency** ω_0 of the circuit is

$$\omega_0 = \frac{1}{\sqrt{LC}} \tag{33.35}$$

The current in a series RLC circuit reaches its maximum value when the frequency of the source equals ω_0—that is, when the "driving" frequency matches the resonance frequency.

Transformers allow for easy changes in alternating voltage. Because energy (and therefore power) are conserved, we can write

$$I_1\,\Delta V_1 = I_2\,\Delta V_2 \tag{33.42}$$

to relate the currents and voltages in the primary and secondary windings of a transformer.

QUESTIONS

1. How can the average value of a current be zero and yet the square root of the average squared current not be zero?

2. What is the time average of the "square-wave" potential shown in Figure Q33.2? What is its rms voltage?

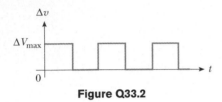

Figure Q33.2

3. Do AC ammeters and voltmeters read maximum, rms, or average values?

4. In the clearest terms you can, explain the statement, "The voltage across an inductor leads the current by 90°."

5. Some fluorescent lights flicker on and off 120 times every second. Explain what causes this. Why can't you see it happening?

6. Why does a capacitor act as a short circuit at high frequencies? Why does it act as an open circuit at low frequencies?

7. Explain how the mnemonic "ELI the ICE man" can be used to recall whether current leads voltage or voltage leads current in *RLC* circuits. Note that E represents emf $\mathcal{E}$.

8. Why is the sum of the maximum voltages across each of the elements in a series *RLC* circuit usually greater than the maximum applied voltage? Doesn't this violate Kirchhoff's loop rule?

9. Does the phase angle depend on frequency? What is the phase angle when the inductive reactance equals the capacitive reactance?

10. In a series *RLC* circuit, what is the possible range of values for the phase angle?

11. If the frequency is doubled in a series *RLC* circuit, what happens to the resistance, the inductive reactance, and the capacitive reactance?

12. Explain why the average power delivered to an *RLC* circuit by the source depends on the phase angle between the current and applied voltage.

13. As shown in Figure 7.5a, a person pulls a vacuum cleaner at speed v across a horizontal floor, exerting on it a force of magnitude F directed upward at an angle θ with the horizontal. At what rate is the person doing work on the cleaner? State as completely as you can the analogy between power in this situation and in an electric circuit.

14. A particular experiment requires a beam of light of very stable intensity. Why would an AC voltage be unsuitable for powering the light source?

15. Do some research to answer these questions: Who invented the metal detector? Why? Did it work?

16. What is the advantage of transmitting power at high voltages?

17. What determines the maximum voltage that can be used on a transmission line?

18. Will a transformer operate if a battery is used for the input voltage across the primary? Explain.

19. Someone argues that high-voltage power lines actually waste more energy. He points out that the rate at which internal energy is produced in a wire is given by $(\Delta V)^2/R$, where R is the resistance of the wire. Therefore, the higher the voltage, the higher the energy waste. What if anything is wrong with his argument?

20. Explain how the quality factor is related to the response characteristics of a radio receiver. Which variable most strongly influences the quality factor?

21. Why are the primary and secondary coils of a transformer wrapped on an iron core that passes through both coils?

22. With reference to Figure Q33.22, explain why the capacitor prevents a DC signal from passing between A and B, yet allows an AC signal to pass from A to B. (The circuits are said to be capacitively coupled.)

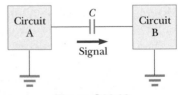

Figure Q33.22

PROBLEMS

1, 2, 3 = straightforward, intermediate, challenging   ☐ = full solution available in the *Student Solutions Manual and Study Guide*

🌀 = coached solution with hints available at http://www.pse6.com   🖥 = computer useful in solving problem

▨ = paired numerical and symbolic problems

Note: Assume all AC voltages and currents are sinusoidal, unless stated otherwise.

Section 33.1   AC Sources
Section 33.2   Resistors in an AC Circuit

1. The rms output voltage of an AC source is 200 V and the operating frequency is 100 Hz. Write the equation giving the output voltage as a function of time.

2. (a) What is the resistance of a lightbulb that uses an average power of 75.0 W when connected to a 60.0-Hz power source having a maximum voltage of 170 V? (b) **What If?** What is the resistance of a 100-W bulb?

3. An AC power supply produces a maximum voltage $\Delta V_{max} = 100$ V. This power supply is connected to a 24.0-Ω resistor, and the current and resistor voltage are measured with an ideal AC ammeter and voltmeter, as shown in Figure P33.3. What does each meter read? Note that an ideal ammeter has zero resistance and that an ideal voltmeter has infinite resistance.

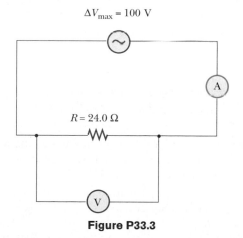

$\Delta V_{max} = 100$ V

$R = 24.0\ \Omega$

Figure P33.3

4. In the simple AC circuit shown in Figure 33.2, $R = 70.0\ \Omega$ and $\Delta v = \Delta V_{max} \sin \omega t$. (a) If $\Delta v_R = 0.250\ \Delta V_{max}$ for the first time at $t = 0.010\ 0$ s, what is the angular frequency of the source? (b) What is the next value of t for which $\Delta v_R = 0.250\ \Delta V_{max}$?

5. The current in the circuit shown in Figure 33.2 equals 60.0% of the peak current at $t = 7.00$ ms. What is the smallest frequency of the source that gives this current?

6. Figure P33.6 shows three lamps connected to a 120-V AC (rms) household supply voltage. Lamps 1 and 2 have 150-W bulbs; lamp 3 has a 100-W bulb. Find the rms current and resistance of each bulb.

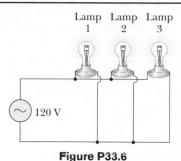

Lamp 1   Lamp 2   Lamp 3

120 V

Figure P33.6

7. An audio amplifier, represented by the AC source and resistor in Figure P33.7, delivers to the speaker alternating voltage at audio frequencies. If the source voltage has an amplitude of 15.0 V, $R = 8.20\ \Omega$, and the speaker is equivalent to a resistance of 10.4 Ω, what is the time-averaged power transferred to it?

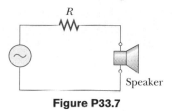

R

Speaker

Figure P33.7

Section 33.3   Inductors in an AC Circuit

8. An inductor is connected to a 20.0-Hz power supply that produces a 50.0-V rms voltage. What inductance is needed to keep the instantaneous current in the circuit below 80.0 mA?

9. In a purely inductive AC circuit, as shown in Figure 33.6, $\Delta V_{max} = 100$ V. (a) The maximum current is 7.50 A at 50.0 Hz. Calculate the inductance L. (b) **What If?** At what angular frequency ω is the maximum current 2.50 A?

10. An inductor has a 54.0-Ω reactance at 60.0 Hz. What is the maximum current if this inductor is connected to a 50.0-Hz source that produces a 100-V rms voltage?

11. 🌀 For the circuit shown in Figure 33.6, $\Delta V_{max} = 80.0$ V, $\omega = 65.0\pi$ rad/s, and $L = 70.0$ mH. Calculate the current in the inductor at $t = 15.5$ ms.

12. A 20.0-mH inductor is connected to a standard electrical outlet ($\Delta V_{rms} = 120$ V; $f = 60.0$ Hz). Determine the energy stored in the inductor at $t = (1/180)$ s, assuming that this energy is zero at $t = 0$.

13. **Review problem.** Determine the maximum magnetic flux through an inductor connected to a standard electrical outlet ($\Delta V_{rms} = 120$ V, $f = 60.0$ Hz).

Section 33.4 Capacitors in an AC Circuit

14. (a) For what frequencies does a 22.0-μF capacitor have a reactance below 175 Ω? (b) **What If?** Over this same frequency range, what is the reactance of a 44.0-μF capacitor?

15. What is the maximum current in a 2.20-μF capacitor when it is connected across (a) a North American electrical outlet having $\Delta V_{rms} = 120$ V, $f = 60.0$ Hz, and (b) **What If?** a European electrical outlet having $\Delta V_{rms} = 240$ V, $f = 50.0$ Hz?

16. A capacitor C is connected to a power supply that operates at a frequency f and produces an rms voltage ΔV. What is the maximum charge that appears on either of the capacitor plates?

17. What maximum current is delivered by an AC source with $\Delta V_{max} = 48.0$ V and $f = 90.0$ Hz when connected across a 3.70-μF capacitor?

18. A 1.00-mF capacitor is connected to a standard electrical outlet ($\Delta V_{rms} = 120$ V; $f = 60.0$ Hz). Determine the current in the capacitor at $t = (1/180)$ s, assuming that at $t = 0$, the energy stored in the capacitor is zero.

Section 33.5 The RLC Series Circuit

19. An inductor ($L = 400$ mH), a capacitor ($C = 4.43\ \mu$F), and a resistor ($R = 500\ \Omega$) are connected in series. A 50.0-Hz AC source produces a peak current of 250 mA in the circuit. (a) Calculate the required peak voltage ΔV_{max}. (b) Determine the phase angle by which the current leads or lags the applied voltage.

20. At what frequency does the inductive reactance of a 57.0-μH inductor equal the capacitive reactance of a 57.0-μF capacitor?

21. A series AC circuit contains the following components: $R = 150\ \Omega$, $L = 250$ mH, $C = 2.00\ \mu$F and a source with $\Delta V_{max} = 210$ V operating at 50.0 Hz. Calculate the (a) inductive reactance, (b) capacitive reactance, (c) impedance, (d) maximum current, and (e) phase angle between current and source voltage.

22. A sinusoidal voltage $\Delta v(t) = (40.0$ V$)\sin(100t)$ is applied to a series RLC circuit with $L = 160$ mH, $C = 99.0\ \mu$F, and $R = 68.0\ \Omega$. (a) What is the impedance of the circuit? (b) What is the maximum current? (c) Determine the numerical values for I_{max}, ω, and ϕ in the equation $i(t) = I_{max}\sin(\omega t - \phi)$.

23. An RLC circuit consists of a 150-Ω resistor, a 21.0-μF capacitor, and a 460-mH inductor, connected in series with a 120-V, 60.0-Hz power supply. (a) What is the phase angle between the current and the applied voltage? (b) Which reaches its maximum earlier, the current or the voltage?

24. Four circuit elements—a capacitor, an inductor, a resistor, and an AC source—are connected together in various ways. First the capacitor is connected to the source, and the rms current is found to be 25.1 mA. The capacitor is disconnected and discharged, and then connected in series with the resistor and the source, making the rms current 15.7 mA. The circuit is disconnected and the capacitor discharged. The capacitor is then connected in series with the inductor and the source, making the rms current 68.2 mA. After the circuit is disconnected and the capacitor discharged, all four circuit elements are connected together in a series loop. What is the rms current in the circuit?

25. A person is working near the secondary of a transformer, as shown in Figure P33.25. The primary voltage is 120 V at 60.0 Hz. The capacitance C_s, which is the stray capacitance between the hand and the secondary winding, is 20.0 pF. Assuming the person has a body resistance to ground $R_b = 50.0$ kΩ, determine the rms voltage across the body. (*Suggestion:* Redraw the circuit with the secondary of the transformer as a simple AC source.)

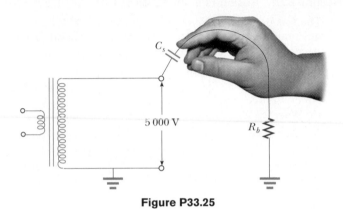

Figure P33.25

26. An AC source with $\Delta V_{max} = 150$ V and $f = 50.0$ Hz is connected between points a and d in Figure P33.26. Calculate the maximum voltages between points (a) a and b, (b) b and c, (c) c and d, and (d) b and d.

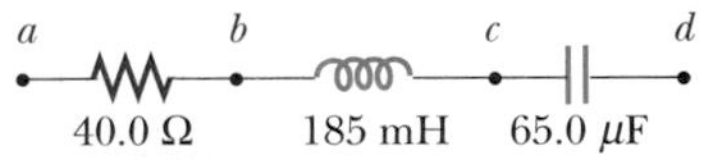

$$a \overset{}{\underset{40.0\ \Omega}{-\!\!\!\bigwedge\!\!\bigwedge\!\!\bigwedge\!\!-}} b \overset{}{\underset{185\ \text{mH}}{-\!\!000\!\!-}} c \overset{}{\underset{65.0\ \mu\text{F}}{-\!\!|\!|\!\!-}} d$$

Figure P33.26 Problems 26 and 68.

27. Draw to scale a phasor diagram showing Z, X_L, X_C, and ϕ for an AC series circuit for which $R = 300\ \Omega$, $C = 11.0\ \mu$F, $L = 0.200$ H, and $f = (500/\pi)$ Hz.

28. In an RLC series circuit that includes a source of alternating current operating at fixed frequency and voltage, the resistance R is equal to the inductive reactance. If the plate separation of the capacitor is reduced to half of its original value, the current in the circuit doubles. Find the initial capacitive reactance in terms of R.

29. A coil of resistance 35.0 Ω and inductance 20.5 H is in series with a capacitor and a 200-V (rms), 100-Hz source. The rms current in the circuit is 4.00 A. (a) Calculate the capacitance in the circuit. (b) What is ΔV_{rms} across the coil?

Section 33.6 Power in an AC Circuit

30. The voltage source in Figure P33.30 has an output of $\Delta V_{rms} = 100$ V at an angular frequency of $\omega = 1\,000$ rad/s. Determine (a) the current in the circuit and (b) the power supplied by the source. (c) Show that the power delivered to the resistor is equal to the power supplied by the source.

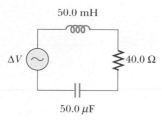

50.0 mH

ΔV 40.0 Ω

50.0 μF

Figure P33.30

31. An AC voltage of the form $\Delta v = (100 \text{ V}) \sin(1\,000t)$ is applied to a series *RLC* circuit. Assume the resistance is 400 Ω, the capacitance is 5.00 μF, and the inductance is 0.500 H. Find the average power delivered to the circuit.

32. A series *RLC* circuit has a resistance of 45.0 Ω and an impedance of 75.0 Ω. What average power is delivered to this circuit when $\Delta V_{\text{rms}} = 210$ V?

33. In a certain series *RLC* circuit, $I_{\text{rms}} = 9.00$ A, $\Delta V_{\text{rms}} = 180$ V, and the current leads the voltage by 37.0°. (a) What is the total resistance of the circuit? (b) Calculate the reactance of the circuit ($X_L - X_C$).

34. Suppose you manage a factory that uses many electric motors. The motors create a large inductive load to the electric power line, as well as a resistive load. The electric company builds an extra-heavy distribution line to supply you with a component of current that is 90° out of phase with the voltage, as well as with current in phase with the voltage. The electric company charges you an extra fee for "reactive volt-amps," in addition to the amount you pay for the energy you use. You can avoid the extra fee by installing a capacitor between the power line and your factory. The following problem models this solution.

In an *RL* circuit, a 120-V (rms), 60.0-Hz source is in series with a 25.0-mH inductor and a 20.0-Ω resistor. What are (a) the rms current and (b) the power factor? (c) What capacitor must be added in series to make the power factor 1? (d) To what value can the supply voltage be reduced, if the power supplied is to be the same as before the capacitor was installed?

35. Suppose power $\mathcal{P}$ is to be transmitted over a distance d at a voltage ΔV with only 1.00% loss. Copper wire of what diameter should be used for each of the two conductors of the transmission line? Assume the current density in the conductors is uniform.

36. A diode is a device that allows current to be carried in only one direction (the direction indicated by the arrowhead in its circuit symbol). Find in terms of ΔV and R the average power delivered to the diode circuit of Figure P33.36.

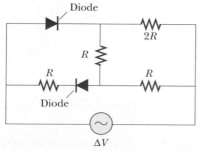

Diode

2R

R

R   R

Diode

ΔV

Figure P33.36

Section 33.7   Resonance in a Series *RLC* Circuit

37. An *RLC* circuit is used in a radio to tune into an FM station broadcasting at 99.7 MHz. The resistance in the circuit is 12.0 Ω, and the inductance is 1.40 μH. What capacitance should be used?

38. The tuning circuit of an AM radio contains an *LC* combination. The inductance is 0.200 mH, and the capacitor is variable, so that the circuit can resonate at any frequency between 550 kHz and 1 650 kHz. Find the range of values required for *C*.

39. A radar transmitter contains an *LC* circuit oscillating at 1.00×10^{10} Hz. (a) What capacitance will resonate with a one-turn loop of inductance 400 pH at this frequency? (b) If the capacitor has square parallel plates separated by 1.00 mm of air, what should the edge length of the plates be? (c) What is the common reactance of the loop and capacitor at resonance?

40. A series *RLC* circuit has components with following values: $L = 20.0$ mH, $C = 100$ nF, $R = 20.0$ Ω, and $\Delta V_{\text{max}} = 100$ V, with $\Delta v = \Delta V_{\text{max}} \sin \omega t$. Find (a) the resonant frequency, (b) the amplitude of the current at the resonant frequency, (c) the Q of the circuit, and (d) the amplitude of the voltage across the inductor at resonance.

41. A 10.0-Ω resistor, 10.0-mH inductor, and 100-μF capacitor are connected in series to a 50.0-V (rms) source having variable frequency. Find the energy that is delivered to the circuit during one period if the operating frequency is twice the resonance frequency.

42. A resistor *R*, inductor *L*, and capacitor *C* are connected in series to an AC source of rms voltage ΔV and variable frequency. Find the energy that is delivered to the circuit during one period if the operating frequency is twice the resonance frequency.

43. Compute the quality factor for the circuits described in Problems 22 and 23. Which circuit has the sharper resonance?

Section 33.8   The Transformer and Power Transmission

44. A step-down transformer is used for recharging the batteries of portable devices such as tape players. The turns ratio inside the transformer is 13:1 and it is used with 120-V (rms) household service. If a particular ideal transformer draws 0.350 A from the house outlet, what are (a) the voltage and (b) the current supplied to a tape player from the transformer? (c) How much power is delivered?

45. A transformer has $N_1 = 350$ turns and $N_2 = 2\,000$ turns. If the input voltage is $\Delta v(t) = (170 \text{ V}) \cos \omega t$, what rms voltage is developed across the secondary coil?

46. A step-up transformer is designed to have an output voltage of 2 200 V (rms) when the primary is connected across a 110-V (rms) source. (a) If the primary winding has 80 turns, how many turns are required on the secondary? (b) If a load resistor across the secondary draws a current of 1.50 A, what is the current in the primary, assuming ideal conditions? (c) **What If?** If the transformer actually has an efficiency of 95.0%, what is the current in the primary when the secondary current is 1.20 A?

47. In the transformer shown in Figure P33.47, the load resistor is 50.0 Ω. The turns ratio $N_1:N_2$ is 5:2, and the source voltage is 80.0 V (rms). If a voltmeter across the load measures 25.0 V (rms), what is the source resistance R_s?

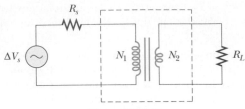

Figure P33.47

48. The secondary voltage of an ignition transformer in a furnace is 10.0 kV. When the primary operates at an rms voltage of 120 V, the primary impedance is 24.0 Ω and the transformer is 90.0% efficient. (a) What turns ratio is required? What are (b) the current in the secondary and (c) the impedance in the secondary?

49. A transmission line that has a resistance per unit length of 4.50×10^{-4} Ω/m is to be used to transmit 5.00 MW over 400 miles $(6.44 \times 10^5$ m). The output voltage of the generator is 4.50 kV. (a) What is the line loss if a transformer is used to step up the voltage to 500 kV? (b) What fraction of the input power is lost to the line under these circumstances? (c) **What If?** What difficulties would be encountered in attempting to transmit the 5.00 MW at the generator voltage of 4.50 kV?

Section 33.9   Rectifiers and Filters

50. One particular plug-in power supply for a radio looks similar to the one shown in Figure 33.23 and is marked with the following information: Input 120 V AC 8 W Output 9 V DC 300 mA. Assume that these values are accurate to two digits. (a) Find the energy efficiency of the device when the radio is operating. (b) At what rate does the device produce wasted energy when the radio is operating? (c) Suppose that the input power to the transformer is 8.0 W when the radio is switched off and that energy costs $0.135/kWh from the electric company. Find the cost of having six such transformers around the house, plugged in for thirty-one days.

51. Consider the filter circuit shown in Figure 33.25a. (a) Show that the ratio of the output voltage to the input voltage is

$$\frac{\Delta V_{out}}{\Delta V_{in}} = \frac{R}{\sqrt{R^2 + \left(\dfrac{1}{\omega C}\right)^2}}$$

(b) What value does this ratio approach as the frequency decreases toward zero? What value does this ratio approach as the frequency increases without limit? (c) At what frequency is the ratio equal to one half?

52. Consider the filter circuit shown in Figure 33.26a. (a) Show that the ratio of the output voltage to the input

voltage is

$$\frac{\Delta V_{out}}{\Delta V_{in}} = \frac{1/\omega C}{\sqrt{R^2 + \left(\dfrac{1}{\omega C}\right)^2}}$$

(b) What value does this ratio approach as the frequency decreases toward zero? What value does this ratio approach as the frequency increases without limit? (c) At what frequency is the ratio equal to one half?

53. The RC high-pass filter shown in Figure 33.25 has a resistance $R = 0.500$ Ω. (a) What capacitance gives an output signal that has half the amplitude of a 300-Hz input signal? (b) What is the ratio $(\Delta V_{out}/\Delta V_{in})$ for a 600-Hz signal? You may use the result of Problem 51.

54. The RC low-pass filter shown in Figure 33.26 has a resistance $R = 90.0$ Ω and a capacitance $C = 8.00$ nF. Calculate the ratio $(\Delta V_{out}/\Delta V_{in})$ for an input frequency of (a) 600 Hz and (b) 600 kHz. You may use the result of Problem 52.

55. The resistor in Figure P33.55 represents the midrange speaker in a three-speaker system. Assume its resistance to be constant at 8.00 Ω. The source represents an audio amplifier producing signals of uniform amplitude $\Delta V_{in} = 10.0$ V at all audio frequencies. The inductor and capacitor are to function as a bandpass filter with $\Delta V_{out}/\Delta V_{in} = 1/2$ at 200 Hz and at 4 000 Hz. (a) Determine the required values of L and C. (b) Find the maximum value of the ratio $\Delta V_{out}/\Delta V_{in}$. (c) Find the frequency f_0 at which the ratio has its maximum value. (d) Find the phase shift between ΔV_{in} and ΔV_{out} at 200 Hz, at f_0, and at 4 000 Hz. (e) Find the average power transferred to the speaker at 200 Hz, at f_0, and at 4 000 Hz. (f) Treating the filter as a resonant circuit, find its quality factor.

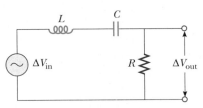

Figure P33.55

Additional Problems

56. Show that the rms value for the sawtooth voltage shown in Figure P33.56 is $\Delta V_{max}/\sqrt{3}$.

Figure P33.56

57. A series RLC circuit consists of an 8.00-Ω resistor, a 5.00-μF capacitor, and a 50.0-mH inductor. A variable

frequency source applies an emf of 400 V (rms) across the combination. Determine the power delivered to the circuit when the frequency is equal to half the resonance frequency.

58. A capacitor, a coil, and two resistors of equal resistance are arranged in an AC circuit, as shown in Figure P33.58. An AC source provides an emf of 20.0 V (rms) at a frequency of 60.0 Hz. When the double-throw switch S is open, as shown in the figure, the rms current is 183 mA. When the switch is closed in position 1, the rms current is 298 mA. When the switch is closed in position 2, the rms current is 137 mA. Determine the values of R, C, and L. Is more than one set of values possible?

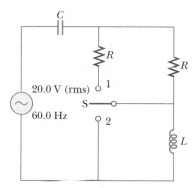

Figure P33.58

59. To determine the inductance of a coil used in a research project, a student first connects the coil to a 12.0-V battery and measures a current of 0.630 A. The student then connects the coil to a 24.0-V (rms), 60.0-Hz generator and measures an rms current of 0.570 A. What is the inductance?

60. **Review problem.** One insulated conductor from a household extension cord has mass per length 19.0 g/m. A section of this conductor is held under tension between two clamps. A subsection is located in a region of magnetic field of magnitude 15.3 mT perpendicular to the length of the cord. The wire carries an AC current of 9.00 A at 60.0 Hz. Determine some combination of values for the distance between the clamps and the tension in the cord so that the cord can vibrate in the lowest-frequency standing-wave vibrational state.

61. In Figure P33.61, find the rms current delivered by the 45.0-V (rms) power supply when (a) the frequency is very large and (b) the frequency is very small.

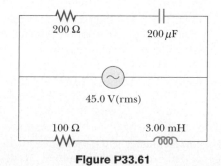

Figure P33.61

62. In the circuit shown in Figure P33.62, assume that all parameters except for C are given. (a) Find the current as a function of time. (b) Find the power delivered to the circuit. (c) Find the current as a function of time after *only* switch 1 is opened. (d) After switch 2 is *also* opened, the current and voltage are in phase. Find the capacitance C. (e) Find the impedance of the circuit when both switches are open. (f) Find the maximum energy stored in the capacitor during oscillations. (g) Find the maximum energy stored in the inductor during oscillations. (h) Now the frequency of the voltage source is doubled. Find the phase difference between the current and the voltage. (i) Find the frequency that makes the inductive reactance half the capacitive reactance.

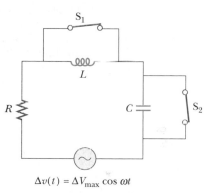

$$\Delta v(t) = \Delta V_{max} \cos \omega t$$

Figure P33.62

63. An 80.0-Ω resistor and a 200-mH inductor are connected in *parallel* across a 100-V (rms), 60.0-Hz source. (a) What is the rms current in the resistor? (b) By what angle does the total current lead or lag behind the voltage?

64. Make an order-of-magnitude estimate of the electric current that the electric company delivers to a town (Figure P33.64) from a remote generating station. State the data you measure or estimate. If you wish, you may consider a suburban bedroom community of 20 000 people.

Eddie Hironaka/Getty Images

Figure P33.64

65. Consider a series RLC circuit having the following circuit parameters: $R = 200\ \Omega$, $L = 663$ mH, and $C = 26.5\ \mu$F. The applied voltage has an amplitude of 50.0 V and a frequency of 60.0 Hz. Find the following amplitudes: (a) The current I_{max}, including its phase constant ϕ relative to the applied voltage Δv, (b) the voltage ΔV_R

across the resistor and its phase relative to the current, (c) the voltage ΔV_C across the capacitor and its phase relative to the current, and (d) the voltage ΔV_L across the inductor and its phase relative to the current.

66. A voltage $\Delta v = (100 \text{ V}) \sin \omega t$ (in SI units) is applied across a series combination of a 2.00-H inductor, a 10.0-μF capacitor, and a 10.0-Ω resistor. (a) Determine the angular frequency ω_0 at which the power delivered to the resistor is a maximum. (b) Calculate the power delivered at that frequency. (c) Determine the two angular frequencies ω_1 and ω_2 at which the power is half the maximum value. [The Q of the circuit is $\omega_0/(\omega_2 - \omega_1)$.]

67. *Impedance matching.* Example 28.2 showed that maximum power is transferred when the internal resistance of a DC source is equal to the resistance of the load. A transformer may be used to provide maximum power transfer between two AC circuits that have different impedances Z1 and Z2, where 1 and 2 are subscripts and the Z's are italic (as in the centered equation). (a) Show that the ratio of turns N_1/N_2 needed to meet this condition is

$$\frac{N_1}{N_2} = \sqrt{\frac{Z_1}{Z_2}}$$

(b) Suppose you want to use a transformer as an impedance-matching device between an audio amplifier that has an output impedance of 8.00 kΩ and a speaker that has an input impedance of 8.00 Ω. What should your N_1/N_2 ratio be?

68. A power supply with $\Delta V_{\text{rms}} = 120$ V is connected between points a and d in Figure P33.26. At what frequency will it deliver a power of 250 W?

69. Figure P33.69a shows a parallel *RLC* circuit, and the corresponding phasor diagram is given in Figure P33.69b. The instantaneous voltages (and rms voltages) across each of the three circuit elements are the same, and each is in phase with the current through the resistor. The currents in C and L lead or lag behind the current in the resistor, as shown in Figure P33.69b. (a) Show that the rms current delivered by the source is

$$I_{\text{rms}} = \Delta V_{\text{rms}} \left[\frac{1}{R^2} + \left(\omega C - \frac{1}{\omega L} \right)^2 \right]^{1/2}$$

(b) Show that the phase angle ϕ between ΔV_{rms} and I_{rms} is

$$\tan \phi = R \left(\frac{1}{X_C} - \frac{1}{X_L} \right)$$

70. An 80.0-Ω resistor, a 200-mH inductor, and a 0.150-μF capacitor are connected *in parallel* across a 120-V (rms) source operating at 374 rad/s. (a) What is the resonant frequency of the circuit? (b) Calculate the rms current in the resistor, inductor, and capacitor. (c) What rms current is delivered by the source? (d) Is the current leading or lagging behind the voltage? By what angle?

71. 🖳 A series *RLC* circuit is operating at 2 000 Hz. At this frequency, $X_L = X_C = 1\,884$ Ω. The resistance of the circuit is 40.0 Ω. (a) Prepare a table showing the values of X_L, X_C, and Z for $f = 300, 600, 800, 1\,000, 1\,500, 2\,000, 3\,000, 4\,000, 6\,000$, and 10 000 Hz. (b) Plot on the same set of axes X_L, X_C, and Z as a function of ln f.

72. 🖳 A series *RLC* circuit in which $R = 1.00$ Ω, $L = 1.00$ mH, and $C = 1.00$ nF is connected to an AC source delivering 1.00 V (rms). Make a precise graph of the power delivered to the circuit as a function of the frequency and verify that the full width of the resonance peak at half-maximum is $R/2\pi L$.

73. 🖳 Suppose the high-pass filter shown in Figure 33.25 has $R = 1\,000$ Ω and $C = 0.050\,0$ μF. (a) At what frequency does $\Delta V_{\text{out}}/\Delta V_{\text{in}} = \frac{1}{2}$? (b) Plot $\log_{10}(\Delta V_{\text{out}}/\Delta V_{\text{in}})$ versus $\log_{10}(f)$ over the frequency range from 1 Hz to 1 MHz. (This log–log plot of gain versus frequency is known as a *Bode plot.*)

Answers to Quick Quizzes

33.1 (a). The phasor in part (a) has the largest projection onto the vertical axis.

33.2 (b). The phasor in part (b) has the smallest-magnitude projection onto the vertical axis.

33.3 (c). The average power is proportional to the rms current, which, as Figure 33.5 shows, is nonzero even though the average current is zero. Condition (a) is valid only for an open circuit, and conditions (b) and (d) can not be true because $i_{\text{av}} = 0$ if the source is sinusoidal.

33.4 (b). For low frequencies, the reactance of the inductor is small so that the current is large. Most of the voltage from the source is across the bulb, so the power delivered to it is large.

33.5 (a). For high frequencies, the reactance of the capacitor is small so that the current is large. Most of the voltage from the source is across the bulb, so the power delivered to it is large.

33.6 (b). For low frequencies, the reactance of the capacitor is large so that very little current exists in the capacitor

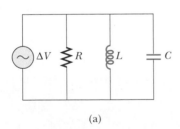

(a)

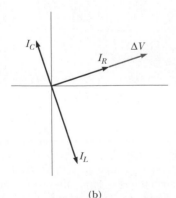

(b)

Figure P33.69

branch. The reactance of the inductor is small so that current exists in the inductor branch and the lightbulb glows. As the frequency increases, the inductive reactance increases and the capacitive reactance decreases. At high frequencies, more current exists in the capacitor branch than the inductor branch and the lightbulb glows more dimly.

33.7 (a) $X_L < X_C$. (b) $X_L = X_C$. (c) $X_L > X_C$.

33.8 (c). The cosine of $-\phi$ is the same as that of $+\phi$, so the $\cos\phi$ factor in Equation 33.31 is the same for both frequencies. The factor ΔV_{rms} is the same because the source voltage is fixed. According to Equation 33.27, changing $+\phi$ to $-\phi$ simply interchanges the values of X_L and X_C. Equation 33.25 tells us that such an interchange does not affect the impedance, so that the current I_{rms} in Equation 33.31 is the same for both frequencies.

33.9 (c). At resonance, $X_L = X_C$. According to Equation 33.25, this gives us $Z = R$.

33.10 (a). The higher the quality factor, the more sensitive the detector. As you can see from Figure 33.19, when $Q = \omega_0/\Delta\omega$ is high, a slight change in the resonance frequency (as might happen when a small piece of metal passes through the portal) causes a large change in current that can be detected easily.

33.11 (a) and (e). The current in an inductive circuit decreases with increasing frequency (see Eq. 33.9). Thus, an inductor connected in series with a woofer blocks high-frequency signals and passes low-frequency signals. The current in a capacitive circuit increases with increasing frequency (see Eq. 33.17). When a capacitor is connected in series with a tweeter, the capacitor blocks low-frequency signals and passes high-frequency signals.

Chapter 34

Electromagnetic Waves

CHAPTER OUTLINE

34.1 Maxwell's Equations and Hertz's Discoveries

34.2 Plane Electromagnetic Waves

34.3 Energy Carried by Electromagnetic Waves

34.4 Momentum and Radiation Pressure

34.5 Production of Electromagnetic Waves by an Antenna

34.6 The Spectrum of Electromagnetic Waves

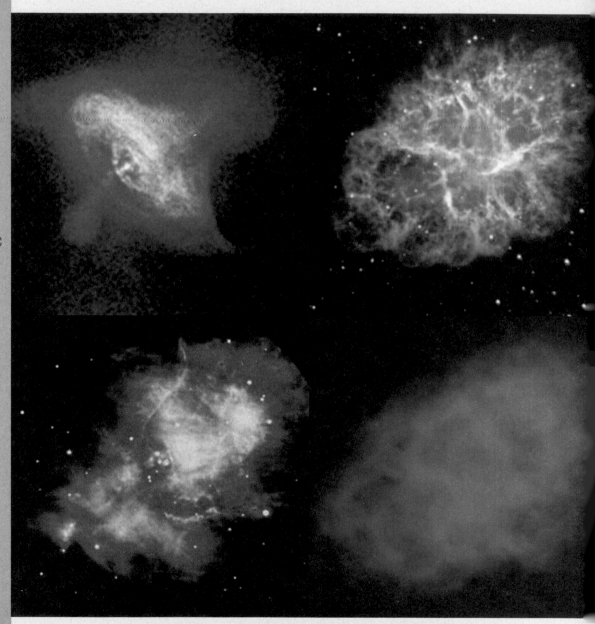

▲ Electromagnetic waves cover a broad spectrum of wavelengths, with waves in various wavelength ranges having distinct properties. These images of the Crab Nebula show different structure for observations made with waves of various wavelengths. The images (clockwise starting from the upper left) were taken with x-rays, visible light, radio waves, and infrared waves. (upper left–NASA/CXC/SAO; upper right–Palomar Observatory; lower right–VLA/NRAO; lower left–WM Keck Observatory)

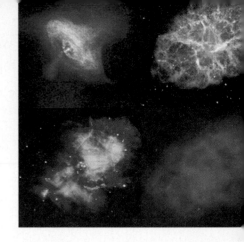

The waves described in Chapters 16, 17, and 18 are mechanical waves. By definition, the propagation of mechanical disturbances—such as sound waves, water waves, and waves on a string—requires the presence of a medium. This chapter is concerned with the properties of electromagnetic waves, which (unlike mechanical waves) can propagate through empty space.

In Section 31.7 we gave a brief description of Maxwell's equations, which form the theoretical basis of all electromagnetic phenomena. The consequences of Maxwell's equations are far-reaching and dramatic. The Ampère–Maxwell law predicts that a time-varying electric field produces a magnetic field, just as Faraday's law tells us that a time-varying magnetic field produces an electric field.

Astonishingly, Maxwell's equations also predict the existence of electromagnetic waves that propagate through space at the speed of light c. This chapter begins with a discussion of how Heinrich Hertz confirmed Maxwell's prediction when he generated and detected electromagnetic waves in 1887. That discovery has led to many practical communication systems, including radio, television, radar, and opto-electronics. On a conceptual level, Maxwell unified the subjects of light and electromagnetism by developing the idea that light is a form of electromagnetic radiation.

Next, we learn how electromagnetic waves are generated by oscillating electric charges. The waves consist of oscillating electric and magnetic fields at right angles to each other and to the direction of wave propagation. Thus, electromagnetic waves are transverse waves. The waves radiated from the oscillating charges can be detected at great distances. Furthermore, electromagnetic waves carry energy and momentum and hence can exert pressure on a surface.

The chapter concludes with a look at the wide range of frequencies covered by electromagnetic waves. For example, radio waves (frequencies of about 10^7 Hz) are electromagnetic waves produced by oscillating currents in a radio tower's transmitting antenna. Light waves are a high-frequency form of electromagnetic radiation (about 10^{14} Hz) produced by oscillating electrons in atoms.

James Clerk Maxwell
Scottish Theoretical Physicist
(1831–1879)

Maxwell developed the electromagnetic theory of light and the kinetic theory of gases, and explained the nature of Saturn's rings and color vision. Maxwell's successful interpretation of the electromagnetic field resulted in the field equations that bear his name. Formidable mathematical ability combined with great insight enabled him to lead the way in the study of electromagnetism and kinetic theory. He died of cancer before he was 50. (*North Wind Picture Archives*)

34.1 Maxwell's Equations and Hertz's Discoveries

In his unified theory of electromagnetism, Maxwell showed that electromagnetic waves are a natural consequence of the fundamental laws expressed in the following four equations (see Section 31.7):

$$\oint \mathbf{E} \cdot d\mathbf{A} = \frac{q}{\epsilon_0} \tag{34.1}$$

$$\oint \mathbf{B} \cdot d\mathbf{A} = 0 \tag{34.2}$$

Maxwell's equations

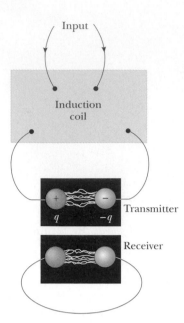

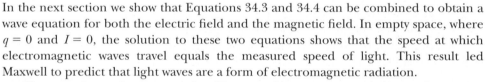

Figure 34.1 Schematic diagram of Hertz's apparatus for generating and detecting electromagnetic waves. The transmitter consists of two spherical electrodes connected to an induction coil, which provides short voltage surges to the spheres, setting up oscillations in the discharge between the electrodes. The receiver is a nearby loop of wire containing a second spark gap.

$$\oint \mathbf{E} \cdot d\mathbf{s} = -\frac{d\Phi_B}{dt} \tag{34.3}$$

$$\oint \mathbf{B} \cdot d\mathbf{s} = \mu_0 I + \mu_0 \epsilon_0 \frac{d\Phi_E}{dt} \tag{34.4}$$

Heinrich Rudolf Hertz

German Physicist (1857–1894)

Hertz made his most important discovery of electromagnetic waves in 1887. After finding that the speed of an electromagnetic wave was the same as that of light, Hertz showed that electromagnetic waves, like light waves, could be reflected, refracted, and diffracted. Hertz died of blood poisoning at the age of 36. During his short life, he made many contributions to science. The hertz, equal to one complete vibration or cycle per second, is named after him. (*Hulton-Deutsch Collection/CORBIS*)

In the next section we show that Equations 34.3 and 34.4 can be combined to obtain a wave equation for both the electric field and the magnetic field. In empty space, where $q = 0$ and $I = 0$, the solution to these two equations shows that the speed at which electromagnetic waves travel equals the measured speed of light. This result led Maxwell to predict that light waves are a form of electromagnetic radiation.

The experimental apparatus that Hertz used to generate and detect electromagnetic waves is shown schematically in Figure 34.1. An induction coil is connected to a transmitter made up of two spherical electrodes separated by a narrow gap. The coil provides short voltage surges to the electrodes, making one positive and the other negative. A spark is generated between the spheres when the electric field near either electrode surpasses the dielectric strength for air (3×10^6 V/m; see Table 26.1). In a strong electric field, the acceleration of free electrons provides them with enough energy to ionize any molecules they strike. This ionization provides more electrons, which can accelerate and cause further ionizations. As the air in the gap is ionized, it becomes a much better conductor, and the discharge between the electrodes exhibits an oscillatory behavior at a very high frequency. From an electric-circuit viewpoint, this is equivalent to an *LC* circuit in which the inductance is that of the coil and the capacitance is due to the spherical electrodes.

Because *L* and *C* are small in Hertz's apparatus, the frequency of oscillation is high, on the order of 100 MHz. (Recall from Eq. 32.22 that $\omega = 1/\sqrt{LC}$ for an *LC* circuit.) Electromagnetic waves are radiated at this frequency as a result of the oscillation (and hence acceleration) of free charges in the transmitter circuit. Hertz was able to detect these waves using a single loop of wire with its own spark gap (the receiver). Such a receiver loop, placed several meters from the transmitter, has its own effective inductance, capacitance, and natural frequency of oscillation. In Hertz's experiment, sparks were induced across the gap of the receiving electrodes when the frequency of the receiver was adjusted to match that of the transmitter. Thus, Hertz demonstrated that the oscillating current induced in the receiver was produced by electromagnetic waves radiated by the transmitter. His experiment is analogous to the mechanical phenomenon in which a tuning fork responds to acoustic vibrations from an identical tuning fork that is oscillating.

Additionally, Hertz showed in a series of experiments that the radiation generated by his spark-gap device exhibited the wave properties of interference, diffraction, reflection, refraction, and polarization, all of which are properties exhibited by light, as we shall see in Part 5 of the text. Thus, it became evident that the radio-frequency waves Hertz was generating had properties similar to those of light waves and differed only in frequency and wavelength. Perhaps his most convincing experiment was the measurement of the speed of this radiation. Waves of known frequency were reflected from a metal sheet and created a standing-wave interference pattern whose nodal points could be detected. The measured distance between the nodal points enabled determination of the wavelength λ. Using the relationship $v = \lambda f$ (Eq. 16.12), Hertz found that v was close to 3×10^8 m/s, the known speed c of visible light.

34.2 Plane Electromagnetic Waves

The properties of electromagnetic waves can be deduced from Maxwell's equations. One approach to deriving these properties is to solve the second-order differential equation obtained from Maxwell's third and fourth equations. A rigorous mathematical treatment of that sort is beyond the scope of this text. To circumvent this problem, we assume that the vectors for the electric field and magnetic field in an electromagnetic wave have a specific space–time behavior that is simple but consistent with Maxwell's equations.

To understand the prediction of electromagnetic waves more fully, let us focus our attention on an electromagnetic wave that travels in the x direction (the *direction of propagation*). In this wave, the electric field **E** is in the y direction, and the magnetic field **B** is in the z direction, as shown in Figure 34.2. Waves such as this one, in which the electric and magnetic fields are restricted to being parallel to a pair of perpendicular axes, are said to be **linearly polarized waves.** Furthermore, we assume that at any point in space, the magnitudes E and B of the fields depend upon x and t only, and not upon the y or z coordinate.

Let us also imagine that the source of the electromagnetic waves is such that a wave radiated from *any* position in the yz plane (not just from the origin as might be suggested by Figure 34.2) propagates in the x direction, and all such waves are emitted in phase. If we define a **ray** as the line along which the wave travels, then all rays for these waves are parallel. This entire collection of waves is often called a **plane wave.** A surface connecting points of equal phase on all waves, which we call a **wave front,** as introduced in Chapter 17, is a geometric plane. In comparison, a point source of radiation sends waves out radially in all directions. A surface connecting points of equal phase for this situation is a sphere, so this is called a **spherical wave.**

We can relate E and B to each other with Equations 34.3 and 34.4. In empty space, where $q = 0$ and $I = 0$, Equation 34.3 remains unchanged and Equation 34.4 becomes

$$\oint \mathbf{B} \cdot d\mathbf{s} = \epsilon_0 \mu_0 \frac{d\Phi_E}{dt} \tag{34.5}$$

Using Equations 34.3 and 34.5 and the plane-wave assumption, we obtain the following differential equations relating E and B. (We shall derive these equations formally later in this section.)

$$\frac{\partial E}{\partial x} = -\frac{\partial B}{\partial t} \tag{34.6}$$

$$\frac{\partial B}{\partial x} = -\mu_0 \epsilon_0 \frac{\partial E}{\partial t} \tag{34.7}$$

Note that the derivatives here are partial derivatives. For example, when we evaluate $\partial E/\partial x$, we assume that t is constant. Likewise, when we evaluate $\partial B/\partial t$, x is held constant. Taking the derivative of Equation 34.6 with respect to x and combining the

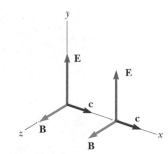

Figure 34.2 An electromagnetic wave traveling at velocity **c** in the positive x direction. The electric field is along the y direction, and the magnetic field is along the z direction. These fields depend only on x and t.

▲ **PITFALL PREVENTION**

34.1 What Is "a" Wave?

A sticky point in these types of discussions is what we mean by a *single* wave. We could define one wave as that which is emitted by a single charged particle. In practice, however, the word *wave* represents both the emission from a *single point* ("wave radiated from any position in the yz plane") and the collection of waves from *all points* on the source ("plane wave"). You should be able to use this term in both ways and to understand its meaning from the context.

result with Equation 34.7, we obtain

$$\frac{\partial^2 E}{\partial x^2} = -\frac{\partial}{\partial x}\left(\frac{\partial B}{\partial t}\right) = -\frac{\partial}{\partial t}\left(\frac{\partial B}{\partial x}\right) = -\frac{\partial}{\partial t}\left(-\mu_0\epsilon_0\frac{\partial E}{\partial t}\right)$$

$$\frac{\partial^2 E}{\partial x^2} = \mu_0\epsilon_0\frac{\partial^2 E}{\partial t^2} \tag{34.8}$$

In the same manner, taking the derivative of Equation 34.7 with respect to x and combining it with Equation 34.6, we obtain

$$\frac{\partial^2 B}{\partial x^2} = \mu_0\epsilon_0\frac{\partial^2 B}{\partial t^2} \tag{34.9}$$

Equations 34.8 and 34.9 both have the form of the general wave equation[1] with the wave speed v replaced by c, where

Speed of electromagnetic waves

$$c = \frac{1}{\sqrt{\mu_0\epsilon_0}} \tag{34.10}$$

Taking $\mu_0 = 4\pi \times 10^{-7}\,\text{T}\cdot\text{m/A}$ and $\epsilon_0 = 8.854\,19 \times 10^{-12}\,\text{C}^2/\text{N}\cdot\text{m}^2$ in Equation 34.10, we find that $c = 2.997\,92 \times 10^8$ m/s. Because this speed is precisely the same as the speed of light in empty space, we are led to believe (correctly) that light is an electromagnetic wave.

The simplest solution to Equations 34.8 and 34.9 is a sinusoidal wave, for which the field magnitudes E and B vary with x and t according to the expressions

Sinusoidal electric and magnetic fields

$$E = E_{\max}\cos(kx - \omega t) \tag{34.11}$$

$$B = B_{\max}\cos(kx - \omega t) \tag{34.12}$$

where $E_{\max}$ and $B_{\max}$ are the maximum values of the fields. The angular wave number is $k = 2\pi/\lambda$, where λ is the wavelength. The angular frequency is $\omega = 2\pi f$, where f is the wave frequency. The ratio ω/k equals the speed of an electromagnetic wave, c:

$$\frac{\omega}{k} = \frac{2\pi f}{2\pi/\lambda} = \lambda f = c$$

where we have used Equation 16.12, $v = c = \lambda f$, which relates the speed, frequency, and wavelength of any continuous wave. Thus, for electromagnetic waves, the wavelength and frequency of these waves are related by

$$\lambda = \frac{c}{f} = \frac{3.00 \times 10^8\,\text{m/s}}{f} \tag{34.13}$$

Figure 34.3a is a pictorial representation, at one instant, of a sinusoidal, linearly polarized plane wave moving in the positive x direction. Figure 34.3b shows how the electric and magnetic field vectors at a fixed location vary with time.

Taking partial derivatives of Equations 34.11 (with respect to x) and 34.12 (with respect to t), we find that

$$\frac{\partial E}{\partial x} = -kE_{\max}\sin(kx - \omega t)$$

$$\frac{\partial B}{\partial t} = \omega B_{\max}\sin(kx - \omega t)$$

[1]   The general wave equation is of the form $(\partial^2 y/\partial x^2) = (1/v^2)(\partial^2 y/\partial t^2)$ where v is the speed of the wave and y is the wave function. The general wave equation was introduced as Equation 16.27, and it would be useful for you to review Section 16.6.

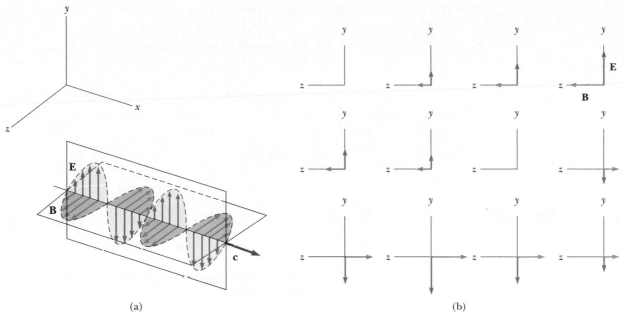

(a)   (b)

Active Figure 34.3 Representation of a sinusoidal, linearly polarized plane electromagnetic wave moving in the positive x direction with velocity **c**. (a) The wave at some instant. Note the sinusoidal variations of E and B with x. (b) A time sequence (starting at the upper left) illustrating the electric and magnetic field vectors at a fixed point in the yz plane, as seen by an observer looking in the negative x direction. The variations of E and B with t are sinusoidal.

At the Active Figures link at http://www.pse6.com, you can observe the wave in part (a) and the variation of the fields in part (b). In addition, you can take a "snapshot" of the wave at an instant of time and investigate the electric and magnetic fields at that instant.

Substituting these results into Equation 34.6, we find that at any instant

$$kE_{max} = \omega B_{max}$$

$$\frac{E_{max}}{B_{max}} = \frac{\omega}{k} = c$$

Using these results together with Equations 34.11 and 34.12, we see that

$$\frac{E_{max}}{B_{max}} = \frac{E}{B} = c \qquad (34.14)$$

That is, **at every instant the ratio of the magnitude of the electric field to the magnitude of the magnetic field in an electromagnetic wave equals the speed of light.**

Finally, note that electromagnetic waves obey the superposition principle (which we discussed in Section 18.1 with respect to mechanical waves) because the differential equations involving E and B are linear equations. For example, we can add two waves with the same frequency and polarization simply by adding the magnitudes of the two electric fields algebraically.

Let us summarize the properties of electromagnetic waves as we have described them:

- The solutions of Maxwell's third and fourth equations are wave-like, with both E and B satisfying a wave equation.

- Electromagnetic waves travel through empty space at the speed of light $c = 1/\sqrt{\epsilon_0 \mu_0}$.

- The components of the electric and magnetic fields of plane electromagnetic waves are perpendicular to each other and perpendicular to the direction of wave propagation. We can summarize the latter property by saying that electromagnetic waves are transverse waves.

▲ **PITFALL PREVENTION**

34.2   E Stronger Than B?

Because the value of c is so large, some students incorrectly interpret Equation 34.14 as meaning that the electric field is much stronger than the magnetic field. Electric and magnetic fields are measured in different units, however, so they cannot be directly compared. In Section 34.3, we find that the electric and magnetic fields contribute equally to the energy of the wave.

Properties of electromagnetic waves

- The magnitudes of **E** and **B** in empty space are related by the expression $E/B = c$.

- Electromagnetic waves obey the principle of superposition.

Quick Quiz 34.1 What is the phase difference between the sinusoidal oscillations of the electric and magnetic fields in Figure 34.3? (a) 180° (b) 90° (c) 0 (d) impossible to determine.

Example 34.1   An Electromagnetic Wave   Interactive

A sinusoidal electromagnetic wave of frequency 40.0 MHz travels in free space in the x direction, as in Figure 34.4.

(A) Determine the wavelength and period of the wave.

Solution Using Equation 34.13 for light waves and given that $f = 40.0$ MHz $= 4.00 \times 10^7$ s^{-1}, we have

$$\lambda = \frac{c}{f} = \frac{3.00 \times 10^8 \text{ m/s}}{4.00 \times 10^7 \text{ s}^{-1}} = \boxed{7.50 \text{ m}}$$

The period T of the wave is the inverse of the frequency:

$$T = \frac{1}{f} = \frac{1}{4.00 \times 10^7 \text{ s}^{-1}} = \boxed{2.50 \times 10^{-8} \text{ s}}$$

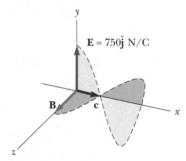

Figure 34.4 (Example 34.1) At some instant, a plane electromagnetic wave moving in the x direction has a maximum electric field of 750 N/C in the positive y direction. The corresponding magnetic field at that point has a magnitude E/c and is in the z direction.

(B) At some point and at some instant, the electric field has its maximum value of 750 N/C and is along the y axis. Calculate the magnitude and direction of the magnetic field at this position and time.

Solution From Equation 34.14 we see that

$$B_{max} = \frac{E_{max}}{c} = \frac{750 \text{ N/C}}{3.00 \times 10^8 \text{ m/s}} = \boxed{2.50 \times 10^{-6} \text{ T}}$$

Because **E** and **B** must be perpendicular to each other and perpendicular to the direction of wave propagation (x in this case), we conclude that **B** is in the z direction.

(C) Write expressions for the space–time variation of the components of the electric and magnetic fields for this wave.

Solution We can apply Equations 34.11 and 34.12 directly:

$$E = E_{max} \cos(kx - \omega t) = (750 \text{ N/C}) \cos(kx - \omega t)$$

$$B = B_{max} \cos(kx - \omega t) = (2.50 \times 10^{-6} \text{ T}) \cos(kx - \omega t)$$

where

$$\omega = 2\pi f = 2\pi(4.00 \times 10^7 \text{ s}^{-1}) = 2.51 \times 10^8 \text{ rad/s}$$

$$k = \frac{2\pi}{\lambda} = \frac{2\pi}{7.50 \text{ m}} = 0.838 \text{ rad/m}$$

  Explore electromagnetic waves of different frequencies at the Interactive Worked Example link at **http://www.pse6.com.**

Derivation of Equations 34.6 and 34.7

To derive Equation 34.6, we start with Faraday's law, Equation 34.3:

$$\oint \mathbf{E} \cdot d\mathbf{s} = -\frac{d\Phi_B}{dt}$$

Let us again assume that the electromagnetic wave is traveling in the x direction, with the electric field **E** in the positive y direction and the magnetic field **B** in the positive z direction.

Consider a rectangle of width dx and height ℓ lying in the xy plane, as shown in Figure 34.5. To apply Equation 34.3, we must first evaluate the line integral of $\mathbf{E} \cdot d\mathbf{s}$ around this rectangle. The contributions from the top and bottom of the rectangle are zero because $\mathbf{E}$ is perpendicular to $d\mathbf{s}$ for these paths. We can express the electric field on the right side of the rectangle as

$$E(x + dx, t) \approx E(x, t) + \frac{dE}{dx}\bigg]_{t\,\text{constant}} dx = E(x, t) + \frac{\partial E}{\partial x} dx$$

while the field on the left side[2] is simply $E(x, t)$. Therefore, the line integral over this rectangle is approximately

$$\oint \mathbf{E} \cdot d\mathbf{s} = [E(x + dx, t)]\ell - [E(x, t)]\ell \approx \ell\left(\frac{\partial E}{\partial x}\right) dx \qquad (34.15)$$

Because the magnetic field is in the z direction, the magnetic flux through the rectangle of area $\ell\,dx$ is approximately $\Phi_B = B\ell\,dx$. (This assumes that dx is very small compared with the wavelength of the wave.) Taking the time derivative of the magnetic flux gives

$$\frac{d\Phi_B}{dt} = \ell\,dx\,\frac{dB}{dt}\bigg]_{x\,\text{constant}} = \ell\,dx\,\frac{\partial B}{\partial t} \qquad (34.16)$$

Substituting Equations 34.15 and 34.16 into Equation 34.3 gives

$$\ell\left(\frac{\partial E}{\partial x}\right) dx = -\ell\,dx\,\frac{\partial B}{\partial t}$$

$$\frac{\partial E}{\partial x} = -\frac{\partial B}{\partial t}$$

This expression is Equation 34.6.

In a similar manner, we can verify Equation 34.7 by starting with Maxwell's fourth equation in empty space (Eq. 34.5). In this case, the line integral of $\mathbf{B} \cdot d\mathbf{s}$ is evaluated around a rectangle lying in the xz plane and having width dx and length ℓ, as in Figure 34.6. Noting that the magnitude of the magnetic field changes from $B(x, t)$ to $B(x + dx, t)$ over the width dx and that the direction in which we take the line integral is as shown in Figure 34.6, the line integral over this rectangle is found to be approximately

$$\oint \mathbf{B} \cdot d\mathbf{s} = [B(x, t)]\ell - [B(x + dx, t)]\ell \approx -\ell\left(\frac{\partial B}{\partial x}\right) dx \qquad (34.17)$$

The electric flux through the rectangle is $\Phi_E = E\ell\,dx$, which, when differentiated with respect to time, gives

$$\frac{\partial \Phi_E}{\partial t} = \ell\,dx\,\frac{\partial E}{\partial t} \qquad (34.18)$$

Substituting Equations 34.17 and 34.18 into Equation 34.5 gives

$$-\ell\left(\frac{\partial B}{\partial x}\right) dx = \mu_0\epsilon_0\,\ell\,dx\left(\frac{\partial E}{\partial t}\right)$$

$$\frac{\partial B}{\partial x} = -\mu_0\epsilon_0\,\frac{\partial E}{\partial t}$$

which is Equation 34.7.

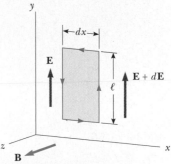

Figure 34.5 At an instant when a plane wave moving in the $+x$ direction passes through a rectangular path of width dx lying in the xy plane, the electric field in the y direction varies from $\mathbf{E}$ to $\mathbf{E} + d\mathbf{E}$. This spatial variation in $\mathbf{E}$ gives rise to a time-varying magnetic field along the z direction, according to Equation 34.6.

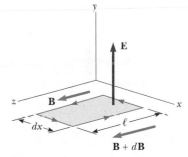

Figure 34.6 At an instant when a plane wave passes through a rectangular path of width dx lying in the xz plane, the magnetic field in the z direction varies from $\mathbf{B}$ to $\mathbf{B} + d\mathbf{B}$. This spatial variation in $\mathbf{B}$ gives rise to a time-varying electric field along the y direction, according to Equation 34.7.

[2]   Because dE/dx in this equation is expressed as the change in E with x at a given instant t, dE/dx is equivalent to the partial derivative $\partial E/\partial x$. Likewise, dB/dt means the change in B with time at a particular position x, so in Equation 34.16 we can replace dB/dt with $\partial B/\partial t$.

34.3 Energy Carried by Electromagnetic Waves

Electromagnetic waves carry energy, and as they propagate through space they can transfer energy to objects placed in their path. The rate of flow of energy in an electromagnetic wave is described by a vector $\mathbf{S}$, called the **Poynting vector,** which is defined by the expression

Poynting vector

$$\mathbf{S} \equiv \frac{1}{\mu_0}\,\mathbf{E} \times \mathbf{B} \tag{34.19}$$

The magnitude of the Poynting vector represents the rate at which energy flows through a unit surface area perpendicular to the direction of wave propagation. Thus, the magnitude of the Poynting vector represents *power per unit area*. The direction of the vector is along the direction of wave propagation (Fig. 34.7). The SI units of the Poynting vector are $J/s \cdot m^2 = W/m^2$.

As an example, let us evaluate the magnitude of $\mathbf{S}$ for a plane electromagnetic wave where $|\mathbf{E} \times \mathbf{B}| = EB$. In this case,

$$S = \frac{EB}{\mu_0} \tag{34.20}$$

Because $B = E/c$, we can also express this as

$$S = \frac{E^2}{\mu_0 c} = \frac{c}{\mu_0}\,B^2$$

These equations for S apply at any instant of time and represent the *instantaneous* rate at which energy is passing through a unit area.

What is of greater interest for a sinusoidal plane electromagnetic wave is the time average of S over one or more cycles, which is called the *wave intensity I.* (We discussed the intensity of sound waves in Chapter 17.) When this average is taken, we obtain an expression involving the time average of $\cos^2(kx - \omega t)$, which equals $\frac{1}{2}$. Hence, the average value of S (in other words, the intensity of the wave) is

Wave intensity

$$I = S_{\text{av}} = \frac{E_{\text{max}}\,B_{\text{max}}}{2\mu_0} = \frac{E_{\text{max}}^2}{2\mu_0 c} = \frac{c}{2\mu_0}\,B_{\text{max}}^2 \tag{34.21}$$

Recall that the energy per unit volume, which is the instantaneous energy density u_E associated with an electric field, is given by Equation 26.13,

$$u_E = \tfrac{1}{2}\epsilon_0 E^2$$

and that the instantaneous energy density u_B associated with a magnetic field is given by Equation 32.14:

$$u_B = \frac{B^2}{2\mu_0}$$

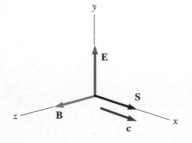

Figure 34.7 The Poynting vector $\mathbf{S}$ for a plane electromagnetic wave is along the direction of wave propagation.

Because E and B vary with time for an electromagnetic wave, the energy densities also vary with time. When we use the relationships $B = E/c$ and $c = 1/\sqrt{\epsilon_0\mu_0}$, the expression for u_B becomes

$$u_B = \frac{(E/c)^2}{2\mu_0} = \frac{\epsilon_0\mu_0}{2\mu_0}E^2 = \tfrac{1}{2}\epsilon_0 E^2$$

Comparing this result with the expression for u_E, we see that

$$u_B = u_E = \tfrac{1}{2}\epsilon_0 E^2 = \frac{B^2}{2\mu_0}$$

That is, **the instantaneous energy density associated with the magnetic field of an electromagnetic wave equals the instantaneous energy density associated with the electric field.** Hence, in a given volume the energy is equally shared by the two fields.

The **total instantaneous energy density** u is equal to the sum of the energy densities associated with the electric and magnetic fields:

$$u = u_F + u_B = \epsilon_0 E^2 = \frac{B^2}{\mu_0}$$

Total instantaneous energy density of an electromagnetic wave

When this total instantaneous energy density is averaged over one or more cycles of an electromagnetic wave, we again obtain a factor of $\tfrac{1}{2}$. Hence, for any electromagnetic wave, the total average energy per unit volume is

$$u_{av} = \epsilon_0 (E^2)_{av} = \tfrac{1}{2}\epsilon_0 E^2_{max} = \frac{B^2_{max}}{2\mu_0} \tag{34.22}$$

Average energy density of an electromagnetic wave

Comparing this result with Equation 34.21 for the average value of S, we see that

$$I = S_{av} = c u_{av} \tag{34.23}$$

In other words, **the intensity of an electromagnetic wave equals the average energy density multiplied by the speed of light.**

Quick Quiz 34.2 An electromagnetic wave propagates in the $-y$ direction. The electric field at a point in space is momentarily oriented in the $+x$ direction. The magnetic field at that point is momentarily oriented in the (a) $-x$ direction (b) $+y$ direction (c) $+z$ direction (d) $-z$ direction.

Quick Quiz 34.3 Which of the following is constant for a plane electromagnetic wave? (a) magnitude of the Poynting vector (b) energy density u_E (c) energy density u_B (d) wave intensity.

Example 34.2 Fields on the Page

Estimate the maximum magnitudes of the electric and magnetic fields of the light that is incident on this page because of the visible light coming from your desk lamp. Treat the bulb as a point source of electromagnetic radiation that is 5% efficient at transforming energy coming in by electrical transmission to energy leaving by visible light.

Solution Recall from Equation 17.7 that the wave intensity I a distance r from a point source is $I = \mathscr{P}_{av}/4\pi r^2$, where $\mathscr{P}_{av}$ is the average power output of the source and $4\pi r^2$ is the area of a sphere of radius r centered on the source. Because

the intensity of an electromagnetic wave is also given by Equation 34.21, we have

$$I = \frac{\mathscr{P}_{av}}{4\pi r^2} = \frac{E^2_{max}}{2\mu_0 c}$$

We must now make some assumptions about numbers to enter in this equation. If we have a 60-W lightbulb, its output at 5% efficiency is approximately 3.0 W by visible light. (The remaining energy transfers out of the bulb by conduction and invisible radiation.) A reasonable distance from the bulb to the page might be 0.30 m. Thus, we have

$$E_{\max} = \sqrt{\frac{\mu_0 c \mathcal{P}_{av}}{2\pi r^2}}$$

$$= \sqrt{\frac{(4\pi \times 10^{-7}\ \text{T·m/A})(3.00 \times 10^8\ \text{m/s})(3.0\ \text{W})}{2\pi(0.30\ \text{m})^2}}$$

$$= \boxed{45\ \text{V/m}}$$

From Equation 34.14,

$$B_{\max} = \frac{E_{\max}}{c} = \frac{45\ \text{V/m}}{3.00 \times 10^8\ \text{m/s}} = \boxed{1.5 \times 10^{-7}\ \text{T}}$$

This value is two orders of magnitude smaller than the Earth's magnetic field, which, unlike the magnetic field in the light wave from your desk lamp, is not oscillating.

34.4 Momentum and Radiation Pressure

Electromagnetic waves transport linear momentum as well as energy. It follows that, as this momentum is absorbed by some surface, pressure is exerted on the surface. We shall assume in this discussion that the electromagnetic wave strikes the surface at normal incidence and transports a total energy U to the surface in a time interval Δt. Maxwell showed that, if the surface absorbs all the incident energy U in this time interval (as does a black body, introduced in Section 20.7), the total momentum **p** transported to the surface has a magnitude

Momentum transported to a perfectly absorbing surface

$$p = \frac{U}{c} \qquad \text{(complete absorption)} \qquad (34.24)$$

The pressure exerted on the surface is defined as force per unit area F/A. Let us combine this with Newton's second law:

$$P = \frac{F}{A} = \frac{1}{A}\frac{dp}{dt}$$

If we now replace p, the momentum transported to the surface by radiation, from Equation 34.24, we have

$$P = \frac{1}{A}\frac{dp}{dt} = \frac{1}{A}\frac{d}{dt}\left(\frac{U}{c}\right) = \frac{1}{c}\frac{(dU/dt)}{A}$$

We recognize $(dU/dt)/A$ as the rate at which energy is arriving at the surface per unit area, which is the magnitude of the Poynting vector. Thus, the radiation pressure P exerted on the perfectly absorbing surface is

Radiation pressure exerted on a perfectly absorbing surface

$$P = \frac{S}{c} \qquad (34.25)$$

If the surface is a perfect reflector (such as a mirror) and incidence is normal, then the momentum transported to the surface in a time interval Δt is twice that given by Equation 34.24. That is, the momentum transferred to the surface by the incoming light is $p = U/c$, and that transferred by the reflected light also is $p = U/c$. Therefore,

$$p = \frac{2U}{c} \qquad \text{(complete reflection)} \qquad (34.26)$$

The momentum delivered to a surface having a reflectivity somewhere between these two extremes has a value between U/c and $2U/c$, depending on the properties of the surface. Finally, the radiation pressure exerted on a perfectly reflecting surface for normal incidence of the wave is[3]

Radiation pressure exerted on a perfectly reflecting surface

$$P = \frac{2S}{c} \qquad (34.27)$$

---

[3] For oblique incidence on a perfectly reflecting surface, the momentum transferred is $(2U\cos\theta)/c$ and the pressure is $P = (2S\cos^2\theta)/c$ where θ is the angle between the normal to the surface and the direction of wave propagation.

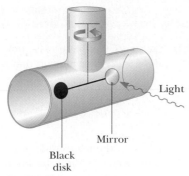

Figure 34.8 An apparatus for measuring the pressure exerted by light. In practice, the system is contained in a high vacuum.

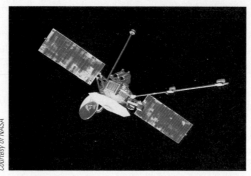

Courtesy of NASA

Figure 34.9 Mariner 10 used its solar panels to "sail on sunlight."

Although radiation pressures are very small (about 5×10^{-6} N/m^2 for direct sunlight), they have been measured with torsion balances such as the one shown in Figure 34.8. A mirror (a perfect reflector) and a black disk (a perfect absorber) are connected by a horizontal rod suspended from a fine fiber. Normal-incidence light striking the black disk is completely absorbed, so all of the momentum of the light is transferred to the disk. Normal-incidence light striking the mirror is totally reflected, and hence the momentum transferred to the mirror is twice as great as that transferred to the disk. The radiation pressure is determined by measuring the angle through which the horizontal connecting rod rotates. The apparatus must be placed in a high vacuum to eliminate the effects of air currents.

NASA is exploring the possibility of *solar sailing* as a low-cost means of sending spacecraft to the planets. Large sheets would experience radiation pressure from sunlight and would be used in much the way canvas sheets are used on earthbound sailboats. In 1973 NASA engineers took advantage of the momentum of the sunlight striking the solar panels of Mariner 10 (Fig. 34.9) to make small course corrections when the spacecraft's fuel supply was running low. (This procedure was carried out when the spacecraft was in the vicinity of the planet Mercury. Would it have worked as well near Pluto?)

Quick Quiz 34.4 To maximize the radiation pressure on the sails of a spacecraft using solar sailing, should the sheets be (a) very black to absorb as much sunlight as possible or (b) very shiny, to reflect as much sunlight as possible?

Quick Quiz 34.5 In an apparatus such as that in Figure 34.8, the disks are illuminated uniformly over their areas. Suppose the black disk is replaced by one with half the radius. Which of the following are different after the disk is replaced? (a) radiation pressure on the disk, (b) radiation force on the disk, (c) radiation momentum delivered to the disk in a given time interval.

Conceptual Example 34.3 Sweeping the Solar System

A great amount of dust exists in interplanetary space. Although in theory these dust particles can vary in size from molecular size to much larger, very little of the dust in our solar system is smaller than about 0.2 μm. Why?

Solution The dust particles are subject to two significant forces—the gravitational force that draws them toward the Sun and the radiation-pressure force that pushes them away from the Sun. The gravitational force is proportional to the cube of the radius of a spherical dust particle because it is proportional to the mass and therefore to the volume $4\pi r^3/3$ of the particle. The radiation pressure is proportional to the square of the radius because it depends on the planar cross section of the particle. For large particles, the gravitational force is greater than the force from radiation pressure. For particles having radii less than about 0.2 μm, the radiation-pressure force is greater than the gravitational force, and as a result these particles are swept out of the Solar System.

Example 34.4   Pressure from a Laser Pointer   `Interactive`

Many people giving presentations use a laser pointer to direct the attention of the audience to information on a screen. If a 3.0-mW pointer creates a spot on a screen that is 2.0 mm in diameter, determine the radiation pressure on a screen that reflects 70% of the light that strikes it. The power 3.0 mW is a time-averaged value.

Solution In conceptualizing this situation, we do not expect the pressure to be very large. We categorize this as a calculation of radiation pressure using something like Equation 34.25 or Equation 34.27, but complicated by the 70% reflection. To analyze the problem, we begin by determining the magnitude of the beam's Poynting vector. We divide the time-averaged power delivered via the electromagnetic wave by the cross-sectional area of the beam:

$$S_{av} = \frac{\mathcal{P}_{av}}{A} = \frac{\mathcal{P}_{av}}{\pi r^2} = \frac{3.0 \times 10^{-3}\ \text{W}}{\pi \left(\dfrac{2.0 \times 10^{-3}\ \text{m}}{2}\right)^2} = 955\ \text{W/m}^2$$

Now we can determine the radiation pressure from the laser beam. Equation 34.27 indicates that a completely reflected beam would apply an average pressure of $P_{av} = 2S_{av}/c$. We can model the actual reflection as follows. Imagine that the surface absorbs the beam, resulting in pressure $P_{av} = S_{av}/c$. Then the surface emits the beam, resulting in additional pressure $P_{av} = S_{av}/c$. If the surface emits only a fraction f of the beam (so that f is the amount of the incident beam reflected), then the pressure due to the emitted beam is $P_{av} = f\,S_{av}/c$. Thus, the total pressure on the surface due to absorption and re-emission (reflection) is

$$P_{av} = \frac{S_{av}}{c} + f\,\frac{S_{av}}{c} = (1 + f)\frac{S_{av}}{c}$$

Notice that if $f = 1$, which represents complete reflection, this equation reduces to Equation 34.27. For a beam that is 70% reflected, the pressure is

$$P_{av} = (1 + 0.70)\,\frac{955\ \text{W/m}^2}{3.0 \times 10^8\ \text{m/s}} = \boxed{5.4 \times 10^{-6}\ \text{N/m}^2}$$

To finalize the example, consider first the magnitude of the Poynting vector. This is about the same as the intensity of sunlight at the Earth's surface. (For this reason, it is not safe to shine the beam of a laser pointer into a person's eyes; that may be more dangerous than looking directly at the Sun.) Note also that the pressure has an extremely small value, as expected. (Recall from Section 14.2 that atmospheric pressure is approximately $10^5\ \text{N/m}^2$.)

What If? What if the laser pointer is moved twice as far away from the screen? Does this affect the radiation pressure on the screen?

Answer Because a laser beam is popularly represented as a beam of light with constant cross section, one might be tempted to claim that the intensity of radiation, and therefore the radiation pressure, would be independent of distance from the screen. However, a laser beam does not have a constant cross section at all distances from the source—there is a small but measurable divergence of the beam. If the laser is moved farther away from the screen, the area of illumination on the screen will increase, decreasing the intensity. In turn, this will reduce the radiation pressure.

In addition, the doubled distance from the screen will result in more loss of energy from the beam due to scattering from air molecules and dust particles as the light travels from the laser to the screen. This will further reduce the radiation pressure.

    At the Interactive Worked Example link at **http://www.pse6.com,** *you can investigate the pressure on the screen for various laser and screen parameters.*

Example 34.5   Solar Energy

As noted in the preceding example, the Sun delivers about $10^3\ \text{W/m}^2$ of energy to the Earth's surface via electromagnetic radiation.

(A) Calculate the total power that is incident on a roof of dimensions 8.00 m × 20.0 m.

Solution We assume that the average magnitude of the Poynting vector for solar radiation at the surface of the Earth is $S_{av} = 1\,000\ \text{W/m}^2$; this represents the power per unit area, or the light intensity. Assuming that the radiation is incident normal to the roof, we obtain

$$\mathcal{P}_{av} = S_{av}A = (1\,000\ \text{W/m}^2)(8.00 \times 20.0\ \text{m}^2)$$

$$= \boxed{1.60 \times 10^5\ \text{W}}$$

(B) Determine the radiation pressure and the radiation force exerted on the roof, assuming that the roof covering is a perfect absorber.

Solution Using Equation 34.25 with $S_{av} = 1\,000\ \text{W/m}^2$, we find that the radiation pressure is

$$P_{av} = \frac{S_{av}}{c} = \frac{1\,000\ \text{W/m}^2}{3.00 \times 10^8\ \text{m/s}} = \boxed{3.33 \times 10^{-6}\ \text{N/m}^2}$$

Because pressure equals force per unit area, this corresponds to a radiation force of

$$F = P_{av}A = (3.33 \times 10^{-6}\ \text{N/m}^2)(160\ \text{m}^2)$$

$$= \boxed{5.33 \times 10^{-4}\ \text{N}}$$

What If? Suppose the energy striking the roof could be captured and used to operate electrical devices in the house. Could the home operate completely from this energy?

Answer The power in part (A) is large compared to the power requirements of a typical home. If this power were maintained for 24 hours per day and the energy could be absorbed and made available to electrical devices, it would provide more than enough energy for the average home. However, solar energy is not easily harnessed, and the prospects for large-scale conversion are not as bright as may appear from this calculation. For example, the efficiency of conversion from solar energy is typically 10% for photovoltaic cells, reducing the available power in part (A) by an order of magnitude. Other considerations reduce the power even further. Depending on location, the radiation will most likely not be incident normal to the roof and, even if it is (in locations near the Equator), this situation exists for only a short time near the middle of the day. No energy is available for about half of each day during the nighttime hours. Furthermore, cloudy days reduce the available energy. Finally, while energy is arriving at a large rate during the middle of the day, some of it must be stored for later use, requiring batteries or other storage devices. The result of these considerations is that complete solar operation of homes is not presently cost-effective for most homes.

34.5 Production of Electromagnetic Waves by an Antenna

Neither stationary charges nor steady currents can produce electromagnetic waves. Whenever the current in a wire changes with time, however, the wire emits electromagnetic radiation. **The fundamental mechanism responsible for this radiation is the acceleration of a charged particle. Whenever a charged particle accelerates, it must radiate energy.**

Let us consider the production of electromagnetic waves by a *half-wave antenna*. In this arrangement, two conducting rods are connected to a source of alternating voltage (such as an *LC* oscillator), as shown in Figure 34.10. The length of each rod is equal to one quarter of the wavelength of the radiation that will be emitted when the oscillator operates at frequency *f*. The oscillator forces charges to accelerate back and forth between the two rods. Figure 34.10 shows the configuration of the electric and magnetic fields at some instant when the current is upward. The electric field lines, due to the separation of charges in the upper and lower portions of the antenna, resemble those of an electric dipole. (As a result, this type of antenna is sometimes called a *dipole antenna*.) Because these charges are continuously oscillating between the two rods, the antenna can be approximated by an oscillating electric dipole. The magnetic field lines, due to the current representing the movement of charges between the ends of the antenna, form concentric circles around the antenna and are perpendicular to the electric field lines at all points. The magnetic field is zero at all points along the axis of the antenna. Furthermore, **E** and **B** are 90° out of phase in time—for example, the current is zero when the charges at the outer ends of the rods are at a maximum.

At the two points where the magnetic field is shown in Figure 34.10, the Poynting vector **S** is directed radially outward. This indicates that energy is flowing away from the antenna at this instant. At later times, the fields and the Poynting vector reverse direction as the current alternates. Because **E** and **B** are 90° out of phase at points near the dipole, the net energy flow is zero. From this, we might conclude (incorrectly) that no energy is radiated by the dipole.

However, we find that energy is indeed radiated. Because the dipole fields fall off as $1/r^3$ (as shown in Example 23.6 for the electric field of a static dipole), they are negligible at great distances from the antenna. At these great distances, something else causes a type of radiation different from that close to the antenna. The source of this radiation is the continuous induction of an electric field by the time-varying magnetic field and the induction of a magnetic field by the time-varying electric field, predicted by Equations 34.3 and 34.4. The electric and magnetic fields produced in this manner are in phase with each other and vary as $1/r$. The result is an outward flow of energy at all times.

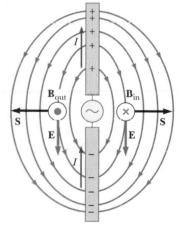

Figure 34.10 A half-wave antenna consists of two metal rods connected to an alternating voltage source. This diagram shows **E** and **B** at an arbitrary instant when the current is upward. Note that the electric field lines resemble those of a dipole (shown in Fig. 23.22).

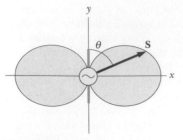

Figure 34.11 Angular dependence of the intensity of radiation produced by an oscillating electric dipole. The distance from the origin to a point on the edge of the gray shape is proportional to the intensity of radiation.

The angular dependence of the radiation intensity produced by a dipole antenna is shown in Figure 34.11. Note that the intensity and the power radiated are a maximum in a plane that is perpendicular to the antenna and passing through its midpoint. Furthermore, the power radiated is zero along the antenna's axis. A mathematical solution to Maxwell's equations for the dipole antenna shows that the intensity of the radiation varies as $(\sin^2 \theta)/r^2$, where θ is measured from the axis of the antenna.

Electromagnetic waves can also induce currents in a receiving antenna. The response of a dipole receiving antenna at a given position is a maximum when the antenna axis is parallel to the electric field at that point and zero when the axis is perpendicular to the electric field.

Quick Quiz 34.6 If the antenna in Figure 34.10 represents the source of a distant radio station, rank the following points in terms of the intensity of the radiation, from greatest to least: (a) a distance d to the right of the antenna (b) a distance $2d$ to the left of the antenna (c) a distance $2d$ in front of the antenna (out of the page) (d) a distance d above the antenna (toward the top of the page).

Quick Quiz 34.7 If the antenna in Figure 34.10 represents the source of a distant radio station, what would be the best orientation for your portable radio antenna located to the right of the figure—(a) up–down along the page, (b) left–right along the page, or (c) perpendicular to the page?

34.6 The Spectrum of Electromagnetic Waves

The various types of electromagnetic waves are listed in Figure 34.12, which shows the **electromagnetic spectrum.** Note the wide ranges of frequencies and wavelengths. No sharp dividing point exists between one type of wave and the next. Remember that **all forms of the various types of radiation are produced by the same phenomenon—accelerating charges.** The names given to the types of waves are simply for convenience in describing the region of the spectrum in which they lie.

Radio waves, whose wavelengths range from more than 10^4 m to about 0.1 m, are the result of charges accelerating through conducting wires. They are generated by such electronic devices as LC oscillators and are used in radio and television communication systems.

Microwaves have wavelengths ranging from approximately 0.3 m to 10^{-4} m and are also generated by electronic devices. Because of their short wavelengths, they are well suited for radar systems and for studying the atomic and molecular properties of matter. Microwave ovens are an interesting domestic application of these waves. It has been suggested that solar energy could be harnessed by beaming microwaves to the Earth from a solar collector in space.

Infrared waves have wavelengths ranging from approximately 10^{-3} m to the longest wavelength of visible light, 7×10^{-7} m. These waves, produced by molecules and room-temperature objects, are readily absorbed by most materials. The infrared (IR) energy absorbed by a substance appears as internal energy because the energy agitates the atoms of the object, increasing their vibrational or translational motion, which results in a temperature increase. Infrared radiation has practical and scientific applications in many areas, including physical therapy, IR photography, and vibrational spectroscopy.

Visible light, the most familiar form of electromagnetic waves, is the part of the electromagnetic spectrum that the human eye can detect. Light is produced by the rearrangement of electrons in atoms and molecules. The various wavelengths of visible

PITFALL PREVENTION

34.6 "Heat Rays"

Infrared rays are often called "heat rays." This is a misnomer. While infrared radiation is used to raise or maintain temperature, as in the case of keeping food warm with "heat lamps" at a fast-food restaurant, all wavelengths of electromagnetic radiation carry energy that can cause the temperature of a system to increase. As an example, consider a potato baking in your microwave oven.

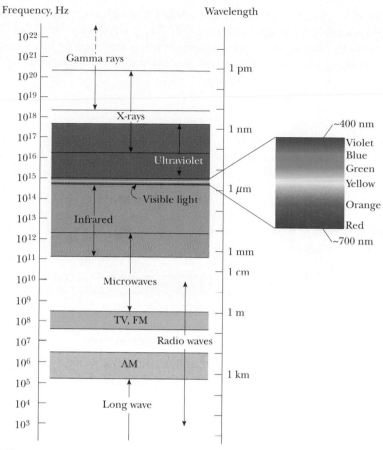

Figure 34.12 The electromagnetic spectrum. Note the overlap between adjacent wave types. The expanded view to the right shows details of the visible spectrum.

Wearing sunglasses that do not block ultraviolet (UV) light is worse for your eyes than wearing no sunglasses. The lenses of any sunglasses absorb some visible light, thus causing the wearer's pupils to dilate. If the glasses do not also block UV light, then more damage may be done to the lens of the eye because of the dilated pupils. If you wear no sunglasses at all, your pupils are contracted, you squint, and much less UV light enters your eyes. High-quality sunglasses block nearly all the eye-damaging UV light.

light, which correspond to different colors, range from red ($\lambda \approx 7 \times 10^{-7}$ m) to violet ($\lambda \approx 4 \times 10^{-7}$ m). The sensitivity of the human eye is a function of wavelength, being a maximum at a wavelength of about 5.5×10^{-7} m. With this in mind, why do you suppose tennis balls often have a yellow-green color?

Ultraviolet waves cover wavelengths ranging from approximately 4×10^{-7} m to 6×10^{-10} m. The Sun is an important source of ultraviolet (UV) light, which is the main cause of sunburn. Sunscreen lotions are transparent to visible light but absorb most UV light. The higher a sunscreen's solar protection factor (SPF), the greater the percentage of UV light absorbed. Ultraviolet rays have also been implicated in the formation of cataracts, a clouding of the lens inside the eye.

Most of the UV light from the Sun is absorbed by ozone (O_3) molecules in the Earth's upper atmosphere, in a layer called the stratosphere. This ozone shield converts lethal high-energy UV radiation to infrared radiation, which in turn warms the stratosphere. Recently, a great deal of controversy has arisen concerning the possible depletion of the protective ozone layer as a result of the chemicals emitted from aerosol spray cans and used as refrigerants.

X-rays have wavelengths in the range from approximately 10^{-8} m to 10^{-12} m. The most common source of x-rays is the stopping of high-energy electrons upon bombarding a metal target. X-rays are used as a diagnostic tool in medicine and as a treatment for certain forms of cancer. Because x-rays damage or destroy living tissues and organisms, care must be taken to avoid unnecessary exposure or overexposure. X-rays are also used in the study of crystal structure because x-ray wavelengths are comparable to the atomic separation distances in solids (about 0.1 nm).

Gamma rays are electromagnetic waves emitted by radioactive nuclei (such as ^{60}Co and ^{137}Cs) and during certain nuclear reactions. High-energy gamma rays are a component of cosmic rays that enter the Earth's atmosphere from space. They have wavelengths ranging from approximately 10^{-10} m to less than 10^{-14} m. They are highly penetrating and produce serious damage when absorbed by living tissues. Consequently, those working near such dangerous radiation must be protected with heavily absorbing materials, such as thick layers of lead.

Quick Quiz 34.8 In many kitchens, a microwave oven is used to cook food. The frequency of the microwaves is on the order of 10^{10} Hz. The wavelengths of these microwaves are on the order of (a) kilometers (b) meters (c) centimeters (d) micrometers.

Quick Quiz 34.9 A radio wave of frequency on the order of 10^5 Hz is used to carry a sound wave with a frequency on the order of 10^3 Hz. The wavelength of this radio wave is on the order of (a) kilometers (b) meters (c) centimeters (d) micrometers.

Example 34.6 A Half-Wave Antenna

A half-wave antenna works on the principle that the optimum length of the antenna is half the wavelength of the radiation being received. What is the optimum length of a car antenna when it receives a signal of frequency 94.7 MHz?

Solution Equation 34.13 tells us that the wavelength of the signal is

$$\lambda = \frac{3.00 \times 10^8 \text{ m/s}}{9.47 \times 10^7 \text{ Hz}} = \boxed{3.16 \text{ m}}$$

Thus, to operate most efficiently, the antenna should have a length of $(3.16 \text{ m})/2 = 1.58$ m. For practical reasons, car antennas are usually one-quarter wavelength in size.

SUMMARY

Take a practice test for this chapter by clicking on the Practice Test link at http://www.pse6.com.

Electromagnetic waves, which are predicted by Maxwell's equations, have the following properties:

- The electric field and the magnetic field each satisfy a wave equation. These two wave equations, which can be obtained from Maxwell's third and fourth equations, are

$$\frac{\partial^2 E}{\partial x^2} = \mu_0 \epsilon_0 \frac{\partial^2 E}{\partial t^2} \qquad (34.8)$$

$$\frac{\partial^2 B}{\partial x^2} = \mu_0 \epsilon_0 \frac{\partial^2 B}{\partial t^2} \qquad (34.9)$$

- The waves travel through a vacuum with the speed of light c, where

$$c = \frac{1}{\sqrt{\mu_0 \epsilon_0}} = 3.00 \times 10^8 \text{ m/s} \qquad (34.10)$$

- The electric and magnetic fields are perpendicular to each other and perpendicular to the direction of wave propagation. (Hence, electromagnetic waves are transverse waves.)

- The instantaneous magnitudes of **E** and **B** in an electromagnetic wave are related by the expression

$$\frac{E}{B} = c \qquad (34.14)$$

- The waves carry energy. The rate of flow of energy crossing a unit area is described by the Poynting vector **S**, where

$$\mathbf{S} \equiv \frac{1}{\mu_0}\,\mathbf{E} \times \mathbf{B} \qquad\qquad (34.19)$$

- Electromagnetic waves carry momentum and hence exert pressure on surfaces. If an electromagnetic wave whose Poynting vector is **S** is completely absorbed by a surface upon which it is normally incident, the radiation pressure on that surface is

$$P = \frac{S}{c} \qquad \text{(complete absorption)} \qquad\qquad (34.25)$$

If the surface totally reflects a normally incident wave, the pressure is doubled.

The electric and magnetic fields of a sinusoidal plane electromagnetic wave propagating in the positive x direction can be written

$$E = E_{max}\cos(kx - \omega t) \qquad\qquad (34.11)$$
$$B = B_{max}\cos(kx - \omega t) \qquad\qquad (34.12)$$

where ω is the angular frequency of the wave and k is the angular wave number. These equations represent special solutions to the wave equations for E and B. The wavelength and frequency of electromagnetic waves are related by

$$\lambda = \frac{c}{f} = \frac{3.00 \times 10^8 \text{ m/s}}{f} \qquad\qquad (34.13)$$

The average value of the Poynting vector for a plane electromagnetic wave has a magnitude

$$S_{av} = \frac{E_{max}\,B_{max}}{2\mu_0} = \frac{E_{max}^2}{2\mu_0 c} = \frac{c}{2\mu_0}\,B_{max}^2 \qquad\qquad (34.21)$$

The intensity of a sinusoidal plane electromagnetic wave equals the average value of the Poynting vector taken over one or more cycles.

The electromagnetic spectrum includes waves covering a broad range of wavelengths, from long radio waves at more than 10^4 m to gamma rays at less than 10^{-14} m.

QUESTIONS

1. Radio stations often advertise "instant news." If they mean that you can hear the news the instant they speak it, is their claim true? About how long would it take for a message to travel across this country by radio waves, assuming that the waves could be detected at this range?

2. Light from the Sun takes approximately 8.3 min to reach the Earth. During this time interval the Earth has continued to rotate on its axis. How far is the actual direction of the Sun from its image in the sky?

3. When light (or other electromagnetic radiation) travels across a given region, what is it that oscillates? What is it that is transported?

4. Do all current-carrying conductors emit electromagnetic waves? Explain.

5. What is the fundamental source of electromagnetic radiation?

6. If a high-frequency current is passed through a solenoid containing a metallic core, the core becomes warm due to induction. Explain why the material rises in temperature in this situation.

7. Does a wire connected to the terminals of a battery emit electromagnetic waves? Explain.

8. If you charge a comb by running it through your hair and then hold the comb next to a bar magnet, do the electric and magnetic fields produced constitute an electromagnetic wave?

9. List as many similarities and differences between sound waves and light waves as you can.

10. Describe the physical significance of the Poynting vector.

11. For a given incident energy of an electromagnetic wave, why is the radiation pressure on a perfectly reflecting surface twice as great as that on a perfect absorbing surface?

12. Before the advent of cable television and satellite dishes, city dwellers often used "rabbit ears" atop their sets (Fig. Q34.12). Certain orientations of the receiving antenna on a television set give better reception than others. Furthermore, the best orientation varies from station to station. Explain.

Figure Q34.12 Questions 12, 14, 15, and 16, and Problem 49. The V-shaped pair of long rods is the VHF antenna and the loop is the UHF antenna.

13. Often when you touch the indoor antenna on a radio or television receiver, the reception instantly improves. Why?

14. Explain how the (dipole) VHF antenna of a television set works. (See Fig. Q34.12.)

15. Explain how the UHF (loop) antenna of a television set works. (See Fig. Q34.12.)

16. Explain why the voltage induced in a UHF (loop) antenna depends on the frequency of the signal, while the voltage in a VHF (dipole) antenna does not. (See Fig. Q34.12.)

17. Electrical engineers often speak of the radiation resistance of an antenna. What do you suppose they mean by this phrase?

18. What does a radio wave do to the charges in the receiving antenna to provide a signal for your car radio?

19. An empty plastic or glass dish being removed from a microwave oven is cool to the touch. How can this be possible? (Assume that your electric bill has been paid.)

20. Why should an infrared photograph of a person look different from a photograph taken with visible light?

21. Suppose that a creature from another planet had eyes that were sensitive to infrared radiation. Describe what the alien would see if it looked around the room you are now in. In particular, what would be bright and what would be dim?

22. A welder must wear protective glasses and clothing to prevent eye damage and sunburn. What does this imply about the nature of the light produced by the welding?

23. A home microwave oven uses electromagnetic waves with a wavelength of about 12.2 cm. Some 2.4-GHz cordless telephones suffer noisy interference when a microwave oven is used nearby. Locate the waves used by both devices on the electromagnetic spectrum. Do you expect them to interfere with each other?

PROBLEMS

1, 2, 3 = straightforward, intermediate, challenging   ☐ = full solution available in the *Student Solutions Manual and Study Guide*

🌐 = coached solution with hints available at http://www.pse6.com   💻 = computer useful in solving problem

▨ = paired numerical and symbolic problems

Section 34.1 Maxwell's Equations and Hertz's Discoveries

Note: Assume that the medium is vacuum unless specified otherwise.

1. A very long, thin rod carries electric charge with the linear density 35.0 nC/m. It lies along the x axis and moves in the x direction at a speed of 15.0 Mm/s. (a) Find the electric field the rod creates at the point (0, 20.0 cm, 0). (b) Find the magnetic field it creates at the same point.

(c) Find the force exerted on an electron at this point, moving with a velocity of $(240\hat{\mathbf{i}})$ Mm/s.

Section 34.2 Plane Electromagnetic Waves

2. (a) The distance to the North Star, Polaris, is approximately 6.44×10^{18} m. If Polaris were to burn out today, in what year would we see it disappear? (b) How long does it take for sunlight to reach the Earth? (c) How long does it take for a microwave radar signal to travel from the Earth to the Moon and back? (d) How long does it take for a radio wave to travel once around the Earth in a great circle, close to the planet's surface? (e) How long does it take for light to reach you from a lightning stroke 10.0 km away?

3. The speed of an electromagnetic wave traveling in a transparent nonmagnetic substance is $v = 1/\sqrt{\kappa \mu_0 \epsilon_0}$, where κ is the dielectric constant of the substance. Determine the speed of light in water, which has a dielectric constant at optical frequencies of 1.78.

4. An electromagnetic wave in vacuum has an electric field amplitude of 220 V/m. Calculate the amplitude of the corresponding magnetic field.

5. 🌐 Figure 34.3 shows a plane electromagnetic sinusoidal wave propagating in the x direction. Suppose that the wavelength is 50.0 m, and the electric field vibrates in the xy plane with an amplitude of 22.0 V/m. Calculate (a) the frequency of the wave and (b) the magnitude and direction of **B** when the electric field has its maximum value in the negative y direction. (c) Write an expression for **B** with the correct unit vector, with numerical values for B_{max}, k, and ω, and with its magnitude in the form

$$B = B_{max} \cos(kx - \omega t)$$

6. Write down expressions for the electric and magnetic fields of a sinusoidal plane electromagnetic wave having a frequency of 3.00 GHz and traveling in the positive x direction. The amplitude of the electric field is 300 V/m.

7. In SI units, the electric field in an electromagnetic wave is described by

$$E_y = 100 \sin(1.00 \times 10^7 x - \omega t)$$

Find (a) the amplitude of the corresponding magnetic field oscillations, (b) the wavelength λ, and (c) the frequency f.

8. Verify by substitution that the following equations are solutions to Equations 34.8 and 34.9, respectively:

$$E = E_{max} \cos(kx - \omega t)$$
$$B = B_{max} \cos(kx - \omega t)$$

9. **Review problem.** A standing-wave interference pattern is set up by radio waves between two metal sheets 2.00 m apart. This is the shortest distance between the plates that will produce a standing-wave pattern. What is the fundamental frequency?

10. A microwave oven is powered by an electron tube called a magnetron, which generates electromagnetic waves of frequency 2.45 GHz. The microwaves enter the oven and are reflected by the walls. The standing-wave pattern produced in the oven can cook food unevenly, with hot spots in the food at antinodes and cool spots at nodes, so a turntable is often used to rotate the food and distribute the energy. If a microwave oven intended for use with a turntable is instead used with a cooking dish in a fixed position, the antinodes can appear as burn marks on foods such as carrot strips or cheese. The separation distance between the burns is measured to be 6 cm ± 5%. From these data, calculate the speed of the microwaves.

Section 34.3   Energy Carried by Electromagnetic Waves

11. How much electromagnetic energy per cubic meter is contained in sunlight, if the intensity of sunlight at the Earth's surface under a fairly clear sky is 1 000 W/m²?

12. An AM radio station broadcasts isotropically (equally in all directions) with an average power of 4.00 kW. A dipole receiving antenna 65.0 cm long is at a location 4.00 miles from the transmitter. Compute the amplitude of the emf that is induced by this signal between the ends of the receiving antenna.

13. What is the average magnitude of the Poynting vector 5.00 miles from a radio transmitter broadcasting isotropically with an average power of 250 kW?

14. A monochromatic light source emits 100 W of electromagnetic power uniformly in all directions. (a) Calculate the average electric-field energy density 1.00 m from the source. (b) Calculate the average magnetic-field energy density at the same distance from the source. (c) Find the wave intensity at this location.

15. 🌐 A community plans to build a facility to convert solar radiation to electrical power. They require 1.00 MW of power, and the system to be installed has an efficiency of 30.0% (that is, 30.0% of the solar energy incident on the surface is converted to useful energy that can power the community). What must be the effective area of a perfectly absorbing surface used in such an installation, assuming sunlight has a constant intensity of 1 000 W/m²?

16. Assuming that the antenna of a 10.0-kW radio station radiates spherical electromagnetic waves, compute the maximum value of the magnetic field 5.00 km from the antenna, and compare this value with the surface magnetic field of the Earth.

17. 🌐 The filament of an incandescent lamp has a 150-Ω resistance and carries a direct current of 1.00 A. The filament is 8.00 cm long and 0.900 mm in radius. (a) Calculate the Poynting vector at the surface of the filament, associated with the static electric field producing the current and the current's static magnetic field. (b) Find the magnitude of the static electric and magnetic fields at the surface of the filament.

18. One of the weapons being considered for the "Star Wars" antimissile system is a laser that could destroy ballistic missiles. When a high-power laser is used in the Earth's atmosphere, the electric field can ionize the air, turning it into a conducting plasma that reflects the laser light. In dry air at 0°C and 1 atm, electric breakdown occurs for fields with amplitudes above about 3.00 MV/m. (a) What laser beam intensity will produce such a field? (b) At this maximum intensity, what power can be delivered in a cylindrical beam of diameter 5.00 mm?

19. In a region of free space the electric field at an instant of time is $\mathbf{E} = (80.0\hat{\mathbf{i}} + 32.0\hat{\mathbf{j}} - 64.0\hat{\mathbf{k}})$ N/C and the magnetic field is $\mathbf{B} = (0.200\hat{\mathbf{i}} + 0.080\,0\hat{\mathbf{j}} + 0.290\hat{\mathbf{k}})$ μT. (a) Show that the two fields are perpendicular to each other. (b) Determine the Poynting vector for these fields.

20. Let us model the electromagnetic wave in a microwave oven as a plane traveling wave moving to the left, with an intensity of 25.0 kW/m². An oven contains two cubical

containers of small mass, each full of water. One has an edge length of 6.00 cm and the other, 12.0 cm. Energy falls perpendicularly on one face of each container. The water in the smaller container absorbs 70.0% of the energy that falls on it. The water in the larger container absorbs 91.0%. (That is, the fraction 0.3 of the incoming microwave energy passes through a 6-cm thickness of water, and the fraction $(0.3)(0.3) = 0.09$ passes through a 12-cm thickness.) Find the temperature change of the water in each container over a time interval of 480 s. Assume that a negligible amount of energy leaves either container by heat.

21. A lightbulb filament has a resistance of 110 Ω. The bulb is plugged into a standard 120-V (rms) outlet, and emits 1.00% of the electric power delivered to it by electromagnetic radiation of frequency f. Assuming that the bulb is covered with a filter that absorbs all other frequencies, find the amplitude of the magnetic field 1.00 m from the bulb.

22. A certain microwave oven contains a magnetron that has an output of 700 W of microwave power for an electrical input power of 1.40 kW. The microwaves are entirely transferred from the magnetron into the oven chamber through a waveguide, which is a metal tube of rectangular cross section with width 6.83 cm and height 3.81 cm. (a) What is the efficiency of the magnetron? (b) Assuming that the food is absorbing all the microwaves produced by the magnetron and that no energy is reflected back into the waveguide, find the direction and magnitude of the Poynting vector, averaged over time, in the waveguide near the entrance to the oven chamber. (c) What is the maximum electric field at this point?

23. High-power lasers in factories are used to cut through cloth and metal (Fig. P34.23). One such laser has a beam diameter of 1.00 mm and generates an electric field

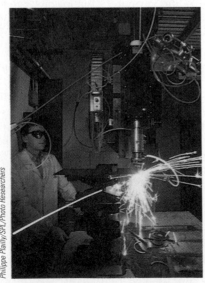

Figure P34.23 A laser cutting device mounted on a robot arm is being used to cut through a metallic plate.

having an amplitude of 0.700 MV/m at the target. Find (a) the amplitude of the magnetic field produced, (b) the intensity of the laser, and (c) the power delivered by the laser.

24. A 10.0-mW laser has a beam diameter of 1.60 mm. (a) What is the intensity of the light, assuming it is uniform across the circular beam? (b) What is the average energy density of the beam?

25. At one location on the Earth, the rms value of the magnetic field caused by solar radiation is 1.80 μT. From this value calculate (a) the rms electric field due to solar radiation, (b) the average energy density of the solar component of electromagnetic radiation at this location, and (c) the average magnitude of the Poynting vector for the Sun's radiation. (d) Compare the value found in part (c) to the value of the solar intensity given in Example 34.5.

Section 34.4 Momentum and Radiation Pressure

26. A 100-mW laser beam is reflected back upon itself by a mirror. Calculate the force on the mirror.

27. A radio wave transmits 25.0 W/m² of power per unit area. A flat surface of area A is perpendicular to the direction of propagation of the wave. Calculate the radiation pressure on it, assuming the surface is a perfect absorber.

28. A possible means of space flight is to place a perfectly reflecting aluminized sheet into orbit around the Earth and then use the light from the Sun to push this "solar sail." Suppose a sail of area 6.00×10^5 m² and mass 6 000 kg is placed in orbit facing the Sun. (a) What force is exerted on the sail? (b) What is the sail's acceleration? (c) How long does it take the sail to reach the Moon, 3.84×10^8 m away? Ignore all gravitational effects, assume that the acceleration calculated in part (b) remains constant, and assume a solar intensity of 1 340 W/m².

29. A 15.0-mW helium–neon laser ($\lambda = 632.8$ nm) emits a beam of circular cross section with a diameter of 2.00 mm. (a) Find the maximum electric field in the beam. (b) What total energy is contained in a 1.00-m length of the beam? (c) Find the momentum carried by a 1.00-m length of the beam.

30. Given that the intensity of solar radiation incident on the upper atmosphere of the Earth is 1 340 W/m², determine (a) the intensity of solar radiation incident on Mars, (b) the total power incident on Mars, and (c) the radiation force that acts on the planet if it absorbs nearly all of the light. (d) Compare this force to the gravitational attraction between Mars and the Sun. (See Table 13.2.)

31. A plane electromagnetic wave has an intensity of 750 W/m². A flat, rectangular surface of dimensions 50.0 cm × 100 cm is placed perpendicular to the direction of the wave. The surface absorbs half of the energy and reflects half. Calculate (a) the total energy absorbed by the surface in 1.00 min and (b) the momentum absorbed in this time.

32. A uniform circular disk of mass 24.0 g and radius 40.0 cm hangs vertically from a fixed, frictionless, horizontal hinge at a point on its circumference. A horizontal beam of electromagnetic radiation with intensity 10.0 MW/m² is incident on the disk in a direction perpendicular to its surface. The disk is perfectly absorbing, and the resulting radiation pressure makes the disk rotate. Find the angle through which the disk rotates as it reaches its new equilibrium position. (Assume that the radiation is *always* perpendicular to the surface of the disk.)

Section 34.5   Production of Electromagnetic Waves by an Antenna

33. Figure 34.10 shows a Hertz antenna (also known as a half-wave antenna, because its length is $\lambda/2$). The antenna is located far enough from the ground that reflections do not significantly affect its radiation pattern. Most AM radio stations, however, use a Marconi antenna, which consists of the top half of a Hertz antenna. The lower end of this (quarter-wave) antenna is connected to Earth ground, and the ground itself serves as the missing lower half. What are the heights of the Marconi antennas for radio stations broadcasting at (a) 560 kHz and (b) 1 600 kHz?

34. Two hand-held radio transceivers with dipole antennas are separated by a large fixed distance. If the transmitting antenna is vertical, what fraction of the maximum received power will appear in the receiving antenna when it is inclined from the vertical by (a) 15.0°? (b) 45.0°? (c) 90.0°?

35. Two radio-transmitting antennas are separated by half the broadcast wavelength and are driven in phase with each other. In which directions are (a) the strongest and (b) the weakest signals radiated?

36. **Review problem.** Accelerating charges radiate electromagnetic waves. Calculate the wavelength of radiation produced by a proton moving in a circle of radius R perpendicular to a magnetic field of magnitude B.

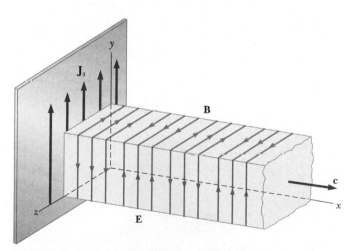

Figure P34.37 Representation of the plane electromagnetic wave radiated by an infinite current sheet lying in the yz plane. The vector **B** is in the z direction, the vector **E** is in the y direction, and the direction of wave motion is along x. Both vectors have magnitudes proportional to $\cos(kx - \omega t)$.

37. A very large flat sheet carries a uniformly distributed electric current with current per unit width J_s. Example 30.6 demonstrated that the current creates a magnetic field on both sides of the sheet, parallel to the sheet and perpendicular to the current, with magnitude $B = \frac{1}{2}\mu_0 J_s$. If the current oscillates in time according to

$$\mathbf{J}_s = J_{max}(\cos \omega t)\hat{\mathbf{j}} = J_{max}[\cos(-\omega t)]\hat{\mathbf{j}},$$

the sheet radiates an electromagnetic wave as shown in Figure P34.37. The magnetic field of the wave is described by the wave function $\mathbf{B} = \frac{1}{2}\mu_0 J_{max}[\cos(kx - \omega t)]\hat{\mathbf{k}}$. (a) Find the wave function for the electric field in the wave. (b) Find the Poynting vector as a function of x and t. (c) Find the intensity of the wave. (d) **What If?** If the sheet is to emit radiation in each direction (normal to the plane of the sheet) with intensity 570 W/m², what maximum value of sinusoidal current density is required?

Section 34.6   The Spectrum of Electromagnetic Waves

38. Classify waves with frequencies of 2 Hz, 2 kHz, 2 MHz, 2 GHz, 2 THz, 2 PHz, 2 EHz, 2 ZHz, and 2 YHz on the electromagnetic spectrum. Classify waves with wavelengths of 2 km, 2 m, 2 mm, 2 μm, 2 nm, 2 pm, 2 fm, and 2 am.

39. The human eye is most sensitive to light having a wavelength of 5.50×10^{-7} m, which is in the green-yellow region of the visible electromagnetic spectrum. What is the frequency of this light?

40. Compute an order-of-magnitude estimate for the frequency of an electromagnetic wave with wavelength equal to (a) your height; (b) the thickness of this sheet of paper. How is each wave classified on the electromagnetic spectrum?

41. What are the wavelengths of electromagnetic waves in free space that have frequencies of (a) 5.00×10^{19} Hz and (b) 4.00×10^9 Hz?

42. Suppose you are located 180 m from a radio transmitter. (a) How many wavelengths are you from the transmitter if the station calls itself 1 150 AM? (The AM band frequencies are in kilohertz.) (b) **What If?** What if this station is 98.1 FM? (The FM band frequencies are in megahertz.)

43. A radar pulse returns to the receiver after a total travel time of 4.00×10^{-4} s. How far away is the object that reflected the wave?

44. *This just in!* An important news announcement is transmitted by radio waves to people sitting next to their radios 100 km from the station, and by sound waves to people sitting across the newsroom, 3.00 m from the newscaster. Who receives the news first? Explain. Take the speed of sound in air to be 343 m/s.

45. The United States Navy has long proposed the construction of extremely low-frequency (ELF) communication systems. Such waves could penetrate the oceans to reach distant submarines. Calculate the length of a quarter-

wavelength antenna for a transmitter generating ELF waves of frequency 75.0 Hz. How practical is this?

46. What are the wavelength ranges in (a) the AM radio band (540–1 600 kHz), and (b) the FM radio band (88.0–108 MHz)?

Additional Problems

47. Assume that the intensity of solar radiation incident on the cloudtops of the Earth is 1 340 W/m². (a) Calculate the total power radiated by the Sun, taking the average Earth–Sun separation to be 1.496×10^{11} m. (b) Determine the maximum values of the electric and magnetic fields in the sunlight at the Earth's location.

48. The intensity of solar radiation at the top of the Earth's atmosphere is 1 340 W/m². Assuming that 60% of the incoming solar energy reaches the Earth's surface and assuming that you absorb 50% of the incident energy, make an order-of-magnitude estimate of the amount of solar energy you absorb in a 60-min sunbath.

49. **Review problem.** In the absence of cable input or a satellite dish, a television set can use a dipole-receiving antenna for VHF channels and a loop antenna for UHF channels (Fig. Q34.12). The UHF antenna produces an emf from the changing magnetic flux through the loop. The TV station broadcasts a signal with a frequency f, and the signal has an electric-field amplitude E_{max} and a magnetic-field amplitude B_{max} at the location of the receiving antenna. (a) Using Faraday's law, derive an expression for the amplitude of the emf that appears in a single-turn circular loop antenna with a radius r, which is small compared with the wavelength of the wave. (b) If the electric field in the signal points vertically, what orientation of the loop gives the best reception?

50. Consider a small, spherical particle of radius r located in space a distance R from the Sun. (a) Show that the ratio F_{rad}/F_{grav} is proportional to $1/r$, where F_{rad} is the force exerted by solar radiation and F_{grav} is the force of gravitational attraction. (b) The result of part (a) means that, for a sufficiently small value of r, the force exerted on the particle by solar radiation exceeds the force of gravitational attraction. Calculate the value of r for which the particle is in equilibrium under the two forces. (Assume that the particle has a perfectly absorbing surface and a mass density of 1.50 g/cm³. Let the particle be located 3.75×10^{11} m from the Sun, and use 214 W/m² as the value of the solar intensity at that point.)

51. A dish antenna having a diameter of 20.0 m receives (at normal incidence) a radio signal from a distant source, as shown in Figure P34.51. The radio signal is a continuous sinusoidal wave with amplitude $E_{max} = 0.200$ μV/m. Assume the antenna absorbs all the radiation that falls on the dish. (a) What is the amplitude of the magnetic field in this wave? (b) What is the intensity of the radiation received by this antenna? (c) What is the power received by the antenna? (d) What force is exerted by the radio waves on the antenna?

Figure P34.51

52. One goal of the Russian space program is to illuminate dark northern cities with sunlight reflected to Earth from a 200-m diameter mirrored surface in orbit. Several smaller prototypes have already been constructed and put into orbit. (a) Assume that sunlight with intensity 1 340 W/m² falls on the mirror nearly perpendicularly and that the atmosphere of the Earth allows 74.6% of the energy of sunlight to pass through it in clear weather. What is the power received by a city when the space mirror is reflecting light to it? (b) The plan is for the reflected sunlight to cover a circle of diameter 8.00 km. What is the intensity of light (the average magnitude of the Poynting vector) received by the city? (c) This intensity is what percentage of the vertical component of sunlight at Saint Petersburg in January, when the sun reaches an angle of 7.00° above the horizon at noon?

53. In 1965, Arno Penzias and Robert Wilson discovered the cosmic microwave radiation left over from the Big Bang expansion of the Universe. Suppose the energy density of this background radiation is 4.00×10^{-14} J/m³. Determine the corresponding electric field amplitude.

54. A hand-held cellular telephone operates in the 860- to 900-MHz band and has a power output of 0.600 W from an antenna 10.0 cm long (Fig. P34.54). (a) Find the average magnitude of the Poynting vector 4.00 cm from the antenna, at the location of a typical person's

Figure P34.54

head. Assume that the antenna emits energy with cylindrical wave fronts. (The actual radiation from antennas follows a more complicated pattern.) (b) The ANSI/IEEE C95.1-1991 maximum exposure standard is 0.57 mW/cm^2 for persons living near cellular telephone base stations, who would be continuously exposed to the radiation. Compare the answer to part (a) with this standard.

55. A linearly polarized microwave of wavelength 1.50 cm is directed along the positive x axis. The electric field vector has a maximum value of 175 V/m and vibrates in the xy plane. (a) Assume that the magnetic field component of the wave can be written in the form $B = B_{\max} \sin(kx - \omega t)$ and give values for $B_{\max}$, k, and ω. Also, determine in which plane the magnetic field vector vibrates. (b) Calculate the average value of the Poynting vector for this wave. (c) What radiation pressure would this wave exert if it were directed at normal incidence onto a perfectly reflecting sheet? (d) What acceleration would be imparted to a 500-g sheet (perfectly reflecting and at normal incidence) with dimensions of 1.00 m $\times$ 0.750 m?

56. The Earth reflects approximately 38.0% of the incident sunlight from its clouds and surface. (a) Given that the intensity of solar radiation is 1 340 W/m², what is the radiation pressure on the Earth, in pascals, at the location where the Sun is straight overhead? (b) Compare this to normal atmospheric pressure at the Earth's surface, which is 101 kPa.

57. An astronaut, stranded in space 10.0 m from his spacecraft and at rest relative to it, has a mass (including equipment) of 110 kg. Because he has a 100-W light source that forms a directed beam, he considers using the beam as a photon rocket to propel himself continuously toward the spacecraft. (a) Calculate how long it takes him to reach the spacecraft by this method. (b) **What If?** Suppose, instead, that he decides to throw the light source away in a direction opposite the spacecraft. If the mass of the light source is 3.00 kg and, after being thrown, it moves at 12.0 m/s relative to the recoiling astronaut, how long does it take for the astronaut to reach the spacecraft?

58. **Review problem.** A 1.00-m-diameter mirror focuses the Sun's rays onto an absorbing plate 2.00 cm in radius, which holds a can containing 1.00 L of water at 20.0°C. (a) If the solar intensity is 1.00 kW/m², what is the intensity on the absorbing plate? (b) What are the maximum magnitudes of the fields **E** and **B**? (c) If 40.0% of the energy is absorbed, how long does it take to bring the water to its boiling point?

59. Lasers have been used to suspend spherical glass beads in the Earth's gravitational field. (a) A black bead has a mass of 1.00 μg and a density of 0.200 g/cm³. Determine the radiation intensity needed to support the bead. (b) If the beam has a radius of 0.200 cm, what is the power required for this laser?

60. Lasers have been used to suspend spherical glass beads in the Earth's gravitational field. (a) A black bead has a mass

m and a density ρ. Determine the radiation intensity needed to support the bead. (b) If the beam has a radius r, what is the power required for this laser?

61. A microwave source produces pulses of 20.0-GHz radiation, with each pulse lasting 1.00 ns. A parabolic reflector with a face area of radius 6.00 cm is used to focus the microwaves into a parallel beam of radiation, as shown in Figure P34.61. The average power during each pulse is 25.0 kW. (a) What is the wavelength of these microwaves? (b) What is the total energy contained in each pulse? (c) Compute the average energy density inside each pulse. (d) Determine the amplitude of the electric and magnetic fields in these microwaves. (e) Assuming this pulsed beam strikes an absorbing surface, compute the force exerted on the surface during the 1.00-ns duration of each pulse.

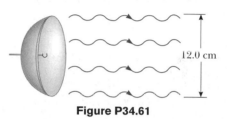

12.0 cm

Figure P34.61

62. The electromagnetic power radiated by a nonrelativistic moving point charge q having an acceleration a is

$$\mathcal{P} = \frac{q^2 a^2}{6 \pi \epsilon_0 c^3}$$

where ϵ_0 is the permittivity of free space and c is the speed of light in vacuum. (a) Show that the right side of this equation has units of watts. (b) An electron is placed in a constant electric field of magnitude 100 N/C. Determine the acceleration of the electron and the electromagnetic power radiated by this electron. (c) **What If?** If a proton is placed in a cyclotron with a radius of 0.500 m and a magnetic field of magnitude 0.350 T, what electromagnetic power does this proton radiate?

63. A thin tungsten filament of length 1.00 m radiates 60.0 W of power in the form of electromagnetic waves. A perfectly absorbing surface in the form of a hollow cylinder of radius 5.00 cm and length 1.00 m is placed concentrically with the filament. Calculate the radiation pressure acting on the cylinder. (Assume that the radiation is emitted in the radial direction, and ignore end effects.)

64. The torsion balance shown in Figure 34.8 is used in an experiment to measure radiation pressure. The suspension fiber exerts an elastic restoring torque. Its torque constant is 1.00×10^{-11} N·m/degree, and the length of the horizontal rod is 6.00 cm. The beam from a 3.00-mW helium–neon laser is incident on the black disk, and the mirror disk is completely shielded.

Calculate the angle between the equilibrium positions of the horizontal bar when the beam is switched from "off" to "on."

65. A "laser cannon" of a spacecraft has a beam of cross-sectional area A. The maximum electric field in the beam is E. The beam is aimed at an asteroid that is initially moving in the direction of the spacecraft. What is the acceleration of the asteroid relative to the spacecraft if the laser beam strikes the asteroid perpendicular to its surface, and the surface is nonreflecting? The mass of the asteroid is m. Ignore the acceleration of the spacecraft.

66. A plane electromagnetic wave varies sinusoidally at 90.0 MHz as it travels along the $+x$ direction. The peak value of the electric field is 2.00 mV/m, and it is directed along the $\pm y$ direction. (a) Find the wavelength, the period, and the maximum value of the magnetic field. (b) Write expressions in SI units for the space and time variations of the electric field and of the magnetic field. Include numerical values and include subscripts to indicate coordinate directions. (c) Find the average power per unit area that this wave carries through space. (d) Find the average energy density in the radiation (in joules per cubic meter). (e) What radiation pressure would this wave exert upon a perfectly reflecting surface at normal incidence?

Note: Section 20.7 introduced electromagnetic radiation as a mode of energy transfer. The following three problems use ideas introduced both there and in the current chapter.

---

67. Eliza is a black cat with four black kittens: Penelope, Rosalita, Sasha, and Timothy. Eliza's mass is 5.50 kg, and each kitten has mass 0.800 kg. One cool night all five sleep snuggled together on a mat, with their bodies forming one hemisphere. (a) Assuming that the purring heap has uniform density 990 kg/m^3, find the radius of the hemisphere. (b) Find the area of its curved surface. (c) Assume the surface temperature is uniformly 31.0°C and the emissivity is 0.970. Find the intensity of radiation emitted by the cats at their curved surface, and (d) the radiated power from this surface. (e) You may think of the emitted electromagnetic wave as having a single predominant frequency (of 31.2 THz). Find the amplitude of the electric field just outside the surface of the cozy pile, and (f) the amplitude of the magnetic field. (g) Are the sleeping cats charged? Are they current-carrying? Are they magnetic? Are they a radiation source? Do they glow in the dark? Give an explanation for your answers so that they do not seem contradictory. (h) **What If?** The next night the kittens all sleep alone, curling up into separate hemispheres like their mother. Find the total radiated power of the family. (For simplicity, we ignore throughout the cats' absorption of radiation from the environment.)

68. **Review problem.** (a) An elderly couple has a solar water heater installed on the roof of their house (Fig. P34.68). The heater consists of a flat closed box with

Figure P34.68

extraordinarily good thermal insulation. Its interior is painted black, and its front face is made of insulating glass. Assume that its emissivity for visible light is 0.900 and its emissivity for infrared light is 0.700. Assume that light from the noon Sun is incident perpendicular to the glass with an intensity of 1 000 W/m^2, and that no water enters or leaves the box. Find the steady-state temperature of the interior of the box. (b) **What If?** The couple builds an identical box with no water tubes. It lies flat on the ground in front of the house. They use it as a cold frame, where they plant seeds in early spring. Assuming the same noon Sun is at an elevation angle of 50.0°, find the steady-state temperature of the interior of this box when its ventilation slots are tightly closed.

69. **Review problem.** The study of Creation suggests a Creator with an inordinate fondness for beetles and for small red stars. A small red star radiates electromagnetic waves with power 6.00×10^{23} W, which is only 0.159% of the luminosity of the Sun. Consider a spherical planet in a circular orbit around this star. Assume the emissivity of the planet is equal for infrared and for visible light. Assume the planet has a uniform surface temperature. Identify the projected area over which the planet absorbs starlight and the radiating area of the planet. If beetles thrive at a temperature of 310 K, what should be the radius of the planet's orbit?

Answers to Quick Quizzes

34.1 (c). Figure 34.3b shows that the **B** and **E** vectors reach their maximum and minimum values at the same time.

34.2 (c). The **B** field must be in the $+z$ direction in order that the Poynting vector be directed along the $-y$ direction.

34.3 (d). The first three choices are instantaneous values and vary in time. The wave intensity is an average over a full cycle.

34.4 (b). To maximize the pressure on the sails, they should be perfectly reflective, so that the pressure is given by Equation 34.27.

34.5 (b), (c). The radiation pressure (a) does not change because pressure is force per unit area. In (b), the smaller disk absorbs less radiation, resulting in a smaller force. For the same reason, the momentum in (c) is reduced.

34.6 (a), (b) = (c), (d). The closest point along the x axis in Figure 34.11 (choice a) will represent the highest intensity. Choices (b) and (c) correspond to points equidistant in different directions. Choice (d) is along the axis of the antenna and the intensity is zero.

34.7 (a). The best orientation is parallel to the transmitting antenna because that is the orientation of the electric field. The electric field moves electrons in the receiving antenna, thus inducing a current that is detected and amplified.

34.8 (c). Either Equation 34.13 or Figure 34.12 can be used to find the order of magnitude of the wavelengths.

34.9 (a). Either Equation 34.13 or Figure 34.12 can be used to find the order of magnitude of the wavelength.

Light and Optics

L ight is basic to almost all life on the Earth. Plants convert the energy transferred by sunlight to chemical energy through photosynthesis. In addition, light is the principal means by which we are able to transmit and receive information to and from objects around us and throughout the Universe.

The nature and properties of light have been a subject of great interest and speculation since ancient times. The Greeks believed that light consisted of tiny particles (corpuscles) that were emitted by a light source and that these particles stimulated the perception of vision upon striking the observer's eye. Newton used this particle theory to explain the reflection and refraction (bending) of light. In 1678, one of Newton's contemporaries, the Dutch scientist Christian Huygens, was able to explain many other properties of light by proposing that light is a wave. In 1801, Thomas Young showed that light beams can interfere with one another, giving strong support to the wave theory. In 1865, Maxwell developed a brilliant theory that electromagnetic waves travel with the speed of light (see Chapter 34). By this time, the wave theory of light seemed to be firmly established.

However, at the beginning of the twentieth century, Max Planck returned to the particle theory of light to explain the radiation emitted by hot objects. Einstein then used the particle theory to explain how electrons are emitted by a metal exposed to light. Today, scientists view light as having a dual nature—that is, light exhibits characteristics of a wave in some situations and characteristics of a particle in other situations.

We shall discuss the particle nature of light in Part 6 of this text, which addresses modern physics. In Chapters 35 through 38, we concentrate on those aspects of light that are best understood through the wave model. First, we discuss the reflection of light at the boundary between two media and the refraction that occurs as light travels from one medium into another. Then, we use these ideas to study reflection and refraction as light forms images due to mirrors and lenses. Next, we describe how the lenses and mirrors used in such instruments as telescopes and microscopes help us view objects not clearly visible to the naked eye. Finally, we study the phenomena of diffraction, polarization, and interference as they apply to light. ■

◀ *The Grand Tetons in western Wyoming are reflected in a smooth lake at sunset. The optical principles that we study in this part of the book will explain the nature of the reflected image of the mountains and why the sky appears red. (David Muench/CORBIS)*

Chapter 35

The Nature of Light and the Laws of Geometric Optics

CHAPTER OUTLINE

35.1 The Nature of Light

35.2 Measurements of the Speed of Light

35.3 The Ray Approximation in Geometric Optics

35.4 Reflection

35.5 Refraction

35.6 Huygens's Principle

35.7 Dispersion and Prisms

35.8 Total Internal Reflection

35.9 Fermat's Principle

▲ *This photograph of a rainbow shows a distinct secondary rainbow with the colors reversed. The appearance of the rainbow depends on three optical phenomena discussed in this chapter—reflection, refraction, and dispersion. (Mark D. Phillips/Photo Researchers, Inc.)*

In this first chapter on optics, we begin by introducing two historical models for light and discussing early methods for measuring the speed of light. Next we study the fundamental phenomena of geometric optics—reflection of light from a surface and refraction as the light crosses the boundary between two media. We will also study the dispersion of light as it refracts into materials, resulting in visual displays such as the rainbow. Finally, we investigate the phenomenon of total internal reflection, which is the basis for the operation of optical fibers and the burgeoning technology of fiber optics.

35.1 The Nature of Light

Before the beginning of the nineteenth century, light was considered to be a stream of particles that either was emitted by the object being viewed or emanated from the eyes of the viewer. Newton, the chief architect of the particle theory of light, held that particles were emitted from a light source and that these particles stimulated the sense of sight upon entering the eye. Using this idea, he was able to explain reflection and refraction.

Most scientists accepted Newton's particle theory. During his lifetime, however, another theory was proposed—one that argued that light might be some sort of wave motion. In 1678, the Dutch physicist and astronomer Christian Huygens showed that a wave theory of light could also explain reflection and refraction.

In 1801, Thomas Young (1773–1829) provided the first clear demonstration of the wave nature of light. Young showed that, under appropriate conditions, light rays interfere with each other. Such behavior could not be explained at that time by a particle theory because there was no conceivable way in which two or more particles could come together and cancel one another. Additional developments during the nineteenth century led to the general acceptance of the wave theory of light, the most important resulting from the work of Maxwell, who in 1873 asserted that light was a form of high-frequency electromagnetic wave. As discussed in Chapter 34, Hertz provided experimental confirmation of Maxwell's theory in 1887 by producing and detecting electromagnetic waves.

Although the wave model and the classical theory of electricity and magnetism were able to explain most known properties of light, they could not explain some subsequent experiments. The most striking of these is the photoelectric effect, also discovered by Hertz: when light strikes a metal surface, electrons are sometimes ejected from the surface. As one example of the difficulties that arose, experiments showed that the kinetic energy of an ejected electron is independent of the light intensity. This finding contradicted the wave theory, which held that a more intense beam of light should add more energy to the electron. An explanation of the photoelectric effect was proposed by Einstein in 1905 in a theory that used the concept of quantization developed by Max Planck (1858–1947) in 1900. The quantization model assumes that the energy of a light wave is present in particles called *photons;* hence, the energy is said to be quantized. According to Einstein's theory, the energy of a photon is proportional to the frequency of the electromagnetic wave:

Energy of a photon

$$E = hf \tag{35.1}$$

where the constant of proportionality $h = 6.63 \times 10^{-34}$ J·s is Planck's constant (see Section 11.6). We will study this theory in Chapter 40.

In view of these developments, light must be regarded as having a dual nature: **Light exhibits the characteristics of a wave in some situations and the characteristics of a particle in other situations.** Light is light, to be sure. However, the question "Is light a wave or a particle?" is inappropriate. Sometimes light acts like a wave, and at other times it acts like a particle. In the next few chapters, we investigate the wave nature of light.

35.2 Measurements of the Speed of Light

Light travels at such a high speed ($c = 3.00 \times 10^8$ m/s) that early attempts to measure its speed were unsuccessful. Galileo attempted to measure the speed of light by positioning two observers in towers separated by approximately 10 km. Each observer carried a shuttered lantern. One observer would open his lantern first, and then the other would open his lantern at the moment he saw the light from the first lantern. Galileo reasoned that, knowing the transit time of the light beams from one lantern to the other, he could obtain the speed. His results were inconclusive. Today, we realize (as Galileo concluded) that it is impossible to measure the speed of light in this manner because the transit time is so much less than the reaction time of the observers.

Roemer's Method

In 1675, the Danish astronomer Ole Roemer (1644–1710) made the first successful estimate of the speed of light. Roemer's technique involved astronomical observations of one of the moons of Jupiter, Io, which has a period of revolution around Jupiter of approximately 42.5 h. The period of revolution of Jupiter around the Sun is about 12 yr; thus, as the Earth moves through 90° around the Sun, Jupiter revolves through only $(1/12)90° = 7.5°$ (Fig. 35.1).

An observer using the orbital motion of Io as a clock would expect the orbit to have a constant period. However, Roemer, after collecting data for more than a year, observed a systematic variation in Io's period. He found that the periods were longer than average when the Earth was receding from Jupiter and shorter than average when the Earth was approaching Jupiter. If Io had a constant period, Roemer should have seen it become eclipsed by Jupiter at a particular instant and should have been able to predict the time of the next eclipse. However, when he checked the time of the second eclipse as the Earth receded from Jupiter, he found that the eclipse was late. If the interval between his observations was three months, then the delay was approximately 600 s. Roemer attributed this variation in period to the fact that the distance between the Earth and Jupiter changed from one observation to the next. In three months (one quarter of the period of revolution of the Earth around the Sun), the light from Jupiter must travel an additional distance equal to the radius of the Earth's orbit.

Using Roemer's data, Huygens estimated the lower limit for the speed of light to be approximately 2.3×10^8 m/s. This experiment is important historically because it demonstrated that light does have a finite speed and gave an estimate of this speed.

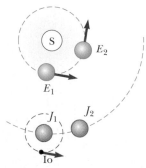

Figure 35.1 Roemer's method for measuring the speed of light. In the time interval during which the Earth travels 90° around the Sun (three months), Jupiter travels only about 7.5° (drawing not to scale).

Fizeau's Method

The first successful method for measuring the speed of light by means of purely terrestrial techniques was developed in 1849 by French physicist Armand H. L. Fizeau (1819–1896). Figure 35.2 represents a simplified diagram of Fizeau's apparatus. The basic procedure is to measure the total time interval during which light travels from some point to a distant mirror and back. If d is the distance between the light source (considered to be at the location of the wheel) and the mirror and if the time interval for one round trip is Δt, then the speed of light is $c = 2d/\Delta t$.

To measure the transit time, Fizeau used a rotating toothed wheel, which converts a continuous beam of light into a series of light pulses. The rotation of such a wheel controls what an observer at the light source sees. For example, if the pulse traveling toward the mirror and passing the opening at point A in Figure 35.2 should return to the wheel at the instant tooth B had rotated into position to cover the return path, the pulse would not reach the observer. At a greater rate of rotation, the opening at point C could move into position to allow the reflected pulse to reach the observer. Knowing the distance d, the number of teeth in the wheel, and the angular speed of the wheel, Fizeau arrived at a value of 3.1×10^8 m/s. Similar measurements made by subsequent investigators yielded more precise values for c, which led to the currently accepted value of 2.9979×10^8 m/s.

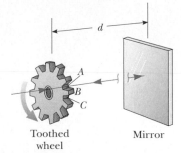

Toothed   Mirror
wheel

Figure 35.2 Fizeau's method for measuring the speed of light using a rotating toothed wheel. The light source is considered to be at the location of the wheel; thus, the distance d is known.

Example 35.1   Measuring the Speed of Light with Fizeau's Wheel

Assume that Fizeau's wheel has 360 teeth and is rotating at 27.5 rev/s when a pulse of light passing through opening A in Figure 35.2 is blocked by tooth B on its return. If the distance to the mirror is 7 500 m, what is the speed of light?

Solution The wheel has 360 teeth, and so it must have 360 openings. Therefore, because the light passes through opening A but is blocked by the tooth immediately adjacent to A, the wheel must rotate through an angular displacement of $(1/720)$ rev in the time interval during which the light pulse

makes its round trip. From the definition of angular speed, that time interval is

$$\Delta t = \frac{\Delta\theta}{\omega} = \frac{(1/720)\ \text{rev}}{27.5\ \text{rev/s}} = 5.05 \times 10^{-5}\ \text{s}$$

Hence, the speed of light calculated from this data is

$$c = \frac{2d}{\Delta t} = \frac{2(7\,500\ \text{m})}{5.05 \times 10^{-5}\ \text{s}} = \boxed{2.97 \times 10^8\ \text{m/s}}$$

35.3 The Ray Approximation in Geometric Optics

The field of **geometric optics** involves the study of the propagation of light, with the assumption that light travels in a fixed direction in a straight line as it passes through a uniform medium and changes its direction when it meets the surface of a different medium or if the optical properties of the medium are nonuniform in either space or time. As we study geometric optics here and in Chapter 36, we use what is called the **ray approximation.** To understand this approximation, first note that the rays of a given wave are straight lines perpendicular to the wave fronts as illustrated in Figure 35.3 for a plane wave. In the ray approximation, we assume that a wave moving through a medium travels in a straight line in the direction of its rays.

If the wave meets a barrier in which there is a circular opening whose diameter is much larger than the wavelength, as in Figure 35.4a, the wave emerging from the opening continues to move in a straight line (apart from some small edge effects); hence, the ray approximation is valid. If the diameter of the opening is on the order of the wavelength, as in Figure 35.4b, the waves spread out from the opening in all directions. This effect is called *diffraction* and will be studied in Chapter 37. Finally, if the opening is much smaller than the wavelength, the opening can be approximated as a point source of waves (Fig. 35.4c). Similar effects are seen when waves encounter an opaque object of dimension d. In this case, when $\lambda \ll d$, the object casts a sharp shadow.

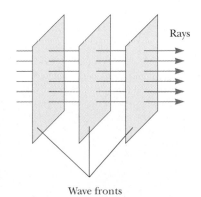

Rays

Wave fronts

Figure 35.3 A plane wave propagating to the right. Note that the rays, which always point in the direction of the wave propagation, are straight lines perpendicular to the wave fronts.

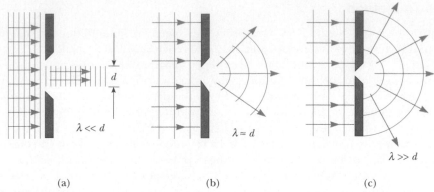

(a)   (b)   (c)

Active Figure 35.4 A plane wave of wavelength λ is incident on a barrier in which there is an opening of diameter d. (a) When $\lambda \ll d$, the rays continue in a straight-line path, and the ray approximation remains valid. (b) When $\lambda \approx d$, the rays spread out after passing through the opening. (c) When $\lambda \gg d$, the opening behaves as a point source emitting spherical waves.

The ray approximation and the assumption that $\lambda \ll d$ are used in this chapter and in Chapter 36, both of which deal with geometric optics. This approximation is very good for the study of mirrors, lenses, prisms, and associated optical instruments, such as telescopes, cameras, and eyeglasses.

35.4 Reflection

When a light ray traveling in one medium encounters a boundary with another medium, part of the incident light is reflected. Figure 35.5a shows several rays of a beam of light incident on a smooth, mirror-like, reflecting surface. The reflected rays are parallel to each other, as indicated in the figure. The direction of a reflected ray is in the plane perpendicular to the reflecting surface that contains the

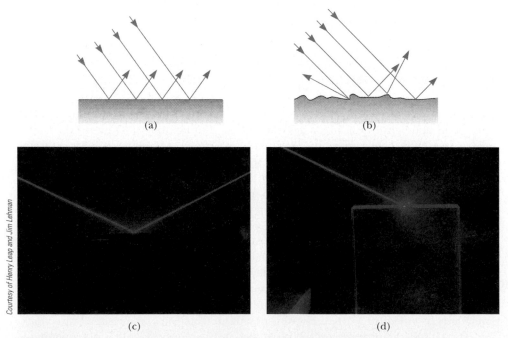

Courtesy of Henry Leap and Jim Lehman

(a)   (b)

(c)   (d)

Figure 35.5 Schematic representation of (a) specular reflection, where the reflected rays are all parallel to each other, and (b) diffuse reflection, where the reflected rays travel in random directions. (c) and (d) Photographs of specular and diffuse reflection using laser light.

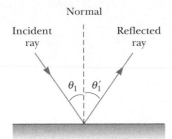

Active Figure 35.6 According to the law of reflection, $\theta_1' = \theta_1$. The incident ray, the reflected ray, and the normal all lie in the same plane.

At the Active Figures link at http://www.pse6.com, *vary the incident angle and see the effect on the reflected ray.*

incident ray. Reflection of light from such a smooth surface is called **specular reflection.** If the reflecting surface is rough, as shown in Figure 35.5b, the surface reflects the rays not as a parallel set but in various directions. Reflection from any rough surface is known as **diffuse reflection.** A surface behaves as a smooth surface as long as the surface variations are much smaller than the wavelength of the incident light.

The difference between these two kinds of reflection explains why it is more difficult to see while driving on a rainy night. If the road is wet, the smooth surface of the water specularly reflects most of your headlight beams away from your car (and perhaps into the eyes of oncoming drivers). When the road is dry, its rough surface diffusely reflects part of your headlight beam back toward you, allowing you to see the highway more clearly. In this book, we concern ourselves only with specular reflection and use the term *reflection* to mean specular reflection.

Consider a light ray traveling in air and incident at an angle on a flat, smooth surface, as shown in Figure 35.6. The incident and reflected rays make angles θ_1 and θ_1', respectively, where the angles are measured between the normal and the rays. (The normal is a line drawn perpendicular to the surface at the point where the incident ray strikes the surface.) Experiments and theory show that **the angle of reflection equals the angle of incidence:**

$$\theta_1' = \theta_1 \qquad (35.2)$$

This relationship is called the **law of reflection.**

> ⚠ **PITFALL PREVENTION**
>
> **35.1 Subscript Notation**
>
> We use the subscript 1 to refer to parameters for the light in the initial medium. When light travels from one medium to another, we use the subscript 2 for the parameters associated with the light in the new medium. In the current discussion, the light stays in the same medium, so we only have to use the subscript 1.

Law of reflection

> **Quick Quiz 35.1** In the movies, you sometimes see an actor looking in a mirror and you can see his face in the mirror. During the filming of this scene, what does the actor see in the mirror? (a) his face (b) your face (c) the director's face (d) the movie camera (e) impossible to determine

---

Example 35.2 The Double-Reflected Light Ray   Interactive

Two mirrors make an angle of 120° with each other, as illustrated in Figure 35.7a. A ray is incident on mirror M_1 at an angle of 65° to the normal. Find the direction of the ray after it is reflected from mirror M_2.

Solution Figure 35.7a helps conceptualize this situation. The incoming ray reflects from the first mirror, and the reflected ray is directed toward the second mirror. Thus, there is a second reflection from this latter mirror. Because the interactions with both mirrors are simple reflections, we categorize this problem as one that will require the law of reflection and some geometry. To analyze the problem, note that from the

law of reflection, we know that the first reflected ray makes an angle of 65° with the normal. Thus, this ray makes an angle of $90° - 65° = 25°$ with the horizontal.

From the triangle made by the first reflected ray and the two mirrors, we see that the first reflected ray makes an angle of 35° with M_2 (because the sum of the interior angles of any triangle is 180°). Therefore, this ray makes an angle of 55° with the normal to M_2. From the law of reflection, the second reflected ray makes an angle of 55° with the normal to M_2.

To finalize the problem, let us explore variations in the angle between the mirrors as follows.

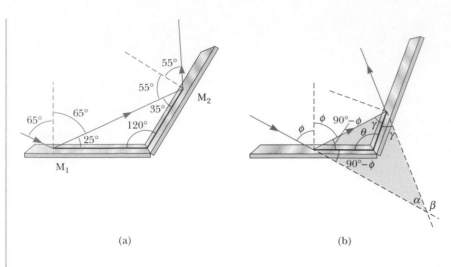

(a)

(b)

Figure 35.7 (Example 35.2) (a) Mirrors M_1 and M_2 make an angle of 120° with each other. (b) The geometry for an arbitrary mirror angle.

What If? If the incoming and outgoing rays in Figure 35.7a are extended behind the mirror, they cross at an angle of 60°, so that the overall change in direction of the light ray is 120°. This is the same as the angle between the mirrors. What if the angle between the mirrors is changed? Is the overall change in the direction of the light ray always equal to the angle between the mirrors?

Answer Making a general statement based on one data point is always a dangerous practice! Let us investigate the change in direction for a general situation. Figure 35.7b shows the mirrors at an arbitrary angle θ and the incoming light ray striking the mirror at an arbitrary angle ϕ with respect to the normal to the mirror surface. In accordance with the law of reflection and the sum of the interior angles of a triangle, the angle γ is $180° - (90° - \phi) - \theta = 90° + \phi - \theta$. Considering the triangle highlighted in blue

in Figure 35.7b, we see that

$$\alpha + 2\gamma + 2(90° - \phi) = 180°$$

$$\alpha = 2(\phi - \gamma)$$

The change in direction of the light ray is angle β, which is $180° - \alpha$:

$$\beta = 180° - \alpha = 180° - 2(\phi - \gamma)$$
$$= 180° - 2[\phi - (90° + \phi - \theta)]$$
$$= 360° - 2\theta$$

Notice that β is not equal to θ. For $\theta = 120°$, we obtain $\beta = 120°$, which happens to be the same as the mirror angle. But this is true only for this special angle between the mirrors. For example, if $\theta = 90°$, we obtain $\beta = 180°$. In this case, the light is reflected straight back to its origin.

 Investigate this reflection situation for various mirror angles at the Interactive Worked Example link at **http://www.pse6.com.**

As discussed in the **What If?** section of the preceding example, if the angle between two mirrors is 90°, the reflected beam returns to the source parallel to its original path. This phenomenon, called *retroreflection*, has many practical applications. If a third mirror is placed perpendicular to the first two, so that the three form the corner of a cube, retroreflection works in three dimensions. In 1969, a panel of many small reflectors was placed on the Moon by the *Apollo 11* astronauts (Fig. 35.8a). A laser beam from the Earth is reflected directly back on itself and its transit time is measured. This information is used to determine the distance to the Moon with an uncertainty of 15 cm. (Imagine how difficult it would be to align a regular flat mirror so that the reflected laser beam would hit a particular location on the Earth!) A more everyday application is found in automobile taillights. Part of the plastic making up the taillight is formed into many tiny cube corners (Fig. 35.8b) so that headlight beams from cars approaching from the rear are reflected back to the drivers. Instead of cube corners, small spherical bumps are sometimes used (Fig. 35.8c). Tiny clear spheres are used in a coating material found on many road signs. Due to retroreflection from these spheres, the stop sign in Figure 35.8d appears much brighter than it would if it were simply a flat, shiny surface reflecting most of the light hitting it away from the highway.

Another practical application of the law of reflection is the digital projection of movies, television shows, and computer presentations. A digital projector makes use

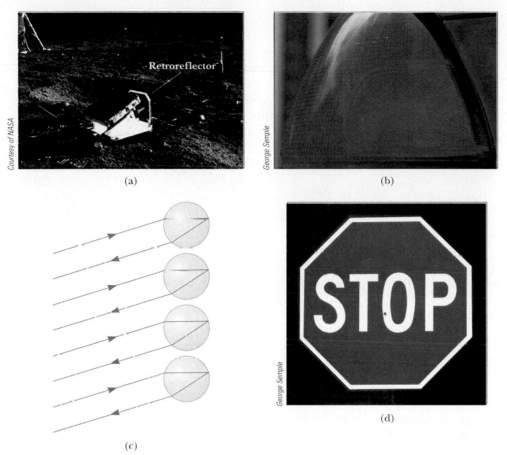

(a)

(b)

(c)

(d)

Figure 35.8 Applications of retroreflection. (a) This panel on the Moon reflects a laser beam directly back to its source on the Earth. (b) An automobile taillight has small retroreflectors that ensure that headlight beams are reflected back toward the car that sent them. (c) A light ray hitting a transparent sphere at the proper position is retroreflected. (d) This stop sign appears to glow in headlight beams because its surface is covered with a layer of many tiny retroreflecting spheres. What would you see if the sign had a mirror-like surface?

(a)

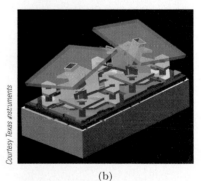

(b)

Figure 35.9 (a) An array of mirrors on the surface of a digital micromirror device. Each mirror has an area of about 16 μm^2. To provide a sense of scale, the leg of an ant appears in the photograph. (b) A close-up view of two single micromirrors. The mirror on the left is "on" and the one on the right is "off."

of an optical semiconductor chip called a *digital micromirror device*. This device contains an array of over one million tiny mirrors (Fig. 35.9a) that can be individually tilted by means of signals to an address electrode underneath the edge of the mirror. Each mirror corresponds to a pixel in the projected image. When the pixel corresponding to a given mirror is to be bright, the mirror is in the "on" position— oriented so as to reflect light from a source illuminating the array to the screen (Fig. 35.9b). When the pixel for this mirror is to be dark, the mirror is "off"—tilted so that the light is reflected away from the screen. The brightness of the pixel is determined by the total time interval during which the mirror is in the "on" position during the display of one image.

Digital movie projectors use three micromirror devices, one for each of the primary colors red, blue, and green, so that movies can be displayed with up to 35 trillion colors. Because information is stored as binary data, a digital movie does not degrade with time as does film. Furthermore, because the movie is entirely in the form of computer software, it can be delivered to theaters by means of satellites, optical discs, or optical fiber networks.

Several movies have been projected digitally to audiences and polls show that 85 percent of the viewers describe the image quality as "excellent." The first all-digital movie, from cinematography to post-production to projection, was *Star Wars Episode II: Attack of the Clones* in 2002.

35.5 Refraction

When a ray of light traveling through a transparent medium encounters a boundary leading into another transparent medium, as shown in Figure 35.10, part of the energy is reflected and part enters the second medium. The ray that enters the second medium is bent at the boundary and is said to be **refracted.** The incident ray, the reflected ray, and the refracted ray all lie in the same plane. The **angle of refraction,** θ_2 in Figure 35.10a, depends on the properties of the two media and on the angle of incidence through the relationship

$$\frac{\sin \theta_2}{\sin \theta_1} = \frac{v_2}{v_1} = \text{constant} \tag{35.3}$$

where v_1 is the speed of light in the first medium and v_2 is the speed of light in the second medium.

The path of a light ray through a refracting surface is reversible. For example, the ray shown in Figure 35.10a travels from point A to point B. If the ray originated at B, it would travel to the left along line BA to reach point A, and the reflected part would point downward and to the left in the glass.

> ### Quick Quiz 35.2 If beam ① is the incoming beam in Figure 35.10b, which of the other four red lines are reflected beams and which are refracted beams?

From Equation 35.3, we can infer that when light moves from a material in which its speed is high to a material in which its speed is lower, as shown in Figure 35.11a, the angle of refraction θ_2 is less than the angle of incidence θ_1, and the ray is bent *toward* the normal. If the ray moves from a material in which light moves slowly to a material in which it moves more rapidly, as illustrated in Figure 35.11b, θ_2 is greater than θ_1, and the ray is bent *away* from the normal.

The behavior of light as it passes from air into another substance and then re-emerges into air is often a source of confusion to students. When light travels in air,

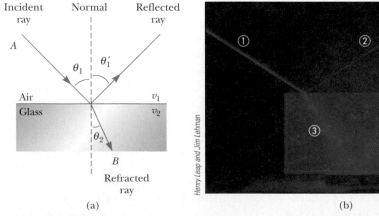

(a)   (b)

At the Active Figures link at http://www.pse6.com, vary the incident angle and see the effect on the reflected and refracted rays.

Active Figure 35.10 (a) A ray obliquely incident on an air–glass interface. The refracted ray is bent toward the normal because $v_2 < v_1$. All rays and the normal lie in the same plane. (b) Light incident on the Lucite block bends both when it enters the block and when it leaves the block.

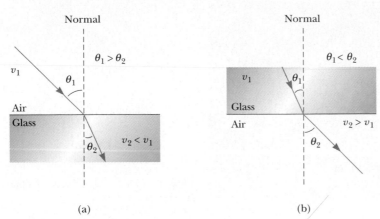

(a)   (b)

Active Figure 35.11 (a) When the light beam moves from air into glass, the light slows down on entering the glass and its path is bent toward the normal. (b) When the beam moves from glass into air, the light speeds up on entering the air and its path is bent away from the normal.

At the Active Figures link at http://www.pse6.com, light passes through three layers of material. You can vary the incident angle and see the effect on the refracted rays for a variety of values of the index of refraction (page 1104) of the three materials.

its speed is 3.00×10^8 m/s, but this speed is reduced to approximately 2×10^8 m/s when the light enters a block of glass. When the light re-emerges into air, its speed instantaneously increases to its original value of 3.00×10^8 m/s. This is far different from what happens, for example, when a bullet is fired through a block of wood. In this case, the speed of the bullet is reduced as it moves through the wood because some of its original energy is used to tear apart the wood fibers. When the bullet enters the air once again, it emerges at the speed it had just before leaving the block of wood.

To see why light behaves as it does, consider Figure 35.12, which represents a beam of light entering a piece of glass from the left. Once inside the glass, the light may encounter an electron bound to an atom, indicated as point A. Let us assume that light is absorbed by the atom; this causes the electron to oscillate (a detail represented by the double-headed vertical arrows). The oscillating electron then acts as an antenna and radiates the beam of light toward an atom at B, where the light is again absorbed. The details of these absorptions and radiations are best explained in terms of quantum mechanics (Chapter 42). For now, it is sufficient to think of light passing from one atom to another through the glass. Although light travels from one glass atom to another at 3.00×10^8 m/s, the absorption and radiation that take place cause the *average* light speed through the material to fall to about 2×10^8 m/s. Once the light emerges into the air, absorption and radiation cease and the speed of the light returns to the original value.

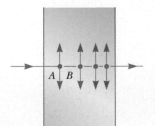

Figure 35.12 Light passing from one atom to another in a medium. The dots are electrons, and the vertical arrows represent their oscillations.

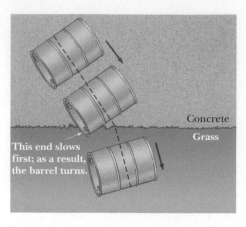

Figure 35.13 Overhead view of a barrel rolling from concrete onto grass.

A mechanical analog of refraction is shown in Figure 35.13. When the left end of the rolling barrel reaches the grass, it slows down, while the right end remains on the concrete and moves at its original speed. This difference in speeds causes the barrel to pivot, and this changes the direction of travel.

Index of Refraction

In general, the speed of light in any material is *less* than its speed in vacuum. In fact, *light travels at its maximum speed in vacuum.* It is convenient to define the **index of refraction** n of a medium to be the ratio

$$n \equiv \frac{\text{speed of light in vacuum}}{\text{speed of light in a medium}} = \frac{c}{v} \tag{35.4}$$

From this definition, we see that the index of refraction is a dimensionless number greater than unity because v is always less than c. Furthermore, n is equal to unity for vacuum. The indices of refraction for various substances are listed in Table 35.1.

As light travels from one medium to another, its frequency does not change but its wavelength does. To see why this is so, consider Figure 35.14. Waves pass an observer at point A in medium 1 with a certain frequency and are

▲ **PITFALL PREVENTION**

35.2 *n* Is Not an Integer Here

We have seen n used several times as an integer, such as in Chapter 18 to indicate the standing wave mode on a string or in an air column. The index of refraction n is *not* an integer.

Index of refraction

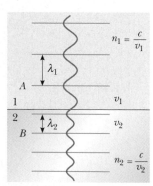

Figure 35.14 As a wave moves from medium 1 to medium 2, its wavelength changes but its frequency remains constant.

Table 35.1

Indices of Refraction[a]			
Substance	**Index of Refraction**	**Substance**	**Index of Refraction**
Solids at 20°C		*Liquids at 20°C*	
Cubic zirconia	2.20	Benzene	1.501
Diamond (C)	2.419	Carbon disulfide	1.628
Fluorite (CaF_2)	1.434	Carbon tetrachloride	1.461
Fused quartz (SiO_2)	1.458	Ethyl alcohol	1.361
Gallium phosphide	3.50	Glycerin	1.473
Glass, crown	1.52	Water	1.333
Glass, flint	1.66		
Ice (H_2O)	1.309	*Gases at 0°C, 1 atm*	
Polystyrene	1.49	Air	1.000 293
Sodium chloride (NaCl)	1.544	Carbon dioxide	1.000 45

[a] All values are for light having a wavelength of 589 nm in vacuum.

incident on the boundary between medium 1 and medium 2. The frequency with which the waves pass an observer at point B in medium 2 must equal the frequency at which they pass point A. If this were not the case, then energy would be piling up at the boundary. Because there is no mechanism for this to occur, the frequency must be a constant as a light ray passes from one medium into another. Therefore, because the relationship $v = f\lambda$ (Eq. 16.12) must be valid in both media and because $f_1 = f_2 = f$, we see that

$$v_1 = f\lambda_1 \quad \text{and} \quad v_2 = f\lambda_2 \quad \text{(35.5)}$$

Because $v_1 \neq v_2$, it follows that $\lambda_1 \neq \lambda_2$.

We can obtain a relationship between index of refraction and wavelength by dividing the first Equation 35.5 by the second and then using Equation 35.4:

$$\frac{\lambda_1}{\lambda_2} = \frac{v_1}{v_2} = \frac{c/n_1}{c/n_2} = \frac{n_2}{n_1} \quad \text{(35.6)}$$

This gives

$$\lambda_1 n_1 = \lambda_2 n_2$$

If medium 1 is vacuum, or for all practical purposes air, then $n_1 = 1$. Hence, it follows from Equation 35.6 that the index of refraction of any medium can be expressed as the ratio

$$n = \frac{\lambda}{\lambda_n} \quad \text{(35.7)}$$

where λ is the wavelength of light in vacuum and λ_n is the wavelength of light in the medium whose index of refraction is n. From Equation 35.7, we see that because $n > 1$, $\lambda_n < \lambda$.

We are now in a position to express Equation 35.3 in an alternative form. If we replace the v_2/v_1 term in Equation 35.3 with n_1/n_2 from Equation 35.6, we obtain

$$n_1 \sin \theta_1 = n_2 \sin \theta_2 \quad \text{(35.8)}$$

The experimental discovery of this relationship is usually credited to Willebrord Snell (1591–1627) and is therefore known as **Snell's law of refraction.** We shall examine this equation further in Sections 35.6 and 35.9.

▲ **PITFALL PREVENTION**

35.3 An Inverse Relationship

The index of refraction is *inversely* proportional to the wave speed. As the wave speed v decreases, the index of refraction n increases. Thus, the higher the index of refraction of a material, the more it *slows down* light from its speed in vacuum. The more the light slows down, the more θ_2 differs from θ_1 in Equation 35.8.

Snell's law of refraction

Quick Quiz 35.3 Light passes from a material with index of refraction 1.3 into one with index of refraction 1.2. Compared to the incident ray, the refracted ray (a) bends toward the normal (b) is undeflected (c) bends away from the normal.

Quick Quiz 35.4 As light from the Sun enters the atmosphere, it refracts due to the small difference between the speeds of light in air and in vacuum. The *optical* length of the day is defined as the time interval between the instant when the top of the Sun is just visibly observed above the horizon to the instant at which the top of the Sun just disappears below the horizon. The *geometric* length of the day is defined as the time interval between the instant when a geometric straight line drawn from the observer to the top of the Sun just clears the horizon to the instant at which this line just dips below the horizon. Which is longer, (a) the optical length of a day, or (b) the geometric length of a day?

Example 35.3   An Index of Refraction Measurement

A beam of light of wavelength 550 nm traveling in air is incident on a slab of transparent material. The incident beam makes an angle of 40.0° with the normal, and the refracted beam makes an angle of 26.0° with the normal. Find the index of refraction of the material.

Solution Using Snell's law of refraction (Eq. 35.8) with these data, and taking $n_1 = 1.00$ for air, we have

$$n_1 \sin \theta_1 = n_2 \sin \theta_2$$

$$n_2 = \frac{n_1 \sin \theta_1}{\sin \theta_2} = (1.00) \frac{\sin 40.0°}{\sin 26.0°}$$

$$= \frac{0.643}{0.438} = \boxed{1.47}$$

From Table 35.1, we see that the material could be fused quartz.

Example 35.4   Angle of Refraction for Glass

A light ray of wavelength 589 nm traveling through air is incident on a smooth, flat slab of crown glass at an angle of 30.0° to the normal, as sketched in Figure 35.15. Find the angle of refraction.

Solution We rearrange Snell's law of refraction to obtain

$$\sin \theta_2 = \frac{n_1}{n_2} \sin \theta_1$$

From Table 35.1, we find that $n_1 = 1.00$ for air and $n_2 = 1.52$ for crown glass. Therefore,

$$\sin \theta_2 = \left(\frac{1.00}{1.52} \right) \sin 30.0° = 0.329$$

$$\theta_2 = \sin^{-1}(0.329) = \boxed{19.2°}$$

Because this is less than the incident angle of 30°, the refracted ray is bent toward the normal, as expected. Its

change in direction is called the *angle of deviation* and is given by $\delta = |\theta_1 - \theta_2| = 30.0° - 19.2° = 10.8°$.

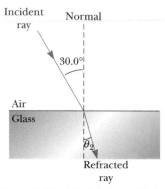

Figure 35.15 (Example 35.4) Refraction of light by glass.

Example 35.5   Laser Light in a Compact Disc

A laser in a compact disc player generates light that has a wavelength of 780 nm in air.

(A) Find the speed of this light once it enters the plastic of a compact disc ($n = 1.55$).

Solution We expect to find a value less than 3.00×10^8 m/s because $n > 1$. We can obtain the speed of light in the plastic by using Equation 35.4:

$$v = \frac{c}{n} = \frac{3.00 \times 10^8 \text{ m/s}}{1.55}$$

$$v = \boxed{1.94 \times 10^8 \text{ m/s}}$$

(B) What is the wavelength of this light in the plastic?

Solution We use Equation 35.7 to calculate the wavelength in plastic, noting that we are given the wavelength in air to be $\lambda = 780$ nm:

$$\lambda_n = \frac{\lambda}{n} = \frac{780 \text{ nm}}{1.55} = \boxed{503 \text{ nm}}$$

Example 35.6   Light Passing Through a Slab   `Interactive`

A light beam passes from medium 1 to medium 2, with the latter medium being a thick slab of material whose index of refraction is n_2 (Fig. 35.16a). Show that the emerging beam is parallel to the incident beam.

Solution First, let us apply Snell's law of refraction to the upper surface:

$$(1) \qquad \sin \theta_2 = \frac{n_1}{n_2} \sin \theta_1$$

Applying this law to the lower surface gives

$$(2) \qquad \sin \theta_3 = \frac{n_2}{n_1} \sin \theta_2$$

Substituting Equation (1) into Equation (2) gives

$$\sin \theta_3 = \frac{n_2}{n_1} \left(\frac{n_1}{n_2} \sin \theta_1 \right) = \sin \theta_1$$

Therefore, $\theta_3 = \theta_1$, and the slab does not alter the direction of the beam. It does, however, offset the beam parallel to itself by the distance d shown in Figure 35.16a.

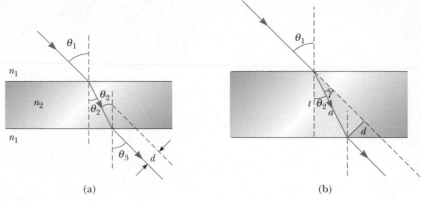

(a)   (b)

Figure 35.16 (Example 35.6) (a) When light passes through a flat slab of material, the emerging beam is parallel to the incident beam, and therefore $\theta_1 = \theta_3$. The dashed line drawn parallel to the ray coming out the bottom of the slab represents the path the light would take if the slab were not there. (b) A magnification of the area of the light path inside the slab.

What If? What if the thickness t of the slab is doubled? Does the offset distance d also double?

Answer Consider the magnification of the area of the light path within the slab in Figure 35.16b. The distance a is the hypotenuse of two right triangles. From the gold triangle, we see

$$a = \frac{t}{\cos \theta_2}$$

and from the blue triangle,

$$d = a \sin \gamma = a \sin(\theta_1 - \theta_2)$$

Combining these equations, we have

$$d = \frac{t}{\cos \theta_2} \sin(\theta_1 - \theta_2)$$

For a given incident angle θ_1, the refracted angle θ_2 is determined solely by the index of refraction, so the offset distance d is proportional to t. If the thickness doubles, so does the offset distance.

 Explore refraction through slabs of various thicknesses at the Interactive Worked Example link at http://www.pse6.com.

35.6 Huygens's Principle

In this section, we develop the laws of reflection and refraction by using a geometric method proposed by Huygens in 1678. **Huygens's principle** is a geometric construction for using knowledge of an earlier wave front to determine the position of a new wave front at some instant. In Huygens's construction,

all points on a given wave front are taken as point sources for the production of spherical secondary waves, called wavelets, which propagate outward through a medium with speeds characteristic of waves in that medium. After some time interval has passed, the new position of the wave front is the surface tangent to the wavelets.

First, consider a plane wave moving through free space, as shown in Figure 35.17a. At $t = 0$, the wave front is indicated by the plane labeled AA'. In Huygens's construction, each point on this wave front is considered a point source. For clarity, only three points on AA' are shown. With these points as sources for the wavelets, we draw circles, each of radius $c\,\Delta t$, where c is the speed of light in vacuum and Δt is some time interval during which the wave propagates. The surface drawn tangent to these wavelets is the plane BB', which is the wave front at a later time, and is parallel to AA'. In a similar manner, Figure 35.17b shows Huygens's construction for a spherical wave.

 PITFALL PREVENTION

35.4 Of What Use Is Huygens's Principle?

At this point, the importance of Huygens's principle may not be evident. Predicting the position of a future wave front may not seem to be very critical. However, we will use Huygens's principle in later chapters to explain additional wave phenomena for light.

Christian Huygens

Dutch Physicist and Astronomer (1629–1695)

Huygens is best known for his contributions to the fields of optics and dynamics. To Huygens, light was a type of vibratory motion, spreading out and producing the sensation of light when impinging on the eye. On the basis of this theory, he deduced the laws of reflection and refraction and explained the phenomenon of double refraction. (*Courtesy of Rijksmuseum voor de Geschiedenis der Natuurwetenschappen and Niels Bohr Library.*)

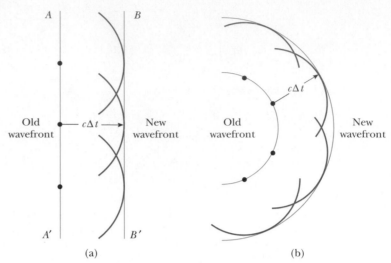

Figure 35.17 Huygens's construction for (a) a plane wave propagating to the right and (b) a spherical wave propagating to the right.

Huygens's Principle Applied to Reflection and Refraction

The laws of reflection and refraction were stated earlier in this chapter without proof. We now derive these laws, using Huygens's principle.

For the law of reflection, refer to Figure 35.18a. The line AB represents a wave front of the incident light just as ray 1 strikes the surface. At this instant, the wave at A sends out a Huygens wavelet (the circular arc centered on A) toward D. At the same time, the wave at B emits a Huygens wavelet (the circular arc centered on B) toward C. Figure 35.18a shows these wavelets after a time interval Δt, after which ray 2 strikes the surface. Because both rays 1 and 2 move with the same speed, we must have $AD = BC = c\,\Delta t$.

The remainder of our analysis depends on geometry, as summarized in Figure 35.18b, in which we isolate the triangles ABC and ADC. Note that these two triangles are congruent because they have the same hypotenuse AC and because $AD = BC$. From Figure 35.18b, we have

$$\cos \gamma = \frac{BC}{AC} \qquad \text{and} \qquad \cos \gamma' = \frac{AD}{AC}$$

where, comparing Figures 35.18a and 35.18b, we see that $\gamma = 90° - \theta_1$ and $\gamma' = 90° - \theta_1'$. Because $AD = BC$, we have

$$\cos \gamma = \cos \gamma'$$

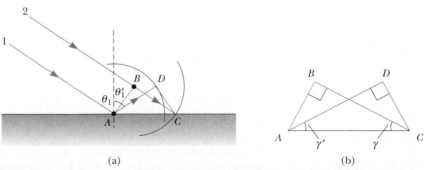

(a)   (b)

Figure 35.18 (a) Huygens's construction for proving the law of reflection. At the instant that ray 1 strikes the surface, it sends out a Huygens wavelet from A and ray 2 sends out a Huygens wavelet from B. We choose a radius of the wavelet to be $c\,\Delta t$, where Δt is the time interval for ray 2 to travel from B to C. (b) Triangle ADC is congruent to triangle ABC.

Therefore,

$$\gamma = \gamma'$$

$$90° - \theta_1 = 90° - \theta_1'$$

and

$$\theta_1 = \theta_1'$$

which is the law of reflection.

Now let us use Huygens's principle and Figure 35.19 to derive Snell's law of refraction. We focus our attention on the instant ray 1 strikes the surface and the subsequent time interval until ray 2 strikes the surface. During this time interval, the wave at A sends out a Huygens wavelet (the arc centered on A) toward D. In the same time interval, the wave at B sends out a Huygens wavelet (the arc centered on B) toward C. Because these two wavelets travel through different media, the radii of the wavelets are different. The radius of the wavelet from A is $AD = v_2 \Delta t$, where v_2 is the wave speed in the second medium. The radius of the wavelet from B is $BC = v_1 \Delta t$, where v_1 is the wave speed in the original medium.

From triangles ABC and ADC, we find that

$$\sin \theta_1 = \frac{BC}{AC} = \frac{v_1 \Delta t}{AC} \quad \text{and} \quad \sin \theta_2 = \frac{AD}{AC} = \frac{v_2 \Delta t}{AC}$$

If we divide the first equation by the second, we obtain

$$\frac{\sin \theta_1}{\sin \theta_2} = \frac{v_1}{v_2}$$

But from Equation 35.4 we know that $v_1 = c/n_1$ and $v_2 = c/n_2$. Therefore,

$$\frac{\sin \theta_1}{\sin \theta_2} = \frac{c/n_1}{c/n_2} = \frac{n_2}{n_1}$$

$$n_1 \sin \theta_1 = n_2 \sin \theta_2$$

which is Snell's law of refraction.

35.7 Dispersion and Prisms

An important property of the index of refraction n is that, for a given material, the index varies with the wavelength of the light passing through the material, as Figure 35.20 shows. This behavior is called **dispersion.** Because n is a function of wavelength, Snell's law of refraction indicates that light of different wavelengths is bent at different angles when incident on a refracting material.

As we see from Figure 35.20, the index of refraction generally decreases with increasing wavelength. This means that violet light bends more than red light does when passing into a refracting material. To understand the effects that dispersion can have on light, consider what happens when light strikes a prism, as shown in Figure 35.21. A ray of single-wavelength light incident on the prism from the left emerges refracted from its original direction of travel by an angle δ, called the **angle of deviation.**

Now suppose that a beam of *white light* (a combination of all visible wavelengths) is incident on a prism, as illustrated in Figure 35.22. The rays that emerge spread out in a series of colors known as the **visible spectrum.** These colors, in order of decreasing wavelength, are red, orange, yellow, green, blue, and violet. Clearly, the angle of deviation δ depends on wavelength. Violet light deviates the most, red the least, and the remaining colors in the visible spectrum fall between these extremes. Newton showed that each color has a particular angle of deviation and that the colors can be recombined to form the original white light.

The dispersion of light into a spectrum is demonstrated most vividly in nature by the formation of a rainbow, which is often seen by an observer positioned between the Sun

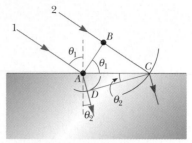

Figure 35.19 Huygens's construction for proving Snell's law of refraction. At the instant that ray 1 strikes the surface, it sends out a Huygens wavelet from A and ray 2 sends out a Huygens wavelet from B. The two wavelets have different radii because they travel in different media.

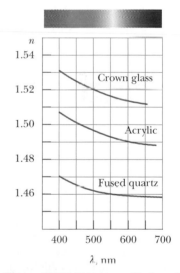

Figure 35.20 Variation of index of refraction with vacuum wavelength for three materials.

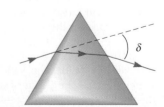

Figure 35.21 A prism refracts a single-wavelength light ray through an angle δ.

David Parker/Science Photo Library/Photo Researchers, Inc.

Figure 35.22 White light enters a glass prism at the upper left. A reflected beam of light comes out of the prism just below the incoming beam. The beam moving toward the lower right shows distinct colors. Different colors are refracted at different angles because the index of refraction of the glass depends on wavelength. Violet light deviates the most; red light deviates the least.

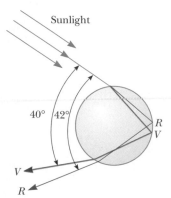

Active Figure 35.23 Path of sunlight through a spherical raindrop. Light following this path contributes to the visible rainbow.

At the Active Figures link at http://www.pse6.com, *you can vary the point at which the sunlight enters the raindrop to verify that the angles shown are the maximum angles.*

▲ **PITFALL PREVENTION**

35.5 A Rainbow of Many Light Rays

Pictorial representations such as Figure 35.23 are subject to misinterpretation. The figure shows one ray of light entering the raindrop and undergoing reflection and refraction, exiting the raindrop in a range of 40° to 42° from the entering ray. This might be interpreted incorrectly as meaning that *all* light entering the raindrop exits in this small range of angles. In reality, light exits the raindrop over a much larger range of angles, from 0° to 42°. A careful analysis of the reflection and refraction from the spherical raindrop shows that the range of 40° to 42° is where the *highest-intensity light* exits the raindrop.

and a rain shower. To understand how a rainbow is formed, consider Figure 35.23. A ray of sunlight (which is white light) passing overhead strikes a drop of water in the atmosphere and is refracted and reflected as follows: It is first refracted at the front surface of the drop, with the violet light deviating the most and the red light the least. At the back surface of the drop, the light is reflected and returns to the front surface, where it again undergoes refraction as it moves from water into air. The rays leave the drop such that the angle between the incident white light and the most intense returning violet ray is 40° and the angle between the white light and the most intense returning red ray is 42°. This small angular difference between the returning rays causes us to see a colored bow.

Now suppose that an observer is viewing a rainbow, as shown in Figure 35.24. If a raindrop high in the sky is being observed, the most intense red light returning from the drop can reach the observer because it is deviated the most, but the most intense violet light passes over the observer because it is deviated the least. Hence, the observer sees this drop as being red. Similarly, a drop lower in the sky would direct the most intense violet light toward the observer and appears to be violet. (The most intense red light from this drop would pass below the eye of the observer and not be

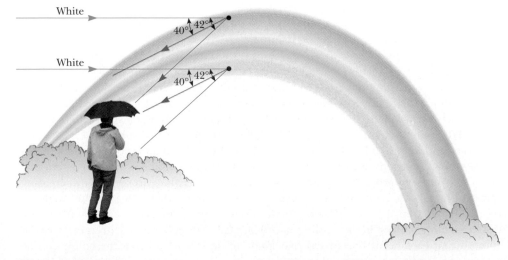

Figure 35.24 The formation of a rainbow seen by an observer standing with the Sun behind his back.

seen.) The most intense light from other colors of the spectrum would reach the observer from raindrops lying between these two extreme positions.

The opening photograph for this chapter shows a *double rainbow*. The secondary rainbow is fainter than the primary rainbow and the colors are reversed. The secondary rainbow arises from light that makes two reflections from the interior surface before exiting the raindrop. In the laboratory, rainbows have been observed in which the light makes over 30 reflections before exiting the water drop. Because each reflection involves some loss of light due to refraction out of the water drop, the intensity of these higher-order rainbows is small compared to the intensity of the primary rainbow.

Quick Quiz 35.5 Lenses in a camera use refraction to form an image on a film. Ideally, you want all the colors in the light from the object being photographed to be refracted by the same amount. Of the materials shown in Figure 35.20, which would you choose for a camera lens? (a) crown glass (b) acrylic (c) fused quartz (d) impossible to determine

---

Example 35.7 Measuring *n* Using a Prism

Although we do not prove it here, the minimum angle of deviation δ_{min} for a prism occurs when the angle of incidence θ_1 is such that the refracted ray inside the prism makes the same angle with the normal to the two prism faces,[1] as shown in Figure 35.25. Obtain an expression for the index of refraction of the prism material.

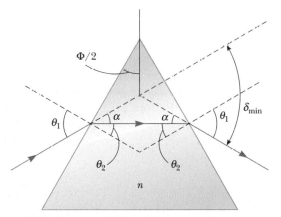

Figure 35.25 (Example 35.7) A light ray passing through a prism at the minimum angle of deviation δ_{min}.

Solution Using the geometry shown in Figure 35.25, we find that $\theta_2 = \Phi/2$, where Φ is the apex angle and

$$\theta_1 = \theta_2 + \alpha = \frac{\Phi}{2} + \frac{\delta_{min}}{2} = \frac{\Phi + \delta_{min}}{2}$$

From Snell's law of refraction, with $n_1 = 1$ because medium 1 is air, we have

$$\sin \theta_1 = n \sin \theta_2$$

$$\sin\left(\frac{\Phi + \delta_{min}}{2}\right) = n \sin(\Phi/2)$$

$$n = \frac{\sin\left(\dfrac{\Phi + \delta_{min}}{2}\right)}{\sin(\Phi/2)} \qquad (35.9)$$

Hence, knowing the apex angle Φ of the prism and measuring δ_{min}, we can calculate the index of refraction of the prism material. Furthermore, we can use a hollow prism to determine the values of n for various liquids filling the prism.

35.8 Total Internal Reflection

An interesting effect called **total internal reflection** can occur when light is directed from a medium having a given index of refraction toward one having a lower index of refraction. Consider a light beam traveling in medium 1 and meeting the boundary between medium 1 and medium 2, where n_1 is greater than n_2 (Fig. 35.26a). Various possible directions of the beam are indicated by rays 1 through 5. The refracted rays are bent away from the normal because n_1 is greater than n_2. At some particular angle of incidence θ_c, called the **critical angle,** the refracted light ray moves parallel to the boundary so that $\theta_2 = 90°$ (Fig. 35.26b).

---

[1] The details of this proof are available in texts on optics.

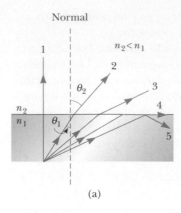

(a)

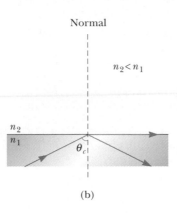

(b)

Active Figure 35.26 (a) Rays travel from a medium of index of refraction n_1 into a medium of index of refraction n_2, where $n_2 < n_1$. As the angle of incidence θ_1 increases, the angle of refraction θ_2 increases until θ_2 is 90° (ray 4). For even larger angles of incidence, total internal reflection occurs (ray 5). (b) The angle of incidence producing an angle of refraction equal to 90° is the critical angle θ_c. At this angle of incidence, all of the energy of the incident light is reflected.

At the Active Figures link at **http://www.pse6.com,** *you can vary the incident angle and see the effect on the refracted ray and the distribution of incident energy between the reflected and refracted rays.*

For angles of incidence greater than θ_c, the beam is entirely reflected at the boundary, as shown by ray 5 in Figure 35.26a. This ray is reflected at the boundary as it strikes the surface. This ray and all those like it obey the law of reflection; that is, for these rays, the angle of incidence equals the angle of reflection.

We can use Snell's law of refraction to find the critical angle. When $\theta_1 = \theta_c$, $\theta_2 = 90°$ and Equation 35.8 gives

$$n_1 \sin \theta_c = n_2 \sin 90° = n_2$$

Critical angle for total internal reflection

$$\sin \theta_c = \frac{n_2}{n_1} \qquad (\text{for } n_1 > n_2) \qquad (35.10)$$

This equation can be used only when n_1 is greater than n_2. That is, **total internal reflection occurs only when light is directed from a medium of a given index of refraction toward a medium of lower index of refraction.** If n_1 were less than n_2, Equation 35.10 would give $\sin \theta_c > 1$; this is a meaningless result because the sine of an angle can never be greater than unity.

The critical angle for total internal reflection is small when n_1 is considerably greater than n_2. For example, the critical angle for a diamond in air is 24°. Any ray inside the diamond that approaches the surface at an angle greater than this is completely reflected back into the crystal. This property, combined with proper faceting, causes diamonds to sparkle. The angles of the facets are cut so that light is "caught" inside the crystal through multiple internal reflections. These multiple reflections give the light a long path through the medium, and substantial dispersion of colors occurs. By the time the light exits through the top surface of the crystal, the rays associated with different colors have been fairly widely separated from one another.

Cubic zirconia also has a high index of refraction and can be made to sparkle very much like a genuine diamond. If a suspect jewel is immersed in corn syrup, the difference in n for the cubic zirconia and that for the syrup is small, and the critical angle is therefore great. This means that more rays escape sooner, and as a result the sparkle completely disappears. A real diamond does not lose all of its sparkle when placed in corn syrup.

Quick Quiz 35.6 In Figure 35.27, five light rays enter a glass prism from the left. How many of these rays undergo total internal reflection at the slanted surface of the prism? (a) 1 (b) 2 (c) 3 (d) 4 (e) 5.

Quick Quiz 35.7 Suppose that the prism in Figure 35.27 can be rotated in the plane of the paper. In order for *all five* rays to experience total internal reflection from the slanted surface, should the prism be rotated (a) clockwise or (b) counterclockwise?

Quick Quiz 35.8 A beam of white light is incident on a crown glass–air interface as shown in Figure 35.26a. The incoming beam is rotated clockwise, so that the incident angle θ increases. Because of dispersion in the glass, some colors of light experience total internal reflection (ray 4 in Figure 35.26a) before other colors, so that the beam refracting out of the glass is no longer white. The last color to refract out of the upper surface is (a) violet (b) green (c) red (d) impossible to determine.

Courtesy of Henry Leap and Jim Lehman

Figure 35.27 (Quick Quiz 35.6 and 35.7) Five nonparallel light rays enter a glass prism from the left.

Example 35.8 A View from the Fish's Eye

Find the critical angle for an air–water boundary. (The index of refraction of water is 1.33.)

Solution We can use Figure 35.26 to solve this problem, with the air above the water having index of refraction n_2 and the water having index of refraction n_1. Applying Equation 35.10, we find that

$$\sin \theta_c = \frac{n_2}{n_1} = \frac{1}{1.33} = 0.752$$

$$\theta_c = 48.8°$$

What If? What if a fish in a still pond looks upward toward the water's surface at different angles relative to the surface, as in Figure 35.28? What does it see?

Answer Because the path of a light ray is reversible, light traveling from medium 2 into medium 1 in Figure 35.26a follows the paths shown, but in the *opposite* direction. A fish looking upward toward the water surface, as in Figure 35.28, can see out of the water if it looks toward the surface at an angle less than the critical angle. Thus, for example, when the fish's line of vision makes an angle of 40° with the normal to the surface, light from above the water reaches the fish's eye. At 48.8°, the critical angle for water, the light has to skim along the water's surface before being refracted to the fish's eye; at this angle, the fish can in principle see the whole shore of the pond. At angles greater than the critical angle, the light reaching the fish comes by means of internal reflection at the surface. Thus, at 60°, the fish sees a reflection of the bottom of the pond.

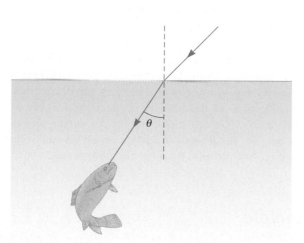

Figure 35.28 (Example 35.8) **What If?** A fish looks upward toward the water surface.

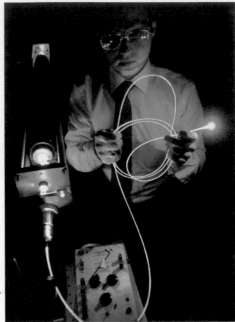

(*Left*) Strands of glass optical fibers are used to carry voice, video, and data signals in telecommunication networks. (*Right*) A bundle of optical fibers is illuminated by a laser.

Figure 35.29 Light travels in a curved transparent rod by multiple internal reflections.

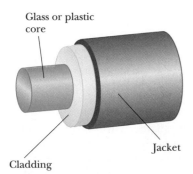

Figure 35.30 The construction of an optical fiber. Light travels in the core, which is surrounded by a cladding and a protective jacket.

Optical Fibers

Another interesting application of total internal reflection is the use of glass or transparent plastic rods to "pipe" light from one place to another. As indicated in Figure 35.29, light is confined to traveling within a rod, even around curves, as the result of successive total internal reflections. Such a light pipe is flexible if thin fibers are used rather than thick rods. A flexible light pipe is called an **optical fiber.** If a bundle of parallel fibers is used to construct an optical transmission line, images can be transferred from one point to another. This technique is used in a sizable industry known as *fiber optics*.

A practical optical fiber consists of a transparent core surrounded by a *cladding*, a material that has a lower index of refraction than the core. The combination may be surrounded by a plastic *jacket* to prevent mechanical damage. Figure 35.30 shows a cutaway view of this construction. Because the index of refraction of the cladding is less than that of the core, light traveling in the core experiences total internal reflection if it arrives at the interface between the core and the cladding at an angle of incidence that exceeds the critical angle. In this case, light "bounces" along the core of the optical fiber, losing very little of its intensity as it travels.

Any loss in intensity in an optical fiber is due essentially to reflections from the two ends and absorption by the fiber material. Optical fiber devices are particularly useful for viewing an object at an inaccessible location. For example, physicians often use such devices to examine internal organs of the body or to perform surgery without making large incisions. Optical fiber cables are replacing copper wiring and coaxial cables for telecommunications because the fibers can carry a much greater volume of telephone calls or other forms of communication than electrical wires can.

35.9 Fermat's Principle

Pierre de Fermat (1601–1665) developed a general principle that can be used to determine the path that light follows as it travels from one point to another. **Fermat's principle** states that **when a light ray travels between any two points, its path is**

the one that requires the smallest time interval. An obvious consequence of this principle is that the paths of light rays traveling in a homogeneous medium are straight lines because a straight line is the shortest distance between two points.

Let us illustrate how Fermat's principle can be used to derive Snell's law of refraction. Suppose that a light ray is to travel from point P in medium 1 to point Q in medium 2 (Fig. 35.31), where P and Q are at perpendicular distances a and b, respectively, from the interface. The speed of light is c/n_1 in medium 1 and c/n_2 in medium 2. Using the geometry of Figure 35.31, and assuming that light leaves P at $t = 0$, we see that the time at which the ray arrives at Q is

$$t = \frac{r_1}{v_1} + \frac{r_2}{v_2} = \frac{\sqrt{a^2 + x^2}}{c/n_1} + \frac{\sqrt{b^2 + (d-x)^2}}{c/n_2} \qquad (35.11)$$

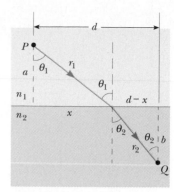

Figure 35.31 Geometry for deriving Snell's law of refraction using Fermat's principle.

To obtain the value of x for which t has its minimum value, we take the derivative of t with respect to x and set the derivative equal to zero:

$$\frac{dt}{dx} = \frac{n_1}{c}\frac{d}{dx}\sqrt{a^2 + x^2} + \frac{n_2}{c}\frac{d}{dx}\sqrt{b^2 + (d-x)^2}$$

$$= \frac{n_1}{c}\left(\tfrac{1}{2}\right)\frac{2x}{(a^2 + x^2)^{1/2}} + \frac{n_2}{c}\left(\tfrac{1}{2}\right)\frac{2(d-x)(-1)}{[b^2 + (d-x)^2]^{1/2}}$$

$$= \frac{n_1 x}{c(a^2 + x^2)^{1/2}} - \frac{n_2(d-x)}{c[b^2 + (d-x)^2]^{1/2}} = 0$$

or

$$\frac{n_1 x}{(a^2 + x^2)^{1/2}} = \frac{n_2(d-x)}{[b^2 + (d-x)^2]^{1/2}} \qquad (35.12)$$

From Figure 35.31,

$$\sin\theta_1 = \frac{x}{(a^2 + x^2)^{1/2}} \qquad \sin\theta_2 = \frac{d-x}{[b^2 + (d-x)^2]^{1/2}}$$

Substituting these expressions into Equation 35.12, we find that

$$n_1 \sin\theta_1 = n_2 \sin\theta_2$$

which is Snell's law of refraction.

This situation is equivalent to the problem of deciding where a lifeguard who can run faster than he can swim should enter the water to help a swimmer in distress. If he enters the water too directly (in other words, at a very small value of θ_1 in Figure 35.31), the distance x is smaller than the value of x that gives the minimum value of the time interval needed for the guard to move from the starting point on the sand to the swimmer. As a result, he spends too little time running and too much time swimming. The guard's optimum location for entering the water so that he can reach the swimmer in the shortest time is at that interface point that gives the value of x that satisfies Equation 35.12.

It is a simple matter to use a similar procedure to derive the law of reflection (see Problem 65).

SUMMARY

In geometric optics, we use the **ray approximation,** in which a wave travels through a uniform medium in straight lines in the direction of the rays.

The **law of reflection** states that for a light ray traveling in air and incident on a smooth surface, the angle of reflection θ_1' equals the angle of incidence θ_1:

$$\theta_1' = \theta_1 \qquad (35.2)$$

Take a practice test for this chapter by clicking on the Practice Test link at http://www.pse6.com.

Light crossing a boundary as it travels from medium 1 to medium 2 is **refracted,** or bent. The angle of refraction θ_2 is defined by the relationship

$$\frac{\sin \theta_2}{\sin \theta_1} = \frac{v_2}{v_1} = \text{constant} \tag{35.3}$$

The **index of refraction** n of a medium is defined by the ratio

$$n \equiv \frac{c}{v} \tag{35.4}$$

where c is the speed of light in a vacuum and v is the speed of light in the medium. In general, n varies with wavelength and is given by

$$n = \frac{\lambda}{\lambda_n} \tag{35.7}$$

where λ is the vacuum wavelength and λ_n is the wavelength in the medium. As light travels from one medium to another, its frequency remains the same.

Snell's law of refraction states that

$$n_1 \sin \theta_1 = n_2 \sin \theta_2 \tag{35.8}$$

where n_1 and n_2 are the indices of refraction in the two media. The incident ray, the reflected ray, the refracted ray, and the normal to the surface all lie in the same plane.

Total internal reflection occurs when light travels from a medium of high index of refraction to one of lower index of refraction. The **critical angle** θ_c for which total internal reflection occurs at an interface is given by

$$\sin \theta_c = \frac{n_2}{n_1} \qquad (\text{for } n_1 > n_2) \tag{35.10}$$

QUESTIONS

1. Light of wavelength λ is incident on a slit of width d. Under what conditions is the ray approximation valid? Under what circumstances does the slit produce enough diffraction to make the ray approximation invalid?

2. Why do astronomers looking at distant galaxies talk about looking backward in time?

3. A solar eclipse occurs when the Moon passes between the Earth and the Sun. Use a diagram to show why some areas of the Earth see a total eclipse, other areas see a partial eclipse, and most areas see no eclipse.

4. The display windows of some department stores are slanted slightly inward at the bottom. This is to decrease the glare from streetlights or the Sun, which would make it difficult for shoppers to see the display inside. Sketch a light ray reflecting from such a window to show how this technique works.

5. You take a child for walks around the neighborhood. She loves to listen to echoes from houses when she shouts or when you clap loudly. A house with a large flat front wall can produce an echo if you stand straight in front of it and reasonably far away. Draw a bird's-eye view of the situation to explain the production of the echo. Shade in the area where you can stand to hear the echo. **What If?** The child helps you to discover that a house with an L-shaped floor plan can produce echoes if you are standing in a wider range of locations. You can be standing at any reasonably distant location from which you can see the inside corner. Explain the echo in this case and draw another diagram for comparison. **What If?** What if the two wings of the house are not perpendicular? Will you and the child, standing close together, hear echoes? **What If?** What if a rectangular house and its garage have a breezeway between them, so that their perpendicular walls do not meet in an inside corner? Will this structure produce strong echoes for people in a wide range of locations? Explain your answers with diagrams.

6. The F-117A stealth fighter (Figure Q35.6) is specifically designed to be a *non*-retroreflector of radar. What aspects of its design help accomplish this? *Suggestion:* Answer the previous question as preparation for this one. Note that the bottom of the plane is flat and that all of the flat exterior panels meet at odd angles.

Figure Q35.6

7. Sound waves have much in common with light waves, including the properties of reflection and refraction. Give examples of these phenomena for sound waves.

8. Does a light ray traveling from one medium into another always bend toward the normal, as shown in Figure 35.10a? Explain.

9. As light travels from one medium to another, does the wavelength of the light change? Does the frequency change? Does the speed change? Explain.

10. A laser beam passing through a nonhomogeneous sugar solution follows a curved path. Explain.

11. A laser beam with vacuum wavelength 632.8 nm is incident from air onto a block of Lucite as shown in Figure 35.10b. The line of sight of the photograph is perpendicular to the plane in which the light moves. Find the speed, frequency, and wavelength of the light in the Lucite.

12. Suppose blue light were used instead of red light in the experiment shown in Figure 35.10b. Would the refracted beam be bent at a larger or smaller angle?

13. The level of water in a clear, colorless glass is easily observed with the naked eye. The level of liquid helium in a clear glass vessel is extremely difficult to see with the naked eye. Explain.

14. In Example 35.6 we saw that light entering a slab with parallel sides will emerge offset, but still parallel to the incoming beam. Our assumption was that the index of refraction of the material did not vary with wavelength. If the slab were made of crown glass (see Fig. 35.20), what would the outgoing beam look like?

15. Explain why a diamond sparkles more than a glass crystal of the same shape and size.

16. Explain why an oar partially in the water appears bent.

17. Total internal reflection is applied in the periscope of a submarine to let the user "see around corners." In this device, two prisms are arranged as shown in Figure Q35.17, so that an incident beam of light follows the path shown. Parallel tilted silvered mirrors could be used, but glass prisms with no silvered surfaces give higher light throughput. Propose a reason for the higher efficiency.

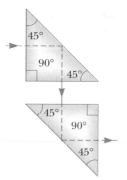

Figure Q35.17

18. Under certain circumstances, sound can be heard over extremely great distances. This frequently happens over a body of water, where the air near the water surface is cooler than the air higher up. Explain how the refraction of sound waves in such a situation could increase the distance over which the sound can be heard.

19. When two colors of light (X and Y) are sent through a glass prism, X is bent more than Y. Which color travels more slowly in the prism?

20. Retroreflection by transparent spheres, mentioned in Section 35.4 in the text, can be observed with dewdrops. To do so, look at the shadow of your head where it falls on dewy grass. Compare your observations to the reactions of two other people: The Renaissance artist Benvenuto Cellini described the phenomenon and his reaction in his *Autobiography*, at the end of Part One. The American philosopher Henry David Thoreau did the same in *Walden*, "Baker Farm," paragraph two. Try to find a person you know who has seen the halo—what did they think?

21. Why does the arc of a rainbow appear with red on top and violet on the bottom?

22. How is it possible that a complete circle of a rainbow can sometimes be seen from an airplane? With a stepladder, a lawn sprinkler, and a sunny day, how can you show the complete circle to children?

23. Is it possible to have total internal reflection for light incident from air on water? Explain.

24. Under what conditions is a mirage formed? On a hot day, what are we seeing when we observe "water on the road"?

PROBLEMS

1, 2, 3 = straightforward, intermediate, challenging   ☐ = full solution available in the *Student Solutions Manual and Study Guide*

🌀 = coached solution with hints available at http://www.pse6.com   💻 = computer useful in solving problem

▦ = paired numerical and symbolic problems

Section 35.1   The Nature of Light
Section 35.2   Measurements of the Speed of Light

1. The *Apollo 11* astronauts set up a panel of efficient corner-cube retroreflectors on the Moon's surface. The speed of light can be found by measuring the time interval required for a laser beam to travel from Earth, reflect from the panel, and return to Earth. If this interval is measured to be 2.51 s, what is the measured speed of light? Take the center-to-center distance from Earth to Moon to be 3.84×10^8 m, and do not ignore the sizes of the Earth and Moon.

2. As a result of his observations, Roemer concluded that eclipses of Io by Jupiter were delayed by 22 min during a 6 month period as the Earth moved from the point in its orbit where it is closest to Jupiter to the diametrically opposite point where it is farthest from Jupiter. Using 1.50×10^8 km as the average radius of the Earth's orbit around the Sun, calculate the speed of light from these data.

3. In an experiment to measure the speed of light using the apparatus of Fizeau (see Fig. 35.2), the distance between light source and mirror was 11.45 km and the wheel had 720 notches. The experimentally determined value of c was 2.998×10^8 m/s. Calculate the minimum angular speed of the wheel for this experiment.

4. Figure P35.4 shows an apparatus used to measure the speed distribution of gas molecules. It consists of two slotted rotating disks separated by a distance d, with the slots displaced by the angle θ. Suppose the speed of light is measured by sending a light beam from the left through this apparatus. (a) Show that a light beam will be seen in the detector (that is, will make it through both slots) only if its speed is given by $c = \omega d/\theta$, where ω is the angular

speed of the disks and θ is measured in radians. (b) What is the measured speed of light if the distance between the two slotted rotating disks is 2.50 m, the slot in the second disk is displaced $1/60$ of one degree from the slot in the first disk, and the disks are rotating at 5 555 rev/s?

Section 35.3   The Ray Approximation in Geometric Optics
Section 35.4   Reflection
Section 35.5   Refraction

Note: You may look up indices of refraction in Table 35.1.

5. A dance hall is built without pillars and with a horizontal ceiling 7.20 m above the floor. A mirror is fastened flat against one section of the ceiling. Following an earthquake, the mirror is in place and unbroken. An engineer makes a quick check of whether the ceiling is sagging by directing a vertical beam of laser light up at the mirror and observing its reflection on the floor. (a) Show that if the mirror has rotated to make an angle ϕ with the horizontal, the normal to the mirror makes an angle ϕ with the vertical. (b) Show that the reflected laser light makes an angle 2ϕ with the vertical. (c) If the reflected laser light makes a spot on the floor 1.40 cm away from the point vertically below the laser, find the angle ϕ.

6. The two mirrors illustrated in Figure P35.6 meet at a right angle. The beam of light in the vertical plane P strikes mirror 1 as shown. (a) Determine the distance the

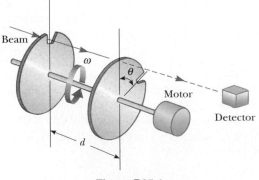

Figure P35.4

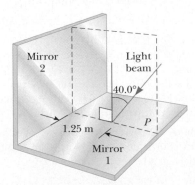

Figure P35.6

reflected light beam travels before striking mirror 2. (b) In what direction does the light beam travel after being reflected from mirror 2?

7. Two flat rectangular mirrors, both perpendicular to a horizontal sheet of paper, are set edge to edge with their reflecting surfaces perpendicular to each other. (a) A light ray in the plane of the paper strikes one of the mirrors at an arbitrary angle of incidence θ_1. Prove that the final direction of the ray, after reflection from both mirrors, is opposite to its initial direction. In a clothing store, such a pair of mirrors shows you an image of yourself as others see you, with no apparent right–left reversal. (b) **What If?** Now assume that the paper is replaced with a third flat mirror, touching edges with the other two and perpendicular to both. The set of three mirrors is called a *corner-cube reflector*. A ray of light is incident from any direction within the octant of space bounded by the reflecting surfaces. Argue that the ray will reflect once from each mirror and that its final direction will be opposite to its original direction. The *Apollo 11* astronauts placed a panel of corner cube retroreflectors on the Moon. Analysis of timing data taken with it reveals that the radius of the Moon's orbit is increasing at the rate of 3.8 cm/yr as it loses kinetic energy because of tidal friction.

8. How many times will the incident beam shown in Figure P35.8 be reflected by each of the parallel mirrors?

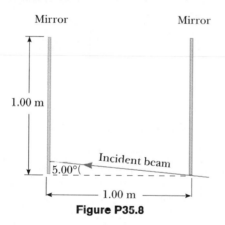

Figure P35.8

9. The distance of a lightbulb from a large plane mirror is twice the distance of a person from the plane mirror. Light from the bulb reaches the person by two paths. It travels to the mirror at an angle of incidence θ, and reflects from the mirror to the person. It also travels directly to the person without reflecting off the mirror. The total distance traveled by the light in the first case is twice the distance traveled by the light in the second case. Find the value of the angle θ.

10. A narrow beam of sodium yellow light, with wavelength 589 nm in vacuum, is incident from air onto a smooth water surface at an angle of incidence of 35.0°. Determine the angle of refraction and the wavelength of the light in water.

11. *Compare this problem with the preceding problem.* A plane sound wave in air at 20°C, with wavelength 589 mm, is incident on a smooth surface of water at 25°C, at an angle of incidence of 3.50°. Determine the angle of refraction for the sound wave and the wavelength of the sound in water.

12. The wavelength of red helium–neon laser light in air is 632.8 nm. (a) What is its frequency? (b) What is its wavelength in glass that has an index of refraction of 1.50? (c) What is its speed in the glass?

13. An underwater scuba diver sees the Sun at an apparent angle of 45.0° above the horizon. What is the actual elevation angle of the Sun above the horizon?

14. A ray of light is incident on a flat surface of a block of crown glass that is surrounded by water. The angle of refraction is 19.6°. Find the angle of reflection.

15. A laser beam is incident at an angle of 30.0° from the vertical onto a solution of corn syrup in water. If the beam is refracted to 19.24° from the vertical, (a) what is the index of refraction of the syrup solution? Suppose the light is red, with vacuum wavelength 632.8 nm. Find its (b) wavelength, (c) frequency, and (d) speed in the solution.

16. Find the speed of light in (a) flint glass, (b) water, and (c) cubic zirconia.

17. A light ray initially in water enters a transparent substance at an angle of incidence of 37.0°, and the transmitted ray is refracted at an angle of 25.0°. Calculate the speed of light in the transparent substance.

18. An opaque cylindrical tank with an open top has a diameter of 3.00 m and is completely filled with water. When the afternoon Sun reaches an angle of 28.0° above the horizon, sunlight ceases to illuminate any part of the bottom of the tank. How deep is the tank?

19. A ray of light strikes a flat block of glass ($n = 1.50$) of thickness 2.00 cm at an angle of 30.0° with the normal. Trace the light beam through the glass, and find the angles of incidence and refraction at each surface.

20. Unpolarized light in vacuum is incident onto a sheet of glass with index of refraction n. The reflected and refracted rays are perpendicular to each other. Find the angle of incidence. This angle is called *Brewster's angle* or the *polarizing angle*. In this situation the reflected light is linearly polarized, with its electric field restricted to be perpendicular to the plane containing the rays and the normal.

21. When the light illustrated in Figure P35.21 passes through the glass block, it is shifted laterally by the distance d. Taking $n = 1.50$, find the value of d.

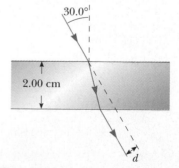

Figure P35.21 Problems 21 and 22.

22. Find the time interval required for the light to pass through the glass block described in the previous problem.

23. The light beam shown in Figure P35.23 makes an angle of 20.0° with the normal line NN' in the linseed oil. Determine the angles θ and θ'. (The index of refraction of linseed oil is 1.48.)

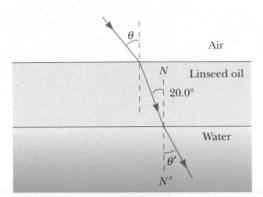

Figure P35.23

24. Three sheets of plastic have unknown indices of refraction. Sheet 1 is placed on top of sheet 2, and a laser beam is directed onto the sheets from above so that it strikes the interface at an angle of 26.5° with the normal. The refracted beam in sheet 2 makes an angle of 31.7° with the normal. The experiment is repeated with sheet 3 on top of sheet 2, and, with the same angle of incidence, the refracted beam makes an angle of 36.7° with the normal. If the experiment is repeated again with sheet 1 on top of sheet 3, what is the expected angle of refraction in sheet 3? Assume the same angle of incidence.

25. When you look through a window, by how much time is the light you see delayed by having to go through glass instead of air? Make an order-of-magnitude estimate on the basis of data you specify. By how many wavelengths is it delayed?

26. Light passes from air into flint glass. (a) What angle of incidence must the light have if the component of its velocity perpendicular to the interface is to remain constant? (b) **What If?** Can the component of velocity parallel to the interface remain constant during refraction?

27. The reflecting surfaces of two intersecting flat mirrors are at an angle θ ($0° < \theta < 90°$), as shown in Figure P35.27. For a light ray that strikes the horizontal mirror, show that the emerging ray will intersect the incident ray at an angle $\beta = 180° - 2\theta$.

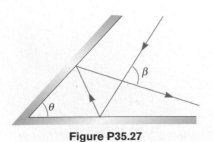

Figure P35.27

Section 35.6 Huygens's Principle

28. The speed of a water wave is described by $v = \sqrt{gd}$, where d is the water depth, assumed to be small compared to the wavelength. Because their speed changes, water waves refract when moving into a region of different depth. Sketch a map of an ocean beach on the eastern side of a landmass. Show contour lines of constant depth under water, assuming reasonably uniform slope. (a) Suppose that waves approach the coast from a storm far away to the north-northeast. Demonstrate that the waves will move nearly perpendicular to the shoreline when they reach the beach. (b) Sketch a map of a coastline with alternating bays and headlands, as suggested in Figure P35.28. Again make a reasonable guess about the shape of contour lines of constant depth. Suppose that waves approach the coast, carrying energy with uniform density along originally straight wavefronts. Show that the energy reaching the coast is concentrated at the headlands and has lower intensity in the bays.

Figure P35.28

Section 35.7 Dispersion and Prisms

Note: The apex angle of a prism is the angle between the surface at which light enters the prism and the second surface the light encounters.

29. A narrow white light beam is incident on a block of fused quartz at an angle of 30.0°. Find the angular width of the light beam inside the quartz.

30. Light of wavelength 700 nm is incident on the face of a fused quartz prism at an angle of 75.0° (with respect to the normal to the surface). The apex angle of the prism is 60.0°. Use the value of n from Figure 35.20 and calculate the angle (a) of refraction at this first surface, (b) of incidence at the second surface, (c) of refraction at the second surface, and (d) between the incident and emerging rays.

31. A prism that has an apex angle of 50.0° is made of cubic zirconia, with $n = 2.20$. What is its angle of minimum deviation?

32. A triangular glass prism with apex angle 60.0° has an index of refraction of 1.50. (a) Show that if its angle of incidence on the first surface is $\theta_1 = 48.6°$, light will pass symmetrically through the prism, as shown in

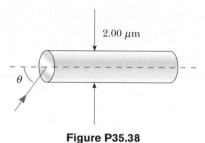

Figure P35.38

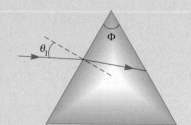

Figure 35.25. (b) Find the angle of deviation δ_{min} for $\theta_1 = 48.6°$. (c) **What If?** Find the angle of deviation if the angle of incidence on the first surface is 45.6°. (d) Find the angle of deviation if $\theta_1 = 51.6°$.

33. A triangular glass prism with apex angle $\Phi = 60.0°$ has an index of refraction $n = 1.50$ (Fig. P35.33). What is the smallest angle of incidence θ_1 for which a light ray can emerge from the other side?

Figure P35.33 Problems 33 and 34.

34. A triangular glass prism with apex angle Φ has index of refraction n. (See Fig. P35.33.) What is the smallest angle of incidence θ_1 for which a light ray can emerge from the other side?

35. The index of refraction for violet light in silica flint glass is 1.66, and that for red light is 1.62. What is the angular dispersion of visible light passing through a prism of apex angle 60.0° if the angle of incidence is 50.0°? (See Fig. P35.35.)

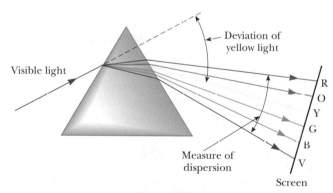

Figure P35.35

Section 35.8 Total Internal Reflection

36. For 589-nm light, calculate the critical angle for the following materials surrounded by air: (a) diamond, (b) flint glass, and (c) ice.

37. Repeat Problem 36 when the materials are surrounded by water.

38. Determine the maximum angle θ for which the light rays incident on the end of the pipe in Figure P35.38 are subject to total internal reflection along the walls of the pipe. Assume that the pipe has an index of refraction of 1.36 and the outside medium is air.

39. Consider a common mirage formed by super-heated air just above a roadway. A truck driver whose eyes are 2.00 m above the road, where $n = 1.000\ 3$, looks forward. She perceives the illusion of a patch of water ahead on the road, where her line of sight makes an angle of 1.20° below the horizontal. Find the index of refraction of the air just above the road surface. (*Suggestion:* Treat this as a problem in total internal reflection.)

40. An optical fiber has index of refraction n and diameter d. It is surrounded by air. Light is sent into the fiber along its axis, as shown in Figure P35.40. (a) Find the smallest outside radius R permitted for a bend in the fiber if no light is to escape. (b) **What If?** Does the result for part (a) predict reasonable behavior as d approaches zero? As n increases? As n approaches 1? (c) Evaluate R assuming the fiber diameter is 100 μm and its index of refraction is 1.40.

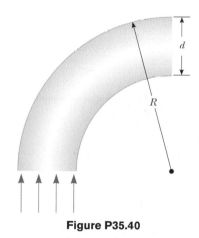

Figure P35.40

41. A large Lucite cube ($n = 1.59$) has a small air bubble (a defect in the casting process) below one surface. When a penny (diameter 1.90 cm) is placed directly over the bubble on the outside of the cube, the bubble cannot be seen by looking down into the cube at any angle. However, when a dime (diameter 1.75 cm) is placed directly over it, the bubble can be seen by looking down into the cube. What is the range of the possible depths of the air bubble beneath the surface?

42. A room contains air in which the speed of sound is 343 m/s. The walls of the room are made of concrete, in which the speed of sound is 1 850 m/s. (a) Find the critical angle for total internal reflection of sound at the concrete–air boundary. (b) In which medium must the sound be traveling in order to undergo total internal reflection? (c) "A bare concrete wall is a highly efficient mirror for sound." Give evidence for or against this statement.

43. In about 1965, engineers at the Toro Company invented a gasoline gauge for small engines, diagrammed in Figure P35.43. The gauge has no moving parts. It consists of a flat slab of transparent plastic fitting vertically into a slot in the cap on the gas tank. None of the plastic has a reflective coating. The plastic projects from the horizontal top down nearly to the bottom of the opaque tank. Its lower edge is cut with facets making angles of 45° with the horizontal. A lawnmower operator looks down from above and sees a boundary between bright and dark on the gauge. The location of the boundary, across the width of the plastic, indicates the quantity of gasoline in the tank. Explain how the gauge works. Explain the design requirements, if any, for the index of refraction of the plastic.

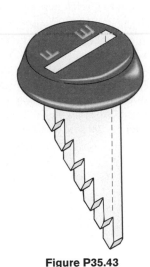

Figure P35.43

Section 35.9 Fermat's Principle

44. 🖥 The shoreline of a lake runs east and west. A swimmer gets into trouble 20.0 m out from shore and 26.0 m to the east of a lifeguard, whose station is 16.0 m in from the shoreline. The lifeguard takes negligible time to accelerate. He can run at 7.00 m/s and swim at 1.40 m/s. To reach the swimmer as quickly as possible, in what direction should the lifeguard start running? You will need to solve a transcendental equation numerically.

Additional Problems

45. Figure P35.45 shows a desk ornament globe containing a photograph. The flat photograph is in air, inside a vertical slot located behind a water-filled compartment having the shape of one half of a cylinder. Suppose you are looking at the center of the photograph and then rotate the globe about a vertical axis. You find that the center of the photograph disappears when you rotate the globe beyond a certain maximum angle (Fig. P35.45b). Account for this phenomenon and calculate the maximum angle. Briefly describe what you would see when you turn the globe beyond this angle.

Courtesy Edwin Lo

(a)   (b)

Figure P35.45

46. A light ray enters the atmosphere of a planet where it descends vertically to the surface a distance h below. The index of refraction where the light enters the atmosphere is 1.000, and it increases linearly to the surface where it has the value n. (a) How long does it take the ray to traverse this path? (b) Compare this to the time interval required in the absence of an atmosphere.

47. A narrow beam of light is incident from air onto the surface of glass with index of refraction 1.56. Find the angle of incidence for which the corresponding angle of refraction is half the angle of incidence. (*Suggestion:* You might want to use the trigonometric identity $\sin 2\theta = 2 \sin \theta \cos \theta$.)

48. 🖥 (a) Consider a horizontal interface between air above and glass of index 1.55 below. Draw a light ray incident from the air at angle of incidence 30.0°. Determine the angles of the reflected and refracted rays and show them on the diagram. (b) **What If?** Suppose instead that the light ray is incident from the glass at angle of incidence 30.0°. Determine the angles of the reflected and refracted rays and show all three rays on a new diagram. (c) For rays incident from the air onto the air–glass surface, determine and tabulate the angles of reflection and refraction for all the angles of incidence at 10.0° intervals from 0° to 90.0°. (d) Do the same for light rays coming up to the interface through the glass.

49. 🌐 A small underwater pool light is 1.00 m below the surface. The light emerging from the water forms a circle on the water surface. What is the diameter of this circle?

50. One technique for measuring the angle of a prism is shown in Figure P35.50. A parallel beam of light is directed on the angle so that parts of the beam reflect from opposite sides. Show that the angular separation of the two reflected beams is given by $B = 2A$.

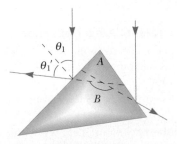

Figure P35.50

51. The walls of a prison cell are perpendicular to the four cardinal compass directions. On the first day of spring, light from the rising Sun enters a rectangular window in

the eastern wall. The light traverses 2.37 m horizontally to shine perpendicularly on the wall opposite the window. A young prisoner observes the patch of light moving across this western wall and for the first time forms his own understanding of the rotation of the Earth. (a) With what speed does the illuminated rectangle move? (b) The prisoner holds a small square mirror flat against the wall at one corner of the rectangle of light. The mirror reflects light back to a spot on the eastern wall close beside the window. How fast does the smaller square of light move across that wall? (c) Seen from a latitude of 40.0° north, the rising Sun moves through the sky along a line making a 50.0° angle with the southeastern horizon. In what direction does the rectangular patch of light on the western wall of the prisoner's cell move? (d) In what direction does the smaller square of light on the eastern wall move?

52. Figure P35.52 shows a top view of a square enclosure. The inner surfaces are plane mirrors. A ray of light enters a small hole in the center of one mirror. (a) At what angle θ must the ray enter in order to exit through the hole after being reflected once by each of the other three mirrors? (b) **What If?** Are there other values of θ for which the ray can exit after multiple reflections? If so, make a sketch of one of the ray's paths.

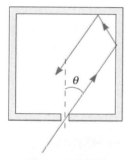

Figure P35.52

53. A hiker stands on an isolated mountain peak near sunset and observes a rainbow caused by water droplets in the air 8.00 km away. The valley is 2.00 km below the mountain peak and entirely flat. What fraction of the complete circular arc of the rainbow is visible to the hiker? (See Fig. 35.24.)

54. A 4.00-m-long pole stands vertically in a lake having a depth of 2.00 m. The Sun is 40.0° above the horizontal. Determine the length of the pole's shadow on the bottom of the lake. Take the index of refraction for water to be 1.33.

55. A laser beam strikes one end of a slab of material, as shown in Figure P35.55. The index of refraction of the slab is 1.48. Determine the number of internal reflections of the beam before it emerges from the opposite end of the slab.

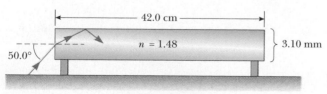

Figure P35.55

56. When light is incident normally on the interface between two transparent optical media, the intensity of the reflected light is given by the expression

$$S_1' = \left(\frac{n_2 - n_1}{n_2 + n_1}\right)^2 S_1$$

In this equation S_1 represents the average magnitude of the Poynting vector in the incident light (the incident intensity), S_1' is the reflected intensity, and n_1 and n_2 are the refractive indices of the two media. (a) What fraction of the incident intensity is reflected for 589-nm light normally incident on an interface between air and crown glass? (b) **What If?** Does it matter in part (a) whether the light is in the air or in the glass as it strikes the interface?

57. Refer to Problem 56 for its description of the reflected intensity of light normally incident on an interface between two transparent media. (a) When light is normally incident on an interface between vacuum and a transparent medium of index n, show that the intensity S_2 of the transmitted light is given by $S_2/S_1 = 4n/(n+1)^2$. (b) Light travels perpendicularly through a diamond slab, surrounded by air, with parallel surfaces of entry and exit. Apply the transmission fraction in part (a) to find the approximate overall transmission through the slab of diamond, as a percentage. Ignore light reflected back and forth within the slab.

58. **What If?** This problem builds upon the results of Problems 56 and 57. Light travels perpendicularly through a diamond slab, surrounded by air, with parallel surfaces of entry and exit. The intensity of the transmitted light is what fraction of the incident intensity? Include the effects of light reflected back and forth inside the slab.

59. The light beam in Figure P35.59 strikes surface 2 at the critical angle. Determine the angle of incidence θ_1.

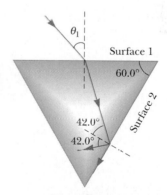

Figure P35.59

60. Builders use a leveling instrument with the beam from a fixed helium–neon laser reflecting in a horizontal plane from a small flat mirror mounted on an accurately vertical rotating shaft. The light is sufficiently bright and the rotation rate is sufficiently high that the reflected light appears as a horizontal line wherever it falls on a wall. (a) Assume the mirror is at the center of a circular grain elevator of radius R. The mirror spins with constant angular velocity ω_m. Find the speed of the spot of laser light on the wall. (b) **What If?** Assume the spinning mirror is at a perpendicular distance d from point O on a flat vertical wall. When the spot of laser light on the wall is at distance x from point O, what is its speed?

61. 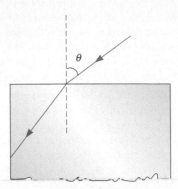 A light ray of wavelength 589 nm is incident at an angle θ on the top surface of a block of polystyrene, as shown in Figure P35.61. (a) Find the maximum value of θ for which the refracted ray undergoes total internal reflection at the left vertical face of the block. **What If?** Repeat the calculation for the case in which the polystyrene block is immersed in (b) water and (c) carbon disulfide.

Figure P35.61

62. Refer to Quick Quiz 35.4. By how much does the duration of an optical day exceed that of a geometric day? Model the Earth's atmosphere as uniform, with index of refraction 1.000 293, a sharply defined upper surface, and depth 8 614 m. Assume that the observer is at the Earth's equator, so that the apparent path of the rising and setting Sun is perpendicular to the horizon.

63. A shallow glass dish is 4.00 cm wide at the bottom, as shown in Figure P35.63. When an observer's eye is placed as shown, the observer sees the edge of the bottom of the empty dish. When this dish is filled with water, the observer sees the center of the bottom of the dish. Find the height of the dish.

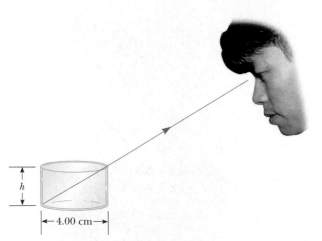

Figure P35.63

64. A ray of light passes from air into water. For its deviation angle $\delta = |\theta_1 - \theta_2|$ to be 10.0°, what must be its angle of incidence?

65. Derive the law of reflection (Eq. 35.2) from Fermat's principle. (See the procedure outlined in Section 35.9 for the derivation of the law of refraction from Fermat's principle.)

66. A material having an index of refraction n is surrounded by a vacuum and is in the shape of a quarter circle of radius R (Fig. P35.66). A light ray parallel to the base of the material is incident from the left at a distance L above the base and emerges from the material at the angle θ. Determine an expression for θ.

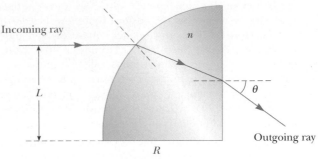

Figure P35.66

67. A transparent cylinder of radius $R = 2.00$ m has a mirrored surface on its right half, as shown in Figure P35.67. A light ray traveling in air is incident on the left side of the cylinder. The incident light ray and exiting light ray are parallel and $d = 2.00$ m. Determine the index of refraction of the material.

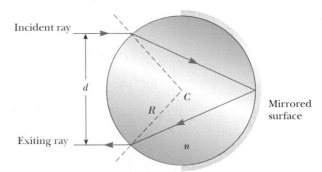

Figure P35.67

68. Suppose that a luminous sphere of radius R_1 (such as the Sun) is surrounded by a uniform atmosphere of radius R_2 and index of refraction n. When the sphere is viewed from a location far away in vacuum, what is its apparent radius? You will need to distinguish between the two cases (a) $R_2 > nR_1$ and (b) $R_2 < nR_1$.

69. A. H. Pfund's method for measuring the index of refraction of glass is illustrated in Figure P35.69. One face of a slab of thickness t is painted white, and a small hole scraped clear at point P serves as a source of diverging rays when the slab is

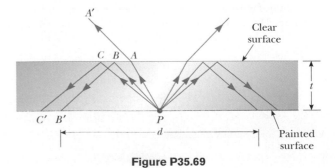

Figure P35.69

illuminated from below. Ray *PBB'* strikes the clear surface at the critical angle and is totally reflected, as are rays such as *PCC'*. Rays such as *PAA'* emerge from the clear surface. On the painted surface there appears a dark circle of diameter d, surrounded by an illuminated region, or halo. (a) Derive an equation for n in terms of the measured quantities d and t. (b) What is the diameter of the dark circle if $n = 1.52$ for a slab 0.600 cm thick? (c) If white light is used, the critical angle depends on color caused by dispersion. Is the inner edge of the white halo tinged with red light or violet light? Explain.

70. A light ray traveling in air is incident on one face of a right-angle prism of index of refraction $n = 1.50$ as shown in Figure P35.70, and the ray follows the path shown in the figure. Assuming $\theta = 60.0°$ and the base of the prism is mirrored, determine the angle ϕ made by the outgoing ray with the normal to the right face of the prism.

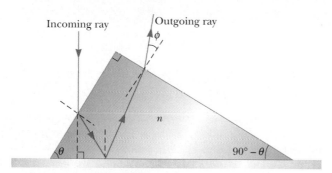

Incoming ray   Outgoing ray
ϕ
n
θ $90° - \theta$
Mirror base
Figure P35.70

71. A light ray enters a rectangular block of plastic at an angle $\theta_1 = 45.0°$ and emerges at an angle $\theta_2 = 76.0°$, as shown in Figure P35.71. (a) Determine the index of refraction of the plastic. (b) If the light ray enters the plastic at a point $L = 50.0$ cm from the bottom edge, how long does it take the light ray to travel through the plastic?

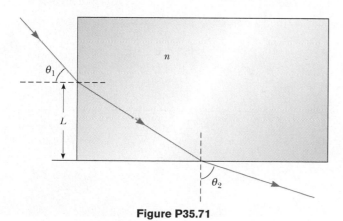

θ_1
n
L
θ_2
Figure P35.71

72. 💻 Students allow a narrow beam of laser light to strike a water surface. They arrange to measure the angle of refraction for selected angles of incidence and record the data shown in the accompanying table. Use the data to verify Snell's law of refraction by plotting the sine of the angle of incidence versus the sine of the angle of refraction. Use the resulting plot to deduce the index of refraction of water.

Angle of Incidence (degrees)	Angle of Refraction (degrees)
10.0	7.5
20.0	15.1
30.0	22.3
40.0	28.7
50.0	35.2
60.0	40.3
70.0	45.3
80.0	47.7

Answers to Quick Quizzes

35.1 (d). The light rays from the actor's face must reflect from the mirror and into the camera. If these light rays are reversed, light from the camera reflects from the mirror into the eyes of the actor.

35.2 Beams ② and ④ are reflected; beams ③ and ⑤ are refracted.

35.3 (c). Because the light is entering a material in which the index of refraction is lower, the speed of light is higher and the light bends away from the normal.

35.4 (a). Due to the refraction of light by air, light rays from the Sun deviate slightly downward toward the surface of the Earth as the light enters the atmosphere. Thus, in the morning, light rays from the upper edge of the Sun arrive at your eyes before the geometric line from your eyes to the top of the Sun clears the horizon. In the evening, light rays from the top of the Sun continue to arrive at your eyes even after the geometric line from your eyes to the top of the Sun dips below the horizon.

35.5 (c). An ideal camera lens would have an index of refraction that does not vary with wavelength so that all colors would be bent through the same angle by the lens. Of the three choices, fused quartz has the least variation in n across the visible spectrum.

35.6 (b). The two bright rays exiting the bottom of the prism on the right in Figure 35.27 result from total internal reflection at the right face of the prism. Notice that there is no refracted light exiting the slanted side for these rays. The light from the other three rays is divided into reflected and refracted parts.

35.7 (b). Counterclockwise rotation of the prism will cause the rays to strike the slanted side of the prism at a larger angle. When all five rays strike at an angle larger than the critical angle, they will all undergo total internal reflection.

35.8 (c). When the outgoing beam approaches the direction parallel to the straight side, the incident angle is approaching the critical angle for total internal reflection. The index of refraction for light at the violet end of the visible spectrum is larger than that at the red end. Thus, as the outgoing beam approaches the straight side, the violet light experiences total internal reflection first, followed by the other colors. The red light is the last to experience total internal reflection.

Chapter 36

Image Formation

CHAPTER OUTLINE

36.1 Images Formed by Flat Mirrors

36.2 Images Formed by Spherical Mirrors

36.3 Images Formed by Refraction

36.4 Thin Lenses

36.5 Lens Aberrations

36.6 The Camera

36.7 The Eye

36.8 The Simple Magnifier

36.9 The Compound Microscope

36.10 The Telescope

▲ The light rays coming from the leaves in the background of this scene did not form a focused image on the film of the camera that took this photograph. Consequently, the background appears very blurry. Light rays passing though the raindrop, however, have been altered so as to form a focused image of the background leaves on the film. In this chapter, we investigate the formation of images as light rays reflect from mirrors and refract through lenses. (Don Hammond/CORBIS)

This chapter is concerned with the images that result when light rays encounter flat and curved surfaces. We find that images can be formed either by reflection or by refraction and that we can design mirrors and lenses to form images with desired characteristics. We continue to use the ray approximation and to assume that light travels in straight lines. Both of these steps lead to valid predictions in the field called *geometric optics*. In subsequent chapters, we shall concern ourselves with interference and diffraction effects—the objects of study in the field of *wave optics*.

36.1 Images Formed by Flat Mirrors

We begin by considering the simplest possible mirror, the flat mirror. Consider a point source of light placed at O in Figure 36.1, a distance p in front of a flat mirror. The distance p is called the **object distance.** Light rays leave the source and are reflected from the mirror. Upon reflection, the rays continue to diverge (spread apart). The dashed lines in Figure 36.1 are extensions of the diverging rays back to a point of intersection at I. The diverging rays appear to the viewer to come from the point I behind the mirror. Point I is called the **image** of the object at O. Regardless of the system under study, we always locate images by extending diverging rays back to a point at which they intersect. **Images are located either at a point from which rays of light *actually* diverge or at a point from which they *appear* to diverge.** Because the rays in Figure 36.1 appear to originate at I, which is a distance q behind the mirror, this is the location of the image. The distance q is called the **image distance.**

Images are classified as **real** or **virtual. A real image is formed when light rays pass through and diverge from the image point; a virtual image is formed when the light rays do not pass through the image point but only appear to diverge from that point.** The image formed by the mirror in Figure 36.1 is virtual. The image of an object seen in a flat mirror is *always* virtual. Real images can be displayed on a

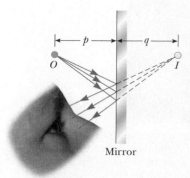

Mirror

Figure 36.1 An image formed by reflection from a flat mirror. The image point I is located behind the mirror a perpendicular distance q from the mirror (the image distance). The image distance has the same magnitude as the object distance p.

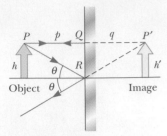

Active Figure 36.2 A geometric construction that is used to locate the image of an object placed in front of a flat mirror. Because the triangles PQR and $P'QR$ are congruent, $|p| = |q|$ and $h = h'$.

 At the Active Figures link at http://www.pse6.com, you can move the object and see the effect on the image.

Lateral magnification

PITFALL PREVENTION

36.1 Magnification Does Not Necessarily Imply Enlargement

For optical elements other than flat mirrors, the magnification defined in Equation 36.1 can result in a number with magnitude larger *or* smaller than 1. Thus, despite the cultural usage of the word *magnification* to mean *enlargement,* the image could be smaller than the object.

---

screen (as at a movie), but virtual images cannot be displayed on a screen. We shall see an example of a real image in Section 36.2.

We can use the simple geometry in Figure 36.2 to examine the properties of the images of extended objects formed by flat mirrors. Even though there are an infinite number of choices of direction in which light rays could leave each point on the object, we need to choose only two rays to determine where an image is formed. One of those rays starts at P, follows a horizontal path to the mirror, and reflects back on itself. The second ray follows the oblique path PR and reflects as shown, according to the law of reflection. An observer in front of the mirror would trace the two reflected rays back to the point at which they appear to have originated, which is point P' behind the mirror. A continuation of this process for points other than P on the object would result in a virtual image (represented by a yellow arrow) behind the mirror. Because triangles PQR and $P'QR$ are congruent, $PQ = P'Q$. We conclude that **the image formed by an object placed in front of a flat mirror is as far behind the mirror as the object is in front of the mirror.**

Geometry also reveals that the object height h equals the image height h'. Let us define **lateral magnification** M of an image as follows:

$$M \equiv \frac{\text{Image height}}{\text{Object height}} = \frac{h'}{h} \tag{36.1}$$

This is a general definition of the lateral magnification for an image from any type of mirror. (This equation is also valid for images formed by lenses, which we study in Section 36.4.) For a flat mirror, $M = 1$ for any image because $h' = h$.

Finally, note that a flat mirror produces an image that has an *apparent* left–right reversal. You can see this reversal by standing in front of a mirror and raising your right hand, as shown in Figure 36.3. The image you see raises its left hand. Likewise, your hair appears to be parted on the side opposite your real part, and a mole on your right cheek appears to be on your left cheek.

This reversal is not *actually* a left–right reversal. Imagine, for example, lying on your left side on the floor, with your body parallel to the mirror surface. Now your head is on the left and your feet are on the right. If you shake your feet, the image does not shake its head! If you raise your right hand, however, the image again raises its left hand. Thus, the mirror again appears to produce a left–right reversal but in the up–down direction!

The reversal is actually a *front–back reversal,* caused by the light rays going forward toward the mirror and then reflecting back from it. An interesting exercise is to stand in front of a mirror while holding an overhead transparency in front of you so that you can read the writing on the transparency. You will also be able to read the writing on the image of the transparency. You may have had a similar experience if you have attached a transparent decal with words on it to the rear window of your car. If the

Figure 36.3 The image in the mirror of a person's right hand is reversed front to back. This makes the right hand appear to be a left hand. Notice that the thumb is on the left side of both real hands and on the left side of the image. That the thumb is not on the right side of the image indicates that there is no left-to-right reversal.

decal can be read from outside the car, you can also read it when looking into your rearview mirror from inside the car.

We conclude that the image that is formed by a flat mirror has the following properties.

- The image is as far behind the mirror as the object is in front.
- The image is unmagnified, virtual, and upright. (By upright we mean that, if the object arrow points upward as in Figure 36.2, so does the image arrow.)
- The image has front–back reversal.

Quick Quiz 36.1 In the overhead view of Figure 36.4, the image of the stone seen by observer 1 is at *C.* At which of the five points *A, B, C, D,* or *E* does observer 2 see the image?

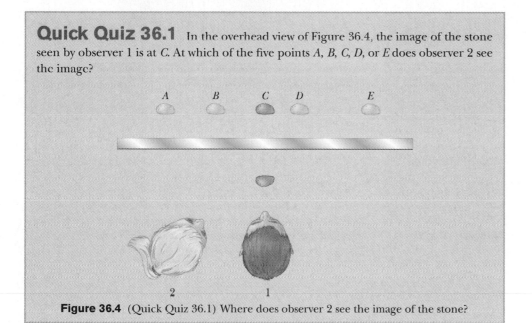

Figure 36.4 (Quick Quiz 36.1) Where does observer 2 see the image of the stone?

Quick Quiz 36.2 You are standing about 2 m away from a mirror. The mirror has water spots on its surface. True or false: It is possible for you to see the water spots and your image both in focus at the same time.

Conceptual Example 36.1   Multiple Images Formed by Two Mirrors

Two flat mirrors are perpendicular to each other, as in Figure 36.5, and an object is placed at point *O.* In this situation, multiple images are formed. Locate the positions of these images.

Solution The image of the object is at I_1 in mirror 1 and at I_2 in mirror 2. In addition, a third image is formed at I_3. This third image is the image of I_1 in mirror 2 or, equivalently, the image of I_2 in mirror 1. That is, the image at I_1 (or I_2) serves as the object for I_3. Note that to form this image at I_3, the rays reflect twice after leaving the object at *O.*

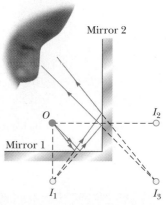

Figure 36.5 (Conceptual Example 36.1) When an object is placed in front of two mutually perpendicular mirrors as shown, three images are formed.

Conceptual Example 36.2   The Levitated Professor

The professor in the box shown in Figure 36.6 appears to be balancing himself on a few fingers, with his feet off the floor. He can maintain this position for a long time, and he appears to defy gravity. How was this illusion created?

Solution This is one of many magicians' optical illusions that make use of a mirror. The box in which the professor stands is a cubical frame that contains a flat vertical mirror positioned in a diagonal plane of the frame. The professor straddles the mirror so that one foot, which you see, is in front of the mirror, and the other foot, which you cannot see, is behind the mirror. When he raises the foot in front of the mirror, the reflection of that foot also rises, so he appears to float in air.

Courtesy of Henry Leap and Jim Lehman

Figure 36.6 (Conceptual Example 36.2) An optical illusion.

Conceptual Example 36.3   The Tilting Rearview Mirror

Most rearview mirrors in cars have a day setting and a night setting. The night setting greatly diminishes the intensity of the image in order that lights from trailing vehicles do not blind the driver. How does such a mirror work?

Solution Figure 36.7 shows a cross-sectional view of a rearview mirror for each setting. The unit consists of a reflective coating on the back of a wedge of glass. In the day setting (Fig. 36.7a), the light from an object behind the car strikes the glass wedge at point 1. Most of the light enters the wedge, refracting as it crosses the front surface, and reflects from the back surface to return to the front surface, where it is refracted again as it re-enters the air as ray *B* (for *bright*). In addition, a small portion of the light is reflected at the front surface of the glass, as indicated by ray *D* (for *dim*).

This dim reflected light is responsible for the image that is observed when the mirror is in the night setting (Fig. 36.7b). In this case, the wedge is rotated so that the path followed by the bright light (ray *B*) does not lead to the eye. Instead, the dim light reflected from the front surface of the wedge travels to the eye, and the brightness of trailing headlights does not become a hazard.

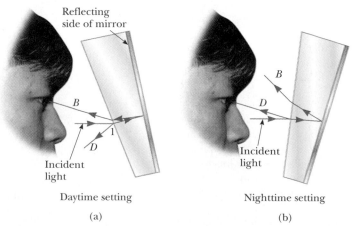

Figure 36.7 (Conceptual Example 36.3) Cross-sectional views of a rearview mirror. (a) With the day setting, the silvered back surface of the mirror reflects a bright ray *B* into the driver's eyes. (b) With the night setting, the glass of the unsilvered front surface of the mirror reflects a dim ray *D* into the driver's eyes.

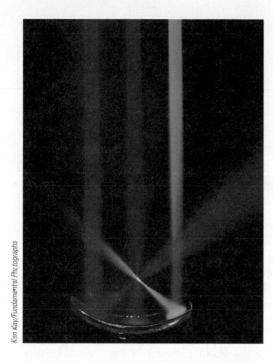

Figure 36.8 Red, blue, and green light rays are reflected by a curved mirror. Note that the three colored beams meet at a point.

36.2 Images Formed by Spherical Mirrors

Concave Mirrors

A **spherical mirror,** as its name implies, has the shape of a section of a sphere. This type of mirror focuses incoming parallel rays to a point, as demonstrated by the colored light rays in Figure 36.8. Figure 36.9a shows a cross section of a spherical mirror, with its surface represented by the solid, curved black line. (The blue band represents the structural support for the mirrored surface, such as a curved piece of glass on which the silvered surface is deposited.) Such a mirror, in which light is reflected from the inner, concave surface, is called a **concave mirror.** The mirror has a radius of curvature R, and its center of curvature is point C. Point V is the center of the spherical section, and a line through C and V is called the **principal axis** of the mirror.

Now consider a point source of light placed at point O in Figure 36.9b, where O is any point on the principal axis to the left of C. Two diverging rays that originate at O are shown. After reflecting from the mirror, these rays converge and cross at the image point I. They then continue to diverge from I as if an object were there. As a result, at point I we have a real image of the light source at O.

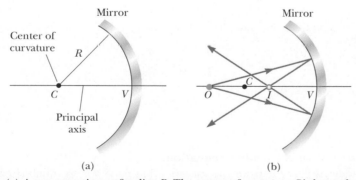

(a)   (b)

Figure 36.9 (a) A concave mirror of radius R. The center of curvature C is located on the principal axis. (b) A point object placed at O in front of a concave spherical mirror of radius R, where O is any point on the principal axis farther than R from the mirror surface, forms a real image at I. If the rays diverge from O at small angles, they all reflect through the same image point.

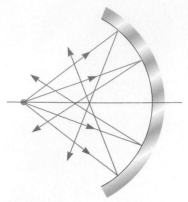

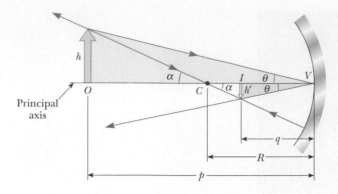

Figure 36.11 The image formed by a spherical concave mirror when the object O lies outside the center of curvature C. This geometric construction is used to derive Equation 36.4.

Figure 36.10 Rays diverging from the object at large angles from the principal axis reflect from a spherical concave mirror to intersect the principal axis at different points, resulting in a blurred image. This condition is called *spherical aberration*.

We shall consider in this section only rays that diverge from the object and make a small angle with the principal axis. Such rays are called **paraxial rays.** All paraxial rays reflect through the image point, as shown in Figure 36.9b. Rays that are far from the principal axis, such as those shown in Figure 36.10, converge to other points on the principal axis, producing a blurred image. This effect, which is called **spherical aberration,** is present to some extent for any spherical mirror and is discussed in Section 36.5.

We can use Figure 36.11 to calculate the image distance q from a knowledge of the object distance p and radius of curvature R. By convention, these distances are measured from point V. Figure 36.11 shows two rays leaving the tip of the object. One of these rays passes through the center of curvature C of the mirror, hitting the mirror perpendicular to the mirror surface and reflecting back on itself. The second ray strikes the mirror at its center (point V) and reflects as shown, obeying the law of reflection. The image of the tip of the arrow is located at the point where these two rays intersect. From the gold right triangle in Figure 36.11, we see that $\tan\theta = h/p$, and from the blue right triangle we see that $\tan\theta = -h'/q$. The negative sign is introduced because the image is inverted, so h' is taken to be negative. Thus, from Equation 36.1 and these results, we find that the magnification of the image is

$$M = \frac{h'}{h} = -\frac{q}{p} \tag{36.2}$$

We also note from the two triangles in Figure 36.11 that have α as one angle that

$$\tan\alpha = \frac{h}{p-R} \quad \text{and} \quad \tan\alpha = -\frac{h'}{R-q}$$

from which we find that

$$\frac{h'}{h} = -\frac{R-q}{p-R} \tag{36.3}$$

If we compare Equations 36.2 and 36.3, we see that

$$\frac{R-q}{p-R} = \frac{q}{p}$$

Simple algebra reduces this to

Mirror equation in terms of radius of curvature

$$\frac{1}{p} + \frac{1}{q} = \frac{2}{R} \tag{36.4}$$

This expression is called the **mirror equation.**

If the object is very far from the mirror—that is, if p is so much greater than R that p can be said to approach infinity—then $1/p \approx 0$, and we see from Equation 36.4 that $q \approx R/2$. That is, when the object is very far from the mirror, the image point is halfway between the center of curvature and the center point on the mirror, as shown in Figure 36.12a. The incoming rays from the object are essentially parallel

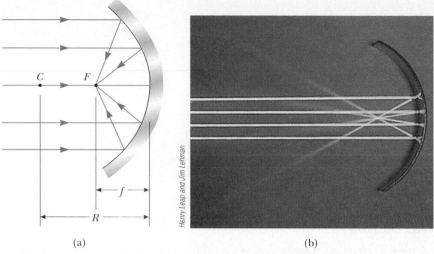

(a)   (b)

Figure 36.12 (a) Light rays from a distant object ($p \rightarrow \infty$) reflect from a concave mirror through the focal point F. In this case, the image distance $q \approx R/2 = f$, where f is the focal length of the mirror. (b) Reflection of parallel rays from a concave mirror.

in this figure because the source is assumed to be very far from the mirror. We call the image point in this special case the **focal point** F and the image distance the **focal length** f, where

$$f = \frac{R}{2} \qquad (36.5)$$

In Figure 36.8, the colored beams are traveling parallel to the principal axis and the mirror reflects all three beams to the focal point. Notice that the point at which the three beams intersect and the colors add is white.

Focal length is a parameter particular to a given mirror and therefore can be used to compare one mirror with another. The mirror equation can be expressed in terms of the focal length:

$$\frac{1}{p} + \frac{1}{q} = \frac{1}{f} \qquad (36.6)$$

Notice that the focal length of a mirror depends only on the curvature of the mirror and not on the material from which the mirror is made. This is because the formation of the image results from rays reflected from the surface of the material. The situation is different for lenses; in that case the light actually passes through the material and the focal length depends on the type of material from which the lens is made.

> ▲ **PITFALL PREVENTION**
>
> **36.2 The *Focal* Point Is Not the *Focus* Point**
>
> The focal point *is usually not* the point at which the light rays focus to form an image. The focal point is determined solely by the curvature of the mirror—it does not depend on the location of the object at all. In general, an image forms at a point different from the focal point of a mirror (or a lens). The *only* exception is when the object is located infinitely far away from the mirror.

Focal length

Mirror equation in terms of focal length

A satellite-dish antenna is a concave reflector for television signals from a satellite in orbit around the Earth. The signals are carried by microwaves that, because the satellite is so far away, are parallel when they arrive at the dish. These waves reflect from the dish and are focused on the receiver at the focal point of the dish.

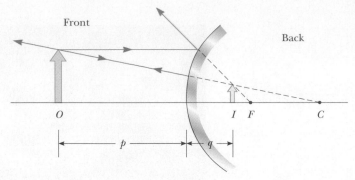

Figure 36.13 Formation of an image by a spherical convex mirror. The image formed by the real object is virtual and upright.

Convex Mirrors

Figure 36.13 shows the formation of an image by a **convex mirror**—that is, one silvered so that light is reflected from the outer, convex surface. This is sometimes called a **diverging mirror** because the rays from any point on an object diverge after reflection as though they were coming from some point behind the mirror. The image in Figure 36.13 is virtual because the reflected rays only appear to originate at the image point, as indicated by the dashed lines. Furthermore, the image is always upright and smaller than the object. This type of mirror is often used in stores to foil shoplifters. A single mirror can be used to survey a large field of view because it forms a smaller image of the interior of the store.

We do not derive any equations for convex spherical mirrors because we can use Equations 36.2, 36.4, and 36.6 for either concave or convex mirrors if we adhere to the following procedure. Let us refer to the region in which light rays move toward the mirror as the *front side* of the mirror, and the other side as the *back side*. For example, in Figures 36.11 and 36.13, the side to the left of the mirrors is the front side, and the side to the right of the mirrors is the back side. Figure 36.14 states the sign conventions for object and image distances, and Table 36.1 summarizes the sign conventions for all quantities.

Ray Diagrams for Mirrors

The positions and sizes of images formed by mirrors can be conveniently determined with *ray diagrams*. These graphical constructions reveal the nature of the image and can be used to check results calculated from the mirror and magnification equations. To draw a ray diagram, we need to know the position of the object and the locations of the mirror's focal point and center of curvature. We then draw three principal rays to locate the image, as shown by the examples in Figure 36.15.

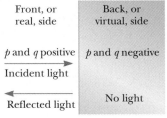

Figure 36.14 Signs of p and q for convex and concave mirrors.

PITFALL PREVENTION

36.3 Watch Your Signs

Success in working mirror problems (as well as problems involving refracting surfaces and thin lenses) is largely determined by proper sign choices when substituting into the equations. The best way to become adept at this is to work a multitude of problems on your own. Watching your instructor or reading the example problems is no substitute for practice.

Table 36.1

Sign Conventions for Mirrors		
Quantity	**Positive When**	**Negative When**
Object location (p)	Object is in front of mirror (real object)	Object is in back of mirror (virtual object)
Image location (q)	Image is in front of mirror (real image)	Image is in back of mirror (virtual image)
Image height (h')	Image is upright	Image is inverted
Focal length (f) and radius (R)	Mirror is concave	Mirror is convex
Magnification (M)	Image is upright	Image is inverted

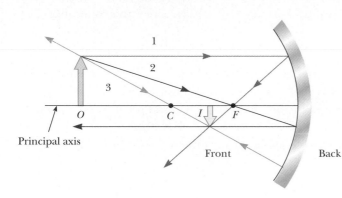

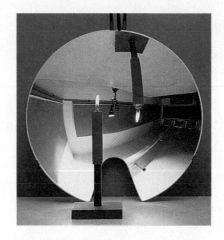

(a)

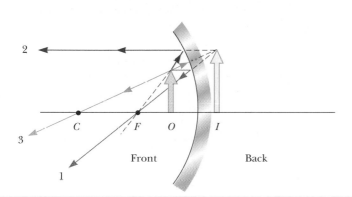

(b)

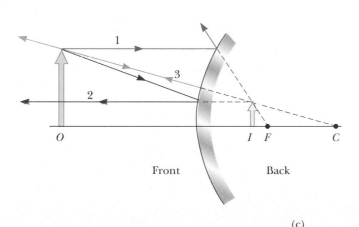

Photos courtesy David Rogers

(c)

Active Figure 36.15 Ray diagrams for spherical mirrors, along with corresponding photographs of the images of candles. (a) When the object is located so that the center of curvature lies between the object and a concave mirror surface, the image is real, inverted, and reduced in size. (b) When the object is located between the focal point and a concave mirror surface, the image is virtual, upright, and enlarged. (c) When the object is in front of a convex mirror, the image is virtual, upright, and reduced in size.

At the Active Figures link at http://www.pse6.com, you can move the objects and change the focal length of the mirrors to see the effect on the images.

36.4 We Are *Choosing* a Small Number of Rays

A *huge* number of light rays leave each point on an object (and pass through each point on an image). In a principal-ray diagram, which displays the characteristics of the image, we choose only a few rays that follow simply stated rules. Locating the image by calculation complements the diagram.

These rays all start from the same object point and are drawn as follows. We may choose any point on the object; here, we choose the top of the object for simplicity. For concave mirrors (see Figs. 36.15a and 36.15b), we draw the following three principal rays:

- Ray 1 is drawn from the top of the object parallel to the principal axis and is reflected through the focal point F.
- Ray 2 is drawn from the top of the object through the focal point and is reflected parallel to the principal axis.
- Ray 3 is drawn from the top of the object through the center of curvature C and is reflected back on itself.

The intersection of any two of these rays locates the image. The third ray serves as a check of the construction. The image point obtained in this fashion must always agree with the value of q calculated from the mirror equation. With concave mirrors, note what happens as the object is moved closer to the mirror. The real, inverted image in Figure 36.15a moves to the left as the object approaches the focal point. When the object is at the focal point, the image is infinitely far to the left. However, when the object lies between the focal point and the mirror surface, as shown in Figure 36.15b, the image is virtual, upright, and enlarged. This latter situation applies when you use a shaving mirror or a makeup mirror, both of which are concave. Your face is closer to the mirror than the focal point, and you see an upright, enlarged image of your face.

For convex mirrors (see Fig. 36.15c), we draw the following three principal rays:

- Ray 1 is drawn from the top of the object parallel to the principal axis and is reflected *away from* the focal point F.
- Ray 2 is drawn from the top of the object toward the focal point on the back side of the mirror and is reflected parallel to the principal axis.
- Ray 3 is drawn from the top of the object toward the center of curvature C on the back side of the mirror and is reflected back on itself.

In a convex mirror, the image of an object is always virtual, upright, and reduced in size as shown in Figure 36.15c. In this case, as the object distance decreases, the virtual image increases in size and moves away from the focal point toward the mirror as the object approaches the mirror. You should construct other diagrams to verify how image position varies with object position.

Figure 36.16 (Quick Quiz 36.4) What type of mirror is this?

Quick Quiz 36.3 You wish to reflect sunlight from a mirror onto some paper under a pile of wood in order to start a fire. Which would be the best choice for the type of mirror? (a) flat (b) concave (c) convex.

Quick Quiz 36.4 Consider the image in the mirror in Figure 36.16. Based on the appearance of this image, you would conclude that (a) the mirror is concave and the image is real. (b) the mirror is concave and the image is virtual. (c) the mirror is convex and the image is real. (d) the mirror is convex and the image is virtual.

Example 36.4 **The Image formed by a Concave Mirror**   Interactive

Assume that a certain spherical mirror has a focal length of +10.0 cm. Locate and describe the image for object distances of

(A) 25.0 cm,

(B) 10.0 cm, and

(C) 5.00 cm.

Solution Because the focal length is positive, we know that this is a concave mirror (see Table 36.1).

(A) This situation is analogous to that in Figure 36.15a; hence, we expect the image to be real. We find the image distance by using Equation 36.6:

$$\frac{1}{p} + \frac{1}{q} = \frac{1}{f}$$

$$\frac{1}{25.0 \text{ cm}} + \frac{1}{q} = \frac{1}{10.0 \text{ cm}}$$

$$q = \boxed{16.7 \text{ cm}}$$

The magnification of the image is given by Equation 36.2:

$$M = -\frac{q}{p} = -\frac{16.7 \text{ cm}}{25.0 \text{ cm}} = -0.668$$

The fact that the absolute value of M is less than unity tells us that the image is smaller than the object, and the negative sign for M tells us that the image is inverted. Because q is positive, the image is located on the front side of the mirror and is real.

(B) When the object distance is 10.0 cm, the object is located at the focal point. Now we find that

$$\frac{1}{10.0 \text{ cm}} + \frac{1}{q} - \frac{1}{10.0 \text{ cm}}$$

$$q = \boxed{\infty}$$

which means that rays originating from an object positioned at the focal point of a mirror are reflected so that the image is formed at an infinite distance from the mirror; that is, the rays travel parallel to one another after reflection. This is the situation in a flashlight, where the bulb filament is placed at the focal point of a reflector, producing a parallel beam of light.

(C) When the object is at $p - 5.00$ cm, it lies halfway between the focal point and the mirror surface, as shown in Figure 36.15b. Thus, we expect a magnified, virtual, upright

image. In this case, the mirror equation gives

$$\frac{1}{5.00 \text{ cm}} + \frac{1}{q} = \frac{1}{10.0 \text{ cm}}$$

$$q = \boxed{-10.0 \text{ cm}}$$

The image is virtual because it is located behind the mirror, as expected. The magnification of the image is

$$M = -\frac{q}{p} = -\left(\frac{-10.0 \text{ cm}}{5.00 \text{ cm}}\right) = +2.00$$

The image is twice as large as the object, and the positive sign for M indicates that the image is upright (see Fig. 36.15b).

What If? Suppose you set up the candle and mirror apparatus illustrated in Figure 36.15a and described in part (A) of the example. While adjusting the apparatus, you accidentally strike the candle with your elbow so that it begins to slide toward the mirror at velocity v_p. How fast does the image of the candle move?

Answer We solve the mirror equation, Equation 36.6, for q:

$$q = \frac{fp}{p - f}$$

Differentiating this equation with respect to time gives us the velocity of the image $v_q = dq/dt$:

$$v_q = \frac{dq}{dt} - \frac{d}{dt}\left(\frac{fp}{p - f}\right) - -\frac{f^2}{(p - f)^2}\frac{dp}{dt} = -\frac{f^2 v_p}{(p - f)^2}$$

For the object position of 25.0 cm in part (A), the velocity of the image is

$$v_q = -\frac{f^2 v_p}{(p - f)^2} = -\frac{(10.0 \text{ cm})^2 v_p}{(25.0 \text{ cm} - 10.0 \text{ cm})^2} = -0.444 \, v_p$$

Thus, the speed of the image is less than that of the object in this case.

We can see two interesting behaviors of this function for v_q. First, note that the velocity is negative regardless of the value of p or f. Thus, if the object moves toward the mirror, the image moves toward the left in Figure 36.15 without regard for the side of the focal point at which the object is located or whether the mirror is concave or convex. Second, in the limit of $p \rightarrow 0$, the velocity v_q approaches $-v_p$. As the object moves very close to the mirror, the mirror looks like a plane mirror, the image is as far behind the mirror as the object is in front, and both the object and the image move with the same speed.

 Investigate the image formed for various object positions and mirror focal lengths at the Interactive Worked Example link at **http://www.pse6.com.**

Example 36.5 **The Image from a Convex Mirror**   Interactive

An anti-shoplifting mirror, as shown in Figure 36.17, shows an image of a woman who is located 3.0 m from the mirror. The focal length of the mirror is −0.25 m. Find

(A) the position of her image and

(B) the magnification of the image.

Solution (A) This situation is depicted in Figure 36.15c. We should expect to find an upright, reduced, virtual image. To find the image position, we use Equation 36.6:

$$\frac{1}{p} + \frac{1}{q} = \frac{1}{f} = \frac{1}{-0.25 \text{ m}}$$

$$\frac{1}{q} = \frac{1}{-0.25 \text{ m}} - \frac{1}{3.0 \text{ m}}$$

$$q = \boxed{-0.23 \text{ m}}$$

The negative value of q indicates that her image is virtual, or behind the mirror, as shown in Figure 36.15c.

(B) The magnification of the image is

$$M = -\frac{q}{p} = -\left(\frac{-0.23 \text{ m}}{3.0 \text{ m}}\right) = \boxed{+0.077}$$

The image is much smaller than the woman, and it is upright because M is positive.

Figure 36.17 (Example 36.5) Convex mirrors, often used for security in department stores, provide wide-angle viewing.

 Investigate the image formed for various object positions and mirror focal lengths at the Interactive Worked Example link at http://www.pse6.com.

36.3 Images Formed by Refraction

In this section we describe how images are formed when light rays are refracted at the boundary between two transparent materials. Consider two transparent media having indices of refraction n_1 and n_2, where the boundary between the two media is a spherical surface of radius R (Fig. 36.18). We assume that the object at O is in the medium for which the index of refraction is n_1. Let us consider the paraxial rays leaving O. As we shall see, all such rays are refracted at the spherical surface and focus at a single point I, the image point.

Figure 36.19 shows a single ray leaving point O and refracting to point I. Snell's law of refraction applied to this ray gives

$$n_1 \sin \theta_1 = n_2 \sin \theta_2$$

Because θ_1 and θ_2 are assumed to be small, we can use the small-angle approximation $\sin \theta \approx \theta$ (with angles in radians) and say that

$$n_1 \theta_1 = n_2 \theta_2$$

Now we use the fact that an exterior angle of any triangle equals the sum of the two opposite interior angles. Applying this rule to triangles OPC and PIC in Figure 36.19 gives

$$\theta_1 = \alpha + \beta$$

$$\beta = \theta_2 + \gamma$$

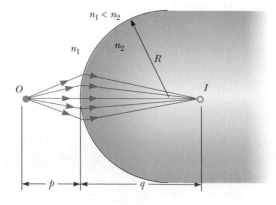

Figure 36.18 An image formed by refraction at a spherical surface. Rays making small angles with the principal axis diverge from a point object at O and are refracted through the image point I.

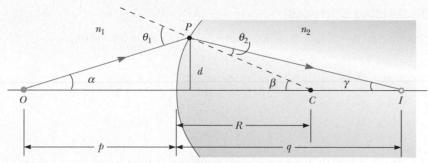

Figure 36.19 Geometry used to derive Equation 36.8, assuming that $n_1 < n_2$.

If we combine all three expressions and eliminate θ_1 and θ_2, we find that

$$n_1\alpha + n_2\gamma = (n_2 - n_1)\beta \tag{36.7}$$

From Figure 36.19, we see three right triangles that have a common vertical leg of length d. For paraxial rays (unlike the relatively large-angle ray shown in Fig. 36.19), the horizontal legs of these triangles are approximately p for the triangle containing angle α, R for the triangle containing angle β, and q for the triangle containing angle γ. In the small-angle approximation, $\tan\theta \approx \theta$, so we can write the approximate relationships from these triangles as follows.

$$\tan\alpha \approx \alpha \approx \frac{d}{p} \qquad \tan\beta \approx \beta \approx \frac{d}{R} \qquad \tan\gamma \approx \gamma \approx \frac{d}{q}$$

We substitute these expressions into Equation 36.7 and divide through by d to give

$$\frac{n_1}{p} + \frac{n_2}{q} = \frac{n_2 - n_1}{R} \tag{36.8}$$

Relation between object and Image distance for a refracting surface

For a fixed object distance p, the image distance q is independent of the angle that the ray makes with the axis. This result tells us that all paraxial rays focus at the same point I.

As with mirrors, we must use a sign convention if we are to apply this equation to a variety of cases. We define the side of the surface in which light rays originate as the front side. The other side is called the back side. Real images are formed by refraction in back of the surface, in contrast with mirrors, where real images are formed in front of the reflecting surface. Because of the difference in location of real images, the refraction sign conventions for q and R are opposite the reflection sign conventions. For example, q and R are both positive in Figure 36.19. The sign conventions for spherical refracting surfaces are summarized in Table 36.2.

We derived Equation 36.8 from an assumption that $n_1 < n_2$ in Figure 36.19. This assumption is not necessary, however. Equation 36.8 is valid regardless of which index of refraction is greater.

Table 36.2

Sign Conventions for Refracting Surfaces		
Quantity	**Positive When**	**Negative When**
Object location (p)	Object is in front of surface (real object)	Object is in back of surface (virtual object)
Image location (q)	Image is in back of surface (real image)	Image is in front of surface (virtual image)
Image height (h')	Image is upright	Image is inverted
Radius (R)	Center of curvature is in back of surface	Center of curvature is in front of surface

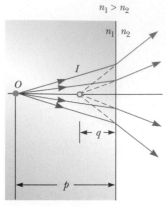

Active Figure 36.20 The image formed by a flat refracting surface is virtual and on the same side of the surface as the object. All rays are assumed to be paraxial.

At the Active Figures link at http://www.pse6.com, *you can move the object to see the effect on the location of the image.*

Flat Refracting Surfaces

If a refracting surface is flat, then R is infinite and Equation 36.8 reduces to

$$\frac{n_1}{p} = -\frac{n_2}{q}$$

$$q = -\frac{n_2}{n_1} p \qquad (36.9)$$

From this expression we see that the sign of q is opposite that of p. Thus, according to Table 36.2, **the image formed by a flat refracting surface is on the same side of the surface as the object.** This is illustrated in Figure 36.20 for the situation in which the object is in the medium of index n_1 and n_1 is greater than n_2. In this case, a virtual image is formed between the object and the surface. If n_1 is less than n_2, the rays in the back side diverge from each other at lesser angles than those in Figure 36.20. As a result, the virtual image is formed to the left of the object.

Quick Quiz 36.5 In Figure 36.18, what happens to the image point I as the object point O is moved to the right from very far away to very close to the refracting surface? (a) It is always to the right of the surface. (b) It is always to the left of the surface. (c) It starts off to the left and at some position of O, I moves to the right of the surface. (d) It starts off to the right and at some position of O, I moves to the left of the surface.

Quick Quiz 36.6 In Figure 36.20, what happens to the image point I as the object point O moves toward the right-hand surface of the material of index of refraction n_1? (a) It always remains between O and the surface, arriving at the surface just as O does. (b) It moves toward the surface more slowly than O so that eventually O passes I. (c) It approaches the surface and then moves to the right of the surface.

Conceptual Example 36.6   Let's Go Scuba Diving!

It is well known that objects viewed under water with the naked eye appear blurred and out of focus. However, a scuba diver using a mask has a clear view of underwater objects. Explain how this works, using the facts that the indices of refraction of the cornea, water, and air are 1.376, 1.333, and 1.00029, respectively.

Solution Because the cornea and water have almost identical indices of refraction, very little refraction occurs

when a person under water views objects with the naked eye. In this case, light rays from an object focus behind the retina, resulting in a blurred image. When a mask is used, the air space between the eye and the mask surface provides the normal amount of refraction at the eye–air interface, and the light from the object focuses on the retina.

Example 36.7   Gaze into the Crystal Ball

A set of coins is embedded in a spherical plastic paperweight having a radius of 3.0 cm. The index of refraction of the plastic is $n_1 = 1.50$. One coin is located 2.0 cm from the edge of the sphere (Fig. 36.21). Find the position of the image of the coin.

Solution Because $n_1 > n_2$, where $n_2 = 1.00$ is the index of refraction for air, the rays originating from the coin are refracted away from the normal at the surface and diverge outward. Hence, the image is formed inside the paperweight and is *virtual*. Applying Equation 36.8 and noting

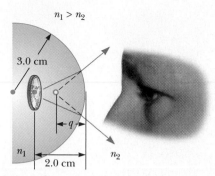

Figure 36.21 (Example 36.7) Light rays from a coin embedded in a plastic sphere form a virtual image between the surface of the object and the sphere surface. Because the object is inside the sphere, the front of the refracting surface is the *interior* of the sphere.

from Table 36.2 that R is negative, we obtain

$$\frac{n_1}{p} + \frac{n_2}{q} = \frac{n_2 - n_1}{R}$$

$$\frac{1.50}{2.0 \text{ cm}} + \frac{1}{q} = \frac{1.00 - 1.50}{-3.0 \text{ cm}}$$

$$q = \boxed{-1.7 \text{ cm}}$$

The negative sign for q indicates that the image is in front of the surface—in other words, in the same medium as the object, as shown in Figure 36.21. Being in the same medium as the object, the image must be virtual. (See Table 36.2.) The coin appears to be closer to the paperweight surface than it actually is.

Example 36.8   The One That Got Away

A small fish is swimming at a depth d below the surface of a pond (Fig. 36.22). What is the apparent depth of the fish, as viewed from directly overhead?

Solution Because the refracting surface is flat, R is infinite. Hence, we can use Equation 36.9 to determine the location of the image with $p = d$. Using the indices of refraction given in Figure 36.22, we obtain

$$q = -\frac{n_2}{n_1} p = -\frac{1.00}{1.33} d = \boxed{-0.752d}$$

Because q is negative, the image is virtual, as indicated by the dashed lines in Figure 36.22. The apparent depth is approximately three-fourths the actual depth.

What If? What if you look more carefully at the fish and measure its apparent *height*, from its upper fin to its lower fin? Is the apparent height h' of the fish different from the actual height h?

Answer Because all points on the fish appear to be fractionally closer to the observer, we would predict that the height would be smaller. If we let the distance d in Figure 36.22 be measured to the top fin and the distance to the bottom fin be $d + h$, then the images of the top and bottom of the fish are located at

$$q_{\text{top}} = -0.752d$$
$$q_{\text{bottom}} = -0.752(d + h)$$

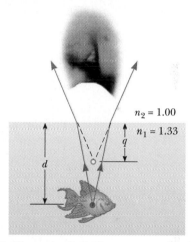

Figure 36.22 (Example 36.8) The apparent depth q of the fish is less than the true depth d. All rays are assumed to be paraxial.

The apparent height h' of the fish is

$$h' = q_{\text{top}} - q_{\text{bottom}} = -0.752d - [-0.752(d + h)]$$
$$= 0.752h$$

and the fish appears to be approximately three-fourths its actual height.

36.4   Thin Lenses

Lenses are commonly used to form images by refraction in optical instruments, such as cameras, telescopes, and microscopes. We can use what we just learned about images formed by refracting surfaces to help us locate the image formed by a lens. We recognize that light passing through a lens experiences refraction at two surfaces. The development we shall follow is based on the notion that **the image**

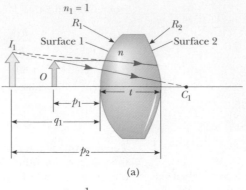

(a)

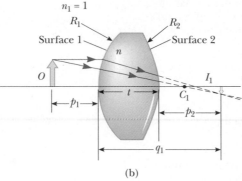

(b)

Figure 36.23 To locate the image formed by a lens, we use the virtual image at I_1 formed by surface 1 as the object for the image formed by surface 2. The point C_1 is the center of curvature of surface 1. (a) The image due to surface 1 is virtual so that I_1 is to the left of the surface. (b) The image due to surface 1 is real so that I_1 is to the right of the surface.

formed by one refracting surface serves as the object for the second surface. We shall analyze a thick lens first and then let the thickness of the lens be approximately zero.

Consider a lens having an index of refraction n and two spherical surfaces with radii of curvature R_1 and R_2, as in Figure 36.23. (Note that R_1 is the radius of curvature of the lens surface that the light from the object reaches first and that R_2 is the radius of curvature of the other surface of the lens.) An object is placed at point O at a distance p_1 in front of surface 1.

Let us begin with the image formed by surface 1. Using Equation 36.8 and assuming that $n_1 = 1$ because the lens is surrounded by air, we find that the image I_1 formed by surface 1 satisfies the equation

$$\frac{1}{p_1} + \frac{n}{q_1} = \frac{n-1}{R_1} \tag{36.10}$$

where q_1 is the position of the image due to surface 1. If the image due to surface 1 is virtual (Fig. 36.23a), q_1 is negative, and it is positive if the image is real (Fig. 36.23b).

Now we apply Equation 36.8 to surface 2, taking $n_1 = n$ and $n_2 = 1$. (We make this switch in index because the light rays approaching surface 2 are *in the material of the lens,* and this material has index n.) Taking p_2 as the object distance for surface 2 and q_2 as the image distance gives

$$\frac{n}{p_2} + \frac{1}{q_2} = \frac{1-n}{R_2} \tag{36.11}$$

We now introduce mathematically the fact that the image formed by the first surface acts as the object for the second surface. We do this by noting from Figure 36.23 that p_2, measured from surface 2, is related to q_1 as follows:

Virtual image from surface 1 (Fig. 36.23a): $p_2 = -q_1 + t$ (q_1 is negative)

Real image from surface 1 (Fig. 36.23b): $p_2 = -q_1 + t$ (q_1 is positive)

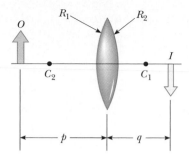

Figure 36.24 Simplified geometry for a thin lens.

where t is the thickness of the lens. For a *thin* lens (one whose thickness is small compared to the radii of curvature), we can neglect t. In this approximation, we see that $p_2 = -q_1$ for either type of image from surface 1. (If the image from surface 1 is real, the image acts as a virtual object, so p_2 is negative.) Hence, Equation 36.11 becomes

$$-\frac{n}{q_1} + \frac{1}{q_2} = \frac{1-n}{R_2}$$

(36.12)

Adding Equations 36.10 and 36.12, we find that

$$\frac{1}{p_1} + \frac{1}{q_2} = (n-1)\left(\frac{1}{R_1} - \frac{1}{R_2}\right)$$

(36.13)

For a thin lens, we can omit the subscripts on p_1 and q_2 in Equation 36.13 and call the object distance p and the image distance q, as in Figure 36.24. Hence, we can write Equation 36.13 in the form

$$\frac{1}{p} + \frac{1}{q} = (n-1)\left(\frac{1}{R_1} - \frac{1}{R_2}\right)$$

(36.14)

This expression relates the image distance q of the image formed by a thin lens to the object distance p and to the lens properties (index of refraction and radii of curvature). It is valid only for paraxial rays and only when the lens thickness is much less than R_1 and R_2.

The **focal length** f of a thin lens is the image distance that corresponds to an infinite object distance, just as with mirrors. Letting p approach ∞ and q approach f in Equation 36.14, we see that the inverse of the focal length for a thin lens is

$$\frac{1}{f} = (n-1)\left(\frac{1}{R_1} - \frac{1}{R_2}\right)$$

(36.15)   **Lens makers' equation**

This relationship is called the **lens makers' equation** because it can be used to determine the values of R_1 and R_2 that are needed for a given index of refraction and a desired focal length f. Conversely, if the index of refraction and the radii of curvature of a lens are given, this equation enables a calculation of the focal length. If the lens is immersed in something other than air, this same equation can be used, with n interpreted as the *ratio* of the index of refraction of the lens material to that of the surrounding fluid.

Using Equation 36.15, we can write Equation 36.14 in a form identical to Equation 36.6 for mirrors:

$$\frac{1}{p} + \frac{1}{q} = \frac{1}{f}$$

(36.16)   **Thin lens equation**

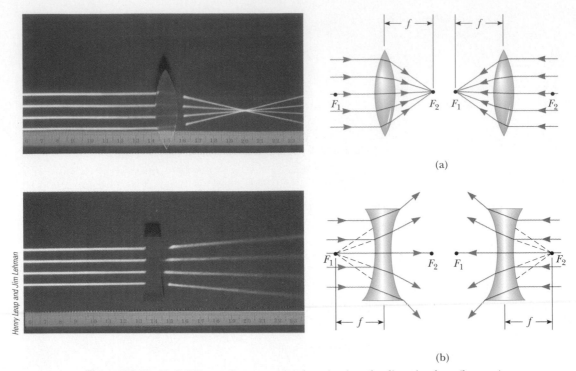

(a)

(b)

Figure 36.25 (*Left*) Effects of a converging lens (top) and a diverging lens (bottom) on parallel rays. (*Right*) Parallel light rays pass through (a) a converging lens and (b) a diverging lens. The focal length is the same for light rays passing through a given lens in either direction. Both focal points F_1 and F_2 are the same distance from the lens.

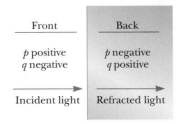

Front	Back
p positive q negative	p negative q positive
Incident light →	Refracted light →

Figure 36.26 A diagram for obtaining the signs of p and q for a thin lens. (This diagram also applies to a refracting surface.)

▲ **PITFALL PREVENTION**

36.5   A Lens Has Two Focal Points but Only One Focal Length

A lens has a focal point on each side, front and back. However, there is only one focal length—each of the two focal points is located the same distance from the lens (Fig, 36.25). This can be seen mathematically by interchanging R_1 and R_2 in Equation 36.15 (and changing the signs of the radii because back and front have been interchanged). As a result, the lens forms an image of an object at the same point if it is turned around. In practice this might not happen, because real lenses are not infinitesimally thin.

This equation, called the **thin lens equation,** can be used to relate the image distance and object distance for a thin lens.

Because light can travel in either direction through a lens, each lens has two focal points, one for light rays passing through in one direction and one for rays passing through in the other direction. This is illustrated in Figure 36.25 for a biconvex lens (two convex surfaces, resulting in a converging lens) and a biconcave lens (two concave surfaces, resulting in a diverging lens).

Figure 36.26 is useful for obtaining the signs of p and q, and Table 36.3 gives the sign conventions for thin lenses. Note that these sign conventions are the same as those for refracting surfaces (see Table 36.2). Applying these rules to a biconvex lens, we see that when $p > f$, the quantities p, q, and R_1 are positive, and R_2 is negative. Therefore, p, q, and f are all positive when a converging lens forms a real image of an object. For a biconcave lens, p and R_2 are positive and q and R_1 are negative, with the result that f is negative.

Table 36.3

Sign Conventions for Thin Lenses		
Quantity	**Positive When**	**Negative When**
Object location (p)	Object is in front of lens (real object)	Object is in back of lens (virtual object)
Image location (q)	Image is in back of lens (real image)	Image is in front of lens (virtual image)
Image height (h')	Image is upright	Image is inverted
R_1 and R_2	Center of curvature is in back of lens	Center of curvature is in front of lens
Focal length (f)	Converging lens	Diverging lens

Henry Leap and Jim Lehman

Various lens shapes are shown in Figure 36.27. Note that a converging lens is thicker at the center than at the edge, whereas a diverging lens is thinner at the center than at the edge.

Magnification of Images

Consider a thin lens through which light rays from an object pass. As with mirrors (Eq. 36.2), we could analyze a geometric construction to show that the lateral magnification of the image is

$$M = \frac{h'}{h} = -\frac{q}{p}$$

From this expression, it follows that when M is positive, the image is upright and on the same side of the lens as the object. When M is negative, the image is inverted and on the side of the lens opposite the object.

Ray Diagrams for Thin Lenses

Ray diagrams are convenient for locating the images formed by thin lenses or systems of lenses. They also help clarify our sign conventions. Figure 36.28 shows such diagrams for three single-lens situations.

To locate the image of a *converging* lens (Fig. 36.28a and b), the following three rays are drawn from the top of the object:

- Ray 1 is drawn parallel to the principal axis. After being refracted by the lens, this ray passes through the focal point on the back side of the lens.
- Ray 2 is drawn through the center of the lens and continues in a straight line.
- Ray 3 is drawn through the focal point on the front side of the lens (or as if coming from the focal point if $p < f$) and emerges from the lens parallel to the principal axis.

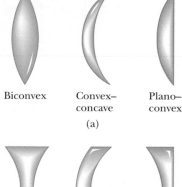
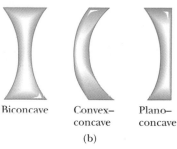

Biconvex   Convex–   Plano–
           concave   convex

(a)

Biconcave   Convex–   Plano–
            concave   concave

(b)

Figure 36.27 Various lens shapes. (a) Converging lenses have a positive focal length and are thickest at the middle. (b) Diverging lenses have a negative focal length and are thickest at the edges.

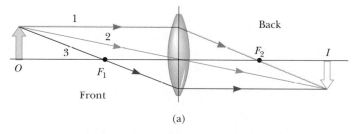

(a)

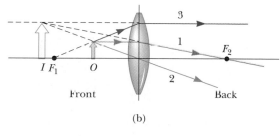

(b)

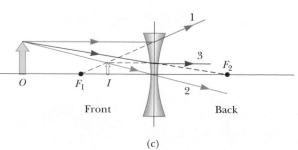

(c)

Active Figure 36.28 Ray diagrams for locating the image formed by a thin lens. (a) When the object is in front of and outside the focal point of a converging lens, the image is real, inverted, and on the back side of the lens. (b) When the object is between the focal point and a converging lens, the image is virtual, upright, larger than the object, and on the front side of the lens. (c) When an object is anywhere in front of a diverging lens, the image is virtual, upright, smaller than the object, and on the front side of the lens.

At the Active Figures link at http://www.pse6.com, you can move the objects and change the focal length of the lenses to see the effect on the images.

To locate the image of a *diverging* lens (Fig. 36.28c), the following three rays are drawn from the top of the object:

- Ray 1 is drawn parallel to the principal axis. After being refracted by the lens, this ray emerges directed away from the focal point on the front side of the lens.
- Ray 2 is drawn through the center of the lens and continues in a straight line.
- Ray 3 is drawn in the direction toward the focal point on the back side of the lens and emerges from the lens parallel to the principal axis.

For the converging lens in Figure 36.28a, where the object is to the left of the focal point ($p > f$), the image is real and inverted. When the object is between the focal point and the lens ($p < f$), as in Figure 36.28b, the image is virtual and upright. For a diverging lens (see Fig. 36.28c), the image is always virtual and upright, regardless of where the object is placed. These geometric constructions are reasonably accurate only if the distance between the rays and the principal axis is much less than the radii of the lens surfaces.

Note that refraction occurs only at the surfaces of the lens. A certain lens design takes advantage of this fact to produce the *Fresnel lens*, a powerful lens without great thickness. Because only the surface curvature is important in the refracting qualities of the lens, material in the middle of a Fresnel lens is removed, as shown in the cross sections of lenses in Figure 36.29. Because the edges of the curved segments cause some distortion, Fresnel lenses are usually used only in situations in which image quality is less important than reduction of weight. A classroom overhead projector often uses a Fresnel lens; the circular edges between segments of the lens can be seen by looking closely at the light projected onto a screen.

Figure 36.29 The Fresnel lens on the left has the same focal length as the thick lens on the right but is made of much less glass.

Quick Quiz 36.7 What is the focal length of a pane of window glass? (a) zero (b) infinity (c) the thickness of the glass (d) impossible to determine

Quick Quiz 36.8 Diving masks often have a lens built into the glass for divers who do not have perfect vision. This allows the individual to dive without the necessity for glasses, because the lenses in the faceplate perform the necessary refraction to provide clear vision. The proper design allows the diver to see clearly with the mask on *both* under water and in the open air. Normal eyeglasses have lenses that are curved on both the front and back surfaces. The lenses in a diving mask should be curved (a) only on the front surface (b) only on the back surface (c) on both the front and back surfaces.

Example 36.9   Images Formed by a Converging Lens   `Interactive`

A converging lens of focal length 10.0 cm forms images of objects placed

(A) 30.0 cm,

(B) 10.0 cm, and

(C) 5.00 cm from the lens.

In each case, construct a ray diagram, find the image distance and describe the image.

Solution

(A) First we construct a ray diagram as shown in Figure 36.30a. The diagram shows that we should expect a real, inverted, smaller image to be formed on the back side of the lens. The thin lens equation, Equation 36.16, can be used to find the image distance:

$$\frac{1}{p} + \frac{1}{q} = \frac{1}{f}$$

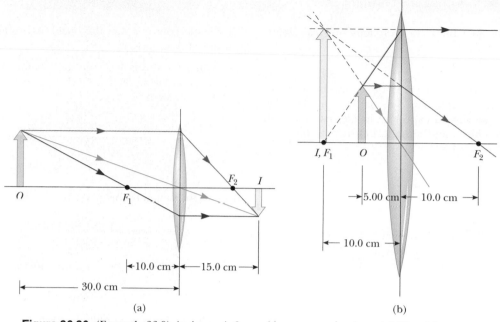

Figure 36.30 (Example 36.9) An image is formed by a converging lens. (a) The object is farther from the lens than the focal point. (b) The object is closer to the lens than the focal point.

$$\frac{1}{30.0 \text{ cm}} + \frac{1}{q} = \frac{1}{10.0 \text{ cm}}$$

$$q = \boxed{+15.0 \text{ cm}}$$

The positive sign for the image distance tells us that the image is indeed real and on the back side of the lens. The magnification of the image is

$$M = -\frac{q}{p} = -\frac{15.0 \text{ cm}}{30.0 \text{ cm}} = \boxed{-0.500}$$

Thus, the image is reduced in height by one half, and the negative sign for M tells us that the image is inverted.

(B) No calculation is necessary for this case because we know that, when the object is placed at the focal point, the image is formed at infinity. This is readily verified by substituting $p = 10.0$ cm into the thin lens equation.

(C) We now move inside the focal point. The ray diagram in Figure 36.30b shows that in this case the lens acts as a magnifying glass; that is, the image is magnified, upright, on the same side of the lens as the object, and virtual. Because the object distance is 5.00 cm, the thin lens equation gives

$$\frac{1}{5.00 \text{ cm}} + \frac{1}{q} = \frac{1}{10.0 \text{ cm}}$$

$$q = \boxed{-10.0 \text{ cm}}$$

and the magnification of the image is

$$M = -\frac{q}{p} = -\left(\frac{-10.0 \text{ cm}}{5.00 \text{ cm}}\right) = \boxed{+2.00}$$

The negative image distance tells us that the image is virtual and formed on the side of the lens from which the light is incident, the front side. The image is enlarged, and the positive sign for M tells us that the image is upright.

What If? What if the object moves right up to the lens surface, so that $p \to 0$? Where is the image?

Answer In this case, because $p \ll R$, where R is either of the radii of the surfaces of the lens, the curvature of the lens can be ignored and it should appear to have the same effect as a plane piece of material. This would suggest that the image is just on the front side of the lens, at $q = 0$. We can verify this mathematically by rearranging the thin lens equation:

$$\frac{1}{q} = \frac{1}{f} - \frac{1}{p}$$

If we let $p \to 0$, the second term on the right becomes very large compared to the first and we can neglect $1/f$. The equation becomes

$$\frac{1}{q} = -\frac{1}{p}$$

$$q = -p = 0$$

Thus, q is on the front side of the lens (because it has the opposite sign as p), and just at the lens surface.

 Investigate the image formed for various object positions and lens focal lengths at the Interactive Worked Example link at **http://www.pse6.com.**

Example 36.10   The Case of a Diverging Lens   **Interactive**

Repeat Example 36.9 for a *diverging* lens of focal length 10.0 cm.

Solution

(A) We begin by constructing a ray diagram as in Figure 36.31a taking the object distance to be 30.0 cm. The diagram shows that we should expect an image that is virtual, smaller than the object, and upright. Let us now apply the thin lens equation with $p = 30.0$ cm:

$$\frac{1}{p} + \frac{1}{q} = \frac{1}{f}$$

$$\frac{1}{30.0 \text{ cm}} + \frac{1}{q} = \frac{1}{-10.0 \text{ cm}}$$

$$q = \boxed{-7.50 \text{ cm}}$$

The magnification of the image is

$$M = -\frac{q}{p} = -\left(\frac{-7.50 \text{ cm}}{30.0 \text{ cm}}\right) = \boxed{+0.250}$$

This result confirms that the image is virtual, smaller than the object, and upright.

(B) When the object is at the focal point, the ray diagram appears as in Figure 36.31b. In the thin lens equation, using $p = 10.0$ cm, we have

$$\frac{1}{10.0 \text{ cm}} + \frac{1}{q} = \frac{1}{-10.0 \text{ cm}}$$

$$q = \boxed{-5.00 \text{ cm}}$$

The magnification of the image is

$$M = -\frac{q}{p} = -\left(\frac{-5.00 \text{ cm}}{10.0 \text{ cm}}\right) = \boxed{+0.500}$$

Notice the difference between this situation and that for a converging lens. For a diverging lens, an object at the focal point does not produce an image infinitely far away.

(C) When the object is inside the focal point, at $p = 5.00$ cm, the ray diagram in Figure 36.31c shows that we expect a virtual image that is smaller than the object and upright. In

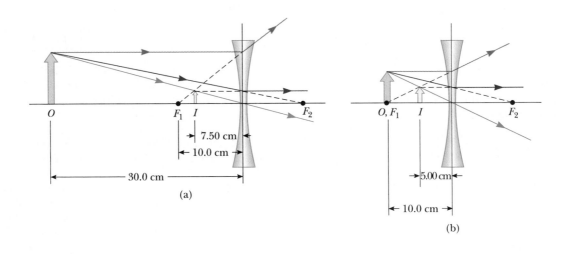

(a)

(b)

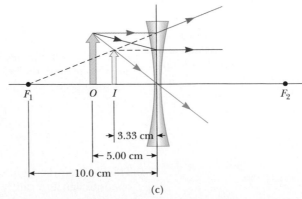

(c)

Figure 36.31 (Example 36.10) An image is formed by a diverging lens. (a) The object is farther from the lens than the focal point. (b) The object is at the focal point. (c) The object is closer to the lens than the focal point.

this case, the thin lens equation gives

$$\frac{1}{5.00 \text{ cm}} + \frac{1}{q} = \frac{1}{-10.0 \text{ cm}}$$

$$q = \boxed{-3.33 \text{ cm}}$$

and the magnification of the image is

$$M = -\left(\frac{-3.33 \text{ cm}}{5.00 \text{ cm}}\right) = \boxed{+0.667}$$

This confirms that the image is virtual, smaller than the object, and upright.

 Investigate the image formed for various object positions and lens focal lengths at the Interactive Worked Example link at **http://www.pse6.com.**

Example 36.11 A Lens Under Water

A converging glass lens ($n = 1.52$) has a focal length of 40.0 cm in air. Find its focal length when it is immersed in water, which has an index of refraction of 1.33.

Solution We can use the lens makers' equation (Eq. 36.15) in both cases, noting that R_1 and R_2 remain the same in air and water:

$$\frac{1}{f_{\text{air}}} = (n - 1)\left(\frac{1}{R_1} - \frac{1}{R_2}\right)$$

$$\frac{1}{f_{\text{water}}} = (n' - 1)\left(\frac{1}{R_1} - \frac{1}{R_2}\right)$$

where n' is the ratio of the index of refraction of glass to that of water: $n' = 1.52/1.33 = 1.14$. Dividing the first

equation by the second gives

$$\frac{f_{\text{water}}}{f_{\text{air}}} = \frac{n - 1}{n' - 1} = \frac{1.52 - 1}{1.14 - 1} = 3.71$$

Because $f_{\text{air}} = 40.0$ cm, we find that

$$f_{\text{water}} = 3.71 f_{\text{air}} = 3.71(40.0 \text{ cm}) = \boxed{148 \text{ cm}}$$

The focal length of any lens is increased by a factor $(n - 1)/(n' - 1)$ when the lens is immersed in a fluid, where n' is the ratio of the index of refraction n of the lens material to that of the fluid.

Combination of Thin Lenses

If two thin lenses are used to form an image, the system can be treated in the following manner. First, the image formed by the first lens is located as if the second lens were not present. Then a ray diagram is drawn for the second lens, with the image formed by the first lens now serving as the object for the second lens. The second image formed is the final image of the system. If the image formed by the first lens lies on the back side of the second lens, then that image is treated as a **virtual object** for the second lens (that is, in the thin lens equation, p is negative). The same procedure can be extended to a system of three or more lenses. Because the magnification due to the second lens is performed on the magnified image due to the first lens, the overall magnification of the image due to the combination of lenses is the product of the individual magnifications.

Let us consider the special case of a system of two lenses of focal lengths f_1 and f_2 in contact with each other. If $p_1 = p$ is the object distance for the combination, application of the thin lens equation (Eq. 36.16) to the first lens gives

$$\frac{1}{p} + \frac{1}{q_1} = \frac{1}{f_1}$$

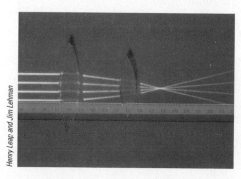

Light from a distant object is brought into focus by two converging lenses.

Henry Leap and Jim Lehman

where q_1 is the image distance for the first lens. Treating this image as the object for the second lens, we see that the object distance for the second lens must be $p_2 = -q_1$. (The distances are the same because the lenses are in contact and assumed to be infinitesimally thin. The object distance is negative because the object is virtual.) Therefore, for the second lens,

$$\frac{1}{p_2} + \frac{1}{q_2} = \frac{1}{f_2}$$

$$-\frac{1}{q_1} + \frac{1}{q} = \frac{1}{f_2}$$

where $q = q_2$ is the final image distance from the second lens, which is the image distance for the combination. Adding the equations for the two lenses eliminates q_1 and gives

$$\frac{1}{p} + \frac{1}{q} = \frac{1}{f_1} + \frac{1}{f_2}$$

If we consider replacing the combination with a single lens that will form an image at the same location, we see that its focal length is related to the individual focal lengths by

Focal length for a combination of two thin lenses in contact

$$\frac{1}{f} = \frac{1}{f_1} + \frac{1}{f_2} \qquad (36.17)$$

Therefore, **two thin lenses in contact with each other are equivalent to a single thin lens having a focal length given by Equation 36.17.**

Example 36.12 Where Is the Final Image? **Interactive**

Two thin converging lenses of focal lengths $f_1 = 10.0$ cm and $f_2 = 20.0$ cm are separated by 20.0 cm, as illustrated in Figure 36.32a. An object is placed 30.0 cm to the left of lens 1. Find the position and the magnification of the final image.

Solution Conceptualize by imagining light rays passing through the first lens and forming a real image (because $p > f$) in the absence of the second lens. Figure 36.32b shows these light rays forming the inverted image I_1. Once the light rays converge to the image point, they do not stop. They continue through the image point and interact with the second lens. The rays leaving the image point behave in the same way as the rays leaving an object. Thus, the image of the first lens serves as the object of the second lens. We categorize this problem as one in which we apply the thin lens equation, but in stepwise fashion to the two lenses.

To analyze the problem, we first draw a ray diagram (Figure 36.32b) showing where the image from the first lens falls and how it acts as the object for the second lens. The location of the image formed by lens 1 is found from the thin lens equation:

$$\frac{1}{p_1} + \frac{1}{q_1} = \frac{1}{f}$$

$$\frac{1}{30.0 \text{ cm}} + \frac{1}{q_1} = \frac{1}{10.0 \text{ cm}}$$

$$q_1 = \boxed{+15.0 \text{ cm}}$$

The magnification of this image is

$$M_1 = -\frac{q_1}{p_1} = -\frac{15.0 \text{ cm}}{30.0 \text{ cm}} = \boxed{-0.500}$$

The image formed by this lens acts as the object for the second lens. Thus, the object distance for the second lens is 20.0 cm − 15.0 cm = 5.00 cm. We again apply the thin lens equation to find the location of the final image:

$$\frac{1}{5.00 \text{ cm}} + \frac{1}{q_2} = \frac{1}{20.0 \text{ cm}}$$

$$q_2 = \boxed{-6.67 \text{ cm}}$$

The magnification of the second image is

$$M_2 = -\frac{q_2}{p_2} = -\frac{(-6.67 \text{ cm})}{5.00 \text{ cm}} = \boxed{+1.33}$$

Thus, the overall magnification of the system is

$$M = M_1 M_2 = (-0.500)(1.33) = \boxed{-0.667}$$

To finalize the problem, note that the negative sign on the overall magnification indicates that the final image is inverted with respect to the initial object. The fact that the absolute value of the magnification is less than one tells us that the final image is smaller than the object. The fact that q_2 is negative tells us that the final image is on the front, or left, side of lens 2. All of these conclusions are consistent with the ray diagram in Figure 36.32b.

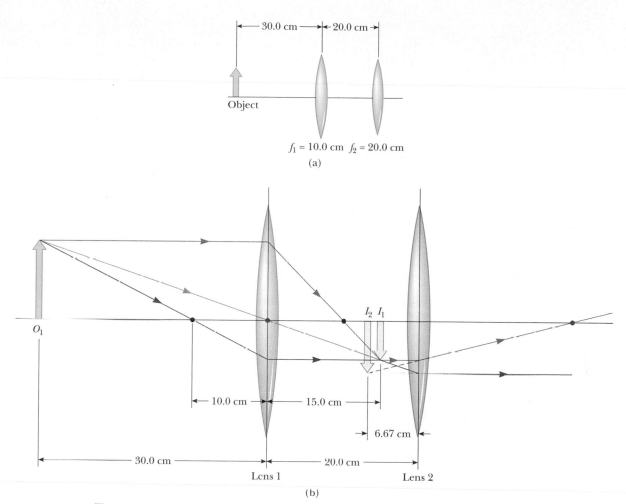

Figure 36.32 (Example 36.12) (a) A combination of two converging lenses. (b) The ray diagram showing the location of the final image due to the combination of lenses. The black dots are the focal points of lens 1 while the red dots are the focal points of lens 2.

What If? Suppose we want to create an upright image with this system of two lenses. How must the second lens be moved in order to achieve this?

Answer Because the object is farther from the first lens than the focal length of that lens, we know that the first image is inverted. Consequently, we need the second lens to invert the image once again so that the final image is upright. An inverted image is only formed by a converging lens if the object is outside the focal point. Thus, the image due to the first lens must be to the left of the focal point of the second lens in Figure 36.32b. To make this happen, we must move the second lens at least as far away from the first lens as the sum $q_1 + f_2 = 15.0$ cm $+ 20.0$ cm $= 35.0$ cm.

 Investigate the image formed by a combination of two lenses at the Interactive Worked Example link at **http://www.pse6.com.**

Conceptual Example 36.13   Watch Your p's and q's!

Use a spreadsheet or a similar tool to create two graphs of image distance as a function of object distance—one for a lens for which the focal length is 10 cm and one for a lens for which the focal length is -10 cm.

Solution The graphs are shown in Figure 36.33. In each graph, a gap occurs where $p = f$, which we shall discuss. Note the similarity in the shapes—a result of the fact that image and object distances for both lenses are related according to the same equation—the thin lens equation.

The curve in the upper right portion of the $f = +10$ cm graph corresponds to an object located on the *front* side of a lens, which we have drawn as the left side of the lens in our previous diagrams. When the object is at positive infinity, a real image forms at the focal point on the back side (the positive side) of the lens, $q = f$. (The incoming rays are parallel in this case.) As the object moves closer to the lens, the image

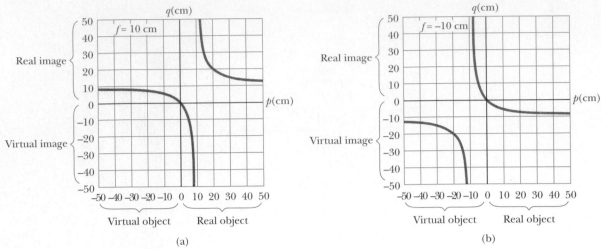

Figure 36.33 (Conceptual Example 36.13) (a) Image position as a function of object position for a lens having a focal length of + 10 cm. (b) Image position as a function of object position for a lens having a focal length of − 10 cm.

moves farther from the lens, corresponding to the upward path of the curve. This continues until the object is located at the focal point on the near side of the lens. At this point, the rays leaving the lens are parallel, making the image infinitely far away. This is described in the graph by the asymptotic approach of the curve to the line $p = f = 10$ cm.

As the object moves inside the focal point, the image becomes virtual and located near $q = -\infty$. We are now following the curve in the lower left portion of Figure 36.33a. As the object moves closer to the lens, the virtual image also moves closer to the lens. As $p \to 0$, the image distance q also approaches 0. Now imagine that we bring the object to the back side of the lens, where $p < 0$. The object is now a virtual object, so it must have been formed by some other lens. For all locations of the virtual object, the image

distance is positive and less than the focal length. The final image is real, and its position approaches the focal point as p becomes more and more negative.

The $f = -10$ cm graph shows that a distant real object forms an image at the focal point on the front side of the lens. As the object approaches the lens, the image remains virtual and moves closer to the lens. But as we continue toward the left end of the p axis, the object becomes virtual. As the position of this virtual object approaches the focal point, the image recedes toward infinity. As we pass the focal point, the image shifts from a location at positive infinity to one at negative infinity. Finally, as the virtual object continues moving away from the lens, the image is virtual, starts moving in from negative infinity, and approaches the focal point.

36.5 Lens Aberrations

Our analysis of mirrors and lenses assumes that rays make small angles with the principal axis and that the lenses are thin. In this simple model, all rays leaving a point source focus at a single point, producing a sharp image. Clearly, this is not always true. When the approximations used in this analysis do not hold, imperfect images are formed.

A precise analysis of image formation requires tracing each ray, using Snell's law at each refracting surface and the law of reflection at each reflecting surface. This procedure shows that the rays from a point object do not focus at a single point, with the result that the image is blurred. The departures of actual images from the ideal predicted by our simplified model are called **aberrations.**

Spherical Aberrations

Spherical aberrations occur because the focal points of rays far from the principal axis of a spherical lens (or mirror) are different from the focal points of rays of the same wavelength passing near the axis. Figure 36.34 illustrates spherical aberration for parallel rays passing through a converging lens. Rays passing through points near the center of

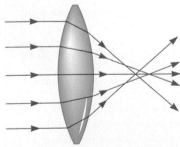

Figure 36.34 Spherical aberration caused by a converging lens. Does a diverging lens cause spherical aberration?

the lens are imaged farther from the lens than rays passing through points near the edges. Figure 36.10 earlier in the chapter showed a similar situation for a spherical mirror.

Many cameras have an adjustable aperture to control light intensity and reduce spherical aberration. (An aperture is an opening that controls the amount of light passing through the lens.) Sharper images are produced as the aperture size is reduced because with a small aperture only the central portion of the lens is exposed to the light; as a result, a greater percentage of the rays are paraxial. At the same time, however, less light passes through the lens. To compensate for this lower light intensity, a longer exposure time is used.

In the case of mirrors, spherical aberration can be minimized through the use of a parabolic reflecting surface rather than a spherical surface. Parabolic surfaces are not used often, however, because those with high-quality optics are very expensive to make. Parallel light rays incident on a parabolic surface focus at a common point, regardless of their distance from the principal axis. Parabolic reflecting surfaces are used in many astronomical telescopes to enhance image quality.

Chromatic Aberrations

The fact that different wavelengths of light refracted by a lens focus at different points gives rise to chromatic aberrations. In Chapter 35 we described how the index of refraction of a material varies with wavelength. For instance, when white light passes through a lens, violet rays are refracted more than red rays (Fig. 36.35). From this we see that the focal length of a lens is greater for red light than for violet light. Other wavelengths (not shown in Fig. 36.35) have focal points intermediate between those of red and violet.

Chromatic aberration for a diverging lens also results in a shorter focal length for violet light than for red light, but on the front side of the lens. Chromatic aberration can be greatly reduced by combining a converging lens made of one type of glass and a diverging lens made of another type of glass.

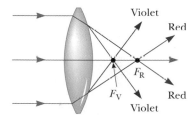

Figure 36.35 Chromatic aberration caused by a converging lens. Rays of different wavelengths focus at different points.

> **Quick Quiz 36.9** A curved mirrored surface can have (a) spherical aberration but not chromatic aberration (b) chromatic aberration but not spherical aberration (c) both spherical aberration and chromatic aberration

36.6   The Camera

The photographic **camera** is a simple optical instrument whose essential features are shown in Figure 36.36. It consists of a light-tight chamber, a converging lens that produces a real image, and a film behind the lens to receive the image. One focuses the camera by varying the distance between lens and film. This is accomplished with an adjustable bellows in antique cameras and with some other mechanical arrangement in contemporary models. For proper focusing—which is necessary for the formation of sharp images—the lens-to-film distance depends on the object distance as well as on the focal length of the lens.

The shutter, positioned behind the lens, is a mechanical device that is opened for selected time intervals, called *exposure times*. One can photograph moving objects by using short exposure times or photograph dark scenes (with low light levels) by using long exposure times. If this adjustment were not available, it would be impossible to take stop-action photographs. For example, a rapidly moving vehicle could move enough in the time interval during which the shutter is open to produce a blurred image. Another major cause of blurred images is the movement of the camera while the shutter is open. To prevent such movement, either short exposure times or a tripod should be used, even for stationary objects. Typical shutter speeds (that is, exposure times) are $(1/30)$s, $(1/60)$s, $(1/125)$s, and $(1/250)$s. For handheld cameras,

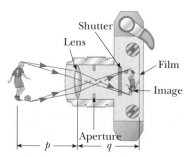

Figure 36.36 Cross-sectional view of a simple camera. Note that in reality, $p \gg q$.

the use of slower speeds can result in blurred images (due to movement), but the use of faster speeds reduces the gathered light intensity. In practice, stationary objects are normally shot with an intermediate shutter speed of $(1/60)$ s.

More expensive cameras have an aperture of adjustable diameter to further control the intensity of the light reaching the film. As noted earlier, when an aperture of small diameter is used, only light from the central portion of the lens reaches the film; in this way spherical aberration is reduced.

The intensity I of the light reaching the film is proportional to the area of the lens. Because this area is proportional to the square of the diameter D, we conclude that I is also proportional to D^2. Light intensity is a measure of the rate at which energy is received by the film per unit area of the image. Because the area of the image is proportional to q^2 and $q \approx f$ (when $p \gg f$, so p can be approximated as infinite), we conclude that the intensity is also proportional to $1/f^2$, and thus $I \propto D^2/f^2$. The brightness of the image formed on the film depends on the light intensity, so we see that the image brightness depends on both the focal length and the diameter of the lens.

The ratio f/D is called the **_f_-number** of a lens:

$$f\text{-number} \equiv \frac{f}{D} \tag{36.18}$$

Hence, the intensity of light incident on the film varies according to the following proportionality:

$$I \propto \frac{1}{(f/D)^2} \propto \frac{1}{(f\text{-number})^2} \tag{36.19}$$

The f-number is often given as a description of the lens "speed." The lower the f-number, the wider the aperture and the higher the rate at which energy from the light exposes the film—thus, a lens with a low f-number is a "fast" lens. The conventional notation for an f-number is "$f/$" followed by the actual number. For example, "$f/4$" means an f-number of 4—it _does not_ mean to divide f by 4! Extremely fast lenses, which have f-numbers as low as approximately $f/1.2$, are expensive because it is very difficult to keep aberrations acceptably small with light rays passing through a large area of the lens. Camera lens systems (that is, combinations of lenses with adjustable apertures) are often marked with multiple f-numbers, usually $f/2.8$, $f/4$, $f/5.6$, $f/8$, $f/11$, and $f/16$. Any one of these settings can be selected by adjusting the aperture, which changes the value of D. Increasing the setting from one f-number to the next higher value (for example, from $f/2.8$ to $f/4$) decreases the area of the aperture by a factor of two. The lowest f-number setting on a camera lens corresponds to a wide-open aperture and the use of the maximum possible lens area.

Simple cameras usually have a fixed focal length and a fixed aperture size, with an f-number of about $f/11$. This high value for the f-number allows for a large **depth of field,** meaning that objects at a wide range of distances from the lens form reasonably sharp images on the film. In other words, the camera does not have to be focused.

Digital cameras are similar to the cameras we have described here except that the light does not form an image on photographic film. The image in a digital camera is formed on a _charge-coupled device_ (CCD), which digitizes the image, turning it into binary code, as we discussed for sound in Section 17.5. (The CCD is described in Section 40.2.) The digital information is then stored on a memory chip for playback on the screen of the camera, or it can be downloaded to a computer and sent to a friend or relative through the Internet.

Quick Quiz 36.10 A camera can be modeled as a simple converging lens that focuses an image on the film, acting as the screen. A camera is initially focused on a distant object. To focus the image of an object close to the camera, the lens must be (a) moved away from the film. (b) left where it is. (c) moved toward the film.

Example 36.14 **Finding the Correct Exposure Time**

The lens of a certain 35-mm camera (where 35 mm is the width of the film strip) has a focal length of 55 mm and a speed (an *f*-number) of *f*/1.8. The correct exposure time for this speed under certain conditions is known to be (1/500) s.

(A) Determine the diameter of the lens.

Solution From Equation 36.18, we find that

$$D = \frac{f}{f\text{-number}} = \frac{55 \text{ mm}}{1.8} = \boxed{31 \text{ mm}}$$

(B) Calculate the correct exposure time if the *f*-number is changed to *f*/4 under the same lighting conditions.

Solution The total light energy hitting the film is proportional to the product of the intensity and the exposure time. If I is the light intensity reaching the film, then in a time interval Δt the energy per unit area received by the film is proportional to $I \Delta t$. Comparing the two situations, we require that $I_1 \Delta t_1 = I_2 \Delta t_2$, where Δt_1 is the correct exposure time for *f*/1.8 and Δt_2 is the correct exposure time for *f*/4. Using this result together with Equation 36.19, we find that

$$\frac{\Delta t_1}{(f_1\text{-number})^2} = \frac{\Delta t_2}{(f_2\text{-number})^2}$$

$$\Delta t_2 = \left(\frac{f_2\text{-number}}{f_1\text{-number}}\right)^2 \Delta t_1 = \left(\frac{4}{1.8}\right)^2 \left(\frac{1}{500}\text{ s}\right) \approx \boxed{\frac{1}{100}\text{ s}}$$

As the aperture size is reduced, exposure time must increase.

36.7 The Eye

Like a camera, a normal eye focuses light and produces a sharp image. However, the mechanisms by which the eye controls the amount of light admitted and adjusts to produce correctly focused images are far more complex, intricate, and effective than those in even the most sophisticated camera. In all respects, the eye is a physiological wonder.

Figure 36.37 shows the basic parts of the human eye. Light entering the eye passes through a transparent structure called the *cornea* (Fig. 36.38), behind which are a clear liquid (the *aqueous humor*), a variable aperture (the *pupil*, which is an opening in the *iris*), and the *crystalline lens*. Most of the refraction occurs at the outer surface of the eye, where the cornea is covered with a film of tears. Relatively little refraction occurs in the crystalline lens because the aqueous humor in contact with the lens has an average index of refraction close to that of the lens. The iris, which is the colored portion

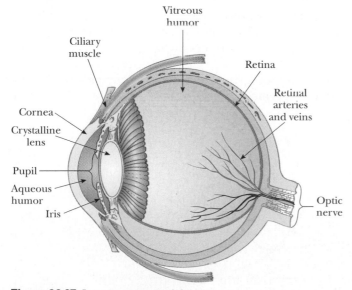

Figure 36.37 Important parts of the eye.

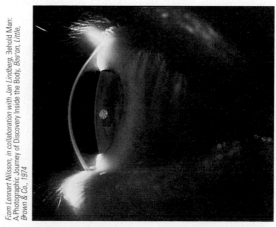

From Lennart Nilsson, in collaboration with Jan Lindberg, Behold Man: A Photographic Journey of Discovery Inside the Body, Boston, Little, Brown & Co., 1974

Figure 36.38 Close-up photograph of the cornea of the human eye.

of the eye, is a muscular diaphragm that controls pupil size. The iris regulates the amount of light entering the eye by dilating the pupil in low-light conditions and contracting the pupil in high-light conditions. The f-number range of the eye is from about $f/2.8$ to $f/16$.

The cornea–lens system focuses light onto the back surface of the eye, the *retina,* which consists of millions of sensitive receptors called *rods* and *cones.* When stimulated by light, these receptors send impulses via the optic nerve to the brain, where an image is perceived. By this process, a distinct image of an object is observed when the image falls on the retina.

The eye focuses on an object by varying the shape of the pliable crystalline lens through an amazing process called **accommodation.** An important component of accommodation is the *ciliary muscle,* which is situated in a circle around the rim of the lens. Thin filaments, called *zonules,* run from this muscle to the edge of the lens. When the eye is focused on a distant object, the ciliary muscle is relaxed, tightening the zonules that attach the muscle to the edge of the lens. The force of the zonules causes the lens to flatten, increasing its focal length. For an object distance of infinity, the focal length of the eye is equal to the fixed distance between lens and retina, about 1.7 cm. The eye focuses on nearby objects by tensing the ciliary muscle, which relaxes the zonules. This action allows the lens to bulge a bit, and its focal length decreases, resulting in the image being focused on the retina. All these lens adjustments take place so swiftly that we are not even aware of the change.

Accommodation is limited in that objects very close to the eye produce blurred images. The **near point** is the closest distance for which the lens can accommodate to focus light on the retina. This distance usually increases with age and has an average value of 25 cm. Typically, at age 10 the near point of the eye is about 18 cm. It increases to about 25 cm at age 20, to 50 cm at age 40, and to 500 cm or greater at age 60. The **far point** of the eye represents the greatest distance for which the lens of the relaxed eye can focus light on the retina. A person with normal vision can see very distant objects and thus has a far point that can be approximated as infinity.

Recall that the light leaving the mirror in Figure 36.8 becomes white where it comes together but then diverges into separate colors again. Because nothing but air exists at the point where the rays cross (and hence nothing exists to cause the colors to separate again), seeing white light as a result of a combination of colors must be a visual illusion. In fact, this is the case. Only three types of color-sensitive cells are present in the retina; they are called red, green, and blue cones because of the peaks of the color ranges to which they respond (Fig. 36.39). If the red and green cones are stimulated simultaneously (as would be the case if yellow light were shining on them), the brain interprets what we see as yellow. If all three types of cones are stimulated by the separate colors red, blue, and green, as in Figure 36.8, we see white. If all three types of cones are stimulated by light that contains *all* colors, such as sunlight, we again see white light.

Color televisions take advantage of this visual illusion by having only red, green, and blue dots on the screen. With specific combinations of brightness in these three

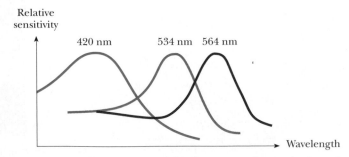

Figure 36.39 Approximate color sensitivity of the three types of cones in the retina.

primary colors, our eyes can be made to see any color in the rainbow. Thus, the yellow lemon you see in a television commercial is not really yellow, it is red and green! The paper on which this page is printed is made of tiny, matted, translucent fibers that scatter light in all directions; the resultant mixture of colors appears white to the eye. Snow, clouds, and white hair are not really white. In fact, there is no such thing as a white pigment. The appearance of these things is a consequence of the scattering of light containing all colors, which we interpret as white.

Conditions of the Eye

When the eye suffers a mismatch between the focusing range of the lens–cornea system and the length of the eye, with the result that light rays from a near object reach the retina before they converge to form an image, as shown in Figure 36.40a, the condition is known as **farsightedness** (or *hyperopia*). A farsighted person can usually see faraway objects clearly but not nearby objects. Although the near point of a normal eye is approximately 25 cm, the near point of a farsighted person is much farther away. The refracting power in the cornea and lens is insufficient to focus the light from all but distant objects satisfactorily. The condition can be corrected by placing a converging lens in front of the eye, as shown in Figure 36.40b. The lens refracts the incoming rays more toward the principal axis before entering the eye, allowing them to converge and focus on the retina.

A person with **nearsightedness** (or *myopia*), another mismatch condition, can focus on nearby objects but not on faraway objects. The far point of the nearsighted eye is not infinity and may be less than 1 m. The maximum focal length of the nearsighted eye is insufficient to produce a sharp image on the retina, and rays from a distant object converge to a focus in front of the retina. They then continue past that point, diverging before they finally reach the retina and causing blurred vision

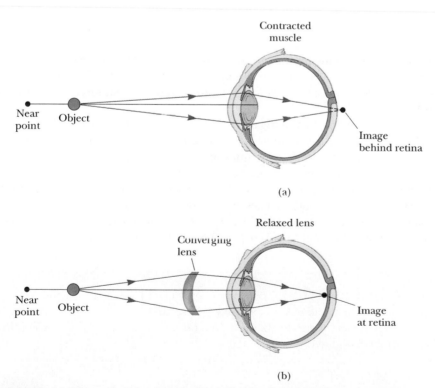

(a)

(b)

Figure 36.40 (a) When a farsighted eye looks at an object located between the near point and the eye, the image point is behind the retina, resulting in blurred vision. The eye muscle contracts to try to bring the object into focus. (b) Farsightedness is corrected with a converging lens.

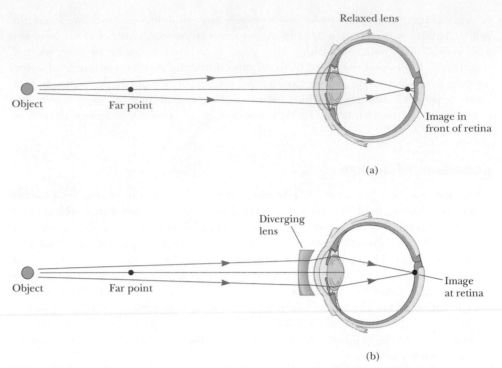

Figure 36.41 (a) When a nearsighted eye looks at an object that lies beyond the eye's far point, the image is formed in front of the retina, resulting in blurred vision. (b) Nearsightedness is corrected with a diverging lens.

(Fig. 36.41a). Nearsightedness can be corrected with a diverging lens, as shown in Figure 36.41b. The lens refracts the rays away from the principal axis before they enter the eye, allowing them to focus on the retina.

Beginning in middle age, most people lose some of their accommodation ability as the ciliary muscle weakens and the lens hardens. Unlike farsightedness, which is a mismatch between focusing power and eye length, **presbyopia** (literally, "old-age vision") is due to a reduction in accommodation ability. The cornea and lens do not have sufficient focusing power to bring nearby objects into focus on the retina. The symptoms are the same as those of farsightedness, and the condition can be corrected with converging lenses.

In the eye defect known as **astigmatism,** light from a point source produces a line image on the retina. This condition arises when either the cornea or the lens or both are not perfectly symmetric. Astigmatism can be corrected with lenses that have different curvatures in two mutually perpendicular directions.

Optometrists and ophthalmologists usually prescribe lenses[1] measured in **diopters:** the **power** P of a lens in diopters equals the inverse of the focal length in meters: $P = 1/f$. For example, a converging lens of focal length $+20$ cm has a power of $+5.0$ diopters, and a diverging lens of focal length -40 cm has a power of -2.5 diopters.

> **Quick Quiz 36.11**   Two campers wish to start a fire during the day. One camper is nearsighted and one is farsighted. Whose glasses should be used to focus the Sun's rays onto some paper to start the fire? (a) either camper (b) the nearsighted camper (c) the farsighted camper.

---

[1]   The word *lens* comes from *lentil,* the name of an Italian legume. (You may have eaten lentil soup.) Early eyeglasses were called "glass lentils" because the biconvex shape of their lenses resembled the shape of a lentil. The first lenses for farsightedness and presbyopia appeared around 1280; concave eyeglasses for correcting nearsightedness did not appear for more than 100 years after that.

Example 36.15   A Case of Nearsightedness

A particular nearsighted person is unable to see objects clearly when they are beyond 2.5 m away (the far point of this particular eye). What should the focal length be in a lens prescribed to correct this problem?

Solution The purpose of the lens in this instance is to "move" an object from infinity to a distance where it can be seen clearly. This is accomplished by having the lens produce an image at the far point. From the thin lens equation, we have

$$\frac{1}{p} + \frac{1}{q} = \frac{1}{\infty} + \frac{1}{-2.5\text{ m}} = \frac{1}{f}$$

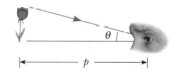

We use a negative sign for the image distance because the image is virtual and in front of the eye. As you should have suspected, the lens must be a diverging lens (one with a negative focal length) to correct nearsightedness.

36.8   The Simple Magnifier

The simple magnifier consists of a single converging lens. As the name implies, this device increases the apparent size of an object.

Suppose an object is viewed at some distance p from the eye, as illustrated in Figure 36.42. The size of the image formed at the retina depends on the angle θ subtended by the object at the eye. As the object moves closer to the eye, θ increases and a larger image is observed. However, an average normal eye cannot focus on an object closer than about 25 cm, the near point (Fig. 36.43a). Therefore, θ is maximum at the near point.

To further increase the apparent angular size of an object, a converging lens can be placed in front of the eye as in Figure 36.43b, with the object located at point O, just inside the focal point of the lens. At this location, the lens forms a virtual, upright, enlarged image. We define **angular magnification** m as the ratio of the angle subtended by an object with a lens in use (angle θ in Fig. 36.43b) to the angle subtended by the object placed at the near point with no lens in use (angle θ_0 in Fig. 36.43a):

$$m \equiv \frac{\theta}{\theta_0} \tag{36.20}$$

The angular magnification is a maximum when the image is at the near point of the eye—that is, when $q = -25$ cm. The object distance corresponding to this image

Figure 36.42 The size of the image formed on the retina depends on the angle θ subtended at the eye.

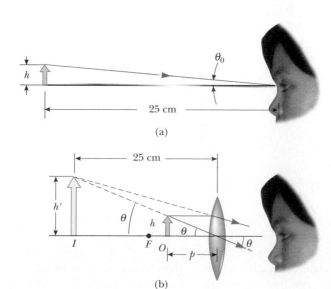

Figure 36.43 (a) An object placed at the near point of the eye ($p = 25$ cm) subtends an angle $\theta_0 \approx h/25$ at the eye. (b) An object placed near the focal point of a converging lens produces a magnified image that subtends an angle $\theta \approx h'/25$ at the eye.

distance can be calculated from the thin lens equation:

$$\frac{1}{p} + \frac{1}{-25 \text{ cm}} = \frac{1}{f}$$

$$p = \frac{25f}{25 + f}$$

where f is the focal length of the magnifier in centimeters. If we make the small-angle approximations

$$\tan \theta_0 \approx \theta_0 \approx \frac{h}{25} \quad \text{and} \quad \tan \theta \approx \theta \approx \frac{h}{p} \quad \text{(36.21)}$$

Equation 36.20 becomes

$$m_{\text{max}} = \frac{\theta}{\theta_0} = \frac{h/p}{h/25} = \frac{25}{p} = \frac{25}{25f/(25 + f)}$$

$$m_{\text{max}} = 1 + \frac{25 \text{ cm}}{f} \quad \text{(36.22)}$$

Although the eye can focus on an image formed anywhere between the near point and infinity, it is most relaxed when the image is at infinity. For the image formed by the magnifying lens to appear at infinity, the object has to be at the focal point of the lens. In this case, Equations 36.21 become

$$\theta_0 \approx \frac{h}{25} \quad \text{and} \quad \theta \approx \frac{h}{f}$$

and the magnification is

$$m_{\text{min}} = \frac{\theta}{\theta_0} = \frac{25 \text{ cm}}{f} \quad \text{(36.23)}$$

A simple magnifier, also called a magnifying glass, is used to view an enlarged image of a portion of a map.

With a single lens, it is possible to obtain angular magnifications up to about 4 without serious aberrations. Magnifications up to about 20 can be achieved by using one or two additional lenses to correct for aberrations.

Example 36.16   Maximum Magnification of a Lens

What is the maximum magnification that is possible with a lens having a focal length of 10 cm, and what is the magnification of this lens when the eye is relaxed?

Solution The maximum magnification occurs when the image is located at the near point of the eye. Under these circumstances, Equation 36.22 gives

$$m_{\text{max}} = 1 + \frac{25 \text{ cm}}{f} = 1 + \frac{25 \text{ cm}}{10 \text{ cm}} = \boxed{3.5}$$

When the eye is relaxed, the image is at infinity. In this case, we use Equation 36.23:

$$m_{\text{min}} = \frac{25 \text{ cm}}{f} = \frac{25 \text{ cm}}{10 \text{ cm}} = \boxed{2.5}$$

36.9   The Compound Microscope

A simple magnifier provides only limited assistance in inspecting minute details of an object. Greater magnification can be achieved by combining two lenses in a device called a **compound microscope,** a schematic diagram of which is shown in Figure 36.44a. It consists of one lens, the *objective,* that has a very short focal length $f_o < 1$ cm and a second lens, the *eyepiece,* that has a focal length f_e of a few

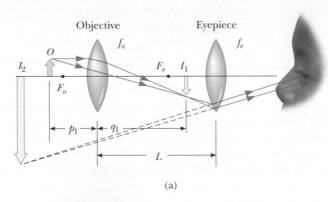

Objective   Eyepiece

(a)

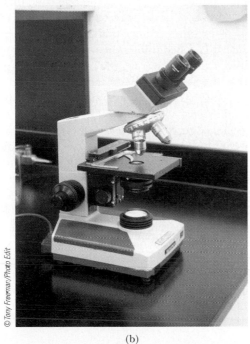

© Tony Freeman/Photo Edit

(b)

Active Figure 36.44 (a) Diagram of a compound microscope, which consists of an objective lens and an eyepiece lens. (b) A compound microscope. The three-objective turret allows the user to choose from several powers of magnification. Combinations of eyepieces with different focal lengths and different objectives can produce a wide range of magnifications.

At the Active Figures link at http://www.pse6.com, *you can adjust the focal lengths of the objective and eyepiece lenses to see the effect on the final image.*

centimeters. The two lenses are separated by a distance L that is much greater than either f_o or f_e. The object, which is placed just outside the focal point of the objective, forms a real, inverted image at I_1, and this image is located at or close to the focal point of the eyepiece. The eyepiece, which serves as a simple magnifier, produces at I_2 a virtual, enlarged image of I_1. The lateral magnification M_1 of the first image is $-q_1/p_1$. Note from Figure 36.44a that q_1 is approximately equal to L and that the object is very close to the focal point of the objective: $p_1 \approx f_o$. Thus, the lateral magnification by the objective is

$$M_o \approx -\frac{L}{f_o}$$

The angular magnification by the eyepiece for an object (corresponding to the image at I_1) placed at the focal point of the eyepiece is, from Equation 36.23,

$$m_e = \frac{25 \text{ cm}}{f_e}$$

The overall magnification of the image formed by a compound microscope is defined as the product of the lateral and angular magnifications:

$$M = M_o m_e = -\frac{L}{f_o}\left(\frac{25 \text{ cm}}{f_e}\right) \tag{36.24}$$

The negative sign indicates that the image is inverted.

The microscope has extended human vision to the point where we can view previously unknown details of incredibly small objects. The capabilities of this instrument have steadily increased with improved techniques for precision grinding of lenses. A question often asked about microscopes is: "If one were extremely patient and careful, would it be possible to construct a microscope that would enable the human eye to see an atom?" The answer is no, as long as light is used to illuminate the object. The reason is that, for an object under an optical microscope (one that uses visible light) to be seen, the object must be at least as large as a wavelength of light. Because the diameter of any atom is many times smaller than the wavelengths of visible light, the mysteries of the atom must be probed using other types of "microscopes."

The ability to use other types of waves to "see" objects also depends on wavelength. We can illustrate this with water waves in a bathtub. Suppose you vibrate your hand in the water until waves having a wavelength of about 15 cm are moving along the surface. If you hold a small object, such as a toothpick, so that it lies in the path of the waves, it does not appreciably disturb the waves; they continue along their path "oblivious" to it. Now suppose you hold a larger object, such as a toy sailboat, in the path of the 15-cm waves. In this case, the waves are considerably disturbed by the object. Because the toothpick is smaller than the wavelength of the waves, the waves do not "see" it. (The intensity of the scattered waves is low.) Because it is about the same size as the wavelength of the waves, however, the boat creates a disturbance. In other words, the object acts as the source of scattered waves that appear to come from it.

Light waves behave in this same general way. The ability of an optical microscope to view an object depends on the size of the object relative to the wavelength of the light used to observe it. Hence, we can never observe atoms with an optical microscope[2] because their dimensions are small (<0.1 nm) relative to the wavelength of the light (<500 nm).

36.10 The Telescope

Two fundamentally different types of **telescopes** exist; both are designed to aid in viewing distant objects, such as the planets in our Solar System. The **refracting telescope** uses a combination of lenses to form an image, and the **reflecting telescope** uses a curved mirror and a lens.

The lens combination shown in Figure 36.45a is that of a refracting telescope. Like the compound microscope, this telescope has an objective and an eyepiece. The two lenses are arranged so that the objective forms a real, inverted image of a distant object very near the focal point of the eyepiece. Because the object is essentially at infinity, this point at which I_1 forms is the focal point of the objective. The eyepiece then forms, at I_2, an enlarged, inverted image of the image at I_1. In order to provide the largest possible magnification, the image distance for the eyepiece is infinite. This means that the light rays exit the eyepiece lens parallel to the principal axis, and the image of the objective lens must form at the focal point of the eyepiece. Hence, the two lenses are separated by a distance $f_o + f_e$, which corresponds to the length of the telescope tube.

---

[2]   Single-molecule near-field optic studies are routinely performed with visible light having wavelengths of about 500 nm. The technique uses very small apertures to produce images having resolutions as small as 10 nm.

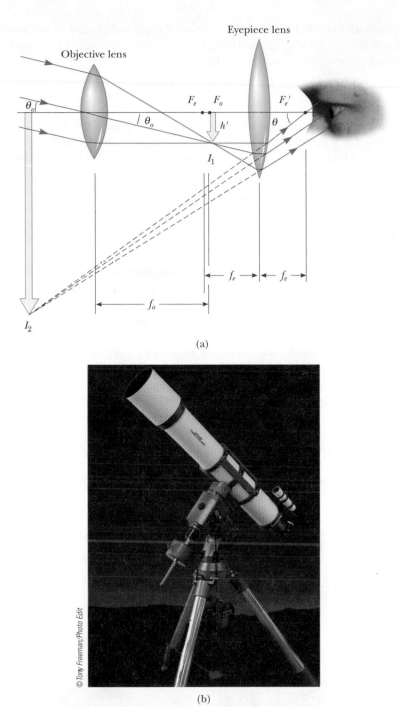

(a)

(b)

Active Figure 36.45 (a) Lens arrangement in a refracting telescope, with the object at infinity. (b) A refracting telescope.

At the Active Figures link at http://www.pse6.com, *you can adjust the focal lengths of the objective and eyepiece lenses to see the effect on the final image.*

The angular magnification of the telescope is given by θ/θ_o, where θ_o is the angle subtended by the object at the objective and θ is the angle subtended by the final image at the viewer's eye. Consider Figure 36.45a, in which the object is a very great distance to the left of the figure. The angle θ_o (to the *left* of the objective) subtended by the object at the objective is the same as the angle (to the *right* of the objective) subtended by the first image at the objective. Thus,

$$\tan \theta_o \approx \theta_o \approx -\frac{h'}{f_o}$$

where the negative sign indicates that the image is inverted.

The angle θ subtended by the final image at the eye is the same as the angle that a ray coming from the tip of I_1 and traveling parallel to the principal axis makes with the principal axis after it passes through the lens. Thus,

$$\tan \theta \approx \theta \approx \frac{h'}{f_e}$$

We have not used a negative sign in this equation because the final image is not inverted; the object creating this final image I_2 is I_1, and both it and I_2 point in the same direction. Hence, the angular magnification of the telescope can be expressed as

$$m = \frac{\theta}{\theta_o} = \frac{h'/f_e}{-h'/f_o} = -\frac{f_o}{f_e} \qquad (36.25)$$

and we see that the angular magnification of a telescope equals the ratio of the objective focal length to the eyepiece focal length. The negative sign indicates that the image is inverted.

When we look through a telescope at such relatively nearby objects as the Moon and the planets, magnification is important. However, individual stars in our galaxy are so far away that they always appear as small points of light no matter how great the magnification. A large research telescope that is used to study very distant objects must have a great diameter to gather as much light as possible. It is difficult and expensive to manufacture large lenses for refracting telescopes. Another difficulty with large lenses is that their weight leads to sagging, which is an additional source of aberration. These problems can be partially overcome by replacing the objective with a concave mirror, which results in a reflecting telescope. Because light is reflected from the mirror and does not pass through a lens, the mirror can have rigid supports on the back side. Such supports eliminate the problem of sagging.

Figure 36.46a shows the design for a typical reflecting telescope. Incoming light rays pass down the barrel of the telescope and are reflected by a parabolic mirror at the base. These rays converge toward point A in the figure, where an image would be formed. However, before this image is formed, a small, flat mirror M reflects the light toward an opening in the side of the tube that passes into an eyepiece. This particular design is said to have a Newtonian focus because Newton developed it. Figure 36.46b

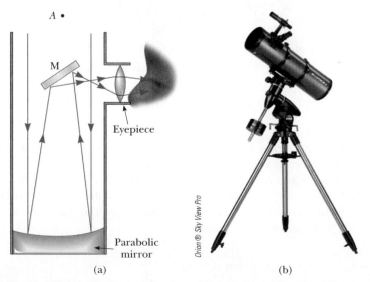

Figure 36.46 (a) A Newtonian-focus reflecting telescope. (b) A reflecting telescope. This type of telescope is shorter than that in Figure 36.45b.

shows such a telescope. Note that in the reflecting telescope the light never passes through glass (except through the small eyepiece). As a result, problems associated with chromatic aberration are virtually eliminated. The reflecting telescope can be made even shorter by orienting the flat mirror so that it reflects the light back toward the objective mirror and the light enters an eyepiece in a hole in the middle of the mirror.

The largest reflecting telescopes in the world are at the Keck Observatory on Mauna Kea, Hawaii. The site includes two telescopes with diameters of 10 m, each containing 36 hexagonally shaped, computer-controlled mirrors that work together to form a large reflecting surface. In contrast, the largest refracting telescope in the world, at the Yerkes Observatory in Williams Bay, Wisconsin, has a diameter of only 1 m.

SUMMARY

Take a practice test for this chapter by clicking on the Practice Test link at http://www.pse6.com.

The **lateral magnification** M of the image due to a mirror or lens is defined as the ratio of the image height h' to the object height h and is equal to the negative of the ratio of the image distance q to the object distance p:

$$M = \frac{h'}{h} = -\frac{q}{p} \qquad (36.1, 36.2)$$

In the paraxial ray approximation, the object distance p and image distance q for a spherical mirror of radius R are related by the **mirror equation:**

$$\frac{1}{p} + \frac{1}{q} = \frac{2}{R} = \frac{1}{f} \qquad (36.4, 36.6)$$

where $f = R/2$ is the **focal length** of the mirror.

An image can be formed by refraction from a spherical surface of radius R. The object and image distances for refraction from such a surface are related by

$$\frac{n_1}{p} + \frac{n_2}{q} = \frac{n_2 - n_1}{R} \qquad (36.8)$$

where the light is incident in the medium for which the index of refraction is n_1 and is refracted in the medium for which the index of refraction is n_2.

The inverse of the **focal length** f of a thin lens surrounded by air is given by the **lens makers' equation:**

$$\frac{1}{f} = (n - 1)\left(\frac{1}{R_1} - \frac{1}{R_2}\right) \qquad (36.15)$$

Converging lenses have positive focal lengths, and **diverging lenses** have negative focal lengths.

For a thin lens, and in the paraxial ray approximation, the object and image distances are related by the **thin lens equation:**

$$\frac{1}{p} + \frac{1}{q} = \frac{1}{f} \qquad (36.16)$$

The ratio of the focal length of a camera lens to the diameter of the lens is called the **f-number** of the lens:

$$f\text{-number} \equiv \frac{f}{D} \qquad (36.18)$$

The intensity of light incident on the film in the camera varies according to:

$$I \propto \frac{1}{(f/D)^2} \propto \frac{1}{(f\text{-number})^2} \qquad (36.19)$$

The maximum magnification of a single lens of focal length f used as a simple magnifier is

$$m_{max} = 1 + \frac{25\text{ cm}}{f} \qquad (36.22)$$

The overall magnification of the image formed by a compound microscope is:

$$M = -\frac{L}{f_o}\left(\frac{25\text{ cm}}{f_e}\right) \qquad (36.24)$$

where f_o and f_e are the focal lengths of the objective and eyepiece lenses, respectively, and L is the distance between the lenses.

The angular magnification of a refracting telescope can be expressed as

$$m = -\frac{f_o}{f_e} \qquad (36.25)$$

where f_o and f_e are the focal lengths of the objective and eyepiece lenses, respectively. The angular magnification of a reflecting telescope is given by the same expression where f_o is the focal length of the objective mirror.

QUESTIONS

1. What is wrong with the caption of the cartoon shown in Figure Q36.1?

© 2003 Sidney Harris

Figure Q36.1 "Most mirrors reverse left and right. This one reverses top and bottom."

2. Consider a concave spherical mirror with a real object. Is the image always inverted? Is the image always real? Give conditions for your answers.

3. Repeat the preceding question for a convex spherical mirror.

4. Do the equations $1/p + 1/q = 1/f$ or $M = -q/p$ apply to the image formed by a flat mirror? Explain your answer.

5. Why does a clear stream, such as a creek, always appear to be shallower than it actually is? By how much is its depth apparently reduced?

6. Consider the image formed by a thin converging lens. Under what conditions is the image (a) inverted, (b) upright, (c) real, (d) virtual, (e) larger than the object, and (f) smaller than the object?

7. Repeat Question 6 for a thin diverging lens.

8. Use the lens makers' equation to verify the sign of the focal length of each of the lenses in Figure 36.27.

9. If a solid cylinder of glass or clear plastic is placed above the words LEAD OXIDE and viewed from above as shown

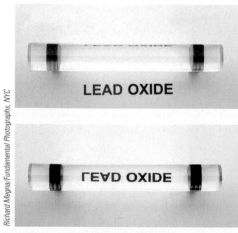

Richard Megna/Fundamental Photographs, NYC

Figure Q36.9

in Figure Q36.9, the LEAD appears inverted but the OXIDE does not. Explain.

10. In Figure 36.28a, assume that the blue object arrow is replaced by one that is much taller than the lens. How many rays from the object will strike the lens? How many principal rays can be drawn in a ray diagram?

11. A zip-lock plastic sandwich bag filled with water can act as a crude converging lens in air. If the bag is filled with air and placed under water, is the effective lens converging or diverging?

12. Explain why a mirror cannot give rise to chromatic aberration.

13. Why do some automobile mirrors have printed on them the statement "Objects in mirror are closer than they appear"? (See Fig. P36.19.)

14. Can a converging lens be made to diverge light if it is placed into a liquid? **What If?** How about a converging mirror?

15. Explain why a fish in a spherical goldfish bowl appears larger than it really is.

16. Why do some emergency vehicles have the symbol ƎƆИAⅬUᗺMA written on the front?

17. A lens forms an image of an object on a screen. What happens to the image if you cover the top half of the lens with paper?

18. Lenses used in eyeglasses, whether converging or diverging, are always designed so that the middle of the lens curves away from the eye, like the center lenses of Figure 36.27a and b. Why?

19. Which glasses in Figure Q36.19 correct nearsightedness and which correct farsightedness?

Figure Q36.19

20. A child tries on either his hyperopic grandfather's or his myopic brother's glasses and complains that "everything looks blurry." Why do the eyes of a person wearing glasses not look blurry? (See Figure Q36.19.)

21. Consider a spherical concave mirror, with the object located to the left of the mirror beyond the focal point. Using ray diagrams, show that the image moves to the left as the object approaches the focal point.

22. In a Jules Verne novel, a piece of ice is shaped to form a magnifying lens to focus sunlight to start a fire. Is this possible?

23. The f-number of a camera is the focal length of the lens divided by its aperture (or diameter). How can the f-number of the lens be changed? How does changing this number affect the required exposure time?

24. A solar furnace can be constructed by using a concave mirror to reflect and focus sunlight into a furnace enclosure. What factors in the design of the reflecting mirror would guarantee very high temperatures?

25. One method for determining the position of an image, either real or virtual, is by means of *parallax*. If a finger or other object is placed at the position of the image, as shown in Figure Q36.25, and the finger and image are viewed simultaneously (the image is viewed through the lens if it is virtual), the finger and image have the same parallax; that is, if they are viewed from different positions, the image will appear to move along with the finger. Use this method to locate the image formed by a lens. Explain why the method works.

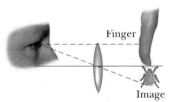

Finger

Image

Figure Q36.25

26. Figure Q36.26 shows a lithograph by M. C. Escher titled *Hand with Reflection Sphere (Self-Portrait in Spherical Mirror)*. Escher had this to say about the work: "The picture

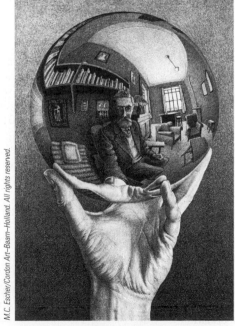

Figure Q36.26

shows a spherical mirror, resting on a left hand. But as a print is the reverse of the original drawing on stone, it was my right hand that you see depicted. (Being left-handed, I needed my left hand to make the drawing.) Such a globe reflection collects almost one's whole surroundings in one disk-shaped image. The whole room, four walls, the floor, and the ceiling, everything, albeit distorted, is compressed into that one small circle. Your own head, or more exactly the point between your eyes, is the absolute center. No matter how you turn or twist yourself, you can't get out of that central point. You are immovably the focus, the unshakable core, of your world." Comment on the accuracy of Escher's description.

27. You can make a corner reflector by placing three flat mirrors in the corner of a room where the ceiling meets the walls. Show that no matter where you are in the room, you can see yourself reflected in the mirrors—upside down.

28. A converging lens of short focal length can take light diverging from a small source and refract it into a beam of parallel rays. A Fresnel lens, as shown in Figure 36.29, is used for this purpose in a lighthouse. A concave mirror can take light diverging from a small source and reflect it into a beam of parallel rays. Is it possible to make a Fresnel mirror? Is this an original idea, or has it already been done? *Suggestion:* Look at the walls and ceiling of an auditorium.

PROBLEMS

1, 2, 3 = straightforward, intermediate, challenging   ☐ = full solution available in the *Student Solutions Manual and Study Guide*

〰〰 = coached solution with hints available at http://www.pse6.com   🖥 = computer useful in solving problem

▨ = paired numerical and symbolic problems

Section 36.1   Images Formed by Flat Mirrors

1. Does your bathroom mirror show you older or younger than you actually are? Compute an order-of-magnitude estimate for the age difference, based on data that you specify.

2. In a church choir loft, two parallel walls are 5.30 m apart. The singers stand against the north wall. The organist faces the south wall, sitting 0.800 m away from it. To enable her to see the choir, a flat mirror 0.600 m wide is mounted on the south wall, straight in front of her. What width of the north wall can she see? *Suggestion:* Draw a top-view diagram to justify your answer.

3. Determine the minimum height of a vertical flat mirror in which a person 5′10″ in height can see his or her full image. (A ray diagram would be helpful.)

4. Two flat mirrors have their reflecting surfaces facing each other, with the edge of one mirror in contact with an edge of the other, so that the angle between the mirrors is α. When an object is placed between the mirrors, a number of images are formed. In general, if the angle α is such that $n\alpha = 360°$, where n is an integer, the number of images formed is $n - 1$. Graphically, find all the image positions for the case $n = 6$ when a point object is between the mirrors (but not on the angle bisector).

5. A person walks into a room with two flat mirrors on opposite walls, which produce multiple images. When the person is located 5.00 ft from the mirror on the left wall and 10.0 ft from the mirror on the right wall, find the distance from the person to the first three images seen in the mirror on the left.

6. A periscope (Figure P36.6) is useful for viewing objects that cannot be seen directly. It finds use in submarines and in watching golf matches or parades from behind a crowd of people. Suppose that the object is a distance p_1 from the upper mirror and that the two flat mirrors are separated by a distance h. (a) What is the distance of the final image from the lower mirror? (b) Is the final image real or virtual? (c) Is it upright or inverted? (d) What is its magnification? (e) Does it appear to be left–right reversed?

Figure P36.6

Section 36.2   Images Formed by Spherical Mirrors

7. A concave spherical mirror has a radius of curvature of 20.0 cm. Find the location of the image for object distances of (a) 40.0 cm, (b) 20.0 cm, and (c) 10.0 cm. For each case, state whether the image is real or virtual and upright or inverted. Find the magnification in each case.

8. At an intersection of hospital hallways, a convex mirror is mounted high on a wall to help people avoid collisions. The mirror has a radius of curvature of 0.550 m. Locate and describe the image of a patient 10.0 m from the mirror. Determine the magnification.

9. 🪐 A spherical convex mirror has a radius of curvature with a magnitude of 40.0 cm. Determine the position of the virtual image and the magnification for object distances of (a) 30.0 cm and (b) 60.0 cm. (c) Are the images upright or inverted?

10. A large church has a niche in one wall. On the floor plan it appears as a semicircular indentation of radius 2.50 m. A worshiper stands on the center line of the niche, 2.00 m out from its deepest point, and whispers a prayer. Where is the sound concentrated after reflection from the back wall of the niche?

11. A concave mirror has a radius of curvature of 60.0 cm. Calculate the image position and magnification of an object placed in front of the mirror at distances of (a) 90.0 cm and (b) 20.0 cm. (c) Draw ray diagrams to obtain the image characteristics in each case.

12. A concave mirror has a focal length of 40.0 cm. Determine the object position for which the resulting image is upright and four times the size of the object.

13. A certain Christmas tree ornament is a silver sphere having a diameter of 8.50 cm. Determine an object location for which the size of the reflected image is three-fourths the size of the object. Use a principal-ray diagram to arrive at a description of the image.

14. (a) A concave mirror forms an inverted image four times larger than the object. Find the focal length of the mirror, assuming the distance between object and image is 0.600 m. (b) A convex mirror forms a virtual image half the size of the object. Assuming the distance between image and object is 20.0 cm, determine the radius of curvature of the mirror.

15. To fit a contact lens to a patient's eye, a *keratometer* can be used to measure the curvature of the front surface of the eye, the cornea. This instrument places an illuminated object of known size at a known distance p from the cornea. The cornea reflects some light from the object, forming an image of the object. The magnification M of the image is measured by using a small viewing telescope that allows comparison of the image formed by the cornea with a second calibrated image projected into the field of view by a prism arrangement. Determine the radius of curvature of the cornea for the case $p = 30.0$ cm and $M = 0.013\ 0$.

16. An object 10.0 cm tall is placed at the zero mark of a meter stick. A spherical mirror located at some point on the meter stick creates an image of the object that is upright, 4.00 cm tall, and located at the 42.0-cm mark of the meter stick. (a) Is the mirror convex or concave? (b) Where is the mirror? (c) What is the mirror's focal length?

17. A spherical mirror is to be used to form, on a screen located 5.00 m from the object, an image five times the size of the object. (a) Describe the type of mirror required. (b) Where should the mirror be positioned relative to the object?

18. A dedicated sports car enthusiast polishes the inside and outside surfaces of a hubcap that is a section of a sphere. When she looks into one side of the hubcap, she sees an image of her face 30.0 cm in back of the hubcap. She then flips the hubcap over and sees another image of her face 10.0 cm in back of the hubcap. (a) How far is her face from the hubcap? (b) What is the radius of curvature of the hubcap?

19. You unconsciously estimate the distance to an object from the angle it subtends in your field of view. This angle θ in radians is related to the linear height of the object h and to the distance d by $\theta = h/d$. Assume that you are driving a car and another car, 1.50 m high, is 24.0 m behind you. (a) Suppose your car has a flat passenger-side rearview mirror, 1.55 m from your eyes. How far from your eyes is the image of the car following you? (b) What angle does the image subtend in your field of view? (c) **What If?** Suppose instead that your car has a convex rearview mirror with a radius of curvature of magnitude 2.00 m (Fig. P36.19). How far from your eyes is the image of the car behind you? (d) What angle does the image subtend at your eyes? (e) Based on its angular size, how far away does the following car appear to be?

THE FAR SIDE® BY GARY LARSON

OBJECTS IN MIRROR ARE CLOSER THAN THEY APPEAR

© 1985 FarWorks, Inc. All Rights Reserved/Dist. by Creators Syndicate

The Far Side® by Gary Larson © 1985 FarWorks, Inc. All Rights Reserved. Used with permission.

Figure P36.19

20. **Review Problem.** A ball is dropped at $t = 0$ from rest 3.00 m directly above the vertex of a concave mirror that has a radius of curvature of 1.00 m and lies in a horizontal plane. (a) Describe the motion of the ball's image in the mirror. (b) At what time do the ball and its image coincide?

Section 36.3   Images Formed by Refraction

21. A cubical block of ice 50.0 cm on a side is placed on a level floor over a speck of dust. Find the location of the image of the speck as viewed from above. The index of refraction of ice is 1.309.

22. A flint glass plate ($n = 1.66$) rests on the bottom of an aquarium tank. The plate is 8.00 cm thick (vertical dimension) and is covered with a layer of water ($n = 1.33$) 12.0 cm deep. Calculate the apparent thickness of the plate as viewed from straight above the water.

23. A glass sphere ($n = 1.50$) with a radius of 15.0 cm has a tiny air bubble 5.00 cm above its center. The sphere is viewed looking down along the extended radius containing the bubble. What is the apparent depth of the bubble below the surface of the sphere?

24. A simple model of the human eye ignores its lens entirely. Most of what the eye does to light happens at the outer surface of the transparent cornea. Assume that this surface has a radius of curvature of 6.00 mm, and assume that the eyeball contains just one fluid with a refractive index of 1.40. Prove that a very distant object will be imaged on the retina, 21.0 mm behind the cornea. Describe the image.

25. One end of a long glass rod ($n = 1.50$) is formed into a convex surface with a radius of curvature of 6.00 cm. An object is located in air along the axis of the rod. Find the image positions corresponding to object distances of (a) 20.0 cm, (b) 10.0 cm, and (c) 3.00 cm from the end of the rod.

26. A transparent sphere of unknown composition is observed to form an image of the Sun on the surface of the sphere opposite the Sun. What is the refractive index of the sphere material?

27. A goldfish is swimming at 2.00 cm/s toward the front wall of a rectangular aquarium. What is the apparent speed of the fish measured by an observer looking in from outside the front wall of the tank? The index of refraction of water is 1.33.

Section 36.4   Thin Lenses

28. A contact lens is made of plastic with an index of refraction of 1.50. The lens has an outer radius of curvature of $+2.00$ cm and an inner radius of curvature of $+2.50$ cm. What is the focal length of the lens?

29. The left face of a biconvex lens has a radius of curvature of magnitude 12.0 cm, and the right face has a radius of curvature of magnitude 18.0 cm. The index of refraction of the glass is 1.44. (a) Calculate the focal length of the lens. (b) **What If?** Calculate the focal length the lens has after it is turned around to interchange the radii of curvature of the two faces.

30. A converging lens has a focal length of 20.0 cm. Locate the image for object distances of (a) 40.0 cm, (b) 20.0 cm, and (c) 10.0 cm. For each case, state whether the image is real or virtual and upright or inverted. Find the magnification in each case.

31. A thin lens has a focal length of 25.0 cm. Locate and describe the image when the object is placed (a) 26.0 cm and (b) 24.0 cm in front of the lens.

32. An object located 32.0 cm in front of a lens forms an image on a screen 8.00 cm behind the lens. (a) Find the focal length of the lens. (b) Determine the magnification. (c) Is the lens converging or diverging?

33. The nickel's image in Figure P36.33 has twice the diameter of the nickel and is 2.84 cm from the lens. Determine the focal length of the lens.

Figure P36.33

34. A person looks at a gem with a jeweler's loupe—a converging lens that has a focal length of 12.5 cm. The loupe forms a virtual image 30.0 cm from the lens. (a) Determine the magnification. Is the image upright or inverted? (b) Construct a ray diagram for this arrangement.

35. Suppose an object has thickness dp so that it extends from object distance p to $p + dp$. Prove that the thickness dq of its image is given by $(-q^2/p^2)\,dp$, so that the longitudinal magnification $dq/dp = -M^2$, where M is the lateral magnification.

36. The projection lens in a certain slide projector is a single thin lens. A slide 24.0 mm high is to be projected so that its image fills a screen 1.80 m high. The slide-to-screen distance is 3.00 m. (a) Determine the focal length of the projection lens. (b) How far from the slide should the lens of the projector be placed in order to form the image on the screen?

37. An object is located 20.0 cm to the left of a diverging lens having a focal length $f = -32.0$ cm. Determine (a) the location and (b) the magnification of the image. (c) Construct a ray diagram for this arrangement.

38. An antelope is at a distance of 20.0 m from a converging lens of focal length 30.0 cm. The lens forms an image of the animal. If the antelope runs away from the lens at a speed of 5.00 m/s, how fast does the image move? Does the image move toward or away from the lens?

39. In some types of optical spectroscopy, such as photoluminescence and Raman spectroscopy, a laser beam exits from a pupil and is focused on a sample to stimulate electromagnetic radiation from the sample. The focusing lens usually has an antireflective coating preventing any light loss. Assume a 100-mW laser is located 4.80 m from the lens, which has a focal length of 7.00 cm. (a) How far from the lens should the sample be located so that an image of the laser exit pupil is formed on the surface of the sample? (b) If the diameter of the laser exit pupil is 5.00 mm, what is the diameter of the light spot on the sample? (c) What is the light intensity at the spot?

40. Figure P36.40 shows a thin glass ($n = 1.50$) converging lens for which the radii of curvature are $R_1 = 15.0$ cm and $R_2 = -12.0$ cm. To the left of the lens is a cube having a face area of 100 cm². The base of the cube is on the axis of the lens, and the right face is 20.0 cm to the left of the lens. (a) Determine the focal length of the lens. (b) Draw the image of the square face formed by the lens. What type of geometric figure is this? (c) Determine the area of the image.

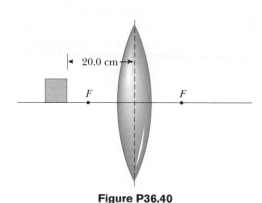

Figure P36.40

41. An object is at a distance d to the left of a flat screen. A converging lens with focal length $f < d/4$ is placed between object and screen. (a) Show that two lens positions exist that form an image on the screen, and determine how far these positions are from the object. (b) How do the two images differ from each other?

42. Figure 36.36 diagrams a cross section of a camera. It has a single lens of focal length 65.0 mm, which is to form an image on the film at the back of the camera. Suppose the position of the lens has been adjusted to focus the image of a distant object. How far and in what direction must the lens be moved to form a sharp image of an object that is 2.00 m away?

43. The South American capybara is the largest rodent on Earth; its body can be 1.20 m long. The smallest rodent is the pygmy mouse found in Texas, with an average body length of 3.60 cm. Assume that a pygmy mouse is observed by looking through a lens placed 20.0 cm from the mouse. The whole image of the mouse is the size of a capybara. Then the lens is moved a certain distance along its axis, and the image of the mouse is the same size as before! How far was the lens moved?

Section 36.5   Lens Aberrations

44. The magnitudes of the radii of curvature are 32.5 cm and 42.5 cm for the two faces of a biconcave lens. The glass has index of refraction 1.53 for violet light and 1.51 for red light. For a very distant object, locate and describe (a) the image formed by violet light, and (b) the image formed by red light.

45. Two rays traveling parallel to the principal axis strike a large plano-convex lens having a refractive index of 1.60 (Fig. P36.45). If the convex face is spherical, a ray near the edge does not pass through the focal point (spherical aberration occurs). Assume this face has a radius of curvature of 20.0 cm and the two rays are at distances $h_1 = 0.500$ cm and $h_2 = 12.0$ cm from the principal axis. Find the difference Δx in the positions where each crosses the principal axis.

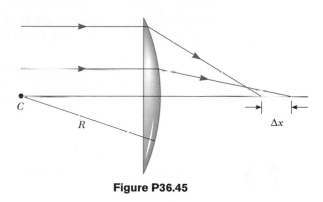

Figure P36.45

Section 36.6   The Camera

46. A camera is being used with a correct exposure at $f/4$ and a shutter speed of $(1/16)$ s. In order to photograph a rapidly moving subject, the shutter speed is changed to $(1/128)$ s. Find the new f-number setting needed to maintain satisfactory exposure.

Section 36.7   The Eye

47. A nearsighted person cannot see objects clearly beyond 25.0 cm (her far point). If she has no astigmatism and contact lenses are prescribed for her, what power and type of lens are required to correct her vision?

48. The accommodation limits for Nearsighted Nick's eyes are 18.0 cm and 80.0 cm. When he wears his glasses, he can see faraway objects clearly. At what minimum distance is he able to see objects clearly?

49. A person sees clearly when he wears eyeglasses that have a power of -4.00 diopters and sit 2.00 cm in front of his eyes. If the person wants to switch to contact lenses, which are placed directly on the eyes, what lens power should be prescribed?

Section 36.8   The Simple Magnifier
Section 36.9   The Compound Microscope
Section 36.10   The Telescope

50. A lens that has a focal length of 5.00 cm is used as a magnifying glass. (a) To obtain maximum magnification, where should the object be placed? (b) What is the magnification?

51. The distance between eyepiece and objective lens in a certain compound microscope is 23.0 cm. The focal length of the eyepiece is 2.50 cm, and that of the objective is 0.400 cm. What is the overall magnification of the microscope?

52. The desired overall magnification of a compound microscope is 140×. The objective alone produces a lateral magnification of 12.0×. Determine the required focal length of the eyepiece.

53. The Yerkes refracting telescope has a 1.00-m diameter objective lens of focal length 20.0 m. Assume it is used with an eyepiece of focal length 2.50 cm. (a) Determine the magnification of the planet Mars as seen through this telescope. (b) Are the Martian polar caps right side up or upside down?

54. Astronomers often take photographs with the objective lens or mirror of a telescope alone, without an eyepiece. (a) Show that the image size h' for this telescope is given by $h' = fh/(f - p)$ where h is the object size, f is the objective focal length, and p is the object distance. (b) **What If?** Simplify the expression in part (a) for the case in which the object distance is much greater than objective focal length. (c) The "wingspan" of the International Space Station is 108.6 m, the overall width of its solar panel configuration. Find the width of the image formed by a telescope objective of focal length 4.00 m when the station is orbiting at an altitude of 407 km.

55. Galileo devised a simple terrestrial telescope that produces an upright image. It consists of a converging objective lens and a diverging eyepiece at opposite ends of the telescope tube. For distant objects, the tube length is equal to the objective focal length minus the absolute value of the eyepiece focal length. (a) Does the user of the telescope see a real or virtual image? (b) Where is the final image? (c) If a telescope is to be constructed with a tube of length 10.0 cm and a magnification of 3.00, what are the focal lengths of the objective and eyepiece?

56. A certain telescope has an objective mirror with an aperture diameter of 200 mm and a focal length of 2 000 mm. It captures the image of a nebula on photographic film at its prime focus with an exposure time of 1.50 min. To produce the same light energy per unit area on the film, what is the required exposure time to photograph the same nebula with a smaller telescope, which has an objective with a diameter of 60.0 mm and a focal length of 900 mm?

Additional Problems

57. The distance between an object and its upright image is 20.0 cm. If the magnification is 0.500, what is the focal length of the lens that is being used to form the image?

58. The distance between an object and its upright image is d. If the magnification is M, what is the focal length of the lens that is being used to form the image?

59. Your friend needs glasses with diverging lenses of focal length −65.0 cm for both eyes. You tell him he looks good

when he doesn't squint, but he is worried about how thick the lenses will be. Assuming the radius of curvature of the first surface is $R_1 = 50.0$ cm and the high-index plastic has a refractive index of 1.66, (a) find the required radius of curvature of the second surface. (b) Assume the lens is ground from a disk 4.00 cm in diameter and 0.100 cm thick at the center. Find the thickness of the plastic at the edge of the lens, measured parallel to the axis. *Suggestion:* Draw a large cross-sectional diagram.

60. A cylindrical rod of glass with index of refraction 1.50 is immersed in water with index 1.33. The diameter of the rod is 9.00 cm. The outer part of each end of the rod has been ground away to form each end into a hemisphere of radius 4.50 cm. The central portion of the rod with straight sides is 75.0 cm long. An object is situated in the water, on the axis of the rod, at a distance of 100 cm from the vertex of the nearer hemisphere. (a) Find the location of the final image formed by refraction at both surfaces. (b) Is the final image real or virtual? Upright or inverted? Enlarged or diminished?

61. A *zoom lens* system is a combination of lenses that produces a variable magnification while maintaining fixed object and image positions. The magnification is varied by moving one or more lenses along the axis. While multiple lenses are used in practice to obtain high-quality images, the effect of zooming in on an object can be demonstrated with a simple two-lens system. An object, two converging lenses, and a screen are mounted on an optical bench. The first lens, which is to the right of the object, has a focal length of 5.00 cm, and the second lens, which is to the right of the first lens, has a focal length of 10.0 cm. The screen is to the right of the second lens. Initially, an object is situated at a distance of 7.50 cm to the left of the first lens, and the image formed on the screen has a magnification of +1.00. (a) Find the distance between the object and the screen. (b) Both lenses are now moved along their common axis, while the object and the screen maintain fixed positions, until the image formed on the screen has a magnification of +3.00. Find the displacement of each lens from its initial position in (a). Can the lenses be displaced in more than one way?

62. The object in Figure P36.62 is midway between the lens and the mirror. The mirror's radius of curvature is 20.0 cm, and the lens has a focal length of −16.7 cm. Considering only the light that leaves the object and travels first toward the mirror, locate the final image formed by this system. Is this image real or virtual? Is it upright or inverted? What is the overall magnification?

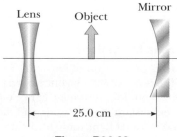

Figure P36.62

63. An object placed 10.0 cm from a concave spherical mirror produces a real image 8.00 cm from the mirror. If the object is moved to a new position 20.0 cm from the mirror, what is the position of the image? Is the latter image real or virtual?

64. In many applications it is necessary to expand or to decrease the diameter of a beam of parallel rays of light. This change can be made by using a converging lens and a diverging lens in combination. Suppose you have a converging lens of focal length 21.0 cm and a diverging lens of focal length − 12.0 cm. How can you arrange these lenses to increase the diameter of a beam of parallel rays? By what factor will the diameter increase?

65. [www] A parallel beam of light enters a glass hemisphere perpendicular to the flat face, as shown in Figure P36.65. The magnitude of the radius is 6.00 cm, and the index of refraction is 1.560. Determine the point at which the beam is focused. (Assume paraxial rays.)

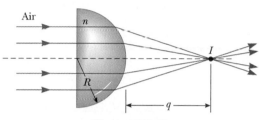

Air
n
I
R
q

Figure P36.65

66. **Review problem.** A spherical lightbulb of diameter 3.20 cm radiates light equally in all directions, with power 4.50 W. (a) Find the light intensity at the surface of the bulb. (b) Find the light intensity 7.20 m away from the center of the bulb. (c) At this 7.20-m distance a lens is set up with its axis pointing toward the bulb. The lens has a circular face with a diameter 15.0 cm and has a focal length of 35.0 cm. Find the diameter of the image of the bulb. (d) Find the light intensity at the image.

67. An object is placed 12.0 cm to the left of a diverging lens of focal length − 6.00 cm. A converging lens of focal length 12.0 cm is placed a distance d to the right of the diverging lens. Find the distance d so that the final image is at infinity. Draw a ray diagram for this case.

68. An observer to the right of the mirror–lens combination shown in Figure P36.68 sees two real images that are the same size and in the same location. One image is upright

and the other is inverted. Both images are 1.50 times larger than the object. The lens has a focal length of 10.0 cm. The lens and mirror are separated by 40.0 cm. Determine the focal length of the mirror. Do not assume that the figure is drawn to scale.

69. [www] The disk of the Sun subtends an angle of 0.533° at the Earth. What are the position and diameter of the solar image formed by a concave spherical mirror with a radius of curvature of 3.00 m?

70. Assume the intensity of sunlight is 1.00 kW/m² at a particular location. A highly reflecting concave mirror is to be pointed toward the Sun to produce a power of at least 350 W at the image. (a) Find the required radius R_a of the circular face area of the mirror. (b) Now suppose the light intensity is to be at least 120 kW/m² at the image. Find the required relationship between R_a and the radius of curvature R of the mirror. The disk of the Sun subtends an angle of 0.533° at the Earth.

71. In a darkened room, a burning candle is placed 1.50 m from a white wall. A lens is placed between candle and wall at a location that causes a larger, inverted image to form on the wall. When the lens is moved 90.0 cm toward the wall, another image of the candle is formed. Find (a) the two object distances that produce the specified images and (b) the focal length of the lens. (c) Characterize the second image.

72. Figure P36.72 shows a thin converging lens for which the radii of curvature are $R_1 = 9.00$ cm and $R_2 = − 11.0$ cm. The lens is in front of a concave spherical mirror with the radius of curvature $R = 8.00$ cm. (a) Assume its focal points F_1 and F_2 are 5.00 cm from the center of the lens. Determine its index of refraction. (b) The lens and mirror are 20.0 cm apart, and an object is placed 8.00 cm to the left of the lens. Determine the position of the final image and its magnification as seen by the eye in the figure. (c) Is the final image inverted or upright? Explain.

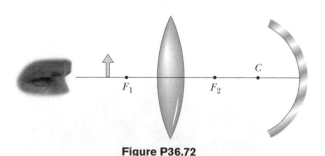

F_1 F_2 C

Figure P36.72

73. A compound microscope has an objective of focal length 0.300 cm and an eyepiece of focal length 2.50 cm. If an object is 3.40 mm from the objective, what is the magnification? (*Suggestion:* Use the lens equation for the objective.)

74. Two converging lenses having focal lengths of 10.0 cm and 20.0 cm are located 50.0 cm apart, as shown in

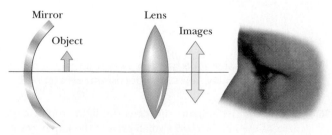

Mirror   Lens
Object   Images

Figure P36.68

Figure P36.74. The final image is to be located between the lenses at the position indicated. (a) How far to the left of the first lens should the object be? (b) What is the overall magnification? (c) Is the final image upright or inverted?

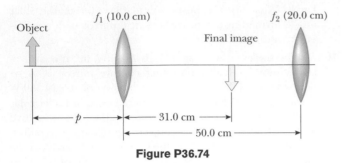

Figure P36.74

75. A cataract-impaired lens in an eye may be surgically removed and replaced by a manufactured lens. The focal length required for the new lens is determined by the lens-to-retina distance, which is measured by a sonar-like device, and by the requirement that the implant provide for correct distant vision. (a) Assuming the distance from lens to retina is 22.4 mm, calculate the power of the implanted lens in diopters. (b) Because no accommodation occurs and the implant allows for correct distant vision, a corrective lens for close work or reading must be used. Assume a reading distance of 33.0 cm and calculate the power of the lens in the reading glasses.

76. A floating strawberry illusion is achieved with two parabolic mirrors, each having a focal length 7.50 cm, facing each other so that their centers are 7.50 cm apart (Fig.

Figure P36.76

P36.76). If a strawberry is placed on the lower mirror, an image of the strawberry is formed at the small opening at the center of the top mirror. Show that the final image is formed at that location and describe its characteristics. (*Note:* A very startling effect is to shine a flashlight beam on this image. Even at a glancing angle, the incoming light beam is seemingly reflected from the image! Do you understand why?)

77. An object 2.00 cm high is placed 40.0 cm to the left of a converging lens having a focal length of 30.0 cm. A diverging lens with a focal length of -20.0 cm is placed 110 cm to the right of the converging lens. (a) Determine the position and magnification of the final image. (b) Is the image upright or inverted? (c) **What If?** Repeat parts (a) and (b) for the case where the second lens is a converging lens having a focal length of $+20.0$ cm.

78. Two lenses made of kinds of glass having different refractive indices n_1 and n_2 are cemented together to form what is called an *optical doublet*. Optical doublets are often used to correct chromatic aberrations in optical devices. The first lens of a doublet has one flat side and one concave side of radius of curvature R. The second lens has two convex sides of radius of curvature R. Show that the doublet can be modeled as a single thin lens with a focal length described by

$$\frac{1}{f} = \frac{2n_2 - n_1 - 1}{R}$$

Answers to Quick Quizzes

36.1 At *C*. A ray traced from the stone to the mirror and then to observer 2 looks like this:

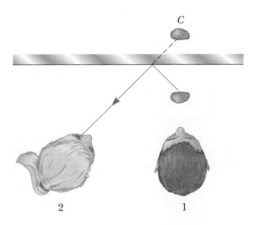

36.2 False. The water spots are 2 m away from you and your image is 4 m away. You cannot focus your eyes on both at the same time.

36.3 (b). A concave mirror will focus the light from a large area of the mirror onto a small area of the paper, resulting in a very high power input to the paper.

36.4 (b). A convex mirror always forms an image with a magnification less than one, so the mirror must be concave. In a concave mirror, only virtual images are upright. This particular photograph is of the Hubble Space Telescope primary mirror.

36.5 (d). When O is far away, the rays refract into the material of index n_2 and converge to form a real image as in Figure 36.18. For certain combinations of R and n_2 as O moves very close to the refracting surface, the incident angle of the rays increases so much that rays are no longer refracted back toward the principal axis. This results in a virtual image as shown below:

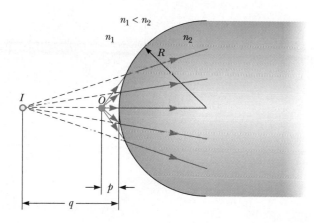

36.6 (a). No matter where O is, the rays refract into the air away from the normal and form a virtual image between O and the surface.

36.7 (b). Because the flat surfaces of the plane have infinite radii of curvature, Equation 36.15 indicates that the focal length is also infinite. Parallel rays striking the plane focus at infinity, which means that they remain parallel after passing through the glass.

36.8 (b). If there is a curve on the front surface, the refraction will differ at that surface when the mask is worn in air and water. In order for there to be no difference in refraction (for normal incidence), the front of the mask should be flat.

36.9 (a). Because the light reflecting from a mirror does not enter the material of the mirror, there is no opportunity for the dispersion of the material to cause chromatic aberration.

36.10 (a). If the object is brought closer to the lens, the image moves farther away from the lens, behind the plane of the film. In order to bring the image back up to the film, the lens is moved toward the object and away from the film.

36.11 (c). The Sun's rays must converge onto the paper. A farsighted person wears converging lenses.

Chapter 37

Interference of Light Waves

CHAPTER OUTLINE

37.1 Conditions for Interference

37.2 Young's Double-Slit Experiment

37.3 Intensity Distribution of the Double-Slit Interference Pattern

37.4 Phasor Addition of Waves

37.5 Change of Phase Due to Reflection

37.6 Interference in Thin Films

37.7 The Michelson Interferometer

▲ The colors in many of a hummingbird's feathers are not due to pigment. The iridescence that makes the brilliant colors that often appear on the throat and belly is due to an interference effect caused by structures in the feathers. The colors will vary with the viewing angle. (RO-MA/Index Stock Imagery)

In the preceding chapter, we used light rays to examine what happens when light passes through a lens or reflects from a mirror. This discussion completed our study of *geometric optics*. Here in Chapter 37 and in the next chapter, we are concerned with *wave optics* or *physical optics*, the study of interference, diffraction, and polarization of light. These phenomena cannot be adequately explained with the ray optics used in Chapters 35 and 36. We now learn how treating light as waves rather than as rays leads to a satisfying description of such phenomena.

37.1 Conditions for Interference

In Chapter 18, we found that the superposition of two mechanical waves can be constructive or destructive. In constructive interference, the amplitude of the resultant wave at a given position or time is greater than that of either individual wave, whereas in destructive interference, the resultant amplitude is less than that of either individual wave. Light waves also interfere with each other. Fundamentally, all interference associated with light waves arises when the electromagnetic fields that constitute the individual waves combine.

If two lightbulbs are placed side by side, no interference effects are observed because the light waves from one bulb are emitted independently of those from the other bulb. The emissions from the two lightbulbs do not maintain a constant phase relationship with each other over time. Light waves from an ordinary source such as a lightbulb undergo random phase changes in time intervals less than a nanosecond. Therefore, the conditions for constructive interference, destructive interference, or some intermediate state are maintained only for such short time intervals. Because the eye cannot follow such rapid changes, no interference effects are observed. Such light sources are said to be **incoherent.**

In order to observe interference in light waves, the following conditions must be met:

As with the hummingbird feathers shown in the opening photograph, the bright colors of peacock feathers are also due to interference. In both types of birds, structures in the feathers split and recombine visible light so that interference occurs for certain colors.

- The sources must be **coherent**—that is, **they must maintain a constant phase with respect to each other.**

- The sources should be **monochromatic**—that is, of a single wavelength.

As an example, single-frequency sound waves emitted by two side-by-side loudspeakers driven by a single amplifier can interfere with each other because the two speakers are coherent—that is, they respond to the amplifier in the same way at the same time.

Conditions for interference

37.2 Young's Double-Slit Experiment

A common method for producing two coherent light sources is to use a monochromatic source to illuminate a barrier containing two small openings (usually in the shape of slits). The light emerging from the two slits is coherent because a single

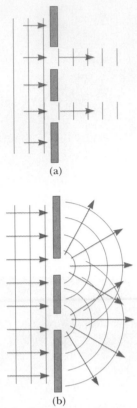

(a)

(b)

At a beach in Tel Aviv, Israel, plane water waves pass through two openings in a breakwall. Notice the diffraction effect—the waves exit the openings with circular wave fronts, as in Figure 37.1b. Notice also how the beach has been shaped by the circular wave fronts.

Courtesy of Sabina Zigman/Benjamin Cardozo High School

Figure 37.1 (a) If light waves did not spread out after passing through the slits, no interference would occur. (b) The light waves from the two slits overlap as they spread out, filling what we expect to be shadowed regions with light and producing interference fringes on a screen placed to the right of the slits.

▲ **PITFALL PREVENTION**

37.1 Interference Patterns Are Not Standing Waves

The interference pattern in Figure 37.2b shows bright and dark regions that appear similar to the antinodes and nodes of a standing-wave pattern on a string (Section 18.3). While both patterns depend on the principle of superposition, here are two major differences: (1) waves on a string propagate in only one dimension while the light-wave interference pattern exists in three dimensions; (2) the standing-wave pattern represents no net energy flow, while there is a net energy flow from the slits to the screen in an interference pattern.

source produces the original light beam and the two slits serve only to separate the original beam into two parts (which, after all, is what is done to the sound signal from the side-by-side loudspeakers at the end of the preceding section). Any random change in the light emitted by the source occurs in both beams at the same time, and as a result interference effects can be observed when the light from the two slits arrives at a viewing screen.

If the light traveled only in its original direction after passing through the slits, as shown in Figure 37.1a, the waves would not overlap and no interference pattern would be seen. Instead, as we have discussed in our treatment of Huygens's principle (Section 35.6), the waves spread out from the slits as shown in Figure 37.1b. In other words, the light deviates from a straight-line path and enters the region that would otherwise be shadowed. As noted in Section 35.3, this divergence of light from its initial line of travel is called **diffraction.**

Interference in light waves from two sources was first demonstrated by Thomas Young in 1801. A schematic diagram of the apparatus that Young used is shown in Figure 37.2a. Plane light waves arrive at a barrier that contains two parallel slits S_1 and S_2. These two slits serve as a pair of coherent light sources because waves emerging from them originate from the same wave front and therefore maintain a constant phase relationship. The light from S_1 and S_2 produces on a viewing screen a visible pattern of bright and dark parallel bands called **fringes** (Fig. 37.2b). When the light from S_1 and that from S_2 both arrive at a point on the screen such that constructive interference occurs at that location, a bright fringe appears. When the light from the two slits combines destructively at any location on the screen, a dark fringe results. Figure 37.3 is a photograph of an interference pattern produced by two coherent vibrating sources in a water tank.

Figure 37.4 shows some of the ways in which two waves can combine at the screen. In Figure 37.4a, the two waves, which leave the two slits in phase, strike the screen at the central point P. Because both waves travel the same distance, they arrive at P in phase. As a result, constructive interference occurs at this location, and a bright fringe is observed. In Figure 37.4b, the two waves also start in phase, but in this case the upper wave has to travel one wavelength farther than the lower

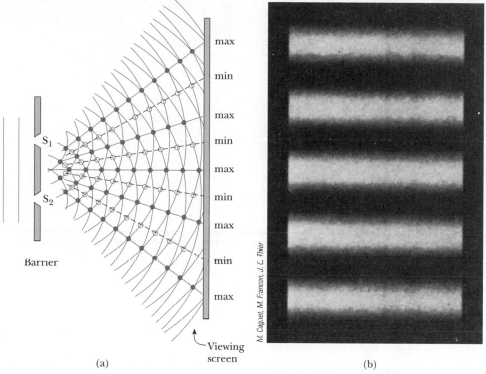

(a)

(b)

M. Cagnet, M. Francon, J. C. Thier

Active Figure 37.2 (a) Schematic diagram of Young's double-slit experiment. Slits S_1 and S_2 behave as coherent sources of light waves that produce an interference pattern on the viewing screen (drawing not to scale). (b) An enlargement of the center of a fringe pattern formed on the viewing screen.

Richard Megna/Fundamental Photographs

Figure 37.3 An interference pattern involving water waves is produced by two vibrating sources at the water's surface. The pattern is analogous to that observed in Young's double-slit experiment. Note the regions of constructive (A) and destructive (B) interference.

At the Active Figures link at http://www.pse6.com, **you can adjust the slit separation and the wavelength of the light to see the effect on the interference pattern.**

wave to reach point Q. Because the upper wave falls behind the lower one by exactly one wavelength, they still arrive in phase at Q, and so a second bright fringe appears at this location. At point R in Figure 37.4c, however, between points P and Q, the upper wave has fallen half a wavelength behind the lower wave. This means that a trough of the lower wave overlaps a crest of the upper wave; this gives rise to destructive interference at point R. For this reason, a dark fringe is observed at this location.

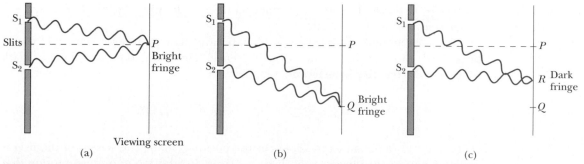

(a)

(b)

(c)

Figure 37.4 (a) Constructive interference occurs at point P when the waves combine. (b) Constructive interference also occurs at point Q. (c) Destructive interference occurs at R when the two waves combine because the upper wave falls half a wavelength behind the lower wave. (All figures not to scale.)

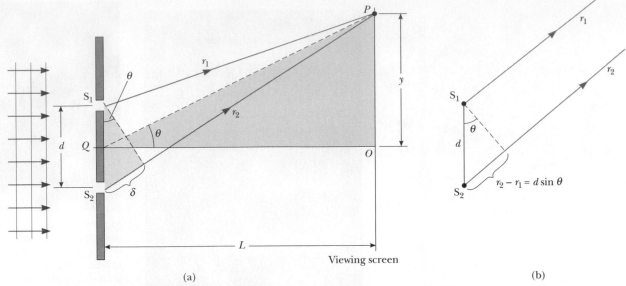

Figure 37.5 (a) Geometric construction for describing Young's double-slit experiment (not to scale). (b) When we assume that r_1 is parallel to r_2, the path difference between the two rays is $r_2 - r_1 = d \sin \theta$. For this approximation to be valid, it is essential that $L \gg d$.

We can describe Young's experiment quantitatively with the help of Figure 37.5. The viewing screen is located a perpendicular distance L from the barrier containing two slits, S_1 and S_2. These slits are separated by a distance d, and the source is monochromatic. To reach any arbitrary point P in the upper half of the screen, a wave from the lower slit must travel farther than a wave from the upper slit by a distance $d \sin \theta$. This distance is called the **path difference** δ (lowercase Greek delta). If we assume that r_1 and r_2 are parallel, which is approximately true if L is much greater than d, then δ is given by

Path difference

$$\delta = r_2 - r_1 = d \sin\theta \tag{37.1}$$

The value of δ determines whether the two waves are in phase when they arrive at point P. If δ is either zero or some integer multiple of the wavelength, then the two waves are in phase at point P and constructive interference results. Therefore, the condition for bright fringes, or **constructive interference,** at point P is

Conditions for constructive interference

$$\delta = d \sin\theta_{\text{bright}} = m\lambda \qquad (m = 0, \pm 1, \pm 2, \ldots) \tag{37.2}$$

The number m is called the **order number.** For constructive interference, the order number is the same as the number of wavelengths that represents the path difference between the waves from the two slits. The central bright fringe at $\theta = 0$ is called the *zeroth-order maximum.* The first maximum on either side, where $m = \pm 1$, is called the *first-order maximum,* and so forth.

When δ is an odd multiple of $\lambda/2$, the two waves arriving at point P are 180° out of phase and give rise to destructive interference. Therefore, the condition for dark fringes, or **destructive interference,** at point P is

Conditions for destructive interference

$$d \sin\theta_{\text{dark}} = (m + \tfrac{1}{2})\lambda \qquad (m = 0, \pm 1, \pm 2, \ldots) \tag{37.3}$$

It is useful to obtain expressions for the positions along the screen of the bright and dark fringes measured vertically from O to P. In addition to our assumption that $L \gg d$, we assume $d \gg \lambda$. These can be valid assumptions because in practice L is often on the order of 1 m, d a fraction of a millimeter, and λ a fraction of a micrometer for visible light. Under these conditions, θ is small; thus, we can use the small angle approximation $\sin\theta \approx \tan\theta$. Then, from triangle OPQ in Figure 37.5a,

we see that

$$y = L \tan\theta \approx L \sin\theta \qquad (37.4)$$

Solving Equation 37.2 for $\sin\theta$ and substituting the result into Equation 37.4, we see that the positions of the bright fringes measured from O are given by the expression

$$y_{\text{bright}} = \frac{\lambda L}{d} m \qquad (m = 0, \pm 1, \pm 2, \ldots) \qquad (37.5)$$

Using Equations 37.3 and 37.4, we find that the dark fringes are located at

$$y_{\text{dark}} = \frac{\lambda L}{d} \left(m + \tfrac{1}{2}\right) \qquad (m = 0, \pm 1, \pm 2, \ldots) \qquad (37.6)$$

As we demonstrate in Example 37.1, Young's double-slit experiment provides a method for measuring the wavelength of light. In fact, Young used this technique to do just that. Additionally, his experiment gave the wave model of light a great deal of credibility. It was inconceivable that particles of light coming through the slits could cancel each other in a way that would explain the dark fringes.

Quick Quiz 37.1 If you were to blow smoke into the space between the barrier and the viewing screen of Figure 37.5a, the smoke would show (a) no evidence of interference between the barrier and the screen (b) evidence of interference everywhere between the barrier and the screen.

Quick Quiz 37.2 In a two-slit interference pattern projected on a screen, the fringes are equally spaced on the screen (a) everywhere (b) only for large angles (c) only for small angles.

Quick Quiz 37.3 Which of the following will cause the fringes in a two-slit interference pattern to move farther apart? (a) decreasing the wavelength of the light (b) decreasing the screen distance L (c) decreasing the slit spacing d (d) immersing the entire apparatus in water.

▲ **PITFALL PREVENTION**

37.2   It May Not Be True That θ Is Small

The approximation $\sin\theta \approx \tan\theta$ is true to three-digit precision only for angles less than about 4°. If you are considering fringes that are far removed from the central fringe, $\tan\theta = y/L$ is still true, but the small-angle approximation may not be valid. In this case, Equations 37.5 and 37.6 cannot be used. These problems can be solved, but the geometry is not as simple.

▲ **PITFALL PREVENTION**

37.3   It May Not Be True That $L \gg d$

Equations 37.2, 37.3, 37.5, and 37.6 were developed under the assumption that $L \gg d$. This is a very common situation, but you are likely to encounter some situations in which this assumption is not valid. In those cases, the geometric construction will be more complicated, but the general approach outlined here will be similar.

Example 37.1   Measuring the Wavelength of a Light Source   `Interactive`

A viewing screen is separated from a double-slit source by 1.2 m. The distance between the two slits is 0.030 mm. The second-order bright fringe ($m = 2$) is 4.5 cm from the center line.

(A) Determine the wavelength of the light.

Solution We can use Equation 37.5, with $m = 2$, $y_{\text{bright}} = 4.5 \times 10^{-2}$ m, $L = 1.2$ m, and $d = 3.0 \times 10^{-5}$ m:

$$\lambda = \frac{y_{\text{bright}} d}{mL} = \frac{(4.5 \times 10^{-2}\ \text{m})(3.0 \times 10^{-5}\ \text{m})}{2(1.2\ \text{m})}$$

$$= 5.6 \times 10^{-7}\ \text{m} = \boxed{560\ \text{nm}}$$

which is in the green range of visible light.

(B) Calculate the distance between adjacent bright fringes.

Solution From Equation 37.5 and the results of part (A), we obtain

$$y_{m+1} - y_m = \frac{\lambda L}{d}(m+1) - \frac{\lambda L}{d} m$$

$$= \frac{\lambda L}{d} = \frac{(5.6 \times 10^{-7}\ \text{m})(1.2\ \text{m})}{3.0 \times 10^{-5}\ \text{m}}$$

$$= 2.2 \times 10^{-2}\ \text{m} = \boxed{2.2\ \text{cm}}$$

 Investigate the double-slit interference pattern at the Interactive Worked Example link at **http://www.pse6.com.**

Example 37.2 Separating Double-Slit Fringes of Two Wavelengths

A light source emits visible light of two wavelengths: $\lambda = 430$ nm and $\lambda' = 510$ nm. The source is used in a double-slit interference experiment in which $L = 1.50$ m and $d = 0.025\,0$ mm. Find the separation distance between the third-order bright fringes.

Solution Using Equation 37.5, with $m = 3$, we find that the fringe positions corresponding to these two wavelengths are

$$y_{bright} = \frac{\lambda L}{d}\,m = 3\,\frac{\lambda L}{d} = 3\,\frac{(430 \times 10^{-9}\text{ m})(1.50\text{ m})}{0.025\,0 \times 10^{-3}\text{ m}}$$

$$= 7.74 \times 10^{-2}\text{ m}$$

$$y'_{bright} = \frac{\lambda' L}{d}\,m = 3\,\frac{\lambda' L}{d} = 3\,\frac{(510 \times 10^{-9}\text{ m})(1.50\text{ m})}{0.025\,0 \times 10^{-3}\text{ m}}$$

$$= 9.18 \times 10^{-2}\text{ m}$$

Hence, the separation distance between the two fringes is

$$\Delta y = 9.18 \times 10^{-2}\text{ m} - 7.74 \times 10^{-2}\text{ m}$$

$$= 1.40 \times 10^{-2}\text{ m} = \boxed{1.40\text{ cm}}$$

What If? What if we examine the entire interference pattern due to the two wavelengths and look for overlapping fringes? Are there any locations on the screen where the bright fringes from the two wavelengths overlap exactly?

Answer We could find such a location by setting the location of any bright fringe due to λ equal to one due to λ',

using Equation 37.5:

$$\frac{\lambda L}{d}\,m = \frac{\lambda' L}{d}\,m'$$

$$\frac{\lambda}{\lambda'} = \frac{m'}{m}$$

Substituting the wavelengths, we have

$$\frac{m'}{m} = \frac{\lambda}{\lambda'} = \frac{430\text{ nm}}{510\text{ nm}} = \frac{43}{51}$$

This might suggest that the 51st bright fringe of the 430-nm light would overlap with the 43rd bright fringe of the 510-nm light. However, if we use Equation 37.5 to find the value of y for these fringes, we find

$$y = 51\,\frac{(430 \times 10^{-9}\text{ m})(1.5\text{ m})}{0.025 \times 10^{-3}\text{ m}} = 1.32\text{ m} = y'$$

This value of y is comparable to L, so that the small-angle approximation used in Equation 37.4 is *not* valid. This suggests that we should not expect Equation 37.5 to give us the correct result. If you use the exact relationship $y = L\tan\theta$, you can show that the bright fringes do indeed overlap when the same condition, $m'/m = \lambda/\lambda'$, is met (see Problem 44). Thus, the 51st fringe of the 430-nm light does overlap with the 43rd fringe of the 510-nm light, but not at the location of 1.32 m. You are asked to find the correct location as part of Problem 44.

37.3 Intensity Distribution of the Double-Slit Interference Pattern

Note that the edges of the bright fringes in Figure 37.2b are not sharp—there is a gradual change from bright to dark. So far we have discussed the locations of only the centers of the bright and dark fringes on a distant screen. Let us now direct our attention to the intensity of the light at other points between the positions of maximum constructive and destructive interference. In other words, we now calculate the distribution of light intensity associated with the double-slit interference pattern.

Again, suppose that the two slits represent coherent sources of sinusoidal waves such that the two waves from the slits have the same angular frequency ω and a constant phase difference ϕ. The total magnitude of the electric field at point P on the screen in Figure 37.6 is the superposition of the two waves. Assuming that the two waves have the same amplitude E_0, we can write the magnitude of the electric field at point P due to each wave separately as

$$E_1 = E_0 \sin \omega t \qquad \text{and} \qquad E_2 = E_0 \sin(\omega t + \phi) \qquad (37.7)$$

Although the waves are in phase at the slits, *their phase difference ϕ at P depends on the path difference* $\delta = r_2 - r_1 = d\sin\theta$. A path difference of λ (for constructive interference) corresponds to a phase difference of 2π rad. A path difference of δ is the same fraction of λ as the phase difference ϕ is of 2π. We can describe this mathematically

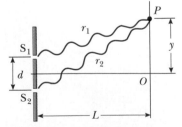

Figure 37.6 Construction for analyzing the double-slit interference pattern. A bright fringe, or intensity maximum, is observed at O.

with the ratio

$$\frac{\delta}{\lambda} = \frac{\phi}{2\pi}$$

which gives us

$$\phi = \frac{2\pi}{\lambda}\,\delta = \frac{2\pi}{\lambda}\,d\sin\theta \qquad (37.8)$$ **Phase difference**

This equation tells us precisely how the phase difference ϕ depends on the angle θ in Figure 37.5.

Using the superposition principle and Equation 37.7, we can obtain the magnitude of the resultant electric field at point P:

$$E_P = E_1 + E_2 = E_0[\sin\omega t + \sin(\omega t + \phi)] \qquad (37.9)$$

To simplify this expression, we use the trigonometric identity

$$\sin A + \sin B = 2\sin\left(\frac{A + B}{2}\right)\cos\left(\frac{A - B}{2}\right)$$

Taking $A = \omega t + \phi$ and $B = \omega t$, we can write Equation 37.9 in the form

$$E_P = 2E_0 \cos\left(\frac{\phi}{2}\right)\sin\left(\omega t + \frac{\phi}{2}\right) \qquad (37.10)$$

This result indicates that the electric field at point P has the same frequency ω as the light at the slits, but that the amplitude of the field is multiplied by the factor $2\cos(\phi/2)$. To check the consistency of this result, note that if $\phi = 0, 2\pi, 4\pi, \ldots$, then the magnitude of the electric field at point P is $2E_0$, corresponding to the condition for maximum constructive interference. These values of ϕ are consistent with Equation 37.2 for constructive interference. Likewise, if $\phi = \pi, 3\pi, 5\pi, \ldots$, then the magnitude of the electric field at point P is zero; this is consistent with Equation 37.3 for total destructive interference.

Finally, to obtain an expression for the light intensity at point P, recall from Section 34.3 that *the intensity of a wave is proportional to the square of the resultant electric field magnitude at that point* (Eq. 34.21). Using Equation 37.10, we can therefore express the light intensity at point P as

$$I \propto E_P^2 = 4E_0^2 \cos^2\left(\frac{\phi}{2}\right)\sin^2\left(\omega t + \frac{\phi}{2}\right)$$

Most light-detecting instruments measure time-averaged light intensity, and the time-averaged value of $\sin^2(\omega t + \phi/2)$ over one cycle is $\frac{1}{2}$. (See Figure 33.5.) Therefore, we can write the average light intensity at point P as

$$I = I_{\max} \cos^2\left(\frac{\phi}{2}\right) \qquad (37.11)$$

where $I_{\max}$ is the maximum intensity on the screen and the expression represents the time average. Substituting the value for ϕ given by Equation 37.8 into this expression, we find that

$$I = I_{\max} \cos^2\left(\frac{\pi d \sin\theta}{\lambda}\right) \qquad (37.12)$$

Alternatively, because $\sin\theta \approx y/L$ for small values of θ in Figure 37.5, we can write Equation 37.12 in the form

$$I \approx I_{\max} \cos^2\left(\frac{\pi d}{\lambda L}\,y\right) \qquad (37.13)$$

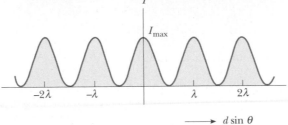

Figure 37.7 Light intensity versus $d \sin \theta$ for a double-slit interference pattern when the screen is far from the two slits ($L \gg d$).

Constructive interference, which produces light intensity maxima, occurs when the quantity $\pi \, dy/\lambda L$ is an integral multiple of π, corresponding to $y = (\lambda L/d)\,m$. This is consistent with Equation 37.5.

A plot of light intensity versus $d \sin \theta$ is given in Figure 37.7. The interference pattern consists of equally spaced fringes of equal intensity. Remember, however, that this result is valid only if the slit-to-screen distance L is much greater than the slit separation, and only for small values of θ.

> **Quick Quiz 37.4** At dark areas in an interference pattern, the light waves have canceled. Thus, there is zero intensity at these regions and, therefore, no energy is arriving. Consequently, when light waves interfere and form an interference pattern, (a) energy conservation is violated because energy disappears in the dark areas (b) energy transferred by the light is transformed to another type of energy in the dark areas (c) the total energy leaving the slits is distributed among light and dark areas and energy is conserved.

37.4 Phasor Addition of Waves

In the preceding section, we combined two waves algebraically to obtain the resultant wave amplitude at some point on a screen. Unfortunately, this analytical procedure becomes cumbersome when we must add several wave amplitudes. Because we shall eventually be interested in combining a large number of waves, we now describe a graphical procedure for this purpose.

Let us again consider a sinusoidal wave whose electric field component is given by

$$E_1 = E_0 \sin \omega t$$

where E_0 is the wave amplitude and ω is the angular frequency. We used phasors in Chapter 33 to analyze AC circuits, and again we find the use of phasors to be valuable

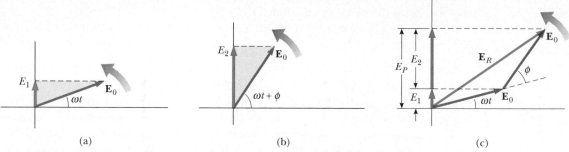

Figure 37.8 (a) Phasor diagram for the wave disturbance $E_1 = E_0 \sin \omega t$. The phasor is a vector of length E_0 rotating counterclockwise. (b) Phasor diagram for the wave $E_2 = E_0 \sin(\omega t + \phi)$. (c) The phasor $\mathbf{E}_R$ represents the combination of the waves in part (a) and (b).

in discussing wave interference. The sinusoidal wave we are discussing can be represented graphically by a phasor of magnitude E_0 rotating about the origin counterclockwise with an angular frequency ω, as in Figure 37.8a. Note that the phasor makes an angle ωt with the horizontal axis. The projection of the phasor on the vertical axis represents E_1, the magnitude of the wave disturbance at some time t. Hence, as the phasor rotates in a circle about the origin, the projection E_1 oscillates along the vertical axis.

Now consider a second sinusoidal wave whose electric field component is given by

$$E_2 = E_0 \sin(\omega t + \phi)$$

This wave has the same amplitude and frequency as E_1, but its phase is ϕ with respect to E_1. The phasor representing E_2 is shown in Figure 37.8b. We can obtain the resultant wave, which is the sum of E_1 and E_2, graphically by redrawing the phasors as shown in Figure 37.8c, in which the tail of the second phasor is placed at the tip of the first. As with vector addition, the resultant phasor $\mathbf{E}_R$ runs from the tail of the first phasor to the tip of the second. Furthermore, $\mathbf{E}_R$ rotates along with the two individual phasors at the same angular frequency ω. The projection of $\mathbf{E}_R$ along the vertical axis equals the sum of the projections of the two other phasors: $E_P = E_1 + E_2$.

It is convenient to construct the phasors at $t = 0$ as in Figure 37.9. From the geometry of one of the right triangles, we see that

$$\cos \alpha = \frac{E_R/2}{E_0}$$

which gives

$$E_R = 2E_0 \cos \alpha$$

Because the sum of the two opposite interior angles equals the exterior angle ϕ, we see that $\alpha = \phi/2$; thus,

$$E_R = 2E_0 \cos\left(\frac{\phi}{2}\right)$$

Hence, the projection of the phasor $\mathbf{E}_R$ along the vertical axis at any time t is

$$E_P = E_R \sin\left(\omega t + \frac{\phi}{2}\right) = 2E_0 \cos(\phi/2) \sin\left(\omega t + \frac{\phi}{2}\right)$$

This is consistent with the result obtained algebraically, Equation 37.10. The resultant phasor has an amplitude $2E_0 \cos(\phi/2)$ and makes an angle $\phi/2$ with the first phasor.

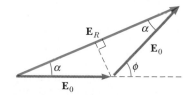

Figure 37.9 A reconstruction of the resultant phasor $\mathbf{E}_R$. From the geometry, note that $\alpha = \phi/2$.

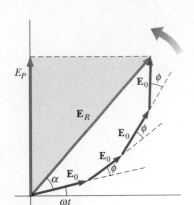

Figure 37.10 The phasor $\mathbf{E}_R$ is the resultant of four phasors of equal amplitude E_0. The phase of $\mathbf{E}_R$ with respect to the first phasor is α. The projection E_P on the vertical axis represents the combination of the four phasors.

Furthermore, the average light intensity at point P, which varies as $E_P{}^2$, is proportional to $\cos^2(\phi/2)$, as described in Equation 37.11.

We can now describe how to obtain the resultant of several waves that have the same frequency:

- Represent the waves by phasors, as shown in Figure 37.10, remembering to maintain the proper phase relationship between one phasor and the next.

- The resultant phasor $\mathbf{E}_R$ is the vector sum of the individual phasors. At each instant, the projection of $\mathbf{E}_R$ along the vertical axis represents the time variation of the resultant wave. The phase angle α of the resultant wave is the angle between $\mathbf{E}_R$ and the first phasor. From Figure 37.10, drawn for four phasors, we see that the resultant wave is given by the expression $E_P = E_R \sin(\omega t + \alpha)$.

Phasor Diagrams for Two Coherent Sources

As an example of the phasor method, consider the interference pattern produced by two coherent sources. Figure 37.11 represents the phasor diagrams for various values of the phase difference ϕ and the corresponding values of the path difference δ, which are obtained from Equation 37.8. The light intensity at a point is a maximum when $\mathbf{E}_R$ is a maximum; this occurs at $\phi = 0, 2\pi, 4\pi, \ldots$. The light intensity at some point is zero when $\mathbf{E}_R$ is zero; this occurs at $\phi = \pi, 3\pi, 5\pi, \ldots$. These results are in complete agreement with the analytical procedure described in the preceding section.

Three-Slit Interference Pattern

Using phasor diagrams, let us analyze the interference pattern caused by three equally spaced slits. We can express the electric field components at a point P on the screen caused by waves from the individual slits as

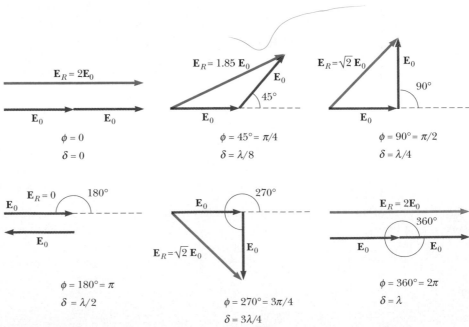

Choose any phase angle at the Active Figures link at http://www.pse6.com and see the resultant phasor.

Active Figure 37.11 Phasor diagrams for a double-slit interference pattern. The resultant phasor $\mathbf{E}_R$ is a maximum when $\phi = 0, 2\pi, 4\pi, \ldots$ and is zero when $\phi = \pi$, $3\pi, 5\pi, \ldots$.

$$E_1 = E_0 \sin \omega t$$

$$E_2 = E_0 \sin(\omega t + \phi)$$

$$E_3 = E_0 \sin(\omega t + 2\phi)$$

where ϕ is the phase difference between waves from adjacent slits. We can obtain the resultant magnitude of the electric field at point P from the phasor diagram in Figure 37.12.

The phasor diagrams for various values of ϕ are shown in Figure 37.13. Note that the resultant magnitude of the electric field at P has a maximum value of $3E_0$, a condition that occurs when $\phi = 0, \pm 2\pi, \pm 4\pi, \ldots$. These points are called *primary maxima*. Such primary maxima occur whenever the three phasors are aligned as shown in Figure 37.13a. We also find secondary maxima of amplitude E_0 occurring between the primary maxima at points where $\phi = \pm \pi, \pm 3\pi, \ldots$. For these points, the wave from one slit exactly cancels that from another slit (Fig. 37.13d). This means that only light from the third slit contributes to the resultant, which consequently has a total amplitude of E_0. Total destructive interference occurs whenever the three phasors form a closed triangle, as shown in Figure 37.13c. These points where $E_R = 0$ correspond to $\phi = \pm 2\pi/3, \pm 4\pi/3, \ldots$. You should be able to construct other phasor diagrams for values of ϕ greater than π.

Figure 37.14 shows multiple-slit interference patterns for a number of configurations. For three slits, note that the primary maxima are nine times more intense than the secondary maxima as measured by the height of the curve. This is because the intensity varies as E_R^2. For N slits, the intensity of the primary maxima is N^2 times greater than that due to a single slit. As the number of slits increases, the primary maxima increase in intensity and become narrower, while the secondary maxima decrease in intensity relative to the primary maxima. Figure 37.14 also shows that as the number of slits increases, the number of secondary maxima also increases. In fact, the number of secondary maxima is always $N - 2$ where N is the number of slits. In Section 38.4 (next chapter), we shall investigate the pattern for a very large number of slits in a device called a *diffraction grating*.

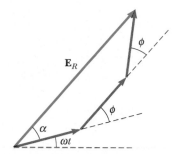

Figure 37.12 Phasor diagram for three equally spaced slits.

Quick Quiz 37.5 Using Figure 37.14 as a model, sketch the interference pattern from six slits.

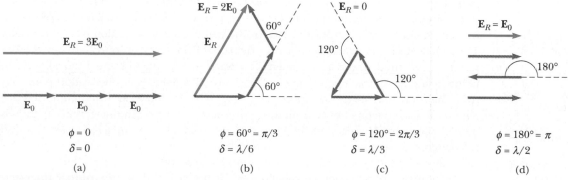

Active Figure 37.13 Phasor diagrams for three equally spaced slits at various values of ϕ. Note from (a) that there are primary maxima of amplitude $3E_0$ and from (d) that there are secondary maxima of amplitude E_0.

Choose any phase angle at the Active Figures link at
http://www.pse6.com *and see the resultant phasor.*

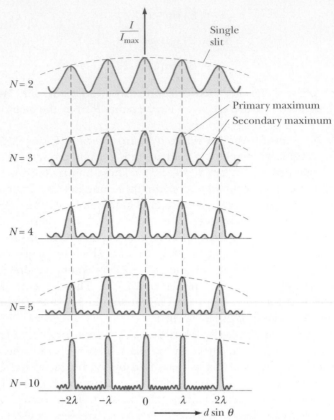

Figure 37.14 Multiple-slit interference patterns. As N, the number of slits, is increased, the primary maxima (the tallest peaks in each graph) become narrower but remain fixed in position and the number of secondary maxima increases. For any value of N, the decrease in intensity in maxima to the left and right of the central maximum, indicated by the blue dashed arcs, is due to *diffraction patterns* from the individual slits, which are discussed in Chapter 38.

37.5   Change of Phase Due to Reflection

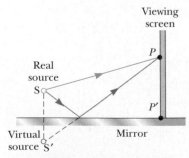

Figure 37.15 Lloyd's mirror. An interference pattern is produced at point P on the screen as a result of the combination of the direct ray (blue) and the reflected ray (brown). The reflected ray undergoes a phase change of 180°.

Young's method for producing two coherent light sources involves illuminating a pair of slits with a single source. Another simple, yet ingenious, arrangement for producing an interference pattern with a single light source is known as *Lloyd's mirror*[1] (Fig. 37.15). A point light source is placed at point S close to a mirror, and a viewing screen is positioned some distance away and perpendicular to the mirror. Light waves can reach point P on the screen either directly from S to P or by the path involving reflection from the mirror. The reflected ray can be treated as a ray originating from a virtual source at point S′. As a result, we can think of this arrangement as a double-slit source with the distance between points S and S′ comparable to length d in Figure 37.5. Hence, at observation points far from the source ($L \gg d$) we expect waves from points S and S′ to form an interference pattern just like the one we see from two real coherent sources. An interference pattern is indeed observed. However, the positions of the dark and bright fringes are reversed relative to the pattern created by two real coherent sources (Young's experiment). This can only occur if the coherent sources at points S and S′ differ in phase by 180°.

To illustrate this further, consider point P', the point where the mirror intersects the screen. This point is equidistant from points S and S′. If path difference alone were responsible for the phase difference, we would see a bright fringe at point P' (because the path difference is zero for this point), corresponding to the central bright fringe of

---

[1]   Developed in 1834 by Humphrey Lloyd (1800–1881), Professor of Natural and Experimental Philosophy, Trinity College, Dublin.

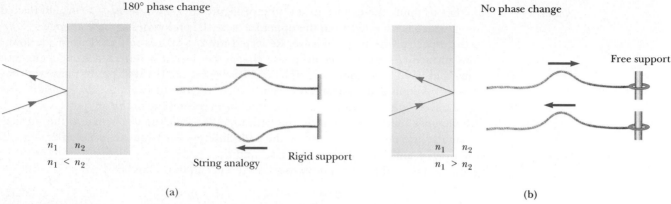

180° phase change   No phase change

n_1 n_2

$n_1 < n_2$ String analogy   Rigid support n_1 n_2

$n_1 > n_2$

Free support

(a)   (b)

Figure 37.16 (a) For $n_1 < n_2$, a light ray traveling in medium 1 when reflected from the surface of medium 2 undergoes a 180° phase change. The same thing happens with a reflected pulse traveling along a string fixed at one end. (b) For $n_1 > n_2$, a light ray traveling in medium 1 undergoes no phase change when reflected from the surface of medium 2. The same is true of a reflected wave pulse on a string whose supported end is free to move.

the two-slit interference pattern. Instead, we observe a dark fringe at point P'. From this, we conclude that a 180° phase change must be produced by reflection from the mirror. In general, **an electromagnetic wave undergoes a phase change of 180° upon reflection from a medium that has a higher index of refraction than the one in which the wave is traveling.**

It is useful to draw an analogy between reflected light waves and the reflections of a transverse wave pulse on a stretched string (Section 16.4). The reflected pulse on a string undergoes a phase change of 180° when reflected from the boundary of a denser medium, but no phase change occurs when the pulse is reflected from the boundary of a less dense medium. Similarly, an electromagnetic wave undergoes a 180° phase change when reflected from a boundary leading to an optically denser medium (defined as a medium with a higher index of refraction), but no phase change occurs when the wave is reflected from a boundary leading to a less dense medium. These rules, summarized in Figure 37.16, can be deduced from Maxwell's equations, but the treatment is beyond the scope of this text.

37.6 Interference in Thin Films

Interference effects are commonly observed in thin films, such as thin layers of oil on water or the thin surface of a soap bubble. The varied colors observed when white light is incident on such films result from the interference of waves reflected from the two surfaces of the film.

Consider a film of uniform thickness t and index of refraction n, as shown in Figure 37.17. Let us assume that the light rays traveling in air are nearly normal to the two surfaces of the film. To determine whether the reflected rays interfere constructively or destructively, we first note the following facts:

- A wave traveling from a medium of index of refraction n_1 toward a medium of index of refraction n_2 undergoes a 180° phase change upon reflection when $n_2 > n_1$ and undergoes no phase change if $n_2 < n_1$.

- The wavelength of light λ_n in a medium whose index of refraction is n (see Section 35.5) is

$$\lambda_n = \frac{\lambda}{n} \tag{37.14}$$

where λ is the wavelength of the light in free space.

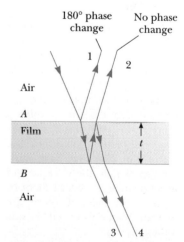

180° phase change   No phase change

1   2

Air

A

Film t

B

Air

3   4

Figure 37.17 Interference in light reflected from a thin film is due to a combination of rays 1 and 2 reflected from the upper and lower surfaces of the film. Rays 3 and 4 lead to interference effects for light transmitted through the film.

Let us apply these rules to the film of Figure 37.17, where $n_{film} > n_{air}$. Reflected ray 1, which is reflected from the upper surface (A), undergoes a phase change of 180° with respect to the incident wave. Reflected ray 2, which is reflected from the lower film surface (B), undergoes no phase change because it is reflected from a medium (air) that has a lower index of refraction. Therefore, ray 1 is 180° out of phase with ray 2, which is equivalent to a path difference of $\lambda_n/2$. However, we must also consider that ray 2 travels an extra distance $2t$ before the waves recombine in the air above surface A. (Remember that we are considering light rays that are close to normal to the surface. If the rays are not close to normal, the path difference is larger than $2t$.) If $2t = \lambda_n/2$, then rays 1 and 2 recombine in phase, and the result is constructive interference. In general, the condition for *constructive* interference in thin films is[2]

$$2t = (m + \tfrac{1}{2})\lambda_n \qquad (m = 0, 1, 2, \ldots) \tag{37.15}$$

This condition takes into account two factors: (1) the difference in path length for the two rays (the term $m\lambda_n$) and (2) the 180° phase change upon reflection (the term $\lambda_n/2$). Because $\lambda_n = \lambda/n$, we can write Equation 37.15 as

Conditions for constructive interference in thin films

$$2nt = (m + \tfrac{1}{2})\lambda \qquad (m = 0, 1, 2, \ldots) \tag{37.16}$$

If the extra distance $2t$ traveled by ray 2 corresponds to a multiple of λ_n, then the two waves combine out of phase, and the result is destructive interference. The general equation for *destructive* interference in thin films is

Conditions for destructive interference in thin films

$$2nt = m\lambda \qquad (m = 0, 1, 2, \ldots) \tag{37.17}$$

The foregoing conditions for constructive and destructive interference are valid when the medium above the top surface of the film is the same as the medium below the bottom surface or, if there are different media above and below the film, the index of refraction of both is less than n. If the film is placed between two different media, one with $n < n_{film}$ and the other with $n > n_{film}$, then the conditions for constructive and destructive interference are reversed. In this case, either there is a phase change of 180° for both ray 1 reflecting from surface A and ray 2 reflecting from surface B, or there is no phase change for either ray; hence, the net change in relative phase due to the reflections is zero.

Rays 3 and 4 in Figure 37.17 lead to interference effects in the light transmitted through the thin film. The analysis of these effects is similar to that of the reflected light. You are asked to explore the transmitted light in Problems 31, 36, and 37.

PITFALL PREVENTION

37.4 Be Careful with Thin Films

Be sure to include *both* effects—path length and phase change—when analyzing an interference pattern resulting from a thin film. The possible phase change is a new feature that we did not need to consider for double-slit interference. Also think carefully about the material on either side of the film. You may have situations in which there is a 180° phase change at *both* surfaces or at *neither* surface, if there are different materials on either side of the film.

Quick Quiz 37.6 In a laboratory accident, you spill two liquids onto water, neither of which mixes with the water. They both form thin films on the water surface. When the films become very thin as they spread, you observe that one film becomes bright and the other dark in reflected light. The film that is dark (a) has an index of refraction higher than that of water (b) has an index of refraction lower than that of water (c) has an index of refraction equal to that of water (d) has an index of refraction lower than that of the bright film.

Quick Quiz 37.7 One microscope slide is placed on top of another with their left edges in contact and a human hair under the right edge of the upper slide. As a result, a wedge of air exists between the slides. An interference pattern results when monochromatic light is incident on the wedge. At the left edges of the slides, there is (a) a dark fringe (b) a bright fringe (c) impossible to determine.

---

[2]   The full interference effect in a thin film requires an analysis of an infinite number of reflections back and forth between the top and bottom surfaces of the film. We focus here only on a single reflection from the bottom of the film, which provides the largest contribution to the interference effect.

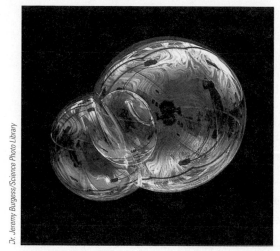

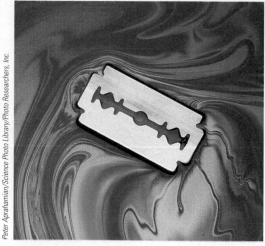

(*Left*) Interference in soap bubbles. The colors are due to interference between light rays reflected from the front and back surfaces of the thin film of soap making up the bubble. The color depends on the thickness of the film, ranging from black where the film is thinnest to magenta where it is thickest. (*Right*) A thin film of oil floating on water displays interference, as shown by the pattern of colors when white light is incident on the film. Variations in film thickness produce the interesting color pattern. The razor blade gives you an idea of the size of the colored bands.

Newton's Rings

Another method for observing interference in light waves is to place a plano-convex lens on top of a flat glass surface, as shown in Figure 37.18a. With this arrangement, the air film between the glass surfaces varies in thickness from zero at the point of contact to some value t at point P. If the radius of curvature R of the lens is much greater than the distance r, and if the system is viewed from above, a pattern of light and dark rings is observed, as shown in Figure 37.18b. These circular fringes, discovered by Newton, are called **Newton's rings.**

The interference effect is due to the combination of ray 1, reflected from the flat plate, with ray 2, reflected from the curved surface of the lens. Ray 1 undergoes a phase change of 180° upon reflection (because it is reflected from a medium of higher index of refraction), whereas ray 2 undergoes no phase change (because it is reflected from a medium of lower refractive index). Hence, the conditions for constructive and destructive interference are given by Equations 37.16 and 37.17, respectively, with $n = 1$ because the film is air.

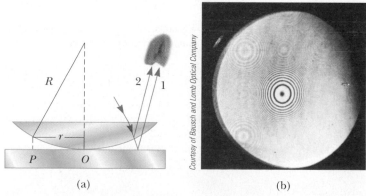

(a)   (b)

Figure 37.18 (a) The combination of rays reflected from the flat plate and the curved lens surface gives rise to an interference pattern known as Newton's rings. (b) Photograph of Newton's rings.

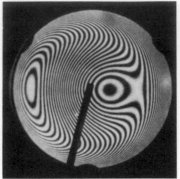

Figure 37.19 This asymmetrical interference pattern indicates imperfections in the lens of a Newton's-rings apparatus.

The contact point at O is dark, as seen in Figure 37.18b, because there is no path difference and the total phase change is due only to the 180° phase change upon reflection.

Using the geometry shown in Figure 37.18a, we can obtain expressions for the radii of the bright and dark bands in terms of the radius of curvature R and wavelength λ. For example, the dark rings have radii given by the expression $r \approx \sqrt{m\lambda R / n}$. The details are left as a problem for you to solve (see Problem 62). We can obtain the wavelength of the light causing the interference pattern by measuring the radii of the rings, provided R is known. Conversely, we can use a known wavelength to obtain R.

One important use of Newton's rings is in the testing of optical lenses. A circular pattern like that pictured in Figure 37.18b is obtained only when the lens is ground to a perfectly symmetric curvature. Variations from such symmetry might produce a pattern like that shown in Figure 37.19. These variations indicate how the lens must be reground and repolished to remove imperfections.

PROBLEM-SOLVING HINTS

Thin-Film Interference

You should keep the following ideas in mind when you work thin-film interference problems:

- Identify the thin film causing the interference.

- The type of interference that occurs is determined by the phase relationship between the portion of the wave reflected at the upper surface of the film and the portion reflected at the lower surface.

- Phase differences between the two portions of the wave have two causes: (1) differences in the distances traveled by the two portions and (2) phase changes that may occur upon reflection.

- When the distance traveled and phase changes upon reflection are both taken into account, the interference is constructive if the equivalent path difference between the two waves is an integral multiple of λ, and it is destructive if the path difference is $\lambda/2$, $3\lambda/2$, $5\lambda/2$, and so forth.

Example 37.3   Interference in a Soap Film

Calculate the minimum thickness of a soap-bubble film that results in constructive interference in the reflected light if the film is illuminated with light whose wavelength in free space is $\lambda = 600$ nm.

Solution The minimum film thickness for constructive interference in the reflected light corresponds to $m = 0$ in Equation 37.16. This gives $2nt = \lambda/2$, or

$$t = \frac{\lambda}{4n} = \frac{600 \text{ nm}}{4(1.33)} = \boxed{113 \text{ nm}}$$

What If? What if the film is twice as thick? Does this situation produce constructive interference?

Answer Using Equation 37.16, we can solve for the thicknesses at which constructive interference will occur:

$$t = (m + \tfrac{1}{2}) \frac{\lambda}{2n} = (2m + 1) \frac{\lambda}{4n} \qquad (m = 0, 1, 2, \ldots)$$

The allowed values of m show that constructive interference will occur for *odd* multiples of the thickness corresponding to $m = 0$, $t = 113$ nm. Thus, constructive interference will *not* occur for a film that is twice as thick.

Example 37.4 Nonreflective Coatings for Solar Cells

Solar cells—devices that generate electricity when exposed to sunlight—are often coated with a transparent, thin film of silicon monoxide (SiO, $n = 1.45$) to minimize reflective losses from the surface. Suppose that a silicon solar cell ($n = 3.5$) is coated with a thin film of silicon monoxide for this purpose (Fig. 37.20). Determine the minimum film thickness that

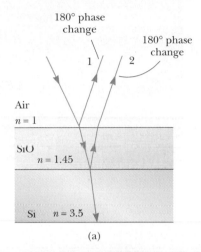

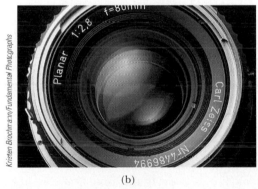

Figure 37.20 (Example 37.4) (a) Reflective losses from a silicon solar cell are minimized by coating the surface of the cell with a thin film of silicon monoxide. (b) The reflected light from a coated camera lens often has a reddish-violet appearance.

produces the least reflection at a wavelength of 550 nm, near the center of the visible spectrum.

Solution Figure 37.20a helps us conceptualize the path of the rays in the SiO film that result in interference in the reflected light. Based on the geometry of the SiO layer, we categorize this as a thin-film interference problem. To analyze the problem, note that the reflected light is a minimum when rays 1 and 2 in Figure 37.20a meet the condition of destructive interference. In this situation, *both* rays undergo a 180° phase change upon reflection—ray 1 from the upper SiO surface and ray 2 from the lower SiO surface. The net change in phase due to reflection is therefore zero, and the condition for a reflection minimum requires a path difference of $\lambda_n/2$, where λ_n is the wavelength of the light in SiO. Hence $2t = \lambda/2n$, where λ is the wavelength in air and n is the index of refraction of SiO. The required thickness is

$$t = \frac{\lambda}{4n} = \frac{550 \text{ nm}}{4(1.45)} = \boxed{94.8 \text{ nm}}$$

To finalize the problem, we can investigate the losses in typical solar cells. A typical uncoated solar cell has reflective losses as high as 30%; a SiO coating can reduce this value to about 10%. This significant decrease in reflective losses increases the cell's efficiency because less reflection means that more sunlight enters the silicon to create charge carriers in the cell. No coating can ever be made perfectly nonreflecting because the required thickness is wavelength-dependent and the incident light covers a wide range of wavelengths.

Glass lenses used in cameras and other optical instruments are usually coated with a transparent thin film to reduce or eliminate unwanted reflection and enhance the transmission of light through the lenses. The camera lens in Figure 37.20b has several coatings (of different thicknesses) to minimize reflection of light waves having wavelengths near the center of the visible spectrum. As a result, the little light that is reflected by the lens has a greater proportion of the far ends of the spectrum and often appears reddish-violet.

 Investigate the interference for various film properties at the Interactive Worked Example link at **http://www.pse6.com.**

Example 37.5 Interference in a Wedge-Shaped Film

A thin, wedge-shaped film of index of refraction n is illuminated with monochromatic light of wavelength λ, as illustrated in Figure 37.21a. Describe the interference pattern observed for this case.

Solution The interference pattern, because it is created by a thin film of variable thickness surrounded by air, is a series of alternating bright and dark parallel fringes. A dark fringe corresponding to destructive interference appears at point O, the apex, because here the upper reflected ray undergoes a 180° phase change while the lower one undergoes no phase change.

According to Equation 37.17, other dark minima appear when $2nt = m\lambda$; thus, $t_1 = \lambda/2n$, $t_2 = \lambda/n$, $t_3 = 3\lambda/2n$, and so on. Similarly, the bright maxima appear at locations where t satisfies Equation 37.16, $2nt = (m + \frac{1}{2})\lambda$, corresponding to thicknesses of $\lambda/4n$, $3\lambda/4n$, $5\lambda/4n$, and so on.

If white light is used, bands of different colors are observed at different points, corresponding to the different wavelengths of light (see Fig. 37.21b). This is why we see different colors in soap bubbles and other films of varying thickness.

Figure 37.21 (Example 37.5) (a) Interference bands in reflected light can be observed by illuminating a wedge-shaped film with monochromatic light. The darker areas correspond to regions where rays tend to cancel each other because of interference effects. (b) Interference in a vertical film of variable thickness. The top of the film appears darkest where the film is thinnest.

Incident light

O

n t_1

t_2

t_3

Richard Megna/Fundamental Photographs

(a)

(b)

37.7 The Michelson Interferometer

The **interferometer,** invented by the American physicist A. A. Michelson (1852–1931), splits a light beam into two parts and then recombines the parts to form an interference pattern. The device can be used to measure wavelengths or other lengths with great precision because a large and precisely measurable displacement of one of the mirrors is related to an exactly countable number of wavelengths of light.

A schematic diagram of the interferometer is shown in Figure 37.22. A ray of light from a monochromatic source is split into two rays by mirror M_0, which is inclined at 45° to the incident light beam. Mirror M_0, called a *beam splitter,* transmits half the light incident on it and reflects the rest. One ray is reflected from M_0 vertically upward toward mirror M_1, and the second ray is transmitted horizontally through M_0 toward mirror M_2. Hence, the two rays travel separate paths L_1 and L_2. After reflecting from M_1 and M_2, the two rays eventually recombine at M_0 to produce an interference pattern, which can be viewed through a telescope.

The interference condition for the two rays is determined by their path length differences. When the two mirrors are exactly perpendicular to each other, the interference pattern is a target pattern of bright and dark circular fringes, similar to Newton's rings. As M_1 is moved, the fringe pattern collapses or expands, depending on the direction in which M_1 is moved. For example, if a dark circle appears at the center of the target pattern (corresponding to destructive interference) and M_1 is then moved a distance $\lambda/4$ toward M_0, the path difference changes by $\lambda/2$. What was a dark circle at the center now becomes a bright circle. As M_1 is moved an additional distance $\lambda/4$ toward M_0, the bright circle becomes a dark circle again. Thus, the fringe pattern shifts by one-half fringe each time M_1 is moved a distance $\lambda/4$. The wavelength of light is then measured by counting the number of fringe shifts for a given displacement of M_1. If the wavelength is accurately known, mirror displacements can be measured to within a fraction of the wavelength.

We will see an important historical use of the Michelson interferometer in our discussion of relativity in Chapter 39. Modern uses include the following two applications.

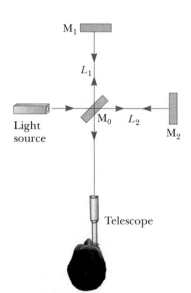

M_1

L_1

Light source

M_0 L_2

M_2

Telescope

Active Figure 37.22 Diagram of the Michelson interferometer. A single ray of light is split into two rays by mirror M_0, which is called a beam splitter. The path difference between the two rays is varied with the adjustable mirror M_1. As M_1 is moved, an interference pattern changes in the field of view.

At the Active Figures link at http://www.pse6.com, move the mirror to see the effect on the interference pattern and use the interferometer to measure the wavelength of light.

Fourier Transform Infrared Spectroscopy (FTIR)

Spectroscopy is the study of the wavelength distribution of radiation from a sample that can be used to identify the characteristics of atoms or molecules in the sample. Infrared spectroscopy is particularly important to organic chemists in analyzing organic molecules. Traditional spectroscopy involves the use of an optical element, such as a prism (Section 35.7) or a diffraction grating (Section 38.4), which spreads out various wavelengths in a complex optical signal from the sample into different angles. In this way, the various wavelengths of radiation and their intensities in the signal can be determined. These types of devices are limited in their resolution and effectiveness because they must be scanned through the various angular deviations of the radiation.

The technique of *Fourier Transform Infrared Spectroscopy* (FTIR) is used to create a higher-resolution spectrum in a time interval of one second that may have required 30 minutes with a standard spectrometer. In this technique, the radiation from a sample enters a Michelson interferometer. The movable mirror is swept through the zero-path-difference condition and the intensity of radiation at the viewing position is recorded. The result is a complex set of data relating light intensity as a function of mirror position, called an *interferogram*. Because there is a relationship between mirror position and light intensity for a given wavelength, the interferogram contains information about all wavelengths in the signal.

In Section 18.8, we discussed Fourier analysis of a waveform. The waveform is a function that contains information about all of the individual frequency components that make up the waveform.[3] Equation 18.16 shows how the waveform is generated from the individual frequency components. Similarly, the interferogram can be analyzed by computer, in a process called a *Fourier transform*, to provide all of the wavelength components. This is the same information generated by traditional spectroscopy, but the resolution of FTIR is much higher.

Laser Interferometer Gravitational-Wave Observatory (LIGO)

Einstein's general theory of relativity (Section 39.10) predicts the existence of *gravitational waves*. These waves propagate from the site of any gravitational disturbance, which could be periodic and predictable, such as the rotation of a double star around a center of mass, or unpredictable, such as the supernova explosion of a massive star.

In Einstein's theory, gravitation is equivalent to a distortion of space. Thus, a gravitational disturbance causes an additional distortion that propagates through space in a manner similar to mechanical or electromagnetic waves. When gravitational waves from a disturbance pass by the Earth, they create a distortion of the local space. The LIGO apparatus is designed to detect this distortion. The apparatus employs a Michelson interferometer that uses laser beams with an effective path length of several kilometers. At the end of an arm of the interferometer, a mirror is mounted on a massive pendulum. When a gravitational wave passes by, the pendulum and the attached mirror move, and the interference pattern due to the laser beams from the two arms changes.

Two sites have been developed in the United States for interferometers in order to allow coincidence studies of gravitational waves. These sites are located in Richland, Washington, and Livingston, Louisiana. Figure 37.23 shows the Washington site. The two arms of the Michelson interferometer are evident in the photograph. Test runs are being performed as of the printing of this book. Cooperation with other gravitational wave detectors, such as VIRGO in Cascina, Italy, will allow detailed studies of gravitational waves.

---

[3]   In acoustics, it is common to talk about the components of a complex signal in terms of frequency. In optics, it is more common to identify the components by wavelength.

LIGO Hanford Observatory

Figure 37.23 The Laser Interferometer Gravitational-Wave Observatory (LIGO) near Richland, Washington. Note the two perpendicular arms of the Michelson interferometer.

SUMMARY

Take a practice test for this chapter by clicking on the Practice Test link at http://www.pse6.com.

Interference in light waves occurs whenever two or more waves overlap at a given point. An interference pattern is observed if (1) the sources are coherent and (2) the sources have identical wavelengths.

In Young's double-slit experiment, two slits S_1 and S_2 separated by a distance d are illuminated by a single-wavelength light source. An interference pattern consisting of bright and dark fringes is observed on a viewing screen. The condition for bright fringes **(constructive interference)** is

$$\delta = d \sin\theta_{\text{bright}} = m\lambda \qquad (m = 0, \pm1, \pm2, \ldots) \tag{37.2}$$

The condition for dark fringes **(destructive interference)** is

$$d \sin\theta_{\text{dark}} = (m + \tfrac{1}{2})\lambda \qquad (m = 0, \pm1, \pm2, \ldots) \tag{37.3}$$

The number m is called the **order number** of the fringe.

The **intensity** at a point in the double-slit interference pattern is

$$I = I_{\max} \cos^2\left(\frac{\pi d \sin\theta}{\lambda}\right) \tag{37.12}$$

where $I_{\max}$ is the maximum intensity on the screen and the expression represents the time average.

A wave traveling from a medium of index of refraction n_1 toward a medium of index of refraction n_2 undergoes a 180° phase change upon reflection when $n_2 > n_1$ and undergoes no phase change when $n_2 < n_1$.

The condition for constructive interference in a film of thickness t and index of refraction n surrounded by air is

$$2nt = (m + \tfrac{1}{2})\lambda \qquad (m = 0, 1, 2, \ldots) \tag{37.16}$$

where λ is the wavelength of the light in free space.

Similarly, the condition for destructive interference in a thin film surrounded by air is

$$2nt = m\lambda \qquad (m = 0, 1, 2, \ldots) \tag{37.17}$$

QUESTIONS

1. What is the necessary condition on the path length difference between two waves that interfere (a) constructively and (b) destructively?

2. Explain why two flashlights held close together do not produce an interference pattern on a distant screen.

3. If Young's double-slit experiment were performed under water, how would the observed interference pattern be affected?

4. In Young's double-slit experiment, why do we use monochromatic light? If white light is used, how would the pattern change?

5. A simple way to observe an interference pattern is to look at a distant light source through a stretched handkerchief or an opened umbrella. Explain how this works.

6. A certain oil film on water appears brightest at the outer regions, where it is thinnest. From this information, what can you say about the index of refraction of oil relative to that of water?

7. As a soap bubble evaporates, it appears black just before it breaks. Explain this phenomenon in terms of the phase changes that occur on reflection from the two surfaces of the soap film.

8. If we are to observe interference in a thin film, why must the film not be very thick (with thickness only on the order of a few wavelengths)?

9. A lens with outer radius of curvature R and index of refraction n rests on a flat glass plate. The combination is illuminated with white light from above and observed from above. Is there a dark spot or a light spot at the center of the lens? What does it mean if the observed rings are noncircular?

10. Why is the lens on a good-quality camera coated with a thin film?

11. Why is it so much easier to perform interference experiments with a laser than with an ordinary light source?

12. Suppose that reflected white light is used to observe a thin transparent coating on glass as the coating material is gradually deposited by evaporation in a vacuum. Describe color changes that might occur during the process of building up the thickness of the coating.

13. In our discussion of thin-film interference, we looked at light *reflecting* from a thin film **What If?** Consider one light ray, the direct ray, which transmits through the film without reflecting. Consider a second ray, the reflected ray, that transmits through the first surface, reflects from the second, reflects again from the first, and then transmits out into the air, parallel to the direct ray. For normal incidence, how thick must the film be, in terms of the wavelength of light, for the outgoing rays to interfere destructively? Is it the same thickness as for reflected destructive interference?

14. Suppose you are watching television by connection to an antenna rather than a cable system. If an airplane flies near your location, you may notice wavering ghost images in the television picture. What might cause this?

PROBLEMS

1, 2, 3 = straightforward, intermediate, challenging   ☐ = full solution available in the *Student Solutions Manual and Study Guide*

🌀 = coached solution with hints available at http://www.pse6.com   💻 = computer useful in solving problem

▨ = paired numerical and symbolic problems

Section 37.1   Conditions for Interference
Section 37.2   Young's Double-Slit Experiment

Note: Problems 8, 9, 10, and 12 in Chapter 18 can be assigned with these sections.

1. A laser beam ($\lambda = 632.8$ nm) is incident on two slits 0.200 mm apart. How far apart are the bright interference fringes on a screen 5.00 m away from the double slits?

2. A Young's interference experiment is performed with monochromatic light. The separation between the slits is 0.500 mm, and the interference pattern on a screen 3.30 m away shows the first side maximum 3.40 mm from the center of the pattern. What is the wavelength?

3. 🌀 Two radio antennas separated by 300 m as shown in Figure P37.3 simultaneously broadcast identical signals at

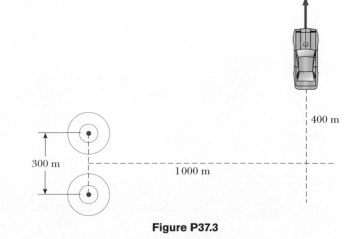

Figure P37.3

the same wavelength. A radio in a car traveling due north receives the signals. (a) If the car is at the position of the second maximum, what is the wavelength of the signals? (b) How much farther must the car travel to encounter the next minimum in reception? (*Note:* Do not use the small-angle approximation in this problem.)

4. In a location where the speed of sound is 354 m/s, a 2 000-Hz sound wave impinges on two slits 30.0 cm apart. (a) At what angle is the first maximum located? (b) **What If?** If the sound wave is replaced by 3.00-cm microwaves, what slit separation gives the same angle for the first maximum? (c) **What If?** If the slit separation is 1.00 μm, what frequency of light gives the same first maximum angle?

5. www Young's double-slit experiment is performed with 589-nm light and a distance of 2.00 m between the slits and the screen. The tenth interference minimum is observed 7.26 mm from the central maximum. Determine the spacing of the slits.

6. The two speakers of a boom box are 35.0 cm apart. A single oscillator makes the speakers vibrate in phase at a frequency of 2.00 kHz. At what angles, measured from the perpendicular bisector of the line joining the speakers, would a distant observer hear maximum sound intensity? Minimum sound intensity? (Take the speed of sound as 340 m/s.)

7. Two narrow, parallel slits separated by 0.250 mm are illuminated by green light (λ = 546.1 nm). The interference pattern is observed on a screen 1.20 m away from the plane of the slits. Calculate the distance (a) from the central maximum to the first bright region on either side of the central maximum and (b) between the first and second dark bands.

8. Light with wavelength 442 nm passes through a double-slit system that has a slit separation d = 0.400 mm. Determine how far away a screen must be placed in order that a dark fringe appear directly opposite both slits, with just one bright fringe between them.

9. A riverside warehouse has two open doors as shown in Figure P37.9. Its walls are lined with sound-absorbing material. A boat on the river sounds its horn. To person A the sound is loud and clear. To person B the sound is barely audible. The principal wavelength of the sound waves is 3.00 m. Assuming person B is at the position of the first minimum, determine the distance between the doors, center to center.

10. Two slits are separated by 0.320 mm. A beam of 500-nm light strikes the slits, producing an interference pattern. Determine the number of maxima observed in the angular range $-30.0° < \theta < 30.0°$.

11. Young's double-slit experiment underlies the *Instrument Landing System* used to guide aircraft to safe landings when the visibility is poor. Although real systems are more complicated than the example described here, they operate on the same principles. A pilot is trying to align her plane with a runway, as suggested in Figure P37.11a. Two radio antennas A_1 and A_2 are positioned adjacent to the runway, separated by 40.0 m. The antennas broadcast unmodulated coherent radio waves at 30.0 MHz. (a) Find the wavelength of the waves. The pilot "locks onto" the strong signal radiated along an interference maximum, and steers the plane to keep the received signal strong. If she has found the central maximum, the plane will have just the right heading to land when it reaches the runway. (b) **What If?** Suppose instead that the plane is flying along the first side maximum (Fig. P37.11b). How far to the side of the runway centerline will the plane be when it is 2.00 km from the antennas? (c) It is possible to tell the pilot she is on the wrong maximum by sending out two signals from each antenna and equipping the aircraft with a two-channel receiver. The ratio of the two frequencies must not be the ratio of small integers (such as 3/4). Explain how this two-frequency system would work, and why it would not necessarily work if the frequencies were related by an integer ratio.

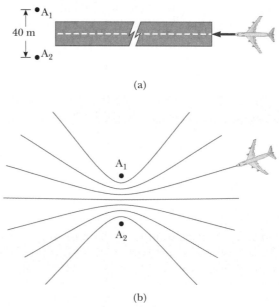

(a)

(b)

Figure P37.11

12. A student holds a laser that emits light of wavelength 633 nm. The beam passes though a pair of slits separated by 0.300 mm, in a glass plate attached to the front of the laser. The beam then falls perpendicularly on a screen, creating an interference pattern on it. The student begins to walk directly toward the screen at 3.00 m/s. The central maximum on the screen is stationary. Find the speed of the first-order maxima on the screen.

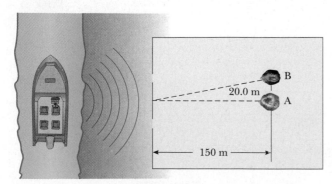

Figure P37.9

13. In Figure 37.5 let $L = 1.20$ m and $d = 0.120$ mm and assume that the slit system is illuminated with monochromatic 500-nm light. Calculate the phase difference between the two wave fronts arriving at P when (a) $\theta = 0.500°$ and (b) $y = 5.00$ mm. (c) What is the value of θ for which the phase difference is 0.333 rad? (d) What is the value of θ for which the path difference is $\lambda/4$?

14. Coherent light rays of wavelength λ strike a pair of slits separated by distance d at an angle θ_1 as shown in Figure P37.14. Assume an interference maximum is formed at an angle θ_2 a great distance from the slits. Show that $d(\sin\theta_2 - \sin\theta_1) = m\lambda$, where m is an integer.

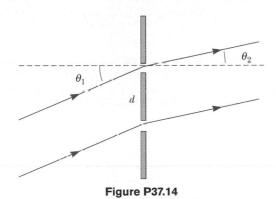

Figure P37.14

15. In a double-slit arrangement of Figure 37.5, $d = 0.150$ mm, $L = 140$ cm, $\lambda = 643$ nm, and $y = 1.80$ cm. (a) What is the path difference δ for the rays from the two slits arriving at P? (b) Express this path difference in terms of λ. (c) Does P correspond to a maximum, a minimum, or an intermediate condition?

Section 37.3 Intensity Distribution of the Double-Slit Interference Pattern

16. The intensity on the screen at a certain point in a double-slit interference pattern is 64.0% of the maximum value. (a) What minimum phase difference (in radians) between sources produces this result? (b) Express this phase difference as a path difference for 486.1-nm light.

17. In Figure 37.5, let $L = 120$ cm and $d = 0.250$ cm. The slits are illuminated with coherent 600-nm light. Calculate the distance y above the central maximum for which the average intensity on the screen is 75.0% of the maximum.

18. Two slits are separated by 0.180 mm. An interference pattern is formed on a screen 80.0 cm away by 656.3-nm light. Calculate the fraction of the maximum intensity 0.600 cm above the central maximum.

19. Two narrow parallel slits separated by 0.850 mm are illuminated by 600-nm light, and the viewing screen is 2.80 m away from the slits. (a) What is the phase difference between the two interfering waves on a screen at a point 2.50 mm from the central bright fringe? (b) What is the ratio of the intensity at this point to the intensity at the center of a bright fringe?

20. Monochromatic coherent light of amplitude E_0 and angular frequency ω passes through three parallel slits each

separated by a distance d from its neighbor. (a) Show that the time-averaged intensity as a function of the angle θ is

$$I(\theta) = I_{max}\left[1 + 2\cos\left(\frac{2\pi d \sin\theta}{\lambda}\right)\right]^2$$

(b) Determine the ratio of the intensities of the primary and secondary maxima.

Section 37.4 Phasor Addition of Waves

Note: Problems 4, 5, and 6 in Chapter 18 can be assigned with this section.

21. Marie Cornu, a physicist at the Polytechnic Institute in Paris, invented phasors in about 1880. This problem helps you to see their utility. Find the amplitude and phase constant of the sum of two waves represented by the expressions

$$E_1 = (12.0 \text{ kN/C}) \sin(15x - 4.5t)$$

and $E_2 = (12.0 \text{ kN/C}) \sin(15x - 4.5t + 70°)$

(a) by using a trigonometric identity (as from Appendix B), and (b) by representing the waves by phasors. (c) Find the amplitude and phase constant of the sum of the three waves represented by

$$E_1 = (12.0 \text{ kN/C}) \sin(15x - 4.5t + 70°),$$

$$E_2 = (15.5 \text{ kN/C}) \sin(15x - 4.5t - 80°),$$

and $E_3 = (17.0 \text{ kN/C}) \sin(15x - 4.5t + 160°)$

22. The electric fields from three coherent sources are described by $E_1 = E_0 \sin \omega t$, $E_2 = E_0 \sin(\omega t + \phi)$, and $E_3 = E_0 \sin(\omega t + 2\phi)$. Let the resultant field be represented by $E_P = E_R \sin(\omega t + \alpha)$. Use phasors to find E_R and α when (a) $\phi = 20.0°$, (b) $\phi = 60.0°$, and (c) $\phi = 120°$. (d) Repeat when $\phi = (3\pi/2)$ rad.

23. Determine the resultant of the two waves given by $E_1 = 6.0 \sin(100\pi t)$ and $E_2 = 8.0 \sin(100\pi t + \pi/2)$.

24. Suppose the slit openings in a Young's double-slit experiment have different sizes so that the electric fields and intensities from each slit are different. With $E_1 = E_{01} \sin(\omega t)$ and $E_2 = E_{02} \sin(\omega t + \phi)$, show that the resultant electric field is $E = E_0 \sin(\omega t + \theta)$, where

$$E_0 = \sqrt{E_{01}^2 + E_{02}^2 + 2E_{01}E_{02}\cos\phi}$$

and

$$\sin\theta = \frac{E_{02}\sin\phi}{E_0}$$

25. Use phasors to find the resultant (magnitude and phase angle) of two fields represented by $E_1 = 12 \sin \omega t$ and $E_2 = 18 \sin(\omega t + 60°)$. (Note that in this case the amplitudes of the two fields are unequal.)

26. Two coherent waves are described by

$$E_1 = E_0 \sin\left(\frac{2\pi x_1}{\lambda} - 2\pi ft + \frac{\pi}{6}\right)$$

$$E_2 = E_0 \sin\left(\frac{2\pi x_2}{\lambda} - 2\pi ft + \frac{\pi}{8}\right)$$

Determine the relationship between x_1 and x_2 that produces constructive interference when the two waves are superposed.

27. When illuminated, four equally spaced parallel slits act as multiple coherent sources, each differing in phase from the adjacent one by an angle ϕ. Use a phasor diagram to determine the smallest value of ϕ for which the resultant of the four waves (assumed to be of equal amplitude) is zero.

28. Sketch a phasor diagram to illustrate the resultant of $E_1 = E_{01} \sin \omega t$ and $E_2 = E_{02} \sin(\omega t + \phi)$, where $E_{02} = 1.50 E_{01}$ and $\pi/6 \leq \phi \leq \pi/3$. Use the sketch and the law of cosines to show that, for two coherent waves, the resultant intensity can be written in the form $I_R = I_1 + I_2 + 2\sqrt{I_1 I_2} \cos\phi$.

29. Consider N coherent sources described as follows: $E_1 = E_0 \sin(\omega t + \phi)$, $E_2 = E_0 \sin(\omega t + 2\phi)$, $E_3 = E_0 \sin(\omega t + 3\phi)$, . . . , $E_N = E_0 \sin(\omega t + N\phi)$. Find the minimum value of ϕ for which $E_R = E_1 + E_2 + E_3 + \cdots + E_N$ is zero.

Section 37.5   Change of Phase Due to Reflection
Section 37.6   Interference in Thin Films

30. A soap bubble ($n = 1.33$) is floating in air. If the thickness of the bubble wall is 115 nm, what is the wavelength of the light that is most strongly reflected?

31. An oil film ($n = 1.45$) floating on water is illuminated by white light at normal incidence. The film is 280 nm thick. Find (a) the color of the light in the visible spectrum most strongly reflected and (b) the color of the light in the spectrum most strongly transmitted. Explain your reasoning.

32. A thin film of oil ($n = 1.25$) is located on a smooth wet pavement. When viewed perpendicular to the pavement, the film reflects most strongly red light at 640 nm and reflects no blue light at 512 nm. How thick is the oil film?

33. A possible means for making an airplane invisible to radar is to coat the plane with an antireflective polymer. If radar waves have a wavelength of 3.00 cm and the index of refraction of the polymer is $n = 1.50$, how thick would you make the coating?

34. A material having an index of refraction of 1.30 is used as an antireflective coating on a piece of glass ($n = 1.50$). What should be the minimum thickness of this film in order to minimize reflection of 500-nm light?

35. A film of MgF_2 ($n = 1.38$) having thickness 1.00×10^{-5} cm is used to coat a camera lens. Are any wavelengths in the visible spectrum intensified in the reflected light?

36. Astronomers observe the chromosphere of the Sun with a filter that passes the red hydrogen spectral line of wavelength 656.3 nm, called the H_α line. The filter consists of a transparent dielectric of thickness d held between two partially aluminized glass plates. The filter is held at a constant temperature. (a) Find the minimum value of d that produces maximum transmission of perpendicular H_α light, if the dielectric has an index of refraction of 1.378. (b) **What If?** If the temperature of the filter increases above the normal value, what happens to the transmitted wavelength? (Its index of refraction does not change significantly.) (c) The dielectric will also pass what near-visible wavelength? One of the glass plates is colored red to absorb this light.

37. A beam of 580-nm light passes through two closely spaced glass plates, as shown in Figure P37.37. For what minimum nonzero value of the plate separation d is the transmitted light bright?

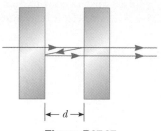

Figure P37.37

38. When a liquid is introduced into the air space between the lens and the plate in a Newton's-rings apparatus, the diameter of the tenth ring changes from 1.50 to 1.31 cm. Find the index of refraction of the liquid.

39. An air wedge is formed between two glass plates separated at one edge by a very fine wire, as shown in Figure P37.39. When the wedge is illuminated from above by 600-nm light and viewed from above, 30 dark fringes are observed. Calculate the radius of the wire.

Figure P37.39 Problems 39 and 40.

40. Two glass plates 10.0 cm long are in contact at one end and separated at the other end by a thread 0.050 0 mm in diameter (Fig. P37.39). Light containing the two wavelengths 400 nm and 600 nm is incident perpendicularly and viewed by reflection. At what distance from the contact point is the next dark fringe?

Section 37.7   The Michelson Interferometer

41. Mirror M_1 in Figure 37.22 is displaced a distance ΔL. During this displacement, 250 fringe reversals (formation of successive dark or bright bands) are counted. The light being used has a wavelength of 632.8 nm. Calculate the displacement ΔL.

42. Monochromatic light is beamed into a Michelson interferometer. The movable mirror is displaced 0.382 mm, causing the interferometer pattern to reproduce itself 1 700 times. Determine the wavelength of the light. What color is it?

43. One leg of a Michelson interferometer contains an evacuated cylinder of length L, having glass plates on each end.

A gas is slowly leaked into the cylinder until a pressure of 1 atm is reached. If N bright fringes pass on the screen when light of wavelength λ is used, what is the index of refraction of the gas?

Additional Problems

44. In the **What If?** section of Example 37.2, it was claimed that overlapping fringes in a two-slit interference pattern for two different wavelengths obey the following relationship even for large values of the angle θ:

$$\frac{\lambda}{\lambda'} = \frac{m'}{m}$$

(a) Prove this assertion. (b) Using the data in Example 37.2, find the value of y on the screen at which the fringes from the two wavelengths first coincide.

45. One radio transmitter A operating at 60.0 MHz is 10.0 m from another similar transmitter B that is 180° out of phase with A. How far must an observer move from A toward B along the line connecting A and B to reach the nearest point where the two beams are in phase?

46. **Review problem.** This problem extends the result of Problem 12 in Chapter 18. Figure P37.46 shows two adjacent vibrating balls dipping into a tank of water. At distant points they produce an interference pattern of water waves, as shown in Figure 37.3. Let λ represent the wavelength of the ripples. Show that the two sources produce a standing wave along the line segment, of length d, between them. In terms of λ and d, find the number of nodes and the number of antinodes in the standing wave. Find the number of zones of constructive and of destructive interference in the interference pattern far away from the sources. Each line of destructive interference springs from a node in the standing wave and each line of constructive interference springs from an antinode.

Courtesy of Central Scientific Company

Figure P37.46

47. Raise your hand and hold it flat. Think of the space between your index finger and your middle finger as one slit, and think of the space between middle finger and ring finger as a second slit. (a) Consider the interference resulting from sending coherent visible light perpendicularly through this pair of openings. Compute an order-of-magnitude estimate for the angle between adjacent zones of constructive interference. (b) To make the angles in the interference pattern easy to measure with a plastic protractor, you should use an electromagnetic wave with frequency of what order of magnitude? How is this wave classified on the electromagnetic spectrum?

48. In a Young's double-slit experiment using light of wavelength λ, a thin piece of Plexiglas having index of refraction n covers one of the slits. If the center point on the screen is a dark spot instead of a bright spot, what is the minimum thickness of the Plexiglas?

49. **Review problem.** A flat piece of glass is held stationary and horizontal above the flat top end of a 10.0-cm-long vertical metal rod that has its lower end rigidly fixed. The thin film of air between the rod and glass is observed to be bright by reflected light when it is illuminated by light of wavelength 500 nm. As the temperature is slowly increased by 25.0°C, the film changes from bright to dark and back to bright 200 times. What is the coefficient of linear expansion of the metal?

50. A certain crude oil has an index of refraction of 1.25. A ship dumps 1.00 m^3 of this oil into the ocean, and the oil spreads into a thin uniform slick. If the film produces a first-order maximum of light of wavelength 500 nm normally incident on it, how much surface area of the ocean does the oil slick cover? Assume that the index of refraction of the ocean water is 1.34.

51. Astronomers observe a 60.0-MHz radio source both directly and by reflection from the sea. If the receiving dish is 20.0 m above sea level, what is the angle of the radio source above the horizon at first maximum?

52. Interference effects are produced at point P on a screen as a result of direct rays from a 500-nm source and reflected rays from the mirror, as shown in Figure P37.52. Assume the source is 100 m to the left of the screen and 1.00 cm above the mirror. Find the distance y to the first dark band above the mirror.

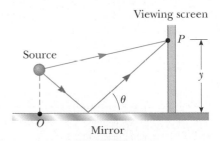

Figure P37.52

53. The waves from a radio station can reach a home receiver by two paths. One is a straight-line path from transmitter to home, a distance of 30.0 km. The second path is by reflection from the ionosphere (a layer of ionized air molecules high in the atmosphere). Assume this reflection takes place at a point midway between receiver and transmitter and that the wavelength broadcast by the radio station is 350 m. Find the minimum height of the ionospheric layer that could produce destructive interference between the direct and reflected beams. (Assume that no phase change occurs on reflection.)

54. Many cells are transparent and colorless. Structures of great interest in biology and medicine can be practically invisible to ordinary microscopy. An *interference microscope* reveals a difference in index of refraction as a shift in interference fringes, to indicate the size and shape of cell structures. The idea is exemplified in the following problem: An air wedge is formed between two glass plates in contact along one edge and slightly separated at the opposite edge. When the plates are illuminated with monochromatic light from above, the reflected light has 85 dark fringes. Calculate the number of dark fringes that appear if water ($n = 1.33$) replaces the air between the plates.

55. Measurements are made of the intensity distribution in a Young's interference pattern (see Fig. 37.7). At a particular value of y, it is found that $I/I_{\max} = 0.810$ when 600-nm light is used. What wavelength of light should be used to reduce the relative intensity at the same location to 64.0% of the maximum intensity?

56. Our discussion of the techniques for determining constructive and destructive interference by reflection from a thin film in air has been confined to rays striking the film at nearly normal incidence. **What If?** Assume that a ray is incident at an angle of 30.0° (relative to the normal) on a film with index of refraction 1.38. Calculate the minimum thickness for constructive interference of sodium light with a wavelength of 590 nm.

57. The condition for constructive interference by reflection from a thin film in air as developed in Section 37.6 assumes nearly normal incidence. **What If?** Show that if the light is incident on the film at a nonzero angle ϕ_1 (relative to the normal), then the condition for constructive interference is $2nt \cos \theta_2 = (m + \frac{1}{2})\lambda$, where θ_2 is the angle of refraction.

58. (a) Both sides of a uniform film that has index of refraction n and thickness d are in contact with air. For normal incidence of light, an intensity minimum is observed in the reflected light at λ_2 and an intensity maximum is observed at λ_1, where $\lambda_1 > \lambda_2$. Assuming that no intensity minima are observed between λ_1 and λ_2, show that the integer m in Equations 37.16 and 37.17 is given by $m = \lambda_1/2(\lambda_1 - \lambda_2)$. (b) Determine the thickness of the film, assuming $n = 1.40$, $\lambda_1 = 500$ nm, and $\lambda_2 = 370$ nm.

59. Figure P37.59 shows a radio-wave transmitter and a receiver separated by a distance d and both a distance h above the ground. The receiver can receive signals both directly from

the transmitter and indirectly from signals that reflect from the ground. Assume that the ground is level between the transmitter and receiver and that a 180° phase shift occurs upon reflection. Determine the longest wavelengths that interfere (a) constructively and (b) destructively.

60. A piece of transparent material having an index of refraction n is cut into the shape of a wedge as shown in Figure P37.60. The angle of the wedge is small. Monochromatic light of wavelength λ is normally incident from above, and viewed from above. Let h represent the height of the wedge and ℓ its width. Show that bright fringes occur at the positions $x = \lambda\ell(m + \frac{1}{2})/2hn$ and dark fringes occur at the positions $x = \lambda\ell m/2hn$, where $m = 0, 1, 2, \ldots$ and x is measured as shown.

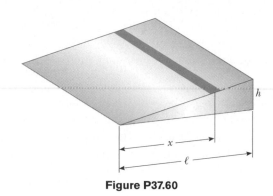

Figure P37.60

61. Consider the double-slit arrangement shown in Figure P37.61, where the slit separation is d and the slit to screen distance is L. A sheet of transparent plastic having an index of refraction n and thickness t is placed over the upper slit. As a result, the central maximum of the interference pattern moves upward a distance y'. Find y'.

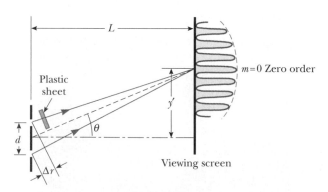

Figure P37.61

62. A plano-convex lens has index of refraction n. The curved side of the lens has radius of curvature R and rests on a flat glass surface of the same index of refraction, with a film of index n_{film} between them, as shown in Fig. 37.18a. The lens is illuminated from above by light of wavelength λ. Show that the dark Newton's rings have radii given approximately by

$$r \approx \sqrt{\frac{m\lambda R}{n_{\text{film}}}}$$

where m is an integer and r is much less than R.

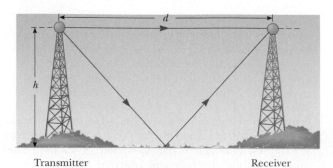

Transmitter　　　　　　　Receiver

Figure P37.59

63. In a Newton's-rings experiment, a plano-convex glass ($n = 1.52$) lens having diameter 10.0 cm is placed on a flat plate as shown in Figure 37.18a. When 650-nm light is incident normally, 55 bright rings are observed with the last one right on the edge of the lens. (a) What is the radius of curvature of the convex surface of the lens? (b) What is the focal length of the lens?

64. A plano-concave lens having index of refraction 1.50 is placed on a flat glass plate, as shown in Figure P37.64. Its curved surface, with radius of curvature 8.00 m, is on the bottom. The lens is illuminated from above with yellow sodium light of wavelength 589 nm, and a series of concentric bright and dark rings is observed by reflection. The interference pattern has a dark spot at the center, surrounded by 50 dark rings, of which the largest is at the outer edge of the lens. (a) What is the thickness of the air layer at the center of the interference pattern? (b) Calculate the radius of the outermost dark ring. (c) Find the focal length of the lens.

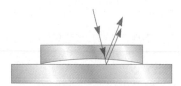

Figure P37.64

65. A plano-convex lens having a radius of curvature of $r = 4.00$ m is placed on a concave glass surface whose radius of curvature is $R = 12.0$ m, as shown in Figure P37.65. Determine the radius of the 100th bright ring, assuming 500-nm light is incident normal to the flat surface of the lens.

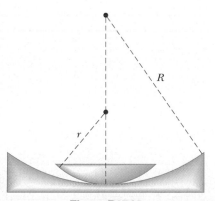

Figure P37.65

66. Use phasor addition to find the resultant amplitude and phase constant when the following three harmonic functions are combined: $E_1 = \sin(\omega t + \pi/6)$, $E_2 = 3.0 \sin(\omega t + 7\pi/2)$, and $E_3 = 6.0 \sin(\omega t + 4\pi/3)$.

67. A soap film ($n = 1.33$) is contained within a rectangular wire frame. The frame is held vertically so that the film drains downward and forms a wedge with flat faces. The thickness of the film at the top is essentially zero. The film is viewed in reflected white light with near-normal incidence, and the first violet ($\lambda = 420$ nm) interference band is observed 3.00 cm from the top edge of the film.

(a) Locate the first red ($\lambda = 680$ nm) interference band. (b) Determine the film thickness at the positions of the violet and red bands. (c) What is the wedge angle of the film?

68. Compact disc (CD) and digital video disc (DVD) players use interference to generate a strong signal from a tiny bump. The depth of a pit is chosen to be one quarter of the wavelength of the laser light used to read the disc. Then light reflected from the pit and light reflected from the adjoining flat differ in path length traveled by one-half wavelength, to interfere destructively at the detector. As the disc rotates, the light intensity drops significantly every time light is reflected from near a pit edge. The space between the leading and trailing edges of a pit determines the time between the fluctuations. The series of time intervals is decoded into a series of zeros and ones that carries the stored information. Assume that infrared light with a wavelength of 780 nm in vacuum is used in a CD player. The disc is coated with plastic having an index of refraction of 1.50. What should be the depth of each pit? A DVD player uses light of a shorter wavelength, and the pit dimensions are correspondingly smaller. This is one factor resulting in greater storage capacity on a DVD compared to a CD.

69. Interference fringes are produced using Lloyd's mirror and a 606-nm source as shown in Figure 37.15. Fringes 1.20 mm apart are formed on a screen 2.00 m from the real source S. Find the vertical distance h of the source above the reflecting surface.

70. Monochromatic light of wavelength 620 nm passes through a very narrow slit S and then strikes a screen in which are two parallel slits, S_1 and S_2, as in Figure P37.70. Slit S_1 is directly in line with S and at a distance of $L = 1.20$ m away from S, whereas S_2 is displaced a distance d to one side. The light is detected at point P on a second screen, equidistant from S_1 and S_2. When either one of the slits S_1 and S_2 is open, equal light intensities are measured at point P. When both are open, the intensity is three times larger. Find the minimum possible value for the slit separation d.

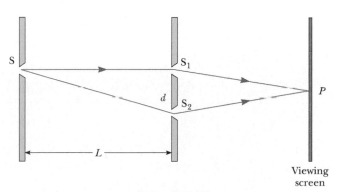

Figure P37.70

71. Slit 1 of a double slit is wider than slit 2, so that the light from 1 has an amplitude 3.00 times that of the light from 2. Show that for this situation, Equation 37.11 is replaced by the equation $I = (4I_{max}/9)(1 + 3 \cos^2 \phi/2)$.

Answers to Quick Quizzes

37.1 (b). The geometrical construction shown in Figure 37.5 is important for developing the mathematical description of interference. It is subject to misinterpretation, however, as it might suggest that the interference can only occur at the position of the screen. A better diagram for this situation is Figure 37.2, which shows *paths* of destructive and constructive interference all the way from the slits to the screen. These paths would be made visible by the smoke.

37.2 (c). Equation 37.5, which shows positions y proportional to order number m, is only valid for small angles.

37.3 (c). Equation 37.5 shows that decreasing λ or L will bring the fringes closer together. Immersing the apparatus in water decreases the wavelength so that the fringes move closer together.

37.4 (c). Conservation of energy cannot be violated. While there is no energy arriving at the location of a dark fringe, there is more energy arriving at the location of a bright fringe than there would be without the double slit.

37.5 The graph is shown in the next column. The width of the primary maxima is slightly narrower than the $N = 5$ primary width but wider than the $N = 10$ primary width. Because $N = 6$, the secondary maxima are $\frac{1}{36}$ as intense as the primary maxima.

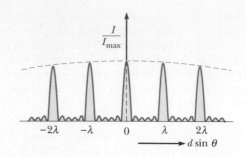

37.6 (a). One of the materials has a higher index of refraction than water, the other lower. For the material with a higher index of refraction, there is a 180° phase shift for the light reflected from the upper surface, but no such phase change from the lower surface, because the index of refraction for water on the other side is lower than that of the film. Thus, the two reflections are out of phase and interfere destructively.

37.7 (a). At the left edge, the air wedge has zero thickness and the only contribution to the interference is the 180° phase shift as the light reflects from the upper surface of the glass slide.

Chapter 38

Diffraction Patterns and Polarization

CHAPTER OUTLINE

38.1 Introduction to Diffraction Patterns

38.2 Diffraction Patterns from Narrow Slits

38.3 Resolution of Single-Slit and Circular Apertures

38.4 The Diffraction Grating

38.5 Diffraction of X-Rays by Crystals

38.6 Polarization of Light Waves

▲ The Hubble Space Telescope does its viewing above the atmosphere and does not suffer from the atmospheric blurring, caused by air turbulence, that plagues ground-based telescopes. Despite this advantage, it does have limitations due to diffraction effects. In this chapter we show how the wave nature of light limits the ability of any optical system to distinguish between closely spaced objects. (©Denis Scott/CORBIS)

Figure 38.1 The diffraction pattern that appears on a screen when light passes through a narrow vertical slit. The pattern consists of a broad central fringe and a series of less intense and narrower side fringes.

When plane light waves pass through a small aperture in an opaque barrier, the aperture acts as if it were a point source of light, with waves entering the shadow region behind the barrier. This phenomenon, known as diffraction, can be described only with a wave model for light, as discussed in Section 35.3. In this chapter, we investigate the features of the *diffraction pattern* that occurs when the light from the aperture is allowed to fall upon a screen.

In Chapter 34, we learned that electromagnetic waves are transverse. That is, the electric and magnetic field vectors associated with electromagnetic waves are perpendicular to the direction of wave propagation. In this chapter, we show that under certain conditions these transverse waves with electric field vectors in all possible transverse directions can be *polarized* in various ways. This means that only certain directions of the electric field vectors are present in the polarized wave.

38.1 Introduction to Diffraction Patterns

In Section 35.3 we discussed the fact that light of wavelength comparable to or larger than the width of a slit spreads out in all forward directions upon passing through the slit. We call this phenomenon *diffraction*. This behavior indicates that light, once it has passed through a narrow slit, spreads beyond the narrow path defined by the slit into regions that would be in shadow if light traveled in straight lines. Other waves, such as sound waves and water waves, also have this property of spreading when passing through apertures or by sharp edges.

We might expect that the light passing through a small opening would simply result in a broad region of light on a screen, due to the spreading of the light as it passes through the opening. We find something more interesting, however. A **diffraction pattern** consisting of light and dark areas is observed, somewhat similar to the interference patterns discussed earlier. For example, when a narrow slit is placed between a distant light source (or a laser beam) and a screen, the light produces a diffraction pattern like that in Figure 38.1. The pattern consists of a broad, intense central band (called the **central maximum**), flanked by a series of narrower, less intense additional bands (called **side maxima** or **secondary maxima**) and a series of intervening dark bands (or **minima**). Figure 38.2 shows a diffraction pattern associated with light passing by the edge of an object. Again we see bright and dark fringes, which is reminiscent of an interference pattern.

Figure 38.3 shows a diffraction pattern associated with the shadow of a penny. A bright spot occurs at the center, and circular fringes extend outward from the shadow's edge. We can explain the central bright spot only by using the wave theory of light, which predicts constructive interference at this point. From the viewpoint of geometric optics (in which light is viewed as rays traveling in straight lines), we expect the center of the shadow to be dark because that part of the viewing screen is completely shielded by the penny.

It is interesting to point out an historical incident that occurred shortly before the central bright spot was first observed. One of the supporters of geometric optics,

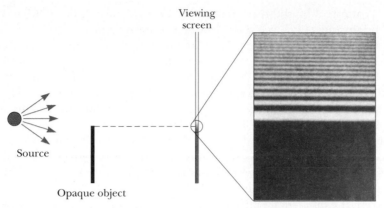

Figure 38.2 Light from a small source passes by the edge of an opaque object and continues on to a screen. A diffraction pattern consisting of bright and dark fringes appears on the screen in the region above the edge of the object.

Figure 38.3 Diffraction pattern created by the illumination of a penny, with the penny positioned midway between screen and light source. Note the bright spot at the center.

Simeon Poisson, argued that if Augustin Fresnel's wave theory of light were valid, then a central bright spot should be observed in the shadow of a circular object illuminated by a point source of light. To Poisson's astonishment, the spot was observed by Dominique Arago shortly thereafter. Thus, Poisson's prediction reinforced the wave theory rather than disproving it.

38.2 Diffraction Patterns from Narrow Slits

Let us consider a common situation, that of light passing through a narrow opening modeled as a slit, and projected onto a screen. To simplify our analysis, we assume that the observing screen is far from the slit, so that the rays reaching the screen are approximately parallel. This can also be achieved experimentally by using a converging lens to focus the parallel rays on a nearby screen. In this model, the pattern on the screen is called a **Fraunhofer diffraction pattern.**[1]

Figure 38.4a shows light entering a single slit from the left and diffracting as it propagates toward a screen. Figure 38.4b is a photograph of a single-slit Fraunhofer

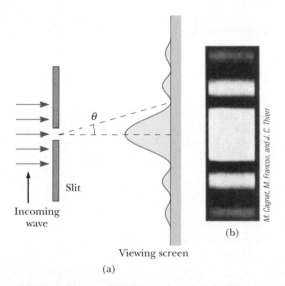

Active Figure 38.4 (a) Fraunhofer diffraction pattern of a single slit. The pattern consists of a central bright fringe flanked by much weaker maxima alternating with dark fringes. (Drawing not to scale.) (b) Photograph of a single-slit Fraunhofer diffraction pattern.

[1]   If the screen is brought close to the slit (and no lens is used), the pattern is a *Fresnel* diffraction pattern. The Fresnel pattern is more difficult to analyze, so we shall restrict our discussion to Fraunhofer diffraction.

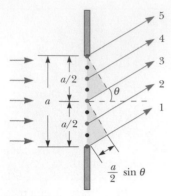

Figure 38.5 Paths of light rays that encounter a narrow slit of width a and diffract toward a screen in the direction described by angle θ. Each portion of the slit acts as a point source of light waves. The path difference between rays 1 and 3, rays 2 and 4, or rays 3 and 5 is $(a/2) \sin \theta$. (Drawing not to scale.)

▲ **PITFALL PREVENTION**

38.2 Similar Equation Warning!

Equation 38.1 has exactly the same form as Equation 37.2, with d, the slit separation, used in Equation 37.2 and a, the slit width, in Equation 38.1. However, Equation 37.2 describes the *bright* regions in a two-slit interference pattern while Equation 38.1 describes the *dark* regions in a single-slit diffraction pattern. Furthermore, $m = 0$ does not represent a dark fringe in the diffraction pattern.

Condition for destructive interference for a single slit

diffraction pattern. A bright fringe is observed along the axis at $\theta = 0$, with alternating dark and bright fringes on each side of the central bright fringe.

Until now, we have assumed that slits are point sources of light. In this section, we abandon that assumption and see how the finite width of slits is the basis for understanding Fraunhofer diffraction. We can deduce some important features of this phenomenon by examining waves coming from various portions of the slit, as shown in Figure 38.5. According to Huygens's principle, **each portion of the slit acts as a source of light waves.** Hence, light from one portion of the slit can interfere with light from another portion, and the resultant light intensity on a viewing screen depends on the direction θ. Based on this analysis, we recognize that a diffraction pattern is actually an interference pattern, in which the different sources of light are different portions of the single slit!

To analyze the diffraction pattern, it is convenient to divide the slit into two halves, as shown in Figure 38.5. Keeping in mind that all the waves are in phase as they leave the slit, consider rays 1 and 3. As these two rays travel toward a viewing screen far to the right of the figure, ray 1 travels farther than ray 3 by an amount equal to the path difference $(a/2)\sin\theta$, where a is the width of the slit. Similarly, the path difference between rays 2 and 4 is also $(a/2) \sin \theta$, as is that between rays 3 and 5. If this path difference is exactly half a wavelength (corresponding to a phase difference of 180°), then the two waves cancel each other and destructive interference results. If this is true for two such rays, then it is true for any two rays that originate at points separated by half the slit width because the phase difference between two such points is 180°. Therefore, waves from the upper half of the slit interfere destructively with waves from the lower half when

$$\frac{a}{2} \sin \theta = \pm \frac{\lambda}{2}$$

or when

$$\sin \theta = \pm \frac{\lambda}{a}$$

If we divide the slit into four equal parts and use similar reasoning, we find that the viewing screen is also dark when

$$\sin \theta = \pm \frac{2\lambda}{a}$$

Likewise, we can divide the slit into six equal parts and show that darkness occurs on the screen when

$$\sin \theta = \pm \frac{3\lambda}{a}$$

Therefore, the general condition for destructive interference is

$$\sin \theta_{\text{dark}} = m \frac{\lambda}{a} \qquad m = \pm 1, \pm 2, \pm 3, \ldots \qquad (38.1)$$

This equation gives the values of θ_{dark} for which the diffraction pattern has zero light intensity—that is, when a dark fringe is formed. However, it tells us nothing about the variation in light intensity along the screen. The general features of the intensity distribution are shown in Figure 38.6. A broad central bright fringe is observed; this fringe is flanked by much weaker bright fringes alternating with dark fringes. The various dark fringes occur at the values of θ_{dark} that satisfy Equation 38.1. Each bright-fringe peak lies approximately halfway between its bordering dark-fringe minima. Note that the central bright maximum is twice as wide as the secondary maxima.

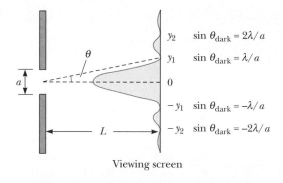

Figure 38.6 Intensity distribution for a Fraunhofer diffraction pattern from a single slit of width a. The positions of two minima on each side of the central maximum are labeled. (Drawing not to scale.)

$y_2 \quad \sin\theta_{dark} = 2\lambda/a$

$y_1 \quad \sin\theta_{dark} = \lambda/a$

0

$-y_1 \quad \sin\theta_{dark} = -\lambda/a$

$-y_2 \quad \sin\theta_{dark} = -2\lambda/a$

Viewing screen

Quick Quiz 38.1 Suppose the slit width in Figure 38.6 is made half as wide. The central bright fringe (a) becomes wider (b) remains the same (c) becomes narrower.

Quick Quiz 38.2 If a classroom door is open slightly, you can hear sounds coming from the hallway. Yet you cannot see what is happening in the hallway. Why is there this difference? (a) Light waves do not diffract through the single slit of the open doorway. (b) Sound waves can pass through the walls, but light waves cannot. (c) The open door is a small slit for sound waves, but a large slit for light waves. (d) The open door is a large slit for sound waves, but a small slit for light waves.

Example 38.1 Where Are the Dark Fringes?   **Interactive**

Light of wavelength 580 nm is incident on a slit having a width of 0.300 mm. The viewing screen is 2.00 m from the slit. Find the positions of the first dark fringes and the width of the central bright fringe.

Solution The problem statement cues us to conceptualize a single-slit diffraction pattern similar to that in Figure 38.6. We categorize this as a straightforward application of our discussion of single-slit diffraction patterns. To analyze the problem, note that the two dark fringes that flank the central bright fringe correspond to $m = \pm 1$ in Equation 38.1. Hence, we find that

$$\sin\theta_{dark} = \pm\frac{\lambda}{a} = \pm\frac{5.80 \times 10^{-7}\text{ m}}{0.300 \times 10^{-3}\text{ m}} = \pm 1.933 \times 10^{-3}$$

From the triangle in Figure 38.6, note that $\tan\theta_{dark} = y_1/L$. Because θ_{dark} is very small, we can use the approximation $\sin\theta_{dark} \approx \tan\theta_{dark}$; thus, $\sin\theta_{dark} \approx y_1/L$. Therefore, the positions of the first minima measured from the central axis are given by

$$y_1 \approx L\sin\theta_{dark} = (2.00\text{ m})(\pm 1.933 \times 10^{-3})$$
$$= \boxed{\pm 3.87 \times 10^{-3}\text{ m}}$$

The positive and negative signs correspond to the dark fringes on either side of the central bright fringe. Hence, the width of the central bright fringe is equal to $2|y_1| = 7.74 \times 10^{-3}\text{ m} = \boxed{7.74\text{ mm}}$. To finalize this problem,

note that this value is much greater than the width of the slit. We finalize further by exploring what happens if we change the slit width.

What If? What if the slit width is increased by an order of magnitude to 3.00 mm? What happens to the diffraction pattern?

Answer Based on Equation 38.1, we expect that the angles at which the dark bands appear will decrease as a increases. Thus, the diffraction pattern narrows. For $a = 3.00$ mm, the sines of the angles θ_{dark} for the $m = \pm 1$ dark fringes are

$$\sin\theta_{dark} = \pm\frac{\lambda}{a} = \pm\frac{5.80 \times 10^{-7}\text{ m}}{3.00 \times 10^{-3}\text{ m}} = \pm 1.933 \times 10^{-4}$$

The positions of the first minima measured from the central axis are given by

$$y_1 \approx L\sin\theta_{dark} = (2.00\text{ m})(\pm 1.933 \times 10^{-4})$$
$$= \pm 3.87 \times 10^{-4}\text{ m}$$

and the width of the central bright fringe is equal to $2|y_1| = 7.74 \times 10^{-4}\text{ m} = 0.774$ mm. Notice that this is *smaller* than the width of the slit.

In general, for large values of a, the various maxima and minima are so closely spaced that only a large central bright area resembling the geometric image of the slit is observed. This is very important in the performance of optical instruments such as telescopes.

Investigate the single-slit diffraction pattern at the Interactive Worked Example link at http://www.pse6.com.

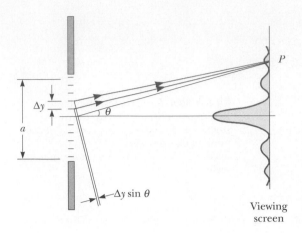

Figure 38.7 Fraunhofer diffraction pattern for a single slit. The light intensity at a distant screen is the resultant of all the incremental electric field magnitudes from zones of width Δy.

Intensity of Single-Slit Diffraction Patterns

We can use phasors to determine the light intensity distribution for a single-slit diffraction pattern. Imagine a slit divided into a large number of small zones, each of width Δy as shown in Figure 38.7. Each zone acts as a source of coherent radiation, and each contributes an incremental electric field of magnitude ΔE at some point on the screen. We obtain the total electric field magnitude E at a point on the screen by summing the contributions from all the zones. The light intensity at this point is proportional to the square of the magnitude of the electric field (Section 37.3).

The incremental electric field magnitudes between adjacent zones are out of phase with one another by an amount $\Delta\beta$, where the phase difference $\Delta\beta$ is related to the path difference $\Delta y \sin\theta$ between adjacent zones by an expression given by an argument similar to that leading to Equation 37.8:

$$\Delta\beta = \frac{2\pi}{\lambda}\,\Delta y \sin\theta \qquad (38.2)$$

To find the magnitude of the total electric field on the screen at any angle θ, we sum the incremental magnitudes ΔE due to each zone. For small values of θ, we can assume that all the ΔE values are the same. It is convenient to use phasor diagrams for various angles, as in Figure 38.8. When $\theta = 0$, all phasors are aligned as in Figure 38.8a because all the waves from the various zones are in phase. In this case, the total electric field at the center of the screen is $E_0 = N\,\Delta E$, where N is the number of zones. The resultant magnitude E_R at some small angle θ is shown in Figure 38.8b, where each phasor differs in phase from an adjacent one by an amount $\Delta\beta$. In this case, E_R is the

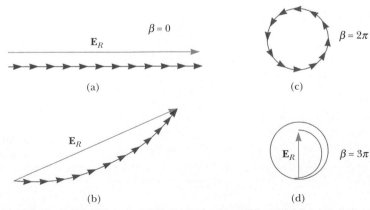

Figure 38.8 Phasor diagrams for obtaining the various maxima and minima of a single-slit diffraction pattern.

vector sum of the incremental magnitudes and hence is given by the length of the chord. Therefore, $E_R < E_0$. The total phase difference β between waves from the top and bottom portions of the slit is

$$\beta = N\,\Delta\beta = \frac{2\pi}{\lambda}\,N\,\Delta y\,\sin\theta = \frac{2\pi}{\lambda}\,a\sin\theta \qquad (38.3)$$

where $a = N\,\Delta y$ is the width of the slit.

As θ increases, the chain of phasors eventually forms the closed path shown in Figure 38.8c. At this point, the vector sum is zero, and so $E_R = 0$, corresponding to the first minimum on the screen. Noting that $\beta = N\,\Delta\beta = 2\pi$ in this situation, we see from Equation 38.3 that

$$2\pi = \frac{2\pi}{\lambda}\,a\sin\theta_{\text{dark}}$$

$$\sin\theta_{\text{dark}} = \frac{\lambda}{a}$$

That is, the first minimum in the diffraction pattern occurs where $\sin\theta_{\text{dark}} = \lambda/a$; this is in agreement with Equation 38.1.

At larger values of θ, the spiral chain of phasors tightens. For example, Figure 38.8d represents the situation corresponding to the second maximum, which occurs when $\beta = 360° + 180° = 540°$ (3π rad). The second minimum (two complete circles, not shown) corresponds to $\beta = 720°$ (4π rad), which satisfies the condition $\sin\theta_{\text{dark}} = 2\lambda/a$.

We can obtain the total electric-field magnitude E_R and light intensity I at any point on the screen in Figure 38.7 by considering the limiting case in which Δy becomes infinitesimal (dy) and N approaches ∞. In this limit, the phasor chains in Figure 38.8 become the curve of Figure 38.9. The arc length of the curve is E_0 because it is the sum of the magnitudes of the phasors (which is the total electric field magnitude at the center of the screen). From this figure, we see that at some angle θ, the resultant electric field magnitude E_R on the screen is equal to the chord length. From the triangle containing the angle $\beta/2$, we see that

$$\sin\frac{\beta}{2} = \frac{E_R/2}{R}$$

where R is the radius of curvature. But the arc length E_0 is equal to the product $R\beta$, where β is measured in radians. Combining this information with the previous expression gives

$$E_R = 2R\sin\frac{\beta}{2} = 2\left(\frac{E_0}{\beta}\right)\sin\frac{\beta}{2} = E_0\left[\frac{\sin(\beta/2)}{\beta/2}\right]$$

Because the resultant light intensity I at a point on the screen is proportional to the square of the magnitude E_R, we find that

$$I = I_{\max}\left[\frac{\sin(\beta/2)}{\beta/2}\right]^2 \qquad (38.4)$$

where $I_{\max}$ is the intensity at $\theta = 0$ (the central maximum). Substituting the expression for β (Eq. 38.3) into Equation 38.4, we have

$$I = I_{\max}\left[\frac{\sin(\pi a\sin\theta/\lambda)}{\pi a\sin\theta/\lambda}\right]^2 \qquad (38.5)$$

From this result, we see that *minima* occur when

$$\frac{\pi a\sin\theta_{\text{dark}}}{\lambda} = m\pi$$

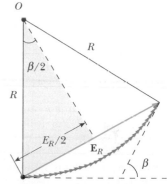

Figure 38.9 Phasor diagram for a large number of coherent sources. All the ends of the phasors lie on the circular arc of radius R. The resultant electric field magnitude E_R equals the length of the chord.

Intensity of a single-slit Fraunhofer diffraction pattern

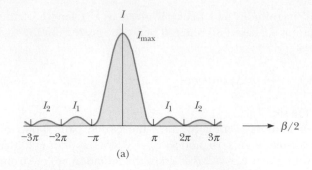

(a)

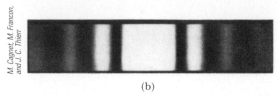

M. Cagnet, M. Francon, and J. C. Thierr

(b)

Figure 38.10 (a) A plot of light intensity I versus $\beta/2$ for the single-slit Fraunhofer diffraction pattern. (b) Photograph of a single-slit Fraunhofer diffraction pattern.

or

Condition for intensity minima for a single slit

$$\sin \theta_{\text{dark}} = m \frac{\lambda}{a} \qquad m = \pm 1, \pm 2, \pm 3, \ldots$$

in agreement with Equation 38.1.

Figure 38.10a represents a plot of Equation 38.4, and Figure 38.10b is a photograph of a single-slit Fraunhofer diffraction pattern. Note that most of the light intensity is concentrated in the central bright fringe.

Example 38.2 Relative Intensities of the Maxima

Find the ratio of the intensities of the secondary maxima to the intensity of the central maximum for the single-slit Fraunhofer diffraction pattern.

Solution To a good approximation, the secondary maxima lie midway between the zero points. From Figure 38.10a, we see that this corresponds to $\beta/2$ values of $3\pi/2$, $5\pi/2$, $7\pi/2$, ... Substituting these values into Equation 38.4 gives for the first two ratios

$$\frac{I_1}{I_{\text{max}}} = \left[\frac{\sin(3\pi/2)}{(3\pi/2)} \right]^2 = \frac{1}{9\pi^2/4} = \boxed{0.045}$$

$$\frac{I_2}{I_{\text{max}}} = \left[\frac{\sin(5\pi/2)}{5\pi/2} \right]^2 = \frac{1}{25\pi^2/4} = \boxed{0.016}$$

That is, the first secondary maxima (the ones adjacent to the central maximum) have an intensity of 4.5% that of the central maximum, and the next secondary maxima have an intensity of 1.6% that of the central maximum.

Intensity of Two-Slit Diffraction Patterns

When more than one slit is present, we must consider not only diffraction patterns due to the individual slits but also the interference patterns due to the waves coming from different slits. Notice the curved dashed lines in Figure 37.14, which indicate a decrease in intensity of the interference maxima as θ increases. This decrease is due to a diffraction pattern. To determine the effects of both two-slit interference and a single-slit diffraction pattern from each slit, we combine Equations 37.12 and 38.5:

$$I = I_{\text{max}} \cos^2 \left(\frac{\pi d \sin \theta}{\lambda} \right) \left[\frac{\sin(\pi a \sin \theta/\lambda)}{\pi a \sin \theta/\lambda} \right]^2 \tag{38.6}$$

Although this expression looks complicated, it merely represents the single-slit diffraction pattern (the factor in square brackets) acting as an "envelope" for a two-slit

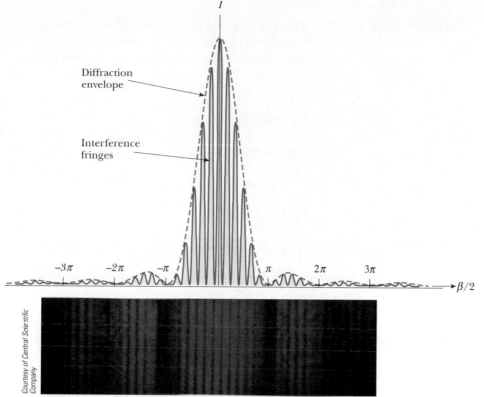

Active Figure 38.11 The combined effects of two-slit and single-slit interference. This is the pattern produced when 650-nm light waves pass through two 3.0-μm slits that are 18 μm apart. Notice how the diffraction pattern acts as an "envelope" and controls the intensity of the regularly spaced interference maxima.

At the Active Figures link at http://www.pse6.com, you can adjust the slit width, slit separation, and the wavelength of the light to see the effect on the interference pattern.

interference pattern (the cosine-squared factor), as shown in Figure 38.11. The broken blue curve in Figure 38.11 represents the factor in square brackets in Equation 38.6. The cosine-squared factor by itself would give a series of peaks all with the same height as the highest peak of the red-brown curve in Figure 38.11. Because of the effect of the square-bracket factor, however, these peaks vary in height as shown.

Equation 37.2 indicates the conditions for interference maxima as $d \sin\theta = m\lambda$, where d is the distance between the two slits. Equation 38.1 specifies that the first diffraction minimum occurs when $a \sin\theta = \lambda$, where a is the slit width. Dividing Equation 37.2 by Equation 38.1 (with $m = 1$) allows us to determine which interference maximum coincides with the first diffraction minimum:

$$\frac{d \sin\theta}{a \sin\theta} = \frac{m\lambda}{\lambda}$$

$$\frac{d}{a} = m \tag{38.7}$$

In Figure 38.11, $d/a = 18\ \mu\text{m}/3.0\ \mu\text{m} = 6$. Therefore, the sixth interference maximum (if we count the central maximum as $m = 0$) is aligned with the first diffraction minimum and cannot be seen.

Quick Quiz 38.3 Using Figure 38.11 as a starting point, make a sketch of the combined diffraction and interference pattern for 650-nm light waves striking two 3.0-μm slits located 9.0 μm apart.

Quick Quiz 38.4 Consider the central peak in the diffraction envelope in Figure 38.11. Suppose the wavelength of the light is changed to 450 nm. What happens to this central peak? (a) The width of the peak decreases and the number of interference fringes it encloses decreases. (b) The width of the peak decreases and the number of interference fringes it encloses increases. (c) The width of the peak decreases and the number of interference fringes it encloses remains the same. (d) The width of the peak increases and the number of interference fringes it encloses decreases. (e) The width of the peak increases and the number of interference fringes it encloses increases. (f) The width of the peak increases and the number of interference fringes it encloses remains the same.

38.3 Resolution of Single-Slit and Circular Apertures

The ability of optical systems to distinguish between closely spaced objects is limited because of the wave nature of light. To understand this difficulty, consider Figure 38.12, which shows two light sources far from a narrow slit of width a. The sources can be two noncoherent point sources S_1 and S_2—for example, they could be two distant stars. If no interference occurred between light passing through different parts of the slit, two distinct bright spots (or images) would be observed on the viewing screen. However, because of such interference, each source is imaged as a bright central region flanked by weaker bright and dark fringes—a diffraction pattern. What is observed on the screen is the sum of two diffraction patterns: one from S_1, and the other from S_2.

If the two sources are far enough apart to keep their central maxima from overlapping as in Figure 38.12a, their images can be distinguished and are said to be *resolved*. If the sources are close together, however, as in Figure 38.12b, the two central maxima overlap, and the images are not resolved. To determine whether two images are resolved, the following condition is often used:

When the central maximum of one image falls on the first minimum of another image, the images are said to be just resolved. This limiting condition of resolution is known as **Rayleigh's criterion.**

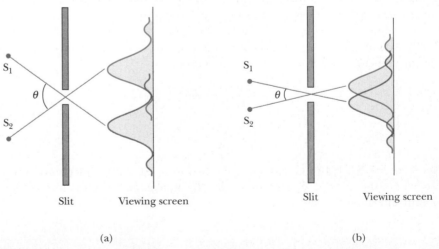

	(a)	(b)

Figure 38.12 Two point sources far from a narrow slit each produce a diffraction pattern. (a) The angle subtended by the sources at the slit is large enough for the diffraction patterns to be distinguishable. (b) The angle subtended by the sources is so small that their diffraction patterns overlap, and the images are not well resolved. (Note that the angles are greatly exaggerated. The drawing is not to scale.)

From Rayleigh's criterion, we can determine the minimum angular separation θ_{min} subtended by the sources at the slit in Figure 38.12 for which the images are just resolved. Equation 38.1 indicates that the first minimum in a single-slit diffraction pattern occurs at the angle for which

$$\sin\theta = \frac{\lambda}{a}$$

where a is the width of the slit. According to Rayleigh's criterion, this expression gives the smallest angular separation for which the two images are resolved. Because $\lambda \ll a$ in most situations, $\sin\theta$ is small, and we can use the approximation $\sin\theta \approx \theta$. Therefore, the limiting angle of resolution for a slit of width a is

$$\theta_{min} = \frac{\lambda}{a} \tag{38.8}$$

where θ_{min} is expressed in radians. Hence, the angle subtended by the two sources at the slit must be greater than λ/a if the images are to be resolved.

Many optical systems use circular apertures rather than slits. The diffraction pattern of a circular aperture, as shown in the lower half of Figure 38.13, consists of a central circular bright disk surrounded by progressively fainter bright and dark rings. Figure 38.13 shows diffraction patterns for three situations in which light from two point sources passes through a circular aperture. When the sources are far apart, their images are well resolved (Fig. 38.13a). When the angular separation of the sources satisfies Rayleigh's criterion, the images are just resolved (Fig. 38.13b). Finally, when the sources are close together, the images are said to be unresolved (Fig. 38.13c).

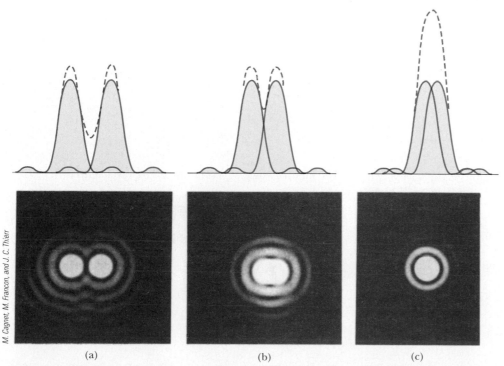

M. Cagnet, M. Francon, and J. C. Thierr

(a)   (b)   (c)

Figure 38.13 Individual diffraction patterns of two point sources (solid curves) and the resultant patterns (dashed curves) for various angular separations of the sources. In each case, the dashed curve is the sum of the two solid curves. (a) The sources are far apart, and the patterns are well resolved. (b) The sources are closer together such that the angular separation just satisfies Rayleigh's criterion, and the patterns are just resolved. (c) The sources are so close together that the patterns are not resolved.

Analysis shows that the limiting angle of resolution of the circular aperture is

Limiting angle of resolution for a circular aperture

$$\theta_{min} = 1.22 \frac{\lambda}{D} \qquad (38.9)$$

where D is the diameter of the aperture. Note that this expression is similar to Equation 38.8 except for the factor 1.22, which arises from a mathematical analysis of diffraction from the circular aperture.

Quick Quiz 38.5 Cat's eyes have pupils that can be modeled as vertical slits. At night, would cats be more successful in resolving (a) headlights on a distant car, or (b) vertically-separated lights on the mast of a distant boat?

Quick Quiz 38.6 Suppose you are observing a binary star with a telescope and are having difficulty resolving the two stars. You decide to use a colored filter to maximize the resolution. (A filter of a given color transmits only that color of light.) What color filter should you choose? (a) blue (b) green (c) yellow (d) red.

Example 38.3   Limiting Resolution of a Microscope

Light of wavelength 589 nm is used to view an object under a microscope. If the aperture of the objective has a diameter of 0.900 cm,

(A) what is the limiting angle of resolution?

Solution Using Equation 38.9, we find that the limiting angle of resolution is

$$\theta_{min} = 1.22 \left(\frac{589 \times 10^{-9} \text{ m}}{0.900 \times 10^{-2} \text{ m}} \right) = \boxed{7.98 \times 10^{-5} \text{ rad}}$$

This means that any two points on the object subtending an angle smaller than this at the objective cannot be distinguished in the image.

(B) If it were possible to use visible light of any wavelength, what would be the maximum limit of resolution for this microscope?

Solution To obtain the smallest limiting angle, we have to use the shortest wavelength available in the visible spectrum. Violet light (400 nm) gives a limiting angle of resolution of

$$\theta_{min} = 1.22 \left(\frac{400 \times 10^{-9} \text{ m}}{0.900 \times 10^{-2} \text{ m}} \right) = \boxed{5.42 \times 10^{-5} \text{ rad}}$$

What If? Suppose that water ($n = 1.33$) fills the space between the object and the objective. What effect does this have on resolving power when 589-nm light is used?

Answer Because light travels more slowly in water, we know that the wavelength of the light in water is smaller than that in vacuum. Based on Equation 38.9, we expect the limiting angle of resolution to be smaller. To find the new value of the limiting angle of resolution, we first calculate the wavelength of the 589-nm light in water using Equation 35.7:

$$\lambda_{water} = \frac{\lambda_{air}}{n_{water}} = \frac{589 \text{ nm}}{1.33} = 443 \text{ nm}$$

The limiting angle of resolution at this wavelength is

$$\theta_{min} = 1.22 \left(\frac{443 \times 10^{-9} \text{ m}}{0.900 \times 10^{-2} \text{ m}} \right) = \boxed{6.00 \times 10^{-5} \text{ rad}}$$

which is indeed smaller than that calculated in part (A).

Example 38.4   Resolution of the Eye

Estimate the limiting angle of resolution for the human eye, assuming its resolution is limited only by diffraction.

Solution Let us choose a wavelength of 500 nm, near the center of the visible spectrum. Although pupil diameter varies from person to person, we estimate a daytime diameter of 2 mm. We use Equation 38.9, taking $\lambda = 500$ nm

and $D = 2$ mm:

$$\theta_{min} = 1.22 \frac{\lambda}{D} = 1.22 \left(\frac{5.00 \times 10^{-7} \text{ m}}{2 \times 10^{-3} \text{ m}} \right)$$

$$\approx 3 \times 10^{-4} \text{ rad} \approx \boxed{1 \text{ min of arc}}$$

We can use this result to determine the minimum separation distance d between two point sources that the eye can distinguish if they are a distance L from the observer (Fig. 38.14). Because θ_{min} is small, we see that

$$\sin \theta_{min} \approx \theta_{min} \approx \frac{d}{L}$$

$$d = L\theta_{min}$$

For example, if the point sources are 25 cm from the eye (the near point), then

$$d = (25 \text{ cm})(3 \times 10^{-4} \text{ rad}) = 8 \times 10^{-3} \text{ cm}$$

This is approximately equal to the thickness of a human hair.

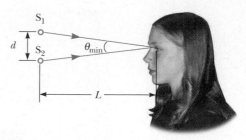

Figure 38.14 (Example 38.4) Two point sources separated by a distance d as observed by the eye.

Example 38.5   Resolution of a Telescope

The Keck telescope at Mauna Kea, Hawaii, has an effective diameter of 10 m. What is its limiting angle of resolution for 600-nm light?

Solution Because $D = 10$ m and $\lambda = 6.00 \times 10^{-7}$ m, Equation 38.9 gives

$$\theta_{min} = 1.22 \frac{\lambda}{D} = 1.22 \left(\frac{6.00 \times 10^{-7} \text{ m}}{10 \text{ m}} \right)$$

$$= 7.3 \times 10^{-8} \text{ rad} \approx \boxed{0.015 \text{ s of arc}}$$

Any two stars that subtend an angle greater than or equal to this value are resolved (if atmospheric conditions are ideal).

The Keck telescope can never reach its diffraction limit because the limiting angle of resolution is always set by atmospheric blurring at optical wavelengths. This seeing limit is usually about 1 s of arc and is never smaller than about 0.1 s of arc. (This is one of the reasons for the superiority of photographs from the Hubble Space Telescope, which views celestial objects from an orbital position above the atmosphere.)

What If? What if we consider radio telescopes? These are much larger in diameter than optical telescopes, but do they have angular resolutions that are better than optical telescopes? For example, the radio telescope at Arecibo, Puerto Rico, has a diameter of 305 m and is designed to detect radio waves of 0.75-m wavelength. How does its resolution compare to that of the Keck telescope?

Answer The increase in diameter might suggest that radio telescopes would have better resolution, but Equation 38.9 shows that θ_{min} depends on *both* diameter and wavelength. Calculating the minimum angle of resolution for the radio telescope, we find

$$\theta_{min} = 1.22 \frac{\lambda}{D} = 1.22 \left(\frac{0.75 \text{ m}}{305 \text{ m}} \right)$$

$$= 3.0 \times 10^{-3} \text{ rad} \approx 10 \text{ min of arc}$$

Notice that this limiting angle of resolution is measured in *minutes* of arc rather than the *seconds* of arc for the optical telescope. Thus, the change in wavelength more than compensates for the increase in diameter, and the limiting angle of resolution for the Arecibo radio telescope is more than 40 000 times larger (that is, *worse*) than the Keck minimum.

As an example of the effects of atmospheric blurring mentioned in Example 38.5, consider telescopic images of Pluto and its moon Charon. Figure 38.15a shows the image taken in 1978 that represents the discovery of Charon. In this photograph taken from an Earth-based telescope, atmospheric turbulence causes the image of Charon to appear only as a bump on the edge of Pluto. In comparison, Figure 38.15b shows a photograph taken with the Hubble Space Telescope. Without the problems of atmospheric turbulence, Pluto and its moon are clearly resolved.

38.4   The Diffraction Grating

The **diffraction grating,** a useful device for analyzing light sources, consists of a large number of equally spaced parallel slits. A *transmission grating* can be made by cutting parallel grooves on a glass plate with a precision ruling machine. The spaces between the grooves are transparent to the light and hence act as separate slits. A *reflection grating* can

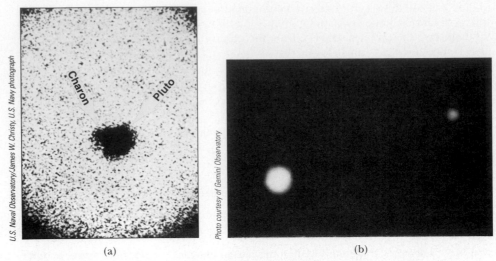

(a)   (b)

Figure 38.15 (a) The photograph on which Charon, the moon of Pluto, was discovered in 1978. From an Earth-based telescope, atmospheric blurring results in Charon appearing only as a subtle bump on the edge of Pluto. (b) A Hubble Space Telescope photo of Pluto and Charon, clearly resolving the two objects.

be made by cutting parallel grooves on the surface of a reflective material. The reflection of light from the spaces between the grooves is specular, and the reflection from the grooves cut into the material is diffuse. Thus, the spaces between the grooves act as parallel sources of reflected light, like the slits in a transmission grating. Current technology can produce gratings that have very small slit spacings. For example, a typical grating ruled with 5 000 grooves/cm has a slit spacing $d = (1/5\,000)$ cm $= 2.00 \times 10^{-4}$ cm.

A section of a diffraction grating is illustrated in Figure 38.16. A plane wave is incident from the left, normal to the plane of the grating. The pattern observed on the

▲ PITFALL PREVENTION

38.3 A Diffraction Grating Is an Interference Grating

As with *diffraction pattern, diffraction grating* is a misnomer, but is deeply entrenched in the language of physics. The diffraction grating depends on diffraction in the same way as the double slit—spreading the light so that light from different slits can interfere. It would be more correct to call it an *interference grating*, but *diffraction grating* is the name in use.

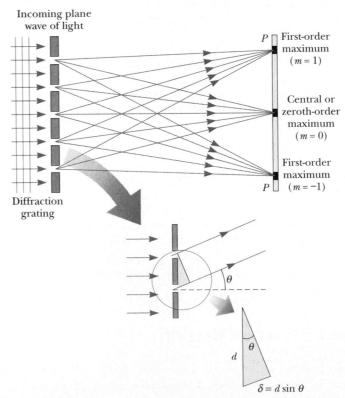

Figure 38.16 Side view of a diffraction grating. The slit separation is d, and the path difference between adjacent slits is $d \sin \theta$.

screen (far to the right of Figure 38.16) is the result of the combined effects of interference and diffraction. Each slit produces diffraction, and the diffracted beams interfere with one another to produce the final pattern.

The waves from all slits are in phase as they leave the slits. However, for some arbitrary direction θ measured from the horizontal, the waves must travel different path lengths before reaching the screen. From Figure 38.16, note that the path difference δ between rays from any two adjacent slits is equal to $d \sin \theta$. If this path difference equals one wavelength or some integral multiple of a wavelength, then waves from all slits are in phase at the screen and a bright fringe is observed. Therefore, the condition for *maxima* in the interference pattern at the angle θ_{bright} is

$$d \sin \theta_{\text{bright}} = m\lambda \qquad m = 0, \pm 1, \pm 2, \pm 3, \ldots \qquad (38.10)$$

Condition for interference maxima for a grating

We can use this expression to calculate the wavelength if we know the grating spacing d and the angle θ_{bright}. If the incident radiation contains several wavelengths, the mth-order maximum for each wavelength occurs at a specific angle. All wavelengths are seen at $\theta = 0$, corresponding to $m = 0$, the zeroth-order maximum. The first-order maximum ($m = 1$) is observed at an angle that satisfies the relationship $\sin \theta_{\text{bright}} = \lambda/d$; the second-order maximum ($m = 2$) is observed at a larger angle θ_{bright}, and so on.

The intensity distribution for a diffraction grating obtained with the use of a monochromatic source is shown in Figure 38.17. Note the sharpness of the principal maxima and the broadness of the dark areas. This is in contrast to the broad bright fringes characteristic of the two-slit interference pattern (see Fig. 37.7). You should also review Figure 37.14, which shows that the width of the intensity maxima decreases as the number of slits increases. Because the principal maxima are so sharp, they are much brighter than two-slit interference maxima.

A schematic drawing of a simple apparatus used to measure angles in a diffraction pattern is shown in Figure 38.18. This apparatus is a *diffraction grating spectrometer*. The light to be analyzed passes through a slit, and a collimated beam of light is incident on the grating. The diffracted light leaves the grating at angles that satisfy Equation 38.10, and a telescope is used to view the image of the slit. The wavelength can be determined by measuring the precise angles at which the images of the slit appear for the various orders.

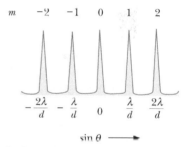

Active Figure 38.17 Intensity versus $\sin \theta$ for a diffraction grating. The zeroth-, first-, and second-order maxima are shown.

At the Active Figures link at http://www.pse6.com, *you can choose the number of slits to be illuminated to see the effect on the interference pattern.*

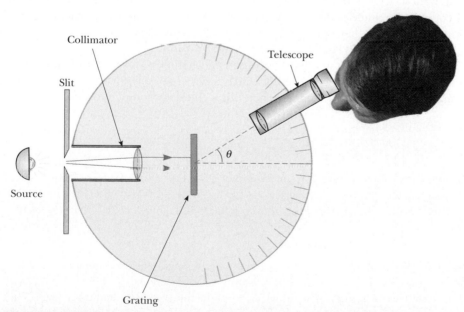

Active Figure 38.18 Diagram of a diffraction grating spectrometer. The collimated beam incident on the grating is spread into its various wavelength components with constructive interference for a particular wavelength occurring at the angles θ_{bright} that satisfy the equation $d \sin \theta_{\text{bright}} = m\lambda$, where $m = 0, 1, 2, \ldots$.

Use the spectrometer at the Active Figures link at http://www.pse6.com *to observe constructive interference for various wavelengths.*

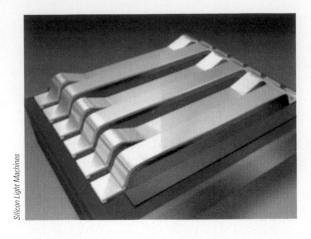

Silicon Light Machines

Figure 38.19 A small portion of a grating light valve. The alternating reflective ribbons at different levels act as a diffraction grating, offering very-high-speed control of the direction of light toward a digital display device.

The spectrometer is a useful tool in *atomic spectroscopy*, in which the light from an atom is analyzed to find the wavelength components. These wavelength components can be used to identify the atom. We will investigate atomic spectra in Chapter 42 of the extended version of this text.

Another application of diffraction gratings is in the recently developed *grating light valve* (GLV), which may compete in the near future in video projection with the digital micromirror devices (DMDs) discussed in Section 35.4. The grating light valve consists of a silicon microchip fitted with an array of parallel silicon nitride ribbons coated with a thin layer of aluminum (Fig. 38.19). Each ribbon is about 20 μm long and about 5 μm wide and is separated from the silicon substrate by an air gap on the order of 100 nm. With no voltage applied, all ribbons are at the same level. In this situation, the array of ribbons acts as a flat surface, specularly reflecting incident light.

When a voltage is applied between a ribbon and the electrode on the silicon substrate, an electric force pulls the ribbon downward, closer to the substrate. Alternate ribbons can be pulled down, while those in between remain in the higher configuration. As a result, the array of ribbons acts as a diffraction grating, such that the constructive interference for a particular wavelength of light can be directed toward a screen or other optical display system. By using three such devices, one each for red, blue, and green light, full-color display is possible.

The GLV tends to be simpler to fabricate and higher in resolution than comparable DMD devices. On the other hand, DMD devices have already made an entry into the market. It will be interesting to watch this technology competition in future years.

Quick Quiz 38.7   If laser light is reflected from a phonograph record or a compact disc, a diffraction pattern appears. This is due to the fact that both devices contain parallel tracks of information that act as a reflection diffraction grating. Which device, (a) record or (b) compact disc, results in diffraction maxima that are farther apart in angle?

Quick Quiz 38.8   Ultraviolet light of wavelength 350 nm is incident on a diffraction grating with slit spacing d and forms an interference pattern on a screen a distance L away. The angular positions θ_{bright} of the interference maxima are large. The locations of the bright fringes are marked on the screen. Now red light of wavelength 700 nm is used with a diffraction grating to form another diffraction pattern on the screen. The bright fringes of this pattern will be located at the marks on the screen if

(a) the screen is moved to a distance $2L$ from the grating (b) the screen is moved to a distance $L/2$ from the grating (c) the grating is replaced with one of slit spacing $2d$ (d) the grating is replaced with one of slit spacing $d/2$ (e) nothing is changed.

Conceptual Example 38.6   A Compact Disc Is a Diffraction Grating

Light reflected from the surface of a compact disc is multicolored, as shown in Figure 38.20. The colors and their intensities depend on the orientation of the disc relative to the eye and relative to the light source. Explain how this works.

Solution The surface of a compact disc has a spiral grooved track (with adjacent grooves having a separation on the order of 1 μm). Thus, the surface acts as a reflection grating. The light reflecting from the regions between these closely spaced grooves interferes constructively only in certain directions that depend on the wavelength and on the direction of the incident light. Any section of the disc serves as a diffraction grating for white light, sending different colors in different directions. The different colors you see when viewing one section change as the light source, the disc, or you move to change the angles of incidence or diffraction.

Figure 38.20 (Conceptual Example 38.6) A compact disc observed under white light. The colors observed in the reflected light and their intensities depend on the orientation of the disc relative to the eye and relative to the light source.

Example 38.7   The Orders of a Diffraction Grating   Interactive

Monochromatic light from a helium–neon laser ($\lambda = 632.8$ nm) is incident normally on a diffraction grating containing 6 000 grooves per centimeter. Find the angles at which the first- and second-order maxima are observed.

Solution First, we must calculate the slit separation, which is equal to the inverse of the number of grooves per centimeter:

$$d = \frac{1}{6\,000} \text{ cm} = 1.667 \times 10^{-4} \text{ cm} = 1\,667 \text{ nm}$$

For the first-order maximum ($m = 1$), we obtain

$$\sin \theta_1 = \frac{\lambda}{d} = \frac{632.8 \text{ nm}}{1\,667 \text{ nm}} = 0.379\,6$$

$$\theta_1 = \boxed{22.31°}$$

For the second-order maximum ($m = 2$), we find

$$\sin \theta_2 = \frac{2\lambda}{d} = \frac{2(632.8 \text{ nm})}{1\,667 \text{ nm}} = 0.759\,2$$

$$\theta_2 = \boxed{49.39°}$$

What If? What if we look for the third-order maximum? Do we find it?

Answer For $m = 3$, we find $\sin \theta_3 = 1.139$. Because $\sin \theta$ cannot exceed unity, this does not represent a realistic solution. Hence, only zeroth-, first-, and second-order maxima are observed for this situation.

 Investigate the interference pattern from a diffraction grating at the Interactive Worked Example link at **http://www.pse6.com.**

Resolving Power of the Diffraction Grating

The diffraction grating is useful for measuring wavelengths accurately. Like the prism, the diffraction grating can be used to separate white light into its wavelength components. Of the two devices, a grating with very small slit separation is more precise if one wants to distinguish two closely spaced wavelengths.

For two nearly equal wavelengths λ_1 and λ_2 between which a diffraction grating can just barely distinguish, the **resolving power** R of the grating is defined as

Resolving power

$$R \equiv \frac{\lambda}{\lambda_2 - \lambda_1} = \frac{\lambda}{\Delta\lambda} \qquad (38.11)$$

where $\lambda = (\lambda_1 + \lambda_2)/2$ and $\Delta\lambda = \lambda_2 - \lambda_1$. Thus, a grating that has a high resolving power can distinguish small differences in wavelength. If N slits of the grating are illuminated, it can be shown that the resolving power in the mth-order diffraction is

Resolving power of a grating

$$R = Nm \qquad (38.12)$$

Thus, resolving power increases with increasing order number and with increasing number of illuminated slits.

Note that $R = 0$ for $m = 0$; this signifies that all wavelengths are indistinguishable for the zeroth-order maximum. However, consider the second-order diffraction pattern ($m = 2$) of a grating that has 5 000 rulings illuminated by the light source. The resolving power of such a grating in second order is $R = 5\,000 \times 2 = 10\,000$. Therefore, for a mean wavelength of, for example, 600 nm, the minimum wavelength separation between two spectral lines that can be just resolved is $\Delta\lambda = \lambda/R = 6.00 \times 10^{-2}$ nm. For the third-order principal maximum, $R = 15\,000$ and $\Delta\lambda = 4.00 \times 10^{-2}$ nm, and so on.

Example 38.8 Resolving Sodium Spectral Lines

When a gaseous element is raised to a very high temperature, the atoms emit radiation having discrete wavelengths. The set of wavelengths for a given element is called its *atomic spectrum* (Chapter 42). Two strong components in the atomic spectrum of sodium have wavelengths of 589.00 nm and 589.59 nm.

(A) What resolving power must a grating have if these wavelengths are to be distinguished?

Solution Using Equation 38.11,

$$R = \frac{\lambda}{\Delta\lambda} = \frac{589.30 \text{ nm}}{589.59 \text{ nm} - 589.00 \text{ nm}} = \frac{589.30}{0.59} = \boxed{999}$$

(B) To resolve these lines in the second-order spectrum, how many slits of the grating must be illuminated?

Solution From Equation 38.12 and the result to part (A), we find that

$$N = \frac{R}{m} = \frac{999}{2} = \boxed{500 \text{ slits}}$$

Application Holography

One interesting application of diffraction gratings is **holography,** the production of three-dimensional images of objects. The physics of holography was developed by Dennis Gabor in 1948, and resulted in the Nobel Prize in physics for Gabor in 1971. The requirement of coherent light for holography, however, delayed the realization of holographic images from Gabor's work until the development of lasers in the 1960s. Figure 38.21 shows a hologram and the three-dimensional character of its image.

Figure 38.22 shows how a hologram is made. Light from the laser is split into two parts by a half-silvered mirror at B. One part of the beam reflects off the object to be photographed and strikes an ordinary photographic film. The other half of the beam is diverged by lens L_2, reflects from mirrors M_1 and M_2, and finally strikes the film. The two beams overlap to form an extremely complicated interference pattern on the film. Such an interference pattern can be produced only if the phase relationship of the two waves is constant throughout the exposure of the film. This condition is met by illuminating the scene with light coming through a pinhole or with coherent laser radiation. The hologram records not only the intensity of the light scattered from the object (as in a conventional photograph), but also the phase difference between the reference beam and the beam scattered from the object. Because of this phase difference, an interference pattern is formed that produces an image in which all three-dimensional information available from the perspective of any point on the hologram is preserved.

In a normal photographic image, a lens is used to focus the image so that each point on the object corresponds to a single point on the film. Notice that there is no lens used in Figure 38.22 to focus the light onto the film. Thus, light from each point on the object reaches *all* points on the film. As a result, each region of the photographic film on which the hologram is recorded contains information about all illuminated points on the object. This leads to a remarkable

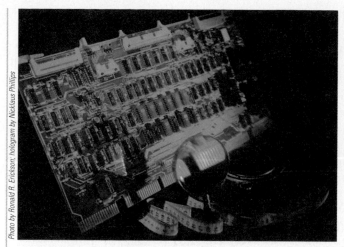
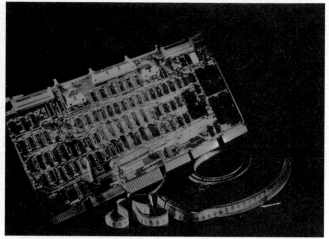

Figure 38.21 In this hologram, a circuit board is shown from two different views. Notice the difference in the appearance of the measuring tape and the view through the magnifying lens.

result—if a small section of the hologram is cut from the film, the complete image can be formed from the small piece! (The quality of the image is reduced, but the entire image is present.)

A hologram is best viewed by allowing coherent light to pass through the developed film as one looks back along the direction from which the beam comes. The interference pattern on the film acts as a diffraction grating. Figure 38.23 shows two rays of light striking the film and passing through. For each ray, the $m = 0$ and $m = \pm 1$ rays in the diffraction pattern are shown emerging from the right side of the film. The $m = +1$ rays converge to

form a real image of the scene, which is not the image that is normally viewed. By extending the light rays corresponding to $m = -1$ back behind the film, we see that there is a virtual image located there, with light coming from it in exactly the same way that light came from the actual object when the film was exposed. This is the image that we see by looking through the holographic film.

Holograms are finding a number of applications. You may have a hologram on your credit card. This is a special type of hologram called a *rainbow hologram*, designed to be viewed in reflected white light.

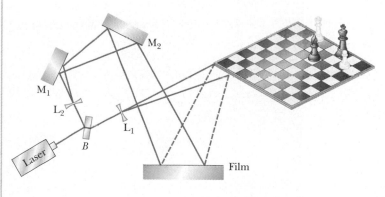

Figure 38.22 Experimental arrangement for producing a hologram.

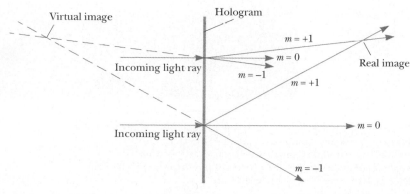

Figure 38.23 Two light rays strike a hologram at normal incidence. For each ray, outgoing rays corresponding to $m = 0$ and $m = \pm 1$ are shown. If the $m = -1$ rays are extended backward, a virtual image of the object photographed in the hologram exists on the front side of the hologram.

38.5   Diffraction of X-Rays by Crystals

In principle, the wavelength of any electromagnetic wave can be determined if a grating of the proper spacing (on the order of λ) is available. X-rays, discovered by Wilhelm Roentgen (1845–1923) in 1895, are electromagnetic waves of very short wavelength (on the order of 0.1 nm). It would be impossible to construct a grating having such a small spacing by the cutting process described at the beginning of Section 38.4. However, the atomic spacing in a solid is known to be about 0.1 nm. In 1913, Max von Laue (1879–1960) suggested that the regular array of atoms in a crystal could act as a three-dimensional diffraction grating for x-rays. Subsequent experiments confirmed this prediction. The diffraction patterns from crystals are complex because of the three-dimensional nature of crystal structure. Nevertheless, x-ray diffraction has proved to be an invaluable technique for elucidating these structures and for understanding the structure of matter.

Figure 38.24 is one experimental arrangement for observing x-ray diffraction from a crystal. A collimated beam of monochromatic x-rays is incident on a crystal. The diffracted beams are very intense in certain directions, corresponding to constructive interference from waves reflected from layers of atoms in the crystal. The diffracted beams, which can be detected by a photographic film, form an array of spots known as a *Laue pattern*, as in Figure 38.25a. One can deduce the crystalline structure by analyzing the positions and intensities of the various spots in the pattern. Fig. 38.25b shows a Laue pattern from a crystalline enzyme, using a wide range of wavelengths so that a swirling pattern results.

The arrangement of atoms in a crystal of sodium chloride (NaCl) is shown in Figure 38.26. Each unit cell (the geometric solid that repeats throughout the crystal) is a cube having an edge length a. A careful examination of the NaCl structure shows that the ions lie in discrete planes (the shaded areas in Fig. 38.26). Now suppose that an incident x-ray beam makes an angle θ with one of the planes, as in Figure 38.27. The beam can be reflected from both the upper plane and the lower one. However,

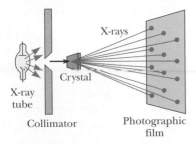

Figure 38.24 Schematic diagram of the technique used to observe the diffraction of x-rays by a crystal. The array of spots formed on the film is called a Laue pattern.

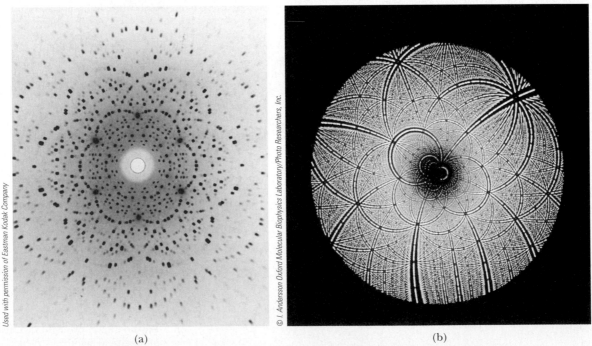

(a)   (b)

Figure 38.25 (a) A Laue pattern of a single crystal of the mineral beryl (beryllium aluminum silicate). Each dot represents a point of constructive interference. (b) A Laue pattern of the enzyme Rubisco, produced with a wide-band x-ray spectrum. This enzyme is present in plants and takes part in the process of photosynthesis. The Laue pattern is used to determine the crystal structure of Rubisco.

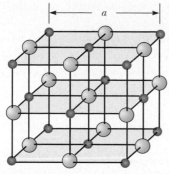

Figure 38.26 Crystalline structure of sodium chloride (NaCl). The blue spheres represent Cl^- ions, and the red spheres represent Na^+ ions. The length of the cube edge is $a = 0.562\ 737$ nm.

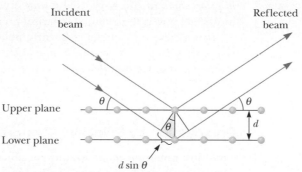

Figure 38.27 A two-dimensional description of the reflection of an x-ray beam from two parallel crystalline planes separated by a distance d. The beam reflected from the lower plane travels farther than the one reflected from the upper plane by a distance $2d \sin \theta$.

the beam reflected from the lower plane travels farther than the beam reflected from the upper plane. The effective path difference is $2d \sin \theta$. The two beams reinforce each other (constructive interference) when this path difference equals some integer multiple of λ. The same is true for reflection from the entire family of parallel planes. Hence, the condition for *constructive* interference (maxima in the reflected beam) is

$$2d \sin \theta = m\lambda \qquad m = 1, 2, 3, \ldots \qquad (38.13)$$

This condition is known as **Bragg's law,** after W. L. Bragg (1890–1971), who first derived the relationship. If the wavelength and diffraction angle are measured, Equation 38.13 can be used to calculate the spacing between atomic planes.

38.6 Polarization of Light Waves

In Chapter 34 we described the transverse nature of light and all other electromagnetic waves. Polarization, discussed in this section, is firm evidence of this transverse nature.

An ordinary beam of light consists of a large number of waves emitted by the atoms of the light source. Each atom produces a wave having some particular orientation of the electric field vector **E**, corresponding to the direction of atomic vibration. The *direction of polarization* of each individual wave is defined to be the direction in which the electric field is vibrating. In Figure 38.28, this direction happens to lie along the y axis. However, an individual electromagnetic wave could have its **E** vector in the yz plane, making any possible angle with the y axis. Because all directions of vibration from a wave source are possible, the resultant electromagnetic wave is a superposition of waves vibrating in many different directions. The result is an **unpolarized** light beam, represented in Figure 38.29a. The direction of wave propagation in this figure is perpendicular to the page. The arrows show a few possible directions of the electric field vectors for the individual waves making up the resultant beam. At any given point and at some instant of time, all these individual electric field vectors add to give one resultant electric field vector.

As noted in Section 34.2, a wave is said to be **linearly polarized** if the resultant electric field **E** vibrates in the same direction *at all times* at a particular point, as shown in Figure 38.29b. (Sometimes, such a wave is described as *plane-polarized*, or simply *polarized.*) The plane formed by **E** and the direction of propagation is called the *plane*

Bragg's law

▲ **PITFALL PREVENTION**

38.4 Different Angles

Notice in Figure 38.27 that the angle θ is measured from the reflecting surface, rather than from the normal, as in the case of the law of reflection in Chapter 35. With slits and diffraction gratings, we also measured the angle θ from the normal to the array of slits. Because of historical tradition, the angle is measured differently in Bragg diffraction, so interpret Equation 38.13 with care.

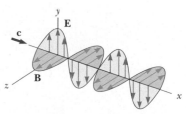

Figure 38.28 Schematic diagram of an electromagnetic wave propagating at velocity **c** in the x direction. The electric field vibrates in the xy plane, and the magnetic field vibrates in the xz plane.

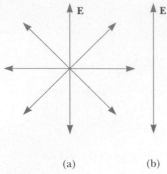

Figure 38.29 (a) A representation of an unpolarized light beam viewed along the direction of propagation (perpendicular to the page). The transverse electric field can vibrate in any direction in the plane of the page with equal probability. (b) A linearly polarized light beam with the electric field vibrating in the vertical direction.

of polarization of the wave. If the wave in Figure 38.28 represents the resultant of all individual waves, the plane of polarization is the *xy* plane.

It is possible to obtain a linearly polarized beam from an unpolarized beam by removing all waves from the beam except those whose electric field vectors oscillate in a single plane. We now discuss four processes for producing polarized light from unpolarized light.

Polarization by Selective Absorption

The most common technique for producing polarized light is to use a material that transmits waves whose electric fields vibrate in a plane parallel to a certain direction and that absorbs waves whose electric fields vibrate in all other directions.

In 1938, E. H. Land (1909–1991) discovered a material, which he called *polaroid*, that polarizes light through selective absorption by oriented molecules. This material is fabricated in thin sheets of long-chain hydrocarbons. The sheets are stretched during manufacture so that the long-chain molecules align. After a sheet is dipped into a solution containing iodine, the molecules become good electrical conductors. However, conduction takes place primarily along the hydrocarbon chains because electrons can move easily only along the chains. As a result, the molecules readily absorb light whose electric field vector is parallel to their length and allow light through whose electric field vector is perpendicular to their length.

It is common to refer to the direction perpendicular to the molecular chains as the *transmission axis*. In an ideal polarizer, all light with **E** parallel to the transmission axis is transmitted, and all light with **E** perpendicular to the transmission axis is absorbed.

Figure 38.30 represents an unpolarized light beam incident on a first polarizing sheet, called the *polarizer*. Because the transmission axis is oriented vertically in the figure, the light transmitted through this sheet is polarized vertically. A second polarizing sheet, called the *analyzer*, intercepts the beam. In Figure 38.30, the analyzer transmission axis is set at an angle θ to the polarizer axis. We call the electric field vector of the first transmitted beam $\mathbf{E}_0$. The component of $\mathbf{E}_0$ perpendicular to the analyzer axis is completely absorbed. The component of $\mathbf{E}_0$ parallel to the analyzer axis, which is allowed through by the analyzer, is $E_0 \cos\theta$. Because the intensity of the transmitted beam varies as the square of its magnitude, we conclude that the intensity of the (polarized) beam transmitted through the analyzer varies as

Malus's law

$$I = I_{\text{max}} \cos^2 \theta \qquad (38.14)$$

Rotate the analyzer at the Active Figures link at http://www.pse6.com, *to see the effect on the transmitted light.*

Unpolarized light

Polarizer

Analyzer

$\mathbf{E}_0$

θ

$\mathbf{E}_0 \cos\theta$

Transmission axis

Polarized light

Active Figure 38.30 Two polarizing sheets whose transmission axes make an angle θ with each other. Only a fraction of the polarized light incident on the analyzer is transmitted through it.

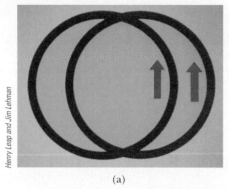
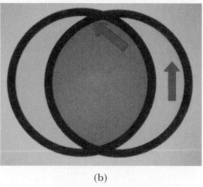
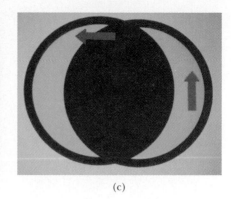

(a)   (b)   (c)

Henry Leap and Jim Lehman

Figure 38.31 The intensity of light transmitted through two polarizers depends on the relative orientation of their transmission axes. (a) The transmitted light has maximum intensity when the transmission axes are aligned with each other. (b) The transmitted light has lesser intensity when the transmission axes are at an angle of 45° with each other. (c) The transmitted light intensity is a minimum when the transmission axes are perpendicular to each other.

where I_{max} is the intensity of the polarized beam incident on the analyzer. This expression, known as **Malus's law,**[2] applies to any two polarizing materials whose transmission axes are at an angle θ to each other. From this expression, we see that the intensity of the transmitted beam is maximum when the transmission axes are parallel ($\theta = 0$ or 180°) and that it is zero (complete absorption by the analyzer) when the transmission axes are perpendicular to each other. This variation in transmitted intensity through a pair of polarizing sheets is illustrated in Figure 38.31.

Polarization by Reflection

When an unpolarized light beam is reflected from a surface, the reflected light may be completely polarized, partially polarized, or unpolarized, depending on the angle of incidence. If the angle of incidence is 0°, the reflected beam is unpolarized. For other angles of incidence, the reflected light is polarized to some extent, and for one particular angle of incidence, the reflected light is completely polarized. Let us now investigate reflection at that special angle.

Suppose that an unpolarized light beam is incident on a surface, as in Figure 38.32a. Each individual electric field vector can be resolved into two components: one parallel to the surface (and perpendicular to the page in Fig. 38.32, represented by the dots), and the other (represented by the brown arrows) perpendicular both to the first component and to the direction of propagation. Thus, the polarization of the entire beam can be described by two electric field components in these directions. It is found that the parallel component reflects more strongly than the perpendicular component, and this results in a partially polarized reflected beam. Furthermore, the refracted beam is also partially polarized.

Now suppose that the angle of incidence θ_1 is varied until the angle between the reflected and refracted beams is 90°, as in Figure 38.32b. At this particular angle of incidence, the reflected beam is completely polarized (with its electric field vector parallel to the surface), and the refracted beam is still only partially polarized. The angle of incidence at which this polarization occurs is called the **polarizing angle** θ_p.

---

2 Named after its discoverer, E. L. Malus (1775–1812). Malus discovered that reflected light was polarized by viewing it through a calcite ($CaCO_3$) crystal.

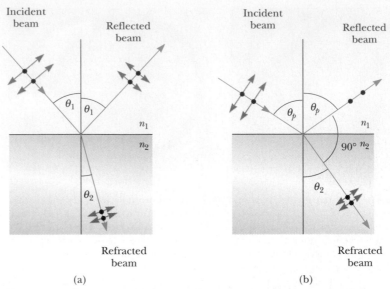

Figure 38.32 (a) When unpolarized light is incident on a reflecting surface, the reflected and refracted beams are partially polarized. (b) The reflected beam is completely polarized when the angle of incidence equals the polarizing angle θ_p, which satisfies the equation $n = \tan\theta_p$. At this incident angle, the reflected and refracted rays are perpendicular to each other.

We can obtain an expression relating the polarizing angle to the index of refraction of the reflecting substance by using Figure 38.32b. From this figure, we see that $\theta_p + 90° + \theta_2 = 180°$; thus $\theta_2 = 90° - \theta_p$. Using Snell's law of refraction (Eq. 35.8) and taking $n_1 = 1.00$ for air and $n_2 = n$, we have

$$n = \frac{\sin\theta_1}{\sin\theta_2} = \frac{\sin\theta_p}{\sin\theta_2}$$

Because $\sin\theta_2 = \sin(90° - \theta_p) = \cos\theta_p$, we can write this expression for n as $n = \sin\theta_p / \cos\theta_p$, which means that

Brewster's law

$$n = \tan\theta_p \tag{38.15}$$

This expression is called **Brewster's law**, and the polarizing angle θ_p is sometimes called **Brewster's angle**, after its discoverer, David Brewster (1781–1868). Because n varies with wavelength for a given substance, Brewster's angle is also a function of wavelength.

We can understand polarization by reflection by imagining that the electric field in the incident light sets electrons at the surface of the material in Figure 38.32b into oscillation. The component directions of oscillation are (1) parallel to the arrows shown on the refracted beam of light and (2) perpendicular to the page. The oscillating electrons act as antennas radiating light with a polarization parallel to the direction of oscillation. For the oscillations in direction (1), there is no radiation in the perpendicular direction, which is along the reflected ray (see the $\theta = 90°$ direction in Figure 34.11). For oscillations in direction (2), the electrons radiate light with a polarization perpendicular to the page (the $\theta = 0$ direction in Figure 34.11). Thus, the light reflected from the surface at this angle is completely polarized parallel to the surface.

Polarization by reflection is a common phenomenon. Sunlight reflected from water, glass, and snow is partially polarized. If the surface is horizontal, the electric field vector of the reflected light has a strong horizontal component. Sunglasses made of polarizing material reduce the glare of reflected light. The transmission axes of the lenses are oriented vertically so that they absorb the strong horizontal component of the reflected light. If you rotate sunglasses through 90 degrees, they are not as effective at blocking the glare from shiny horizontal surfaces.

Polarization by Double Refraction

Solids can be classified on the basis of internal structure. Those in which the atoms are arranged in a specific order are called *crystalline;* the NaCl structure of Figure 38.26 is just one example of a crystalline solid. Those solids in which the atoms are distributed randomly are called *amorphous.* When light travels through an amorphous material, such as glass, it travels with a speed that is the same in all directions. That is, glass has a single index of refraction. In certain crystalline materials, however, such as calcite and quartz, the speed of light is not the same in all directions. Such materials are characterized by two indices of refraction. Hence, they are often referred to as **double-refracting** or **birefringent** materials.

Upon entering a calcite crystal, unpolarized light splits into two plane-polarized rays that travel with different velocities, corresponding to two angles of refraction, as shown in Figure 38.33. The two rays are polarized in two mutually perpendicular directions, as indicated by the dots and arrows. One ray, called the **ordinary (O) ray,** is characterized by an index of refraction n_O that is the same in all directions. This means that if one could place a point source of light inside the crystal, as in Figure 38.34, the ordinary waves would spread out from the source as spheres.

The second plane-polarized ray, called the **extraordinary (E) ray,** travels with different speeds in different directions and hence is characterized by an index of refraction n_E that varies with the direction of propagation. Consider again the point source within a birefringent material, as in Figure 38.34. The source sends out an extraordinary wave having wave fronts that are elliptical in cross section. Note from Figure 38.34 that there is one direction, called the **optic axis,** along which the ordinary and extraordinary rays have the same speed, corresponding to the direction for which $n_O = n_E$. The difference in speed for the two rays is a maximum in the direction perpendicular to the optic axis. For example, in calcite, $n_O = 1.658$ at a wavelength of 589.3 nm, and n_E varies from 1.658 along the optic axis to 1.486 perpendicular to the optic axis. Values for n_O and n_E for various double-refracting crystals are given in Table 38.1.

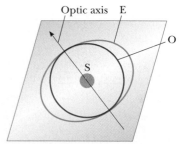

Figure 38.34 A point source S inside a double-refracting crystal produces a spherical wave front corresponding to the ordinary ray and an elliptical wave front corresponding to the extraordinary ray. The two waves propagate with the same velocity along the optic axis.

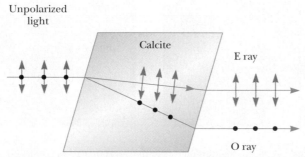

Figure 38.33 Unpolarized light incident on a calcite crystal splits into an ordinary (O) ray and an extraordinary (E) ray. These two rays are polarized in mutually perpendicular directions. (Drawing not to scale.)

Table 38.1

Indices of Refraction for Some Double-Refracting Crystals at a Wavelength of 589.3 nm			
Crystal	n_O	n_E	n_O/n_E
Calcite ($CaCO_3$)	1.658	1.486	1.116
Quartz (SiO_2)	1.544	1.553	0.994
Sodium nitrate ($NaNO_3$)	1.587	1.336	1.188
Sodium sulfite ($NaSO_3$)	1.565	1.515	1.033
Zinc chloride ($ZnCl_2$)	1.687	1.713	0.985
Zinc sulfide (ZnS)	2.356	2.378	0.991

Henry Leap and Jim Lehman

Figure 38.35 A calcite crystal produces a double image because it is a birefringent (double-refracting) material.

If we place a piece of calcite on a sheet of paper and then look through the crystal at any writing on the paper, we see two images, as shown in Figure 38.35. As can be seen from Figure 38.33, these two images correspond to one formed by the ordinary ray and one formed by the extraordinary ray. If the two images are viewed through a sheet of rotating polarizing glass, they alternately appear and disappear because the ordinary and extraordinary rays are plane-polarized along mutually perpendicular directions.

Some materials, such as glass and plastic, become birefringent when stressed. Suppose that an unstressed piece of plastic is placed between a polarizer and an analyzer so that light passes from polarizer to plastic to analyzer. When the plastic is unstressed and the analyzer axis is perpendicular to the polarizer axis, none of the polarized light passes through the analyzer. In other words, the unstressed plastic has no effect on the light passing through it. If the plastic is stressed, however, regions of greatest stress become birefringent and the polarization of the light passing through the plastic changes. Hence, a series of bright and dark bands is observed in the transmitted light, with the bright bands corresponding to regions of greatest stress.

Engineers often use this technique, called *optical stress analysis,* in designing structures ranging from bridges to small tools. They build a plastic model and analyze it under different load conditions to determine regions of potential weakness and failure under stress. Some examples of plastic models under stress are shown in Figure 38.36.

Polarization by Scattering

When light is incident on any material, the electrons in the material can absorb and reradiate part of the light. Such absorption and reradiation of light by electrons in the gas molecules that make up air is what causes sunlight reaching an observer on the Earth to be partially polarized. You can observe this effect—called **scattering**—by looking

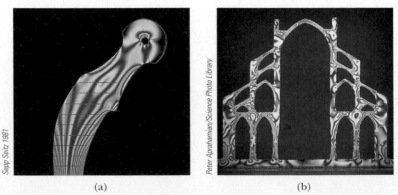

Sepp Seitz 1981

Peter Aprahamian/Science Photo Library

(a)   (b)

Figure 38.36 (a) Strain distribution in a plastic model of a hip replacement used in a medical research laboratory. The pattern is produced when the plastic model is viewed between a polarizer and analyzer oriented perpendicular to each other. (b) A plastic model of an arch structure under load conditions observed between perpendicular polarizers. Such patterns are useful in the optimal design of architectural components.

directly up at the sky through a pair of sunglasses whose lenses are made of polarizing material. Less light passes through at certain orientations of the lenses than at others.

Figure 38.37 illustrates how sunlight becomes polarized when it is scattered. The phenomenon is similar to that creating completely polarized light upon reflection from a surface at Brewster's angle. An unpolarized beam of sunlight traveling in the horizontal direction (parallel to the ground) strikes a molecule of one of the gases that make up air, setting the electrons of the molecule into vibration. These vibrating charges act like the vibrating charges in an antenna. The horizontal component of the electric field vector in the incident wave results in a horizontal component of the vibration of the charges, and the vertical component of the vector results in a vertical component of vibration. If the observer in Figure 38.37 is looking straight up (perpendicular to the original direction of propagation of the light), the vertical oscillations of the charges send no radiation toward the observer. Thus, the observer sees light that is completely polarized in the horizontal direction, as indicated by the brown arrows. If the observer looks in other directions, the light is partially polarized in the horizontal direction.

Some phenomena involving the scattering of light in the atmosphere can be understood as follows. When light of various wavelengths λ is incident on gas molecules of diameter d, where $d \ll \lambda$, the relative intensity of the scattered light varies as $1/\lambda^4$. The condition $d \ll \lambda$ is satisfied for scattering from oxygen (O_2) and nitrogen (N_2) molecules in the atmosphere, whose diameters are about 0.2 nm. Hence, short wavelengths (blue light) are scattered more efficiently than long wavelengths (red light). Therefore, when sunlight is scattered by gas molecules in the air, the short-wavelength radiation (blue) is scattered more intensely than the long-wavelength radiation (red).

When you look up into the sky in a direction that is not toward the Sun, you see the scattered light, which is predominantly blue; hence, you see a blue sky. If you look toward the west at sunset (or toward the east at sunrise), you are looking in a direction toward the Sun and are seeing light that has passed through a large distance of air. Most of the blue light has been scattered by the air between you and the Sun. The light that survives this trip through the air to you has had much of its blue component scattered and is thus heavily weighted toward the red end of the spectrum; as a result, you see the red and orange colors of sunset.

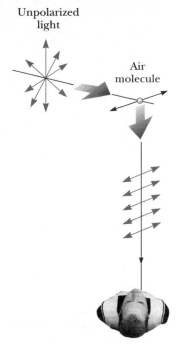

Figure 38.37 The scattering of unpolarized sunlight by air molecules. The scattered light traveling perpendicular to the incident light is plane-polarized because the vertical vibrations of the charges in the air molecule send no light in this direction.

On the right side of this photograph is a view from the side of the freeway (cars and a truck are visible at the left) of a rocket launch from Vandenburg Air Force Base, California. The trail left by the rocket shows the effects of scattering of light by air molecules. The lower portion of the trail appears red, due to the scattering of wavelengths at the violet end of the spectrum as the light from the Sun travels through a large portion of the atmosphere to light up the trail. The upper portion of the trail is illuminated by light that has traveled through much less atmosphere and appears white.

Optical Activity

Many important applications of polarized light involve materials that display **optical activity.** A material is said to be optically active if it rotates the plane of polarization of any light transmitted through the material. The angle through which the light is rotated by a specific material depends on the length of the path through the material and on concentration if the material is in solution. One optically active material is a solution of the common sugar dextrose. A standard method for determining the concentration of sugar solutions is to measure the rotation produced by a fixed length of the solution.

Molecular asymmetry determines whether a material is optically active. For example, some proteins are optically active because of their spiral shape.

The liquid crystal displays found in most calculators have their optical activity changed by the application of electric potential across different parts of the display. Try using a pair of polarizing sunglasses to investigate the polarization used in the display of your calculator.

Quick Quiz 38.9 A polarizer for microwaves can be made as a grid of parallel metal wires about a centimeter apart. Is the electric field vector for microwaves transmitted through this polarizer (a) parallel or (b) perpendicular to the metal wires?

Quick Quiz 38.10 You are walking down a long hallway that has many light fixtures in the ceiling and a very shiny, newly waxed floor. In the floor, you see reflections of every light fixture. Now you put on sunglasses that are polarized. Some of the reflections of the light fixtures can no longer be seen (Try this!) The reflections that disappear are those (a) nearest to you (b) farthest from you (c) at an intermediate distance from you.

SUMMARY

Take a practice test for this chapter by clicking on the Practice Test link at http://www.pse6.com.

Diffraction is the deviation of light from a straight-line path when the light passes through an aperture or around an obstacle. Diffraction is due to the wave nature of light.

The **Fraunhofer diffraction pattern** produced by a single slit of width a on a distant screen consists of a central bright fringe and alternating bright and dark fringes of much lower intensities. The angles θ_{dark} at which the diffraction pattern has zero intensity, corresponding to destructive interference, are given by

$$\sin \theta_{\text{dark}} = m \frac{\lambda}{a} \qquad m = \pm 1, \pm 2, \pm 3, \ldots \tag{38.1}$$

The intensity I of a single-slit diffraction pattern as a function of angle θ is given by the expression

$$I = I_{\text{max}} \left[\frac{\sin(\beta/2)}{\beta/2} \right]^2 \tag{38.4}$$

where $\beta = (2\pi a \sin \theta)/\lambda$ and I_{max} is the intensity at $\theta = 0$.

Rayleigh's criterion, which is a limiting condition of resolution, states that two images formed by an aperture are just distinguishable if the central maximum of the diffraction pattern for one image falls on the first minimum of the diffraction pattern for the other image. The limiting angle of resolution for a slit of width a is $\theta_{\text{min}} = \lambda/a$, and the limiting angle of resolution for a circular aperture of diameter D is $\theta_{\text{min}} = 1.22\lambda/D$.

A **diffraction grating** consists of a large number of equally spaced, identical slits. The condition for intensity maxima in the interference pattern of a diffraction grating for normal incidence is

$$d \sin\theta_{\text{bright}} = m\lambda \qquad m = 0, \pm 1, \pm 2, \pm 3, \ldots \qquad (38.10)$$

where d is the spacing between adjacent slits and m is the order number of the diffraction pattern. The resolving power of a diffraction grating in the mth order of the diffraction pattern is

$$R = Nm \qquad (38.12)$$

where N is the number of lines in the grating that are illuminated.

When polarized light of intensity I_{max} is emitted by a polarizer and then incident on an analyzer, the light transmitted through the analyzer has an intensity equal to $I_{\text{max}} \cos^2\theta$, where θ is the angle between the polarizer and analyzer transmission axes.

In general, reflected light is partially polarized. However, reflected light is completely polarized when the angle of incidence is such that the angle between the reflected and refracted beams is 90°. This angle of incidence, called the **polarizing angle** θ_p, satisfies **Brewster's law:**

$$n = \tan\theta_p \qquad (38.15)$$

where n is the index of refraction of the reflecting medium.

QUESTIONS

1. Why can you hear around corners, but not see around corners?

2. Holding your hand at arm's length, you can readily block sunlight from reaching your eyes. Why can you not block sound from reaching your ears this way?

3. Observe the shadow of your book when it is held a few inches above a table with a small lamp several feet above the book. Why is the shadow somewhat fuzzy at the edges?

4. Knowing that radio waves travel at the speed of light and that a typical AM radio frequency is 1 000 kHz while an FM radio frequency might be 100 MHz, estimate the wavelengths of typical AM and FM radio signals. Use this information to explain why AM radio stations can fade out when you drive your car through a short tunnel or underpass, when FM radio stations do not.

5. Describe the change in width of the central maximum of the single-slit diffraction pattern as the width of the slit is made narrower.

6. John William Strutt, Lord Rayleigh (1842–1919), is known as the last person to understand all of physics and all of mathematics. He invented an improved foghorn. To warn ships of a coastline, a foghorn should radiate sound in a wide horizontal sheet over the ocean's surface. It should not waste energy by broadcasting sound upward. It should not emit sound downward, because the water in front of the foghorn would reflect that sound upward. Rayleigh's foghorn trumpet is shown in Figure Q38.6. Is it installed in the correct orientation? Decide whether the long dimension of the rectangular opening should be horizontal or vertical, and argue for your decision.

Figure Q38.6

7. Featured in the motion picture $M*A*S*H$ (20th Century Fox, Aspen Productions, 1970) is a loudspeaker mounted on an exterior wall of an Army barracks. It has an approximately rectangular aperture. Its design can be thought of as based on Lord Rayleigh's foghorn trumpet, described in Question 6. Borrow or rent a copy of the movie, sketch the orientation of the loudspeaker, decide whether it is installed in the correct orientation, and argue for your decision.

8. Assuming that the headlights of a car are point sources, estimate the maximum distance from an observer to the car at which the headlights are distinguishable from each other.

9. A laser beam is incident at a shallow angle on a machinist's ruler that has a finely calibrated scale. The engraved rulings on the scale give rise to a diffraction pattern on a screen. Discuss how you can use this technique to obtain a measure of the wavelength of the laser light.

10. When you receive a chest x-ray at a hospital, the rays pass through a series of parallel ribs in your chest. Do the ribs act as a diffraction grating for x-rays?

11. Certain sunglasses use a polarizing material to reduce the intensity of light reflected from shiny surfaces. What orientation of polarization should the material have to be most effective?

12. During the "day" on the Moon (when the Sun is visible), you see a black sky and the stars can be clearly seen. During the day on the Earth, you see a blue sky with no stars. Account for this difference.

13. You can make the path of a light beam visible by placing dust in the air (perhaps by clapping two blackboard erasers in the path of the light beam). Explain why you can see the beam under these circumstances. In general, when is light visible?

14. Is light from the sky polarized? Why is it that clouds seen through Polaroid glasses stand out in bold contrast to the sky?

15. If a coin is glued to a glass sheet and this arrangement is held in front of a laser beam, the projected shadow has diffraction rings around its edge and a bright spot in the center. How is this possible?

16. How could the index of refraction of a flat piece of dark obsidian glass be determined?

17. A laser produces a beam a few millimeters wide, with uniform intensity across its width. A hair is stretched vertically across the front of the laser to cross the beam. How is the diffraction pattern it produces on a distant screen related to that of a vertical slit equal in width to the hair? How could you determine the width of the hair from measurements of its diffraction pattern?

18. A radio station serves listeners in a city to the northeast of its broadcast site. It broadcasts from three adjacent towers on a mountain ridge, along a line running east and west. Show that by introducing time delays among the signals the individual towers radiate, the station can maximize net intensity in the direction toward the city (and in the opposite direction) and minimize the signal transmitted in other directions. The towers together are said to form a *phased array*.

PROBLEMS

Section 38.2 Diffraction Patterns from Narrow Slits

1. Helium–neon laser light (λ = 632.8 nm) is sent through a 0.300-mm-wide single slit. What is the width of the central maximum on a screen 1.00 m from the slit?

2. A beam of green light is diffracted by a slit of width 0.550 mm. The diffraction pattern forms on a wall 2.06 m beyond the slit. The distance between the positions of zero intensity on both sides of the central bright fringe is 4.10 mm. Calculate the wavelength of the laser light.

3. 🌐 A screen is placed 50.0 cm from a single slit, which is illuminated with 690-nm light. If the distance between the first and third minima in the diffraction pattern is 3.00 mm, what is the width of the slit?

4. Coherent microwaves of wavelength 5.00 cm enter a long, narrow window in a building otherwise essentially opaque to the microwaves. If the window is 36.0 cm wide, what is the distance from the central maximum to the first-order minimum along a wall 6.50 m from the window?

5. Sound with a frequency 650 Hz from a distant source passes through a doorway 1.10 m wide in a sound-absorbing wall. Find the number and approximate directions of the diffraction-maximum beams radiated into the space beyond.

6. Light of wavelength 587.5 nm illuminates a single slit 0.750 mm in width. (a) At what distance from the slit should a screen be located if the first minimum in the diffraction pattern is to be 0.850 mm from the center of the principal maximum? (b) What is the width of the central maximum?

7. A beam of laser light of wavelength 632.8 nm has a circular cross section 2.00 mm in diameter. A rectangular aperture is to be placed in the center of the beam so that, when the light falls perpendicularly on a wall 4.50 m away, the central maximum fills a rectangle 110 mm wide and 6.00 mm high. The dimensions are measured between the minima bracketing the central maximum. Find the required width and height of the aperture.

8. **What If?** Assume the light in Figure 38.5 strikes the single slit at an angle β from the perpendicular direction. Show that Equation 38.1, the condition for destructive interference, must be modified to read

$$\sin \theta_{\text{dark}} = m\left(\frac{\lambda}{a}\right) - \sin \beta$$

9. A diffraction pattern is formed on a screen 120 cm away from a 0.400-mm-wide slit. Monochromatic 546.1-nm light is used. Calculate the fractional intensity I/I_{max} at a point on the screen 4.10 mm from the center of the principal maximum.

10. Coherent light of wavelength 501.5 nm is sent through two parallel slits in a large flat wall. Each slit is 0.700 μm wide. Their centers are 2.80 μm apart. The light then falls on a semicylindrical screen, with its axis at the midline between the slits. (a) Predict the direction of each interference maximum on the screen, as an angle away from the bisector of the line joining the slits. (b) Describe the pattern of light on the screen, specifying the number of bright fringes and the location of each. (c) Find the intensity of light on the screen at the center of each bright fringe, expressed as a fraction of the light intensity I_{max} at the center of the pattern.

Section 38.3   Resolution of Single-Slit and Circular Apertures

11. The pupil of a cat's eye narrows to a vertical slit of width 0.500 mm in daylight. What is the angular resolution for horizontally separated mice? Assume that the average wavelength of the light is 500 nm.

12. Find the radius a star image forms on the retina of the eye if the aperture diameter (the pupil) at night is 0.700 cm and the length of the eye is 3.00 cm. Assume the representative wavelength of starlight in the eye is 500 nm.

13. A helium–neon laser emits light that has a wavelength of 632.8 nm. The circular aperture through which the beam emerges has a diameter of 0.500 cm. Estimate the diameter of the beam 10.0 km from the laser.

14. You are vacationing in a Wonderland populated by friendly elves and a cannibalistic Cyclops that devours physics students. The elves and the Cyclops look precisely alike (everyone wears loose jeans and sweatshirts) except that each elf has two eyes, about 10.0 cm apart, and the Cyclops—you guessed it—has only one eye of about the same size as an elf's. The elves and the Cyclops are constantly at war with each other, so they rarely sleep and all have red eyes, predominantly reflecting light with a wavelength of 660 nm. From what maximum distance can you distinguish between a friendly elf and the predatory Cyclops? The air in Wonderland is always clear. Dilated with fear, your pupils have a diameter of 7.00 mm.

15. Narrow, parallel, glowing gas-filled tubes in a variety of colors form block letters to spell out the name of a night club. Adjacent tubes are all 2.80 cm apart. The tubes forming one letter are filled with neon and radiate predominantly red light with a wavelength of 640 nm. For another letter, the tubes emit predominantly violet light at 440 nm. The pupil of a dark-adapted viewer's eye is 5.20 mm in diameter. If she is in a certain range of distances away, the viewer can resolve the separate tubes of one color but not the other. Which color is easier to resolve? The viewer's distance must be in what range for her to resolve the tubes of only one color?

16. On the night of April 18, 1775, a signal was sent from the steeple of Old North Church in Boston to Paul Revere, who was 1.80 mi away: "One if by land, two if by sea." At what minimum separation did the sexton have to set the lanterns for Revere to receive the correct message about the approaching British? Assume that the patriot's pupils had a diameter of 4.00 mm at night and that the lantern light had a predominant wavelength of 580 nm.

17. The Impressionist painter Georges Seurat created paintings with an enormous number of dots of pure pigment, each of which was approximately 2.00 mm in diameter. The idea was to have colors such as red and green next to each other to form a scintillating canvas (Fig. P38.17). Outside what distance would one be unable to discern individual dots on the canvas? (Assume that $\lambda = 500$ nm and that the pupil diameter is 4.00 mm.)

Figure P38.17 *Sunday Afternoon on the Island of La Grande Jatte*, by Georges Seurat.

18. A binary star system in the constellation Orion has an angular interstellar separation of 1.00×10^{-5} rad. If $\lambda = 500$ nm, what is the smallest diameter the telescope can have to just resolve the two stars?

19. A spy satellite can consist essentially of a large-diameter concave mirror forming an image on a digital-camera detector and sending the picture to a ground receiver by radio waves. In effect, it is an astronomical telescope in orbit, looking down instead of up. Can a spy satellite read a license plate? Can it read the date on a dime? Argue for your answers by making an order-of-magnitude calculation, specifying the data you estimate.

20. A circular radar antenna on a Coast Guard ship has a diameter of 2.10 m and radiates at a frequency of 15.0 GHz. Two small boats are located 9.00 km away from the ship. How close together could the boats be and still be detected as two objects?

21. Grote Reber was a pioneer in radio astronomy. He constructed a radio telescope with a 10.0-m-diameter receiving dish. What was the telescope's angular resolution for 2.00-m radio waves?

22. When Mars is nearest the Earth, the distance separating the two planets is 88.6×10^6 km. Mars is viewed through a telescope whose mirror has a diameter of 30.0 cm. (a) If the wavelength of the light is 590 nm, what is the angular resolution of the telescope? (b) What is the smallest distance that can be resolved between two points on Mars?

Section 38.4   The Diffraction Grating

> *Note:* In the following problems, assume that the light is incident normally on the gratings.

23. White light is spread out into its spectral components by a diffraction grating. If the grating has 2 000 grooves per centimeter, at what angle does red light of wavelength 640 nm appear in first order?

24. Light from an argon laser strikes a diffraction grating that has 5 310 grooves per centimeter. The central and first-order principal maxima are separated by 0.488 m on a wall 1.72 m from the grating. Determine the wavelength of the laser light.

25. The hydrogen spectrum has a red line at 656 nm and a blue line at 434 nm. What are the angular separations between these two spectral lines obtained with a diffraction grating that has 4 500 grooves/cm?

26. A helium–neon laser ($\lambda = 632.8$ nm) is used to calibrate a diffraction grating. If the first-order maximum occurs at 20.5°, what is the spacing between adjacent grooves in the grating?

27. Three discrete spectral lines occur at angles of 10.09°, 13.71°, and 14.77° in the first-order spectrum of a grating spectrometer. (a) If the grating has 3 660 slits/cm, what are the wavelengths of the light? (b) At what angles are these lines found in the second-order spectrum?

28. Show that, whenever white light is passed through a diffraction grating of any spacing size, the violet end of the continuous visible spectrum in third order always overlaps with red light at the other end of the second-order spectrum.

29. A diffraction grating of width 4.00 cm has been ruled with 3 000 grooves/cm. (a) What is the resolving power of this grating in the first three orders? (b) If two monochromatic waves incident on this grating have a mean wavelength of 400 nm, what is their wavelength separation if they are just resolved in the third order?

30. The laser in a CD player must precisely follow the spiral track, along which the distance between one loop of the spiral and the next is only about 1.25 μm. A feedback mechanism lets the player know if the laser drifts off the track, so that the player can steer it back again. Figure P38.30 shows how a diffraction grating is used to provide information to keep the beam on track. The laser light passes through a diffraction grating just before it reaches the disk. The strong central maximum of the diffraction pattern is used to read the information in the track of pits. The two first-order side maxima are used for steering. The grating is designed so that the first-order maxima fall on the flat surfaces on both sides of the information track. Both side beams are reflected into their own detectors. As long as both beams are reflecting from smooth nonpitted surfaces, they are detected with constant high intensity. If the main beam wanders off the track, however, one of the side beams will begin to strike pits on the information track and the reflected light will diminish. This change is used with an electronic circuit to guide the beam back to

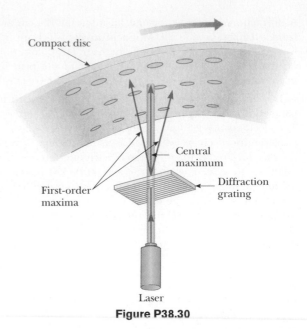

Figure P38.30

the desired location. Assume that the laser light has a wavelength of 780 nm and that the diffraction grating is positioned 6.90 μm from the disk. Assume that the first-order beams are to fall on the disk 0.400 μm on either side of the information track. What should be the number of grooves per millimeter in the grating?

31. A source emits 531.62-nm and 531.81-nm light. (a) What minimum number of grooves is required for a grating that resolves the two wavelengths in the first-order spectrum? (b) Determine the slit spacing for a grating 1.32 cm wide that has the required minimum number of grooves.

32. A diffraction grating has 4 200 rulings/cm. On a screen 2.00 m from the grating, it is found that for a particular order m, the maxima corresponding to two closely spaced wavelengths of sodium (589.0 nm and 589.6 nm) are separated by 1.59 mm. Determine the value of m.

33. A grating with 250 grooves/mm is used with an incandescent light source. Assume the visible spectrum to range in wavelength from 400 to 700 nm. In how many orders can one see (a) the entire visible spectrum and (b) the short-wavelength region?

34. A wide beam of laser light with a wavelength of 632.8 nm is directed through several narrow parallel slits, separated by 1.20 mm, and falls on a sheet of photographic film 1.40 m away. The exposure time is chosen so that the film stays unexposed everywhere except at the central region of each bright fringe. (a) Find the distance between these interference maxima. The film is printed as a transparency—it is opaque everywhere except at the exposed lines. Next, the same beam of laser light is directed through the transparency and allowed to fall on a screen 1.40 m beyond. (b) Argue that several narrow parallel bright regions, separated by 1.20 mm, will appear on the screen, as real images of the original slits. If at last the screen is removed, light will diverge from the images of the original slits with the same reconstructed wave fronts as the original slits produced. (*Suggestion:* You may find it useful to draw diagrams similar to Figure 38.16. A train of thought like this, at a soccer game, led Dennis Gabor to the invention of holography.)

Section 38.5   Diffraction of X-Rays by Crystals

35. Potassium iodide (KI) has the same crystalline structure as NaCl, with atomic planes separated by 0.353 nm. A monochromatic x-ray beam shows a first-order diffraction maximum when the grazing angle is 7.60°. Calculate the x-ray wavelength.

36. A wavelength of 0.129 nm characterizes K_α x-rays from zinc. When a beam of these x-rays is incident on the surface of a crystal whose structure is similar to that of NaCl, a first-order maximum is observed at 8.15°. Calculate the interplanar spacing based on this information.

37. If the interplanar spacing of NaCl is 0.281 nm, what is the predicted angle at which 0.140-nm x-rays are diffracted in a first-order maximum?

38. The first-order diffraction maximum is observed at 12.6° for a crystal in which the interplanar spacing is 0.240 nm. How many other orders can be observed?

39. In water of uniform depth, a wide pier is supported on pilings in several parallel rows 2.80 m apart. Ocean waves of uniform wavelength roll in, moving in a direction that makes an angle of 80.0° with the rows of posts. Find the three longest wavelengths of waves that will be strongly reflected by the pilings.

Section 38.6   Polarization of Light Waves

> Problem 34 in Chapter 34 can be assigned with this section.

40. Unpolarized light passes through two polaroid sheets. The axis of the first is vertical, and that of the second is at 30.0° to the vertical. What fraction of the incident light is transmitted?

41. Plane-polarized light is incident on a single polarizing disk with the direction of E_0 parallel to the direction of the transmission axis. Through what angle should the disk be rotated so that the intensity in the transmitted beam is reduced by a factor of (a) 3.00, (b) 5.00, (c) 10.0?

42. Three polarizing disks whose planes are parallel are centered on a common axis. The direction of the transmission axis in each case is shown in Figure P38.42 relative to the

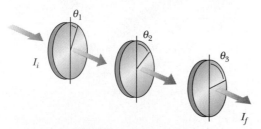

Figure P38.42 Problems 42 and 48.

common vertical direction. A plane-polarized beam of light with E_0 parallel to the vertical reference direction is incident from the left on the first disk with intensity $I_i = 10.0$ units (arbitrary). Calculate the transmitted intensity I_f when (a) $\theta_1 = 20.0°$, $\theta_2 = 40.0°$, and $\theta_3 = 60.0°$; (b) $\theta_1 = 0°$, $\theta_2 = 30.0°$, and $\theta_3 = 60.0°$.

43. The angle of incidence of a light beam onto a reflecting surface is continuously variable. The reflected ray is found to be completely polarized when the angle of incidence is 48.0°. What is the index of refraction of the reflecting material?

44. Review Problem. (a) A transparent plate with index of refraction n_2 is immersed in a medium with index n_1. Light traveling in the surrounding medium strikes the top surface of the plate at Brewster's angle. Show that if and only if the surfaces of the plate are parallel, the refracted light will strike the bottom surface of the plate at Brewster's angle for that interface. (b) **What If?** Instead of a plate, consider a prism of index of refraction n_2 separating media of different refractive indices n_1 and n_3. Is there one particular apex angle between the surfaces of the prism for which light can fall on both of its surfaces at Brewster's angle as it passes through the prism? If so, determine it.

45. The critical angle for total internal reflection for sapphire surrounded by air is 34.4°. Calculate the polarizing angle for sapphire.

46. For a particular transparent medium surrounded by air, show that the critical angle for total internal reflection and the polarizing angle are related by $\cot \theta_p = \sin \theta_c$.

47. How far above the horizon is the Moon when its image reflected in calm water is completely polarized? ($n_{water} = 1.33$)

Additional Problems

48. In Figure P38.42, suppose that the transmission axes of the left and right polarizing disks are perpendicular to each other. Also, let the center disk be rotated on the common axis with an angular speed ω. Show that if unpolarized light is incident on the left disk with an intensity I_{max}, the intensity of the beam emerging from the right disk is

$$I = \tfrac{1}{16} I_{max} (1 - \cos 4\omega t)$$

This means that the intensity of the emerging beam is modulated at a rate that is four times the rate of rotation of the center disk. [*Suggestion:* Use the trigonometric identities $\cos^2 \theta = (1 + \cos 2\theta)/2$ and $\sin^2 \theta = (1 - \cos 2\theta)/2$, and recall that $\theta = \omega t$.]

49. You want to rotate the plane of polarization of a polarized light beam by 45.0° with a maximum intensity reduction of 10.0%. (a) How many sheets of perfect polarizers do you need to achieve your goal? (b) What is the angle between adjacent polarizers?

50. Figure P38.50 shows a megaphone in use. Construct a theoretical description of how a megaphone works. You may assume that the sound of your voice radiates just through

Figure P38.50

the opening of your mouth. Most of the information in speech is carried not in a signal at the fundamental frequency, but in noises and in harmonics, with frequencies of a few thousand hertz. Does your theory allow any prediction that is simple to test?

51. Light from a helium–neon laser ($\lambda = 632.8$ nm) is incident on a single slit. What is the maximum width of the slit for which no diffraction minima are observed?

52. What are the approximate dimensions of the smallest object on Earth that astronauts can resolve by eye when they are orbiting 250 km above the Earth? Assume that $\lambda = 500$ nm and that a pupil diameter is 5.00 mm.

53. **Review problem.** A beam of 541-nm light is incident on a diffraction grating that has 400 grooves/mm. (a) Determine the angle of the second-order ray. (b) **What If?** If the entire apparatus is immersed in water, what is the new second-order angle of diffraction? (c) Show that the two diffracted rays of parts (a) and (b) are related through the law of refraction.

54. The *Very Large Array* (VLA) is a set of 27 radio telescope dishes in Caton and Socorro Counties, New Mexico (Fig. P38.54). The antennas can be moved apart on railroad tracks, and their combined signals give the resolving power of a synthetic aperture 36.0 km in diameter. (a) If the detectors are tuned to a frequency of 1.40 GHz, what is the angular resolution of the VLA? (b) Clouds of hydrogen radiate at this frequency. What must be the separation distance of two clouds 26 000 lightyears away at the center of the galaxy, if they are to be resolved? (c) **What If?** As the telescope looks up, a circling hawk looks down. Find the

angular resolution of the hawk's eye. Assume that the hawk is most sensitive to green light having a wavelength of 500 nm and that it has a pupil of diameter 12.0 mm. (d) A mouse is on the ground 30.0 m below. By what distance must the mouse's whiskers be separated if the hawk can resolve them?

55. A 750-nm light beam hits the flat surface of a certain liquid, and the beam is split into a reflected ray and a refracted ray. If the reflected ray is completely polarized at 36.0°, what is the wavelength of the refracted ray?

56. Iridescent peacock feathers are shown in Figure P38.56a. The surface of one microscopic barbule is composed of transparent keratin that supports rods of dark brown melanin in a regular lattice, represented in Figure P38.56b. (Your fingernails are made of keratin, and melanin is the dark pigment giving color to human skin.) In a portion of the feather that can appear turquoise, assume that the melanin rods are uniformly separated by 0.25 μm, with air between them. (a) Explain how this structure can appear blue-green when it contains no blue or green pigment.

(a)

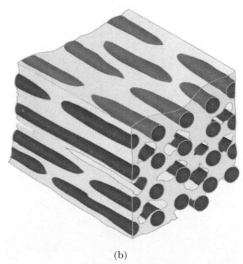

(b)

Figure P38.56 (a) Iridescent peacock feathers. (b) Microscopic section of a feather showing dark melanin rods in a pale keratin matrix.

Figure P38.54 Some of the radio telescope dishes in the Very Large Array.

(b) Explain how it can also appear violet if light falls on it in a different direction. (c) Explain how it can present different colors to your two eyes at the same time—a characteristic of iridescence. (d) A compact disc can appear to be any color of the rainbow. Explain why this portion of the feather cannot appear yellow or red. (e) What could be different about the array of melanin rods in a portion of the feather that does appear to be red?

57. Light of wavelength 500 nm is incident normally on a diffraction grating. If the third-order maximum of the diffraction pattern is observed at 32.0°, (a) what is the number of rulings per centimeter for the grating? (b) Determine the total number of primary maxima that can be observed in this situation.

58. Light strikes a water surface at the polarizing angle. The part of the beam refracted into the water strikes a submerged glass slab (index of refraction, 1.50), as shown in Figure P38.58. The light reflected from the upper surface of the slab is completely polarized. Find the angle between the water surface and the glass slab.

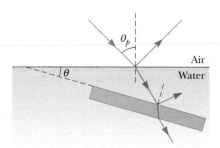

Figure P38.58

59. A beam of bright red light of wavelength 654 nm passes through a diffraction grating. Enclosing the space beyond the grating is a large screen forming one half of a cylinder centered on the grating, with its axis parallel to the slits in the grating. Fifteen bright spots appear on the screen. Find the maximum and minimum possible values for the slit separation in the diffraction grating.

60. A *pinhole camera* has a small circular aperture of diameter D. Light from distant objects passes through the aperture into an otherwise dark box, falling on a screen located a distance L away. If D is too large, the display on the screen will be fuzzy, because a bright point in the field of view will send light onto a circle of diameter slightly larger than D. On the other hand, if D is too small, diffraction will blur the display on the screen. The screen shows a reasonably sharp image if the diameter of the central disk of the diffraction pattern, specified by Equation 38.9, is equal to D at the screen. (a) Show that for monochromatic light with plane wave fronts and $L \gg D$, the condition for a sharp view is fulfilled if $D^2 = 2.44\lambda L$. (b) Find the optimum pinhole diameter for 500-nm light projected onto a screen 15.0 cm away.

61. An American standard television picture is composed of about 485 horizontal lines of varying light intensity. Assume that your ability to resolve the lines is limited only by the Rayleigh criterion and that the pupils of your eyes are 5.00 mm in diameter. Calculate the ratio of minimum viewing distance to the vertical dimension of the picture such that you will not be able to resolve the lines. Assume

that the average wavelength of the light coming from the screen is 550 nm.

62. (a) Light traveling in a medium of index of refraction n_1 is incident at an angle θ on the surface of a medium of index n_2. The angle between reflected and refracted rays is β. Show that

$$\tan \theta = \frac{n_2 \sin \beta}{n_1 - n_2 \cos \beta}$$

(*Suggestion:* Use the identity $\sin(A + B) = \sin A \cos B + \cos A \sin B$.) (b) **What If?** Show that this expression for $\tan \theta$ reduces to Brewster's law when $\beta = 90°$, $n_1 = 1$, and $n_2 = n$.

63. Suppose that the single slit in Figure 38.6 is 6.00 cm wide and in front of a microwave source operating at 7.50 GHz. (a) Calculate the angle subtended by the first minimum in the diffraction pattern. (b) What is the relative intensity I/I_{max} at $\theta = 15.0°$? (c) Assume that two such sources, separated laterally by 20.0 cm, are behind the slit. What must the maximum distance between the plane of the sources and the slit be if the diffraction patterns are to be resolved? (In this case, the approximation $\sin \theta \approx \tan \theta$ is not valid because of the relatively small value of a/λ.)

64. Two polarizing sheets are placed together with their transmission axes crossed so that no light is transmitted. A third sheet is inserted between them with its transmission axis at an angle of 45.0° with respect to each of the other axes. Find the fraction of incident unpolarized light intensity transmitted by the three-sheet combination. (Assume each polarizing sheet is ideal.)

65. Two wavelengths λ and $\lambda + \Delta\lambda$ (with $\Delta\lambda \ll \lambda$) are incident on a diffraction grating. Show that the angular separation between the spectral lines in the mth-order spectrum is

$$\Delta\theta = \frac{\Delta\lambda}{\sqrt{(d/m)^2 - \lambda^2}}$$

where d is the slit spacing and m is the order number.

66. Two closely spaced wavelengths of light are incident on a diffraction grating. (a) Starting with Equation 38.10, show that the angular dispersion of the grating is given by

$$\frac{d\theta}{d\lambda} = \frac{m}{d \cos \theta}$$

(b) A square grating 2.00 cm on each side containing 8 000 equally spaced slits is used to analyze the spectrum of mercury. Two closely spaced lines emitted by this element have wavelengths of 579.065 nm and 576.959 nm. What is the angular separation of these two wavelengths in the second-order spectrum?

67. The scale of a map is a number of kilometers per centimeter, specifying the distance on the ground that any distance on the map represents. The scale of a spectrum is its *dispersion*, a number of nanometers per centimeter, which specifies the change in wavelength that a distance across the spectrum represents. One must know the dispersion in order to compare one spectrum with another and to make a measurement of (for example) a Doppler shift. Let y represent the position relative to the center of a diffraction pattern projected onto a flat screen at distance L by a diffraction grating with slit spacing d. The dispersion is $d\lambda/dy$.

(a) Prove that the dispersion is given by

$$\frac{d\lambda}{dy} = \frac{L^2 d}{m(L^2 + y^2)^{3/2}}$$

(b) Calculate the dispersion in first order for light with a mean wavelength of 550 nm, analyzed with a grating having 8 000 rulings/cm, and projected onto a screen 2.40 m away.

68. Derive Equation 38.12 for the resolving power of a grating, $R = Nm$, where N is the number of slits illuminated and m is the order in the diffraction pattern. Remember that Rayleigh's criterion (Section 38.3) states that two wavelengths will be resolved when the principal maximum for one falls on the first minimum for the other.

69. Figure P38.69a is a three-dimensional sketch of a birefringent crystal. The dotted lines illustrate how a thin parallel-faced slab of material could be cut from the larger specimen with the optic axis of the crystal parallel to the faces of the plate. A section cut from the crystal in this manner is known as a *retardation plate*. When a beam of light is incident on the plate perpendicular to the direction of the optic axis, as shown in Figure P38.69b, the O ray and the E ray travel along a single straight line, but with different speeds. (a) Let the thickness of the plate be d and show that the phase difference between the O ray and the E ray is

$$\theta = \frac{2\pi d}{\lambda} |n_O - n_E|$$

where λ is the wavelength in air. (b) In a particular case the incident light has a wavelength of 550 nm. Find the minimum value of d for a quartz plate for which $\theta = \pi/2$. Such a plate is called a *quarter-wave plate*. (Use values of n_O and n_E from Table 38.1.)

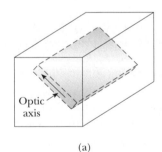

(a)

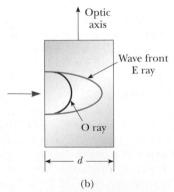

(b)

Figure P38.69

70. How much diffraction spreading does a light beam undergo? One quantitative answer is the *full width at half*

maximum of the central maximum of the single-slit Fraunhofer diffraction pattern. You can evaluate this angle of spreading in this problem and in the next. (a) In Equation 38.4, define $\beta/2 = \phi$ and show that, at the point where $I = 0.5I_{max}$, we must have $\sin \phi = \phi/\sqrt{2}$. (b) Let $y_1 = \sin \phi$ and $y_2 = \phi/\sqrt{2}$. Plot y_1 and y_2 on the same set of axes over a range from $\phi = 1$ rad to $\phi = \pi/2$ rad. Determine ϕ from the point of intersection of the two curves. (c) Then show that, if the fraction λ/a is not large, the angular full width at half maximum of the central diffraction maximum is $\Delta\theta = 0.886\,\lambda/a$.

71. Another method to solve the equation $\phi = \sqrt{2}\sin\phi$ in Problem 70 is to guess a first value of ϕ, use a computer or calculator to see how nearly it fits, and continue to update your estimate until the equation balances. How many steps (iterations) does this take?

72. In the diffraction pattern of a single slit, described by the equation

$$I_\theta = I_{max}\frac{\sin^2(\beta/2)}{(\beta/2)^2}$$

with $\beta = (2\pi a \sin\theta)/\lambda$, the central maximum is at $\beta = 0$ and the side maxima are *approximately* at $\beta/2 = (m + \frac{1}{2})\pi$ for $m = 1, 2, 3, \ldots$. Determine more precisely (a) the location of the first side maximum, where $m = 1$, and (b) the location of the second side maximum. Observe in Figure 38.10a that the graph of intensity versus $\beta/2$ has a horizontal tangent at maxima and also at minima. You will need to solve a transcendental equation.

73. Light of wavelength 632.8 nm illuminates a single slit, and a diffraction pattern is formed on a screen 1.00 m from the slit. Using the data in the table below, plot relative intensity versus position. Choose an appropriate value for the slit width a and on the same graph used for the experimental data, plot the theoretical expression for the relative intensity

$$\frac{I_\theta}{I_{max}} = \frac{\sin^2(\beta/2)}{(\beta/2)^2}$$

What value of a gives the best fit of theory and experiment?

Relative Intensity	Position Relative to Central Maximum (mm)
1.00	0
0.95	0.8
0.80	1.6
0.60	2.4
0.39	3.2
0.21	4.0
0.079	4.8
0.014	5.6
0.003	6.5
0.015	7.3
0.036	8.1
0.047	8.9
0.043	9.7
0.029	10.5
0.013	11.3
0.002	12.1
0.000 3	12.9
0.005	13.7

0.012	14.5
0.016	15.3
0.015	16.1
0.010	16.9
0.004 4	17.7
0.000 6	18.5
0.000 3	19.3
0.003	20.2

Answers to Quick Quizzes

38.1 (a). Equation 38.1 shows that a decrease in a results in an increase in the angles at which the dark fringes appear.

38.2 (c). The space between the slightly open door and the doorframe acts as a single slit. Sound waves have wavelengths that are larger than the opening and so are diffracted and spread throughout the room you are in. Because light wavelengths are much smaller than the slit width, they experience negligible diffraction. As a result, you must have a direct line of sight to detect the light waves.

38.3 The situation is like that depicted in Figure 38.11 except that now the slits are only half as far apart. The diffraction pattern is the same, but the interference pattern is stretched out because d is smaller. Because $d/a = 3$, the $m = 3$ interference maximum coincides with the first diffraction minimum. Your sketch should look like the figure below.

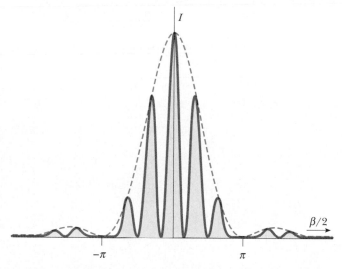

38.4 (c). In Equation 38.7, the ratio d/a is independent of wavelength, so the number of interference fringes in the central diffraction pattern peak remains the same. Equation 38.1 tells us that a decrease in wavelength causes a decrease in the width of the central peak.

38.5 (b). The effective slit width in the vertical direction of the cat's eye is larger than that in the horizontal direction. Thus, the eye has more resolving power for lights separated in the vertical direction and would be more effective at resolving the mast lights on the boat.

38.6 (a). We would like to reduce the minimum angular separation for two objects below the angle subtended by the two stars in the binary system. We can do that by reducing the wavelength of the light—this in essence makes the aperture larger, relative to the light wavelength, increasing the resolving power. Thus, we should choose a blue filter.

38.7 (b). The tracks of information on a compact disc are much closer together than on a phonograph record. As a result, the diffraction maxima from the compact disc will be farther apart than those from the record.

38.8 (c). With the doubled wavelength, the pattern will be wider. Choices (a) and (d) make the pattern even wider. From Equation 38.10, we see that choice (b) causes $\sin \theta$ to be twice as large. Because we cannot use the small angle approximation, however, a doubling of $\sin \theta$ is not the same as a doubling of θ, which would translate to a doubling of the position of a maximum along the screen. If we only consider small-angle maxima, choice (b) would work, but it does not work in the large-angle case.

38.9 (b). Electric field vectors parallel to the metal wires cause electrons in the metal to oscillate parallel to the wires. Thus, the energy from the waves with these electric field vectors is transferred to the metal by accelerating these electrons and is eventually transformed to internal energy through the resistance of the metal. Waves with electric-field vectors perpendicular to the metal wires pass through because they are not able to accelerate electrons in the wires.

38.10 (c). At some intermediate distance, the light rays from the fixtures will strike the floor at Brewster's angle and reflect to your eyes. Because this light is polarized horizontally, it will not pass through your polarized sunglasses. Tilting your head to the side will cause the reflections to reappear.

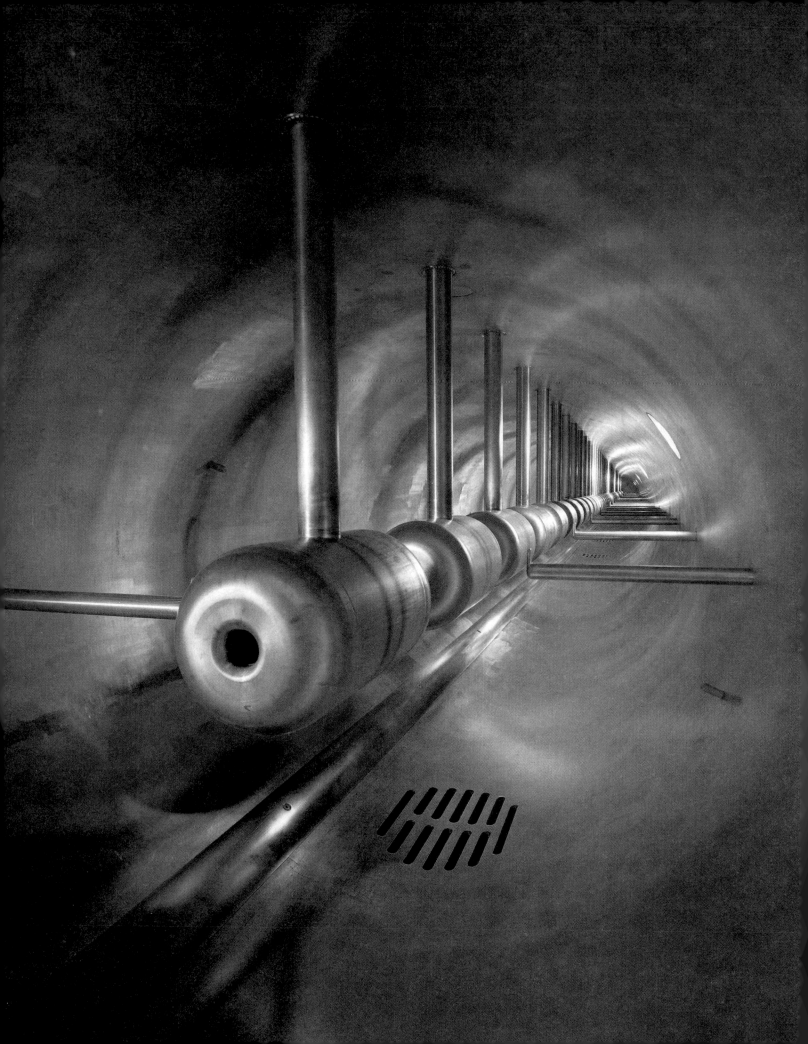

Modern Physics

A t the end of the nineteenth century, many scientists believed that they had learned most of what there was to know about physics. Newton's laws of motion and his theory of universal gravitation, Maxwell's theoretical work in unifying electricity and magnetism, the laws of thermodynamics and kinetic theory, and the principles of optics were highly successful in explaining a variety of phenomena.

As the nineteenth century turned to the twentieth, however, a major revolution shook the world of physics. In 1900 Planck provided the basic ideas that led to the formulation of the quantum theory, and in 1905 Einstein formulated his brilliant special theory of relativity. The excitement of the times is captured in Einstein's own words: "It was a marvelous time to be alive." Both ideas were to have a profound effect on our understanding of nature. Within a few decades, these two theories inspired new developments and theories in the fields of atomic physics, nuclear physics, and condensed-matter physics.

In Chapter 39 we introduce the special theory of relativity. The theory provides us with a new and deeper view of physical laws. Although the concepts underlying this theory often violate our common sense, the theory correctly predicts the results of experiments involving speeds near the speed of light. In the extended version of this textbook, *Physics for Scientists and Engineers with Modern Physics,* we cover the basic concepts of quantum mechanics and their application to atomic and molecular physics, and we introduce solid-state physics, nuclear physics, particle physics, and cosmology.

You should keep in mind that, although the physics that was developed during the twentieth century has led to a multitude of important technological achievements, the story is still incomplete. Discoveries will continue to evolve during our lifetimes, and many of these discoveries will deepen or refine our understanding of nature and the world around us. It is still a "marvelous time to be alive." ∎

◄ *A portion of the accelerator tunnel at Fermilab, near Chicago, Illinois. The tunnel is circular and 1.9 km in diameter. Using electric and magnetic fields, protons and antiprotons are accelerated to speeds close to that of light and then allowed to collide head-on, in order to investigate the production of new particles. (Fermilab Photo)*

Chapter 39

Relativity

CHAPTER OUTLINE

39.1 The Principle of Galilean Relativity

39.2 The Michelson–Morley Experiment

39.3 Einstein's Principle of Relativity

39.4 Consequences of the Special Theory of Relativity

39.5 The Lorentz Transformation Equations

39.6 The Lorentz Velocity Transformation Equations

39.7 Relativistic Linear Momentum and the Relativistic Form of Newton's Laws

39.8 Relativistic Energy

39.9 Mass and Energy

39.10 The General Theory of Relativity

▲ Standing on the shoulders of a giant. *David Serway, son of one of the authors, watches over his children, Nathan and Kaitlyn, as they frolic in the arms of Albert Einstein at the Einstein memorial in Washington, D.C. It is well known that Einstein, the principal architect of relativity, was very fond of children. (Emily Serway)*

Our everyday experiences and observations have to do with objects that move at speeds much less than the speed of light. Newtonian mechanics was formulated by observing and describing the motion of such objects, and this formalism is very successful in describing a wide range of phenomena that occur at low speeds. However, it fails to describe properly the motion of objects whose speeds approach that of light.

Experimentally, the predictions of Newtonian theory can be tested at high speeds by accelerating electrons or other charged particles through a large electric potential difference. For example, it is possible to accelerate an electron to a speed of $0.99c$ (where c is the speed of light) by using a potential difference of several million volts. According to Newtonian mechanics, if the potential difference is increased by a factor of 4, the electron's kinetic energy is four times greater and its speed should double to $1.98c$. However, experiments show that the speed of the electron—as well as the speed of any other object in the Universe—always remains less than the speed of light, regardless of the size of the accelerating voltage. Because it places no upper limit on speed, Newtonian mechanics is contrary to modern experimental results and is clearly a limited theory.

In 1905, at the age of only 26, Einstein published his special theory of relativity. Regarding the theory, Einstein wrote:

> The relativity theory arose from necessity, from serious and deep contradictions in the old theory from which there seemed no escape. The strength of the new theory lies in the consistency and simplicity with which it solves all these difficulties . . . .[1]

Although Einstein made many other important contributions to science, the special theory of relativity alone represents one of the greatest intellectual achievements of all time. With this theory, experimental observations can be correctly predicted over the range of speeds from $v = 0$ to speeds approaching the speed of light. At low speeds, Einstein's theory reduces to Newtonian mechanics as a limiting situation. It is important to recognize that Einstein was working on electromagnetism when he developed the special theory of relativity. He was convinced that Maxwell's equations were correct, and in order to reconcile them with one of his postulates, he was forced into the revolutionary notion of assuming that space and time are not absolute.

This chapter gives an introduction to the special theory of relativity, with emphasis on some of its consequences. The special theory covers phenomena such as the slowing down of moving clocks and the contraction of moving lengths. We also discuss the relativistic forms of momentum and energy.

In addition to its well-known and essential role in theoretical physics, the special theory of relativity has practical applications, including the design of nuclear power plants and modern global positioning system (GPS) units. These devices do not work if designed in accordance with nonrelativistic principles.

---

[1] A. Einstein and L. Infeld, *The Evolution of Physics*, New York, Simon and Schuster, 1961.

39.1   The Principle of Galilean Relativity

To describe a physical event, we must establish a frame of reference. You should recall from Chapter 5 that an inertial frame of reference is one in which an object is observed to have no acceleration when no forces act on it. Furthermore, any system moving with constant velocity with respect to an inertial frame must also be in an inertial frame.

There is no absolute inertial reference frame. This means that the results of an experiment performed in a vehicle moving with uniform velocity will be identical to the results of the same experiment performed in a stationary vehicle. The formal statement of this result is called the **principle of Galilean relativity:**

Principle of Galilean relativity   The laws of mechanics must be the same in all inertial frames of reference.

Let us consider an observation that illustrates the equivalence of the laws of mechanics in different inertial frames. A pickup truck moves with a constant velocity, as shown in Figure 39.1a. If a passenger in the truck throws a ball straight up, and if air effects are neglected, the passenger observes that the ball moves in a vertical path. The motion of the ball appears to be precisely the same as if the ball were thrown by a person at rest on the Earth. The law of universal gravitation and the equations of motion under constant acceleration are obeyed whether the truck is at rest or in uniform motion.

Both observers agree on the laws of physics—they each throw a ball straight up and it rises and falls back into their hand. What about the *path* of the ball thrown by the observer in the truck? Do the observers agree on the path? The observer on the ground sees the path of the ball as a parabola, as illustrated in Figure 39.1b, while, as mentioned earlier, the observer in the truck sees the ball move in a vertical path. Furthermore, according to the observer on the ground, the ball has a horizontal component of velocity equal to the velocity of the truck. Although the two observers disagree on certain aspects of the situation, they agree on the validity of Newton's laws and on such classical principles as conservation of energy and conservation of linear momentum. This agreement implies that no mechanical experiment can detect any difference between the two inertial frames. The only thing that can be detected is the relative motion of one frame with respect to the other.

> **Quick Quiz 39.1** Which observer in Figure 39.1 sees the ball's *correct* path? (a) the observer in the truck (b) the observer on the ground (c) both observers.

(a)   (b)

Figure 39.1 (a) The observer in the truck sees the ball move in a vertical path when thrown upward. (b) The Earth observer sees the path of the ball as a parabola.

Suppose that some physical phenomenon, which we call an *event*, occurs and is observed by an observer at rest in an inertial reference frame. The event's location and time of occurrence can be specified by the four coordinates (x, y, z, t). We would like to be able to transform these coordinates from those of an observer in one inertial frame to those of another observer in a frame moving with uniform relative velocity compared to the first frame. When we say an observer is "in a frame," we mean that the observer is at rest with respect to the origin of that frame.

Consider two inertial frames S and S′ (Fig. 39.2). The frame S′ moves with a constant velocity **v** along the common x and x' axes, where **v** is measured relative to S. We assume that the origins of S and S′ coincide at $t = 0$ and that an event occurs at point P in space at some instant of time. An observer in S describes the event with space–time coordinates (x, y, z, t), whereas an observer in S′ uses the coordinates (x', y', z', t') to describe the same event. As we see from the geometry in Figure 39.2, the relationships among these various coordinates can be written

$$x' = x - vt \qquad y' = y \qquad z' = z \qquad t' = t \tag{39.1}$$

These equations are the **Galilean space–time transformation equations.** Note that time is assumed to be the same in both inertial frames. That is, within the framework of classical mechanics, all clocks run at the same rate, regardless of their velocity, so that the time at which an event occurs for an observer in S is the same as the time for the same event in S′. Consequently, the time interval between two successive events should be the same for both observers. Although this assumption may seem obvious, it turns out to be incorrect in situations where v is comparable to the speed of light.

Now suppose that a particle moves through a displacement of magnitude dx along the x axis in a time interval dt as measured by an observer in S. It follows from Equations 39.1 that the corresponding displacement dx' measured by an observer in S′ is $dx' = dx - v\, dt$, where frame S′ is moving with speed v in the x direction relative to frame S. Because $dt = dt'$, we find that

$$\frac{dx'}{dt'} = \frac{dx}{dt} - v$$

or

$$u'_x = u_x - v \tag{39.2}$$

where u_x and u'_x are the x components of the velocity of the particle measured by observers in S and S′, respectively. (We use the symbol **u** for particle velocity rather than **v**, which is used for the relative velocity of two reference frames.) This is the **Galilean velocity transformation equation.** It is consistent with our intuitive notion of time and space as well as with our discussions in Section 4.6. As we shall soon see, however, it leads to serious contradictions when applied to electromagnetic waves.

Quick Quiz 39.2 A baseball pitcher with a 90-mi/h fastball throws a ball while standing on a railroad flatcar moving at 110 mi/h. The ball is thrown in the same direction as that of the velocity of the train. Applying the Galilean velocity transformation equation, the speed of the ball relative to the Earth is (a) 90 mi/h (b) 110 mi/h (c) 20 mi/h (d) 200 mi/h (e) impossible to determine.

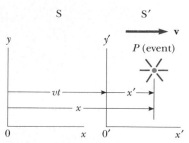

Figure 39.2 An event occurs at a point P. The event is seen by two observers in inertial frames S and S′, where S′ moves with a velocity **v** relative to S.

Galilean transformation equations

▲ **PITFALL PREVENTION**

39.1 The Relationship Between the S and S′ Frames

Many of the mathematical representations in this chapter are true *only* for the specified relationship between the S and S′ frames. The x and x' axes coincide, except that their origins are different. The y and y' axes (and the z and z' axes), are parallel, but do not coincide due to the displacement of the origin of S′ with respect to that of S. We choose the time $t = 0$ to be the instant at which the origins of the two coordinate systems coincide. If the S′ frame is moving in the positive x direction relative to S, v is positive; otherwise it is negative.

The Speed of Light

It is quite natural to ask whether the principle of Galilean relativity also applies to electricity, magnetism, and optics. Experiments indicate that the answer is no. Recall from Chapter 34 that Maxwell showed that the speed of light in free space is $c = 3.00 \times 10^8$ m/s. Physicists of the late 1800s thought that light waves moved through a medium called the *ether* and that the speed of light was c only in a special, absolute frame

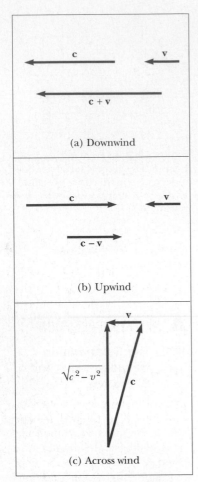

Figure 39.3 If the velocity of the ether wind relative to the Earth is **v** and the velocity of light relative to the ether is **c**, then the speed of light relative to the Earth is (a) $c + v$ in the downwind direction, (b) $c - v$ in the upwind direction, and (c) $(c^2 - v^2)^{1/2}$ in the direction perpendicular to the wind.

at rest with respect to the ether. The Galilean velocity transformation equation was expected to hold for observations of light made by an observer in any frame moving at speed v relative to the absolute ether frame. That is, if light travels along the x axis and an observer moves with velocity **v** along the x axis, the observer will measure the light to have speed $c \pm v$, depending on the directions of travel of the observer and the light.

Because the existence of a preferred, absolute ether frame would show that light was similar to other classical waves and that Newtonian ideas of an absolute frame were true, considerable importance was attached to establishing the existence of the ether frame. Prior to the late 1800s, experiments involving light traveling in media moving at the highest laboratory speeds attainable at that time were not capable of detecting differences as small as that between c and $c \pm v$. Starting in about 1880, scientists decided to use the Earth as the moving frame in an attempt to improve their chances of detecting these small changes in the speed of light.

As observers fixed on the Earth, we can take the view that we are stationary and that the absolute ether frame containing the medium for light propagation moves past us with speed v. Determining the speed of light under these circumstances is just like determining the speed of an aircraft traveling in a moving air current, or wind; consequently, we speak of an "ether wind" blowing through our apparatus fixed to the Earth.

A direct method for detecting an ether wind would use an apparatus fixed to the Earth to measure the ether wind's influence on the speed of light. If v is the speed of the ether relative to the Earth, then light should have its maximum speed $c + v$ when propagating downwind, as in Figure 39.3a. Likewise, the speed of light should have its minimum value $c - v$ when the light is propagating upwind, as in Figure 39.3b, and an intermediate value $(c^2 - v^2)^{1/2}$ in the direction perpendicular to the ether wind, as in Figure 39.3c. If the Sun is assumed to be at rest in the ether, then the velocity of the ether wind would be equal to the orbital velocity of the Earth around the Sun, which has a magnitude of approximately 3×10^4 m/s. Because $c = 3 \times 10^8$ m/s, it is necessary to detect a change in speed of about 1 part in 10^4 for measurements in the upwind or downwind directions. However, while such a change is experimentally measurable, all attempts to detect such changes and establish the existence of the ether wind (and hence the absolute frame) proved futile! We explore the classic experimental search for the ether in Section 39.2.

The principle of Galilean relativity refers only to the laws of mechanics. If it is assumed that the laws of electricity and magnetism are the same in all inertial frames, a paradox concerning the speed of light immediately arises. We can understand this by recognizing that Maxwell's equations seem to imply that the speed of light always has the fixed value 3.00×10^8 m/s in all inertial frames, a result in direct contradiction to what is expected based on the Galilean velocity transformation equation. According to Galilean relativity, the speed of light should *not* be the same in all inertial frames.

To resolve this contradiction in theories, we must conclude that either (1) the laws of electricity and magnetism are not the same in all inertial frames or (2) the Galilean velocity transformation equation is incorrect. If we assume the first alternative, then a preferred reference frame in which the speed of light has the value c must exist and the measured speed must be greater or less than this value in any other reference frame, in accordance with the Galilean velocity transformation equation. If we assume the second alternative, then we are forced to abandon the notions of absolute time and absolute length that form the basis of the Galilean space–time transformation equations.

39.2 The Michelson–Morley Experiment

The most famous experiment designed to detect small changes in the speed of light was first performed in 1881 by Albert A. Michelson (see Section 37.7) and later repeated under various conditions by Michelson and Edward W. Morley (1838–1923). We state at the outset that the outcome of the experiment contradicted the ether hypothesis.

The experiment was designed to determine the velocity of the Earth relative to that of the hypothetical ether. The experimental tool used was the Michelson interferometer, which was discussed in Section 37.7 and is shown again in Figure 39.4. Arm 2 is aligned along the direction of the Earth's motion through space. The Earth moving through the ether at speed v is equivalent to the ether flowing past the Earth in the opposite direction with speed v. This ether wind blowing in the direction opposite the direction of Earth's motion should cause the speed of light measured in the Earth frame to be $c - v$ as the light approaches mirror M_2 and $c + v$ after reflection, where c is the speed of light in the ether frame.

The two light beams reflect from M_1 and M_2 and recombine, and an interference pattern is formed, as discussed in Section 37.7. The interference pattern is observed while the interferometer is rotated through an angle of 90°. This rotation interchanges the speed of the ether wind between the arms of the interferometer. The rotation should cause the fringe pattern to shift slightly but measurably. Measurements failed, however, to show any change in the interference pattern! The Michelson–Morley experiment was repeated at different times of the year when the ether wind was expected to change direction and magnitude, but the results were always the same: **no fringe shift of the magnitude required was ever observed.**[2]

The negative results of the Michelson–Morley experiment not only contradicted the ether hypothesis but also showed that it was impossible to measure the absolute velocity of the Earth with respect to the ether frame. However, Einstein offered a postulate for his special theory of relativity that places quite a different interpretation on these null results. In later years, when more was known about the nature of light, the idea of an ether that permeates all of space was abandoned. **Light is now understood to be an electromagnetic wave, which requires no medium for its propagation.** As a result, the idea of an ether in which these waves travel became unnecessary.

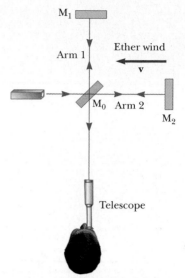

Active Figure 39.4 According to the ether wind theory, the speed of light should be $c - v$ as the beam approaches mirror M_2 and $c + v$ after reflection.

At the Active Figures link at http://www.pse6.com, *you can adjust the speed of the ether wind to see the effect on the light beams if there were an ether.*

Details of the Michelson–Morley Experiment

To understand the outcome of the Michelson–Morley experiment, let us assume that the two arms of the interferometer in Figure 39.4 are of equal length L. We shall analyze the situation as if there were an ether wind, because that is what Michelson and Morley expected to find. As noted above, the speed of the light beam along arm 2 should be $c - v$ as the beam approaches M_2 and $c + v$ after the beam is reflected. Thus, the time interval for travel to the right is $L/(c - v)$, and the time interval for travel to the left is $L/(c + v)$. The total time interval for the round trip along arm 2 is

$$\Delta t_{\text{arm 2}} = \frac{L}{c + v} + \frac{L}{c - v} = \frac{2Lc}{c^2 - v^2} = \frac{2L}{c}\left(1 - \frac{v^2}{c^2}\right)^{-1}$$

Now consider the light beam traveling along arm 1, perpendicular to the ether wind. Because the speed of the beam relative to the Earth is $(c^2 - v^2)^{1/2}$ in this case (see Fig. 39.3), the time interval for travel for each half of the trip is $L/(c^2 - v^2)^{1/2}$, and the total time interval for the round trip is

$$\Delta t_{\text{arm 1}} = \frac{2L}{(c^2 - v^2)^{1/2}} = \frac{2L}{c}\left(1 - \frac{v^2}{c^2}\right)^{-1/2}$$

Thus, the time difference Δt between the horizontal round trip (arm 2) and the vertical round trip (arm 1) is

$$\Delta t = \Delta t_{\text{arm 2}} - \Delta t_{\text{arm 1}} = \frac{2L}{c}\left[\left(1 - \frac{v^2}{c^2}\right)^{-1} - \left(1 - \frac{v^2}{c^2}\right)^{-1/2}\right]$$

[2]   From an Earth observer's point of view, changes in the Earth's speed and direction of motion in the course of a year are viewed as ether wind shifts. Even if the speed of the Earth with respect to the ether were zero at some time, six months later the speed of the Earth would be 60 km/s with respect to the ether, and as a result a fringe shift should be noticed. No shift has ever been observed, however.

Because $v^2/c^2 \ll 1$, we can simplify this expression by using the following binomial expansion after dropping all terms higher than second order:

$$(1 - x)^n \approx 1 - nx \qquad \text{(for } x \ll 1\text{)}$$

In our case, $x = v^2/c^2$, and we find that

$$\Delta t = \Delta t_{\text{arm 2}} - \Delta t_{\text{arm 1}} \approx \frac{Lv^2}{c^3} \qquad (39.3)$$

This time difference between the two instants at which the reflected beams arrive at the viewing telescope gives rise to a phase difference between the beams, producing an interference pattern when they combine at the position of the telescope. A shift in the interference pattern should be detected when the interferometer is rotated through 90° in a horizontal plane, so that the two beams exchange roles. This rotation results in a time difference twice that given by Equation 39.3. Thus, the path difference that corresponds to this time difference is

$$\Delta d = c\,(2\,\Delta t) = \frac{2Lv^2}{c^2}$$

Because a change in path length of one wavelength corresponds to a shift of one fringe, the corresponding fringe shift is equal to this path difference divided by the wavelength of the light:

$$\text{Shift} = \frac{2Lv^2}{\lambda c^2} \qquad (39.4)$$

In the experiments by Michelson and Morley, each light beam was reflected by mirrors many times to give an effective path length L of approximately 11 m. Using this value and taking v to be equal to 3.0×10^4 m/s, the speed of the Earth around the Sun, we obtain a path difference of

$$\Delta d = \frac{2(11 \text{ m})(3.0 \times 10^4 \text{ m/s})^2}{(3.0 \times 10^8 \text{ m/s})^2} = 2.2 \times 10^{-7} \text{ m}$$

This extra travel distance should produce a noticeable shift in the fringe pattern. Specifically, using 500-nm light, we expect a fringe shift for rotation through 90° of

$$\text{Shift} = \frac{\Delta d}{\lambda} = \frac{2.2 \times 10^{-7} \text{ m}}{5.0 \times 10^{-7} \text{ m}} \approx 0.44$$

The instrument used by Michelson and Morley could detect shifts as small as 0.01 fringe. However, **it detected no shift whatsoever in the fringe pattern.** Since then, the experiment has been repeated many times by different scientists under a wide variety of conditions, and no fringe shift has ever been detected. Thus, it was concluded that the motion of the Earth with respect to the postulated ether cannot be detected.

Many efforts were made to explain the null results of the Michelson–Morley experiment and to save the ether frame concept and the Galilean velocity transformation equation for light. All proposals resulting from these efforts have been shown to be wrong. No experiment in the history of physics received such valiant efforts to explain the absence of an expected result as did the Michelson–Morley experiment. The stage was set for Einstein, who solved the problem in 1905 with his special theory of relativity.

39.3   Einstein's Principle of Relativity

In the previous section we noted the impossibility of measuring the speed of the ether with respect to the Earth and the failure of the Galilean velocity transformation equation in the case of light. Einstein proposed a theory that boldly removed these difficulties and at the same time completely altered our notion of space and time.[3] He based his special theory of relativity on two postulates:

1. **The principle of relativity:** The laws of physics must be the same in all inertial reference frames.
2. **The constancy of the speed of light:** The speed of light in vacuum has the same value, $c = 3.00 \times 10^8$ m/s, in all inertial frames, regardless of the velocity of the observer or the velocity of the source emitting the light.

The first postulate asserts that *all* the laws of physics—those dealing with mechanics, electricity and magnetism, optics, thermodynamics, and so on—are the same in all reference frames moving with constant velocity relative to one another. This postulate is a sweeping generalization of the principle of Galilean relativity, which refers only to the laws of mechanics. From an experimental point of view, Einstein's principle of relativity means that any kind of experiment (measuring the speed of light, for example) performed in a laboratory at rest must give the same result when performed in a laboratory moving at a constant velocity with respect to the first one. Hence, no preferred inertial reference frame exists, and it is impossible to detect absolute motion.

Note that postulate 2 is required by postulate 1: if the speed of light were not the same in all inertial frames, measurements of different speeds would make it possible to distinguish between inertial frames; as a result, a preferred, absolute frame could be identified, in contradiction to postulate 1.

Although the Michelson–Morley experiment was performed before Einstein published his work on relativity, it is not clear whether or not Einstein was aware of the details of the experiment. Nonetheless, the null result of the experiment can be readily understood within the framework of Einstein's theory. According to his principle of relativity, the premises of the Michelson–Morley experiment were incorrect. In the process of trying to explain the expected results, we stated that when light traveled against the ether wind its speed was $c - v$, in accordance with the Galilean velocity transformation equation. However, if the state of motion of the observer or of the source has no influence on the value found for the speed of light, one always measures the value to be c. Likewise, the light makes the return trip after reflection from the mirror at speed c, not at speed $c + v$. Thus, the motion of the Earth does not influence the fringe pattern observed in the Michelson–Morley experiment, and a null result should be expected.

If we accept Einstein's theory of relativity, we must conclude that relative motion is unimportant when measuring the speed of light. At the same time, we shall see that we must alter our common-sense notion of space and time and be prepared for some surprising consequences. It may help as you read the pages ahead to keep in mind that our common-sense ideas are based on a lifetime of everyday experiences and not on observations of objects moving at hundreds of thousands of kilometers per second. Thus, these results will seem strange, but that is only because we have no experience with them.

39.4 Consequences of the Special Theory of Relativity

Before we discuss the consequences of Einstein's special theory of relativity, we must first understand how an observer located in an inertial reference frame describes an event. As mentioned earlier, an event is an occurrence describable by three space

Albert Einstein
German-American Physicist
(1879–1955)

Einstein, one of the greatest physicists of all times, was born in Ulm, Germany. In 1905, at the age of 26, he published four scientific papers that revolutionized physics. Two of these papers were concerned with what is now considered his most important contribution: the special theory of relativity.

In 1916, Einstein published his work on the general theory of relativity. The most dramatic prediction of this theory is the degree to which light is deflected by a gravitational field. Measurements made by astronomers on bright stars in the vicinity of the eclipsed Sun in 1919 confirmed Einstein's prediction, and as a result Einstein became a world celebrity.

Einstein was deeply disturbed by the development of quantum mechanics in the 1920s despite his own role as a scientific revolutionary. In particular, he could never accept the probabilistic view of events in nature that is a central feature of quantum theory. The last few decades of his life were devoted to an unsuccessful search for a unified theory that would combine gravitation and electromagnetism. *(AIP Niels Bohr Library)*

---

[3]   A. Einstein, "On the Electrodynamics of Moving Bodies," *Ann. Physik* 17:891, 1905. For an English translation of this article and other publications by Einstein, see the book by H. Lorentz, A. Einstein, H. Minkowski, and H. Weyl, *The Principle of Relativity*, Dover, 1958.

coordinates and one time coordinate. Observers in different inertial frames will describe the same event with coordinates that have different values.

As we examine some of the consequences of relativity in the remainder of this section, we restrict our discussion to the concepts of simultaneity, time intervals, and lengths, all three of which are quite different in relativistic mechanics from what they are in Newtonian mechanics. For example, in relativistic mechanics the distance between two points and the time interval between two events depend on the frame of reference in which they are measured. That is, **in relativistic mechanics there is no such thing as an absolute length or absolute time interval.** Furthermore, **events at different locations that are observed to occur simultaneously in one frame are not necessarily observed to be simultaneous in another frame moving uniformly with respect to the first.**

Simultaneity and the Relativity of Time

A basic premise of Newtonian mechanics is that a universal time scale exists that is the same for all observers. In fact, Newton wrote that "Absolute, true, and mathematical time, of itself, and from its own nature, flows equably without relation to anything external." Thus, Newton and his followers simply took simultaneity for granted. In his special theory of relativity, Einstein abandoned this assumption.

Einstein devised the following thought experiment to illustrate this point. A boxcar moves with uniform velocity, and two lightning bolts strike its ends, as illustrated in Figure 39.5a, leaving marks on the boxcar and on the ground. The marks on the boxcar are labeled A' and B', and those on the ground are labeled A and B. An observer O' moving with the boxcar is midway between A' and B', and a ground observer O is midway between A and B. The events recorded by the observers are the striking of the boxcar by the two lightning bolts.

The light signals emitted from A and B at the instant at which the two bolts strike reach observer O at the same time, as indicated in Figure 39.5b. This observer realizes that the signals have traveled at the same speed over equal distances, and so rightly concludes that the events at A and B occurred simultaneously. Now consider the same events as viewed by observer O'. By the time the signals have reached observer O, observer O' has moved as indicated in Figure 39.5b. Thus, the signal from B' has already swept past O', but the signal from A' has not yet reached O'. In other words, O' sees the signal from B' before seeing the signal from A'. According to Einstein, *the two observers must find that light travels at the same speed.* Therefore, observer O' concludes that the lightning strikes the front of the boxcar before it strikes the back.

This thought experiment clearly demonstrates that the two events that appear to be simultaneous to observer O do not appear to be simultaneous to observer O'.

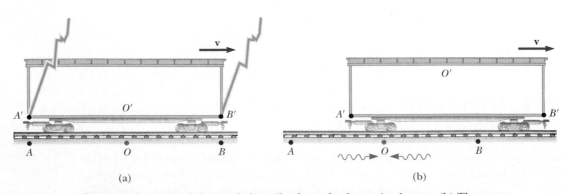

(a)   (b)

Figure 39.5 (a) Two lightning bolts strike the ends of a moving boxcar. (b) The events appear to be simultaneous to the stationary observer O, standing midway between A and B. The events do not appear to be simultaneous to observer O', who claims that the front of the car is struck before the rear. Note that in (b) the leftward-traveling light signal has already passed O' but the rightward-traveling signal has not yet reached O'.

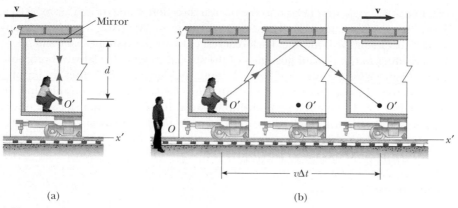

(a)   (b)   (c)

Active Figure 39.6 (a) A mirror is fixed to a moving vehicle, and a light pulse is sent out by observer O' at rest in the vehicle. (b) Relative to a stationary observer O standing alongside the vehicle, the mirror and O' move with a speed v. Note that what observer O measures for the distance the pulse travels is greater than $2d$. (c) The right triangle for calculating the relationship between Δt and Δt_p.

At the Active Figures link at http://www.pse6.com, you can observe the bouncing of the light pulse for various speeds of the train.

In other words,

two events that are simultaneous in one reference frame are in general not simultaneous in a second frame moving relative to the first. That is, simultaneity is not an absolute concept but rather one that depends on the state of motion of the observer.

Einstein's thought experiment demonstrates that two observers can disagree on the simultaneity of two events. **This disagreement, however, depends on the transit time of light to the observers and, therefore, does *not* demonstrate the deeper meaning of relativity.** In relativistic analyses of high-speed situations, relativity shows that simultaneity is relative even when the transit time is subtracted out. In fact, all of the relativistic effects that we will discuss from here on will assume that we are ignoring differences caused by the transit time of light to the observers.

Time Dilation

We can illustrate the fact that observers in different inertial frames can measure different time intervals between a pair of events by considering a vehicle moving to the right with a speed v, such as the boxcar shown in Figure 39.6a. A mirror is fixed to the ceiling of the vehicle, and observer O' at rest in the frame attached to the vehicle holds a flashlight a distance d below the mirror. At some instant, the flashlight emits a pulse of light directed toward the mirror (event 1), and at some later time after reflecting from the mirror, the pulse arrives back at the flashlight (event 2). Observer O' carries a clock and uses it to measure the time interval Δt_p between these two events. (The subscript p stands for *proper*, as we shall see in a moment.) Because the light pulse has a speed c, the time interval required for the pulse to travel from O' to the mirror and back is

$$\Delta t_p = \frac{\text{distance traveled}}{\text{speed}} = \frac{2d}{c} \tag{39.5}$$

Now consider the same pair of events as viewed by observer O in a second frame, as shown in Figure 39.6b. According to this observer, the mirror and flashlight are moving to the right with a speed v, and as a result the sequence of events appears entirely different. By the time the light from the flashlight reaches the mirror, the mirror has moved to the right a distance $v\,\Delta t/2$, where Δt is the time interval required for the light to travel from O' to the mirror and back to O' as measured by O. In other words, O concludes that, because of the motion of the vehicle, if the light is to hit the mirror, it must leave the flashlight at an angle with respect to the vertical direction. Comparing Figure 39.6a and b, we see that the light must travel farther in (b) than in (a). (Note that neither observer "knows" that he or she is moving. Each is at rest in his or her own inertial frame.)

According to the second postulate of the special theory of relativity, both observers must measure c for the speed of light. Because the light travels farther according to O, it follows that the time interval Δt measured by O is longer than the time interval Δt_p measured by O'. To obtain a relationship between these two time intervals, it is convenient to use the right triangle shown in Figure 39.6c. The Pythagorean theorem gives

$$\left(\frac{c\,\Delta t}{2}\right)^2 = \left(\frac{v\,\Delta t}{2}\right)^2 + d^2$$

Solving for Δt gives

$$\Delta t = \frac{2d}{\sqrt{c^2 - v^2}} = \frac{2d}{c\sqrt{1 - \dfrac{v^2}{c^2}}} \tag{39.6}$$

Because $\Delta t_p = 2d/c$, we can express this result as

Time dilation

$$\Delta t = \frac{\Delta t_p}{\sqrt{1 - \dfrac{v^2}{c^2}}} = \gamma\,\Delta t_p \tag{39.7}$$

where

$$\gamma = \frac{1}{\sqrt{1 - \dfrac{v^2}{c^2}}} \tag{39.8}$$

Because γ is always greater than unity, this result says that **the time interval Δt measured by an observer moving with respect to a clock is longer than the time interval Δt_p measured by an observer at rest with respect to the clock.** This effect is known as **time dilation.**

We can see that time dilation is not observed in our everyday lives by considering the factor γ. This factor deviates significantly from a value of 1 only for very high speeds, as shown in Figure 39.7 and Table 39.1. For example, for a speed of $0.1c$, the value of γ is 1.005. Thus, there is a time dilation of only 0.5% at one-tenth the speed of light. Speeds that we encounter on an everyday basis are far slower than this, so we do not see time dilation in normal situations.

The time interval Δt_p in Equations 39.5 and 39.7 is called the **proper time interval.** (In German, Einstein used the term *Eigenzeit*, which means "own-time.") In

Table 39.1

Approximate Values for γ at Various Speeds	
v/c	γ
0.001 0	1.000 000 5
0.010	1.000 05
0.10	1.005
0.20	1.021
0.30	1.048
0.40	1.091
0.50	1.155
0.60	1.250
0.70	1.400
0.80	1.667
0.90	2.294
0.92	2.552
0.94	2.931
0.96	3.571
0.98	5.025
0.99	7.089
0.995	10.01
0.999	22.37

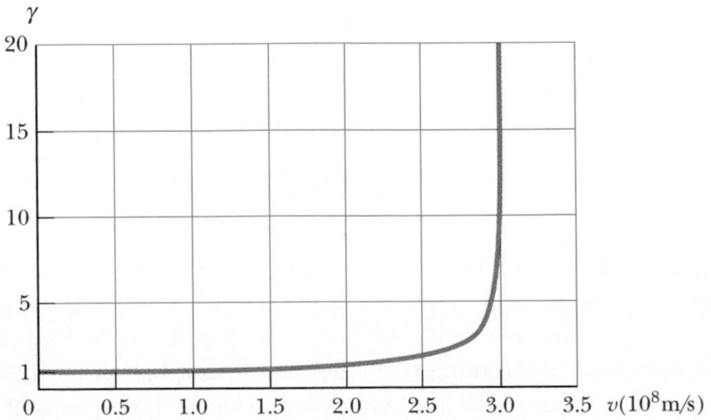

Figure 39.7 Graph of γ versus v. As the speed approaches that of light, γ increases rapidly.

general, **the proper time interval is the time interval between two events measured by an observer who sees the events occur at the same point in space.**

If a clock is moving with respect to you, the time interval between ticks of the moving clock is observed to be longer than the time interval between ticks of an identical clock in your reference frame. Thus, it is often said that a moving clock is measured to run more slowly than a clock in your reference frame by a factor γ. This is true for mechanical clocks as well as for the light clock just described. We can generalize this result by stating that all physical processes, including chemical and biological ones, are measured to slow down when those processes occur in a frame moving with respect to the observer. For example, the heartbeat of an astronaut moving through space would keep time with a clock inside the spacecraft. Both the astronaut's clock and heartbeat would be measured to slow down according to an observer on Earth comparing time intervals with his own clock (although the astronaut would have no sensation of life slowing down in the spacecraft).

> **Quick Quiz 39.3** Suppose the observer O' on the train in Figure 39.6 aims her flashlight at the far wall of the boxcar and turns it on and off, sending a pulse of light toward the far wall. Both O' and O measure the time interval between when the pulse leaves the flashlight and it hits the far wall. Which observer measures the proper time interval between these two events? (a) O' (b) O (c) both observers (d) neither observer.

> **Quick Quiz 39.4** A crew watches a movie that is two hours long in a spacecraft that is moving at high speed through space. Will an Earthbound observer, who is watching the movie through a powerful telescope, measure the duration of the movie to be (a) longer than, (b) shorter than, or (c) equal to two hours?

Strange as it may seem, time dilation is a verifiable phenomenon. An experiment reported by Hafele and Keating provided direct evidence of time dilation.[4] Time intervals measured with four cesium atomic clocks in jet flight were compared with time intervals measured by Earth-based reference atomic clocks. In order to compare these results with theory, many factors had to be considered, including periods of speeding up and slowing down relative to the Earth, variations in direction of travel, and the fact that the gravitational field experienced by the flying clocks was weaker than that experienced by the Earth-based clock. The results were in good agreement with the predictions of the special theory of relativity and can be explained in terms of the relative motion between the Earth and the jet aircraft. In their paper, Hafele and Keating stated that "Relative to the atomic time scale of the U.S. Naval Observatory, the flying clocks lost 59 ± 10 ns during the eastward trip and gained 273 ± 7 ns during the westward trip. . . . These results provide an unambiguous empirical resolution of the famous clock paradox with macroscopic clocks."

Another interesting example of time dilation involves the observation of *muons*, unstable elementary particles that have a charge equal to that of the electron and a mass 207 times that of the electron. (We will study the muon and other particles in Chapter 46.) Muons can be produced by the collision of cosmic radiation with atoms high in the atmosphere. Slow-moving muons in the laboratory have a lifetime which is measured to be the proper time interval $\Delta t_p = 2.2\ \mu s$. If we assume that the speed of atmospheric muons is close to the speed of light, we find that these particles can travel a distance of approximately $(3.0 \times 10^8\ \text{m/s})(2.2 \times 10^{-6}\ \text{s}) \approx 6.6 \times 10^2\ \text{m}$ before they decay (Fig. 39.8a). Hence, they are unlikely to reach the surface of the Earth from

---

[4] J. C. Hafele and R. E. Keating, "Around the World Atomic Clocks: Relativistic Time Gains Observed," *Science*, 177:168, 1972.

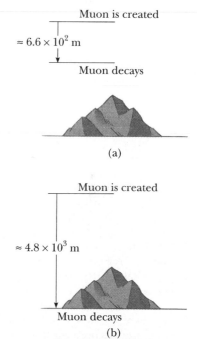

Figure 39.8 (a) Without relativistic considerations, muons created in the atmosphere and traveling downward with a speed of $0.99c$ travel only about 6.6×10^2 m before decaying with an average lifetime of $2.2\ \mu s$. Thus, very few muons reach the surface of the Earth. (b) With relativistic considerations, the muon's lifetime is dilated according to an observer on Earth. As a result, according to this observer, the muon can travel about 4.8×10^3 m before decaying. This results in many of them arriving at the surface.

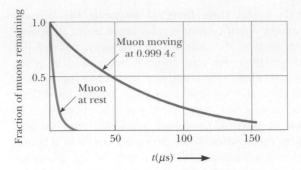

Figure 39.9 Decay curves for muons at rest and for muons traveling at a speed of 0.999 4c.

high in the atmosphere where they are produced. However, experiments show that a large number of muons *do* reach the surface. The phenomenon of time dilation explains this effect. As measured by an observer on Earth, the muons have a dilated lifetime equal to $\gamma \Delta t_p$. For example, for $v = 0.99c$, $\gamma \approx 7.1$ and $\gamma \Delta t_p \approx 16$ μs. Hence, the average distance traveled by the muons in this time as measured by an observer on Earth is approximately $(0.99)(3.0 \times 10^8 \text{ m/s})(16 \times 10^{-6} \text{ s}) \approx 4.8 \times 10^3$ m, as indicated in Figure 39.8b.

In 1976, at the laboratory of the European Council for Nuclear Research (CERN) in Geneva, muons injected into a large storage ring reached speeds of approximately 0.999 4c. Electrons produced by the decaying muons were detected by counters around the ring, enabling scientists to measure the decay rate and hence the muon lifetime. The lifetime of the moving muons was measured to be approximately 30 times as long as that of the stationary muon (Fig. 39.9), in agreement with the prediction of relativity to within two parts in a thousand.

Example 39.1 What Is the Period of the Pendulum?

The period of a pendulum is measured to be 3.00 s in the reference frame of the pendulum. What is the period when measured by an observer moving at a speed of 0.950c relative to the pendulum?

Solution To conceptualize this problem, let us change frames of reference. Instead of the observer moving at 0.950c, we can take the equivalent point of view that the observer is at rest and the pendulum is moving at 0.950c past the stationary observer. Hence, the pendulum is an example of a clock moving at high speed with respect to an observer and we can categorize this problem as one involving time dilation.

To analyze the problem, note that the proper time interval, measured in the rest frame of the pendulum, is $\Delta t_p = 3.00$ s. Because a clock moving with respect to an observer is measured to run more slowly than a stationary clock by a factor γ, Equation 39.7 gives

$$\Delta t = \gamma \Delta t_p = \frac{1}{\sqrt{1 - \dfrac{(0.950c)^2}{c^2}}} \Delta t_p = \frac{1}{\sqrt{1 - 0.902}} \Delta t_p$$

$$= (3.20)(3.00 \text{ s}) = \boxed{9.60 \text{ s}}$$

To finalize this problem, we see that indeed a moving pendulum is measured to take longer to complete a period

than a pendulum at rest does. The period increases by a factor of $\gamma = 3.20$. We see that this is consistent with Table 39.1, where this value lies between those for γ for $v/c = 0.94$ and $v/c = 0.96$.

What If? What if we increase the speed of the observer by 5.00%? Does the dilated time interval increase by 5.00%?

Answer Based on the highly nonlinear behavior of γ as a function of v in Figure 39.7, we would guess that the increase in Δt would be different from 5.00%. Increasing v by 5.00% gives us

$$v_{\text{new}} = (1.050\ 0)(0.950c) = 0.997\ 5c$$

(Because γ varies so rapidly with v when v is this large, we will keep one additional significant figure until the final answer.) If we perform the time dilation calculation again, we find that

$$\Delta t = \gamma \Delta t_p = \frac{1}{\sqrt{1 - \dfrac{(0.997\ 5c)^2}{c^2}}} \Delta t_p = \frac{1}{\sqrt{1 - 0.995\ 0}} \Delta t_p$$

$$= (14.15)(3.00 \text{ s}) = 42.5 \text{ s}$$

Thus, the 5.00% increase in speed has caused over a 300% increase in the dilated time!

Example 39.2   How Long Was Your Trip?

Suppose you are driving your car on a business trip and are traveling at 30 m/s. Your boss, who is waiting at your destination, expects the trip to take 5.0 h. When you arrive late, your excuse is that your car clock registered the passage of 5.0 h but that you were driving fast and so your clock ran more slowly than your boss's clock. If your car clock actually did indicate a 5.0-h trip, how much time passed on your boss's clock, which was at rest on the Earth?

Solution   We begin by calculating γ from Equation 39.8:

$$\gamma = \frac{1}{\sqrt{1 - \dfrac{v^2}{c^2}}} = \frac{1}{\sqrt{1 - \dfrac{(3 \times 10^1 \text{ m/s})^2}{(3 \times 10^8 \text{ m/s})^2}}}$$

$$= \frac{1}{\sqrt{1 - 10^{-14}}}$$

If you try to determine this value on your calculator, you will probably obtain $\gamma = 1$. However, if we perform a binomial expansion, we can more precisely determine the value as

$$\gamma = (1 - 10^{-14})^{-1/2} \approx 1 + \tfrac{1}{2}(10^{-14}) = 1 + 5.0 \times 10^{-15}$$

This result indicates that at typical automobile speeds, γ is not much different from 1.

Applying Equation 39.7, we find Δt, the time interval measured by your boss, to be

$$\Delta t = \gamma \, \Delta t_p = (1 + 5.0 \times 10^{-15})(5.0 \text{ h})$$

$$= 5.0 \text{ h} + 2.5 \times 10^{-14} \text{ h} = \boxed{5.0 \text{ h} + 0.09 \text{ ns}}$$

Your boss's clock would be only 0.09 ns ahead of your car clock. You might want to think of another excuse!

The Twin Paradox

An intriguing consequence of time dilation is the so-called *twin paradox* (Fig. 39.10). Consider an experiment involving a set of twins named Speedo and Goslo. When they are 20 yr old, Speedo, the more adventuresome of the two, sets out on an epic journey to Planet X, located 20 ly from the Earth. (Note that 1 lightyear (ly) is the distance light travels through free space in 1 year.) Furthermore, Speedo's spacecraft is capable of reaching a speed of $0.95c$ relative to the inertial frame of his twin brother back home. After reaching Planet X, Speedo becomes homesick and immediately returns to the Earth at the same speed $0.95c$. Upon his return, Speedo is shocked to discover that Goslo has aged 42 yr and is now 62 yr old. Speedo, on the other hand, has aged only 13 yr.

At this point, it is fair to raise the following question—which twin is the traveler and which is really younger as a result of this experiment? From Goslo's frame of reference, he was at rest while his brother traveled at a high speed away from him and then came back. According to Speedo, however, he himself remained stationary while Goslo and the Earth raced away from him and then headed back. This leads to an apparent

Speedo   Goslo   Speedo   Goslo
        (a)   (b)

Figure 39.10   (a) As one twin leaves his brother on the Earth, both are the same age. (b) When Speedo returns from his journey to Planet X, he is younger than his twin Goslo.

contradiction due to the apparent symmetry of the observations. Which twin has developed signs of excess aging?

The situation in our current problem is actually not symmetrical. To resolve this apparent paradox, recall that the special theory of relativity describes observations made in inertial frames of reference moving relative to each other. Speedo, the space traveler, must experience a series of accelerations during his journey because he must fire his rocket engines to slow down and start moving back toward Earth. As a result, his speed is not always uniform, and consequently he is not in an inertial frame. Therefore, there is no paradox—only Goslo, who is always in a single inertial frame, can make correct predictions based on special relativity. During each passing year noted by Goslo, slightly less than 4 months elapses for Speedo.

Only Goslo, who is in a single inertial frame, can apply the simple time-dilation formula to Speedo's trip. Thus, Goslo finds that instead of aging 42 yr, Speedo ages only $(1 - v^2/c^2)^{1/2}(42 \text{ yr}) = 13 \text{ yr}$. Thus, according to Goslo, Speedo spends 6.5 yr traveling to Planet X and 6.5 yr returning, for a total travel time of 13 yr, in agreement with our earlier statement.

Quick Quiz 39.5 Suppose astronauts are paid according to the amount of time they spend traveling in space. After a long voyage traveling at a speed approaching c, would a crew rather be paid according to (a) an Earth-based clock, (b) their spacecraft's clock, or (c) either clock?

Length Contraction

▲ **PITFALL PREVENTION**

39.4 The Proper Length

As with the proper time interval, it is *very* important in relativistic calculations to correctly identify the observer who measures the proper length. The proper length between two points in space is always the length measured by an observer at rest with respect to the points. Often the proper time interval and the proper length are *not* measured by the same observer.

The measured distance between two points also depends on the frame of reference. **The proper length L_p of an object is the length measured by someone at rest relative to the object.** The length of an object measured by someone in a reference frame that is moving with respect to the object is always less than the proper length. This effect is known as **length contraction.**

Consider a spacecraft traveling with a speed v from one star to another. There are two observers: one on the Earth and the other in the spacecraft. The observer at rest on the Earth (and also assumed to be at rest with respect to the two stars) measures the distance between the stars to be the proper length L_p. According to this observer, the time interval required for the spacecraft to complete the voyage is $\Delta t = L_p/v$. The passages of the two stars by the spacecraft occur at the same position for the space traveler. Thus, the space traveler measures the proper time interval Δt_p. Because of time dilation, the proper time interval is related to the Earth-measured time interval by $\Delta t_p = \Delta t/\gamma$. Because the space traveler reaches the second star in the time Δt_p, he or she concludes that the distance L between the stars is

$$L = v \Delta t_p = v \frac{\Delta t}{\gamma}$$

Because the proper length is $L_p = v \Delta t$, we see that

Length contraction

$$L = \frac{L_p}{\gamma} = L_p \sqrt{1 - \frac{v^2}{c^2}} \tag{39.9}$$

where $\sqrt{1 - v^2/c^2}$ is a factor less than unity. **If an object has a proper length L_p when it is measured by an observer at rest with respect to the object, then when it moves with speed v in a direction parallel to its length, its length L is measured to be shorter according to $L = L_p \sqrt{1 - v^2/c^2} = L_p/\gamma$.**

For example, suppose that a meter stick moves past a stationary Earth observer with speed v, as in Figure 39.11. The length of the stick as measured by an observer in a frame attached to the stick is the proper length L_p shown in Figure 39.11a. The length of the stick L measured by the Earth observer is shorter than L_p by the factor $(1 - v^2/c^2)^{1/2}$. Note that **length contraction takes place only along the direction of motion.**

The proper length and the proper time interval are defined differently. The proper length is measured by an observer for whom the end points of the length remain fixed in space. The proper time interval is measured by someone for whom the two events take place at the same position in space. As an example of this point, let us return to the decaying muons moving at speeds close to the speed of light. An observer in the muon's reference frame would measure the proper lifetime, while an Earth-based observer would measure the proper length (the distance from creation to decay in Figure 39.8). In the muon's reference frame, there is no time dilation but the distance of travel to the surface is observed to be shorter when measured in this frame. Likewise, in the Earth observer's reference frame, there is time dilation, but the distance of travel is measured to be the proper length. Thus, when calculations on the muon are performed in both frames, the outcome of the experiment in one frame is the same as the outcome in the other frame—more muons reach the surface than would be predicted without relativistic effects.

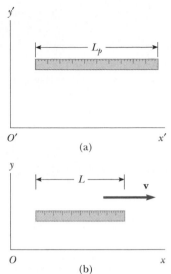

Active Figure 39.11 (a) A meter stick measured by an observer in a frame attached to the stick (that is, both have the same velocity) has its proper length L_p. (b) The stick measured by an observer in a frame in which the stick has a velocity **v** relative to the frame is measured to be shorter than its proper length L_p by a factor $(1 - v^2/c^2)^{1/2}$.

At the Active Figures link at http://www.pse6.com, you can view the meter stick from the points of view of two observers to compare the measured length of the stick.

> **Quick Quiz 39.6** You are packing for a trip to another star. During the journey, you will be traveling at $0.99c$. You are trying to decide whether you should buy smaller sizes of your clothing, because you will be thinner on your trip, due to length contraction. Also, you are considering saving money by reserving a smaller cabin to sleep in, because you will be shorter when you lie down. Should you (a) buy smaller sizes of clothing, (b) reserve a smaller cabin, (c) do neither of these, or (d) do both of these?

> **Quick Quiz 39.7** You are observing a spacecraft moving away from you. You measure it to be shorter than when it was at rest on the ground next to you. You also see a clock through the spacecraft window, and you observe that the passage of time on the clock is measured to be slower than that of the watch on your wrist. Compared to when the spacecraft was on the ground, what do you measure if the spacecraft turns around and comes *toward* you at the same speed? (a) The spacecraft is measured to be longer and the clock runs faster. (b) The spacecraft is measured to be longer and the clock runs slower. (c) The spacecraft is measured to be shorter and the clock runs faster. (d) The spacecraft is measured to be shorter and the clock runs slower.

Space–Time Graphs

It is sometimes helpful to make a *space–time graph*, in which ct is the ordinate and position x is the abscissa. The twin paradox is displayed in such a graph in Figure 39.12

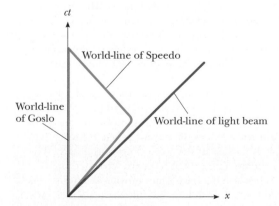

Figure 39.12 The twin paradox on a space–time graph. The twin who stays on the Earth has a world-line along the ct axis. The path of the traveling twin through space–time is represented by a world-line that changes direction.

from the point of view of Goslo. A path through space–time is called a **world-line.** At the origin, the world-lines of Speedo and Goslo coincide because the twins are in the same location at the same time. After Speedo leaves on his trip, his world-line diverges from that of his brother. Goslo's world-line is vertical because he remains fixed in location. At their reunion, the two world-lines again come together. Note that it would be impossible for Speedo to have a world-line that crossed the path of a light beam that left the Earth when he did. To do so would require him to have a speed greater than c (not possible, as shown in Sections 39.6 and 39.7).

World-lines for light beams are diagonal lines on space–time graphs, typically drawn at 45° to the right or left of vertical (assuming that the x and ct axes have the same scales), depending on whether the light beam is traveling in the direction of increasing or decreasing x. These two world-lines mean that all possible future events for Goslo and Speedo lie within two 45° lines extending from the origin. Either twin's presence at an event outside this "light cone" would require that twin to move at a speed greater than c, which we have said is not possible. Also, the only past events that Goslo and Speedo could have experienced occurred within two similar 45° world-lines that approach the origin from below the x axis.

Example 39.3   The Contraction of a Spacecraft

A spacecraft is measured to be 120.0 m long and 20.0 m in diameter while at rest relative to an observer. If this spacecraft now flies by the observer with a speed of $0.99c$, what length and diameter does the observer measure?

Solution From Equation 39.9, the length measured by the

observer is

$$L = L_p \sqrt{1 - \frac{v^2}{c^2}} = (120.0 \text{ m}) \sqrt{1 - \frac{(0.99c)^2}{c^2}} = \boxed{17 \text{ m}}$$

The diameter measured by the observer is still 20.0 m because the diameter is a dimension perpendicular to the motion and length contraction occurs only along the direction of motion.

Example 39.4   The Pole-in-the-Barn Paradox   Interactive

The twin paradox, discussed earlier, is a classic "paradox" in relativity. Another classic "paradox" is this: Suppose a runner moving at $0.75c$ carries a horizontal pole 15 m long toward a barn that is 10 m long. The barn has front and rear doors. An observer on the ground can instantly and simultaneously open and close the two doors by remote control. When the runner and the pole are inside the barn, the ground observer closes and then opens both doors so that the runner and pole are momentarily captured inside the barn and then proceed to exit the barn from the back door. Do both the runner and the ground observer agree that the runner makes it safely through the barn?

Solution From our everyday experience, we would be surprised to see a 15-m pole fit inside a 10-m barn. But the pole is in motion with respect to the ground observer, who measures the pole to be contracted to a length L_{pole}, where

$$L_{\text{pole}} = L_p \sqrt{1 - \frac{v^2}{c^2}} = (15 \text{ m}) \sqrt{1 - (0.75)^2} = 9.9 \text{ m}$$

Thus, the ground observer measures the pole to be slightly shorter than the barn and there is no problem with momentarily capturing the pole inside it. The "paradox" arises when we consider the runner's point of view. The runner

sees the barn contracted to

$$L_{\text{barn}} = L_p \sqrt{1 - \frac{v^2}{c^2}} = (10 \text{ m}) \sqrt{1 - (0.75)^2} = \boxed{6.6 \text{ m}}$$

Because the pole is in the rest frame of the runner, the runner measures it to have its proper length of 15 m. How can a 15-m pole fit inside a 6.6-m barn? While this is the classic question that is often asked, this is not the question we have asked, because it is not the important question. We asked *if the runner can make it safely through the barn.*

The resolution of the "paradox" lies in the relativity of simultaneity. The closing of the two doors is measured to be simultaneous by the ground observer. Because the doors are at different positions, however, they do not close simultaneously as measured by the runner. The rear door closes and then opens first, allowing the leading edge of the pole to exit. The front door of the barn does not close until the trailing edge of the pole passes by.

We can analyze this using a space-time graph. Figure 39.13a is a space–time graph from the ground observer's point of view. We choose $x = 0$ as the position of the front door of the barn and $t = 0$ as the instant at which the leading end of the pole is located at the front door of the barn. The world-lines for the two ends of the barn are separated by 10 m and are vertical because the barn is not moving relative to this observer. For the pole, we follow two tilted world-lines, one

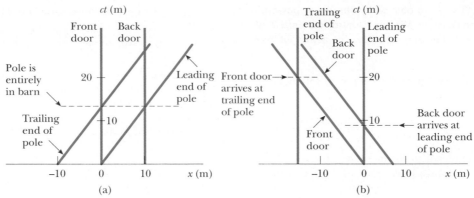

Figure 39.13 (Example 39.4) Space–time graphs for the pole-in-the-barn paradox. (a) From the ground observer's point of view, the world-lines for the front and back doors of the barn are vertical lines. The world-lines for the ends of the pole are tilted and are 9.9 m apart horizontally. The front door of the barn is at $x = 0$, and the leading end of the pole enters the front door at $t = 0$. The entire pole is inside the barn at the time indicated by the dashed line. (b) From the runner's point of view, the world-lines for the ends of the pole are vertical. The barn is moving in the negative direction, so the world-lines for the front and back doors are tilted to the left. The leading end of the pole exits the back door before the trailing end arrives at the front door.

for each end of the moving pole. These world-lines are 9.9 m apart horizontally, which is the contracted length seen by the ground observer. As seen in Figure 39.13a, at one instant, the pole is entirely within the barn.

Figure 39.13b shows the space–time graph according to the runner. Here, the world-lines for the pole are separated by 15 m and are vertical because the pole is at rest in the runner's frame of reference. The barn is hurtling *toward* the runner, so the world-lines for the front and rear doors of the barn are tilted in the opposite direction compared to Figure 39.13a. The world-lines for the barn are separated by 6.6 m, the contracted length as seen by the runner. Notice that the front of the pole leaves the rear door of the barn long before the back of the pole enters the barn. Thus, the opening of the rear door occurs before the closing of the front door.

From the ground observer's point of view, the time at which the trailing end of the pole enters the barn is found from

$$\Delta t = t - 0 = t = \frac{\Delta x}{v} = \frac{9.9 \text{ m}}{0.75c} = \frac{13.2 \text{ m}}{c}$$

Thus, the pole should be completely inside the barn at a time corresponding to $ct = 13.2$ m. This is consistent with the point on the ct axis in Figure 39.13a where the pole is inside the barn.

From the runner's point of view, the time at which the leading end of the pole leaves the barn is found from

$$\Delta t = t - 0 = t = \frac{\Delta x}{v} = \frac{6.6 \text{ m}}{0.75c} = \frac{8.8 \text{ m}}{c}$$

leading to $ct = 8.8$ m. This is consistent with the point on the ct axis in Figure 39.13b where the back door of the barn arrives at the leading end of the pole. Finally, the time at which the trailing end of the pole enters the front door of the barn is found from

$$\Delta t = t - 0 = t = \frac{\Delta x}{v} = \frac{15 \text{ m}}{0.75c} = \frac{20 \text{ m}}{c}$$

This gives $ct = 20$ m, which agrees with the instant shown in Figure 39.13b.

 Investigate the pole-in-the-barn paradox at the Interactive Worked Example link at **http://www.pse6.com.**

Example 39.5 A Voyage to Sirius

An astronaut takes a trip to Sirius, which is located a distance of 8 lightyears from the Earth. The astronaut measures the time of the one-way journey to be 6 yr. If the spaceship moves at a constant speed of $0.8c$, how can the 8-ly distance be reconciled with the 6-yr trip time measured by the astronaut?

Solution The distance of 8 ly represents the proper length from the Earth to Sirius measured by an observer seeing both objects nearly at rest. The astronaut sees Sirius approaching her at $0.8c$ but also sees the distance

contracted to

$$\frac{8 \text{ ly}}{\gamma} = (8 \text{ ly}) \sqrt{1 - \frac{v^2}{c^2}} = (8 \text{ ly}) \sqrt{1 - \frac{(0.8c)^2}{c^2}} = 5 \text{ ly}$$

Thus, the travel time measured on her clock is

$$\Delta t = \frac{d}{v} = \frac{5 \text{ ly}}{0.8c} = 6 \text{ yr}$$

Note that we have used the value for the speed of light as $c = 1$ ly/yr.

What If? What if this trip is observed with a very powerful telescope by a technician in Mission Control on Earth? At what time will this technician *see* that the astronaut has arrived at Sirius?

Answer The time interval that the technician will measure for the astronaut to arrive is

$$\Delta t = \frac{d}{v} = \frac{8 \text{ ly}}{0.8c} = 10 \text{ yr}$$

In order for the technician to *see* the arrival, the light from the scene of the arrival must travel back to Earth and enter

the telescope. This will require a time interval of

$$\Delta t = \frac{d}{v} = \frac{8 \text{ ly}}{c} = 8 \text{ yr}$$

Thus, the technician sees the arrival after 10 yr + 8 yr = 18 yr. Notice that if the astronaut immediately turns around and comes back home, she arrives, according to the technician, 20 years after leaving, only 2 years after he *saw* her arrive! In addition, she would have aged by only 12 years.

The Relativistic Doppler Effect

Another important consequence of time dilation is the shift in frequency found for light emitted by atoms in motion as opposed to light emitted by atoms at rest. This phenomenon, known as the Doppler effect, was introduced in Chapter 17 as it pertains to sound waves. In the case of sound, the motion of the source with respect to the medium of propagation can be distinguished from the motion of the observer with respect to the medium. Light waves must be analyzed differently, however, because they require no medium of propagation, and no method exists for distinguishing the motion of a light source from the motion of the observer.

If a light source and an observer approach each other with a relative speed v, the frequency f_{obs} measured by the observer is

$$f_{obs} = \frac{\sqrt{1 + v/c}}{\sqrt{1 - v/c}} f_{source} \tag{39.10}$$

where f_{source} is the frequency of the source measured in its rest frame. Note that this relativistic Doppler shift equation, unlike the Doppler shift equation for sound, depends only on the relative speed v of the source and observer and holds for relative speeds as great as c. As you might expect, the equation predicts that $f_{obs} > f_{source}$ when the source and observer approach each other. We obtain the expression for the case in which the source and observer recede from each other by substituting negative values for v in Equation 39.10.

The most spectacular and dramatic use of the relativistic Doppler effect is the measurement of shifts in the frequency of light emitted by a moving astronomical object such as a galaxy. Light emitted by atoms and normally found in the extreme violet region of the spectrum is shifted toward the red end of the spectrum for atoms in other galaxies—indicating that these galaxies are *receding* from us. The American astronomer Edwin Hubble (1889–1953) performed extensive measurements of this *red shift* to confirm that most galaxies are moving away from us, indicating that the Universe is expanding.

39.5 The Lorentz Transformation Equations

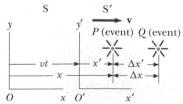

Figure 39.14 Events occur at points P and Q and are observed by an observer at rest in the S frame and another in the S′ frame, which is moving to the right with a speed v.

Suppose an event that occurs at some point P is reported by two observers, one at rest in a frame S and another in a frame S′ that is moving to the right with speed v as in Figure 39.14. The observer in S reports the event with space–time coordinates (x, y, z, t), while the observer in S′ reports the same event using the coordinates (x', y', z', t'). If two events occur at P and Q, Equation 39.1 predicts that $\Delta x = \Delta x'$, that is, the distance between the two points in space at which the events occur does not depend on motion of the observer. Because this is contradictory to the notion of length contraction, the Galilean transformation is not valid when v approaches the speed of light. In this section, we state the correct transformation equations that apply for all speeds in the range $0 \le v < c$.

The equations that are valid for all speeds and enable us to transform coordinates from S to S′ are the **Lorentz transformation equations:**

$$x' = \gamma(x - vt) \qquad y' = y \qquad z' = z \qquad t' = \gamma\left(t - \frac{v}{c^2}x\right) \tag{39.11}$$

Lorentz transformation for S → S′

These transformation equations were developed by Hendrik A. Lorentz (1853–1928) in 1890 in connection with electromagnetism. However, it was Einstein who recognized their physical significance and took the bold step of interpreting them within the framework of the special theory of relativity.

Note the difference between the Galilean and Lorentz time equations. In the Galilean case, $t = t'$, but in the Lorentz case the value for t' assigned to an event by an observer O' in the S′ frame in Figure 39.14 depends both on the time t and on the coordinate x as measured by an observer O in the S frame. This is consistent with the notion that an event is characterized by four space–time coordinates (x, y, z, t). In other words, in relativity, space and time are *not* separate concepts but rather are closely interwoven with each other.

If we wish to transform coordinates in the S′ frame to coordinates in the S frame, we simply replace v by $-v$ and interchange the primed and unprimed coordinates in Equations 39.11:

$$x = \gamma(x' + vt') \qquad y = y' \qquad z = z' \qquad t = \gamma\left(t' + \frac{v}{c^2}x'\right) \tag{39.12}$$

Inverse Lorentz transformation for S′ → S

When $v \ll c$, the Lorentz transformation equations should reduce to the Galilean equations. To verify this, note that as v approaches zero, $v/c \ll 1$; thus, $\gamma \to 1$, and Equations 39.11 reduce to the Galilean space–time transformation equations:

$$x' = x - vt \qquad y' = y \qquad z' = z \qquad t' = t$$

In many situations, we would like to know the difference in coordinates between two events or the time interval between two events as seen by observers O and O'. We can accomplish this by writing the Lorentz equations in a form suitable for describing pairs of events. From Equations 39.11 and 39.12, we can express the differences between the four variables x, x', t, and t' in the form

$$\left.\begin{aligned} \Delta x' &= \gamma(\Delta x - v\,\Delta t) \\ \Delta t' &= \gamma\left(\Delta t - \frac{v}{c^2}\,\Delta x\right) \end{aligned}\right\} \text{S} \to \text{S}' \tag{39.13}$$

$$\left.\begin{aligned} \Delta x &= \gamma(\Delta x' + v\,\Delta t') \\ \Delta t &= \gamma\left(\Delta t' + \frac{v}{c^2}\,\Delta x'\right) \end{aligned}\right\} \text{S}' \to \text{S} \tag{39.14}$$

where $\Delta x' = x'_2 - x'_1$ and $\Delta t' = t'_2 - t'_1$ are the differences measured by observer O' and $\Delta x = x_2 - x_1$ and $\Delta t = t_2 - t_1$ are the differences measured by observer O. (We have not included the expressions for relating the y and z coordinates because they are unaffected by motion along the x direction.[5])

---

[5] Although relative motion of the two frames along the x axis does not change the y and z coordinates of an object, it does change the y and z velocity components of an object moving in either frame, as noted in Section 39.6.

Example 39.6 **Simultaneity and Time Dilation Revisited**

Use the Lorentz transformation equations in difference form to show that

(A) simultaneity is not an absolute concept and that

(B) a moving clock is measured to run more slowly than a clock that is at rest with respect to an observer.

Solution **(A)** Suppose that two events are simultaneous and separated in space such that $\Delta t' = 0$ and $\Delta x' \neq 0$ according to an observer O' who is moving with speed v relative to O. From the expression for Δt given in Equation 39.14, we see that in this case the time interval Δt measured by observer O is $\Delta t = \gamma v \, \Delta x'/c^2$. That is, the

time interval for the same two events as measured by O is nonzero, and so the events do not appear to be simultaneous to O.

(B) Suppose that observer O' carries a clock that he uses to measure a time interval $\Delta t'$. He finds that two events occur at the same place in his reference frame ($\Delta x' = 0$) but at different times ($\Delta t' \neq 0$). Observer O' is moving with speed v relative to O, who measures the time interval between the events to be Δt. In this situation, the expression for Δt given in Equation 39.14 becomes $\Delta t = \gamma \, \Delta t'$. This is the equation for time dilation found earlier (Eq. 39.7), where $\Delta t' = \Delta t_p$ is the proper time measured by the clock carried by observer O'. Thus, O measures the moving clock to run slow.

39.6 The Lorentz Velocity Transformation Equations

Suppose two observers in relative motion with respect to each other are both observing the motion of an object. Previously, we defined an event as occurring at an instant of time. Now, we wish to interpret the "event" as the motion of the object. We know that the Galilean velocity transformation (Eq. 39.2) is valid for low speeds. How do the observers' measurements of the velocity of the object relate to each other if the speed of the object is close to that of light? Once again S′ is our frame moving at a speed v relative to S. Suppose that an object has a velocity component u_x' measured in the S′ frame, where

$$u_x' = \frac{dx'}{dt'} \tag{39.15}$$

Using Equation 39.11, we have

$$dx' = \gamma(dx - v \, dt)$$

$$dt' = \gamma\left(dt - \frac{v}{c^2} \, dx\right)$$

Substituting these values into Equation 39.15 gives

$$u_x' = \frac{dx'}{dt'} = \frac{dx - v \, dt}{dt - \dfrac{v}{c^2} \, dx} = \frac{\dfrac{dx}{dt} - v}{1 - \dfrac{v}{c^2} \dfrac{dx}{dt}}$$

But dx/dt is just the velocity component u_x of the object measured by an observer in S, and so this expression becomes

Lorentz velocity transformation for S → S′

$$u_x' = \frac{u_x - v}{1 - \dfrac{u_x v}{c^2}} \tag{39.16}$$

If the object has velocity components along the y and z axes, the components as measured by an observer in S′ are

$$u_y' = \frac{u_y}{\gamma\left(1 - \dfrac{u_x v}{c^2}\right)} \quad \text{and} \quad u_z' = \frac{u_z}{\gamma\left(1 - \dfrac{u_x v}{c^2}\right)} \tag{39.17}$$

Note that u'_y and u'_z do not contain the parameter v in the numerator because the relative velocity is along the x axis.

When v is much smaller than c (the nonrelativistic case), the denominator of Equation 39.16 approaches unity, and so $u'_x \approx u_x - v$, which is the Galilean velocity transformation equation. In another extreme, when $u_x = c$, Equation 39.16 becomes

$$ u'_x = \frac{c - v}{1 - \dfrac{cv}{c^2}} = \frac{c\left(1 - \dfrac{v}{c}\right)}{1 - \dfrac{v}{c}} = c $$

From this result, we see that a speed measured as c by an observer in S is also measured as c by an observer in S′—independent of the relative motion of S and S′. Note that this conclusion is consistent with Einstein's second postulate—that the speed of light must be c relative to all inertial reference frames. Furthermore, we find that the speed of an object can never be measured as larger than c. That is, the speed of light is the ultimate speed. We return to this point later.

To obtain u_x in terms of u'_x, we replace v by $-v$ in Equation 39.16 and interchange the roles of u_x and u'_x:

$$ u_x = \frac{u'_x + v}{1 + \dfrac{u'_x v}{c^2}} \tag{39.18} $$

Quick Quiz 39.8 You are driving on a freeway at a relativistic speed. Straight ahead of you, a technician standing on the ground turns on a searchlight and a beam of light moves exactly vertically upward, as seen by the technician. As you observe the beam of light, you measure the magnitude of the vertical component of its velocity as (a) equal to c (b) greater than c (c) less than c.

Quick Quiz 39.9 Consider the situation in Quick Quiz 39.8 again. If the technician aims the searchlight directly at you instead of upward, you measure the magnitude of the horizontal component of its velocity as (a) equal to c (b) greater than c (c) less than c.

Example 39.7 Relative Velocity of Two Spacecraft

Two spacecraft A and B are moving in opposite directions, as shown in Figure 39.15. An observer on the Earth measures the speed of craft A to be $0.750c$ and the speed of craft B to be $0.850c$. Find the velocity of craft B as observed by the crew on craft A.

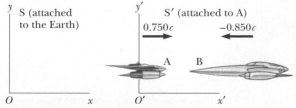

Figure 39.15 (Example 39.7) Two spacecraft A and B move in opposite directions. The speed of B relative to A is *less* than c and is obtained from the relativistic velocity transformation equation.

Solution To conceptualize this problem, we carefully identify the observers and the event. The two observers are on the Earth and on spacecraft A. The event is the motion of spacecraft B. Because the problem asks to find an observed velocity, we categorize this problem as one requiring the Lorentz velocity transformation. To analyze the problem, we note that the Earth observer makes two measurements, one of each spacecraft. We identify this observer as being at rest in the S frame. Because the velocity of spacecraft B is what we wish to measure, we identify the speed u_x as $-0.850c$. The velocity of spacecraft A is also the velocity of the observer at rest in the S′ frame, which is attached to the spacecraft, relative to the observer at rest in S. Thus, $v = 0.750c$. Now we can obtain the velocity u'_x of craft B relative to craft A by using Equation 39.16:

$$u'_x = \frac{u_x - v}{1 - \dfrac{u_x v}{c^2}} = \frac{-0.850c - 0.750c}{1 - \dfrac{(-0.850c)(0.750c)}{c^2}}$$

$$= \boxed{-0.977c}$$

To finalize this problem, note that the negative sign indicates that craft B is moving in the negative x direction as observed by the crew on craft A. Is this consistent with your expectation from Figure 39.15? Note that the speed is less than c. That is, an object whose speed is less than c in one frame of reference must have a speed less than c in any other frame. (If the Galilean velocity transformation equation were used in

this example, we would find that $u'_x = u_x - v = -0.850c - 0.750c = -1.60c$, which is impossible. The Galilean transformation equation does not work in relativistic situations.)

What If? What if the two spacecraft pass each other? Now what is their relative speed?

Answer The calculation using Equation 39.16 involves only the velocities of the two spacecraft and does not depend on their locations. After they pass each other, they have the same velocities, so the velocity of craft B as observed by the crew on craft A is the same, $-0.977c$. The only difference after they pass is that B is receding from A whereas it was approaching A before it passed.

Example 39.8   The Speeding Motorcycle

Imagine a motorcycle moving with a speed $0.80c$ past a stationary observer, as shown in Figure 39.16. If the rider

tosses a ball in the forward direction with a speed of $0.70c$ relative to himself, what is the speed of the ball relative to the stationary observer?

Solution The speed of the motorcycle relative to the stationary observer is $v = 0.80c$. The speed of the ball in the frame of reference of the motorcyclist is $u'_x = 0.70c$. Therefore, the speed u_x of the ball relative to the stationary observer is

$$u_x = \frac{u'_x + v}{1 + \dfrac{u'_x v}{c^2}} = \frac{0.70c + 0.80c}{1 + \dfrac{(0.70c)(0.80c)}{c^2}} = \boxed{0.96c}$$

Figure 39.16 (Example 39.8) A motorcyclist moves past a stationary observer with a speed of $0.80c$ and throws a ball in the direction of motion with a speed of $0.70c$ relative to himself.

Example 39.9   Relativistic Leaders of the Pack   `Interactive`

Two motorcycle pack leaders named David and Emily are racing at relativistic speeds along perpendicular paths, as shown in Figure 39.17. How fast does Emily recede as seen by David over his right shoulder?

Solution Figure 39.17 represents the situation as seen by a police officer at rest in frame S, who observes the

following:

David: $u_x = 0.75c$ $u_y = 0$

Emily: $u_x = 0$ $u_y = -0.90c$

To calculate Emily's speed of recession as seen by David, we take S′ to move along with David and then calculate u'_x and

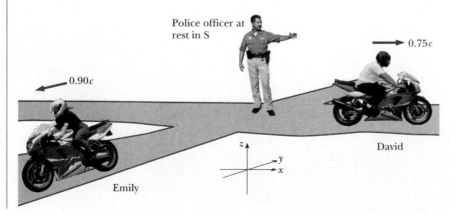

Figure 39.17 (Example 39.9) David moves to the east with a speed $0.75c$ relative to the police officer, and Emily travels south at a speed $0.90c$ relative to the officer.

u'_y for Emily using Equations 39.16 and 39.17:

$$u'_x = \frac{u_x - v}{1 - \dfrac{u_x v}{c^2}} = \frac{0 - 0.75c}{1 - \dfrac{(0)(0.75c)}{c^2}} = -0.75c$$

$$u'_y = \frac{u_y}{\gamma\left(1 - \dfrac{u_x v}{c^2}\right)} = \frac{\sqrt{1 - \dfrac{(0.75c)^2}{c^2}}\,(-0.90c)}{\left(1 - \dfrac{(0)(0.75c)}{c^2}\right)}$$

$$= -0.60c$$

Thus, the speed of Emily as observed by David is

$$u' = \sqrt{(u'_x)^2 + (u'_y)^2} = \sqrt{(-0.75c)^2 + (-0.60c)^2}$$

$$= \boxed{0.96c}$$

Note that this speed is less than c, as required by the special theory of relativity.

 Investigate this situation with various speeds of David and Emily at the Interactive Worked Example link at **http://www.pse6.com.**

39.7 Relativistic Linear Momentum and the Relativistic Form of Newton's Laws

We have seen that in order to describe properly the motion of particles within the framework of the special theory of relativity, we must replace the Galilean transformation equations by the Lorentz transformation equations. Because the laws of physics must remain unchanged under the Lorentz transformation, we must generalize Newton's laws and the definitions of linear momentum and energy to conform to the Lorentz transformation equations and the principle of relativity. These generalized definitions should reduce to the classical (nonrelativistic) definitions for $v \ll c$.

First, recall that the law of conservation of linear momentum states that when two particles (or objects that can be modeled as particles) collide, the total momentum of the isolated system of the two particles remains constant. Suppose that we observe this collision in a reference frame S and confirm that the momentum of the system is conserved. Now imagine that the momenta of the particles are measured by an observer in a second reference frame S′ moving with velocity **v** relative to the first frame. Using the Lorentz velocity transformation equation and the classical definition of linear momentum, **p** = *m***u** (where **u** is the velocity of a particle), we find that linear momentum is *not* measured to be conserved by the observer in S′. However, because the laws of physics are the same in all inertial frames, linear momentum of the system must be conserved in all frames. We have a contradiction. In view of this contradiction and assuming that the Lorentz velocity transformation equation is correct, we must modify the definition of linear momentum to satisfy the following conditions:

- The linear momentum of an isolated system must be conserved in all collisions.
- The relativistic value calculated for the linear momentum **p** of a particle must approach the classical value *m***u** as **u** approaches zero.

For any particle, the correct relativistic equation for linear momentum that satisfies these conditions is

$$\mathbf{p} \equiv \frac{m\mathbf{u}}{\sqrt{1 - \dfrac{u^2}{c^2}}} = \gamma m\mathbf{u} \qquad (39.19)$$

where **u** is the velocity of the particle and m is the mass of the particle. When u is much less than c, $\gamma = (1 - u^2/c^2)^{-1/2}$ approaches unity and **p** approaches m**u**. Therefore,

▲ **PITFALL PREVENTION**

39.6   Watch Out for "Relativistic Mass"

Some older treatments of relativity maintained the conservation of momentum principle at high speeds by using a model in which the mass of a particle increases with speed. You might still encounter this notion of "relativistic mass" in your outside reading, especially in older books. Be aware that this notion is no longer widely accepted and mass is considered as *invariant*, independent of speed. The mass of an object in all frames is considered to be the mass as measured by an observer at rest with respect to the object.

Definition of relativistic linear momentum

The speed of light is the speed limit of the Universe. It is the maximum possible speed for energy transfer and for information transfer. Any object with mass must move at a lower speed.

the relativistic equation for $\mathbf{p}$ does indeed reduce to the classical expression when u is much smaller than c.

The relativistic force $\mathbf{F}$ acting on a particle whose linear momentum is $\mathbf{p}$ is defined as

$$\mathbf{F} \equiv \frac{d\mathbf{p}}{dt} \tag{39.20}$$

where $\mathbf{p}$ is given by Equation 39.19. This expression, which is the relativistic form of Newton's second law, is reasonable because it preserves classical mechanics in the limit of low velocities and is consistent with conservation of linear momentum for an isolated system ($\mathbf{F} = 0$) both relativistically and classically.

It is left as an end-of-chapter problem (Problem 69) to show that under relativistic conditions, the acceleration $\mathbf{a}$ of a particle decreases under the action of a constant force, in which case $a \propto (1 - u^2/c^2)^{3/2}$. From this proportionality, we see that as the particle's speed approaches c, the acceleration caused by any finite force approaches zero. Hence, it is impossible to accelerate a particle from rest to a speed $u \geq c$. This argument shows that the speed of light is the ultimate speed, as noted at the end of the preceding section.

Example 39.10   Linear Momentum of an Electron

An electron, which has a mass of 9.11×10^{-31} kg, moves with a speed of $0.750c$. Find its relativistic momentum and compare this value with the momentum calculated from the classical expression.

Solution Using Equation 39.19 with $u = 0.750c$, we have

$$p = \frac{m_e u}{\sqrt{1 - \dfrac{u^2}{c^2}}}$$

$$p = \frac{(9.11 \times 10^{-31}\ \text{kg})(0.750)(3.00 \times 10^8\ \text{m/s})}{\sqrt{1 - \dfrac{(0.750c)^2}{c^2}}}$$

$$p = \boxed{3.10 \times 10^{-22}\ \text{kg} \cdot \text{m/s}}$$

The classical expression (used incorrectly here) gives

$$p_{\text{classical}} = m_e u = 2.05 \times 10^{-22}\ \text{kg} \cdot \text{m/s}$$

Hence, the correct relativistic result is 50% greater than the classical result!

39.8 Relativistic Energy

We have seen that the definition of linear momentum requires generalization to make it compatible with Einstein's postulates. This implies that most likely the definition of kinetic energy must also be modified.

To derive the relativistic form of the work–kinetic energy theorem, let us imagine a particle moving in one dimension along the x axis. A force in the x direction causes the momentum of the particle to change according to Equation 39.20. The work done by the force F on the particle is

$$W = \int_{x_1}^{x_2} F\,dx = \int_{x_1}^{x_2} \frac{dp}{dt}\,dx \tag{39.21}$$

In order to perform this integration and find the work done on the particle and the relativistic kinetic energy as a function of u, we first evaluate dp/dt:

$$\frac{dp}{dt} = \frac{d}{dt}\frac{mu}{\sqrt{1 - \dfrac{u^2}{c^2}}} = \frac{m(du/dt)}{\left(1 - \dfrac{u^2}{c^2}\right)^{3/2}}$$

Substituting this expression for dp/dt and $dx = u\,dt$ into Equation 39.21 gives

$$W = \int_0^t \frac{m(du/dt)\,u\,dt}{\left(1 - \dfrac{u^2}{c^2}\right)^{3/2}} = m \int_0^u \frac{u}{\left(1 - \dfrac{u^2}{c^2}\right)^{3/2}}\,du$$

where we use the limits 0 and u in the integral because the integration variable has been changed from t to u. We assume that the particle is accelerated from rest to some final speed u. Evaluating the integral, we find that

$$W = \frac{mc^2}{\sqrt{1 - \dfrac{u^2}{c^2}}} - mc^2 \tag{39.22}$$

Recall from Chapter 7 that the work done by a force acting on a system consisting of a single particle equals the change in kinetic energy of the particle. Because we assumed that the initial speed of the particle is zero, we know that its initial kinetic energy is zero. We therefore conclude that the work W in Equation 39.22 is equivalent to the relativistic kinetic energy K:

$$K = \frac{mc^2}{\sqrt{1 - \dfrac{u^2}{c^2}}} - mc^2 = \gamma mc^2 - mc^2 = (\gamma - 1)\,mc^2 \tag{39.23}$$

Relativistic kinetic energy

This equation is routinely confirmed by experiments using high-energy particle accelerators.

At low speeds, where $u/c \ll 1$, Equation 39.23 should reduce to the classical expression $K = \frac{1}{2}mu^2$. We can check this by using the binomial expansion $(1 - \beta^2)^{-1/2} \approx 1 + \frac{1}{2}\beta^2 + \cdots$ for $\beta \ll 1$, where the higher-order powers of β are neglected in the expansion. (In treatments of relativity, β is a common symbol used to represent u/c or v/c.) In our case, $\beta = u/c$, so that

$$\gamma = \frac{1}{\sqrt{1 - \dfrac{u^2}{c^2}}} = \left(1 - \frac{u^2}{c^2}\right)^{-1/2} \approx 1 + \frac{1}{2}\frac{u^2}{c^2}$$

Substituting this into Equation 39.23 gives

$$K \approx \left[\left(1 + \frac{1}{2}\frac{u^2}{c^2}\right) - 1\right] mc^2 = \frac{1}{2}mu^2 \qquad \text{(for } u/c \ll 1\text{)}$$

which is the classical expression for kinetic energy. A graph comparing the relativistic and nonrelativistic expressions is given in Figure 39.18. In the relativistic case, the particle speed never exceeds c, regardless of the kinetic energy. The two curves are in good agreement when $u \ll c$.

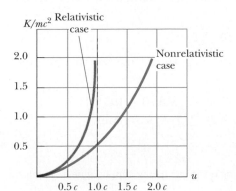

Figure 39.18 A graph comparing relativistic and nonrelativistic kinetic energy of a moving particle. The energies are plotted as a function of particle speed u. In the relativistic case, u is always less than c.

The constant term mc^2 in Equation 39.23, which is independent of the speed of the particle, is called the **rest energy** E_R of the particle:

Rest energy

$$E_R = mc^2 \qquad (39.24)$$

The term γmc^2, which does depend on the particle speed, is therefore the sum of the kinetic and rest energies. We define γmc^2 to be the **total energy** E:

$$\text{Total energy} = \text{kinetic energy} + \text{rest energy}$$

$$E = K + mc^2 \qquad (39.25)$$

or

Total energy of a relativistic particle

$$E = \frac{mc^2}{\sqrt{1 - \dfrac{u^2}{c^2}}} = \gamma mc^2 \qquad (39.26)$$

The relationship $E = K + mc^2$ shows that **mass is a form of energy,** where c^2 in the rest energy term is just a constant conversion factor. This expression also shows that a small mass corresponds to an enormous amount of energy, a concept fundamental to nuclear and elementary-particle physics.

In many situations, the linear momentum or energy of a particle is measured rather than its speed. It is therefore useful to have an expression relating the total energy E to the relativistic linear momentum p. This is accomplished by using the expressions $E = \gamma mc^2$ and $p = \gamma mu$. By squaring these equations and subtracting, we can eliminate u (Problem 43). The result, after some algebra, is[6]

Energy–momentum relationship for a relativistic particle

$$E^2 = p^2c^2 + (mc^2)^2 \qquad (39.27)$$

When the particle is at rest, $p = 0$ and so $E = E_R = mc^2$.

In Section 35.1, we introduced the concept of a particle of light, called a **photon.** For particles that have zero mass, such as photons, we set $m = 0$ in Equation 39.27 and find that

$$E = pc \qquad (39.28)$$

This equation is an exact expression relating total energy and linear momentum for photons, which always travel at the speed of light (in vacuum).

Finally, note that because the mass m of a particle is independent of its motion, m must have the same value in all reference frames. For this reason, m is often called the **invariant mass.** On the other hand, because the total energy and linear momentum of a particle both depend on velocity, these quantities depend on the reference frame in which they are measured.

When we are dealing with subatomic particles, it is convenient to express their energy in electron volts (Section 25.1) because the particles are usually given this energy by acceleration through a potential difference. The conversion factor, as you recall from Equation 25.5, is

$$1 \text{ eV} = 1.60 \times 10^{-19} \text{ J}$$

For example, the mass of an electron is 9.11×10^{-31} kg. Hence, the rest energy of the electron is

$$m_e c^2 = (9.11 \times 10^{-31} \text{ kg})(3.00 \times 10^8 \text{ m/s})^2 = 8.20 \times 10^{-14} \text{ J}$$
$$= (8.20 \times 10^{-14} \text{ J})(1 \text{ eV}/1.60 \times 10^{-19} \text{ J}) = 0.511 \text{ MeV}$$

---

[6]   One way to remember this relationship is to draw a right triangle having a hypotenuse of length E and legs of lengths pc and mc^2.

Quick Quiz 39.10 The following *pairs* of energies represent the rest energy and total energy of three different particles: particle 1: E, $2E$; particle 2: E, $3E$; particle 3: $2E$, $4E$. Rank the particles, from greatest to least, according to their (a) mass; (b) kinetic energy; (c) speed.

Example 39.11   The Energy of a Speedy Electron

An electron in a television picture tube typically moves with a speed $u = 0.250c$. Find its total energy and kinetic energy in electron volts.

Solution Using the fact that the rest energy of the electron is 0.511 MeV together with Equation 39.26, we have

$$E = \frac{m_e c^2}{\sqrt{1 - \dfrac{u^2}{c^2}}} = \frac{0.511 \text{ MeV}}{\sqrt{1 - \dfrac{(0.250c)^2}{c^2}}}$$

$$= 1.03(0.511 \text{ MeV}) = \boxed{0.528 \text{ MeV}}$$

This is 3% greater than the rest energy.

We obtain the kinetic energy by subtracting the rest energy from the total energy:

$$K = E - m_e c^2 = 0.528 \text{ MeV} - 0.511 \text{ MeV}$$

$$= \boxed{0.017 \text{ MeV}}$$

Example 39.12   The Energy of a Speedy Proton

(A) Find the rest energy of a proton in electron volts.

Solution Using Equation 39.24,

$$E_R = m_p c^2 = (1.67 \times 10^{-27} \text{ kg})(3.00 \times 10^8 \text{ m/s})^2$$

$$= (1.50 \times 10^{-10} \text{ J})\left(\frac{1.00 \text{ eV}}{1.60 \times 10^{-19} \text{ J}}\right)$$

$$= \boxed{938 \text{ MeV}}$$

(B) If the total energy of a proton is three times its rest energy, what is the speed of the proton?

Solution Equation 39.26 gives

$$E = 3m_p c^2 = \frac{m_p c^2}{\sqrt{1 - \dfrac{u^2}{c^2}}}$$

$$3 = \frac{1}{\sqrt{1 - \dfrac{u^2}{c^2}}}$$

Solving for u gives

$$\left(1 - \frac{u^2}{c^2}\right) = \frac{1}{9}$$

$$\frac{u^2}{c^2} = \frac{8}{9}$$

$$u = \frac{\sqrt{8}}{3}c = 0.943c = \boxed{2.83 \times 10^8 \text{ m/s}}$$

(C) Determine the kinetic energy of the proton in electron volts.

Solution From Equation 39.25,

$$K = E - m_p c^2 = 3m_p c^2 - m_p c^2 = 2m_p c^2$$

Because $m_p c^2 = 938$ MeV, we see that $K = \boxed{1\,880 \text{ MeV}}$.

(D) What is the proton's momentum?

Solution We can use Equation 39.27 to calculate the momentum with $E = 3m_p c^2$:

$$E^2 = p^2 c^2 + (m_p c^2)^2 = (3m_p c^2)^2$$

$$p^2 c^2 = 9(m_p c^2)^2 - (m_p c^2)^2 = 8(m_p c^2)^2$$

$$p = \sqrt{8}\,\frac{m_p c^2}{c} = \sqrt{8}\,\frac{(938 \text{ MeV})}{c} = \boxed{2\,650 \text{ MeV}/c}$$

The unit of momentum is written MeV/c, which is a common unit in particle physics.

What If? In classical physics, if the momentum of a particle doubles, the kinetic energy increases by a factor of 4. What happens to the kinetic energy of the speedy proton in this example if its momentum doubles?

Answer Based on what we have seen so far in relativity, it is likely that you would predict that its kinetic energy does not increase by a factor of 4. If the momentum doubles, the new momentum is

$$p_{\text{new}} = 2\left(\sqrt{8}\,\frac{m_p c^2}{c}\right) = 4\sqrt{2}\,\frac{m_p c^2}{c}$$

Using Equation 39.27, we find the square of the new total energy:

$$E_{\text{new}}^2 = p_{\text{new}}^2 c^2 + (m_p c^2)^2$$

$$E^2_{\text{new}} = \left(4\sqrt{2}\ \frac{m_p c^2}{c}\right)^2 c^2 + (m_p c^2)^2 = 33(m_p c^2)^2$$

$$E_{\text{new}} = \sqrt{33}(m_p c^2) = 5.7 m_p c^2$$

Now, using Equation 39.25, we find the new kinetic energy:

$$K_{\text{new}} = E_{\text{new}} - m_p c^2 = 5.7 m_p c^2 - m_p c^2 = 4.7 m_p c^2$$

Notice that this is only 2.35 times as large as the kinetic energy we found in part (C), not four times as large. In general, the factor by which the kinetic energy increases if the momentum doubles will depend on the initial momentum, but will approach 4 as the momentum approaches zero. In this latter situation, classical physics correctly describes the situation.

39.9   Mass and Energy

Equation 39.26, $E = \gamma m c^2$, which represents the total energy of a particle, suggests that even when a particle is at rest ($\gamma = 1$) it still possesses enormous energy through its mass. The clearest experimental proof of the equivalence of mass and energy occurs in nuclear and elementary particle interactions in which the conversion of mass into kinetic energy takes place. Because of this, in relativistic situations, we cannot use the principle of conservation of energy as it was outlined in Chapters 7 and 8. We must include rest energy as another form of energy storage.

This concept is important in atomic and nuclear processes, in which the change in mass is a relatively large fraction of the initial mass. For example, in a conventional nuclear reactor, the uranium nucleus undergoes *fission*, a reaction that results in several lighter fragments having considerable kinetic energy. In the case of ^{235}U, which is used as fuel in nuclear power plants, the fragments are two lighter nuclei and a few neutrons. The total mass of the fragments is less than that of the ^{235}U by an amount Δm. The corresponding energy Δmc^2 associated with this mass difference is exactly equal to the total kinetic energy of the fragments. The kinetic energy is absorbed as the fragments move through water, raising the internal energy of the water. This internal energy is used to produce steam for the generation of electrical power.

Next, consider a basic *fusion* reaction in which two deuterium atoms combine to form one helium atom. The decrease in mass that results from the creation of one helium atom from two deuterium atoms is $\Delta m = 4.25 \times 10^{-29}$ kg. Hence, the corresponding energy that results from one fusion reaction is $\Delta mc^2 = 3.83 \times 10^{-12}$ J = 23.9 MeV. To appreciate the magnitude of this result, if only 1 g of deuterium is converted to helium, the energy released is on the order of 10^{12} J! At the year 2003 cost of electrical energy, this would be worth about $30 000. We shall present more details of these nuclear processes in Chapter 45 of the extended version of this textbook.

Example 39.13   Mass Change in a Radioactive Decay

The ^{216}Po nucleus is unstable and exhibits radioactivity (Chapter 44). It decays to ^{212}Pb by emitting an alpha particle, which is a helium nucleus, ^{4}He. Find

(A) the mass change in this decay and

(B) the energy that this represents.

Solution Using values in Table A.3, we see that the initial and final masses are

$$m_i = m(^{216}\text{Po}) = 216.001\,905\text{ u}$$

$$m_f = m(^{212}\text{Pb}) + m(^4\text{He}) = 211.991\,888\text{ u} + 4.002\,603\text{ u}$$

$$= 215.994\,491\text{ u}$$

Thus, the mass change is

$$\Delta m = 216.001\,905\text{ u} - 215.994\,491\text{ u} = 0.007\,414\text{ u}$$

$$= \boxed{1.23 \times 10^{-29}\text{ kg}}$$

(B) The energy associated with this mass change is

$$E = \Delta mc^2 = (1.23 \times 10^{-29}\text{ kg})(3.00 \times 10^8\text{ m/s})^2$$

$$= 1.11 \times 10^{-12}\text{ J} = \boxed{6.92\text{ MeV}}$$

This energy appears as the kinetic energy of the alpha particle and the ^{212}Pb nucleus after the decay.

39.10   The General Theory of Relativity

Up to this point, we have sidestepped a curious puzzle. Mass has two seemingly different properties: a *gravitational attraction* for other masses and an *inertial* property that represents a resistance to acceleration. To designate these two attributes, we use the subscripts g and i and write

$$\text{Gravitational property} \qquad F_g = m_g g$$

$$\text{Inertial property} \qquad \sum F = m_i a$$

The value for the gravitational constant G was chosen to make the magnitudes of m_g and m_i numerically equal. Regardless of how G is chosen, however, the strict proportionality of m_g and m_i has been established experimentally to an extremely high degree: a few parts in 10^{12}. Thus, it appears that gravitational mass and inertial mass may indeed be exactly proportional.

But why? They seem to involve two entirely different concepts: a force of mutual gravitational attraction between two masses, and the resistance of a single mass to being accelerated. This question, which puzzled Newton and many other physicists over the years, was answered by Einstein in 1916 when he published his theory of gravitation, known as the *general theory of relativity*. Because it is a mathematically complex theory, we offer merely a hint of its elegance and insight.

In Einstein's view, the dual behavior of mass was evidence for a very intimate and basic connection between the two behaviors. He pointed out that no mechanical experiment (such as dropping an object) could distinguish between the two situations illustrated in Figures 39.19a and 39.19b. In Figure 39.19a, a person is standing in an elevator on the surface of a planet, and feels pressed into the floor, due to the gravitational force. In Figure 39.19b, the person is in an elevator in empty space accelerating upward with $a = g$. The person feels pressed into the floor with the same force as in Figure 39.19a. In each case, an object released by the observer undergoes a downward acceleration of magnitude g relative to the floor. In Figure 39.19a, the person is in an inertial frame in a gravitational field. In Figure 39.19b, the person is in a noninertial frame accelerating in gravity-free space. Einstein's claim is that these two situations are completely equivalent.

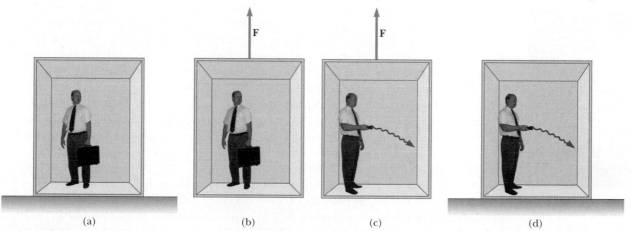

(a)   (b)   (c)   (d)

Figure 39.19 (a) The observer is at rest in a uniform gravitational field **g**, directed downward. (b) The observer is in a region where gravity is negligible, but the frame is accelerated by an external force **F** that produces an acceleration **g** directed upward. According to Einstein, the frames of reference in parts (a) and (b) are equivalent in every way. No local experiment can distinguish any difference between the two frames. (c) In the accelerating frame, a ray of light would appear to bend downward due to the acceleration of the elevator. (d) If parts (a) and (b) are truly equivalent, as Einstein proposed, then part (c) suggests that a ray of light would bend downward in a gravitational field.

Einstein carried this idea further and proposed that *no* experiment, mechanical or otherwise, could distinguish between the two cases. This extension to include all phenomena (not just mechanical ones) has interesting consequences. For example, suppose that a light pulse is sent horizontally across an elevator that is accelerating upward in empty space, as in Figure 39.19c. From the point of view of an observer in an inertial frame outside of the elevator, the light travels in a straight line while the floor of the elevator accelerates upward. According to the observer on the elevator, however, the trajectory of the light pulse bends downward as the floor of the elevator (and the observer) accelerates upward. Therefore, based on the equality of parts (a) and (b) of the figure for all phenomena, Einstein proposed that **a beam of light should also be bent downward by a gravitational field,** as in Figure 39.19d. Experiments have verified the effect, although the bending is small. A laser aimed at the horizon falls less than 1 cm after traveling 6 000 km. (No such bending is predicted in Newton's theory of gravitation.)

The two postulates of Einstein's **general theory of relativity** are

Postulates of the general theory of relativity

- All the laws of nature have the same form for observers in any frame of reference, whether accelerated or not.

- The results of experiments based on the laws of physics, as measured by an observer at rest in an inertial reference frame in a uniform gravitational field, are equivalent to those measured by an observer at rest in a uniformly accelerated reference frame in gravity-free space.

One interesting effect predicted by the general theory is that time is altered by gravity. A clock in the presence of gravity runs slower than one located where gravity is negligible. Consequently, the frequencies of radiation emitted by atoms in the presence of a strong gravitational field are *red-shifted* to lower frequencies when compared with the same emissions in the presence of a weak field. This gravitational red shift has been detected in spectral lines emitted by atoms in massive stars. It has also been verified on the Earth by comparing the frequencies of gamma rays emitted from nuclei separated vertically by about 20 m.

The second postulate suggests that a gravitational field may be "transformed away" at any point if we choose an appropriate accelerated frame of reference—a freely falling one. Einstein developed an ingenious method of describing the acceleration necessary to make the gravitational field "disappear." He specified a concept, the *curvature of space–time*, that describes the gravitational effect at every point. In fact, the curvature of space–time completely replaces Newton's gravitational theory. According to Einstein, there is no such thing as a gravitational force. Rather, the presence of a mass causes a curvature of space–time in the vicinity of the mass, and this curvature dictates the space–time path that all freely moving objects must follow. In 1979, John Wheeler summarized Einstein's general theory of relativity in a single sentence: "Space tells matter how to move and matter tells space how to curve."

As an example of the effects of curved space–time, imagine two travelers moving on parallel paths a few meters apart on the surface of the Earth and maintaining an exact northward heading along two longitude lines. As they observe each other near the equator, they will claim that their paths are exactly parallel. As they approach the North Pole, however, they notice that they are moving closer together, and they will actually meet at the North Pole. Thus, they will claim that they moved along parallel paths, but moved toward each other, *as if there were an attractive force between them.* They will make this conclusion based on their everyday experience of moving on flat surfaces. From our mental representation, however, we realize that they are walking on a curved surface, and it is the geometry of the curved surface that causes them to converge, rather than an attractive force. In a similar way, general relativity replaces the notion of forces with the movement of objects through curved space–time.

One prediction of the general theory of relativity is that a light ray passing near the Sun should be deflected in the curved space–time created by the Sun's mass. This prediction was confirmed when astronomers detected the bending of starlight near the

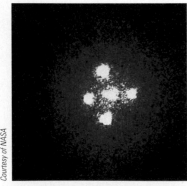

Einstein's cross. The four bright spots are images of the same galaxy that have been bent around a massive object located between the galaxy and the Earth. The massive object acts like a lens, causing the rays of light that were diverging from the distant galaxy to converge on the Earth. (If the intervening massive object had a uniform mass distribution, we would see a bright ring instead of four spots.)

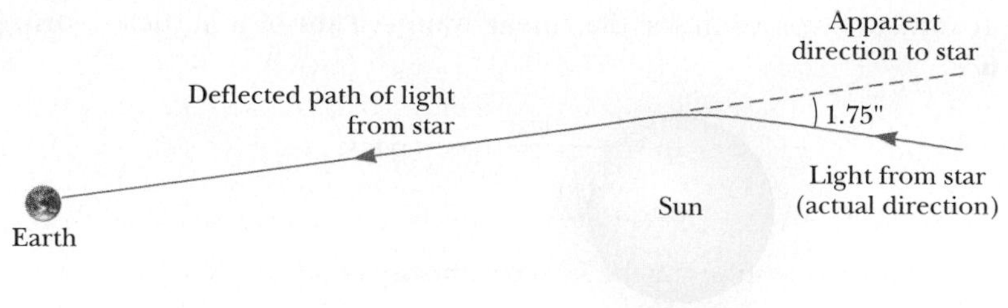

Figure 39.20 Deflection of starlight passing near the Sun. Because of this effect, the Sun or some other remote object can act as a *gravitational lens*. In his general theory of relativity, Einstein calculated that starlight just grazing the Sun's surface should be deflected by an angle of 1.75 s of arc.

Sun during a total solar eclipse that occurred shortly after World War I (Fig. 39.20). When this discovery was announced, Einstein became an international celebrity.

If the concentration of mass becomes very great, as is believed to occur when a large star exhausts its nuclear fuel and collapses to a very small volume, a **black hole** may form. Here, the curvature of space–time is so extreme that, within a certain distance from the center of the black hole, all matter and light become trapped, as discussed in Section 13.7.

SUMMARY

The two basic postulates of the special theory of relativity are

- The laws of physics must be the same in all inertial reference frames.

- The speed of light in vacuum has the same value, $c = 3.00 \times 10^8$ m/s, in all inertial frames, regardless of the velocity of the observer or the velocity of the source emitting the light.

Take a practice test for this chapter by clicking on the Practice Test link at http://www.pse6.com.

Three consequences of the special theory of relativity are

- Events that are measured to be simultaneous for one observer are not necessarily measured to be simultaneous for another observer who is in motion relative to the first.

- Clocks in motion relative to an observer are measured to run slower by a factor $\gamma = (1 - v^2/c^2)^{-1/2}$. This phenomenon is known as **time dilation.**

- The length of objects in motion are measured to be contracted in the direction of motion by a factor $1/\gamma = (1 - v^2/c^2)^{1/2}$. This phenomenon is known as **length contraction.**

To satisfy the postulates of special relativity, the Galilean transformation equations must be replaced by the **Lorentz transformation equations:**

$$x' = \gamma(x - vt) \qquad y' = y \qquad z' = z \qquad t' = \gamma\left(t - \frac{v}{c^2}\,x\right) \qquad (39.11)$$

where $\gamma = (1 - v^2/c^2)^{-1/2}$ and the S′ frame moves in the x direction relative to the S frame.

The relativistic form of the **velocity transformation equation** is

$$u_x' = \frac{u_x - v}{1 - \dfrac{u_x v}{c^2}} \qquad (39.16)$$

where u_x is the speed of an object as measured in the S frame and u_x' is its speed measured in the S′ frame.

The relativistic expression for the **linear momentum** of a particle moving with a velocity **u** is

$$\mathbf{p} \equiv \frac{m\mathbf{u}}{\sqrt{1 - \dfrac{u^2}{c^2}}} = \gamma m\mathbf{u} \tag{39.19}$$

The relativistic expression for the **kinetic energy** of a particle is

$$K = \frac{mc^2}{\sqrt{1 - \dfrac{u^2}{c^2}}} - mc^2 = (\gamma - 1)mc^2 \tag{39.23}$$

The constant term mc^2 in Equation 39.23 is called the **rest energy** E_R of the particle:

$$E_R = mc^2 \tag{39.24}$$

The total energy E of a particle is given by

$$E = \frac{mc^2}{\sqrt{1 - \dfrac{u^2}{c^2}}} = \gamma mc^2 \tag{39.26}$$

The relativistic linear momentum of a particle is related to its total energy through the equation

$$E^2 = p^2c^2 + (mc^2)^2 \tag{39.27}$$

QUESTIONS

1. What two speed measurements do two observers in relative motion always agree on?

2. A spacecraft with the shape of a sphere moves past an observer on Earth with a speed $0.5c$. What shape does the observer measure for the spacecraft as it moves past?

3. The speed of light in water is 230 Mm/s. Suppose an electron is moving through water at 250 Mm/s. Does this violate the principle of relativity?

4. Two identical clocks are synchronized. One is then put in orbit directed eastward around the Earth while the other remains on the Earth. Which clock runs slower? When the moving clock returns to the Earth, are the two still synchronized?

5. Explain why it is necessary, when defining the length of a rod, to specify that the positions of the ends of the rod are to be measured simultaneously.

6. A train is approaching you at very high speed as you stand next to the tracks. Just as an observer on the train passes you, you both begin to play the same Beethoven symphony on portable compact disc players. (a) According to you, whose CD player finishes the symphony first? (b) **What If?** According to the observer on the train, whose CD player finishes the symphony first? (c) Whose CD player really finishes the symphony first?

7. List some ways our day-to-day lives would change if the speed of light were only 50 m/s.

8. Does saying that a moving clock runs slower than a stationary one imply that something is physically unusual about the moving clock?

9. How is acceleration indicated on a space–time graph?

10. A particle is moving at a speed less than $c/2$. If the speed of the particle is doubled, what happens to its momentum?

11. Give a physical argument that shows that it is impossible to accelerate an object of mass m to the speed of light, even with a continuous force acting on it.

12. The upper limit of the speed of an electron is the speed of light c. Does that mean that the momentum of the electron has an upper limit?

13. Because mass is a measure of energy, can we conclude that the mass of a compressed spring is greater than the mass of the same spring when it is not compressed?

14. It is said that Einstein, in his teenage years, asked the question, "What would I see in a mirror if I carried it in my hands and ran at the speed of light?" How would you answer this question?

15. Some distant astronomical objects, called quasars, are receding from us at half the speed of light (or greater). What is the speed of the light we receive from these quasars?

16. Photons of light have zero mass. How is it possible that they have momentum?

17. "Newtonian mechanics correctly describes objects moving at ordinary speeds and relativistic mechanics correctly describes objects moving very fast." "Relativistic mechanics must make a smooth transition as it reduces to Newtonian mechanics in a case where the speed of an object becomes small compared to the speed of light." Argue for or against each of these two statements.

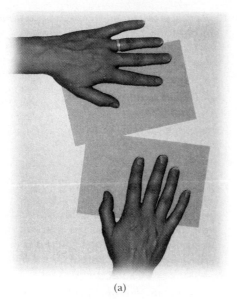

(a)

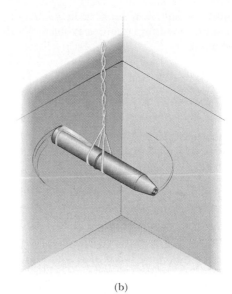

(b)

Figure Q39.18

18. Two cards have straight edges. Suppose that the top edge of one card crosses the bottom edge of another card at a small angle, as in Figure Q39.18a. A person slides the cards together at a moderately high speed. In what direction does the intersection point of the edges move? Show that it can move at a speed greater than the speed of light.

A small flashlight is suspended in a horizontal plane and set into rapid rotation. Show that the spot of light it produces on a distant screen can move across the screen at a speed greater than the speed of light. (If you use a laser pointer, as in Figure Q39.18b, make sure the direct laser light cannot enter a person's eyes.) Argue that these experiments do not invalidate the principle that no material, no energy, and no information can move faster than light moves in a vacuum.

19. Describe how the results of Example 39.7 would change if, instead of fast space vehicles, two ordinary cars were approaching each other at highway speeds.

20. Two objects are identical except that one is hotter than the other. Compare how they respond to identical forces.

21. With regard to reference frames, how does general relativity differ from special relativity?

22. Two identical clocks are in the same house, one upstairs in a bedroom, and the other downstairs in the kitchen. Which clock runs more slowly? Explain.

23. *A thought experiment.* Imagine ants living on a merry-go-round turning at relativistic speed, which is their two-dimensional world. From measurements on small circles they are thoroughly familiar with the number π. When they measure the circumference of their world, and divide it by the diameter, they expect to calculate the number $\pi = 3.141\ 59.\ .\ .\ .$ We see the merry-go-round turning at relativistic speed. From our point of view, the ants' measuring rods on the circumference are experiencing length contraction in the tangential direction; hence the ants will need some extra rods to fill that entire distance. The rods measuring the diameter, however, do not contract, because their motion is perpendicular to their lengths. As a result, the computed ratio does not agree with the number π. If you were an ant, you would say that the rest of the universe is spinning in circles, and your disk is stationary. What possible explanation can you then give for the discrepancy, in light of the general theory of relativity?

PROBLEMS

1, 2, 3 = straightforward, intermediate, challenging ☐ = full solution available in the *Student Solutions Manual and Study Guide*

🌐 = coached solution with hints available at http://www.pse6.com 🖥 = computer useful in solving problem

▨ = paired numerical and symbolic problems

Section 39.1 The Principle of Galilean Relativity

1. A 2 000-kg car moving at 20.0 m/s collides and locks together with a 1 500-kg car at rest at a stop sign. Show that momentum is conserved in a reference frame moving at 10.0 m/s in the direction of the moving car.

2. A ball is thrown at 20.0 m/s inside a boxcar moving along the tracks at 40.0 m/s. What is the speed of the ball relative to the ground if the ball is thrown (a) forward (b) backward (c) out the side door?

3. In a laboratory frame of reference, an observer notes that Newton's second law is valid. Show that it is also valid for an observer moving at a constant speed, small compared with the speed of light, relative to the laboratory frame.

4. Show that Newton's second law is *not* valid in a reference frame moving past the laboratory frame of Problem 3 with a constant acceleration.

Section 39.2 The Michelson–Morley Experiment
Section 39.3 Einstein's Principle of Relativity
Section 39.4 Consequences of the Special Theory of Relativity

Problem 43 in Chapter 4 can be assigned with this section.

5. How fast must a meter stick be moving if its length is measured to shrink to 0.500 m?

6. At what speed does a clock move if it is measured to run at a rate that is half the rate of a clock at rest with respect to an observer?

7. An astronaut is traveling in a space vehicle that has a speed of $0.500c$ relative to the Earth. The astronaut measures her pulse rate at 75.0 beats per minute. Signals generated by the astronaut's pulse are radioed to Earth when the vehicle is moving in a direction perpendicular to the line that connects the vehicle with an observer on the Earth. (a) What pulse rate does the Earth observer measure? (b) **What If?** What would be the pulse rate if the speed of the space vehicle were increased to $0.990c$?

8. An astronomer on Earth observes a meteoroid in the southern sky approaching the Earth at a speed of $0.800c$. At the time of its discovery the meteoroid is 20.0 ly from the Earth. Calculate (a) the time interval required for the meteoroid to reach the Earth as measured by the Earth-bound astronomer, (b) this time interval as measured by a tourist on the meteoroid, and (c) the distance to the Earth as measured by the tourist.

9. An atomic clock moves at 1 000 km/h for 1.00 h as measured by an identical clock on the Earth. How many nanoseconds slow will the moving clock be compared with the Earth clock, at the end of the 1.00-h interval?

10. A muon formed high in the Earth's atmosphere travels at speed $v = 0.990c$ for a distance of 4.60 km before it decays into an electron, a neutrino, and an antineutrino $(\mu^- \rightarrow e^- + \nu + \overline{\nu})$. (a) How long does the muon live, as measured in its reference frame? (b) How far does the Earth travel, as measured in the frame of the muon?

11. A spacecraft with a proper length of 300 m takes 0.750 μs to pass an Earth observer. Determine the speed of the spacecraft as measured by the Earth observer.

12. (a) An object of proper length L_p takes a time interval Δt to pass an Earth observer. Determine the speed of the object as measured by the Earth observer. (b) A column of tanks, 300 m long, takes 75.0 s to pass a child waiting at a street corner on her way to school. Determine the speed of the armored vehicles. (c) Show that the answer to part (a) includes the answer to Problem 11 as a special case, and includes the answer to part (b) as another special case.

13. **Review problem.** In 1963 Mercury astronaut Gordon Cooper orbited the Earth 22 times. The press stated that for each orbit he aged 2 millionths of a second less than he would have if he had remained on the Earth. (a) Assuming that he was 160 km above the Earth in a circular orbit, determine the time difference between someone on the Earth and the orbiting astronaut for the 22 orbits. You will need to use the approximation $\sqrt{1 - x} \approx 1 - x/2$, for small x. (b) Did the press report accurate information? Explain.

14. For what value of v does $\gamma = 1.010\,0$? Observe that for speeds lower than this value, time dilation and length contraction are effects amounting to less than 1%.

15. A friend passes by you in a spacecraft traveling at a high speed. He tells you that his craft is 20.0 m long and that the identically constructed craft you are sitting in is 19.0 m long. According to your observations, (a) how long is your spacecraft, (b) how long is your friend's craft, and (c) what is the speed of your friend's craft?

16. The identical twins Speedo and Goslo join a migration from the Earth to Planet X. It is 20.0 ly away in a reference frame in which both planets are at rest. The twins, of the same age, depart at the same time on different spacecraft. Speedo's craft travels steadily at $0.950c$, and Goslo's at $0.750c$. Calculate the age difference between the twins after Goslo's spacecraft lands on Planet X. Which twin is the older?

17. An interstellar space probe is launched from the Earth. After a brief period of acceleration it moves with a constant velocity, with a magnitude of 70.0% of the speed of light. Its nuclear-powered batteries supply the energy to keep its data transmitter active continuously. The batteries have a lifetime of 15.0 yr as measured in a rest frame. (a) How long do the batteries on the space probe last as measured by Mission Control on the Earth? (b) How far is the probe from the Earth when its batteries fail, as measured by Mission Control? (c) How far is the probe from the Earth when its batteries fail, as measured by its built-in trip odometer? (d) For what total time interval after launch are data received from the probe by Mission Control? Note that radio waves travel at the speed of light and fill the space between the probe and the Earth at the time of battery failure.

18. **Review problem.** An alien civilization occupies a brown dwarf, nearly stationary relative to the Sun, several lightyears away. The extraterrestrials have come to love original broadcasts of *I Love Lucy*, on our television channel 2, at carrier frequency 57.0 MHz. Their line of sight to us is in the plane of the Earth's orbit. Find the difference between the highest and lowest frequencies they receive due to the Earth's orbital motion around the Sun.

19. Police radar detects the speed of a car (Fig. P39.19) as follows. Microwaves of a precisely known frequency are broadcast toward the car. The moving car reflects the microwaves with a Doppler shift. The reflected waves are received and combined with an attenuated version of the transmitted wave. Beats occur between the two microwave signals. The beat frequency is measured. (a) For an electromagnetic wave reflected back to its source from a mirror approaching at speed v, show that the reflected

wave has frequency

$$f = f_{source} \frac{c + v}{c - v}$$

where f_{source} is the source frequency. (b) When v is much less than c, the beat frequency is much smaller than the transmitted frequency. In this case use the approximation $f + f_{source} \approx 2 f_{source}$ and show that the beat frequency can be written as $f_{beat} = 2v/\lambda$. (c) What beat frequency is measured for a car speed of 30.0 m/s if the microwaves have frequency 10.0 GHz? (d) If the beat frequency measurement is accurate to ± 5 Hz, how accurate is the velocity measurement?

Figure P39.19

20. *The red shift.* A light source recedes from an observer with a speed v_{source} that is small compared with c. (a) Show that the fractional shift in the measured wavelength is given by the approximate expression

$$\frac{\Delta\lambda}{\lambda} \approx \frac{v_{source}}{c}$$

This phenomenon is known as the red shift, because the visible light is shifted toward the red. (b) Spectroscopic measurements of light at $\lambda = 397$ nm coming from a galaxy in Ursa Major reveal a red shift of 20.0 nm. What is the recessional speed of the galaxy?

21. A physicist drives through a stop light. When he is pulled over, he tells the police officer that the Doppler shift made the red light of wavelength 650 nm appear green to him, with a wavelength of 520 nm. The police officer writes out a traffic citation for speeding. How fast was the physicist traveling, according to his own testimony?

Section 39.5 The Lorentz Transformation Equations

22. Suzanne observes two light pulses to be emitted from the same location, but separated in time by 3.00 μs. Mark sees the emission of the same two pulses separated in time by 9.00 μs. (a) How fast is Mark moving relative to Suzanne? (b) According to Mark, what is the separation in space of the two pulses?

23. A moving rod is observed to have a length of 2.00 m and to be oriented at an angle of 30.0° with respect to the direction of motion, as shown in Figure P39.23. The rod has a speed of 0.995c. (a) What is the proper length of the rod? (b) What is the orientation angle in the proper frame?

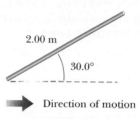

2.00 m

30.0°

Direction of motion

Figure P39.23

24. An observer in reference frame S sees two events as simultaneous. Event A occurs at the point (50.0 m, 0, 0) at the instant 9:00:00 Universal time, 15 January 2004. Event B occurs at the point (150 m, 0, 0) at the same moment. A second observer, moving past with a velocity of $0.800c\hat{\mathbf{i}}$, also observes the two events. In her reference frame S', which event occurred first and what time interval elapsed between the events?

25. A red light flashes at position $x_R = 3.00$ m and time $t_R = 1.00 \times 10^{-9}$ s, and a blue light flashes at $x_B = 5.00$ m and $t_B = 9.00 \times 10^{-9}$ s, all measured in the S reference frame. Reference frame S' has its origin at the same point as S at $t = t' = 0$; frame S' moves uniformly to the right. Both flashes are observed to occur at the same place in S'. (a) Find the relative speed between S and S'. (b) Find the location of the two flashes in frame S'. (c) At what time does the red flash occur in the S' frame?

Section 39.6 The Lorentz Velocity Transformation Equations

26. A Klingon spacecraft moves away from the Earth at a speed of 0.800c (Fig. P39.26). The starship *Enterprise* pursues at a speed of 0.900c relative to the Earth. Observers on the Earth see the *Enterprise* overtaking the Klingon craft at a relative speed of 0.100c. With what speed is the *Enterprise* overtaking the Klingon craft as seen by the crew of the *Enterprise*?

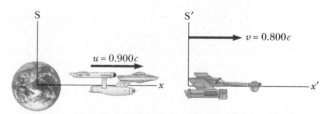

S

S'

$v = 0.800c$

$u = 0.900c$

x

x'

Figure P39.26

27. Two jets of material from the center of a radio galaxy are ejected in opposite directions. Both jets move at 0.750c

relative to the galaxy. Determine the speed of one jet relative to the other.

28. A spacecraft is launched from the surface of the Earth with a velocity of $0.600c$ at an angle of $50.0°$ above the horizontal positive x axis. Another spacecraft is moving past, with a velocity of $0.700c$ in the negative x direction. Determine the magnitude and direction of the velocity of the first spacecraft as measured by the pilot of the second spacecraft.

Section 39.7 Relativistic Linear Momentum and the Relativistic Form of Newton's Laws

29. Calculate the momentum of an electron moving with a speed of (a) $0.010\ 0c$, (b) $0.500c$, and (c) $0.900c$.

30. The nonrelativistic expression for the momentum of a particle, $p = mu$, agrees with experiment if $u \ll c$. For what speed does the use of this equation give an error in the momentum of (a) 1.00% and (b) 10.0%?

31. A golf ball travels with a speed of 90.0 m/s. By what fraction does its relativistic momentum magnitude p differ from its classical value mu? That is, find the ratio $(p - mu)/mu$.

32. Show that the speed of an object having momentum of magnitude p and mass m is

$$u = \frac{c}{\sqrt{1 + (mc/p)^2}}$$

33. An unstable particle at rest breaks into two fragments of unequal mass. The mass of the first fragment is 2.50×10^{-28} kg, and that of the other is 1.67×10^{-27} kg. If the lighter fragment has a speed of $0.893c$ after the breakup, what is the speed of the heavier fragment?

Section 39.8 Relativistic Energy

34. Determine the energy required to accelerate an electron from (a) $0.500c$ to $0.900c$ and (b) $0.900c$ to $0.990c$.

35. A proton in a high-energy accelerator moves with a speed of $c/2$. Use the work–kinetic energy theorem to find the work required to increase its speed to (a) $0.750c$ and (b) $0.995c$.

36. Show that, for any object moving at less than one-tenth the speed of light, the relativistic kinetic energy agrees with the result of the classical equation $K = \frac{1}{2} mu^2$ to within less than 1%. Thus for most purposes, the classical equation is good enough to describe these objects, whose motion we call *nonrelativistic*.

37. Find the momentum of a proton in MeV/c units assuming its total energy is twice its rest energy.

38. Find the kinetic energy of a 78.0-kg spacecraft launched out of the solar system with speed 106 km/s by using (a) the classical equation $K = \frac{1}{2} mu^2$. (b) **What If?** Calculate its kinetic energy using the relativistic equation.

39. A proton moves at $0.950c$. Calculate its (a) rest energy, (b) total energy, and (c) kinetic energy.

40. A cube of steel has a volume of 1.00 cm^3 and a mass of 8.00 g when at rest on the Earth. If this cube is now given a speed $u = 0.900c$, what is its density as measured by a

stationary observer? Note that relativistic density is defined as $E_R/c^2 V$.

41. An unstable particle with a mass of 3.34×10^{-27} kg is initially at rest. The particle decays into two fragments that fly off along the x axis with velocity components $0.987c$ and $-0.868c$. Find the masses of the fragments. (*Suggestion:* Conserve both energy and momentum.)

42. An object having mass 900 kg and traveling at speed $0.850c$ collides with a stationary object having mass $1\ 400$ kg. The two objects stick together. Find (a) the speed and (b) the mass of the composite object.

43. Show that the energy–momentum relationship $E^2 = p^2c^2 + (mc^2)^2$ follows from the expressions $E = \gamma mc^2$ and $p = \gamma mu$.

44. In a typical color television picture tube, the electrons are accelerated through a potential difference of $25\ 000$ V. (a) What speed do the electrons have when they strike the screen? (b) What is their kinetic energy in joules?

45. Consider electrons accelerated to an energy of 20.0 GeV in the 3.00-km-long Stanford Linear Accelerator. (a) What is the γ factor for the electrons? (b) What is their speed? (c) How long does the accelerator appear to them?

46. Compact high-power lasers can produce a 2.00-J light pulse of duration 100 fs, focused to a spot 1 μm in diameter. (See Mourou and Umstader, "Extreme Light," *Scientific American*, May 2002, page 81.) The electric field in the light accelerates electrons in the target material to near the speed of light. (a) What is the average power of the laser during the pulse? (b) How many electrons can be accelerated to $0.999\ 9c$ if $0.010\ 0\%$ of the pulse energy is converted into energy of electron motion?

47. A pion at rest $(m_\pi = 273m_e)$ decays to a muon $(m_\mu = 207m_e)$ and an antineutrino $(m_{\bar{\nu}} \approx 0)$. The reaction is written $\pi^- \rightarrow \mu^- + \bar{\nu}$. Find the kinetic energy of the muon and the energy of the antineutrino in electron volts. (*Suggestion:* Conserve both energy and momentum.)

48. According to observer A, two objects of equal mass and moving along the x axis collide head on and stick to each other. Before the collision, this observer measures that object 1 moves to the right with a speed of $3c/4$, while object 2 moves to the left with the same speed. According to observer B, however, object 1 is initially at rest. (a) Determine the speed of object 2 as seen by observer B. (b) Compare the total initial energy of the system in the two frames of reference.

Section 39.9 Mass and Energy

49. Make an order-of-magnitude estimate of the ratio of mass increase to the original mass of a flag, as you run it up a flagpole. In your solution explain what quantities you take as data and the values you estimate or measure for them.

50. When 1.00 g of hydrogen combines with 8.00 g of oxygen, 9.00 g of water is formed. During this chemical reaction, 2.86×10^5 J of energy is released. How much mass do the constituents of this reaction lose? Is the loss of mass likely to be detectable?

51. In a nuclear power plant the fuel rods last 3 yr before they are replaced. If a plant with rated thermal power 1.00 GW

operates at 80.0% capacity for 3.00 yr, what is the loss of mass of the fuel?

52. Review problem. The total volume of water in the oceans is approximately 1.40×10^9 km³. The density of sea water is 1 030 kg/m³, and the specific heat of the water is 4 186 J/(kg·°C). Find the increase in mass of the oceans produced by an increase in temperature of 10.0°C.

53. The power output of the Sun is 3.77×10^{26} W. How much mass is converted to energy in the Sun each second?

54. A gamma ray (a high-energy photon) can produce an electron (e^-) and a positron (e^+) when it enters the electric field of a heavy nucleus: $\gamma \rightarrow e^+ + e^-$. What minimum gamma-ray energy is required to accomplish this task? (*Note:* The masses of the electron and the positron are equal.)

Section 39.10   The General Theory of Relativity

55. An Earth satellite used in the global positioning system moves in a circular orbit with period 11 h 58 min. (a) Determine the radius of its orbit. (b) Determine its speed. (c) The satellite contains an oscillator producing the principal nonmilitary GPS signal. Its frequency is 1 575.42 MHz in the reference frame of the satellite. When it is received on the Earth's surface, what is the fractional change in this frequency due to time dilation, as described by special relativity? (d) The gravitational blue shift of the frequency according to general relativity is a separate effect. The magnitude of that fractional change is given by

$$\frac{\Delta f}{f} = \frac{\Delta U_g}{mc^2}$$

where ΔU_g is the change in gravitational potential energy of an object–Earth system when the object of mass m is moved between the two points at which the signal is observed. Calculate this fractional change in frequency. (e) What is the overall fractional change in frequency? Superposed on both of these relativistic effects is a Doppler shift that is generally much larger. It can be a red shift or a blue shift, depending on the motion of a particular satellite relative to a GPS receiver (Fig. P39.55).

Figure P39.55 This global positioning system (GPS) receiver incorporates relativistically corrected time calculations in its analysis of signals it receives from orbiting satellites. This allows the unit to determine its position on the Earth's surface to within a few meters. If these corrections were not made, the location error would be about 1 km.

Additional Problems

56. An astronaut wishes to visit the Andromeda galaxy, making a one-way trip that will take 30.0 yr in the spacecraft's frame of reference. Assume that the galaxy is 2.00×10^6 ly away and that the astronaut's speed is constant. (a) How fast must he travel relative to the Earth? (b) What will be the kinetic energy of his 1 000-metric-ton spacecraft? (c) What is the cost of this energy if it is purchased at a typical consumer price for electric energy: $0.130/kWh?

57. The cosmic rays of highest energy are protons that have kinetic energy on the order of 10^{13} MeV. (a) How long would it take a proton of this energy to travel across the Milky Way galaxy, having a diameter $\sim 10^5$ ly, as measured in the proton's frame? (b) From the point of view of the proton, how many kilometers across is the galaxy?

58. An electron has a speed of $0.750c$. (a) Find the speed of a proton that has the same kinetic energy as the electron. (b) **What If?** Find the speed of a proton that has the same momentum as the electron.

59. Ted and Mary are playing a game of catch in frame S′, which is moving at $0.600c$ with respect to frame S, while Jim, at rest in frame S, watches the action (Fig. P39.59). Ted throws the ball to Mary at $0.800c$ (according to Ted) and their separation (measured in S′) is 1.80×10^{12} m. (a) According to Mary, how fast is the ball moving? (b) According to Mary, how long does it take the ball to reach her? (c) According to Jim, how far apart are Ted and Mary, and how fast is the ball moving? (d) According to Jim, how long does it take the ball to reach Mary?

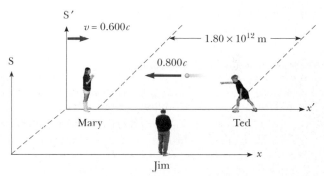

Figure P39.59

60. A rechargeable AA battery with a mass of 25.0 g can supply a power of 1.20 W for 50.0 min. (a) What is the difference in mass between a charged and an uncharged battery? (b) What fraction of the total mass is this mass difference?

61. The net nuclear fusion reaction inside the Sun can be written as $4\,^1H \rightarrow\,^4He + \Delta E$. The rest energy of each hydrogen atom is 938.78 MeV and the rest energy of the helium-4 atom is 3 728.4 MeV. Calculate the percentage of the starting mass that is transformed to other forms of energy.

62. An object disintegrates into two fragments. One of the fragments has mass 1.00 MeV/c^2 and momentum 1.75 MeV/c in the positive x direction. The other fragment has mass 1.50 MeV/c^2 and momentum 2.00 MeV/c in the positive y direction. Find (a) the mass and (b) the speed of the original object.

63. An alien spaceship traveling at $0.600c$ toward the Earth launches a landing craft with an advance guard of purchasing agents and physics teachers. The lander travels in the same direction with a speed of $0.800c$ relative to the mother ship. As observed on the Earth, the spaceship is 0.200 ly from the Earth when the lander is launched. (a) What speed do the Earth observers measure for the approaching lander? (b) What is the distance to the Earth at the time of lander launch, as observed by the aliens? (c) How long does it take the lander to reach the Earth as observed by the aliens on the mother ship? (d) If the lander has a mass of 4.00×10^5 kg, what is its kinetic energy as observed in the Earth reference frame?

64. A physics professor on the Earth gives an exam to her students, who are in a spacecraft traveling at speed v relative to the Earth. The moment the craft passes the professor, she signals the start of the exam. She wishes her students to have a time interval T_0 (spacecraft time) to complete the exam. Show that she should wait a time interval (Earth time) of

$$T = T_0 \sqrt{\frac{1 - v/c}{1 + v/c}}$$

before sending a light signal telling them to stop. (*Suggestion:* Remember that it takes some time for the second light signal to travel from the professor to the students.)

65. Spacecraft I, containing students taking a physics exam, approaches the Earth with a speed of $0.600c$ (relative to the Earth), while spacecraft II, containing professors proctoring the exam, moves at $0.280c$ (relative to the Earth) directly toward the students. If the professors stop the exam after 50.0 min have passed on their clock, how long does the exam last as measured by (a) the students (b) an observer on the Earth?

66. Energy reaches the upper atmosphere of the Earth from the Sun at the rate of 1.79×10^{17} W. If all of this energy were absorbed by the Earth and not re-emitted, how much would the mass of the Earth increase in 1.00 yr?

67. A supertrain (proper length 100 m) travels at a speed of $0.950c$ as it passes through a tunnel (proper length 50.0 m). As seen by a trackside observer, is the train ever completely within the tunnel? If so, with how much space to spare?

68. Imagine that the entire Sun collapses to a sphere of radius R_g such that the work required to remove a small mass m from the surface would be equal to its rest energy mc^2. This radius is called the *gravitational radius* for the Sun. Find R_g. (It is believed that the ultimate fate of very massive stars is to collapse beyond their gravitational radii into black holes.)

69. A particle with electric charge q moves along a straight line in a uniform electric field $\mathbf{E}$ with a speed of u. The electric force exerted on the charge is $q\mathbf{E}$. The motion and the electric field are both in the x direction. (a) Show that the acceleration of the particle in the x direction is given by

$$a = \frac{du}{dt} = \frac{qE}{m} \left(1 - \frac{u^2}{c^2}\right)^{3/2}$$

(b) Discuss the significance of the dependence of the acceleration on the speed. (c) **What If?** If the particle starts from rest at $x = 0$ at $t = 0$, how would you proceed to find the speed of the particle and its position at time t?

70. An observer in a coasting spacecraft moves toward a mirror at speed v relative to the reference frame labeled by S in Figure P39.70. The mirror is stationary with respect to S. A light pulse emitted by the spacecraft travels toward the mirror and is reflected back to the craft. The front of the craft is a distance d from the mirror (as measured by observers in S) at the moment the light pulse leaves the craft. What is the total travel time of the pulse as measured by observers in (a) the S frame and (b) the front of the spacecraft?

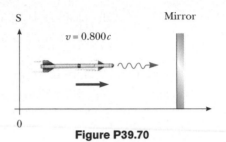

Figure P39.70

71. The creation and study of new elementary particles is an important part of contemporary physics. Especially interesting is the discovery of a very massive particle. To create a particle of mass M requires an energy Mc^2. With enough energy, an exotic particle can be created by allowing a fast moving particle of ordinary matter, such as a proton, to collide with a similar target particle. Let us consider a perfectly inelastic collision between two protons: an incident proton with mass m_p, kinetic energy K, and momentum magnitude p joins with an originally stationary target proton to form a single product particle of mass M. You might think that the creation of a new product particle, nine times more massive than in a previous experiment, would require just nine times more energy for the incident proton. Unfortunately not all of the kinetic energy of the incoming proton is available to create the product particle, since conservation of momentum requires that after the collision the system as a whole still must have some kinetic energy. Only a fraction of the energy of the incident particle is thus available to create a new particle. You will determine how the energy available for particle creation depends on the energy of the moving proton. Show that the energy available to create a product particle is given by

$$Mc^2 = 2m_p c^2 \sqrt{1 + \frac{K}{2m_p c^2}}$$

From this result, when the kinetic energy K of the incident proton is large compared to its rest energy $m_p c^2$, we see that M approaches $(2m_p K)^{1/2}/c$. Thus if the energy of the incoming proton is increased by a factor of nine, the mass you can create increases only by a factor of three. This disappointing result is the main reason that most modern accelerators, such as those at CERN (in Europe), at Fermilab (near Chicago), at SLAC (at Stanford), and at DESY (in Germany), use *colliding beams*. Here the total momentum of a pair of interacting particles can be zero. The center of mass can be at rest after the collision, so in principle all of the initial kinetic energy can be used for particle creation, according to

$$Mc^2 = 2mc^2 + K = 2mc^2\left(1 + \frac{K}{2mc^2}\right)$$

where K is the total kinetic energy of two identical colliding particles. Here if $K \gg mc^2$, we have M directly proportional to K, as we would desire. These machines are difficult to build and to operate, but they open new vistas in physics.

72. A particle of mass m moving along the x axis with a velocity component $+u$ collides head-on and sticks to a particle of mass $m/3$ moving along the x axis with the velocity component $-u$. What is the mass M of the resulting particle?

73. A rod of length L_0 moving with a speed v along the horizontal direction makes an angle θ_0 with respect to the x' axis. (a) Show that the length of the rod as measured by a stationary observer is $L = L_0[1 - (v^2/c^2)\cos^2\theta_0]^{1/2}$. (b) Show that the angle that the rod makes with the x axis is given by $\tan\theta = \gamma\tan\theta_0$. These results show that the rod is both contracted and rotated. (Take the lower end of the rod to be at the origin of the primed coordinate system.)

74. Suppose our Sun is about to explode. In an effort to escape, we depart in a spacecraft at $v = 0.800c$ and head toward the star Tau Ceti, 12.0 ly away. When we reach the midpoint of our journey from the Earth, we see our Sun explode and, unfortunately, at the same instant we see Tau Ceti explode as well. (a) In the spacecraft's frame of reference, should we conclude that the two explosions occurred simultaneously? If not, which occurred first? (b) **What If?** In a frame of reference in which the Sun and Tau Ceti are at rest, did they explode simultaneously? If not, which exploded first?

75. A ^{57}Fe nucleus at rest emits a 14.0-keV photon. Use conservation of energy and momentum to deduce the kinetic energy of the recoiling nucleus in electron volts. (Use $Mc^2 = 8.60 \times 10^{-9}$ J for the final state of the ^{57}Fe nucleus.)

76. 🖳 Prepare a graph of the relativistic kinetic energy and the classical kinetic energy, both as a function of speed, for an object with a mass of your choice. At what speed does the classical kinetic energy underestimate the experimental value by 1%? by 5%? by 50%?

Answers to Quick Quizzes

39.1 (c). While the observers' measurements differ, both are correct.

39.2 (d). The Galilean velocity transformation gives us $u_x = u'_x + v = 110$ mi/h $+ 90$ mi/h $= 200$ mi/h.

39.3 (d). The two events (the pulse leaving the flashlight and the pulse hitting the far wall) take place at different locations for both observers, so neither measures the proper time interval.

39.4 (a). The two events are the beginning and the end of the movie, both of which take place at rest with respect to the spacecraft crew. Thus, the crew measures the proper time interval of 2 h. Any observer in motion with respect to the spacecraft, which includes the observer on Earth, will measure a longer time interval due to time dilation.

39.5 (a). If their on-duty time is based on clocks that remain on the Earth, they will have larger paychecks. A shorter time interval will have passed for the astronauts in their frame of reference than for their employer back on the Earth.

39.6 (c). Both your body and your sleeping cabin are at rest in your reference frame; thus, they will have their proper length according to you. There will be no change in measured lengths of objects, including yourself, within your spacecraft.

39.7 (d). Time dilation and length contraction depend only on the relative speed of one observer relative to another, not on whether the observers are receding or approaching each other.

39.8 (c). Because of your motion toward the source of the light, the light beam has a horizontal component of velocity as measured by you. The magnitude of the vector sum of the horizontal and vertical component vectors must be equal to c, so the magnitude of the vertical component must be smaller than c.

39.9 (a). In this case, there is only a horizontal component of the velocity of the light, and you must measure a speed of c.

39.10 (a) $m_3 > m_2 = m_1$; the rest energy of particle 3 is $2E$, while it is E for particles 1 and 2. (b) $K_3 = K_2 > K_1$; the kinetic energy is the difference between the total energy and the rest energy. The kinetic energy is $4E - 2E = 2E$ for particle 3, $3E - E = 2E$ for particle 2, and $2E - E = E$ for particle 1. (c) $u_2 > u_3 = u_1$; from Equation 39.26, $E = \gamma E_R$. Solving this for the square of the particle speed u, we find $u^2 = c^2(1 - (E_R/E)^2)$. Thus, the particle with the smallest ratio of rest energy to total energy will have the largest speed. Particles 1 and 3 have the same ratio as each other, and the ratio of particle 2 is smaller.

Chapter 40

Introduction to Quantum Physics

CHAPTER OUTLINE

40.1 Blackbody Radiation and Planck's Hypothesis

40.2 The Photoelectric Effect

40.3 The Compton Effect

40.4 Photons and Electromagnetic Waves

40.5 The Wave Properties of Particles

40.6 The Quantum Particle

40.7 The Double–Slit Experiment Revisited

40.8 The Uncertainty Principle

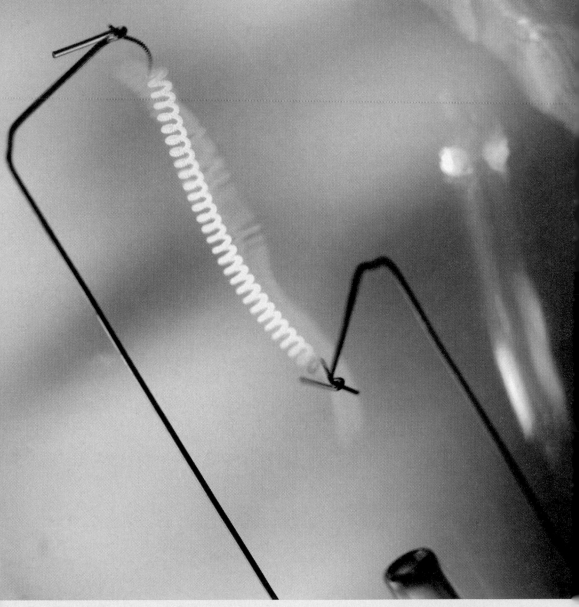

▲ *This lightbulb filament glows with an orange color. Why? Classical physics is unable to explain the experimentally observed wavelength distribution of electromagnetic radiation from a hot object. A theory proposed in 1900 and describing the radiation from such objects represents the dawn of quantum physics. (Steve Cole/Getty Images)*

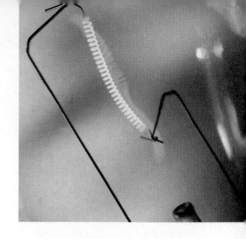

In the preceding chapter, we discussed the fact that Newtonian mechanics must be replaced by Einstein's special theory of relativity when dealing with particle speeds comparable to the speed of light. As the twentieth century progressed, many experimental and theoretical problems were resolved by the special theory of relativity. There were many other problems, however, for which neither relativity nor classical physics could provide a theoretical answer. Attempts to apply the laws of classical physics to explain the behavior of matter on the atomic scale were consistently unsuccessful. For example, the emission of discrete wavelengths of light from atoms in a high-temperature gas could not be explained within the framework of classical physics. While models based on classical physics predicted light frequencies with the correct order of magnitude, they predicted that the frequencies would be continuous rather than discrete.

As physicists sought new ways to solve these puzzles, another revolution took place in physics between 1900 and 1930. A new theory called *quantum mechanics* was highly successful in explaining the behavior of particles of microscopic size. Like the special theory of relativity, the quantum theory requires a modification of our ideas concerning the physical world.

The first explanation of a phenomenon using quantum theory was introduced by Max Planck, but most of the subsequent mathematical developments and interpretations were made by a number of other distinguished physicists, including Einstein, Bohr, de Broglie, Schrödinger, Heisenberg, Born, and Dirac. Despite the great success of the quantum theory, Einstein frequently played the role of its critic, especially with regard to the manner in which the theory was interpreted.

Because an extensive study of quantum theory is beyond the scope of this book, this chapter is simply an introduction to its underlying principles.

40.1 Blackbody Radiation and Planck's Hypothesis

An object at any temperature emits **thermal radiation** from its surface, as discussed in Section 20.7. The characteristics of this radiation depend on the temperature and properties of the surface. Careful study shows that the thermal radiation it emits consists of a continuous distribution of wavelengths from all portions of the electromagnetic spectrum. If the object is at room temperature, the wavelengths of thermal radiation are mainly in the infrared region, and hence the radiation is not detected by the eye. As the surface temperature of the object increases, the object eventually begins to glow visibly red. At sufficiently high temperatures, the glowing object appears white, as in the hot tungsten filament of a lightbulb.

From a classical viewpoint, thermal radiation originates from accelerated charged particles in the atoms near the surface of the object; those charged particles emit radiation much as small antennas do. The thermally agitated accelerating particles can have a distribution of energies, which accounts for the continuous spectrum of radiation emitted by the object. By the end of the nineteenth century, however, it became apparent that the classical theory of thermal radiation was inadequate. The basic problem was in understanding the observed distribution of wavelengths in the radiation emitted by a black

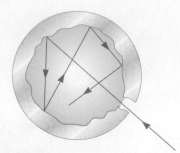

Figure 40.1 The opening to the cavity inside a hollow object is a good approximation of a black body. Light entering the small opening strikes the interior walls, where some is absorbed and some is reflected at a random angle. The cavity walls re-radiate at wavelengths corresponding to their temperature, producing standing waves in the cavity. Some of the energy from these standing waves can leave through the opening.

body. As defined in Section 20.7, a **black body** is an ideal system that absorbs all radiation incident on it. The electromagnetic radiation emitted by the black body is called **blackbody radiation.**

A good approximation of a black body is a small hole leading to the inside of a hollow object, as shown in Figure 40.1. Any radiation incident on the hole from outside the cavity enters the hole and is absorbed or reflected a number of times from the walls of the cavity; hence the hole acts as a perfect absorber. The nature of the radiation leaving the cavity through the hole depends only on the temperature of the cavity walls and not on the material of which the walls are made. The spaces between lumps of hot charcoal (Fig. 40.2) emit light that is very much like blackbody radiation.

The radiation emitted by oscillators in the cavity walls experiences boundary conditions—it reflects from the walls of the cavity. Consequently, standing electromagnetic waves are established within the three-dimensional interior of the cavity. Many standing-wave modes are possible, and the distribution of the energy in the cavity among these modes determines the wavelength distribution of the radiation leaving the cavity through the hole.

The wavelength distribution of radiation from cavities was studied extensively in the late nineteenth century. Figure 40.3 shows how the intensity of blackbody radiation is found to vary with temperature and wavelength. The following two consistent experimental findings were seen as especially significant:

1. The total power of the emitted radiation increases with temperature.

We discussed this briefly in Chapter 20, where we introduced **Stefan's law,**

Stefan's law

$$\mathcal{P} = \sigma A e T^4 \qquad (40.1)$$

where $\mathcal{P}$ is the power in watts radiated from the surface of an object, σ is the Stefan–Boltzmann constant, equal to 5.670×10^{-8} W/m$^2 \cdot$ K^4, A is the surface area of the object in square meters, e is the emissivity of the surface, and T is the surface temperature in kelvins. For a black body, the emissivity is $e = 1$ exactly. Recalling that $I \equiv \mathcal{P}/A$ is the definition of intensity and that $e = 1$ for a black body, we can write Stefan's law in the form $I = \sigma T^4$ at the surface of the object.

Figure 40.2 The glow emanating from the spaces between these hot charcoal briquettes is, to a close approximation, blackbody radiation. The color of the light depends only upon the temperature of the briquettes.

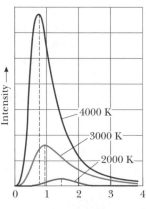

Active Figure 40.3 Intensity of blackbody radiation versus wavelength at three temperatures. Note that the amount of radiation emitted (the area under a curve) increases with increasing temperature. The visible range of wavelengths is between 0.4 μm and 0.7 μm. Thus, the 4 000-K curve has a peak that is near the visible range and represents an object that would glow with a yellowish-white appearance. At about 6 000 K, the peak is in the center of the visible wavelengths and the object appears white.

✎ *At the Active Figures link at* http://www.pse6.com, *you can adjust the temperature of the black body and study the radiation emitted from it.*

2. **The peak of the wavelength distribution shifts to shorter wavelengths as the temperature increases.**

This behavior was found to be described by the following relationship, called **Wien's displacement law:**

$$\lambda_{max} T = 2.898 \times 10^{-3} \, \text{m} \cdot \text{K} \qquad (40.2)$$

Wien's displacement law

where λ_{max} is the wavelength at which the curve peaks and T is the absolute temperature of the surface of the object emitting the radiation. The wavelength at the peak of the curve is inversely proportional to the absolute temperature; that is, as the temperature increases, the peak is "displaced" to shorter wavelengths (Fig. 40.3).

This second finding (Wien's displacement law) is consistent with the behavior of the object mentioned at the beginning of this section. At room temperature, it does not appear to glow (peak in the infrared, $\lambda > 0.7 \, \mu$m). At higher temperature it glows red (peak in the near infrared with some radiation at the red end of the visible spectrum, $\lambda \approx 0.7 \, \mu$m), and at still higher temperature it glows white (peak in the visible, $0.4 \, \mu$m $< \lambda < 0.7 \, \mu$m).

Quick Quiz 40.1 Figure 40.4 shows two stars in the constellation Orion. Betelgeuse appears to glow red, while Rigel looks blue in color. Which star has a higher surface temperature? (a) Betelgeuse (b) Rigel (c) They both have the same surface temperature. (d) Impossible to determine.

Figure 40.4 (Quick Quiz 40.1) Which star is hotter?

A successful theory for blackbody radiation must predict the shape of the curves in Figure 40.3, the temperature dependence expressed in Stefan's law, and the shift of the peak with temperature described by Wien's displacement law. Early attempts to use classical ideas to explain the shapes of the curves in Figure 40.3 failed.

We consider one of these early attempts here. To describe the distribution of energy from a black body, it is useful to define $I(\lambda, T) \, d\lambda$ to be the intensity, or power per unit area, emitted in the wavelength interval $d\lambda$. The result of a calculation based on a classical theory of blackbody radiation known as the **Rayleigh–Jeans law** is

$$I(\lambda, T) = \frac{2\pi c k_B T}{\lambda^4} \qquad (40.3)$$

Rayleigh–Jeans law

where k_B is Boltzmann's constant. The black body is modeled as the hole leading into a cavity supporting many modes of oscillation of the electromagnetic field caused by accelerated charges in the cavity walls, resulting in the emission of electromagnetic waves at all wavelengths. In the classical theory used to derive Equation 40.3, the average energy for each wavelength of the standing wave modes is assumed to be proportional to $k_B T$, based on the theorem of equipartition of energy that we discussed in Section 21.1.

An experimental plot of the blackbody radiation spectrum is shown in Figure 40.5, together with the theoretical prediction of the Rayleigh–Jeans law. At long wavelengths, the Rayleigh–Jeans law is in reasonable agreement with experimental data, but at short wavelengths major disagreement is apparent.

Note that as λ approaches zero, the function $I(\lambda, T)$ given by Equation 40.3 approaches infinity. Hence, according to classical theory, not only should short wavelengths predominate in a blackbody spectrum, but also the energy emitted by any black body should become infinite in the limit of zero wavelength. In contrast to this prediction, the experimental data plotted in Figure 40.5 show that as λ approaches zero, $I(\lambda, T)$ also approaches zero. This mismatch of theory and experiment was so disconcerting that scientists called it the *ultraviolet catastrophe*. (This "catastrophe"—infinite energy—occurs as the wavelength approaches zero; the word "ultraviolet" was applied because ultraviolet wavelengths are short.)

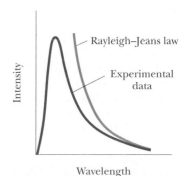

Figure 40.5 Comparison of experimental results and the curve predicted by the Rayleigh–Jeans law for the distribution of blackbody radiation.

Max Planck
German Physicist (1858–1947)

Planck introduced the concept of "quantum of action" (Planck's constant, h) in an attempt to explain the spectral distribution of blackbody radiation, which laid the foundations for quantum theory. In 1918 he was awarded the Nobel Prize for this discovery of the quantized nature of energy. (©*Bettmann/CORBIS*)

In 1900, Max Planck developed a theory of blackbody radiation that leads to an equation for $I(\lambda, T)$ that is in complete agreement with experimental results at all wavelengths. Planck assumed that the cavity radiation came from atomic oscillators in the cavity walls in Figure 40.1. Planck made two bold and controversial assumptions concerning the nature of the oscillators in the cavity walls:

- The energy of an oscillator can have only certain *discrete* values E_n:

$$E_n = nhf \qquad (40.4)$$

where n is a positive integer called a **quantum number,**[1] f is the frequency of oscillation, and h is a parameter that Planck introduced and is now called **Planck's constant.** Because the energy of each oscillator can have only discrete values given by Equation 40.4, we say the energy is **quantized.** Each discrete energy value corresponds to a different **quantum state,** represented by the quantum number n. When the oscillator is in the $n = 1$ quantum state, its energy is hf; when it is in the $n = 2$ quantum state, its energy is $2hf$; and so on.

- The oscillators emit or absorb energy when making a transition from one quantum state to another. The entire energy difference between the initial and final states in the transition is emitted or absorbed as a single quantum of radiation. If the transition is from one state to a lower adjacent state—say, from the $n = 3$ state to the $n = 2$ state—Equation 40.4 shows that the amount of energy emitted by the oscillator is

$$E = hf \qquad (40.5)$$

An oscillator emits or absorbs energy only when it changes quantum states. If it remains in one quantum state, no energy is absorbed or emitted. Figure 40.6 is an **energy-level diagram** showing the quantized energy levels and allowed transitions proposed by Planck. This is an important semigraphical representation that is used often in quantum physics.[2] The vertical axis is linear in energy, and the allowed energy levels are represented as horizontal lines. The quantized system can have only the energies represented by the horizontal lines.

The key point in Planck's theory is the radical assumption of quantized energy states. This development was a clear deviation from classical physics and marked the birth of the quantum theory.

In the Rayleigh–Jeans model, the average energy associated with a particular wavelength of standing waves in the cavity is the same for all wavelengths and equal to $k_B T$. Planck used the same classical ideas as in the Rayleigh–Jeans model to arrive at the energy density as a product of constants and the average energy for a given wavelength,

▲ **PITFALL PREVENTION**

40.2 n Is Again an Integer

In the preceding chapters on optics, we used the symbol n for the index of refraction, which was not an integer. Here we are again using n as we did in Chapter 18 to indicate the standing-wave mode on a string or in an air column. In quantum physics, n is often used as an integer quantum number to identify a particular quantum state of a system.

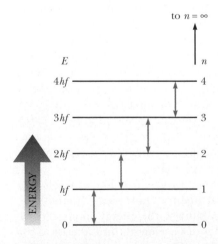

Figure 40.6 Allowed energy levels for an oscillator with frequency f. Allowed transitions are indicated by the double-headed arrows.

---

[1] A quantum number is generally an integer (although half–integer quantum numbers can occur) that describes an allowed state of a system, like the values of n describing the normal modes of oscillation of a string fixed at both ends, as discussed in Section 18.3.

[2] We first saw an energy-level diagram in Section 21.4.

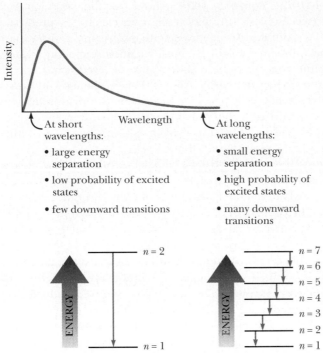

Active Figure 40.7 In Planck's model, the average energy associated with a given wavelength is the product of the energy of a transition and a factor related to the probability of the transition occurring. As the energy levels move farther apart at shorter wavelengths (higher energy), the probability of excitation decreases, as does the probability of a transition from the excited state.

Investigate the energy levels and observe the emission of radiation of different wavelengths at the Active Figures link at http://www.pse6.com.

but the average energy is not given by the equipartition theorem. The average energy of a wave is the average energy difference between levels of the oscillator, *weighted according to the probability of the wave being emitted.* This weighting is based on the occupation of higher-energy states as described by the Boltzmann distribution law, which we presented in Section 21.5. According to this law, the probability of a state being occupied is proportional to the factor $e^{-E/k_B T}$ where E is the energy of the state.

At low frequencies, the energy levels are close together, as on the right in Figure 40.7, and many of the energy states are excited because the Boltzmann factor $e^{-E/k_B T}$ is relatively large for these states. Thus, there are many contributions to the outgoing radiation, although each contribution has very low energy. Now, consider high-frequency radiation, that is, radiation with short wavelength. In order to obtain this radiation, the allowed energies are very far apart, as on the left in Figure 40.7. The probability of thermal agitation exciting these high energy levels is small because of the small value of the Boltzmann factor for large values of E. At high frequencies, the low probability of excitation results in very little contribution to the total energy, even though each quantum is of large energy. This "turns the curve over" and brings it down to zero again at short wavelengths.

Using this approach, Planck was able to generate a theoretical expression for the wavelength distribution that agreed remarkably well with the experimental curves in Figure 40.3:

$$I(\lambda, \, T) = \frac{2\pi hc^2}{\lambda^5(e^{hc/\lambda k_B T} - 1)} \tag{40.6}$$

Planck's wavelength distribution function

This function includes the parameter h, which Planck adjusted so that his curve matched the experimental data at all wavelengths. The value of this parameter is found to be independent of the material of which the black body is made and independent of the temperature—it is a fundamental constant of nature. The value of h, Planck's constant, which we saw briefly first in Chapter 11 and again in Chapter 35, is

$$h = 6.626 \times 10^{-34} \, \text{J} \cdot \text{s} \tag{40.7}$$

Planck's constant

At long wavelengths, Equation 40.6 reduces to the Rayleigh–Jeans expression, Equation 40.3 (see Problem 12), and at short wavelengths it predicts an exponential decrease in $I(\lambda, T)$ with decreasing wavelength, in agreement with experimental results.

When Planck presented his theory, most scientists (including Planck!) did not consider the quantum concept to be realistic. They believed it was a mathematical trick that happened to predict the correct results. Hence, Planck and others continued to search for a more "rational" explanation of blackbody radiation. However, subsequent developments showed that a theory based on the quantum concept (rather than on classical concepts) had to be used to explain not only blackbody radiation but also a number of other phenomena at the atomic level.

In 1905, Einstein rederived Planck's results by assuming that the cavity oscillations of the electromagnetic field were themselves quantized. In other words, he proposed that quantization is a fundamental property of light and other electromagnetic radiation. This led to the concept of photons as will be discussed in Section 40.2. Critical to the success of the quantum or photon theory was the relation between energy and frequency, which classical theory completely failed to predict.

You may have had your body temperature measured at the doctor's office by an *ear thermometer,* which can read your temperature in a matter of seconds (Fig. 40.8). This type of thermometer measures the amount of infrared radiation emitted by the eardrum in a fraction of a second. It then converts the amount of radiation into a temperature reading. This thermometer is very sensitive because temperature is raised to the fourth power in Stefan's law. Suppose that you have a fever of 1°C above normal. Because absolute temperatures are found by adding 273 to Celsius temperatures, the ratio of your fever temperature to normal body temperature of 37°C is

$$\frac{T_{\text{fever}}}{T_{\text{normal}}} = \frac{38°\text{C} + 273°\text{C}}{37°\text{C} + 273°\text{C}} = 1.0032$$

This is only a 0.32% increase in temperature. The increase in radiated power, however, is proportional to the fourth power of temperature, so

$$\frac{\mathcal{P}_{\text{fever}}}{\mathcal{P}_{\text{normal}}} = \left(\frac{38°\text{C} + 273°\text{C}}{37°\text{C} + 273°\text{C}}\right)^4 = 1.013$$

This is a 1.3% increase in radiated power, which is easily measured by modern infrared radiation sensors.

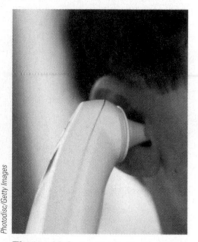

Figure 40.8 An ear thermometer measures a patient's temperature by detecting the intensity of infrared radiation leaving the eardrum.

Example 40.1 Thermal Radiation from Different Objects

Find the peak wavelength of the blackbody radiation emitted by each of the following:

(A) The human body when the skin temperature is 35°C

(B) The tungsten filament of a lightbulb, which operates at 2 000 K

(C) The Sun, which has a surface temperature of about 5 800 K

Solution **(A)** From Wien's displacement law (Eq. 40.2), we have $\lambda_{\max} T = 2.898 \times 10^{-3}\ \text{m} \cdot \text{K}$. Solving for $\lambda_{\max}$ and noting that 35°C corresponds to an absolute temperature of 308 K, we have

$$\lambda_{\max} = \frac{2.898 \times 10^{-3}\ \text{m} \cdot \text{K}}{308\ \text{K}} = \boxed{9.4\ \mu\text{m}}$$

This radiation is in the infrared region of the spectrum and is invisible to the human eye. Some animals (pit vipers, for instance) are able to detect radiation of this wavelength and therefore can locate warm-blooded prey even in the dark.

(B) Following the same procedure as in part (A), we find

$$\lambda_{\max} = \frac{2.898 \times 10^{-3}\ \text{m} \cdot \text{K}}{2\,000\ \text{K}} = \boxed{1.4\ \mu\text{m}}$$

This also is in the infrared, meaning that most of the energy emitted by a lightbulb is not visible to us.

(C) Again following the same procedure, we have

$$\lambda_{\max} = \frac{2.898 \times 10^{-3}\ \text{m} \cdot \text{K}}{5\,800\ \text{K}} = \boxed{0.50\ \mu\text{m}}$$

This is near the center of the visible spectrum, about the color of a yellow-green tennis ball. Because it is the most prevalent color in sunlight, our eyes have evolved to be most sensitive to light of approximately this wavelength.

Example 40.2   The Quantized Oscillator

A 2.0-kg block is attached to a massless spring that has a force constant of $k = 25$ N/m. The spring is stretched 0.40 m from its equilibrium position and released.

(A) Find the total energy of the system and the frequency of oscillation according to classical calculations.

Solution Because of our study of oscillating blocks in Chapter 15, this problem is easy to conceptualize. The phrase "according to classical calculations" tells us that we should categorize this part of the problem as a classical analysis of the oscillator. To analyze the problem, we know that the total energy of a simple harmonic oscillator having an amplitude A is $\frac{1}{2}kA^2$ (Eq. 15.21). Therefore,

$$E = \tfrac{1}{2}kA^2 = \tfrac{1}{2}(25 \text{ N/m})(0.40 \text{ m})^2 = \boxed{2.0 \text{ J}}$$

The frequency of oscillation is (Eq. 15.14)

$$f = \frac{1}{2\pi}\sqrt{\frac{k}{m}} = \frac{1}{2\pi}\sqrt{\frac{25 \text{ N/m}}{2.0 \text{ kg}}} = \boxed{0.56 \text{ Hz}}$$

(B) Assuming that the energy is quantized, find the quantum number n for the system oscillating with this amplitude.

Solution This part of the problem is categorized as a quantum analysis of the oscillator. To analyze the problem, we note that the energy of the oscillator is quantized according to Equation 40.4. Thus,

$$E_n = nhf = n(6.626 \times 10^{-34} \text{ J·s})(0.56 \text{ Hz}) = 2.0 \text{ J}$$

$$n = \boxed{5.4 \times 10^{33}}$$

To finalize this problem, note that 5.4×10^{33} is a very large quantum number, which is typical for macroscopic systems. Changes from one quantum state to another appear as smooth transitions rather than discontinuous changes as is the case for a microscopic system. Changes between quantum states for the oscillator are explored in the following **What If?**

What If? Suppose the oscillator makes a transition from the $n = 5.4 \times 10^{33}$ state to the state corresponding to $n = 5.4 \times 10^{33} - 1$. By how much does the energy of the oscillator change in this one-quantum change?

Answer From Equation 40.5, the energy difference between states differing in n by 1 is

$$E = hf = (6.626 \times 10^{-34} \text{ J·s})(0.56 \text{ Hz}) = 3.7 \times 10^{-34} \text{ J}$$

This energy change due to a one-quantum change is fractionally equal to 3.7×10^{-34} J/2.0 J, or about one part in 10^{34}! It is such a small fraction of the total energy of the oscillator that we cannot detect it. Thus, even though the energy of a macroscopic spring–block system is quantized and does indeed decrease by small quantum jumps, our senses perceive the decrease as continuous. Quantum effects become important and measurable only on the submicroscopic level of atoms and molecules.

40.2 The Photoelectric Effect

Blackbody radiation was historically the first phenomenon to be explained with a quantum model. In the latter part of the nineteenth century, at the same time that data were taken on thermal radiation, experiments showed that light incident on certain metallic surfaces causes electrons to be emitted from those surfaces. This phenomenon, which we first met in Section 35.1, is known as the **photoelectric effect,** and the emitted electrons are called **photoelectrons.**[3]

Figure 40.9 is a diagram of an apparatus for studying the photoelectric effect. An evacuated glass or quartz tube contains a metallic plate E connected to the negative terminal of a battery and another metallic plate C that is connected to the positive terminal of the battery. When the tube is kept in the dark, the ammeter reads zero, indicating no current in the circuit. However, when plate E is illuminated by light having an appropriate wavelength, a current is detected by the ammeter, indicating a flow of charges across the gap between plates E and C. This current arises from photoelectrons emitted from the negative plate (the emitter) and collected at the positive plate (the collector).

Figure 40.10 is a plot of photoelectric current versus potential difference ΔV applied between plates E and C for two light intensities. At large values of ΔV, the current

---

[3] Photoelectrons are not different from other electrons. They are given this name solely because of their ejection from a metal by light in the photoelectric effect.

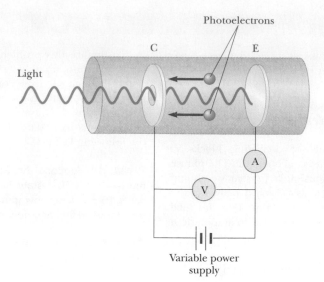

Photoelectrons

C   E

Light

Active Figure 40.9 A circuit diagram for studying the photoelectric effect. When light strikes the plate E (the emitter), photoelectrons are ejected from the plate. Electrons moving from plate E to plate C (the collector) constitute a current in the circuit.

A

V

Variable power supply

Current

High intensity

Low intensity

$-\Delta V_s$ Applied voltage

Active Figure 40.10 Photoelectric current versus applied potential difference for two light intensities. The current increases with intensity but reaches a saturation level for large values of ΔV. At voltages equal to or more negative than $-\Delta V_s$, where ΔV_s is the stopping potential, the current is zero.

reaches a maximum value; all of the electrons emitted from E are collected at C, and the current cannot increase further. In addition, the maximum current increases as the intensity of the incident light increases, as you might expect—more electrons are ejected by the higher-intensity light. Finally, when ΔV is negative—that is, when the battery in the circuit is reversed to make plate E positive and plate C negative—the current drops because many of the photoelectrons emitted from E are repelled by the now negative plate C. In this situation, only those photoelectrons having a kinetic energy greater than $e|\Delta V|$ reach plate C, where e is the magnitude of the charge on the electron. When ΔV is equal to or more negative than $-\Delta V_s$, where ΔV_s is the **stopping potential,** no photoelectrons reach C and the current is zero.

Let us consider the combination of the electric field between the plates and an electron ejected from plate E to be an isolated system. Suppose this electron stops just as it reaches plate C. Because the system of the electron and the electric field between the plates is isolated, the total mechanical energy of the system must be conserved:

$$K_1 + U_1 = K_2 + U_2$$

where configuration 1 refers to the instant that the electron leaves the metal with kinetic energy K_1 and configuration 2 is when the electron stops just before touching plate C. If we define the electric potential energy of the system in configuration 1 to be zero, we have

$$K_1 + 0 = 0 + (-e)(-\Delta V)$$

Now suppose we increase the potential difference ΔV in the negative direction just until no current flows. In this case the electron that stops just before reaching plate C will have the maximum possible kinetic energy upon leaving the metal surface, and ΔV will be the stopping potential ΔV_s. The previous equation can then be written as

$$K_{max} = e\,\Delta V_s \qquad (40.8)$$

This equation allows us to measure K_{max} experimentally by determining the voltage ΔV_s at which the current drops to zero.

Below we list several features of the photoelectric effect. For each feature, we compare the predictions made by a classical approach, using the wave model for light, with the experimental results.

1. Dependence of photoelectron kinetic energy on light intensity

   Classical prediction: Electrons should absorb energy continuously from the electromagnetic waves. As the light intensity incident on a metal is increased, energy should be transferred into the metal at a higher rate and the electrons should be ejected with more kinetic energy.

Experimental result: The maximum kinetic energy of photoelectrons is *independent* of light intensity. This is shown in Figure 40.10 by the fact that both curves fall to zero at the *same* negative voltage. According to Equation 40.8, the maximum kinetic energy is proportional to the stopping potential.

2. Time interval between incidence of light and ejection of photoelectrons

Classical prediction: At low light intensities, a measurable time interval should pass between the instant the light is turned on and the time an electron is ejected from the metal. This time interval is required for the electron to absorb the incident radiation before it acquires enough energy to escape from the metal.

Experimental result: Electrons are emitted from the surface of the metal almost *instantaneously* (less than 10^{-9} s after the surface is illuminated), even at very low light intensities.

3. Dependence of ejection of electrons on light frequency

Classical prediction: Electrons should be ejected from the metal at any incident light frequency, as long as the light intensity is high enough, because energy is transferred to the metal regardless of the incident light frequency.

Experimental result: No electrons are emitted if the incident light frequency falls below some **cutoff frequency** f_c, whose value is characteristic of the material being illuminated. No electrons are ejected below this cutoff frequency *regardless* of the light intensity.

4. Dependence of photoelectron kinetic energy on light frequency

Classical prediction: There should be *no* relationship between the frequency of the light and the electron kinetic energy. The kinetic energy should be related to the intensity of the light.

Experimental result: The maximum kinetic energy of the photoelectrons increases with increasing light frequency.

Notice that experimental results contradict *all four* classical predictions. A successful explanation of the photoelectric effect was given by Einstein in 1905, the same year he published his special theory of relativity. As part of a general paper on electromagnetic radiation, for which he received a Nobel prize in 1921, Einstein extended Planck's concept of quantization to electromagnetic waves, as mentioned in Section 40.1. He assumed that light (or any other electromagnetic wave) of frequency f can be considered a stream of quanta, regardless of the source of the radiation. Today we call these quanta **photons.** Each photon has an energy E, given by Equation 40.5, $E = hf$. Each photon moves in a vacuum at the speed of light c, where $c = 3.00 \times 10^8$ m/s.

Quick Quiz 40.2 While standing outdoors one evening, you are exposed to the following four types of electromagnetic radiation: yellow light from a sodium street lamp, radio waves from an AM radio station, radio waves from an FM radio station, and microwaves from an antenna of a communications system. Rank these types of waves in terms of increasing photon energy, lowest first.

In Einstein's model of the photoelectric effect, a photon of the incident light gives *all* its energy hf to a *single* electron in the metal. Thus, the absorption of energy by the electrons is not a continuous process as envisioned in the wave model, but rather a discontinuous process in which energy is delivered to the electrons in discrete bundles. The energy transfer is accomplished via a one photon–one electron event.[4]

Electrons ejected from the surface of the metal and not making collisions with other metal atoms before escaping possess the maximum kinetic energy K_{max}. Accord-

---

[4]   In principle, two photons could combine to provide an electron with their combined energy. This is highly improbable, however, without the high intensity of radiation available from very strong lasers.

ing to Einstein, the maximum kinetic energy for these liberated electrons is

Photoelectric effect equation

$$K_{max} = hf - \phi \tag{40.9}$$

where ϕ is called the **work function** of the metal. **The work function represents the minimum energy with which an electron is bound in the metal** and is on the order of a few electron volts. Table 40.1 lists work functions for various metals.

We can understand Equation 40.9 by rearranging it as follows:

$$K_{max} + \phi = hf$$

In this form, we see that Einstein's equation is Equation 7.17 applied to the system of the electron and the metal. K_{max} is the change in kinetic energy of the electron, assuming it begins at rest, ϕ is the change in potential energy of the system, assuming we define the potential energy when the electron is within the metal as zero, and hf is the transfer of energy into the system by electromagnetic radiation.

Using the photon model of light, one can explain the observed features of the photoelectric effect that could not be understood using classical concepts:

1. Dependence of photoelectron kinetic energy on light intensity

   Equation 40.9 shows that K_{max} is independent of the light intensity. The maximum kinetic energy of any one electron, which equals $hf - \phi$, depends only on the light frequency and the work function. If the light intensity is doubled, the number of photons arriving per unit time is doubled, which doubles the rate at which photoelectrons are emitted. The maximum kinetic energy of any one photoelectron, however, is unchanged.

2. Time interval between incidence of light and ejection of photoelectrons

   Near-instantaneous emission of electrons is consistent with the photon model of light. The incident energy appears in small packets, and there is a one-to-one interaction between photons and electrons. Even if the incident light has very low intensity, there may be very few photons arriving per unit time interval, but each one can have sufficient energy to eject an electron immediately.

3. Dependence of ejection of electrons on light frequency

   Failure to observe the photoelectric effect below a certain cutoff frequency follows from the fact that the photon must have energy greater than the work function ϕ in order to eject an electron. If the energy of an incoming photon does not satisfy this requirement, an electron cannot be ejected from the surface, regardless of light intensity.

4. Dependence of photoelectron kinetic energy on light frequency

   That K_{max} increases with increasing frequency is easily understood with Equation 40.9.

Einstein's model predicts a linear relationship (Eq. 40.9) between the maximum electron kinetic energy K_{max} and the light frequency f. Experimental observation of a linear relationship between K_{max} and f would be a final confirmation of Einstein's theory. Indeed, such a linear relationship is observed, as sketched in Figure 40.11, and the slope of the lines in such a plot is Planck's constant h. The intercept on the horizontal axis gives the cutoff frequency below which no photoelectrons are emitted, regardless of light intensity. The cutoff frequency is related to the work function through the relationship $f_c = \phi/h$. The cutoff frequency corresponds to a **cutoff wavelength** λ_c, where

Cutoff wavelength

$$\lambda_c = \frac{c}{f_c} = \frac{c}{\phi/h} = \frac{hc}{\phi} \tag{40.10}$$

and c is the speed of light. Wavelengths greater than λ_c incident on a material having a work function ϕ do not result in the emission of photoelectrons.

The combination hc occurs often when relating the energy of a photon to its wavelength. A common shortcut to use in solving problems is to express this

Table 40.1

Work Functions of Selected Metals[a]	
Metal	**ϕ (eV)**
Na	2.46
Al	4.08
Cu	4.70
Zn	4.31
Ag	4.73
Pt	6.35
Pb	4.14
Fe	4.50

[a] Values are typical for metals listed. Actual values may vary depending on whether the metal is a single crystal or polycrystalline. Values may also depend on the face from which electrons are ejected from crystalline metals. Furthermore, different experimental procedures may produce differing values.

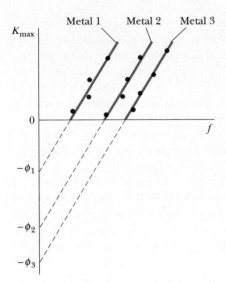

Active Figure 40.11 A plot of K_{max} of photoelectrons versus frequency of incident light in a typical photoelectric effect experiment. Photons with frequency less than the cutoff frequency for a given metal do not have sufficient energy to eject an electron from the metal.

At the Active Figures link at http://www.pse6.com, you can sweep through the frequency range and observe the curve for different target metals.

combination in useful units according to the following approximation:

$$hc = 1\ 240\ eV \cdot nm$$

One of the first practical uses of the photoelectric effect was as the detector in a light meter of a camera. Light reflected from the object to be photographed strikes a photoelectric surface in the meter, causing it to emit photoelectrons that then pass through a sensitive ammeter. The magnitude of the current in the ammeter depends on the light intensity.

The phototube, another early application of the photoelectric effect, acts much like a switch in an electric circuit. It produces a current in the circuit when light of sufficiently high frequency falls on a metal plate in the phototube, but produces no current in the dark. Phototubes were used in burglar alarms and in the detection of the soundtrack on motion picture film. Modern semiconductor devices have now replaced older devices based on the photoelectric effect.

The photoelectric effect is used today in the operation of photomultiplier tubes. Figure 40.12 shows the structure of such a device. A photon striking the photocathode ejects an electron by means of the photoelectric effect. This electron is accelerated across the potential difference between the photocathode and the first *dynode*, shown as being at $+200$ V relative to the photocathode in Figure 40.12. This high-energy

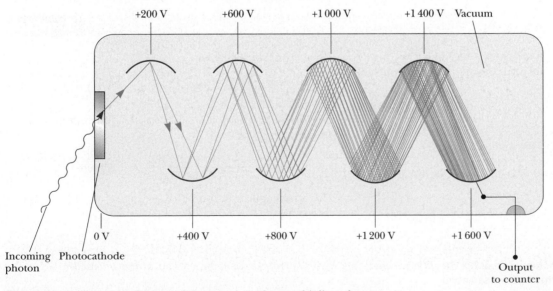

Figure 40.12 The multiplication of electrons in a photomultiplier tube.

electron strikes the dynode and ejects several more electrons. This process is repeated through a series of dynodes at ever higher potentials until an electrical pulse is produced as millions of electrons strike the last dynode. This is why the tube is called a *multiplier*—one photon at the input has resulted in millions of electrons at the output.

The photomultiplier tube is used in nuclear detectors to detect photons produced by the interaction of energetic charged particles or gamma rays with certain materials. It is also used in astronomy in a technique called *photoelectric photometry*. In this technique, the light collected by a telescope from a single star is allowed to fall on a photomultiplier tube for a time interval. The tube measures the total light energy during the time interval, which can then be converted to a luminosity of the star.

The photomultiplier tube is being replaced in many astronomical observations with a *charge-coupled device* (CCD), which is the same device that is used in a digital camera. In this device, an array of pixels is formed on the silicon surface of an integrated circuit (Section 43.7). When the surface is exposed to light from an astronomical scene through a telescope or a terrestrial scene through a digital camera, electrons generated by the photoelectric effect are caught in "traps" beneath the surface. The number of electrons is related to the intensity of the light striking the surface. A signal processor measures the number of electrons associated with each pixel and converts this information into a digital code that a computer can use to reconstruct and display the scene.

The *electron bombardment CCD camera* allows higher sensitivity than a conventional CCD. In this device, electrons ejected from a photocathode by the photoelectric effect are accelerated through a high voltage before striking a CCD array. The higher energy of the electrons results in a very sensitive detector of low-intensity radiation.

The explanation of the photoelectric effect with a quantum model, combined with Planck's quantum model for blackbody radiation, laid a strong foundation for further investigation into quantum physics. In the next section, we present a third experimental result that provides further strong evidence of the quantum nature of light.

Quick Quiz 40.3 Consider one of the curves in Figure 40.10. Suppose the intensity of the incident light is held fixed but its frequency is increased. The stopping potential in Figure 40.10 (a) remains fixed (b) moves to the right (c) moves to the left.

Quick Quiz 40.4 Suppose classical physicists had come up with the idea of predicting the appearance of a plot of K_{max} versus f, as in Figure 40.11. Draw a graph of what their expected plot would look like, based on the wave model for light.

Example 40.3   The Photoelectric Effect for Sodium   Interactive

A sodium surface is illuminated with light having a wavelength of 300 nm. The work function for sodium metal is 2.46 eV. Find

(A) the maximum kinetic energy of the ejected photoelectrons and

(B) the cutoff wavelength for sodium.

Solution (A) The energy of each photon in the illuminating light beam is

$$E = hf = \frac{hc}{\lambda} = \frac{1\,240 \text{ eV} \cdot \text{nm}}{300 \text{ nm}} = 4.13 \text{ eV}$$

Using Equation 40.9 gives

$$K_{max} = hf - \phi = 4.13 \text{ eV} - 2.46 \text{ eV} = \boxed{1.67 \text{ eV}}$$

(B) We can calculate the cutoff wavelength from Equation 40.10:

$$\lambda_c = \frac{hc}{\phi} = \frac{1\,240 \text{ eV} \cdot \text{nm}}{2.46 \text{ eV}} = \boxed{504 \text{ nm}}$$

This wavelength is in the yellow-green region of the visible spectrum.

 Investigate** the photoelectric effect for different materials and different wavelengths of light at the Interactive Worked Example link at **http://www.pse6.com.

40.3 The Compton Effect

In 1919, Einstein concluded that a photon of energy E travels in a single direction and carries a momentum equal to $E/c = hf/c$. In 1923, Arthur Holly Compton (1892–1962) and Peter Debye (1884–1966) independently carried Einstein's idea of photon momentum further.

Prior to 1922, Compton and his co-workers had accumulated evidence showing that the classical wave theory of light failed to explain the scattering of x-rays from electrons. According to classical theory, electromagnetic waves of frequency f_0 incident on electrons should have two effects, as shown in Figure 40.13a: (1) radiation pressure (see Section 34.4) should cause the electrons to accelerate in the direction of propagation of the waves, and (2) the oscillating electric field of the incident radiation should set the electrons into oscillation at the apparent frequency f', where f' is the frequency in the frame of the moving electrons. This apparent frequency f' is different from the frequency f_0 of the incident radiation because of the Doppler effect (see Section 17.4). Each electron first absorbs radiation as a moving particle and then re-radiates as a moving particle, thereby exhibiting two Doppler shifts in the frequency of radiation.

Because different electrons will move at different speeds after the interaction, depending on the amount of energy absorbed from the electromagnetic waves, the scattered wave frequency at a given angle to the incoming radiation should show a

Arthur Holly Compton
American Physicist
(1892–1962)

Compton was born in Wooster, Ohio, and attended Wooster College and Princeton University. He became the director of the laboratory at the University of Chicago, where experimental work concerned with sustained nuclear chain reactions was conducted. This work was of central importance to the construction of the first nuclear weapon. His discovery of the Compton effect led to his sharing of the 1927 Nobel Prize in physics with Charles Wilson. (*Courtesy of AIP Niels Bohr Library*)

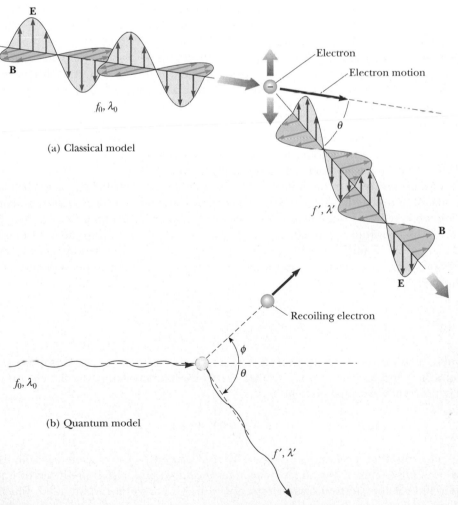

(a) Classical model

(b) Quantum model

Figure 40.13 X-ray scattering from an electron: (a) the classical model; (b) the quantum model.

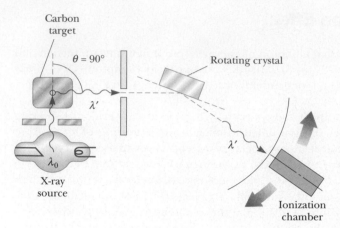

Figure 40.14 Schematic diagram of Compton's apparatus. The wavelength was measured with a rotating crystal spectrometer using graphite (carbon) as the target.

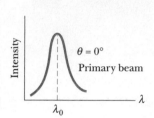

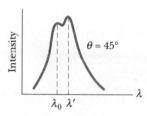

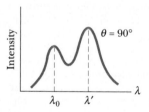

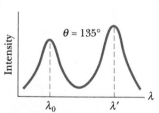

Figure 40.15 Scattered x-ray intensity versus wavelength for Compton scattering at $\theta = 0°$, $45°$, $90°$, and $135°$.

distribution of Doppler-shifted values. Contrary to this prediction, Compton's experiments showed that, at a given angle, only *one* frequency of radiation is observed. Compton and his co-workers realized that they could explain these experiments by treating photons not as waves but rather as point-like particles having energy hf and momentum hf/c, and by assuming that the energy and momentum of the isolated system of the colliding photon–electron pair are conserved. Compton adopted a particle model for something that was well known as a wave, and today this scattering phenomenon is known as the **Compton effect.** Figure 40.13b shows the quantum picture of the collision between an individual x-ray photon and an electron.

The second difference between the classical and quantum models is also shown in Figure 40.13b. In the classical model, the electron is pushed along the direction of propagation of the incident x-ray by radiation pressure. In the quantum model, the electron is scattered through an angle ϕ with respect to this direction, as if this were a billiard-ball type of collision. (The symbol ϕ used here is an angle and not to be confused with the work function, which was discussed in the preceding section.)

Figure 40.14 is a schematic diagram of the apparatus used by Compton. The x-rays, scattered from a graphite target, were analyzed with a rotating crystal spectrometer, and the intensity was measured with an ionization chamber that generated a current proportional to the intensity. The incident beam consisted of monochromatic x-rays of wavelength $\lambda_0 = 0.071$ nm. The experimental intensity-versus-wavelength plots observed by Compton for four scattering angles (corresponding to θ in Fig. 40.13) are shown in Figure 40.15. The graphs for the three nonzero angles show two peaks, one at λ_0 and one at $\lambda' > \lambda_0$. The shifted peak at λ' is caused by the scattering of x-rays from free electrons, and it was predicted by Compton to depend on scattering angle as

Compton shift equation

$$\lambda' - \lambda_0 = \frac{h}{m_e c} (1 - \cos\theta) \tag{40.11}$$

where m_e is the mass of the electron. This expression is known as the **Compton shift equation,** and the factor $h/m_e c$ is called the **Compton wavelength** of the electron. It has a currently accepted value of

Compton wavelength

$$\frac{h}{m_e c} = 0.002\ 43 \text{ nm}$$

The unshifted peak at λ_0 in Figure 40.15 is caused by x-rays scattered from electrons tightly bound to the target atoms. This unshifted peak also is predicted by Equation 40.11 if the electron mass is replaced with the mass of a carbon atom, which is about 23 000 times the mass of the electron. Thus, there is a wavelength shift for scattering from an electron bound to an atom, but it is so small that it was undetectable in Compton's experiment.

Compton's measurements were in excellent agreement with the predictions of Equation 40.11. It is fair to say that these results were the first that really convinced many physicists of the fundamental validity of quantum theory!

Quick Quiz 40.5 Note that for any given scattering angle θ, Equation 40.11 gives the same value for the Compton shift for any wavelength. Keeping this in mind, for which of the following types of radiation is the fractional shift in wavelength at a given scattering angle the largest? (a) radio waves (b) microwaves (c) visible light (d) x-rays.

Derivation of the Compton Shift Equation

We can derive the Compton shift equation by assuming that the photon behaves like a particle and collides elastically with a free electron initially at rest, as shown in Figure 40.13b. In this model, the photon is treated as a particle having energy $E = hf = hc/\lambda$ and zero rest energy. In the scattering process, the total energy and total linear momentum of the system must be conserved. Applying the principle of conservation of energy to this process gives

$$\frac{hc}{\lambda_0} = \frac{hc}{\lambda'} + K_e$$

where hc/λ_0 is the energy of the incident photon, hc/λ' is the energy of the scattered photon, and K_e is the kinetic energy of the recoiling electron. Because the electron may recoil at a speed comparable to that of light, we must use the relativistic expression $K_e = (\gamma - 1)m_ec^2$ (Eq. 39.23). Therefore,

$$\frac{hc}{\lambda_0} = \frac{hc}{\lambda'} + (\gamma - 1)m_ec^2 \tag{40.12}$$

where $\gamma = 1/\sqrt{1 - (v^2/c^2)}$.

Next, we apply the law of conservation of momentum to this collision, noting that the x and y components of momentum are each conserved independently. Figure 40.16 shows the initial momentum of the incident photon and the vector sum of the final momenta of the recoiling electron and the scattered photon in the collision of Figure 40.13b. Equation 39.28 shows that the momentum of a photon has a magnitude $p = E/c$, and we know from Equation 40.5 that $E = hf$. Therefore, $p = hf/c$. Substituting λf for c (Eq. 16.12) in this expression gives us $p = h/\lambda$. Because the relativistic expression for the momentum of the recoiling electron is $p_e = \gamma m_e v$ (Eq. 39.19), we obtain the following expressions for the x and y components of linear momentum, where the angles are as described in Figure 40.16:

$$x\text{ component:} \quad \frac{h}{\lambda_0} = \frac{h}{\lambda'}\cos\theta + \gamma m_e v\cos\phi \tag{40.13}$$

$$y\text{ component:} \quad 0 = \frac{h}{\lambda'}\sin\theta - \gamma m_e v\sin\phi \tag{40.14}$$

By eliminating v and ϕ from Equations 40.12 to 40.14, we obtain a single expression that relates the remaining three variables (λ', λ_0, and θ). After some algebra (see Problem 58), we obtain the Compton shift equation:

$$\lambda' - \lambda_0 = \frac{h}{m_ec}(1 - \cos\theta)$$

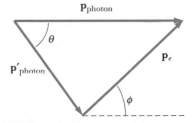

Figure 40.16 Momentum vectors for the Compton scattering event shown in Figure 40.13b.

Example 40.4   **Compton Scattering at 45°**   Interactive

X-rays of wavelength $\lambda_0 = 0.200\,000$ nm are scattered from a block of material. The scattered x-rays are observed at an angle of 45.0° to the incident beam. Calculate their wavelength.

Solution The shift in wavelength of the scattered x-rays is given by Equation 40.11:

$$\lambda' - \lambda_0 = \frac{h}{m_e c}\,(1 - \cos\theta)$$

$$\lambda' = \lambda_0 + \frac{6.626 \times 10^{-34}\,\mathrm{J\cdot s}}{(9.11 \times 10^{-31}\,\mathrm{kg})(3.00 \times 10^8\,\mathrm{m/s})}$$

$$\times\,(1 - \cos 45.0°)$$

$$= 0.200\,000 \times 10^{-9}\,\mathrm{m} + 7.10 \times 10^{-13}\,\mathrm{m}$$

$$= \boxed{0.200\,710\ \mathrm{nm}}$$

What If? What if we move our detector so that scattered x-rays are detected at an angle larger than 45°? Does the wavelength of the scattered x-rays increase or decrease as the angle θ increases?

Answer In Equation 40.11, if the angle θ increases, the term $\cos\theta$ decreases. Consequently, the factor $(1 - \cos\theta)$ increases. Therefore, the magnitude of the Compton shift increases. Because the shift is defined as the scattered wavelength minus the incident wavelength, the scattered wavelength increases.

We could also apply an energy argument to achieve this same result. As the scattering angle increases, more energy is transferred from the incident photon to the electron. As a result, the energy of the scattered photon decreases with increasing scattering angle. Because $E = hf$, the frequency of the scattered photon decreases and, because $\lambda = c/f$, the wavelength increases.

 Study Compton scattering for different angles at the Interactive Worked Example link at **http://www.pse6.com.**

40.4   Photons and Electromagnetic Waves

Phenomena such as the photoelectric effect and the Compton effect offer ironclad evidence that when light (or other forms of electromagnetic radiation) and matter interact, the light behaves as if it were composed of particles having energy hf and momentum h/λ. How can light be considered a photon (in other words, a particle) when we know it is a wave? On the one hand, we describe light in terms of photons having energy and momentum. On the other hand, we recognize that light and other electromagnetic waves exhibit interference and diffraction effects, which are consistent only with a wave interpretation.

Which model is correct? Is light a wave or a particle? The answer depends on the phenomenon being observed. Some experiments can be explained either better or solely with the photon model, whereas others are explained better or solely with the wave model. The end result is that **we must accept both models and admit that the true nature of light is not describable in terms of any single classical picture.** However, you should recognize that the same light beam that can eject photoelectrons from a metal (meaning that the beam consists of photons) can also be diffracted by a grating (meaning that the beam is a wave). In other words, **the particle model and the wave model of light complement each other.**

The success of the particle model of light in explaining the photoelectric effect and the Compton effect raises many other questions. If light is a particle, what is the meaning of the "frequency" and "wavelength" of the particle, and which of these two properties determines its energy and momentum? Is light *simultaneously* a wave and a particle? Although photons have no rest energy (a nonobservable quantity because a photon cannot be at rest!), is there a simple expression for the *effective mass* of a moving photon? If photons have effective mass, do they experience gravitational attraction? What is the spatial extent of a photon, and how does an electron absorb or scatter one photon? Some of these questions can be answered, but others demand a view of atomic processes that is too pictorial and literal. Many of these questions stem from classical analogies such as colliding billiard balls and water waves breaking on a shore. Quantum mechanics gives light a more fluid and flexible nature by treating the particle model and wave model of light as both necessary and complementary. Neither model can be used exclusively to describe all properties of light. A complete understanding of the observed behavior of light can be attained only if the two models are combined in a complementary manner. Before discussing this combination in more detail, we now turn our attention from electromagnetic waves to the behavior of entities that we have called particles.

40.5 The Wave Properties of Particles

Students introduced to the dual nature of light often find the concept difficult to accept. In the world around us, we are accustomed to regarding such things as baseballs solely as particles and other things such as sound waves solely as forms of wave motion. Every large-scale observation can be interpreted by considering either a wave explanation or a particle explanation, but in the world of photons and electrons, such distinctions are not as sharply drawn. Even more disconcerting is the fact that, under certain conditions, the things we unambiguously call "particles" exhibit wave characteristics!

In his 1923 doctoral dissertation, Louis de Broglie postulated that **because photons have both wave and particle characteristics, perhaps all forms of matter have both properties.** This was a highly revolutionary idea with no experimental confirmation at that time. According to de Broglie, electrons, just like light, have a dual particle–wave nature.

In Section 40.3, we found that the momentum of a photon can be expressed as

$$p = \frac{h}{\lambda}$$

From this equation we see that the photon wavelength can be specified by its momentum: $\lambda = h/p$. De Broglie suggested that material particles of momentum p have a characteristic wavelength that is given by the same expression, $\lambda = h/p$. Because the magnitude of the momentum of a particle of mass m and speed v is $p = mv$, the **de Broglie wavelength** of that particle is[5]

$$\lambda = \frac{h}{p} = \frac{h}{mv} \qquad\qquad (40.15)$$

de Broglie wavelength

Furthermore, in analogy with photons, de Broglie postulated that particles obey the Einstein relation $E = hf$, where E is the total energy of the particle. Then the frequency of a particle is

$$f = \frac{E}{h} \qquad\qquad (40.16)$$

The dual nature of matter is apparent in these last two equations because each contains both particle concepts (mv and E) and wave concepts (λ and f). The fact that these relationships are established experimentally for photons makes the de Broglie hypothesis somewhat easier to accept.

The Davisson–Germer Experiment

De Broglie's proposal in 1923 that matter exhibits both wave and particle properties was regarded as pure speculation. If particles such as electrons had wave properties, then under the correct conditions they should exhibit diffraction effects. Only three years later, C. J. Davisson (1881–1958) and L. H. Germer (1896–1971) succeeded in measuring the wavelength of electrons. Their important discovery provided the first experimental confirmation of the matter waves proposed by de Broglie.

Louis de Broglie
French Physicist (1892–1987)

De Broglie was born in Dieppe, France. At the Sorbonne in Paris, he studied history in preparation for what he hoped to be a career in the diplomatic service. The world of science is lucky that he changed his career path to become a theoretical physicist. De Broglie was awarded the Nobel Prize in 1929 for his prediction of the wave nature of electrons. (*AIP Niels Bohr Library*)

---

[5] The de Broglie wavelength for a particle moving at *any* speed v is $\lambda = h/\gamma mv$ where $\gamma = [1 - (v^2/c^2)]^{-1/2}$.

Interestingly, the intent of the initial Davisson–Germer experiment was not to confirm the de Broglie hypothesis. In fact, their discovery was made by accident (as is often the case). The experiment involved the scattering of low-energy electrons (about 54 eV) from a nickel target in a vacuum. During one experiment, the nickel surface was badly oxidized because of an accidental break in the vacuum system. After the target was heated in a flowing stream of hydrogen to remove the oxide coating, electrons scattered by it exhibited intensity maxima and minima at specific angles. The experimenters finally realized that the nickel had formed large crystalline regions upon heating and that the regularly spaced planes of atoms in these regions served as a diffraction grating for electron matter waves. (See the discussion of diffraction by crystals in Section 38.5.)

Shortly thereafter, Davisson and Germer performed more extensive diffraction measurements on electrons scattered from single-crystal targets. Their results showed conclusively the wave nature of electrons and confirmed the de Broglie relationship $p = h/\lambda$. In the same year, G. P. Thomson (1892–1975) of Scotland also observed electron diffraction patterns by passing electrons through very thin gold foils. Diffraction patterns have since been observed in the scattering of helium atoms, hydrogen atoms, and neutrons. Hence, the universal nature of matter waves has been established in various ways.

The problem of understanding the dual nature of matter and radiation is conceptually difficult because the two models seem to contradict each other. This problem as it applies to light was discussed earlier. The **principle of complementarity** states that the **wave and particle models of either matter or radiation complement each other.** Neither model can be used exclusively to describe matter or radiation adequately. Because humans tend to generate mental images based on their experiences from the everyday world (baseballs, water waves, and so forth), we use both descriptions in a complementary manner to explain any given set of data from the quantum world.

▲ **PITFALL PREVENTION**

40.3 What's Waving?

If particles have wave properties, what's waving? You are familiar with waves on strings, which are very concrete. Sound waves are more abstract, but you are likely comfortable with them. Electromagnetic waves are even more abstract, but at least they can be described in terms of physical variables and electric and magnetic fields. In contrast, waves associated with particles are completely abstract and cannot be associated with a physical variable. In Chapter 41, we will describe the wave associated with a particle in terms of probability.

Quick Quiz 40.6 An electron and a proton both moving at nonrelativistic speeds have the same de Broglie wavelength. Which of the following are also the same for the two particles? (a) speed (b) kinetic energy (c) momentum (d) frequency.

Quick Quiz 40.7 We have discussed two wavelengths associated with the electron—the Compton wavelength and the de Broglie wavelength. Which is an actual *physical* wavelength associated with the electron? (a) the Compton wavelength (b) the de Broglie wavelength (c) both wavelengths (d) neither wavelength.

Example 40.5 The Wavelength of an Electron

Calculate the de Broglie wavelength for an electron ($m_e = 9.11 \times 10^{-31}$ kg) moving at 1.00×10^7 m/s.

Solution Equation 40.15 gives

$$\lambda = \frac{h}{m_e v} = \frac{6.63 \times 10^{-34}\,\text{J}\cdot\text{s}}{(9.11 \times 10^{-31}\,\text{kg})(1.00 \times 10^7\,\text{m/s})}$$

$$= 7.28 \times 10^{-11}\,\text{m}$$

Example 40.6 The Wavelength of a Rock

A rock of mass 50 g is thrown with a speed of 40 m/s. What is its de Broglie wavelength?

Solution From Equation 40.15 we have

$$\lambda = \frac{h}{mv} = \frac{6.63 \times 10^{-34}\,\text{J}\cdot\text{s}}{(50 \times 10^{-3}\,\text{kg})(40\,\text{m/s})}$$

$$\lambda = 3.32 \times 10^{-34}\,\text{m}$$

This wavelength is much smaller than any aperture through which the rock could possibly pass. This means that we could not observe diffraction effects, and as a result the wave properties of large-scale objects cannot be observed.

Example 40.7   An Accelerated Charged Particle

A particle of charge q and mass m has been accelerated from rest to a nonrelativistic speed through a potential difference ΔV. Find an expression for its de Broglie wavelength.

Solution When a charged particle is accelerated from rest through a potential difference ΔV, its gain in kinetic energy $\frac{1}{2}mv^2$ must equal the loss in potential energy $q \Delta V$ of the charge–field system:

$$\tfrac{1}{2}mv^2 = q \Delta V$$

Because $p = mv$, we can express this equation in the form

$$\frac{p^2}{2m} = q \Delta V$$

$$p = \sqrt{2mq \Delta V}$$

Substituting this expression for p into Equation 40.15 gives

$$\lambda = \frac{h}{p} = \frac{h}{\sqrt{2mq \Delta V}}$$

The Electron Microscope

A practical device that relies on the wave characteristics of electrons is the **electron microscope.** A *transmission* electron microscope, used for viewing flat, thin samples, is shown in Figure 40.17. In many respects it is similar to an optical microscope, but the electron microscope has a much greater resolving power because it can accelerate electrons to very high kinetic energies, giving them very short wavelengths. No microscope can resolve details that are significantly smaller than the wavelength of the waves used to illuminate the object. Typically, the wavelengths of electrons are about 100 times shorter than those

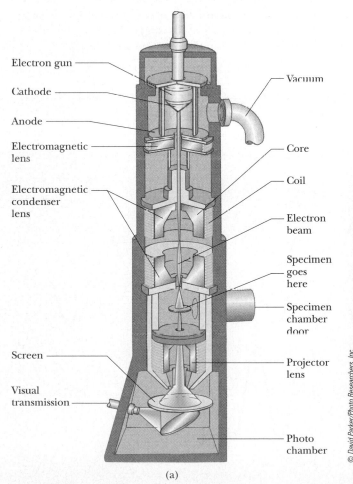

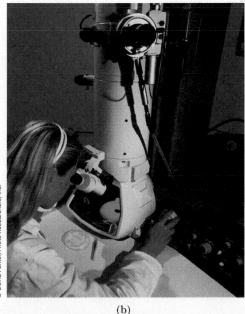

© David Parker/Photo Researchers, Inc.

(a)   (b)

Figure 40.17 (a) Diagram of a transmission electron microscope for viewing a thinly sectioned sample. The "lenses" that control the electron beam are magnetic deflection coils. (b) An electron microscope.

Figure 40.18 A color-enhanced electron microscope photograph shows significant detail of a storage mite, *Lepidoglyphus destructor*. The mite is so small, with a maximum length of 0.75 mm, that ordinary microscopes do not reveal minute anatomical details.

of the visible light used in optical microscopes. As a result, an electron microscope with ideal lenses would be able to distinguish details about 100 times smaller than those distinguished by an optical microscope. (Electromagnetic radiation of the same wavelength as the electrons in an electron microscope is in the x-ray region of the spectrum.)

The electron beam in an electron microscope is controlled by electrostatic or magnetic deflection, which acts on the electrons to focus the beam and form an image. Rather than examining the image through an eyepiece as in an optical microscope, the viewer looks at an image formed on a monitor or other type of display screen. Figure 40.18 shows the amazing detail available with an electron microscope.

40.6 The Quantum Particle

The discussions presented in previous sections may be quite disturbing, because we considered the particle and wave models to be distinct in the past. The notion that both light and material particles have both particle and wave properties does not fit with this distinction. We have experimental evidence, however, that this is just what we must accept. The recognition of this dual nature leads to a new model, the **quantum particle.** In this model, entities have both particle and wave characteristics, and we must choose one appropriate behavior—particle or wave—in order to understand a particular phenomenon.

In this section, we shall explore this model in a way that might make you more comfortable with this idea. We shall do this by demonstrating that we can construct from waves an entity that exhibits properties of a particle.

Let us first recall some characteristics of ideal particles and ideal waves. An ideal particle has zero size. Thus, an essential feature of a particle is that it is *localized* in space. An ideal wave has a single frequency and is infinitely long, as suggested by Figure 40.19a.

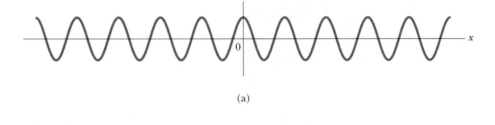

(a)

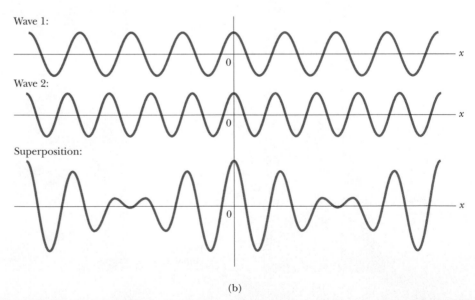

Wave 1:

Wave 2:

Superposition:

(b)

Figure 40.19 (a) An idealized wave of an exact single frequency is the same throughout space and time. (b) If two ideal waves with slightly different frequencies are combined, beats result (Section 18.7). The regions of space at which there is constructive interference are different from those at which there is destructive interference.

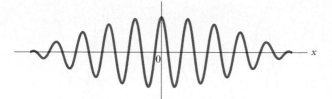

Active Figure 40.20 If a large number of waves are combined, the result is a wave packet, which represents a particle.

Choose the number of waves to add together and observe the resulting wave packet at the Active Figures link at http://www.pse6.com.

Thus, an ideal wave is *unlocalized* in space. Let us show that we can build a localized entity from infinitely long waves. Imagine drawing one wave along the x axis, with one of its crests located at $x = 0$, as at the top of Figure 40.19b. Now draw a second wave, of the same amplitude but a different frequency, with one of its crests also at $x = 0$. The result of the superposition of these two waves is a *beat*, as the waves are alternately in phase and out of phase. (Beats were discussed in Section 18.7.) The lowest curve in Figure 40.19b shows the results of superposing these two waves.

Notice that we have already introduced some localization by superposing the two waves. A single wave has the same amplitude everywhere in space—no point in space is any different from any other point. By adding a second wave, however, there is something different about the in-phase points compared to the out-of-phase points.

Now imagine that more and more waves are added to our original two, each new wave having a new frequency. Each new wave is added so that one of its crests is at $x = 0$. The result is that all of the waves add constructively at $x = 0$. When we consider a large number of waves, the probability of a positive value of a wave function at any point $x \neq 0$ is equal to the probability of a negative value, and there is destructive interference *everywhere* except near $x = 0$, where we superposed all of the crests. The result of this is shown in Figure 40.20. The small region of constructive interference is called a **wave packet.** This is a localized region of space that is different from all other regions. We can identify the wave packet as a particle—it has the localized nature of what we have come to recognize as a particle! The location of the wave packet corresponds to the position of the particle.

The localized nature of this entity is the *only* characteristic of a particle that was generated with this process. We have not addressed how the wave packet might achieve such particle characteristics as mass, electric charge, spin, etc. Thus, you may not be completely convinced that we have built a particle. As further evidence that the wave packet can represent the particle, let us show that the wave packet has another characteristic of a particle.

We return to our combination of two waves in order to make the mathematical representation simple. Consider two waves with equal amplitudes but different frequencies f_1 and f_2. We can represent the waves mathematically as

$$y_1 = A\cos(k_1 x - \omega_1 t) \qquad \text{and} \qquad y_2 = A\cos(k_2 x - \omega_2 t)$$

where, as in Chapter 16, $\omega = 2\pi f$ and $k = 2\pi/\lambda$. Using the superposition principle, we add the waves:

$$y = y_1 + y_2 = A\cos(k_1 x - \omega_1 t) + A\cos(k_2 x - \omega_2 t)$$

It is convenient to write this in a form that uses the trigonometric identity

$$\cos a + \cos b = 2\cos\left(\frac{a - b}{2}\right)\cos\left(\frac{a + b}{2}\right)$$

Letting $a = k_1 x - \omega_1 t$ and $b = k_2 x - \omega_2 t$, we find that

$$y = 2A\cos\left(\frac{(k_1 x - \omega_1 t) - (k_2 x - \omega_2 t)}{2}\right)\cos\left(\frac{(k_1 x - \omega_1 t) + (k_2 x - \omega_2 t)}{2}\right)$$

$$= \left[2A\cos\left(\frac{\Delta k}{2}x - \frac{\Delta\omega}{2}t\right)\right]\cos\left(\frac{k_1 + k_2}{2}x - \frac{\omega_1 + \omega_2}{2}t\right) \qquad (40.17)$$

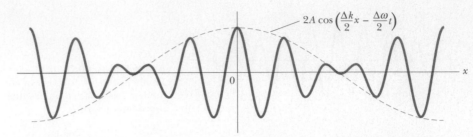

Observe the movement of the waves and the envelope at the Active Figures link at http://www.pse6.com.

Active Figure 40.21 The beat pattern of Figure 40.19b, with an envelope function (blue curve) superimposed.

where $\Delta k = k_1 - k_2$ and $\Delta\omega = \omega_1 - \omega_2$. The second cosine factor represents a wave with a wave number and frequency that are equal to the averages of the values for the individual waves.

The factor in square brackets represents the envelope of the wave, as shown by the blue curve in Figure 40.21. Notice that this factor also has the mathematical form of a wave. **This envelope of the combination can travel through space with a different speed than the individual waves.** As an extreme example of this possibility, imagine combining two identical waves moving in opposite directions. The two waves move with the same speed, but the envelope has a speed of *zero*, because we have built a standing wave, which we studied in Section 18.2.

For an individual wave, the speed is given by Equation 16.11,

Phase speed of a wave in a wave packet

$$v_{\text{phase}} = \frac{\omega}{k} \qquad (40.18)$$

This speed is called the **phase speed** because it is the rate of advance of a crest on a single wave, which is a point of fixed phase. This equation can be interpreted as follows: the phase speed of a wave is the ratio of the coefficient of the time variable t to the coefficient of the space variable x in the equation for the wave, $y = A\cos(kx - \omega t)$.

The factor in brackets in Equation 40.17 is of the form of a wave, so it moves with a speed given by this same ratio:

$$v_g = \frac{\text{coefficient of time variable } t}{\text{coefficient of space variable } x} = \frac{(\Delta\omega/2)}{(\Delta k/2)} = \frac{\Delta\omega}{\Delta k}$$

The subscript g on the speed indicates that this is commonly called the **group speed,** or the speed of the wave packet (the *group* of waves) that we have built. We have generated this expression for a simple addition of two waves. For a superposition of a very large number of waves to form a wave packet, this ratio becomes a derivative:

Group speed of a wave packet

$$v_g = \frac{d\omega}{dk} \qquad (40.19)$$

Let us multiply the numerator and the denominator by $\hbar$, where $\hbar = h/2\pi$:

$$v_g = \frac{\hbar\, d\omega}{\hbar\, dk} = \frac{d(\hbar\omega)}{d(\hbar k)} \qquad (40.20)$$

We look at the terms in the parentheses in the numerator and denominator of Equation 40.20 separately. For the numerator,

$$\hbar\omega = \frac{h}{2\pi}\,(2\pi f) = hf = E$$

For the denominator,

$$\hbar k = \frac{h}{2\pi}\left(\frac{2\pi}{\lambda}\right) = \frac{h}{\lambda} = p$$

Thus, Equation 40.20 can be written as

$$v_g = \frac{d(\hbar\omega)}{d(\hbar k)} = \frac{dE}{dp}$$ (40.21)

Because we are exploring the possibility that the envelope of the combined waves represents the particle, consider a free particle moving with a speed u that is small compared to that of light. The energy of the particle is its kinetic energy:

$$E = \tfrac{1}{2}mu^2 = \frac{p^2}{2m}$$

Differentiating this equation with respect to p gives

$$v_g = \frac{dE}{dp} = \frac{d}{dp}\left(\frac{p^2}{2m}\right) = \frac{1}{2m}(2p) = u$$ (40.22)

Thus, the group speed of the wave packet is identical to the speed of the particle that it is modeled to represent! This gives us further confidence that the wave packet is a reasonable way to build a particle.

Quick Quiz 40.8 As an analogy to wave packets, consider an "automobile packet" that occurs near the scene of an accident on a freeway. The phase speed is analogous to the speed of individual automobiles as they move through the backup caused by the accident. The group speed can be identified as the speed of the leading edge of the packet of cars. For the automobile packet, (a) the group speed is the same as the phase speed. (b) the group speed is less than the phase speed. (c) the group speed is greater than the phase speed.

Quick Quiz 40.9 As another analogy to wave packets, consider a "runner packet" that occurs at the start of a footrace of length L. As the runners begin the race, the packet of runners spreads in size as the faster runners outpace the slower runners. The phase speed is the speed of a single runner, while we can identify the group speed v_g as the speed with which the average position of the entire packet of runners moves. The time interval for the winning runner to run the race is (a) greater than L/v_g. (b) equal to L/v_g. (c) less than L/v_g.

40.7 The Double-Slit Experiment Revisited

Wave–particle duality is now a firmly accepted concept reinforced by experimental results including those of the Davisson–Germer experiment. However, as with the postulates of special relativity, this concept often leads to clashes with familiar thought patterns we hold from everyday experience.

One way to crystallize our ideas about the electron's wave–particle duality is to consider an experiment in which electrons are fired at a double slit. Consider a parallel beam of mono-energetic electrons that is incident on a double slit as in Figure 40.22. We shall assume that the slit widths are small compared to the electron wavelength, so that we need not worry about diffraction maxima and minima, as we discussed for light in Section 38.2. An electron detector is positioned far from the slits at a distance much greater than d, the separation distance of the slits. If the detector collects electrons for a long enough time, one finds a typical wave interference pattern for the counts per minute, or probability of arrival of electrons. Such an interference pattern would not be expected if the electrons behaved as classical particles. This is clear evidence that electrons are interfering—a distinct wave-like behavior.

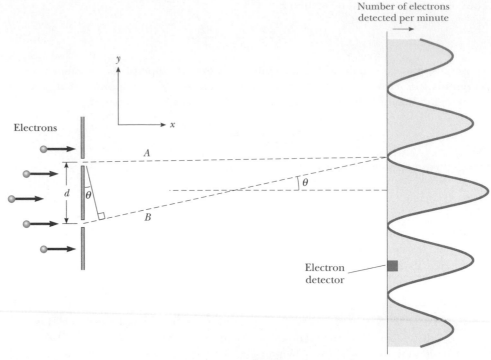

From E. R. Huggins, Physics I, New York, W. A. Benjamin, 1968

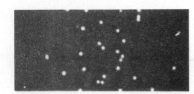

(a) After 28 electrons

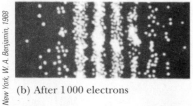

(b) After 1 000 electrons

(c) After 10 000 electrons

From C. Jönsson, Zeitschrift fur Physik 161:454, 1961; used with permission

(d) Two-slit electron pattern

Active Figure 40.23 (a), (b), (c) Computer-simulated interference patterns for a beam of electrons incident on a double slit. (d) Photograph of a double-slit interference pattern produced by electrons.

At the Active Figures link at http://www.pse6.com, *you can watch the interference pattern develop over time and see how it is destroyed by the action of keeping track of which slit an electron goes through.*

Figure 40.22 Electron diffraction. The slit separation d is much greater than the individual slit widths and much less than the distance between the slit and the detector. The electron detector is movable along the y direction in the drawing and so can detect electrons diffracted at different values of θ. The detector acts like the "viewing screen" of Young's double-slit experiment with light discussed in Chapter 37.

If the experiment is carried out at lower electron beam intensities, the interference pattern is still observed if the time interval for the measurement is sufficiently long. This is illustrated by the computer-simulated patterns in Figure 40.23. Note that the interference pattern becomes clearer as the number of electrons reaching the screen increases.

If one imagines a single electron in the beam reaching one of the slits, standard wave theory can be used to find the angular separation θ between the central probability maximum and its neighboring minimum. The minimum occurs when the path length difference between A and B is half a wavelength, or when

$$d \sin \theta = \frac{\lambda}{2}$$

Because an electron's wavelength is given by $\lambda = h/p$, we see that for small θ,

$$\sin \theta \approx \theta = \frac{h}{2pd}$$

Thus, the dual nature of the electron is clearly shown in this experiment: **the electrons are detected as particles at a localized spot at some instant of time, but the probability of arrival at that spot is determined by finding the intensity of two interfering waves.**

Let us now look at this experiment from another point of view. If one slit is covered during the experiment, a symmetric curve peaked around the center of the open slit is observed—the central maximum of the single-slit diffraction pattern. Plots of the counts per minute (probability of arrival of electrons) with the lower or upper slit closed are shown as blue curves in the central part of Figure 40.24.

If another experiment is now performed with slit 2 blocked half of the time and then slit 1 blocked during the remaining time, the accumulated pattern of counts/min shown by the blue curve on the right side of Figure 40.24 is completely different from

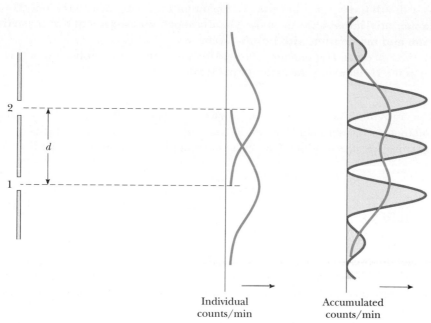

Figure 40.24 Accumulated results (blue) of the two-slit electron diffraction experiment with each slit closed half the time. The result with both slits open is shown in red.

the case with both slits open (red curve). There is no longer a maximum probability of arrival of an electron at $\theta = 0$. In fact, **the interference pattern has been lost and the accumulated result is simply the sum of the individual results.** When only one slit is open at a time, we know that the electron has the same localizability and indivisibility at the slits as we measure at the detector, because the electron clearly goes through slit 1 or slit 2. Thus, the total must be analyzed as the sum of those electrons that come through slit 1, and those that come through slit 2.

When both slits are open, it is tempting to assume that the electron goes through either slit 1 or slit 2, and that the counts/min are again given by the combination of the single-slit distributions. However, the experimental results indicated by the red interference pattern in Figure 40.24 contradict this assumption. Hence, our assumption that the electron is localized and goes through only one slit when both slits are open must be wrong (a painful conclusion!).

In order to interpret these results, we are forced to conclude that **an electron interacts with both slits simultaneously.** If we attempt to determine experimentally which slit the electron goes through, the act of measuring destroys the interference pattern. It is impossible to determine which slit the electron goes through. In effect, **we can say only that the electron passes through both slits!** The same arguments apply to photons.

If we restrict ourselves to a pure particle model, it is an uncomfortable notion that the electron can be present at both slits at once. From the quantum particle model, however, the particle can be considered to be built from waves that exist throughout space. Thus, the wave components of the electron are present at both slits at the same time, and this model leads to a more comfortable interpretation of this experiment.

40.8 The Uncertainty Principle

Whenever one measures the position or velocity of a particle at any instant, experimental uncertainties are built into the measurements. According to classical mechanics, there is no fundamental barrier to an ultimate refinement of the apparatus or experimental procedures. In other words, it is possible, in principle, to make such measurements with arbitrarily small uncertainty. Quantum theory predicts, however, that **it is fundamentally impossible to make simultaneous measurements of a particle's position and momentum with infinite accuracy.**

In 1927, Werner Heisenberg (1901–1976) introduced this notion, which is now known as the **Heisenberg uncertainty principle:**

If a measurement of the position of a particle is made with uncertainty Δx and a simultaneous measurement of its x component of momentum is made with uncertainty Δp_x, the product of the two uncertainties can never be smaller than $\hbar/2$:

Heisenberg uncertainty principle

$$\Delta x \, \Delta p_x \geq \frac{\hbar}{2} \qquad (40.23)$$

That is, **it is physically impossible to measure simultaneously the exact position and exact momentum of a particle.** Heisenberg was careful to point out that the inescapable uncertainties Δx and Δp_x do not arise from imperfections in practical measuring instruments. Rather, **the uncertainties arise from the quantum structure of matter.**

To understand the uncertainty principle suppose the wavelength of a particle is known *exactly*. According to the de Broglie relation, $\lambda = h/p$, we would therefore know the momentum to be precisely $p = h/\lambda$.

In reality, a single-wavelength wave would exist throughout space. Any region along this wave is the same as any other region (Fig. 40.19a). If we were to ask, "Where is the particle that this wave represents?", there is no special location in space along the wave that could be identified with the particle—all points along the wave are the same. Therefore, we have *infinite* uncertainty in the position of the particle—we know nothing about its location. Perfect knowledge of the particle's momentum has cost us all information about its location.

In comparison, now consider a particle whose momentum is uncertain, so that it has a range of possible values of momentum. According to the de Broglie relation, this results in a range of wavelengths. Thus, the particle is not represented by a single wavelength, but a combination of wavelengths within this range. This combination forms a wave packet, as we discussed in Section 40.6 and illustrated in Figure 40.20. If we are asked to determine the location of the particle, we can only say that it is somewhere in the region defined by the wave packet—because there is a distinct difference between this region and the rest of space. Thus, by losing some information about the momentum of the particle, we have gained information about its position.

If we were to lose *all* information about the momentum, then we would be adding together waves of all possible wavelengths. This would result in a wave packet of zero length. Thus, if we know nothing about the momentum, we know exactly where the particle is.

The mathematical form of the uncertainty principle states that the product of the uncertainties in position and momentum is always larger than some minimum value. This value can be calculated from the types of arguments discussed above, which result in the value of $\hbar/2$ in Equation 40.23.

Another form of the uncertainty principle can be generated by reconsidering Figure 40.20. Imagine that the horizontal axis is time rather than spatial position x. We can then make the same arguments that we made about knowledge of wavelength and position in the time domain. The corresponding variables would be frequency and time. Because frequency is related to the energy of the particle by $E = hf$, the uncertainty principle in this form is

$$\Delta E \, \Delta t \geq \frac{\hbar}{2} \qquad (40.24)$$

Werner Heisenberg
German Theoretical Physicist (1901–1976)

Heisenberg obtained his Ph.D. in 1923 at the University of Munich. While other physicists tried to develop physical models of quantum phenomena, Heisenberg developed an abstract mathematical model called *matrix mechanics*. The more widely accepted physical models were shown to be equivalent to matrix mechanics. Heisenberg made many other significant contributions to physics, including his famous uncertainty principle for which he received a Nobel Prize in 1932, the prediction of two forms of molecular hydrogen, and theoretical models of the nucleus. (*Courtesy of the University of Hamburg*)

This form of the uncertainty principle suggests that energy conservation can appear to be violated by an amount ΔE as long as it is only for a short time interval Δt consistent with Equation 40.24. We shall use this notion to estimate the rest energies of particles in Chapter 46.

> ## Quick Quiz 40.10
> The location of a particle is measured and specified as being exactly at $x = 0$, with *zero* uncertainty in the x direction. How does this affect the uncertainty of its velocity component in the y direction? (a) It does not affect it. (b) It makes it infinite. (c) It makes it zero.
>
> ## Quick Quiz 40.11
> A quantum argument for the phenomenon of diffraction of light claims that photons passing through a narrow slit have been localized to the width of the slit. Because we have gained information about their position, they must have a larger uncertainty in momentum along the plane of the screen in which the slit is cut. Thus, the photons gain momentum perpendicular to their original direction of propagation and spread out, forming on a screen a bright area that is wider than the slit. Suppose we are observing diffraction of light and suddenly Planck's constant drops to half its previous value. This quantum argument for diffraction would claim that (a) the bright area on the screen is unchanged (b) the bright area on the screen becomes wider (c) the bright area on the screen becomes narrower.

▲ **PITFALL PREVENTION**

40.4 The Uncertainty Principle

Some students incorrectly interpret the uncertainty principle as meaning that a measurement interferes with the system. For example, if an electron is observed in a hypothetical experiment using an optical microscope, the photon used to see the electron collides with it and makes it move, giving it an uncertainty in momentum. This is *not* the idea of the uncertainty principle. The uncertainty principle is independent of the measurement process and is grounded in the wave nature of matter.

Example 40.8 Locating an Electron

The speed of an electron is measured to be 5.00×10^3 m/s to an accuracy of 0.003 00%. Find the minimum uncertainty in determining the position of this electron.

Solution Assuming that the electron is moving along the x axis, the x component of the momentum of the electron is

$$p_x = mv_x = (9.11 \times 10^{-31} \text{ kg})(5.00 \times 10^3 \text{ m/s})$$
$$- 4.56 \times 10^{-27} \text{ kg} \cdot \text{m/s}$$

The uncertainty in p_x is 0.003 00% of this value:

$$\Delta p_x = (0.000\ 030\ 0)(4.56 \times 10^{-27} \text{ kg} \cdot \text{m/s})$$
$$= 1.37 \times 10^{-31} \text{ kg} \cdot \text{m/s}$$

We can now calculate the minimum uncertainty in position by using this value of Δp_x and Equation 40.23:

$$\Delta x\, \Delta p_x \geq \frac{\hbar}{2}$$

$$\Delta x \geq \frac{\hbar}{2\, \Delta p_x} = \frac{1.05 \times 10^{-34} \text{ J} \cdot \text{s}}{2(1.37 \times 10^{-31} \text{ kg} \cdot \text{m/s})}$$

$$= \boxed{0.383 \text{ mm}}$$

Example 40.9 The Line Width of Atomic Emissions

Atoms have quantized energy levels similar to those of Planck's oscillators, although the energy levels of an atom are generally not evenly spaced. When an atom makes a transition between states, energy is emitted in the form of a photon. Although an excited atom can radiate at any time from $t = 0$ to $t = \infty$, the average time interval after excitation during which an atom radiates is called the **lifetime** τ. If $\tau = 1.0 \times 10^{-8}$ s, use the uncertainty principle to compute the line width Δf produced by this finite lifetime.

Solution We use $\Delta E\, \Delta t \geq \hbar/2$, where $\Delta E = h\, \Delta f$ and $\Delta t = 1.0 \times 10^{-8}$ s is the lifetime. Thus, the minimum value of

Δf is

$$\Delta f = \frac{\Delta E}{h} = \frac{\hbar}{2h\, \Delta t} = \frac{1}{4\pi\, \Delta t} = \frac{1}{4\pi(1.0 \times 10^{-8} \text{ s})}$$

$$= \boxed{8.0 \times 10^6 \text{ Hz}}$$

Note that ΔE is the uncertainty in the energy of the excited atom. It is also the uncertainty in the energy of the photon emitted by an atom in this state.

What If? What if this same lifetime were associated with a transition that emits a radio wave rather than a visible light

wave from an atom? Is the fractional line width $\Delta f/f$ larger or smaller than for the visible light?

Answer Because we are assuming the same lifetime for both transitions, Δf is independent of the frequency of radiation. Radio waves have lower frequencies than light waves, so the ratio $\Delta f/f$ will be larger for the radio waves. Assuming a light-wave frequency f of 6.00×10^{14} Hz, the fractional line width is

$$\frac{\Delta f}{f} = \frac{8.0 \times 10^6 \text{ Hz}}{6.00 \times 10^{14} \text{ Hz}} = 1.3 \times 10^{-8}$$

This narrow fractional line width can be measured with a sensitive interferometer. Usually, however, temperature and pressure effects overshadow the natural line width and broaden the line through mechanisms associated with the Doppler effect and collisions.

Assuming a radio-wave frequency f of 94.7×10^6 Hz, the fractional line width is

$$\frac{\Delta f}{f} = \frac{8.0 \times 10^6 \text{ Hz}}{94.7 \times 10^6 \text{ Hz}} = 8.4 \times 10^{-2}$$

Thus, for the radio wave, this same absolute line width corresponds to a fractional line width of over 8%.

SUMMARY

Take a practice test for this chapter by clicking on the Practice Test link at http://www.pse6.com.

The characteristics of **blackbody radiation** cannot be explained using classical concepts. Planck introduced the quantum concept and Planck's constant when he assumed that atomic oscillators existing only in discrete energy states were responsible for this radiation. In Planck's model, radiation is emitted in single quantized packets whenever an oscillator makes a transition between discrete energy states. Einstein successfully extended Planck's quantum hypothesis to the standing waves of electromagnetic radiation in a cavity used in the blackbody radiation model.

The **photoelectric effect** is a process whereby electrons are ejected from a metal surface when light is incident on that surface. In Einstein's model, light is viewed as a stream of particles, or **photons,** each having energy $E = hf$, where f is the frequency and h is Planck's constant. The maximum kinetic energy of the ejected photoelectron is

$$K_{max} = hf - \phi \tag{40.9}$$

where ϕ is the **work function** of the metal.

X-rays are scattered at various angles by electrons in a target. In such a scattering event, a shift in wavelength is observed for the scattered x-rays, and the phenomenon is known as the **Compton effect.** Classical physics does not predict the correct behavior in this effect. If the x-ray is treated as a photon, conservation of energy and linear momentum applied to the photon–electron collisions yields for the Compton shift:

$$\lambda' - \lambda_0 = \frac{h}{m_e c} (1 - \cos \theta) \tag{40.11}$$

where m_e is the mass of the electron, c is the speed of light, and θ is the scattering angle.

Light has a dual nature in that it has both wave and particle characteristics. Some experiments can be explained either better or solely by the particle model, whereas others can be explained either better or solely by the wave model.

Every object of mass m and momentum $p = mv$ has wave properties, with a **de Broglie wavelength** given by:

$$\lambda = \frac{h}{p} = \frac{h}{mv} \tag{40.15}$$

By combining a large number of waves, a single region of constructive interference, called a **wave packet,** can be created. The wave packet carries the characteristic of localization like a particle, but has wave properties because it is built from waves. For an individual wave in the wave packet, the **phase speed** is

$$v_{phase} = \frac{\omega}{k} \tag{40.18}$$

For the wave packet as a whole, the **group speed** is

$$v_g = \frac{d\omega}{dk} \qquad (40.19)$$

For a wave packet representing a particle, the group speed can be shown to be the same as the speed of the particle.

The **Heisenberg uncertainty principle** states that if a measurement of position is made with uncertainty Δx and a simultaneous measurement of linear momentum is made with uncertainty Δp_x, the product of the two uncertainties is restricted to

$$\Delta x \, \Delta p_x \geq \frac{\hbar}{2} \qquad (40.23)$$

Another form of the uncertainty principle relates measurements of energy and time:

$$\Delta E \, \Delta t \geq \frac{\hbar}{2} \qquad (40.24)$$

QUESTIONS

1. What assumptions did Planck make in dealing with the problem of blackbody radiation? Discuss the consequences of these assumptions.

2. The classical model of blackbody radiation given by the Rayleigh–Jeans law has two major flaws. Identify them and explain how Planck's law deals with them.

3. All objects radiate energy. Why, then, are we not able to see all objects in a dark room?

4. The brightest star in the constellation Lyra is the bluish star Vega, and the brightest star in Boötes is the reddish star Arcturus. How do you account for the difference in color of the two stars?

5. If the photoelectric effect is observed for one metal, can you conclude that the effect will also be observed for another metal under the same conditions? Explain.

6. What does the slope of the line in Figure 40.11 represent? What does the y intercept represent? How would such graphs for different metals compare with one another?

7. Why does the existence of a cutoff frequency in the photoelectric effect favor a particle theory for light over a wave theory?

8. In the photoelectric effect, explain why the stopping potential depends on the frequency of light but not on the intensity.

9. A student claims that she is going to eject electrons from a piece of metal by placing a radio transmitter antenna adjacent to the metal and sending a strong AM radio signal into the antenna. The work function of a metal is typically a few electron volts. Will this work?

10. Suppose the photoelectric effect occurs in a gaseous target rather than a solid plate. Will photoelectrons be produced at all frequencies of the incident photon? Explain.

11. Which has more energy, a photon of ultraviolet radiation or a photon of yellow light?

12. Which is more likely to cause sunburn (because more energy is absorbed by individual molecules in skin cells): (a) infrared light, (b) visible light, or (c) ultraviolet light?

13. How does the Compton effect differ from the photoelectric effect?

14. What assumptions did Compton make in dealing with the scattering of a photon from an electron?

15. An x-ray photon is scattered by an electron. What happens to the frequency of the scattered photon relative to that of the incident photon?

16. Suppose a photograph were made of a person's face using only a few photons. Would the result be simply a very faint image of the face? Explain your answer.

17. Is light a wave or a particle? Support your answer by citing specific experimental evidence.

18. Is an electron a wave or a particle? Support your answer by citing some experimental results.

19. Why was the demonstration of electron diffraction by Davisson and Germer an important experiment?

20. An electron and a proton are accelerated from rest through the same potential difference. Which particle has the longer wavelength?

21. If matter has a wave nature, why is this wave-like characteristic not observable in our daily experiences?

22. Why is it impossible to measure simultaneously with infinite accuracy the position and speed of a particle?

23. In describing the passage of electrons through a slit and arriving at a screen, the physicist Richard Feynman said that "electrons arrive in lumps, like particles, but the probability of arrival of these lumps is determined as the intensity of the waves would be. It is in this sense that the electron behaves sometimes like a particle and sometimes like a wave." Elaborate on this point in your own words. (For a further discussion of this point, see R. Feynman, *The Character of Physical Law*, Cambridge, MA, MIT Press, 1980, Chapter 6.)

24. Why is an electron microscope more suitable than an optical microscope for "seeing" objects less than 1 μm in size?

25. Shown in the photographs on pages 1176 and 1177, *iridescence* is the phenomenon that gives shining colors to the feathers of peacocks, hummingbirds, resplendent quetzals,

and even ducks and grackles. Without pigments, it colors Morpho butterflies, Urania moths, some beetles and flies, rainbow trout, and mother-of-pearl in abalone shells. Iridescent colors change as you turn an object. They can look different to your two eyes, so that the object appears to have metallic luster. Iridescent colors were first described in print not by an artist or biologist, but by the physicist Isaac Newton. They are produced by a wide variety of intricate structures in different species—Problem 56 in Chapter 38 describes those in a peacock feather. These structures were all unknown until the invention of the electron microscope. Explain why light microscopes cannot reveal them.

26. *Blacker than black, brighter than white.* (a) Take a large, closed, empty cardboard box. Cut a slot a few millimeters wide in one side. Use black pens, markers, and black material to make some stripes next to the slot, as shown in Figure Q40.26a. Inspect them with care and choose which is blackest—the figure may not show enough contrast to reveal which it is. Explain why it is blackest. (b) Locate an intri-

cately shaped compact fluorescent light fixture. Look at it through dark glasses and describe where it appears brightest. Explain why it is brightest there. Figure Q40.26b shows two such light fixtures held near each other. *Suggestion:* Gustav Kirchhoff, professor at Heidelberg and master of the obvious, gave the same answer to part (a) as you likely will. His answer to part (b) would begin like this: When electromagnetic radiation falls on its surface, an object reflects some fraction r of the energy and absorbs the rest. Whether the fraction reflected is 0.8 or 0.001, the fraction absorbed is $a = 1 - r$. Suppose the object and its surroundings are at the same temperature. The energy the object absorbs joins its fund of internal energy, but the second law of thermodynamics implies that the absorbed energy cannot raise the object's temperature. It does not produce a temperature increase because the object's energy budget has one more term: energy radiated . . . . You still have to make the observations and answer questions (a) and (b), but you can incorporate some of Kirchhoff's ideas into your answer if you wish.

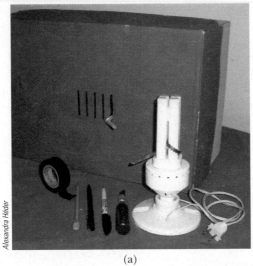

(a)   (b)

Figure Q40.26

PROBLEMS

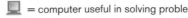

Section 40.1 Blackbody Radiation and Planck's Hypothesis

1. The human eye is most sensitive to 560-nm light. What is the temperature of a black body that would radiate most intensely at this wavelength?

2. (a) Lightning produces a maximum air temperature on the order of 10^4 K, whereas (b) a nuclear explosion produces a temperature on the order of 10^7 K. Use Wien's displacement law to find the order of magnitude of the wavelength of the thermally produced photons radiated

with greatest intensity by each of these sources. Name the part of the electromagnetic spectrum where you would expect each to radiate most strongly.

3. A black body at 7 500 K consists of an opening of diameter 0.050 0 mm, looking into an oven. Find the number of photons per second escaping the hole and having wavelengths between 500 nm and 501 nm.

4. Consider a black body of surface area 20.0 cm^2 and temperature 5 000 K. (a) How much power does it radiate? (b) At what wavelength does it radiate most intensely? Find

the spectral power per wavelength at (c) this wavelength and at wavelengths of (d) 1.00 nm (an x- or γ ray), (e) 5.00 nm (ultraviolet light or an x-ray), (f) 400 nm (at the boundary between UV and visible light), (g) 700 nm (at the boundary between visible and infrared light), (h) 1.00 mm (infrared light or a microwave) and (i) 10.0 cm (a microwave or radio wave). (j) About how much power does the object radiate as visible light?

5. The radius of our Sun is 6.96×10^8 m, and its total power output is 3.77×10^{26} W. (a) Assuming that the Sun's surface emits as a black body, calculate its surface temperature. (b) Using the result of part (a), find λ_{max} for the Sun.

6. A sodium-vapor lamp has a power output of 10.0 W. Using 589.3 nm as the average wavelength of this source, calculate the number of photons emitted per second.

7. 🚀 Calculate the energy, in electron volts, of a photon whose frequency is (a) 620 THz, (b) 3.10 GHz, (c) 46.0 MHz. (d) Determine the corresponding wavelengths for these photons and state the classification of each on the electromagnetic spectrum.

8. The average threshold of dark-adapted (scotopic) vision is 4.00×10^{-11} W/m^2 at a central wavelength of 500 nm. If light having this intensity and wavelength enters the eye and the pupil is open to its maximum diameter of 8.50 mm, how many photons per second enter the eye?

9. An FM radio transmitter has a power output of 150 kW and operates at a frequency of 99.7 MHz. How many photons per second does the transmitter emit?

10. A simple pendulum has a length of 1.00 m and a mass of 1.00 kg. The amplitude of oscillations of the pendulum is 3.00 cm. Estimate the quantum number for the pendulum.

11. **Review problem.** A star moving away from the Earth at $0.280c$ emits radiation that we measure to be most intense at the wavelength 500 nm. Determine the surface temperature of this star.

12. Show that at long wavelengths, Planck's radiation law (Equation 40.6) reduces to the Rayleigh–Jeans law (Equation 40.3).

Section 40.2 The Photoelectric Effect

13. Molybdenum has a work function of 4.20 eV. (a) Find the cutoff wavelength and cutoff frequency for the photoelectric effect. (b) What is the stopping potential if the incident light has a wavelength of 180 nm?

14. Electrons are ejected from a metallic surface with speeds ranging up to 4.60×10^5 m/s when light with a wavelength of 625 nm is used. (a) What is the work function of the surface? (b) What is the cutoff frequency for this surface?

15. Lithium, beryllium, and mercury have work functions of 2.30 eV, 3.90 eV, and 4.50 eV, respectively. Light with a wavelength of 400 nm is incident on each of these metals. Determine (a) which metals exhibit the photoelectric effect and (b) the maximum kinetic energy for the photoelectrons in each case.

16. A student studying the photoelectric effect from two different metals records the following information: (i) the stopping potential for photoelectrons released from metal 1 is 1.48 V larger than that for metal 2, and (ii) the threshold frequency for metal 1 is 40.0% smaller than that for metal 2. Determine the work function for each metal.

17. Two light sources are used in a photoelectric experiment to determine the work function for a particular metal surface. When green light from a mercury lamp ($\lambda = 546.1$ nm) is used, a stopping potential of 0.376 V reduces the photocurrent to zero. (a) Based on this measurement, what is the work function for this metal? (b) What stopping potential would be observed when using the yellow light from a helium discharge tube ($\lambda = 587.5$ nm)?

18. From the scattering of sunlight, Thomson calculated the classical radius of the electron as having a value of 2.82×10^{-15} m. Sunlight with an intensity of 500 W/m^2 falls on a disk with this radius. Calculate the time interval required to accumulate 1.00 eV of energy. Assume that light is a classical wave and that the light striking the disk is completely absorbed. How does your result compare with the observation that photoelectrons are emitted promptly (within 10^{-9} s)?

19. **Review problem.** An isolated copper sphere of radius 5.00 cm, initially uncharged, is illuminated by ultraviolet light of wavelength 200 nm. What charge will the photoelectric effect induce on the sphere? The work function for copper is 4.70 eV.

20. **Review problem.** A light source emitting radiation at 7.00×10^{14} Hz is incapable of ejecting photoelectrons from a certain metal. In an attempt to use this source to eject photoelectrons from the metal, the source is given a velocity toward the metal. (a) Explain how this procedure produces photoelectrons. (b) When the speed of the light source is equal to $0.280c$, photoelectrons just begin to be ejected from the metal. What is the work function of the metal? (c) When the speed of the light source is increased to $0.900c$, determine the maximum kinetic energy of the photoelectrons.

Section 40.3 The Compton Effect

21. Calculate the energy and momentum of a photon of wavelength 700 nm.

22. X-rays having an energy of 300 keV undergo Compton scattering from a target. The scattered rays are detected at 37.0° relative to the incident rays. Find (a) the Compton shift at this angle, (b) the energy of the scattered x-ray, and (c) the energy of the recoiling electron.

23. 🚀 A 0.001 60-nm photon scatters from a free electron. For what (photon) scattering angle does the recoiling electron have kinetic energy equal to the energy of the scattered photon?

24. A 0.110-nm photon collides with a stationary electron. After the collision, the electron moves forward and the photon recoils backward. Find the momentum and the kinetic energy of the electron.

25. A 0.880-MeV photon is scattered by a free electron initially at rest such that the scattering angle of the scattered electron is equal to that of the scattered photon ($\theta = \phi$ in Fig. 40.13b). (a) Determine the angles θ and ϕ. (b) Determine the energy and momentum of the scattered photon. (c) Determine the kinetic energy and momentum of the scattered electron.

26. A photon having energy E_0 is scattered by a free electron initially at rest such that the scattering angle of the scattered electron is equal to that of the scattered photon ($\theta = \phi$ in Fig. 40.13b). (a) Determine the angles θ and ϕ. (b) Determine the energy and momentum of the scattered photon. (c) Determine the kinetic energy and momentum of the scattered electron.

27. In a Compton scattering experiment, an x-ray photon scatters through an angle of 17.4° from a free electron that is initially at rest. The electron recoils with a speed of 2 180 km/s. Calculate (a) the wavelength of the incident photon and (b) the angle through which the electron scatters.

28. A 0.700-MeV photon scatters off a free electron such that the scattering angle of the photon is twice the scattering angle of the electron (Fig. P40.28). Determine (a) the scattering angle for the electron and (b) the final speed of the electron.

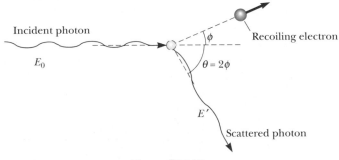

Figure P40.28

29. A photon having wavelength λ scatters off a free electron at A (Fig. P40.29) producing a second photon having wavelength λ'. This photon then scatters off another free electron at B, producing a third photon having wavelength λ'' and moving in a direction directly opposite the original photon as shown in Figure P40.29. Determine the numerical value of $\Delta\lambda = \lambda'' - \lambda$.

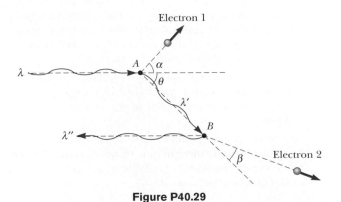

Figure P40.29

30. Find the maximum fractional energy loss for a 0.511-MeV gamma ray that is Compton scattered from a free (a) electron (b) proton.

Section 40.4   Photons and Electromagnetic Waves

31. An electromagnetic wave is called *ionizing radiation* if its photon energy is larger than say 10.0 eV, so that a single photon has enough energy to break apart an atom. With reference to Figure 34.12, identify what regions of the electromagnetic spectrum fit this definition of ionizing radiation and what does not.

32. **Review problem.** A helium–neon laser produces a beam of diameter 1.75 mm, delivering 2.00×10^{18} photons/s. Each photon has a wavelength of 633 nm. (a) Calculate the amplitudes of the electric and magnetic fields inside the beam. (b) If the beam shines perpendicularly onto a perfectly reflecting surface, what force does it exert on the surface? (c) If the beam is absorbed by a block of ice at 0°C for 1.50 h, what mass of ice is melted?

Section 40.5   The Wave Properties of Particles

33. Calculate the de Broglie wavelength for a proton moving with a speed of 1.00×10^6 m/s.

34. Calculate the de Broglie wavelength for an electron that has kinetic energy (a) 50.0 eV and (b) 50.0 keV.

35. (a) An electron has a kinetic energy of 3.00 eV. Find its wavelength. (b) **What If?** A photon has energy 3.00 eV. Find its wavelength.

36. (a) Show that the wavelength of a nonrelativistic neutron is

$$\lambda = \frac{2.86 \times 10^{-11}}{\sqrt{K_n}} \text{ m}$$

where K_n is the kinetic energy of the neutron in electron volts. (b) What is the wavelength of a 1.00-keV neutron?

37. The nucleus of an atom is on the order of 10^{-14} m in diameter. For an electron to be confined to a nucleus, its de Broglie wavelength would have to be on this order of magnitude or smaller. (a) What would be the kinetic energy of an electron confined to this region? (b) Given that typical binding energies of electrons in atoms are measured to be on the order of a few eV, would you expect to find an electron in a nucleus? Explain.

38. In the Davisson–Germer experiment, 54.0-eV electrons were diffracted from a nickel lattice. If the first maximum in the diffraction pattern was observed at $\phi = 50.0°$ (Fig. P40.38),

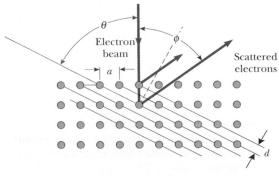

Figure P40.38

what was the lattice spacing a between the vertical rows of atoms in the figure? (It is not the same as the spacing between the horizontal rows of atoms.)

39. (a) Show that the frequency f and wavelength λ of a freely moving particle are related by the expression

$$\left(\frac{f}{c}\right)^2 = \frac{1}{\lambda^2} + \frac{1}{\lambda_C{}^2}$$

where $\lambda_C = h/mc$ is the Compton wavelength of the particle. (b) Is it ever possible for a particle having nonzero mass to have the same wavelength *and* frequency as a photon? Explain.

40. A photon has an energy equal to the kinetic energy of a particle moving with a speed of $0.900c$. (a) Calculate the ratio of the wavelength of the photon to the wavelength of the particle. (b) What would this ratio be for a particle having a speed of $0.001\,00c$? (c) **What If?** What value does the ratio of the two wavelengths approach at high particle speeds? (d) At low particle speeds?

41. The resolving power of a microscope depends on the wavelength used. If one wished to "see" an atom, a resolution of approximately 1.00×10^{-11} m would be required. (a) If electrons are used (in an electron microscope), what minimum kinetic energy is required for the electrons? (b) **What If?** If photons are used, what minimum photon energy is needed to obtain the required resolution?

42. After learning about de Broglie's hypothesis that particles of momentum p have wave characteristics with wavelength $\lambda = h/p$, an 80.0-kg student has grown concerned about being diffracted when passing through a 75.0-cm-wide doorway. Assume that significant diffraction occurs when the width of the diffraction aperture is less that 10.0 times the wavelength of the wave being diffracted. (a) Determine the maximum speed at which the student can pass through the doorway in order to be significantly diffracted. (b) With that speed, how long will it take the student to pass through the doorway if it is in a wall 15.0 cm thick? Compare your result to the currently accepted age of the Universe, which is 4×10^{17} s. (c) Should this student worry about being diffracted?

Section 40.6 The Quantum Particle

43. Consider a freely moving quantum particle with mass m and speed u. Its energy is $E = K = \frac{1}{2}mu^2$. Determine the phase speed of the quantum wave representing the particle and show that it is different from the speed at which the particle transports mass and energy.

44. For a free relativistic quantum particle moving with speed v, the total energy is $E = hf = \hbar\omega = \sqrt{p^2c^2 + m^2c^4}$ and the momentum is $p = h/\lambda = \hbar k = \gamma mv$. For the quantum wave representing the particle, the group speed is $v_g = d\omega/dk$. Prove that the group speed of the wave is the same as the speed of the particle.

Section 40.7 The Double-Slit Experiment Revisited

45. Neutrons traveling at 0.400 m/s are directed through a pair of slits having a 1.00-mm separation. An array of detectors is placed 10.0 m from the slits. (a) What is the de Broglie wavelength of the neutrons? (b) How far off axis is the first zero-intensity point on the detector array? (c) When a neutron reaches a detector, can we say which slit the neutron passed through? Explain.

46. A modified oscilloscope is used to perform an electron interference experiment. Electrons are incident on a pair of narrow slits $0.060\,0$ μm apart. The bright bands in the interference pattern are separated by 0.400 mm on a screen 20.0 cm from the slits. Determine the potential difference through which the electrons were accelerated to give this pattern.

47. In a certain vacuum tube, electrons evaporate from a hot cathode at a slow, steady rate and accelerate from rest through a potential difference of 45.0 V. Then they travel 28.0 cm as they pass through an array of slits and fall on a screen to produce an interference pattern. If the beam current is below a certain value, only one electron at a time will be in flight in the tube. What is this value? In this situation, the interference pattern still appears, showing that each individual electron can interfere with itself.

Section 40.8 The Uncertainty Principle

48. Suppose Fuzzy, a quantum–mechanical duck, lives in a world in which $h = 2\pi$ J · s. Fuzzy has a mass of 2.00 kg and is initially known to be within a pond 1.00 m wide. (a) What is the minimum uncertainty in the component of his velocity parallel to the width of the pond? (b) Assuming that this uncertainty in speed prevails for 5.00 s, determine the uncertainty in his position after this time interval.

49. An electron ($m_e = 9.11 \times 10^{-31}$ kg) and a bullet ($m = 0.020\,0$ kg) each have a velocity of magnitude of 500 m/s, accurate to within 0.010 0%. Within what limits could we determine the position of the objects along the direction of the velocity?

50. An air rifle is used to shoot 1.00-g particles at 100 m/s through a hole of diameter 2.00 mm. How far from the rifle must an observer be in order to see the beam spread by 1.00 cm because of the uncertainty principle? Compare this answer with the diameter of the visible Universe (2×10^{26} m).

51. Use the uncertainty principle to show that if an electron were confined inside an atomic nucleus of diameter 2×10^{-15} m, it would have to be moving relativistically, while a proton confined to the same nucleus can be moving nonrelativistically.

52. (a) Show that the kinetic energy of a nonrelativistic particle can be written in terms of its momentum as $K = p^2/2m$. (b) Use the results of (a) to find the minimum kinetic energy of a proton confined within a nucleus having a diameter of 1.00×10^{-15} m.

53. A woman on a ladder drops small pellets toward a point target on the floor. (a) Show that, according to the uncertainty principle, the average miss distance must be at least

$$\Delta x_f = \left(\frac{2\hbar}{m}\right)^{1/2}\left(\frac{2H}{g}\right)^{1/4}$$

where H is the initial height of each pellet above the floor and m is the mass of each pellet. Assume that the spread in impact points is given by $\Delta x_f = \Delta x_i + (\Delta v_x)t$. (b) If $H = 2.00$ m and $m = 0.500$ g, what is Δx_f?

Additional Problems

54. Figure P40.54 shows the stopping potential versus the incident photon frequency for the photoelectric effect for sodium. Use the graph to find (a) the work function, (b) the ratio h/e, and (c) the cutoff wavelength. The data are taken from R. A. Millikan, *Phys. Rev.* 7:362 (1916).

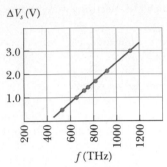

Figure P40.54

55. The following table shows data obtained in a photoelectric experiment. (a) Using these data, make a graph similar to Figure 40.11 that plots as a straight line. From the graph, determine (b) an experimental value for Planck's constant (in joule-seconds) and (c) the work function (in electron volts) for the surface. (Two significant figures for each answer are sufficient.)

Wavelength (nm)	Maximum Kinetic Energy of Photoelectrons (eV)
588	0.67
505	0.98
445	1.35
399	1.63

56. Review problem. Photons of wavelength λ are incident on a metal. The most energetic electrons ejected from the metal are bent into a circular arc of radius R by a magnetic field having a magnitude B. What is the work function of the metal?

57. A 200-MeV photon is scattered at $40.0°$ by a free proton initially at rest. (a) Find the energy (in MeV) of the scattered photon. (b) What kinetic energy (in MeV) does the proton acquire?

58. Derive the equation for the Compton shift (Eq. 40.11) from Equations 40.12, 40.13, and 40.14.

59. Show that a photon cannot transfer all of its energy to a free electron. (*Suggestion:* Note that system energy and momentum must be conserved.)

60. Show that the speed of a particle having de Broglie wavelength λ and Compton wavelength $\lambda_C = h/(mc)$ is

$$v = \frac{c}{\sqrt{1 + (\lambda/\lambda_C)^2}}$$

61. The total power per unit area radiated by a black body at a temperature T is the area under the $I(\lambda, T)$-versus-λ curve, as shown in Figure 40.3. (a) Show that this power per unit area is

$$\int_0^\infty I(\lambda, T) \, d\lambda = \sigma T^4$$

where $I(\lambda, T)$ is given by Planck's radiation law and σ is a constant independent of T. This result is known as Stefan's law. (See Section 20.7.) To carry out the integration, you should make the change of variable $x = hc/\lambda k_B T$ and use the fact that

$$\int_0^\infty \frac{x^3 \, dx}{e^x - 1} = \frac{\pi^4}{15}$$

(b) Show that the Stefan–Boltzmann constant σ has the value

$$\sigma = \frac{2\pi^5 k_B^4}{15 c^2 h^3} = 5.67 \times 10^{-8} \text{ W/m}^2 \cdot \text{K}^4$$

62. Derive Wien's displacement law from Planck's law. Proceed as follows. In Figure 40.3 note that the wavelength at which a black body radiates with greatest intensity is the wavelength for which the graph of $I(\lambda, T)$ versus λ has a horizontal tangent. From Equation 40.6 evaluate the derivative $dI/d\lambda$. Set it equal to zero. Solve the resulting transcendental equation numerically to prove $hc/\lambda_{max} k_B T = 4.965 \ldots$, or $\lambda_{max} T = hc/4.965 \, k_B$. Evaluate the constant as precisely as possible and compare it with Wien's experimental value.

63. The spectral distribution function $I(\lambda, T)$ for an ideal black body at absolute temperature T is shown in Figure P40.63. (a) Show that the percentage of the total power radiated per unit area in the range $0 \leq \lambda \leq \lambda_{max}$ is

$$\frac{A}{A + B} = 1 - \frac{15}{\pi^4} \int_0^{4.965} \frac{x^3}{e^x - 1} \, dx$$

independent of the value of T. (b) Using numerical integration, show that this ratio is approximately 1/4.

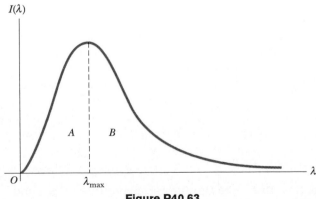

Figure P40.63

64. The neutron has a mass of 1.67×10^{-27} kg. Neutrons emitted in nuclear reactions can be slowed down via collisions with matter. They are referred to as thermal neutrons once they come into thermal equilibrium with their surroundings. The average kinetic energy $(3k_BT/2)$ of a thermal neutron is approximately 0.04 eV. Calculate the de Broglie wavelength of a neutron with a kinetic energy of 0.040 0 eV. How does it compare with the characteristic atomic spacing in a crystal? Would you expect thermal neutrons to exhibit diffraction effects when scattered by a crystal?

65. Show that the ratio of the Compton wavelength λ_C to the de Broglie wavelength $\lambda = h/p$ for a relativistic electron is

$$\frac{\lambda_C}{\lambda} = \left[\left(\frac{E}{m_ec^2}\right)^2 - 1\right]^{1/2}$$

where E is the total energy of the electron and m_e is its mass.

66. Johnny Jumper's favorite trick is to step out of his 16th-story window and fall 50.0 m into a pool. A news reporter takes a picture of 75.0-kg Johnny just before he makes a splash, using an exposure time of 5.00 ms. Find (a) Johnny's de Broglie wavelength at this moment, (b) the uncertainty of his kinetic energy measurement during such a period of time, and (c) the percent error caused by such an uncertainty.

67. A π^0 meson is an unstable particle produced in high-energy particle collisions. Its rest energy is about 135 MeV, and it exists for an average lifetime of only 8.70×10^{-17} s before decaying into two gamma rays. Using the uncertainty principle, estimate the fractional uncertainty $\Delta m/m$ in its mass determination.

68. A photon of initial energy E_0 undergoes Compton scattering at an angle θ from a free electron (mass m_e) initially at rest. Using relativistic equations for energy and momentum conservation, derive the following relationship for the final energy E' of the scattered photon:

$$E' = E_0\left[1 + \left(\frac{E_0}{m_ec^2}\right)(1 - \cos\theta)\right]^{-1}.$$

69. **Review problem.** Consider an extension of Young's double-slit experiment performed with photons. Think of Figure 40.24 as a top view looking down on the apparatus. The viewing screen can be a large flat array of *charge-coupled detectors*. Each cell in the array registers individual photons with high efficiency, so we can see where individual photons strike the screen in real time. We cover slit 1 with a polarizer with its transmission axis horizontal, and slit 2 with a polarizer with vertical transmission axis. Any one photon is either absorbed by a polarizing filter or allowed to pass through. The photons that come through a polarizer have their electric field oscillating in the plane defined by their direction of motion and the filter axis. Now we place another large sheet of polarizing material just in front of the screen. For experimental trial 1, we make the transmission axis of this third polarizer horizontal. This choice in effect blocks slit 2. After many photons have been sent through the apparatus, their distribution on the viewing screen is shown by the lower blue curve in the middle of Figure 40.24. For trial 2, we turn the polarizer at the screen to make its transmission axis vertical. Then the screen receives photons only by way of slit 2, and their distribution is shown as the upper blue curve. For trial 3, we temporarily remove

the third sheet of polarizing material. Then the interference pattern shown by the red curve on the right in Figure 40.24 appears. (a) Is the light arriving at the screen to form the interference pattern polarized? Explain your answer. (b) Next, in trial 4 we replace the large square of polarizing material in front of the screen and set its transmission axis to 45°, halfway between horizontal and vertical. What appears on the screen? (c) Suppose we repeat all of trials 1 through 4 with very low light intensity, so that only one photon is present in the apparatus at a time. What are the results now? (d) We go back to high light intensity for convenience and in trial 5 make the large square of polarizer turn slowly and steadily about a rotation axis through its center and perpendicular to its area. What appears on the screen? (e) **What If?** At last, we go back to very low light intensity and replace the large square sheet of polarizing plastic with a flat layer of liquid crystal, to which we can apply an electric field in either a horizontal or a vertical direction. With the applied field we can very rapidly switch the liquid crystal to transmit only photons with horizontal electric field, to act as a polarizer with a vertical transmission axis, or to transmit all photons with high efficiency. We keep track of photons as they are emitted individually by the source. For each photon we wait until it has passed through the pair of slits. Then we quickly choose the setting of the liquid crystal and make that photon encounter a horizontal polarizer, a vertical polarizer, or no polarizer before it arrives at the detector array. We can alternate among the conditions we earlier set up in trials 1, 2, and 3. We keep track of our settings of the liquid crystal and sort out how photons behave under the different conditions, to end up with full sets of data for all three of those trials. What are the results?

70. A photon with wavelength λ_0 moves toward a free electron that is moving with speed u in the same direction as the photon (Fig. P40.70a). The photon scatters at an angle θ (Fig. P40.70b). Show that the wavelength of the scattered photon is

$$\lambda' = \lambda_0\left(\frac{1 - (u/c)\cos\theta}{1 - (u/c)}\right) + \frac{h}{m_ec}\sqrt{\frac{1 + (u/c)}{1 - (u/c)}}(1 - \cos\theta)$$

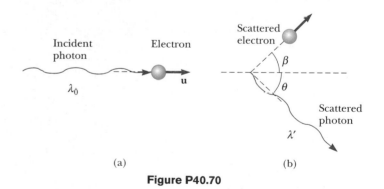

(a)   (b)

Figure P40.70

Answers to Quick Quizzes

40.1 (b). A very hot star will have its peak in the blackbody intensity distribution curve at wavelengths shorter than the visible. As a result, more blue light is emitted than red light.

40.2 AM radio, FM radio, microwaves, sodium light. The order of photon energy will be the same as the order of frequency. See Figure 34.12 for a pictorial representation of electromagnetic radiation in order of frequency.

40.3 (c). When the frequency is increased, the photons each carry more energy, so a stopping potential larger in magnitude is required for the current to fall to zero.

40.4 Classical physics predicts that light of sufficient intensity causes emission of photoelectrons, independent of frequency and without a cutoff frequency. Also, the greater the intensity, the larger the maximum kinetic energy of the electrons, with some time delay in emission at low intensities. Thus the classical expectation (which did not match experiment) yields a graph that looks like this:

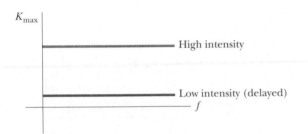

40.5 (d). The shift $\Delta\lambda$ is independent of λ. Thus, the largest fractional shift will correspond to the smallest wavelength.

40.6 (c). According to Equation 40.15, two particles with the same de Broglie wavelength will have the same momentum $p = mv$. If the electron and proton have the same momentum, they cannot have the same speed (a) because of the difference in their masses. For the same reason, because $K = p^2/2m$, they cannot have the same kinetic energy (b). Because the particles have different kinetic energies, Equation 40.16 tells us that the particles do not have the same frequency (d).

40.7 (b). The Compton wavelength (Section 40.3) is a combination of constants and has no relation to the motion of the electron. The de Broglie wavelength (Eq. 40.15) is associated with the motion of the electron through its momentum.

40.8 (b). The group speed is zero because the leading edge of the packet remains fixed at the location of the accident.

40.9 (c). The phase speed of the winning runner is larger than the average speed of all the runners, so the time interval for the winning runner is less than L/v_g.

40.10 (a). The uncertainty principle relates uncertainty in position and velocity along the same axis. The zero uncertainty in position along the x axis results in infinite uncertainty in its velocity component in the x direction, but it is unrelated to the y direction.

40.11 (c). According to the uncertainty principle, if Planck's constant is smaller, the uncertainty in momentum can be smaller and the momentum perpendicular to the original direction of propagation would be smaller. Note that classical wave theory does not include Planck's constant, so it would predict no effect.

ROSE IS ROSE reprinted by permission of United Feature Syndicate, Inc.

Chapter 41

Quantum Mechanics

CHAPTER OUTLINE

41.1 An Interpretation of Quantum Mechanics

41.2 A Particle in a Box

41.3 The Particle Under Boundary Conditions

41.4 The Schrödinger Equation

41.5 A Particle in a Well of Finite Height

41.6 Tunneling Through a Potential Energy Barrier

41.7 The Scanning Tunneling Microscope

41.8 The Simple Harmonic Oscillator

▲ A quantum corral *shows two aspects of current technological advances in physics. The first aspect involves control over individual atoms. This corral is formed by positioning iron atoms in a stadium-shaped ring on a copper surface. The second aspect is the ability to image the individual atoms with a scanning tunneling microscope. The corral can be used to study the quantized states of electrons trapped in a small region. (Courtesy of IBM Research, Almaden Research Center. Unauthorized use prohibited.)*

In this chapter we introduce *quantum mechanics,* an extremely successful theory for explaining the behavior of microscopic particles. This theory, developed from 1925 to 1926 by Erwin Schrödinger, Werner Heisenberg, and others, enables us to understand a host of phenomena involving atoms, molecules, nuclei, and solids. The discussion in this chapter follows from the introductory quantum concepts that were developed in Chapter 40 and incorporates some of the features of waves under boundary conditions that were explored in Chapter 18. We shall also discuss practical applications, including the scanning tunneling microscope and nanoscale devices that may be used in future quantum computers. Finally, we shall return to the simple harmonic oscillator that was introduced in Chapter 15 and examine it from a quantum mechanical point of view. We shall find a strong connection between this discussion and Planck's model for blackbody radiation, which represented the beginning of our study of quantum physics in Chapter 40.

41.1 An Interpretation of Quantum Mechanics

The preceding chapter introduced some new and strange ideas. In particular, we concluded on the basis of experimental evidence that both matter and electromagnetic radiation are sometimes best described as particles and sometimes as waves, depending on the phenomenon being observed. We can improve our understanding of quantum physics by making another conceptual connection between particles and waves using the notion of probability, a term that was introduced in Chapter 40.

We begin by discussing electromagnetic radiation from the particle point of view. The probability per unit volume of finding a photon in a given region of space at an instant of time is proportional to the number N of photons per unit volume at that time:

$$\frac{\text{Probability}}{V} \propto \frac{N}{V}$$

The number of photons per unit volume is proportional to the intensity of the radiation:

$$\frac{N}{V} \propto I$$

Now, we form a connection between the particle model and the wave model by recalling that the intensity of electromagnetic radiation is proportional to the square of the electric field amplitude E for the electromagnetic wave (Equation 34.21):

$$I \propto E^2$$

Equating the beginning and the end of this string of proportionalities, we have

$$\frac{\text{Probability}}{V} \propto E^2 \tag{41.1}$$

Thus, for electromagnetic radiation, the probability per unit volume of finding a particle associated with this radiation (the photon) is proportional to the square of the amplitude of the associated electromagnetic wave.

Recognizing the wave–particle duality of both electromagnetic radiation and matter, we should suspect a parallel proportionality for a material particle—the probability per unit volume of finding the particle is proportional to the square of the amplitude of a wave representing the particle. In the previous chapter, we learned that there is a de Broglie wave associated with every particle. The amplitude of the de Broglie wave associated with a particle is not a measurable quantity. (This is because the wave function representing a particle is generally a complex function, as we discuss below.) In contrast, the electric field is a measurable quantity for an electromagnetic wave. The matter analog to Equation 41.1 relates the square of the amplitude of the wave to the probability per unit volume of finding the particle. As a result, we simply call the amplitude of the wave associated with the particle the **probability amplitude,** or the **wave function,** and give it the symbol Ψ.

In general, the complete wave function Ψ for a system depends on the positions of all the particles in the system and on time, and therefore can be written $\Psi(\mathbf{r}_1, \mathbf{r}_2, \mathbf{r}_3, \dots, \mathbf{r}_j, \dots, t)$, where $\mathbf{r}_j$ is the position vector of the jth particle in the system. For many systems of interest, including all of those that we shall study in this text, the wave function Ψ is mathematically separable in space and time and can be written as a product of a space function ψ for one particle of the system and a complex time function[1]:

$$\Psi(\mathbf{r}_1, \mathbf{r}_2, \mathbf{r}_3, \dots, \mathbf{r}_j, \dots, t) = \psi(\mathbf{r}_j)\,e^{-i\omega t} \qquad (41.2)$$

Space- and time-dependent wave function Ψ

where $\omega(= 2\pi f)$ is the angular frequency of the wave function and $i = \sqrt{-1}$.

For any system in which the potential energy is time-independent and depends only on the positions of particles within the system, the important information about the system is contained within the space part of the wave function. The time part is simply the factor $e^{-i\omega t}$. Therefore, the understanding of ψ is the critical aspect of a given problem.

The wave function ψ is often complex-valued. The absolute square $|\psi|^2 = \psi^*\psi$, where ψ^* is the complex conjugate[2] of ψ, is always real and positive, and is proportional to the probability per unit volume of finding a particle at a given point at some instant. The wave function contains within it all the information that can be known about the particle.

This probabilistic interpretation of the wave function was first suggested by Max Born (1882–1970) in 1928. In 1926, Erwin Schrödinger (1887–1961) proposed a wave equation that describes the manner in which the wave function changes in space and time. The *Schrödinger wave equation*, which we shall examine in Section 41.4, represents a key element in the theory of quantum mechanics.

The concepts of quantum mechanics, strange as they sometimes may seem, developed from classical ideas. In fact, when the techniques of quantum mechanics are applied to macroscopic systems, the results are essentially identical to those of classical physics. This blending of the two approaches occurs when the de Broglie wavelength is small compared with the dimensions of the system. The situation is similar to the agreement between relativistic mechanics and classical mechanics when $v \ll c$.

 PITFALL PREVENTION

41.1 The Wave Function Belongs to a System

The common language in quantum mechanics is to associate a wave function with a particle. The wave function, however, is determined by the particle *and* its interaction with its environment, so it more rightfully belongs to a system. In many cases, the particle is the only part of the system that experiences a change, which is why the common language has developed. You will see examples in the future in which it is more proper to think of the system wave function rather than the particle wave function.

---

[1]   The standard form of a complex number is $a + ib$. The notation $e^{i\theta}$ is equivalent to the standard form as follows:

$$e^{i\theta} = \cos\theta + i\sin\theta$$

Thus, the notation $e^{-i\omega t}$ in Equation 41.2 is equivalent to $\cos(-\omega t) + i\sin(-\omega t) = \cos\omega t - i\sin\omega t$.

[2]   For a complex number $z = a + ib$, the complex conjugate is found by changing i to $-i$: $z^* = a - ib$. The product of a complex number and its complex conjugate is always real and positive: $z^*z = (a - ib)(a + ib) = a^2 - (ib)^2 = a^2 - (i)^2 b^2 = a^2 + b^2$.

In Section 40.5, we found that the de Broglie equation relates the momentum of a particle to its wavelength through the relation $p = h/\lambda$. If an ideal free particle has a precisely known momentum p_x, its wave function is an uninterrupted sinusoidal wave of wavelength $\lambda = h/p_x$, and the particle has equal probability of being at any point along the x axis (Fig. 40.19a). The wave function ψ for such a free particle moving along the x axis can be written as

Wave function for a free particle

$$\psi(x) = Ae^{ikx} \tag{41.3}$$

where $k = 2\pi/\lambda$ is the angular wave number (Eq. 16.8) of the wave representing the particle and A is a constant amplitude.[3]

Although we cannot measure ψ, we can measure the real quantity $|\psi|^2$, which can be interpreted as follows. If ψ represents a single particle, then $|\psi|^2$—called the **probability density**—is the relative probability per unit volume that the particle will be found at any given point in the volume. This interpretation can also be stated in the following manner. If dV is a small volume element surrounding some point, then the probability of finding the particle in that volume element is $|\psi|^2 \, dV$.

Probability density $|\psi|^2$

One-Dimensional Wave Functions and Expectation Values

In this section we deal only with one-dimensional systems, where the particle must be located along the x axis, so the probability $|\psi|^2 \, dV$ is modified to become $|\psi|^2 \, dx$. In this case, the probability $P(x) \, dx$ that the particle will be found in the infinitesimal interval dx around the point x is

$$P(x) \, dx = |\psi|^2 \, dx \tag{41.4}$$

Although it is not possible to specify the position of a particle with complete certainty, it is possible through $|\psi|^2$ to specify the probability of observing it in a region surrounding a given point x. **The probability of finding the particle in the arbitrary interval $a \le x \le b$ is**

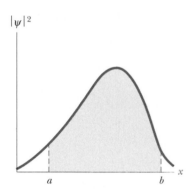

$|\psi|^2$

a b

x

Figure 41.1 The probability of a particle being in the interval $a \le x \le b$ is the area under the probability density curve from a to b.

$$P_{ab} = \int_a^b |\psi|^2 \, dx \tag{41.5}$$

The probability P_{ab} is the area under the curve of $|\psi|^2$ versus x between the points $x = a$ and $x = b$, as in Figure 41.1.

Experimentally, there is a finite probability of finding a particle in an interval near some point at some instant. The value of that probability must lie between the limits 0 and 1. For example, if the probability is 0.30, there is a 30% chance of finding the particle in the interval.

Because the particle must be somewhere along the x axis, the sum of the probabilities over all values of x must be 1:

Normalization condition on ψ

$$\int_{-\infty}^{\infty} |\psi|^2 \, dx = 1 \tag{41.6}$$

Any wave function satisfying Equation 41.6 is said to be **normalized.** Normalization is simply a statement that the particle exists at some point in space.

The wave function ψ satisfies a wave equation, just as the electric field associated with an electromagnetic wave satisfies a wave equation that follows from Maxwell's

---

[3]   For the free particle, the full wave function, based on Equation 41.2, is

$$\Psi(x, t) = Ae^{ikx}e^{-i\omega t} = Ae^{i(kx - \omega t)} = A[\cos(kx - \omega t) + i\sin(kx - \omega t)]$$

The real part of this wave function has the same form as the waves that we added together to form wave packets in Section 40.6.

equations. The wave equation satisfied by ψ is the Schrödinger equation (Section 41.4), and ψ can be computed from it. Although ψ is not a measurable quantity, all the measurable quantities of a particle, such as its energy and momentum, can be derived from a knowledge of ψ. For example, once the wave function for a particle is known, it is possible to calculate the average position at which you would expect to find the particle after many measurements. This average position is called the **expectation value** of x and is defined by the equation

$$\langle x \rangle \equiv \int_{-\infty}^{\infty} \psi^* x \psi \, dx \qquad (41.7)$$

Expectation value for position x

(Brackets, $\langle \, . \, . \, . \, \rangle$, are used to denote expectation values.) Furthermore, one can find the expectation value of any function $f(x)$ associated with the particle by using the following equation[4]:

$$\langle f(x) \rangle \equiv \int_{-\infty}^{\infty} \psi^* f(x) \psi \, dx \qquad (41.8)$$

Expectation value for a function $f(x)$

Quick Quiz 41.1 Consider the wave function for the free particle, Equation 41.3. At what value of x is the particle most likely to be found at a given time? (a) at $x = 0$ (b) at small nonzero values of x (c) at large values of x (d) anywhere along the x axis.

Example 41.1   A Wave Function for a Particle

Consider a particle whose wave function is given by the following expression:

$$\psi(x) = Ae^{-ax^2}$$

This wave function is graphed in Figure 41.2.

(A) What is the value of A if this wave function is normalized?

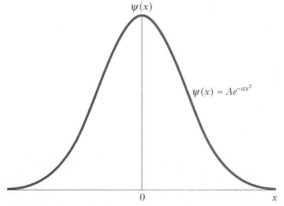

$\psi(x)$

$\psi(x) = Ae^{-ax^2}$

0 x

Figure 41.2 (Example 41.1) A symmetric wave function for a particle, given by $\psi(x) = Ae^{-ax^2}$.

Solution We use the normalization condition, Equation 41.6:

$$\int_{-\infty}^{\infty} |\psi|^2 \, dx = \int_{-\infty}^{\infty} (Ae^{-ax^2})^2 \, dx = A^2 \int_{-\infty}^{\infty} e^{-2ax^2} \, dx = 1$$

The integral can be expressed as the sum of two integrals:

$$A^2 \int_{-\infty}^{\infty} e^{-2ax^2} \, dx = A^2 \left(\int_{0}^{\infty} e^{-2ax^2} \, dx + \int_{-\infty}^{0} e^{-2ax^2} \, dx \right) = 1$$

Because of the x^2 factor in the exponential function, these two integrals have the same value whether x is positive or negative. Thus,

$$(1) \qquad 2A^2 \int_{0}^{\infty} e^{-2ax^2} \, dx = 1$$

We evaluate the integral with the help of Table B.6 in the Appendix:

$$\int_{0}^{\infty} e^{-2ax^2} \, dx = \frac{1}{2} \sqrt{\frac{\pi}{2a}}$$

Thus, Equation (1) becomes

$$2A^2 \left(\frac{1}{2} \sqrt{\frac{\pi}{2a}} \right) = 1$$

---

[4]   Expectation values are analogous to "weighted averages," in which each possible value of a function is multiplied by the probability of the occurrence of that value before summing over all possible values. We write the expectation value as $\int_{-\infty}^{\infty} \psi^* f(x) \psi \, dx$ rather than $\int_{-\infty}^{\infty} f(x) |\psi|^2 \, dx$ because in more advanced treatments of quantum mechanics, $f(x)$ may be represented by an operator (such as a derivative) rather than a simple multiplicative function. In these situations, the operator is applied only to ψ and not to ψ^*.

$$A = \left(\frac{2a}{\pi}\right)^{1/4}$$

(B) What is the expectation value of x for this particle?

Solution We evaluate the expectation value using Equation 41.7:

$$\langle x \rangle \equiv \int_{-\infty}^{\infty} \psi^* x \psi \, dx = \int_{-\infty}^{\infty} (Ae^{-ax^2}) \, x \, (Ae^{-ax^2}) \, dx$$

$$= A^2 \int_{-\infty}^{\infty} xe^{-2ax^2} \, dx$$

As in part (A), we express the integral as a sum of two integrals,

$$(2) \qquad \langle x \rangle = A^2 \left(\int_0^{\infty} xe^{-2ax^2} \, dx + \int_{-\infty}^{\infty} xe^{-2ax^2} \, dx \right)$$

If we change the integration variable from x to $-x$ in the second integral, it becomes

$$\int_{-\infty}^{0} xe^{-2ax^2} \, dx = \int_{\infty}^{0} -xe^{-2ax^2} (-dx) = \int_{\infty}^{0} xe^{-2ax^2} \, dx$$

Reversing the order of the limits introduces a negative sign:

$$\int_{\infty}^{0} xe^{-2ax^2} \, dx = -\int_0^{\infty} xe^{-2ax^2} \, dx$$

This is the negative of the first integral in Equation (2), so

$$\langle x \rangle = \boxed{0}$$

Given the symmetry of the wave function around $x = 0$ in Figure 41.2, it is not a surprising result that the average position of the particle is at $x = 0$.

In Section 41.8, we show that the wave function that we have studied in this example represents the lowest-energy state of the quantum harmonic oscillator. We shall explore more details of this system in that section.

We close this section by noting the important mathematical features of a physically reasonable wave function $\psi(x)$ for a system.

- $\psi(x)$ may be a complex function or a real function, depending on the system;
- $\psi(x)$ must be defined at all points in space and be single-valued;
- $\psi(x)$ must be normalized according to Equation 41.6;
- $\psi(x)$ must be continuous in space—there must be no discontinuous jumps in the value of the wave function at any point.[5]

41.2   A Particle in a Box

In this section, we shall apply some of the ideas we have developed to a simple physical problem—a particle confined to a one-dimensional region of space, called the *particle-in-a-box* problem (even though the "box" is one-dimensional!). From a classical viewpoint, if a particle is bouncing elastically back and forth along the x axis between two impenetrable walls separated by a distance L, as in Figure 41.3a, its motion is easy to describe. If the speed of the particle is v, the magnitude of its momentum mv remains constant, as does its kinetic energy. Classical physics places no restrictions on the values of a particle's momentum and energy. The quantum-mechanical approach to this problem is quite different and requires that we find the appropriate wave function consistent with the conditions of the situation.

Because the walls are impenetrable, there is zero probability of finding the particle outside the box, so the wave function $\psi(x)$ must be zero for $x < 0$ and for $x > L$. As we pointed out at the end of Section 41.1, a wave function must be continuous in space. Thus, if ψ is zero outside the walls, it must also be zero *at* the walls; that is, $\psi(0) = 0$ and $\psi(L) = 0$. Only those wave functions that satisfy these boundary conditions are allowed.

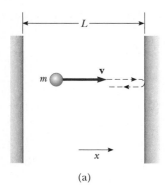

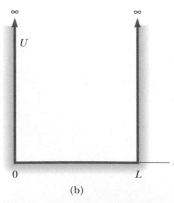

(a)

(b)

Figure 41.3 (a) A particle of mass m and velocity **v**, confined to bouncing between two impenetrable walls separated by a distance L. (b) The potential-energy function for the system.

---

[5] If the wave function were not continuous at a point, the derivative of the wave function at that point would be infinite. This leads to difficulties in the Schrödinger equation, for which the wave function is a solution as discussed in Section 41.4.

Figure 41.3b, a graphical representation of the particle-in-a-box problem, shows the potential energy of the particle–environment system as a function of the position of the particle. As long as the particle is inside the box, the potential energy of the system does not depend on the location of the particle and we can choose its value to be zero. Outside the box, we must ensure that the wave function is zero. We can do this by defining the potential energy of the system as infinitely large if the particle were outside the box. Therefore, the only way a particle could be outside the box is if the system has an infinite amount of energy, which is impossible.

The wave function for a particle in the box can be expressed as a real sinusoidal function[6]:

$$\psi(x) = A \sin\left(\frac{2\pi x}{\lambda}\right) \tag{41.9}$$

where λ is the de Broglie wavelength associated with the particle. This wave function must satisfy boundary conditions at the walls. The boundary condition at $x = 0$ is satisfied already because the sine function is zero when $x = 0$. For the boundary condition at $x = L$, we have

$$\psi(L) = 0 = A \sin\left(\frac{2\pi L}{\lambda}\right)$$

which can only be true if

$$\frac{2\pi L}{\lambda} = n\pi \quad \longrightarrow \quad \lambda = \frac{2L}{n} \tag{41.10}$$

where $n = 1, 2, 3, \ldots$. Thus, only certain wavelengths for the particle are allowed! Each of the allowed wavelengths corresponds to a quantum state for the system, and n is the quantum number. Expressing the wave function in terms of the quantum number n, we have

$$\psi(x) = A \sin\left(\frac{2\pi x}{\lambda}\right) = A \sin\left(\frac{2\pi x}{2L/n}\right) = A \sin\left(\frac{n\pi x}{L}\right) \tag{41.11}$$

Wave functions for a particle in a box

Figures 41.4a and b are graphical representations of ψ versus x and $|\psi|^2$ versus x for $n = 1, 2$, and 3 for the particle in a box.[7] Note that although ψ can be positive or negative, $|\psi|^2$ is always positive. Because $|\psi|^2$ represents a probability density, a negative value for $|\psi|^2$ would be meaningless.

Further inspection of Figure 41.4b shows that $|\psi|^2$ is zero at the boundaries, satisfying our boundary conditions. In addition, $|\psi|^2$ is zero at other points, depending on the value of n. For $n = 2$, $|\psi|^2 = 0$ at $x = L/2$; for $n = 3$, $|\psi|^2 = 0$ at $x = L/3$ and $x = 2L/3$. The number of zero points increases by one each time the quantum number increases by one.

Because the wavelengths of the particle are restricted by the condition $\lambda = 2L/n$, the magnitude of the momentum of the particle is also restricted to specific values, which we can find from the expression for the de Broglie wavelength, Equation 40.15:

$$p = \frac{h}{\lambda} = \frac{h}{2L/n} = \frac{nh}{2L}$$

---

[6] We show this explicitly in Section 41.4.

[7] Note that $n = 0$ is not allowed because, according to Equation 41.11, the wave function would be $\psi = 0$. This is not a physically reasonable wave function; for example, it cannot be normalized because $\int_{-\infty}^{\infty} |\psi|^2 \, dx = 0$, but Equation 41.6 tells us that this integral must equal 1.

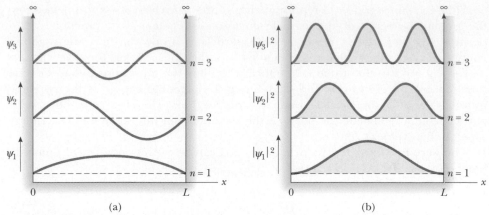

(a)   (b)

Active Figure 41.4 The first three allowed states for a particle confined to a one-dimensional box. The states are shown superimposed on the potential energy function of Figure 41.3b. (a) The wave functions ψ for $n = 1, 2$, and 3. (b) The probability densities $|\psi|^2$ for $n = 1, 2$, and 3. The wave functions and probability densities are plotted vertically from separate axes that are offset vertically for clarity. The positions of these axes on the potential-energy function suggest the relative energies of the states, but the positions are not shown to scale.

We have chosen the potential energy of the system to be zero when the particle is inside the box. Therefore, the allowed values of the energy of the system, which is simply the kinetic energy of the particle, are

$$E_n = \tfrac{1}{2}mv^2 = \frac{p^2}{2m} = \frac{(nh/2L)^2}{2m}$$

**Quantized energies for a
particle in a box**

$$E_n = \left(\frac{h^2}{8mL^2}\right)n^2 \qquad n = 1, 2, 3, \ldots \qquad (41.12)$$

As we see from this expression, **the energy of the particle is quantized.** The lowest allowed energy corresponds to the **ground state,** which is the lowest energy state for any system. For the particle in a box, the ground state corresponds to $n = 1$, for which $E_1 = h^2/8mL^2$. Because $E_n = n^2E_1$, the **excited states** corresponding to $n = 2, 3, 4, \ldots$ have energies given by $4E_1, 9E_1, 16E_1, \ldots$.

Figure 41.5 is an energy-level diagram describing the energy values of the allowed states. Note that the state $n = 0$, for which E would be equal to zero, is *not* allowed. (See footnote 7 on page 1327.) This means that **according to quantum mechanics, the particle can never be at rest.** The smallest energy it can have, corresponding to $n = 1$, is called the **ground-state energy.** This result contradicts the classical viewpoint, in which $E = 0$ is an acceptable state, as are *all* positive values of E.

n

4 —————————— $E_4 = 16E_1$

3 —————————— $E_3 = 9E_1$

2 —————————— $E_2 = 4E_1$

1 —————————— E_1
- - - - - - - - - - - $E = 0$

Ground-state energy > 0

Active Figure 41.5 Energy-level diagram for a particle confined to a one-dimensional box of length L. The lowest allowed energy is $E_1 = h^2/8mL^2$.

Quick Quiz 41.2 Redraw Figure 41.4b, the probability of finding a particle at a particular location in a box, on the basis of classical mechanics instead of quantum mechanics.

Quick Quiz 41.3 A particle is in a box of length L. Suddenly, the length of the box is increased to $2L$. What happens to the energy levels shown in Figure 41.5? (a) Nothing—they are unaffected. (b) They move farther apart. (c) They move closer together.

Example 41.2 A Bound Electron

An electron is confined between two impenetrable walls 0.200 nm apart. Determine the energy levels for the states $n = 1, 2,$ and 3.

Solution For the state $n = 1$, Equation 41.12 gives

$$E_1 = \frac{h^2}{8m_e L^2} = \frac{(6.63 \times 10^{-34}\,\text{J}\cdot\text{s})^2}{8(9.11 \times 10^{-31}\,\text{kg})(2.00 \times 10^{-10}\,\text{m})^2}$$

$$E_1 = 1.51 \times 10^{-18}\,\text{J} = \boxed{9.42\,\text{eV}}$$

For $n = 2$ and $n = 3$, $E_2 = 4E_1 = \boxed{37.7\,\text{eV}}$ and $E_3 = 9E_1 = \boxed{84.8\,\text{eV}}$. Although this is a rather primitive model, it can be used to estimate the energy of an electron trapped in a vacant crystal site.

 Investigate the energy levels of various particles trapped in a one-dimensional box at the Interactive Worked Example link at http://www.pse6.com

Example 41.3 Energy Quantization for a Macroscopic Object

A 0.500-kg baseball is confined between two rigid walls of a stadium that can be modeled as a box of length 100 m. Calculate the minimum speed of the baseball.

Solution The minimum speed corresponds to the state for which $n = 1$. Using Equation 41.12 with $n = 1$ gives the ground-state energy:

$$E_1 = \frac{h^2}{8mL^2} = \frac{(6.63 \times 10^{-34}\,\text{J}\cdot\text{s})^2}{8(0.500\,\text{kg})(100\,\text{m})^2} = 1.10 \times 10^{-71}\,\text{J}$$

Because $E = K = \frac{1}{2}mv^2$, we have

$$v = \left(\frac{2K}{m}\right)^{1/2} = \left[\frac{2(1.10 \times 10^{-71}\,\text{J})}{0.500\,\text{kg}}\right]^{1/2}$$

$$= \boxed{6.63 \times 10^{-36}\,\text{m/s}}$$

This speed is so small that the object can be considered to be at rest, which is what one would expect for the minimum speed of a macroscopic object.

What If? What if a sharp line drive is hit so that the baseball is moving with a speed of 150 m/s? What is the quantum number of the state in which the baseball now resides?

Answer We expect the quantum number to be very large because the baseball is a macroscopic object. The kinetic energy of the baseball is

$$\tfrac{1}{2}mv^2 = \tfrac{1}{2}(0.500\,\text{kg})(150\,\text{m/s})^2 = 5.62 \times 10^3\,\text{J}$$

From Equation 41.12, we find the quantum number n:

$$n = \sqrt{\frac{8mL^2 E_n}{h^2}} = \sqrt{\frac{8(0.500\,\text{kg})(100\,\text{m})^2(5.62 \times 10^3\,\text{J})}{(6.63 \times 10^{-34}\,\text{J}\cdot\text{s})^2}}$$

$$= 2.26 \times 10^{37}$$

This is a tremendously large quantum number. Thus, as the baseball pushes air out of the way, hits the ground, and rolls to a stop, it moves through more than 10^{37} quantum states. These states are so close together in energy that we do not see the transitions from one state to the next. We see what appears to be a smooth variation in the speed of the ball. The quantum nature of the universe is simply not evident in the motion of macroscopic objects.

Example 41.4 Model of an Atom

An atom can be viewed as several electrons moving around a positively charged nucleus, where the electrons are subject mainly to the electrical attraction of the nucleus. (This attraction is partially "screened" by the inner-core electrons and is therefore diminished.) Figure 41.6 represents the potential energy of an atom as a function of r, the distance between the electron and the nucleus.

(A) Use the simple model of a particle in a box to *estimate* the energy (in electron volts) required to raise an atom from the state $n = 1$ to the state $n = 2$, assuming that the atom has a radius of 0.100 nm and that the moving electron carries the energy that has been added to the atom.

Solution Using Equation 41.12 and taking the length L of the box to be 0.200 nm (the diameter of the atom) and $m = m_e = 9.11 \times 10^{-31}$ kg, we find that

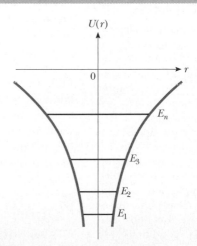

Figure 41.6 (Example 41.4) Model of potential energy versus separation distance r for an atom. The allowed energy levels of the atom are shown superimposed on the potential energy function.

$$E_n = \left(\frac{h^2}{8m_eL^2}\right)n^2$$

$$= \frac{(6.63 \times 10^{-34}\,\text{J}\cdot\text{s})^2}{8(9.11 \times 10^{-31}\,\text{kg})(2.00 \times 10^{-10}\,\text{m})^2}\,n^2$$

$$= (1.51 \times 10^{-18})\,n^2\,\text{J} = 9.4n^2\,\text{eV}$$

Hence, the energy difference between the states $n = 1$ and $n = 2$ is

$$\Delta E = E_2 - E_1 = 9.42(2)^2\,\text{eV} - 9.42(1)^2\,\text{eV} = \boxed{28.3\,\text{eV}}$$

(B) Atoms may be excited to higher energy states by absorbing incoming radiation. Calculate the wavelength of the photon that would cause the transition from the $n = 1$ state to the $n = 2$ state.

Solution Using the fact that $\Delta E = hf = hc/\lambda$, we obtain

$$\lambda = \frac{hc}{\Delta E} = \frac{1\,240\,\text{eV}\cdot\text{nm}}{(28.3\,\text{eV})} = \boxed{43.8\,\text{nm}}$$

This wavelength is in the far ultraviolet region, and it is interesting to note that the result is roughly correct. This oversimplified model gives a reasonable order-of-magnitude estimate for transitions between lowest-lying levels of the atom, but the estimate becomes progressively less valid for higher-energy transitions.

 PITFALL PREVENTION

41.2 Reminder: Energy Belongs to a System

We often refer to the energy of a particle in commonly used language. As in Pitfall Prevention 41.1, we are actually describing the energy of the *system* of the particle and whatever environment is establishing the impenetrable walls. For the particle in a box, (Section 41.2) the only type of energy is kinetic energy belonging to the particle—this is the origin of the common description.

 PITFALL PREVENTION

41.3 Quantum States Are Not Necessarily Standing Waves

While there are many similarities between the quantization of states for quantum systems and the quantization of frequencies for waves on a string (Chapter 18), quantum states are not necessarily standing waves. There may be no stationary "nodes" and no sinusoidal shape associated with a quantum wave function under boundary conditions. Systems more complicated than the particle in a box will have more complicated wave functions, and some boundary conditions will not lead to zeroes of the wave function at fixed points.

41.3 The Particle Under Boundary Conditions

The allowed wavelengths for the wave function of a particle in a box (Eq. 41.10) are identical in form to the allowed wavelengths for mechanical waves on a string fixed at both ends (Eq. 18.6, page 554). In both the string wave and the quantum particle wave, we applied **boundary conditions** to determine the allowed states of the system. For the string fixed at both ends, the boundary conditions were that the elements of the string at the boundaries must have zero displacement. For the particle in a box, the probability amplitude at the boundaries must be zero. In both cases, this results in quantized wavelengths. In the case of the vibrating string, wavelength is related to the frequency, so we have a set of harmonics, or quantized frequencies, given by Equation 18.8. In the case of the particle in a box, we also have quantized frequencies. We can go further in the quantum case, however, because the frequency of a particle is related to its energy through $E = hf$, and we generate a set of quantized *energies*.

In the model of a particle under boundary conditions, **an interaction of a particle with its environment represents one or more boundary conditions and, if the interaction restricts the particle to a finite region of space, results in quantization of the energy of the system.** Because particles have wave-like characteristics, the allowed quantum states of a system are those in which the boundary conditions on the wave function representing the system are satisfied.

In general, boundary conditions are related to the coordinates describing the problem. For the particle in a box, we require a zero value of the wave function at two values of x. In the case of a three-dimensional system such as the hydrogen atom to be discussed in Chapter 42, the problem is best presented in *spherical coordinates*. These are an extension of the plane polar coordinates introduced in Section 3.1, and consist of a radial coordinate r and two angular coordinates. Boundary conditions on the radial part of the wave function, related to r, require that the wave function must approach zero as $r \rightarrow \infty$ (so that the wave function can be normalized) and remain finite as $r \rightarrow 0$. A boundary condition on the angular part of the wave function is that adding 2π radians to the angle must return the wave function to the same value, because an addition of 2π results in the same angular position. The generation of the wave function and application of the boundary conditions for the hydrogen atom are beyond the scope of this book. However, we shall examine the behavior of some of the hydrogen-atom wave functions in Chapter 42.

Quick Quiz 41.4 Which of the following will exhibit quantized energy levels? (a) an atom in a crystal (b) an electron and a proton in a hydrogen atom (c) a proton in the nucleus of a heavy atom (d) all of the above (e) none of the above.

41.4 The Schrödinger Equation

In Section 34.2, we discussed a wave equation for electromagnetic radiation. The waves associated with particles also satisfy a wave equation. We might guess that the wave equation for material particles is different from that associated with photons due to the fact that material particles have a nonzero rest energy. The appropriate wave equation was developed by Schrödinger in 1926. In analyzing the behavior of a quantum system, the approach is to determine a solution to this equation and then apply the appropriate boundary conditions to the solution. The solution yields the allowed wave functions and energy levels of the system under consideration. Proper manipulation of the wave function then enables one to calculate all measurable features of the system.

The Schrödinger equation as it applies to a particle of mass m confined to moving along the x axis and interacting with its environment through a potential energy function $U(x)$ is

$$-\frac{\hbar^2}{2m}\frac{d^2\psi}{dx^2} + U\psi = E\psi \qquad (41.13)$$

Time-independent Schrödinger equation

where E is a constant equal to the total energy of the system (the particle and its environment). Because this equation is independent of time, it is commonly referred to as the **time-independent Schrödinger equation.** (We shall not discuss the time-dependent Schrödinger equation in this text.)

The Schrödinger equation is consistent with the concept of mechanical energy of a system. Problem 15 shows, both for a free particle and a particle in a box, that the first term in the Schrödinger equation reduces to the kinetic energy of the particle multiplied by the wave function. Therefore, Equation 41.13 tells us that the total energy is the sum of the kinetic energy and the potential energy, and that the total energy is a constant: $K + U = E =$ constant.

In principle, if the potential energy function U for the system is known, one can solve Equation 41.13 and obtain the wave functions and energies for the allowed states of the system. Because U may vary with position, it may be necessary to obtain solutions to the equation for different regions of the x axis. In the process, the wave function must be physically reasonable, as noted at the end of Section 41.1. Solutions to the Schrödinger equation in different regions must join smoothly at the boundaries—we require that $\psi(x)$ be *continuous.* In order that $\psi(x)$ obey the normalization condition, we require that $\psi(x)$ approach zero as x approaches $\pm\infty$. Finally, $d\psi/dx$ must also be continuous for finite values of the potential energy.[8]

The task of solving the Schrödinger equation may be very difficult, depending on the form of the potential energy function. As it turns out, the Schrödinger equation has been extremely successful in explaining the behavior of atomic and nuclear systems, whereas classical physics has failed to explain this behavior. Furthermore, when quantum mechanics is applied to macroscopic objects, the results agree with classical physics.

Erwin Schrödinger
Austrian Theoretical Physicist (1887–1961)

Schrödinger is best known as one of the creators of quantum mechanics. His approach to quantum mechanics was demonstrated to be mathematically equivalent to the more abstract matrix mechanics developed by Heisenberg. Schrödinger also produced important papers in the fields of statistical mechanics, color vision, and general relativity. (*AIP Emilio Segrè Visual Archives*)

The Particle in a Box Revisited

To see how the Schrödinger equation is applied to a problem, let us return to our particle in a one-dimensional box of length L (see Fig. 41.3) and analyze it with the Schrödinger equation. Figure 41.3b is the potential-energy diagram that describes this problem. A potential-energy diagram such as this is a useful representation for understanding and solving problems with the Schrödinger equation.

---

[8] If $d\psi/dx$ were not continuous, we would not be able to evaluate $d^2\psi/dx^2$ in Equation 41.13 at the point of discontinuity.

Because of the shape of the curve in Figure 41.3b, the particle in a box is sometimes said to be in a **square well,**[9] where a **well** is an upward-facing region of the curve in a potential-energy diagram. (A downward-facing region is called a *barrier*, which we shall investigate in Section 41.6.) Figure 41.3b shows an infinite square well.

In the region $0 < x < L$, where $U = 0$, we can express the Schrödinger equation in the form

$$\frac{d^2\psi}{dx^2} = -\frac{2mE}{\hbar^2}\psi = -k^2\psi \tag{41.14}$$

where

$$k = \frac{\sqrt{2mE}}{\hbar}$$

The solution to Equation 41.14 is a function ψ whose second derivative is the negative of the same function multiplied by a constant k^2. Both the sine and cosine functions satisfy this requirement. Therefore, the most general solution to the equation is a linear combination of both solutions:

$$\psi(x) = A \sin kx + B \cos kx$$

where A and B are constants that are determined by the boundary and normalization conditions.

Our first boundary condition is that $\psi(0) = 0$:

$$\psi(0) = A \sin 0 + B \cos 0 = 0 + B = 0$$

which means that $B = 0$. Thus, our solution reduces to

$$\psi(x) = A \sin kx$$

The second boundary condition, $\psi(L) = 0$, when applied to the reduced solution, gives

$$\psi(L) = A \sin kL = 0$$

This equation could be satisfied by setting $A = 0$, but this means that $\psi = 0$ everywhere, which is not a valid wave function. We can satisfy our boundary condition only if kL is an integral multiple of π, that is, if $kL = n\pi$, where n is an integer. Because $k = \sqrt{2mE}/\hbar$, we have

$$kL = \frac{\sqrt{2mE}}{\hbar}L = n\pi$$

Each value of the integer n corresponds to a quantized energy that we call E_n. Solving for the allowed energies E_n gives

$$E_n = \left(\frac{h^2}{8mL^2}\right)n^2 \tag{41.15}$$

which are identical to the allowed energies in Equation 41.12.

Substituting the values of k in the wave function, the allowed wave functions $\psi_n(x)$ are given by

$$\psi_n(x) = A \sin\left(\frac{n\pi x}{L}\right) \tag{41.16}$$

This is the wave function (Equation 41.11) that we used in our initial discussion of the particle in a box.

---

[9] It is called a square well even if it has a rectangular shape in a potential energy diagram.

Normalizing this wave function shows that $A = \sqrt{2/L}$. (See Problem 17.) Therefore, the normalized wave function is

$$\psi_n(x) = \sqrt{\frac{2}{L}} \sin\left(\frac{n\pi x}{L}\right) \qquad (41.17)$$

Normalized wave function for a particle in a box

Quick Quiz 41.5 Consider an electron, a proton, and an alpha particle (a helium nucleus), each trapped separately in identical infinite square wells. Which particle corresponds to the highest ground-state energy? (a) the electron (b) the proton (c) the alpha particle. (d) The ground-state energy is the same in all three cases.

Quick Quiz 41.6 Consider the three particles in Quick Quiz 41.5 again. Which particle has the longest wavelength when the system is in the ground state? (a) the electron (b) the proton (c) the alpha particle. (d) All three particles have the same wavelength.

Example 41.5 The Expectation Values for the Particle in a Box

A particle of mass m is confined to a one dimensional box between $x = 0$ and $x = L$. Find the expectation value of the position x of the particle in the ground state.

Solution Using Equation 41.7 and the wave function in Equation 41.17, with $n = 1$, we can express the expectation value for x as follows:

$$(1) \qquad \langle x \rangle = \int_{-\infty}^{\infty} \psi_1 * x \psi_1 \, dx = \int_0^L x \left[\sqrt{\frac{2}{L}} \sin\left(\frac{\pi x}{L}\right) \right]^2 dx$$

$$= \frac{2}{L} \int_0^L x \sin^2\left(\frac{\pi x}{L}\right) dx$$

Note that we reduced the limits on the integral to 0 to L because the value of the wave function is zero elsewhere. Evaluating the integral by consulting an integral table or by mathematical integration[10] gives

$$(2) \qquad \langle x \rangle = \frac{2}{L}\left[\frac{x^2}{4} - \frac{x \sin\left(2\dfrac{\pi x}{L}\right)}{4\dfrac{\pi}{L}} - \frac{\cos\left(2\dfrac{\pi x}{L}\right)}{8\left(\dfrac{\pi}{L}\right)^2} \right]_0^L$$

$$= \frac{2}{L}\left[\frac{L^2}{4} \right] = \frac{L}{2}$$

This result shows that the expectation value of x is right at the center of the box, which we would expect from the symmetry of the square of the wave function (the probability density) about the center (Fig. 41.4b).

What If? What if we consider states of higher n? How does the expectation value of x depend on n?

Answer Because the squares of all wave functions are symmetric about the midpoint, we would not expect the expectation value to depend on n. If we write Equation (1) with a general wave function, we have

$$\langle x \rangle = \int_{-\infty}^{\infty} \psi_n * x \psi_n \, dx = \int_0^L x \left[\sqrt{\frac{2}{L}} \sin\left(\frac{n\pi x}{L}\right) \right]^2 dx$$

$$= \frac{2}{L} \int_0^L x \sin^2\left(\frac{n\pi x}{L}\right) dx$$

which leads to a new expression for Equation (2).

$$\langle x \rangle = \frac{2}{L}\left[\frac{x^2}{4} - \frac{x \sin\left(2\dfrac{n\pi x}{L}\right)}{4\dfrac{n\pi}{L}} - \frac{\cos\left(2\dfrac{n\pi x}{L}\right)}{8\left(\dfrac{n\pi}{L}\right)^2} \right]_0^L$$

$$= \frac{2}{L}\left[\frac{L^2}{4} \right] = \frac{L}{2}$$

The presence of n in the wave function has not affected the expectation value, as we suspected.

The $n = 2$ wave function in Figure 41.4b has a value of zero at the midpoint of the box. How can the expectation value of the particle be at a position at which the particle has zero probability of existing? Remember that the expectation value is the *average* position. Therefore, the particle is as likely to be found to the right of the midpoint as to the left, so that its average position is right at the midpoint, even though its probability of being there is zero. As an analogy, consider a group of students for whom the average final examination score is 50%. There is no requirement that some student achieve a score of exactly 50% for the average of all students to be 50%.

---

[10] To integrate this function, first replace $\sin^2(\pi x/L)$ with $\frac{1}{2}(1 - \cos 2\pi x/L)$ (Table B.3 in Appendix B). This will allow $\langle x \rangle$ to be expressed as two integrals. The second integral can then be evaluated by partial integration (Section B.7 in Appendix B).

41.5 A Particle in a Well of Finite Height

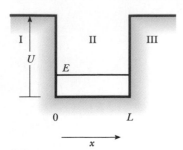

Figure 41.7 Potential energy diagram of a well of finite height U and length L. A particle is trapped in a well. The total energy E of the particle–well system is less than U.

Now consider a particle in a *finite* potential well, that is, a system having a potential energy that is zero when the particle is in the region $0 < x < L$ and a finite value U when the particle is outside this region, as in Figure 41.7. The system consisting of the particle and its environment determines the shape of the well. If the total energy E of the system is less than U, classically the particle is permanently bound in the potential well. If the particle were outside the well, its kinetic energy would have to be negative, an impossibility. However, according to quantum mechanics, a finite probability exists that the particle can be found outside the well even if $E < U$. That is, the wave function ψ is generally nonzero outside the well—regions I and III in Figure 41.7—so the probability density $|\psi|^2$ is also nonzero in these regions. While this may be an uncomfortable notion, the uncertainty principle tells us that the energy of the system is uncertain. This allows the particle to be outside the well as long as the apparent violation of conservation of energy does not exist in any measurable way.

In region II, where $U = 0$, the allowed wave functions are again sinusoidal because they represent solutions of Equation 41.14. However, the boundary conditions no longer require that ψ be zero at the ends of the well, as was the case with the infinite square well.

The Schrödinger equation for regions I and III may be written

$$\frac{d^2\psi}{dx^2} = \frac{2m(U - E)}{\hbar^2}\psi \qquad (41.18)$$

Because $U > E$, the coefficient of ψ on the right-hand side is necessarily positive. Therefore, we can express Equation 41.18 in the form

$$\frac{d^2\psi}{dx^2} = C^2\psi \qquad (41.19)$$

where $C^2 = 2m(U - E)/\hbar^2$ is a positive constant in regions I and III. As you can verify by substitution, the general solution of Equation 41.19 is

$$\psi = Ae^{Cx} + Be^{-Cx}$$

where A and B are constants.

We can use this general solution as a starting point for determining the appropriate solution for regions I and III. The function we choose for our solution must remain finite over the entire region under consideration. In region I, where $x < 0$, we must rule out the term Be^{-Cx}. In other words, we must require that $B = 0$ in region I to avoid an infinite value for ψ for large negative values of x. Likewise, in region III, where $x > L$, we must rule out the term Ae^{Cx}; this is accomplished by taking $A = 0$ in this region. This choice avoids an infinite value for ψ for large positive x values. Hence, the solutions in regions I and III are

$$\psi_{\text{I}} = Ae^{Cx} \qquad \text{for } x < 0$$

$$\psi_{\text{III}} = Be^{-Cx} \qquad \text{for } x > L$$

In region II the wave function is sinusoidal and has the general form

$$\psi_{\text{II}}(x) = F\sin kx + G\cos kx$$

where F and G are constants.

These results show that the wave functions outside the potential well (where classical physics forbids the presence of the particle) decay exponentially with distance. At large negative x values, ψ_{I} approaches zero; at large positive x values, ψ_{III} approaches zero. These functions, together with the sinusoidal solution in region II, are shown in

particle is either reflected or transmitted, we require that $T + R = 1$. An approximate expression for the transmission coefficient that is obtained when $T \ll 1$ (a very wide barrier or a very high barrier, that is, $U \gg E$) is

$$T \approx e^{-2CL} \tag{41.20}$$

where

$$C = \frac{\sqrt{2m(U - E)}}{\hbar} \tag{41.21}$$

This quantum model of barrier penetration and specifically Equation 41.20 show that T can be nonzero. The fact that the phenomenon of tunneling is observed experimentally provides further confidence in the principles of quantum physics.

Quick Quiz 41.8 Which of the following changes would increase the probability of transmission of a particle through a potential barrier? (You may choose more than one answer.) (a) decreasing the width of the barrier (b) increasing the width of the barrier (c) decreasing the height of the barrier (d) increasing the height of the barrier (e) decreasing the kinetic energy of the incident particle (f) increasing the kinetic energy of the incident particle.

Example 41.6 **Transmission Coefficient for an Electron**   `Interactive`

A 30-eV electron is incident on a square barrier of height 40 eV. What is the probability that the electron will tunnel through the barrier if its width is

(A) 1.0 nm?

(B) 0.10 nm?

Solution (A) In this situation, the quantity $U - E$ has the value

$$U - E = (40 \text{ eV} - 30 \text{ eV}) = 10 \text{ eV} = 1.6 \times 10^{-18} \text{ J}$$

Using Equations 41.20 and 41.21, we find that

$$2CL = 2 \frac{\sqrt{2(9.11 \times 10^{-31} \text{ kg})(1.6 \times 10^{-18} \text{ J})}}{1.054 \times 10^{-34} \text{ J} \cdot \text{s}} (1.0 \times 10^{-9} \text{ m})$$

$$= 32.4$$

Thus, the probability of tunneling through the barrier is

$$T \approx e^{-2CL} = e^{-32.4} = \boxed{8.5 \times 10^{-15}}$$

The electron has about 1 chance in 10^{14} of tunneling through the barrier.

(B) For $L = 0.10$ nm, $2CL = 3.24$, and

$$T \approx e^{-2CL} = e^{-3.24} = \boxed{0.039}$$

Now the electron has a much higher probability (3.9%) of penetrating the barrier. Thus, reducing the width of the barrier by only one order of magnitude increases the probability of tunneling by about 12 orders of magnitude!

 Investigate the tunneling of particles through barriers at the Interactive Worked Example link at **http://www.pse6.com**

Some Applications of Tunneling

As we have seen, tunneling is a quantum phenomenon, a manifestation of the wave nature of matter. Many examples exist (on the atomic and nuclear scales) for which tunneling is very important.

- **Alpha decay.** One form of radioactive decay is the emission of alpha particles (the nuclei of helium atoms) by unstable, heavy nuclei (Chapter 44). In order for an alpha particle to escape from the nucleus, it must penetrate a barrier whose height is several times larger than the energy of the nucleus–alpha particle system. The barrier is due to a combination of the attractive nuclear force (discussed in

Chapter 44) and the Coulomb repulsion (discussed in detail in Chapter 23) between the alpha particle and the rest of the nucleus. Occasionally an alpha particle tunnels through the barrier, which explains the basic mechanism for this type of decay and the large variations in the mean lifetimes of various radioactive nuclei.

- **Nuclear fusion.** The basic reaction that powers the Sun and, indirectly, almost everything else in the solar system, is fusion, which we will study in Chapter 45. In one step of the process that occurs at the core of the Sun, protons must approach each other to within such a small distance that they fuse to form a deuterium nucleus. (See Section 45.4.) According to classical physics, these protons cannot overcome and penetrate the barrier caused by their mutual electrical repulsion. Quantum-mechanically, however, the protons are able to tunnel through the barrier and fuse together.

- **Scanning tunneling microscopes,** discussed in Section 41.8.

As a final example of an application of tunneling, let us expand on the quantum-dot discussion in Section 41.5 by exploring the **resonant tunneling device.** Figure 41.10a shows the physical construction of such a device. The island of gallium arsenide is a quantum dot located between two barriers formed from the thin extensions of aluminum arsenide. Figure 41.10b shows the potential barriers that are encountered by electrons incident from the left as well as the quantized energy levels in the quantum dot. This situation differs from the one shown in Figure 41.9 in that there are quantized energy levels on the right of the first barrier. In Figure 41.9, an electron that tunnels through the barrier is considered a free particle and can have any energy. In contrast, in Figure 41.10b, as the electrons encounter the first barrier they have no

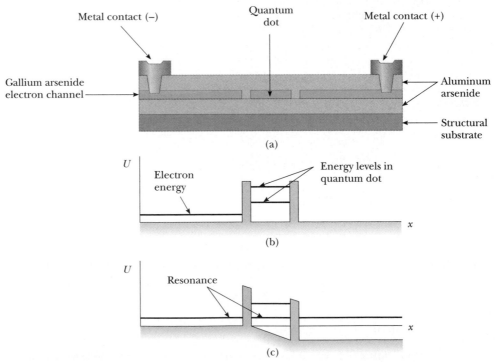

(a)

(b)

(c)

Active Figure 41.10 (a) The physical structure of a resonant tunneling device. Electrons travel in the gallium arsenide semiconductor and strike the barrier of the quantum dot from the left. (b) A potential energy diagram showing the double barrier representing the walls of the quantum dot. With no voltage applied to the device, none of the quantized states in the quantum dot is resonant with the incident electron energy, and the current is zero. (c) A voltage is applied across the device. The distortion of the potential energy curve causes one of the states in the quantum dot to resonate with the incident electron energy. The electrons are now able to tunnel through the barrier and produce a current in the device.

Vary the voltage across the resonant tunneling device at the Active Figures link at http://www.pse6.com.

energy levels available on the right side of the barrier, which greatly reduces the probability of tunneling.

Figure 41.10c shows the effect of applying a voltage—the potential decreases with position as we move to the right across the device. The deformation of the potential barrier results in an energy level in the quantum dot coinciding with the energy of the incident electrons. This "resonance" of energies gives the device its name. When the voltage is applied, the probability of tunneling increases tremendously and the device carries current. In this manner, the device can be used as a very fast switch on a nanotechnological scale.

Figure 41.11a shows the addition of a gate electrode at the top of the device over the quantum dot. This electrode turns the device into a **resonant tunneling transistor.** The basic function of a transistor is amplification—to convert a small varying voltage into a large varying voltage. Figure 41.11b, representing the potential-energy diagram for the tunneling transistor, has a slope at the bottom of the quantum dot due to the differing voltages at the source and drain electrodes. In this configuration, there is no resonance between the electron energies outside the quantum dot and the quantized energies within the dot. By applying a small voltage to the gate electrode as in Figure 41.11c, the quantized energies can be brought into resonance with the electron energy outside the well, and resonant tunneling occurs. The current that results causes a voltage across an external resistor that is much larger than that of the gate voltage; hence, the device amplifies the input signal to the gate electrode.

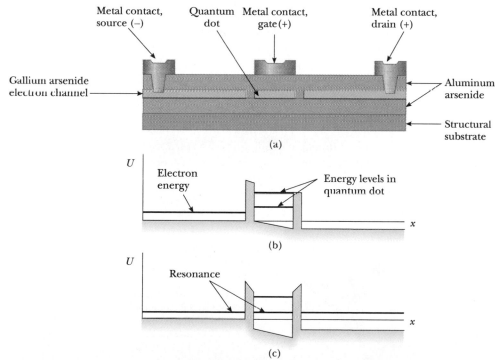

Figure 41.11 (a) The addition of a gate electrode to the structure in Figure 41.10 converts it to a resonant tunneling transistor. Electrons travel in the gallium arsenide semiconductor and strike the barrier of the quantum dot from the left. (b) A potential energy diagram showing the double barrier representing the walls of the quantum dot. With a low voltage applied across the device, none of the quantized states in the quantum dot is resonant with the incident electron energy, and the current is zero. (c) A voltage is applied to the gate electrode. The potential in the region of the quantum dot drops, along with the quantized energy levels. For very small changes in the gate voltage, the quantized level goes in and out of resonance with the electron energy, causing a large change in the current in the device and the voltage across an external resistor.

41.7 The Scanning Tunneling Microscope[11]

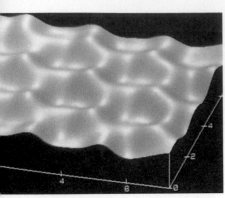

Figure 41.12 The surface of graphite as "viewed" with a scanning tunneling microscope. This type of microscope enables scientists to see details with a lateral resolution of about 0.2 nm and a vertical resolution of 0.001 nm. The contours seen here represent the ring-like arrangement of individual carbon atoms on the crystal surface.

One of the basic phenomena of quantum mechanics—tunneling—is at the heart of a very practical device, the scanning tunneling microscope (STM), which enables us to obtain highly detailed images of surfaces at resolutions comparable to the size of a *single atom*. Figure 41.12, showing the surface of a piece of graphite, demonstrates what the STM can do. What makes this image so remarkable is that its resolution is about 0.2 nm. For an optical microscope, the resolution is limited by the wavelength of the light used to make the image. Thus, an optical microscope has a resolution no better than 200 nm, about half the wavelength of visible light, and so could never show the detail displayed in Figure 41.12. An ideal electron microscope (Section 40.5) could have a resolution of 0.2 nm by using electron waves of this wavelength, given by the de Broglie expression $\lambda = h/p$. The linear momentum p of an electron required to give this wavelength is 10 000 eV/c, corresponding to an electron speed of 2% of the speed of light. Electrons traveling at this speed would penetrate into the interior of the graphite in Figure 41.12 and thus could not give us information about individual surface atoms.

The STM achieves its very fine resolution by using the basic idea shown in Figure 41.13. An electrically conducting probe with a very sharp tip is brought near the surface to be studied. The empty space between tip and surface represents the "barrier" we have been discussing, and the tip and surface are the two walls of the "potential well." Because electrons obey quantum rules rather than Newtonian rules, they can "tunnel" across the barrier of empty space. If a voltage is applied between surface and tip, electrons in the atoms of the surface material can be made to tunnel preferentially from surface to tip to produce a tunneling current. In this way the tip samples the distribution of electrons just above the surface.

Because of the nature of tunneling, the STM is very sensitive to the distance z from tip to surface—in other words, to the thickness of the barrier (see Example 41.6). The reason is that in the empty space between tip and surface, the electron wave function falls off exponentially (see Fig. 41.9, region II) with a decay length on the order of 0.1 nm; that is, the wave function decreases by $1/e$ over that distance. For distances $z > 1$ nm (that is, beyond a few atomic diameters), essentially no tunneling takes place. This exponential behavior causes the current of electrons tunneling from surface to tip to depend very strongly on z. This sensitivity is the basis of the operation of the STM. By monitoring the tunneling current as the tip is scanned over the surface, scientists obtain a sensitive measure of the topography of the electron distribution on the

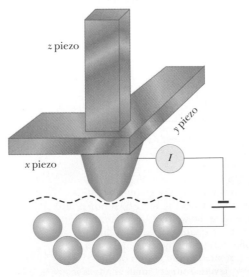

Figure 41.13 Schematic view of an STM. The tip, shown as a rounded cone, is mounted on a piezoelectric *xyz* scanner. (A piezoelectric crystal develops a voltage that is related to the amount of deformation of the crystal.) A scan of the tip over the sample can reveal contours of the surface down to the atomic level. An STM image is composed of a series of scans displaced laterally from one another. (*Based on a drawing from P. K. Hansma, V. B. Elings, O. Marti, and C. Bracker, Science 242:209, 1988.*) © 1988 by the AAAS.

[11] An earlier version of this section was written by Roger A. Freedman and Paul K. Hansma, University of California, Santa Barbara.

surface. The result of this scan is used to make images like that in Figure 41.12. In this way the STM can measure the height of surface features to within 0.001 nm, approximately 1/100 of an atomic diameter!

You can see just how sensitive the STM is by examining Figure 41.12. Of the six carbon atoms in each ring, three appear lower than the other three. In fact, all six atoms are at the same height, but all have slightly different electron distributions. The three atoms that appear lower are bonded to other carbon atoms directly beneath them in the underlying atomic layer; as a result, their electron distributions, which are responsible for the bonding, extend downward beneath the surface. The atoms in the surface layer that appear higher do not lie directly over subsurface atoms and hence are not bonded to any underlying atoms. For these higher-appearing atoms, the electron distribution extends upward into the space above the surface. This extra electron density is what makes these atoms appear higher in Figure 41.12, because what the STM maps is the topography of the electron distribution.

The STM has one serious limitation: Its operation depends on the electrical conductivity of the sample and the tip. Unfortunately, most materials are not electrically conductive at their surfaces. Even metals, which are usually excellent electrical conductors, are covered with nonconductive oxides. A newer microscope, the atomic force microscope (AFM), overcomes this limitation.

41.8 The Simple Harmonic Oscillator

Let us use the understanding of quantum mechanics developed in this chapter to consider again the subject of blackbody radiation, discussed in Section 40.1. In Planck's analysis, standing waves in a cavity are established by radiation from vibrating charges in the walls of the cavity. These sources of radiation have allowed energy levels equally spaced in energy (Figure 40.6). Let us see if this is consistent with a quantum treatment of the vibrating charges as simple harmonic oscillators.

We begin by considering a particle that is subject to a linear restoring force $F = -kx$, where x is the position of the particle relative to equilibrium ($x = 0$) and k is the force constant. The classical motion of a particle subject to such a force is simple harmonic motion, which was discussed in Chapter 15. The potential energy of the system is, from Equation 15.20,

$$U = \tfrac{1}{2}kx^2 = \tfrac{1}{2}m\omega^2 x^2$$

where the angular frequency of vibration is $\omega = \sqrt{k/m}$. Classically, if the particle is displaced from its equilibrium position and released, it oscillates between the points $x = -A$ and $x = A$, where A is the amplitude of the motion. Furthermore, its total energy E is, from Equation 15.21,

$$E = K + U = \tfrac{1}{2}kA^2 = \tfrac{1}{2}m\omega^2 A^2$$

In the classical model, any value of E is allowed, including $E = 0$, which is the total energy when the particle is at rest at $x = 0$.

The Schrödinger equation for this problem is obtained by substituting $U = \tfrac{1}{2}m\omega^2 x^2$ into Equation 41.13:

$$-\frac{\hbar^2}{2m}\frac{d^2\psi}{dx^2} + \tfrac{1}{2}m\omega^2 x^2\psi = E\psi \qquad (41.22)$$

The mathematical technique for solving this equation is beyond the level of this text. However, it is instructive to guess at a solution. We take as our guess the following wave function:

$$\psi = Be^{-Cx^2} \qquad (41.23)$$

Substituting this function into Equation 41.22, we find that it is a satisfactory solution to the Schrödinger equation, provided that

$$C = \frac{m\omega}{2\hbar} \quad \text{and} \quad E = \tfrac{1}{2}\hbar\omega$$

It turns out that the solution we have guessed corresponds to the ground state of the system, which has an energy $\tfrac{1}{2}\hbar\omega$. Because $C = m\omega/2\hbar$ it follows from Equation 41.23 that the wave function for this state is

Wave function for the ground state of a simple harmonic oscillator

$$\psi = Be^{-(m\omega/2\hbar)x^2} \tag{41.24}$$

where B is a constant to be determined from the normalization condition. This is only one solution to Equation 41.22. The remaining solutions that describe the excited states are more complicated, but all solutions include the exponential factor e^{-Cx^2}.

The energy levels of a harmonic oscillator are quantized, as we would expect because the oscillating particle is bound to stay near $x = 0$. The energy for an arbitrary quantum number n is

Quantized energies for a simple harmonic oscillator

$$E_n = (n + \tfrac{1}{2})\hbar\omega \qquad n = 0, 1, 2, \ldots \tag{41.25}$$

The state $n = 0$ corresponds to the ground state, whose energy is $E_0 = \tfrac{1}{2}\hbar\omega$; the state $n = 1$ corresponds to the first excited state, whose energy is $E_1 = \tfrac{3}{2}\hbar\omega$, and so on. The energy-level diagram for this system is shown in Figure 41.14. Note that the separations

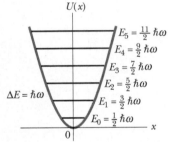

Figure 41.14 Energy-level diagram for a simple harmonic oscillator, superimposed on the potential energy function. The levels are equally spaced, with separation $\hbar\omega$. The ground-state energy is $E_0 = \tfrac{1}{2}\hbar\omega$.

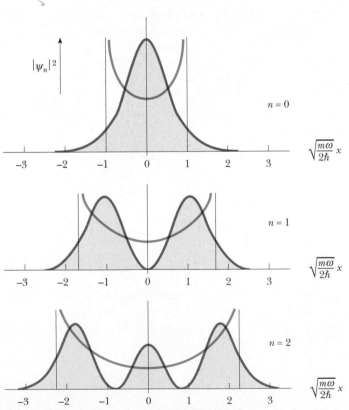

Figure 41.15 The brown curves represent probability densities $|\psi|^2$ for the first three states of a quantum simple harmonic oscillator. The blue curves represent classical probability densities corresponding to the same energies. (*From C. W. Sherwin*, Introduction to Quantum Mechanics, *New York, Holt, Rinehart and Winston, 1959. Used with permission.*)

between adjacent levels are equal and are given by

$$\Delta E = \hbar \omega \qquad (41.26)$$

The brown curves in Figure 41.15 indicate the probability densities $|\psi|^2$ for the first three states of a simple harmonic oscillator. The blue curves represent the classical probability densities that correspond to the same energy and are provided for comparison. We see that as n increases, the agreement between the classical and quantum-mechanical results improves, as expected.

Notice that the energy levels for the harmonic oscillator in Figure 41.14 are equally spaced, just as Planck proposed for the oscillators in the walls of the cavity that was used in the model for blackbody radiation in Section 40.1. Planck's Equation 40.4 for the energy levels of the oscillators differs from Equation 41.25 only in the term $\frac{1}{2}$ added to n. This additional term does not affect the energy emitted in a transition, given by Equation 40.5, which is equivalent to Equation 41.26. The fact that Planck generated these concepts without the benefit of the Schrödinger equation is testimony to his genius.

Quick Quiz 41.9 For a particle undergoing simple harmonic motion in the $n = 0$ state, the most probable value of x for the particle according to quantum mechanics is (a) $x = 0$ (b) $x = \pm A$. (c) All values of x are equally likely.

Example 41.7   Molar Specific Heat of Hydrogen Gas

In Figure 21.7 (page 652), which shows the molar specific heat of hydrogen as a function of temperature, vibration does not contribute to the molar specific heat at room temperature. Explain why this is so, modeling the hydrogen molecule as a simple harmonic oscillator. The effective spring constant for the bond in the hydrogen molecule is 573 N/m.

Solution We conceptualize this problem by imagining the only mode of vibration available to a diatomic molecule. This mode (shown in Figure 21.6c) consists of the two atoms always moving in opposite directions. We categorize this as a quantum harmonic oscillator problem, with the molecule as a two-particle system. We can analyze the motion of the particles relative to the center of mass by considering a single particle with reduced mass μ given by

$$\mu = \frac{m_1 m_2}{m_1 + m_2}$$

where m_1 and m_2 are the masses of the two particles. (See Problem 37.) For the hydrogen molecule, the masses of the two particles are the same, so $\mu = \frac{1}{2}m$, where m is the mass of a hydrogen atom.

Let us assume that the molecule is in its vibrational ground state and calculate the energy necessary to excite the molecule to its first excited vibrational state. From Equation 41.26, this energy is

$$\Delta E = \hbar \omega = \hbar \sqrt{\frac{k}{\mu}}$$

where we have used Equation 15.9 for the angular frequency of the oscillator and have substituted the reduced mass μ for

the single equivalent particle. Evaluating this numerically, we find

$$\Delta E = \hbar \sqrt{\frac{k}{\frac{1}{2}m}} = \hbar \sqrt{\frac{2k}{m}}$$

$$= (1.055 \times 10^{-34}\,\text{J}\cdot\text{s})\sqrt{\frac{2(573\,\text{N/m})}{1.67 \times 10^{-27}\,\text{kg}}}$$

$$= 8.74 \times 10^{-20}\,\text{J}$$

From Equation 21.4, we know that the average translational kinetic energy of molecules in a gas is given by $\frac{3}{2}k_B T$, where k_B is Boltzmann's constant. Let us find the temperature at which the average molecular translational kinetic energy is equal to that required to excite the first vibrational state of the molecule:

$$\tfrac{3}{2}k_B T = 8.74 \times 10^{-20}\,\text{J}$$

$$T = \tfrac{2}{3}\left(\frac{8.74 \times 10^{-20}\,\text{J}}{k_B}\right) = \tfrac{2}{3}\left(\frac{8.74 \times 10^{-20}\,\text{J}}{1.38 \times 10^{-23}\,\text{J/K}}\right)$$

$$= 4.22 \times 10^3\,\text{K}$$

Thus, the temperature of the gas must be more than 4 000 K in order for the translational kinetic energy to be comparable to the energy required to excite the first vibrational state. This excitation energy must come from collisions between molecules, so if the molecules do not have sufficient translational kinetic energy, they will not be excited to the first vibrational state and vibration will not contribute to the molar specific heat. This explains why the curve in Figure 21.7 does not rise to a value corresponding to the contribution of vibration until the hydrogen gas has been raised to thousands of kelvins.

To finalize this problem, we see from Figure 21.7 that rotational energy levels must be more closely spaced in energy than vibrational levels because they are excited at a lower temperature than the vibrational levels. The translational energy levels are those of a particle in a three-dimensional box, where the box is the container holding the gas. These levels are given by an expression similar to Equation 41.12. Because the box is macroscopic in size, L is very large, and the energy levels are very close together. In fact, they are so close together that translational energy levels are excited at a fraction of a kelvin.

SUMMARY

Take a practice test for this chapter by clicking on the Practice Test link at http://www.pse6.com.

In quantum mechanics, a particle in a system can be represented by a wave function $\psi(x, y, z)$. The probability per unit volume (or probability density) that a particle will be found at a point is $|\psi|^2 = \psi^*\psi$, where ψ^* is the complex conjugate of ψ. If the particle is confined to moving along the x axis, then the probability that it is located in an interval dx is $|\psi|^2 dx$. Furthermore, the sum of all these probabilities over all values of x must be 1:

$$\int_{-\infty}^{\infty} |\psi|^2 \, dx = 1 \qquad (41.6)$$

This is called the **normalization condition.**

The measured position x of the particle, averaged over many trials, is called the **expectation value** of x and is defined by

$$\langle x \rangle \equiv \int_{-\infty}^{\infty} \psi^* x \psi \, dx \qquad (41.7)$$

If a particle of mass m is confined to moving in a one-dimensional box of length L whose walls are impenetrable, we require that ψ be zero at the walls and outside the box. The wave functions for this system are given by

$$\psi(x) = A \sin\left(\frac{n\pi x}{L}\right) \qquad n = 1, 2, 3, \ldots \qquad (41.11)$$

where A is the maximum value of ψ. The particle has a well-defined wavelength λ with values such that $L = n\lambda/2$. The allowed states of a particle in a box have quantized energies given by

$$E_n = \left(\frac{h^2}{8mL^2}\right) n^2 \qquad n = 1, 2, 3, \ldots \qquad (41.12)$$

The wave function for a system must satisfy the **Schrödinger equation.** The time-independent Schrödinger equation for a particle confined to moving along the x axis is

$$-\frac{\hbar^2}{2m} \frac{d^2\psi}{dx^2} + U\psi = E\psi \qquad (41.13)$$

where E is the total energy of the system and U is the potential energy.

The approach of quantum mechanics is to solve Equation 41.13 for ψ and E, given the potential energy $U(x)$ for the system. In doing so, we must place the following restrictions on $\psi(x)$: (1) $\psi(x)$ must be continuous, (2) $\psi(x)$ must approach zero as x approaches $\pm\infty$, (3) $\psi(x)$ must be single-valued, and (4) $d\psi/dx$ must be continuous for all finite values of $U(x)$.

QUESTIONS

1. What is the significance of the wave function ψ?

2. For a particle in a box, the probability density at certain points is zero, as seen in Figure 41.4b. Does this imply that the particle cannot move across these points? Explain.

3. Discuss the relationship between ground-state energy and the uncertainty principle.

4. As a particle of energy E is reflected from a potential barrier of height U, where $E < U$, how does the amplitude

of the reflected wave change as the barrier height is reduced?

5. A philosopher once said that "it is necessary for the very existence of science that the same conditions always produce the same results." In view of what has been discussed in this chapter, present an argument showing that this statement is false. How might the statement be reworded to make it true?

6. In quantum mechanics it is possible for the energy E of a particle to be less than the potential energy, but classically this is not possible. Explain.

7. Consider two square wells of the same length, one with finite walls and the other with infinite walls. Compare the energy and momentum of a particle trapped in the finite well with the energy and momentum of an identical particle in the infinite well.

8. Why is it impossible for the lowest-energy state of a harmonic oscillator to be zero?

9. What is the Schrödinger equation? How is it useful in describing quantum phenomena?

PROBLEMS

1, 2, 3 = straightforward, intermediate, challenging   ☐ = full solution available in the *Student Solutions Manual and Study Guide*

🪐 = coached solution with hints available at http://www.pse6.com   💻 = computer useful in solving problem

▨ = paired numerical and symbolic problems

Section 41.1   An Interpretation of Quantum Mechanics

1. A free electron has a wave function

$$\psi(x) = Ae^{i(5.00 \times 10^{10}x)}$$

where x is in meters. Find (a) its de Broglie wavelength, (b) its momentum, and (c) its kinetic energy in electron volts.

2. The wave function for a particle is

$$\psi(x) = \sqrt{\frac{a}{\pi(x^2 + a^2)}}$$

for $a > 0$ and $-\infty < x < +\infty$. Determine the probability that the particle is located somewhere between $x = -a$ and $x = +a$.

Section 41.2   A Particle in a Box

3. An electron is confined to a one-dimensional region in which its ground-state ($n = 1$) energy is 2.00 eV. (a) What is the length of the region? (b) How much energy is required to promote the electron to its first excited state?

4. An electron that has an energy of approximately 6 eV moves between rigid walls 1.00 nm apart. Find (a) the quantum number n for the energy state that the electron occupies and (b) the precise energy of the electron.

5. 🪐 An electron is contained in a one-dimensional box of length 0.100 nm. (a) Draw an energy-level diagram for the electron for levels up to $n = 4$. (b) Find the wavelengths of all photons that can be emitted by the electron in making downward transitions that could eventually carry it from the $n = 4$ state to the $n = 1$ state.

6. An alpha particle in a nucleus can be modeled as a particle moving in a box of length 1.00×10^{-14} m (the approximate diameter of a nucleus). Using this model, estimate the energy and momentum of an alpha particle in its lowest energy state. ($m_\alpha = 4 \times 1.66 \times 10^{-27}$ kg)

7. A ruby laser emits 694.3-nm light. Assume light of this wavelength is due to a transition of an electron in a box from its $n = 2$ state to its $n = 1$ state. Find the length of the box.

8. A laser emits light of wavelength λ. Assume this light is due to a transition of an electron in a box from its $n = 2$ state to its $n = 1$ state. Find the length of the box.

9. The nuclear potential energy that binds protons and neutrons in a nucleus is often approximated by a square well. Imagine a proton confined in an infinitely high square well of length 10.0 fm, a typical nuclear diameter. Calculate the wavelength and energy associated with the photon emitted when the proton moves from the $n = 2$ state to the ground state. In what region of the electromagnetic spectrum does this wavelength belong?

10. A proton is confined to move in a one-dimensional box of length 0.200 nm. (a) Find the lowest possible energy of the proton. (b) **What If?** What is the lowest possible energy of an electron confined to the same box? (c) How do you account for the great difference in your results for (a) and (b)?

11. Use the particle-in-a-box model to calculate the first three energy levels of a neutron trapped in a nucleus of diameter 20.0 fm. Do the energy-level differences have a realistic order of magnitude?

12. A photon with wavelength λ is absorbed by an electron confined to a box. As a result, the electron moves from state $n = 1$ to $n = 4$. (a) Find the length of the box. (b) What is the wavelength of the photon emitted in the transition of that electron from the state $n = 4$ to the state $n = 2$?

Section 41.3   The Particle Under Boundary Conditions
Section 41.4   The Schrödinger Equation

13. Show that the wave function $\psi = Ae^{i(kx - \omega t)}$ is a solution to the Schrödinger equation (Eq. 41.13) where $k = 2\pi/\lambda$ and $U = 0$.

14. The wave function of a particle is given by

$$\psi(x) = A \cos(kx) + B \sin(kx)$$

where A, B, and k are constants. Show that ψ is a solution of the Schrödinger equation (Eq. 41.13), assuming the particle is free $(U = 0)$, and find the corresponding energy E of the particle.

15. Prove that the first term in the Schrödinger equation, $-(\hbar^2/2m)(d^2\psi/dx^2)$, reduces to the kinetic energy of the particle multiplied by the wave function (a) for a freely moving particle, with the wave function given by Equation 41.3 and (b) for a particle in a box, with the wave function given by Equation 41.17.

16. A particle in an infinitely deep square well has a wave function given by

$$\psi_2(x) = \sqrt{\frac{2}{L}} \sin\left(\frac{2\pi x}{L}\right)$$

for $0 \le x \le L$ and zero otherwise. (a) Determine the expectation value of x. (b) Determine the probability of finding the particle near $L/2$, by calculating the probability that the particle lies in the range $0.490L \le x \le 0.510L$. (c) **What If?** Determine the probability of finding the particle near $L/4$, by calculating the probability that the particle lies in the range $0.240L \le x \le 0.260L$. (d) Argue that the result of part (a) does not contradict the results of parts (b) and (c).

17. The wave function for a particle confined to moving in a one-dimensional box is

$$\psi(x) = A \sin\left(\frac{n\pi x}{L}\right)$$

Use the normalization condition on ψ to show that

$$A = \sqrt{\frac{2}{L}}$$

Suggestion: Because the box length is L, the wave function is zero for $x < 0$ and for $x > L$, so the normalization condition (Eq. 41.6) reduces to

$$\int_0^L |\psi|^2 \, dx = 1$$

18. The wave function of an electron is

$$\psi(x) = \sqrt{\frac{2}{L}} \sin\left(\frac{2\pi x}{L}\right)$$

Calculate the probability of finding the electron between $x = 0$ and $x = L/4$.

19. An electron in an infinitely deep square well has a wave function that is given by

$$\psi_2(x) = \sqrt{\frac{2}{L}} \sin\left(\frac{2\pi x}{L}\right)$$

for $0 \le x \le L$ and is zero otherwise. What are the most probable positions of the electron?

20. ⌨ A particle is in the $n = 1$ state of an infinite square well with walls at $x = 0$ and $x = L$. Let ℓ be an arbitrary value of x between $x = 0$ and $x = L$. (a) Find an expres-

sion for the probability, as a function of ℓ, that the particle will be found between $x = 0$ and $x = \ell$. (b) Sketch the probability as a function of ℓ/L. Choose values of ℓ/L ranging from 0 to 1.00 in steps of 0.100. (c) Find the value of ℓ for which the probability of finding the particle between $x = 0$ and $x = \ell$ is twice the probability of finding the particle between $x = \ell$ and $x = L$. You can solve the transcendental equation for ℓ/L numerically.

21. A particle in an infinite square well has a wave function that is given by

$$\psi_1(x) = \sqrt{\frac{2}{L}} \sin\left(\frac{\pi x}{L}\right)$$

for $0 \le x \le L$ and is zero otherwise. (a) Determine the probability of finding the particle between $x = 0$ and $x = L/3$. (b) Use the result of this calculation and symmetry arguments to find the probability of finding the particle between $x = L/3$ and $x = 2L/3$. Do not re-evaluate the integral. (c) **What If?** Compare the result of part (a) with the classical probability.

22. Consider a particle moving in a one-dimensional box for which the walls are at $x = -L/2$ and $x = L/2$. (a) Write the wave functions and probability densities for $n = 1$, $n = 2$, and $n = 3$. (b) Sketch the wave functions and probability densities. (*Suggestion:* Make an analogy to the case of a particle in a box for which the walls are at $x = 0$ and $x = L$.)

23. A particle of mass m moves in a potential well of length $2L$. Its potential energy is infinite for $x < -L$ and for $x > +L$. Inside the region $-L < x < L$, its potential energy is given by

$$U(x) = \frac{-\hbar^2 x^2}{mL^2(L^2 - x^2)}$$

In addition, the particle is in a stationary state that is described by the wave function $\psi(x) = A(1 - x^2/L^2)$ for $-L < x < +L$, and by $\psi(x) = 0$ elsewhere. (a) Determine the energy of the particle in terms of $\hbar$, m, and L. (*Suggestion:* Use the Schrödinger equation, Eq. 41.13.) (b) Show that $A = (15/16L)^{1/2}$. (c) Determine the probability that the particle is located between $x = -L/3$ and $x = +L/3$.

24. In a region of space, a particle with zero total energy has a wave function

$$\psi(x) = Axe^{-x^2/L^2}$$

(a) Find the potential energy U as a function of x. (b) Make a sketch of $U(x)$ versus x.

Section 41.5 A Particle in a Well of Finite Height

25. Suppose a particle is trapped in its ground state in a box that has infinitely high walls (Fig. 41.4). Now suppose the left-hand wall is suddenly lowered to a finite height and width. (a) Qualitatively sketch the wave function for the particle a short time later. (b) If the box has a length L, what is the wavelength of the wave that penetrates the left-hand wall?

26. Sketch the wave function $\psi(x)$ and the probability density $|\psi(x)|^2$ for the $n = 4$ state of a particle in a finite potential well. (See Fig. 41.8.)

Section 41.6   Tunneling Through a Potential Energy Barrier

27. An electron with kinetic energy $E = 5.00$ eV is incident on a barrier with thickness $L = 0.200$ nm and height $U = 10.0$ eV (Fig. P41.27). What is the probability that the electron (a) will tunnel through the barrier? (b) will be reflected?

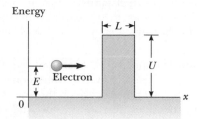

Energy

Figure P41.27 Problems 27 and 28.

28. An electron having total energy $E = 4.50$ eV approaches a rectangular energy barrier with $U = 5.00$ eV and $L = 950$ pm as shown in Figure P41.27. Classically, the electron cannot pass through the barrier because $E < U$. However, quantum-mechanically the probability of tunneling is not zero. Calculate this probability, which is the transmission coefficient.

29. What If? In Problem 28, by how much would the width L of the potential barrier have to be increased for the chance of an incident 4.50-eV electron tunneling through the barrier to be one in a million?

30. An electron has a kinetic energy of 12.0 eV. The electron is incident upon a rectangular barrier of height 20.0 eV and thickness 1.00 nm. By what factor would the electron's probability of tunneling through the barrier increase assuming that the electron absorbs all the energy of a photon with wavelength 546 nm (green light)?

Section 41.7   The Scanning Tunneling Microscope

31. A scanning tunneling microscope (STM) can precisely determine the depths of surface features because the current through its tip is very sensitive to differences in the width of the gap between the tip and the sample surface. Assume that in this direction the electron wave function falls off exponentially with a decay length of 0.100 nm; that is, with $C = 10.0$/nm. Determine the ratio of the current when the STM tip is 0.500 nm above a surface feature to the current when the tip is 0.515 nm above the surface.

32. The design criterion for a typical scanning tunneling microscope specifies that it must be able to detect, on the sample below its tip, surface features that differ in height by only 0.002 00 nm. What percentage change in electron transmission must the electronics of the STM be able to detect, to achieve this resolution? Assume that the electron transmission coefficient is e^{-2CL} with $C = 10.0$/nm.

Section 41.8   The Simple Harmonic Oscillator

Note: Problem 43 in Chapter 16 can be assigned with this section.

33. Show that Equation 41.24 is a solution of Equation 41.22 with energy $E = \frac{1}{2}\hbar\omega$.

34. A one-dimensional harmonic oscillator wave function is

$$\psi = Axe^{-bx^2}$$

(a) Show that ψ satisfies Equation 41.22. (b) Find b and the total energy E. (c) Is this a ground state or a first excited state?

35. A quantum simple harmonic oscillator consists of an electron bound by a restoring force proportional to its position relative to a certain equilibrium point. The proportionality constant is 8.99 N/m. What is the longest wavelength of light that can excite the oscillator?

36. (a) Normalize the wave function for the ground state of a simple harmonic oscillator. That is, apply Equation 41.6 to Equation 41.24 and find the required value for the constant B, in terms of m, ω, and fundamental constants. (b) Determine the probability of finding the oscillator in a narrow interval $-\delta/2 < x < \delta/2$ around its equilibrium position.

37. Two particles with masses m_1 and m_2 are joined by a light spring of force constant k. They vibrate along a straight line with their center of mass fixed. (a) Show that the total energy

$$\tfrac{1}{2}m_1v_1^2 + \tfrac{1}{2}m_2v_2^2 + \tfrac{1}{2}kx^2$$

can be written as $\frac{1}{2}\mu v^2 + \frac{1}{2}kx^2$ where $v = |v_1| + |v_2|$ is the *relative* speed of the particles and $\mu = m_1m_2/(m_1 + m_2)$ is the reduced mass of the system. This result demonstrates that the pair of freely vibrating particles can be precisely modeled as a single particle vibrating on one end of a spring that has its other end fixed. (b) Differentiate the equation

$$\tfrac{1}{2}\mu v^2 + \tfrac{1}{2}kx^2 = \text{constant}$$

with respect to x. Proceed to show that the system executes simple harmonic motion. Find its frequency.

38. The total energy of a particle–spring system in which the particle moves with simple harmonic motion along the x axis is

$$E = \frac{p_x^2}{2m} + \frac{kx^2}{2}$$

where p_x is the momentum of the particle and k is the spring constant. (a) Using the uncertainty principle, show that this expression can also be written

$$E \geq \frac{p_x^2}{2m} + \frac{k\hbar^2}{8p_x^2}$$

(b) Show that the minimum energy of the harmonic oscillator is

$$E_{\min} = K + U = \tfrac{1}{4}\hbar\sqrt{\frac{k}{m}} + \frac{\hbar\omega}{4} = \frac{\hbar\omega}{2}$$

Additional Problems

39. Keeping a constant speed of 0.8 m/s, a marble rolls back and forth across a shoebox. Make an order-of-magnitude estimate of the probability of its escaping through the wall of the box by quantum tunneling. State the quantities you take as data and the values you measure or estimate for them.

40. A particle of mass 2.00×10^{-28} kg is confined to a one-dimensional box of length 1.00×10^{-10} m. For $n = 1$, what are (a) the particle's wavelength, (b) its momentum, and (c) its ground-state energy?

41. An electron is represented by the time-independent wave function

$$\psi(x) = \begin{cases} Ae^{-\alpha x} & \text{for } x > 0 \\ Ae^{+\alpha x} & \text{for } x < 0 \end{cases}$$

(a) Sketch the wave function as a function of x. (b) Sketch the probability density representing the likelihood that the electron is found between x and $x + dx$. (c) Argue that this can be a physically reasonable wave function. (d) Normalize the wave function. (e) Determine the probability of finding the electron somewhere in the range

$$x_1 = -\frac{1}{2\alpha} \quad \text{to} \quad x_2 = \frac{1}{2\alpha}$$

42. Particles incident from the left are confronted with a step in potential energy shown in Figure P41.42. Located at $x = 0$, the step has a height U. The particles have energy $E > U$. Classically, we would expect all of the particles to continue on, although with reduced speed. According to quantum mechanics, a fraction of the particles are reflected at the barrier. (a) Prove that the reflection coefficient R for this case is

$$R = \frac{(k_1 - k_2)^2}{(k_1 + k_2)^2}$$

where $k_1 = 2\pi/\lambda_1$ and $k_2 = 2\pi/\lambda_2$ are the wave numbers for the incident and transmitted particles. Proceed as follows. Show that the wave function $\psi_1 = Ae^{ik_1 x} + Be^{-ik_1 x}$ satisfies the Schrödinger equation in region 1, for $x < 0$. Here $Ae^{ik_1 x}$ represents the incident beam and $Be^{-ik_1 x}$ represents the reflected particles. Show that $\psi_2 = Ce^{ik_2 x}$ satisfies the Schrödinger equation in region 2, for $x > 0$. Impose the boundary conditions $\psi_1 = \psi_2$ and $d\psi_1/dx = d\psi_2/dx$ at $x = 0$, to find the relationship between B and A. Then evaluate $R = B^2/A^2$. (b) A particle that has kinetic energy $E = 7.00$ eV is incident from a region where the potential energy is zero onto one in which $U = 5.00$ eV. Find its probability of being reflected and its probability of being transmitted.

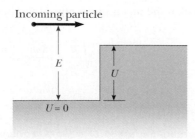

Incoming particle

E

U

$U = 0$

Figure P41.42 Problems 42 and 43.

43. Particles incident from the left are confronted with a step in potential energy shown in Figure P41.42. The step has a height U, and the particles have energy $E = 2U$. Classically, all the particles would pass into the region of higher potential energy at the right. However, according to quantum mechanics, a fraction of the particles are reflected at the barrier. Use the result of Problem 42 to determine the fraction of the incident particles that are reflected. (This situation is analogous to the partial reflection and transmission of light striking an interface between two different media.)

44. An electron is trapped in a quantum dot. The quantum dot may be modeled as a one-dimensional, rigid-walled box of length 1.00 nm. (a) Sketch the wave functions and probability densities for the $n = 1$ and $n = 2$ states. (b) For the $n = 1$ state, calculate the probability of finding the electron between $x_1 = 0.150$ nm and $x_2 = 0.350$ nm, where $x = 0$ is the left side of the box. (c) Repeat part (b) for the $n = 2$ state. (d) Calculate the energies in electron volts of the $n = 1$ and $n = 2$ states. *Suggestion:* For parts (b) and (c), use Equation 41.5 and note that

$$\int \sin^2 ax \, dx = \frac{1}{2}x - \frac{1}{4a} \sin 2ax$$

45. An atom in an excited state 1.80 eV above the ground state remains in that excited state 2.00 μs before moving to the ground state. Find (a) the frequency and (b) the wavelength of the emitted photon. (c) Find the approximate uncertainty in energy of the photon.

46. An electron is confined to move in the xy plane in a rectangle whose dimensions are L_x and L_y. That is, the electron is trapped in a two-dimensional potential well having lengths of L_x and L_y. In this situation, the allowed energies of the electron depend on two quantum numbers n_x and n_y. The allowed energies are given by

$$E = \frac{h^2}{8m_e}\left(\frac{n_x^2}{L_x^2} + \frac{n_y^2}{L_y^2}\right)$$

(a) Assuming $L_x = L_y = L$, find the energies of the lowest four energy levels for the electron. (b) Construct an energy-level diagram for the electron, and determine the energy difference between the second excited state and the ground state.

47. For a particle described by a wave function $\psi(x)$, the expectation value of a physical quantity $f(x)$ associated with the particle is defined by

$$\langle f(x) \rangle \equiv \int_{-\infty}^{\infty} \psi^* f(x) \psi \, dx$$

For a particle in a one-dimensional box extending from $x = 0$ to $x = L$, show that

$$\langle x^2 \rangle = \frac{L^2}{3} - \frac{L^2}{2n^2\pi^2}$$

48. A particle is described by the wave function

$$\psi(x) = \begin{cases} A\cos\left(\dfrac{2\pi x}{L}\right) & \text{for } -\dfrac{L}{4} \le x \le \dfrac{L}{4} \\ 0 & \text{for other values of } x \end{cases}$$

(a) Determine the normalization constant A. (b) What is the probability that the particle will be found between $x = 0$ and $x = L/8$ if its position is measured? (*Suggestion:* Use Eq. 41.5.)

49. A particle has a wave function

$$\psi(x) = \begin{cases} \sqrt{\dfrac{2}{a}}\, e^{-x/a} & \text{for } x > 0 \\ \\ 0 & \text{for } x < 0 \end{cases}$$

(a) Find and sketch the probability density. (b) Find the probability that the particle will be at any point where $x < 0$. (c) Show that ψ is normalized, and then find the probability that the particle will be found between $x = 0$ and $x = a$.

50. A particle of mass m is placed in a one-dimensional box of length L. **What If?** Assume the box is so small that the particle's motion is *relativistic*, so that $K = p^2/2m$ is not valid. (a) Derive an expression for the kinetic energy levels of the particle. (b) Assume the particle is an electron in a box of length $L = 1.00 \times 10^{-12}$ m. Find its lowest possible kinetic energy. By what percent is the nonrelativistic equation in error? (*Suggestion:* See Eq. 39.23.)

51. Consider a "crystal" consisting of two nuclei and two electrons as shown in Figure P41.51. (a) Taking into account all the pairs of interactions, find the potential energy of the system as a function of d. (b) Assuming the electrons to be restricted to a one-dimensional box of length $3d$, find the minimum kinetic energy of the two electrons. (c) Find the value of d for which the total energy is a minimum. (d) Compare this value of d with the spacing of atoms in lithium, which has a density of 0.530 g/cm^3 and an atomic mass of 7 u. (This type of calculation can be used to estimate the density of crystals and certain stars.)

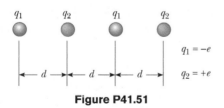

$q_1 = -e$
$q_2 = +e$

Figure P41.51

52. *The simple harmonic oscillator excited.* The wave function

$$\psi(x) = Bxe^{-(m\omega/2\hbar)x^2}$$

is a solution to the simple harmonic oscillator problem. (a) Find the energy of this state. (b) At what position are you least likely to find the particle? (c) At what positions are you most likely to find the particle? (d) Determine the value of B required to normalize the wave function. (e) **What If?** Determine the classical probability of finding the particle in an interval of small width δ centered at the position $x = 2(\hbar/m\omega)^{1/2}$. (f) What is the actual probability of finding the particle in this interval?

53. *Normalization of wave functions.* (a) Find the normalization constant A for a wave function made up of the two lowest states of a particle in a box:

$$\psi(x) = A\left[\sin\left(\frac{\pi x}{L}\right) + 4\sin\left(\frac{2\pi x}{L}\right)\right]$$

(b) A particle is described in the space $-a \le x \le a$ by the wave function

$$\psi(x) = A\cos\left(\frac{\pi x}{2a}\right) + B\sin\left(\frac{\pi x}{a}\right)$$

Determine the relationship between the values of A and B required for normalization. (*Suggestion:* Use the identity $\sin 2\theta = 2\sin\theta\cos\theta$.)

54. The normalized wave functions for the ground state, $\psi_0(x)$, and the first excited state, $\psi_1(x)$, of a quantum harmonic oscillator are

$$\psi_0(x) = \left(\frac{a}{\pi}\right)^{1/4} e^{-ax^2/2} \qquad \psi_1(x) = \left(\frac{4a^3}{\pi}\right)^{1/4} xe^{-ax^2/2}$$

where $a = m\omega/\hbar$. A mixed state, $\psi_{01}(x)$, is constructed from these states:

$$\psi_{01}(x) = \frac{1}{\sqrt{2}}[\psi_0(x) + \psi_1(x)]$$

The symbol $\langle q \rangle_s$ denotes the expectation value of the quantity q for the state $\psi_s(x)$. Calculate the following expectation values: (a) $\langle x \rangle_0$ (b) $\langle x \rangle_1$ (c) $\langle x \rangle_{01}$.

55. A two-slit electron diffraction experiment is done with slits of *unequal* widths. When only slit 1 is open, the number of electrons reaching the screen per second is 25.0 times the number of electrons reaching the screen per second when only slit 2 is open. When both slits are open, an interference pattern results in which the destructive interference is not complete. Find the ratio of the probability of an electron arriving at an interference maximum to the probability of an electron arriving at an adjacent interference minimum. (*Suggestion:* Use the superposition principle.)

Answers to Quick Quizzes

41.1 (d). The probability density for this wave function is $|\psi|^2 = \psi^*\psi = (Ae^{-ikx})(Ae^{ikx}) = A^2$, which is independent of x. Consequently, the particle is equally likely to be found at any value of x. This is consistent with the uncertainty principle—if the wavelength is known precisely (based on a specific value of k in Equation 41.3), we have no knowledge of the position of the particle.

41.2 Classically, we expect the particle to bounce back and forth between the two walls at constant speed. Thus, we are as likely to find it on the left side of the box as in the middle, on the right side, or anywhere else inside the box. Our graph of probability density versus x would therefore be a horizontal line, with a total area under the line of unity:

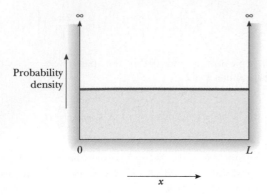

41.3 (c). According to Equation 41.12, if L is increased, all quantized energies become smaller. Thus, the energy levels move closer together. As L becomes macroscopic, the energy levels are so close together that we do not observe the quantized behavior.

41.4 (d). The particles in all three parts (a), (b), and (c) are part of a bound system.

41.5 (a). In Equation 41.15, we set $n = 1$ for the ground-state energy and see that the energy is inversely proportional to the particle mass.

41.6 (d). The wavelength is determined by the length L of the well.

41.7 (c). The longer wavelength results in a smaller value of the momentum from the de Broglie relationship. If the momentum of the particle is decreased, the energy also decreases.

41.8 (a), (c), (f). Decreasing the barrier height and increasing the particle energy both reduce the value of C in Equation 41.21, increasing the transmission coefficient in Equation 41.20. Decreasing the width L of the barrier increases the transmission coefficient in Equation 41.20.

41.9 (a). The ground state represents the largest deviation from classical behavior. Classically, the most probable values of x are $x = \pm A$ because the particle is moving the slowest near these points. As the top graph in Figure 41.15 shows, quantum mechanics predicts that the maximum probability density in the ground state is at $x = 0$.

Chapter 42

Atomic Physics

▲ *Much of the luminous matter in the Universe is hydrogen. In this photograph of the central portion of the Orion Nebula, the colors come from transitions between quantized states in hydrogen atoms. (NASA/GRIN)*

CHAPTER OUTLINE

42.1 Atomic Spectra of Gases

42.2 Early Models of the Atom

42.3 Bohr's Model of the Hydrogen Atom

42.4 The Quantum Model of the Hydrogen Atom

42.5 The Wave Functions for Hydrogen

42.6 Physical Interpretation of the Quantum Numbers

42.7 The Exclusion Principle and the Periodic Table

42.8 More on Atomic Spectra: Visible and X-Ray

42.9 Spontaneous and Stimulated Transitions

42.10 Lasers

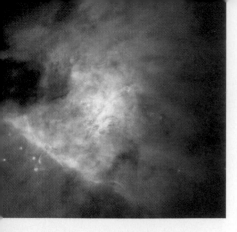

In Chapter 41 we introduced some of the basic concepts and techniques used in quantum mechanics, along with their applications to various one-dimensional systems. In this chapter, we apply quantum mechanics to atomic systems. A large portion of the chapter is focused on the application of quantum mechanics to the study of the hydrogen atom. Understanding the hydrogen atom, the simplest atomic system, is important for several reasons:

- The hydrogen atom is the only atomic system that can be solved exactly.

- Much of what was learned in the twentieth century about the hydrogen atom, with its single electron, can be extended to such single-electron ions as He^+ and Li^{2+}.

- The hydrogen atom proved to be an ideal system for performing precise tests of theory against experiment and for improving our overall understanding of atomic structure.

- The quantum numbers that are used to characterize the allowed states of hydrogen can also be used to investigate more complex atoms, and such a description enables us to understand the periodic table of the elements. This understanding is one of the greatest triumphs of quantum mechanics.

- The basic ideas about atomic structure must be well understood before we attempt to deal with the complexities of molecular structures and the electronic structure of solids.

The full mathematical solution of the Schrödinger equation applied to the hydrogen atom gives a complete and beautiful description of the atom's properties. However, because the mathematical procedures that are involved are beyond the scope of this text, many of the details are omitted. The solutions for some states of hydrogen are discussed, together with the quantum numbers used to characterize various allowed states. We also discuss the physical significance of the quantum numbers and the effect of a magnetic field on certain quantum states.

A new physical idea, the *exclusion principle,* is presented in this chapter. This principle is extremely important for understanding the properties of multielectron atoms and the arrangement of elements in the periodic table. In fact, the implications of the exclusion principle are almost as far-reaching as those of the Schrödinger equation.

Finally, we apply our knowledge of atomic structure to describe the mechanisms involved in the production of x-rays and in the operation of a laser.

42.1 Atomic Spectra of Gases

As pointed out in Section 40.1, all objects emit thermal radiation characterized by a continuous distribution of wavelengths. In sharp contrast to this continuous-distribution spectrum is the discrete **line spectrum** observed when a low-pressure gas is subject to an electric discharge. (Electric discharge occurs when the gas is subjected to a potential difference that creates an electric field greater than the dielectric strength of the gas.) Observation and analysis of these spectral lines is called **emission spectroscopy.**

When the light from a gas discharge is examined using a spectrometer (see Fig. 38.18), it is found to consist of a few bright lines of color on a generally dark background. This discrete line spectrum contrasts sharply with the continuous rainbow of colors seen when a glowing solid is viewed through the same instrument. Figure 42.1a shows that the wavelengths contained in a given line spectrum are characteristic of the element emitting the light. The simplest line spectrum is that for atomic hydrogen, and we describe this spectrum in detail. Other atoms exhibit completely different line spectra. Because no two elements have the same line spectrum, this phenomenon represents a practical and sensitive technique for identifying the elements present in unknown samples.

Another form of spectroscopy very useful in analyzing substances is **absorption spectroscopy.** An absorption spectrum is obtained by passing white light from a continuous source through a gas or a dilute solution of the element being analyzed. The absorption spectrum consists of a series of dark lines superimposed on the continuous spectrum of the light source, as shown in Figure 42.1b for atomic hydrogen.

The absorption spectrum of an element has many practical applications. For example, the continuous spectrum of radiation emitted by the Sun must pass through the cooler gases of the solar atmosphere and through the Earth's atmosphere. The various absorption lines observed in the solar spectrum have been used to identify elements in the solar atmosphere. In early studies of the solar spectrum, experimenters found some lines that did not correspond to any known element. A new element had been discovered! The new element was named helium, after the Greek

> ▲ **PITFALL PREVENTION**
>
> **42.1   Why Lines?**
>
> The phrase "spectral lines" is used very often when discussing the radiation from atoms. Lines are seen because the light passes through a long and very narrow slit before being separated by wavelength. You will see many references to these "lines" in both physics and chemistry.

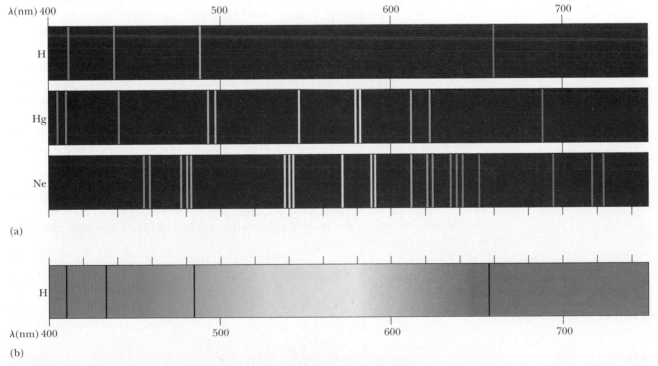

Figure 42.1 (a) Emission line spectra for hydrogen, mercury, and neon. (b) The absorption spectrum for hydrogen. Note that the dark absorption lines occur at the same wavelengths as the hydrogen emission lines in (a). *(K. W. Whitten, R. E. Davis, M. L. Peck, and G. G. Stanley, General Chemistry, 7th ed., Belmont, CA, Brooks/Cole, 2004.)*

Figure 42.2 This street in Las Vegas is filled with neon lights of varying bright colors.

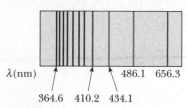

Figure 42.3 The Balmer series of spectral lines for atomic hydrogen, with several lines marked with the wavelength in nanometers. The line labeled 364.6 is the shortest-wavelength line and is in the ultraviolet region of the electromagnetic spectrum. The other labeled lines are in the visible region.

Balmer series

Lyman series

Paschen series

Brackett series

word for Sun, *helios.* Helium was subsequently isolated from subterranean gas on the Earth.

Scientists are able to examine the light from stars other than our Sun in this fashion, but elements other than those present on the Earth have never been detected. Absorption spectroscopy has also been useful in analyzing heavy-metal contamination of the food chain. For example, the first determination of high levels of mercury in tuna was made with the use of atomic absorption spectroscopy.

The discrete emissions of light from gas discharges are used in "neon" signs such as those in Figure 42.2. Neon, the first gas used in these types of signs and the gas after which they are named, emits strongly in the red region. As a result, a glass tube filled with neon gas emits bright red light when an applied voltage causes a continuous discharge. Early signs used different gases to provide different colors, although the brightness of these signs was generally very low. Many present-day "neon" signs contain mercury vapor, which emits strongly in the ultraviolet range of the electromagnetic spectrum. The inside of the glass tube is coated with a material that emits a particular color when it absorbs ultraviolet radiation from the mercury. The color of the light from the tube is due to the particular material chosen. A fluorescent light operates in the same manner, with a white-emitting material coating the inside of the glass tube.

From 1860 to 1885, scientists accumulated a great deal of data on atomic emissions using spectroscopic measurements. In 1885, a Swiss school teacher, Johann Jacob Balmer (1825–1898), found an empirical equation that correctly predicted the wavelengths of four visible emission lines of hydrogen: H_α (red), H_β (green), H_γ (blue), and H_δ (violet). Figure 42.3 shows these and other lines (in the ultraviolet) in the emission spectrum of hydrogen. The complete set of lines is called the **Balmer series.** The four visible lines occur at the wavelengths 656.3 nm, 486.1 nm, 434.1 nm, and 410.2 nm. The wavelengths of these lines can be described by the following equation, which is a modification made by Johannes Rydberg (1854–1919) of Balmer's original equation:

$$\frac{1}{\lambda} = R_H\left(\frac{1}{2^2} - \frac{1}{n^2}\right) \qquad n = 3, 4, 5, \ldots \tag{42.1}$$

where R_H is a constant now called the **Rydberg constant.** If the wavelength is in meters, R_H has the value $1.097\,373\,2 \times 10^7$ m^{-1}. In Balmer's original equation, the integer values of n varied from 3 to 6 to give the four visible lines from 656.3 nm (red) down to 410.2 nm (violet). Rydberg's modification shown in Equation 42.1 describes the ultraviolet spectral lines in the Balmer series if n is carried out beyond $n = 6$. The **series limit** is the shortest wavelength in the series and corresponds to $n \rightarrow \infty$, with a wavelength of 364.6 nm, as in Figure 42.3. The measured spectral lines agree with the empirical equation, Equation 42.1, to within 0.1%.

Other lines in the spectrum of hydrogen were found following Balmer's discovery. These spectra are called the Lyman, Paschen, and Brackett series after their discoverers. The wavelengths of the lines in these series can be calculated through the use of the following empirical equations:

$$\frac{1}{\lambda} = R_H\left(1 - \frac{1}{n^2}\right) \qquad n = 2, 3, 4, \ldots \tag{42.2}$$

$$\frac{1}{\lambda} = R_H\left(\frac{1}{3^2} - \frac{1}{n^2}\right) \qquad n = 4, 5, 6, \ldots \tag{42.3}$$

$$\frac{1}{\lambda} = R_H\left(\frac{1}{4^2} - \frac{1}{n^2}\right) \qquad n = 5, 6, 7, \ldots \tag{42.4}$$

No theoretical basis existed for these equations; they simply worked. Note that the same constant R_H appears in each equation. In Section 42.3, we discuss the remarkable achievement of a theory for the hydrogen atom that provided an explanation for these equations.

42.2 Early Models of the Atom

The model of the atom in the days of Newton was a tiny, hard, indestructible sphere. Although this model provided a good basis for the kinetic theory of gases, new models had to be devised when experiments revealed the electrical nature of atoms. In 1897, J. J. Thomson established the charge-to-mass ratio for electrons. (See Figure 29.25, page 912.) The following year, he suggested a model that describes the atom as a region in which positive charge is spread out in space with electrons embedded throughout the region, much like the seeds in a watermelon or raisins in thick pudding (Fig. 42.4). The atom as a whole would then be electrically neutral.

In 1911, Ernest Rutherford (1871–1937) and his students Hans Geiger and Ernest Marsden performed a critical experiment that showed that Thomson's model could not be correct. In this experiment, a beam of positively charged alpha particles (helium nuclei) was projected into a thin metallic foil, such as the target in Figure 42.5a. Most of the particles passed through the foil as if it were empty space. But some of the results of the experiment were astounding. Many of the particles deflected from their original direction of travel were scattered through *large* angles. Some particles were even deflected backward, completely reversing their direction of travel! When Geiger informed Rutherford that some alpha particles were scattered backward, Rutherford wrote, "It was quite the most incredible event that has ever happened to me in my life. It was almost as incredible as if you fired a 15-inch [artillery] shell at a piece of tissue paper and it came back and hit you."

Such large deflections were not expected on the basis of Thomson's model. According to that model, the positive charge of an atom in the foil is spread out over such a great volume (the entire atom) that there is no concentration of positive charge strong enough to cause any large-angle deflections of the positively charged alpha particles. Furthermore, the electrons are so much less massive than the alpha particles that they would not cause large-angle scattering either. Rutherford explained his astonishing results by developing a new atomic model, one that assumed that the positive charge in the atom was concentrated in a region that was small relative to the size of the atom. He called this concentration of positive charge the **nucleus** of the atom. Any electrons belonging to the atom were assumed to be in the relatively large volume outside the nucleus. To explain why these electrons were not pulled into the nucleus by the attractive electric force, Rutherford modeled them as moving in orbits around the nucleus in the same manner as the planets orbit the Sun (Fig. 42.5b). For this reason, this model is often referred to as the planetary model of the atom.

Two basic difficulties exist with Rutherford's planetary model. As we saw in Section 42.1, an atom emits (and absorbs) certain characteristic frequencies of electromagnetic radiation and no others; the Rutherford model cannot explain this phenomenon. A

Joseph John Thomson
English physicist (1856–1940)

The recipient of a Nobel Prize in physics in 1906, Thomson is usually considered the discoverer of the electron. He opened up the field of subatomic particle physics with his extensive work on the deflection of cathode rays (electrons) in an electric field. (*Stock Montage, Inc.*)

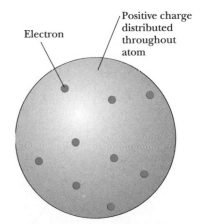

Figure 42.4 Thomson's model of the atom: negatively charged electrons in a volume of continuous positive charge.

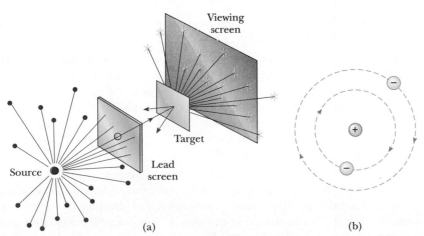

(a)   (b)

Figure 42.5 (a) Rutherford's technique for observing the scattering of alpha particles from a thin foil target. The source is a naturally occurring radioactive substance, such as radium. (b) Rutherford's planetary model of the atom.

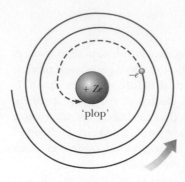

Figure 42.6 The classical model of the nuclear atom. Because the accelerating electron radiates energy, the orbit decays until the electron falls into the nucleus.

second difficulty is that Rutherford's electrons are undergoing a centripetal acceleration. According to Maxwell's theory of electromagnetism, centripetally accelerated charges revolving with frequency f should radiate electromagnetic waves of frequency f. Unfortunately, this classical model leads to a prediction of self-destruction when applied to the atom. As the electron radiates, energy is carried away from the atom, the radius of the electron's orbit steadily decreases, and its frequency of revolution increases. This would lead to an ever-increasing frequency of emitted radiation and an ultimate collapse of the atom as the electron plunges into the nucleus (Fig. 42.6).

42.3   Bohr's Model of the Hydrogen Atom

Given the situation described at the end of Section 42.2, the stage was set for Niels Bohr in 1913 when he presented a new model of the hydrogen atom that circumvented the erroneous deductions of Rutherford's planetary model. Bohr applied Planck's ideas of quantized energy levels (Section 40.1) to orbiting atomic electrons. Bohr's theory was historically important to the development of quantum physics, and it appears to explain the four spectral line series described by Equations 42.1 through 42.4. Though his model is now considered obsolete and has been completely replaced by a probabilistic quantum-mechanical theory, we can use his model to develop the notions of energy quantization and angular momentum quantization as applied to atomic-sized systems. We can also use the model to calculate certain properties of the hydrogen atom and then compare these predictions to experimental observations.

Bohr combined ideas from Planck's original quantum theory, Einstein's concept of the photon, Rutherford's planetary model of the atom, and Newtonian mechanics to arrive at a semiclassical model based on some revolutionary postulates. The basic ideas of the Bohr theory as it applies to the hydrogen atom are as follows:

1. The electron moves in circular orbits around the proton under the influence of the electric force of attraction, as shown in Figure 42.7.

2. Only certain electron orbits are stable. When in one of these **stationary states,** as Bohr called them, the electron does not emit energy in the form of radiation. Hence, the total energy of the atom remains constant, and classical mechanics can be used to describe the electron's motion. Note that this representation claims that the centripetally accelerated electron does not continuously emit radiation, losing energy and eventually spiraling into the nucleus, as predicted by classical physics in the form of Rutherford's planetary model.

3. Radiation is emitted by the atom when the electron makes a transition from a more energetic initial orbit to a lower-energy orbit. This transition cannot be visualized or treated classically. In particular, the frequency f of the photon emitted in the transition is related to the change in the atom's energy and *is independent of the frequency of the electron's orbital motion.* The frequency of the emitted radiation is found from the energy-conservation expression

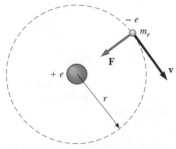

Figure 42.7 Diagram representing Bohr's model of the hydrogen atom. The orbiting electron is allowed to be only in specific orbits of discrete radii.

$$E_i - E_f = hf \qquad (42.5)$$

where E_i is the energy of the initial state, E_f is the energy of the final state, and $E_i > E_f$. In addition, energy of an incident photon can be absorbed by the atom, but only if the photon has an energy that exactly matches the difference in energy between an allowed state of the atom and its existing state upon incidence of the photon. Upon absorption, the photon disappears and the atom makes a transition to the higher-energy state.

4. The size of an allowed electron orbit is determined by a condition imposed on the electron's orbital angular momentum: the allowed orbits are those for which the electron's orbital angular momentum about the nucleus is quantized and equal to an integral multiple of $\hbar = h/2\pi$,

$$m_e vr = n\hbar \qquad n = 1, 2, 3, \ldots \qquad (42.6)$$

Note that assumption 3 implies the existence of a characteristic discrete emission line spectrum *and also* a corresponding absorption line spectrum of the kind shown in Figure 42.1a and b for hydrogen. Using these four assumptions, we can calculate the allowed energy levels and emission wavelengths of the hydrogen atom. We can find the electric potential energy of the system shown in Figure 42.7 from Equation 25.13, $U = k_e q_1 q_2/r = -k_e e^2/r$, where k_e is the Coulomb constant and the negative sign arises from the charge $-e$ on the electron. Thus, the total energy of the atom, which consists of the kinetic energy of the electron and the potential energy of the system, is

$$E = K + U = \tfrac{1}{2} m_e v^2 - k_e \frac{e^2}{r} \qquad (42.7)$$

Applying Newton's second law to the electron, we see that the electric force $k_e e^2/r^2$ exerted on the electron must equal the product of its mass and its centripetal acceleration ($a_c = v^2/r$):

$$\frac{k_e e^2}{r^2} = \frac{m_e v^2}{r}$$

From this expression, we see that the kinetic energy of the electron is

$$K = \tfrac{1}{2} m_e v^2 = \frac{k_e e^2}{2r} \qquad (42.8)$$

Substituting this value of K into Equation 42.7, we find that the total energy of the atom is[1]

$$E = -\frac{k_e e^2}{2r} \qquad (42.9)$$

Note that the total energy is *negative*, indicating a bound electron–proton system. This means that energy in the amount of $k_e e^2/2r$ must be added to the atom to remove the electron and make the total energy of the system zero.

We can obtain an expression for r, the radius of the allowed orbits, by solving Equations 42.6 and 42.8 for v^2 and equating the results:

$$v^2 = \frac{n^2 \hbar^2}{m_e^2 r^2} = \frac{k_e e^2}{m_e r}$$

$$r_n = \frac{n^2 \hbar^2}{m_e k_e e^2} \qquad n = 1, 2, 3, \ldots \qquad (42.10)$$

This equation shows that the radii of the allowed orbits have discrete values—they are quantized. The result is based on the *assumption* that the electron can exist only in certain allowed orbits determined by the integer n.

Niels Bohr
Danish Physicist (1885–1962)

Bohr was an active participant in the early development of quantum mechanics and provided much of its philosophical framework. During the 1920s and 1930s, he headed the Institute for Advanced Studies in Copenhagen. The institute was a magnet for many of the world's best physicists and provided a forum for the exchange of ideas. When Bohr visited the United States in 1939 to attend a scientific conference, he brought news that the fission of uranium had been observed by Hahn and Strassman in Berlin. The results were the foundations of the nuclear weapon developed in the United States during World War II. Bohr was awarded the 1922 Nobel Prize in physics for his investigation of the structure of atoms and the radiation emanating from them. (*Photo courtesy of AIP Niels Bohr Library, Margarethe Bohr Collection*)

---

[1]   Compare Equation 42.9 to its gravitational counterpart, Equation 13.18.

The orbit with the smallest radius, called the **Bohr radius** a_0, corresponds to $n = 1$ and has the value

Bohr radius

$$a_0 = \frac{\hbar^2}{m_e k_e e^2} = 0.052\,9 \text{ nm} \tag{42.11}$$

We obtain a general expression for the radius of any orbit in the hydrogen atom by substituting Equation 42.11 into Equation 42.10:

Radii of Bohr orbits in hydrogen

$$r_n = n^2 a_0 = n^2(0.052\,9 \text{ nm}) \tag{42.12}$$

Bohr's theory predicts a value for the radius of a hydrogen atom on the right order of magnitude, based on experimental measurements. This result was a striking triumph for Bohr's theory. The first three Bohr orbits are shown to scale in Figure 42.8.

The quantization of orbit radii immediately leads to energy quantization. We can see this by substituting $r_n = n^2 a_0$ into Equation 42.9, obtaining for the allowed energy levels

Allowed energies of the Bohr hydrogen atom

$$E_n = -\frac{k_e e^2}{2a_0}\left(\frac{1}{n^2}\right) \qquad n = 1, 2, 3, \ldots \tag{42.13}$$

Inserting numerical values into this expression, we find

$$E_n = -\frac{13.606}{n^2} \text{ eV} \qquad n = 1, 2, 3, \ldots \tag{42.14}$$

Only energies satisfying this equation are permitted. The lowest allowed energy level, the ground state, has $n = 1$ and energy $E_1 = -13.606$ eV. The next energy level, the first excited state, has $n = 2$ and energy $E_2 = E_1/2^2 = -3.401$ eV. Figure 42.9 is an energy-level diagram showing the energies of these discrete energy states and the corresponding quantum numbers n. The uppermost level, corresponding to $n = \infty$ (or $r = \infty$) and $E = 0$, represents the state for which the electron is removed from the atom.

The minimum energy required to ionize the atom in its ground state (that is, to completely remove an electron from the proton's influence) is called the **ionization energy**. As can be seen from Figure 42.9, the ionization energy for hydrogen in the ground state, based on Bohr's calculation, is 13.6 eV. This constituted another major

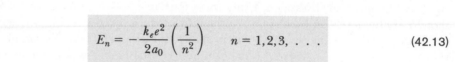

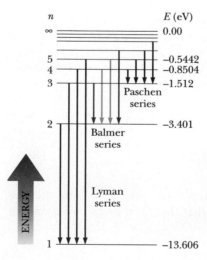

Active Figure 42.8 The first three circular orbits predicted by the Bohr model of the hydrogen atom.

At the Active Figures link at http://www.pse6.com, choose the initial and final states of the hydrogen atom and observe the transitions in this figure and in Figure 42.9.

At the Active Figures link at http://www.pse6.com, choose the initial and final states of the hydrogen atom and observe the transitions in this figure and in Figure 42.8.

Active Figure 42.9 An energy-level diagram for the hydrogen atom. Quantum numbers are given on the left and energies (in electron volts) on the right. Vertical arrows represent the four lowest-energy transitions for each of the spectral series shown. The colored arrows for the Balmer series indicate that this series results in visible light.

achievement for the Bohr theory because the ionization energy for hydrogen had already been measured to be 13.6 eV.

Equations 42.5 and 42.13 can be used to calculate the frequency of the photon emitted when the electron makes a transition from an outer orbit to an inner orbit:

$$f = \frac{E_i - E_f}{h} = \frac{k_e e^2}{2 a_0 h}\left(\frac{1}{n_f{}^2} - \frac{1}{n_i{}^2}\right)$$ (42.15)

Frequency of a photon emitted from hydrogen

Because the quantity measured experimentally is wavelength, it is convenient to use $c = f\lambda$ to express Equation 42.15 in terms of wavelength:

$$\frac{1}{\lambda} = \frac{f}{c} = \frac{k_e e^2}{2 a_0 h c}\left(\frac{1}{n_f{}^2} - \frac{1}{n_i{}^2}\right)$$ (42.16)

The remarkable fact is that this expression, which is purely theoretical, is *identical* to the general form of the empirical relationships discovered by Balmer and Rydberg and given by Equations 42.1 to 42.4,

$$\frac{1}{\lambda} = R_H\left(\frac{1}{n_f{}^2} - \frac{1}{n_i{}^2}\right)$$ (42.17)

provided the constant $k_e e^2 / 2 a_0 h c$ is equal to the experimentally determined Rydberg constant $R_H = 1.097\,373\,2 \times 10^7$ m^{-1}. Soon after Bohr demonstrated that these two quantities agree to within approximately 1%, this work was recognized as the crowning achievement of his new quantum theory of the hydrogen atom. Furthermore, Bohr showed that all of the spectral series for hydrogen have a natural interpretation in his theory. Figure 42.9 shows the origin of these spectral series as transitions between energy levels.

Bohr immediately extended his model for hydrogen to other elements in which all but one electron had been removed. These systems have the same structure as the hydrogen atom except that the nuclear charge is larger. Ionized elements such as He$^+$, Li^{2+}, and Be^{3+} were suspected to exist in hot stellar atmospheres, where atomic collisions frequently have enough energy to completely remove one or more atomic electrons. Bohr showed that many mysterious lines observed in the spectra of the Sun and several other stars could not be due to hydrogen but were correctly predicted by his theory if attributed to singly ionized helium. In general, to describe a single electron orbiting a fixed nucleus of charge $+Ze$, where Z is the atomic number of the element (the number of protons in the nucleus), Bohr's theory gives

$$r_n = (n^2)\frac{a_0}{Z}$$ (42.18)

$$E_n = -\frac{k_e e^2}{2 a_0}\left(\frac{Z^2}{n^2}\right) \qquad n = 1, 2, 3, \ldots$$ (42.19)

While the Bohr theory was triumphant in its agreement with some experimental results on the hydrogen atom, it suffered from some difficulties. One of the first indications that the Bohr theory needed to be modified arose when improved spectroscopic techniques were used to examine the spectral lines of hydrogen. It was found that many of the lines in the Balmer and other series were not single lines at all. Instead, each was a group of lines spaced very close together. An additional difficulty arose when it was observed that, in some situations, certain single spectral lines were split into three closely spaced lines when the atoms were placed in a strong magnetic field. Efforts to explain these and other deviations from the Bohr model led to modifications in the theory and ultimately to a replacement theory that we will discuss in Section 42.4.

 PITFALL PREVENTION

42.2 The Bohr Model Is Great, But . . .

The Bohr model correctly predicts the ionization energy for hydrogen, but it cannot account for the spectra of more complex atoms and is unable to predict many subtle spectral details of hydrogen and other simple atoms. The notion of electrons in well-defined orbits around the nucleus is *not* consistent with current models of the hydrogen atom.

Bohr's Correspondence Principle

In our study of relativity, we found that Newtonian mechanics is a special case of relativistic mechanics and is usable only when v is much less than c. Similarly, **quantum physics agrees with classical physics when the difference between quantized levels becomes vanishingly small.** This principle, first set forth by Bohr, is called the **correspondence principle.**[2]

For example, consider an electron orbiting the hydrogen atom with $n > 10\,000$. For such large values of n, the energy differences between adjacent levels approach zero, and therefore the levels are nearly continuous. Consequently, the classical model is reasonably accurate in describing the system for large values of n. According to the classical picture, the frequency of the light emitted by the atom is equal to the frequency of revolution of the electron in its orbit about the nucleus. Calculations show that for $n > 10\,000$, this frequency is different from that predicted by quantum mechanics by less than 0.015%.

Quick Quiz 42.1   A hydrogen atom is in its ground state. Incident on the atom are many photons each having an energy of 10.5 eV. The result is that (a) the atom is excited to a higher allowed state (b) the atom is ionized (c) the photons pass by the atom without interaction.

Quick Quiz 42.2   When electrons collide with an atom, they can transfer some or all of their energy to the atom. Suppose a hydrogen atom in its ground state is struck by many electrons each having a kinetic energy of 10.5 eV. The result is that (a) the atom is excited to a higher allowed state (b) the atom is ionized (c) the electrons pass by the atom without interaction.

Quick Quiz 42.3   A hydrogen atom makes a transition from the $n = 3$ level to the $n = 2$ level. It then makes a transition from the $n = 2$ level to the $n = 1$ level. Which transition results in emission of the longest-wavelength photon? (a) the first transition (b) the second transition (c) neither, because the wavelengths are the same for both transitions.

Example 42.1   Spectral Lines from the Star ξ-Puppis

Some mysterious lines observed in 1896 in the emission spectrum of the star ξ-Puppis (ξ is the Greek letter xi) fit the empirical equation

$$\frac{1}{\lambda} = R_H \left(\frac{1}{(n_f/2)^2} - \frac{1}{(n_i/2)^2} \right)$$

Show that these lines can be explained by the Bohr theory as originating from He$^+$.

Solution   The ion He$^+$ has $Z = 2$. Thus, the allowed energy levels are given by Equation 42.19 as

$$E_n = -\frac{k_e e^2}{2a_0} \left(\frac{4}{n^2} \right)$$

Similar to Equation 42.15, we find

$$f = \frac{E_i - E_f}{h} = \frac{k_e e^2}{2a_0 h} \left(\frac{4}{n_f{}^2} - \frac{4}{n_i{}^2} \right)$$

$$f = \frac{k_e e^2}{2a_0 h} \left(\frac{1}{(n_f/2)^2} - \frac{1}{(n_i/2)^2} \right)$$

$$\frac{1}{\lambda} = \frac{f}{c} = \frac{k_e e^2}{2a_0 hc} \left(\frac{1}{(n_f/2)^2} - \frac{1}{(n_i/2)^2} \right)$$

This is the desired solution when we recognize that $R_H \equiv k_e e^2/2a_0 hc$. (See text discussion immediately following Eq. 42.17.)

---

[2]   The correspondence principle is in reality the starting point for Bohr's postulate 4 on angular momentum quantization. To see how postulate 4 arises from the correspondence principle, see pages 353–356 of J. W. Jewett, *Physics Begins With Another M . . . Mysteries, Magic, Myth, and Modern Physics,* Allyn & Bacon, Boston, 1996.

Example 42.2   Electronic Transitions in Hydrogen

(A) The electron in a hydrogen atom makes a transition from the $n = 2$ energy level to the ground level ($n = 1$). Find the wavelength and frequency of the emitted photon.

Solution We can use Equation 42.17 directly to obtain λ, with $n_i = 2$ and $n_f = 1$:

$$\frac{1}{\lambda} = R_H \left(\frac{1}{n_f{}^2} - \frac{1}{n_i{}^2} \right)$$

$$\frac{1}{\lambda} = R_H \left(\frac{1}{1^2} - \frac{1}{2^2} \right) = \frac{3R_H}{4}$$

$$\lambda = \frac{4}{3R_H} = \frac{4}{3(1.097 \times 10^7 \text{ m}^{-1})}$$

$$= 1.215 \times 10^{-7} \text{ m} = \boxed{121.5 \text{ nm}} \quad \text{(ultraviolet)}$$

Because $c = f\lambda$, the frequency of the photon is

$$f = \frac{c}{\lambda} = \frac{3.00 \times 10^8 \text{ m/s}}{1.215 \times 10^{-7} \text{ m}} = \boxed{2.47 \times 10^{15} \text{ Hz}}$$

(B) In interstellar space, highly excited hydrogen atoms called Rydberg atoms have been observed. Find the wavelength to which radio astronomers must tune to detect signals from electrons dropping from the $n = 273$ level to $n = 272$.

Solution We can again use Equation 42.17, this time with $n_i = 273$ and $n_f = 272$:

$$\frac{1}{\lambda} = R_H \left(\frac{1}{n_f{}^2} - \frac{1}{n_i{}^2} \right)$$

$$\lambda = \boxed{0.922 \text{ m}}$$

(C) What is the radius of the electron orbit for a Rydberg atom for which $n = 273$?

Solution Using Equation 42.12, we find

$$r_{273} = (273)^2 (0.052\ 9 \text{ nm}) = \boxed{3.94 \ \mu\text{m}}$$

This is large enough that the atom is on the verge of becoming macroscopic!

(D) How fast is the electron moving in a Rydberg atom for which $n = 273$?

Solution Using Equation 42.8, we find

$$v = \sqrt{\frac{k_e e^2}{m_e r}} = \sqrt{\frac{(8.99 \times 10^9 \text{ N} \cdot \text{m}^2/\text{C}^2)(1.60 \times 10^{-19} \text{ C})^2}{(9.11 \times 10^{-31} \text{ kg})(3.94 \times 10^{-6} \text{ m})}}$$

$$= \boxed{8.01 \times 10^3 \text{ m/s}}$$

What If? What if radiation from the Rydberg atom in part **(B)** is treated classically? What is the wavelength of radiation emitted by the atom in the $n = 273$ level?

Answer Classically, the frequency of the emitted radiation is that of the rotation of the electron around the nucleus. We can calculate this using the period defined in Equation 4.16:

$$f = \frac{1}{T} = \frac{v}{2\pi r}$$

We have the radius and speed from parts (C) and (D), so

$$f = \frac{v}{2\pi r} = \frac{8.01 \times 10^3 \text{ m/s}}{2\pi (3.94 \times 10^{-6} \text{ m})} = 3.24 \times 10^8 \text{ Hz}$$

The wavelength of the radiation is

$$\lambda = \frac{c}{f} = \frac{3.00 \times 10^8 \text{ m/s}}{3.24 \times 10^8 \text{ Hz}} = 0.926 \text{ m}$$

This is less than half of one percent different from the wavelength calculated in part (B). As indicated in the discussion of Bohr's correspondence principle, this difference becomes even smaller for higher values of n.

 Investigate transitions of the atom between states at the Interactive Worked Example link at **http://www.pse6.com.**

42.4   The Quantum Model of the Hydrogen Atom

In the preceding section we described how the Bohr model views the electron as a particle orbiting the nucleus in nonradiating, quantized energy levels. This approach leads to an analysis that combines both classical and quantum concepts. While the model demonstrates excellent agreement with some experimental results, it cannot explain others such as the "splitting" of spectral lines, as noted in Section 42.3. These difficulties are removed when a full quantum model involving the Schrödinger equation is used to describe the hydrogen atom.

The potential energy function for the hydrogen atom is

$$U(r) = -k_e \frac{e^2}{r} \qquad (42.20)$$

where $k_e = 8.99 \times 10^9$ N·m^2/C^2 is the Coulomb constant and r is the radial distance from the proton (situated at $r = 0$) to the electron.

The formal procedure for solving the problem of the hydrogen atom is to substitute $U(r)$ into the Schrödinger equation and find appropriate solutions to the equation, as we did for the particle in a box in Chapter 41. The present problem is more complicated, however, because it is three-dimensional and because U depends on the radial coordinate r. If the time-independent Schrödinger equation (Equation 41.13) is extended to three-dimensional rectangular coordinates, the result is:

$$-\frac{\hbar^2}{2m} \left(\frac{\partial^2 \psi}{\partial x^2} + \frac{\partial^2 \psi}{\partial y^2} + \frac{\partial^2 \psi}{\partial z^2} \right) + U\psi = E\psi$$

It is easier to solve this equation for the hydrogen atom if rectangular coordinates are converted to spherical polar coordinates, an extension of the plane polar coordinates we introduced in Section 3.1. In spherical polar coordinates, a point in space is represented by the three variables r, θ, and ϕ where r is the radial distance from the origin, $r = \sqrt{x^2 + y^2 + z^2}$. With the point represented at the end of a position vector **r** as shown in Figure 42.10, the angular coordinate θ specifies its angular position relative to the z axis. Once that position vector is projected onto the xy plane, the angular coordinate ϕ specifies the projection's (and therefore the point's) angular position relative to the x axis.

The conversion of the three-dimensional time-independent Schrödinger equation for $\psi(x, y, z)$ to the equivalent form for $\psi(r, \theta, \phi)$ is straightforward but very tedious, and we will omit the details.[3] As we did in Chapter 41 when we separated the time dependence from the space dependence in solutions to the one-dimensional Schrödinger equation, in this case we can separate the three space variables by writing $\psi(r, \theta, \phi)$ as a product of functions of each single variable:

$$\psi(r, \theta, \phi) = R(r) f(\theta) g(\phi)$$

In this way, the three-dimensional partial differential equation can be transformed into three separate ordinary differential equations: one for $R(r)$, one for $f(\theta)$, and one for $g(\phi)$. Each of these functions is subject to boundary conditions. For example, $R(r)$ must remain finite as $r \to 0$ and $r \to \infty$, and $g(\phi)$ must have the same value as $g(\phi + 2\pi)$.

When the full set of boundary conditions is applied to all three functions, we are led to three different quantum numbers for each allowed state of the hydrogen atom. These are restricted to integer values and correspond to the three independent degrees of freedom (three space dimensions).

The first quantum number, associated with the radial function $R(r)$ of the full wave function, is called the **principal quantum number** and is assigned the symbol n. The radial wave equation will lead to functions giving the probability of finding the electron at a certain radial distance from the nucleus. In Section 42.5, we will display two of these radial wave functions. Note that the potential-energy function given in Equation 42.20 depends *only* on the radial coordinate r and not on either of the angular coordinates. The energies of the allowed states for the hydrogen atom are found to be

$$E_n = -\left(\frac{k_e e^2}{2a_0} \right) \frac{1}{n^2} = -\frac{13.606 \text{ eV}}{n^2} \qquad n = 1, 2, 3, \ldots \qquad (42.21)$$

This result is in exact agreement with that obtained in the Bohr theory (Eqs. 42.13 and 42.14).

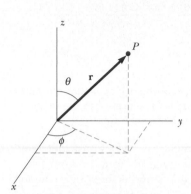

Figure 42.10 A point P in space is located by means of a position vector **r**. In Cartesian coordinates, the components of this vector are x, y, and z. In spherical polar coordinates, the point is described by r, the distance from the origin, θ, the angle between **r** and the z axis, and ϕ, the angle between the x axis and a projection of **r** onto the xy plane.

▲ **PITFALL PREVENTION**

42.3 Energy Depends on n Only for Hydrogen

This statement that the energy depends only on the quantum number n is only true for the hydrogen atom. For more complicated atoms, we will use the same quantum numbers that we develop here for hydrogen. The energy levels for these atoms will depend primarily on n, but will also depend to a lesser degree on other quantum numbers.

Allowed energies of the quantum hydrogen atom

---

[3] Descriptions of the solutions to the Schrödinger equation for the hydrogen atom are available in modern physics textbooks, such as R. A. Serway, C. Moses, and C. A. Moyer, *Modern Physics*, 2nd ed., Saunders College Publishing, 1998.

The **orbital quantum number**, symbolized ℓ, is associated with the orbital angular momentum of the electron, as is the **orbital magnetic quantum number** m_ℓ. Both ℓ and m_ℓ are integers. We will expand our discussion of these two quantum numbers in Section 42.6, and also introduce a fourth (non-integral) quantum number, resulting from a relativistic treatment of the hydrogen atom. Each set of appropriate values for these four quantum numbers corresponds to an allowed state of the atom.

The application of boundary conditions on the three parts of the full wave function leads to important relationships among the three quantum numbers, as well as certain restrictions on their values:

> The values of n can range from 1 to ∞.
>
> The values of ℓ can range from 0 to $n - 1$.
>
> The values of m_ℓ can range from $-\ell$ to ℓ.

Restrictions on the values of hydrogen-atom quantum numbers

For example, if $n = 1$, only $\ell = 0$ and $m_\ell = 0$ are permitted. If $n = 2$, ℓ may be 0 or 1; if $\ell = 0$, then $m_\ell = 0$; but if $\ell = 1$, then m_ℓ may be 1, 0, or -1. Table 42.1 summarizes the rules for determining the allowed values of ℓ and m_ℓ for a given n.

For historical reasons, **all states having the same principal quantum number are said to form a shell.** Shells are identified by the letters K, L, M, . . . , which designate the states for which $n = 1, 2, 3, ...$ Likewise, **all states having the same values of n and ℓ are said to form a subshell.** The letters[4] $s, p, d, f, g, h, . . .$ are used to designate the subshells for which $\ell = 0, 1, 2, 3, ...$ For example, the state designated by $3p$ has the quantum numbers $n = 3$ and $\ell = 1$; the $2s$ state has the quantum numbers $n = 2$ and $\ell = 0$. These notations are summarized in Table 42.2.

PITFALL PREVENTION

42.4   Quantum Numbers Describe a System

It is common to assign the quantum numbers to an electron. Remember, however, that these quantum numbers arise from the Schrödinger equation, which involves a potential energy function for the *system* of the electron and the nucleus. Thus, it is more *proper* to assign the quantum numbers to the atom, but it is more *popular* to assign them to an electron. We will follow this latter usage because it is so common.

Table 42.1

Three Quantum Numbers for the Hydrogen Atom			
Quantum Number	Name	Allowed Values	Number of Allowed States
n	Principal quantum number	$1, 2, 3, . . .$	Any number
ℓ	Orbital quantum number	$0, 1, 2, . . . , n - 1$	n
m_ℓ	Orbital magnetic quantum number	$-\ell, -\ell + 1, . . . , 0,$ $. . . , \ell - 1, \ell$	$2\ell + 1$

Table 42.2

Atomic Shell and Subshell Notations			
n	Shell Symbol	ℓ	Subshell Symbol
1	K	0	s
2	L	1	p
3	M	2	d
4	N	3	f
5	O	4	g
6	P	5	h
.		.	
.		.	
.		.	

[4]   The first four of these letters come from early classifications of spectral lines: sharp, principal, diffuse, and fundamental. The remaining letters are in alphabetical order.

States that violate the rules given in Table 42.1 do not exist. (They do not satisfy the boundary conditions on the wave function.) For instance, the $2d$ state, which would have $n = 2$ and $\ell = 2$, cannot exist because the highest allowed value of ℓ is $n - 1$, which in this case is 1. Thus, for $n = 2$, the $2s$ and $2p$ states are allowed but $2d$, $2f$, . . . are not. For $n = 3$, the allowed subshells are $3s$, $3p$, and $3d$.

Quick Quiz 42.4 How many possible subshells are there for the $n = 4$ level of hydrogen? (a) 5 (b) 4 (c) 3 (d) 2 (e) 1.

Quick Quiz 42.5 When the principal quantum number is $n = 5$, how many different values of (a) ℓ and (b) m_ℓ are possible?

Quick Quiz 42.6 In the hydrogen atom, the quantum number n can increase without limit. True or False: because of this, the frequency of possible spectral lines from hydrogen also increases without limit.

Example 42.3 The $n = 2$ Level of Hydrogen

For a hydrogen atom, determine the number of allowed states corresponding to the principal quantum number $n = 2$, and calculate the energies of these states.

Solution When $n = 2$, ℓ can be 0 or 1. If $\ell = 0$, the only value that m_ℓ can have is 0; for $\ell = 1$, m_ℓ can be -1, 0, or 1. Hence, we have one state, designated as the $2s$ state, that is associated with the quantum numbers $n = 2$, $\ell = 0$, and $m_\ell = 0$, and three states, designated as $2p$ states, for which

the quantum numbers are $n = 2$, $\ell = 1$, $m_\ell = -1$; $n = 2$, $\ell = 1$, $m_\ell = 0$; and $n = 2$, $\ell = 1$, $m_\ell = 1$.

Because all four of these states have the same principal quantum number $n = 2$, they all have the same energy, according to Equation 42.21:

$$E_2 = -\frac{13.606 \text{ eV}}{2^2} = -3.401 \text{ eV}$$

42.5 The Wave Functions for Hydrogen

Because the potential energy of the hydrogen atom depends only on the radial distance r between nucleus and electron, some of the allowed states for this atom can be represented by wave functions that depend only on r. (For these states, $f(\theta)$ and $g(\phi)$ are constants.) The simplest wave function for hydrogen is the one that describes the $1s$ state and is designated $\psi_{1s}(r)$:

Wave function for hydrogen in its ground state

$$\psi_{1s}(r) = \frac{1}{\sqrt{\pi a_0^3}} e^{-r/a_0} \tag{42.22}$$

where a_0 is the Bohr radius. (In Problem 21, you can verify that this function satisfies the Schrödinger equation.) Note that ψ_{1s} approaches zero as r approaches ∞ and is normalized as presented (see Eq. 41.6). Furthermore, because ψ_{1s} depends only on r, it is *spherically symmetric*. This symmetry exists for all s states.

Recall that the probability of finding a particle in any region is equal to an integral of the probability density $|\psi|^2$ for the particle over the region. The probability density for the $1s$ state is

$$|\psi_{1s}|^2 = \left(\frac{1}{\pi a_0^3}\right) e^{-2r/a_0} \tag{42.23}$$

Because we imagine the nucleus to be fixed in space at $r = 0$, we can assign this probability density to the question of locating the electron. The probability of finding the electron in a volume element dV is $|\psi|^2\,dV$. It is convenient to define the *radial probability density function* $P(r)$ as the probability per unit radial length of finding the electron in a spherical shell of radius r and thickness dr. Thus, $P(r)\,dr$ is the probability of finding the electron in this shell. The volume dV of such an infinitesimally thin shell equals its surface area $4\pi r^2$ multiplied by the shell thickness dr (Fig. 42.11), so we can write this probability as

$$P(r)\,dr = |\psi|^2\,dV = |\psi|^2\,4\pi r^2\,dr$$

Thus, the radial probability density function is

$$P(r) = 4\pi r^2\,|\psi|^2 \tag{42.24}$$

Substituting Equation 42.23 into Equation 42.24 gives the radial probability density function for the hydrogen atom in its ground state:

$$P_{1s}(r) = \left(\frac{4r^2}{a_0^3}\right) e^{-2r/a_0} \tag{42.25}$$

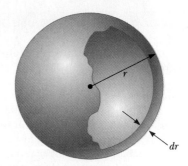

Figure 42.11 A spherical shell of radius r and thickness dr has a volume equal to $4\pi r^2 dr$.

Radial probability density for the 1s state of hydrogen

A plot of the function $P_{1s}(r)$ versus r is presented in Figure 42.12a. The peak of the curve corresponds to the most probable value of r for this particular state. We show in Example 42.4 that this peak occurs at the Bohr radius, the radial position of the electron when the hydrogen atom is in its ground state (in the Bohr theory)! This agreement, along with that of the energies given by Equation 42.21, is remarkable because the Bohr theory and the full quantum theory arrive at these results from completely different starting points.

We also find that the average value of r for the ground state of hydrogen is $\frac{3}{2}a_0$, which is 50% greater than the most probable value. The reason the average value is so much greater is the asymmetry in the radial probability density function (Fig. 42.12a), which has more area to the right of the peak. According to quantum mechanics, the atom has no sharply defined boundary as suggested by the Bohr theory. The probability distribution in Figure 42.12a suggests that the charge of the electron is extended throughout a diffuse region of space, commonly referred to as an *electron cloud*. Figure 42.12b shows the probability density of the electron in a hydrogen atom in the 1s state as a function of position in the xy plane. The darkest portion of the distribution appears at $r = a_0$, corresponding to the most probable value of r for the electron.

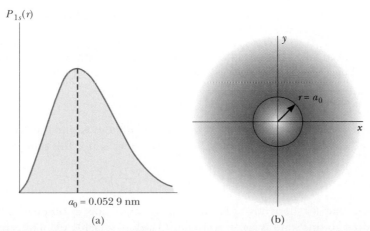

(a)   (b)

Figure 42.12 (a) The probability of finding the electron as a function of distance from the nucleus for the hydrogen atom in the 1s (ground) state. Note that the probability has its maximum value when r equals the Bohr radius a_0. (b) The cross section in the xy plane of the spherical electronic charge distribution for the hydrogen atom in its 1s state.

Example 42.4 The Ground State of Hydrogen

Calculate the most probable value of r for an electron in the ground state of the hydrogen atom.

Solution We conceptualize a hydrogen atom as having a single electron and proton. Because the statement of the problem asks for the "most probable value of r," we categorize this as a problem in which we use the quantum approach. (In the Bohr atom, the electron moves in an orbit with an *exact* value of r.) Thus, our conceptualization should include the electron cloud image of the electron rather than the well-defined orbits of the Bohr model. To analyze the problem, we note that the most probable value of r corresponds to the peak of the plot of $P_{1s}(r)$ versus r. Because the slope of the curve at this point is zero, we can evaluate the most probable value of r by setting $dP_{1s}/dr = 0$ and solving for r. Using Equation 42.25, we obtain

$$\frac{dP_{1s}}{dr} = \frac{d}{dr}\left[\left(\frac{4r^2}{a_0^{\,3}}\right)e^{-2r/a_0}\right] = 0$$

Carrying out the derivative operation and simplifying the expression, we obtain

$$e^{-2r/a_0}\frac{d}{dr}\left(r^2\right) + r^2\frac{d}{dr}\left(e^{-2r/a_0}\right) = 0$$

$$2re^{-2r/a_0} + r^2\left(-2/a_0\right)e^{-2r/a_0} = 0$$

$$(1) \qquad 2r[1 - (r/a_0)]e^{-2r/a_0} = 0$$

This expression is satisfied if

$$1 - \frac{r}{a_0} = 0$$

$$r = \boxed{a_0}$$

The most probable value of r is the Bohr radius!

To finalize the problem, note that Equation (1) is also satisfied at $r = 0$ and as $r \to \infty$. These are points of *minimum* probability, which is equal to zero, as seen in Figure 42.12a.

What If? What if you were asked for the *average* value of r for the electron in the ground state rather than the most probable value?

Answer From the discussion before Example 42.4, we expect the value to be $\frac{3}{2}a_0$. The average value of r is the same as the expectation value for r. From Equation 41.7,

$$r_{av} = \langle r \rangle = \int_0^\infty rP(r)\,dr = \int_0^\infty r\left[\left(\frac{4r^2}{a_0^{\,3}}\right)e^{-2r/a_0}\right]dr$$

$$= \left(\frac{4}{a_0^{\,3}}\right)\int_0^\infty r^3 e^{-2r/a_0}\,dr$$

We can evaluate the integral with the help of the first integral listed in Table B.6 in Appendix B:

$$r_{av} = \left(\frac{4}{a_0^{\,3}}\right)\int_0^\infty r^3 e^{-2r/a_0}\,dr = \left(\frac{4}{a_0^{\,3}}\right)\left(\frac{3!}{(2/a_0)^4}\right) = \frac{3}{2}\,a_0$$

Example 42.5 Probabilities for the Electron in Hydrogen

Calculate the probability that the electron in the ground state of hydrogen will be found outside the first Bohr radius.

Solution The probability is found by integrating the radial probability density function $P_{1s}(r)$ for this state from the Bohr radius a_0 to ∞. Using Equation 42.25, we obtain

$$P = \int_{a_0}^\infty P_{1s}(r)\,dr = \frac{4}{a_0^{\,3}}\int_{a_0}^\infty r^2 e^{-2r/a_0}\,dr$$

We can put the integral in dimensionless form by changing variables from r to $z = 2r/a_0$. Noting that $z = 2$ when $r = a_0$

and that $dr = (a_0/2)\,dz$, we obtain

$$P = \frac{1}{2}\int_2^\infty z^2 e^{-z}\,dz = -\frac{1}{2}(z^2 + 2z + 2)e^{-z}\Big|_2^\infty$$

where we have used partial integration (see Appendix B.7). Evaluating between the limits, we have

$$P = 5e^{-2} = \boxed{0.677 \text{ or } 67.7\%}$$

This probability is larger than 50%, which is consistent with the fact that, as shown in Example 42.4, the average value of r is *greater* than the most probable value of r, the Bohr radius.

The next-simplest wave function for the hydrogen atom is the one corresponding to the $2s$ state ($n = 2$, $\ell = 0$). The normalized wave function for this state is

Wave function for hydrogen in the 2s state

$$\psi_{2s}(r) = \frac{1}{4\sqrt{2\pi}}\left(\frac{1}{a_0}\right)^{3/2}\left(2 - \frac{r}{a_0}\right)e^{-r/2a_0} \qquad (42.26)$$

Again we see that ψ_{2s} depends only on r and is spherically symmetric. The energy corresponding to this state is $E_2 = -(13.606/4)\text{ eV} = -3.401\text{ eV}$. This energy level represents the first excited state of hydrogen. A plot of the radial probability density function for this state in comparison to the $1s$ state is shown in Figure 42.13. The plot for the $2s$ state has two peaks. In this case, the most probable value corresponds

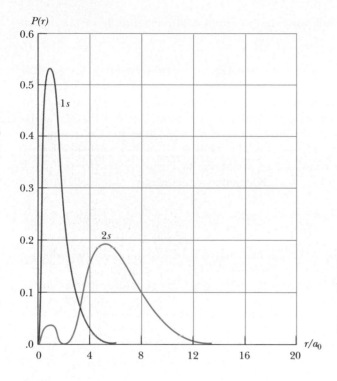

Active Figure 42.13 The radial probability density function versus r/a_0 for the $1s$ and $2s$ states of the hydrogen atom. *(From E. U. Condon and G. H. Shortley,* The Theory of Atomic Spectra, *Cambridge, England, Cambridge University Press, 1953. Used with permission.)*

At the Active Figures link at http://www.pse6.com, choose values of r/a_0 and find the probability that the electron is located between two values.

to that value of r that has the highest value of P ($\approx 5a_0$). An electron in the $2s$ state would be much farther from the nucleus (on the average) than an electron in the $1s$ state.

42.6 Physical Interpretation of the Quantum Numbers

The energy of a particular state in the hydrogen atom depends on the principal quantum number n (Eq. 42.21). Now let us see what the other quantum numbers contribute to our atomic model.

The Orbital Quantum Number ℓ

We begin this discussion by returning briefly to the Bohr model of the atom. If the electron moves in a circle of radius r, the magnitude of its angular momentum relative to the center of the circle is $L = m_e v r$. The direction of **L** is perpendicular to the plane of the circle and is given by a right-hand rule. According to classical physics, the magnitude L of the orbital angular momentum can have any value. However, the Bohr model of hydrogen postulates that the magnitude of the angular momentum of the electron is restricted to multiples of $\hbar$; that is, $m_e v r = n\hbar$. This model must be modified because it predicts (incorrectly) that the ground state of hydrogen has one unit of angular momentum. Furthermore, if L is taken to be zero in the Bohr model, we are forced to accept a picture of the electron as a particle oscillating along a straight line through the nucleus, a physically unacceptable situation.

These difficulties are resolved with the quantum-mechanical model of the atom, although we must give up the mental representation of an electron orbiting in a well-defined circular path. Despite the absence of this representation, the atom does indeed possess an angular momentum and we still call it orbital angular momentum. According to quantum mechanics, an atom in a state whose principal quantum number is n can take on the following *discrete* values of the magnitude of the orbital angular momentum[5]:

Allowed values of L

$$L = \sqrt{\ell(\ell + 1)}\,\hbar \qquad \ell = 0, 1, 2, \ldots, n - 1 \qquad (42.27)$$

Given these allowed values of ℓ, we see that $L = 0$ (corresponding to $\ell = 0$) is an acceptable value of the magnitude of the angular momentum. The fact that L can be zero in this model serves to point out the inherent difficulties in any attempt to describe results based on quantum mechanics in terms of a purely particle-like (classical) model. In the quantum-mechanical interpretation, the electron cloud for the $L = 0$ state is spherically symmetric and has no fundamental rotation axis.

Example 42.6   Calculating L for a p State

Calculate the magnitude of the orbital angular momentum of an electron in a p state of hydrogen.

Solution Because we know that $\hbar = 1.054 \times 10^{-34}\,\text{J}\cdot\text{s}$, we can use Equation 42.27 to calculate L. With $\ell = 1$ for a p state (Table 42.2), we have

$$L = \sqrt{1(1 + 1)}\,\hbar = \sqrt{2}\,\hbar = \boxed{1.49 \times 10^{-34}\,\text{J}\cdot\text{s}}$$

This number is extremely small relative to, say, the orbital angular momentum of the Earth orbiting the Sun, which is approximately $2.7 \times 10^{40}\,\text{J}\cdot\text{s}$. The quantum number that describes L for macroscopic objects such as the Earth is so large that the separation between adjacent states cannot be measured. Once again, the correspondence principle is upheld.

The Orbital Magnetic Quantum Number m_ℓ

Because angular momentum is a vector, its direction must be specified. Recall from Chapter 29 that a current loop has a corresponding magnetic moment $\boldsymbol{\mu} = I\mathbf{A}$ (Eq. 29.10), where I is the current in the loop and $\mathbf{A}$ is a vector perpendicular to the loop whose magnitude is the area of the loop. Such a moment placed in a magnetic field $\mathbf{B}$ interacts with the field. Suppose a weak magnetic field applied along the z axis defines a direction in space. According to classical physics, the energy of the loop–field system depends on the direction of the magnetic moment of the loop with respect to the magnetic field, as described by Equation 29.12, $U = -\boldsymbol{\mu} \cdot \mathbf{B}$. Any energy between $-\mu B$ and $+\mu B$ is allowed by classical physics.

In the Bohr theory, the circulating electron represents a current loop. In the quantum-mechanical approach to the hydrogen atom, we abandon the circular orbit viewpoint of the Bohr theory, but the atom still possesses an orbital angular momentum. Thus, there is some sense of rotation of the electron around the nucleus, so that a magnetic moment is present due to this angular momentum.

As mentioned in Section 42.3, spectral lines from some atoms are observed to split into groups of three closely spaced lines when the atoms are placed in a magnetic field. Suppose we place the hydrogen atom in a magnetic field. According to quantum mechanics, **there are discrete directions allowed for the magnetic moment vector $\boldsymbol{\mu}$ with respect to the magnetic field vector B.** This is very different from the situation in classical physics, in which all directions are allowed.

Because the magnetic moment $\boldsymbol{\mu}$ of the atom can be related[6] to the angular momentum vector $\mathbf{L}$, the discrete directions of $\boldsymbol{\mu}$ translate to the fact that the direction of $\mathbf{L}$ is quantized. This quantization means that L_z (the projection of $\mathbf{L}$ along the z axis) can have

---

[5]   Equation 42.27 is a direct result of the mathematical solution of the Schrödinger equation and the application of angular boundary conditions. This development, however, is beyond the scope of this text and will not be presented.

[6]   See Equation 30.25 for this relationship, as derived from a classical viewpoint. Quantum mechanics arrives at the same result.

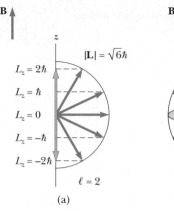
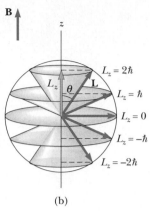

Figure 42.14 A vector model for $\ell = 2$. (a) The allowed projections of the orbital angular momentum **L**. (b) The orbital angular momentum vector **L** lies on the surface of a cone and precesses about the z axis when a magnetic field **B** is applied in the z direction.

only discrete values. The orbital magnetic quantum number m_ℓ specifies the allowed values of the z component of the orbital angular momentum according to the expression

$$L_z = m_\ell \hbar \qquad (42.28)$$

Allowed values of L_z

The quantization of the possible orientations of **L** with respect to an external magnetic field is often referred to as **space quantization.**

Let us look at the possible orientations of **L** for a given value of ℓ. Recall that m_ℓ can have values ranging from $-\ell$ to ℓ. If $\ell = 0$, then $m_\ell = 0$ and $L_z = 0$. If $\ell = 1$, then the possible values of m_ℓ are -1, 0, and 1, so that L_z may be $-\hbar$, 0, or $\hbar$. If $\ell = 2$, then m_ℓ can be -2, -1, 0, 1, or 2, corresponding to L_z values of $-2\hbar$, $-\hbar$, 0, $\hbar$, or $2\hbar$, and so on.

Figure 42.14a shows what is commonly called a **vector model** that describes space quantization for the case $\ell = 2$. Note that **L** can never be aligned parallel or antiparallel to **B** because L_z must be less than the total angular momentum L. For L_z to be zero, **L** must be perpendicular to **B**.

It can be shown that **L** does not point in one specific direction, even though its z component is fixed. If **L** were known exactly, then all three components L_x, L_y, and L_z would be specified, which is inconsistent with the uncertainty principle. How can the magnitude and z component of a vector be specified, but the vector not be completely specified? To answer this, we can imagine that **L** must lie anywhere on the surface of a cone that makes an angle θ with the z axis, as shown in Figure 42.14b. From the figure, we see that θ is also quantized and that its values are specified through the relationship

$$\cos \theta = \frac{L_z}{|\mathbf{L}|} = \frac{m_\ell}{\sqrt{\ell(\ell + 1)}} \qquad (42.29)$$

Note that m_ℓ is never greater than ℓ, and therefore θ can never be zero.

If the atom is placed in a magnetic field, the energy $U = -\boldsymbol{\mu} \cdot \mathbf{B}$ is additional energy for the atom beyond that described in Equation 42.21. Because the directions of $\boldsymbol{\mu}$ are quantized, there are discrete total energies for the atom corresponding to different values of m_ℓ. Figure 42.15a shows a transition between two atomic levels in the absence of a magnetic field. In Figure 42.15b, a magnetic field is applied, and the upper level, with $\ell = 1$, splits into three levels corresponding to the different directions of $\boldsymbol{\mu}$. There are now three possible transitions from the $\ell = 1$ subshell to the $\ell = 0$ subshell. Thus, in a collection of atoms, the single spectral line in Figure 42.15a will split into three spectral lines when the atoms are placed in a magnetic field. This phenomenon is called the *Zeeman effect*.

The Zeeman effect can be used to measure extraterrestrial magnetic fields. For example, the splitting of spectral lines in light from hydrogen atoms in the surface of the Sun can be used to calculate the magnitude of the magnetic field at that location. The Zeeman effect is one of many phenomena that cannot be explained with the Bohr model but are successfully explained by the quantum model of the atom.

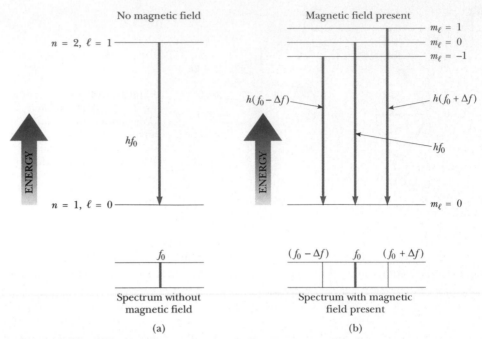

Figure 42.15 The Zeeman effect. (a) Energy levels for the ground and first excited states of a hydrogen atom. When **B** = 0, only a single spectral line at f_0 is observed. (b) When the atom is immersed in a magnetic field **B**, the state with $\ell = 1$ splits into three states. Atoms in the excited states decay to the ground state with the emission of photons with three different energies, giving rise to emission lines at f_0, $f_0 + \Delta f$, and $f_0 - \Delta f$, where Δf is the frequency shift of the emission caused by the magnetic field.

Quick Quiz 42.7   Sketch a vector model (shown in Figure 42.14 for $\ell = 2$) for $\ell = 1$.

Example 42.7   Space Quantization for Hydrogen   `Interactive`

Consider the hydrogen atom in the $\ell = 3$ state. Calculate the magnitude of **L**, the allowed values of L_z, and the corresponding angles θ that $|\mathbf{L}|$ makes with the z axis.

Solution We can calculate the magnitude of the orbital angular momentum using Equation 42.27:

$$L = \sqrt{\ell(\ell + 1)}\, \hbar = \sqrt{3(3 + 1)}\, \hbar = \boxed{2\sqrt{3}\, \hbar}$$

The allowed values of L_z can be calculated using $L_z = m_\ell \hbar$ with $m_\ell = -3, -2, -1, 0, 1, 2,$ and 3:

$$L_z = \boxed{-3\hbar, -2\hbar, -\hbar, 0, \hbar, 2\hbar, 3\hbar}$$

Finally, we calculate the allowed values of θ using Equation 42.29:

$$\cos\theta = \frac{m_\ell}{2\sqrt{3}}$$

Substituting the allowed values of m_ℓ gives

$$\cos\theta = \pm 0.866,\ \pm 0.577,\ \pm 0.289,\ \text{and } 0$$

$$\theta = \boxed{30.0°,\ 54.8°,\ 73.2°,\ 90.0°,\ 107°,\ 125°,\ \text{and } 150°}$$

What If? What if the value of ℓ is an arbitrary integer? For an arbitrary value of ℓ, how many values of m_ℓ are allowed?

Answer For a given value of ℓ, the values of m_ℓ range from $-\ell$ to $+\ell$ in steps of 1. Thus, there are 2ℓ nonzero values of m_ℓ (specifically, $\pm 1, \pm 2, \ldots, \pm \ell$). In addition, one more value of $m_\ell = 0$ is possible. This makes a total of $2\ell + 1$ values of m_ℓ. This result is critical in the understanding of the results of the Stern–Gerlach experiment described below with regard to spin.

 At the Interactive Worked Example link at http://www.pse6.com, *practice evaluating the angular momentum for various quantum states of the hydrogen atom.*

The Spin Magnetic Quantum Number m_s

The three quantum numbers n, ℓ, and m_ℓ that we have discussed so far are generated by applying boundary conditions to solutions of the Schrödinger equation, and we can assign a physical interpretation to each of the quantum numbers. Let us now consider **electron spin,** which does *not* come from the Schrödinger equation.

In Example 42.3, we found four quantum states corresponding to $n = 2$. In reality, however, eight such states occur. The additional four states can be explained by requiring a fourth quantum number for each state—the **spin magnetic quantum number** m_s.

The need for this new quantum number arises because of an unusual feature that is observed in the spectra of certain gases, such as sodium vapor. Close examination of one prominent line in the emission spectrum of sodium reveals that the line is, in fact, two closely spaced lines called a *doublet.*[7] The wavelengths of these lines occur in the yellow region of the electromagnetic spectrum at 589.0 nm and 589.6 nm. In 1925, when this doublet was first observed, it could not be explained with the existing atomic theory. To resolve this dilemma, Samuel Goudsmit (1902–1978) and George Uhlenbeck (1900–1988), following a suggestion made by the Austrian physicist Wolfgang Pauli (1900–1958), proposed the spin quantum number.

To describe this new quantum number, it is convenient (but technically incorrect) to think of the electron as spinning about its axis as it orbits the nucleus, as described in Section 30.8. Only two directions exist for the electron spin, as illustrated in Figure 42.16. If the direction of spin is as shown in Figure 42.16a, the electron is said to have *spin up*. If the direction of spin is as shown in Figure 42.16b, the electron is said to have *spin down*. In the presence of a magnetic field, the energy of the electron is slightly different for the two spin directions, and this energy difference accounts for the sodium doublet.

The classical description of electron spin—as resulting from a spinning electron—is incorrect. More recent theory indicates that the electron is a point particle, without spatial extent. Thus, the electron cannot be considered to be spinning. Despite this conceptual difficulty, all experimental evidence supports the idea that an electron does have some intrinsic angular momentum that can be described by the spin magnetic quantum number. Paul Dirac (1902–1984) showed that this fourth quantum number originates in the relativistic properties of the electron.

In 1921, Otto Stern (1888–1969) and Walter Gerlach (1889–1979) performed an experiment that demonstrated space quantization. However, their results were not in quantitative agreement with the atomic theory that existed at that time. In their experiment, a beam of silver atoms sent through a nonuniform magnetic field was split into two discrete components (Fig. 42.17). They repeated the experiment using other atoms, and in each case the beam split into two or more components. The classical argument is as follows: If the z direction is chosen to be the direction of the maximum nonuniformity of **B**, the net magnetic force on the atoms is along the z axis and is proportional to the component of the magnetic moment $\boldsymbol{\mu}$ of the atom in the z direction. Classically, $\boldsymbol{\mu}$ can have any orientation, so the deflected beam should be spread out continuously. According to quantum mechanics, however, the deflected beam has an integral number of discrete components, and the number of components determines the number of possible values of μ_z. Therefore, because the Stern–Gerlach experiment showed split beams, space quantization was at least qualitatively verified.

For the moment, let us assume that the magnetic moment $\boldsymbol{\mu}$ of the atom is due to the orbital angular momentum. Because μ_z is proportional to m_ℓ, the number of possible values of μ_z is $2\ell + 1$, as found in the **What If?** section of Example 42.7. Furthermore, because ℓ is an integer, the number of values of μ_z is always odd. This prediction is not consistent with Stern and Gerlach's observation of two components (an *even*

Wolfgang Pauli and Niels Bohr watch a spinning top. The spin of the electron is analogous to the spin of the top but different in many ways.

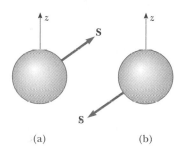

Figure 42.16 The spin of an electron can be either (a) up or (b) down relative to a specified z axis. The spin can never be aligned with the axis.

▲ **PITFALL PREVENTION**

42.5 The Electron Is Not Spinning

While the concept of a spinning electron is conceptually useful, it should not be taken literally. The spin of the Earth is a physical rotation. Electron spin is a purely quantum effect that gives the electron an angular momentum as if it were physically spinning.

---

[7] This is a Zeeman effect for spin and is identical in nature to the Zeeman effect for orbital angular momentum discussed before Example 42.7. The magnetic field for this Zeeman effect is internal to the atom and arises due to the relative motion of the electron and the nucleus.

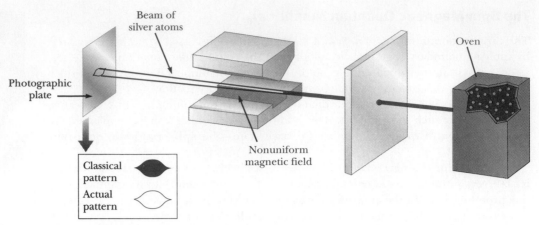

Photographic plate

Beam of silver atoms

Oven

Nonuniform magnetic field

Classical pattern

Actual pattern

Figure 42.17 The technique used by Stern and Gerlach to verify space quantization. A beam of silver atoms is split in two by a nonuniform magnetic field.

number) in the deflected beam of silver atoms. Hence, we are forced to conclude that either quantum mechanics is incorrect or the model is in need of refinement.

In 1927, Phipps and Taylor repeated the Stern–Gerlach experiment using a beam of hydrogen atoms. This experiment was important because it involved an atom containing a single electron in its ground state, for which the quantum theory makes reliable predictions. Recall that $\ell = 0$ for hydrogen in its ground state, so $m_\ell = 0$. Therefore, we would not expect the beam to be deflected by the field because the magnetic moment $\boldsymbol{\mu}$ of the atom is zero. However, the beam in the Phipps–Taylor experiment was again split into two components. On the basis of that result, we can come to only one conclusion: something other than the electron's orbital motion is contributing to the atomic magnetic moment.

As we learned earlier, Goudsmit and Uhlenbeck had proposed that the electron has an intrinsic angular momentum, spin, apart from its orbital angular momentum. In other words, the total angular momentum of the electron in a particular electronic state contains both an orbital contribution **L** and a spin contribution **S**. The Phipps–Taylor result confirmed the hypothesis of Goudsmit and Uhlenbeck.

In 1929, Dirac used the relativistic form of the total energy of a system to solve the relativistic wave equation for the electron in a potential well. His analysis confirmed the fundamental nature of electron spin. (Spin, like mass and charge, is an *intrinsic* property of a particle, independent of its surroundings.) Furthermore, the analysis showed that electron spin[8] can be described by a single quantum number s, whose value can be only $s = \frac{1}{2}$. The spin angular momentum of the electron *never changes*. This notion contradicts classical laws, which dictate that a rotating charge slows down in the presence of an applied magnetic field because of the Faraday emf that accompanies the changing field. Furthermore, if the electron is viewed as a spinning ball of charge subject to classical laws, parts of it near its surface would be rotating with speeds exceeding the speed of light. Thus, the classical picture must not be pressed too far; ultimately, the spinning electron is a quantum entity defying any simple classical description.

Because spin is a form of angular momentum, it must follow the same quantum rules as orbital angular momentum. In accordance with Equation 42.27, the magnitude of the **spin angular momentum S** for the electron is

Magnitude of the spin angular momentum of an electron

$$S = \sqrt{s(s + 1)}\,\hbar = \frac{\sqrt{3}}{2}\,\hbar \qquad (42.30)$$

8 Physicists often use the word *spin* when referring to the spin angular momentum quantum number. For example, it is common to make the statement "The electron has a spin of one half."

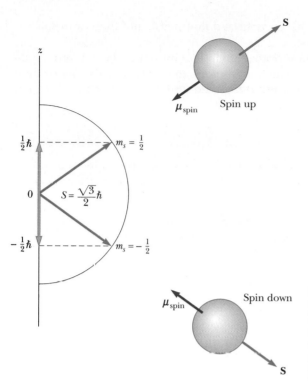

Figure 42.18 Spin angular momentum **S** exhibits space quantization. This figure shows the two allowed orientations of the spin angular momentum vector **S** and the spin magnetic moment $\boldsymbol{\mu}_{\text{spin}}$ for a spin-$\frac{1}{2}$ particle, such as the electron.

Like orbital angular momentum **L**, spin angular momentum **S** is quantized in space, as described in Figure 42.18. It can have two orientations relative to a z axis, specified by the spin magnetic quantum number $m_s = \pm \frac{1}{2}$. Similar to Equation 42.28 for orbital angular momentum, the z component of spin angular momentum is

$$S_z = m_s \hbar = \pm \tfrac{1}{2} \hbar \qquad (42.31)$$

Allowed values of S_z

The two values $\pm \hbar/2$ for S_z correspond to the two possible orientations for **S** shown in Figure 42.18. The value $m_s = +\frac{1}{2}$ refers to the spin-up case, and $m_s = -\frac{1}{2}$ refers to the spin-down case. Notice that Equations 42.30 and 42.31 do not allow the spin vector to lie along the z axis. The actual direction of **S** is at a relatively large angle with respect to the z axis as shown in Figures 42.16 and 42.18.

The spin magnetic moment $\boldsymbol{\mu}_{\text{spin}}$ of the electron is related to its spin angular momentum **S** by the expression

$$\boldsymbol{\mu}_{\text{spin}} = -\frac{e}{m_e} \mathbf{S} \qquad (42.32)$$

where e is the electronic charge and m_e is the mass of the electron. Because $S_z = \pm \frac{1}{2} \hbar$, the z component of the spin magnetic moment can have the values

$$\mu_{\text{spin},z} = \pm \frac{e \hbar}{2 m_e} \qquad (42.33)$$

As we learned in Section 30.8, the quantity $e\hbar/2m_e$ is the Bohr magneton $\mu_{\text{B}} = 9.27 \times 10^{-24}$ J/T. Note that the ratio of magnetic moment to angular momentum is twice as great for spin angular momentum (Eq. 42.32) as it is for orbital angular momentum (Eq. 30.25). The factor of 2 is explained in a relativistic treatment first carried out by Dirac.

Today physicists explain the Stern–Gerlach experiment as follows. The observed magnetic moments for both silver and hydrogen are due to spin angular momentum only, with no contribution from orbital angular momentum. A single-electron atom such as hydrogen has its electron spin quantized in the magnetic field in such a way that the z component of spin angular momentum is either $\frac{1}{2}\hbar$ or $-\frac{1}{2}\hbar$, corresponding to $m_s = \pm\frac{1}{2}$. Electrons with spin $+\frac{1}{2}$ are deflected downward, and those with spin $-\frac{1}{2}$ are deflected upward.

The Stern–Gerlach experiment provided two important results. First, it verified the concept of space quantization. Second, it showed that spin angular momentum exists—even though this property was not recognized until four years after the experiments were performed.

Example 42.8 Putting Some Spin on Hydrogen

For a hydrogen atom, determine the quantum numbers associated with the possible states that correspond to the principal quantum number $n = 2$.

Solution With the addition of the spin quantum number, we have the possibilities given in the accompanying table.

n	ℓ	m_ℓ	m_s	Subshell	Shell	Number of States in Subshell
2	0	0	$\frac{1}{2}$			
2	0	0	$-\frac{1}{2}$	$2s$	L	2
2	1	1	$\frac{1}{2}$			
2	1	1	$-\frac{1}{2}$			
2	1	0	$\frac{1}{2}$			
2	1	0	$-\frac{1}{2}$	$2p$	L	6
2	1	-1	$\frac{1}{2}$			
2	1	-1	$-\frac{1}{2}$			

42.7 The Exclusion Principle and the Periodic Table

We have found that the state of a hydrogen atom is specified by four quantum numbers: n, ℓ, m_ℓ, and m_s. As it turns out, the number of states available to other atoms may also be predicted by this same set of quantum numbers. In fact, these four quantum numbers can be used to describe all the electronic states of an atom regardless of the number of electrons in its structure.

For the present discussion of atoms with many electrons, it is often easiest to assign the quantum numbers to the electrons in the atom as opposed to the entire atom. An obvious question that arises here is: "How many electrons can be in a particular quantum state?" Pauli answered this important question in 1925, in a statement known as the **exclusion principle:**

> No two electrons can ever be in the same quantum state; therefore, no two electrons in the same atom can have the same set of quantum numbers.

If this principle were not valid, an atom could radiate energy until every electron in the atom is in the lowest possible energy state, and the chemical behavior of the elements would be grossly modified. Nature as we know it would not exist!

In reality, we can view the electronic structure of complex atoms as a succession of filled levels increasing in energy. As a general rule, the order of filling of an atom's subshells is as follows. Once a subshell is filled, the next electron goes into the lowest-energy vacant subshell. We can understand this behavior by recognizing that if the atom were not in the lowest energy state available to it, it would radiate energy until it reached this state.

 PITFALL PREVENTION

42.6 The Exclusion Principle Is More General

A more general form of the exclusion principle will be discussed in Chapter 46. It states that no two *fermions* can be in the same quantum state. Fermions are particles with half-integral spin ($\frac{1}{2}$, $\frac{3}{2}$, $\frac{5}{2}$, etc.).

Table 42.3

Allowed Quantum States for an Atom up to $n = 3$													
n	1	2			3								
ℓ	0	0	1		0	1			2				
m_ℓ	0	0	1	0	−1	0	1	0	−1	2	1	0	−1	−2
m_s	⇅	⇅	⇅	⇅	⇅	⇅	⇅	⇅	⇅	⇅	⇅	⇅	⇅	⇅

Wolfgang Pauli
Austrian Theoretical Physicist (1900–1958)

An extremely talented theoretician who made important contributions in many areas of modern physics, Pauli gained public recognition at the age of 21 with a masterful review article on relativity that is still considered one of the finest and most comprehensive introductions to the subject. His other major contributions were the discovery of the exclusion principle, the explanation of the connection between particle spin and statistics, theories of relativistic quantum electrodynamics, the neutrino hypothesis, and the hypothesis of nuclear spin. (*CERN, courtesy of AIP Emilio Segré Visual Archive*)

Before we discuss the electronic configuration of various elements, it is convenient to define an *orbital* as the atomic state characterized by the quantum numbers n, ℓ, and m_ℓ. From the exclusion principle, we see that **only two electrons can be present in any orbital.** One of these electrons has a spin magnetic quantum number $m_s = +\frac{1}{2}$, and the other has $m_s = -\frac{1}{2}$. Because each orbital is limited to two electrons, the number of electrons that can occupy the various shells is also limited.

Table 42.3 shows the allowed quantum states for an atom up to $n = 3$. The arrows pointing upward indicate an electron described by $m_s = +\frac{1}{2}$, and those pointing downward indicate that $m_s = -\frac{1}{2}$. The $n = 1$ shell can accommodate only two electrons because $m_\ell = 0$ means that only one orbital is allowed. (The three quantum numbers describing this orbital are $n = 1$, $\ell = 0$, and $m_\ell = 0$.) The $n = 2$ shell has two subshells, one for $\ell = 0$ and one for $\ell = 1$. The $\ell = 0$ subshell is limited to two electrons because $m_\ell = 0$. The $\ell = 1$ subshell has three allowed orbitals, corresponding to $m_\ell = 1$, 0, and -1. Because each orbital can accommodate two electrons, the $\ell = 1$ subshell can hold six electrons. Thus, the $n = 2$ shell can contain eight electrons, as shown in Example 42.8. The $n = 3$ shell has three subshells ($\ell = 0, 1, 2$) and nine orbitals, accommodating up to 18 electrons. In general, each shell can accommodate up to $2n^2$ electrons.

The exclusion principle can be illustrated by an examination of the electronic arrangement in a few of the lighter atoms. The **atomic number** Z of any element is the number of protons in the nucleus of an atom of that element. A neutral atom of that element has Z electrons. Hydrogen ($Z = 1$) has only one electron—which, in the ground state of the atom, can be described by either of two sets of quantum numbers n, ℓ, m_ℓ, m_s: $1, 0, 0, \frac{1}{2}$ or $1, 0, 0, -\frac{1}{2}$. This electronic configuration is often written $1s^1$. The notation $1s$ refers to a state for which $n = 1$ and $\ell = 0$, and the superscript indicates that one electron is present in the s subshell.

Neutral helium ($Z = 2$) has two electrons. In the ground state, their quantum numbers are $1, 0, 0, \frac{1}{2}$ and $1, 0, 0, -\frac{1}{2}$. No other possible combinations of quantum numbers exist for this level, and we say that the K shell is filled. This electronic configuration is written $1s^2$.

Neutral lithium ($Z = 3$) has three electrons. In the ground state, two of these are in the $1s$ subshell. The third is in the $2s$ subshell because this subshell is slightly lower in energy than the $2p$ subshell.[9] Hence, the electronic configuration for lithium is $1s^2 2s^1$.

The electronic configurations of lithium and the next several elements are provided in Figure 42.19. The electronic configuration of beryllium ($Z = 4$), with its four electrons, is $1s^2 2s^2$, and boron ($Z = 5$) has a configuration of $1s^2 2s^2 2p^1$. The $2p$ electron in boron may be described by any of the six equally probable sets of quantum numbers listed in

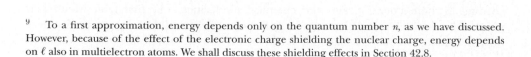

[9]   To a first approximation, energy depends only on the quantum number n, as we have discussed. However, because of the effect of the electronic charge shielding the nuclear charge, energy depends on ℓ also in multielectron atoms. We shall discuss these shielding effects in Section 42.8.

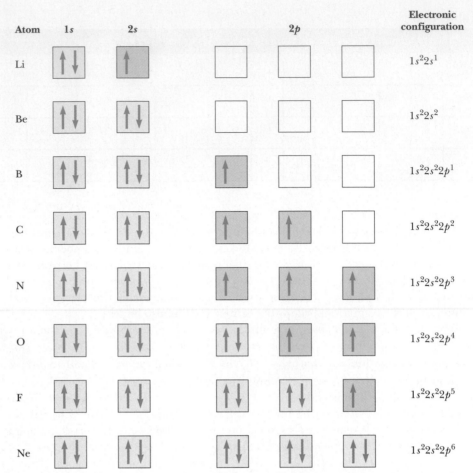

Atom	1s	2s	2p			Electronic configuration
Li	↑↓	↑				$1s^2 2s^1$
Be	↑↓	↑↓				$1s^2 2s^2$
B	↑↓	↑↓	↑			$1s^2 2s^2 2p^1$
C	↑↓	↑↓	↑	↑		$1s^2 2s^2 2p^2$
N	↑↓	↑↓	↑	↑	↑	$1s^2 2s^2 2p^3$
O	↑↓	↑↓	↑↓	↑	↑	$1s^2 2s^2 2p^4$
F	↑↓	↑↓	↑↓	↑↓	↑	$1s^2 2s^2 2p^5$
Ne	↑↓	↑↓	↑↓	↑↓	↑↓	$1s^2 2s^2 2p^6$

Figure 42.19 The filling of electronic states must obey both the exclusion principle and Hund's rule.

Example 42.8. In Figure 42.19, we show this electron in the leftmost $2p$ box with spin up, but it is equally likely to be in any $2p$ box with spin either up or down.

Carbon ($Z = 6$) has six electrons, giving rise to a question concerning how to assign the two $2p$ electrons. Do they go into the same orbital with paired spins (↑↓), or do they occupy different orbitals with unpaired spins (↑↑)? Experimental data show that the most stable configuration (that is, the one that is energetically preferred) is the latter, in which the spins are unpaired. Hence, the two $2p$ electrons in carbon and the three $2p$ electrons in nitrogen ($Z = 7$) have unpaired spins, as Figure 42.19 shows. The general rule that governs such situations, called **Hund's rule,** states that

> when an atom has orbitals of equal energy, the order in which they are filled by electrons is such that a maximum number of electrons have unpaired spins.

Some exceptions to this rule occur in elements having subshells that are close to being filled or half-filled.

A complete list of electronic configurations is provided in Table 42.4. In 1871, long before quantum mechanics was developed, the Russian chemist Dmitri Mendeleev (1834–1907) made an early attempt at finding some order among the chemical elements. He was trying to organize the elements for the table of contents of a book he was writing. He arranged the atoms in a table similar to that shown in Appendix C, according to their atomic masses and chemical similarities. The first table Mendeleev proposed contained many blank spaces, and he boldly stated that the gaps were there only because the elements had not yet been discovered. By noting the columns in

Table 42.4

Electronic Configuration of the Elements			
Atomic Number Z	**Symbol**	**Ground-State Configuration**	**Ionization Energy (eV)**
1	H	$1s^1$	13.595
2	He	$1s^2$	24.581
3	Li	[He] $2s^1$	5.39
4	Be	$2s^2$	9.320
5	B	$2s^2 2p^1$	8.296
6	C	$2s^2 2p^2$	11.256
7	N	$2s^2 2p^3$	14.545
8	O	$2s^2 2p^4$	13.614
9	F	$2s^2 2p^5$	17.418
10	Nc	$2s^2 2p^6$	21.559
11	Na	[Ne] $3s^1$	5.138
12	Mg	$3s^2$	7.644
13	Al	$3s^2 3p^1$	5.984
14	Si	$3s^2 3p^2$	8.149
15	P	$3s^2 3p^3$	10.484
16	S	$3s^2 3p^4$	10.357
17	Cl	$3s^2 3p^5$	13.01
18	Ar	$3s^2 3p^6$	15.755
19	K	[Ar] $4s^1$	4.339
20	Ca	$4s^2$	6.111
21	Sc	$3d^1 4s^2$	6.54
22	Ti	$3d^2 4s^2$	6.83
23	V	$3d^3 4s^2$	6.74
24	Cr	$3d^5 4s^1$	6.76
25	Mn	$3d^5 4s^2$	7.432
26	Fe	$3d^6 4s^2$	7.87
27	Co	$3d^7 4s^2$	7.86
28	Ni	$3d^8 4s^2$	7.633
29	Cu	$3d^{10} 4s^1$	7.724
30	Zn	$3d^{10} 4s^2$	9.391
31	Ga	$3d^{10} 4s^2 4p^1$	6.00
32	Ge	$3d^{10} 4s^2 4p^2$	7.88
33	As	$3d^{10} 4s^2 4p^3$	9.81
34	Se	$3d^{10} 4s^2 4p^4$	9.75
35	Br	$3d^{10} 4s^2 4p^5$	11.84
36	Kr	$3d^{10} 4s^2 4p^6$	13.996
37	Rb	[Kr] $5s^1$	4.176
38	Sr	$5s^2$	5.692
39	Y	$4d^1 5s^2$	6.377
40	Zr	$4d^2 5s^2$	
41	Nb	$4d^4 5s^1$	6.881
42	Mo	$4d^5 5s^1$	7.10
43	Tc	$4d^6 5s^1$	7.228
44	Ru	$4d^7 5s^1$	7.365
45	Rh	$4d^8 5s^1$	7.461
46	Pd	$4d^{10}$	8.33
47	Ag	$4d^{10} 5s^1$	7.574
48	Cd	$4d^{10} 5s^2$	8.991

continued

Table 42.4

Electronic Configuration of the Elements *continued*			
Atomic Number Z	**Symbol**	**Ground-State Configuration**	**Ionization Energy (eV)**
49	In	$4d^{10}5s^25p^1$	
50	Sn	$4d^{10}5s^25p^2$	7.342
51	Sb	$4d^{10}5s^25p^3$	8.639
52	Te	$4d^{10}5s^25p^4$	9.01
53	I	$4d^{10}5s^25p^5$	10.454
54	Xe	$4d^{10}5s^25p^6$	12.127
55	Cs	[Xe] $6s^1$	3.893
56	Ba	$6s^2$	5.210
57	La	$5d^16s^2$	5.61
58	Ce	$4f^15d^16s^2$	6.54
59	Pr	$4f^36s^2$	5.48
60	Nd	$4f^46s^2$	5.51
61	Pm	$4f^56s^2$	
62	Fm	$4f^66s^2$	5.6
63	Eu	$4f^76s^2$	5.67
64	Gd	$4f^75d^16s^2$	6.16
65	Tb	$4f^96s^2$	6.74
66	Dy	$4f^{10}6s^2$	
67	Ho	$4f^{11}6s^2$	
68	Er	$4f^{12}6s^2$	
69	Tm	$4f^{13}6s^2$	
70	Yb	$4f^{14}6s^2$	6.22
71	Lu	$4f^{14}5d^16s^2$	6.15
72	Hf	$4f^{14}5d^26s^2$	7.0
73	Ta	$4f^{14}5d^36s^2$	7.88
74	W	$4f^{14}5d^46s^2$	7.98
75	Re	$4f^{14}5d^56s^2$	7.87
76	Os	$4f^{14}5d^66s^2$	8.7
77	Ir	$4f^{14}5d^76s^2$	9.2
78	Pt	$4f^{14}5d^96s^1$	8.88
79	Au	[Xe, $4f^{14}5d^{10}$] $6s^1$	9.22
80	Hg	$6s^2$	10.434

which some missing elements should be located, he was able to make rough predictions about their chemical properties. Within 20 years of this announcement, most of these elements were indeed discovered.

The elements in the **periodic table** (Appendix C) are arranged so that all those in a column have similar chemical properties. For example, consider the elements in the last column, which are all gases at room temperature: He (helium), Ne (neon), Ar (argon), Kr (krypton), Xe (xenon), and Rn (radon). The outstanding characteristic of all these elements is that they do not normally take part in chemical reactions—that is, they do not join with other atoms to form molecules. They are therefore called *inert gases*.

We can partially understand this behavior by looking at the electronic configurations in Table 42.4. The chemical behavior of an element depends on the outermost shell that contains electrons. The electronic configuration for helium is $1s^2$—the $n = 1$ shell (which is the outermost shell because it is the only shell) is filled. Additionally, the energy of the atom in this configuration is considerably lower than the energy for the configuration in which an electron is in the next available level, the $2s$ subshell. Next, look at the electronic configuration for neon, $1s^22s^22p^6$. Again, the outermost

Table 42.4

Electronic Configuration of the Elements *continued*			
Atomic Number Z	**Symbol**	**Ground-State Configuration**	**Ionization Energy (eV)**
81	Tl	$6s^26p^1$	6.106
82	Pb	$6s^26p^2$	7.415
83	Bi	$6s^26p^3$	7.287
84	Po	$6s^26p^4$	8.43
85	At	$6s^26p^5$	
86	Rn	$6s^26p^6$	10.745
87	Fr	[Rn] $7s^1$	
88	Ra	$7s^2$	5.277
89	Ac	$6d^17s^2$	6.9
90	Th	$6d^27s^2$	
91	Pa	$5f^26d^17s^2$	
92	U	$5f^36d^17s^2$	4.0
93	Np	$5f^46d^17s^2$	
94	Pu	$5f^67s^2$	
95	Am	$5f^77s^2$	
96	Cm	$5f^76d^17s^2$	
97	Bk	$5f^97s^2$	
98	Cf	$5f^{10}7s^2$	
99	Es	$5f^{11}7s^2$	
100	Fm	$5f^{12}7s^2$	
101	Md	$5f^{13}7s^2$	
102	No	$5f^{14}7s^2$	
103	Lr	$5f^{14}7s^27p^1$	
104	Rf	$5f^{14}6d^27s^2$	
105	Db	$5f^{14}6d^37s^2$	
106	Sg	$5f^{14}6d^47s^2$	
107	Bh	$5f^{14}6d^57s^2$	
108	Hs	$5f^{14}6d^67s^2$	
109	Mt	$5f^{14}6d^77s^2$	
110	Ds	$5f^{14}6d^97s^1$	

Note: The bracket notation is used as a shorthand method to avoid repetition in indicating inner-shell electrons. Thus, [He] represents $1s^2$, [Ne] represents $1s^22s^22p^6$, [Ar] represents $1s^22s^22p^63s^23p^6$, and so on. Configurations for elements above $Z = 102$ are tentative.

shell ($n = 2$ in this case) is filled, and a wide gap in energy occurs between the filled $2p$ subshell and the next available one, the $3s$ subshell. Argon has the configuration $1s^22s^22p^63s^23p^6$. Here, it is only the $3p$ subshell that is filled, but again a wide gap in energy occurs between the filled $3p$ subshell and the next available one, the $3d$ subshell. This pattern continues through all the inert gases. Krypton has a filled $4p$ subshell, xenon a filled $5p$ subshell, and radon a filled $6p$ subshell.

If we consider the column to the left of the inert gases in the periodic table, we find a group of elements called the *halogens*: fluorine, chlorine, bromine, iodine, and astatine. At room temperature, fluorine and chlorine are gases, bromine is a liquid, and iodine and astatine are solids. In each of these atoms, the outer subshell is one electron short of being filled. As a result, the halogens are chemically very active, readily accepting an electron from another atom to form a closed shell. The halogens tend to form strong ionic bonds with atoms at the other side of the periodic table. (We will discuss ionic bonds in Chapter 43.) In a halogen lightbulb, bromine or iodine atoms combine with tungsten atoms evaporated from the filament and return them to the filament, resulting in a longer-lasting bulb. In addition, the filament can

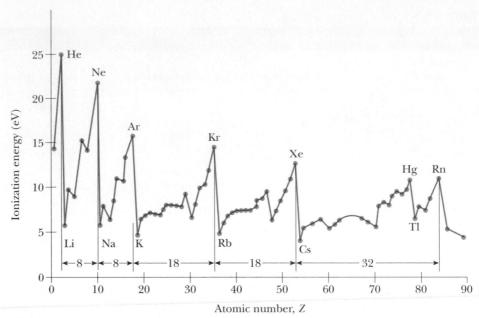

Figure 42.20 Ionization energy of the elements versus atomic number.

be operated at a higher temperature than in ordinary bulbs, giving a brighter and whiter light.

At the left side of the periodic table, the Group I elements consist of hydrogen and the *alkali metals*, lithium, sodium, potassium, rubidium, cesium, and francium. Each of these atoms contains one electron in a subshell outside of a closed subshell. Thus, these elements easily form positive ions because the lone electron is bound with a relatively low energy and is easily removed. Because of this, the alkali metal atoms are chemically active and form very strong bonds with halogen atoms. For example, table salt, NaCl, is a combination of an alkali metal and a halogen. Because the outer electron is weakly bound, pure alkali metals tend to be good electrical conductors, although, because of their high chemical activity, pure alkali metals are not generally found in nature.

It is interesting to plot ionization energy (see page 1358) versus atomic number Z, as in Figure 42.20. Note the pattern of $\Delta Z = 2, 8, 8, 18, 18, 32$ for the various peaks. This pattern follows from the exclusion principle and helps explain why the elements repeat their chemical properties in groups. For example, the peaks at $Z = 2, 10, 18$, and 36 correspond to the inert gases helium, neon, argon, and krypton, which, as we have mentioned, all have filled outermost shells. These elements have relatively high ionization energies and similar chemical behavior.

Quick Quiz 42.8 Rank the energy necessary to remove the outermost electron from the following three elements, smallest to largest: helium, neon, argon.

42.8 More on Atomic Spectra: Visible and X-Ray

In Section 42.1 we discussed the observation and early interpretation of visible spectral lines from gases. These spectral lines have their origin in transitions between quantized atomic states. We shall investigate these transitions more deeply in these final three sections of this chapter.

A modified energy-level diagram for hydrogen is shown in Figure 42.21. In this diagram, the allowed values of ℓ for each shell are separated horizontally. Figure 42.21 shows only those states up to $\ell = 2$; the shells from $n = 4$ upward would have more sets

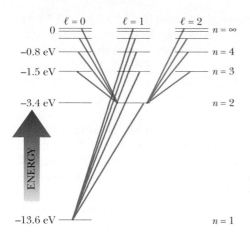

Figure 42.21 Some allowed electronic transitions for hydrogen, represented by the colored lines. These transitions must obey the selection rule $\Delta \ell = \pm 1$.

of states to the right which are not shown. We find that transitions for which ℓ does not change are very unlikely to occur and are called *forbidden transitions*. (Such transitions actually can occur, but their probability is very low relative to the probability of "allowed" transitions.) The various diagonal lines represent allowed transitions between stationary states. Whenever an atom makes a transition from a higher energy state to a lower one, a photon of light is emitted. The frequency of this photon is $f = \Delta E / h$, where ΔE is the energy difference between the two states and h is Planck's constant. The **selection rules** for the *allowed transitions* are

$$\Delta \ell = \pm 1 \qquad \text{and} \qquad \Delta m_\ell = 0, \pm 1 \tag{42.34}$$

Selection rules for allowed atomic transitions

Because the orbital angular momentum of an atom changes when a photon is emitted or absorbed (that is, as a result of a transition between states) and because angular momentum of the atom–photon system must be conserved, we conclude that **the photon involved in the process must carry angular momentum.** In fact, the photon has an angular momentum equivalent to that of a particle having a spin of 1. Therefore, a photon has energy, linear momentum, and angular momentum.

Recall from Equation 42.19 that the allowed energies for one-electron atoms and ions, such as hydrogen and He$^+$, are

$$E_n = -\frac{k_e e^2}{2a_0}\left(\frac{Z^2}{n^2}\right) = -\frac{13.6Z^2}{n^2}\ \text{eV} \tag{42.35}$$

This equation was developed from the Bohr theory, but it serves as a good first approximation in quantum theory as well. For multielectron atoms, the positive nuclear charge Ze is largely shielded by the negative charge of the inner-shell electrons. Therefore, the outer electrons interact with a net charge that is smaller than the nuclear charge. The expression for the allowed energies for multielectron atoms has the same form as Equation 42.35 with Z replaced by an effective atomic number Z_{eff}:

$$E_n = -\frac{13.6Z_{\text{eff}}^2}{n^2}\ \text{eV} \tag{42.36}$$

where Z_{eff} depends on n and ℓ.

X-Ray Spectra

X-rays are emitted when high-energy electrons or any other charged particles bombard a metal target. The x-ray spectrum typically consists of a broad continuous band containing a series of sharp lines, as shown in Figure 42.22. In Section 34.5, we mentioned the fact that an accelerated electric charge emits electromagnetic radiation. The x-rays we see in Figure 42.22 are the result of the slowing down of high-energy electrons as they strike the target. It may take several interactions with the atoms of the target before the

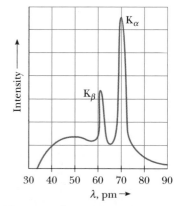

Figure 42.22 The x-ray spectrum of a metal target consists of a broad continuous spectrum (*bremsstrahlung*) plus a number of sharp lines, which are due to *characteristic x-rays*. The data shown were obtained when 37-keV electrons bombarded a molybdenum target.

electron loses all of its kinetic energy. (We explored this idea in Quick Quiz 42.2.) The amount of kinetic energy lost in any given interaction can vary from zero up to the entire kinetic energy of the electron. Thus, the wavelength of radiation from these interactions lies in a continuous range from some minimum value up to infinity. It is this general slowing down of the electrons that provides the continuous curve in Figure 42.22, which shows the cutoff of x-rays below a minimum wavelength value that depends on the kinetic energy of the incoming electrons. X-ray radiation with its origin in the slowing down of electrons is called **bremsstrahlung,** the German word for "braking radiation."

The discrete lines in Figure 42.22, called **characteristic x-rays** and discovered in 1908, have a different origin. Their origin remained unexplained until the details of atomic structure were understood. The first step in the production of characteristic x-rays occurs when a bombarding electron collides with a target atom. The electron must have sufficient energy to remove an inner-shell electron from the atom. The vacancy created in the shell is filled when an electron in a higher level drops down into the level containing the vacancy. The time interval for this to happen is very short, less than 10^{-9} s. As usual, this transition is accompanied by the emission of a photon whose energy equals the difference in energy between the two levels. Typically, the energy of such transitions is greater than 1 000 eV, and the emitted x-ray photons have wavelengths in the range of 0.01 nm to 1 nm.

Let us assume that the incoming electron has dislodged an atomic electron from the innermost shell—the K shell. If the vacancy is filled by an electron dropping from the next higher shell—the L shell—the photon emitted has an energy corresponding to the K_α characteristic x-ray line on the curve of Figure 42.22. In this notation, K refers to the final level of the electron and the subscript α, as the *first* letter of the Greek alphabet, refers to the initial level as the *first* one above the final level. If the vacancy in the K shell is filled by an electron dropping from the M shell, the K_β line in Figure 42.22 is produced.

Other characteristic x-ray lines are formed when electrons drop from upper levels to vacancies other than those in the K shell. For example, L lines are produced when vacancies in the L shell are filled by electrons dropping from higher shells. An L_α line is produced as an electron drops from the M shell to the L shell, and an L_β line is produced by a transition from the N shell to the L shell.

Although multielectron atoms cannot be analyzed exactly with either the Bohr model or the Schrödinger equation, we can apply our knowledge of Gauss's law from Chapter 24 to make some surprisingly accurate estimates of expected x-ray energies and wavelengths. Consider an atom of atomic number Z in which one of the two electrons in the K shell has been ejected. Imagine that a gaussian sphere is drawn just inside the most probable radius of the L electrons. The electric field at the position of the L electrons is a combination of the fields created by the nucleus, the single K electron, the other L electrons, and the outer electrons. The wave functions of the outer electrons are such that the electrons have a very high probability of being farther from the nucleus than the L electrons are. Therefore, they are much more likely to be outside the gaussian surface than inside and, on the average, do not contribute significantly to the electric field at the position of the L electrons. The effective charge inside the gaussian surface is the positive nuclear charge and one negative charge due to the single K electron. If we ignore the interactions between L electrons, a single L electron behaves as if it experiences an electric field due to a charge $(Z - 1)e$ enclosed by the gaussian surface. The nuclear charge is shielded by the electron in the K shell such that Z_{eff} in Equation 42.36 is $Z - 1$. For higher-level shells, the nuclear charge is shielded by electrons in all of the inner shells.

We can now use Equation 42.36 to estimate the energy associated with an electron in the L shell:

$$E_{\text{L}} = -(Z - 1)^2 \frac{13.6 \text{ eV}}{2^2}$$

After the atom makes the transition, there are two electrons in the K shell. Using a similar argument for a gaussian surface drawn just inside the most probable radius for a K electron, we find that the energy associated with one of these electrons is approximately that of a one-electron atom with the nuclear charge reduced by the negative charge of the other electron. That is,

$$E_K \approx -(Z-1)^2 (13.6 \text{ eV}) \tag{42.37}$$

As Example 42.9 shows, the energy of the atom with an electron in an M shell can be estimated in a similar fashion. Taking the energy difference between the initial and final levels, we can then calculate the energy and wavelength of the emitted photon.

In 1914, Henry G. J. Moseley (1887–1915) plotted the Z values for a number of elements versus $\sqrt{1/\lambda}$, where λ is the wavelength of the K_α line of each element. He found that the plot is a straight line, as in Figure 42.23. This is consistent with rough calculations of the energy levels given by Equation 42.37. From this plot, Moseley determined the Z values of elements that had not yet been discovered and produced a periodic table in excellent agreement with the known chemical properties of the elements. Until that experiment, atomic numbers had been merely placeholders for the elements that appeared in the periodic table, the elements being ordered according to mass.

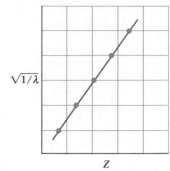

Figure 42.23 A Moseley plot of $\sqrt{1/\lambda}$ versus Z, where λ is the wavelength of the K_α x-ray line of the element of atomic number Z.

Quick Quiz 42.9 What are the initial and final shells for an M_β line in an x-ray spectrum?

Quick Quiz 42.10 In an x-ray tube, as you increase the energy of the electrons striking the metal target, the wavelengths of the characteristic x-rays (a) increase (b) decrease (c) do not change.

Quick Quiz 42.11 True or false: it is possible for an x-ray spectrum to show the continuous spectrum of x-rays without the presence of the characteristic x-rays.

Example 42.9   Estimating the Energy of an X-Ray

Estimate the energy of the characteristic x-ray emitted from a tungsten target when an electron drops from an M shell ($n = 3$ state) to a vacancy in the K shell ($n = 1$ state).

Solution The atomic number for tungsten is $Z = 74$. Using Equation 42.37, we see that the energy associated with the electron in the K shell is approximately

$$E_K \approx -(74-1)^2 (13.6 \text{ eV}) = -7.2 \times 10^4 \text{ eV}$$

An electron in the M shell is subject to an effective nuclear charge that depends on the number of electrons in the $n = 1$ and $n = 2$ states, which shield the nucleus. Because there are eight electrons in the $n = 2$ state and one electron in the $n = 1$ state, nine electrons shield the nucleus, and so

$Z_{\text{eff}} = Z - 9$. Hence, the energy of the M shell, following Equation 42.36, is approximately

$$E_M \approx -\frac{(13.6 \text{ eV})(74-9)^2}{(3)^2} \approx -6.4 \times 10^3 \text{ eV}$$

Therefore, the emitted x-ray has an energy equal to $hf = E_M - E_K \approx -6.4 \times 10^3 \text{ eV} - (-7.2 \times 10^4 \text{ eV}) \approx 6.6 \times 10^4 \text{ eV} = \boxed{66 \text{ keV}}.$

Consultation of x-ray tables shows that the M–K transition energies in tungsten vary from 66.9 keV to 67.7 keV, where the range of energies is due to slightly different energy values for states of different ℓ. Thus, our estimate differs from the midpoint of this experimentally measured range by about 2%.

42.9   Spontaneous and Stimulated Transitions

We have seen that an atom absorbs and emits electromagnetic radiation only at frequencies that correspond to the energy differences between allowed states. Let us now examine more details of these processes. Consider an atom having the allowed energy levels labeled E_1, E_2, E_3, ... When radiation is incident on the atom, only those photons whose energy hf matches the energy separation ΔE between two

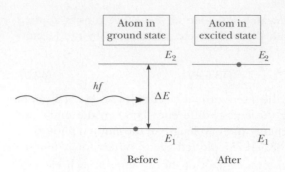

Adjust the energy difference between states at the Active Figures link at http://www.pse6.com and observe stimulated absorption.

Active Figure 42.24 Stimulated absorption of a photon. The blue dot represents an electron. The electron is transferred from the ground state to the excited state when the atom absorbs a photon of energy $hf = E_2 - E_1$.

energy levels can be absorbed by the atom, as represented in Figure 42.24. This process is called **stimulated absorption** because the photon stimulates the atom to make the upward transition. At ordinary temperatures, most of the atoms in a sample are in the ground state. If a vessel containing many atoms of a gaseous element is illuminated with radiation of all possible photon frequencies (that is, a continuous spectrum), only those photons having energy $E_2 - E_1$, $E_3 - E_1$, $E_4 - E_1$, and so on are absorbed by the atoms. As a result of this absorption, some of the atoms are raised to excited states.

Once an atom is in an excited state, the excited atom can make a transition back to a lower energy level, emitting a photon in the process, as in Figure 42.25. This process is known as **spontaneous emission** because it happens naturally, without requiring an event to trigger the transition. Typically, an atom remains in an excited state for only about 10^{-8} s.

In addition to spontaneous emission, **stimulated emission** occurs. Suppose an atom is in an excited state E_2, as in Figure 42.26. If the excited state is a *metastable state*—that is, if its lifetime is much longer than the typical 10^{-8} s lifetime of excited states—then the time interval until spontaneous emission occurs will be relatively long. Let us imagine that during that interval a photon of energy $hf = E_2 - E_1$ is incident on the atom. One possibility is that the photon energy will be sufficient for the photon to ionize the atom. Another possibility is that the interaction between the incoming photon and the atom will cause the atom to return to the ground state[10] and thereby emit a second photon with energy $hf = E_2 - E_1$. In this process the incident photon is not absorbed; thus, after the stimulated emission, two photons with identical energy exist—the incident photon and the emitted photon. The two are in phase and travel in the same direction—an important consideration in lasers, which we shall discuss in the next section.

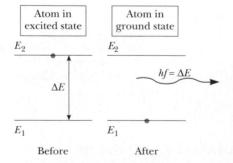

Adjust the energy difference between states at the Active Figures link at http://www.pse6.com and observe spontaneous emission.

Active Figure 42.25 Spontaneous emission of a photon by an atom that is initially in the excited state E_2. When the atom falls to the ground state, it emits a photon of energy $hf = E_2 - E_1$.

---

[10] This is fundamentally a *resonance* phenomenon. The incoming photon is considered to be driving the system of the atom. Because the driving frequency matches that associated with a transition between states—one of the natural frequencies of the atom—there is a large response: the atom makes the transition.

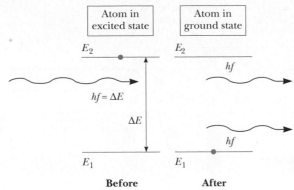

Active Figure 42.26 Stimulated emission of a photon by an incoming photon of energy $hf = E_2 - E_1$. Initially, the atom is in the excited state. The incoming photon stimulates the atom to emit a second photon of energy $hf = E_2 - E_1$.

Adjust the energy difference between states at the Active Figures link at http://www.pse6.com and observe stimulated emission.

42.10 Lasers

In this section, we shall explore the nature of laser light and a variety of applications of lasers in our technological society. The primary properties of laser light that make it useful in these technological applications are these:

- Laser light is coherent. The individual rays of light in a laser beam maintain a fixed phase relationship with each other, resulting in no destructive interference.
- Laser light is monochromatic. Light in a laser beam has a very narrow range of wavelengths.
- Laser light has a small angle of divergence. The beam spreads out very little, even over large distances.

In order to understand the origin of these properties, let us combine our knowledge of atomic energy levels from this chapter with some special requirements for the atoms that emit laser light.

We have described how an incident photon can cause atomic energy transitions either upward (stimulated absorption) or downward (stimulated emission). The two processes are equally probable. When light is incident on a collection of atoms, a net absorption of energy usually occurs because, when the system is in thermal equilibrium, many more atoms are in the ground state than in excited states. However, if the situation can be inverted so that more atoms are in an excited state than in the ground state, a net emission of photons can result. Such a condition is called **population inversion.**

This, in fact, is the fundamental principle involved in the operation of a **laser**—an acronym for *l*ight *a*mplification by *s*timulated *e*mission of *r*adiation. The full name indicates one of the requirements for laser light—the process of stimulated emission must occur in order to achieve laser action.

Suppose an atom is in the excited state E_2, as in Figure 42.26, and a photon with energy $hf = E_2 - E_1$ is incident on it. As described in Section 42.9, the incoming photon can stimulate the excited atom to return to the ground state and thereby emit a second photon having the same energy hf and traveling in the same direction. The incident photon is not absorbed, so after the stimulated emission, there are two identical photons—the incident photon and the emitted photon. The emitted photon is in phase with the incident photon. These photons can stimulate other atoms to emit photons in a chain of similar processes. The many photons produced in this fashion are the source of the intense, coherent light in a laser.

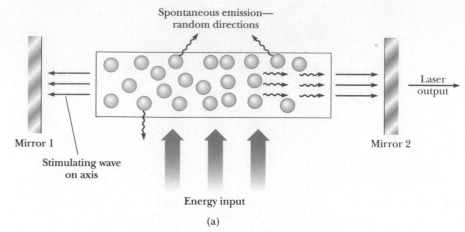

(a)

Figure 42.27 Schematic diagram of a laser design. The tube contains the atoms that are the active medium. An external source of energy (for example, an optical or electrical device) "pumps" the atoms to excited states. The parallel end mirrors confine the photons to the tube, but mirror 2 is only partially reflective.

In order that the stimulated emission result in laser light, we must have a buildup of photons in the system. The following three conditions must be satisfied in order to achieve this buildup:

- The system must be in a state of population inversion—there must be more atoms in an excited state than in the ground state. This must be true because the number of photons emitted must be greater than the number absorbed.

- The excited state of the system must be a *metastable state*, which means its lifetime must be long compared with the usually short lifetimes of excited states, which are typically 10^{-8} s. In this case, the population inversion can be established and stimulated emission is likely to occur before spontaneous emission.

- The emitted photons must be confined in the system long enough to enable them to stimulate further emission from other excited atoms. This is achieved by using reflecting mirrors at the ends of the system. One end is made totally reflecting, and the other is partially reflecting. A fraction of the light intensity passes through the partially reflecting end, forming the beam of laser light (Fig. 42.27).

One device that exhibits stimulated emission of radiation is the helium–neon gas laser. Figure 42.28 is an energy-level diagram for the neon atom in this system. The mixture of helium and neon is confined to a glass tube that is sealed at the ends by mirrors. A voltage applied across the tube causes electrons to sweep through the tube, colliding with the atoms of the gases and raising them into excited states. Neon atoms are excited to state $E_3{}^*$ through this process (the asterisk * indicates a metastable state) and also as a result of collisions with excited helium atoms. Stimulated emission occurs, causing neon atoms to make transitions to state E_2. Neighboring excited atoms are also stimulated. This results in the production of coherent light at a wavelength of 632.8 nm.

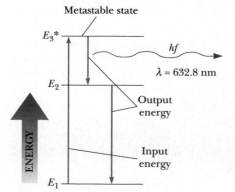

Figure 42.28 Energy-level diagram for a neon atom in a helium–neon laser. The atom emits 632.8-nm photons through stimulated emission in the transition $E_3{}^* - E_2$. This is the source of coherent light in the laser.

Applications

Since the development of the first laser in 1960, tremendous growth has occurred in laser technology. Lasers that cover wavelengths in the infrared, visible, and ultraviolet regions are now available. Applications include surgical "welding" of detached retinas, precision surveying and length measurement, precision cutting of metals and other materials (such as the fabric in Figure 42.29), and telephone communication along optical fibers. These and other applications are possible because of the unique characteristics of laser light. In addition to being highly monochromatic, laser light is also highly directional and can be sharply focused to produce regions of extremely intense light energy (with energy densities 10^{12} times that in the flame of a typical cutting torch).

Lasers are used in precision long-range distance measurement (range finding). In recent years it has become important, for astronomical and geophysical purposes, to measure as precisely as possible the distances from various points on the surface of the Earth to a point on the Moon's surface. To facilitate this, the Apollo astronauts set up a 0.5-m square of reflector prisms on the Moon, which enables laser pulses directed from an Earth station to be retroreflected to the same station (see Fig. 35.8a). Using the known speed of light and the measured round-trip travel time of a 1-ns pulse, the Earth–Moon distance can be determined to a precision of better than 10 cm.

Medical applications use the fact that various laser wavelengths can be absorbed in specific biological tissues. For example, certain laser procedures have greatly reduced blindness in glaucoma and diabetes patients. Glaucoma is a widespread eye condition characterized by a high fluid pressure in the eye, a condition that can lead to destruction of the optic nerve. A simple laser operation (iridectomy) can "burn" open a tiny hole in a clogged membrane, relieving the destructive pressure. A serious side effect of diabetes is neovascularization, the proliferation of weak blood vessels, which often leak blood. When this occurs in the retina, vision deteriorates (diabetic retinopathy) and finally is destroyed. It is now possible to direct the green light from an argon ion laser through the clear eye lens and eye fluid, focus on the retina edges, and photocoagulate the leaky vessels. Even people who have only minor vision defects such as nearsightedness are benefiting from the use of lasers to reshape the cornea, changing its focal length and reducing the need for eyeglasses.

Laser surgery is now an everyday occurrence at hospitals around the world. Infrared light at 10 μm from a carbon dioxide laser can cut through muscle tissue, primarily by vaporizing the water contained in cellular material. Laser power of about 100 W is required in this technique. The advantage of the "laser knife" over conventional

Philippe Plailly/SPL/Photo Researchers, Inc.

Figure 42.29 This robot carrying laser scissors, which can cut up to 50 layers of fabric at a time, is one of the many applications of laser technology.

methods is that laser radiation cuts tissue and coagulates blood at the same time, leading to a substantial reduction in blood loss. In addition, the technique virtually eliminates cell migration, an important consideration when tumors are being removed.

A laser beam can be trapped in fine optical-fiber light guides (endoscopes) by means of total internal reflection. An endoscope can be introduced through natural orifices, conducted around internal organs, and directed to specific interior body locations, eliminating the need for invasive surgery. For example, bleeding in the gastrointestinal tract can be optically cauterized by endoscopes inserted through the mouth.

In biological and medical research, it is often important to isolate and collect unusual cells for study and growth. A laser cell separator exploits the fact that specific cells can be tagged with fluorescent dyes. All cells are then dropped from a tiny charged nozzle and laser-scanned for the dye tag. If triggered by the correct light-emitting tag, a small voltage applied to parallel plates deflects the falling electrically charged cell into a collection beaker.

An exciting area of research and technological applications arose in the 1990s with the development of *laser trapping* of atoms. One scheme, called *optical molasses* and developed by Steven Chu, of Stanford University, and his colleagues, involves focusing six laser beams onto a small region in which atoms are to be trapped. Each pair of lasers is along one of the *x, y,* and *z* axes and emits light in opposite directions (Fig. 42.30). The frequency of the laser light is tuned to be just below the absorption frequency of the subject atom. Imagine that an atom has been placed into the trap region and moves along the positive *x* axis toward the laser which is emitting light toward it (the rightmost laser in Figure 42.30). Because the atom is moving, the light from the laser appears Doppler-shifted upward in frequency in the reference frame of the atom. This creates a match between the Doppler-shifted laser frequency and the absorption frequency of the atom, and the atom absorbs photons.[11] The momentum carried by these photons results in the atom being pushed back to the center of the trap. By incorporating six lasers, the atoms are pushed back into the trap regardless of which way they move along any axis.

In 1986, Chu developed *optical tweezers,* a device that uses a single tightly focused laser beam to trap and manipulate small particles. In combination with microscopes, optical tweezers have opened up many new possibilities for biologists. Optical tweezers

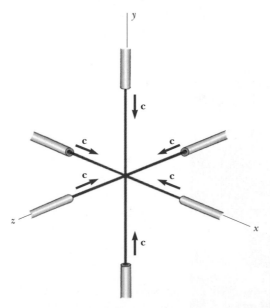

Figure 42.30 An optical trap for atoms is formed at the intersection point of six counterpropagating laser beams along mutually perpendicular axes. The frequency of the laser light is tuned to be just below that for absorption by the trapped atoms. If an atom moves away from the trap, it absorbs the Doppler-shifted laser light, and the momentum of the light pushes the atom back into the trap.

[11]   The laser light traveling in the same direction as the atom is Doppler-shifted further downward in frequency, so that there is no absorption. Thus, the atom is not pushed out of the trap by the diametrically opposed laser.

Figure 42.31 A staff member of the National Institute of Standards and Technology views a sample of trapped sodium atoms (the small yellow dot in the center of the vacuum chamber) cooled to a temperature of less than 1 mK.

have been used to manipulate live bacteria without damage, move chromosomes within a cell nucleus, and measure the elastic properties of a single DNA molecule. Steven Chu shared the 1997 Nobel Prize in physics with two of his colleagues for the development of the techniques of optical trapping.

An extension of laser trapping, *laser cooling*, is possible because the normal high speeds of the atoms are reduced when they are restricted to the region of the trap. As a result, the temperature of the collection of atoms can be reduced to a few microkelvins. The technique of laser cooling allows scientists to study the behavior of atoms at extremely low temperatures (Figure 42.31).

SUMMARY

The wavelengths of spectral lines from hydrogen, called the **Balmer series,** can be described by the following equation:

$$\frac{1}{\lambda} = R_{\mathrm{H}} \left(\frac{1}{2^2} - \frac{1}{n^2} \right) \qquad n = 3, 4, 5, \ldots \qquad (42.1)$$

where R_{H} is the **Rydberg constant.** The spectral lines corresponding to values of n from 3 to 6 are in the visible range of the electromagnetic spectrum. Values of n higher than 6 correspond to spectral lines in the ultraviolet region of the spectrum.

The Bohr model of the atom is successful in describing the spectra of atomic hydrogen and hydrogen-like ions. One of the basic assumptions of the model is that the electron can exist only in discrete orbits such that the angular momentum of the electron $m_e v r$ is an integral multiple of $h/2\pi = \hbar$. When we assume circular orbits and a simple Coulomb attraction between electron and proton, the energies of the quantum states for hydrogen are calculated to be

$$E_n = -\frac{k_e e^2}{2a_0} \left(\frac{1}{n^2} \right) \qquad n = 1, 2, 3, \ldots \qquad (42.13)$$

where k_e is the Coulomb constant, e is the electronic charge, n is an integer called the **quantum number,** and $a_0 = 0.052\,9$ nm is the **Bohr radius.**

If the electron in a hydrogen atom makes a transition from an orbit whose quantum number is n_i to one whose quantum number is n_f, where $n_f < n_i$, a photon is emitted by the atom, and the frequency of this photon is

$$f = \frac{k_e e^2}{2a_0 h} \left(\frac{1}{n_f{}^2} - \frac{1}{n_i{}^2} \right) \qquad (42.15)$$

Take a practice test for this chapter by clicking on the Practice Test link at http://www.pse6.com.

Quantum mechanics can be applied to the hydrogen atom by the use of the potential energy function $U(r) = -k_e e^2/r$ in the Schrödinger equation. The solution to this equation yields wave functions for allowed states and allowed energies:

$$E_n = -\left(\frac{k_e e^2}{2a_0}\right)\frac{1}{n^2} = -\frac{13.606 \text{ eV}}{n^2} \qquad n = 1, 2, 3, \ldots \qquad (42.21)$$

where n is the **principal quantum number.** The allowed wave functions depend on three quantum numbers: n, ℓ, and m_ℓ, where ℓ is the **orbital quantum number** and m_ℓ is the **orbital magnetic quantum number.** The restrictions on the quantum numbers are

$$n = 1, 2, 3, \ldots$$
$$\ell = 0, 1, 2, \ldots, n-1$$
$$m_\ell = -\ell, -\ell + 1, \ldots, \ell - 1, \ell$$

All states having the same principal quantum number n form a **shell,** identified by the letters K, L, M, . . . (corresponding to $n = 1, 2, 3, \ldots$). All states having the same values of n and ℓ form a **subshell,** designated by the letters $s, p, d, f, \ldots$ (corresponding to $\ell = 0, 1, 2, 3, \ldots$).

An atom in a state characterized by a specific value of n can have the following values of L, the magnitude of the atom's orbital angular momentum $\mathbf{L}$:

$$L = \sqrt{\ell(\ell+1)}\,\hbar \qquad (42.27)$$

The allowed values of the projection of $\mathbf{L}$ along the z axis are

$$L_z = m_\ell \hbar \qquad (42.28)$$

Only discrete values of L_z are allowed, as determined by the restrictions on m_ℓ. This quantization of L_z is referred to as **space quantization.**

The electron has an intrinsic angular momentum called the **spin angular momentum.** That is, the total angular momentum of an electron in an atom has two contributions, one arising from the spin of the electron ($\mathbf{S}$) and one arising from the orbital motion of the electron ($\mathbf{L}$). Electron spin can be described by a single quantum number $s = \frac{1}{2}$. To completely describe a quantum state, it is necessary to include a fourth quantum number m_s, called the **spin magnetic quantum number.** This quantum number can have only two values, $\pm\frac{1}{2}$. The magnitude of the spin angular momentum is

$$S = \frac{\sqrt{3}}{2}\,\hbar \qquad (42.30)$$

and the z component of $\mathbf{S}$ is

$$S_z = m_s \hbar = \pm\frac{1}{2}\hbar \qquad (42.31)$$

That is, the spin angular momentum is also quantized in space, as specified by the spin magnetic quantum number $m_s = \pm\frac{1}{2}$.

The magnetic moment $\boldsymbol{\mu}_{\text{spin}}$ associated with the spin angular momentum of an electron is

$$\boldsymbol{\mu}_{\text{spin}} = -\frac{e}{m_e}\mathbf{S} \qquad (42.32)$$

The z component of $\boldsymbol{\mu}_{\text{spin}}$ can have the values

$$\mu_{\text{spin},z} = \pm\frac{e\hbar}{2m_e} \qquad (42.33)$$

The **exclusion principle** states that **no two electrons in an atom can be in the same quantum state.** In other words, no two electrons can have the same set of quantum numbers n, ℓ, m_ℓ, and m_s. Using this principle, the electronic configurations of the elements can be determined. This serves as a basis for understanding atomic structure and the chemical properties of the elements. The allowed electronic transitions between any two levels in an atom are governed by the **selection rules**

$$\Delta\ell = \pm 1 \quad \text{and} \quad \Delta m_\ell = 0, \pm 1 \tag{42.34}$$

The x-ray spectrum of a metal target consists of a set of sharp characteristic lines superimposed on a broad continuous spectrum. **Bremsstrahlung** is x-radiation with its origin in the slowing down of high-energy electrons as they encounter the target. **Characteristic x-rays** are emitted by atoms when an electron undergoes a transition from an outer shell to a vacancy in an inner shell.

Atomic transitions can be described with three processes: **stimulated absorption,** in which an incoming photon raises the atom to a higher energy state; **spontaneous emission**, in which the atom makes a transition to a lower energy state, emitting a photon; and **stimulated emission,** in which an incident photon causes an excited atom to make a downward transition, emitting a photon identical to the incident one.

QUESTIONS

1. Does the light emitted by a neon sign constitute a continuous spectrum or only a few colors? Defend your answer.

2. The Bohr theory of the hydrogen atom is based upon several assumptions. Discuss these assumptions and their significance. Do any of them contradict classical physics?

3. Suppose that the electron in the hydrogen atom obeyed classical mechanics rather than quantum mechanics. Why should such a hypothetical atom emit a continuous spectrum rather than the observed line spectrum?

4. Can a hydrogen atom in the ground state absorb a photon of energy (a) less than 13.6 eV and (b) greater than 13.6 eV?

5. Explain why, in the Bohr model, the total energy of the atom is negative.

6. Let $-E$ represent the energy of a hydrogen atom. What is the kinetic energy of the electron? What is the potential energy of the atom?

7. According to Bohr's model of the hydrogen atom, what is the uncertainty in the radial coordinate of the electron? What is the uncertainty in the radial component of the velocity of the electron? In what way does the model violate the uncertainty principle?

8. Why are three quantum numbers needed to describe the state of a one-electron atom (ignoring spin)?

9. Compare the Bohr theory and the Schrödinger treatment of the hydrogen atom. Comment on the total energy and orbital angular momentum.

10. Discuss why the term *electron cloud* is used to describe the electronic arrangement in the quantum-mechanical model of an atom.

11. Why is the direction of the orbital angular momentum of an electron opposite to that of its magnetic moment?

12. Could the Stern–Gerlach experiment be performed with ions rather than neutral atoms? Explain.

13. Why is a nonuniform magnetic field used in the Stern–Gerlach experiment?

14. Discuss some of the consequences of the exclusion principle.

15. Describe some experiments that support the conclusion that the spin magnetic quantum number for electrons can only have the values $\pm\frac{1}{2}$.

16. Why do lithium, potassium, and sodium exhibit similar chemical properties?

17. Explain why a photon must have a spin of 1.

18. An energy of about 21 eV is required to excite an electron in a helium atom from the $1s$ state to the $2s$ state. The same transition for the He$^+$ ion requires approximately twice as much energy. Explain.

19. The absorption or emission spectrum of a gas consists of lines that broaden as the density of gas molecules increases. Why do you suppose this occurs?

20. How is it possible that electrons, whose positions are described by a probability distribution around a nucleus, can exist in states of *definite* energy (e.g., $1s$, $2p$, $3d$, . . .)?

21. It is easy to understand how two electrons (one spin up, one spin down) can fill the $1s$ shell for a helium atom. How is it possible that eight more electrons can fit into the $2s$, $2p$ level to complete the $1s^2 2s^2 2p^6$ shell for a neon atom?

22. In 1914, Henry Moseley was able to define the atomic number of an element from its characteristic x-ray spectrum. How was this possible? (*Suggestion:* See Figure 42.23.)

23. Does the intensity of light from a laser fall off as $1/r^2$?

24. Why is stimulated emission so important in the operation of a laser?

25. (a) "As soon as I define a particular direction as the z axis, precisely one half of the electrons in this part of the Universe have their magnetic moment vectors oriented at 54.735 61° to that axis, and all the rest have their magnetic moments at 125.264 39°." Argue for or against this statement. (b) "The Universe is not simply stranger than we suppose; it is stranger than we *can* suppose." Argue for or against this statement.

PROBLEMS

1, 2, 3 = straightforward, intermediate, challenging   ☐ = full solution available in the *Student Solutions Manual and Study Guide*

🌀 = coached solution with hints available at http://www.pse6.com   💻 = computer useful in solving problem

▨ = paired numerical and symbolic problems

Section 42.1   Atomic Spectra of Gases

1. (a) What value of n_i is associated with the 94.96-nm spectral line in the Lyman series of hydrogen? (b) **What If?** Could this wavelength be associated with the Paschen or Balmer series?

2. (a) Compute the shortest wavelength in each of these hydrogen spectral series: Lyman, Balmer, Paschen, and Brackett. (b) Compute the energy (in electron volts) of the highest-energy photon produced in each series.

Section 42.2   Early Models of the Atom

3. 🌀 According to classical physics, a charge e moving with an acceleration a radiates at a rate

$$\frac{dE}{dt} = -\frac{1}{6\pi\epsilon_0}\frac{e^2 a^2}{c^3}$$

(a) Show that an electron in a classical hydrogen atom (see Fig. 42.6) spirals into the nucleus at a rate

$$\frac{dr}{dt} = -\frac{e^4}{12\pi^2\epsilon_0^2 r^2 m_e^2 c^3}$$

(b) Find the time interval over which the electron will reach $r = 0$, starting from $r_0 = 2.00 \times 10^{-10}$ m.

4. In the Rutherford scattering experiment, 4.00-MeV alpha particles (^{4}He nuclei containing 2 protons and 2 neutrons) scatter off gold nuclei (containing 79 protons and 118 neutrons). Assume that a particular alpha particle makes a direct head-on collision with the gold nucleus and scatters backward at 180°. Determine (a) the distance of closest approach of the alpha particle to the gold nucleus, and (b) the maximum force exerted on the alpha particle. Assume that the gold nucleus remains fixed throughout the entire process.

Section 42.3   Bohr's Model of the Hydrogen Atom

5. For a hydrogen atom in its ground state, use the Bohr model to compute (a) the orbital speed of the electron, (b) the kinetic energy of the electron, and (c) the electric potential energy of the atom.

6. Four possible transitions for a hydrogen atom are as follows:

(i) $n_i = 2$; $n_f = 5$ (ii) $n_i = 5$; $n_f = 3$

(iii) $n_i = 7$; $n_f = 4$ (iv) $n_i = 4$; $n_f = 7$

(a) In which transition is light of the shortest wavelength emitted? (b) In which transition does the atom gain the most energy? (c) In which transition(s) does the atom lose energy?

7. 🌀 A hydrogen atom is in its first excited state ($n = 2$). Using the Bohr theory of the atom, calculate (a) the radius of the orbit, (b) the linear momentum of the electron, (c) the angular momentum of the electron, (d) the kinetic energy of the electron, (e) the potential energy of the system, and (f) the total energy of the system.

8. How much energy is required to ionize hydrogen (a) when it is in the ground state? (b) when it is in the state for which $n = 3$?

9. A photon is emitted as a hydrogen atom undergoes a transition from the $n = 6$ state to the $n = 2$ state. Calculate (a) the energy, (b) the wavelength, and (c) the frequency of the emitted photon.

10. Show that the speed of the electron in the nth Bohr orbit in hydrogen is given by

$$v_n = \frac{k_e e^2}{n\hbar}$$

11. Two hydrogen atoms collide head-on and end up with zero kinetic energy. Each atom then emits light with a wavelength of 121.6 nm ($n = 2$ to $n = 1$ transition). At what speed were the atoms moving before the collision?

12. A monochromatic beam of light is absorbed by a collection of ground-state hydrogen atoms in such a way that six different wavelengths are observed when the hydrogen relaxes back to the ground state. What is the wavelength of the incident beam?

13. (a) Construct an energy-level diagram for the He$^+$ ion, for which $Z = 2$. (b) What is the ionization energy for He$^+$?

14. In a hot star, because of the high temperature, an atom can absorb sufficient energy to remove several electrons from the atom. Consider such a multiply ionized atom with a single remaining electron. The ion produces a series of spectral lines as described by the Bohr model. The series corresponds to electronic transitions that terminate in the same final state. The longest and shortest wavelengths of the series are 63.3 nm and 22.8 nm, respectively. (a) What is the ion? (b) Find the wavelengths of the next three spectral lines nearest to the line of longest wavelength.

15. (a) Calculate the angular momentum of the Moon due to its orbital motion about the Earth. In your calculation, use 3.84×10^8 m as the average Earth–Moon distance and 2.36×10^6 s as the period of the Moon in its orbit. (b) Assume the Moon's angular momentum is described by Bohr's assumption $mvr = n\hbar$. Determine the corresponding quantum number. (c) By what fraction would the Earth–Moon distance have to be increased to raise the quantum number by 1?

Section 42.4   The Quantum Model of the Hydrogen Atom

16. A general expression for the energy levels of one-electron atoms and ions is

$$E_n = -\frac{\mu k_e^2 q_1^2 q_2^2}{2\hbar^2 n^2}$$

where k_e is the Coulomb constant, q_1 and q_2 are the charges of the electron and the nucleus, and μ is the reduced mass, given by $\mu = m_1 m_2/(m_1 + m_2)$. The wavelength for the $n = 3$ to $n = 2$ transition of the hydrogen atom is 656.3 nm (visible red light). **What If?** What are the wavelengths for this same transition in (a) positronium, which consists of an electron and a positron, and (b) singly ionized helium? (*Note:* A positron is a positively charged electron.)

17. An electron of momentum p is at a distance r from a stationary proton. The electron has kinetic energy $K = p^2/2m_e$. The atom has potential energy $U = -k_e e^2/r$, and total energy $E = K + U$. If the electron is bound to the proton to form a hydrogen atom, its average position is at the proton, but the uncertainty in its position is approximately equal to the radius r of its orbit. The electron's average vector momentum is zero, but its average squared momentum is approximately equal to the squared uncertainty in its momentum, as given by the uncertainty principle. Treating the atom as a one-dimensional system, (a) estimate the uncertainty in the electron's momentum in terms of r. (b) Estimate the electron's kinetic, potential, and total energies in terms of r. (c) The actual value of r is the one that *minimizes the total energy*, resulting in a stable atom. Find that value of r and the resulting total energy. Compare your answer with the predictions of the Bohr theory.

Section 42.5   The Wave Functions for Hydrogen

18. Plot the wave function $\psi_{1s}(r)$ (see Eq. 42.22) and the radial probability density function $P_{1s}(r)$ (see Eq. 42.25) for hydrogen. Let r range from 0 to $1.5a_0$, where a_0 is the Bohr radius.

19. The ground-state wave function for the electron in a hydrogen atom is

$$\psi(r) = \frac{1}{\sqrt{\pi a_0^3}} \, e^{-r/a_0}$$

where r is the radial coordinate of the electron and a_0 is the Bohr radius. (a) Show that the wave function as given is normalized. (b) Find the probability of locating the electron between $r_1 = a_0/2$ and $r_2 = 3a_0/2$.

20. The wave function for an electron in the $2p$ state of hydrogen is

$$\psi_{2p} = \frac{1}{\sqrt{3}(2a_0)^{3/2}} \frac{r}{a_0} e^{-r/2a_0}$$

What is the most likely distance from the nucleus to find an electron in the $2p$ state?

21.   For a spherically symmetric state of a hydrogen atom, the Schrödinger equation in spherical coordinates is

$$-\frac{\hbar^2}{2m}\left(\frac{d^2\psi}{dr^2} + \frac{2}{r}\frac{d\psi}{dr}\right) - \frac{k_e e^2}{r}\psi = E\psi$$

Show that the $1s$ wave function for an electron in hydrogen,

$$\psi(r) = \frac{1}{\sqrt{\pi a_0^3}} \, e^{-r/a_0}$$

satisfies the Schrödinger equation.

22. In an experiment, electrons are fired at a sample of neutral hydrogen atoms and observations are made of how the incident particles scatter. A large set of trials can be thought of as containing 1 000 observations of the electron in the ground state of a hydrogen atom being momentarily at a distance $a_0/2$ from the nucleus. How many times is the atomic electron observed at a distance $2a_0$ from the nucleus in this set of trials?

Section 42.6   Physical Interpretation of the Quantum Numbers

23. List the possible sets of quantum numbers for electrons in (a) the $3d$ subshell and (b) the $3p$ subshell.

24. Calculate the angular momentum for an electron in (a) the $4d$ state and (b) the $6f$ state.

25. If an electron has orbital angular momentum equal to 4.714×10^{-34} J·s, what is the orbital quantum number for the state of the electron?

26. A hydrogen atom is in its fifth excited state, with principal quantum number 6. The atom emits a photon with a wavelength of 1 090 nm. Determine the maximum possible orbital angular momentum of the electron after emission.

27.   How many sets of quantum numbers are possible for an electron for which (a) $n = 1$, (b) $n = 2$, (c) $n = 3$, (d) $n = 4$, and (e) $n = 5$? Check your results to show that they agree with the general rule that the number of sets of quantum numbers for a shell is equal to $2n^2$.

28. Find all possible values of L, L_z, and θ for an electron in a $3d$ state of hydrogen.

29. (a) Find the mass density of a proton, modeling it as a solid sphere of radius 1.00×10^{-15} m. (b) **What If?** Consider a classical model of an electron as a solid sphere with the same density as the proton. Find its radius. (c) Imagine that this electron possesses spin angular momentum $I\omega = \hbar/2$ because of classical rotation about the z axis. Determine the speed of a point on the equator of the electron and (d) compare this speed to the speed of light.

30. An electron is in the N shell. Determine the maximum value the z component of its angular momentum could have.

31. The ρ-meson has a charge of $-e$, a spin quantum number of 1, and a mass 1 507 times that of the electron. **What If?** Imagine that the electrons in atoms were replaced by ρ-mesons. List the possible sets of quantum numbers for ρ-mesons in the $3d$ subshell.

Section 42.7   The Exclusion Principle and the Periodic Table

32. (a) Write out the electronic configuration for the ground state of oxygen ($Z = 8$). (b) Write out a set of possible values for the quantum numbers n, ℓ, m_ℓ, and m_s for each electron in oxygen.

33. As we go down the periodic table, which subshell is filled first, the $3d$ or the $4s$ subshell? Which electronic configuration has a lower energy: $[Ar]3d^4 4s^2$ or $[Ar]3d^5 4s^1$? Which has the greater number of unpaired spins? Identify this element and discuss Hund's rule in this case. (*Note:* The notation [Ar] represents the filled configuration for argon.)

34. Devise a table similar to that shown in Figure 42.19 for atoms containing 11 through 19 electrons. Use Hund's rule and educated guesswork.

35. A certain element has its outermost electron in a $3p$ state. It has valence $+3$, since it has 3 more electrons than a certain noble gas. What element is it?

36. Two electrons in the same atom both have $n = 3$ and $\ell = 1$. (a) List the quantum numbers for the possible states of the atom. (b) **What If?** How many states would be possible if the exclusion principle were inoperative?

37. (a) Scanning through Table 42.4 in order of increasing atomic number, note that the electrons fill the subshells in such a way that those subshells with the lowest values of $n + \ell$ are filled first. If two subshells have the same value of $n + \ell$, the one with the lower value of n is filled first. Using these two rules, write the order in which the subshells are filled through $n + \ell = 7$. (b) Predict the chemical valence for the elements that have atomic numbers 15, 47, and 86, and compare your predictions with the actual valences (which may be found in a chemistry text).

38. For a neutral atom of element 110, what would be the probable ground-state electronic configuration?

39. Review problem. For an electron with magnetic moment $\boldsymbol{\mu}_S$ in a magnetic field **B**, Section 29.3 showed the following. The electron can be in a higher energy state with the z component of its magnetic moment opposite to the field, or in a lower energy state with the z component of its magnetic moment in the direction of the field. The difference in energy between the two states is $2\mu_B B$.

Under high resolution, many spectral lines are observed to be doublets. The most famous of these are the two yellow lines in the spectrum of sodium (the D lines), with wavelengths of 588.995 nm and 589.592 nm. Their existence was explained in 1925 by Goudsmit and Uhlenbeck, who postulated that an electron has intrinsic spin angular momentum. When the sodium atom is excited with its outermost electron in a $3p$ state, the orbital motion of the outermost electron creates a magnetic field. The atom's energy is somewhat different depending on whether the electron is itself spin-up or spin-down in this field. Then the photon energy the atom radiates as it falls back into its ground state depends on the energy of the excited state. Calculate the magnitude of the internal magnetic field mediating this so-called *spin-orbit coupling*.

Section 42.8 More on Atomic Spectra: Visible and X-Ray

40. (a) Determine the possible values of the quantum numbers ℓ and m_ℓ for the He^+ ion in the state corresponding to $n = 3$. (b) What is the energy of this state?

41. If you wish to produce 10.0-nm x-rays in the laboratory, what is the minimum voltage you must use in accelerating the electrons?

42 In x-ray production, electrons are accelerated through a high voltage ΔV and then decelerated by striking a target. Show that the shortest wavelength of an x-ray that can be produced is

$$\lambda_{min} = \frac{1\ 240\ \text{nm} \cdot \text{V}}{\Delta V}$$

43. Use the method illustrated in Example 42.9 to calculate the wavelength of the x-ray emitted from a molybdenum target ($Z = 42$) when an electron moves from the L shell ($n = 2$) to the K shell ($n = 1$).

44. The K series of the discrete x-ray spectrum of tungsten contains wavelengths of 0.018 5 nm, 0.020 9 nm, and 0.021 5 nm. The K-shell ionization energy is 69.5 keV. Determine the ionization energies of the L, M, and N shells. Draw a diagram of the transitions.

45. The wavelength of characteristic x-rays in the K_β line is 0.152 nm. Determine the material in the target.

Section 42.9 Spontaneous and Stimulated Transitions
Section 42.10 Lasers

46. Figure P42.46 shows portions of the energy-level diagrams of the helium and neon atoms. An electrical discharge excites the He atom from its ground state to its excited state of 20.61 eV. The excited He atom collides with a Ne atom in its ground state and excites this atom to the state at 20.66 eV. Lasing action takes place for electron transitions from E_3^* to E_2 in the Ne atoms. From the data in the figure, show that the wavelength of the red He–Ne laser light is approximately 633 nm.

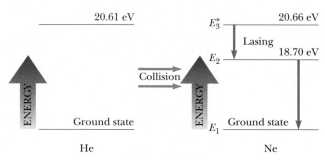

Figure P42.46

47. The carbon dioxide laser is one of the most powerful developed. The energy difference between the two laser levels is 0.117 eV. Determine the frequency and wavelength of the radiation emitted by this laser. In what portion of the electromagnetic spectrum is this radiation?

48. A Nd:YAG laser used in eye surgery emits a 3.00-mJ pulse in 1.00 ns, focused to a spot 30.0 μm in diameter on the retina. (a) Find (in SI units) the power per unit area at the retina. (This quantity is called the *irradiance* in the optics industry.) (b) What energy is delivered to an area of molecular size, taken as a circular area 0.600 nm in diameter?

49. A ruby laser delivers a 10.0-ns pulse of 1.00 MW average power. If the photons have a wavelength of 694.3 nm, how many are contained in the pulse?

50. The number N of atoms in a particular state is called the population of that state. This number depends on the energy of that state and the temperature. In thermal equilibrium the population of atoms in a state of energy E_n is given by a Boltzmann distribution expression

$$N = N_g e^{-(E_n - E_g)/k_B T}$$

where T is the absolute temperature and N_g is the population of the ground state, of energy E_g. For simplicity, we assume that each energy level has only one quantum state associated with it. (a) Before the power is switched on, the neon atoms in a laser are in thermal equilibrium at 27.0°C. Find the equilibrium ratio of the populations of the states $E_3{}^*$ and E_2 shown in Figure 42.28. Lasers operate by a clever artificial production of a "population inversion" between the upper and lower atomic energy states involved in the lasing transition. This means that more atoms are in the upper excited state then in the lower one. Consider the helium–neon laser transition at 632.8 nm. Assume that 2% more atoms occur in the upper state than in the lower. (b) To demonstrate how unnatural such a situation is, find the temperature for which the Boltzmann distribution describes a 2.00% population inversion. (c) Why does such a situation not occur naturally?

51. **Review problem.** A helium–neon laser can produce a green laser beam instead of red. Refer to Figure 42.28, which omits some energy levels between E_2 and E_1. After a population inversion is established, neon atoms will make a variety of downward transitions in falling from the state labeled $E_3{}^*$ down eventually to level E_1. The atoms will emit both red light with a wavelength of 632.8 nm and also green light with a wavelength of 543 nm in a competing transition. Assume the atoms are in a cavity between mirrors designed to reflect the green light with high efficiency but to allow the red light to leave the cavity immediately. Then stimulated emission can lead to the buildup of a collimated beam of green light between the mirrors having a greater intensity than does the red light. A small fraction of the green light is permitted to escape by transmission through one mirror, to constitute the radiated laser beam. The mirrors forming the resonant cavity can be made of layers of silicon dioxide and titanium dioxide. (a) How thick a layer of silicon dioxide, between layers of titanium dioxide, would minimize reflection of the red light? (b) What should be the thickness of a similar but separate layer of silicon dioxide to maximize reflection of the green light?

Additional Problems

52. As the Earth moves around the Sun, its orbits are quantized. (a) Follow the steps of Bohr's analysis of the hydrogen atom to show that the allowed radii of the Earth's orbit are given by

$$r = \frac{n^2 \hbar^2}{G M_S M_E{}^2}$$

where M_S is the mass of the Sun, M_E is the mass of the Earth, and n is an integer quantum number. (b) Calculate the numerical value of n. (c) Find the distance between the orbit for quantum number n and the next orbit out from the Sun corresponding to the quantum number $n + 1$. Discuss the significance of your results.

53. **LENINGRAD, 1930**—Four years after publication of the Schrödinger equation, Lev Davidovich Landau, age 23,

solved the equation for a charged particle moving in a uniform magnetic field. A single electron moving perpendicular to a field **B** can be considered as a model atom without a nucleus, or as the irreducible quantum limit of the cyclotron. Landau proved that its energy is quantized in uniform steps of $e\hbar B/m_e$. **HARVARD, 1999**—Gerald Gabrielse traps a single electron in an evacuated centimeter-size metal can cooled to a temperature of 80 mK. In a magnetic field of magnitude 5.26 T, the electron circulates for hours in its lowest energy level, generating a measurable signal as it moves. (a) Evaluate the size of a quantum jump in the electron's energy. (b) For comparison, evaluate $k_B T$ as a measure of the energy available to the electron in blackbody radiation from the walls of its container. (c) Microwave radiation can be introduced to excite the electron. Calculate the frequency and wavelength of the photon that the electron absorbs as it jumps to its second energy level. Measurement of the resonant absorption frequency verifies the theory and permits precise determination of properties of the electron.

54. Example 42.4 calculates the most probable value and the average value for the radial coordinate r of the electron in the ground state of a hydrogen atom. **What If?** For comparison with these modal and mean values, find the median value of r. Proceed as follows. (a) Derive an expression for the probability, as a function of r, that the electron in the ground state of hydrogen will be found outside a sphere of radius r centered on the nucleus. (b) Make a graph of the probability as a function of r/a_0. Choose values of r/a_0 ranging from 0 to 4.00 in steps of 0.250. (c) Find the value of r for which the probability of finding the electron outside a sphere of radius r is equal to the probability of finding the electron inside this sphere. You must solve a transcendental equation numerically, and your graph is a good starting point.

55. The positron is the antiparticle to the electron. It has the same mass and a positive electric charge of the same magnitude as that of the electron. Positronium is a hydrogen-like atom consisting of a positron and an electron revolving around each other. Using the Bohr model, find the allowed distances between the two particles and the allowed energies of the system.

56. **Review problem.** (a) How much energy is required to cause an electron in hydrogen to move from the $n = 1$ state to the $n = 2$ state? (b) Suppose the electron gains this energy through collisions among hydrogen atoms at a high temperature. At what temperature would the average atomic kinetic energy $3k_B T/2$, where k_B is the Boltzmann constant, be great enough to excite the electron?

57. *An example of the correspondence principle.* Use Bohr's model of the hydrogen atom to show that when the electron moves from the state n to the state $n - 1$, the frequency of the emitted light is

$$f = \left(\frac{2\pi^2 m_e k_e{}^2 e^4}{h^3 n^2} \right) \frac{2n - 1}{(n - 1)^2}$$

Show that as $n \to \infty$, this expression varies as $1/n^3$ and reduces to the classical frequency one expects the atom to emit. (*Suggestion:* To calculate the classical frequency, note that the frequency of revolution is $v/2\pi r$, where r is given by Eq. 42.10.)

58. Astronomers observe a series of spectral lines in the light from a distant galaxy. On the hypothesis that the lines form the Lyman series for a (new?) one-electron atom, they start to construct the energy-level diagram shown in Figure P42.58, which gives the wavelengths of the first four lines and the short-wavelength limit of this series. Based on this information, calculate (a) the energies of the ground state and first four excited states for this one-electron atom, and (b) the wavelengths of the first three lines and the short-wavelength limit in the Balmer series for this atom. (c) Show that the wavelengths of the first four lines and the short wavelength limit of the Lyman series for the hydrogen atom are all 60.0% of the wavelengths for the Lyman series in the one-electron atom described in part (b). (d) Based on this observation, explain why this atom could be hydrogen.

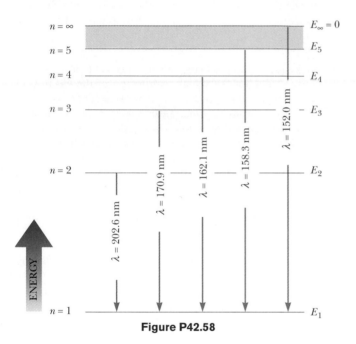

Figure P42.58

59. Suppose a hydrogen atom is in the $2s$ state, with its wave function given by Equation 42.26. Taking $r = a_0$, calculate values for (a) $\psi_{2s}(a_0)$, (b) $|\psi_{2s}(a_0)|^2$, and (c) $P_{2s}(a_0)$.

60. The states of matter are solid, liquid, gas, and plasma. Plasma can be described as a gas of charged particles, or a gas of ionized atoms. Most of the matter in the Solar System is plasma (throughout the interior of the Sun). In fact, most of the matter in the Universe is plasma; so is a candle flame. Use the information in Figure 42.20 to make an order-of-magnitude estimate for the temperature to which a typical chemical element must be raised to turn into plasma by ionizing most of the atoms in a sample. Explain your reasoning.

61. A pulsed ruby laser emits light at 694.3 nm. For a 14.0-ps pulse containing 3.00 J of energy, find (a) the physical length of the pulse as it travels through space and (b) the number of photons in it. (c) The beam has a circular cross section of diameter 0.600 cm. Find the number of photons per cubic millimeter.

62. A pulsed laser emits light of wavelength λ. For a pulse of duration Δt having energy E, find (a) the physical length of the pulse as it travels through space and (b) the number of photons in it. (c) The beam has a circular cross section having diameter d. Find the number of photons per unit volume.

63. Assume that three identical uncharged particles of mass m and spin $1/2$ are contained in a one-dimensional box of length L. What is the ground-state energy of this system?

64. The force on a magnetic moment μ_z in a nonuniform magnetic field B_z is given by $F_z = \mu_z (dB_z / dz)$. If a beam of silver atoms travels a horizontal distance of 1.00 m through such a field and each atom has a speed of 100 m/s, how strong must be the field gradient dB_z / dz in order to deflect the beam 1.00 mm?

65. (a) Show that the most probable radial position for an electron in the $2s$ state of hydrogen is $r = 5.236a_0$. (b) Show that the wave function given by Equation 42.26 is normalized.

66. Suppose the ionization energy of an atom is 4.10 eV. In the spectrum of this same atom, we observe emission lines with wavelengths 310 nm, 400 nm, and 1 377.8 nm. Use this information to construct the energy-level diagram with the fewest levels. Assume that the higher levels are closer together.

67. An electron in chromium moves from the $n = 2$ state to the $n = 1$ state without emitting a photon. Instead, the excess energy is transferred to an outer electron (one in the $n = 4$ state), which is then ejected by the atom. (This is called an Auger [pronounced 'ohjay'] process, and the ejected electron is referred to as an Auger electron.) Use the Bohr theory to find the kinetic energy of the Auger electron.

68. In interstellar space, atomic hydrogen produces the sharp spectral line called the 21-cm radiation, which astronomers find most helpful in detecting clouds of hydrogen between stars. This radiation is useful because it is the only signal cold hydrogen emits and because interstellar dust that obscures visible light is transparent to these radio waves. The radiation is not generated by an electron transition between energy states characterized by different values of n. Instead, in the ground state ($n = 1$), the electron and proton spins may be parallel or antiparallel, with a resultant slight difference in these energy states. (a) Which condition has the higher energy? (b) More precisely, the line has wavelength 21.11 cm. What is the energy difference between the states? (c) The average lifetime in the excited state is about 10^7 yr. Calculate the associated uncertainty in energy of the excited energy level.

69. For hydrogen in the $1s$ state, what is the probability of finding the electron farther than $2.50a_0$ from the nucleus?

70. All atoms have the same size, to an order of magnitude. (a) To show this, estimate the diameters for aluminum (with molar mass 27.0 g/mol and density 2.70 g/cm^3) and uranium (molar mass 238 g/mol and density 18.9 g/cm^3). (b) What do the results imply about the wave functions for inner-shell electrons as we progress to higher and higher atomic mass atoms? (*Suggestion:* The molar volume is approximately $D^3 N_A$, where D is the atomic diameter and N_A is Avogadro's number.)

71. In the technique known as electron spin resonance (ESR), a sample containing unpaired electrons is placed in a magnetic field. Consider the simplest situation, in which only one electron is present and therefore only two energy states are possible, corresponding to $m_s = \pm\frac{1}{2}$. In ESR, the absorption of a photon causes the electron's spin magnetic moment to flip from the lower energy state to the higher energy state. According to Section 29.3, the change in energy is $2\mu_B B$. (The lower energy state corresponds to the case where the z component of the magnetic moment $\boldsymbol{\mu}_{spin}$ is aligned with the magnetic field, and the higher energy state is the case where the z component of $\boldsymbol{\mu}_{spin}$ is aligned opposite to the field.) What is the photon frequency required to excite an ESR transition in a 0.350-T magnetic field?

72. Show that the wave function for an electron in the $2s$ state in hydrogen

$$\psi_{2s}(r) = \frac{1}{4\sqrt{2\pi}}\left(\frac{1}{a_0}\right)^{3/2}\left(2 - \frac{r}{a_0}\right)e^{-r/2a_0}$$

satisfies the spherically symmetric Schrödinger equation given in Problem 21.

73. **Review problem.** Steven Chu, Claude Cohen-Tannoudji, and William Phillips received the 1997 Nobel Prize in physics for "the development of methods to cool and trap atoms with laser light." One part of their work was with a beam of atoms (mass $\sim 10^{-25}$ kg) that move at a speed on the order of 1 km/s, similar to the speed of molecules in air at room temperature. An intense laser light beam tuned to a visible atomic transition (assume 500 nm) is directed straight into the atomic beam. That is, the atomic beam and light beam are traveling in opposite directions. An atom in the ground state immediately absorbs a photon. Total system momentum is conserved in the absorption process. After a lifetime on the order of 10^{-8} s, the excited atom radiates by spontaneous emission. It has an equal probability of emitting a photon in any direction. Thus, the average "recoil" of the atom is zero over many absorption and emission cycles. (a) Estimate the average deceleration of the atomic beam. (b) What is the order of magnitude of the distance over which the atoms in the beam will be brought to a halt?

74. Find the average (expectation) value of $1/r$ in the $1s$ state of hydrogen. It is given by

$$\langle 1/r \rangle = \int_{\text{all space}} |\psi|^2\,(1/r)\,dV = \int_0^\infty P(r)\,(1/r)4\pi r^2\,dr$$

Is the result equal to the inverse of the average value of r?

Answers to Quick Quizzes

42.1 (c). Because the energy of 10.5 eV does not correspond to raising the atom from the ground state to an allowed excited state, there is no interaction between the photon and the atom.

42.2 (a), (c). As the electrons strike the atom, they can give up any amount of energy between 0 and 10.5 eV, unlike the photons in Quick Quiz 42.1, which must give up all of their energy in one interaction. Thus, those electrons that undergo the appropriate collision with the atom can transfer 13.606 eV − 3.401 eV = 10.205 eV to the atom and excite it to the $n = 2$ state. Those electrons that do not make the appropriate collision will transfer only enough kinetic energy to the atom as a whole to satisfy conservation of momentum in the collision, without raising the atom to an excited state.

42.3 (a). The longest-wavelength photon is associated with the lowest energy transition, which is $n = 3$ to $n = 2$.

42.4 (b). The number of subshells is the same as the number of allowed values of ℓ. The allowed values of ℓ for $n = 4$ are $\ell = 0, 1, 2,$ and 3, so there are four subshells.

42.5 (a). Five values (0, 1, 2, 3, 4) of ℓ and (b) nine different values (−4, −3, −2, −1, 0, 1, 2, 3, 4) of m_ℓ as follows:

ℓ	m_ℓ
0	0
1	$-1, 0, 1$
2	$-2, -1, 0, 1, 2$
3	$-3, -2, -1, 0, 1, 2, 3$
4	$-4, -3, -2, -1, 0, 1, 2, 3, 4$

42.6 False. If the energy of the hydrogen atom were proportional to n (or any power of n), the energy would become infinite as n grows to infinity. But the energy of the atom is *inversely* proportional to n^2. Thus, as n increases to very large values, the energy of the atom approaches zero from the negative side. As a result, the maximum frequency of emitted radiation approaches a value determined by the difference in energy between zero and the (negative) energy of the ground state.

42.7 The vector model for $\ell = 1$ is shown below:

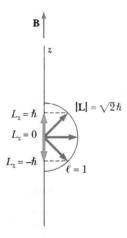

42.8 Argon, neon, helium. The higher the value of Z, the closer to zero is the energy associated with the outermost electron and the smaller is the ionization energy.

42.9 Final: M. Initial: O (because the subscript β indicates that the initial shell is the second shell higher than M).

42.10 (c). The wavelengths of the characteristic x-rays are determined by the separation between energy levels in the atoms of the target, which is unrelated to the energy with which electrons are fired at the target. The only dependence is that the incoming electrons must have enough energy to eject an atomic electron from an inner shell.

42.11 True. If the electrons arrive at the target with very low energy, atomic electrons cannot be ejected and characteristic x-rays do not appear. Because the incoming electrons experience accelerations, the continuous spectrum appears.

Chapter 43

Molecules and Solids

CHAPTER OUTLINE

43.1 Molecular Bonds

43.2 Energy States and Spectra of Molecules

43.3 Bonding in Solids

43.4 Free-Electron Theory of Metals

43.5 Band Theory of Solids

43.6 Electrical Conduction in Metals, Insulators, and Semiconductors

43.7 Semiconductor Devices

43.8 Superconductivity

▲ An understanding of the physics of solids has led to the technology of integrated circuits, found in countless electronic devices used by consumers in today's society. In this photograph, the microchip sitting on a fingertip contains millions of electrical components. (Bruce Dale/Getty Images)

The most random atomic arrangement, that of a gas, was well understood in the 1800s, as we discussed in Chapter 21. In a crystalline solid, the atoms are not randomly arranged, but form a regular array. The symmetry of the arrangement of atoms both stimulated and allowed rapid progress in the field of solid-state physics in the twentieth century. Recently, our understanding of liquids and amorphous solids has advanced. (In an amorphous solid, the atoms do not form a regular array.) The recent interest in the physics of low-cost amorphous materials has been driven by their use in such devices as solar cells, memory elements, and fiber-optic waveguides.

In this chapter, we begin by studying the aggregates of atoms known as molecules. We describe the bonding mechanisms in molecules, the various modes of molecular excitation, and the radiation emitted or absorbed by molecules. We then take the next logical step and show how molecules combine to form solids. Then, by examining their energy-level structure, we explain the differences between insulating, conducting, semiconducting, and superconducting materials. The chapter also includes discussions of semiconducting junctions and several semiconductor devices.

43.1   Molecular Bonds

The bonding mechanisms in a molecule are fundamentally due to electric forces between atoms (or ions), but these forces are more complex than the simple Coulomb attraction or repulsion between single charges that we studied in Chapter 23. If the atoms in a molecule are considered as a system, the forces between components of the system are related to a potential energy function. A stable molecule would be expected at a configuration for which the potential energy function has its minimum value. (See Section 8.6.)

A potential energy function that can be used to model a molecule should account for two known features of molecular bonding:

- The force between atoms is repulsive at very small separation distance. When two atoms are brought close to each other, some of their electron shells overlap, resulting in repulsion between the shells. This repulsion is partly electrostatic in origin and partly the result of the exclusion principle. Because all electrons must obey the exclusion principle, some electrons in the overlapping shells are forced into higher energy states, and the system energy increases, as if a repulsive force existed between the atoms.

- At somewhat larger separations, the force between atoms is attractive. If this were not true, the atoms in a molecule would not be bound together. For many molecules, the attractive force is due to dipole–dipole interaction between charge distributions within the atoms of the molecule. We found in Example 23.6 that an electric dipole possesses an electric field, despite the fact that it carries no net charge. The electric fields of two dipoles will interact, resulting in a force between the dipoles.

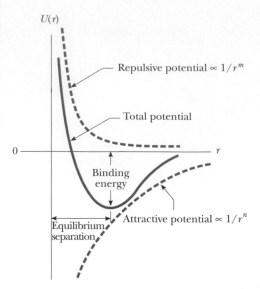

Figure 43.1 Total potential energy as a function of internuclear separation distance for a system of two atoms.

Taking into account these two features, the potential energy for a system of two atoms can be represented by an expression of the form

$$U(r) = -\frac{A}{r^n} + \frac{B}{r^m} \tag{43.1}$$

where r is the internuclear separation distance between the two atoms, and n and m are small integers. The parameter A is associated with the attractive force and B with the repulsive force. Example 8.11 on page 237 gives one common model for such a potential energy function, the Lennard–Jones potential.

Potential energy versus internuclear separation distance for a two-atom system is graphed in Figure 43.1. At large separation distances between the two atoms, the slope of the curve is positive, corresponding to a net attractive force. At the equilibrium separation distance, the attractive and repulsive forces just balance. At this point the potential energy has its minimum value, and the slope of the curve is zero.

A complete description of the bonding mechanisms in molecules is highly complex because bonding involves the mutual interactions of many particles. In this section, therefore, we discuss only some simplified models: ionic bonding, covalent bonding, van der Waals bonding, and hydrogen bonding.

Ionic Bonding

When two atoms combine in such a way that one or more outer electrons are transferred from one atom to the other, the bond formed is called an **ionic bond.** Ionic bonds are fundamentally caused by the Coulomb attraction between oppositely charged ions.

A familiar example of an ionically bonded solid is sodium chloride, NaCl, which is common table salt. Sodium, which has the electronic configuration $1s^2 2s^2 2p^6 3s^1$, is ionized relatively easily, giving up its $3s$ electron to form a Na^+ ion. The energy required to ionize the atom to form Na^+ is 5.1 eV. Chlorine, which has the electronic configuration $1s^2 2s^2 2p^5$, is one electron short of the filled-shell structure of argon. If we compare the energy of a system of a free electron and a Cl atom to one in which the electron joins the atom to make the Cl^- ion, we find that the energy of the ion is lower. When the electron makes a transition from the $E = 0$ state to the negative energy state associated with the available shell in the atom, energy is released. This amount of energy is called the **electron affinity** of the atom. For chlorine, the electron affinity is 3.7 eV. Therefore, the energy required to form Na^+ and Cl^- from isolated atoms is $5.1 - 3.7 = 1.4$ eV. It costs 5.1 eV to remove the electron from the Na atom but you gain 3.7 eV of it back when that electron is allowed to join with the Cl atom.

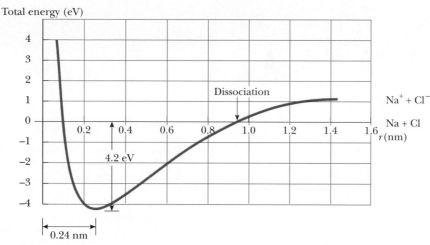

Figure 43.2 Total energy versus internuclear separation distance for Na^+ and Cl^- ions. The horizontal axis is labeled Na + Cl because we define zero energy as that for the system of neutral sodium and chlorine *atoms*. The asymptote of the curve for large values of r is marked $Na^+ + Cl^-$ because that is the energy of the system of sodium and chloride *ions*.

Now imagine that these two charged ions interact with one another to form a NaCl "molecule."[1] The total energy of the NaCl molecule versus internuclear separation distance is graphed in Figure 43.2. At very large separation distances, the energy of the system of ions is 1.4 eV, as calculated above. The total energy has a minimum value of -4.2 eV at the equilibrium separation distance, which is about 0.24 nm. This means that the energy required to break the Na^+–Cl^- bond and form neutral sodium and chlorine atoms, called the **dissociation energy,** is 4.2 eV. Notice that the energy of the molecule is lower than that of the system of two neutral atoms. Consequently, we say that it is **energetically favorable** for the molecule to form—if a lower energy state of a system exists, the system will tend to emit energy in order to achieve this lower energy state. The system of neutral sodium and chlorine atoms can reduce its total energy by transferring energy out of the system (by electromagnetic radiation, for example) and forming the NaCl molecule.

Covalent Bonding

A **covalent bond** between two atoms is one in which electrons supplied by either one or both atoms are shared by the two atoms. Many diatomic molecules, such as H_2, F_2, and CO, owe their stability to covalent bonds. We will describe covalent bonds by means of atomic wave functions, focusing on the bonds between hydrogen atoms, for which we saw the ground-state wave function in Chapter 42:

$$\psi_{1s}(r) = \frac{1}{\sqrt{\pi a_0^3}} e^{-r/a_0}$$

This wave function is graphed in Figure 43.3a for two hydrogen atoms that are far apart. Notice that there is very little overlap of the wave functions $\psi_1(r)$ for atom 1, located at $r = 0$, and $\psi_2(r)$ for atom 2, located some distance away. Suppose now we bring the two atoms close together. As this happens, their wave functions overlap and form the compound wave function $\psi_1(r) + \psi_2(r)$ shown in Figure 43.3b. Notice that the probability amplitude is larger between the atoms than it is on either side of the combination of atoms. As a result, the probability is higher that the electrons associated with the atoms will be located between the atoms than on the outer regions of the system. Consequently, the average position of negative charge in the system is halfway

 PITFALL PREVENTION

43.1 Ionic and Covalent Bonds

In practice, these descriptions of ionic and covalent bonds represent extreme ends of a spectrum of bonds involving electron transfer. In a real bond, the electron may not be *completely* transferred as in an ionic bond or *equally* shared as in a covalent bond. Thus, real bonds lie somewhere between these extremes.

---

[1]   NaCl does not exist as an isolated molecule. In the solid state, NaCl forms a crystalline array of ions as described in Section 43.3. In the liquid state, or in solution with water, the Na^+ and Cl^- ions dissociate and are free to move relative to each other.

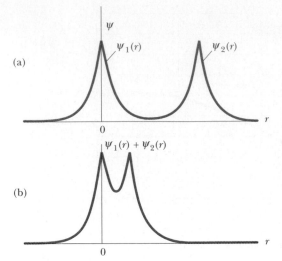

Active Figure 43.3 Ground-state wave functions $\psi_1(r)$ and $\psi_2(r)$ for two atoms making a covalent bond. (a) The atoms are far apart and their wave functions overlap minimally. (b) The atoms are close together, forming a composite wave function $\psi_1(r) + \psi_2(r)$ for the system. The probability amplitude for an electron to be between the atoms is high.

between the atoms. This can be modeled as if there were a fixed negative charge between the atoms, exerting attractive Coulomb forces on both nuclei. The result is an overall attractive force between the atoms, resulting in a covalent bond.

Because of the exclusion principle, the two electrons in the ground state of H_2 must have antiparallel spins. Also because of the exclusion principle, if a third H atom is brought near the H_2 molecule, the third electron would have to occupy a higher energy level, which is an energetically unfavorable situation. For this reason, the H_3 molecule is not stable and does not form.

Van der Waals Bonding

Ionic and covalent bonds occur between atoms to form molecules or ionic solids, so they can be described as bonds *within* molecules. Two additional types of bonds, van der Waals bonds and hydrogen bonds, can occur *between* molecules.

We might expect that two neutral molecules would not interact by means of the electric force because they each have zero net charge. We find, however, that they are attracted to each other by weak electrostatic forces called **van der Waals forces.** Likewise, atoms that do not form ionic or covalent bonds are attracted to each other by van der Waals forces. Inert gases, for example, because of their filled shell structure, do not generally form molecules. Because of van der Waals forces, however, at sufficiently low temperatures at which thermal excitations are negligible, inert gases first condense to liquids and then solidify (with the exception of helium, which does not solidify at atmospheric pressure).

The van der Waals force is due to the fact that, while being electrically neutral, a molecule has a charge distribution with positive and negative centers at different positions in the molecule. As a result, the molecule may act as an electric dipole. Because of the dipole electric fields, two molecules can interact such that there is an attractive force between them.

There are three types of van der Waals forces. The first type, called the *dipole–dipole force,* is an interaction between two molecules each having a permanent electric dipole moment—for example, polar molecules such as HCl have permanent electric dipole moments and attract other polar molecules.

The second type, the *dipole–induced dipole force,* results when a polar molecule having a permanent electric dipole moment induces a dipole moment in a nonpolar molecule. In this case, the electric field of the polar molecule creates the dipole moment in the nonpolar molecule, which then results in an attractive force between the molecules.

The third type is called the *dispersion force,* an attractive force that occurs between two nonpolar molecules. In this case, the interaction results from the fact that, although the average dipole moment of a nonpolar molecule is zero, the average of the square of the dipole moment is nonzero because of charge fluctuations. Two non-polar molecules near each other tend to have dipole moments that are correlated in time so as to produce an attractive van der Waals force.

Hydrogen Bonding

Because hydrogen has only one electron, it is expected to form a covalent bond with only one other atom within a molecule. A hydrogen atom in a given molecule can also form a second type of bond between molecules called a **hydrogen bond.** Let us use the water molecule H_2O as an example. In the two covalent bonds in this molecule, the electrons from the hydrogen atoms are more likely to be found near the oxygen atom than near the hydrogen atoms. This leaves essentially bare protons at the positions of the hydrogen atoms. This unshielded positive charge can be attracted to the negative end of another polar molecule. Because the proton is unshielded by electrons, the negative end of the other molecule can come very close to the proton to form a bond that is strong enough to form a solid crystalline structure, such as that of ice. The bonds within a water molecule are covalent, but the bonds between water molecules in ice are hydrogen bonds.

The hydrogen bond is relatively weak compared with other chemical bonds—it can be broken with an input energy of about 0.1 eV. Because of this, ice melts at the low temperature of 0°C. Despite the fact that this bond is very weak, however, hydrogen bonding is a critical mechanism responsible for the linking of biological molecules and polymers. For example, in the case of the DNA (deoxyribonucleic acid) molecule, which has a double-helix structure (Fig. 43.4), hydrogen bonds formed by the sharing of a proton between two atoms create linkages between the turns of the helix.

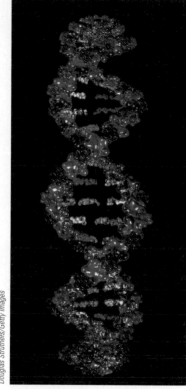

Douglas Struthers/Getty Images

Figure 43.4 DNA molecules are held together by hydrogen bonds.

> **Quick Quiz 43.1** For each of the following, identify the most likely type of bonding that will occur between the atoms or between the molecules. Choose from the following list: ionic, covalent, van der Waals, hydrogen. (a) atoms of krypton (b) potassium and chlorine atoms (c) hydrogen fluoride (HF) molecules (d) chlorine and oxygen atoms in a hypochlorite ion (ClO^-).

43.2   Energy States and Spectra of Molecules

As in the case of atoms, we can study the structure and properties of molecules by examining the radiation they emit or absorb. Before we describe these processes, it is important to understand the various ways of exciting a molecule.

Consider an individual molecule in the gaseous phase of a substance. The energy of the molecule can be divided into four categories: (1) electronic energy, due to the interactions between the molecule's electrons and nuclei; (2) translational energy, due to the motion of the molecule's center of mass through space; (3) rotational energy, due to the rotation of the molecule about its center of mass; and (4) vibrational energy, due to the vibration of the molecule's constituent atoms:

$$E = E_{el} + E_{trans} + E_{rot} + E_{vib}$$

Total energy of a molecule

We explored the role of translational, rotational, and vibrational energy of molecules in determining the molar specific heats of gases in Sections 21.2 and 21.4. Because the translational energy is unrelated to internal structure, this molecular energy is unimportant in interpreting molecular spectra. The electronic energy of a molecule is very complex because it involves the interaction of many charged particles, but various techniques have been developed to approximate its values. While the electronic energies can be studied, significant information about a molecule can be determined by analyzing its rotational and vibrational energy states, which give spectral lines in the microwave and infrared regions of the electromagnetic spectrum, respectively.

Rotational Motion of Molecules

Let us consider the rotation of a molecule around its center of mass, confining our discussion to the diatomic case (Fig. 43.5a) but noting that the same ideas can be extended to polyatomic molecules. A diatomic molecule aligned along an x axis has only two rotational degrees of freedom, corresponding to rotations about the y and z axes. If ω is the angular frequency of rotation about one of these axes, the rotational kinetic energy of the molecule about that axis can be expressed in the form

$$E_{\text{rot}} = \tfrac{1}{2} I \omega^2 \tag{43.2}$$

In this equation, I is the moment of inertia of the molecule, given by

Moment of inertia for a diatomic molecule

$$I = \left(\frac{m_1 m_2}{m_1 + m_2}\right) r^2 = \mu r^2 \tag{43.3}$$

where m_1 and m_2 are the masses of the atoms that form the molecule, r is the atomic separation, and μ is the **reduced mass** of the molecule (see Example 41.7):

Reduced mass of a diatomic molecule

$$\mu = \frac{m_1 m_2}{m_1 + m_2} \tag{43.4}$$

The magnitude of the molecule's angular momentum is $L = I\omega$, which classically can have any value. Quantum mechanics, however, restricts the molecule to certain quantized rotational frequencies such that the angular momentum of the molecule has the values[2]

Allowed values of rotational angular momentum

$$L = \sqrt{J(J+1)}\, \hbar \qquad J = 0, 1, 2, \ldots \tag{43.5}$$

where J is an integer called the **rotational quantum number.** Combining Equations 43.5 and 43.2, we obtain an expression for the allowed values of the rotational

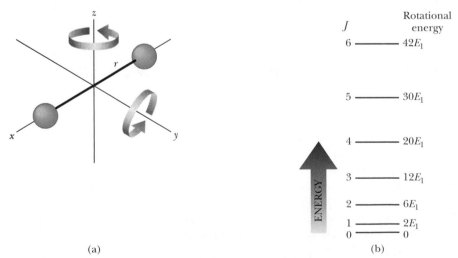

At the Active Figures link at http://www.pse6.com, you can adjust the distance between the atoms and choose the initial rotational energy state of the molecule. Then observe transitions of the molecule to lower energy states.

(a)   (b)

Active Figure 43.5 (a) A diatomic molecule oriented along the x axis has two rotational degrees of freedom, corresponding to rotation about the y and z axes. (b) Allowed rotational energies of a diatomic molecule calculated with Equation 43.6, where $E_1 = \hbar^2/2I$.

[2] Equation 43.5 is similar to Equation 42.27 for orbital angular momentum in an atom. In fact, the relationship between the magnitude of the angular momentum of a system and the associated quantum number is the same as it is in these equations for *any* system that exhibits rotation, as long as the potential energy function for the system is spherically symmetric.

kinetic energy of the molecule:

$$E_{rot} = \tfrac{1}{2} I\omega^2 = \frac{1}{2I}(I\omega)^2 = \frac{L^2}{2I} = \frac{\left(\sqrt{J(J+1)}\,\hbar\right)^2}{2I}$$

$$E_{rot} = \frac{\hbar^2}{2I} J(J+1) \qquad J = 0, 1, 2, \ldots \qquad (43.6)$$

Allowed values of rotational energy

This result shows that **the rotational energy of the molecule is quantized and depends on its moment of inertia.** The allowed rotational energies of a diatomic molecule are plotted in Figure 43.5b. As the quantum number J goes up, the states become farther apart, as displayed earlier for rotational energy levels in Figure 21.8.

For most molecules, transitions between adjacent rotational energy levels result in radiation that lies in the microwave range of frequencies ($f \sim 10^{11}$ Hz). When a molecule absorbs a microwave photon, the molecule jumps from a lower rotational energy level to a higher one. The allowed rotational transitions of linear molecules are regulated by the selection rule $\Delta J = \pm 1$. Given this selection rule, all absorption lines in the spectrum of a linear molecule correspond to energy separations equal to $E_J - E_{J-1}$, where $J = 1, 2, 3, \ldots$. From Equation 43.6, we see that the energies of the allowed transitions are given by the condition

$$\Delta E = E_J - E_{J-1} = \frac{\hbar^2}{2I}[J(J+1) - (J-1)J]$$

$$\Delta E = \frac{\hbar^2}{I}J = \frac{h^2}{4\pi^2 I}J \qquad J = 1, 2, 3, \ldots \qquad (43.7)$$

Energy separation between adjacent rotational levels

where J is the rotational quantum number of the higher energy state. Because $\Delta E = hf$, where f is the frequency of the absorbed photon, we see that the allowed frequency for the transition $J = 0$ to $J = 1$ is $f_1 = h/4\pi^2 I$. The frequency corresponding to the $J = 1$ to $J = 2$ transition is $2f_1$, and so on. These predictions are in excellent agreement with the observed frequencies.

The wavelengths and frequencies for the microwave absorption spectrum of the carbon monoxide molecule are given in Table 43.1. From these data, we can evaluate the moment of inertia and bond length of the molecule (Example 43.1).

Quick Quiz 43.2 A gas of diatomic molecules is absorbing electromagnetic radiation over a wide range of frequencies. Molecule 1 is in the $J = 0$ rotation state and makes a transition to the $J = 1$ state. Molecule 2 is in the $J = 2$ state and makes a transition to the $J = 3$ state. The ratio of the frequency of the photon that excited molecule 2 to that of the photon that excited molecule 1 is (a) 1 (b) 2 (c) 3 (d) 4 (e) impossible to determine.

Table 43.1

Several Rotational Transitions of the CO Molecule		
Rotational Transition	**Wavelength of Absorbed Photon (m)**	**Frequency of Absorbed Photon (Hz)**
$J = 0 \rightarrow J = 1$	2.60×10^{-3}	1.15×10^{11}
$J = 1 \rightarrow J = 2$	1.30×10^{-3}	2.30×10^{11}
$J = 2 \rightarrow J = 3$	8.77×10^{-4}	3.46×10^{11}
$J = 3 \rightarrow J = 4$	6.50×10^{-4}	4.61×10^{11}

From G. M. Barrows, *The Structure of Molecules,* New York, W. A. Benjamin, 1963.

Example 43.1   Rotation of the CO Molecule   Interactive

The $J = 0$ to $J = 1$ rotational transition of the CO molecule occurs at a frequency of 1.15×10^{11} Hz.

(A) Use this information to calculate the moment of inertia of the molecule.

Solution From Equation 43.7, we see that the energy difference between the $J = 0$ and $J = 1$ rotational levels is $h^2/4\pi^2 I$. Equating this ΔE value to the energy of the absorbed photon, we have

$$\Delta E = \frac{h^2}{4\pi^2 I} = hf$$

Solving for I gives

$$I = \frac{h}{4\pi^2 f} = \frac{6.626 \times 10^{-34}\,\text{J}\cdot\text{s}}{4\pi^2 (1.15 \times 10^{11}\,\text{s}^{-1})}$$

$$= \boxed{1.46 \times 10^{-46}\,\text{kg}\cdot\text{m}^2}$$

(B) Calculate the bond length of the molecule.

Solution We can use Equation 43.3 to calculate the bond length, but we first need to know the value for the reduced mass μ of the CO molecule:

$$\mu = \frac{m_1 m_2}{m_1 + m_2} = \frac{(12\,\text{u})(16\,\text{u})}{12\,\text{u} + 16\,\text{u}} = 6.86\,\text{u}$$

$$\mu = (6.86\,\text{u})\left(\frac{1.66 \times 10^{-27}\,\text{kg}}{1\,\text{u}}\right) = 1.14 \times 10^{-26}\,\text{kg}$$

where we have used the fact that $1\,\text{u} = 1.66 \times 10^{-27}$ kg.

Substituting this value and the result of part (A) into Equation 43.3 and solving for r, we obtain

$$r = \sqrt{\frac{I}{\mu}} = \sqrt{\frac{1.46 \times 10^{-46}\,\text{kg}\cdot\text{m}^2}{1.14 \times 10^{-26}\,\text{kg}}}$$

$$= 1.13 \times 10^{-10}\,\text{m} = \boxed{0.113\,\text{nm}}$$

What If? What if another photon of frequency 1.15×10^{11} Hz is incident on the CO molecule while it is in the $J = 1$ state? What happens?

Answer Because the rotational quantum states are not equally spaced in energy, the $J = 1$ to $J = 2$ transition does not have the same energy as the $J = 0$ to $J = 1$ transition. Thus, the molecule will *not* be excited to the $J = 2$ state. Two possibilities are available. First, the photon could pass by the molecule with no interaction. Second, the photon could induce a stimulated emission, similar to that for atoms and discussed in Section 42.9. In this case, the molecule makes a transition back to the $J = 0$ state, and the original photon and a second identical photon leave the scene of the interaction.

 At the Interactive Worked Example link at **http://www.pse6.com,** *practice evaluating the moment of inertia and the bond length from spectroscopic data of diatomic molecules.*

Vibrational Motion of Molecules

If we consider a molecule to be a flexible structure in which the atoms are bonded together by "effective springs" as shown in Figure 43.6a, we can model the molecule as a simple harmonic oscillator, as long as the atoms in the molecule are not too far from their equilibrium positions. Figure 43.6b shows a plot of potential energy versus atomic separation for the molecule in Figure 43.6a, where r_0 is the equilibrium atomic separation. For separations close to r_0, the shape of the potential energy curve closely

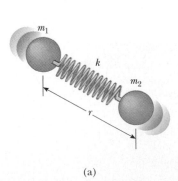

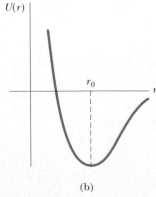

 At the Active Figures link at http://www.pse6.com, *adjust the spring constant and choose the initial vibrational energy state of the molecule. Then observe transitions of the molecule to lower energy states in Figure 43.7.*

(a)   (b)

Active Figure 43.6 (a) Effective-spring model of a diatomic molecule. The vibration is along the molecular axis. (b) Plot of the potential energy of a diatomic molecule versus atomic separation distance, where r_0 is the equilibrium separation distance of the atoms. Compare with Figure 15.12, page 464.

resembles a parabola. Recall from Section 15.3 that the potential energy function for a simple harmonic oscillator is indeed parabolic, varying as the square of the displacement from equilibrium. (See Equation 15.20.)

According to classical mechanics, the frequency of vibration for the system shown in Figure 43.6a is given by Equation 15.14:

$$f = \frac{1}{2\pi} \sqrt{\frac{k}{\mu}} \tag{43.8}$$

where k is the effective spring constant and μ is the reduced mass given by Equation 43.4.

Quantum mechanics predicts that a molecule will vibrate in quantized states as described in Section 41.8. The vibrational motion and quantized vibrational energy can be altered if the molecule acquires energy of the proper value to cause a transition between quantized vibrational states. As discussed in Section 41.8, the allowed vibrational energies are

$$E_{\text{vib}} = (v + \tfrac{1}{2})hf \qquad v = 0, 1, 2, \dots \tag{43.9}$$

where v is an integer called the **vibrational quantum number.** (We used n in Section 41.8 for a general harmonic oscillator, but v is often used for the quantum number when discussing molecular vibrations.) If the system is in the lowest vibrational state, for which $v = 0$, its ground-state energy is $\tfrac{1}{2}hf$. The accompanying vibration is always present, even if the molecule is not excited. In the first excited state, $v = 1$ and the vibrational energy is $\tfrac{3}{2}hf$, and so on.

Substituting Equation 43.8 into Equation 43.9, we obtain the following expression for the allowed vibrational energies:

$$E_{\text{vib}} = (v + \tfrac{1}{2})\frac{h}{2\pi} \sqrt{\frac{k}{\mu}} \qquad v = 0, 1, 2, \dots \tag{43.10}$$

The selection rule for the allowed vibrational transitions is $\Delta v = \pm 1$. From Equation 43.10, we see that the energy difference between any two successive vibrational levels is

$$\Delta E_{\text{vib}} = \frac{h}{2\pi} \sqrt{\frac{k}{\mu}} = hf \tag{43.11}$$

where f is the frequency of the oscillator.

The vibrational energies of a diatomic molecule are plotted in Figure 43.7. At ordinary temperatures, most molecules have vibrational energies corresponding to the $v = 0$ state because the spacing between vibrational states is much greater than $k_B T$ where k_B is Boltzmann's constant and T is the temperature. The molecules are not thermally excited into higher states, as discussed in Example 41.7.

Transitions between vibrational levels are caused by absorption of photons in the infrared region of the spectrum. The photon frequencies corresponding to the $v = 0$ to $v = 1$ transition are listed in Table 43.2 for several diatomic molecules, together with the force constants of the effective springs holding the molecules together. The latter values were calculated using Equation 43.11. The "stiffness" of a bond can be measured by the size of the effective force constant k.

Quick Quiz 43.3 A gas of diatomic molecules is absorbing electromagnetic radiation over a wide range of frequencies. Molecule 1, initially in the $v = 0$ vibrational state, makes a transition to the $v = 1$ state. Molecule 2, initially in the $v = 2$ state, makes a transition to the $v = 3$ state. The ratio of the frequency of the photon that excited molecule 2 to that of the photon that excited molecule 1 is (a) 1 (b) 2 (c) 3 (d) 4 (e) impossible to determine.

Allowed values of vibrational energy

v	Vibrational energy
5	$\frac{11}{2} hf$
4	$\frac{9}{2} hf$
3	$\frac{7}{2} hf$
2	$\frac{5}{2} hf$
1	$\frac{3}{2} hf$
0	$\frac{1}{2} hf$

ΔE_{vib}

ENERGY

Active Figure 43.7 Allowed vibrational energies of a diatomic molecule, where f is the frequency of vibration of the molecule, given by Equation 43.8. The spacings between adjacent vibrational levels are equal if the molecule behaves as a harmonic oscillator.

At the Active Figures link at http://www.pse6.com, adjust the spring constant in Figure 43.6 and choose the initial vibrational energy state of the molecule. Then observe transitions of the molecule to lower energy states.

Table 43.2

Photon Frequency and Effective Spring Force Constant for $v = 0$ to $v = 1$ Transition in Some Diatomic Molecules		
Molecule	**Photon Frequency (Hz)**	**Force Constant k (N/m)**
HF	8.72×10^{13}	970
HCl	8.66×10^{13}	480
HBr	7.68×10^{13}	410
HI	6.69×10^{13}	320
CO	6.42×10^{13}	1 850
NO	5.63×10^{13}	1 530

From G. M. Barrows, *The Structure of Molecules,* New York, W. A. Benjamin, 1963. The k values were calculated from Equation 43.11.

Example 43.2 Vibration of the CO Molecule Interactive

The frequency of the photon that causes the $v = 0$ to $v = 1$ transition in the CO molecule is 6.42×10^{13} Hz. We will ignore any changes in the rotational energy for this example.

(A) Calculate the force constant k for this molecule.

Solution We can use Equation 43.11 and the value $\mu = 1.14 \times 10^{-26}$ kg that we calculated in Example 43.1b:

$$\frac{h}{2\pi}\sqrt{\frac{k}{\mu}} = hf$$

$$k = 4\pi^2 \mu f^2$$

$$= 4\pi^2 (1.14 \times 10^{-26} \text{ kg})(6.42 \times 10^{13} \text{ s}^{-1})^2$$

$$= \boxed{1.85 \times 10^3 \text{ N/m}}$$

(B) What is the maximum classical amplitude of vibration for this molecule in the $v = 0$ vibrational state?

Solution The maximum elastic potential energy in the molecule is $\frac{1}{2}kA^2$, where A is the classical amplitude of

vibration. Equating this maximum energy to the vibrational energy given by Equation 43.10 with $v = 0$, we have

$$\frac{1}{2}kA^2 = \frac{h}{4\pi}\sqrt{\frac{k}{\mu}}$$

$$A = \sqrt{\frac{h}{2\pi}}\left(\frac{1}{\mu k}\right)^{1/4}$$

Substituting the value for k from part (A) and the value for μ gives

$$A = \sqrt{\frac{6.626 \times 10^{-34} \text{ J} \cdot \text{s}}{2\pi}}$$

$$\times \left(\frac{1}{(1.14 \times 10^{-26} \text{ kg})(1.85 \times 10^3 \text{ N/m})}\right)^{1/4}$$

$$= 4.79 \times 10^{-12} \text{ m} = \boxed{0.004\ 79 \text{ nm}}$$

Comparing this result with the bond length of 0.113 nm we calculated in Example 43.1b, we see that the classical amplitude of vibration is about 4% of the bond length.

 At the Interactive Worked Example link at http://www.pse6.com, *practice evaluating the effective force constant and the classical vibration amplitude from spectroscopic data of diatomic molecules.*

Molecular Spectra

In general, a molecule vibrates and rotates simultaneously. To a first approximation, these motions are independent of each other, and so the total energy of the molecule for these motions is the sum of Equations 43.6 and 43.9:

$$E = (v + \tfrac{1}{2})hf + \frac{\hbar^2}{2I}J(J + 1) \tag{43.12}$$

The energy levels of any molecule can be calculated from this expression, and each level is indexed by the two quantum numbers v and J. From these calculations, an energy-level diagram like the one shown in Figure 43.8a can be constructed. For each allowed value of the vibrational quantum number v, there is a complete set of rotational

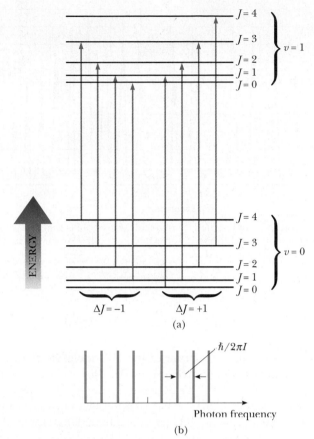

Active Figure 43.8 (a) Absorptive transitions between the $v = 0$ and $v = 1$ vibrational states of a diatomic molecule. The transitions obey the selection rule $\Delta J = \pm 1$ and fall into two sequences, those for $\Delta J = +1$ and those for $\Delta J = -1$. The transition energies are given by Equations 43.13 and 43.14. Compare the energy levels in this figure with those in Figure 21.8 on page 653. (b) Expected lines in the absorption spectrum of a molecule. The lines to the right of the center mark correspond to transitions in which J changes by $+1$; the lines to the left of the center mark correspond to transitions for which J changes by -1. These same lines appear in the emission spectrum.

levels corresponding to $J = 0, 1, 2, \ldots$. The energy separation between successive rotational levels is much smaller than the separation between successive vibrational levels. As noted earlier, most molecules at ordinary temperatures are in the $v = 0$ vibrational state; these molecules can be in various rotational states, as Figure 43.8a shows.

When a molecule absorbs a photon with the appropriate energy, the vibrational quantum number v increases by one unit while the rotational quantum number J either increases or decreases by one unit, as can be seen in Figure 43.8. Therefore, the molecular absorption spectrum consists of two groups of lines: one group to the right of center and satisfying the selection rules $\Delta J = +1$ and $\Delta v = +1$, and the other group to the left of center and satisfying the selection rules $\Delta J = -1$ and $\Delta v = +1$.

The energies of the absorbed photons can be calculated from Equation 43.12:

$$\Delta E = hf + \frac{\hbar^2}{I}(J + 1) \qquad J = 0, 1, 2, \ldots \qquad (\Delta J = +1) \qquad (43.13)$$

$$\Delta E = hf - \frac{\hbar^2}{I}J \qquad J = 1, 2, 3, \ldots \qquad (\Delta J = -1) \qquad (43.14)$$

where J is the rotational quantum number of the *initial* state. Equation 43.13 generates the series of equally spaced lines *higher* than the frequency f, whereas Equation 43.14 generates the series *lower* than this frequency. Adjacent lines are separated in frequency by the fundamental unit $\hbar/2\pi I$. Figure 43.8b shows the expected frequencies in the absorption spectrum of the molecule; these same frequencies appear in the emission spectrum.

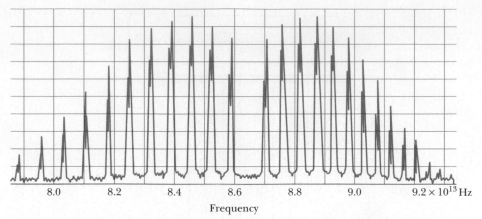

Figure 43.9 Absorption spectrum of the HCl molecule. Each line is split into a doublet because the sample contains two chlorine isotopes that have different masses and therefore different moments of inertia.

The experimental absorption spectrum of the HCl molecule shown in Figure 43.9 follows this pattern very well and reinforces our model. However, one peculiarity is apparent: each line is split into a doublet. This doubling occurs because two chlorine isotopes (see Section 44.1) were present in the sample used to obtain this spectrum. Because the isotopes have different masses, the two HCl molecules have different values of I.

The intensity of the spectral lines in Figure 43.9 follows an interesting pattern, rising first as one moves away from the central gap (at about 8.65×10^{13} Hz, corresponding to the forbidden $J = 0$ to $J = 0$ transition) and then falling. This intensity is determined by a product of two functions of J. The first function corresponds to the number of available states for a given value of J. This function is $2J + 1$, corresponding to the number of values of m_J, the molecular rotation analog to m_ℓ for atomic states. For example, the $J = 2$ state has five substates with five values of m_J ($m_J = -2, -1, 0, 1, 2$), while the $J = 1$ state has only three substates ($m_J = -1, 0, 1$). Thus, on the average and without regard for the second function described below, $\frac{5}{3}$ as many molecules will make the transition from the $J = 2$ state as from the $J = 1$ state.

The second function determining the envelope of the intensity of the spectral lines is the Boltzmann factor, introduced in Section 21.5. The number of molecules in an excited rotational state is given by

$$n = n_0 e^{-\hbar^2 J(J+1)/(2Ik_B T)}$$

where n_0 is the number of molecules in the $J = 0$ state.

Multiplying these factors together indicates that the intensity of spectral lines should be described by a function of J as follows:

Intensity variation in the vibration-rotation spectrum of a molecule

$$I \propto (2J + 1)e^{-\hbar^2 J(J+1)/(2Ik_B T)} \tag{43.15}$$

The factor $(2J + 1)$ increases with J while the exponential second factor decreases. The product of the two factors gives a behavior that describes very well the envelope of the spectral lines in Figure 43.9.

The excitation of rotational and vibrational energy levels is an important consideration in current models for the phenomenon of global warming. For CO_2 molecules, most of the absorption lines are in the infrared portion of the spectrum. Thus, visible light from the Sun is not absorbed by atmospheric CO_2 but instead strikes the Earth's surface, warming it. In turn, the surface of the Earth, being at a much lower temperature than the Sun, emits thermal radiation that peaks in the infrared portion of the electromagnetic spectrum (Section 40.1). This infrared radiation is absorbed by the CO_2 molecules in the air instead of radiating out into space. Thus, atmospheric CO_2 acts like a one-way valve for energy from the Sun and is responsible, along with some other atmospheric molecules, for raising the temperature of the Earth's surface above its value in the absence of an atmosphere. This phenomenon is commonly called the "greenhouse effect." The burning of fossil fuels in today's industrialized society adds more CO_2 to the atmosphere. Many scientists fear that this will increase the absorption of infrared radiation, raising the Earth's temperature further, and may cause substantial climatic changes.

Quick Quiz 43.4 In Figure 43.9, there is a gap between the two sets of peaks. This is because (a) there is no pair of energy levels that are separated by an energy corresponding to this frequency (b) the transition between energy levels that are separated by an energy corresponding to this frequency is not allowed (c) molecules cannot rotate and vibrate simultaneously.

Quick Quiz 43.5 A spectrum like that in Figure 43.9 is taken when the HCl gas is at a much higher temperature. Compared to Figure 43.9, in this new spectrum, the absorption peak with the highest intensity (a) is the same as the one in Figure 43.9 (b) is farther from the gap than in Figure 43.9 (c) is closer to the gap than in Figure 43.9.

Conceptual Example 43.3   Comparing Figures 43.8 and 43.9

In Figure 43.8a, the transitions indicated correspond to spectral lines that are equally spaced, as shown in Figure 43.8b. The actual spectrum in Figure 43.9, however, shows lines that move closer together as the frequency increases. Why does the spacing of the actual spectral lines differ from the diagram in Figure 43.8?

Solution In Figure 43.8, we modeled the rotating diatomic molecule as a rigid object (Chapter 10). In reality, however, as the molecule rotates faster and faster, the effective spring in Figure 43.6a stretches, in order to provide the increased force associated with the larger centripetal acceleration of each atom. As the molecule stretches along its length, its moment of inertia I increases. Thus, in Equation 43.12, the rotational part of the energy expression has an extra dependence on J in the moment of inertia I. Because the increasing moment of inertia is in the denominator, as J increases, the energies do not increase as rapidly with J as indicated in Equation 43.12. With each higher energy level being lower than indicated by Equation 43.12, the energy associated with a transition to that level is smaller, as is the frequency of the absorbed photon. This destroys the even spacing of the spectral lines and gives the uneven spacing seen in Figure 43.9.

43.3 Bonding in Solids

A crystalline solid consists of a large number of atoms arranged in a regular array, forming a periodic structure. Two of the bonding mechanisms described in Section 43.1–ionic and covalent–are appropriate for describing bonds in solids. For example, the ions in the NaCl crystal are ionically bonded, as already noted, and the carbon atoms in the crystal that we call diamond form covalent bonds with one another. The metallic bond described at the end of this section is responsible for the cohesion of copper, silver, sodium, and other solid metals.

Ionic Solids

Many crystals are formed by ionic bonding, in which the dominant interaction between ions is through the Coulomb force. Consider the NaCl crystal in Figure 43.10. Each Na^+ ion has six nearest-neighbor Cl^- ions, and each Cl^- ion has six nearest-neighbor Na^+ ions. Each Na^+ ion is attracted to its six Cl^- neighbors. The corresponding potential

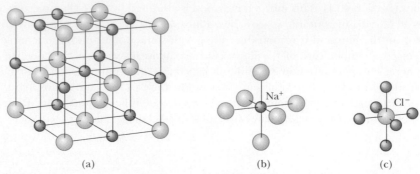

Figure 43.10 (a) Crystalline structure of NaCl. (b) Each positive sodium ion (orange spheres) is surrounded by six negative chloride ions (blue spheres). (c) Each chloride ion is surrounded by six sodium ions.

energy is $-6k_e e^2/r$, where k_e is the Coulomb constant and r is the separation distance between each Na^+ and Cl^-. In addition, there are 12 next nearest-neighbor Na^+ ions at a distance of $\sqrt{2}\,r$ from the Na^+ ion, and these 12 positive ions exert weaker repulsive forces on the central Na^+. Furthermore, beyond these 12 Na^+ ions are more Cl^- ions that exert an attractive force, and so on. The net effect of all these interactions is a resultant negative electric potential energy

$$U_{\text{attractive}} = -\alpha k_e \frac{e^2}{r} \tag{43.16}$$

where α is a dimensionless number known as the **Madelung constant.** The value of α depends only on the particular crystalline structure of the solid. For example, $\alpha = 1.747\,6$ for the NaCl structure. When the constituent ions of a crystal are brought close together, a repulsive force exists because of electrostatic forces and the exclusion principle, as discussed in Section 43.1. The potential energy term B/r^m in Equation 43.1 accounts for this repulsive force. We do not include neighbors other than nearest neighbors here because the repulsive forces occur only for ions that are very close together. (Electron shells must overlap for exclusion-principle effects to become important.) Therefore, we can express the total potential energy of the crystal as

$$U_{\text{total}} = -\alpha k_e \frac{e^2}{r} + \frac{B}{r^m} \tag{43.17}$$

where m in this expression is some small integer.

A plot of total potential energy versus ion separation distance is shown in Figure 43.11. The potential energy has its minimum value U_0 at the equilibrium separation, when $r = r_0$. It is left as a problem for you (Problem 53) to show that

$$U_0 = -\alpha k_e \frac{e^2}{r_0}\left(1 - \frac{1}{m}\right) \tag{43.18}$$

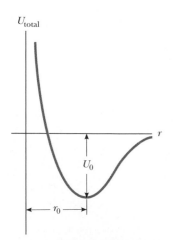

Figure 43.11 Total potential energy versus ion separation distance for an ionic solid, where U_0 is the ionic cohesive energy and r_0 is the equilibrium separation distance between ions.

This minimum energy U_0 is called the **ionic cohesive energy** of the solid, and its absolute value represents the energy required to separate the solid into a collection of isolated positive and negative ions. Its value for NaCl is -7.84 eV per ion pair.

To calculate the **atomic cohesive energy,** which is the binding energy relative to the energy of the neutral atoms, we must add 5.14 eV to the ionic cohesive energy value to account for the transition from Na^+ to Na, and we must subtract 3.61 eV to account for the conversion of Cl^- to Cl. Therefore, the atomic cohesive energy of NaCl is

$$-7.84 \text{ eV} + 5.14 \text{ eV} - 3.61 \text{ eV} = -6.31 \text{ eV}$$

In other words, it requires 6.31 eV of energy per ion pair to separate the solid into isolated neutral atoms of Na and Cl.

Ionic crystals have the following general properties:

- They form relatively stable, hard crystals.

- They are poor electrical conductors because they contain no free electrons. This means that each electron in the solid is bound tightly to one of the ions, so it is not sufficiently mobile to carry current.

- They have high melting points. (The melting point of NaCl is 801°C.)

- They are transparent to visible radiation but absorb strongly in the infrared region. No visible light is absorbed because the shells formed by the electrons in ionic solids are so tightly bound that visible radiation does not possess sufficient energy to promote electrons to the next allowed shell. Infrared radiation is absorbed strongly because the vibrations of the ions have natural resonant frequencies in the low-energy infrared region.

- Many are quite soluble in polar liquids, such as water. The polar solvent molecules exert an attractive electric force on the charged ions, which breaks the ionic bonds and dissolves the solid.

Covalent Solids

Solid carbon, in the form of diamond, is a crystal whose atoms are covalently bonded. Because atomic carbon has the electronic configuration $1s^2 2s^2 2p^2$, it is four electrons short of filling its $n = 2$ shell, which can accommodate eight electrons. Hence, two carbon atoms have a strong attraction for each other, with a cohesive energy of 7.37 eV.

In the diamond structure, each carbon atom is covalently bonded to four other carbon atoms located at four corners of a cube, as shown in Figure 43.12a. To form such a configuration of bonds, one 2s electron of each atom must be promoted to the $2p$ subshell so that the electronic configuration becomes $1s^2 2s^1 2p^3$, which corresponds to a half-filled p subshell. The promotion of this electron requires an energy input of about 4 eV.

The crystalline structure of diamond is shown in Figure 43.12b. Note that each carbon atom forms covalent bonds with four nearest-neighbor atoms. The basic structure of diamond is called tetrahedral (each carbon atom is at the center of a regular tetrahedron), and the angle between the bonds is 109.5°. Other crystals such as silicon and germanium have the same structure.

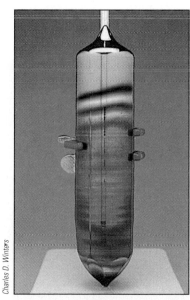

A cylinder of nearly pure crystalline silicon (Si), approximately 25 cm long. Such crystals are cut into wafers and processed to make various semiconductor devices.

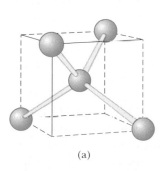

(a)

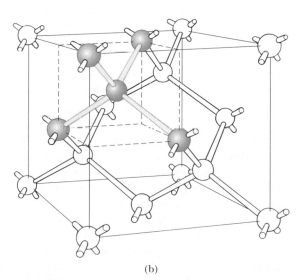

(b)

Figure 43.12 (a) Each carbon atom in a diamond crystal is covalently bonded to four other carbon atoms and forms a tetrahedral structure. (b) The crystal structure of diamond, showing the tetrahedral bond arrangement.

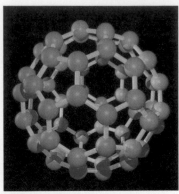

Figure 43.13 Computer rendering of a "buckyball," short for the molecule buckminsterfullerene. These nearly spherical molecular structures that look like soccer balls were named for the inventor of the geodesic dome. This form of carbon, C_{60}, was discovered by astrophysicists while investigating the carbon gas that exists between stars. Scientists are actively studying the properties and potential uses of buckminsterfullerene and related molecules.

Table 43.3

Atomic Cohesive Energies of Some Covalent Solids	
Solid	Cohesive Energy (eV per ion pair)
C (diamond)	7.37
Si	4.63
Ge	3.85
InAs	5.70
SiC	6.15
ZnS	6.32
CuCl	9.24

Carbon is interesting in that it can form several different types of structures. In addition to the diamond structure, it forms graphite, with completely different properties. In this form, the carbon atoms form flat layers with hexagonal arrays of atoms. There is a very weak interaction between the layers. This allows the layers to be removed easily under friction, as occurs in the graphite used in pencil lead.

Carbon atoms can also form a large hollow structure; in this case, the compound is called **buckminsterfullerene** after the famous architect R. Buckminster Fuller, who invented the geodesic dome. The unique shape of this molecule (Fig. 43.13) provides a "cage" to hold other atoms or molecules. Related structures, called "buckytubes" because of their long, narrow cylindrical arrangements of carbon atoms, may provide the basis for extremely strong, yet lightweight materials.

The atomic cohesive energies of some covalent solids are given in Table 43.3. The large energies account for the hardness of covalent solids. Diamond is particularly hard and has an extremely high melting point (about 4 000 K). Covalently bonded solids are usually very hard, have high bond energies and high melting points, and are good electrical insulators.

Metallic Solids

Metallic bonds are generally weaker than ionic or covalent bonds. The outer electrons in the atoms of a metal are relatively free to move throughout the material, and the number of such mobile electrons in a metal is large. The metallic structure can be viewed as a "sea" or a "gas" of nearly free electrons surrounding a lattice of positive ions (Fig. 43.14). The bonding mechanism in a metal is the attractive force between the entire collection of positive ions and the electron gas. Metals have a cohesive energy in the range of 1 to 3 eV per atom, which is less than the cohesive energies of ionic or covalent solids.

Light interacts strongly with the free electrons in metals. Hence, visible light is absorbed and re-emitted quite close to the surface of a metal, which accounts for the shiny nature of metal surfaces. In addition to the high electrical conductivity of metals produced by the free electrons, the nondirectional nature of the metallic bond allows many different types of metal atoms to be dissolved in a host metal in varying amounts. The resulting *solid solutions*, or *alloys*, may be designed to have particular properties, such as tensile strength, ductility, electrical and thermal conductivity, and resistance to corrosion. Such properties are usually controllable and in many cases predictable.

Because the bonding in metals is between all of the electrons and all of the positive ions, metals tend to bend when stressed. This is in contrast to nonmetallic solids, which tend to fracture when stressed. This latter property is due to the fact that bonding in nonmetallic solids is primarily with nearest-neighbor ions or atoms. When the distortion causes sufficient stress between some set of nearest neighbors, fracture occurs.

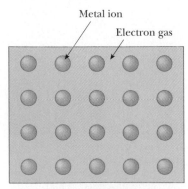

Metal ion

Electron gas

Figure 43.14 Highly schematic diagram of a metal. The blue area represents the electron gas, and the orange circles represent the positive metal ions.

43.4   Free-Electron Theory of Metals

In the remaining sections of this chapter, we discuss various aspects of electrical conduction in solids. This important area of study has led to many technological developments, a few of which we will discuss.

In Section 27.3 we described a classical free-electron theory of electrical conduction in metals that led to Ohm's law. According to this theory, a metal is modeled as a classical gas of conduction electrons moving through a fixed lattice of ion cores. Although this theory predicts the correct functional form of Ohm's law, it does not predict the correct values of electrical and thermal conductivities.

This section introduces a quantum-based free-electron theory of metals, which remedies the shortcomings of the classical model by taking into account the wave nature of the electrons. In this model, one imagines that the outer-shell electrons are free to move through the metal but are trapped within a three-dimensional box formed by the metal surfaces. Therefore, each electron is represented as a particle in a box. As we discussed in Section 41.2, particles in a box are restricted to quantized energy levels.

Statistical physics can be applied to a collection of particles in an effort to relate microscopic properties to macroscopic properties, as we saw with kinetic theory of gases in Chapter 21. In the case of electrons, it is necessary to use *quantum statistics,* with the requirement that each state of the system can be occupied by only two electrons (one with spin up and the other with spin down) as a consequence of the exclusion principle. The probability that a particular state having energy E is occupied by one of the electrons in a solid is given by

$$f(E) = \frac{1}{e^{(E-E_F)/k_B T} + 1} \qquad (43.19)$$

Fermi–Dirac distribution function

where $f(E)$ is called the **Fermi–Dirac distribution function** and E_F is called the **Fermi energy.** A plot of $f(E)$ versus E at $T = 0$ K is shown in Figure 43.15a. Note that $f(E) = 1$ for $E < E_F$ and $f(E) = 0$ for $E > E_F$. That is, at 0 K all states having energies less than the Fermi energy are occupied, and all states having energies greater than the Fermi energy are vacant. A plot of $f(E)$ versus E at some temperature $T > 0$ K is shown in Figure 43.15b. This curve shows that as T increases, the distribution rounds off slightly, with states near and below E_F losing population and states near and above E_F gaining population, due to thermal excitation. The Fermi energy E_F also depends on temperature, but the dependence is weak in metals.

Let us now follow up on our discussion of the particle in a box in Chapter 41 in order to generalize the results to a three-dimensional box. We found that if a particle of mass m is confined to move in a one-dimensional box of length L, the allowed states

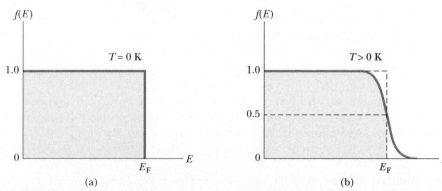

Active Figure 43.15 Plot of the Fermi–Dirac distribution function $f(E)$ versus energy at (a) $T = 0$ K and (b) $T > 0$ K. The energy E_F is the Fermi energy.

At the *Active Figures* link at http://www.pse6.com, *adjust the temperature and observe the effect on the Fermi–Dirac distribution function.*

have quantized energy levels given by Equation 41.12:

$$E_n = \frac{h^2}{8mL^2}n^2 = \frac{\hbar^2\pi^2}{2mL^2}n^2 \qquad n = 1, 2, 3, \ldots$$

Now imagine a piece of metal in the shape of a solid cube of sides L and volume L^3, and let us focus on one electron that is free to move anywhere in this volume. Thus, the electron is modeled as a particle in a three-dimensional box. In this model, we require that $\psi(x, y, z) = 0$ at the boundaries of the metal. It can be shown (see Problem 34) that the energy for such an electron is

$$E = \frac{\hbar^2\pi^2}{2m_e L^2}(n_x^2 + n_y^2 + n_z^2) \tag{43.20}$$

where m_e is the mass of the electron and n_x, n_y, and n_z are quantum numbers. As we expect, the energies are quantized, and each allowed value of the energy is characterized by this set of three quantum numbers (one for each degree of freedom) and the spin quantum number m_s. For example, the ground state, corresponding to $n_x = n_y = n_z = 1$, has an energy equal to $3\hbar^2\pi^2/2m_e L^2$, and can be occupied by two electrons, corresponding to spin-up and spin-down.

Because of the macroscopic size L of the box, the energy levels for the electrons are very close together. As a result, we can treat the quantum numbers as continuous variables. Under this assumption, the number of allowed states per unit volume that have energies between E and $E + dE$ is

$$g(E)\,dE = \frac{8\sqrt{2}\pi m_e^{3/2}}{h^3}E^{1/2}\,dE \tag{43.21}$$

(See Example 43.5 on page 1417.) The function $g(E)$ is called the **density-of-states function.**

For a metal in thermal equilibrium, the number of electrons per unit volume $N(E)\,dE$ that have energy between E and $E + dE$ is equal to the product of the number of allowed states and the probability that a state is occupied, that is, $N(E)\,dE = f(E)g(E)\,dE$:

$$N(E)\,dE = \frac{8\sqrt{2}\pi m_e^{3/2}}{h^3}\frac{E^{1/2}\,dE}{e^{(E-E_F)/k_B T} + 1} \tag{43.22}$$

Plots of $N(E)$ versus E for two temperatures are given in Figure 43.16.

If n_e is the total number of electrons per unit volume, we require that

$$n_e = \int_0^\infty N(E)\,dE = \frac{8\sqrt{2}\pi m_e^{3/2}}{h^3}\int_0^\infty \frac{E^{1/2}\,dE}{e^{(E-E_F)/k_B T} + 1} \tag{43.23}$$

We can use this condition to calculate the Fermi energy. At $T = 0$ K, the Fermi-Dirac distribution function $f(E) = 1$ for $E < E_F$ and $f(E) = 0$ for $E > E_F$. Therefore, at $T = 0$ K, Equation 43.23 becomes

$$n_e = \frac{8\sqrt{2}\pi m_e^{3/2}}{h^3}\int_0^{E_F} E^{1/2}\,dE = \frac{2}{3}\frac{8\sqrt{2}\pi m_e^{3/2}}{h^3}E_F^{3/2} \tag{43.24}$$

Solving for the Fermi energy at 0 K, we obtain

$$E_F(0) = \frac{h^2}{2m_e}\left(\frac{3n_e}{8\pi}\right)^{2/3} \tag{43.25}$$

Fermi energy at $T = 0$ K

The order of magnitude of the Fermi energy for metals is about 5 eV. Representative values for various metals are given in Table 43.4. It is left as a problem (Problem 33) to show that the average energy of a free electron in a metal at 0 K is

$$E_{av} = \frac{3}{5}E_F \tag{43.26}$$

In summary, we can consider a metal to be a system comprising a very large number of energy levels available to the free electrons. These electrons fill these levels in accordance with the Pauli exclusion principle, beginning with $E = 0$ and ending with E_F. At $T = 0$ K, all levels below the Fermi energy are filled and all levels above the

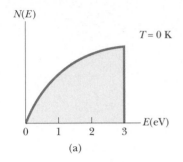

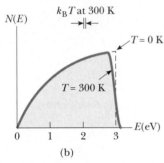

Figure 43.16 Plot of the electron distribution function versus energy in a metal at (a) $T = 0$ K and (b) $T = 300$ K. The Fermi energy E_F is 3 eV.

Table 43.4

Calculated Values of the Fermi Energy for Metals at 300 K Based on the Free-Electron Theory		
Metal	**Electron Concentration (m^{-3})**	**Fermi Energy (eV)**
Li	4.70×10^{28}	4.72
Na	2.65×10^{28}	3.23
K	1.40×10^{28}	2.12
Cu	8.49×10^{28}	7.05
Ag	5.85×10^{28}	5.48
Au	5.90×10^{28}	5.53

Fermi energy are empty. At 300 K, a very small fraction of the free electrons are excited above the Fermi energy. In Section 43.6, we will combine our understanding of the Fermi energy with the information that we study in Section 43.5 on band theory of solids to explain the high electrical conductivity of metals relative to other materials in the solid state.

Quick Quiz 43.6 In a piece of silver, free-electron energy levels are measured near 2 eV and near 6 eV. (The Fermi energy for silver is 5.48 eV.) Near which of these energies are the energy levels closer together? (a) 2 eV (b) 6 eV (c) The spacing of energy levels is the same near both energies.

Quick Quiz 43.7 Consider the situation in Quick Quiz 43.6 again. Near which of these energies are there more electrons occupying quantum states? (a) 2 eV (b) 6 eV (c) The number of electrons is the same near both energies.

Example 43.4 The Fermi Energy of Gold

Each atom of gold (Au) contributes one free electron to the metal. Compute the Fermi energy for gold.

Solution The concentration of free electrons in gold is 5.90×10^{28} m^{-3} (see Table 43.4). Substitution of this value into Equation 43.25 gives

$$E_F(0) = \frac{h^2}{2m_e}\left(\frac{3n_e}{8\pi}\right)^{2/3}$$

$$= \frac{(6.626 \times 10^{-34}\,\text{J}\cdot\text{s})^2}{2(9.11 \times 10^{-31}\,\text{kg})}\left(\frac{3(5.90 \times 10^{28}\,\text{m}^{-3})}{8\pi}\right)^{2/3}$$

$$= 8.85 \times 10^{-19}\,\text{J} = \boxed{5.53\ \text{eV}}$$

Example 43.5 Deriving Equation 43.21

Based on the allowed states of a particle in a three-dimensional box, derive Equation 43.21.

Solution We begin by conceptualizing a particle confined to a three-dimensional box, subject to boundary conditions in three dimensions. We categorize this problem as that of a quantum system in which the energies of the particle are quantized. As noted previously, the allowed states of the particle in a three-dimensional box are described by three quantum numbers n_x, n_y, and n_z. To analyze the problem, imagine a three-dimensional quantum number space whose axes represent n_x, n_y, and n_z. The allowed states in this space can be represented as dots located at integral values of the three quantum numbers, as in Figure 43.17.

Equation 43.20 can be written as

$$(1)\qquad n_x{}^2 + n_y{}^2 + n_z{}^2 = \frac{E}{E_0} = n^2$$

where

$$(2)\qquad E_0 = \frac{\hbar^2\pi^2}{2m_e L^2}$$

and $n = (E/E_0)^{1/2}$. In this space, Equation (1) is the equation of a sphere of radius n. Thus, the number of allowed states having energies between E and $E + dE$ is equal to the number of points in a spherical shell of radius n and thickness dn. The "volume" of this shell is

$$(3)\qquad G(E)\, dE = \tfrac{1}{8}(4\pi n^2\, dn) = \tfrac{1}{2}\pi n^2\, dn$$

where we have taken one eighth of the total volume because we are restricted to the octant of a three-dimensional space in which all three quantum numbers are positive. We use $G(E)$ here to represent the total number of states, while

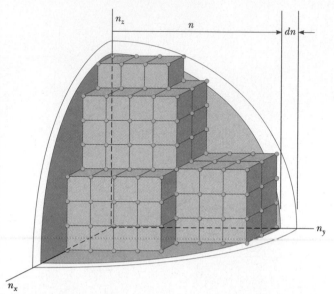

Figure 43.17 (Example 43.5) The allowed states of particles in a three-dimensional box can be represented by dots (blue circles) in a quantum number space. This is not traditional space in which a location is specified by coordinates x, y, and z. It is a space in which allowed states can be specified by coordinates representing the quantum numbers. The dots representing the allowed states are located at integer values of n_x, n_y, and n_z and are therefore at the corners of cubes with sides of "length" 1. The number of allowed states having energies between E and $E + dE$ corresponds to the number of dots in the spherical shell of radius n and thickness dn.

$g(E)$ represents the number of states per unit volume V (in normal space). Thus, $g(E) = G(E)/V$.

Now, we replace n in Equation (3) with its equivalent in terms of E using the relation $n^2 = E/E_0$ from Equation (1):

$$G(E)\, dE = \tfrac{1}{2}\, \pi \left(\frac{E}{E_0}\right) d\left[\left(\frac{E}{E_0}\right)^{1/2}\right]$$

$$= \tfrac{1}{2}\, \pi \left(\frac{E}{E_0}\right) E_0^{-1/2} \tfrac{1}{2} E^{-1/2}\, dE = \tfrac{1}{4}\, \pi E_0^{-3/2} E^{1/2}\, dE$$

Substituting for E_0 from Equation (2) gives

$$G(E)\, dE = \tfrac{1}{4}\, \pi \left(\frac{\hbar^2 \pi^2}{2 m_e L^2}\right)^{-3/2} E^{1/2}\, dE$$

$$= \frac{\sqrt{2}}{2} \frac{m_e^{3/2} L^3}{\pi^2 \hbar^3} E^{1/2}\, dE$$

Recognizing L^3 as the volume V of the cubical box in normal space, and recalling that $g(E) = G(E)/V$, we find

$$g(E)\, dE = \frac{G(E)}{V}\, dE = \frac{\sqrt{2}}{2} \frac{m_e^{3/2}}{\pi^2 \hbar^3} E^{1/2}\, dE$$

Substituting $\hbar = h/2\pi$ gives

$$g(E)\, dE = \frac{4\sqrt{2}\, \pi m_e^{3/2}}{h^3} E^{1/2}\, dE$$

Finally, we multiply by 2 for the two possible spin states in each particle-in-a-box state:

$$g(E)\, dE = \frac{8\sqrt{2}\, \pi m_e^{3/2}}{h^3} E^{1/2}\, dE$$

To finalize this problem, note that this is Equation 43.21, which is what we set out to derive.

43.5 Band Theory of Solids

In the preceding section, we modeled the electrons in a metal as particles free to move around inside a three-dimensional box and we paid no attention to the influence of the parent atoms. In this section, we make the model more sophisticated by incorporating the contribution of the parent atoms that form the crystal.

Recall from Section 41.1 that the probability density $|\psi|^2$ for a system is physically significant while the probability amplitude ψ is not. Let us consider as an example an atom that has a single s electron outside of a closed shell. Both of the following wave functions are valid for such an atom with atomic number Z:

$$\psi_s^+ (r) = +Af(r)e^{-Zr/na_0} \qquad \psi_s^- (r) = -Af(r)e^{-Zr/na_0}$$

where A is the normalization constant and $f(r)$ is a function[3] of r that varies with the value of n. Choosing either of these wave functions leads to the same value of $|\psi|^2$, so both choices are equivalent. A difference arises, however, when we combine two atoms.

If two identical atoms are very far apart, they do not interact and their electronic energy levels can be considered to be those of isolated atoms. Suppose that the two atoms

---

[3]   The functions $f(r)$ are called *Laguerre polynomials*. They can be found in the quantum treatment of the hydrogen atom in modern physics textbooks.

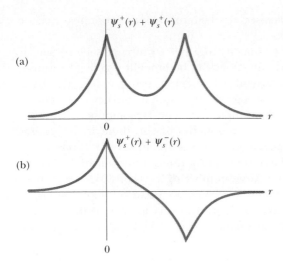

Figure 43.18 The wave functions of two atoms combine to form a composite wave function for the two-atom system when the atoms are close together. (a) Two atoms with wave functions $\psi_s^+(r)$ combine. (b) Two atoms with wave functions $\psi_s^+(r)$ and $\psi_s^-(r)$ combine.

are sodium, each having a lone $3s$ electron that is in a well-defined quantum state. As the two sodium atoms are brought closer together, their wave functions begin to overlap, as we discussed for covalent bonding in Section 43.1. The properties of the combined system differ depending on whether we combine two atoms with wave functions $\psi_s^+(r)$, as in Figure 43.18a, or one with wave function $\psi_s^+(r)$ and the other with $\psi_s^-(r)$, as in Figure 43.18b. (The choice of two atoms with wave function $\psi_s^-(r)$ is physically equivalent to that with two positive wave functions, so we do not consider it separately.) When two wave functions $\psi_s^+(r)$ are combined, the result is a composite wave function in which the probability amplitudes add between the atoms. If $\psi_s^+(r)$ combines with $\psi_s^-(r)$, however, the wave functions between the nuclei subtract. Thus, the composite probability amplitudes for the two possibilities are different. **These two possible combinations of wave functions represent two possible states of the two-atom system.** We interpret these curves as representing the probability amplitude of finding an electron. The positive–positive curve shows some probability of finding the electron at the midpoint between the atoms. The positive–negative function shows no such probability. A state with a high probability of an electron *between* two positive nuclei (resulting in a covalent binding force from the Coulomb attraction of the electron on both nuclei) must have a different energy than a state with a high probability of the electron being elsewhere! Thus, the states are *split* into two energy levels due to the two ways of combining the wave functions. The energy difference is relatively small, so the two states are close together on an energy scale.

Figure 43.19a shows this splitting effect as a function of separation distance. For large separations r, the electron clouds do not overlap and there is no splitting. As the

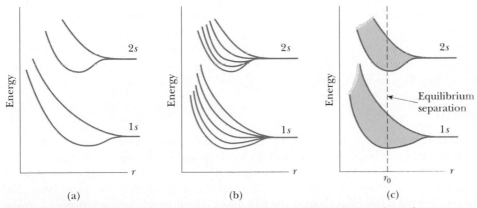

Figure 43.19 (a) Splitting of the $1s$ and $2s$ levels when two sodium atoms are brought together. (b) Splitting of the $1s$ and $2s$ levels when five sodium atoms are brought together. (c) Formation of energy bands when a large number of sodium atoms are assembled to form a solid.

atoms are brought closer so that r decreases, the electron clouds overlap so that we need to consider the system of two atoms.

When a large number of atoms are brought together to form a solid, a similar phenomenon occurs. The individual wave functions can be brought together in various combinations of $\psi_s^+(r)$ and $\psi_s^-(r)$, each possible combination corresponding to a different energy. As the atoms are brought close together, the various isolated-atom energy levels split into multiple energy levels for the composite system. This splitting in levels for five atoms in close proximity is shown in Figure 43.19b. In this case, there are five energy levels corresponding to five different combinations of isolated-atom wave functions.

If we extend this argument to the large number of atoms found in solids (on the order of 10^{23} atoms per cm^3), we obtain a large number of levels of varying energy so closely spaced that they may be regarded as a continuous **band** of energy levels, as shown in Figure 43.19c. In the case of sodium, it is customary to refer to the continuous distributions of allowed energy levels as s bands because the bands originate from the s levels of the individual sodium atoms.

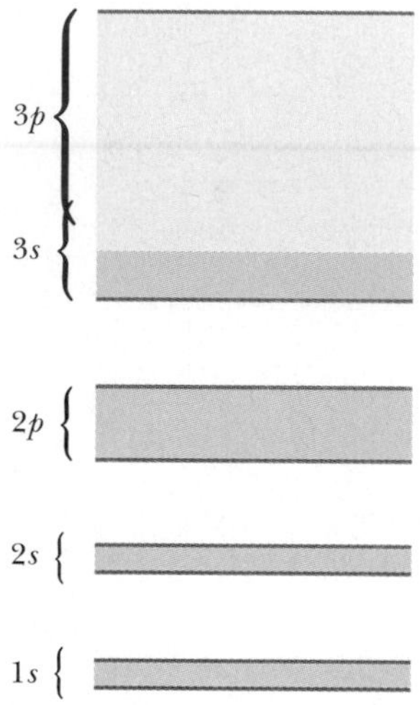

In general, a crystalline solid has a large number of allowed energy bands that arise from the various atomic energy levels. Figure 43.20 shows the allowed energy bands of sodium. Note that energy gaps, corresponding to *forbidden energies,* occur between the allowed bands. In addition, some bands exhibit sufficient spreading in energy that there is an overlap between bands arising from different quantum states ($3s$ and $3p$).

The $1s$, $2s$, and $2p$ bands of sodium are each full of electrons, as indicated by the blue-shaded areas in Figure 43.20. This is because the $1s$, $2s$, and $2p$ states of each atom are full. An energy level in which the orbital angular momentum is ℓ can hold $2(2\ell + 1)$ electrons. The factor 2 arises from the two possible electron spin orientations, and the factor $2\ell + 1$ corresponds to the number of possible orientations of the orbital angular momentum. The capacity of each band for a system of N atoms is $2(2\ell + 1)N$ electrons. Therefore, the $1s$ and $2s$ bands each contain $2N$ electrons ($\ell = 0$), and the $2p$ band contains $6N$ electrons ($\ell = 1$). Because sodium has only one $3s$ electron and there are a total of N atoms in the solid, the $3s$ band contains only N electrons and is partially full, as indicated by the blue coloring in Figure 43.20. The $3p$ band, which is the higher region of the overlapping bands, is completely empty (all gold in the figure).

Band theory allows us to build simple models in order to understand the behavior of conductors, insulators, and semiconductors, as well as that of semiconductor devices, as we see in the following sections.

Figure 43.20 Energy bands of a sodium crystal. Note the energy gaps (white regions) between the allowed bands; electrons cannot occupy states that lie in these gaps. Blue represents energy bands occupied by the sodium electrons when the atom is in its ground state. Gold represents energy bands that are empty.

43.6 Electrical Conduction in Metals, Insulators, and Semiconductors

Good electrical conductors contain a high density of free charge carriers, and the density of free charge carriers in insulators is nearly zero. **Semiconductors** are a class of technologically important materials in which charge-carrier densities are intermediate between those of insulators and those of conductors. In this section, we discuss the mechanisms of conduction in these three classes of materials in terms of a model based on energy bands.

Metals

If a material is to be a good electrical conductor, the charge carriers in the material must be free to move in response to an applied electric field. Let us consider the electrons in a metal as the charge carriers we shall investigate. The motion of the electrons in response to an electric field represents an increase in energy of the system of the metal lattice and the free electrons corresponding to the additional kinetic energy of the moving electrons. Therefore, when an electric field is applied to a conductor, electrons must move upward to an available higher energy state on an energy-level diagram.

Figure 43.21 shows a half-filled band in a metal at $T = 0$ K, where the blue region represents levels filled with electrons. Because electrons obey Fermi–Dirac

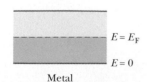

Figure 43.21 Half-filled band of a metal, an electrical conductor. At $T = 0$ K, the Fermi energy lies in the middle of the band.

statistics, all levels below the Fermi energy are filled with electrons and all levels above the Fermi energy are empty. The Fermi energy lies in the band at the highest filled state. At temperatures slightly greater than 0 K, some electrons are thermally excited to levels above E_F, but overall there is little change from the 0 K case. However, **if a potential difference is applied to the metal, electrons having energies near the Fermi energy require only a small amount of additional energy from the applied field to reach nearby empty energy states above the Fermi energy.** Therefore, electrons in a metal experiencing only a small applied electric field are free to move because there are many empty levels available close to the occupied energy levels. The model of metals based on band theory shows that metals are excellent electrical conductors.

Insulators

Now consider the two outermost energy bands of a material in which the lower band is filled with electrons and the higher band is empty at 0 K (Fig. 43.22). The lower, filled band is called the **valence band,** and the upper, empty band is the **conduction band.** (The conduction band is the one that is partially filled in a metal.) It is common to refer to the energy separation between the valence and conduction bands as the **energy gap** E_g of the material. The Fermi energy lies somewhere in the energy gap,[4] as shown in Figure 43.22.

Suppose a material has a relatively large energy gap, for example, about 5 eV. At 300 K (room temperature), $k_B T = 0.025$ eV, which is much smaller than the energy gap. At such temperatures, the Fermi–Dirac distribution predicts that very few electrons are thermally excited into the conduction band. There are no available states that lie close in energy and into which electrons can move upward in order to account for the extra kinetic energy associated with motion through the material in response to an electric field. Consequently, the electrons do not move—the material is an insulator. Although an insulator has many vacant states in its conduction band that can accept electrons, these states are separated from the filled states by a large energy gap. Only a few electrons occupy these states, so the overall electrical conductivity of insulators is very small.

Semiconductors

Semiconductors have the same type of band structure as an insulator but the energy gap is much smaller—on the order of 1 eV. Table 43.5 shows the energy gaps for some representative materials. The band structure of a semiconductor is shown in Figure 43.23. Because the Fermi level is located near the middle of the gap for a semiconductor and because E_g is small, appreciable numbers of electrons are thermally excited from the valence band to the conduction band. There are many empty levels above the thermally filled levels in the conduction band; therefore, a small applied potential difference can easily raise the energy of the electrons in the conduction band, resulting in a moderate current.

At $T = 0$ K, all electrons in these materials are in the valence band, and no energy is available to excite them across the energy gap. Thus, semiconductors are poor conductors at very low temperatures. Because the thermal excitation of electrons across the narrow gap is more probable at higher temperatures, the conductivity of semiconductors increases rapidly with temperature. This contrasts sharply with the conductivity of metals, which decreases slowly with increasing temperature.

Charge carriers in a semiconductor can be negative or positive, or both. When an electron moves from the valence band into the conduction band, it leaves behind a vacant site, called a **hole,** in the otherwise filled valence band. This hole

---

[4] We defined the Fermi energy as the energy of the highest filled state at $T = 0$. This might suggest that the Fermi energy should be at the top of the valence band in Figure 43.22. A more sophisticated general treatment of the Fermi energy, however, shows that it is located at that energy at which the probability of occupation is $\frac{1}{2}$ (see Fig. 43.15b). According to this definition, the Fermi energy lies in the energy gap between the bands.

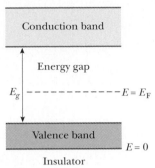

Figure 43.22 An electrical insulator at $T = 0$ K has a filled valence band and an empty conduction band. The Fermi level lies somewhere between these bands in the region known as the energy gap.

Table 43.5

Energy-Gap Values for Some Semiconductors[a]		
	E_g (eV)	
Crystal	**0 K**	**300 K**
Si	1.17	1.14
Ge	0.744	0.67
InP	1.42	1.35
GaP	2.32	2.26
GaAs	1.52	1.43
CdS	2.582	2.42
CdTe	1.607	1.45
ZnO	3.436	3.2
ZnS	3.91	3.6

[a] From C. Kittel, *Introduction to Solid State Physics*, 5th ed., New York, John Wiley & Sons, 1976.

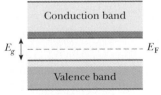

Figure 43.23 Band structure of a semiconductor at ordinary temperatures ($T \approx 300$ K). The energy gap is much smaller than in an insulator, and some electrons from the valence band occupy states in the conduction band.

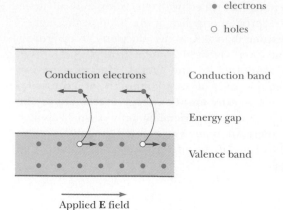

Figure 43.24 Movement of charges (holes and electrons) in an intrinsic semiconductor. The electrons move in the direction opposite the direction of the external electric field, and the holes move in the direction of the field.

(electron-deficient site) acts as a charge carrier in the sense that a free electron from a nearby site can transfer into the hole. Whenever an electron does so, it creates a new hole at the site it abandoned. Thus, the net effect can be viewed as the hole migrating through the material in the direction opposite the direction of electron movement. The hole behaves as if it were a particle with a positive charge $+e$.

A pure semiconductor crystal containing only one element or one compound is called an **intrinsic semiconductor.** In such a semiconductor, there are equal numbers of conduction electrons and holes. Such combinations of charges are called **electron–hole pairs.** In the presence of an external electric field, the holes move in the direction of the field and the conduction electrons move in the direction opposite the field (Fig. 43.24). Because the electrons and holes have opposite signs, both of these motions correspond to a current in the same direction.

Quick Quiz 43.8 Consider the data on three materials in the table below:

Material	Conduction Band	E_g
A	Empty	1.2 eV
B	Half full	1.2 eV
C	Empty	8.0 eV

Identify these materials as a conductor, an insulator, or a semiconductor.

Doped Semiconductors

When impurities are added to a semiconductor, both the band structure of the semiconductor and its resistivity are modified. The process of adding impurities, called **doping,** is important in controlling the conductivity of semiconductors. For example, when an atom containing five outer-shell electrons, such as arsenic, is added to a Group IV semiconductor, four of the electrons form covalent bonds with atoms of the semiconductor and one is left over (Fig. 43.25a). This extra electron is nearly free of its parent atom and can be modeled as having an energy level that lies in the energy gap, just below the conduction band (Fig. 43.25b). Such a pentavalent atom in effect donates an electron to the structure and hence is referred to as a **donor atom.** Because the spacing between the energy level of the electron of the donor atom and the bottom of the conduction band is very small (typically, about 0.05 eV), only a small amount of thermal excitation is needed to cause this electron to move into the conduction band. (Recall that the average energy of an electron at room temperature is about $k_BT \approx 0.025$ eV.) Semiconductors doped with

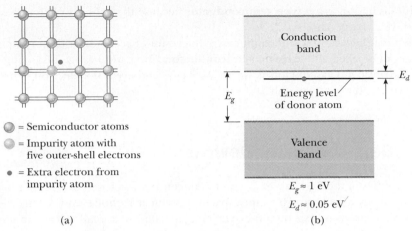

$E_g \approx 1$ eV

$E_d \approx 0.05$ eV

(a)   (b)

Figure 43.25 (a) Two-dimensional representation of a semiconductor consisting of Group IV atoms (gray) and an impurity atom (yellow) that has five outer-shell electrons. Each double line between atoms represents a covalent bond in which two electrons are shared. (b) Energy-band diagram for a semiconductor in which the nearly free electron of the impurity atom lies in the energy gap, just below the bottom of the conduction band. A small amount of energy can excite the electron into the conduction band.

donor atoms are called **n-type semiconductors** because the majority of charge carriers are electrons, which are **n**egatively charged.

If a Group IV semiconductor is doped with atoms containing three outer-shell electrons, such as indium and aluminum, the three electrons form covalent bonds with neighboring semiconductor atoms, leaving an electron deficiency—a hole—where the fourth bond would be if an impurity-atom electron were available to form it (Fig. 43.26a). This can be modeled by placing an energy level in the energy gap, just above the valence band, as in Figure 43.26b. An electron from the valence band has enough energy at room temperature to fill this impurity level, leaving behind a hole in the valence band. This hole can carry current in the presence of an electric field. Because a trivalent atom in effect accepts an electron from the valence band, such impurities are referred to as **acceptor atoms.** A semiconductor doped with trivalent (acceptor)

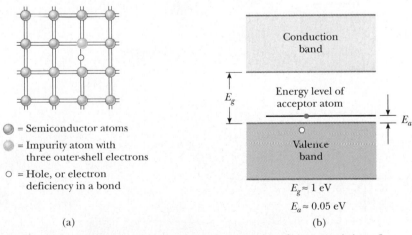

$E_g \approx 1$ eV

$E_a \approx 0.05$ eV

(a)   (b)

Figure 43.26 (a) Two-dimensional representation of a semiconductor consisting of Group IV atoms (gray) and an impurity atom (yellow) having three outer-shell electrons. The single line between the impurity atom and the semiconductor atom below it represents the fact that there is only one electron shared in this bond. (b) Energy-band diagram for a semiconductor in which the hole resulting from the trivalent impurity atom lies in the energy gap, just above the top of the valence band. This diagram shows an electron excited into the energy level of the acceptor atom, leaving a hole in the valence band.

impurities is known as a **p-type semiconductor** because the majority of charge carriers are *p*ositively charged holes.

When conduction in a semiconductor is the result of acceptor or donor impurities, the material is called an **extrinsic semiconductor.** The typical range of doping densities for extrinsic semiconductors is 10^{13} to 10^{19} cm^{-3}, whereas the electron density in a typical semiconductor is roughly 10^{21} cm^{-3}.

43.7 Semiconductor Devices

The electronics of the first half of the twentieth century was based on vacuum tubes, in which electrons pass through empty space between a cathode and an anode. The number of electrons in a vacuum tube can be controlled by a small voltage applied to a grid electrode, which repels electrons in their path to the anode. We have seen vacuum tubes in Figure 23.27 (the cathode-ray tube), Figure 29.5 (television picture tube), Figure 29.20 (circular electron beam), Figure 29.25 (Thomson's apparatus for measuring e/m_e for the electron), and Figure 40.9 (photoelectric effect apparatus).

The transistor was invented in 1948, leading to a shift away from vacuum tubes and toward semiconductors as the basis of electronic devices. This phase of electronics has been underway for several decades. As discussed in Chapter 41, there may be a new phase of electronics in the near future using nanotechnological devices employing quantum dots and other nanoscale structures.

In this section, we discuss electronic devices based on semiconductors, which are still in wide use and will be for many years to come.

The Junction Diode

A fundamental unit of a semiconductor device is formed when a *p*-type semiconductor is joined to an *n*-type semiconductor to form a **p–n junction.** A **junction diode** is a device that is based on a single *p–n* junction. The role of a diode of any type is to pass current in one direction but not the other. Thus, it acts as a one-way valve for current.

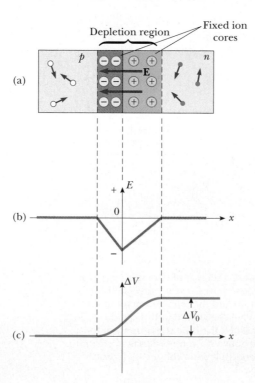

Figure 43.27 (a) Physical arrangement of a *p–n* junction. (b) Internal electric field versus *x* for the *p–n* junction. (c) Internal electric potential difference ΔV versus *x* for the *p–n* junction. The potential difference ΔV_0 represents the potential difference across the junction in the absence of an applied electric field.

The p–n junction is shown in Figure 43.27a. We can identify three distinct regions: a p region, an n region, and a small area that extends several micrometers to either side of the interface, called a *depletion region*.

The depletion region may be visualized as arising when the two halves of the junction are brought together. The mobile n-side donor electrons nearest the junction (deep-blue area in Fig. 43.27a) diffuse to the p side and fill holes located there, leaving behind immobile positive ions. While this happens, we can model the holes that are being filled as diffusing to the n side, leaving behind a region (brown area in Fig. 43.27a) of fixed negative ions.

Because the two sides of the depletion region each carry a net charge, an internal electric field on the order of 10^4 to 10^6 V/cm exists in the depletion region (see Fig. 43.27b). This field produces an electric force on any remaining mobile charge carriers that sweeps them out of the depletion region. The depletion region is so named because it is depleted of mobile charge carriers. This internal electric field creates an internal potential difference ΔV_0 that prevents further diffusion of holes and electrons across the junction and thereby ensures zero current in the junction when no potential difference is applied.

The operation of the junction as a diode is easiest to understand in terms of the potential-difference graph shown in Figure 43.27c. If a voltage ΔV is applied to the junction such that the p side is connected to the positive terminal of a voltage source, as shown in Figure 43.28a, the internal potential difference ΔV_0 across the junction is decreased; the decrease results in a current that increases exponentially with increasing forward voltage, or *forward bias*. For *reverse bias* (where the n side of the junction is connected to the positive terminal of a voltage source), the internal potential difference ΔV_0 increases with increasing reverse bias; the increase results in a very small reverse current that quickly reaches a saturation value I_0. The current–voltage relationship for an ideal diode is

$$I = I_0 \, (e^{e\Delta V/k_\mathrm{B}T} - 1) \tag{43.27}$$

where the first e is the base of the natural logarithm, the second e represents the magnitude of the electron charge, k_B is Boltzmann's constant, and T is the absolute temperature. Figure 43.28b shows a circuit diagram for a diode under forward bias and Figure 43.28c shows an I–ΔV plot characteristic of a real p–n junction, demonstrating the diode behavior.

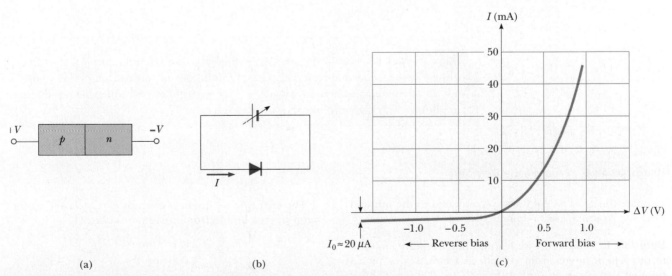

(a)   (b)   (c)

Figure 43.28 (a) A p–n junction under forward bias. (b) The circuit diagram for a diode under forward bias, showing a battery with an adjustable voltage. Both positive and negative voltages can be applied to the diode to study its nonlinear behavior. (c) The characteristic curve for a real p–n junction.

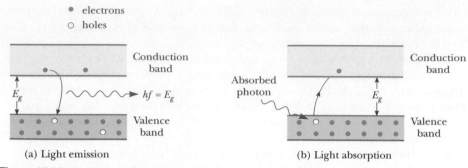

Figure 43.29 (a) Light emission from a semiconductor. (b) Light absorption by a semiconductor.

Light-Emitting and Light-Absorbing Diodes

Light-emitting diodes (LEDs) and semiconductor lasers are common examples of devices that depend on the behavior of semiconductors. LEDs are commonly used in traffic signals, in electronic displays, and as indicator lights for electronic equipment. Semiconductor lasers are often used for pointers in presentations and in compact disc and DVD playback equipment.

Light emission and absorption in semiconductors is similar to light emission and absorption by gaseous atoms, except that in the discussion of semiconductors, we must incorporate the concept of energy bands rather than the discrete energy levels in single atoms. As shown in Figure 43.29a, an electron excited electrically into the conduction band can easily recombine with a hole (especially if the electron is injected into a p region). As this recombination takes place, a photon of energy E_g is emitted. With proper design of the semiconductor and the associated plastic envelope or mirrors, the light from a large number of these transitions will serve as the source of an LED or a semiconductor laser.

Conversely, an electron in the valence band may absorb an incoming photon of light and be promoted to the conduction band, leaving a hole behind (Fig. 43.29b). This absorbed energy can be used to operate an electrical circuit. One device that operates on this principle is the photovoltaic solar cell, which appears in many handheld calculators. Arrays of solar cells are used in generating electric power in space vehicles and in remote areas on the Earth.

Quick Quiz 43.9 Figure 43.29a shows a photon of frequency $f = E_g/h$ being emitted from a semiconductor. In addition to photons of this frequency, (a) photons with no other frequencies are emitted (b) photons with lower frequencies are emitted (c) photons with higher frequencies are emitted.

Example 43.6   Where's the Remote?

Estimate the band gap of the semiconductor in the infrared LED of a typical television remote control.

Solution In Chapter 34 we learned that the wavelength of infrared light ranges from 700 nm to 1 mm. Let us pick a number that is easy to work with, such as 1 000 nm. (This is not a bad estimate. Remote controls typically operate in the range of 880 to 950 nm.)

The energy of a photon is given by $E = hc/\lambda$, and so the energy of the photons from the remote control is

$$E = \frac{hc}{\lambda} = \frac{1\ 240\ \text{eV} \cdot \text{nm}}{1\ 000\ \text{nm}} = \boxed{1.2\ \text{eV}}$$

This corresponds to an energy gap E_g of approximately 1.2 eV in the LED's semiconductor.

The Transistor

The invention of the transistor by John Bardeen (1908–1991), Walter Brattain (1902–1987), and William Shockley (1910–1989) in 1948 totally revolutionized the world of electronics. For this work, these three men shared a Nobel Prize in physics in 1956. By 1960, the transistor had replaced the vacuum tube in many electronic applications. The advent of the transistor created a multitrillion-dollar industry that produces such popular devices as portable radios, CD players, handheld calculators, computers, television receivers, wireless telephones, and electronic games.

A **junction transistor** consists of a semiconducting material in which a very narrow n region is sandwiched between two p regions or a p region is sandwiched between two n regions. In either case, the transistor is formed from two p–n junctions. These types of transistors were used widely in the early days of semiconductor electronics.

During the 1960s, the electronics industry converted many electronic applications from the junction transistor to the **field-effect transistor,** which is much easier to manufacture and just as effective. Figure 43.30a shows the structure of a very common device, the **MOSFET,** or **metal-oxide-semiconductor field-effect transistor.** You are likely using millions of MOSFET devices when you are working on your computer.

There are three metal connections (the M in MOSFET) to the transistor—the *source, drain,* and *gate*. The source and drain are connected to n-type semiconductor regions (the S in MOSFET) at either end of the structure. These regions are connected by a narrow channel of additional n-type material, the n channel. The source and drain regions and the n channel are embedded in a p-type substrate material. This forms a depletion region, as in the junction diode, along the bottom of the n channel. (Depletion regions also exist at the junctions underneath the source and drain regions, but we will ignore these because the operation of the device depends primarily on the behavior in the channel.)

The gate is separated from the n channel by a layer of insulating silicon dioxide (the O in MOSFET, for oxide). Thus, it does not make electrical contact with the rest of the semiconducting material.

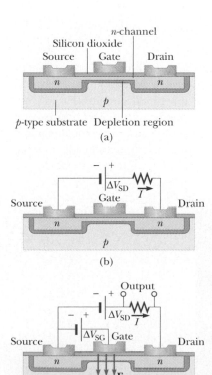

Figure 43.30 (a) The structure of a metal-oxide-semiconductor field-effect transistor (MOSFET). (b) A source–drain voltage is applied, with the result that current exists in the circuit. (c) A gate voltage is applied. The gate voltage can be used to control the source–drain current, so that the MOSFET acts as an amplifier.

Imagine that a voltage source ΔV_{SD} is applied across the source and drain as shown in Figure 43.30b. With this situation, electrons will flow through the upper region of the n channel. Electrons cannot flow through the depletion region in the lower part of the n channel because this region is depleted of charge carriers. Now a second voltage ΔV_{SG} is applied across the source and gate as in Figure 43.30c. The positive potential on the gate electrode results in an electric field below the gate that is directed downward in the n channel. (This is the field in "field-effect.") This electric field exerts upward forces on electrons in the region below the gate, causing them to move into the n channel. This causes the depletion region to become smaller, widening the area through which there is current between the top of the n channel and the depletion region. As the area becomes wider, the value of the current increases.

If a varying voltage, such as that generated from music stored on a compact disc, is applied to the gate, the area through which the source–drain current exists will vary in size according to the varying gate voltage. A small variation in gate voltage results in a large variation in current and a correspondingly large voltage across the resistor in Figure 43.30c. Therefore, the MOSFET acts as a voltage amplifier. A circuit consisting of a chain of such transistors can result in a very small initial signal from a microphone being amplified enough to drive the powerful speakers at an outdoor concert.

The MOSFET can be used as a switch by reversing the potential difference ΔV_{SG} in Figure 43.30c. In this case, increasing the voltage causes the n channel to decrease in size. If this voltage is large enough to completely block the area through which electrons pass, the current falls to zero. This on–off behavior is used in many applications.

The MOSFET is used widely in current electronic equipment. As mentioned in Chapter 41, a new era of transistors based on quantum dots and other nanotechnologies may await us in the future.

The Integrated Circuit

Invented independently by Jack Kilby (b. 1923, Nobel Prize, 2000) at Texas Instruments in late 1958 and by Robert Noyce (b. 1927) at Fairchild Camera and Instrument in early 1959, the integrated circuit has been justly called "the most remarkable technology ever to hit mankind." Kilby's first device is shown in Figure 43.31. Integrated circuits have indeed started a "second industrial revolution" and are found at the heart of computers, watches, cameras, automobiles, aircraft, robots, space vehicles, and all sorts of communication and switching networks.

In simplest terms, an **integrated circuit** is a collection of interconnected transistors, diodes, resistors, and capacitors fabricated on a single piece of silicon known as a *chip*. Contemporary electronic devices often contain many integrated circuits (Fig. 43.32). State-of-the-art chips easily contain several million components within a 1-cm^2 area with the number of components per square inch having increased steadily since the integrated circuit was invented. Figure 43.33a illustrates the dramatic advances made in chip technology since Intel introduced the first microprocessor in 1971. Figure 43.33b is a

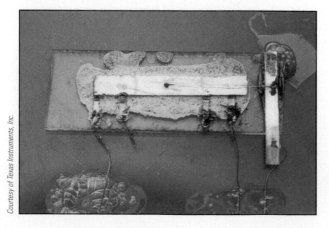

Courtesy of Texas Instruments, Inc.

Figure 43.31 Jack Kilby's first integrated circuit, tested on September 12, 1958.

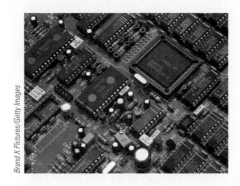

Figure 43.32 Integrated circuits are prevalent in many electronic devices. All of the flat circuit elements with black-topped surfaces in this photograph are integrated circuits.

graph of the logarithm of the number of transistors in a chip as a function of the year in which the chip was introduced. Because this growth follows an approximately straight line, we conclude that that the growth is exponential. "Moore's law," proposed in 1965 by Gordon Moore, a co-founder of Intel, claims that the number of transistors per square inch on integrated circuits should double every 18 months. The doubling time in Figure 43.33b is longer than 18 months because this graph shows only the total number of transistors and does not take into account the shrinking size of the integrated circuit over the years. When the density of transistors per square centimeter is graphed, the results are similar to those predicted by Moore's law.

Integrated circuits were invented partly to solve the interconnection problem spawned by the transistor. In the era of vacuum tubes, power and size considerations of individual components set modest limits on the number of components that could be interconnected in a given circuit. With the advent of the tiny, low-power, highly reliable transistor, design limits on the number of components disappeared and were replaced by the problem of wiring together hundreds of thousands of components. The magnitude of this problem can be appreciated when we consider that second-generation computers (consisting of discrete transistors rather than integrated circuits) contained several hundred thousand components requiring more than a million joints that had to be hand-soldered and tested.

In addition to solving the interconnection problem, integrated circuits possess the advantages of miniaturization and fast response, two attributes critical for high-speed computers. The fast response results from the miniaturization and close packing of components, because the response time of a circuit depends on the time interval required for electrical signals traveling at the speed of light to pass from one component to another.

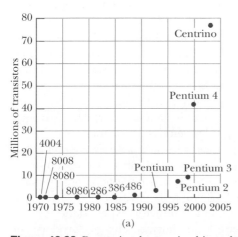

(a)

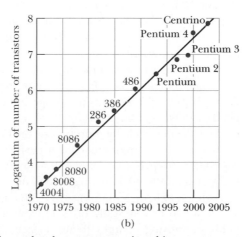

(b)

Figure 43.33 Dramatic advances in chip technology related to computer microchips manufactured by Intel. (a) A plot of the number of transistors on a single computer chip versus year of manufacture. (b) A plot of the logarithm of the number of transistors in part (a). The fact that this plot is approximately a straight line shows that the growth is exponential.

Table 43.6

Critical Temperatures for Various Superconductors	
Material	T_c (K)
Zn	0.88
Al	1.19
Sn	3.72
Hg	4.15
Pb	7.18
Nb	9.46
Nb_3Sn	18.05
Nb_3Ge	23.2
$YBa_2Cu_3O_7$	92
Bi–Sr–Ca–Cu–O	105
Tl–Ba–Ca–Cu–O	125
$HgBa_2Ca_2Cu_3O_8$	134

43.8 Superconductivity

We learned in Section 27.5 that there is a class of metals and compounds known as **superconductors** whose electrical resistance decreases to virtually zero below a certain temperature T_c called the *critical temperature* (Table 43.6). Let us now look at these amazing materials in greater detail, using what we have just learned about the properties of solids to help us understand the behavior of superconductors.

Let us start by examining the Meissner effect, described in Section 30.8 as the exclusion of magnetic flux from the interior of superconductors. Simple arguments based on the laws of electricity and magnetism can be used to show that the magnetic field inside a superconductor cannot change with time. According to Equation 27.8, $R = \Delta V/I$, and because the potential difference ΔV across a conductor is proportional to the electric field inside the conductor, we see that the electric field is proportional to the resistance of the conductor. Thus, because $R = 0$ for a superconductor at or below its critical temperature, *the electric field in its interior must be zero.* Now recall that Faraday's law of induction can be expressed in the form shown in Equation 31.9:

$$\oint \mathbf{E} \cdot d\mathbf{s} = -\frac{d\Phi_B}{dt} \tag{43.28}$$

That is, the line integral of the electric field around any closed loop is equal to the negative rate of change in the magnetic flux Φ_B through the loop. Because $\mathbf{E}$ is zero everywhere inside the superconductor, the integral over any closed path inside the superconductor is zero. Hence, $d\Phi_B/dt = 0$; this tells us that **the magnetic flux in the superconductor cannot change.** From this information, we conclude that $B\,(=\Phi_B/A)$ must remain constant inside the superconductor.

Before 1933, it was assumed that superconductivity was a manifestation of perfect conductivity. If a perfect conductor is cooled below its critical temperature in the presence of an applied magnetic field, the field should be trapped in the interior of the conductor even after the external field is removed. In addition, the final state of the perfect conductor should depend on which occurs first, the application of the field or cooling the material below T_c. If the field is applied after the material has been cooled, the field should be expelled from the superconductor. If the field is applied before the material is cooled, the field should not be expelled once the material has been cooled. In 1933, however, W. Hans Meissner and Robert Ochsenfeld discovered that, when a metal becomes superconducting in the presence of a weak magnetic field, the field is expelled. Thus, the same final state $\mathbf{B} = 0$ is achieved whether the field is applied before or after the material is cooled below its critical temperature.

The Meissner effect is illustrated in Figure 43.34 for a superconducting material in the shape of a long cylinder. Note that the field penetrates the cylinder when its temperature is greater than T_c (Fig. 43.34a). As the temperature is lowered to below T_c, however, the field lines are spontaneously expelled from the interior of the superconductor (Fig. 43.34b). Thus, a superconductor is more than a perfect conductor (resistivity $\rho = 0$); it is also a perfect diamagnet ($\mathbf{B} = 0$).The property that $\mathbf{B} = 0$ in the interior of a superconductor is as fundamental as the property of zero resistance. If the magnitude of the applied magnetic field exceeds a critical value B_c, defined as the value of B that destroys a material's superconducting properties, the field again penetrates the sample.

Because a superconductor is a perfect diamagnet having a *negative* magnetic susceptibility, it repels a permanent magnet. In fact, one can perform a demonstration of the Meissner effect by floating a small permanent magnet above a superconductor and achieving magnetic levitation, as seen in Figure 30.35 on page 952.

Recall from our study of electricity that a good conductor expels static electric fields by moving charges to its surface. In effect, the surface charges produce an electric field that exactly cancels the externally applied field inside the conductor. In a similar manner, a superconductor expels magnetic fields by forming surface currents. To see why this happens, consider again the superconductor shown in Figure 43.34. Let us assume that the sample is initially at a temperature $T > T_c$ as

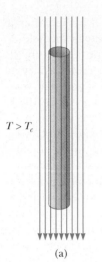
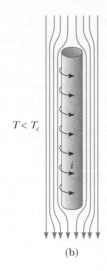

$T > T_c$

$T < T_c$

i

(a)

(b)

Figure 43.34 A superconductor in the form of a long cylinder in the presence of an external magnetic field. (a) At temperatures above T_c, the field lines penetrate the cylinder because it is in its normal state. (b) When the cylinder is cooled to $T < T_c$ and becomes superconducting, magnetic flux is excluded from its interior by the induction of surface currents.

illustrated in Figure 43.34a, so that the magnetic field penetrates the cylinder. As the cylinder is cooled to a temperature $T < T_c$, the field is expelled, as shown in Figure 43.34b. Surface currents induced on the superconductor's surface produce a magnetic field that exactly cancels the externally applied field inside the superconductor. As you would expect, the surface currents disappear when the external magnetic field is removed.

A successful theory for superconductivity in metals was published in 1957 by J. Bardeen, L. N. Cooper, and J. R. Schrieffer, and is generally called BCS theory, based on the first letters of their last names. This theory led to a Nobel Prize in physics for the three scientists in 1972. In this theory, two electrons can interact via distortions in the array of lattice ions so that there is a net attractive force between the electrons.[5] As a result, the two electrons are bound into an entity called a *Cooper pair*. The Cooper pair behaves like a particle with integral spin. Particles with integral spin are called *bosons*. (As noted in Pitfall Prevention 42.6, *fermions* make up another class of particles, those with half-integral spin.) An important feature of bosons is that they do not obey the Pauli exclusion principle. Consequently, at very low temperatures, it is possible for all bosons in a collection of such particles to be in the lowest quantum state. As a result, the entire collection of Cooper pairs in the metal is described by a single wave function. Above the energy level associated with this wave function is an energy gap equal to the binding energy of a Cooper pair. Under the action of an applied electric field, the Cooper pairs experience an electric force and move through the metal. A random scattering event of a Cooper pair from a lattice ion would represent resistance to the electric current. Such a collision would change the energy of the Cooper pair because some energy would be transferred to the lattice ion. But there are no available energy levels below that of the Cooper pair (it is already in the lowest state) and none available above, because of the energy gap. As a result, collisions do not occur and there is no resistance to the movement of Cooper pairs.

An important development in physics that elicited much excitement in the scientific community was the discovery of high-temperature copper oxide–based superconductors. The excitement began with a 1986 publication by J. Georg Bednorz (b. 1950) and K. Alex Müller (b. 1927), scientists at the IBM Zurich Research Laboratory in Switzerland. In their seminal paper,[6] Bednorz and Müller reported strong evidence for superconductivity at 30 K in an oxide of barium, lanthanum, and copper. They were awarded

[5]   A highly simplified explanation of this attraction between electrons is as follows. Around one electron, the attractive Coulomb force causes surrounding positively charged lattice ions to move inward slightly toward the electron. As a result, there is a higher concentration of positive charge in this region than elsewhere in the lattice. A second electron is attracted to the higher concentration of positive charge.

[6]   J. G. Bednorz and K. A. Müller, *Z. Phys. B* 64:189, 1986.

the Nobel Prize for physics in 1987 for their remarkable discovery. Shortly thereafter, a new family of compounds was open for investigation, and research activity in the field of superconductivity proceeded vigorously. In early 1987, groups at the University of Alabama at Huntsville and the University of Houston announced superconductivity at about 92 K in an oxide of yttrium, barium, and copper ($YBa_2Cu_3O_7$). Later that year, teams of scientists from Japan and the United States reported superconductivity at 105 K in an oxide of bismuth, strontium, calcium, and copper. More recently, scientists have reported superconductivity at temperatures as high as 150 K in an oxide containing mercury. Today, one cannot rule out the possibility of room-temperature superconductivity, and the mechanisms responsible for the behavior of high-temperature superconductors are still under investigation. The search for novel superconducting materials continues both for scientific reasons and because practical applications become more probable and widespread as the critical temperature is raised.

While BCS theory was very successful in explaining superconductivity in metals, there is currently no widely accepted theory for high-temperature superconductivity. This remains an area of active research.

SUMMARY

Take a practice test for this chapter by clicking on the Practice Test link at http://www.pse6.com.

Two or more atoms combine to form molecules because of a net attractive force between the atoms. The mechanisms responsible for molecular bonding can be classified as follows:

- **Ionic bonds** form primarily because of the Coulomb attraction between oppositely charged ions. Sodium chloride (NaCl) is one example.

- **Covalent bonds** form when the constituent atoms of a molecule share electrons. For example, the two electrons of the H_2 molecule are equally shared between the two nuclei.

- **Van der Waals bonds** are weak electrostatic bonds between molecules or between atoms that do not form ionic or covalent bonds. These bonds are responsible for the condensation of inert gas atoms and nonpolar molecules into the liquid phase.

- **Hydrogen bonds** form between the center of positive charge in a polar molecule that includes one or more hydrogen atoms and the center of negative charge in another polar molecule.

The energy of a gas molecule consists of contributions from the electronic energy in the bonds and from the translational, rotational, and vibrational motions of the molecule.

The allowed values of the rotational energy of a diatomic molecule are

$$E_{\text{rot}} = \frac{\hbar^2}{2I} J(J + 1) \qquad J = 0, 1, 2, \ldots \tag{43.6}$$

where I is the moment of inertia of the molecule and J is an integer called the **rotational quantum number.** The selection rule for transitions between rotational states is given by $\Delta J = \pm 1$.

The allowed values of the vibrational energy of a diatomic molecule are

$$E_{\text{vib}} = (v + \tfrac{1}{2}) \frac{h}{2\pi} \sqrt{\frac{k}{\mu}} \qquad v = 0, 1, 2, \ldots \tag{43.10}$$

where v is the **vibrational quantum number,** k is the force constant of the "effective spring" bonding the molecule, and μ is the **reduced mass** of the molecule. The selection rule for allowed vibrational transitions is $\Delta v = \pm 1$, and the energy difference between any two adjacent levels is the same regardless of which two levels are involved.

Bonding mechanisms in solids can be classified in a manner similar to the schemes for molecules. For example, the Na$^+$ and Cl$^-$ ions in NaCl form **ionic bonds,** while the carbon atoms in diamond form **covalent bonds.** The **metallic bond** is characterized by a net attractive force between positive ion cores and the mobile free electrons of a metal.

In the **free-electron theory of metals,** the free electrons fill the quantized levels in accordance with the Pauli exclusion principle. The number of states per unit volume available to the conduction electrons having energies between E and $E + dE$ is

$$N(E)\,dE = \frac{8\sqrt{2}\,\pi m_e^{3/2}}{h^3}\;\frac{E^{1/2}\,dE}{e^{(E-E_{\mathrm{F}})/k_{\mathrm{B}}T} + 1} \tag{43.22}$$

where E_{F} is the **Fermi energy.** At $T = 0$ K, all levels below E_{F} are filled, all levels above E_{F} are empty, and

$$E_{\mathrm{F}}(0) = \frac{h^2}{2m_e}\left(\frac{3n_e}{8\pi}\right)^{2/3} \tag{43.25}$$

where n_e is the total number of conduction electrons per unit volume. Only those electrons having energies near E_{F} can contribute to the electrical conductivity of the metal.

In a crystalline solid, the energy levels of the system form a set of **bands.** Electrons occupy the lowest energy states, with no more than one electron per state. Energy gaps are present between the bands of allowed states.

A **semiconductor** is a material having an energy gap of approximately 1 eV and a valence band that is filled at $T = 0$ K. Because of the small energy gap, a significant number of electrons can be thermally excited from the valence band into the conduction band. The band structures and electrical properties of a Group IV semiconductor can be modified by the addition of either donor atoms containing five outer-shell electrons (such as arsenic) or acceptor atoms containing three outer-shell electrons (such as indium). A semiconductor **doped** with donor impurity atoms is called an *n*-type **semiconductor,** and one doped with acceptor impurity atoms is called a *p*-type **semiconductor.** The energy levels of these impurity atoms fall within the energy gap of the material.

QUESTIONS

Note: Questions 14 and 15 in Chapter 27 can be assigned wih this chapter.

1. Discuss the three major forms of excitation of a molecule (other than translational motion) and the relative energies associated with these three forms.

2. Explain the role of the Pauli exclusion principle in describing the electrical properties of metals.

3. Discuss the properties of a material that determine whether it is a good electrical insulator or a good conductor.

4. Table 43.5 shows that the energy gaps for semiconductors decrease with increasing temperature. What do you suppose accounts for this behavior?

5. The resistivity of metals increases with increasing temperature, whereas the resistivity of an intrinsic semiconductor decreases with increasing temperature. Explain.

6. Discuss the differences in the band structures of metals, insulators, and semiconductors. How does the band-

structure model enable you to better understand the electrical properties of these materials?

7. Discuss models for the different types of bonds that form stable molecules.

8. Discuss the electrical, physical, and optical properties of ionically bonded solids. Compare your expectations with tabulated properties for such solids.

9. Discuss the electrical and physical properties of covalently bonded solids. Compare your expectations with tabulated properties for such solids.

10. Discuss the electrical and physical properties of metals.

11. When a photon is absorbed by a semiconductor, an electron–hole pair is created. Give a physical explanation of this statement using the energy-band model as the basis for your description.

12. Pentavalent atoms such as arsenic are donor atoms in a semiconductor such as silicon, while trivalent atoms such as indium are acceptors. Inspect the periodic table in Appendix C, and determine what other elements might make good donors or acceptors.

13. What are the essential assumptions made in the free-electron theory of metals? How does the energy-band model differ from the free-electron theory in describing the properties of metals?

14. How do the vibrational and rotational levels of heavy hydrogen (D_2) molecules compare with those of H_2 molecules?

15. Which is easier to excite in a diatomic molecule, rotational or vibrational motion?

16. The energies of photons of visible light range between 1.8 and 3.1 eV. Does this explain why silicon, with an energy gap of 1.1 eV (see Table 43.5), appears opaque, whereas diamond, with an energy gap of 5.5 eV, appears transparent?

17. How can the analysis of the rotational spectrum of a molecule lead to an estimate of the size of that molecule?

PROBLEMS

1, 2, 3 = straightforward, intermediate, challenging ☐ = full solution available in the *Student Solutions Manual and Study Guide*

🌀 = coached solution with hints available at http://www.pse6.com 💻 = computer useful in solving problem

▨ = paired numerical and symbolic problems

Section 43.1 Molecular Bonds

1. 🌀 **Review problem.** A K^+ ion and a Cl^- ion are separated by a distance of 5.00×10^{-10} m. Assuming the two ions act like point charges, determine (a) the force each ion exerts on the other and (b) the potential energy of the two-ion system in electron volts.

2. Potassium chloride is an ionically bonded molecule, sold as a salt substitute for use in a low-sodium diet. The electron affinity of chlorine is 3.6 eV. An energy input of 0.7 eV is required to form separate K^+ and Cl^- ions from separate K and Cl atoms. What is the ionization energy of K?

3. One description of the potential energy of a diatomic molecule is given by the Lennard–Jones potential,

$$U = \frac{A}{r^{12}} - \frac{B}{r^6}$$

where A and B are constants. Find, in terms of A and B, (a) the value r_0 at which the energy is a minimum and (b) the energy E required to break up a diatomic molecule. (c) Evaluate r_0 in meters and E in electron volts for the H_2 molecule. In your calculations, take $A = 0.124 \times 10^{-120}$ eV·m^{12} and $B = 1.488 \times 10^{-60}$ eV·m^6. (*Note:* Although this potential is widely used for modeling, it is known to have serious defects. For example, its behavior at both small and large values of r is significantly in error.)

4. Potassium iodide can be taken as a medicine to reduce radiation dosage to the thyroid gland before or after exposure to radioactive iodine. In the potassium iodide molecule, assume that the K and I atoms bond ionically by the transfer of one electron from K to I. (a) The ionization energy of K is 4.34 eV, and the electron affinity of I is 3.06 eV. What energy is needed to transfer an electron from K to I, to form K^+ and I^- ions from neutral atoms? This is sometimes called the activation energy E_a. (b) A model potential energy function for the KI molecule is the Lennard–Jones potential:

$$U(r) = 4\epsilon \left[\left(\frac{\sigma}{r} \right)^{12} - \left(\frac{\sigma}{r} \right)^6 \right] + E_a$$

where r is the internuclear separation distance, and σ and ϵ are adjustable parameters. The E_a term is added to ensure the correct asymptotic behavior at large r. At the equilibrium separation distance, $r = r_0 = 0.305$ nm, $U(r)$ is a minimum, and $dU/dr = 0$. Now $U(r_0)$ is the negative of the dissociation energy: $U(r_0) = -3.37$ eV. Evaluate σ and ϵ. (c) Calculate the force needed to break up a KI molecule. (d) Calculate the force constant for small oscillations about $r = r_0$. (*Suggestion:* Set $r = r_0 + s$ where $s/r_0 \ll 1$, and expand $U(r)$ in powers of s/r_0 up to second-order terms.)

5. A van der Waals dispersion force between helium atoms produces a very shallow potential well, with a depth on the order of 1 meV. At about what temperature would you expect helium to condense?

Section 43.2 Energy States and Spectra of Molecules

6. The cesium iodide (CsI) molecule has an atomic separation of 0.127 nm. (a) Determine the energy of the lowest excited rotational state and the frequency of the photon absorbed in the $J = 0$ to $J = 1$ transition. (b) **What If?** What would be the fractional change in this frequency if the estimate of the atomic separation is off by 10%?

7. 🌀 An HCl molecule is excited to its first rotational energy level, corresponding to $J = 1$. If the distance between its nuclei is 0.127 5 nm, what is the angular speed of the molecule about its center of mass?

8. The CO molecule makes a transition from the $J = 1$ to the $J = 2$ rotational state when it absorbs a photon of frequency 2.30×10^{11} Hz. Find the moment of inertia of this molecule from these data.

9. A diatomic molecule consists of two atoms having masses m_1 and m_2 and separated by a distance r. Show that the moment of inertia about an axis through the center of mass of the molecule is given by Equation 43.3, $I = \mu r^2$.

10. (a) Calculate the moment of inertia of an NaCl molecule about its center of mass. The atoms are separated by a

distance $r = 0.28$ nm. (b) Calculate the wavelength of radiation emitted when an NaCl molecule undergoes a transition from the $J = 2$ state to the $J = 1$ state.

11. The rotational spectrum of the HCl molecule contains lines with wavelengths of 0.060 4, 0.069 0, 0.080 4, 0.096 4, and 0.120 4 mm. What is the moment of inertia of the molecule?

12. Use the data in Table 43.2 to calculate the minimum amplitude of vibration for (a) the HI molecule and (b) the HF molecule. Which has the weaker bond?

13. [icon] Taking the effective force constant of a vibrating HCl molecule as $k = 480$ N/m, find the energy difference between the ground state and the first excited vibrational level.

14. The nuclei of the O_2 molecule are separated by 1.20×10^{-10} m. The mass of each oxygen atom in the molecule is 2.66×10^{-26} kg. (a) Determine the rotational energies of an oxygen molecule in electron volts for the levels corresponding to $J = 0$, 1, and 2. (b) The effective force constant k between the atoms in the oxygen molecule is 1 177 N/m. Determine the vibrational energies (in electron volts) corresponding to $v = 0$, 1, and 2.

15. Figure P43.15 is a model of a benzene molecule. All atoms lie in a plane, and the carbon atoms form a regular hexagon, as do the hydrogen atoms. The carbon atoms are 0.110 nm apart center-to-center. Determine the allowed energies of rotation about an axis perpendicular to the plane of the paper through the center point O. Hydrogen and carbon atoms have masses of 1.67×10^{-27} kg and 1.99×10^{-26} kg, respectively.

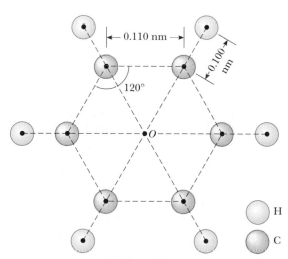

Figure P43.15

16. Calculate the longest wavelength in the rotational spectrum of HCl. Take the Cl atom to be the isotope ^{35}Cl. The equilibrium separation of the H and Cl atoms is 0.127 46 nm. The atomic mass of the H atom is 1.007 825 u, and that of the ^{35}Cl atom is 34.968 853 u. (b) **What If?** Repeat the calculation in (a), but take the Cl atom to be the isotope ^{37}Cl, which has atomic mass 36.965 903 u. The equilibrium separation distance is the same as in (a). (c) Naturally occurring chlorine contains approximately three parts of ^{35}Cl to one part of ^{37}Cl. Because of the two different Cl

masses, each line in the microwave rotational spectrum of HCl is split into a doublet. Calculate the doublet separation for the longest wavelength. (Figure 43.9 shows the doublets in the infrared vibrational spectrum).

17. If the CO molecule were rigid, the rotational transition into what J state would absorb the same wavelength photon as the 0 to 1 vibrational transition? (Use information given in Table 43.2.)

18. Calculate the moment of inertia of an HCl molecule from its infrared absorption spectrum shown in Figure 43.9.

19. An H_2 molecule is in its vibrational and rotational ground states. It absorbs a photon of wavelength 2.211 2 μm and jumps to the $v = 1$, $J = 1$ energy level. It then drops to the $v = 0$, $J = 2$ energy level, while emitting a photon of wavelength 2.405 4 μm. Calculate (a) the moment of inertia of the H_2 molecule about an axis through its center of mass and perpendicular to the H–H bond (b) the vibrational frequency of the H_2 molecule (c) the equilibrium separation distance for this molecule.

20. Photons of what frequencies can be spontaneously emitted by CO molecules in the state with $v = 1$ and $J = 0$?

21. Most of the mass of an atom is in its nucleus. Model the mass distribution in a diatomic molecule as two spheres, each of radius 2.00×10^{-15} m and mass 1.00×10^{-26} kg, located at points along the x axis in Figure 43.5a, and separated by 2.00×10^{-10} m. Rotation about the axis joining the nuclei in the diatomic molecule is ordinarily ignored because the first excited state would have an energy that is too high to access. To see why, calculate the ratio of the energy of the first excited state for rotation about the x axis to the energy of the first excited state for rotation about the y axis.

Section 43.3   Bonding in Solids

22. Use a magnifying glass to look at the table salt that comes out of a salt shaker. Compare what you see to Figure 43.10(a). The distance between a sodium ion and a nearest-neighbor chlorine ion is 0.261 nm. (a) Make an order-of-magnitude estimate of the number N of atoms in a typical grain of salt. (b) **What If?** Suppose that you had a number of grains of salt equal to this number N. What would be the volume of this quantity of salt?

23. Use Equation 43.18 to calculate the ionic cohesive energy for NaCl. Take $\alpha = 1.747$ 6, $r_0 = 0.281$ nm, and $m = 8$.

24. The distance between the K^+ and Cl^- ions in a KCl crystal is 0.314 nm. Calculate the distances from one K^+ ion to its nearest-neighbor K^+ ions, to its second-nearest-neighbor K^+ ions, and to its third-nearest-neighbor K^+ ions.

25. Consider a one-dimensional chain of alternating positive and negative ions. Show that the potential energy associated with one of the ions and its interactions with the rest of this hypothetical crystal is

$$U(r) = -k_e \alpha \frac{e^2}{r}$$

where the Madelung constant is $\alpha = 2 \ln 2$ and r is the interionic spacing. [*Suggestion:* Use the series expansion for $\ln(1 + x)$.]

Section 43.4   Free-Electron Theory of Metals

Section 43.5   Band Theory of Solids

26. Show that Equation 43.25 can be expressed as $E_F = (3.65 \times 10^{-19})\, n_e^{2/3}$ eV where E_F is in electron volts when n_e is in electrons per cubic meter.

27. The Fermi energy for silver is 5.48 eV. Silver has a density of 10.6×10^3 kg/m^3 and an atomic mass of 108. Use this information to show that silver has one free electron per atom.

28. (a) Find the typical speed of a conduction electron in copper, taking its kinetic energy as equal to the Fermi energy, 7.05 eV. (b) How does this compare with a drift speed of 0.1 mm/s?

29. Sodium is a monovalent metal having a density of 0.971 g/cm^3 and a molar mass of 23.0 g/mol. Use this information to calculate (a) the density of charge carriers and (b) the Fermi energy.

30. When solid silver starts to melt, what is the approximate fraction of the conduction electrons that are thermally excited above the Fermi level?

31. Calculate the energy of a conduction electron in silver at 800 K, assuming the probability of finding an electron in that state is 0.950. The Fermi energy is 5.48 eV at this temperature.

32. Consider a cube of gold 1.00 mm on an edge. Calculate the approximate number of conduction electrons in this cube whose energies lie in the range 4.000 to 4.025 eV.

33. Show that the average kinetic energy of a conduction electron in a metal at 0 K is $E_{av} = \frac{3}{5}E_F$. (*Suggestion:* In general, the average kinetic energy is

$$E_{av} = \frac{1}{n_e} \int E\, N(E)\, dE$$

where n_e is the density of particles, $N(E)\, dE$ is given by Equation 43.22, and the integral is over all possible values of the energy.)

34. **Review problem.** An electron moves in a three-dimensional box of edge length L and volume L^3. The wave function of the particle is $\psi = A \sin(k_x x) \sin(k_y y) \sin(k_z z)$. Show that its energy is given by Equation 43.20,

$$E = \frac{\hbar^2 \pi^2}{2m_e L^2}\, (n_x^2 + n_y^2 + n_z^2)$$

where the quantum numbers (n_x, n_y, n_z) are integers ≥ 1. (*Suggestions:* The Schrödinger equation in three dimensions may be written

$$\frac{\hbar^2}{2m}\left(\frac{\partial^2 \psi}{\partial x^2} + \frac{\partial^2 \psi}{\partial y^2} + \frac{\partial^2 \psi}{\partial z^2}\right) = (U - E)\psi$$

To confine the electron inside the box, take $U = 0$ inside and $U = \infty$ outside.)

35. (a) Consider a system of electrons confined to a three-dimensional box. Calculate the ratio of the number of allowed energy levels at 8.50 eV to the number at 7.00 eV. (b) **What If?** Copper has a Fermi energy of 7.0 eV at 300 K. Calculate the ratio of the number of occupied levels at an energy of 8.50 eV to the number at the Fermi energy. Compare your answer with that obtained in part (a).

Section 43.6   Electrical Conduction in Metals, Insulators, and Semiconductors

36. The energy gap for silicon at 300 K is 1.14 eV. (a) Find the lowest-frequency photon that will promote an electron from the valence band to the conduction band. (b) What is the wavelength of this photon?

37. Light from a hydrogen discharge tube is incident on a CdS crystal. Which spectral lines from the Balmer series are absorbed and which are transmitted?

38. A light-emitting diode (LED) made of the semiconductor GaAsP emits red light ($\lambda = 650$ nm). Determine the energy-band gap E_g in the semiconductor.

39. Most solar radiation has a wavelength of 1 μm or less. What energy gap should the material in a solar cell have in order to absorb this radiation? Is silicon appropriate (see Table 43.5)?

40. Assume you are to build a scientific instrument that is thermally isolated from its surroundings, but such that you can use an external laser to raise the temperature of a target inside it. (It might be a calorimeter, but these design criteria could apply to other devices as well.) Since you know that diamond is transparent and a good thermal insulator, you decide to use a diamond window in the apparatus. Diamond has an energy gap of 5.5 eV between its valence and conduction bands. What is the shortest laser wavelength you can use to warm the sample inside?

41. **Review problem.** Silicon is a semiconductor widely used in computer chips and other electronic devices. Its most important properties result from doping it with impurities in order to control its electrical conductivity. Phosphorus, which is adjacent to silicon in the periodic table, has five outer valence electrons as compared to four for silicon. When a phosphorus atom is substituted for a silicon atom in a crystal, four of the phosphorus valence electrons form bonds with neighboring atoms and the remaining electron is much more loosely bound. You can model the electron as free to move through the crystal lattice. The phosphorus nucleus has one more positive charge than does the silicon nucleus, however, so the extra electron provided by the phosphorus atom is attracted to this single nuclear charge $+ e$. The energy levels of the extra electron are similar to those of the electron in the Bohr hydrogen atom with two important exceptions. First, the Coulomb attraction between the electron and the positive charge on the phosphorus nucleus is reduced by a factor of $1/\kappa$ from what it would be in free space (see Eq. 26.21), so the orbit radii are greatly increased. Here κ is the dielectric constant of the crystal, with a value of 11.7 in silicon. Second, the influence of the periodic electric potential of the lattice causes the electron to move as if it had an effective mass m^*, quite different from the mass m_e of a free electron. One can use the Bohr model of hydrogen to obtain fairly accurate values for the allowed energy levels of the extra electron. These energy levels, called donor states, play an important role in semiconductor devices. Assume that $m^* = 0.220\, m_e$. Calculate the energy and the radius for an extra electron in the first Bohr orbit around a donor atom in silicon.

Section 43.7   Semiconductor Devices

> *Note:* Problem 74 in Chapter 27 can be assigned with this section.

42. For what value of the bias voltage ΔV in Equation 43.27 does (a) $I = 9.00I_0$? (b) **(What If?)** $I = -0.900I_0$? Assume $T = 300$ K.

43. The diode shown in Figure 43.28 is connected in series with a battery and a 150-Ω resistor. What battery emf is required for a current of 25.0 mA?

44. You put a diode in a microelectronic circuit to protect the system in case an untrained person installs the battery backward. In the correct forward-bias situation, the current is 200 mA with a potential difference of 100 mV across the diode at room temperature (300 K). If the battery were reversed, what would be the magnitude of the current through the diode?

45. A diode, a resistor, and a battery are connected in a series circuit. The diode is at a temperature for which $k_B T = 25.0$ meV, and the saturation value I_0 of the current is 1.00 μA. The resistance of the resistor is 745 Ω, and the battery maintains a constant potential difference between its terminals of 2.42 V. (a) Find graphically the current in the loop. Proceed as follows. On the same axes, draw graphs of the diode current I_D and the current in the wire I_W versus the voltage across the diode ΔV. Choose values of ΔV ranging from 0 to 0.250 V in steps of 0.025 V. Determine the value of ΔV at the intersection of the two graph lines, and calculate the corresponding currents I_D and I_W. Do they agree? (b) Find the ohmic resistance of the diode, which is defined as the ratio $\Delta V/I_D$. (c) Find the dynamic resistance of the diode, which is defined as the derivative $d(\Delta V)/dI_D$.

Section 43.8   Superconductivity

> *Note:* Problem 28 in Chapter 30 and Problems 76 through 79 in Chapter 32 can be assigned wih this section.

46. A thin rod of superconducting material 2.50 cm long is placed into a 0.540-T magnetic field with its cylindrical axis along the magnetic field lines. (a) Sketch the directions of the applied field and the induced surface current. (b) Find the magnitude of the surface current on the curved surface of the rod.

47. Determine the current generated in a superconducting ring of niobium metal 2.00 cm in diameter when a 0.020 0-T magnetic field directed perpendicular to the ring is suddenly decreased to zero. The inductance of the ring is 3.10×10^{-8} H.

48. *A convincing demonstration of zero resistance.* A direct and relatively simple demonstration of zero DC resistance can be carried out using the four-point probe method. The probe shown in Figure P43.48 consists of a disk of $YBa_2Cu_3O_7$ (a high-T_c superconductor) to which four wires are attached by indium solder or some other suitable contact material. Current is maintained through the sample by applying a DC voltage between points a and b, and it is measured with a DC ammeter. The current can be varied with the variable resistance R. The potential difference ΔV_{cd} between c and d is measured with a digital voltmeter. When the probe is immersed in liquid nitrogen, the sample quickly cools to 77 K, below the critical temperature of the material, 92 K. The current remains approximately constant, but ΔV_{cd} drops *abruptly to zero.* (a) Explain this observation on the basis of what you know about superconductors. (b) The data in Table P43.48 represent actual values of ΔV_{cd} for different values of I taken on the sample at room temperature. A 6-V battery in series with a variable resistor R supplied the current. The values of R ranged from 10 Ω to 100 Ω. The data are from one author's laboratory. Make an I-ΔV plot of the data, and determine whether the sample behaves in a linear manner. From the data obtain a value for the DC resistance of the sample at room temperature. (c) At room temperature it is found that $\Delta V_{cd} = 2.234$ mV for $I = 100.3$ mA, but after the sample is cooled to 77 K, $\Delta V_{cd} = 0$ and $I = 98.1$ mA. What do you think might cause the slight decrease in current?

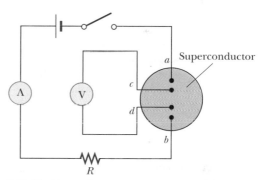

Figure P43.48 Circuit diagram used in the four-point probe measurement of the DC resistance of a sample. A DC digital ammeter is used to measure the current, and the potential difference between c and d is measured with a DC digital voltmeter. Note that there is no voltage source in the inner loop circuit where ΔV_{cd} is measured.

Table P43.48

Current Versus Potential Difference ΔV_{cd} Measured in a Bulk Ceramic Sample of $YBa_2Cu_3O_{7-\delta}$ at Room Temerature	
I (mA)	ΔV_{cd} (mV)
57.8	1.356
61.5	1.441
68.3	1.602
76.8	1.802
87.5	2.053
102.2	2.398
123.7	2.904
155	3.61

Additional Problems

49. As you will learn in Chapter 44, carbon-14 (^{14}C) is an isotope of carbon. It has the same chemical properties and electronic structure as the much more abundant isotope carbon-12 (^{12}C) but has different nuclear properties. Its mass is 14 u, greater because it has two extra neutrons in its nucleus. Assume that the CO molecular potential is the same for both isotopes of carbon, and that the tables and examples in Section 43.2 refer to carbon monoxide with carbon-12 atoms. (a) What is the vibrational frequency of ^{14}CO? (b) What is the moment of inertia of ^{14}CO? (c) What wavelengths of light can be absorbed by ^{14}CO in the ($v = 0, J = 10$) state that will cause it to end up in the $v = 1$ level?

50. The effective spring constant associated with bonding in the N_2 molecule is $2\,297$ N/m. The nitrogen atoms each have a mass of 2.32×10^{-26} kg, and their nuclei are 0.120 nm apart. Assume that the molecule is rigid and in the ground vibrational state. Calculate the J value of the rotational state that has the same energy as the first excited vibrational state.

51. The hydrogen molecule comes apart (dissociates) when it is excited internally by 4.5 eV. Assuming that this molecule behaves like a harmonic oscillator having classical angular frequency $\omega = 8.28 \times 10^{14}$ rad/s, find the highest vibrational quantum number for a state below the 4.5-eV dissociation energy.

52. Under pressure, liquid helium can solidify as each atom bonds with four others, and each bond has an average energy of 1.74×10^{-23} J. Find the latent heat of fusion for helium in joules per gram. (The molar mass of He is 4.00 g/mol.)

53. Show that the ionic cohesive energy of an ionically bonded solid is given by Equation 43.18. (*Suggestion:* Start with Equation 43.17, and note that $dU/dr = 0$ at $r = r_0$.)

54. The dissociation energy of ground-state molecular hydrogen is 4.48 eV, while it only takes 3.96 eV to dissociate it when it starts in the first excited vibrational state with $J = 0$. Using this information, determine the depth of the H_2 molecular potential-energy function.

55. A particle moves in one-dimensional motion through a field for which the potential energy of the particle–field system is

$$U(x) = \frac{A}{x^3} - \frac{B}{x}$$

where $A = 0.150$ eV·nm^3 and $B = 3.68$ eV·nm. The general shape of this function is shown in Figure 43.11, where x replaces r. (a) Find the static equilibrium position x_0 of the particle. (b) Determine the depth U_0 of this potential well. (c) In moving along the x axis, what maximum force toward the negative x direction does the particle experience?

56. A particle of mass m moves in one-dimensional motion through a field for which the potential energy of the particle–field system is

$$U(x) = \frac{A}{x^3} - \frac{B}{x}$$

where A and B are constants with appropriate units. The general shape of this function is shown in Figure 43.11, where x replaces r. (a) Find the static equilibrium position x_0 of the particle in terms of m, A, and B. (b) Determine the depth U_0 of this potential well. (c) In moving along the x axis, what maximum force toward the negative x direction does the particle experience?

57. As an alternative to Equation 43.1, another useful model for the potential energy of a diatomic molecule is the Morse potential

$$U(r) = B[e^{-a(r-r_0)} - 1]^2$$

where B, a, and r_0 are parameters used to adjust the shape of the potential and its depth. (a) What is the equilibrium separation of the nuclei? (b) What is the depth of the potential well, i.e., the difference in energy between the potential's minimum value and its asymptote as r approaches infinity? (c) If μ is the reduced mass of the system of two nuclei, what is the vibrational frequency of the diatomic molecule in its ground state? (Assume that the potential is nearly parabolic about the well minimum). (d) What amount of energy needs to be supplied to the ground-state molecule to separate the two nuclei to infinity?

58. The Fermi-Dirac distribution function can be written as

$$f(E) = \frac{1}{e^{(E-E_F)/k_B T} + 1} = \frac{1}{e^{(E/E_F - 1)T_F/T} + 1}$$

where T_F is the *Fermi temperature*, defined according to

$$k_B T_F \equiv E_F$$

Write a spreadsheet to calculate and plot $f(E)$ versus E/E_F at a fixed temperature T. Examine the curves obtained for $T = 0.1T_F$, $0.2T_F$, and $0.5T_F$.

59. The Madelung constant for sodium chloride may be found by summing an infinite alternating series of terms giving the electric potential energy between a Na$^+$ ion and its six nearest Cl$^-$ neighbors, its twelve next-nearest Na$^+$ neighbors, and so on (Fig. 43.10a). (a) From this expression, show that the first three terms of the series yield $\alpha = 2.13$ for the NaCl structure. (b) **What If?** Does this series converge rapidly? Calculate the fourth term as a check.

Answers to Quick Quizzes

43.1 (a) van der Waals (b) ionic (c) hydrogen (d) covalent

43.2 (c). Equation 43.7 shows that the energy spacing is proportional to J, the quantum number of the higher state in the transition. Because the frequency of the absorbed photon is proportional to the energy separation of the states, the frequencies are in the same ratio as the energy separations.

43.3 (a). This is similar to Quick Quiz 43.2, except that the vibrational states are all separated by the same energy difference.

43.4 (b). The $J = 0$ to $J = 0$ transition is forbidden according to the selection rules.

43.5 (b). At a higher temperature, transitions from higher excited states will be more likely, so that peaks farther from the gap will be more intense. In Equation 43.15, at a higher temperature, the exponential factor does not decrease as rapidly with J if the temperature is higher.

43.6 (b). In Equation 43.21, the density of states increases with energy. Thus, at the higher energy of 6 eV, the states are closer together.

43.7 (a). Because the energy of 6 eV is higher than the Fermi energy, there are essentially no electrons in states near 6 eV.

43.8 A: semiconductor; B: conductor; C: insulator

43.9 (c). Photons can be emitted when electrons make a transition from anywhere in the conduction band above its bottom to anywhere in the valence band below its top. The lowest-energy transition, resulting in the lowest-frequency photon, is that shown in Figure 43.29a.

Chapter 44

Nuclear Structure

CHAPTER OUTLINE

44.1 Some Properties of Nuclei

44.2 Nuclear Binding Energy

44.3 Nuclear Models

44.4 Radioactivity

44.5 The Decay Processes

44.6 Natural Radioactivity

44.7 Nuclear Reactions

44.8 Nuclear Magnetic Resonance and Magnetic Resonance Imaging

▲ The Ice Man, discovered in 1991 when an Italian glacier melted enough to expose his remains. His possessions, particularly his tools, have shed light on the way people lived in the Bronze Age. Radioactivity was used to determine how long ago this person lived. (Paul Hanny/Gamma Liaison)

The year 1896 marks the birth of nuclear physics when the French physicist Henri Becquerel (1852–1908) discovered radioactivity in uranium compounds. This discovery prompted scientists to investigate the details of radioactivity and, ultimately, the structure of the nucleus. Pioneering work by Rutherford showed that the radiation emitted from radioactive substances is of three types: alpha, beta, and gamma rays, classified according to the nature of their electric charge and their ability to penetrate matter and ionize air. Later experiments showed that alpha rays are helium nuclei, beta rays are electrons, and gamma rays are high-energy photons.

In 1911, Rutherford, Geiger, and Marsden performed the alpha-particle scattering experiments described in Section 42.2. These experiments established that (a) the nucleus of an atom can be regarded as essentially a point mass and point charge and (b) most of the atomic mass is contained in the nucleus. Subsequent studies revealed the presence of a new type of force, the short-range nuclear force, which is predominant at particle separation distances less than approximately 10^{-14} m and is zero for large distances.

Other milestones in the development of nuclear physics include

- The observation of nuclear reactions in 1930 by Cockroft and Walton using artificially accelerated nuclei;
- The discovery of the neutron in 1932 by Chadwick and the conclusion that neutrons make up about half of the nucleus;
- The discovery of artificial radioactivity in 1933 by Irène and Frédéric Joliot-Curie;
- The discovery of nuclear fission in 1938 by Hahn and Strassmann;
- The development of the first controlled fission reactor in 1942 by Fermi and his collaborators.

In this chapter we discuss the properties and structure of the atomic nucleus. We start by describing the basic properties of nuclei, followed by a discussion of nuclear forces and binding energy, nuclear models, and the phenomenon of radioactivity. We then discuss various processes by which nuclei decay, and describe nuclear reactions.

44.1 Some Properties of Nuclei

All nuclei are composed of two types of particles: protons and neutrons. The only exception is the ordinary hydrogen nucleus, which is a single proton. We describe the atomic nucleus by the number of protons and neutrons it contains, using the following quantities:

- The **atomic number** Z, which equals the number of protons in the nucleus (sometimes called the *charge number*);

▲ **PITFALL PREVENTION**

44.1 Mass Number Is Not Atomic Mass

The mass number A should not be confused with the atomic mass, defined in Section 1.3. Mass number is an integer specific to an isotope and has no units—it is simply a count of the number of nucleons. Atomic mass is generally not an integer, because it is an average of the masses of the naturally occurring isotopes of a given element, and has units of u.

- The **neutron number** N, which equals the number of neutrons in the nucleus;

- The **mass number** $A = Z + N$, which equals the number of **nucleons** (neutrons plus protons) in the nucleus.

In representing nuclei, it is convenient to use the symbol $^A_Z X$ to show how many protons and neutrons are present, where X represents the chemical symbol of the element. For example, $^{56}_{26}Fe$ (iron) has mass number 56 and atomic number 26; therefore, it contains 26 protons and 30 neutrons. When no confusion is likely to arise, we omit the subscript Z because the chemical symbol can always be used to determine Z. Thus, $^{56}_{26}Fe$ is the same as ^{56}Fe and can also be expressed as "iron-56."

The nuclei of all atoms of a particular element contain the same number of protons but often contain different numbers of neutrons. Nuclei that are related in this way are called **isotopes.** The isotopes of an element have the same Z value but different N and A values.

The natural abundance of isotopes can differ substantially. For example $^{11}_6C$, $^{12}_6C$, $^{13}_6C$, and $^{14}_6C$ are four isotopes of carbon. The natural abundance of the $^{12}_6C$ isotope is approximately 98.9%, whereas that of the $^{13}_6C$ isotope is only about 1.1%. Some isotopes, such as $^{11}_6C$ and $^{14}_6C$, do not occur naturally but can be produced by nuclear reactions in the laboratory or by cosmic rays.

Even the simplest element, hydrogen, has isotopes: ^1_1H, the ordinary hydrogen nucleus; ^2_1H, deuterium; and ^3_1H, tritium.

Quick Quiz 44.1 Consider the following three nuclei: ^{12}C, ^{13}N, ^{14}O. These nuclei have the same (a) number of protons (b) number of neutrons (c) number of nucleons.

Quick Quiz 44.2 Consider the following three nuclei: ^{12}N, ^{13}N, ^{14}N. These nuclei have the same (a) number of protons (b) number of neutrons (c) number of nucleons.

Quick Quiz 44.3 Consider the following three nuclei: ^{14}C, ^{14}N, ^{14}O. These nuclei have the same (a) number of protons (b) number of neutrons (c) number of nucleons.

Charge and Mass

The proton carries a single positive charge e, equal in magnitude to the charge $-e$ on the electron ($e = 1.6 \times 10^{-19}$ C). The neutron is electrically neutral, as its name implies. Because the neutron has no charge, it was difficult to detect with early experimental apparatus and techniques. Today we can detect neutrons relatively easily with devices such as plastic scintillators.

Nuclear masses can be measured with great precision with the use of a mass spectrometer (see Section 29.5) and by the analysis of nuclear reactions. The proton is approximately 1 836 times as massive as the electron, and the masses of the proton and the neutron are almost equal. The atomic mass unit u, first introduced in Section 1.3, is defined in such a way that the mass of one atom of the isotope ^{12}C is exactly 12 u, where 1 u = $1.660\,539 \times 10^{-27}$ kg. According to this definition, the proton and

Table 44.1

Masses of Selected Particles in Various Units			
	Mass		
Particle	**kg**	**u**	**MeV/c^2**
Proton	$1.672\,62 \times 10^{-27}$	1.007 276	938.28
Neutron	$1.674\,93 \times 10^{-27}$	1.008 665	939.57
Electron	$9.109\,39 \times 10^{-31}$	$5.48\,579 \times 10^{-4}$	0.510 999
$^1_1$H atom	$1.673\,53 \times 10^{-27}$	1.007 825	938.783
$^4_2$He nucleus	$6.644\,66 \times 10^{-27}$	4.001 506	3 727.38
$^{12}_6$C atom	$1.992\,65 \times 10^{-27}$	12.000 000	11 177.9

neutron each have a mass of approximately 1 u, and the electron has a mass that is only a small fraction of this value:

$$\text{Mass of proton} = 1.007\ 276\ \text{u}$$

$$\text{Mass of neutron} = 1.008\ 665\ \text{u}$$

$$\text{Mass of electron} = 0.000\ 548\ 6\ \text{u}$$

One might wonder how six protons and six neutrons, each having a mass larger than 1 u, can be combined with six electrons to form a carbon-12 atom having a mass of exactly 12 u. The bound system of ^{12}C has a lower rest energy (Section 39.8) than that of six separate protons and six separate neutrons. According to Equation 39.24, $E_R = mc^2$, this lower rest energy corresponds to a smaller mass for the bound system. The extra mass of the separated particles is equal to the binding energy when the particles are combined to form the nucleus. We shall discuss this point in more detail in Section 44.3.

Because the rest energy of a particle is given by $E_R = mc^2$, it is often convenient to express the atomic mass unit in terms of its *rest energy equivalent*. For one atomic mass unit, we have

$$E_R = mc^2 = (1.660\ 539 \times 10^{-27}\ \text{kg})(2.997\ 92 \times 10^8\ \text{m/s})^2 = 931.494\ \text{MeV}$$

where we have used the conversion $1\ \text{eV} = 1.602\ 176 \times 10^{-19}\ \text{J}$.

Based on the rest-energy expression in Equation 39.24, nuclear physicists often express mass in terms of the unit MeV/c^2. The masses of several nuclei and atoms are given in Table 44.1. The masses and some other properties of selected isotopes are provided in Appendix A.3.

Example 44.1   The Atomic Mass Unit

Use Avogadro's number to show that $1\ \text{u} = 1.66 \times 10^{-27}$ kg.

Solution From the definition of the mole given in Section 19.5, we know that exactly 12 g $(= 1\ \text{mol})$ of ^{12}C contains Avogadro's number of atoms, where $N_A = 6.02 \times 10^{23}$ atoms/mol. Thus, the mass of one carbon atom is

$$\text{Mass of one } {}^{12}\text{C atom} = \frac{0.012\ \text{kg}}{6.02 \times 10^{23}\ \text{atoms}}$$

$$= 1.99 \times 10^{-26}\ \text{kg}$$

Because one atom of ^{12}C is defined to have a mass of 12.0 u, we find that

$$1\ \text{u} = \frac{1.99 \times 10^{-26}\ \text{kg}}{12.0} = 1.66 \times 10^{-27}\ \text{kg}$$

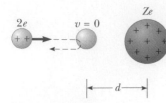

Active Figure 44.1 An alpha particle on a head-on collision course with a nucleus of charge Ze. Because of the Coulomb repulsion between the charges of the same sign, the alpha particle approaches to a distance d from the nucleus, called the distance of closest approach.

At the Active Figures link at http://www.pse6.com, adjust the atomic number of the target nucleus and the kinetic energy of the alpha particle. Then observe the approach of the alpha particle toward the nucleus.

Figure 44.2 A nucleus can be modeled as a cluster of tightly packed spheres, where each sphere is a nucleon.

Nuclear radius

The Size and Structure of Nuclei

In Rutherford's scattering experiments, positively charged nuclei of helium atoms (alpha particles) were directed at a thin piece of metallic foil. As the alpha particles moved through the foil, they often passed near a metal nucleus. Because of the positive charge on both the incident particles and the nuclei, the particles were deflected from their straight-line paths by the Coulomb repulsive force. Some particles were even deflected straight backward! These particles apparently were moving directly toward a nucleus, on a head-on collision course.

Rutherford used conservation of energy for an isolated system to find an expression for the separation distance d at which an alpha particle approaching a nucleus head-on is turned around by Coulomb repulsion. In such a head-on collision, he reasoned, the kinetic energy of the incoming particle must be converted completely to electric potential energy of the alpha-particle–nucleus system when the particle stops momentarily at the point of closest approach before moving back along the same path (Fig. 44.1). If we equate the initial and final energies of the system, we have

$$\frac{1}{2}mv^2 = k_e\frac{q_1q_2}{r} = k_e\frac{(2e)(Ze)}{d}$$

where m is the mass of the alpha particle and Z is the atomic number of the target nucleus. Solving for d, the distance of closest approach, we obtain

$$d = \frac{4k_e Z e^2}{mv^2}$$

From this expression, Rutherford found that the alpha particles approached nuclei to within 3.2×10^{-14} m when the foil was made of gold. Thus, the radius of the gold nucleus must be less than this value. For silver atoms, the distance of closest approach was found to be 2×10^{-14} m. From the results of his scattering experiments, Rutherford concluded that the positive charge in an atom is concentrated in a small sphere, which he called the nucleus, whose radius is no greater than about 10^{-14} m.

Because such small lengths are common in nuclear physics, an often-used convenient length unit is the femtometer (fm), which is sometimes called the **fermi** and is defined as

$$1 \text{ fm} \equiv 10^{-15} \text{ m}$$

In the early 1920s it was known that the nucleus of an atom contains Z protons and has a mass nearly equivalent to that of A protons, where on the average $A \approx 2Z$ for lighter nuclei ($Z \leq 20$) and $A > 2Z$ for heavier nuclei. To account for the nuclear mass, Rutherford proposed that each nucleus must also contain $A - Z$ neutral particles that he called neutrons. In 1932, the British physicist James Chadwick (1891–1974) discovered the neutron and was awarded the Nobel Prize for this important work.

Since the time of Rutherford's scattering experiments, a multitude of other experiments have shown that most nuclei are approximately spherical and have an average radius given by

$$r = r_0 A^{1/3} \tag{44.1}$$

where r_0 is a constant equal to 1.2×10^{-15} m and A is the mass number. Because the volume of a sphere is proportional to the cube of its radius, it follows from Equation 44.1 that the volume of a nucleus (assumed to be spherical) is directly proportional to A, the total number of nucleons. This proportionality suggests that *all nuclei have nearly the same density*. When nucleons combine to form a nucleus, they combine as though they were tightly packed spheres (Fig. 44.2). This fact has led to an analogy between the nucleus and a drop of liquid, in which the density of the drop is independent of its size. We shall discuss the liquid-drop model of the nucleus in Section 44.3.

Example 44.2 The Volume and Density of a Nucleus

For a nucleus of mass number A, find

(A) an approximate expression for the mass of the nucleus,

(B) an expression for the volume of this nucleus in terms of A, and

(C) a numerical value for its density.

Solution (A) The mass of the proton is approximately equal to that of the neutron. Thus, if the mass of one of these particles is m, the mass of the nucleus is approximately Am.

(B) Assuming the nucleus is spherical and using Equation 44.1, we find that the volume is

$$V_{\text{nucleus}} = \tfrac{4}{3}\pi r^3 = \boxed{\tfrac{4}{3}\pi r_0^3 A}$$

(C) The nuclear density is

$$\rho = \frac{m_{\text{nucleus}}}{V_{\text{nucleus}}} = \frac{Am}{\tfrac{4}{3}\pi r_0^3 A} = \frac{3m}{4\pi r_0^3}$$

Taking $r_0 = 1.2 \times 10^{-15}$ m and $m = 1.67 \times 10^{-27}$ kg, we find that

$$\rho = \frac{3(1.67 \times 10^{-27}\text{ kg})}{4\pi(1.2 \times 10^{-15}\text{ m})^3} = \boxed{2.3 \times 10^{17}\text{ kg/m}^3}$$

The nuclear density is approximately 2.3×10^{14} times as great as the density of water ($\rho_{\text{water}} = 1.0 \times 10^3$ kg/m^3).

What If? What if the Earth were compressed until it had this density? How large would it be?

Answer Because this density is so large, we predict that an Earth of this density would be very small. Using Equation 1.1 and the mass of the Earth, we can find the volume of the compressed Earth:

$$V = \frac{M_E}{\rho} = \frac{5.98 \times 10^{24}\text{ kg}}{2.3 \times 10^{17}\text{ kg/m}^3} = 2.6 \times 10^7\text{ m}^3$$

From this volume we find the radius:

$$V = \tfrac{4}{3}\pi r^3$$

$$r = \left(\frac{3}{4}\frac{V}{\pi}\right)^{1/3} = \left[\frac{3}{4}\left(\frac{2.6 \times 10^7\text{ m}^3}{\pi}\right)\right]^{1/3} = 1.8 \times 10^2\text{ m}$$

This is indeed a small Earth!

Nuclear Stability

Because the nucleus is viewed as a closely packed collection of protons and neutrons, you might be surprised that it can exist. Because charges of the same sign (the protons) in proximity exert very large repulsive Coulomb forces on each other, these forces should cause the nucleus to fly apart. Because this doesn't happen, there must be a counteracting attractive force. The **nuclear force** is a very-short-range (about 2 fm) attractive force that acts between all nuclear particles. The protons attract each other by means of the nuclear force, and, at the same time, they repel each other through the Coulomb force. The nuclear force also acts between pairs of neutrons and between neutrons and protons. The nuclear force dominates the Coulomb repulsive force within the nucleus (at short ranges), so that stable nuclei can exist.

The nuclear force is independent of charge. In other words, the forces associated with the proton–proton, proton–neutron, and neutron–neutron interactions are the same, apart from the additional repulsive Coulomb force for the proton–proton interaction.

Evidence for the limited range of nuclear forces comes from scattering experiments and from studies of nuclear binding energies. The short range of the nuclear force is shown in the neutron–proton (n–p) potential energy plot of Figure 44.3a obtained by scattering neutrons from a target containing hydrogen. The depth of the n–p potential energy well is 40 to 50 MeV, and there is a strong repulsive component that prevents the nucleons from approaching much closer than 0.4 fm.

The nuclear force does not affect electrons, enabling energetic electrons to serve as point-like probes of nuclei. The charge independence of the nuclear force also means that the main difference between the n–p and p–p interactions is that the p–p potential energy consists of a *superposition* of nuclear and Coulomb interactions as shown in Figure 44.3b. At distances less than 2 fm, both p–p and n–p potential energies are nearly identical, but for distances greater than this, the p–p potential has a positive energy barrier with a maximum at 4 fm.

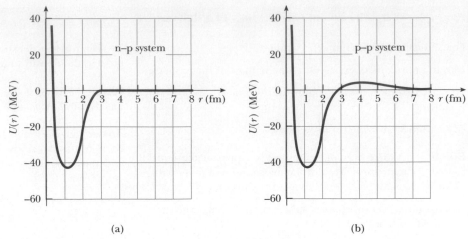

(a)   (b)

Figure 44.3 (a) Potential energy versus separation distance for a neutron–proton system. (b) Potential energy versus separation distance for a proton–proton system. The difference in the two curves is due to the large Coulomb repulsion in the case of the proton–proton interaction. The height of the peak for the proton–proton curve has been exaggerated by a factor of 10 in order to display the difference in the curves on this scale.

The general features of the nuclear force responsible for stable nuclei have been revealed in a wide variety of experiments and are summarized as follows:

- The nuclear force is attractive and is the strongest force in nature.

- It is a short-range force that falls to zero value when the separation between nucleons exceeds several fermis.

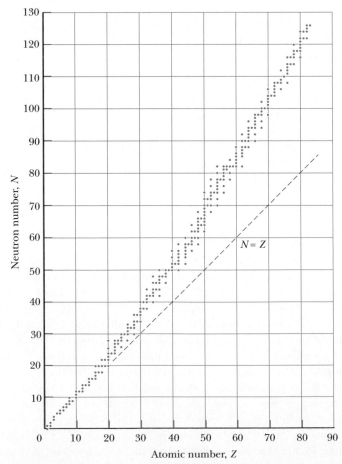

Figure 44.4 Neutron number N versus atomic number Z for stable nuclei (blue dots). These nuclei lie in a narrow band called the line of stability. The dashed line corresponds to the condition $N = Z$.

- The magnitude of the nuclear force depends on the relative spin orientations of the nucleons.

- Scattering experiments and other indirect evidence show that the nuclear force is independent of the charge of the interacting nucleons.

The existence of the nuclear force results in approximately 270 stable nuclei; hundreds of other nuclei have been observed, but they are unstable. A plot of neutron number N versus atomic number Z for a number of stable nuclei is given in Figure 44.4. The stable nuclei are represented by the blue dots, which lie in a narrow range called the *line of stability*. Note that the light stable nuclei contain an equal number of protons and neutrons; that is, $N = Z$. Also, note that in heavy stable nuclei the number of neutrons exceeds the number of protons—above $Z = 20$, the line of stability deviates upward from the line representing $N = Z$. This can be understood by recognizing that, as the number of protons increases, the strength of the Coulomb force increases, which tends to break the nucleus apart. As a result, more neutrons are needed to keep the nucleus stable because neutrons experience only the attractive nuclear force. Eventually, the repulsive Coulomb forces between protons cannot be compensated by the addition of more neutrons. This occurs when $Z = 83$, meaning that elements that contain more than 83 protons do not have stable nuclei.

44.2 Nuclear Binding Energy

As mentioned in the discussion of ^{12}C in the preceding section, the total mass of a nucleus is less than the sum of the masses of its individual nucleons. Because mass is a measure of energy, **the total energy of the bound system (the nucleus) is less than the combined energy of the separated nucleons.** This difference in energy is called the **binding energy** of the nucleus and can be interpreted as the energy that must be added to a nucleus to break it apart into its components. Therefore, in order to separate a nucleus into protons and neutrons, energy must be delivered to the system.

Conservation of energy and the Einstein mass–energy equivalence relationship show that the binding energy E_b of any nucleus of mass M_A is

$$E_b = (Zm_p + Nm_n - M_A) \times 931.494 \text{ MeV/u} \qquad (44.2)$$

Binding energy of a nucleus

where m_p is the mass of the proton, m_n is the mass of the neutron, and the masses are all expressed in atomic mass units. In practice, it is often more convenient to use the mass of neutral atoms (nuclear mass plus mass of electrons) in computing binding energy because mass spectrometers generally measure atomic masses.[1]

A plot of binding energy per nucleon E_b/A as a function of mass number A for various stable nuclei is shown in Figure 44.5. Note that the curve in Figure 44.5 peaks in the vicinity of $A = 60$. That is, nuclei having mass numbers either greater or less than 60 are not as strongly bound as those near the middle of the periodic table. The higher values of binding energy per nucleon near $A = 60$ imply that energy is released when a heavy nucleus splits, or *fissions*, into two lighter nuclei. Energy is released in fission because the nucleons in each product nucleus are more tightly bound to one another than are the nucleons in the original nucleus. The important process of fission and a second important process of *fusion*, in which energy is released as light nuclei combine, are considered in detail in Chapter 45.

Another important feature of Figure 44.5 is that the binding energy per nucleon is approximately constant at around 8 MeV per nucleon for all nuclei with $A > 50$. For these nuclei, the nuclear forces are said to be *saturated*, meaning that, in the closely packed structure shown in Figure 44.2, a particular nucleon can form attractive bonds with only a limited number of other nucleons.

 PITFALL PREVENTION

44.2 Binding Energy

When separate nucleons are combined to form a nucleus, the energy of the system is reduced. Thus, the change in energy is negative. The absolute value of this change is called the binding energy. This difference in sign may be a source of confusion. For example, an *increase* in binding energy corresponds to a *decrease* in the energy of the system.

---

[1] It is possible to use atomic masses rather than nuclear masses because electron masses cancel in the calculations.

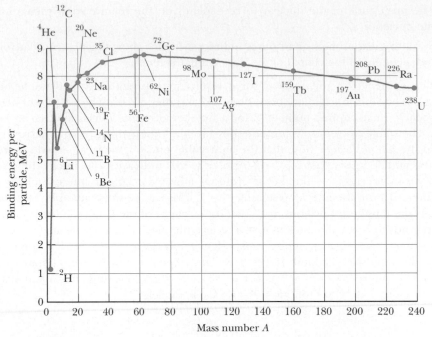

Figure 44.5 Binding energy per nucleon versus mass number for nuclei that lie along the line of stability in Figure 44.4. Some representative nuclei appear as blue dots with labels. (Nuclei to the right of ^{208}Pb are unstable. The curve represents the binding energy for the most stable isotopes.)

Figure 44.5 provides insight into fundamental questions about the origin of the chemical elements. In the early life of the Universe, the only elements that existed were hydrogen and helium. Clouds of cosmic gas coalesced under gravitational forces to form stars. As a star ages, it produces heavier elements from the lighter elements contained within it, beginning by fusing hydrogen atoms to form helium. This process continues as the star becomes older, generating atoms having larger and larger atomic numbers, up through the isotope of iron having $A = 56$, which is near the peak of the curve shown in Figure 44.5.

The nuclide $^{62}_{28}$Ni has the largest binding energy per nucleon of 8.794 5 MeV per nucleon. It takes additional energy to create elements with mass numbers larger than 62 because of their lower binding energies per nucleon. This energy comes from the supernova explosion that occurs at the end of some large stars' lives. Thus, all the heavy atoms in your body were produced from the explosions of ancient stars. You are literally made of stardust!

44.3   Nuclear Models

The details of the nuclear force are still an area of active research. Several nuclear models have been proposed, and these are useful in understanding general features of nuclear experimental data and the mechanisms responsible for binding energy. Two such models are discussed below. The liquid-drop model provides good agreement with the observed behavior of nuclear binding energy, and the shell model predicts the existence of stable isotopes.

Liquid-Drop Model

In 1936, Bohr proposed treating nucleons like molecules in a drop of liquid. In this **liquid-drop model,** the nucleons interact strongly with one another and undergo frequent collisions as they jiggle around within the nucleus. This jiggling motion is analogous to the thermally agitated motion of molecules in a drop of liquid.

Four major effects influence the binding energy of the nucleus in the liquid-drop model:

- **The volume effect.** Figure 44.5 shows that, for $A > 50$, the binding energy per nucleon is approximately constant. This indicates that the nuclear force on a given nucleon is due only to a few nearest neighbors and not to all the other nucleons in the nucleus. On the average, then, the binding energy associated with the nuclear force for each nucleon is the same in all nuclei—that associated with an interaction with a few neighbors. This tells us that the total binding energy of the nucleus is proportional to A and therefore proportional to the nuclear volume. The contribution to the binding energy of the entire nucleus is $C_1 A$, where C_1 is an adjustable constant that we can determine by fitting the prediction of the model to experimental results.

- **The surface effect.** Because nucleons on the surface of the drop have fewer neighbors than those in the interior, surface nucleons reduce the binding energy by an amount proportional to their number. Because the number of surface nucleons is proportional to the surface area $4\pi r^2$ of the nucleus, and because $r^2 \propto A^{2/3}$ (Eq. 44.1), the surface term can be expressed as $-C_2 A^{2/3}$, where C_2 is a second adjustable constant.

- **The Coulomb repulsion effect.** Each proton repels every other proton in the nucleus. The corresponding potential energy per pair of interacting protons is $k_e e^2 / r$, where k_e is the Coulomb constant. The total electric potential energy is equivalent to the work required to assemble Z protons, initially infinitely far apart, into a sphere of volume V. This energy is proportional to the number of proton pairs $Z(Z - 1)/2$ and inversely proportional to the nuclear radius. Consequently, the reduction in binding energy that results from the Coulomb effect is $-C_3 Z(Z - 1)/A^{1/3}$, where C_3 is yet another adjustable constant.

- **The symmetry effect.** Another effect that lowers the binding energy is related to the symmetry of the nucleus in terms of values of N and Z. For small values of A, stable nuclei tend to have $N \approx Z$. Any large asymmetry between N and Z for light nuclei reduces the binding energy and makes the nucleus less stable. For larger A, the value of N for stable nuclei is naturally larger than Z. This effect can be described by a binding-energy term of the form $-C_4 (N - Z)^2/A$, where C_4 is another adjustable constant.[2] For small A, any large asymmetry between values of N and Z makes this term relatively large and reduces the binding energy. For large A, the A in the denominator of this term reduces its value so that it has little effect on the overall binding energy.

Adding these contributions, we arrive at an expression for the total binding energy

$$E_b = C_1 A - C_2 A^{2/3} - C_3 \frac{Z(Z - 1)}{A^{1/3}} \quad C_4 \frac{(N - Z)^2}{A} \tag{44.3}$$

Semiempirical binding-energy formula

This equation, often referred to as the **semiempirical binding-energy formula**, contains four constants that are adjusted to fit the theoretical expression to experimental data. For nuclei having $A \geq 15$, the constants have the values

$$C_1 = 15.7 \text{ MeV} \qquad C_2 = 17.8 \text{ MeV}$$
$$C_3 = 0.71 \text{ MeV} \qquad C_4 = 23.6 \text{ MeV}$$

---

[2] The liquid-drop model *describes* the fact that heavy nuclei have $N > Z$. The shell model, as we shall see shortly, *explains* why this is true with a physical argument.

Equation 44.3, together with these constants, fits the known nuclear mass values very well. However, the liquid-drop model does not account for some finer details of nuclear structure, such as stability rules and angular momentum. On the other hand, it does provide a qualitative description of nuclear fission, as we shall discuss in Chapter 45.

Example 44.3   Applying the Semiempirical Binding-Energy Formula

The nucleus ^{64}Zn has a tabulated binding energy of 559.09 MeV. Use the semiempirical binding-energy formula to generate a theoretical estimate of the binding energy for this nucleus.

Solution For this nucleus, we have $Z = 30$, $N = 34$, and $A = 64$. The four terms of the semiempirical binding-energy formula have the following values:

$$C_1 A = (15.7 \text{ MeV})(64) = 1\,005 \text{ MeV}$$

$$C_2 A^{2/3} = (17.8 \text{ MeV})(64)^{2/3} = 285 \text{ MeV}$$

$$C_3 \frac{Z(Z-1)}{A^{1/3}} = (0.71 \text{ MeV}) \frac{(30)(29)}{(64)^{1/3}} = 154 \text{ MeV}$$

$$C_4 \frac{(N-Z)^2}{A} = (23.6 \text{ MeV}) \frac{(34-30)^2}{64} = 5.90 \text{ MeV}$$

Substituting these values into Equation 44.3, we find

$$E_b = 1\,005 \text{ MeV} - 285 \text{ MeV} - 154 \text{ MeV} - 5.90 \text{ MeV}$$

$$= \boxed{560 \text{ MeV}}$$

This value differs from the tabulated value by less than 0.2%. Notice how the sizes of the terms decrease from the first to the fourth term. The fourth term is particularly small, especially for this nucleus, which does not have an excessive number of neutrons.

The Shell Model

The liquid-drop model describes the general behavior of nuclear binding energies relatively well. However, when the binding energies are studied more closely, we find the following features:

- Most stable nuclei have an even value of A. Furthermore, only eight stable nuclei have odd values for both Z and N.

- Figure 44.6 shows a graph of the difference between the binding energy per nucleon calculated by Equation 44.3 and the measured binding energy. There is evidence for regularly spaced peaks in the data that are not described by the semiempirical binding-energy formula. The peaks occur at values of N or Z that have become known as **magic numbers:**

Magic numbers

$$Z \text{ or } N = 2, 8, 20, 28, 50, 82 \tag{44.4}$$

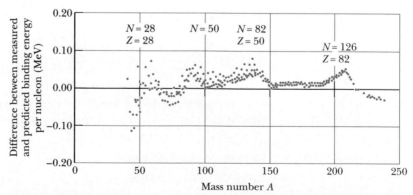

Figure 44.6 The difference between measured binding energies and those calculated from the liquid-drop model as a function of A. The appearance of regular peaks in the experimental data suggests behavior that is not predicted in the liquid-drop model. (Adapted from R. A. Dunlap, *The Physics of Nuclei and Particles*, Brooks/Cole, Belmont CA, 2004.)

- High-precision studies of nuclear radii show deviations from the simple expression in Equation 44.1. Graphs of experimental data show peaks in the curve of radius versus N at values of N equal to the magic numbers.

- We defined isotopes in Section 44.1. A group of *isotones* is a collection of nuclei having the same value of N and varying values of Z. When the number of stable isotones is graphed as function of N, there are peaks in the graph—again at the magic numbers in Equation 44.4.

- Several other nuclear measurements show anomalous behavior at the magic numbers.[3]

These peaks in graphs of experimental data are reminiscent of the peaks in Figure 42.20 for the ionization energy of atoms, which arose because of the shell structure of the atom. The **shell model** of the nucleus, also called the **independent-particle model,** was developed independently in 1949 and 1950 by Maria Goeppert-Mayer and Hans Jensen, both German scientists, who shared the 1963 Nobel Prize in physics. In this model, each nucleon is assumed to exist in a shell, similar to an atomic shell for an electron. The nucleons exist in quantized energy states, and there are few collisions between nucleons. Obviously, the assumptions of this model differ greatly from those made in the liquid-drop model.

The quantized states occupied by the nucleons can be described by a set of quantum numbers. Because both the proton and the neutron have spin $\frac{1}{2}$, the exclusion principle can be applied to describe the allowed states (as we did for electrons in Chapter 42). That is, each state can contain only two protons (or two neutrons) having *opposite* spins (Fig. 44.7). The protons have a set of allowed states, and these states differ from those of the neutrons because the two species move in different potential wells. The proton energy levels are farther apart than the neutron levels because the protons experience a superposition of the Coulomb force and the nuclear force, whereas the neutrons experience only the nuclear force.

One factor influencing the observed characteristics of nuclear ground states is *nuclear spin–orbit* effects. The spin–orbit interaction between the spin of an electron and its orbital motion in an atom gives rise to the sodium doublet discussed in Section 42.6 and is magnetic in origin. In contrast, the spin–orbit effect for nucleons in a nucleus is due to the nuclear force. It is much stronger than in the atomic case, and it has opposite sign. When these effects are taken into account, the shell model is able to account for the observed magic numbers.

The shell model helps us understand why nuclei containing an even number of protons and neutrons are more stable than other nuclei. (There are 160 stable even–even isotopes.) Any particular state is filled when it contains two protons (or two neutrons) having opposite spins. An extra proton or neutron can be added to the nucleus only at

Maria Goeppert-Mayer
German Scientist (1906–1972)

Goeppert-Mayer was born and educated in Germany. She is best known for her development of the shell model (independent-particle model) of the nucleus, published in 1950. A similar model was simultaneously developed by Hans Jensen, another German scientist. Goeppert-Mayer and Jensen were awarded the Nobel Prize in physics in 1963 for their extraordinary work in understanding the structure of the nucleus. (*Courtesy of Louise Barker/AIP Niels Bohr Library*)

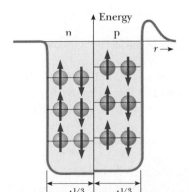

Figure 44.7 A square potential well containing 12 nucleons. The orange circles represent protons, and the green circles represent neutrons. The energy levels for the protons are slightly higher than those for the neutrons because of the electric potential energy associated with the system of protons. The difference in the levels increases as Z increases. Note that only two nucleons having opposite spins can occupy a given level, as required by the exclusion principle.

[3] For further details, see Chapter 5 of R. A. Dunlap, *The Physics of Nuclei and Particles*, Brooks/Cole, 2004.

the expense of increasing the energy of the nucleus. This increase in energy leads to a nucleus that is less stable than the original nucleus. A careful inspection of the stable nuclei shows that the majority have a special stability when their nucleons combine in pairs, which results in a total angular momentum of zero.

The shell model also helps us to understand why nuclei tend to have more neutrons than protons. As in Figure 44.7, the proton energy levels are higher than those for neutrons, due to the extra energy associated with Coulomb repulsion. This effect becomes more pronounced as Z increases. As a result, as Z increases and higher states are filled, a proton level for a given quantum number will be much higher in energy than the neutron level for the same quantum number. In fact, it will be even higher in energy than neutron levels for higher quantum numbers. As a result, it is more energetically favorable for the nucleus to form with neutrons in the lower energy levels rather than protons in the higher energy levels, so that the number of neutrons is greater than the number of protons.

More sophisticated models of the nucleus have been developed and continue to be developed. For example, the *collective model* combines features of the liquid-drop and shell models. The development of theoretical models of the nucleus continues to be an active area of research.

44.4 Radioactivity

In 1896, Henri Becquerel accidentally discovered that uranyl potassium sulfate crystals emit an invisible radiation that can darken a photographic plate when the plate is covered to exclude light. After a series of experiments, he concluded that the radiation emitted by the crystals was of a new type, one that requires no external stimulation and was so penetrating that it could darken protected photographic plates and ionize gases. This process of spontaneous emission of radiation by uranium was soon to be called **radioactivity.**

Subsequent experiments by other scientists showed that other substances were more powerfully radioactive. The most significant investigations of this type were conducted by Marie and Pierre Curie. After several years of careful and laborious chemical separation processes on tons of pitchblende, a radioactive ore, the Curies reported the discovery of two previously unknown elements, both radioactive. These were named polonium and radium. Additional experiments, including Rutherford's famous work on alpha-particle scattering, suggested that radioactivity is the result of the *decay*, or disintegration, of unstable nuclei.

Three types of radioactive decay occur in radioactive substances: alpha (α) decay, in which the emitted particles are ^{4}He nuclei; beta (β) decay, in which the emitted particles are either electrons or positrons; and gamma (γ) decay, in which the emitted "rays" are high-energy photons. A **positron** is a particle like the electron in all respects except that the positron has a charge of $+e$. (The positron is the *antimatter twin* of the electron; see Section 46.2.) The symbol e^- is used to designate an electron, and e^+ designates a positron.

It is possible to distinguish among these three forms of radiation by using the scheme described in Figure 44.8. The radiation from a variety of radioactive samples is directed into a region in which there is a magnetic field. The radiation beam splits into three components, two bending in opposite directions and the third experiencing no change in direction. From this simple observation, we can conclude that the radiation of the undeflected beam carries no charge (the gamma ray), the component deflected upward corresponds to positively charged particles (alpha particles), and the component deflected downward corresponds to negatively charged particles (e^-). If the beam includes a positron (e^+), it is deflected upward like the alpha particle but follows a different trajectory due to its smaller mass.

The three types of radiation have quite different penetrating powers. Alpha particles barely penetrate a sheet of paper, beta particles (electrons and positrons) can penetrate a few millimeters of aluminum, and gamma rays can penetrate several centimeters of lead.

Marie Curie
Polish Scientist (1867–1934)

In 1903, Marie Curie shared the Nobel Prize in physics with her husband Pierre and with Becquerel for their studies of radioactive substances. In 1911 she was awarded a Nobel Prize in chemistry for the discovery of radium and polonium. She died of leukemia caused by years of exposure to radioactive substances. "I persist in believing that the ideas that then guided us are the only ones which can lead to true social progress. We cannot hope to build a better world without improving the individual. Toward this end, each of us must work toward his own highest development, accepting at the same time his share of responsibility in the general life of humanity." (*FPG International*)

▲ PITFALL PREVENTION

44.3 Rays or Particles?

Early in the history of nuclear physics, the term *radiation* was used to describe the emanations from radioactive nuclei. We now know that alpha radiation and beta radiation involve the emission of particles with nonzero rest energy. Even though these are not examples of electromagnetic radiation, the use of the term *radiation* for all three types of emission is deeply entrenched in our language and in the physics community.

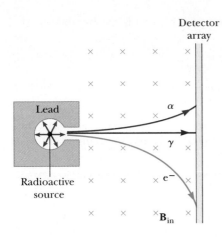

Figure 44.8 The radiation from radioactive sources can be separated into three components by using a magnetic field to deflect the charged particles. The detector array at the right records the events. The gamma ray is not deflected by the magnetic field.

The decay process is probabilistic in nature and can be described with statistical calculations for a radioactive substance of macroscopic size, so that the number of radioactive nuclei is large. For large numbers of radioactive nuclei, the rate at which a particular decay process occurs in a sample is proportional to the number of radioactive nuclei present (that is, the number of nuclei that have not yet decayed). If N is the number of undecayed radioactive nuclei present at some instant, the rate of change of N is

$$\frac{dN}{dt} = -\lambda N \tag{44.5}$$

where λ, called the **decay constant,** is the probability of decay per nucleus per second. The negative sign indicates that dN/dt is negative; that is, N decreases in time.

Equation 44.5 can be written in the form

$$\frac{dN}{N} = -\lambda\, dt$$

which, upon integration, gives

$$N = N_0 e^{-\lambda t} \tag{44.6}$$

where the constant N_0 represents the number of undecayed radioactive nuclei at $t = 0$. Equation 44.6 shows that the number of undecayed radioactive nuclei in a sample decreases exponentially with time. The plot of N versus t shown in Figure 44.9 illustrates the exponential nature of the decay.

▲ **PITFALL PREVENTION**

44.4 Notation Warning

In Section 44.1, we introduced the symbol N as an integer representing the number of neutrons in a nucleus. In the present discussion, the symbol N represents the number of undecayed nuclei in a radioactive sample remaining after some time interval. As you read further, be sure to consider the context to determine the appropriate meaning for the symbol N.

Exponential behavior of the number of undecayed nuclei

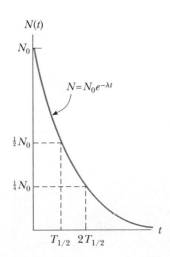

Active Figure 44.9 Plot of the exponential decay law for radioactive nuclei. The vertical axis represents the number of undecayed radioactive nuclei present at any time t, and the horizontal axis is time. The time $T_{1/2}$ is the half-life of the sample.

At the Active Figures link at http://www.pse6.com, observe the decay curves for nuclei with varying half-lives.

The **decay rate** R, which is the number of decays per second, can be obtained by combining Equations 44.5 and 44.6:

Exponential behavior of the decay rate

$$R = \left| \frac{dN}{dt} \right| = \lambda N = N_0 \lambda e^{-\lambda t} = R_0 e^{-\lambda t} \qquad (44.7)$$

where $R_0 = N_0 \lambda$ is the decay rate at $t = 0$. The decay rate R of a sample is often referred to as its **activity.** Note that both N and R decrease exponentially with time.

Another parameter useful in characterizing nuclear decay is the **half-life** $T_{1/2}$:

▲ **PITFALL PREVENTION**

44.5 Half-Life

It is *not* true that all of the original nuclei have decayed after two half-lives! In one half-life, *half of those nuclei that are left* will decay.

The **half-life** of a radioactive substance is the time interval during which half of a given number of radioactive nuclei decay.

To find an expression for the half-life, we first set $N = N_0/2$ and $t = T_{1/2}$ in Equation 44.6 to give

$$\frac{N_0}{2} = N_0 e^{-\lambda T_{1/2}}$$

Canceling the N_0 factors and then taking the reciprocal of both sides, we obtain $e^{\lambda T_{1/2}} = 2$. Taking the natural logarithm of both sides gives

Half-life

$$T_{1/2} = \frac{\ln 2}{\lambda} = \frac{0.693}{\lambda} \qquad (44.8)$$

This is a convenient expression relating the half-life $T_{1/2}$ to the decay constant λ. After a time interval equal to one half-life, there are $N_0/2$ radioactive nuclei remaining (by definition); after two half-lives, half of these remaining nuclei have decayed and $N_0/4$ radioactive nuclei are left; after three half-lives, $N_0/8$ are left, and so on. In general, after n half-lives, the number of undecayed radioactive nuclei remaining is $N_0/2^n$.

A frequently used unit of activity is the **curie** (Ci), defined as

The curie

$$1 \text{ Ci} \equiv 3.7 \times 10^{10} \text{ decays/s}$$

This value was originally selected because it is the approximate activity of 1 g of radium. The SI unit of activity is the **becquerel** (Bq):

The becquerel

$$1 \text{ Bq} \equiv 1 \text{ decay/s}$$

Therefore, $1 \text{ Ci} = 3.7 \times 10^{10} \text{ Bq}$. The curie is a rather large unit, and the more frequently used activity units are the millicurie and the microcurie.

Quick Quiz 44.4 On your birthday, you measure the activity of a sample of ^{210}Bi, which has a half-life of 5.01 days. The activity you measure is 1.000 μCi. What is the activity of this sample on your next birthday? (a) 1.000 μCi (b) 0 (c) ~ 0.2 μCi (d) ~ 0.01 μCi (e) $\sim 10^{-22}$ μCi

Quick Quiz 44.5 Suppose you have a pure radioactive material with a half-life of $T_{1/2}$. You begin with N_0 undecayed nuclei of the material at $t = 0$. At $t = \frac{1}{2}T_{1/2}$, how many of the nuclei *have decayed*? (a) $\frac{1}{4}N_0$ (b) $\frac{1}{2}N_0$ (c) $\frac{3}{4}N_0$ (d) $0.707N_0$ (e) $0.293N_0$

Example 44.4 How Many Nuclei Are Left?

The isotope carbon-14, $^{14}_{6}$C, is radioactive and has a half-life of 5 730 years. If you start with a sample of 1 000 carbon-14 nuclei, how many will still be undecayed in 22 920 years?

Solution In 5 730 years, half the sample will have decayed, leaving 500 carbon-14 nuclei remaining. In another 5 730 years (for a total time interval of 11 460 years), the number will be reduced to 250 nuclei. After another 5 730 years (total time interval 17 190 years), 125 remain. Finally, after four half-lives (22 920 years), only about 62 remain.

What If? What if you were able to perform this experiment? How would the results compare to our predictions?

Answer The numbers found in the example represent ideal circumstances. As we have mentioned, radioactive decay is actually a probabilistic statistical process and accurate statistical predictions are only possible with a very large number of atoms. The original sample in this example contained only 1 000 nuclei, certainly not a very large number. Thus, if you counted the number remaining, for example, after one half-life for this small sample, it probably would not be exactly 500.

Example 44.5 The Activity of Radium

Interactive

The half-life of the radioactive nucleus radium-226, $^{226}_{88}$Ra, is 1.6×10^3 yr.

(A) What is the decay constant λ of this nucleus?

Solution We can calculate λ using Equation 44.8 and the fact that

$$T_{1/2} = 1.6 \times 10^3 \text{ yr} \left(\frac{3.16 \times 10^7 \text{ s}}{1 \text{ yr}} \right)$$

$$= 5.0 \times 10^{10} \text{ s}$$

Therefore,

$$\lambda = \frac{0.693}{T_{1/2}} = \frac{0.693}{5.0 \times 10^{10} \text{ s}} = \boxed{1.4 \times 10^{-11} \text{ s}^{-1}}$$

Note that this result is also the probability that any single $^{226}_{88}$Ra nucleus will decay in a time interval of one second.

(B) If a sample contains 3.0×10^{16} $^{226}_{88}$Ra nuclei at $t = 0$, determine its activity in curies at this time.

Solution By definition (Eq. 44.7) R_0, the activity at $t = 0$, is λN_0, where N_0 is the number of radioactive nuclei present

at $t = 0$. With $N_0 = 3.0 \times 10^{16}$, we have

$$R_0 = \lambda N_0 = (1.4 \times 10^{-11} \text{ s}^{-1})(3.0 \times 10^{16})$$

$$= (4.2 \times 10^5 \text{ Bq}) \left(\frac{1 \text{ Ci}}{3.7 \times 10^{10} \text{ Bq}} \right)$$

$$= \boxed{11 \ \mu\text{Ci}}$$

(C) What is the activity after the sample is 2.0×10^3 yr old?

Solution We use Equation 44.7 and the fact that the age of the sample is 2.0×10^3 yr $= 6.3 \times 10^{10}$ s:

$$R = R_0 e^{-\lambda t}$$

$$= (4.2 \times 10^5 \text{ Bq}) e^{-(1.4 \times 10^{-11} \text{ s}^{-1})(6.3 \times 10^{10} \text{ s})}$$

$$= 1.7 \times 10^5 \text{ Bq} \left(\frac{1 \text{ Ci}}{3.7 \times 10^{10} \text{ Bq}} \right)$$

$$= \boxed{4.7 \ \mu\text{Ci}}$$

 At the Interactive Worked Example link at http://www.pse6.com, *practice evaluating the parameters for radioactive decay of various isotopes of radium.*

Example 44.6 The Activity of Carbon

At time $t = 0$, a radioactive sample contains 3.50 μg of pure $^{11}_{6}$C, which has a half-life of 20.4 min.

(A) Determine the number N_0 of nuclei in the sample at $t = 0$.

Solution The molar mass of $^{11}_{6}$C is approximately 11.0 g/mol, and so 11.0 g contains Avogadro's number (6.02×10^{23}) of nuclei. Therefore, 3.50 μg contains N_0 nuclei, where

$$\frac{N_0}{6.02 \times 10^{23} \text{ nuclei/mol}} = \frac{3.50 \times 10^{-6} \text{ g}}{11.0 \text{ g/mol}}$$

$$N_0 = \boxed{1.92 \times 10^{17} \text{ nuclei}}$$

(B) What is the activity of the sample initially and after 8.00 h?

Solution With $T_{1/2} = 20.4$ min $= 1 224$ s, the decay constant is

$$\lambda = \frac{0.693}{T_{1/2}} = \frac{0.693}{1 224 \text{ s}} = 5.66 \times 10^{-4} \text{ s}^{-1}$$

Therefore, the initial activity of the sample is

$$R_0 = \lambda N_0 = (5.66 \times 10^{-4} \text{ s}^{-1})(1.92 \times 10^{17})$$

$$= \boxed{1.09 \times 10^{14} \text{ Bq}}$$

We use Equation 44.7 to find the activity at $t = 8.00$ h $= 2.88 \times 10^4$ s:

$$R = R_0 e^{-\lambda t} = (1.09 \times 10^{14}\,\text{Bq})\,e^{-(5.66 \times 10^{-4}\,\text{s}^{-1})(2.88 \times 10^4\,\text{s})}$$

$$= \boxed{9.08 \times 10^6\,\text{Bq}}$$

A listing of activity versus time for this situation is given in Table 44.2.

Table 44.2

Activity Versus Time for the Sample Described in Example 44.6

t (h)	R (Bq)
0	1.09×10^{14}
1	1.42×10^{13}
2	1.85×10^{12}
3	2.41×10^{11}
4	3.15×10^{10}
5	4.10×10^{9}
6	5.34×10^{8}
7	6.97×10^{7}
8	9.08×10^{6}

Example 44.7 A Radioactive Isotope of Iodine

A sample of the isotope ^{131}I, which has a half-life of 8.04 days, has an activity of 5.0 mCi at the time of shipment. Upon receipt in a medical laboratory, the activity is 4.2 mCi. How much time has elapsed between the two measurements?

Solution To conceptualize this problem, consider that the sample is continuously decaying as it is in transit. The decrease in the activity is 16% during the time interval between shipment and receipt, so we expect the elapsed time to be less than the half-life of 8.04 d. The stated activity corresponds to many decays per second, so that N is large and we can categorize this problem as one in which we can use our statistical analysis of radioactivity. To analyze the problem we use Equation 44.7 in the form

$$\frac{R}{R_0} = e^{-\lambda t}$$

where the sample is shipped at $t = 0$, at which time the activity is R_0. Taking the natural logarithm of each side, we have

$$\ln\left(\frac{R}{R_0}\right) = -\lambda t$$

$$(1) \qquad t = -\frac{1}{\lambda}\ln\left(\frac{R}{R_0}\right)$$

To find λ, we use Equation 44.8:

$$(2) \qquad \lambda = \frac{0.693}{T_{1/2}} = \frac{0.693}{8.04\,\text{d}} = 8.62 \times 10^{-2}\,\text{d}^{-1}$$

Substituting Equation (2) into Equation (1) gives

$$t = -\left(\frac{1}{8.62 \times 10^{-2}\,\text{d}^{-1}}\right)\ln\left(\frac{4.2\,\text{mCi}}{5.0\,\text{mCi}}\right) = \boxed{2.0\,\text{d}}$$

To finalize this problem, note that this is indeed less than the half-life, as we expected. This example demonstrates the difficulty in shipping radioactive samples with short half-lives. If the shipment were to be delayed by several days, only a small fraction of the sample might remain upon receipt. This difficulty can be addressed by shipping a combination of isotopes in which the desired isotope is the product of a decay occurring within the sample. It is possible for the desired isotope to be in *equilibrium*, in which case it is created at the same rate as it decays. Therefore, the amount of the desired isotope remains constant during the shipping process. Upon receipt, the desired isotope can be separated from the rest of the sample and its decay from the initial activity begins upon receipt rather than upon shipment.

44.5 The Decay Processes

As we stated in the preceding section, a radioactive nucleus spontaneously decays by one of three processes: alpha decay, beta decay, or gamma decay. Figure 44.10 shows a close-up view of a portion of Figure 44.4 from $Z = 65$ to $Z = 80$. The blue circles are the stable nuclei seen in Figure 44.4. In addition, unstable nuclei above and below the line of stability for each value of Z are shown. Above the line of stability, the red circles show unstable nuclei that are neutron-rich and undergo a beta decay process in which an electron is emitted. Below the blue circles are green circles corresponding to proton-rich unstable nuclei that primarily undergo a beta-decay process in which a positron is emitted or a competing process called electron capture. Beta decay and electron capture are described in more detail below. Further below the line of stability (with a few exceptions) are yellow circles that represent very proton-rich nuclei for which the primary decay mechanism is alpha decay, which we will discuss first.

Alpha Decay

A nucleus emitting an alpha particle ($^4_2$He) loses two protons and two neutrons. Therefore, the atomic number Z decreases by 2, the mass number A decreases by 4, and the neutron number decreases by 2. The decay can be written

$$^A_Z X \longrightarrow \; ^{A-4}_{Z-2}Y + \; ^4_2He \tag{44.9}$$

where X is called the **parent nucleus** and Y the **daughter nucleus.** As a general rule in any decay expression such as this, (1) the sum of the mass numbers A must be the same on both sides of the decay and (2) the sum of the atomic numbers Z must be the same on both sides of the decay. As examples, ^{238}U and ^{226}Ra are both alpha emitters and decay according to the schemes

$$^{238}_{92}U \longrightarrow \; ^{234}_{90}Th + \; ^4_2He \tag{44.10}$$

$$^{226}_{88}Ra \longrightarrow \; ^{222}_{86}Rn + \; ^4_2He \tag{44.11}$$

The decay of ^{226}Ra is shown in Figure 44.11.

When the nucleus of one element changes into the nucleus of another, as happens in alpha decay, the process is called **spontaneous decay.** In any spontaneous decay, relativistic energy and momentum of the isolated parent nucleus must be conserved. If we call M_X the mass of the parent nucleus, M_Y the mass of the daughter nucleus, and M_α the mass of the alpha particle, we can define the **disintegration energy** Q of the system as

$$Q = (M_X - M_Y - M_\alpha)\,c^2 \tag{44.12}$$

The energy Q is in joules when the masses are in kilograms and c is the speed of light, 3.00×10^8 m/s. However, when the masses are expressed in the more convenient unit u, Q can be calculated in MeV using the expression

$$Q = (M_X - M_Y - M_\alpha) \times 931.494 \text{ MeV/u} \tag{44.13}$$

The disintegration energy Q appears in the form of kinetic energy in the daughter nucleus and the alpha particle, and is sometimes referred to as the Q value of the nuclear decay. In the case of the ^{226}Ra decay described in Figure 44.11, if the parent nucleus is at rest before the decay, the total kinetic energy of the products is 4.87 MeV. Most of this kinetic energy is associated with the alpha particle because this particle is much less massive than the daughter nucleus ^{222}Rn. That is, because momentum must be conserved, the lighter alpha particle recoils with a much higher speed than does the daughter nucleus. Generally, less massive particles carry off most of the energy in nuclear decays.

Experimental observations of alpha-particle energies show a number of discrete energies rather than a single energy. This is due to the fact that the daughter nucleus may

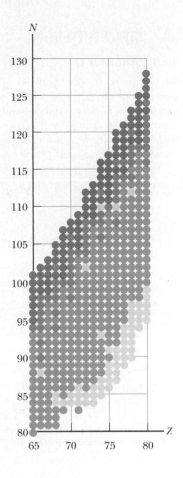

- ● Beta (electron)
- ● Stable
- ● Beta (positron) or electron capture
- ● Alpha

Active Figure 44.10 A close-up view of the line of stability in Figure 44.4 from $Z = 65$ to $Z = 80$. The blue dots represent stable nuclei as in Figure 44.4. The other colored dots represent unstable isotopes above and below the line of stability, with the color of the dot indicating the primary means of decay.

Study the decay modes and decay energies by clicking on any of the colored dots at the Active Figures link at http://www.pse6.com.

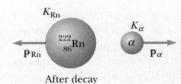

Active Figure 44.11 The alpha decay of radium-226. The radium nucleus is initially at rest. After the decay, the radon nucleus has kinetic energy K_{Rn} and momentum $\mathbf{p}_{Rn}$, and the alpha particle has kinetic energy K_α and momentum $\mathbf{p}_\alpha$.

At the Active Figures link at http://www.pse6.com, observe the decay of radium-226. For a large number of decays, observe the development of the graph in Figure 44.14b.

44.6 Another Q

We have seen the symbol Q before, but this is a brand new meaning for this symbol—the disintegration energy. In this context, it is neither heat nor charge, for which we have used Q before.

be left in an excited quantum state after the decay. As a result, not all of the disintegration energy is available as kinetic energy of the alpha particle and daughter nucleus. The emission of an alpha particle is followed by one or more gamma-ray photons (see below) as the excited nucleus decays to the ground state. The observed discrete alpha-particle energies represent evidence of the quantized nature of the nucleus and allow a determination of the energies of the quantum states.

Finally, it is interesting to note that if one assumed that ^{238}U (or any other alpha emitter) decayed by emitting either a proton or a neutron, the mass of the decay products would exceed that of the parent nucleus, corresponding to a negative Q value. A negative Q value indicates that such a proposed decay does not occur spontaneously.

Quick Quiz 44.6 Which of the following is the correct daughter nucleus associated with the alpha decay of $^{157}_{72}$Hf? (a) $^{153}_{72}$Hf (b) $^{153}_{70}$Yb (c) $^{157}_{70}$Yb

---

Example 44.8 The Energy Liberated when Radium Decays

The ^{226}Ra nucleus undergoes alpha decay according to Equation 44.11. Calculate the Q value for this process. Take the masses to be 226.025 403 u for ^{226}Ra, 222.017 570 u for ^{222}Rn, and 4.002 603 u for $^{4}_{2}$He, as found in Table A.3.

Solution We may add 88 electrons to both sides of the decay expression in Equation 44.11. The differences in electron binding energies are negligible when compared with the Q value for the nuclear decay process. Then, we may use the masses of neutral atoms in Equation 44.13 to see that

$Q = (M_X - M_Y - M_\alpha) \times 931.494 \text{ MeV/u}$

$= (226.025\,403 \text{ u} - 222.017\,570 \text{ u} - 4.002\,603 \text{ u})$
$\times 931.494 \text{ MeV/u}$

$= (0.005\,230 \text{ u}) \times 931.494 \text{ MeV/u} = \boxed{4.87 \text{ MeV}}$

What If? Suppose we measured the kinetic energy of the alpha particle from this decay. Would we measure 4.87 MeV?

Answer The value of Q that we have calculated is the disintegration energy, not the kinetic energy of the alpha particle. The energy of 4.87 MeV includes the kinetic energy of both the alpha particle and the daughter nucleus after the decay. Thus, the kinetic energy of the alpha particle must be *less* than 4.87 MeV.

Let us determine this kinetic energy mathematically. The parent nucleus is an isolated system that decays into an alpha particle and a daughter nucleus. Thus, momentum must be conserved for the system. The initial momentum is zero, so

$$(1) \qquad 0 = M_Y v_Y - M_\alpha v_\alpha$$

where we have explicitly indicated with the minus sign that the particles move in opposite directions, so we can interpret v_Y and v_α as speeds rather than velocity components.

The disintegration energy is equal to the sum of the kinetic energies of the alpha particle and the daughter nucleus (assuming that the daughter nucleus is left in the ground state):

$$(2) \qquad Q = \tfrac{1}{2} M_\alpha v_\alpha^2 + \tfrac{1}{2} M_Y v_Y^2$$

If we solve Equation (1) for v_Y and substitute into Equation (2), we find

$$Q = \tfrac{1}{2} M_\alpha v_\alpha^2 + \tfrac{1}{2} M_Y \left(\frac{M_\alpha v_\alpha}{M_Y}\right)^2 = \tfrac{1}{2} M_\alpha v_\alpha^2 \left(1 + \frac{M_\alpha}{M_Y}\right)$$

$$= K_\alpha \left(\frac{M_Y + M_\alpha}{M_Y}\right)$$

Thus, in general, the kinetic energy of the alpha particle is related to the disintegration energy according to

$$K_\alpha = Q \left(\frac{M_Y}{M_Y + M_\alpha}\right)$$

For the specific decay of ^{226}Ra that we are exploring in this example, we have

$$K_\alpha = (4.87 \text{ MeV}) \left(\frac{222}{222 + 4}\right) = 4.78 \text{ MeV}$$

To understand the mechanism of alpha decay, let us model the parent nucleus as a system consisting of (1) the alpha particle, already formed as an entity within the nucleus, and (2) the daughter nucleus that will result when the alpha particle is emitted. Figure 44.12 shows a plot of potential energy versus separation distance r between the alpha particle and the daughter nucleus, where the distance marked R is the range of the nuclear force. The curve represents the combined effects of (1) the repulsive Coulomb force, which gives the positive part of the curve for $r > R$, and (2) the attractive nuclear force, which causes the curve to be negative for $r < R$. As we saw in Example 44.8, a typical disintegration energy Q is about 5 MeV, which is the approximate kinetic energy of the alpha particle, represented by the lower dashed line in Figure 44.12.

According to classical physics, the alpha particle is trapped in a potential well. How, then, does it ever escape from the nucleus? The answer to this question was first provided by George Gamow (1904–1968) in 1928 and independently by R. W. Gurney and E. U. Condon in 1929, using quantum mechanics. In Section 41.6, we discussed the view of quantum mechanics that there is always some probability that a particle can tunnel through a barrier. This is exactly how we can describe alpha decay—the alpha particle tunnels through the barrier in Figure 44.12, escaping the nucleus. Furthermore, this model agrees with the observation that higher-energy alpha particles come from nuclei with shorter half-lives. For higher-energy alpha particles in Figure 44.12, the barrier is narrower and the probability is higher that tunneling will occur. The higher probability translates to a shorter half-life.

As an example, consider the decays of ^{238}U and ^{226}Ra in Equations 44.10 and 44.11, along with the corresponding half-lives and alpha-particle energies:

$$^{238}\text{U}: \quad T_{1/2} = 4.47 \times 10^9 \,\text{yr} \quad K_\alpha = 4.20 \,\text{MeV}$$

$$^{226}\text{Ra}: \quad T_{1/2} = 1.60 \times 10^3 \,\text{yr} \quad K_\alpha = 4.78 \,\text{MeV}$$

Notice that a relatively small difference in alpha-particle energy is associated with a tremendous difference of six orders of magnitude in the half-life. The origin of this effect can be seen as follows. First, in Figure 44.12, notice that the curve below an alpha-particle energy of 5 MeV has a slope with a relatively small magnitude. Thus, a small difference in energy on the vertical axis has a relatively large effect on the width of the potential barrier. Second, recall Equation 41.20, which describes the exponential dependence of the probability of transmission on the barrier width. These two factors combine to give the very sensitive relationship between half-life and alpha-particle energy that the data above suggest.

The Smoke Detector. A life-saving application of alpha decay is in the household smoke detector, shown in Figure 44.13. Most of the common ones use a radioactive material. The detector consists of an ionization chamber, a sensitive current detector, and an alarm. A weak radioactive source (usually $^{241}_{95}\text{Am}$) ionizes the air in the chamber of the

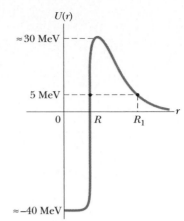

Figure 44.12 Potential energy versus separation distance for a system consisting of an alpha particle and a daughter nucleus. Classically, the energy of the alpha particle is not sufficiently large to overcome the energy barrier, and so the particle should not be able to escape from the nucleus. In reality, the alpha particle does escape by tunneling through the barrier.

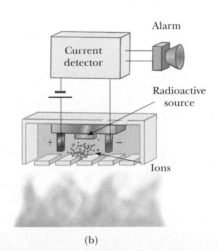

(a)   (b)

Figure 44.13 (a) A smoke detector uses alpha decay to determine whether smoke is in the air. The alpha source is in the black cylinder at the right. (b) Smoke entering the chamber reduces the detected current, causing the alarm to sound. © *1998 by the AAAS.*

detector, creating charged particles. A voltage is maintained between the plates inside the chamber, setting up a small but detectable current in the external circuit due to the ions acting as charge carriers between the plates. As long as the current is maintained, the alarm is deactivated. However, if smoke drifts into the chamber, the ions become attached to the smoke particles. These heavier particles do not drift as readily as do the lighter ions, which causes a decrease in the detector current. The external circuit senses this decrease in current and sets off the alarm.

Beta Decay

When a radioactive nucleus undergoes beta decay, the daughter nucleus contains the same number of nucleons as the parent nucleus but the atomic number is changed by 1, which means that the number of protons changes:

$$\ _{Z}^{A}X \ \longrightarrow \ _{Z+1}^{A}Y + e^{-} \qquad \text{(incomplete expression)} \qquad (44.14)$$

$$\ _{Z}^{A}X \ \longrightarrow \ _{Z-1}^{A}Y + e^{+} \qquad \text{(incomplete expression)} \qquad (44.15)$$

where, as we mentioned in Section 44.4, e^{-} is used to designate an electron and e^{+} designates a positron, with *beta particle* being the general term referring to either. *Beta decay is not described completely by these expressions.* We shall give reasons for this shortly.

As with alpha decay, the nucleon number and total charge are both conserved in beta decays. From the fact that A does not change but Z does, we conclude that in beta decay, either a neutron changes to a proton (Eq. 44.14) or a proton changes to a neutron (Eq. 44.15). Note that the electron or positron emitted in these decays is not present beforehand in the nucleus; it is created in the process of the decay from the rest energy of the decaying nucleus. Two typical beta-decay processes are

$$\ _{6}^{14}C \ \longrightarrow \ _{7}^{14}N + e^{-} \qquad \text{(incomplete expression)} \qquad (44.16)$$

$$\ _{7}^{12}N \ \longrightarrow \ _{6}^{12}C + e^{+} \qquad \text{(incomplete expression)} \qquad (44.17)$$

Let us consider the energy of the system undergoing beta decay before and after the decay. As with alpha decay, energy of the isolated system must be conserved. Experimentally, it is found that beta particles from a single type of nucleus are emitted over a continuous range of energies (Fig. 44.14a), as opposed to alpha decay, in which the alpha particles are emitted with discrete energies (Fig. 44.14b). The kinetic energy of the system after the decay is equal to the decrease in mass-energy of the system, that is, the Q value. However, because all decaying nuclei in the sample have the same initial mass, *the Q value must be the same for each decay.* In view of this, why do the emitted particles have the range of kinetic energies shown in Figure 44.14a? The law of conservation of energy seems to be violated! And it becomes worse: further analysis of the decay processes

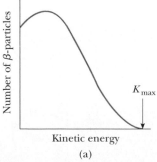
(a)

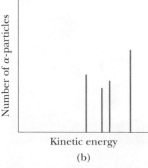
(b)

Active Figure 44.14 (a) Distribution of beta particle energies in a typical beta decay. All energies are observed up to a maximum value. (b) In contrast, the energies of alpha particles from an alpha decay are discrete.

described by Equations 44.14 and 44.15 shows that the laws of conservation of both angular momentum (spin) and linear momentum are also violated!

After a great deal of experimental and theoretical study, Pauli in 1930 proposed that a third particle must be present in the decay products to carry away the "missing" energy and momentum. Fermi later named this particle the **neutrino** (little neutral one) because it had to be electrically neutral and have little or no mass. Although it eluded detection for many years, the neutrino (symbol ν, Greek nu) was finally detected experimentally in 1956 by Frederick Reines, who received the Nobel Prize for this work in 1995. It has the following properties:

- It has zero electric charge.

Properties of the neutrino

- Its mass is either zero (in which case it travels at the speed of light) or very small; there is much recent persuasive experimental evidence that suggests that the neutrino mass is not zero. Current experiments place the upper bound of the mass of the neutrino at about 7 eV/c^2.

- It has a spin of $\frac{1}{2}$, which allows the law of conservation of angular momentum to be satisfied in beta decay.

- It interacts very weakly with matter and is therefore very difficult to detect.

We can now write the beta-decay processes (Eqs. 44.14 and 44.15) in their correct and complete form:

$$\,^{A}_{Z}X \longrightarrow \,^{A}_{Z+1}Y + e^{-} + \overline{\nu} \qquad \text{(complete expression)} \qquad \textbf{(44.18)}$$

Beta decay processes

$$\,^{A}_{Z}X \longrightarrow \,^{A}_{Z-1}Y + e^{+} + \nu \qquad \text{(complete expression)} \qquad \textbf{(44.19)}$$

as well as those for carbon-14 and nitrogen-12 (Eqs. 44.16 and 44.17):

$$\,^{14}_{6}C \longrightarrow \,^{14}_{7}N + e^{-} + \overline{\nu} \qquad \text{(complete expression)} \qquad \textbf{(44.20)}$$

$$\,^{12}_{7}N \longrightarrow \,^{12}_{6}C + e^{+} + \nu \qquad \text{(complete expression)} \qquad \textbf{(44.21)}$$

where the symbol $\overline{\nu}$ represents the **antineutrino,** the antiparticle to the neutrino. We shall discuss antiparticles further in Chapter 46. For now, it suffices to say that **a neutrino is emitted in positron decay and an antineutrino is emitted in electron decay.** As with alpha decay, the decays listed above are analyzed by applying conservation laws, but relativistic expressions must be used for beta particles because their kinetic energy is large (typically 1 MeV) compared with their rest energy of 0.511 MeV. Figure 44.15 shows a pictorial representation of the decays described by Equations 44.20 and 44.21.

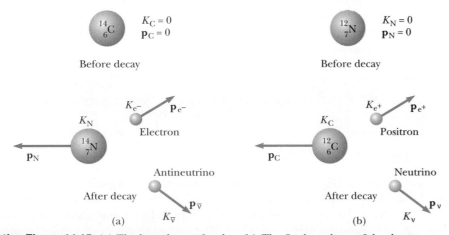

Active Figure 44.15 (a) The beta decay of carbon-14. The final products of the decay are the nitrogen-14 nucleus, an electron, and an antineutrino. (b) The beta decay of nitrogen-12. The final products of the decay are the carbon-12 nucleus, a positron, and a neutrino.

At the Active Figures link at http://www.pse6.com, observe the decay of carbon-14. For a large number of decays, observe the development of the graph in Figure 44.14a.

In Equation 44.18, the number of protons has increased by one and the number of neutrons has decreased by one. We can write the fundamental process of e^- decay in terms of a neutron changing into a proton as follows:

$$n \longrightarrow p + e^- + \bar{\nu} \tag{44.22}$$

The electron and the antineutrino are ejected from the nucleus, with the net result that there is one more proton and one fewer neutron, consistent with the changes in Z and $A - Z$. A similar process occurs in e^+ decay, with a proton changing into a neutron, a positron, and a neutrino. This latter process can only occur within the nucleus, with the result that the nuclear mass decreases. It cannot occur for an isolated proton because its mass is less than that of the neutron.

A process that competes with e^+ decay is **electron capture.** This occurs when a parent nucleus captures one of its own orbital electrons and emits a neutrino. The final product after decay is a nucleus whose charge is $Z - 1$:

$$^A_Z X + \, ^0_{-1}e \longrightarrow \, ^A_{Z-1}Y + \nu \tag{44.23}$$

In most cases, it is a K-shell electron that is captured, and for this reason the process is referred to as **K capture.** One example is the capture of an electron by $^7_4$Be:

$$^7_4 Be + \, ^0_{-1}e \longrightarrow \, ^7_3Li + \nu$$

Because the neutrino is very difficult to detect, electron capture is usually observed by the x-rays given off as higher-shell electrons cascade downward to fill the vacancy created in the K shell.

Finally, we specify Q values for the beta-decay processes. The Q values for e^- decay and electron capture are given by $Q = (M_X - M_Y)c^2$, where M_X and M_Y are the masses of neutral atoms. The Q values for e^+ decay are given by $Q = (M_X - M_Y - 2m_e)c^2$. The extra term $-2m_ec^2$ in this expression is due to the fact that the atomic number of the parent decreases by one when the daughter is formed. To form a neutral atom, the daughter atom sheds one electron. Thus, the final products are the daughter atom, the shed electron, and the ejected positron. In e^- decay, the parent nucleus experiences an increase in atomic number and must add an electron in order for the atom to become neutral. If we imagine the neutral parent atom and an electron that combine with the daughter as the initial system and the final system as the neutral daughter atom and the beta-ejected electron, there is a free electron in the system both before and after the decay. Thus, in subtracting the initial and final mass-energies of the system, this electron mass cancels.

These relationships are useful in determining whether or not a process is energetically possible. For example, the expression for proposed e^+ decay for a particular parent nucleus may turn out to be negative. In this case, this decay will not occur. The expression for electron capture for this parent nucleus, however, may give a positive number, so that electron capture can occur even though e^+ decay is not possible. This is the case for the decay of $^7_4$Be shown above.

> **Quick Quiz 44.7** Which of the following is the correct daughter nucleus associated with the beta decay of $^{184}_{72}$Hf? (a) $^{183}_{72}$Hf (b) $^{183}_{73}$Ta (c) $^{184}_{73}$Ta

Carbon Dating

The beta decay of ^{14}C (Eq. 44.20) is commonly used to date organic samples. Cosmic rays in the upper atmosphere cause nuclear reactions (Section 44.7) that create ^{14}C. The ratio of ^{14}C to ^{12}C in the carbon dioxide molecules of our atmosphere has a constant value of approximately 1.3×10^{-12}. The carbon atoms in all living organisms have this same ^{14}C/^{12}C ratio because the organisms continuously exchange carbon dioxide with their surroundings. When an organism dies, however, it no longer absorbs ^{14}C from the atmosphere, and so the ^{14}C/^{12}C ratio decreases as the ^{14}C decays with a

---

Electron capture

 PITFALL PREVENTION

44.7 Mass Number of the Electron

An alternative notation for an electron is the symbol $_{-1}^{0}e$. This does not imply that the electron has zero rest energy. The mass of the electron is so much smaller than that of the lightest nucleon, however, that we approximate it as zero in the context of nuclear decays and reactions.

half-life of 5 730 yr. It is therefore possible to measure the age of a material by measuring its ^{14}C activity. Using this technique, scientists have been able to identify samples of wood, charcoal, bone, and shell as having lived from 1 000 to 25 000 years ago. This knowledge has helped us reconstruct the history of living organisms—including humans—during this time span.

A particularly interesting example is the dating of the Dead Sea Scrolls. This group of manuscripts was discovered by a shepherd in 1947. Translation showed them to be religious documents, including most of the books of the Old Testament. Because of their historical and religious significance, scholars wanted to know their age. Carbon dating applied to the material in which they were wrapped established their age at approximately 1 950 yr.

Conceptual Example 44.9   The Age of Ice Man

In 1991, a German tourist discovered the well-preserved remains of a man, now called the "Ice Man," trapped in a glacier in the Italian Alps. (See the chapter opening photograph.) Radioactive dating with ^{14}C revealed that this person was alive about 5 300 years ago. Why did scientists date a sample of the Ice Man using ^{14}C rather than ^{11}C, which is a beta emitter having a half life of 20.4 min?

Solution Because ^{14}C has a half-life of 5 730 yr, the fraction of ^{14}C nuclei remaining after one half-life is high enough to allow accurate measurements of changes in the sample's activity. Because ^{11}C has a very short half-life, it is not useful—its activity decreases to a vanishingly small value over the age of the sample, making it impossible to detect.

An isotope used to date a sample must be present in a known amount in the sample when it is formed. As a general rule, the isotope chosen to date a sample should also have a half-life that is on the same order of magnitude as the age of the sample. If the half-life is much less than the age of the sample, there won't be enough activity left to measure because almost all of the original radioactive nuclei will have decayed. If the half-life is much greater than the age of the sample, the amount of decay that has taken place since the sample died will be too small to measure. For example, if you have a specimen estimated to have died 50 years ago, neither ^{14}C (5 730 yr) nor ^{11}C (20 min) is suitable. If you know your sample contains hydrogen, however, you can measure the activity of ^{3}H (tritium), a beta emitter that has a half-life of 12.3 yr.

Example 44.10   Radioactive Dating

A piece of charcoal containing 25.0 g of carbon is found in some ruins of an ancient city. The sample shows a ^{14}C activity R of 250 decays/min. How long has the tree from which this charcoal came been dead?

Solution First, let us calculate the decay constant for ^{14}C, which has a half-life of 5 730 yr:

$$\lambda = \frac{0.693}{T_{1/2}} = \frac{0.693}{(5\ 730\ \text{yr})(3.16 \times 10^7\ \text{s/yr})}$$
$$= 3.83 \times 10^{-12}\ \text{s}^{-1}$$

The number of ^{14}C nuclei can be calculated in two steps. First, the number of ^{12}C nuclei in 25.0 g of carbon is

$$N(^{12}\text{C}) = \frac{6.02 \times 10^{23}\ \text{nuclei/mol}}{12.0\ \text{g/mol}}(25.0\ \text{g})$$
$$= 1.25 \times 10^{24}\ \text{nuclei}$$

Knowing that the ratio of ^{14}C to ^{12}C in the live sample was 1.3×10^{-12}, we see that the number of ^{14}C nuclei in 25.0 g *before* decay was

$$N_0(^{14}\text{C}) = (1.3 \times 10^{-12})(1.25 \times 10^{24}) = 1.6 \times 10^{12}\ \text{nuclei}$$

Hence, the initial activity of the sample was

$$R_0 = N_0\lambda = (1.6 \times 10^{12}\ \text{nuclei})(3.83 \times 10^{-12}\ \text{s}^{-1})$$
$$= 6.13\ \text{decays/s} = 368\ \text{decays/min}$$

We now use Equation 44.7, which relates the activity R at any time t to the initial activity R_0:

$$R = R_0 e^{-\lambda t}$$

$$e^{-\lambda t} = \frac{R}{R_0}$$

Using $R = 250$ decays/min and $R_0 = 368$ decays/min, we calculate t by taking the natural logarithm of both sides of this expression:

$$-\lambda t = \ln\left(\frac{R}{R_0}\right) = \ln\left(\frac{250}{368}\right) = -0.39$$

$$t = \frac{0.39}{\lambda} = \frac{0.39}{3.83 \times 10^{-12}\ \text{s}^{-1}}$$

$$= 1.0 \times 10^{11}\ \text{s} = \boxed{3.2 \times 10^3\ \text{yr}}$$

 Practice using carbon dating on samples at the Interactive Worked Example link at http://www.pse6.com.

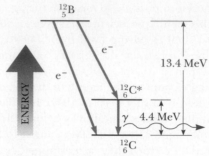

Figure 44.16 An energy-level diagram showing the initial nuclear state of a ^{12}B nucleus and two possible lower-energy states of the ^{12}C nucleus. The beta decay of the ^{12}B nucleus can result in either of two situations: the ^{12}C nucleus is in the ground state or in the excited state, in which case the nucleus is denoted as ^{12}C*. In the latter case, the beta decay to ^{12}C* is followed by a gamma decay to ^{12}C as the excited nucleus makes a transition to the ground state.

Gamma Decay

Very often, a nucleus that undergoes radioactive decay is left in an excited energy state. The nucleus can then undergo a second decay to a lower energy state, perhaps to the ground state, by emitting a high-energy photon:

Gamma decay

$$^{A}_{Z}X^{*} \longrightarrow {}^{A}_{Z}X + \gamma \tag{44.24}$$

where X* indicates a nucleus in an excited state. The typical half-life of an excited nuclear state is 10^{-10} s. Photons emitted in such a de-excitation process are called gamma rays. Such photons have very high energy (1 MeV–1 GeV) relative to the energy of visible light (about 1 eV). Recall from Section 42.3 that the energy of a photon emitted or absorbed by an atom equals the difference in energy between the two electronic states involved in the transition. Similarly, a gamma-ray photon has an energy hf that equals the energy difference ΔE between two nuclear energy levels. When a nucleus decays by emitting a gamma ray, the only change in the nucleus is that it ends up in a lower energy state. There are no changes in Z, N, or A.

A nucleus may reach an excited state as the result of a violent collision with another particle. However, it is more common for a nucleus to be in an excited state after it has undergone alpha or beta decay. The following sequence of events represents a typical situation in which gamma decay occurs:

$$^{12}_{5}B \longrightarrow {}^{12}_{6}C^{*} + e^{-} + \bar{\nu} \tag{44.25}$$

$$^{12}_{6}C^{*} \longrightarrow {}^{12}_{6}C + \gamma \tag{44.26}$$

Figure 44.16 shows the decay scheme for ^{12}B, which undergoes beta decay to either of two levels of ^{12}C. It can either (1) decay directly to the ground state of ^{12}C by emitting a 13.4-MeV electron or (2) undergo beta decay to an excited state of ^{12}C* followed by gamma decay to the ground state. The latter process results in the emission of a 9.0-MeV electron and a 4.4-MeV photon.

The various pathways by which a radioactive nucleus can undergo decay are summarized in Table 44.3.

Table 44.3

Various Decay Pathways	
Alpha decay	$^{A}_{Z}X \rightarrow {}^{A-4}_{Z-2}Y + {}^{4}_{2}He$
Beta decay (e^{-})	$^{A}_{Z}X \rightarrow {}_{Z+1}^{A}Y + e^{-} + \bar{\nu}$
Beta decay (e^{+})	$^{A}_{Z}X \rightarrow {}_{Z-1}^{A}Y + e^{+} + \nu$
Electron capture	$^{A}_{Z}X + e^{-} \rightarrow {}_{Z-1}^{A}Y + \nu$
Gamma decay	$^{A}_{Z}X^{*} \rightarrow {}^{A}_{Z}X + \gamma$

Table 44.4

The Four Radioactive Series			
Series	Starting Isotope	Half-Life (years)	Stable End Product
Uranium ⎫	$^{238}_{92}U$	4.47×10^9	$^{206}_{82}Pb$
Actinium ⎬ Natural	$^{235}_{92}U$	7.04×10^8	$^{207}_{82}Pb$
Thorium ⎭	$^{232}_{90}Th$	1.41×10^{10}	$^{208}_{82}Pb$
Neptunium	$^{237}_{93}Np$	2.14×10^6	$^{209}_{83}Bi$

44.6 Natural Radioactivity

Radioactive nuclei are generally classified into two groups: (1) unstable nuclei found in nature, which give rise to **natural radioactivity,** and (2) unstable nuclei produced in the laboratory through nuclear reactions, which exhibit **artificial radioactivity.**

As Table 44.4 shows, there are three series of naturally occurring radioactive nuclei. Each series starts with a specific long-lived radioactive isotope whose half-life exceeds that of any of its unstable descendants. The three natural series begin with the isotopes ^{238}U, ^{235}U, and ^{232}Th, and the corresponding stable end products are three isotopes of lead: ^{206}Pb, ^{207}Pb, and ^{208}Pb. The fourth series in Table 44.4 begins with ^{237}Np and has as its stable end product ^{209}Bi. The element ^{237}Np is a *transuranic* element (one having an atomic number greater than that of uranium) not found in nature. This element has a half-life of "only" 2.14×10^6 years.

Figure 44.17 shows the successive decays for the ^{232}Th series. Note that ^{232}Th first undergoes alpha decay to ^{228}Ra. Next, ^{228}Ra undergoes two successive beta decays to ^{228}Th. The series continues and finally branches when it reaches ^{212}Bi. At this point, there are two decay possibilities. The end of the decay series is the stable isotope ^{208}Pb. The sequence shown in Figure 44.17 is characterized by a mass-number decrease of either 4 (for alpha decays) or 0 (for beta or gamma decays). The two uranium series are more complex than the ^{232}Th series. Also, there are several naturally occurring radioactive isotopes, such as ^{14}C and ^{40}K, which are not part of any decay series.

Because of these radioactive series, our environment is constantly replenished with radioactive elements that would otherwise have disappeared long ago. For example, because the Solar System is approximately 5×10^9 years old, the supply of ^{226}Ra (whose half-life is only 1 600 years) would have been depleted by radioactive decay long ago if it were not for the radioactive series starting with ^{238}U.

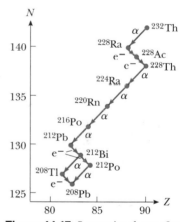

Figure 44.17 Successive decays for the ^{232}Th series.

44.7 Nuclear Reactions

We have studied radioactivity, which is a spontaneous process in which the structure of a nucleus changes. It is also possible to stimulate changes in the structure of nuclei by bombarding them with energetic particles. Such collisions, which change the identity of the target nuclei, are called **nuclear reactions.** Rutherford was the first to observe them, in 1919, using naturally occurring radioactive sources for the bombarding particles. Since then, thousands of nuclear reactions have been observed following the development of charged-particle accelerators in the 1930s. With today's advanced technology in particle accelerators and particle detectors, it is possible to achieve particle energies of at least 1 000 GeV = 1 TeV. These high-energy particles are used to create new particles whose properties are helping to solve the mysteries of the nucleus.

Consider a reaction in which a target nucleus X is bombarded by a particle a, resulting in a daughter nucleus Y and an outgoing particle b:

$$a + X \longrightarrow Y + b \tag{44.27}$$

Nuclear reaction

Sometimes this reaction is written in the more compact form

$$X(a, b)Y$$

In Section 44.5, the Q value, or disintegration energy, of a radioactive decay was defined as the mass-energy transformed to kinetic energy as a result of the decay process. Likewise, we define the **reaction energy** Q associated with a nuclear reaction as *the total change in mass-energy resulting from the reaction*:

Reaction energy Q

$$Q = (M_a + M_X - M_Y - M_b)c^2 \tag{44.28}$$

As an example, consider the reaction ^{7}Li (p, α)^{4}He. The notation p indicates a proton, which is a hydrogen nucleus. Thus, we can write this reaction in the expanded form

$$^1_1\text{H} + ^7_3\text{Li} \longrightarrow ^4_2\text{He} + ^4_2\text{He}$$

The Q value for this reaction is 17.3 MeV. A reaction such as this, for which Q is positive, is called **exothermic.** A reaction for which Q is negative is called **endothermic.** In order to satisfy conservation of momentum, an endothermic reaction does not occur unless the bombarding particle has a kinetic energy greater than Q. (See problem 56.) The minimum energy necessary for such a reaction to occur is called the **threshold energy.**

Nuclear reactions must obey the law of conservation of linear momentum. Generally the only force acting on the interacting particles is their mutual force of interaction; that is, there are no external accelerating electric fields present near the colliding particles. Thus, the system is considered to be isolated for purposes of momentum conservation.

If particles a and b in a nuclear reaction are identical, so that X and Y are also necessarily identical, the reaction is called a **scattering event.** If the kinetic energy of the system (a and X) before the event is the same as that of the system (b and Y) after the event, it is classified as *elastic scattering*. If the kinetic energies of the system before and after the event are not the same, the reaction is described as *inelastic scattering*. In this case, the difference in energy is accounted for by the fact that the target nucleus has been raised to an excited state by the event. The final system now consists of b and an excited nucleus Y*, and eventually will become b, Y, and γ, where γ is the gamma-ray photon that is emitted when the system returns to the ground state. This elastic and inelastic terminology is identical to that used in describing collisions between macroscopic objects (Section 9.3).

Measured Q values for a number of nuclear reactions involving light nuclei are given in Table 44.5.

In addition to energy and momentum, the total charge and total number of nucleons must be conserved in any nuclear reaction. For example, consider the reaction ^{19}F(p, α)^{16}O, which has a Q value of 8.11 MeV. We can show this reaction more completely as

$$^1_1\text{H} + ^{19}_9\text{F} \longrightarrow ^{16}_8\text{O} + ^4_2\text{He} \tag{44.29}$$

The total number of nucleons before the reaction $(1 + 19 = 20)$ is equal to the total number after the reaction $(16 + 4 = 20)$. Furthermore, the total charge is the same before $(1 + 9)$ and after $(8 + 2)$ the reaction.

Example 44.11 Bombarding Fluorine

Verify that the Q value for the reaction in Equation 44.29 is 8.11 MeV.

Solution Using Table A.3 in the Appendix, we can evaluate the atomic masses in the expression for the Q value:

$$Q = (M_a + M_X - M_Y - M_b) \times 931.494 \text{ MeV/u}$$
$$= (1.007\,825 + 18.998\,403 \text{ u} - 15.994\,915 \text{ u}$$
$$- 4.002\,603 \text{ u}) \times 931.494 \text{ MeV/u}$$
$$= (0.008\,710 \text{ u}) \times 931.494 \text{ MeV/u} = \boxed{8.11 \text{ MeV}}$$

Table 44.5

Q Values for Nuclear Reactions Involving Light Nuclei	
Reaction[a]	**Q Value (MeV)**
$^2\text{H}(n, \gamma)^3\text{H}$	6.257
$^2\text{H}(d, p)^3\text{H}$	4.033
$^6\text{Li}(p, \alpha)^3\text{He}$	4.019
$^6\text{Li}(d, p)^7\text{Li}$	5.025
$^7\text{Li}(p, n)^7\text{Be}$	-1.644
$^7\text{Li}(p, \alpha)^4\text{He}$	17.347
$^9\text{Be}(n, \gamma)^{10}\text{Be}$	6.812
$^9\text{Be}(\gamma, n)^8\text{Be}$	-1.665
$^9\text{Be}(d, p)^{10}\text{Be}$	4.588
$^9\text{Be}(p, \alpha)^6\text{Li}$	2.125
$^{10}\text{B}(n, \alpha)^7\text{Li}$	2.789
$^{10}\text{B}(p, \alpha)^7\text{Be}$	1.145
$^{12}\text{C}(n, \gamma)^{13}\text{C}$	4.946
$^{13}\text{C}(p, n)^{13}\text{N}$	-3.003
$^{14}\text{N}(n, p)^{14}\text{C}$	0.626
$^{14}\text{N}(n, \gamma)^{15}\text{N}$	10.833
$^{18}\text{O}(p, n)^{18}\text{F}$	-2.438
$^{19}\text{F}(p, \alpha)^{16}\text{O}$	8.114

[a] The symbols n, p, d, α, and γ denote the neutron, proton, deuteron, alpha particle, and photon, respectively.

44.8 Nuclear Magnetic Resonance and Magnetic Resonance Imaging

In this section, we shall describe an important application of nuclear physics in medicine called magnetic resonance imaging. In order to understand this application, we must first discuss the spin angular momentum of the nucleus. This discussion will have parallels with the discussion of spin for atomic electrons.

In Chapter 42, we discussed the fact that the electron has an intrinsic angular momentum, which we called spin. Nuclei also have spin because their component particles—neutrons and protons—each have spin $\frac{1}{2}$, as well as orbital angular momentum within the nucleus. All types of angular momentum obey the quantum rules that we outlined for orbital and spin angular momentum in Chapter 42. In particular, there are two quantum numbers associated with the angular momentum that determine the allowed values of the magnitude of the angular momentum vector and its direction in space. The magnitude of the nuclear angular momentum is $\sqrt{I(I + 1)}\,\hbar$, where I is called the **nuclear spin quantum number** and may be an integer or a half-integer, depending on how the individual proton and neutron spins combine. The maximum value of the z component of the spin angular momentum vector is $I\hbar$. Figure 44.18 is a vector model (see Section 42.6) illustrating the possible orientations of the nuclear spin vector and its projections along the z axis for the case in which $I = \frac{3}{2}$.

Nuclear spin has an associated, corresponding nuclear magnetic moment, similar to that of the electron. The spin magnetic moment of a nucleus is measured in terms of the **nuclear magneton** μ_n, a unit of moment defined as

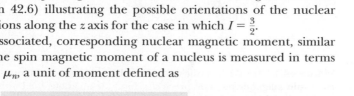

$$\mu_n = \frac{e\hbar}{2m_p} = 5.05 \times 10^{-27} \text{ J/T} \qquad (44.30)$$

Nuclear magneton

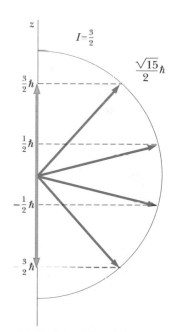

Figure 44.18 A vector model showing possible orientations of the nuclear spin angular momentum vector and its projections along the z axis for the case $I = \frac{3}{2}$.

where m_p is the mass of the proton. This definition is analogous to that of the Bohr magneton μ_B, which corresponds to the spin magnetic moment of a free electron (see Section 42.6). Note that μ_n is smaller than μ_B ($= 9.274 \times 10^{-24}$ J/T) by a factor of 1 836 because of the large difference between the proton mass and the electron mass.

The magnetic moment of a free proton is $2.792\ 8\mu_n$. Unfortunately, there is no general theory of nuclear magnetism that explains this value. The neutron also has a magnetic moment, which has a value of $-1.913\ 5\mu_n$. The negative sign indicates that this moment is opposite the spin angular momentum of the neutron. The existence of a magnetic moment for the neutron is surprising in view of the fact that the neutron is uncharged. This suggests that the neutron is not a fundamental particle but rather has an underlying structure consisting of charged constituents. We shall explore this structure in Chapter 46.

Quick Quiz 44.8 Which of the following do you expect *not* to vary substantially among different isotopes of an element? (a) atomic mass number, (b) nuclear spin magnetic moment, (c) chemical properties?

Nuclear magnetic moments, as well as electronic magnetic moments, precess when placed in an external magnetic field. This is due to the fact that a magnetic torque is applied to the spin angular momentum, so that precession occurs just as it does for a rotating top in a gravitational field, as discussed in Section 11.5. The frequency at which a magnetic moment precesses, called the **Larmor precessional frequency** ω_p, is directly proportional to the magnitude of the magnetic field. This is described schematically in Figure 44.19a, in which the external magnetic field is along the z axis. For example, the Larmor frequency of a proton in a 1-T magnetic field is 42.577 MHz. The potential energy of a magnetic dipole moment $\boldsymbol{\mu}$ in an external magnetic field $\mathbf{B}$ is given by $-\boldsymbol{\mu} \cdot \mathbf{B}$ (Eq. 29.12). When the magnetic moment $\boldsymbol{\mu}$ is lined up with the field as closely as quantum physics allows, the potential energy of the dipole moment in the field has its minimum value E_{min}. When $\boldsymbol{\mu}$ is as antiparallel to the field as possible, the potential energy has its maximum value E_{max}. In general, there are other energy states between these values corresponding to the quantized directions of the magnetic moment with respect to the field. For a nucleus with spin $\frac{1}{2}$, there are only two allowed states, with energies E_{min} and E_{max}. These two energy states are shown in Figure 44.19b.

It is possible to observe transitions between these two spin states using a technique called **NMR**, for **nuclear magnetic resonance.** A constant magnetic field ($\mathbf{B}$ in

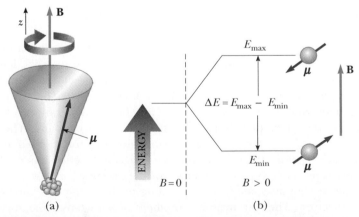

(a) (b)

Figure 44.19 (a) When a nucleus is placed in an external magnetic field $\mathbf{B}$, the nuclear spin magnetic moment precesses about the magnetic field with a frequency proportional to the magnitude of the field. (b) A nucleus with spin $\frac{1}{2}$ can occupy one of two energy states when placed in an external magnetic field. The lower energy state E_{min} corresponds to the case where the spin is aligned with the field as much as possible according to quantum mechanics, and the higher energy state E_{max} corresponds to the case where the spin is opposite the field as much as possible.

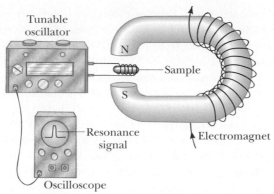

Tunable oscillator

N

Sample

S

Resonance signal

Electromagnet

Oscilloscope

Figure 44.20 Experimental arrangement for nuclear magnetic resonance. The radio-frequency magnetic field created by the coil surrounding the sample and provided by the variable-frequency oscillator is perpendicular to the constant magnetic field created by the electromagnet. When the nuclei in the sample meet the resonance condition, the nuclei absorb energy from the radio-frequency field of the coil, and this absorption changes the characteristics of the circuit in which the coil is included. Most modern NMR spectrometers use superconducting magnets at fixed field strengths and operate at frequencies of approximately 200 MHz.

Fig. 44.19a) is introduced to define a z axis. A second, weaker, oscillating magnetic field is then applied perpendicular to **B**. When the frequency of the oscillating field is adjusted to match the Larmor precessional frequency of the nuclei in the sample, a torque that acts on the precessing magnetic moments causes them to "flip" between the two spin states shown in Figure 44.19b. These transitions result in a net absorption of energy by the nuclei, an absorption that can be detected electronically.

A diagram of the apparatus used in nuclear magnetic resonance is illustrated in Figure 44.20. The energy absorbed by the nuclei is supplied by the generator producing the oscillating magnetic field. Nuclear magnetic resonance and a related technique called *electron spin resonance* are extremely important methods for studying nuclear and atomic systems and the ways in which these systems interact with their surroundings.

A widely used medical diagnostic technique called **MRI,** for **magnetic resonance imaging,** is based on nuclear magnetic resonance. Because nearly two thirds of the atoms in the human body are hydrogen (which gives a strong signal), MRI works exceptionally well for viewing internal tissues. The patient is placed inside a large solenoid that supplies a magnetic field that is constant in time but whose magnitude varies spatially across the body. Because of the variation in the field, protons in different parts of the body precess at different frequencies, so the resonance signal can be used to provide information about the positions of the protons. A computer is used to analyze the position information to provide data for constructing a final image. An MRI scan showing incredible detail in internal body structure is shown in Figure 44.21.

The main advantage of MRI over other imaging techniques is that it causes minimal cellular damage. The photons associated with the radio-frequency signals used in MRI

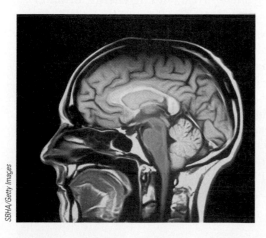

Figure 44.21 • A color-enhanced MRI scan of a human brain.

have energies of only about 10^{-7} eV. Because molecular bond strengths are much larger (approximately 1 eV), the radio-frequency radiation causes little cellular damage. In comparison, x-rays have energies ranging from 10^4 to 10^6 eV and can cause considerable cellular damage. Thus, despite some individuals' fears of the word *nuclear* associated with MRI, the radio-frequency radiation involved is overwhelmingly safer than the x-rays that these individuals might accept more readily! A disadvantage of MRI is that the equipment required to conduct the procedure is very expensive so MRI images are costly.

The magnetic field produced by the solenoid is sufficient to lift a car, and the radio signal is about the same magnitude as that from a small commercial broadcasting station! While MRI is inherently safe in normal use, the strong magnetic field of the solenoid requires diligent care to make sure that no ferromagnetic materials are located in the room near the MRI apparatus. Several accidents have occurred, such as a 2000 incident in which a gun was pulled from a police officer's hand and discharged upon striking the machine.

SUMMARY

Take a practice test for this chapter by clicking on the Practice Test link at http://www.pse6.com.

A nucleus is represented by the symbol ${}_{Z}^{A}X$, where A is the **mass number** (the total number of nucleons), and Z is the **atomic number** (the total number of protons). The total number of neutrons in a nucleus is the **neutron number** N, where $A = N + Z$. Nuclei having the same Z value but different A and N values are **isotopes** of each other. Assuming that nuclei are spherical, their radius is given by

$$r = r_0 A^{1/3} \tag{44.1}$$

where $r_0 = 1.2$ fm.

Nuclei are stable because of the **nuclear force** between nucleons. This short-range force dominates the Coulomb repulsive force at distances of less than about 2 fm and is independent of charge. Light stable nuclei have equal numbers of protons and neutrons. Heavy stable nuclei have more neutrons than protons. The most stable nuclei have Z and N values that are both even.

The difference between the sum of the masses of a group of separate nucleons and the mass of the compound nucleus containing these nucleons, when multiplied by c^2, gives the **binding energy** E_b of the nucleus. We can calculate the binding energy of the nucleus of mass M_A using the expression

$$E_b = (Zm_p + Nm_n - M_A) \times 931.494 \text{ MeV/u} \tag{44.2}$$

where m_p is the mass of the proton and m_n is the mass of the neutron.

The **liquid-drop model** of nuclear structure treats the nucleons as molecules in a drop of liquid. The four main contributions influencing binding energy are the volume effect, the surface effect, the Coulomb repulsion effect, and the symmetry effect. Summing such contributions results in the **semiempirical binding-energy formula:**

$$E_b = C_1 A - C_2 A^{2/3} - C_3 \frac{Z(Z - 1)}{A^{1/3}} - C_4 \frac{(N - Z)^2}{A} \tag{44.3}$$

The **shell model,** or **independent-particle model,** assumes that each nucleon exists in a shell and can only have discrete energy values. The stability of certain nuclei can be explained with this model.

A radioactive substance decays by **alpha decay, beta decay,** or **gamma decay.** An alpha particle is the ^{4}He nucleus; a beta particle is either an electron (e^-) or a positron (e^+); a gamma particle is a high-energy photon.

If a radioactive material contains N_0 radioactive nuclei at $t = 0$, the number N of nuclei remaining after a time t has elapsed is

$$N = N_0 e^{-\lambda t} \tag{44.6}$$

where λ is the **decay constant,** a number equal to the probability per second that a nucleus will decay. The **decay rate,** or **activity,** of a radioactive substance is

$$R = \left| \frac{dN}{dt} \right| = R_0 e^{-\lambda t} \tag{44.7}$$

where $R_0 = N_0 \lambda$ is the activity at $t = 0$. The **half-life** $T_{1/2}$ is defined as the time interval required for half of a given number of radioactive nuclei to decay, where

$$T_{1/2} = \frac{0.693}{\lambda} \tag{44.8}$$

In alpha decay, a helium nucleus is ejected from the parent nucleus with a discrete set of kinetic energies. A nucleus undergoing beta decay emits either an electron (e^-) and an antineutrino ($\bar{\nu}$) or a positron (e^+) and a neutrino (ν). The electron or positron is ejected with a range of energies. In **electron capture,** the nucleus of an atom absorbs one of its own electrons and emits a neutrino. In gamma decay, a nucleus in an excited state decays to its ground state and emits a gamma ray.

Nuclear reactions can occur when a target nucleus X is bombarded by a particle a, resulting in a daughter nucleus Y and an outgoing particle b:

$$a + X \longrightarrow Y + b \tag{44.27}$$

The mass–energy conversion in such a reaction, called the **reaction energy** Q, is

$$Q = (M_a + M_X - M_Y - M_b) c^2 \tag{44.28}$$

Nuclei have an intrinsic spin angular momentum of magnitude $\sqrt{I(I + 1)}\,\hbar$, where I is the **nuclear spin quantum number.** The magnetic moment of a nucleus is measured in terms of the **nuclear magneton** μ_n, where

$$\mu_n \equiv \frac{e\hbar}{2m_p} = 5.05 \times 10^{-27}\,\text{J/T} \tag{44.30}$$

When a nuclear spin magnetic moment is placed in an external magnetic field, it precesses about the field with a frequency (the **Larmor precessional frequency**) that is proportional to the magnitude of the field.

QUESTIONS

1. In Rutherford's experiment, assume that an alpha particle is headed directly toward the nucleus of an atom. Why doesn't the alpha particle make physical contact with the nucleus?

2. How many protons are there in the nucleus $^{222}_{86}\text{Rn}$? How many neutrons? How many orbiting electrons are there in the neutral atom?

3. Element X has several isotopes. What do these isotopes have in common? How do they differ?

4. Why are very heavy nuclei unstable?

5. Explain why nuclei that are well off the line of stability in Figure 44.4 tend to be unstable.

6. Consider two heavy nuclei X and Y having similar mass numbers. If X has the higher binding energy, which nucleus tends to be more unstable?

7. Why do nearly all the naturally occurring isotopes lie above the $N = Z$ line in Figure 44.4?

8. Discuss the differences between the liquid-drop model and the shell model of the nucleus.

9. Would the liquid-drop or shell model be more appropriate to predict the behavior of a nucleus in a fission reaction? Which would be more successful in predicting the magnetic moment of a given nucleus? Which could better explain the gamma-ray spectrum of an excited nucleus?

10. From Table A.3, identify the four stable nuclei that have magic numbers in both Z and N.

11. If a nucleus has a half-life of 1 year, does this mean it will be completely decayed after 2 years? Explain.

12. "If no more people were to be born, the law of population growth would strongly resemble the radioactive decay law." Discuss this statement.

13. Two samples of the same radioactive nuclide are prepared. Sample A has twice the initial activity of sample B. How does the half-life of A compare with the half-life of B? After each has passed through five half-lives, what is the ratio of their activities?

14. What fraction of a radioactive sample has decayed after two half-lives have elapsed?

15. Explain why the half-lives for radioactive nuclei are essentially independent of temperature.

16. The radioactive nucleus $^{226}_{88}$Ra has a half-life of approximately 1.6×10^3 years. Being that the Solar System is about 5 billion years old, why do we still find this nucleus in nature (Figure Q44.16)?

©Richard Megna/Fundamental Photographs

Figure Q44.16 Paint on the hands and numbers of this older watch contains minute amounts of natural radium mixed with a phosphorescent material. The radioactive decay of radium causes the phosphor to glow continuously.

17. A free neutron undergoes beta decay with a half-life of about 10 min. Can a free proton undergo a similar decay?

18. What is the difference between a neutrino and a photon?

19. Pick any beta-decay process and show that the neutrino must have zero charge.

20. Use Equations 44.20 to 44.22 to explain why the neutrino must have a spin of $\frac{1}{2}$.

21. If a nucleus such as ^{226}Ra initially at rest undergoes alpha decay, which has more kinetic energy after the decay, the alpha particle or the daughter nucleus?

22. Use the analogy of a bullet and a rifle to explain why the recoiling nucleus carries off only a very small fraction of the disintegration energy in the alpha decay of a nucleus.

23. Can a nucleus emit alpha particles that have different energies? Explain.

24. Explain why many heavy nuclei undergo alpha decay but do not spontaneously emit neutrons or protons.

25. If an alpha particle and an electron have the same kinetic energy, which undergoes the greater deflection when passed through a magnetic field?

26. If film is kept in a wooden box, alpha particles from a radioactive source outside the box cannot expose the film but beta particles can. Explain.

27. Does the reaction energy of a nuclear reaction represent the quantity (final mass − initial mass)c^2, or does it represent the quantity (initial mass − final mass)c^2?

28. Explain how you can carbon-date the age of a sample.

29. Suppose it could be shown that the cosmic-ray intensity at the Earth's surface was much greater 10 000 years ago. How would this difference affect what we accept as valid carbon-dated values of the age of ancient samples of once-living matter?

30. Why is carbon dating unable to provide accurate measurements of the ages of very old material?

31. Explain the main differences between alpha, beta, and gamma rays.

32. How many values of I_z are possible for $I = 5/2$? for $I = 3$?

33. In nuclear magnetic resonance, how does increasing the value of the constant magnetic field change the frequency of the radio-frequency field that excites a particular transition?

34. Do all natural events have causes? Is the Universe intelligible? Give reasons for your answers. *Note:* You may wish to consider again Question 21 in Chapter 6, on whether the future is determinate.

PROBLEMS

1, 2, 3 = straightforward, intermediate, challenging   ☐ = full solution available in the *Student Solutions Manual and Study Guide*

🌐 = coached solution with hints available at http://www.pse6.com   🖥 = computer useful in solving problem

▨ = paired numerical and symbolic problems

Note: Atomic masses are listed in Table A.3 in Appendix A.

Section 44.1   Some Properties of Nuclei

1. What is the order of magnitude of the number of protons in your body? of the number of neutrons? of the number of electrons?

2. **Review problem.** Singly ionized carbon is accelerated through 1 000 V and passed into a mass spectrometer to determine the isotopes present (see Chapter 29). The magnitude of the magnetic field in the spectrometer is 0.200 T. (a) Determine the orbit radii for the ^{12}C and the ^{13}C isotopes as they pass through the field. (b) Show that the ratio of radii may be written in the form

$$\frac{r_1}{r_2} = \sqrt{\frac{m_1}{m_2}}$$

and verify that your radii in part (a) agree with this.

3. **Review problem.** An alpha particle ($Z = 2$, mass 6.64×10^{-27} kg) approaches to within 1.00×10^{-14} m of a carbon nucleus ($Z = 6$). What are (a) the maximum Coulomb

force on the alpha particle, (b) the acceleration of the alpha particle at this point, and (c) the potential energy of the alpha particle–nucleus system when the alpha particle is at this point?

4. In a Rutherford scattering experiment, alpha particles having kinetic energy of 7.70 MeV are fired toward a gold nucleus. (a) Use energy conservation to determine the distance of closest approach between the alpha particle and gold nucleus. Assume the nucleus remains at rest. (b) **What If?** Calculate the de Broglie wavelength for the 7.70-MeV alpha particle and compare it to the distance obtained in part (a). (c) Based on this comparison, why is it proper to treat the alpha particle as a particle and not as a wave in the Rutherford scattering experiment?

5. (a) Use energy methods to calculate the distance of closest approach for a head-on collision between an alpha particle having an initial energy of 0.500 MeV and a gold nucleus (^{197}Au) at rest. (Assume the gold nucleus remains at rest during the collision.) (b) What minimum initial speed must the alpha particle have to get as close as 300 fm?

6. How much energy (in MeV units) must an alpha particle have to reach the surface of a gold nucleus ($Z = 79$, $A = 197$)? Assume the gold nucleus remains stationary.

7. Find the radius of (a) a nucleus of $^4_2$He and (b) a nucleus of $^{238}_{92}$U.

8. Find the nucleus that has a radius approximately equal to half the radius of uranium $^{238}_{92}$U.

9. A star ending its life with a mass of two times the mass of the Sun is expected to collapse, combining its protons and electrons to form a neutron star. Such a star could be thought of as a gigantic atomic nucleus. If a star of mass $2 \times 1.99 \times 10^{30}$ kg collapsed into neutrons ($m_n = 1.67 \times 10^{-27}$ kg), what would its radius be? (Assume that $r = r_0 A^{1/3}$.)

10. **Review problem.** What would be the gravitational force between two golf balls (each with a 4.30-cm diameter), 1.00 m apart, if they were made of nuclear matter?

11. From Table A.3, identify the stable nuclei that correspond to the magic numbers given by Equation 44.4.

12. Consider the stable isotopes in Table A.3 with abundances over 25.0%. That is, select the nuclei that are not radioactive and that occur in more than 25.0% of the atoms of each element in nature. Count the number of these nuclei that are (a) even Z, even N; (b) even Z, odd N; (c) odd Z, even N; and (d) odd Z, odd N.

13. Nucleus 1 has eight times as many protons as nucleus 2, five times as many neutrons, and six times as many nucleons as nucleus 2. Nucleus 1 has four more neutrons than protons. (a) What are the two nuclei? (b) Is each nucleus stable? If not, what is the minimum number of neutrons that must be added to, or removed from, each unstable nucleus to make it stable?

Section 44.2 Nuclear Binding Energy

14. Calculate the binding energy per nucleon for (a) ^{2}H, (b) ^{4}He, (c) ^{56}Fe, and (d) ^{238}U.

15. The iron isotope ^{56}Fe is near the peak of the stability curve. This is why iron is generally prominent in the spectrum of the Sun and stars. Show that ^{56}Fe has a higher binding energy per nucleon than its neighbors ^{55}Mn and ^{59}Co. Compare your results with Figure 44.5.

16. Two nuclei having the same mass number are known as *isobars*. Calculate the difference in binding energy per nucleon for the isobars $^{23}_{11}$Na and $^{23}_{12}$Mg. How do you account for the difference?

17. Nuclei having the same mass numbers are called *isobars*. The isotope $^{139}_{57}$La is stable. A radioactive isobar, $^{139}_{59}$Pr, is located below the line of stable nuclei in Figure 44.4 and decays by e^+ emission. Another radioactive isobar of $^{139}_{57}$La, $^{139}_{55}$Cs, decays by e^- emission and is located above the line of stable nuclei in Figure 44.4. (a) Which of these three isobars has the highest neutron-to-proton ratio? (b) Which has the greatest binding energy per nucleon? (c) Which do you expect to be heavier, $^{139}_{59}$Pr or $^{139}_{55}$Cs?

18. The energy required to construct a uniformly charged sphere of total charge Q and radius R is $U = 3k_e Q^2/5R$, where k_e is the Coulomb constant (see Problem 71). Assume that a ^{40}Ca nucleus contains 20 protons uniformly distributed in a spherical volume. (a) How much energy is required to counter their electrical repulsion according to the above equation? (*Suggestion:* First calculate the radius of a ^{40}Ca nucleus.) (b) Calculate the binding energy of ^{40}Ca. (c) Explain what you can conclude from comparing the result of part (b) and that of part (a).

19. A pair of nuclei for which $Z_1 = N_2$ and $Z_2 = N_1$ are called *mirror isobars* (the atomic and neutron numbers are interchanged). Binding-energy measurements on these nuclei can be used to obtain evidence of the charge independence of nuclear forces (that is, proton–proton, proton–neutron, and neutron–neutron nuclear forces are equal). Calculate the difference in binding energy for the two mirror isobars $^{15}_8$O and $^{15}_7$N. The electric repulsion among eight protons rather than seven accounts for the difference.

20. Calculate the minimum energy required to remove a neutron from the $^{43}_{20}$Ca nucleus.

Section 44.3 Nuclear Models

21. Using the graph in Figure 44.5, estimate how much energy is released when a nucleus of mass number 200 fissions into two nuclei each of mass number 100.

22. (a) In the liquid-drop model of nuclear structure, why does the surface-effect term $- C_2 A^{2/3}$ have a negative sign? (b) **What If?** The binding energy of the nucleus increases as the volume-to-surface ratio increases. Calculate this ratio for both spherical and cubical shapes, and explain which is more plausible for nuclei.

23. (a) Use the semiempirical binding-energy formula to compute the binding energy for $^{56}_{26}$Fe. (b) What percentage is contributed to the binding energy by each of the four terms?

Section 44.4 Radioactivity

24. The half life of ^{131}I is 8.04 days. On a certain day, the activity of an iodine-131 sample is 6.40 mCi. What is its activity 40.2 days later?

25. A sample of radioactive material contains 1.00×10^{15} atoms and has an activity of 6.00×10^{11} Bq. What is its half-life?

26. Determine the activity of 1.00 g of ^{60}Co. The half-life of ^{60}Co is 5.27 yr.

27. 🪐 A freshly prepared sample of a certain radioactive isotope has an activity of 10.0 mCi. After 4.00 h, its activity is 8.00 mCi. (a) Find the decay constant and half-life. (b) How many atoms of the isotope were contained in the freshly prepared sample? (c) What is the sample's activity 30.0 h after it is prepared?

28. How much time elapses before 90.0% of the radioactivity of a sample of $^{72}_{33}$As disappears, as measured by its activity? The half-life of $^{72}_{33}$As is 26 h.

29. The radioactive isotope ^{198}Au has a half-life of 64.8 h. A sample containing this isotope has an initial activity ($t = 0$) of 40.0 μCi. Calculate the number of nuclei that decay in the time interval between $t_1 = 10.0$ h and $t_2 = 12.0$ h.

30. A radioactive nucleus has half-life $T_{1/2}$. A sample containing these nuclei has initial activity R_0. Calculate the number of nuclei that decay during the interval between the times t_1 and t_2.

31. In an experiment on the transport of nutrients in the root structure of a plant, two radioactive nuclides X and Y are used. Initially 2.50 times more nuclei of type X are present than of type Y. Just three days later there are 4.20 times more nuclei of type X than of type Y. Isotope Y has a half-life of 1.60 d. What is the half-life of isotope X?

32. (a) The daughter nucleus formed in radioactive decay is often radioactive. Let N_{10} represent the number of parent nuclei at time $t = 0$, $N_1(t)$ the number of parent nuclei at time t, and λ_1 the decay constant of the parent. Suppose the number of daughter nuclei at time $t = 0$ is zero, let $N_2(t)$ be the number of daughter nuclei at time t, and let λ_2 be the decay constant of the daughter. Show that $N_2(t)$ satisfies the differential equation

$$\frac{dN_2}{dt} = \lambda_1 N_1 - \lambda_2 N_2$$

(b) Verify by substitution that this differential equation has the solution

$$N_2(t) = \frac{N_{10}\lambda_1}{\lambda_1 - \lambda_2}\left(e^{-\lambda_2 t} - e^{-\lambda_1 t}\right)$$

This equation is the law of successive radioactive decays. (c) ^{218}Po decays into ^{214}Pb with a half-life of 3.10 min, and ^{214}Pb decays into ^{214}Bi with a half-life of 26.8 min. On the same axes, plot graphs of $N_1(t)$ for ^{218}Po and $N_2(t)$ for ^{214}Pb. Let $N_{10} = 1\,000$ nuclei, and choose values of t from 0 to 36 min in 2-min intervals. The curve for ^{214}Pb at first rises to a maximum and then starts to decay. At what instant t_m is the number of ^{214}Pb nuclei a maximum? (d) By applying the condition for a maximum $dN_2/dt = 0$, derive a symbolic equation for t_m in terms of λ_1 and λ_2. Does the value obtained in (c) agree with this equation?

Section 44.5 The Decay Processes

33. Find the energy released in the alpha decay

$$^{238}_{92}\text{U} \longrightarrow {}^{234}_{90}\text{Th} + {}^{4}_{2}\text{He}$$

You will find Table A.3 useful.

34. Identify the missing nuclide or particle (X):
(a) X $\rightarrow$ $^{65}_{28}$Ni + γ
(b) $^{215}_{84}$Po $\rightarrow$ X + α
(c) X $\rightarrow$ $^{55}_{26}$Fe + e^+ + ν
(d) $^{109}_{48}$Cd + X $\rightarrow$ $^{109}_{47}$Ag + ν
(e) $^{14}_{7}$N + $^{4}_{2}$He $\rightarrow$ X + $^{17}_{8}$O

35. A living specimen in equilibrium with the atmosphere contains one atom of ^{14}C (half-life = 5 730 yr) for every 7.7×10^{11} stable carbon atoms. An archeological sample of wood (cellulose, $C_{12}H_{22}O_{11}$) contains 21.0 mg of carbon. When the sample is placed inside a shielded beta counter with 88.0% counting efficiency, 837 counts are accumulated in one week. Assuming that the cosmic-ray flux and the Earth's atmosphere have not changed appreciably since the sample was formed, find the age of the sample.

36. A certain African artifact is found to have a carbon-14 activity of (0.12 ± 0.01) Bq per gram of carbon. Assume the uncertainty is negligible in the half-life of ^{14}C (5 730 yr) and in the activity of atmospheric carbon (0.25 Bq per gram). The age of the object lies within what range?

37. A ^{3}H nucleus beta decays into ^{3}He by creating an electron and an antineutrino according to the reaction

$$^{3}_{1}\text{H} \longrightarrow {}^{3}_{2}\text{He} + e^- + \bar{\nu}$$

The symbols in this reaction refer to nuclei. Write the reaction referring to neutral atoms by adding one electron to both sides. Then use Table A.3 to determine the total energy released in this reaction.

38. Determine which decays can occur spontaneously:
(a) $^{40}_{20}$Ca $\rightarrow$ e^+ + $^{40}_{19}$K
(b) $^{98}_{44}$Ru $\rightarrow$ $^{4}_{2}$He + $^{94}_{42}$Mo
(c) $^{144}_{60}$Nd $\rightarrow$ $^{4}_{2}$He + $^{140}_{58}$Ce

39. The nucleus $^{15}_{8}$O decays by electron capture. The nuclear reaction is written

$$^{15}_{8}\text{O} + e^- \longrightarrow {}^{15}_{7}\text{N} + \nu$$

(a) Write the process going on for a single particle within the nucleus. (b) Write the decay process referring to neutral atoms. (c) Determine the energy of the neutrino. Disregard the daughter's recoil.

Section 44.6 Natural Radioactivity

40. A rock sample contains traces of ^{238}U, ^{235}U, ^{232}Th, ^{208}Pb, ^{207}Pb, and ^{206}Pb. Careful analysis shows that the ratio of the amount of ^{238}U to ^{206}Pb is 1.164. (a) Assume that the rock originally contained no lead, and determine the age of the rock. (b) What should be the ratios of ^{235}U to ^{207}Pb and of ^{232}Th to ^{208}Pb so that they would yield the same age for the rock? Ignore the minute amounts of the intermediate decay products in the decay chains. Note that this form of multiple dating gives reliable geological dates.

41. Enter the correct isotope symbol in each open square in Figure P44.41, which shows the sequences of decays in the natural radioactive series starting with the long-lived isotope uranium-235 and ending with the stable nucleus lead-207.

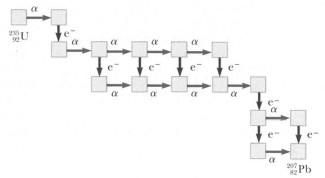

Figure P44.41

42. *Indoor air pollution.* Uranium is naturally present in rock and soil. At one step in its series of radioactive decays, ^{238}U produces the chemically inert gas radon-222, with a half-life of 3.82 days. The radon seeps out of the ground to mix into the atmosphere, typically making open air radioactive with activity 0.3 pCi/L. In homes ^{222}Rn can be a serious pollutant, accumulating to reach much higher activities in enclosed spaces. If the radon radioactivity exceeds 4 pCi/L, the Environmental Protection Agency suggests taking action to reduce it, such as by reducing infiltration of air from the ground. (a) Convert the activity 4 pCi/L to units of becquerel per cubic meter. (b) How many ^{222}Rn atoms are in one cubic meter of air displaying this activity? (c) What fraction of the mass of the air does the radon constitute?

43. Two radioactive samples consist of nuclei X in one sample, and nuclei Y in the other. The two samples have the same initial activity. After 0.685 h, the activity of the sample containing nuclei X is 1.04 times the activity of the other sample. The half-life of nucleus X exceeds the half-life of nucleus Y by 77.2 h. (a) Find the half-life of each nucleus. (b) Use Table A.3 to identify the two nuclei. (c) Nucleus X decays to nucleus Y by an alpha-decay chain. How many alpha-decay events are there in the chain?

44. The most common isotope of radon is ^{222}Rn, which has half-life 3.82 days. (a) What fraction of the nuclei that were on Earth one week ago are now undecayed? (b) What fraction of those that existed one year ago? (c) In view of these results, explain why radon remains a problem, contributing significantly to our background radiation exposure.

Section 44.7   Nuclear Reactions

45. The reaction $^{27}_{13}$Al$(\alpha, \text{n})^{30}_{15}$P, achieved in 1934, was the first known in which the product nucleus is radioactive. Calculate the Q value of this reaction.

46. Identify the unknown nuclei and particles X and X′ in the following nuclear reactions:
    (a) X + $^{4}_{2}$He $\rightarrow$ $^{24}_{12}$Mg + $^{1}_{0}$n
    (b) $^{235}_{92}$U + $^{1}_{0}$n $\rightarrow$ $^{90}_{38}$Sr + X + 2$^{1}_{0}$n
    (c) 2$^{1}_{1}$H $\rightarrow$ $^{2}_{1}$H + X + X′

47. Natural gold has only one isotope, $^{197}_{79}$Au. If natural gold is irradiated by a flux of slow neutrons, electrons are emitted. (a) Write the reaction equation. (b) Calculate the maximum energy of the emitted electrons.

48. A beam of 6.61-MeV protons is incident on a target of $^{27}_{13}$Al. Those that collide produce the reaction

$$\text{p} + {}^{27}_{13}\text{Al} \longrightarrow {}^{27}_{14}\text{Si} + \text{n}$$

($^{27}_{14}$Si has mass 26.986 705 u.) Ignoring any recoil of the product nucleus, determine the kinetic energy of the emerging neutrons.

49. Using the mass of ^{9}Be from Table A.3 and Q values of appropriate reactions from Table 44.5, calculate the masses of ^{8}Be and ^{10}Be in atomic mass units to four decimal places.

50. (a) Suppose $^{10}_{5}$B is struck by an alpha particle, releasing a proton and a product nucleus in the reaction. What is the product nucleus? (b) An alpha particle and a product nucleus are produced when $^{13}_{6}$C is struck by a proton. What is the product nucleus?

51. Determine the Q value associated with the spontaneous fission of ^{236}U into the fragments ^{90}Rb and ^{143}Cs, which have mass 89.914 809 u and 142.927 330 u, respectively. The masses of the other particles involved in the reaction are given in Table A.3.

Section 44.8   Nuclear Magnetic Resonance and Magnetic Resonance Imaging

52. Construct a diagram like that of Figure 44.18 for the cases when I equals (a) 5/2 and (b) 4.

53. The radio frequency at which a nucleus displays resonance absorption between spin states is called the Larmor precessional frequency and is given by

$$f = \frac{\Delta E}{h} = \frac{2\mu B}{h}$$

Calculate the Larmor frequency for (a) free neutrons in a magnetic field of 1.00 T, (b) free protons in a magnetic field of 1.00 T, and (c) free protons in the earth's magnetic field at a location where the magnitude of the field is 50.0 μT.

Additional Problems

54. As part of his discovery of the neutron in 1932, James Chadwick determined the mass of the newly identified particle by firing a beam of fast neutrons, all having the same speed, at two different targets and measuring the maximum recoil speeds of the target nuclei. The maximum speeds arise when an elastic head-on collision occurs between a neutron and a stationary target nucleus. (a) Represent the masses and final speeds of the two target nuclei as m_1, v_1, m_2, and v_2, and assume Newtonian mechanics

applies. Show that the neutron mass can be calculated from the equation

$$m_n = \frac{m_1 v_1 - m_2 v_2}{v_2 - v_1}$$

(b) Chadwick directed a beam of neutrons (produced from a nuclear reaction) on paraffin, which contains hydrogen. The maximum speed of the protons ejected was found to be 3.3×10^7 m/s. Because the velocity of the neutrons could not be determined directly, a second experiment was performed using neutrons from the same source and nitrogen nuclei as the target. The maximum recoil speed of the nitrogen nuclei was found to be 4.7×10^6 m/s. The masses of a proton and a nitrogen nucleus were taken as 1 u and 14 u, respectively. What was Chadwick's value for the neutron mass?

55. (a) One method of producing neutrons for experimental use is bombardment of light nuclei with alpha particles. In the method used by Chadwick in 1932, alpha particles emitted by polonium are incident on beryllium nuclei:

$$^4_2\text{He} + ^9_4\text{Be} \longrightarrow ^{12}_6\text{C} + ^1_0\text{n}$$

What is the Q value? (b) Neutrons are also often produced by small-particle accelerators. In one design, deuterons accelerated in a Van de Graaff generator bombard other deuterium nuclei:

$$^2_1\text{H} + ^2_1\text{H} \longrightarrow ^3_2\text{He} + ^1_0\text{n}$$

Is this reaction exothermic or endothermic? Calculate its Q value.

56. When the nuclear reaction represented by Equation 44.27 is endothermic, the reaction energy Q is negative. For the reaction to proceed, the incoming particle must have a minimum energy called the threshold energy, E_{th}. Some fraction of the energy of the incident particle is transferred to the compound nucleus to conserve momentum. Therefore, E_{th} must be greater than Q. (a) Show that

$$E_{th} = -Q\left(1 + \frac{M_a}{M_X}\right)$$

(b) Calculate the threshold energy of the incident alpha particle in the reaction

$$^4_2\text{He} + ^{14}_7\text{N} \longrightarrow ^{17}_8\text{O} + ^1_1\text{H}$$

57. One method of producing neutrons for experimental use is to bombard ^7_3Li with protons. The neutrons are emitted according to the reaction

$$^1_1\text{H} + ^7_3\text{Li} \longrightarrow ^7_4\text{Be} + ^1_0\text{n}$$

What is the minimum kinetic energy the incident proton must have if this reaction is to occur? You may use the result of Problem 56.

58. A byproduct of some fission reactors is the isotope $^{239}_{94}\text{Pu}$, an alpha emitter having a half-life of 24 120 yr:

$$^{239}_{94}\text{Pu} \longrightarrow ^{235}_{92}\text{U} + \alpha$$

Consider a sample of 1.00 kg of pure $^{239}_{94}\text{Pu}$ at $t = 0$. Calculate (a) the number of $^{239}_{94}\text{Pu}$ nuclei present at $t = 0$ and

(b) the initial activity in the sample. (c) **What If?** How long does the sample have to be stored if a "safe" activity level is 0.100 Bq?

59. The atomic mass of ^{57}Co is 56.936 296 u. (a) Can ^{57}Co decay by e^+ emission? Explain. (b) **What If?** Can ^{14}C decay by e^- emission? Explain. (c) If either answer is yes, what is the range of kinetic energies available for the beta particle?

60. (a) Find the radius of the $^{12}_6\text{C}$ nucleus. (b) Find the force of repulsion between a proton at the surface of a $^{12}_6\text{C}$ nucleus and the remaining five protons. (c) How much work (in MeV) has to be done to overcome this electric repulsion to put the last proton into the nucleus? (d) Repeat (a), (b), and (c) for $^{238}_{92}\text{U}$.

61. (a) Why is the beta decay $p \rightarrow n + e^+ + \nu$ forbidden for a free proton? (b) **What If?** Why is the same reaction possible if the proton is bound in a nucleus? For example, the following reaction occurs:

$$^{13}_7\text{N} \longrightarrow ^{13}_6\text{C} + e^+ + \nu$$

(c) How much energy is released in the reaction given in (b)? *Suggestion:* Add seven electrons to both sides of the reaction to write it for neutral atoms. You may use the masses $m(e^+) = 0.000\,549$ u, $M(^{13}\text{C}) = 13.003\,355$ u, $M(^{13}\text{N}) = 13.005\,739$ u.

62. 🖥 The activity of a radioactive sample was measured over 12 h, with the net count rates shown in the table.

Time(h)	Counting Rate (counts/min)
1.00	3 100
2.00	2 450
4.00	1 480
6.00	910
8.00	545
10.0	330
12.0	200

(a) Plot the logarithm of counting rate as a function of time. (b) Determine the decay constant and half-life of the radioactive nuclei in the sample. (c) What counting rate would you expect for the sample at $t = 0$? (d) Assuming the efficiency of the counting instrument to be 10.0%, calculate the number of radioactive atoms in the sample at $t = 0$.

63. The ^{145}Pm nucleus decays by alpha emission. (a) Determine the daughter nucleus. (b) Using the values given in Table A.3, determine the energy released in this decay. (c) What fraction of this energy is carried away by the alpha particle when the recoil of the daughter is taken into account?

64. When, after a reaction or disturbance of any kind, a nucleus is left in an excited state, it can return to its normal (ground) state by emission of a gamma-ray photon (or several photons). This process is illustrated by Equation 44.24. The emitting nucleus must recoil to conserve both energy and momentum. (a) Show that the recoil energy of

the nucleus is

$$E_r = \frac{(\Delta E)^2}{2Mc^2}$$

where ΔE is the difference in energy between the excited and ground states of a nucleus of mass M. (b) Calculate the recoil energy of the ^{57}Fe nucleus when it decays by gamma emission from the 14.4-keV excited state. For this calculation, take the mass to be 57 u. (*Suggestions:* When writing the equation for conservation of energy, use $(Mv)^2/2M$ for the kinetic energy of the recoiling nucleus. Also, assume that $hf \ll Mc^2$ and use the binomial expansion.)

65. After the sudden release of radioactivity from the Chernobyl nuclear reactor accident in 1986, the radioactivity of milk in Poland rose to 2 000 Bq/L due to iodine-131 present in the grass eaten by dairy cattle. Radioactive iodine, with half-life 8.04 days, is particularly hazardous, because the thyroid gland concentrates iodine. The Chernobyl accident caused a measurable increase in thyroid cancers among children in Belarus. (a) For comparison, find the activity of milk due to potassium. Assume that one liter of milk contains 2.00 g of potassium, of which 0.011 7% is the isotope ^{40}K with a half-life 1.28×10^9 yr. (b) After what elapsed time would the activity due to iodine fall below that due to potassium?

66. Europeans named a certain direction in the sky as between the horns of Taurus the Bull. On the day they named as July 4, 1054 A.D., a brilliant light appeared there. Europeans left no surviving record of the supernova, which could be seen in daylight for some days. As it faded it remained visible for years, dimming for a time with the 77.1-day half-life of the radioactive cobalt-56 that had been created in the explosion. (a) The remains of the star now form the Crab nebula (see page 1066). In it, the cobalt-56 has now decreased to what fraction of its original activity? (b) Suppose that an American, of the people called the Anasazi, made a charcoal drawing of the supernova. The carbon-14 in the charcoal has now decayed to what fraction of its original activity?

67. A theory of nuclear astrophysics proposes that all the elements heavier than iron are formed in supernova explosions ending the lives of massive stars. Assume that at the time of the explosion the amounts of ^{235}U and ^{238}U were equal. How long ago did the star(s) explode that released the elements that formed our Earth? The present ^{235}U/^{238}U ratio is 0.007 25. The half-lives of ^{235}U and ^{238}U are 0.704×10^9 yr and 4.47×10^9 yr.

68. After determining that the Sun has existed for hundreds of millions of years, but before the discovery of nuclear physics, scientists could not explain why the Sun has continued to burn for such a long time. For example, if it were a coal fire, it would have burned up in about 3 000 yr. Assume that the Sun, whose mass is 1.99×10^{30} kg, originally consisted entirely of hydrogen and that its total power output is 3.77×10^{26} W. (a) If the energy-generating mechanism of the Sun is the fusion of hydrogen into helium via the net reaction

$$4(^1_1\text{H}) + 2(\text{e}^-) \longrightarrow ^4_2\text{He} + 2\nu + \gamma$$

calculate the energy (in joules) given off by this reaction. (b) Determine how many hydrogen atoms constitute the Sun. Take the mass of one hydrogen atom to be 1.67×10^{-27} kg. (c) If the total power output remains constant, after what time interval will all the hydrogen be converted into helium, making the Sun die? The actual projected lifetime of the Sun is about 10 billion years, because only the hydrogen in a relatively small core is available as a fuel. Only in the core are temperatures and densities high enough for the fusion reaction to be self-sustaining.

69. **Review problem.** Consider the Bohr model of the hydrogen atom, with the electron in the ground state. The magnetic field at the nucleus produced by the orbiting electron has a value of 12.5 T. (See Chapter 30, Problem 1.) The proton can have its magnetic moment aligned in either of two directions perpendicular to the plane of the electron's orbit. Because of the interaction of the proton's magnetic moment with the electron's magnetic field, there will be a difference in energy between the states with the two different orientations of the proton's magnetic moment. Find that energy difference in eV.

70. Many radioisotopes have important industrial, medical, and research applications. One of these is ^{60}Co, which has a half-life of 5.27 yr and decays by the emission of a beta particle (energy 0.31 MeV) and two gamma photons (energies 1.17 MeV and 1.33 MeV). A scientist wishes to prepare a ^{60}Co sealed source that will have an activity of 10.0 Ci after 30.0 months of use. (a) What is the initial mass of ^{60}Co required? (b) At what rate will the source emit energy after 30.0 months?

71. **Review problem.** Consider a model of the nucleus in which the positive charge (Ze) is uniformly distributed throughout a sphere of radius R. By integrating the energy density $\frac{1}{2}\epsilon_0 E^2$ over all space, show that the electric potential energy may be written

$$U = \frac{3Z^2 e^2}{20\pi\epsilon_0 R} = \frac{3k_e Z^2 e^2}{5R}$$

72. The ground state of $^{93}_{43}$Tc (molar mass 92.910 2 g/mol) decays by electron capture and e^+ emission to energy levels of the daughter (molar mass 92.906 8 g/mol in ground state) at 2.44 MeV, 2.03 MeV, 1.48 MeV, and 1.35 MeV. (a) For which of these levels are electron capture and e^+ decay allowed? (b) Identify the daughter and sketch the decay scheme, assuming all excited states de-excite by direct γ decay to the ground state.

73. Free neutrons have a characteristic half-life of 10.4 min. What fraction of a group of free neutrons with kinetic energy 0.040 0 eV will decay before traveling a distance of 10.0 km?

74. In a piece of rock from the Moon, the ^{87}Rb content is assayed to be 1.82×10^{10} atoms per gram of material, and the ^{87}Sr content is found to be 1.07×10^9 atoms per gram. (a) Calculate the age of the rock. (b) **What If?** Could the material in the rock actually be much older? What assumption is implicit in using the radioactive dating method? (The relevant decay is $^{87}\text{Rb} \rightarrow ^{87}\text{Sr} + \text{e}^- + \overline{\nu}$. The half-life of the decay is 4.75×10^{10} yr.)

75. *Student determination of the half-life of* ^{137}Ba. The radioactive barium isotope ^{137}Ba has a relatively short half-life and can be easily extracted from a solution containing its parent cesium (^{137}Cs). This barium isotope is commonly used in an undergraduate laboratory exercise for demonstrating the radioactive decay law. Undergraduate students using modest experimental equipment took the data presented in Figure P44.75. Determine the half-life for the decay of ^{137}Ba using their data.

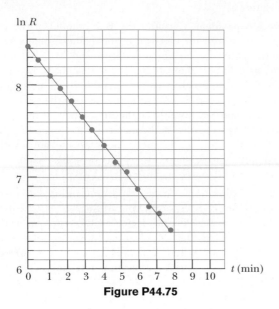

Figure P44.75

Answers to Quick Quizzes

44.1 (b). The value of $N = A - Z$ is the same for all three nuclei.

44.2 (a). The value of Z is the same for all three nuclei because they are all nuclei of nitrogen.

44.3 (c). The value of A is the same for all three nuclei, as seen by the unchanging pre-superscript.

44.4 (e). A year of 365 days is equivalent to 365 d/5.01 d $\approx$ 73 half-lives. Thus, the activity will be reduced after one year to approximately $(1/2)^{73}$ (1.000 μCi) $\sim 10^{-22}$ μCi.

44.5 (e). The time we are interested in is half of a half-life. Thus, the number of *remaining* nuclei is $(\frac{1}{2})^{1/2}N_0 = N_0/\sqrt{2} = 0.707N_0$. The number of nuclei that *have decayed* is $N_0 - 0.707N_0 = 0.293N_0$.

44.6 (b). In alpha decay, the atomic number decreases by two and the atomic mass number decreases by four.

44.7 (c). In e$^-$ decay, the atomic number increases by one and the atomic mass number stays fixed. None of the choices is consistent with e$^+$ decay, so we assume that the decay must be by e$^-$.

44.8 (c). Isotopes of a given element correspond to nuclei with different numbers of neutrons. This results in different atomic mass numbers of the nucleus—and different magnetic moments, because the neutron, despite being uncharged, has a magnetic moment. The chemical behavior, however, is governed by the electrons. All isotopes of a given element have the same number of electrons and, therefore, the same chemical behavior.

Chapter 45

Applications of Nuclear Physics

CHAPTER OUTLINE

45.1 Interactions Involving Neutrons

45.2 Nuclear Fission

45.3 Nuclear Reactors

45.4 Nuclear Fusion

45.5 Radiation Damage

45.6 Radiation Detectors

45.7 Uses of Radiation

▲ The San Onofre Nuclear Generating Station, south of San Clemente, California, is one of dozens of nuclear power plants around the world that provide energy from uranium. These plants operate via a nuclear process called fission, while plants based on a second process, fusion, are years in the future. (Tony Freeman/Index Stock Imagery)

In this chapter, we study two means for deriving energy from nuclear reactions: fission, in which a large nucleus splits into two smaller nuclei, and fusion, in which two small nuclei fuse to form a larger one. In both cases, the released energy can be used either constructively (as in the production of electric power) or destructively (as in nuclear weapons). We also examine the ways in which radiation interacts with matter and look at several devices used to detect radiation. The chapter concludes with a discussion of some industrial and biological applications of radiation.

45.1 Interactions Involving Neutrons

Nuclear fission is the process that occurs in a nuclear reactor and ultimately results in energy supplied to a community by electrical transmission. Nuclear fusion is an area of active research but has not yet been commercially developed for the supply of energy. We will discuss fission first and then explore fusion in Section 45.4.

To understand nuclear fission and the physics of nuclear reactors, we must first understand how neutrons interact with nuclei. Because of their charge neutrality, neutrons are not subject to Coulomb forces and as a result do not interact electrically with electrons or the nucleus. Therefore, neutrons can easily penetrate deep into an atom and collide with the nucleus.

A fast neutron (energy greater than about 1 MeV) traveling through matter undergoes many collisions with nuclei. In each collision, the neutron gives up some of its kinetic energy to a nucleus. For some materials and for fast neutrons, elastic collisions dominate. Materials for which this occurs are called **moderators** because they slow down (or moderate) the originally energetic neutrons very effectively. Moderator nuclei should be of low mass so that more kinetic energy is transferred to them in elastic collisions. For this reason, materials that are abundant in hydrogen, such as paraffin and water, are good moderators for neutrons.

Sooner or later, most neutrons bombarding a moderator become **thermal neutrons,** which means they are in thermal equilibrium with the moderator material. Their average kinetic energy at room temperature is, from Equation 21.4,

$$K_{av} = \tfrac{3}{2}k_{B}T \approx \tfrac{3}{2}k_{B}(300 \text{ K}) \approx 0.04 \text{ eV}$$

which corresponds to a neutron root-mean-square speed of about 2 800 m/s. Thermal neutrons have a distribution of speeds, just as the molecules in a container of gas do (see Chapter 21). High-energy neutrons, those with energy of several MeV, *thermalize* (that is, their average energy reaches K_{av}) in less than 1 ms when they are incident on a moderator.

Once the neutrons have thermalized and the energy of a particular neutron is sufficiently low, there is a high probability that the neutron will be captured by a nucleus, an event that is accompanied by the emission of a gamma ray. This **neutron capture** reaction can be written

Neutron capture reaction

$$^{1}_{0}\text{n} + ^{A}_{Z}\text{X} \longrightarrow ^{A+1}_{Z}\text{X}^{*} \longrightarrow ^{A+1}_{Z}\text{X} + \gamma \qquad (45.1)$$

Once the neutron is captured, the nucleus $^{A+1}_{\ \ Z}X^*$ is in an excited state for a very short time before it undergoes gamma decay. The product nucleus $^{A+1}_{\ \ Z}X$ is usually radioactive and decays by beta emission.

The neutron-capture rate for neutrons passing through any sample depends on the type of atoms in the sample and on the energy of the incident neutrons. The interaction of neutrons with matter increases with decreasing neutron energy because a slow neutron spends more time in the vicinity of target nuclei.

45.2 Nuclear Fission

As we mentioned in Section 44.2, nuclear **fission** occurs when a heavy nucleus, such as ^{235}U, splits into two smaller nuclei. Fission is initiated when a heavy nucleus captures a thermal neutron, as described by the first step in Equation 45.1. The absorption of the neutron creates a nucleus that is unstable and can change to a lower-energy configuration by splitting into two smaller nuclei. In such a reaction, the combined mass of the daughter nuclei is less than the mass of the parent nucleus, and the difference in mass is called the **mass defect.** Multiplying the mass defect by c^2 gives the numerical value of the released energy. Energy is released because the binding energy per nucleon of the daughter nuclei is about 1 MeV greater than that of the parent nucleus (see Fig. 44.5).

Nuclear fission was first observed in 1938 by Otto Hahn (1879–1968) and Fritz Strassman (1902–1980) following some basic studies by Fermi. After bombarding uranium with neutrons, Hahn and Strassman discovered among the reaction products two medium-mass elements, barium and lanthanum. Shortly thereafter, Lise Meitner (1878–1968) and her nephew Otto Frisch (1904–1979) explained what had happened. The uranium nucleus had split into two nearly equal fragments plus several neutrons after absorbing a neutron. Such an occurrence was of considerable interest to physicists attempting to understand the nucleus, but it was to have even more far-reaching consequences. Measurements showed that about 200 MeV of energy was released in each fission event, and this fact was to affect the course of history.

The fission of ^{235}U by thermal neutrons can be represented by the reaction

$$^{1}_{0}n + {}^{235}_{92}U \longrightarrow {}^{236}_{92}U^* \longrightarrow X + Y + \text{neutrons} \qquad (45.2)$$

where $^{236}U^*$ is an intermediate excited state that lasts only for about 10^{-12} s before splitting into medium-mass nuclei X and Y, which are called **fission fragments.** In any fission reaction, there are many combinations of X and Y that satisfy the requirements of conservation of energy and charge. In the case of uranium, for example, there are about 90 daughter nuclei that can be formed.

Fission also results in the production of several neutrons, typically two or three. On the average, about 2.5 neutrons are released per event. A typical fission reaction for uranium is

$$^{1}_{0}n + {}^{235}_{92}U \longrightarrow {}^{141}_{56}Ba + {}^{92}_{36}Kr + 3({}^{1}_{0}n) \qquad (45.3)$$

Figure 45.1 is a graph of the distribution of fission products versus mass number A. The most probable products have mass numbers $A \approx 140$ and $A \approx 95$. Suppose these products are $^{140}_{53}I$ (with 87 neutrons) and $^{95}_{39}Y$ (with 56 neutrons). If these nuclei are located on the graph of Figure 44.4, it is seen that both are well above the line of stability. These fragments, because they are very unstable owing to their unusually high number of neutrons, almost instantaneously release two or three neutrons.

The breakup of the uranium nucleus can be compared to what happens to a drop of water when excess energy is added to it. (Recall the liquid-drop model of the nucleus described in Section 44.3.) Initially, all the atoms in the drop have some energy, but this is not enough to break up the drop. Imagine adding energy to a drop

▲ **PITFALL PREVENTION**

45.1 Binding Energy Reminder

Remember from Chapter 44 that binding energy is the absolute value of the system energy and is related to the system mass. Therefore, when considering Figure 44.5, imagine flipping it upside down for a curve representing system mass. In a fission reaction, the system mass decreases. This decrease in mass appears in the system as kinetic energy of the fission products.

General fission reaction for ^{235}U

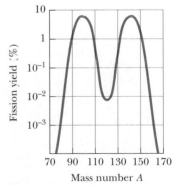

Figure 45.1 Distribution of fission products versus mass number for the fission of ^{235}U bombarded with thermal neutrons. Note that the vertical axis is logarithmic.

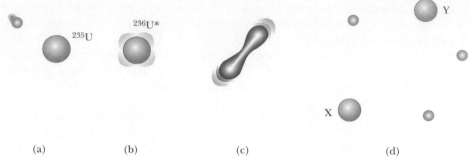

Figure 45.2 A nuclear fission event as described by the liquid-drop model of the nucleus. (a) A slow neutron approaches a ^{235}U nucleus. (b) The neutron is absorbed by the ^{235}U nucleus, changing it to $^{236}U^*$, that is the ^{236}U nucleus in an excited state. (c) The nucleus deforms and oscillates like a liquid drop. (d) The nucleus undergoes fission, resulting in two lighter nuclei X, and Y, and several neutrons.

of water floating in empty space so that it begins to elongate and compress (as in Fig. 45.2c). If enough energy is added, the drop vibrates until the amplitude of vibration becomes large enough to cause it to break. In the uranium nucleus, a similar process occurs, as described in the four steps of Figure 45.2.

Quick Quiz 45.1 Which of the following is a possible set of products for a $^{235}_{92}U$ fission reaction in which two neutrons are released? (a) $^{141}_{56}Ba$, $^{93}_{36}Kr$ (b) $^{142}_{55}Cs$, $^{93}_{37}Rb$ (c) $^{139}_{54}Xe$, $^{96}_{38}Sr$ (d) $^{140}_{53}I$, $^{93}_{38}Sr$

Let us estimate the disintegration energy Q released in a typical fission process. From Figure 44.5 we see that the binding energy per nucleon is about 7.2 MeV for heavy nuclei ($A \approx 240$) and about 8.2 MeV for nuclei of intermediate mass. This means that the nuclei in fission fragments are more tightly bound and therefore have less combined mass than the parent nucleus. This decrease in mass appears as released energy when fission occurs. The amount of energy released is approximately 8.2 MeV − 7.2 MeV = 1 MeV per nucleon. Because there are a total of 235 nucleons in $^{235}_{92}U$, we estimate that the energy released per fission event is

$$Q \approx (235 \text{ nucleons})(8.2 \text{ MeV/nucleon} - 7.2 \text{ MeV/nucleon}) = 235 \text{ MeV}$$

This is a large amount of energy relative to the amount released in chemical processes. For example, the energy released in the combustion of one molecule of octane used in gasoline engines is about one millionth of the energy released in a single fission event!

Quick Quiz 45.2 When a nucleus undergoes fission, the two daughter nuclei are generally radioactive. By which process are they most likely to decay? (a) alpha decay (b) beta decay (e^-) (c) beta decay (e^+).

Quick Quiz 45.3 Which of the following are possible fission reactions?

(a) $\quad ^1_0n + ^{235}_{92}U \longrightarrow ^{140}_{54}Xe + ^{94}_{38}Sr + 2\,(^1_0n)$

(b) $\quad ^1_0n + ^{235}_{92}U \longrightarrow ^{132}_{50}Sn + ^{101}_{42}Mo + 3\,(^1_0n)$

(c) $\quad ^1_0n + ^{239}_{94}Pu \longrightarrow ^{137}_{53}I + ^{97}_{41}Nb + 3\,(^1_0n)$

Example 45.1   The Energy Released in the Fission of ^{235}U

Calculate the energy released when 1.00 kg of ^{235}U fissions, taking the disintegration energy per event to be $Q = 208$ MeV.

Solution We need to know the number of nuclei in 1.00 kg of uranium. Because $A = 235$ for uranium, we know that the molar mass of this isotope is 235 g/mol. Therefore, the number of nuclei in our sample is

$$N = \left(\frac{6.02 \times 10^{23} \text{ nuclei/mol}}{235 \text{ g/mol}} \right)(1.00 \times 10^3 \text{ g})$$

$$= 2.56 \times 10^{24} \text{ nuclei}$$

Hence, the total disintegration energy is

$$E = NQ = (2.56 \times 10^{24} \text{ nuclei})(208 \text{ MeV/nucleus})$$

$$= 5.32 \times 10^{26} \text{ MeV}$$

Let us convert this energy to kWh:

$$E = (5.32 \times 10^{26} \text{ MeV}) \left(\frac{1.60 \times 10^{-13} \text{ J}}{\text{MeV}} \right) \left(\frac{1 \text{ kWh}}{3.60 \times 10^6 \text{ J}} \right)$$

$$= 2.36 \times 10^7 \text{ kWh}$$

This is enough energy to keep a 100-W lightbulb burning for 30 000 years! If the available fission energy in 1 kg of ^{235}U were suddenly released, it would be equivalent to detonating about 20 000 tons of TNT.

45.3   Nuclear Reactors

In the preceding section, we learned that, when ^{235}U fissions, an average of 2.5 neutrons are emitted per event. These neutrons can trigger other nuclei to fission, with the possibility of a chain reaction (Fig. 45.3). Calculations show that if the chain reaction is not controlled (that is, if it does not proceed slowly), it can result in a violent explosion, with the

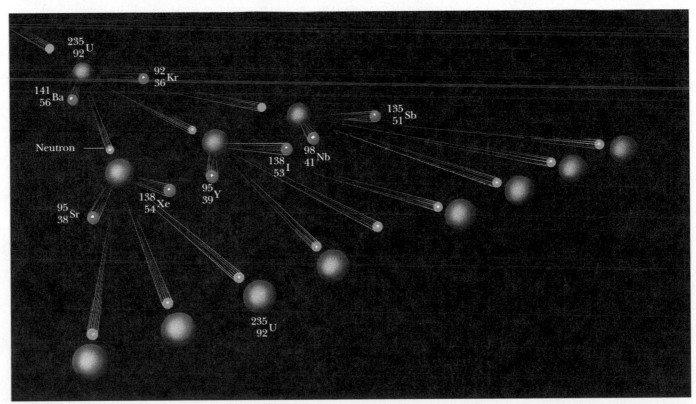

Active Figure 45.3 A nuclear chain reaction initiated by the capture of a neutron.

Observe the chain reaction at the Active Figures link at http://www.pse6.com.

Figure 45.4 Artist's rendition of the world's first nuclear reactor. Because of wartime secrecy, there are few photographs of the completed reactor, which was composed of layers of moderating graphite interspersed with uranium. A self-sustained chain reaction was first achieved on December 2, 1942. Word of the success was telephoned immediately to Washington with this message: "The Italian navigator has landed in the New World and found the natives very friendly." The historic event took place in an improvised laboratory in the racquet court under the stands of the University of Chicago's Stagg Field, and the Italian navigator was Enrico Fermi.

Enrico Fermi
Italian Physicist (1901–1954)

Fermi was awarded the Nobel Prize in physics in 1938 for producing transuranic elements by neutron irradiation and for his discovery of nuclear reactions brought about by thermal neutrons. He made many other outstanding contributions to physics, including his theory of beta decay, the free-electron theory of metals, and the development of the world's first fission reactor in 1942. Fermi was truly a gifted theoretical and experimental physicist. He was also well known for his ability to present physics in a clear and exciting manner. "Whatever Nature has in store for mankind, unpleasant as it may be, men must accept, for ignorance is never better than knowledge." (*National Accelerator Laboratory*)

release of an enormous amount of energy. When the reaction is controlled, however, the energy released can be put to constructive use. In the United States, for example, nearly 20% of the electricity generated each year comes from nuclear power plants, and nuclear power is used extensively in many countries, including France, Japan, and Germany.

A nuclear reactor is a system designed to maintain what is called a **self-sustained chain reaction.** This important process was first achieved in 1942 by Enrico Fermi and his team at the University of Chicago, with naturally occurring uranium as the fuel.[1] In the first nuclear reactor (Fig. 45.4), Fermi placed bricks of graphite (carbon) between the fuel elements. Carbon nuclei are about 12 times more massive than neutrons, but after several collisions with carbon nuclei, a neutron is slowed sufficiently to increase its likelihood of fission with ^{235}U. In this design, carbon is the moderator; most modern reactors use water as the moderator.

Most reactors in operation today also use uranium as fuel. However, naturally occurring uranium contains only about 0.7% of the ^{235}U isotope, with the remaining 99.3% being ^{238}U. This fact is important to the operation of a reactor because ^{238}U almost never fissions. Instead, it tends to absorb neutrons without a subsequent fission event, producing neptunium and plutonium. For this reason, reactor fuels must be artificially *enriched* to contain at least a few percent ^{235}U.

In the process of slowing down, neutrons may be captured by nuclei that do not fission. The most common event of this type is neutron capture by ^{238}U, which constitutes more than 90% of the uranium in the enriched fuel elements. The probability of neutron capture by ^{238}U is very high when the neutrons have high kinetic energies and very low when they have low kinetic energies. Thus, the slowing down of the neutrons by the moderator serves the secondary purpose of making them available for reaction with ^{235}U and decreasing their chances of being captured by ^{238}U.

To achieve a self-sustained chain reaction, an average of one neutron emitted in each ^{235}U fission must be captured by another ^{235}U nucleus and cause that nucleus to undergo fission. A useful parameter for describing the level of reactor operation is the **reproduction constant** K, defined as **the average number of neutrons from each**

[1]   Although Fermi's reactor was the first manufactured nuclear reactor, there is evidence that a natural fission reaction may have sustained itself for perhaps hundreds of thousands of years in a deposit of uranium in Gabon, West Africa. See G. Cowan, "A Natural Fission Reactor," *Scientific American* 235(5):36, 1976.

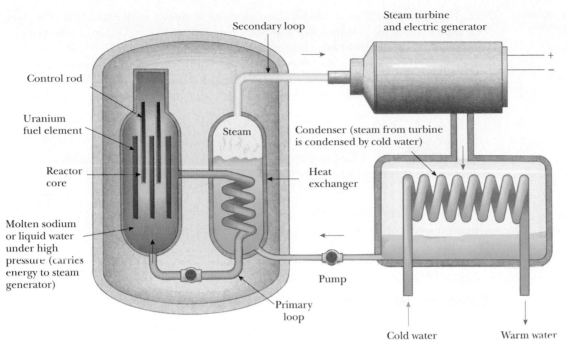

Figure 45.5 Main components of a pressurized-water nuclear reactor.

fission event that cause another fission event. As we have seen, K has an average value of 2.5 in the uncontrolled fission of uranium.

A self-sustained chain reaction is achieved when $K = 1$. Under this condition, the reactor is said to be **critical.** When $K < 1$, the reactor is subcritical and the reaction dies out. When $K > 1$, the reactor is supercritical and a runaway reaction occurs. In a nuclear reactor used to furnish power to a utility company, it is necessary to maintain a value of K close to 1. If K rises above this value, the internal energy produced in the reaction could melt the reactor.

Several types of reactor systems allow the kinetic energy of fission fragments to be transformed to other types of energy and eventually transferred out of the reactor plant by electrical transmission. The most common reactor in use in the United States is the pressurized-water reactor (Fig. 45.5). We shall examine this type because its main parts are common to all reactor designs. Fission events in the uranium **fuel elements** in the reactor core raise the temperature of the water contained in the primary (closed) loop, which is maintained at high pressure to keep the water from boiling. (This water also serves as the moderator to slow down the neutrons released in the fission events with energy of about 2 MeV.) The hot water is pumped through a heat exchanger, where the internal energy of the water is transferred by conduction to the water contained in the secondary loop. The hot water in the secondary loop is converted to steam, which does work to drive a turbine–generator system to create electric power. The water in the secondary loop is isolated from the water in the primary loop to avoid contamination of the secondary water and the steam by radioactive nuclei from the reactor core.

In any reactor, a fraction of the neutrons produced in fission leak out of the uranium fuel elements before inducing other fission events. If the fraction leaking out is too large, the reactor will not operate. The percentage lost is large if the fuel elements are very small because leakage is a function of the ratio of surface area to volume. Therefore, a critical feature of the reactor design is an optimal surface area–to–volume ratio of the fuel elements.

Control of Power Level

Safety is of critical importance in the operation of a nuclear reactor. The reproduction constant K must not be allowed to rise above 1, lest a runaway reaction occur. Consequently, reactor design must include a means of controlling the value of K.

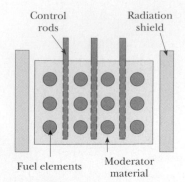

Control rods   Radiation shield

Fuel elements   Moderator material

Figure 45.6 Cross section of a reactor core showing the control rods, fuel elements containing enriched fuel, and moderating material, all surrounded by a radiation shield.

The basic design of a nuclear reactor core is shown in Figure 45.6. The fuel elements consist of uranium that has been enriched in the ^{235}U isotope. To control the power level, **control rods** are inserted into the reactor core. These rods are made of materials, such as cadmium, that are very efficient in absorbing neutrons. By adjusting the number and position of the control rods in the reactor core, the K value can be varied and any power level within the design range of the reactor can be achieved.

Quick Quiz 45.4 Which of the following has the function of slowing down fast neutrons? (a) control rods (b) moderator (c) fuel elements.

Quick Quiz 45.5 To reduce the value of the reproduction constant K, do you (a) push the control rods deeper into the core or (b) pull the control rods farther out of the core?

Safety and Waste Disposal

The 1979 near-disaster at a nuclear power plant at Three Mile Island in Pennsylvania and the 1986 accident at the Chernobyl reactor in Ukraine rightfully focused attention on reactor safety. The Three Mile Island accident was the result of inadequate control-room instrumentation and poor emergency-response training. There were no injuries or detectable health impacts from the event, even though more than one third of the fuel melted.

This unfortunately was not the case at Chernobyl, where the activity of the materials released immediately after the accident totaled approximately 1.2×10^{19} Bq and resulted in the evacuation of 135 000 people. Thirty individuals died during the accident or shortly thereafter, and data from the Ukraine Radiological Institute suggest that more than 2 500 deaths could be attributed to the Chernobyl accident. In the period 1986–1997 there was a tenfold increase in the number of children contracting thyroid cancer from the ingestion of radioactive iodine in milk from cows that ate contaminated grass. One conclusion of an international conference studying the Ukraine accident was that the main causes of the Chernobyl accident were the coincidence of severe deficiencies in the reactor physical design and a violation of safety procedures. Most of these deficiencies have been addressed at plants of similar design in Russia and neighboring countries of the former Soviet Union.

Commercial reactors achieve safety through careful design and rigid operating protocol, and it is only when these variables are compromised that reactors pose a danger. Radiation exposure and the potential health risks associated with such exposure are controlled by three layers of containment. The fuel and radioactive fission products are contained inside the reactor vessel. Should this vessel rupture, the reactor building acts as a second containment structure to prevent radioactive material from contaminating the environment. Finally, the reactor facilities must be in a remote location to protect the general public from exposure should radiation escape the reactor building.

A continuing concern about nuclear fission reactors is the safe disposal of radioactive material when the reactor core is replaced. This waste material contains long-lived, highly radioactive isotopes and must be stored over long time intervals in such a way that there is no chance of environmental contamination. At present, sealing radioactive wastes in waterproof containers and burying them in deep geologic repositories seems to be the most promising solution.

Transport of reactor fuel and reactor wastes poses additional safety risks. Accidents during transport of nuclear fuel could expose the public to harmful levels of radiation. The Department of Energy requires stringent crash tests of all containers used to transport nuclear materials. Container manufacturers must demonstrate that their containers will not rupture even in high-speed collisions.

Despite these risks, there are advantages to the use of nuclear power to be weighed against the risks. For example, nuclear power plants do not produce air pollution and greenhouse gases as do fossil fuel plants, and the supply of uranium on the Earth is predicted to last longer than the supply of fossil fuels. For each source of energy, whether nuclear, hydroelectric, fossil fuel, wind, or solar, the risks must be weighed against the benefits and the availability of the energy source.

45.4   Nuclear Fusion

In Chapter 44 we found that the binding energy for light nuclei ($A < 20$) is much smaller than the binding energy for heavier nuclei. This suggests a process that is the reverse of fission. As we mentioned in Section 39.9, when two light nuclei combine to form a heavier nucleus, the process is called nuclear **fusion.** Because the mass of the final nucleus is less than the combined masses of the original nuclei, there is a loss of mass accompanied by a release of energy.

Two examples of such energy-liberating fusion reactions are as follows:

$$^1_1H + ^1_1H \longrightarrow ^2_1H + e^+ + \nu$$

$$^1_1H + ^2_1H \longrightarrow ^3_2He + \gamma$$

These reactions occur in the core of a star and are responsible for the outpouring of energy from the star. This second reaction is followed by either hydrogen–helium fusion or helium–helium fusion:

$$^1_1H + ^3_2He \longrightarrow ^4_2He + e^+ + \nu$$

$$^3_2He + ^3_2He \longrightarrow ^4_2He + ^1_1H + ^1_1H$$

These are the basic reactions in the **proton–proton cycle,** believed to be one of the basic cycles by which energy is generated in the Sun and other stars that contain an abundance of hydrogen. Most of the energy production takes place in the Sun's interior, where the temperature is approximately 1.5×10^7 K. Because such high temperatures are required to drive these reactions, they are called **thermonuclear fusion reactions.** All of the reactions in the proton–proton cycle are exothermic. An overview of the cycle is that four protons combine to form an alpha particle and two positrons.

▲ **PITFALL PREVENTION**

45.2   Fission and Fusion

The words *fission* and *fusion* sound similar, but they correspond to different processes. Consider the binding-energy curve in Figure 44.5. There are two directions from which you can approach the peak of the curve so that energy is released: combining two light nuclei—fusion—or separating a heavy nucleus into two lighter nuclei—fission.

Quick Quiz 45.6 In the core of a star, hydrogen nuclei combine in fusion reactions. Once the hydrogen has been exhausted, fusion of helium nuclei can occur. Once the helium is used up, if the star is sufficiently massive, fusion of heavier and heavier nuclei can occur. Consider fusion reactions involving two nuclei with the same value of A. For these types of reactions, which of the following values of A are impossible? (a) 12 (b) 20 (c) 28 (d) 64

Example 45.2   Energy Released in Fusion

Find the total energy released in the fusion reactions in the proton–proton cycle.

Solution As mentioned earlier, the net result of the proton–proton cycle is to fuse four protons to form an alpha particle. The initial mass of four protons is, using atomic masses from Table A.3,

$$4(1.007\ 825\ u) = 4.031\ 300\ u$$

The change in mass of the system is this value minus the mass of the resultant alpha particle:

$$4.031\ 300\ u - 4.002\ 603\ u = 0.028\ 697\ u$$

Now we convert this mass change into energy units:

$$E = 0.028\ 697\ u \times 931.494\ \text{MeV/u} = \boxed{26.7\ \text{MeV}}$$

Terrestrial Fusion Reactions

The enormous amount of energy released in fusion reactions suggests the possibility of harnessing this energy for useful purposes. A great deal of effort is currently under way to develop a sustained and controllable thermonuclear reactor—a fusion power reactor. Controlled fusion is often called the ultimate energy source because of the availability of its fuel source: water. For example, if deuterium were used as the fuel, 0.12 g of it could be extracted from 1 gal of water at a cost of about four cents. This amount of deuterium would release about 10^{10} J if all nuclei underwent fusion. By comparison, a gallon of gasoline releases about 10^8 J upon burning, and costs far more than four cents.

An additional advantage of fusion reactors is that comparatively few radioactive byproducts are formed. For the proton–proton cycle, for instance, the end product is safe, nonradioactive helium. Unfortunately, a thermonuclear reactor that can deliver a net power output spread out over a reasonable time interval is not yet a reality, and many difficulties must be resolved before a successful device is constructed.

The Sun's energy is based, in part, upon a set of reactions in which hydrogen is converted to helium. However, the proton–proton interaction is not suitable for use in a fusion reactor because the event requires very high temperatures and densities. The process works in the Sun only because of the extremely high density of protons in the Sun's interior.

The reactions that appear most promising for a fusion power reactor involve deuterium ($^2_1$H) and tritium ($^3_1$H):

$$^2_1\text{H} + {}^2_1\text{H} \longrightarrow {}^3_2\text{He} + {}^1_0\text{n} \qquad Q = 3.27 \text{ MeV}$$

$$^2_1\text{H} + {}^2_1\text{H} \longrightarrow {}^3_1\text{H} + {}^1_1\text{H} \qquad Q = 4.03 \text{ MeV} \qquad (45.4)$$

$$^2_1\text{H} + {}^3_1\text{H} \longrightarrow {}^4_2\text{He} + {}^1_0\text{n} \qquad Q = 17.59 \text{ MeV}$$

As noted earlier, deuterium is available in almost unlimited quantities from our lakes and oceans and is very inexpensive to extract. Tritium, however, is radioactive ($T_{1/2} = 12.3$ yr) and undergoes beta decay to ^{3}He. For this reason, tritium does not occur naturally to any great extent and must be artificially produced.

One of the major problems in obtaining energy from nuclear fusion is that the Coulomb repulsive force between two nuclei, which carry positive charges, must be overcome before they can fuse. Potential energy as a function of the separation distance between two deuterons (deuterium nuclei, each having charge $+ e$) is shown in Figure 45.7. The potential energy is positive in the region $r > R$, where the Coulomb repulsive force dominates ($R \approx 1$ fm), and negative in the region $r < R$, where the nuclear force dominates. The fundamental problem then is to give the two nuclei enough kinetic energy to overcome this repulsive force. This can be accomplished by raising the fuel to extremely high temperatures (to about 10^8 K, far greater than the interior temperature of the Sun). At these high temperatures, the atoms are ionized and the system consists of a collection of electrons and nuclei, commonly referred to as a *plasma*.

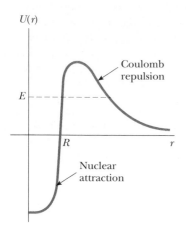

Figure 45.7 Potential energy as a function of separation distance between two deuterons. The Coulomb repulsive force is dominant at long range, and the nuclear force is dominant at short range, where R is on the order of 1 fm. If we neglect tunneling, to undergo fusion the two deuterons require an energy E greater than the height of the barrier.

Example 45.3   The Fusion of Two Deuterons · Interactive

The separation distance between two deuterons must be about 1.0×10^{-14} m in order for the nuclear force to overcome the repulsive Coulomb force.

(A) Calculate the height of the potential barrier due to the repulsive force.

Solution Conceptualize this problem by imagining moving two deuterons toward each other. As they move closer together, the Coulomb repulsion force becomes stronger.

Work must be done to push against this force, and this work appears in the system of two deuterons as electric potential energy. We categorize this problem as one involving the electric potential energy of a system of two charged particles. To analyze the problem, recall that the potential energy associated with two charges separated by a distance r is, from Equation 25.13,

$$U = k_e \frac{q_1 q_2}{r}$$

where k_e is the Coulomb constant. For the case of two deuterons, $q_1 = q_2 = + e$, so that

$$U = k_e \frac{e^2}{r} = (8.99 \times 10^9 \text{ N} \cdot \text{m}^2/\text{C}^2) \frac{(1.60 \times 10^{-19} \text{ C})^2}{1.0 \times 10^{-14} \text{ m}}$$

$$= 2.3 \times 10^{-14} \text{ J} = \boxed{0.14 \text{ MeV}}$$

(B) Estimate the temperature required for a deuteron to overcome the potential barrier, assuming an energy of $\frac{3}{2} k_B T$ per deuteron (where k_B is Boltzmann's constant).

Solution Because the total Coulomb energy of the pair is 0.14 MeV, the Coulomb energy per deuteron is 0.07 MeV = 1.1×10^{-14} J. Setting this energy equal to the average energy per deuteron gives

$$\tfrac{3}{2} k_B T = 1.1 \times 10^{-14} \text{ J}$$

Solving for T gives

$$T = \frac{2(1.1 \times 10^{-14} \text{ J})}{3(1.38 \times 10^{-23} \text{ J/K})} = \boxed{5.3 \times 10^8 \text{ K}}$$

To finalize this problem, note that this calculated temperature is too high because the particles in the plasma have a Maxwellian speed distribution (Section 21.6), and therefore some fusion reactions are caused by particles in the high-energy tail of this distribution. Furthermore, even those particles that do not have enough energy to overcome the barrier have some probability of tunneling through. When these effects are taken into account, a temperature of "only" 4×10^8 K appears adequate to fuse two deuterons in a plasma.

(C) Find the energy released in the deuterium–deuterium reaction

$$_1^2\text{H} + {_1^2}\text{H} \longrightarrow {_1^3}\text{H} + {_1^1}\text{H}$$

Solution The mass of a single deuterium atom is equal to 2.014 102 u. Thus, the total mass before the reaction is 4.028 204 u. After the reaction, the sum of the masses is 3.016 049 u + 1.007 825 u = 4.023 874 u. The excess mass is 0.004 33 u, equivalent to an energy of

$$0.004\,33 \text{ u} \times 931.494 \text{ MeV/u} = \boxed{4.03 \text{ MeV}}$$

Notice that this energy value is consistent with that already given in Equation 45.4.

What If? Suppose that the tritium resulting from the reaction in part (C) reacts with another deuterium in the reaction

$$_1^2\text{H} + {_1^3}\text{H} \longrightarrow {_2^4}\text{He} + {_0^1}\text{n}$$

How much energy is released in the sequence of two reactions?

Answer The overall effect of the sequence of two reactions is that three deuterium nuclei have combined to form a helium nucleus, a hydrogen nucleus, and a neutron. The initial mass is 3(2.014 102 u) = 6.042 306 u. After the reaction, the sum of the masses is 4.002 603 u + 1.007 825 u + 1.008 665 = 6.019 093 u. The excess mass is 0.023 213 u, equivalent to an energy of 21.6 MeV. Notice that this value is the sum of the Q values for the second and third reactions in Equation 45.4.

 At the Interactive Worked Example link at **http://www.pse6.com,** *practice evaluating parameters for a number of different fusion reactions.*

The temperature at which the power generation rate in any fusion reaction exceeds the loss rate is called the **critical ignition temperature** T_{ignit}. This temperature for the deuterium–deuterium (D–D) reaction is 4×10^8 K. From the relationship $E \approx \frac{3}{2} k_B T$, the ignition temperature is equivalent to approximately 52 keV. The critical ignition temperature for the deuterium-tritium (D–T) reaction is about 4.5×10^7 K, or only 6 keV. A plot of the power $\mathcal{P}_{\text{gen}}$ generated by fusion versus temperature for the two reactions is shown in Figure 45.8. The green straight line represents the power $\mathcal{P}_{\text{lost}}$ lost via the radiation mechanism known as bremsstrahlung (Section 42.8). In this principal mechanism of energy loss, radiation (primarily x-rays) is emitted as the result of electron–ion collisions within the plasma. The intersections of the $\mathcal{P}_{\text{lost}}$ line with the $\mathcal{P}_{\text{gen}}$ curves give the critical ignition temperatures.

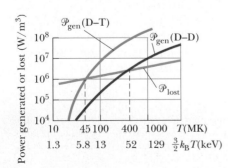

Figure 45.8 Power generated versus temperature for deuterium–deuterium (D–D) and deuterium–tritium (D–T) fusion. The green line represents power lost as a function of temperature. When the generation rate exceeds the loss rate, ignition takes place.

In addition to the high-temperature requirements, there are two other critical parameters that determine whether or not a thermonuclear reactor is successful: the **ion density** n and **confinement time** τ, which is the time interval during which energy injected into the plasma remains within the plasma. The British physicist J. D. Lawson has shown that the ion density and confinement time must both be large enough to ensure that more fusion energy is released than the amount required to raise the temperature of the plasma. A graph of the value of $n\tau$ necessary to achieve a net energy output for the D–T and D–D reactions at different temperatures is shown in Figure 45.9. The product $n\tau$ is referred to as the **Lawson number** of a reaction. In particular, **Lawson's criterion** states that a net energy output is possible for values of $n\tau$ that meet the following conditions:

Lawson's criterion

$$n\tau \geq 10^{14} \text{ s/cm}^3 \quad \text{(D–T)}$$
$$n\tau \geq 10^{16} \text{ s/cm}^3 \quad \text{(D–D)} \tag{45.5}$$

These values represent the minima of the curves in Figure 45.9.

Lawson's criterion was arrived at by comparing the energy required to raise the temperature of a given plasma with the energy generated by the fusion process.[2] The energy E_{in} required to raise the temperature of the plasma is proportional to the ion density n, which we can express as $E_{in} = C_1 n$, where C_1 is some constant. The energy generated by the fusion process is proportional to $n^2\tau$, or $E_{gen} = C_2 n^2 \tau$. This may be understood by realizing that the fusion energy released is proportional to both the rate at which interacting ions collide ($\propto n^2$) and the confinement time τ. Net energy is produced when $E_{gen} > E_{in}$. When the constants C_1 and C_2 are calculated for different reactions, the condition that $E_{gen} \geq E_{in}$ leads to Lawson's criterion.

In summary, the two basic requirements of a successful thermonuclear power reactor are

- The plasma temperature must be very high—about 4.5×10^7 K for the D–T reaction and 4×10^8 K for the D–D reaction.

- To meet Lawson's criterion, the product $n\tau$ must be large. For a given value of n, the probability of fusion between two particles increases as τ increases. For a given value of τ, the collision rate between nuclei increases as n increases.

Current efforts are aimed at meeting Lawson's criterion at temperatures exceeding T_{ignit}. Although the minimum required plasma densities have been achieved, the problem of confinement time is more difficult. How can a plasma be confined at 10^8 K and achieve a Lawson number of 10^{14} s/cm³? The two basic techniques under investigation are magnetic confinement and inertial confinement.

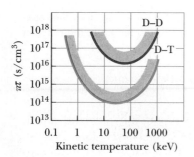

Figure 45.9 The Lawson number $n\tau$ at which net energy output is possible versus temperature for the D–T and D–D fusion reactions. The regions above the colored curves represent favorable conditions for fusion.

Magnetic Confinement

Many fusion-related plasma experiments use **magnetic confinement** to contain the plasma. A toroidal device called a **tokamak,** first developed in Russia, is shown in Figure 45.10a. A combination of two magnetic fields is used to confine and stabilize the plasma: (1) a strong toroidal field produced by the current in the toroidal windings surrounding a donut-shaped vacuum chamber and (2) a weaker "poloidal" field produced by the toroidal current. In addition to confining the plasma, the toroidal current is used to raise its temperature. The resultant helical magnetic field lines spiral around the plasma and keep it from touching the walls of the vacuum chamber. (If the plasma touches the walls, its temperature is reduced and heavy impurities sputtered from the walls "poison" it and lead to large power losses.)

---

[2] Lawson's criterion neglects the energy needed to set up the strong magnetic field used to confine the hot plasma in a magnetic confinement approach. This energy is expected to be about 20 times greater than the energy required to raise the temperature of the plasma. For this reason, it is necessary either to have a magnetic energy recovery system or to use superconducting magnets.

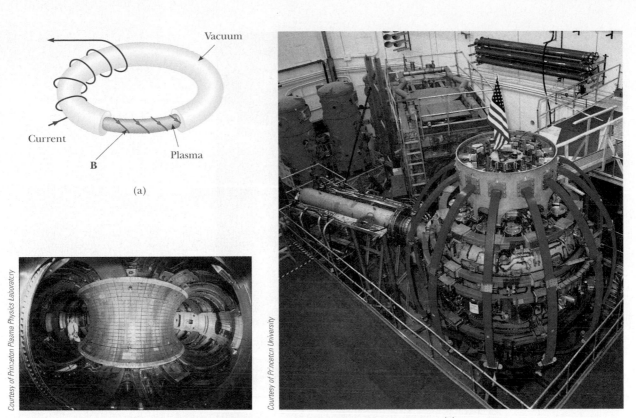

Figure 45.10 (a) Diagram of a tokamak used in the magnetic confinement scheme.
(b) Interior view of the closed Tokamak Fusion Test Reactor (TFTR) vacuum vessel at
the Princeton Plasma Physics Laboratory. (c) The National Spherical Torus Experiment
(NSTX) that began operation in March 1999.

One of the major breakthroughs in magnetic confinement in the 1980s was in the
area of auxiliary energy input to reach ignition temperatures. Experiments have shown
that injecting a beam of energetic neutral particles into the plasma is a very efficient
method of raising it to ignition temperatures. Radio-frequency energy input will probably be needed for reactor-size plasmas.

When it was in operation, the Tokamak Fusion Test Reactor (TFTR, Fig. 45.10b) at
Princeton reported central ion temperatures of 510 million degrees Celsius, more than
30 times hotter than the center of the Sun. The $n\tau$ values in the TFTR for the D–T
reaction were well above 10^{13} s/cm^3 and close to the value required by Lawson's criterion. In 1991, reaction rates of 6×10^{17} D–T fusions per second were reached in the
JET tokamak at Abington, England.

One of the new generation of fusion experiments is the National Spherical Torus
Experiment (NSTX) shown in Figure 45.10c. Rather than the donut-shaped plasma of
a tokamak, the NSTX produces a spherical plasma that has a hole through its center.
The major advantage of the spherical configuration is its ability to confine the plasma
at a higher pressure in a given magnetic field. This approach could lead to development of smaller, more economical fusion reactors.

An international collaborative effort involving Canada, Europe, Japan, and Russia
is currently under way to build a fusion reactor called ITER (International Thermonuclear Experimental Reactor). China and the United States began to participate in program activities in early 2003. This facility will address the remaining technological and
scientific issues concerning the feasibility of fusion power. The design is completed
(Fig. 45.11), and site and construction negotiations are under way. As of the printing
of this text, four possible sites are proposed—Cadarache, France; Clarington, Canada;
Rokkasho-mura, Japan; and Vandellòs, Spain. If the planned device works as expected,
the Lawson number for ITER will be about six times greater than the current record
holder, the JT-60U tokamak in Japan. ITER will produce 1.5 GW of power, and the

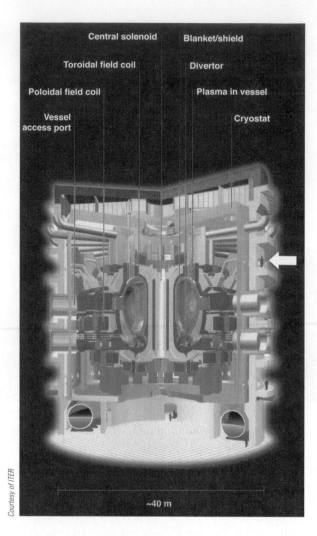

Central solenoid

Toroidal field coil

Poloidal field coil

Vessel access port

Blanket/shield

Divertor

Plasma in vessel

Cryostat

~40 m

Courtesy of ITER

Figure 45.11 Cutaway diagram of the ITER (International Thermonuclear Experimental Reactor). Note the size of the reactor relative to that of a person *(arrow)*.

energy content of the alpha particles inside the reactor will be so intense that they will sustain the fusion reaction, allowing the auxiliary energy sources to be turned off once the reaction is initiated.

Example 45.4 Inside a Fusion Reactor

In 1998, the JT-60U tokamak in Japan was operated with a D–T plasma density of 4.8×10^{13} cm^{-3} at a temperature (in energy units) of 24.1 keV. It was able to confine this plasma inside a magnetic field for 1.1 s.

(A) Does this meet Lawson's criterion?

Solution Equation 45.5 says that for a D–T plasma, the Lawson number $n\tau$ must be greater than 10^{14} s/cm^3. For the JT-60U,

$$n\tau = (4.8 \times 10^{13} \text{ cm}^{-3})(1.1 \text{ s}) = 5.3 \times 10^{13} \text{ s/cm}^3$$

which is close to meeting Lawson's criterion. In fact, scientists recorded a power gain of 1.25, indicating the reactor was operating slightly past the break-even point and was producing more energy than it required to maintain the plasma.

(B) How does the plasma density compare with the density of atoms in an ideal gas when the gas is at room temperature and pressure?

Solution The density of atoms in a sample of ideal gas is given by N_A/V_{mol}, where N_A is Avogadro's number and V_{mol} is the molar volume of an ideal gas under standard conditions, 2.24×10^{-2} m^3/mol. Thus, the density of the gas is

$$\frac{N_A}{V_{mol}} = \frac{6.02 \times 10^{23} \text{ atoms/mol}}{2.24 \times 10^{-2} \text{ m}^3/\text{mol}} = 2.7 \times 10^{25} \text{ atoms/m}^3$$

This is more than 500 000 times greater than the plasma density in the reactor.

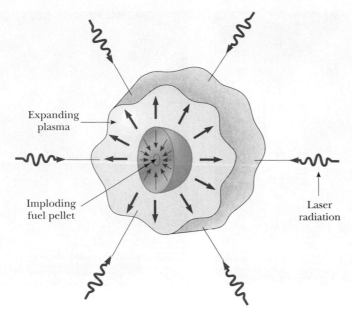

Figure 45.12 In inertial confinement, a D–T fuel pellet fuses when struck by several high-intensity laser beams simultaneously.

Inertial Confinement

The second technique for confining a plasma is called **inertial confinement** and makes use of a D–T target that has a very high particle density. In this scheme, the confinement time is very short (typically 10^{-11} to 10^{-9} s), so that, because of their own inertia, the particles do not have a chance to move appreciably from their initial positions. Thus, Lawson's criterion can be satisfied by combining a high particle density with a short confinement time.

Laser fusion is the most common form of inertial confinement. A small D–T pellet, about 1 mm in diameter, is struck simultaneously by several focused, high-intensity laser beams, resulting in a large pulse of input energy that causes the surface of the fuel pellet to evaporate (Fig. 45.12). The escaping particles exert a third-law reaction force on the core of the pellet, resulting in a strong, inwardly moving compressive shock wave. This shock wave increases the pressure and density of the core and produces a corresponding increase in temperature. When the temperature of the core reaches ignition temperature, fusion reactions occur.

Two of the leading laser fusion laboratories in the United States are the Omega facility at the University of Rochester in New York and the Nova facility at Lawrence Livermore National Laboratory in California. The Omega facility focuses 24 laser beams on the target, and the Nova facility employs 10 beams. Figure 45.13a shows the target chamber at Nova, and Figure 45.13b shows the tiny, spherical D–T pellets used. Nova is capable of injecting a power of 2×10^{14} W into a 0.5-mm D–T pellet and has achieved values of $n\tau \approx 5 \times 10^{14}$ s/cm^3 and ion temperatures of 5.0 keV. These values are close to those required for D–T ignition. This steady progress has led the U.S. Department of Energy and other groups to plan a national facility that will involve a laser fusion device with an input energy in the 5- to 10-MJ range.

Fusion Reactor Design

In the D–T fusion reaction

$$^2_1\text{H} + {}^3_1\text{H} \longrightarrow {}^4_2\text{He} + {}^1_0\text{n} \qquad Q = 17.59 \text{ MeV}$$

the alpha particle carries 20% of the energy and the neutron carries 80%, or about 14 MeV. The alpha particles, because they are charged, are primarily absorbed by the

Courtesy of University of California Lawrence Livermore National Laboratory and the U.S. Department of Energy

Courtesy of Los Alamos National Laboratory

(a) (b)

Figure 45.13 (a) The target chamber of the Nova Laser Facility at Lawrence Livermore Laboratory. (b) Spherical plastic target shells used to contain the D–T fuel, shown clustered on a quarter. The shells have very smooth surfaces and are about 100 nm thick.

plasma, causing the plasma's temperature to increase. In contrast, the 14-MeV neutrons, being electrically neutral, pass through the plasma and are absorbed by a surrounding blanket material, where their large kinetic energy is extracted and used to generate electric power. A diagram of the deuterium—tritium fusion reaction is shown in Figure 45.14.

One scheme is to use molten lithium metal as the neutron-absorbing material and to circulate the lithium in a closed heat-exchange loop to produce steam and drive turbines as in a conventional power plant. Figure 45.15 shows a diagram of such a reactor. It is estimated that a blanket of lithium about 1 m thick will capture nearly 100% of the neutrons from the fusion of a small D–T pellet.

The capture of neutrons by lithium is described by the reaction

$$\frac{1}{0}n + \frac{6}{3}Li \longrightarrow \frac{3}{1}H + \frac{4}{2}He$$

where the kinetic energies of the charged tritium $\frac{3}{1}H$ and alpha particle are converted to internal energy in the molten lithium. An extra advantage of using lithium as the energy-transfer medium is that the tritium produced can be separated from the lithium and returned as fuel to the reactor.

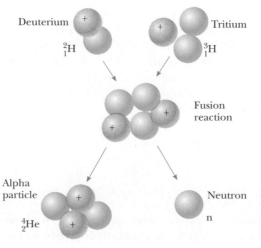

At the Active Figures link at http://www.pse6.com, choose to observe several fusion reactions and measure the energy released.

Active Figure 45.14 Deuterium–tritium fusion. Eighty percent of the energy released is in the 14-MeV neutron.

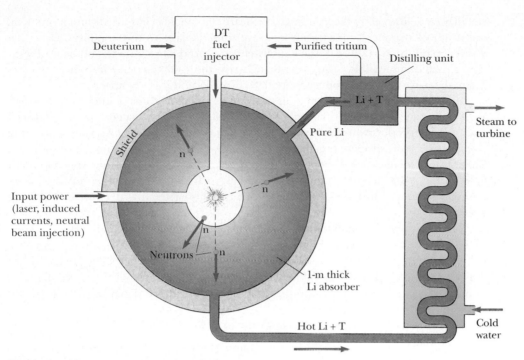

Figure 45.15 Diagram of a fusion reactor.

Advantages and Problems of Fusion

If fusion power can ever be harnessed, it will offer several advantages over fission-generated power: (1) low cost and abundance of fuel (deuterium), (2) impossibility of runaway accidents, and (3) decreased radiation hazard. Some of the anticipated problems and disadvantages include (1) scarcity of lithium, (2) limited supply of helium, which is needed for cooling the superconducting magnets used to produce strong confining fields, and (3) structural damage and induced radioactivity caused by neutron bombardment. If such problems and the engineering design factors can be resolved, nuclear fusion may become a feasible source of energy by the middle of the twenty-first century.

45.5 Radiation Damage

In Chapter 34 we learned that electromagnetic radiation is all around us in the form of radio waves, microwaves, light waves, and so on. In this section, we turn to forms of radiation that can cause severe damage as they pass through matter. These include radiation resulting from radioactive processes and radiation in the form of energetic particles such as neutrons and protons. It is these forms that we refer to here and in the following two sections when we use the word *radiation*.

The degree and type of damage depend upon several factors, including the type and energy of the radiation and the properties of the matter. The metals used in nuclear reactor structures can be severely weakened by high fluxes of energetic neutrons because these high fluxes often lead to metal fatigue. The damage in such situations is in the form of atomic displacements, often resulting in major alterations in the properties of the material.

Radiation damage in biological organisms is primarily due to ionization effects in cells. The normal operation of a cell may be disrupted when highly reactive ions are formed as the result of ionizing radiation. For example, hydrogen and the hydroxyl radical OH^- produced from water molecules can induce chemical reactions that may break bonds in proteins and other vital molecules. Furthermore, the ionizing radiation may affect vital molecules directly by removing electrons from their structure. Large doses of radiation are especially dangerous because damage to a great number of molecules in a cell may cause the cell to die. Although the death of a single cell is usually not a problem, the death of many cells may result in irreversible damage to the organism. Cells that divide rapidly, such as those of the digestive tract, reproductive organs, and hair follicles, are especially susceptible. Also, cells that survive the radiation may become defective. These defective cells can produce more defective cells and lead to cancer.

In biological systems, it is common to separate radiation damage into two categories: somatic damage and genetic damage. *Somatic damage* is that associated with any body cell except the reproductive cells. Somatic damage can lead to cancer or seriously alter the characteristics of specific organisms. *Genetic damage* affects only reproductive cells. Damage to the genes in reproductive cells can lead to defective offspring. It is important to be aware of the effect of diagnostic treatments, such as x-rays and other forms of radiation exposure, and to balance the significant benefits of treatment with the damaging effects.

Damage caused by radiation also depends on the radiation's penetrating power. Alpha particles cause extensive damage, but penetrate only to a shallow depth in a material due to the strong interaction with other charged particles. Neutrons do not interact via the electric force and hence penetrate deeper, causing significant damage. Gamma rays are high-energy photons that can cause severe damage, but often pass through matter without interaction.

Several units have been used historically to quantify the amount, or dose, of any radiation that interacts with a substance.

The **roentgen** (R) is that amount of ionizing radiation that produces an electric charge of 3.33×10^{-10} C in 1 cm^3 of air under standard conditions.

Equivalently, the roentgen is that amount of radiation that increases the energy of 1 kg of air by 8.76×10^{-3} J.

For most applications, the roentgen has been replaced by the rad (an acronym for *radiation absorbed dose*):

One **rad** is that amount of radiation that increases the energy of 1 kg of absorbing material by 1×10^{-2} J.

Although the rad is a perfectly good physical unit, it is not the best unit for measuring the degree of biological damage produced by radiation because damage depends not only on the dose but also on the type of the radiation. For example, a given dose of alpha particles causes about ten times more biological damage than an equal dose of x-rays. The **RBE** (relative biological effectiveness) factor for a given type of radiation is **the number of rads of x-radiation or gamma radiation that produces the same biological damage as 1 rad of the radiation being used.** The RBE factors for different types of radiation are given in Table 45.1. The values are only approximate because

Table 45.1

RBE[a] Factors for Several Types of Radiation	
Radiation	**RBE Factor**
X-rays and gamma rays	1.0
Beta particles	1.0–1.7
Alpha particles	10–20
Thermal neutrons	4–5
Fast neutrons and protons	10
Heavy ions	20

[a] RBE = relative biological effectiveness.

Table 45.2

Units for Radiation Dosage					
Quantity	SI Unit	Symbol	Relation to Other SI units	Older Unit	Conversion
Absorbed dose	gray	Gy	$= 1 \text{ J/kg}$	rad	1 Gy = 100 rad
Dose equivalent	sievert	Sv	$= 1 \text{ J/kg}$	rem	1 Sv = 100 rem

they vary with particle energy and with the form of the damage. The RBE factor should be considered only a first-approximation guide to the actual effects of radiation.

Finally, the **rem** (radiation equivalent in man) is the product of the dose in rad and the RBE factor:

$$\text{Dose in rem} \equiv \text{dose in rad} \times \text{RBE} \qquad (45.6)$$

▶ Radiation dose in rem

According to this definition, 1 rem of any two types of radiation produces the same amount of biological damage. From Table 45.1, we see that a dose of 1 rad of fast neutrons represents an effective dose of 10 rem, but 1 rad of gamma radiation is equivalent to a dose of only 1 rem.

Low-level radiation from natural sources, such as cosmic rays and radioactive rocks and soil, delivers to each of us a dose of about 0.13 rem/yr; this radiation is called *background radiation*. Background radiation varies with geography, with the main factors being altitude (exposure to cosmic rays) and geology (radon gas released by some rock formations, deposits of naturally radioactive minerals).

The upper limit of radiation dose rate recommended by the U.S. government (apart from background radiation) is about 0.5 rem/yr. Many occupations involve much higher radiation exposures, and so an upper limit of 5 rem/yr has been set for combined whole-body exposure. Higher upper limits are permissible for certain parts of the body, such as the hands and the forearms. A dose of 400 to 500 rem results in a mortality rate of about 50% (which means that half the people exposed to this radiation level die). The most dangerous form of exposure for most people is either ingestion or inhalation of radioactive isotopes, especially isotopes of those elements the body retains and concentrates, such as ^{90}Sr.

This discussion has focused on measurements of radiation dosage in units such as rads and rems because these units are still widely used. These units, however, have been formally replaced with new SI units. The rad has been replaced with the *gray* (Gy), equal to 100 rad. The rem has been replaced with the *sievert* (Sv), equal to 100 rem. Table 45.2 summarizes the older and the current SI units of radiation dosage.

45.6 Radiation Detectors

Particles passing through matter interact with the matter in several ways. The particle can ionize atoms, scatter from atoms, be absorbed by atoms, etc. Radiation detectors exploit these interactions to allow a measurement of the particle's energy, momentum, or charge, and sometimes the very existence of the particle if it is otherwise difficult to detect. Various devices have been developed for detecting radiation. These devices are used for a variety of purposes, including medical diagnoses, radioactive dating measurements, measuring background radiation, and measuring the mass, energy, and momentum of particles created in high-energy nuclear reactions.

In the early part of the twentieth century, detectors were much simpler than those used today. We will discuss three of these early detectors first. A **photographic emulsion** is the simplest example of a detector. A charged particle ionizes the atoms in an

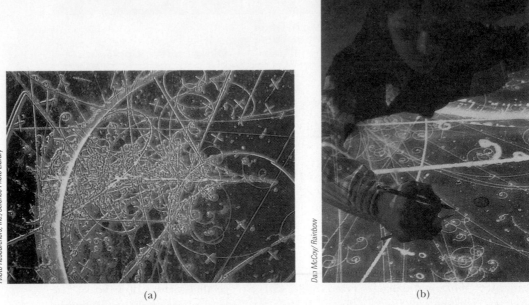

(a)   (b)

Figure 45.16 (a) Artificially colored bubble-chamber photograph showing tracks of particles that have passed through the chamber. (b) This research scientist is studying a photograph of particle tracks made in a bubble chamber at Fermilab. The curved tracks are produced by charged particles moving through the chamber in the presence of an applied magnetic field. Negatively charged particles deflect in one direction, while positively charged particles deflect in the opposite direction.

emulsion layer. The path of the particle corresponds to a family of points at which chemical changes have occurred in the emulsion. When the emulsion is developed, the particle's track becomes visible. A **cloud chamber** contains a gas that has been supercooled to just below its usual condensation point. An energetic particle passing through ionizes the gas along the particle's path. The ions serve as centers for condensation of the supercooled gas. The track of the particle can be seen with the naked eye and can be photographed. A magnetic field can be applied to determine the charges of the particles, as well as their momentum and energy. A device called a **bubble chamber,** invented in 1952 by D. Glaser, uses a liquid (usually liquid hydrogen) maintained near its boiling point. Ions produced by incoming charged particles leave bubble tracks, which can be photographed (Fig. 45.16). Because the density of the detecting medium in a bubble chamber is much higher than the density of the gas in a cloud chamber, the bubble chamber has a much higher sensitivity.

More contemporary detectors involve more sophisticated processes. In an **ion chamber** (Fig. 45.17), electron–ion pairs are generated as radiation passes through a gas and produces an electrical signal. Two plates in the chamber are connected to a voltage supply and thereby maintained at different electric potentials. The positive plate attracts the electrons, and the negative plate attracts positive ions, causing a current pulse that is proportional to the number of electron–ion pairs produced when a particle passes through the chamber. When an ion chamber is used both to detect the presence of a particle and to measure its energy, it is called a **proportional counter.**

The **Geiger counter** (Fig. 45.18) is perhaps the most common form of ion chamber used to detect radioactivity. It can be considered the prototype of all counters that use the ionization of a medium as the basic detection process. It consists of a thin wire electrode aligned along the central axis of a cylindrical metallic tube filled with a gas at low pressure. The wire is maintained at a high positive electric potential (about 10^3 V) relative to the tube. When a high-energy particle resulting, for example, from a radioactive decay enters the tube through a thin window at one end, some of the gas atoms are ionized. The electrons removed from these atoms are attracted toward the

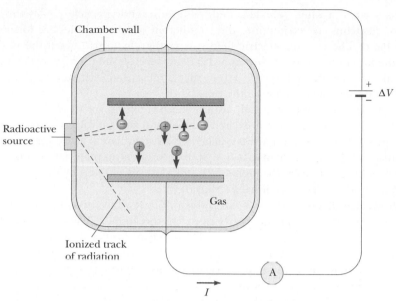

Figure 45.17 Simplified diagram of an ion chamber. The radioactive source creates electrons and positive ions that are collected by the charged plates. The current set up in the external circuit is proportional to a radioactive particle's kinetic energy if the particle stops in the chamber.

wire electrode, and in the process they ionize other atoms in their path. This sequential ionization results in an *avalanche* of electrons that produces a current pulse. After the pulse has been amplified, it can either be used to trigger an electronic counter or be delivered to a loudspeaker that clicks each time a particle is detected. Although a Geiger counter easily detects the presence of a particle, the energy lost by the particle in the counter is *not* proportional to the current pulse produced. Thus, a Geiger counter cannot be used to measure the energy of a particle.

A **semiconductor-diode detector** is essentially a reverse bias p–n junction. Recall from Section 43.7 that a p–n junction passes current readily when forward-biased and

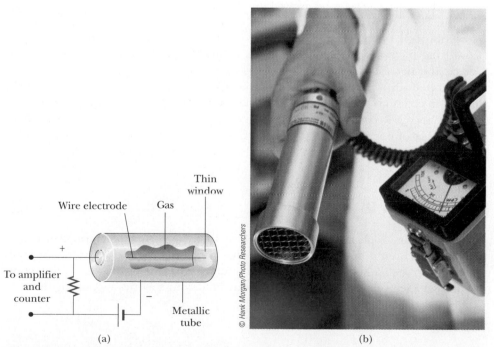

Figure 45.18 (a) Diagram of a Geiger counter. The voltage between the wire electrode and the metallic tube is usually about 1 000 V. (b) A scientist uses a Geiger counter to make a measurement.

prohibits a current when reverse-biased. As an energetic particle passes through the junction, electrons are excited into the conduction band and holes are formed in the valence band. The internal electric field sweeps the electrons toward the positive (n) side of the junction and the holes toward the negative (p) side. This movement of electrons and holes creates a pulse of current that is measured with an electronic counter. In a typical device, the duration of the pulse is 10^{-8} s.

A **scintillation counter** usually uses a solid or liquid material whose atoms are easily excited by radiation. The excited atoms then emit photons when they return to their ground state. Common materials used as scintillators are transparent crystals of sodium iodide and certain plastics. If the scintillator material is attached to a photomultiplier (PM) tube (Section 40.2), the photons emitted by the scintillator can be detected and an electrical signal produced.

Both the scintillator and the semiconductor-diode detector are much more sensitive than a Geiger counter, mainly because of the higher density of the detecting medium. Both measure the total energy deposited in the detector, which can be very useful in particle identification. In addition, if the particle stops in the detector, they can both be used to measure the total particle energy.

Track detectors are devices used to view the tracks of charged particles directly. High-energy particles produced in particle accelerators may have energies ranging from 10^9 to 10^{12} eV. Therefore, they often cannot be stopped and cannot have their energy measured with the detectors already mentioned. Instead, the energy and momentum of these energetic particles are found from the curvature of their path in a magnetic field of known magnitude and direction.

A **spark chamber** is a counting device that consists of an array of conducting parallel plates and is capable of recording a three-dimensional track record. Even-numbered plates are grounded, and odd-numbered plates are maintained at a high electric potential (about 10 kV). The spaces between the plates contain an inert gas at atmospheric pressure. When a charged particle passes through the chamber, gas atoms are ionized, resulting in a current surge and visible sparks along the particle path. These sparks may be photographed or electronically detected and the data sent to a computer for path reconstruction and determination of particle mass, momentum, and energy.

Newer versions of the spark chamber have been developed. A **drift chamber** has thousands of high-voltage wires arrayed through the space of the detector, which is filled with gas. The result is an array of thousands of Geiger counters. When a charged particle passes through the detector, it ionizes gas molecules, and the ejected electrons drift toward the high-voltage wires, creating an electrical signal upon arrival. A computer detects the signals and reconstructs the path through the detector. A large-volume, sophisticated drift chamber that has provided significant results in studying particles formed in collisions of atoms is the Solenoidal Tracker at RHIC (STAR). (The acronym RHIC stands for Relativistic Heavy Ion Collider, a facility at Brookhaven National Laboratory that began operation in 2000.) This type of drift chamber is called a **time projection chamber.** A photograph of the STAR detector is shown in Figure 45.19.

Courtesy of Brookhaven National Laboratory/ RHIC-STAR

Figure 45.19 The STAR detector at the Relativistic Heavy Ion Collider at Brookhaven National Laboratory.

45.7   Uses of Radiation

Nuclear physics applications are extremely widespread in manufacturing, medicine, and biology. Even a brief discussion of all the possibilities would fill an entire book, and to keep such a book up to date would require frequent revisions. In this section, we present a few of these applications and the underlying theories supporting them.

Tracing

Radioactive tracers are used to track chemicals participating in various reactions. One of the most valuable uses of radioactive tracers is in medicine. For example, iodine, a nutrient needed by the human body, is obtained largely through the intake of iodized

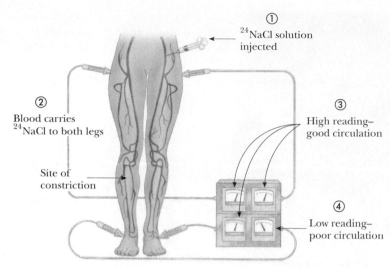

Figure 45.20 A tracer technique for determining the condition of the human circulatory system.

salt and seafood. To evaluate the performance of the thyroid, the patient drinks a very small amount of radioactive sodium iodide containing ^{131}I, an artificially produced isotope of iodine (the natural, nonradioactive isotope is ^{127}I). The amount of iodine in the thyroid gland is determined as a function of time by measuring the radiation intensity at the neck area. How much of the isotope ^{131}I remains in the thyroid is a measure of how well that gland is functioning.

A second medical application is indicated in Figure 45.20. A solution containing radioactive sodium is injected into a vein in the leg, and the time at which the radioisotope arrives at another part of the body is detected with a radiation counter. The elapsed time is a good indication of the presence or absence of constrictions in the circulatory system.

Tracers are also useful in agricultural research. Suppose the best method of fertilizing a plant is to be determined. A certain element in a fertilizer, such as nitrogen, can be *tagged* (identified) with one of its radioactive isotopes. The fertilizer is then sprayed on one group of plants, sprinkled on the ground for a second group, and raked into the soil for a third. A Geiger counter is then used to track the nitrogen through the three groups.

Tracing techniques are as wide ranging as human ingenuity can devise. Present applications range from checking how teeth absorb fluoride to monitoring how cleansers contaminate food-processing equipment to studying deterioration inside an automobile engine. In the latter case, a radioactive material is used in the manufacture of the piston rings, and the oil is checked for radioactivity to determine the amount of wear on the rings.

Materials Analysis

For centuries, a standard method of identifying the elements in a sample of material has been chemical analysis, which involves determining how the material reacts with various chemicals. A second method is spectral analysis, which uses the fact that, when excited, each element emits its own characteristic set of electromagnetic wavelengths. These methods are now supplemented by a third technique, **neutron activation analysis.** Both chemical and spectral methods have the disadvantage that a fairly large sample of the material must be destroyed for the analysis. In addition, extremely small quantities of an element may go undetected by either method. Neutron activation analysis has an advantage over the other two methods in both respects.

When a material is irradiated with neutrons, nuclei in the material absorb the neutrons and are changed to different isotopes, most of which are radioactive. For

AP/Wide World Photos

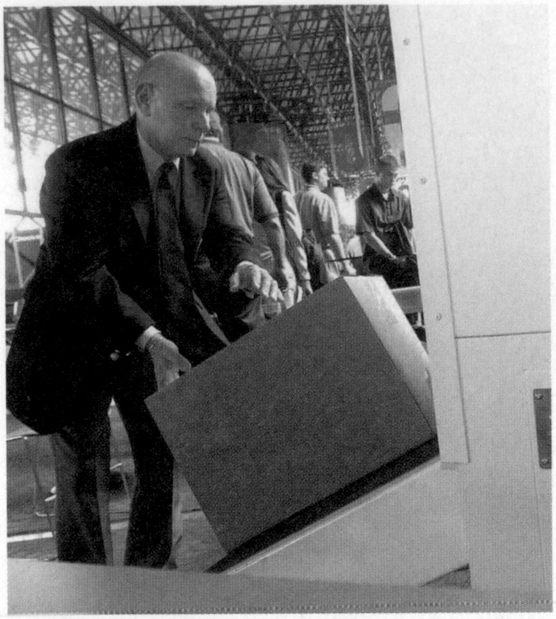

Figure 45.21 A bomb detector will irradiate this package with neutrons. If there are hidden explosives inside, chemicals within the explosive materials become radioactive and can be easily detected. The half-life of the resulting radiation is so short that there is no danger to the security personnel removing the package for further inspection.

example, ^{65}Cu absorbs a neutron to become ^{66}Cu, which undergoes beta decay:

$$^{1}_{0}\text{n} + ^{65}_{29}\text{Cu} \longrightarrow ^{66}_{29}\text{Cu} \longrightarrow ^{66}_{30}\text{Zn} + \text{e}^{-} + \bar{\nu}$$

The presence of the copper can be deduced because it is known that ^{66}Cu has a half-life of 5.1 min and decays with the emission of beta particles having maximum energies of 2.63 and 1.59 MeV. Also emitted in the decay of ^{66}Cu is a 1.04-MeV gamma ray. By examining the radiation emitted by a substance after it has been exposed to neutron irradiation, one can detect extremely small amounts of an element in that substance.

Neutron activation analysis is used routinely in a number of industries—for example, in commercial aviation for the checking of airline luggage for hidden explosives (Fig. 45.21). The following nonroutine use is of interest. Napoleon died on the island of St. Helena in 1821, supposedly of natural causes. Over the years, suspicion has existed that his death was not all that natural. After his death, his head was shaved and locks of his hair were sold as souvenirs. In 1961, the amount of arsenic in a sample of this hair was measured by neutron activation analysis, and an unusually large quantity of arsenic was found. (Activation analysis is so sensitive that very small pieces of a single hair could be analyzed.) Results showed that the arsenic was fed to him irregularly. In fact, the arsenic concentration pattern corresponded to the fluctuations in the severity of Napoleon's illness as determined from historical records.

Art historians use neutron activation analysis to detect forgeries. The pigments used in paints have changed throughout history, and old and new pigments react differently to neutron activation. The method can even reveal hidden works of art behind existing paintings because an older, hidden layer of paint reacts differently than the surface layer to neutron activation.

Martin Dohrn/Science Photo Library/ Photo Researchers

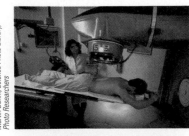

Figure 45.22 This large machine is being set to deliver a dose of radiation from ^{60}Co in an effort to destroy a cancerous tumor. Cancer cells are especially susceptible to this type of therapy because they tend to divide more often than cells of healthy tissue nearby.

Radiation Therapy

Radiation causes the most damage to rapidly dividing cells. Therefore, it is useful in cancer treatment because tumor cells divide extremely rapidly. Several mechanisms can be used to deliver radiation to a tumor. In some cases, a narrow beam of x-rays or radiation from a source such as ^{60}Co is used, as shown in Figure 45.22. In other situations, thin radioactive needles called *seeds* are implanted in the cancerous tissue. The radioactive isotope ^{131}I is used to treat cancer of the thyroid.

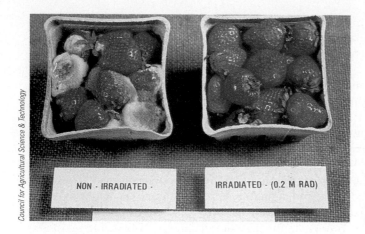

Figure 45.23 The strawberries on the left are untreated and have become moldy. The unspoiled strawberries on the right have been irradiated. The radiation has killed or incapacitated the mold spores that have spoiled the strawberries on the left.

Food Preservation

Radiation is finding increasing use as a means of preserving food because exposure to high levels of radiation can destroy or incapacitate bacteria and mold spores (Fig. 45.23). Techniques include exposing foods to gamma rays, high-energy electron beams, or x-rays. Food preserved this way can be placed in a sealed container (to keep out new spoiling agents) and stored for long periods of time. There is no evidence of adverse effect on the taste or nutritional value of food from irradiation. The safety of irradiated foods has been endorsed by the World Health Organization (WHO), the Centers for Disease Control and Prevention (CDC), the U.S. Department of Agriculture (USDA), and the Food and Drug Administration (FDA).

SUMMARY

The probability that neutrons are captured as they move through matter generally increases with decreasing neutron energy. A **thermal neutron** is a slow-moving neutron that has a high probability of being captured by a nucleus in a **neutron capture event:**

$$ {}^{1}_{0}n + {}^{A}_{Z}X \longrightarrow {}^{A+1}_{Z}X^* \longrightarrow {}^{A+1}_{Z}X + \gamma \tag{45.1} $$

where ${}^{A+1}_{Z}X^*$ is an excited intermediate nucleus that rapidly emits a photon.

Nuclear fission occurs when a very heavy nucleus, such as ^{235}U, splits into two smaller **fission fragments.** Thermal neutrons can create fission in ^{235}U:

$$ {}^{1}_{0}n + {}^{235}_{92}U \longrightarrow {}^{236}_{92}U^* \longrightarrow X + Y + \text{neutrons} \tag{45.2} $$

where X and Y are the fission fragments and ^{236}U* is an intermediate excited state. On the average, 2.5 neutrons are released per fission event. The fragments then undergo a series of beta and gamma decays to various stable isotopes. The energy released per fission event is about 200 MeV.

The **reproduction constant** K is the average number of neutrons released from each fission event that cause another event. In a fission reactor, it is necessary to maintain $K \approx 1$. The value of K is affected by such factors as reactor geometry, mean neutron energy, and probability of neutron capture. Neutron energies are reduced with a moderator material that slows down energetic neutrons and therefore increases the probability of neutron capture by other ^{235}U nuclei. The power level of the reactor is adjusted with control rods made of a material that is very efficient in absorbing neutrons.

In **nuclear fusion,** two light nuclei fuse to form a heavier nucleus and release energy. The major obstacle in obtaining useful energy from fusion is the large Coulomb repulsive force between the charged nuclei at small separation distances. Sufficient energy must be supplied to the particles to overcome this Coulomb barrier. The

Take a practice test for this chapter by clicking on the Practice Test link at http://www.pse6.com.

temperature required to produce fusion is on the order of 10^8 K, and at this temperature all matter occurs as a plasma.

In a fusion reactor, the plasma temperature must reach the **critical ignition temperature,** the temperature at which the power generated by the fusion reactions exceeds the power lost in the system. The most promising fusion reaction is the D–T reaction, which has a critical ignition temperature of approximately 4.5×10^7 K. Two critical parameters in fusion reactor design are **ion density** n and **confinement time** τ, the time interval during which the interacting particles must be maintained at $T > T_{\text{ignit}}$. **Lawson's criterion** states that for the D–T reaction, $n\tau \geq 10^{14}$ s/cm^3.

Energy absorbed in a substance due to radiation is measured with a unit called the **rad,** where one rad is that amount of radiation that increases the energy of 1 kg of absorbing material by 1×10^{-2} J. The radiation dose that is related to the amount of biological damage in an organism is measured by a unit called the **rem** (radiation equivalent in man) and is the product of the dose in rad and the RBE factor:

$$\text{Dose in rem} \equiv \text{dose in rad} \times \text{RBE} \tag{45.6}$$

where the **RBE** (relative biological effectiveness) factor for a given type of radiation is the number of rads of x-radiation or gamma radiation that produces the same biological damage as 1 rad of the radiation being used.

The rad and the rem have been formally replaced with new SI units. The rad has been replaced with the *gray* (Gy), equal to 100 rad. The rem has been replaced with the *sievert* (Sv), equal to 100 rem.

QUESTIONS

1. Explain the function of a moderator in a fission reactor.

2. Why is water a better shield against neutrons than lead or steel?

3. If a nucleus captures a slow-moving neutron, the product is left in a highly excited state, with an energy approximately 8 MeV above the ground state. Explain the source of the excitation energy.

4. Discuss the advantages and disadvantages of fission reactors from the point of view of safety, pollution, and resources. Make a comparison with power generated from the burning of fossil fuels.

5. Why would a fusion reactor produce less radioactive waste than a fission reactor?

6. Lawson's criterion states that the product of ion density and confinement time must exceed a certain number before a break-even fusion reaction can occur. Why should these two parameters determine the outcome?

7. Why is the temperature required for the D–T fusion less than that needed for the D–D fusion? Estimate the relative importance of Coulomb repulsion and nuclear attraction in each case.

8. What factors make a fusion reaction difficult to achieve?

9. Discuss the similarities and differences between fusion and fission.

10. Discuss the advantages and disadvantages of fusion power from the viewpoint of safety, pollution, and resources.

11. Discuss three major problems associated with the development of a controlled fusion reactor.

12. Describe two techniques being pursued in an effort to obtain power from nuclear fusion.

13. If two radioactive samples have the same activity measured in curies, will they necessarily create the same damage to a medium? Explain.

14. Why should a radiologist be extremely cautious about x-ray doses when treating pregnant women?

15. The design of a PM tube (Figure 40.12) might suggest that any number of dynodes may be used to amplify a weak signal. What factors do you suppose would limit the amplification in this device?

16. *And swift, and swift past comprehension*
   Turn round Earth's beauty and her might.
   The heavens blaze in alternation
   With deep and chill and rainy night.
   In mighty currents foams the ocean
   Up from the rocks' abyssal base,
   With rock and sea torn into motion
   In ever-swift celestial race.
   And tempests bluster in a contest
   From sea to land, from land to sea.
   In rage they forge a chain around us
   Of deepest meaning, energy.
   There flames a lightning disaster
   Before the thunder, in its way.
   But all Your servants honor, Master,
   The gentle order of Your day.
   Johann Wolfgang von Goethe wrote this song of the archangels in *Faust* half a century before the law of conservation of energy was recognized. Students often find it useful to think of a list of several "forms of energy," from kinetic to nuclear. Argue for or against the view that these lines of poetry make an obvious or oblique reference to every form of energy and energy transfer.

PROBLEMS

1, **2**, **3** = straightforward, intermediate, challenging   ☐ = full solution available in the *Student Solutions Manual and Study Guide*

🪐 = coached solution with hints available at http://www.pse6.com   💻 = computer useful in solving problem

▨ = paired numerical and symbolic problems

Section 45.2   Nuclear Fission

> *Note:* Problem 53 in Chapter 25 and Problems 21, 51, and 68 in Chapter 44 can be assigned with this section.

1. Burning one metric ton (1 000 kg) of coal can yield an energy of 3.30×10^{10} J. Fission of one nucleus of uranium-235 yields an average of about 208 MeV. What mass of uranium produces the same energy as a ton of coal?

2. Find the energy released in the fission reaction

$$^1_0n + ^{235}_{92}U \longrightarrow ^{98}_{40}Zr + ^{135}_{52}Te + 3(^1_0n)$$

The atomic masses of the fission products are: $^{98}_{40}Zr$, 97.912 7 u; $^{135}_{52}Te$, 134.916 5 u.

3. Strontium-90 is a particularly dangerous fission product of ^{235}U because it is radioactive and it substitutes for calcium in bones. What other direct fission products would accompany it in the neutron-induced fission of ^{235}U? (*Note:* This reaction may release two, three, or four free neutrons.)

4. List the nuclear reactions required to produce ^{239}Pu from ^{238}U under fast neutron bombardment.

5. 🪐 List the nuclear reactions required to produce ^{233}U from ^{232}Th under fast neutron bombardment.

6. (a) The following fission reaction is typical of those occurring in a nuclear electric generating station:

$$^1_0n + ^{235}_{92}U \longrightarrow ^{141}_{56}Ba + ^{92}_{36}Kr + 3(^1_0n)$$

Find the energy released. The required masses are

$$M(^1_0n) = 1.008\ 665 \text{ u}$$

$$M(^{235}_{92}U) = 235.043\ 923 \text{ u}$$

$$M(^{141}_{56}Ba) = 140.914\ 4 \text{ u}$$

$$M(^{92}_{36}Kr) = 91.926\ 2 \text{ u}$$

(b) What fraction of the initial mass of the system is transformed?

7. A reaction that has been considered as a source of energy is the absorption of a proton by a boron-11 nucleus to produce three alpha particles:

$$^1_1H + ^{11}_5B \longrightarrow 3(^4_2He)$$

This is an attractive possibility because boron is easily obtained from the Earth's crust. A disadvantage is that the protons and boron nuclei must have large kinetic energies in order for the reaction to take place. This is in contrast to the initiation of uranium fission by slow neutrons. (a) How much energy is released in each reaction? (b) Why must the reactant particles have high kinetic energies?

8. A typical nuclear fission power plant produces about 1.00 GW of electrical power. Assume that the plant has an overall efficiency of 40.0% and that each fission produces 200 MeV of energy. Calculate the mass of ^{235}U consumed each day.

9. **Review problem.** Suppose enriched uranium containing 3.40% of the fissionable isotope $^{235}_{92}U$ is used as fuel for a ship. The water exerts an average friction force of magnitude 1.00×10^5 N on the ship. How far can the ship travel per kilogram of fuel? Assume that the energy released per fission event is 208 MeV and that the ship's engine has an efficiency of 20.0%.

Section 45.3   Nuclear Reactors

10. To minimize neutron leakage from a reactor, the surface area–to-volume ratio should be a minimum. For a given volume V, calculate this ratio for (a) a sphere, (b) a cube, and (c) a parallelepiped of dimensions $a \times a \times 2a$. (d) Which of these shapes would have minimum leakage? Which would have maximum leakage?

11. 🪐 It has been estimated that on the order of 10^9 tons of natural uranium is available at concentrations exceeding 100 parts per million, of which 0.7% is the fissionable isotope ^{235}U. Assume that all the world's energy use (7×10^{12} J/s) were supplied by ^{235}U fission in conventional nuclear reactors, releasing 208 MeV for each reaction. How long would the supply last? The estimate of uranium supply is taken from K. S. Deffeyes and I. D. MacGregor, "World Uranium Resources," *Scientific American* **242**(1):66, 1980.

12. If the reproduction constant is 1.000 25 for a chain reaction in a fission reactor and the average time interval between successive fissions is 1.20 ms, by what factor will the reaction rate increase in one minute?

13. A large nuclear power reactor produces about 3 000 MW of power in its core. Three months after a reactor is shut down, the core power from radioactive byproducts is 10.0 MW. Assuming that each emission delivers 1.00 MeV of energy to the power, find the activity in becquerels three months after the reactor is shut down.

Section 45.4   Nuclear Fusion

14. (a) Consider a fusion generator built to create 3.00 GW of power. Determine the rate of fuel burning in grams per hour if the D–T reaction is used. (b) Do the same for the D–D reaction assuming that the reaction products are split evenly between (n, ^3He) and (p, ^3H).

15. Two nuclei having atomic numbers Z_1 and Z_2 approach each other with a total energy E. (a) Suppose they will

spontaneously fuse if they approach within a distance of 1.00×10^{-14} m. Find the minimum value of E required to produce fusion, in terms of Z_1 and Z_2. (b) Evaluate the minimum energy for fusion for the D–D and D–T reactions (the first and third reactions in Eq. 45.4).

16. Review problem. Consider the deuterium–tritium fusion reaction with the tritium nucleus at rest:

$$^2_1\text{H} + ^3_1\text{H} \longrightarrow ^4_2\text{He} + ^1_0\text{n}$$

(a) Suppose that the reactant nuclei will spontaneously fuse if their surfaces touch. From Equation 44.1, determine the required distance of closest approach between their centers. (b) What is the electric potential energy (in eV) at this distance? (c) Suppose the deuteron is fired straight at an originally stationary tritium nucleus with just enough energy to reach the required distance of closest approach. What is the common speed of the deuterium and tritium nuclei as they touch, in terms of the initial deuteron speed v_i? (*Suggestion:* At this point, the two nuclei have a common velocity equal to the center-of-mass velocity.) (d) Use energy methods to find the minimum initial deuteron energy required to achieve fusion. (e) Why does the fusion reaction actually occur at much lower deuteron energies than that calculated in (d)?

17. To understand why plasma containment is necessary, consider the rate at which an unconfined plasma would be lost. (a) Estimate the rms speed of deuterons in a plasma at 4.00×10^8 K. (b) **What If?** Estimate the order of magnitude of the time interval during which such a plasma would remain in a 10-cm cube if no steps were taken to contain it.

18. Of all the hydrogen in the oceans, 0.030 0% of the mass is deuterium. The oceans have a volume of 317 million mi^3. (a) If nuclear fusion were controlled and all the deuterium in the oceans were fused to ^4_2He, how many joules of energy would be released? (b) **What If?** World power consumption is about 7.00×10^{12} W. If consumption were 100 times greater, how many years would the energy calculated in part (a) last?

19. It has been suggested that fusion reactors are safe from explosion because there is never enough energy in the plasma to do much damage. (a) In 1992, the TFTR reactor achieved an ion temperature of 4.0×10^8 K, an ion density of 2.0×10^{13} cm^{-3}, and a confinement time of 1.4 s. Calculate the amount of energy stored in the plasma of the TFTR reactor. (b) How many kilograms of water could be boiled away by this much energy? (The plasma volume of the TFTR reactor is about 50 m^3.)

20. Review problem. To confine a stable plasma, the magnetic energy density in the magnetic field (Eq. 32.14) must exceed the pressure $2nk_BT$ of the plasma by a factor of at least 10. In the following, assume a confinement time $\tau = 1.00$ s. (a) Using Lawson's criterion, determine the ion density required for the D–T reaction. (b) From the ignition-temperature criterion, determine the required plasma pressure. (c) Determine the magnitude of the magnetic field required to contain the plasma.

21. Find the number of ^6Li and the number of ^7Li nuclei present in 2.00 kg of lithium. (The natural abundance of ^6Li is 7.5%; the remainder is ^7Li.)

22. One old prediction for the future was to have a fusion reactor supply energy to dissociate the molecules in garbage into separate atoms and then to ionize the atoms. This material could be put through a giant mass spectrometer, so that trash would be a new source of isotopically pure elements—the mine of the future. Assuming an average atomic mass of 56 and an average charge of 26 (a high estimate, considering all the organic materials), at a beam current of 1.00 MA, how long would it take to process 1.00 metric ton of trash?

Section 45.5 Radiation Damage

23. A building has become accidentally contaminated with radioactivity. The longest-lived material in the building is strontium-90. ($^{90}_{38}\text{Sr}$ has an atomic mass 89.907 7 u, and its half-life is 29.1 yr. It is particularly dangerous because it substitutes for calcium in bones.) Assume that the building initially contained 5.00 kg of this substance uniformly distributed throughout the building (a very unlikely situation) and that the safe level is defined as less than 10.0 decays/min (to be small in comparison to background radiation). How long will the building be unsafe?

24. Review problem. A particular radioactive source produces 100 mrad of 2-MeV gamma rays per hour at a distance of 1.00 m. (a) How long could a person stand at this distance before accumulating an intolerable dose of 1 rem? (b) **What If?** Assuming the radioactive source is a point source, at what distance would a person receive a dose of 10.0 mrad/h?

25. Assume that an x-ray technician takes an average of eight x-rays per day and receives a dose of 5 rem/yr as a result. (a) Estimate the dose in rem per photograph taken. (b) How does the technician's exposure compare with low-level background radiation?

26. When gamma rays are incident on matter, the intensity of the gamma rays passing through the material varies with depth x as $I(x) = I_0e^{-\mu x}$, where μ is the absorption coefficient and I_0 is the intensity of the radiation at the surface of the material. For 0.400-MeV gamma rays in lead, the absorption coefficient is 1.59 cm^{-1}. (a) Determine the "half-thickness" for lead—that is, the thickness of lead that would absorb half the incident gamma rays. (b) What thickness will reduce the radiation by a factor of 10^4?

27. A "clever" technician decides to warm some water for his coffee with an x-ray machine. If the machine produces 10.0 rad/s, how long will it take to raise the temperature of a cup of water by 50.0°C?

28. Review problem. The danger to the body from a high dose of gamma rays is not due to the amount of energy absorbed but occurs because of the ionizing nature of the radiation. To illustrate this, calculate the rise in body temperature that would result if a "lethal" dose of 1 000 rad were absorbed strictly as internal energy. Take the specific heat of living tissue as 4 186 J/kg·°C.

29. Technetium-99 is used in certain medical diagnostic procedures. Assume 1.00×10^{-8} g of ^{99}Tc is injected into a 60.0-kg patient and half of the 0.140-MeV gamma rays are absorbed in the body. Determine the total radiation dose received by the patient.

30. Strontium-90 from the testing of nuclear bombs can still be found in the atmosphere. Each decay of ^{90}Sr releases 1.1 MeV of energy into the bones of a person who has had strontium replace the calcium. Assume a 70.0-kg person receives 1.00 μg of ^{90}Sr from contaminated milk. Calculate the absorbed dose rate (in J/kg) in one year. Take the half-life of ^{90}Sr to be 29.1 yr.

Section 45.6   Radiation Detectors

31. In a Geiger tube, the voltage between the electrodes is typically 1.00 kV and the current pulse discharges a 5.00-pF capacitor. (a) What is the energy amplification of this device for a 0.500-MeV electron? (b) How many electrons participate in the avalanche caused by the single initial electron?

32. Assume a photomultiplier tube (Figure 40.12) has seven dynodes with potentials of 100, 200, 300, . . . , 700 V. The average energy required to free an electron from the dynode surface is 10.0 eV. Assume that just one electron is incident and that the tube functions with 100% efficiency. (a) How many electrons are freed at the first dynode? (b) How many electrons are collected at the last dynode? (c) What is the energy available to the counter for each electron?

33. (a) Your grandmother recounts to you how, as young children, your father, aunts, and uncles made the screen door slam continually as they ran between the house and the back yard. The time interval between one slam and the next varied randomly, but the average slamming rate stayed constant at 38.0/h from dawn to dusk every summer day. If the slamming rate suddenly dropped to zero, the children would have found a nest of baby field mice or gotten into some other mischief requiring adult intervention. How long after the last screen-door slam would a prudent and attentive parent wait before leaving her or his work to see about the children? Explain your reasoning. (b) A student wishes to measure the half-life of a radioactive substance, using a small sample. Consecutive clicks of her Geiger counter are randomly spaced in time. The counter registers 372 counts during one 5.00-min interval, and 337 counts during the next 5.00 min. The average background rate is 15 counts per minute. Find the most probable value for the half-life. (c) Estimate the uncertainty in the half-life determination. Explain your reasoning.

Section 45.7   Uses of Radiation

34. During the manufacture of a steel engine component, radioactive iron (^{59}Fe) is included in the total mass of 0.200 kg. The component is placed in a test engine when the activity due to this isotope is 20.0 μCi. After a 1 000-h test period, some of the lubricating oil is removed from the engine and found to contain enough ^{59}Fe to produce 800 disintegrations/min/L of oil. The total volume of oil in the engine is 6.50 L. Calculate the total mass worn from the engine component per hour of operation. (The half-life of ^{59}Fe is 45.1 d.)

35. At some time in your past or future, you may find yourself in a hospital to have a PET scan. The acronym stands for *positron-emission tomography*. In the procedure, a radioactive element that undergoes e^+ decay is introduced into your body. The equipment detects the gamma rays that result from pair annihilation when the emitted positron encounters an electron in your body's tissue. Suppose you receive an injection of glucose that contains on the order of 10^{10} atoms of ^{14}O. Assume that the oxygen is uniformly distributed through 2 L of blood after 5 min. What will be the order of magnitude of the activity of the oxygen atoms in 1 cm^3 of the blood?

36. You want to find out how many atoms of the isotope ^{65}Cu are in a small sample of material. You bombard the sample with neutrons to ensure that on the order of 1% of these copper nuclei absorb a neutron. After activation you turn off the neutron flux, and then use a highly efficient detector to monitor the gamma radiation that comes out of the sample. Assume that half of the ^{66}Cu nuclei emit a 1.04-MeV gamma ray in their decay. (The other half of the activated nuclei decay directly to the ground state of ^{66}Ni.) If after 10 min (two half-lives) you have detected 10^4 MeV of photon energy at 1.04 MeV, (a) about how many ^{65}Cu atoms are in the sample? (b) Assume the sample contains natural copper. Refer to the isotopic abundances listed in Table A.3 and estimate the total mass of copper in the sample.

37. *Neutron activation analysis* is a method for chemical analysis at the level of isotopes. When a sample is irradiated by neutrons, radioactive atoms are produced continuously and then decay according to their characteristic half-lives. (a) Assume that one species of radioactive nuclei is produced at a constant rate R and that its decay is described by the conventional radioactive decay law. If irradiation begins at time $t = 0$, show that the number of radioactive atoms accumulated at time t is

$$N = \frac{R}{\lambda}\,(1 - e^{-\lambda t})$$

(b) What is the maximum number of radioactive atoms that can be produced?

Additional Problems

38. The nuclear bomb dropped on Hiroshima on August 6, 1945, released 5×10^{13} J of energy, equivalent to that from 12 000 tons of TNT. The fission of one $^{235}_{92}$U nucleus releases an average of 208 MeV. Estimate (a) the number of nuclei fissioned, and (b) the mass of this $^{235}_{92}$U.

39. Carbon detonations are powerful nuclear reactions that temporarily tear apart the cores inside massive stars late in their lives. These blasts are produced by carbon fusion, which requires a temperature of about 6×10^8 K to overcome the strong Coulomb repulsion between carbon nuclei. (a) Estimate the repulsive energy barrier to fusion, using the temperature required for carbon fusion. (In other words, what is the average kinetic energy of a carbon nucleus at 6×10^8 K?) (b) Calculate the energy (in MeV) released in each of these "carbon-burning" reactions:

$$^{12}C + {}^{12}C \longrightarrow {}^{20}Ne + {}^{4}He$$
$$^{12}C + {}^{12}C \longrightarrow {}^{24}Mg + \gamma$$

(c) Calculate the energy (in kWh) given off when 2.00 kg of carbon completely fuses according to the first reaction.

40. Review problem. Consider a nucleus at rest, which then spontaneously splits into two fragments of masses m_1 and m_2. Show that the fraction of the total kinetic energy that is carried by fragment m_1 is

$$\frac{K_1}{K_{\text{tot}}} = \frac{m_2}{m_1 + m_2}$$

and the fraction carried by m_2 is

$$\frac{K_2}{K_{\text{tot}}} = \frac{m_1}{m_1 + m_2}$$

assuming relativistic corrections can be ignored. (*Note:* If the parent nucleus was moving before the decay, then the fission products still divide the kinetic energy as shown, as long as all velocities are measured in the center-of-mass frame of reference, in which the total momentum of the system is zero.)

41. A stationary $^{236}_{92}$U nucleus fissions spontaneously into two primary fragments, $^{87}_{35}$Br and $^{149}_{57}$La. (a) Calculate the disintegration energy. The required atomic masses are 86.920 711 u for $^{87}_{35}$Br, 148.934 370 u for $^{149}_{57}$La, and 236.045 562 u for $^{236}_{92}$U. (b) How is the disintegration energy split between the two primary fragments? You may use the result of Problem 40. (c) Calculate the speed of each fragment immediately after the fission.

42. Compare the fractional energy loss in a typical ^{235}U fission reaction with the fractional energy loss in D–T fusion.

43. The half-life of tritium is 12.3 yr. If the TFTR fusion reactor contained 50.0 m^3 of tritium at a density equal to 2.00×10^{14} ions/cm^3, how many curies of tritium were in the plasma? Compare this value with a fission inventory (the estimated supply of fissionable material) of 4×10^{10} Ci.

44. Review problem. A very slow neutron (with speed approximately equal to zero) can initiate the reaction

$$^1_0\text{n} + ^{10}_5\text{B} \longrightarrow ^7_3\text{Li} + ^4_2\text{He}$$

The alpha particle moves away with speed 9.25×10^6 m/s. Calculate the kinetic energy of the lithium nucleus. Use nonrelativistic equations.

45. Review problem. A nuclear power plant operates by using the energy released in nuclear fission to convert 20°C water into 400°C steam. How much water could theoretically be converted to steam by the complete fissioning of 1.00 g of ^{235}U at 200 MeV/fission?

46. Review problem. A nuclear power plant operates by using the energy released in nuclear fission to convert liquid water at T_c into steam at T_h. How much water could theoretically be converted to steam by the complete fissioning of a mass m of ^{235}U at 200 MeV/fission?

47. About 1 of every 3 300 water molecules contains one deuterium atom. (a) If all the deuterium nuclei in 1 L of water are fused in pairs according to the D–D fusion reaction $^2\text{H} + ^2\text{H} \rightarrow ^3\text{He} + \text{n} + 3.27$ MeV, how much energy in joules is liberated? (b) **What If?** Burning gasoline produces about 3.40×10^7 J/L. Compare the energy obtainable from the fusion of the deuterium in a liter of water with the energy liberated from the burning of a liter of gasoline.

48. Review problem. The first nuclear bomb was a fissioning mass of plutonium-239 exploded in the Trinity test, before dawn on July 16, 1945, at Alamogordo, New Mexico. Enrico Fermi was 14 km away, lying on the ground facing away from the bomb. After the whole sky had flashed with unbelievable brightness, Fermi stood up and began dropping bits of paper to the ground. They first fell at his feet in the calm and silent air. As the shock wave passed, about 40 s after the explosion, the paper then in flight jumped about 5 cm away from ground zero. (a) Assume that the shock wave in air propagated equally in all directions without absorption. Find the change in volume of a sphere of radius 14 km as it expands by 5 cm. (b) Find the work $P\Delta V$ done by the air in this sphere on the next layer of air farther from the center. (c) Assume the shock wave carried on the order of one tenth of the energy of the explosion. Make an order-of-magnitude estimate of the bomb yield. (d) One ton of exploding trinitrotoluene (TNT) releases 4.2 GJ of energy. What was the order of magnitude of the energy of the Trinity test in equivalent tons of TNT? The dawn revealed the mushroom cloud. Fermi's immediate knowledge of the bomb yield agreed with that determined days later by analysis of elaborate measurements.

49. A certain nuclear plant generates internal energy at a rate of 3.065 GW and transfers energy out of the plant by electrical transmission at a rate of 1.000 GW. Of the wasted energy, 3.0% is ejected to the atmosphere and the remainder is passed into a river. A state law requires that the river water be warmed by no more than 3.50°C when it is returned to the river. (a) Determine the amount of cooling water necessary (in kg/h and m^3/h) to cool the plant. (b) Assume fission generates 7.80×10^{10} J/g of ^{235}U. Determine the rate of fuel burning (in kg/h) of ^{235}U.

50. The alpha-emitter polonium-210 ($^{210}_{84}$Po) is used in a nuclear energy source on a spacecraft (Fig. P45.50). Determine the initial power output of the source. Assume that it

Figure P45.50 The *Pioneer 10* spacecraft leaves the Solar System. It carries radioactive power supplies at the ends of two booms. Solar panels would not work in this region far from the Sun.

contains 0.155 kg of ^{210}Po and that the efficiency for conversion of radioactive decay energy to energy transferred by electrical transmission is 1.00%.

51. Natural uranium must be processed to produce uranium enriched in ^{235}U for bombs and power plants. The processing yields a large quantity of nearly pure ^{238}U as a byproduct, called "depleted uranium." Because of its high mass density, it is used in armor-piercing artillery shells. (a) Find the edge dimension of a 70.0-kg cube of ^{238}U. (Refer to Table 1.5.) (b) The isotope ^{238}U has a long half-life of 4.47×10^9 yr. As soon as one nucleus decays, it begins a relatively rapid series of 14 steps that together constitute the net reaction

$$^{238}_{92}\text{U} \longrightarrow 8(^4_2\text{He}) + 6(^{\ 0}_{-1}\text{e}) + ^{206}_{82}\text{Pb} + 6\bar{\nu} + Q_{net}$$

Find the net decay energy. (Refer to Table A.3.) (c) Argue that a radioactive sample with decay rate R and decay energy Q has power output $\mathcal{P} = QR$. (d) Consider an artillery shell with a jacket of 70.0 kg of ^{238}U. Find its power output due to the radioactivity of the uranium and its daughters. Assume the shell is old enough that the daughters have reached steady-state amounts. Express the power in joules per year. (e) **What If?** A 17-year-old soldier of mass 70.0 kg works in an arsenal where many such artillery shells are stored. Assume his radiation exposure is limited to 5.00 rem per year. Find the rate at which he can absorb energy of radiation, in joules per year. Assume an average RBE factor of 1.10.

52. A 2.0-MeV neutron is emitted in a fission reactor. If it loses half its kinetic energy in each collision with a moderator atom, how many collisions must it undergo in order to become a thermal neutron, with energy 0.039 eV?

53. Assuming that a deuteron and a triton are at rest when they fuse according to the reaction $^2_1\text{H} + ^3_1\text{H} \rightarrow ^4_2\text{He} + ^1_0\text{n} + 17.6$ MeV, determine the kinetic energy acquired by the neutron.

54. A sealed capsule containing the radiopharmaceutical phosphorus-32 ($^{32}_{15}$P), an e^- emitter, is implanted into a patient's tumor. The average kinetic energy of the beta particles is 700 keV. The initial activity is 5.22 MBq. Determine the absorbed dose during a 10.0-day period. Assume the beta particles are completely absorbed in 100 g of tissue. (*Suggestion:* Find the number of beta particles emitted.)

55. (a) Calculate the energy (in kilowatt-hours) released if 1.00 kg of ^{239}Pu undergoes complete fission and the energy released per fission event is 200 MeV. (b) Calculate the energy (in electron volts) released in the deuterium–tritium fusion reaction

$$^2_1\text{H} + ^3_1\text{H} \longrightarrow ^4_2\text{He} + ^1_0\text{n}$$

(c) Calculate the energy (in kilowatt-hours) released if 1.00 kg of deuterium undergoes fusion according to this reaction. (d) **What If?** Calculate the energy (in kilowatt-hours) released by the combustion of 1.00 kg of coal if each $C + O_2 \rightarrow CO_2$ reaction yields 4.20 eV. (e) List advantages and disadvantages of each of these methods of energy generation.

56. The Sun radiates energy at the rate of 3.77×10^{26} W. Suppose that the net reaction

$$4(^1_1\text{H}) + 2(^{\ 0}_{-1}\text{e}) \longrightarrow ^4_2\text{He} + 2\nu + \gamma$$

accounts for all the energy released. Calculate the number of protons fused per second.

57. Consider the two nuclear reactions

(I) $A + B \longrightarrow C + E$
(II) $C + D \longrightarrow F + G$

(a) Show that the net disintegration energy for these two reactions ($Q_{net} = Q_I + Q_{II}$) is identical to the disintegration energy for the net reaction

$$A + B + D \longrightarrow E + F + G$$

(b) One chain of reactions in the proton–proton cycle in the Sun's core is

$$^1_1\text{H} + ^1_1\text{H} \longrightarrow ^2_1\text{H} + ^{\ 0}_{+1}\text{e} + \nu$$
$$^{\ 0}_{+1}\text{e} + ^{\ 0}_{-1}\text{e} \longrightarrow 2\gamma$$
$$^1_1\text{H} + ^2_1\text{H} \longrightarrow ^3_2\text{He} + \gamma$$
$$^1_1\text{H} + ^3_2\text{He} \longrightarrow ^4_2\text{He} + ^{\ 0}_{+1}\text{e} + \nu$$
$$^{\ 0}_{+1}\text{e} + ^{\ 0}_{-1}\text{e} \longrightarrow 2\gamma$$

Based on part (a), what is Q_{net} for this sequence?

58. Suppose the target in a laser fusion reactor is a sphere of solid hydrogen that has a diameter of 1.50×10^{-4} m and a density of 0.200 g/cm^3. Also assume that half of the nuclei are ^{2}H and half are ^{3}H. (a) If 1.00% of a 200-kJ laser pulse is delivered to this sphere, what temperature does the sphere reach? (b) If all of the hydrogen "burns" according to the D–T reaction, how many joules of energy are released?

59. In addition to the proton–proton cycle described in the chapter text, the carbon cycle, first proposed by Hans Bethe in 1939, is another cycle by which energy is released in stars as hydrogen is converted to helium. The carbon cycle requires higher temperatures than the proton–proton cycle. The series of reactions is

$$^{12}\text{C} + ^1\text{H} \longrightarrow ^{13}\text{N} + \gamma$$
$$^{13}\text{N} \longrightarrow ^{13}\text{C} + e^+ + \nu$$
$$e^+ + e^- \longrightarrow 2\gamma$$
$$^{13}\text{C} + ^1\text{H} \longrightarrow ^{14}\text{N} + \gamma$$
$$^{14}\text{N} + ^1\text{H} \longrightarrow ^{15}\text{O} + \gamma$$
$$^{15}\text{O} \longrightarrow ^{15}\text{N} + e^+ + \nu$$
$$e^+ + e^- \longrightarrow 2\gamma$$
$$^{15}\text{N} + ^1\text{H} \longrightarrow ^{12}\text{C} + ^4\text{He}$$

(a) If the proton–proton cycle requires a temperature of 1.5×10^7 K, estimate by proportion the temperature required for the carbon cycle. (b) Calculate the Q value for each step in the carbon cycle and the overall energy released. (c) Do you think the energy carried off by the neutrinos is deposited in the star? Explain.

60. When photons pass through matter, the intensity I of the beam (measured in watts per square meter) decreases exponentially according to

$$I = I_0 e^{-\mu x}$$

where I_0 is the intensity of the incident beam and I is the intensity of the beam that just passed through a thickness

x of material. The constant μ is known as the *linear absorption coefficient*, and its value depends on the absorbing material and the wavelength of the photon beam. This wavelength (or energy) dependence allows us to filter out unwanted wavelengths from a broad-spectrum x-ray beam. (a) Two x-ray beams of wavelengths λ_1 and λ_2 and equal incident intensities pass through the same metal plate. Show that the ratio of the emergent beam intensities is

$$\frac{I_2}{I_1} = e^{-(\mu_2 - \mu_1)x}$$

(b) Compute the ratio of intensities emerging from an aluminum plate 1.00 mm thick if the incident beam contains equal intensities of 50 pm and 100 pm x-rays. The values of μ for aluminum at these two wavelengths are $\mu_1 = 5.4$ cm^{-1} at 50 pm and $\mu_2 = 41.0$ cm^{-1} at 100 pm. (c) Repeat for an aluminum plate 10.0 mm thick.

61. *To build a bomb.* (a) At time $t = 0$ a sample of uranium is exposed to a neutron source that causes N_0 nuclei to undergo fission. The sample is in a supercritical state, with a reproduction constant $K > 1$. A chain reaction occurs which proliferates fission throughout the mass of uranium. The chain reaction can be thought of as a succession of *generations*. The N_0 fissions produced initially are the zeroth generation of fissions. From this generation, N_0K neutrons go off to produce fission of new uranium nuclei. The N_0K fissions that occur subsequently are the first generation of fissions, and from this generation, N_0K^2 neutrons go in search of uranium nuclei in which to cause fission. The subsequent N_0K^2 fissions are the second generation of fissions. This process can continue until all the uranium nuclei have fissioned. Show that the cumulative total of fissions N that have occurred up to and including the nth generation after the zeroth generation, is given by

$$N = N_0 \left(\frac{K^{n+1} - 1}{K - 1} \right)$$

(b) Consider a hypothetical uranium bomb made from 5.50 kg of isotopically pure ^{235}U. The chain reaction has a reproduction constant of 1.10, and starts with a zeroth generation of 1.00×10^{20} fissions. The average time interval between one fission generation and the next is 10.0 ns. How long after the zeroth generation does it take the uranium in this bomb to fission completely? (c) Assume that the bulk modulus of uranium is 150 GPa. Find the speed of sound in uranium. You may ignore the density difference between ^{235}U and natural uranium. (d) Find the time interval required for a compressional wave to cross the radius of a 5.50-kg sphere of uranium. This time interval indicates how quickly the motion of explosion begins. (e) Fission must occur in a time interval that is short compared to that in part (d), for otherwise most of the uranium will disperse in small chunks without having fissioned. Can the bomb considered in part (b) release the explosive energy of all of its uranium? If so, how much energy does it release, in equivalent tons of TNT? Assume that one ton of TNT releases 4.20 GJ and that each uranium fission releases 200 MeV of energy.

Answers to Quick Quizzes

45.1 (a). This is the only set of products of the four that satisfies the conditions that the atomic numbers add to 92 and the atomic mass numbers add to 234 (allowing for two neutrons to make up the total of 236).

45.2 (b). According to Figure 44.4, the ratio N/Z increases with increasing Z. As a result, when a heavy nucleus fissions to two lighter nuclei, the lighter nuclei tend to have too many neutrons for the nucleus to be stable. Beta decay in which electrons are ejected decreases the number of neutrons and increases the number of protons in order to stabilize the nucleus.

45.3 (a) and (b). In both these cases the Z and A values balance on the two sides of the equations. In reaction (c), $Z_{\text{left}} = Z_{\text{right}}$ but $A_{\text{left}} \neq A_{\text{right}}$.

45.4 (b). Collisions of the neutrons with moderator nuclei slow them down.

45.5 (a). To reduce the value of K, more neutrons need to be absorbed, so a larger volume of the control rods must be inside the reactor core.

45.6 (d). Figure 44.5 shows that the curve representing the binding energy per nucleon peaks at $A \approx 60$. Consequently, combining two nuclei with equal values of $A > 60$ results in an increase in mass, so that a fusion reaction will not occur.

Chapter 46

Particle Physics and Cosmology

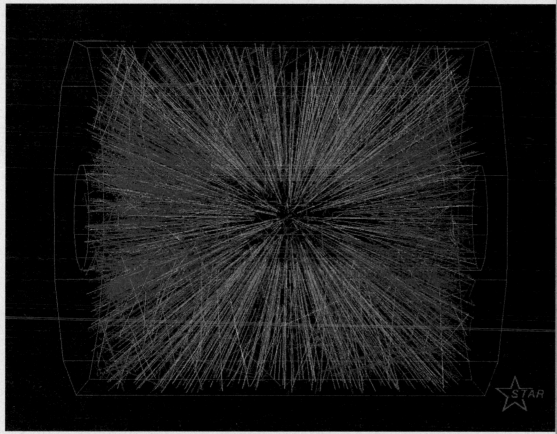

▲ *A shower of particle tracks from a head-on collision of gold nuclei, each moving with energy 100 GeV. This collision occurred at the Relativistic Heavy Ion Collider (RHIC) at Brookhaven National Laboratory and was recorded with the STAR (Solenoidal Tracker at RHIC) detector. The tracks represent many fundamental particles arising from the energy of the collision. (Courtesy of Brookhaven National Laboratory/RHIC-STAR)*

CHAPTER OUTLINE

46.1 The Fundamental Forces in Nature

46.2 Positrons and Other Antiparticles

46.3 Mesons and the Beginning of Particle Physics

46.4 Classification of Particles

46.5 Conservation Laws

46.6 Strange Particles and Strangeness

46.7 Making Particles and Measuring Their Properties

46.8 Finding Patterns in the Particles

46.9 Quarks

46.10 Multicolored Quarks

46.11 The Standard Model

46.12 The Cosmic Connection

46.13 Problems and Perspectives

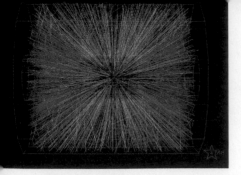

The word *atom* comes from the Greek *atomos,* which means "indivisible." The early Greeks believed that atoms were the indivisible constituents of matter; that is, they regarded them as elementary particles. After 1932 physicists viewed all matter as consisting of three constituent particles: electrons, protons, and neutrons. Beginning in the 1940s, many "new" particles were discovered in experiments involving high-energy collisions between known particles. The new particles are characteristically very unstable and have very short half-lives, ranging between 10^{-6} s and 10^{-23} s. So far, more than 300 of them have been catalogued.

Until the 1960s, physicists were bewildered by the great number and variety of subatomic particles that were being discovered. They wondered whether the particles had no systematic relationship connecting them, or whether a pattern was emerging that would provide a better understanding of the elaborate structure in the subatomic world. The fact that the neutron has a magnetic moment despite having zero electric charge (Section 44.8) suggests an underlying structure to the neutron. The fact that the periodic table explains how over a hundred elements can be formed from three types of particles (electrons, protons, and neutrons) suggests that perhaps there is a means of forming over 300 subatomic particles from a small number of basic building blocks.

Recall Figure 1.2, in which we illustrated the various levels of structure in matter. We studied the atomic structure of matter in Chapter 42. In Chapter 44, we delved deeper and investigated the substructure of the atom by describing the structure of the nucleus. As mentioned in Section 1.2, the protons and neutrons in the nucleus, and a host of other exotic particles, are now known to be composed of six different varieties of particles called *quarks.* In this concluding chapter, we examine the current theory of elementary particles, in which all matter is constructed from only two families of particles, quarks and leptons. We also discuss how clarifications of such models might help scientists understand the birth and evolution of the Universe.

46.1   The Fundamental Forces in Nature

As noted in Section 5.1, all natural phenomena can be described by four fundamental forces acting between particles. In order of decreasing strength, they are the nuclear force, the electromagnetic force, the weak force, and the gravitational force.

The nuclear force (Chapter 44) is an attractive force between nucleons. It is very short-range and is negligible for separation distances between nucleons greater than about 10^{-15} m (about the size of the nucleus). The electromagnetic force, which binds atoms and molecules together to form ordinary matter, has about 10^{-2} times the strength of the nuclear force. It is a long-range force that decreases in magnitude as the inverse square of the separation between interacting particles. The weak force is a short-range force that tends to produce instability in certain nuclei. It is responsible for decay processes, and its strength is only about 10^{-5} times that of the nuclear force. Finally, the gravitational force is a long-range force that has a strength of only about

Table 46.1

Particle Interactions				
Interaction	Relative Strength	Range of Force	Mediating Field Particle	Mass of Field Particle (GeV/c^2)
Nuclear	1	Short (≈ 1 fm)	Gluon	0
Electromagnetic	10^{-2}	∞	Photon	0
Weak	10^{-5}	Short ($\approx 10^{-3}$ fm)	$W^{\pm}$, Z^0 bosons	80.4, 80.4, 91.2
Gravitational	10^{-39}	∞	Graviton	0

10^{-39} times that of the nuclear force. Although this familiar interaction is the force that holds the planets, stars, and galaxies together, its effect on elementary particles is negligible.

In Section 13.5, we discussed the difficulty that early scientists had with the notion of the gravitational force acting at a distance, with no physical contact between the interacting objects. To resolve this difficulty, the concept of the gravitational field was introduced. Similarly, in Chapter 23, we introduced the electric field to describe the electric force acting between charged objects, followed by a discussion of the magnetic field in Chapter 29. In modern physics, the nature of the interaction between particles is carried a step further. These interactions are described in terms of the exchange of entities called **field particles** or **exchange particles.** Field particles are also called **gauge bosons.**[1] The interacting particles continuously emit and absorb field particles. The emission of a field particle by one particle and its absorption by another manifests as a force between the two interacting particles. In the case of the electromagnetic interaction, for instance, the field particles are photons. In the language of modern physics, the electromagnetic force is said to be *mediated* by photons, and photons are the field particles of the electromagnetic field. Likewise, the nuclear force is mediated by field particles called *gluons* (so called because they act as a "glue" to bond the nucleons together). The weak force is mediated by field particles called *W* and *Z bosons,* and the gravitational force is proposed to be mediated by field particles called *gravitons.* These interactions, their ranges, and their relative strengths are summarized in Table 46.1.

46.2   Positrons and Other Antiparticles

In the 1920s, Paul Dirac developed a relativistic quantum-mechanical description of the electron that successfully explained the origin of the electron's spin and its magnetic moment. His theory had one major problem, however: its relativistic wave equation required solutions corresponding to negative energy states, and if negative energy states existed, an electron in a state of positive energy would be expected to make a rapid transition to one of these states, emitting a photon in the process.

Dirac circumvented this difficulty by postulating that all negative energy states are filled. The electrons occupying these negative energy states are collectively called the *Dirac sea.* Electrons in the Dirac sea are not directly observable because the Pauli exclusion principle does not allow them to react to external forces—there are no available states to which an electron can make a transition in response to an external force. Therefore, an electron in such a state acts as an isolated system, unless an interaction

Paul Adrien Maurice Dirac
British Physicist (1902–1984)

Dirac was instrumental in the understanding of antimatter and the unification of quantum mechanics and relativity. He made many contributions to the development of quantum physics and cosmology. Dirac won the Nobel Prize for physics in 1933. *(Courtesy of AIP Emilio Segré Visual Archives)*

---

[1]   The word *bosons* suggests that the field particles have integral spin, as discussed in Section 43.8. The word *gauge* comes from *gauge theory,* which is a sophisticated mathematical analysis that is beyond the scope of this book.

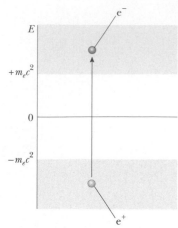

Figure 46.1 Dirac's model for the existence of antielectrons (positrons). The states lower in energy than $-m_e c^2$ are filled with electrons (the Dirac sea). One of these electrons can make a transition out of its state only if it is provided with energy equal to or larger than $2m_e c^2$. This leaves a vacancy in the Dirac sea, which can behave as a particle identical to the electron except for its positive charge.

▲ **PITFALL PREVENTION**

46.1 Antiparticles

An antiparticle is not identified solely on the basis of opposite charge; even neutral particles have antiparticles, which are defined in terms of other properties, such as spin.

with the environment is strong enough to excite the electron to a positive energy state. Such an excitation causes one of the negative energy states to be vacant, as in Figure 46.1, leaving a hole in the sea of filled states. *The hole can react to external forces and is observable.* The hole reacts in a way similar to that of the electron, except that it has a positive charge—it is the *antiparticle* to the electron.

This theory strongly suggested that *for every particle an antiparticle exists*, not only for fermions such as electrons but also for bosons. This has subsequently been verified for *all* particles known today. The antiparticle for a charged particle has the same mass as the particle but opposite charge. For example, the electron's antiparticle (the *positron* mentioned in Section 44.4) has a rest energy of 0.511 MeV and a positive charge of $+1.60 \times 10^{-19}$ C.

Carl Anderson (1905–1991) observed the positron experimentally in 1932, and in 1936 he was awarded a Nobel Prize for his achievement. Anderson discovered the positron while examining tracks created in a cloud chamber by electron-like particles of positive charge. (These early experiments used cosmic rays—mostly energetic protons passing through interstellar space—to initiate high-energy reactions on the order of several GeV.) To discriminate between positive and negative charges, Anderson placed the cloud chamber in a magnetic field, causing moving charges to follow curved paths. He noted that some of the electron-like tracks deflected in a direction corresponding to a positively charged particle.

Since Anderson's discovery, positrons have been observed in a number of experiments. A common source of positrons is **pair production.** In this process, a gamma-ray photon with sufficiently high energy interacts with a nucleus, and an electron–positron pair is created from the photon. (The presence of the nucleus allows the principle of conservation of momentum to be satisfied.) Because the total rest energy of the electron–positron pair is $2m_e c^2 = 1.02$ MeV (where m_e is the mass of the electron), the photon must have at least this much energy to create an electron–positron pair. Therefore, the energy of a photon is converted to rest energy of the electron and positron in accordance with Einstein's relationship $E_R = mc^2$. If the gamma-ray photon has energy in excess of the rest energy of the electron–positron pair, the excess appears as kinetic energy of the two particles. Figure 46.2 shows early observations of tracks of electron–positron pairs in a bubble chamber created by 300-MeV gamma rays striking a lead sheet.

Quick Quiz 46.1 Given the identification of the particles in Figure 46.2b, what is the direction of the external magnetic field in Figure 46.2a? (a) into the page (b) out of the page (c) impossible to determine.

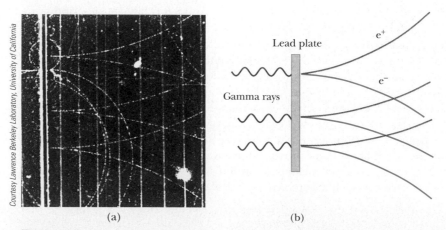

Figure 46.2 (a) Bubble-chamber tracks of electron–positron pairs produced by 300-MeV gamma rays striking a lead sheet. (b) The pertinent pair-production events. The positrons deflect upward and the electrons downward in an applied magnetic field.

The reverse process can also occur. Under the proper conditions, an electron and a positron can annihilate each other to produce two gamma-ray photons that have a combined energy of at least 1.02 MeV:

$$e^- + e^+ \longrightarrow 2\gamma$$

Because the initial momentum of the electron–positron system is approximately zero, the two gamma rays travel in opposite directions after the annihilation, satisfying the principle of conservation of momentum for the system.

Practically every known elementary particle has a distinct antiparticle. Among the exceptions are the photon and the neutral pion (π^0—see Section 46.3). Following the construction of high-energy accelerators in the 1950s, many other antiparticles were revealed. These included the antiproton, discovered by Emilio Segré (1905–1989) and Owen Chamberlain (b. 1920) in 1955, and the antineutron,[2] discovered shortly thereafter.

Electron–positron annihilation is used in the medical diagnostic technique called *positron emission tomography* (PET). The patient is injected with a glucose solution containing a radioactive substance that decays by positron emission, and the material is carried by the blood throughout the body. A positron emitted during a decay event in one of the radioactive nuclei in the glucose solution annihilates with an electron in the surrounding tissue, resulting in two gamma-ray photons emitted in opposite directions. A gamma detector surrounding the patient pinpoints the source of the photons and, with the assistance of a computer, displays an image of the sites at which the glucose accumulates. (Glucose is metabolized rapidly in cancerous tumors and accumulates at those sites, providing a strong signal for a PET detector system.) The images from a PET scan can indicate a wide variety of disorders in the brain, including Alzheimer's disease (Fig. 46.3). In addition, because glucose metabolizes more rapidly in active areas of the brain, a PET scan can indicate which areas of the brain are involved in the activities in which the patient is engaging at the time of the scan, such as language use, music, and vision.

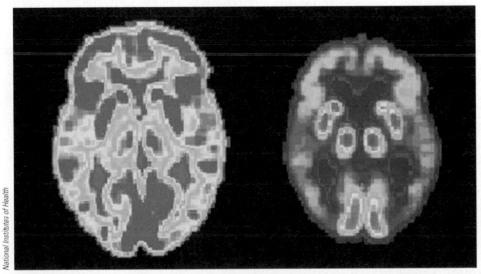

National Institutes of Health

Figure 46.3 PET scans of the brain of a healthy older person *(left)* and that of a patient suffering from Alzheimer's disease *(right)*. Lighter regions contain higher concentrations of radioactive glucose, indicating higher metabolism rates and therefore increased brain activity.

---

[2] Antiparticles of charged particles have the opposite charge. Antiparticles for uncharged particles, such as the neutron, are a little more difficult to describe. One basic process that can detect the existence of an antiparticle is pair annihilation. For example, a neutron and an antineutron can annihilate to form two gamma rays. Because the photon and the neutral pion do not have distinct antiparticles, we do not observe pair annihilation with either of these particles.

46.3 Mesons and the Beginning of Particle Physics

Physicists in the mid-1930s had a fairly simple view of the structure of matter. The building blocks were the proton, the electron, and the neutron. Three other particles were either known or postulated at the time: the photon, the neutrino, and the positron. Together these six particles were considered the fundamental constituents of matter. With this simple picture, however, no one was able to answer the following important question: in view of the fact that the protons in any nucleus should strongly repel one another due to their charges of the same sign, what is the nature of the force that holds the nucleus together? Scientists recognized that this mysterious force must be much stronger than anything encountered in nature up to that time. This is the nuclear force discussed in Section 44.1 and examined in historical perspective in the following paragraphs.

The first theory to explain the nature of the nuclear force was proposed in 1935 by the Japanese physicist Hideki Yukawa—an effort that earned him a Nobel Prize in physics in 1949. To understand Yukawa's theory, recall the introduction of field particles in Section 46.1, which stated that each fundamental force is mediated by a field particle exchanged between the interacting particles. Yukawa used this idea to explain the nuclear force, proposing the existence of a new particle whose exchange between nucleons in the nucleus causes the nuclear force. He established that the range of the force is inversely proportional to the mass of this particle and predicted the mass to be about 200 times the mass of the electron. (Yukawa's predicted particle is *not* the gluon mentioned in Section 46.1, which is massless and is today considered to be the field particle for the nuclear force.) Because the new particle would have a mass between that of the electron and that of the proton, it was called a **meson** (from the Greek *meso*, "middle").

In efforts to substantiate Yukawa's predictions, physicists began experimental searches for the meson by studying cosmic rays entering the Earth's atmosphere. In 1937, Carl Anderson and his collaborators discovered a particle of mass $106\ \text{MeV}/c^2$, about 207 times the mass of the electron. This was thought to be Yukawa's meson. However, subsequent experiments showed that the particle interacted very weakly with matter and hence could not be the field particle for the nuclear force. That puzzling situation inspired several theoreticians to propose two mesons having slightly different masses equal to about 200 times that of the electron—one having been discovered by Anderson and the other, still undiscovered, predicted by Yukawa. This idea was confirmed in 1947 with the discovery of the **pi meson** (π), or simply **pion.** The particle discovered by Anderson in 1937, the one initially thought to be Yukawa's meson, is not really a meson. (We shall discuss the characteristics of mesons in Section 46.4.) Instead, it takes part in the weak and electromagnetic interactions only and is now called the **muon** (μ).

The pion comes in three varieties, corresponding to three charge states: π^+, π^-, and π^0. The π^+ and π^- particles (π^- is the antiparticle of π^+) each have a mass of $139.6\ \text{MeV}/c^2$, and the π^0 mass is $135.0\ \text{MeV}/c^2$. Two muons exist: μ^- and its antiparticle μ^+.

Pions and muons are very unstable particles. For example, the π^-, which has a mean lifetime of 2.6×10^{-8} s, decays to a muon and an antineutrino.[3] The muon, which has a mean lifetime of $2.2\ \mu\text{s}$, then decays to an electron, a neutrino, and an antineutrino:

$$\pi^- \longrightarrow \mu^- + \bar{\nu}$$

$$\mu^- \longrightarrow e^- + \nu + \bar{\nu} \tag{46.1}$$

Hideki Yukawa

**Japanese Physicist
(1907–1981)**

Yukawa was awarded the Nobel Prize in 1949 for predicting the existence of mesons. This photograph of him at work was taken in 1950 in his office at Columbia University. Yukawa came to Columbia in 1949 after spending the early part of his career in Japan. *(UPI/Corbis-Bettman)*

---

[3] The antineutrino is another zero-charge particle for which the identification of the antiparticle is more difficult than that for a charged particle. Although the details are beyond the scope of this text, the neutrino and antineutrino can be differentiated by means of the relationship between the linear momentum and the spin angular momentum of the particles.

Note that for chargeless particles (as well as some charged particles, such as the proton), a bar over the symbol indicates an antiparticle, as in beta decay (see Section 44.5). Other antiparticles, such as e^+ and μ^+, use a different notation.

The interaction between two particles can be represented in a simple diagram called a **Feynman diagram,** developed by the American physicist Richard P. Feynman. Figure 46.4 is such a diagram for the electromagnetic interaction between two electrons. A Feynman diagram is a qualitative graph of time on the vertical axis versus space on the horizontal axis. It is qualitative in the sense that the actual values of time and space are not important, but the overall appearance of the graph provides a representation of the process.

In the simple case of the electron–electron interaction in Figure 46.4, a photon (the field particle) mediates the electromagnetic force between the electrons. Imagining time as the vertical axis, notice that the entire interaction is represented in the diagram as occurring at a single point in time. Therefore, the paths of the electrons appear to undergo a discontinuous change in direction at the moment of interaction. This is different from the *actual* paths, which would be curved due to the continuous exchange of large numbers of field particles. This is another aspect of the qualitative nature of Feynman diagrams.

In the electron–electron interaction, the photon, which transfers energy and momentum from one electron to the other, is called a *virtual photon* because it vanishes during the interaction without having been detected. In Chapter 40, we discussed the fact that a photon has energy $E = hf$, where f is its frequency. Consequently, for a system of two electrons initially at rest, the system has energy $2m_e c^2$ before a virtual photon is released and energy $2m_e c^2 + hf$ after the virtual photon is released (plus any kinetic energy of the electron resulting from the emission of the photon). Is this a violation of the law of conservation of energy for an isolated system? No; this process does *not* violate the law of conservation of energy because the virtual photon has a very short lifetime Δt that makes the uncertainty in the energy $\Delta E \approx \hbar/2\Delta t$ of the system consisting of two electrons and the photon greater than the photon energy. Therefore, within the constraints of the uncertainty principle, the energy of the system is conserved.

Now consider a pion mediating the nuclear force between a proton and a neutron, as in Yukawa's model (Fig. 46.5a). The rest energy E_R of a pion of mass m_π is given by Einstein's equation $E_R = m_\pi c^2$. As with the photon in Figure 46.4, in order to conserve energy, the uncertainty in the system energy must be greater than the rest energy of the pion: $\Delta E > E_R$. The existence of the pion would violate the law of conservation of energy if the particle existed for a time interval greater than $\Delta t \approx \hbar/2E_R$ (from the uncertainty principle), where E_R is the rest energy of the pion and Δt is the time interval required for the pion to transfer from one nucleon to the other. Therefore,

$$\Delta t \approx \frac{\hbar}{2E_R} = \frac{\hbar}{2m_\pi c^2} \qquad (46.2)$$

Figure 46.4 Feynman diagram representing a photon mediating the electromagnetic force between two electrons.

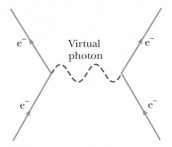

Richard Feynman
**American Physicist
(1918–1988)**

Inspired by Dirac, Feynman developed quantum electrodynamics, the theory of the interaction of light and matter on a relativistic and quantum basis. Feynman won the Nobel Prize for physics in 1965. The prize was shared by Feynman, Julian Schwinger, and Sin Itiro Tomonaga. Early in his career, he was a leading member of the team developing the first nuclear weapon in the Manhattan Project. Toward the end of his career, he worked on the commission investigating the 1986 *Challenger* tragedy and demonstrated the effects of cold temperatures on the rubber O-rings used in the space shuttle. *(© Shelly Gazin/CORBIS)*

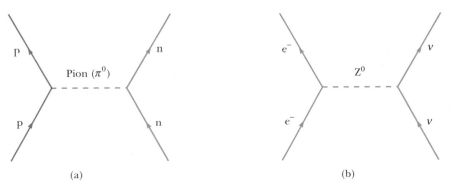

(a)   (b)

Figure 46.5 (a) Feynman diagram representing a proton and a neutron interacting via the nuclear force with a neutral pion mediating the force. (This is *not* the current model for nucleon interaction.) (b) Feynman diagram for an electron and a neutrino interacting via the weak force, with a Z^0 boson mediating the force.

Because the pion cannot travel faster than the speed of light, the maximum distance d it can travel in a time interval Δt is $c\Delta t$. Therefore,

$$d = c\Delta t \approx \frac{\hbar}{2m_\pi c} \tag{46.3}$$

From Table 46.1, we know that the range of the nuclear force is approximately 1×10^{-15} m. Using this value for d in Equation 46.3, we estimate the rest energy of the pion to be

$$m_\pi c^2 \approx \frac{\hbar c}{2d} = \frac{(1.05 \times 10^{-34}\,\text{J}\cdot\text{s})(3.00 \times 10^8\,\text{m/s})}{2(1 \times 10^{-15}\,\text{m})}$$

$$= 1.6 \times 10^{-11}\,\text{J} \approx 100\,\text{MeV}$$

Because this is the same order of magnitude as the observed rest energies of the pions, we have some confidence in the field-particle model.

The concept we have just described is quite revolutionary. In effect, it says that a system of two nucleons can change into two nucleons plus a pion, as long as it returns to its original state in a very short time interval. (Remember that this is the older historical model, which assumes that the pion is the field particle for the nuclear force; keep in mind that the gluon is the actual field particle in current models.) Physicists often say that a nucleon undergoes *fluctuations* as it emits and absorbs field particles. These fluctuations are a consequence of a combination of quantum mechanics (through the uncertainty principle) and special relativity (through Einstein's energy–mass relationship $E_R = mc^2$).

This section has dealt with the field particles that were originally proposed to mediate the nuclear force (pions) and those that mediate the electromagnetic force (photons). The graviton, the field particle for the gravitational force, has yet to be observed. $W^\pm$ and Z^0 particles, which mediate the weak force, were discovered in 1983 by the Italian physicist Carlo Rubbia (b. 1934) and his associates, using a proton–antiproton collider. Rubbia and Simon van der Meer (b. 1925), both at CERN,[4] shared the 1984 Nobel Prize in physics for the discovery of the $W^\pm$ and Z^0 particles and the development of the proton–antiproton collider. Figure 46.5b shows a Feynman diagram for a weak interaction mediated by a Z^0 boson.

▲ **PITFALL PREVENTION**

46.2 The Nuclear Force and the Strong Force

The nuclear force that we discussed in Chapter 44 was originally called the strong force. Once the quark theory (Section 46.9) was established, however, the phrase *strong force* was reserved for the force between quarks. We shall follow this convention—the strong force is between quarks and the nuclear force is between nucleons. The nuclear force is a secondary result of the strong force, as we discuss in Section 46.10. Because of this historical development of the names for these forces, other books sometimes refer to the nuclear force as the strong force.

46.4 Classification of Particles

All particles other than field particles can be classified into two broad categories, *hadrons* and *leptons*. The criterion for separating these particles into categories is whether or not they interact via the strong force. The nuclear force between nucleons in a nucleus is a particular manifestation of the strong force, but we will use the term *strong force* in general to refer to any interaction between particles made up of quarks. (For more detail on quarks and the strong force, see Section 46.9.) Table 46.2 provides a summary of the properties of hadrons and leptons.

Hadrons

Particles that interact through the strong force (as well as through the other fundamental forces) are called **hadrons.** The two classes of hadrons, *mesons* and *baryons*, are distinguished by their masses and spins.

---

[4] CERN was originally the Conseil Européen pour la Recherche Nucléaire—European Organization for Nuclear Research; the name has been altered to European Laboratory for Particle Physics, but the CERN acronym has been retained.

Table 46.2

Some Particles and Their Properties

Category	Particle Name	Symbol	Anti-particle	Mass (MeV/c^2)	B	L_e	L_μ	L_τ	S	Lifetime(s)	Principal Decay Modes[a]
Leptons	Electron	e^-	e^+	0.511	0	+1	0	0	0	Stable	
	Electron–neutrino	ν_e	$\overline{\nu}_e$	< 7 eV/c^2	0	+1	0	0	0	Stable	
	Muon	μ^-	μ^+	105.7	0	0	+1	0	0	2.20×10^{-6}	$e^- \overline{\nu}_e \nu_\mu$
	Muon–neutrino	ν_μ	$\overline{\nu}_\mu$	< 0.3	0	0	+1	0	0	Stable	
	Tau	τ^-	τ^+	1 784	0	0	0	+1	0	$< 4 \times 10^{-13}$	$\mu^- \overline{\nu}_\mu \nu_\tau$, $e^- \overline{\nu}_e \nu_\tau$
	Tau–neutrino	ν_τ	$\overline{\nu}_\tau$	< 30	0	0	0	+1	0	Stable	
Hadrons											
Mesons	Pion	π^+	π^-	139.6	0	0	0	0	0	2.60×10^{-8}	$\mu^+ \nu_\mu$
		π^0	Self	135.0	0	0	0	0	0	0.83×10^{-16}	2γ
	Kaon	K^+	K^-	493.7	0	0	0	0	+1	1.24×10^{-8}	$\mu^+ \nu_\mu$, $\pi^+ \pi^0$
		K_s^0	$\overline{K}_s^0$	497.7	0	0	0	0	+1	0.89×10^{-10}	$\pi^+ \pi^-$, $2\pi^0$
		K_L^0	$\overline{K}_L^0$	497.7	0	0	0	0	+1	5.2×10^{-8}	$\pi^\pm e^\mp \overline{\nu}_e$, $3\pi^0$ $\pi^\pm \mu^\mp \overline{\nu}_\mu$
	Eta	η	Self	548.8	0	0	0	0	0	$< 10^{-18}$	2γ, $3\pi^0$
		η'	Self	958	0	0	0	0	0	2.2×10^{-21}	$\eta \pi^+ \pi^-$
Baryons	Proton	p	$\overline{\text{p}}$	938.3	+1	0	0	0	0	Stable	
	Neutron	n	$\overline{\text{n}}$	939.6	+1	0	0	0	0	614	$\text{p}e^- \overline{\nu}_e$
	Lambda	Λ^0	$\overline{\Lambda}^0$	1 115.6	+1	0	0	0	−1	2.6×10^{-10}	$\text{p}\pi^-$, $\text{n}\pi^0$
	Sigma	Σ^+	$\overline{\Sigma}^-$	1 189.4	+1	0	0	0	−1	0.80×10^{-10}	$\text{p}\pi^0$, $\text{n}\pi^+$
		Σ^0	$\overline{\Sigma}^0$	1 192.5	+1	0	0	0	−1	6×10^{-20}	$\Lambda^0 \gamma$
		Σ^-	$\overline{\Sigma}^+$	1 197.3	+1	0	0	0	−1	1.5×10^{-10}	$\text{n}\pi^-$
	Delta	Δ^{++}	$\overline{\Delta}^{--}$	1 230	+1	0	0	0	0	6×10^{-24}	$\text{p}\pi^+$
		Δ^+	$\overline{\Delta}^-$	1 231	+1	0	0	0	0	6×10^{-24}	$\text{p}\pi^0$, $\text{n}\pi^+$
		Δ^0	$\overline{\Delta}^0$	1 232	+1	0	0	0	0	6×10^{-24}	$\text{n}\pi^0$, $\text{p}\pi^-$
		Δ^-	$\overline{\Delta}^+$	1 234	+1	0	0	0	0	6×10^{-24}	$\text{n}\pi^-$
	Xi	Ξ^0	$\overline{\Xi}^0$	1 315	+1	0	0	0	−2	2.9×10^{-10}	$\Lambda^0 \pi^0$
		Ξ^-	Ξ^+	1 321	+1	0	0	0	−2	1.64×10^{-10}	$\Lambda^0 \pi^-$
	Omega	Ω^-	Ω^+	1 672	+1	0	0	0	−3	0.82×10^{-10}	$\Xi^- \pi^0$, $\Xi^0 \pi^-$, $\Lambda^0 K^-$

a Notations in this column such as $\text{p}\pi^-$, $\text{n}\pi^0$ mean two possible decay modes. In this case, the two possible decays are $\Lambda^0 \rightarrow \text{p} + \pi^-$ and $\Lambda^0 \rightarrow \text{n} + \pi^0$.

Mesons all have zero or integer spin (0 or 1). As indicated in Section 46.3, the name comes from the expectation that Yukawa's proposed meson mass would lie between the masses of the electron and the proton. Several meson masses do lie in this range, although mesons having masses greater than that of the proton have been found to exist.

All mesons are known to decay finally into electrons, positrons, neutrinos, and photons. The pions are the lightest known mesons; they have masses of about 1.4×10^2 MeV/c^2 and all three pions—π^+, π^-, and π^0—have a spin of 0. (This indicates that the particle discovered by Anderson in 1937, the muon, is not a meson; the muon has spin $\frac{1}{2}$. It belongs in the *lepton* classification, described on page 1520.)

Baryons, the second class of hadrons, have masses equal to or greater than the proton mass (the name *baryon* means "heavy" in Greek), and their spin is always a half-integer value ($\frac{1}{2}$ or $\frac{3}{2}$). Protons and neutrons are baryons, as are many other particles. With the exception of the proton, all baryons decay, in such a way that the end products include a proton. For example, the baryon called the Ξ hyperon (Greek capital xi) decays to the Λ^0 baryon (Greek capital lambda) in about 10^{-10} s. The Λ^0 then decays to a proton and a π^- in approximately 3×10^{-10} s.

Today it is believed that hadrons are not elementary particles but are composed of more elementary units called quarks, per Section 46.9.

Leptons

Leptons (from the Greek *leptos,* meaning "small" or "light") are a group of particles that do not interact by means of the strong force. All leptons have spin $\frac{1}{2}$. Unlike hadrons, which have size and structure, leptons appear to be truly elementary, meaning that they have no structure and are point-like.

Quite unlike the case with hadrons, the number of known leptons is small. Currently, scientists believe that only six leptons exist—the electron, the muon, the tau, and a neutrino associated with each: e^-, μ^-, τ^-, ν_e, ν_μ, ν_τ. The tau lepton, discovered in 1975, has a mass about twice that of the proton. Direct experimental evidence for the neutrino associated with the tau was announced by the Fermi National Accelerator Laboratory (Fermilab) in July, 2000. Each of the six leptons has an antiparticle.

Current studies indicate that neutrinos have a small but nonzero mass. If they do have mass, then they cannot travel at the speed of light. Also, because so many neutrinos exist, their combined mass may be sufficient to cause all the matter in the Universe to eventually collapse into a single point, which might then explode and create a completely new Universe! We shall discuss this possibility in more detail in Section 46.12.

46.5 Conservation Laws

In general, the laws of conservation of energy, linear momentum, angular momentum, and electric charge provide us with a set of rules that all processes must follow. In Chapter 44 we learned that conservation laws are important for understanding why certain radioactive decays and nuclear reactions occur and others do not. In the study of elementary particles, a number of additional conservation laws are important. Although the two described here have no theoretical foundation, they are supported by abundant empirical evidence.

Baryon Number

Experimental results tell us that whenever a baryon is created in a decay or nuclear reaction, an antibaryon is also created. This scheme can be quantified by assigning every particle a quantum number, the **baryon number,** as follows: $B = +1$ for all baryons, $B = -1$ for all antibaryons, and $B = 0$ for all other particles. (See Table 46.2.) Therefore, the **law of conservation of baryon number** states that **whenever a nuclear reaction or decay occurs, the sum of the baryon numbers before the process must equal the sum of the baryon numbers after the process.**

Conservation of baryon number

If baryon number is absolutely conserved, the proton must be absolutely stable. For example, a decay of the proton to a positron and a neutral pion would satisfy conservation of energy, momentum, and electric charge. However, such a decay has never been observed. The law of conservation of baryon number would be consistent with the absence of this decay, as the proposed decay would involve the loss of a baryon. At the present, all we can say is that protons have a half-life of at least 10^{33} years (the estimated age of the Universe is only 10^{10} years), from experimental observations as pointed out in Example 46.2. Some recent theories, however, predict that the proton is unstable. According to this theory, baryon number is not absolutely conserved.

Quick Quiz 46.2 Consider the following decay: $n \rightarrow \pi^+ + \pi^- + \mu^+ + \mu^-$. What conservation laws are violated by this decay? (a) energy (b) electric charge (c) baryon number (d) angular momentum (e) no conservation laws.

Quick Quiz 46.3 Consider the following decay: $n \rightarrow p + \pi^-$. What conservation laws are violated by this decay? (a) energy (b) electric charge (c) baryon number (d) angular momentum (e) no conservation laws.

Example 46.1   Checking Baryon Numbers

Use the law of conservation of baryon number to determine whether the following reactions can occur:

(A) $p + n \rightarrow p + p + n + \bar{p}$

(B) $p + n \rightarrow p + p + \bar{p}$

Solution (A) The left side of the equation gives a total baryon number of $1 + 1 = 2$. The right side gives a total baryon number of $1 + 1 + 1 + (-1) = 2$. Therefore, baryon number is conserved and the reaction can occur (provided the initial particles have sufficient kinetic energy that energy conservation is satisfied). (B) The left side of the equation gives a total baryon number of $1 + 1 = 2$. However, the right side gives $1 + 1 + (-1) = 1$. Because baryon number is not conserved, the reaction cannot occur.

Example 46.2   Detecting Proton Decay

Interactive

Measurements taken at the Super Kamiokande neutrino detection facility (Fig. 46.6) indicate that the half-life of protons is at least 10^{33} years.

(A) Estimate how long we would have to watch, on average, to see a proton in a glass of water decay.

Solution To conceptualize the problem, imagine the number of protons in a glass of water. Although this number is huge, we know that the probability of a single proton undergoing decay is small, so we would expect to wait a long time before observing a decay. Because a half-life is provided in the problem, we categorize this problem as one in which we can apply our statistical analysis techniques from Section 44.4. To analyze the problem, let us estimate that a glass contains about 250 g of water. The number of molecules of water is

$$\frac{(250 \text{ g})(6.02 \times 10^{23} \text{ molecules/mol})}{18 \text{ g/mol}}$$

$$= 8.4 \times 10^{24} \text{ molecules}$$

Each water molecule contains one proton in each of its two hydrogen atoms plus eight protons in its oxygen atom, for a total of ten. Therefore, there are 8.4×10^{25} protons in the glass of water. The decay constant (Section 44.4) is given by Equation 44.8:

$$\lambda - \frac{0.693}{T_{1/2}} = \frac{0.693}{10^{33} \text{ yr}} = 6.9 \times 10^{-34} \text{ yr}^{-1}$$

This is the probability that *one* proton will decay in one year. The probability that *any* proton in our glass of water will decay in the one-year interval is (Eqs. 44.5 and 44.7)

$$R = (8.4 \times 10^{25})(6.9 \times 10^{-34} \text{ yr}^{-1}) = 5.8 \times 10^{-8} \text{ yr}^{-1}$$

To finalize this part of the problem, note that we have to watch our glass of water for $1/R \approx$ 17 million years! This indeed is a long time, as we suspected.

(B) The Super Kamiokande neutrino facility contains 50 000 metric tons of water. Estimate the average time interval between detected proton decays in this much water if the half-life of a proton is 10^{33} yr.

Courtesy of KRR [Institute for Cosmic Ray Research], University of Tokyo

Figure 46.6 (Example 46.2) This detector at the Super Kamiokande neutrino facility in Japan is used to study photons and neutrinos. It holds 50 000 metric tons of highly purified water and 13 000 photomultipliers. The photograph was taken while the detector was being filled. Technicians use a raft to clean the photodetectors before they are submerged.

Solution We find the ratio of the number of molecules in 50 000 metric tons of water to that in the glass of water in part (A), which will be the same as the ratio of masses:

$$\frac{N_{\text{Kamiokande}}}{N_{\text{glass}}} = \frac{m_{\text{Kamiokande}}}{m_{\text{glass}}}$$

$$= \frac{50\,000\,\text{metric ton}}{250\,\text{g}}\left(\frac{1\,000\,\text{kg}}{1\,\text{metric ton}}\right)\left(\frac{1\,000\,\text{g}}{1\,\text{kg}}\right)$$

$$= 2.0 \times 10^8$$

$$N_{\text{Kamiokande}} = (2.0 \times 10^8)N_{\text{glass}}$$

$$= (2.0 \times 10^8)(8.4 \times 10^{24}\,\text{molecules})$$

$$= 1.7 \times 10^{33}\,\text{molecules}$$

Each of these molecules contains ten protons. The probability that one of these protons will decay in one year is

$$R = (10)(1.7 \times 10^{33})(6.9 \times 10^{-34}\,\text{yr}^{-1}) \approx 12\,\text{yr}^{-1}$$

Note that the average time interval between decays is about one twelfth of a year, or approximately one month.

This is much shorter than the time interval in part (A), due to the tremendous amount of water in the detector facility.

 Practice the statistics of proton decay at the Interactive Worked Example link at **http://www.pse6.com.**

Lepton Number

We have three conservation laws involving lepton numbers, one for each variety of lepton. The **law of conservation of electron lepton number** states that whenever a nuclear reaction or decay occurs, **the sum of the electron lepton numbers before the process must equal the sum of the electron lepton numbers after the process.**

Conservation of lepton number

The electron and the electron neutrino are assigned an electron lepton number $L_e = +1$ and the antileptons e^+ and $\bar{\nu}_e$ are assigned an electron lepton number $L_e = -1$. All other particles have $L_e = 0$. For example, consider the decay of the neutron:

$$\text{n} \longrightarrow \text{p} + e^- + \bar{\nu}_e$$

Before the decay, the electron lepton number is $L_e = 0$; after the decay, it is $0 + 1 + (-1) = 0$. Therefore, electron lepton number is conserved. (Baryon number must also be conserved, of course, and it is: before the decay, $B = +1$, and after the decay $B = +1 + 0 + 0 = +1$.)

Similarly, when a decay involves muons, the muon lepton number L_μ is conserved. The μ^- and the ν_μ are assigned a muon lepton number $L_\mu = +1$ and the antimuons μ^+ and $\bar{\nu}_\mu$ are assigned a muon lepton number $L_\mu = -1$. All other particles have $L_\mu = 0$.

Finally, tau lepton number L_τ is conserved with similar assignments made for the tau lepton, its neutrino, and their two antiparticles.

Quick Quiz 46.4 Consider the following decay: $\pi^0 \rightarrow \mu^- + e^+ + \nu_\mu$. What conservation laws are violated by this decay? (a) energy (b) angular momentum (c) electric charge (d) baryon number (e) electron lepton number (f) muon lepton number (g) tau lepton number (h) no conservation laws.

Quick Quiz 46.5 Suppose a claim is made that the decay of the neutron is given by $\text{n} \rightarrow \text{p} + e^-$. What conservation laws are violated by this decay? (a) energy (b) angular momentum (c) electric charge (d) baryon number (e) electron lepton number (f) muon lepton number (g) tau lepton number (h) no conservation laws.

Quick Quiz 46.6 A student claims to have observed a decay of an electron into two electron neutrinos, traveling in opposite directions. What conservation laws would be violated by this decay? (a) energy (b) angular momentum (c) electric charge (d) baryon number (e) electron lepton number (f) muon lepton number (g) tau lepton number (h) no conservation laws.

Example 46.3 **Checking Lepton Numbers**

Use the law of conservation of lepton numbers to determine which of the following decay schemes can occur:

(A) $\mu^- \rightarrow e^- + \bar{\nu}_e + \nu_\mu$

(B) $\pi^+ \rightarrow \mu^+ + \nu_\mu + \nu_e$

Solution (A) Because this decay involves a muon and an electron, L_μ and L_e must both be conserved. Before

the decay, $L_\mu = +1$ and $L_e = 0$. After the decay, $L_\mu = 0 + 0 + 1 = +1$ and $L_e = +1 + (-1) + 0 = 0$. Therefore, both numbers are conserved, and on this basis the decay is possible.

(B) Before the decay, $L_\mu = 0$ and $L_e = 0$. After the decay, $L_\mu = -1 + 1 + 0 = 0$, but $L_e = 0 + 0 + 1 = 1$. Therefore, the decay is not possible because electron lepton number is not conserved.

46.6 Strange Particles and Strangeness

Many particles discovered in the 1950s were produced by the interaction of pions with protons and neutrons in the atmosphere. A group of these—the kaon (K), lambda (Λ), and sigma (Σ) particles—exhibited unusual properties both as they were created and as they decayed and hence were called *strange particles*.

One unusual property of strange particles is that they are always produced in pairs. For example, when a pion collides with a proton, a highly probable result is the production of two neutral strange particles (Fig. 46.7):

$$\pi^- + p \longrightarrow K^0 + \Lambda^0$$

However, the reaction $\pi^- + p \rightarrow K^0 + n^0$, where only one of the final particles is strange, never occurs, even though no known conservation laws would be violated and even though the energy of the pion is sufficient to initiate the reaction.

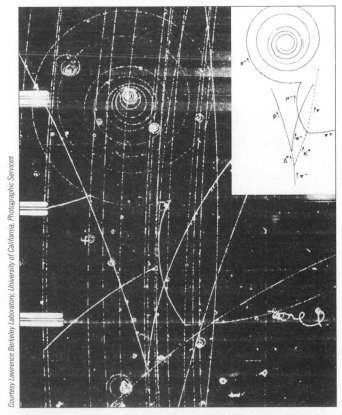

Courtesy Lawrence Berkeley Laboratory, University of California, Photographic Services

Figure 46.7 This bubble-chamber photograph shows many events, and the inset is a drawing of identified tracks. The strange particles Λ^0 and K^0 are formed at the bottom as a π^- particle interacts with a proton in the reaction $\pi^- + p \rightarrow \Lambda^0 + K^0$. (Note that the neutral particles leave no tracks, as indicated by the dashed lines in the inset.) The Λ^0 then decays in the reaction $\Lambda^0 \rightarrow \pi^- + p$ and the K^0 in the reaction $K^0 \rightarrow \pi^+ + \mu^- + \bar{\nu}_\mu$.

The second peculiar feature of strange particles is that, although they are produced in reactions involving the strong interaction at a high rate, they do not decay into particles that interact via the strong force at a high rate. Instead, they decay very slowly, which is characteristic of the weak interaction. Their half-lives are in the range 10^{-10} s to 10^{-8} s, whereas most other particles that interact via the strong force have much shorter lifetimes on the order of 10^{-23} s.

To explain these unusual properties of strange particles, a new quantum number S, called **strangeness,** was introduced, together with a conservation law. The strangeness numbers for some particles are given in Table 46.2. The production of strange particles in pairs is explained by assigning $S = +1$ to one of the particles, $S = -1$ to the other, and $S = 0$ to all nonstrange particles. The **law of conservation of strangeness** states that **in a nuclear reaction or decay that occurs via the strong force, strangeness is conserved, that is, the sum of the strangeness numbers before the process must equal the sum of the strangeness numbers after the process. In processes that occur via the weak interaction, strangeness may not be conserved.**

Conservation of strangeness number

The low decay rate of strange particles can be explained by assuming that the strong and electromagnetic interactions obey the law of conservation of strangeness but the weak interaction does not. Because the decay of a strange particle involves the loss of one strange particle, it violates strangeness conservation and hence proceeds slowly via the weak interaction.

Example 46.4   Is Strangeness Conserved?

(A) Use the law of strangeness conservation to determine whether the reaction $\pi^0 + n \rightarrow K^+ + \Sigma^-$ occurs.

Solution From Table 46.2, we see that the initial strangeness is $S = 0 + 0 = 0$. Because the strangeness of the K^+ is $S = +1$ and the strangeness of the Σ^- is $S = -1$, the strangeness of the final products is $+1 - 1 = 0$. Therefore, strangeness is conserved, and the reaction is allowed.

(B) Show that the reaction $\pi^- + p \rightarrow \pi^- + \Sigma^+$ does not conserve strangeness.

Solution Before: $S = 0 + 0 = 0$; after: $S = 0 + (-1) = -1$. Therefore, strangeness is not conserved.

46.7   Making Particles and Measuring Their Properties

The bewildering array of entries in Table 46.2 leaves one yearning for firm ground. It is natural to wonder about an entry, for example, which shows a particle (Σ^0) which exists for 10^{-20} s and has a mass of $1\,192.5$ MeV/c^2. How is it possible to detect a particle that exists for only 10^{-20} s? In this section we answer such questions and explain how elementary particles are produced and their properties measured.

Most elementary particles are unstable and are created in nature only rarely, in cosmic ray showers. In the laboratory, however, great numbers of these particles are created in controlled collisions between high-energy particles and a suitable target. The incident particles must have very high energy, and it takes a relatively long time interval for electromagnetic fields to accelerate particles to high energies. Therefore, stable charged particles such as electrons or protons generally make up the incident beam. In addition, targets must be simple and stable, and the simplest target, hydrogen, serves nicely as both a target (the proton) and a detector.

Figure 46.7 documents a typical event in which a bubble chamber served as both target source and detector. Many parallel tracks of negative pions are visible entering the photograph from the bottom. As the labels in the inset drawing show, one of the pions has hit a stationary proton in the hydrogen and produced two strange particles, Λ^0 and K^0, according to the reaction

$$\pi^- + p \longrightarrow \Lambda^0 + K^0$$

Neither neutral strange particle leaves a track, but their subsequent decay into charged particles can be seen in Figure 46.7. A magnetic field directed into the plane of the page causes the track of each charged particle to curve, and from the measured curvature we can determine the particle's charge and linear momentum. If the mass and momentum of the incident particle are known, we can then usually calculate the product particle's mass, kinetic energy, and speed from the laws of conservation of momentum and energy. Finally, combining a product particle's speed with the length of the track it leaves, we can calculate the particle's lifetime. Figure 46.7 shows that sometimes we can use this lifetime technique even for a neutral particle, which leaves no track. As long as the beginning and end of the missing track are known, as well as the particle speed, we can infer the missing track length and so determine the lifetime of the neutral particle.

Resonance Particles

Using drift chambers or other modern detectors, one can measure decay track lengths as short as 10^{-6} m. This means that lifetimes as short as 10^{-16} s can be measured for high-energy particles traveling at about the speed of light. We arrive at this result by assuming that a decaying particle travels 1 μm at a speed of $0.99c$ in the reference frame of the laboratory, yielding a lifetime of $\Delta t_{\text{lab}} = 1 \times 10^{-6}$ m$/0.99c \approx 3.4 \times 10^{-15}$ s. This is not our final result, however, because we must account for the relativistic effects of time dilation. Because the proper lifetime Δt_p measured in the decaying particle's reference frame is shorter than the laboratory-frame value Δt_{lab} by a factor of $\sqrt{1 - (v^2/c^2)}$ (see Eq. 39.7), we can calculate the proper lifetime as follows:

$$\Delta t_p = \Delta t_{\text{lab}} \sqrt{1 - \frac{v^2}{c^2}} = (3.4 \times 10^{-15} \text{ s}) \sqrt{1 - \frac{(0.99c)^2}{c^2}} = 4.8 \times 10^{-16} \text{ s}$$

Unfortunately, even with Einstein's help, the best answer we can obtain with the track-length method is several orders of magnitude away from lifetimes of 10^{-20} s. How, then, can we detect the presence of particles that exist for time intervals such as 10^{-20} s? For such short-lived particles, known as **resonance particles,** all we can do is infer their masses, their lifetimes, and their very existence from data on their decay products.

Let us consider this detection process in detail by examining the case of the resonance particle called the delta plus (Δ^+), which has a mass of 1 231 MeV$/c^2$ and a lifetime of about 6×10^{-24} s. This particle is produced in the reaction

$$e^- + p \longrightarrow e^- + \Delta^+ \tag{46.4}$$

followed in 6×10^{-24} s by the decay

$$\Delta^+ \longrightarrow \pi^+ + n \tag{46.5}$$

Because the Δ^+ lifetime is so short, the particle leaves no measurable track in a drift chamber. It might therefore seem impossible to distinguish the reactions given in Equations 46.4 and 46.5 from the reaction

$$e^- + p \longrightarrow e^- + \pi^+ + n \tag{46.6}$$

in which the reactants of Equation 46.4 decay directly to e^-, π^+, and n with no intermediate step in which a Δ^+ is produced. Distinguishing between these two possibilities is not impossible, however. If a Δ^+ particle exists, it has a distinct rest energy, which must come from the kinetic energy of the incoming particles. If we imagine firing electrons with increasing kinetic energy at protons, eventually we will provide enough energy to the system to create the Δ^+ particle. This is very similar to firing photons of increasing energy at an atom until you provide them with enough energy to excite the atom to a higher quantum state. In fact, the Δ^+ particle is an excited state of the proton, which we can understand via the quark theory discussed in Section 46.9. After the Δ^+ particle is formed, its rest energy becomes the energies of the outgoing pion and neutron. Equation 39.27 can be solved for the rest energy of the Δ^+ particle in terms of its kinetic energy and linear momentum:

$$(m_{\Delta^+}c^2)^2 = E_{\Delta^+}^2 - p_{\Delta^+}^2 c^2 = E_{\Delta^+}^2 - (\mathbf{p}_{\Delta^+})^2 c^2$$

When the Δ^+ particle decays into a pion and a neutron, conservation of energy and momentum require that

$$E_{\Delta^+} = E_{\pi^+} + E_n \qquad \mathbf{p}_{\Delta^+} = \mathbf{p}_{\pi^+} + \mathbf{p}_n$$

Therefore, the rest energy of the Δ^+ particle can be expressed in terms of the energies and momenta of the outgoing particles, which can all be measured in the drift-chamber computer reconstruction:

$$(m_{\Delta^+}c^2)^2 = (E_{\pi^+} + E_n)^2 - (\mathbf{p}_{\pi^+} + \mathbf{p}_n)^2 c^2$$

Any pions and neutrons that come from the decay of a Δ^+ particle must have energies and momenta that combine in this equation to give the rest energy of the Δ^+ particle. Pions and neutrons coming from the reaction of Equation 46.6 will have a variety of energies and momenta with no particular pattern because the energy of the reactants can divide up in many ways among the three outgoing particles in this reaction. At the energy at which the rest energy of the Δ^+ particle can be created, many reactions occur, as evidenced by the proper combinations of energy and momentum already described.

To show the existence of the Δ^+ particle, we analyze a large number of events in which a π^+ and a neutron are produced. Then the number of events in a given energy range is plotted versus energy. Following this procedure, we obtain a slowly varying curve that has a sharp peak superimposed on it. The peak represents the incident electron energy at which the rest energy of the Δ^+ particle was created, revealing the existence of the particle.

Figure 46.8 is an experimental plot for the Δ^+ particle. The dashed broad curve was produced by direct events in which no Δ^+ was created (see Eq. 46.6). The sharp peak near 1 230 MeV was produced by all the events in which a Δ^+ was formed and decayed to a pion and a neutron. Therefore, the rest energy of the Δ^+ particle is near 1 230 MeV. Peaks corresponding to two resonance particles with masses greater than that of the Δ^+ particle can also be seen in Figure 46.8.

Graphs such as Figure 46.8 can tell us not only the mass of a short-lived particle but also its lifetime. The width of the resonance peak and the uncertainty relation $\Delta E \Delta t \approx \hbar/2$ are used to infer the lifetime Δt of the particle. The measured width of

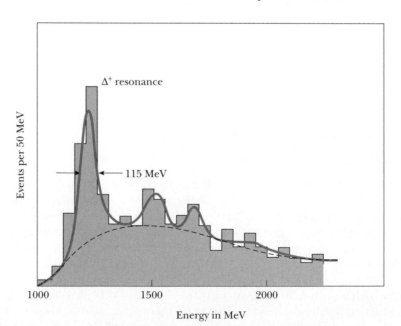

Figure 46.8 Experimental evidence for the existence of the Δ^+ particle. The sharp peak near 1 230 MeV was produced by events in which a Δ^+ formed and promptly decayed to a π^+ and a neutron.

115 MeV in Figure 46.8 leads to a lifetime of 6×10^{-24} s for the Δ^+ particle. In this incredibly short lifetime, a Δ^+ particle moving at $0.99c$ travels only 10^{-14} m, which is about ten nuclear diameters.

46.8 Finding Patterns in the Particles

One of the tools scientists use is the detection of patterns in data, patterns that contribute to our understanding of nature. One of the best examples of the use of this tool is the development of the periodic table, which provides a fundamental understanding of the chemical behavior of the elements. As mentioned in the introduction, the periodic table explains how more than 100 elements can be formed from three particles—the electron, the proton, and the neutron. The table of nuclides, part of which is shown in Table A.3, contains hundreds of nuclides, but all can be built from protons and neutrons.

The number of observed particles and resonances observed by particle physicists is also in the hundreds. Is it possible that a small number of entities exist from which all of these can be built? Taking a hint from the success of the periodic table and the table of nuclides, let us explore the historical search for patterns among the particles.

Many classification schemes have been proposed for grouping particles into families. Consider, for instance, the baryons listed in Table 46.2 that have spins of $\frac{1}{2}$: p, n, Λ^0, Σ^+, Σ^0, Σ^-, Ξ^0, and Ξ^-. If we plot strangeness versus charge for these baryons using a sloping coordinate system, as in Figure 46.9a, we observe a fascinating pattern: six of the baryons form a hexagon, and the remaining two are at the hexagon's center.

As a second example, consider the following nine spin-zero mesons listed in Table 46.2: π^+, π^0, π^-, K^+, K^0, K^-, η, η', and the antiparticle $\overline{K}^0$. Figure 46.9b is a plot of strangeness versus charge for this family. Again, a hexagonal pattern emerges. In this case, each particle on the perimeter of the hexagon lies opposite its antiparticle, and the remaining three (which form their own antiparticles) are at the center of the hexagon. These and related symmetric patterns were developed independently in 1961 by Murray Gell-Mann and Yuval Ne'eman (b. 1925). Gell-Mann called the patterns the **eightfold way,** after the eightfold path to nirvana in Buddhism.

Groups of baryons and mesons can be displayed in many other symmetric patterns within the framework of the eightfold way. For example, the family of spin-$\frac{3}{2}$ baryons known in 1961 contains nine particles arranged in a pattern like that of the pins in a bowling alley, as in Figure 46.10. (The particles Σ^{*+}, Σ^{*0}, Σ^{*-}, Ξ^{*0}, and Ξ^{*-} are excited states of the particles Σ^+, Σ^0, Σ^-, Ξ^0, and Ξ^-. In these higher-energy states, the spins

Murray Gell-Mann
American Physicist (b. 1929)

Murray Gell-Mann was awarded the Nobel Prize in 1969 for his theoretical studies dealing with subatomic particles. *(Courtesy of Michael R. Dressler)*

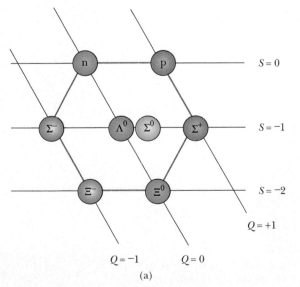

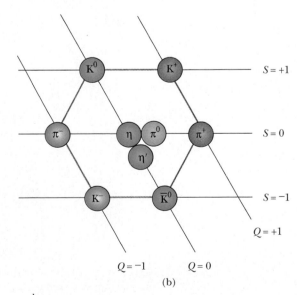

Figure 46.9 (a) The hexagonal eightfold-way pattern for the eight spin-$\frac{1}{2}$ baryons. This strangeness-versus-charge plot uses a sloping axis for charge number Q and a horizontal axis for strangeness S. (b) The eightfold-way pattern for the nine spin-zero mesons.

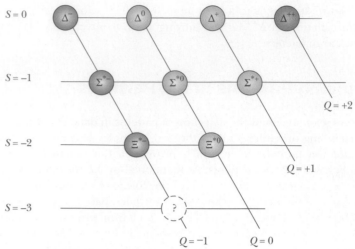

Figure 46.10 The pattern for the higher-mass, spin-$\frac{3}{2}$ baryons known at the time the pattern was proposed. The three Σ^* and two Ξ^* particles are excited states of the corresponding spin-$\frac{1}{2}$ particles in Figure 46.9. These excited states have higher mass and spin $\frac{3}{2}$. The absence of a particle in the bottom position was evidence of a new particle yet to be discovered, the Ω^-.

of the three quarks (see Section 46.9) making up the particle are aligned so that the total spin of the particle is $\frac{3}{2}$.) When this pattern was proposed, an empty spot occurred in it (at the bottom position), corresponding to a particle that had never been observed. Gell-Mann predicted that the missing particle, which he called the omega minus (Ω^-), should have spin $\frac{3}{2}$, charge -1, strangeness -3, and rest energy of approximately $1\,680$ MeV. Shortly thereafter, in 1964, scientists at the Brookhaven National Laboratory found the missing particle through careful analyses of bubble-chamber photographs (Fig. 46.11) and confirmed all its predicted properties.

The prediction of the missing particle in the eightfold way has much in common with the prediction of missing elements in the periodic table. Whenever a vacancy occurs in an organized pattern of information, experimentalists have a guide for their investigations.

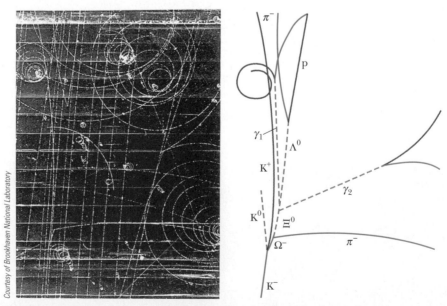

Courtesy of Brookhaven National Laboratory

Figure 46.11 Discovery of the Ω^- particle. The photograph on the left shows the original bubble-chamber tracks. The drawing on the right isolates the tracks of the important events. The K^- particle at the bottom collides with a proton to produce the first detected Ω^- particle plus a K^0 and a K^+.

46.9 Quarks

As we have noted, leptons appear to be truly elementary particles because there are only a few types of them, and experiments indicate that they have no measurable size or internal structure. Hadrons, on the other hand, are complex particles having size and structure. The existence of the strangeness–charge patterns of the eightfold way suggests that hadrons have substructure. Furthermore, we know that hundreds of types of hadrons exist and that many of them decay into other hadrons.

The Original Quark Model

In 1963 Gell-Mann and George Zweig (b. 1937) independently proposed a model for the substructure of hadrons. According to their model, all hadrons are composed of two or three elementary constituents called **quarks.** (Gell-Mann borrowed the word *quark* from the passage "Three quarks for Muster Mark" in James Joyce's *Finnegans Wake.* In Zweig's model, he called the constituents "aces.") The model has three types of quarks, designated by the symbols u, d, and s. These are given the arbitrary names **up, down,** and **strange.** The various types of quarks are called **flavors.** Figure 46.12 is a pictorial representation of the quark compositions of several hadrons.

An unusual property of quarks is that they carry a fractional electronic charge. The u, d, and s quarks have charges of $+2e/3$, $-e/3$, and $-e/3$, respectively, where e is the elementary charge 1.60×10^{-19} C. These and other properties of quarks and anti-quarks are given in Table 46.3. Notice that quarks have spin $\frac{1}{2}$, which means that all quarks are fermions, defined as any particle having half-integral spin, as pointed out in Section 43.8. As Table 46.3 shows, associated with each quark is an antiquark of opposite charge, baryon number, and strangeness.

The compositions of all hadrons known when Gell-Mann and Zweig presented their model can be completely specified by three simple rules:

- A meson consists of one quark and one antiquark, giving it a baryon number of 0, as required.
- A baryon consists of three quarks.
- An antibaryon consists of three antiquarks.

The theory put forth by Gell-Mann and Zweig is referred to as the *original quark model.*

> **Quick Quiz 46.7** Using a coordinate system like that in Figure 46.9, draw an eightfold-way diagram for the three quarks in the original quark model.

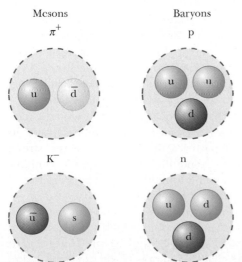

Mesons   Baryons

π^+ p

K⁻   n

Active Figure 46.12 Quark composition of two mesons and two baryons.

At the Active Figures link at **http://www.pse6.com,** *observe the quark composition for the mesons and baryons in Tables 46.4 and 46.5.*

Table 46.3

Properties of Quarks and Antiquarks								
Quarks								
Name	**Symbol**	**Spin**	**Charge**	**Baryon Number**	**Strangeness**	**Charm**	**Bottomness**	**Topness**
Up	u	$\frac{1}{2}$	$+\frac{2}{3}e$	$\frac{1}{3}$	0	0	0	0
Down	d	$\frac{1}{2}$	$-\frac{1}{3}e$	$\frac{1}{3}$	0	0	0	0
Strange	s	$\frac{1}{2}$	$-\frac{1}{3}e$	$\frac{1}{3}$	-1	0	0	0
Charmed	c	$\frac{1}{2}$	$+\frac{2}{3}e$	$\frac{1}{3}$	0	$+1$	0	0
Bottom	b	$\frac{1}{2}$	$-\frac{1}{3}e$	$\frac{1}{3}$	0	0	$+1$	0
Top	t	$\frac{1}{2}$	$+\frac{2}{3}e$	$\frac{1}{3}$	0	0	0	$+1$
Antiquarks								
Name	**Symbol**	**Spin**	**Charge**	**Baryon Number**	**Strangeness**	**Charm**	**Bottomness**	**Topness**
Anti-up	$\overline{u}$	$\frac{1}{2}$	$-\frac{2}{3}e$	$-\frac{1}{3}$	0	0	0	0
Anti-down	$\overline{d}$	$\frac{1}{2}$	$+\frac{1}{3}e$	$-\frac{1}{3}$	0	0	0	0
Anti-strange	$\overline{s}$	$\frac{1}{2}$	$+\frac{1}{3}e$	$-\frac{1}{3}$	$+1$	0	0	0
Anti-charmed	$\overline{c}$	$\frac{1}{2}$	$-\frac{2}{3}e$	$-\frac{1}{3}$	0	-1	0	0
Anti-bottom	$\overline{b}$	$\frac{1}{2}$	$+\frac{1}{3}e$	$-\frac{1}{3}$	0	0	-1	0
Anti-top	$\overline{t}$	$\frac{1}{2}$	$-\frac{2}{3}e$	$-\frac{1}{3}$	0	0	0	-1

Charm and Other Developments

Although the original quark model was highly successful in classifying particles into families, some discrepancies occurred between its predictions and certain experimental decay rates. Consequently, several physicists proposed a fourth quark flavor in 1967. They argued that if four types of leptons exist (as was thought at the time), then there should also be four flavors of quarks because of an underlying symmetry in nature. The fourth quark, designated c, was assigned a property called **charm.** A *charmed* quark has charge $+2e/3$, just as the up quark does, but its charm distinguishes it from the other three quarks. This introduces a new quantum number C, representing charm. The new quark has charm $C = +1$, its antiquark has charm of $C = -1$, and all other quarks have $C = 0$. Charm, like strangeness, is conserved in strong and electromagnetic interactions but not in weak interactions.

Evidence that the charmed quark exists began to accumulate in 1974, when a heavy meson called the J/Ψ particle (or simply Ψ, uppercase Greek psi) was discovered independently by two groups, one led by Burton Richter (b. 1931) at the Stanford Linear Accelerator (SLAC), and the other led by Samuel Ting (b. 1936) at the Brookhaven National Laboratory. In 1976 Richter and Ting were awarded a Nobel Prize in physics for this work. The J/Ψ particle does not fit into the three-quark model; instead, it has properties of a combination of the proposed charmed quark and its antiquark ($c\overline{c}$). It is much more massive than the other known mesons ($\sim 3\,100$ MeV/c^2), and its lifetime is much longer than the lifetimes of particles that interact via the strong force. Soon, related mesons were discovered, corresponding to such quark combinations as $\overline{c}d$ and $c\overline{d}$, all of which have great masses and long lifetimes. The existence of these new mesons provided firm evidence for the fourth quark flavor.

In 1975, researchers at Stanford University reported strong evidence for the tau (τ) lepton, mass $1\,784$ MeV/c^2. This was the fifth type of lepton, which led physicists to propose that more flavors of quarks might exist, on the basis of symmetry arguments similar to those leading to the proposal of the charmed quark. These proposals led to more elaborate quark models and the prediction of two new quarks, **top** (t) and

Table 46.4

Quark[a] Composition of Mesons									
		Antiquarks							
		$\bar{b}$		$\bar{c}$		$\bar{s}$		$\bar{d}$	$\bar{u}$

Quarks		$\bar{b}$		$\bar{c}$		$\bar{s}$		$\bar{d}$		$\bar{u}$	
	b	Υ	$(b\bar{b})$	B_c^-	$(\bar{c}b)$	$\bar{B}_s{}^0$	$(\bar{s}b)$	$\bar{B}_d{}^0$	$(\bar{d}b)$	B^-	$(\bar{u}b)$
	c	B_c^+	$(\bar{b}c)$	J/Ψ	$(\bar{c}c)$	D_s^+	$(\bar{s}c)$	D^+	$(\bar{d}c)$	D^0	$(\bar{u}c)$
	s	$B_s{}^0$	$(\bar{b}s)$	D_s^-	$(\bar{c}s)$	η, η'	$(\bar{s}s)$	$\bar{K}^0$	$(\bar{d}s)$	K^-	$(\bar{u}s)$
	d	$B_d{}^0$	$(\bar{b}d)$	D^-	$(\bar{c}d)$	K^0	$(\bar{s}d)$	π^0, η, η'	$(\bar{d}d)$	π^-	$(\bar{u}d)$
	u	B^+	$(\bar{b}u)$	$\bar{D}^0$	$(\bar{c}u)$	K^+	$(\bar{s}u)$	π^+	$(\bar{d}u)$	π^0, η, η'	$(\bar{u}u)$

[a] The top quark does not form mesons because it decays too quickly.

bottom (b). (Some physicists prefer *truth* and *beauty*.) To distinguish these quarks from the others, quantum numbers called *topness* and *bottomness* (with allowed values $+1$, 0, -1) were assigned to all quarks and antiquarks (see Table 46.3). In 1977, researchers at the Fermi National Laboratory, under the direction of Leon Lederman (b. 1922), reported the discovery of a very massive new meson Υ^- (Greek capital upsilon), whose composition is considered to be $b\bar{b}$, providing evidence for the bottom quark. In March 1995, researchers at Fermilab announced the discovery of the top quark (supposedly the last of the quarks to be found), which has a mass of 173 GeV/c^2.

Table 46.4 lists the quark compositions of mesons formed from the up, down, strange, charmed, and bottom quarks. Table 46.5 shows the quark combinations for the baryons listed in Table 46.2. Note that only two flavors of quarks, u and d, are contained in all hadrons encountered in ordinary matter (protons and neutrons).

You are probably wondering whether the discoveries of elementary particles will ever end. How many "building blocks" of matter really exist? At the present, physicists believe that the elementary particles in nature are six quarks and six leptons, together with their antiparticles, and the four field particles listed in Table 46.1. Table 46.6 lists the rest energies and charges of the quarks and leptons.

Table 46.5

Quark Composition of Several Baryons[a]	
Particle	**Quark Composition**
p	uud
n	udd
Λ^0	uds
Σ^+	uus
Σ^0	uds
Σ^-	dds
Δ^{++}	uuu
Δ^+	uud
Δ^0	udd
Δ^-	ddd
Ξ^0	uss
Ξ^-	dss
Ω^-	sss

[a] Some baryons have the same quark composition, such as the p and the Δ^+ and the n and the Δ^0. In these cases, the Δ particles are considered to be excited states of the proton and neutron.

Table 46.6

The Elementary Particles and Their Rest Energies and Charges		
Particle	**Rest Energy**	**Charge**
Quarks		
u	360 MeV	$+\frac{2}{3}e$
d	360 MeV	$-\frac{1}{3}e$
s	540 MeV	$-\frac{1}{3}e$
c	1 500 MeV	$+\frac{2}{3}e$
b	5 GeV	$-\frac{1}{3}e$
t	173 GeV	$+\frac{2}{3}e$
Leptons		
e^-	511 keV	$-e$
μ^-	105.7 MeV	$-e$
τ^-	1 784 MeV	$-e$
ν_e	<7 eV	0
ν_μ	<0.3 MeV	0
ν_τ	<30 MeV	0

Despite extensive experimental effort, no isolated quark has ever been observed. Physicists now believe that at ordinary temperatures quarks are permanently confined inside ordinary particles because of an exceptionally strong force that prevents them from escaping, called (appropriately) the **strong force**[5] (which we introduced at the beginning of Section 46.4 and we discuss further in Section 46.10). This force increases with separation distance, similar to the force exerted by a stretched spring. Current efforts are underway to form a **quark–gluon plasma,** a state of matter in which the quarks are freed from neutrons and protons. In 2000, scientists at CERN announced evidence for a quark–gluon plasma formed by colliding lead nuclei. Experiments continue at CERN as well as at the Relativistic Heavy Ion Collider (RHIC) at Brookhaven to verify the production of a quark–gluon plasma.

Quick Quiz 46.8 Doubly charged baryons, such as the Δ^{++}, are known to exist. True or false: doubly charged mesons also exist.

Quick Quiz 46.9 Imagine hypothetical super-heavy baryons with the following quark combinations: (i) ccc (ii) bbb (iii) ttt. Rank these baryons according to (a) mass, from smallest to largest (b) spin (c) electric charge and (d) baryon number.

46.10   Multicolored Quarks

Shortly after the concept of quarks was proposed, scientists recognized that certain particles had quark compositions that violated the exclusion principle. In Section 42.7, we applied the exclusion principle to electrons in atoms. The principle is more general, however, and applies to all particles with half-integral spin ($\frac{1}{2}$, $\frac{3}{2}$, etc.), which we collectively call fermions. Because all quarks are fermions having spin $\frac{1}{2}$, they are expected to follow the exclusion principle. One example of a particle that appears to violate the exclusion principle is the Ω^- (sss) baryon, which contains three strange quarks having parallel spins, giving it a total spin of $\frac{3}{2}$. All three quarks have the same spin quantum number, in violation of the exclusion principle. Other examples of baryons made up of identical quarks having parallel spins are the Δ^{++} (uuu) and the Δ^- (ddd).

To resolve this problem, it was suggested that quarks possess an additional property called **color charge.** This property is similar in many respects to electric charge except that it occurs in six varieties rather than two. The colors assigned to quarks are red, green, and blue, and antiquarks have the colors antired, antigreen, and antiblue. Therefore, the colors red, green, and blue serve as the "quantum numbers" for the color of the quark. To satisfy the exclusion principle, the three quarks in any baryon must all have different colors. The three colors "neutralize" to white. A quark and an antiquark in a meson must be of a color and the corresponding anticolor and will consequently neutralize to white, similar to the way electric charges + and − neutralize to zero net charge. The result is that baryons and mesons are always colorless (or white). Therefore, the apparent violation of the exclusion principle in the Ω^- baryon is removed because the three quarks in the particle have different colors.

Note that the new property of color increases the number of quarks by a factor of three, since each of the six quarks comes in three colors. Although the concept of

▲ **PITFALL PREVENTION**

46.3   Color Charge Is Not Really Color

The description of color for a quark has nothing to do with visual sensation from light. It is simply a convenient name for a property that is analogous to electric charge.

[5]   As a reminder, the original meaning of the term *strong force* was the short-range attractive force between nucleons, which we have called the *nuclear force.* The nuclear force between nucleons is a secondary effect of the strong force between quarks.

color in the quark model was originally conceived to satisfy the exclusion principle, it also provided a better theory for explaining certain experimental results. For example, the modified theory correctly predicts the lifetime of the π^0 meson.

The theory of how quarks interact with each other is called **quantum chromo-dynamics,** or QCD, to parallel the name *quantum electrodynamics* (the theory of the electrical interaction between light and matter). In QCD, each quark is said to carry a color charge, in analogy to electric charge. The strong force between quarks is often called the **color force.** Therefore, the terms *strong force* and *color force* are used interchangeably.

In Section 46.1 we stated that the nuclear interaction between hadrons is mediated by massless field particles called **gluons.** As we have mentioned, the nuclear force is actually a secondary effect of the strong force between quarks. The gluons are the mediators of the strong force. When a quark emits or absorbs a gluon, the quark's color may change. For example, a blue quark that emits a gluon may become a red quark, and a red quark that absorbs this gluon becomes a blue quark.

The color force between quarks is analogous to the electric force between charges: particles with the same color repel, and those with opposite colors attract. Therefore, two green quarks repel each other, but a green quark is attracted to an antigreen quark. The attraction between quarks of opposite color to form a meson ($q\bar{q}$) is indicated in Figure 46.13a. Differently colored quarks also attract one another, although with less intensity than the oppositely colored quark and antiquark. For example, a cluster of red, blue, and green quarks all attract one another to form a baryon, as in Figure 46.13b. Therefore, every baryon contains three quarks of three different colors.

Although the nuclear force between two colorless hadrons is negligible at large separations, the net strong force between their constituent quarks is not exactly zero at small separations. This residual strong force is the nuclear force that binds protons and neutrons to form nuclei. It is similar to the force between two electric dipoles. Each dipole is electrically neutral. An electric field surrounds the dipoles, however, because of the separation of the positive and negative charges (see Section 23.6). As a result, an electric interaction occurs between the dipoles that is weaker than the force between single charges. In Section 43.1, we explored how this interaction results in the Van der Waals force between neutral molecules.

According to QCD, a more basic explanation of the nuclear force can be given in terms of quarks and gluons. Figure 46.14a shows the nuclear interaction between a neutron and a proton by means of Yukawa's pion, in this case a π^-. This drawing differs from Figure 46.5a, in which the field particle is a π^0—there is no transfer of

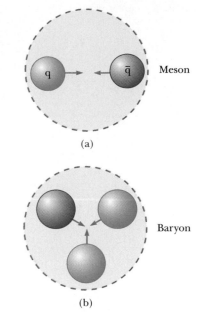

Figure 46.13 (a) A green quark is attracted to an antigreen quark. This forms a meson whose quark structure is ($q\bar{q}$). (b) Three quarks of different colors attract each other to form a baryon.

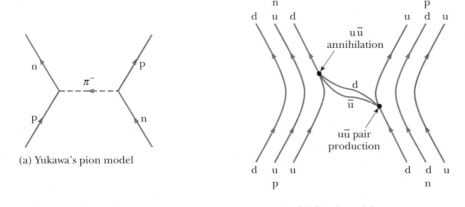

(a) Yukawa's pion model

(b) Quark model

Figure 46.14 (a) A nuclear interaction between a proton and a neutron explained in terms of Yukawa's pion-exchange model. Because the pion carries charge, the proton and neutron switch identities. (b) The same interaction, explained in terms of quarks and gluons. Note that the exchanged $\bar{u}d$ quark pair makes up a π^- meson.

charge from one nucleon to the other in Figure 46.5a. In Figure 46.14a, the charged pion carries charge from one nucleon to the other, so the nucleons change identities—the proton becomes a neutron and the neutron becomes a proton.

Let us look at the same interaction from the viewpoint of the quark model, shown in Figure 46.14b. In this Feynman diagram, the proton and neutron are represented by their quark constituents. Each quark in the neutron and proton is continuously emitting and absorbing gluons. The energy of a gluon can result in the creation of quark–antiquark pairs. This is similar to the creation of electron–positron pairs in pair production, which we investigated in Section 46.2. When the neutron and proton approach to within 1 fm of each other, these gluons and quarks can be exchanged between the two nucleons, and such exchanges produce the nuclear force. Figure 46.14b depicts one possibility for the process shown in Figure 46.14a. A down quark in the neutron on the right emits a gluon. The energy of the gluon is then transformed to create a $u\bar{u}$ pair. The u quark stays within the nucleon (which has now changed to a proton), and the recoiling d quark and the $\bar{u}$ antiquark are transmitted to the proton on the left side of the diagram. Here the $\bar{u}$ annihilates a u quark within the proton (with the creation of a gluon), and the d is captured. The net effect is to change a u quark to a d quark, and the proton on the left has changed to a neutron.

As the d quark and $\bar{u}$ antiquark in Figure 46.14b transfer between the nucleons, the d and $\bar{u}$ exchange gluons with each other and can be considered to be bound to each other by means of the strong force. If we look back at Table 46.4, we see that this combination is a π^-—Yukawa's field particle! Therefore, the quark model of interactions between nucleons is consistent with the pion-exchange model.

46.11 The Standard Model

Scientists now believe that there are three classifications of truly elementary particles: leptons, quarks, and field particles. These three particles are further classified as either fermions or bosons. Note that quarks and leptons have spin $\frac{1}{2}$ and hence are fermions, while the field particles have integral spin of 1 or higher and are bosons.

Recall from Section 46.1 that the weak force is believed to be mediated by the W^+, W^-, and Z^0 bosons. These particles are said to have *weak charge,* just as quarks have color charge. Therefore, each elementary particle can have mass, electric charge, color charge, and weak charge. Of course, one or more of these could be zero.

In 1979, Sheldon Glashow (b. 1932), Abdus Salam (1926–1996), and Steven Weinberg (b. 1933) won a Nobel Prize in physics for developing a theory that unifies the electromagnetic and weak interactions. This **electroweak theory** postulates that the weak and electromagnetic interactions have the same strength when the particles involved have very high energies. The two interactions are viewed as different manifestations of a single unifying electroweak interaction. The theory makes many concrete predictions, but perhaps the most spectacular is the prediction of the masses of the W and Z particles at about 82 GeV/c^2 and 93 GeV/c^2, respectively. These predictions are close to the masses in Table 46.1 determined by experiment.

The combination of the electroweak theory and QCD for the strong interaction is referred to in high-energy physics as the **Standard Model.** Although the details of the Standard Model are complex, its essential ingredients can be summarized with the help of Figure 46.15. (The Standard Model does not include the gravitational force at present; however, we include gravity in Figure 46.15 because physicists hope to eventually incorporate this force into a unified theory.) This diagram shows that quarks participate in all of the fundamental forces and that leptons participate in all except the strong force.

The Standard Model does not answer all questions. A major question that is still unanswered is why, of the two mediators of the electroweak interaction, the photon has no mass but the W and Z bosons do. Because of this mass difference, the elec-

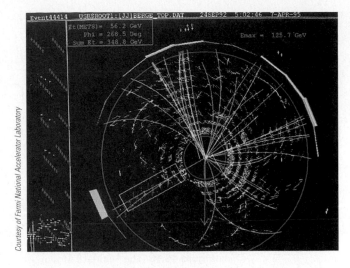

Courtesy of Fermi National Accelerator Laboratory

Figure 46.17 Computers at Fermilab create a pictorial representation such as this of the paths of particles after a collision.

46.12 The Cosmic Connection

In this section we describe one of the most fascinating theories in all of science—the Big Bang theory of the creation of the Universe—and the experimental evidence that supports it. This theory of cosmology states that the Universe had a beginning and, furthermore, that the beginning was so cataclysmic that it is impossible to look back beyond it. According to this theory, the Universe erupted from an infinitely dense singularity about 10 billion years ago.[6] The first few minutes after the Big Bang saw such extremely high energy that it is believed that all four interactions of physics were unified and all matter was contained in a quark–gluon plasma.

The evolution of the four fundamental forces from the Big Bang to the present is shown in Figure 46.18. During the first 10^{-43} s (the ultrahot epoch, $T \sim 10^{32}$ K), it is

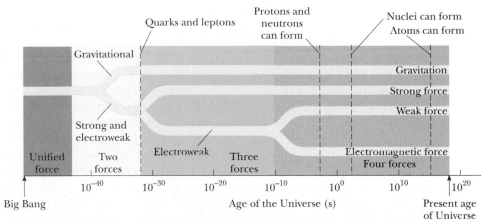

Figure 46.18 A brief history of the Universe from the Big Bang to the present. The four forces became distinguishable during the first nanosecond. Following this, all the quarks combined to form particles that interact via the nuclear force. However, the leptons remained separate and to this day exist as individual, observable particles.

---

[6]   An updated estimate for the age of the Universe of between 11.2 billion and 20 billion years was published shortly before this book went to press: L. M. Krauss and B. Chaboyer, "Age Estimates of Globular Clusters in the Milky Way: Constraints on Cosmology," *Science,* January 3, 2003; 299: pp. 65–69. Data released from the Wilkinson Microwave Anisotropy Probe in February 2003 pinpoints the age as 13.7 ± 0.2 billion years ("WMAP Spacecraft Maps the Entire Cosmic Microwave Sky with Unprecedented Precision", *Physics Today*, April 2003; 56(4), pp. 21–24.)

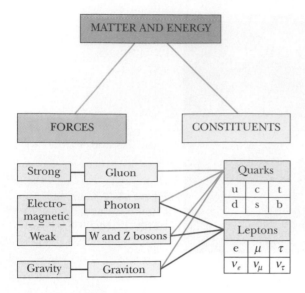

Figure 46.15 The Standard Model of particle physics.

tromagnetic and weak forces are quite distinct at low energies but become similar at very high energies—when the rest energy is negligible relative to the total energy. The behavior as one goes from high to low energies is called *symmetry breaking* because the forces are similar, or symmetric, at high energies but are very different at low energies. The nonzero rest energies of the W and Z bosons raise the question of the origin of particle masses. To resolve this problem, a hypothetical particle called the **Higgs boson,** which provides a mechanism for breaking the electroweak symmetry, has been proposed. The Standard Model modified to include the Higgs mechanism provides a logically consistent explanation of the massive nature of the W and Z bosons. Unfortunately, the Higgs boson has not yet been found, but physicists know that its rest energy should be less than 1 TeV. In order to determine whether the Higgs boson exists, two quarks each having at least 1 TeV of energy must collide. However, calculations show that such a collision requires injecting 40 TeV of energy within the volume of a proton.

Because of the limited energy available in conventional accelerators using fixed targets, it is necessary to employ colliding-beam accelerators called **colliders.** The concept of colliders is straightforward. Particles that have equal masses and equal kinetic energies, traveling in opposite directions in an accelerator ring, collide head-on to produce the required reaction and form new particles. Because the total momentum of the interacting particles is zero, all of their kinetic energy is available for the reaction. The Large Electron–Positron (LEP) Collider at CERN (Fig. 46.16) and the Stanford Linear Collider collide both electrons and positrons. The Super Proton Synchrotron at CERN accelerates protons and antiprotons to energies of 270 GeV. The world's highest-energy proton accelerator, the Tevatron at the Fermi National Laboratory in Illinois, produces protons at almost 1 000 GeV (1 TeV). The Relativistic Heavy Ion Collider (RHIC) at Brookhaven National Laboratory will collide heavy ions to search for the quark–gluon plasma, as discussed earlier. CERN expects a 2007 completion date for the Large Hadron Collider (LHC), a proton–proton collider that will provide a center-of-mass energy of 14 TeV and enable exploration of Higgs-boson physics. The accelerator will be constructed in the same 27-km circumference tunnel now housing the LEP Collider, and many countries will participate in the project.

In addition to increasing energies in modern accelerators, detection techniques have become increasingly sophisticated. We saw simple bubble-chamber photographs earlier in this chapter that required hours of analysis by hand. Figure 46.17 shows a modern detection display of particle tracks after a reaction; the tracks are analyzed rapidly by computer. The chapter opening photograph shows a complex set of tracks from a collision of gold nuclei.

Courtesy of CERN

Figure 46.16 A view from inside the Large Electron–Positron (LEP) Collider tunnel, which is 27 km in circumference.

presumed that the strong, electroweak, and gravitational forces were joined to form a completely unified force. In the first 10^{-35} s following the Big Bang (the hot epoch, $T \sim 10^{29}$ K), symmetry breaking occurred for gravity while the strong and electroweak forces remained unified. This was a period when particle energies were so great ($> 10^{16}$ GeV) that very massive particles as well as quarks, leptons, and their antiparticles existed. Then, after 10^{-35} s, the Universe rapidly expanded and cooled (the warm epoch, $T \sim 10^{29}$ to 10^{15} K) and the strong and electroweak forces parted company. As the Universe continued to cool, the electroweak force split into the weak force and the electromagnetic force about 10^{-10} s after the Big Bang.

After a few minutes, protons and neutrons condensed out of the plasma. For half an hour the Universe underwent thermonuclear detonation, exploding as a hydrogen bomb and producing most of the helium nuclei that now exist. The Universe continued to expand, and its temperature dropped. Until about 700 000 years after the Big Bang, the Universe was dominated by radiation. Energetic radiation prevented matter from forming single hydrogen atoms because collisions would instantly ionize any atoms that happened to form. Photons experienced continuous Compton scattering from the vast numbers of free electrons, resulting in a Universe that was opaque to radiation. By the time the Universe was about 700 000 years old, it had expanded and cooled to about 3 000 K, and protons could bind to electrons to form neutral hydrogen atoms. Because of the quantized energies of the atoms, far more wavelengths of radiation were not absorbed by atoms than were absorbed, and the Universe suddenly became transparent to photons. Radiation no longer dominated the Universe, and clumps of neutral matter steadily grew—first atoms, then molecules, gas clouds, stars, and finally galaxies.

Observation of Radiation from the Primordial Fireball

In 1965, Arno A. Penzias (b. 1933) and Robert W. Wilson (b. 1936) of Bell Laboratories were testing a sensitive microwave receiver and made an amazing discovery. A pesky signal producing a faint background hiss was interfering with their satellite communications experiments. In spite of their valiant efforts, the signal remained. Ultimately, it became clear that they were detecting microwave background radiation (at a wavelength of 7.35 cm), which represented the leftover "glow" from the Big Bang.

The microwave horn that served as their receiving antenna is shown in Figure 46.19. The intensity of the detected signal remained unchanged as the antenna was

Figure 46.19 Robert W. Wilson *(left)* and Arno A. Penzias with the Bell Telephone Laboratories horn-reflector antenna.

pointed in different directions. The fact that the radiation had equal strengths in all directions suggested that the entire Universe was the source of this radiation. Evicting a flock of pigeons from the 20-ft horn and cooling the microwave detector both failed to remove the signal. Through a casual conversation, Penzias and Wilson discovered that a group at Princeton had predicted the residual radiation from the Big Bang and were planning an experiment to attempt to confirm the theory. The excitement in the scientific community was high when Penzias and Wilson announced that they had already observed an excess microwave background compatible with a 3-K blackbody source, which was consistent with the predicted temperature of the Universe at this time after the Big Bang.

Because Penzias and Wilson made their measurements at a single wavelength, they did not completely confirm the radiation as 3-K blackbody radiation. Subsequent experiments by other groups added intensity data at different wavelengths, as shown in Figure 46.20. The results confirm that the radiation is that of a black body at 2.7 K. This figure is perhaps the most clear-cut evidence for the Big Bang theory. The 1978 Nobel Prize in physics was awarded to Penzias and Wilson for this most important discovery.

The discovery of the cosmic background radiation brought with it a problem, however—the radiation was too uniform. Scientists believed that slight fluctuations in this background had to occur to act as nucleation sites for the formation of the galaxies and other objects we now see in the sky. In 1989, NASA launched a satellite called COBE (KOH-bee), for Cosmic Background Explorer, to study this radiation in greater detail. In 1992, George Smoot (b. 1945) at the Lawrence Berkeley Laboratory found, on the basis of the data collected, that the background was not perfectly uniform but instead contained irregularities that corresponded to temperature variations of 0.000 3 K. In 2000, the BOOMERANG experiment was carried out by an international team of scientists from Italy, the U.K., the U.S., Canada, and France. This team used instruments flying in a balloon over Antarctica to generate data with over 25 times the angular resolution of the COBE data. These data clearly show a peak in fluctuation intensity 300 000 years after the Big Bang. The Wilkinson Microwave Anisotropy Probe, launched in June 2001, exhibited data in early 2003 that allowed observation of temperature differences in the cosmos in the microkelvin range.

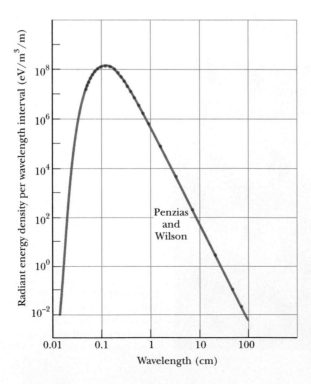

Figure 46.20 Theoretical blackbody (brown curve) and measured radiation spectra (blue points) of the Big Bang. Most of the data were collected from the Cosmic Background Explorer (COBE) satellite. The datum of Penzias and Wilson is indicated.

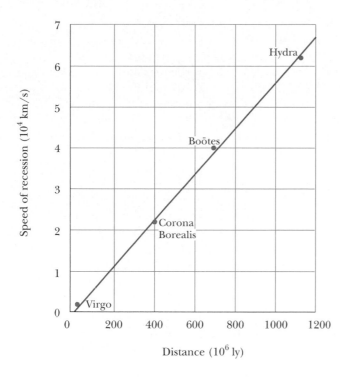

Figure 46.21 Hubble's law: a plot of speed of recession versus distance for four galaxies.

Other Evidence for an Expanding Universe

The Big Bang theory of cosmology predicts that the Universe is expanding. Most of the key discoveries supporting the theory of an expanding Universe were made in the twentieth century. Vesto Melvin Slipher (1875–1969), an American astronomer, reported in 1912 that most nebulae are receding from the Earth at speeds up to several million miles per hour. Slipher was one of the first scientists to use Doppler shifts (see Section 17.4) in spectral lines to measure velocities.

In the late 1920s, Edwin P. Hubble (1889–1953) made the bold assertion that the whole Universe is expanding. From 1928 to 1936, until they reached the limits of the 100-inch telescope, Hubble and Milton Humason (1891–1972) worked at Mount Wilson in California to prove this assertion. The results of that work and of its continuation with the use of a 200-inch telescope in the 1940s showed that the speeds at which galaxies are receding from the Earth increase in direct proportion to their distance R from us (Fig. 46.21). This linear relationship, known as **Hubble's law,** may be written

$$v = HR \qquad\qquad (46.7) \qquad \textbf{Hubble's law}$$

where H, called the **Hubble constant,** has the approximate value

$$H \approx 17 \times 10^{-3}\ \text{m/s} \cdot \text{ly}$$

Example 46.5 Recession of a Quasar

A quasar is an object that appears similar to a star and is very distant from the Earth. Its speed can be determined from Doppler-shift measurements in the light it emits. A certain quasar recedes from the Earth at a speed of $0.55c$. How far away is it?

Solution We can find the distance through Hubble's law:

$$R = \frac{v}{H} = \frac{(0.55)(3.00 \times 10^8\ \text{m/s})}{17 \times 10^{-3}\ \text{m/s} \cdot \text{ly}}$$

$$= \;\; 9.7 \times 10^9\ \text{ly}$$

What If? Suppose we assume that the quasar has moved at this speed ever since the Big Bang. With this assumption, estimate the age of the Universe.

Answer We approximate the distance from Earth to the quasar as the distance that the quasar has moved from the singularity since the Big Bang. We can then find the time interval from a calculation as performed in Chapter 2: $\Delta t = \Delta x / v = R/v = 1/H \approx 18$ billion years, which is in approximate agreement with other calculations.

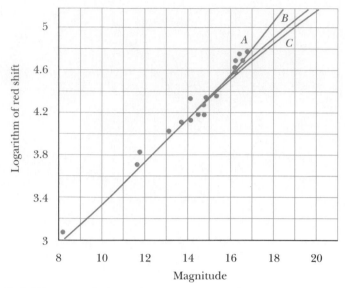

Figure 46.22 Red shift, or speed of recession, versus magnitude (which is related to brightness) of 18 faint galaxy clusters. Significant scatter of the data occurs, so the extrapolation of the curve to the upper right is uncertain. Curve A is the trend suggested by the six faintest clusters. Curve C corresponds to a Universe having a constant rate of expansion. If more data are taken and the complete set of data indicates a curve that falls between B and C, the expansion will slow but never stop. If the data fall to the left of B, expansion will eventually stop and the Universe will begin to contract.

Will the Universe Expand Forever?

In the 1950s and 1960s, Allan R. Sandage (b. 1926) used the 200-inch telescope at Mount Palomar to measure the speeds of galaxies at distances of up to 6 billion lightyears away from the Earth. These measurements showed that these very distant galaxies were moving about 10 000 km/s faster than Hubble's law predicted. According to this result, the Universe must have been expanding more rapidly 1 billion years ago, and consequently we conclude from these data that the expansion rate is slowing[7] (Fig. 46.22). Today, astronomers and physicists are trying to determine the rate of expansion. If the average mass density of the Universe is less than some critical value, the galaxies will slow in their outward rush but still escape to infinity. If the average density exceeds the critical value, the expansion will eventually stop and contraction will begin, possibly leading to a superdense state followed by another expansion. In this scenario, we have an oscillating Universe.

Example 46.6   The Critical Density of the Universe

(A) Starting from energy conservation, derive an expression for the critical mass density of the Universe ρ_c in terms of the Hubble constant H and the universal gravitational constant G.

Solution Figure 46.23 shows a large section of the Universe, contained within a sphere of radius R. The total mass of the galaxies in this volume is M. A galaxy of mass $m \ll M$ that has a speed v at a distance R from the center of the sphere will escape to infinity (at which its speed will approach zero) if the sum of its kinetic energy and the gravitational potential energy

of the system—the galaxy plus the rest of the Universe—is zero at any time. The Universe may be infinite in spatial extent, but Gauss's law for gravitation (Problem 73 in Chapter 24) implies that only the mass M inside the sphere contributes to the gravitational potential energy of the galaxy:

$$E_{\text{total}} = 0 = K + U = \tfrac{1}{2}mv^2 - \frac{GmM}{R}$$

We substitute for the mass M contained within the sphere the product of the critical density and the volume of the

[7]   The data at large distances have large observational uncertainties and may be systematically in error from effects such as abnormal brightness in the most distant visible clusters.

sphere:

$$\frac{1}{2}mv^2 = \frac{Gm(\frac{4}{3}\pi R^3 \rho_c)}{R}$$

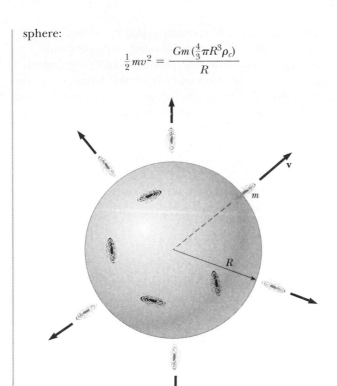

Figure 46.23 (Example 46.6) The galaxy marked with mass *m* is escaping from a large cluster of galaxies contained within a spherical volume of radius *R*. Only the mass within *R* slows the galaxy.

Solving for the critical density gives

$$\rho_c = \frac{3v^2}{8\pi GR^2}$$

From Hubble's law, the ratio of *v* to *R* is $v/R = H$, so this expression becomes

$$\rho_c = \frac{3H^2}{8\pi G}$$

(B) Estimate a numerical value for the critical density in grams per cubic centimeter.

Solution Using $H = 17 \times 10^{-3}$ m/s·ly where 1 ly = 9.46 $\times 10^{15}$ m, we find for the critical density

$$\rho_c = \frac{3H^2}{8\pi G} = \frac{3(17 \times 10^{-3}\,\text{m/s·ly})^2}{8\pi(6.67 \times 10^{-11}\,\text{N·m}^2/\text{kg}^2)}$$

$$\times \left(\frac{1\,\text{ly}}{9.46 \times 10^{15}\,\text{m}}\right)^2$$

$$= 6 \times 10^{-27}\,\text{kg/m}^3 = \boxed{6 \times 10^{-30}\,\text{g/cm}^3}$$

Because the mass of a hydrogen atom is 1.67×10^{-24} g, this value of ρ_c corresponds to 3×10^{-6} hydrogen atoms per cubic centimeter or 3 atoms per cubic meter.

Missing Mass in the Universe?

The luminous matter in galaxies averages out to a Universe density of 5×10^{-33} g/cm³. The radiation in the Universe has a mass equivalent of approximately 2% of the luminous matter. The total mass of all nonluminous matter (such as interstellar gas and black holes) may be estimated from the speeds of galaxies orbiting each other in a cluster. The higher the galaxy speeds, the more mass in the cluster. Measurements on the Coma cluster of galaxies indicate, surprisingly, that the amount of nonluminous matter is 20 to 30 times the amount of luminous matter present in stars and luminous gas clouds. Yet even this large invisible component of *dark matter,* if extrapolated to the Universe as a whole, leaves the observed mass density a factor of 10 less than ρ_c. The deficit, called *missing mass,* has been the subject of intense theoretical and experimental work, with exotic particles such as axions, photinos, and superstring particles suggested as candidates for the missing mass. Some researchers have made the more mundane proposal that the missing mass is present in neutrinos. In fact, neutrinos are so abundant that a tiny neutrino rest energy on the order of only 20 eV would furnish the missing mass and "close" the Universe. Current experiments designed to measure the rest energy of the neutrino will have an impact on predictions for the future of the Universe.

Mysterious Energy in the Universe?

A surprising twist in the story of the Universe arose in 1998 with the observation of a class of supernovae that have a fixed absolute brightness. By combining the apparent brightness and the redshift of light from these explosions, their distance and speed of recession of the Earth can be determined. These observations led to the conclusion that the expansion of the Universe is not slowing down, but it is accelerating! Observations by other groups also led to the same interpretation.

In order to explain this acceleration, physicists have proposed *dark energy*, which is energy possessed by the vacuum of space. In the early life of the Universe, gravity dominated over the dark energy. As the Universe expanded and the gravitational force between galaxies became smaller because of the great distances between them, the dark energy became more important. The dark energy results in an effective repulsive force that causes the expansion rate to increase.[8]

Although we have some degree of certainty about the beginning of the Universe, we are uncertain about how the story will end. Will the Universe keep on expanding forever, or will it someday collapse and then expand again, perhaps in an endless series of oscillations? Results and answers to these questions remain inconclusive, and the exciting controversy continues.

46.13 Problems and Perspectives

While particle physicists have been exploring the realm of the very small, cosmologists have been exploring cosmic history back to the first microsecond of the Big Bang. Observation of the events that occur when two particles collide in an accelerator is essential for reconstructing the early moments in cosmic history. For this reason, perhaps the key to understanding the early Universe is to first understand the world of elementary particles. Cosmologists and physicists now find that they have many common goals and are joining hands in an attempt to understand the physical world at its most fundamental level.

Our understanding of physics at short distances is far from complete. Particle physics is faced with many questions. Why does so little antimatter exist in the Universe? Is it possible to unify the strong and electroweak theories in a logical and consistent manner? Why do quarks and leptons form three similar but distinct families? Are muons the same as electrons apart from their difference in mass, or do they have other subtle differences that have not been detected? Why are some particles charged and others neutral? Why do quarks carry a fractional charge? What determines the masses of the elementary constituents of matter? Can isolated quarks exist?

An important and obvious question that remains is whether leptons and quarks have an underlying structure. If they do, we can envision an infinite number of deeper structure levels. However, if leptons and quarks are indeed the ultimate constituents of matter, as physicists today tend to believe, we should be able to construct a final theory of the structure of matter, just as Einstein dreamed of doing. This theory, whimsically called the Theory of Everything, is a combination of the Standard Model and a quantum theory of gravity.

String Theory—a New Perspective

Let us briefly discuss one current effort at answering some of these questions by proposing a new perspective on particles. While reading this book, you may recall starting off with the particle model and doing quite a bit of physics with it. In Part 2, we introduced the wave model, and there was more physics to be investigated via the properties of waves. We used a wave model for light in Part 5. Early in Part 6, however, we saw the need to return to the particle model for light. Furthermore, we found that

[8] For an update on dark energy, see S. Perlmutter, "Supernovae, Dark Energy, and the Accelerating Universe", *Physics Today*, April 2003; 56(4), pp. 53–60.

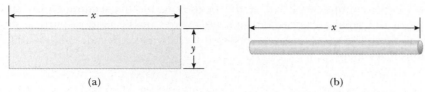

Figure 46.24 (a) A piece of paper is cut into a rectangular shape. As a rectangle, the shape has two dimensions. (b) The paper is rolled up into a soda straw. From far away, it appears to be one-dimensional. The curled-up second dimension is not visible when viewed from a distance large compared to the diameter of the straw.

material particles had wave-like characteristics. The quantum particle model discussed in Chapter 40 allowed us to build particles out of waves, suggesting that a wave is the fundamental entity. In the current chapter, however, we introduced elementary particles as the fundamental entities. It seems as if we cannot make up our mind! In this final section, we shall discuss a current research effort to build particles out of waves and vibrations.

String theory is an effort to unify the four fundamental forces by modeling all particles as various vibrational modes of a single entity—an incredibly small string. The typical length of such a string is on the order of 10^{-35} m, called the **Planck length.** We have seen quantized modes before—the frequencies of vibrating guitar strings in Chapter 18 and the quantized energy levels of atoms in Chapter 42. In string theory, each quantized mode of vibration of the string corresponds to a different elementary particle in the Standard Model.

One of the complicating factors in string theory is that it requires space–time to have ten dimensions. Despite the theoretical and conceptual difficulties in dealing with ten dimensions, string theory holds promise in incorporating gravity with the other forces. Four of the ten dimensions are visible to us—three space dimensions and one time dimension—and the other six are said to be *compactified*. That is, the six dimensions are curled up so tightly that they are not visible in the macroscopic world.

As an analogy, consider a soda straw. We can build a soda straw by cutting a rectangular piece of paper (Fig. 46.24a), which clearly has two dimensions, and rolling it up into a small tube (Fig. 46.24b). From far away, the soda straw looks like a one-dimensional straight line. The second dimension has been curled up and is not visible. String theory claims that six space–time dimensions are curled up in an analogous way, with the curling being on the size of the Planck length—impossible to see from our viewpoint.

Another complicating factor with string theory is that it is difficult for string theorists to guide experimentalists as to what to look for in an experiment. The Planck length is so small that direct experimentation on strings is impossible. Until the theory has been further developed, string theorists are restricted to applying the theory to known results and testing for consistency.

One of the predictions of string theory, called **supersymmetry** (SUSY), suggests that every elementary particle has a superpartner that has not yet been observed. It is believed that supersymmetry is a broken symmetry (like the broken electroweak symmetry at low energies) and the masses of the superpartners are above our current capabilities of detection by accelerators. Some theorists claim that the mass of superpartners is the missing mass discussed in Section 46.12. Keeping with the whimsical trend in naming particles and their properties, superpartners are given names such as the *squark* (the superpartner to a quark), the *selectron* (electron), and the *gluinos* (gluon).

Other theorists are working on **M-theory**, which is an eleven-dimensional theory based on membranes rather than strings. In a way reminiscent of the correspondence principle, M-theory is claimed to reduce to string theory if one compactifies from eleven dimensions to ten dimensions.

The questions that we listed at the beginning of this section go on and on. Because of the rapid advances and new discoveries in the field of particle physics, many of these questions may be resolved in the next decade while other new questions may emerge.

SUMMARY

Take a practice test for this chapter by clicking on the Practice Test link at http://www.pse6.com.

Before quark theory was developed, the four fundamental forces in nature were identified as nuclear, electromagnetic, weak, and gravitational. All the interactions in which these forces take part are mediated by **field particles.** The electromagnetic interaction is mediated by the photon; the weak interaction is mediated by the $W^{\pm}$ and Z^0 bosons; the gravitational interaction is mediated by gravitons; the nuclear interaction is mediated by gluons.

A charged particle and its **antiparticle** have the same mass but opposite charge, and other properties will have opposite values, such as lepton number and baryon number. It is possible to produce particle–antiparticle pairs in nuclear reactions if the available energy is greater than $2mc^2$, where m is the mass of the particle (or antiparticle).

Particles other than field particles are classified as hadrons or leptons. **Hadrons** interact via all four fundamental forces. They have size and structure, and are not elementary particles. There are two types—**baryons** and **mesons.** Baryons, which generally are the most massive particles, have nonzero **baryon number** and a spin of $\frac{1}{2}$ or $\frac{3}{2}$. Mesons have baryon number zero and either zero or integral spin.

Leptons have no structure or size and are considered truly elementary. They interact only via the weak, gravitational, and electromagnetic forces. Six types of leptons exist: the electron e^-, the muon μ^-, the tau τ^-; and their neutrinos ν_e, ν_μ, and ν_τ.

In all reactions and decays, quantities such as energy, linear momentum, angular momentum, electric charge, baryon number, and lepton number are strictly conserved. Certain particles have properties called **strangeness** and **charm.** These unusual properties are conserved in all decays and nuclear reactions except those that occur via the weak force.

Theorists in elementary particle physics have postulated that all hadrons are composed of smaller units known as **quarks,** and experimental evidence agrees with this model. Quarks have fractional electric charge and come in six **flavors:** up (u), down (d), strange (s), charmed (c), top (t), and bottom (b). Each baryon contains three quarks, and each meson contains one quark and one antiquark.

According to the theory of **quantum chromodynamics,** quarks have a property called **color,** and the force between quarks is referred to as the **strong force** or the **color force.** The strong force is now considered to be a fundamental force. The nuclear force, which was originally considered to be fundamental, is now understood to be a secondary effect of the strong force, due to gluon exchanges between hadrons.

The electromagnetic and weak forces are now considered to be manifestations of a single force called the **electroweak force.** The combination of quantum chromodynamics and the electroweak theory is called the **Standard Model.**

The background microwave radiation discovered by Penzias and Wilson strongly suggests that the Universe started with a Big Bang 11 to 20 billion years ago. The background radiation is equivalent to that of a black body at 3 K. Various astronomical measurements strongly suggest that the Universe is expanding. According to **Hubble's law,** distant galaxies are receding from the Earth at a speed $v = HR$, where R is the distance from the Earth to the galaxy and H is the **Hubble constant,** $H \approx 17 \times 10^{-3}$ m/s·ly.

QUESTIONS

1. Name the four fundamental interactions and the field particle that mediates each.

2. When an electron and a positron meet at low speeds in free space, why are *two* 0.511-MeV gamma rays produced, rather than one gamma ray with an energy of 1.02 MeV?

3. Describe the quark model of hadrons, including the properties of quarks.

4. What are the differences between hadrons and leptons?

5. Describe the properties of baryons and mesons and the important differences between them.

6. Particles known as resonances have very short lifetimes, on the order of 10^{-23} s. From this information, would you guess that they are hadrons or leptons? Explain.

7. Kaons all decay into final states that contain no protons or neutrons. What is the baryon number of kaons?

8. The Ξ^0 particle decays by the weak interaction according to the decay mode $\Xi^0 \rightarrow \Lambda^0 + \pi^0$. Would you expect this decay to be fast or slow? Explain.

9. Identify the particle decays in Table 46.2 that occur by the weak interaction. Justify your answers.

10. Identify the particle decays in Table 46.2 that occur by the electromagnetic interaction. Justify your answers.

11. Two protons in a nucleus interact via the nuclear interaction. Are they also subject to the weak interaction?

12. Discuss the following conservation laws: energy, linear momentum, angular momentum, electric charge, baryon number, lepton number, and strangeness. Are all of these laws based on fundamental properties of nature? Explain.

13. An antibaryon interacts with a meson. Can a baryon be produced in such an interaction? Explain.

14. Describe the essential features of the Standard Model of particle physics.

15. How many quarks are in each of the following: (a) a baryon, (b) an antibaryon, (c) a meson, (d) an antimeson? How do you account for the fact that baryons have half-integral spins while mesons have spins of 0 or 1? (*Note:* Quarks have spin $\frac{1}{2}$.)

16. In the theory of quantum chromodynamics, quarks come in three colors. How would you justify the statement that "all baryons and mesons are colorless"?

17. Which baryon did Murray Gell-Mann predict in 1961? What is the quark composition of this particle?

18. What is the quark composition of the Ξ^- particle? (See Table 46.5.)

19. The W and Z bosons were first produced at CERN in 1983 by causing a beam of protons and a beam of antiprotons to meet at high energy. Why was this an important discovery?

20. How did Edwin Hubble determine in 1928 that the Universe is expanding?

21. Neutral atoms did not exist until hundreds of thousands of years after the Big Bang. Why?

22. **Review question.** A girl and her grandmother grind corn while the woman tells the girl stories about what is most important. A boy keeps crows away from ripening corn while his grandfather sits in the shade and explains to him the Universe and his place in it. What the children do not understand this year they will better understand next year. Now you must take the part of the adults. State the most general, most fundamental, most universal truths that you know. If you need to repeat someone else's ideas, get the best version of those ideas you can, and state your source. If there is something you do not understand, make a plan to understand it better within the next year.

PROBLEMS

1, 2, 3 = straightforward, intermediate, challenging ☐ = full solution available in the *Student Solutions Manual and Study Guide*

= coached solution with hints available at http://www.pse6.com 💻 = computer useful in solving problem

= paired numerical and symbolic problems

Section 46.1 The Fundamental Forces in Nature
Section 46.2 Positrons and Other Antiparticles

1. A photon produces a proton–antiproton pair according to the reaction $\gamma \rightarrow p + \bar{p}$. What is the minimum possible frequency of the photon? What is its wavelength?

2. Two photons are produced when a proton and antiproton annihilate each other. In the reference frame in which the center of mass of the proton–antiproton system is stationary, what are the minimum frequency and corresponding wavelength of each photon?

3. A photon with an energy $E_\gamma = 2.09$ GeV creates a proton–antiproton pair in which the proton has a kinetic energy of 95.0 MeV. What is the kinetic energy of the antiproton? ($m_p c^2 = 938.3$ MeV.)

Section 46.3 Mesons and the Beginning of Particle Physics

4. Occasionally, high-energy muons collide with electrons and produce two neutrinos according to the reaction $\mu^+ + e^- \rightarrow 2\nu$. What kind of neutrinos are these?

5. One of the mediators of the weak interaction is the Z^0 boson, with mass 93 GeV/c^2. Use this information to find the order of magnitude of the range of the weak interaction.

6. Calculate the range of the force that might be produced by the virtual exchange of a proton.

7. A neutral pion at rest decays into two photons according to

$$\pi^0 \longrightarrow \gamma + \gamma$$

Find the energy, momentum, and frequency of each photon.

8. When a high-energy proton or pion traveling near the speed of light collides with a nucleus, it travels an average distance of 3×10^{-15} m before interacting. From this information, find the order of magnitude of the time interval required for the strong interaction to occur.

9. A free neutron beta decays by creating a proton, an electron and an antineutrino according to the reaction $n \rightarrow p + e^- + \bar{\nu}$. **What If?** Imagine that a free neutron decays by creating a proton and electron according to the reaction

$$n \longrightarrow p + e^-$$

and assume that the neutron is initially at rest in the laboratory. (a) Determine the energy released in this reaction. (b) Determine the speeds of the proton and electron after the reaction. (Energy and momentum are conserved in the reaction.) (c) Is either of these particles moving at a relativistic speed? Explain.

Section 46.4 Classification of Particles

10. Identify the unknown particle on the left side of the following reaction:

$$? + p \longrightarrow n + \mu^+$$

11. Name one possible decay mode (see Table 46.2) for Ω^+, $\overline{K_S^0}$, $\overline{\Lambda^0}$ and $\bar{n}$.

Section 46.5 Conservation Laws

12. Each of the following reactions is forbidden. Determine a conservation law that is violated for each reaction.
(a) $p + \bar{p} \rightarrow \mu^+ + e^-$
(b) $\pi^- + p \rightarrow p + \pi^+$
(c) $p + p \rightarrow p + \pi^+$
(d) $p + p \rightarrow p + p + n$
(e) $\gamma + p \rightarrow n + \pi^0$

13. (a) Show that baryon number and charge are conserved in the following reactions of a pion with a proton.

$$\pi^+ + p \longrightarrow K^+ + \Sigma^+ \quad (1)$$

$$\pi^+ + p \longrightarrow \pi^+ + \Sigma^+ \quad (2)$$

(b) The first reaction is observed, but the second never occurs. Explain.

14. The first of the following two reactions can occur, but the second cannot. Explain.

$$K_S^0 \longrightarrow \pi^+ + \pi^- \quad \text{(can occur)}$$

$$\Lambda^0 \longrightarrow \pi^+ + \pi^- \quad \text{(cannot occur)}$$

15. The following reactions or decays involve one or more neutrinos. In each case, supply the missing neutrino $(\nu_e, \nu_\mu, \text{ or } \nu_\tau)$ or antineutrino.
(a) $\pi^- \rightarrow \mu^- + ?$
(b) $K^+ \rightarrow \mu^+ + ?$
(c) $? + p \rightarrow n + e^+$
(d) $? + n \rightarrow p + e^-$
(e) $? + n \rightarrow p + \mu^-$
(f) $\mu^- \rightarrow e^- + ? + ?$

16. A K_S^0 particle at rest decays into a π^+ and a π^-. What will be the speed of each of the pions? The mass of the K_S^0 is 497.7 MeV/c^2, and the mass of each π is 139.6 MeV/c^2.

17. Determine which of the following reactions can occur. For those that cannot occur, determine the conservation law (or laws) violated:
(a) $p \rightarrow \pi^+ + \pi^0$
(b) $p + p \rightarrow p + p + \pi^0$
(c) $p + p \rightarrow p + \pi^+$
(d) $\pi^+ \rightarrow \mu^+ + \nu_\mu$
(e) $n \rightarrow p + e^- + \bar{\nu}_e$
(f) $\pi^+ \rightarrow \mu^+ + n$

18. (a) Show that the proton-decay reaction

$$p \longrightarrow e^+ + \gamma$$

cannot occur, because it violates conservation of baryon number. (b) **What If?** Imagine that this reaction does occur, and that the proton is initially at rest. Determine the energy and momentum of the positron and photon after the reaction. (*Suggestion:* Recall that energy and momentum must be conserved in the reaction.) (c) Determine the speed of the positron after the reaction.

19. Determine the type of neutrino or antineutrino involved in each of the following processes:
(a) $\pi^+ \rightarrow \pi^0 + e^+ + ?$
(b) $? + p \rightarrow \mu^- + p + \pi^+$
(c) $\Lambda^0 \rightarrow p + \mu^- + ?$
(d) $\tau^+ \rightarrow \mu^+ + ? + ?$

Section 46.6 Strange Particles and Strangeness

20. The neutral meson ρ^0 decays by the strong interaction into two pions: $\rho^0 \rightarrow \pi^+ + \pi^-$, half-life 10^{-23} s. The neutral kaon also decays into two pions: $K_S^0 \rightarrow \pi^+ + \pi^-$, half-life 10^{-10} s. How do you explain the difference in half-lives?

21. Determine whether or not strangeness is conserved in the following decays and reactions.
(a) $\Lambda^0 \rightarrow p + \pi^-$
(b) $\pi^- + p \rightarrow \Lambda^0 + K^0$
(c) $\bar{p} + p \rightarrow \overline{\Lambda^0} + \Lambda^0$
(d) $\pi^- + p \rightarrow \pi^- + \Sigma^+$
(e) $\Xi^- \rightarrow \Lambda^0 + \pi^-$
(f) $\Xi^0 \rightarrow p + \pi^-$

22. For each of the following forbidden decays, determine which conservation law is violated:
(a) $\mu^- \rightarrow e^- + \gamma$
(b) $n \rightarrow p + e^- + \nu_e$
(c) $\Lambda^0 \rightarrow p + \pi^0$

(d) $p \rightarrow e^+ + \pi^0$

(e) $\bar{\Xi}^0 \rightarrow n + \pi^0$

23. Which of the following processes are allowed by the strong interaction, the electromagnetic interaction, the weak interaction, or no interaction at all?

(a) $\pi^- + p \rightarrow 2\eta$

(b) $K^- + n \rightarrow \Lambda^0 + \pi^-$

(c) $K^- \rightarrow \pi^- + \pi^0$

(d) $\Omega^- \rightarrow \Xi^- + \pi^0$

(e) $\eta \rightarrow 2\gamma$

24. Identify the conserved quantities in the following processes:

(a) $\Xi^- \rightarrow \Lambda^0 + \mu^- + \nu_\mu$

(b) $K_S^0 \rightarrow 2\pi^0$

(c) $K^- + p \rightarrow \Sigma^0 + n$

(d) $\Sigma^0 \rightarrow \Lambda^0 + \gamma$

(e) $e^+ + e^- \rightarrow \mu^+ + \mu^-$

(f) $\bar{p} + n \rightarrow \bar{\Lambda}^0 + \Sigma^-$

25. Fill in the missing particle. Assume that (a) occurs via the strong interaction and (b) and (c) involve the weak interaction.

(a) $K^+ + p \rightarrow ? + p$

(b) $\Omega^- \rightarrow ? + \pi^-$

(c) $K^+ \rightarrow ? + \mu^+ + \nu_\mu$

Section 46.7 Making Elementary Particles and Measuring their Properties

26. The particle decay $\Sigma^+ \rightarrow \pi^+ + n$ is observed in a bubble chamber. Figure P46.26 represents the curved tracks of the particles Σ^+ and π^+, and the invisible track of the neutron, in the presence of a uniform magnetic field of 1.15 T directed out of the page. The measured radii of curvature are 1.99 m for the Σ^+ particle and 0.580 m for the π^1 particle. (a) Find the momenta of the Σ^+ and the π^+ particles, in units of MeV/c. (b) The angle between the momenta of the Σ^+ and the π^+ particles at the moment of decay is 64.5°. Find the momentum of the neutron. (c) Calculate the total energy of the π^+ particle, and of the neutron, from their known masses $(m_\pi = 139.6 \text{ MeV}/c^2, \quad m_n = 939.6 \text{ MeV}/c^2)$ and the relativistic energy–momentum relation. What is the total

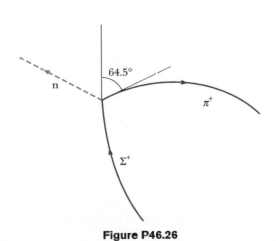

Figure P46.26

energy of the Σ^+ particle? (d) Calculate the mass and speed of the Σ^+ particle.

27. If a K_S^0 meson at rest decays in 0.900×10^{-10} s, how far will a K_S^0 meson travel if it is moving at $0.960c$?

28. A particle of mass m_1 is fired at a stationary particle of mass m_2, and a reaction takes place in which new particles are created out of the incident kinetic energy. Taken together, the product particles have total mass m_3. The minimum kinetic energy that the bombarding particle must have in order to induce the reaction is called the threshold energy. At this energy, the kinetic energy of the products is a minimum, so that the fraction of the incident kinetic energy that is available to create new particles is a maximum. This occurs when all the product particles have the same velocity, so that the particles have no kinetic energy of motion relative to one another. (a) By using conservation of relativistic energy and momentum, and the relativistic energy-momentum relation, show that the threshold energy is given by

$$K_{min} = \frac{\left[m_3^2 - (m_1 + m_2)^2\right]c^2}{2m_2}$$

Calculate the threshold energy for each of the following reactions:

(b) $p + p \rightarrow p + p + p + \bar{p}$

  (One of the initial protons is at rest. Antiprotons are produced.)

(c) $\pi^- + p \rightarrow K^0 + \Lambda^0$

  (The proton is at rest. Strange particles are produced.)

(d) $p + p \rightarrow p + p + \pi^0$

  (One of the initial protons is at rest. Pions are produced.)

(e) $p + \bar{p} \rightarrow Z^0$

  (One of the initial particles is at rest. Z^0 particles (mass 91.2 GeV/c^2) are produced.)

Section 46.8 Finding Patterns in the Particles
Section 46.9 Quarks
Section 46.10 Multicolored Quarks
Section 46.11 The Standard Model

Note: Problem 71 in Chapter 39 can be assigned with Section 46.11.

29. (a) Find the number of electrons and the number of each species of quarks in 1 L of water. (b) Make an order-of-magnitude estimate of the number of each kind of fundamental matter particle in your body. State your assumptions and the quantities you take as data.

30. The quark composition of the proton is uud, and that of the neutron is udd. Show that in each case the charge, baryon number, and strangeness of the particle equal, respectively, the sums of these numbers for the quark constituents.

31. **What If?** Imagine that binding energies could be ignored. Find the masses of the u and d quarks from the masses of the proton and neutron.

32. The quark compositions of the K^0 and Λ^0 particles are $\bar{s}d$ and uds, respectively. Show that the charge, baryon number, and strangeness of these particles equal, respectively, the sums of these numbers for the quark constituents.

33. Analyze each reaction in terms of constituent quarks:
    (a) $\pi^- + p \rightarrow K^0 + \Lambda^0$
    (b) $\pi^+ + p \rightarrow K^+ + \Sigma^+$
    (c) $K^- + p \rightarrow K^+ + K^0 + \Omega^-$
    (d) $p + p \rightarrow K^0 + p + \pi^+ + ?$

    In the last reaction, identify the mystery particle.

34. The text states that the reaction $\pi^- + p \rightarrow K^0 + \Lambda^0$ occurs with high probability, whereas the reaction $\pi^- + p \rightarrow K^0 + n$ never occurs. Analyze these reactions at the quark level. Show that the first reaction conserves the total number of each type of quark, and the second reaction does not.

35. A Σ^0 particle traveling through matter strikes a proton; then a Σ^+ and a gamma ray emerge, as well as a third particle. Use the quark model of each to determine the identity of the third particle.

36. Identify the particles corresponding to the quark combinations (a) suu, (b) $\bar{u}d$, (c)$\bar{s}d$, and (d) ssd,

37. What is the electrical charge of the baryons with the quark compositions (a) $\bar{u}\,\bar{u}\,\bar{d}$ and (b) $\bar{u}\,\bar{d}\,d$? What are these baryons called?

Section 46.12 The Cosmic Connection

> *Note:* Problem 20 in Chapter 39 can be assigned with this section.

38. **Review problem**. Refer to Section 39.4. Prove that the Doppler shift in wavelength of electromagnetic waves is described by

$$\lambda' = \lambda \sqrt{\frac{1 + v/c}{1 - v/c}}$$

where λ' is the wavelength measured by an observer moving at speed v away from a source radiating waves of wavelength λ.

39. A distant quasar is moving away from Earth at such high speed that the blue 434-nm H_γ line of hydrogen is observed at 510 nm, in the green portion of the spectrum (Fig. P46.39). (a) How fast is the quasar receding? You may use the result of Problem 38. (b) Edwin Hubble discovered that all objects outside the local group of galaxies are moving away from us, with speeds proportional to their distances. Hubble's law is expressed as $v = HR$, where Hubble's constant has the approximate value $H = 17 \times 10^{-3}$ m/s · ly. Determine the distance from Earth to this quasar.

40. The various spectral lines observed in the light from a distant quasar have longer wavelengths λ_n' than the wavelengths λ_n measured in light from a stationary source. Here n is an index taking different values for different spectral lines. The fractional change in wavelength toward the red is the same for all spectral lines. That is, the redshift parameter Z defined by

$$Z = \frac{\lambda_n' - \lambda_n}{\lambda_n}$$

is common to all spectral lines for one object. In terms of Z, determine (a) the speed of recession of the quasar and (b) the distance from Earth to this quasar. Use the result of Problem 38 and Hubble's law.

41. Using Hubble's law, find the wavelength of the 590-nm sodium line emitted from galaxies (a) 2.00×10^6 ly away from Earth, (b) 2.00×10^8 ly away, and (c) 2.00×10^9 ly away. You may use the result of Problem 38.

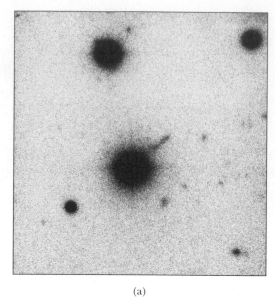

(a)

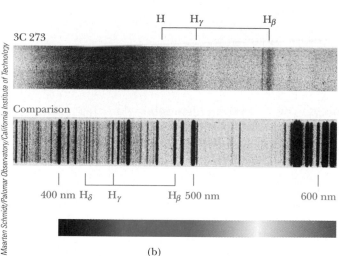

(b)

Figure P46.39 (a) Image of the quasar 3C273. (b) Spectrum of the quasar above a comparison spectrum emitted by stationary hydrogen and helium atoms. Both parts of the figure are printed as black-and-white photographic negatives to reveal detail.

Maarten Schmidt/Palomar Observatory/California Institute of Technology

42. Review problem. The cosmic background radiation is blackbody radiation from a source at a temperature of 2.73 K. (a) Use Wien's law to determine the wavelength at which this radiation has its maximum intensity. (b) In what part of the electromagnetic spectrum is the peak of the distribution?

43. Review problem. Use Stefan's law to find the intensity of the cosmic background radiation emitted by the fireball of the Big Bang at a temperature of 2.73 K.

44. It is mostly your roommate's fault. Nosy astronomers have discovered enough junk and clutter in your dorm room to constitute the missing mass required to close the Universe. After observing your floor, closet, bed, and computer files, they extrapolate to slobs in other galaxies and calculate the average density of the observable Universe as $1.20\rho_c$. How many times larger will the Universe become before it begins to collapse? That is, by what factor will the distance between remote galaxies increase in the future?

45. The early Universe was dense with gamma-ray photons of energy $\sim k_B T$ and at such a high temperature that protons and antiprotons were created by the process $\gamma \rightarrow p + \bar{p}$ as rapidly as they annihilated each other. As the Universe cooled in adiabatic expansion, its temperature fell below a certain value, and proton pair production became rare. At that time slightly more protons than antiprotons existed, and essentially all of the protons in the Universe today date from that time. (a) Estimate the order of magnitude of the temperature of the Universe when protons condensed out. (b) Estimate the order of magnitude of the temperature of the Universe when electrons condensed out.

46. If the average density of the Universe is small compared to the critical density, the expansion of the Universe described by Hubble's law proceeds with speeds that are nearly constant over time. (a) Prove that in this case the age of the Universe is given by the inverse of Hubble's constant. (b) Calculate $1/H$ and express it in years.

47. Assume that the average density of the Universe is equal to the critical density. (a) Prove that the age of the Universe is given by $2/3H$. (b) Calculate $2/3H$ and express it in years.

48. Hubble's law can be stated in vector form as $\mathbf{v} = H\mathbf{R}$: Outside the local group of galaxies, all objects are moving away from us with velocities proportional to their displacements from us. In this form, it sounds as if our location in the Universe is specially privileged. Prove that Hubble's law would be equally true for an observer elsewhere in the Universe. Proceed as follows. Assume that we are at the origin of coordinates, that one galaxy cluster is at location $\mathbf{R}_1$ and has velocity $\mathbf{v}_1 = H\mathbf{R}_1$ relative to us, and that another galaxy cluster has radius vector $\mathbf{R}_2$ and velocity $\mathbf{v}_2 = H\mathbf{R}_2$. Suppose the speeds are nonrelativistic. Consider the frame of reference of an observer in the first of these galaxy clusters. Show that our velocity relative to her, together with the displacement vector of our galaxy cluster from hers, satisfies Hubble's law. Show that the displacement and velocity of cluster 2 relative to cluster 1 satisfy Hubble's law.

Section 46.13   Problems and Perspectives

49. Classical general relativity views the structure of space–time as deterministic and well defined down to arbitrarily small distances. On the other hand, quantum general relativity forbids distances smaller than the Planck length given by $L = (\hbar G/c^3)^{1/2}$. (a) Calculate the value of the Planck length. The quantum limitation suggests that after the Big Bang, when all the presently observable section of the Universe was contained within a point-like singularity, nothing could be observed until that singularity grew larger than the Planck length. Because the size of the singularity grew at the speed of light, we can infer that no observations were possible during the time interval required for light to travel the Planck length. (b) Calculate this time interval, known as the Planck time T, and compare it with the ultrahot epoch mentioned in the text. (c) Does this suggest we may never know what happened between the time $t = 0$ and the time $t = T$?

Additional Problems

50. Review problem. Supernova Shelton 1987A, located about 170 000 ly from the Earth, is estimated to have emitted a burst of neutrinos carrying energy $\sim 10^{46}$ J (Fig. P46.50).

Anglo-Australian Telescope Board

Figure P46.50 (Problems 50 and 51) The giant star Sanduleak −69° 202 in the "before" picture became Supernova Shelton 1987A in the "after" picture.

Suppose the average neutrino energy was 6 MeV and your body presented cross-sectional area 5 000 cm^2. To an order of magnitude, how many of these neutrinos passed through you?

51. The most recent naked-eye supernova was Supernova Shelton 1987A (Fig. P46.50). It was 170 000 ly away in the next galaxy to ours, the Large Magellanic Cloud. About 3 h before its optical brightening was noticed, two continuously running neutrino detection experiments simultaneously registered the first neutrinos from an identified source other than the Sun. The Irvine–Michigan–Brookhaven experiment in a salt mine in Ohio registered 8 neutrinos over a 6-s period, and the Kamiokande II experiment in a zinc mine in Japan counted 11 neutrinos in 13 s. (Because the supernova is far south in the sky, these neutrinos entered the detectors from below. They passed through the Earth before they were by chance absorbed by nuclei in the detectors.) The neutrino energies were between about 8 MeV and 40 MeV. If neutrinos have no mass, then neutrinos of all energies should travel together at the speed of light—the data are consistent with this possibility. The arrival times could show scatter simply because neutrinos were created at different moments as the core of the star collapsed into a neutron star. If neutrinos have nonzero mass, then lower-energy neutrinos should move comparatively slowly. The data are consistent with a 10-MeV neutrino requiring at most about 10 s more than a photon would require to travel from the supernova to us. Find the upper limit that this observation sets on the mass of a neutrino. (Other evidence sets an even tighter limit.)

52. Name at least one conservation law that prevents each of the following reactions: (a) $\pi^- + p \rightarrow \Sigma^+ + \pi^0$ (b) $\mu^- \rightarrow \pi^- + \nu_e$ (c) $p \rightarrow \pi^+ + \pi^+ + \pi^-$

53. The energy flux carried by neutrinos from the Sun is estimated to be on the order of 0.4 W/m^2 at Earth's surface. Estimate the fractional mass loss of the Sun over 10^9 yr due to the emission of neutrinos. (The mass of the Sun is 2×10^{30} kg. The Earth–Sun distance is 1.5×10^{11} m.)

54. Two protons approach each other head-on, each with 70.4 MeV of kinetic energy, and engage in a reaction in which a proton and positive pion emerge at rest. What third particle, obviously uncharged and therefore difficult to detect, must have been created?

55. A rocket engine for space travel using photon drive and matter–antimatter annihilation has been suggested. Suppose the fuel for a short-duration burn consists of N protons and N antiprotons, each with mass m. (a) Assume all of the fuel is annihilated to produce photons. When the photons are ejected from the rocket, what momentum can be imparted to it? (b) **What If?** If half of the protons and antiprotons annihilate each other and the energy released is used to eject the remaining particles, what momentum could be given to the rocket? (c) Which scheme results in the greater change in speed for the rocket?

56. The nuclear force can be attributed to the exchange of an elementary particle between protons and neutrons when they are sufficiently close. Take the range of the nuclear force as approximately 1.4×10^{-15} m. (a) Use the uncertainty principle $\Delta E \Delta t \geq \hbar/2$ to estimate the mass of the elementary particle. Assume that it moves at nearly the speed of light. (b) Using Table 46.2, identify the particle.

57. Determine the kinetic energies of the proton and pion resulting from the decay of a Λ^0 at rest:

$$\Lambda^0 \longrightarrow p + \pi^-$$

58. A gamma-ray photon strikes a stationary electron. Determine the minimum gamma-ray energy to make this reaction occur:

$$\gamma + e^- \longrightarrow e^- + e^- + e^+$$

59. An unstable particle, initially at rest, decays into a proton (rest energy 938.3 MeV) and a negative pion (rest energy 139.6 MeV). A uniform magnetic field of 0.250 T exists perpendicular to the velocities of the created particles. The radius of curvature of each track is found to be 1.33 m. What is the mass of the original unstable particle?

60. A Σ^0 particle at rest decays according to

$$\Sigma^0 \longrightarrow \Lambda^0 + \gamma$$

Find the gamma-ray energy.

61. Two protons approach each other with velocities of equal magnitude in opposite directions. What is the minimum kinetic energy of each of the protons if they are to produce a π^+ meson at rest in the following reaction?

$$p + p \longrightarrow p + n + \pi^+$$

62. A π-meson at rest decays according to $\pi^- \rightarrow \mu^- + \bar{\nu}_\mu$. What is the energy carried off by the neutrino? (Assume that the neutrino has no mass and moves off with the speed of light. Take $m_\pi c^2 = 139.6$ MeV and $m_\mu c^2 = 105.7$ MeV.)

63. **Review problem.** Use the Boltzmann distribution function $e^{-E/k_B T}$ to calculate the temperature at which 1.00% of a population of photons will have energy greater than 1.00 eV. The energy required to excite an atom is on the order of 1 eV. Therefore as the temperature of the Universe fell below the value you calculate, neutral atoms could form from plasma, and the Universe became transparent. The cosmic background radiation represents our vastly red-shifted view of the opaque fireball of the Big Bang as it was at this time and temperature. The fireball surrounds us; we are embers.

64. What processes are described by the Feynman diagrams in Figure P46.64? What is the exchanged particle in each process?

(a)

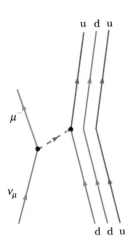

(b)

Figure P46.64

65. Identify the mediators for the two interactions described in the Feynman diagrams shown in Figure P46.65.

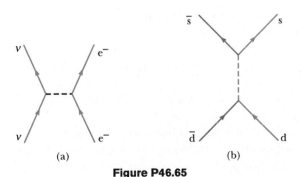

(a)   (b)

Figure P46.65

66. The cosmic rays of highest energy are mostly protons, accelerated by unknown sources. Their spectrum shows a cutoff at an energy on the order of 10^{20} eV. Above that energy, a proton will interact with a photon of cosmic

microwave background radiation to produce mesons, for example according to

$$p + \gamma \longrightarrow p + \pi^0$$

Demonstrate this fact by taking the following steps: (a) Find the minimum photon energy required to produce this reaction in the reference frame where the total momentum of the photon–proton system is zero. The reaction was observed experimentally in the 1950s with photons of a few hundred MeV. (b) Use Wien's displacement law to find the wavelength of a photon at the peak of the blackbody spectrum of the primordial microwave background radiation, with a temperature of 2.73 K. (c) Find the energy of this photon. (d) Consider the reaction in part (a) in a moving reference frame so that the photon is the same as that in part (c). Calculate the energy of the proton in this frame, which represents the Earth reference frame.

Answers to Quick Quizzes

46.1 (a). The right-hand rule for the positive particle tells you that this is the direction that leads to a force directed toward the center of curvature of the path.

46.2 (c), (d). There is a baryon, the neutron, on the left of the reaction, but no baryon on the right. Therefore, baryon number is not conserved. The neutron has spin $\frac{1}{2}$. On the right side of the reaction, the pions each have integral spin, and the combination of two muons must also have integral spin. Therefore, the total spin of the particles on the right-hand side must be integral and angular momentum is not conserved.

46.3 (a). The sum of the proton and pion masses is larger than the mass of the neutron, so energy conservation is violated.

46.4 (b), (e), (f). The pion on the left has integral spin while the three spin-$\frac{1}{2}$ leptons on the right must result in a total spin that is half-integral. Therefore, angular momentum is not conserved. Electron lepton number is zero on the left and -1 on the right. There are no muons on the left, but a muon and its neutrino on the right (both with $L_\mu = +1$). Therefore, muon lepton number is not conserved.

46.5 (b), (e). There is one spin-$\frac{1}{2}$ particle on the left and two on the right, so angular momentum is not conserved. There are no leptons on the left and an electron on the right, so electron lepton number is not conserved.

46.6 (b), (c), (e). Angular momentum is not conserved because there is one spin-$\frac{1}{2}$ particle before the decay and two spin-$\frac{1}{2}$ particles afterward. Electric charge is not conserved because the negative charge on the electron disappears. Electron lepton number is not conserved, because there is an electron with $L_e = 1$ before the decay and two neutrinos, each with $L_e = 1$, afterward.

46.7 The diagram would look this:

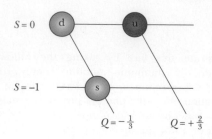

46.8 False. Because the charges on quarks are $+2e/3$ and $-e/3$, the maximum possible charge of a combination of a quark and an antiquark is $\pm e$.

46.9 (a) (i), (ii), (iii). The order of increasing quark mass is c, b, t. (b) (i) = (ii) = (iii). All three baryons are combinations of three spin-$\frac{1}{2}$ quarks. (c) (i) = (iii) > (ii). The total charge of ccc and ttt is $2e$, while the total charge of bbb is $-e$. (d) (i) = (ii) = (iii). All three particles have a baryon number of 1.

"Particles, particles, particles."

© 2003 Sidney Harris

The Meaning of Success

To earn the respect of intelligent people and to win the affection of children;

To appreciate the beauty in nature and all that surrounds us;

To seek out and nurture the best in others;

To give the gift of yourself to others without the slightest thought of return, for it is in giving that we receive;

To have accomplished a task, whether it be saving a lost soul, healing a sick child, writing a book, or risking your life for a friend;

To have celebrated and laughed with great joy and enthusiasm and sung with exultation;

To have hope even in times of despair, for as long as you have hope, you have life;

To love and be loved;

To be understood and to understand;

To know that even one life has breathed easier because you have lived;

This is the meaning of success.

—Ralph Waldo Emerson
Modified by Ray Serway, December 1989

Table A.1

Conversion Factors

Length

	m	cm	km	in.	ft	mi
1 meter	1	10^2	10^{-3}	39.37	3.281	6.214×10^{-4}
1 centimeter	10^{-2}	1	10^{-5}	0.393 7	3.281×10^{-2}	6.214×10^{-6}
1 kilometer	10^3	10^5	1	3.937×10^4	3.281×10^3	0.621 4
1 inch	2.540×10^{-2}	2.540	2.540×10^{-5}	1	8.333×10^{-2}	1.578×10^{-5}
1 foot	0.304 8	30.48	3.048×10^{-4}	12	1	1.894×10^{-4}
1 mile	1 609	1.609×10^5	1.609	6.336×10^4	5 280	1

Mass

	kg	g	slug	u
1 kilogram	1	10^3	6.852×10^{-2}	6.024×10^{26}
1 gram	10^{-3}	1	6.852×10^{-5}	6.024×10^{23}
1 slug	14.59	1.459×10^4	1	8.789×10^{27}
1 atomic mass unit	1.660×10^{-27}	1.660×10^{-24}	1.137×10^{-28}	1

Note: 1 metric ton = 1 000 kg.

Time

	s	min	h	day	yr
1 second	1	1.667×10^{-2}	2.778×10^{-4}	1.157×10^{-5}	3.169×10^{-8}
1 minute	60	1	1.667×10^{-2}	6.994×10^{-4}	1.901×10^{-6}
1 hour	3 600	60	1	4.167×10^{-2}	1.141×10^{-4}
1 day	8.640×10^4	1 440	24	1	2.738×10^{-5}
1 year	3.156×10^7	5.259×10^5	8.766×10^3	365.2	1

Speed

	m/s	cm/s	ft/s	mi/h
1 meter per second	1	10^2	3.281	2.237
1 centimeter per second	10^{-2}	1	3.281×10^{-2}	2.237×10^{-2}
1 foot per second	0.304 8	30.48	1	0.681 8
1 mile per hour	0.447 0	44.70	1.467	1

Note: 1 mi/min = 60 mi/h = 88 ft/s.

Force

	N	lb
1 newton	1	0.224 8
1 pound	4.448	1

continued

Table A.1

Conversion Factors *continued*

Work, Energy, Heat

	J	ft·lb	eV
1 joule	1	0.737 6	6.242×10^{18}
1 foot-pound	1.356	1	8.464×10^{18}
1 electron volt	1.602×10^{-19}	1.182×10^{-19}	1
1 calorie	4.186	3.087	2.613×10^{19}
1 British thermal unit	1.055×10^{3}	7.779×10^{2}	6.585×10^{21}
1 kilowatt hour	3.600×10^{6}	2.655×10^{6}	2.247×10^{25}

	cal	Btu	kWh
1 joule	0.238 9	9.481×10^{-4}	2.778×10^{-7}
1 foot-pound	0.323 9	1.285×10^{-3}	3.766×10^{-7}
1 electron volt	3.827×10^{-20}	1.519×10^{-22}	4.450×10^{-26}
1 calorie	1	3.968×10^{-3}	1.163×10^{-6}
1 British thermal unit	2.520×10^{2}	1	2.930×10^{-4}
1 kilowatt hour	8.601×10^{5}	3.413×10^{2}	1

Pressure

	Pa	atm
1 pascal	1	9.869×10^{-6}
1 atmosphere	1.013×10^{5}	1
1 centimeter mercury[a]	1.333×10^{3}	1.316×10^{-2}
1 pound per square inch	6.895×10^{3}	6.805×10^{-2}
1 pound per square foot	47.88	4.725×10^{-4}

	cm Hg	lb/in.^{2}	lb/ft^2
1 pascal	7.501×10^{-4}	1.450×10^{-4}	2.089×10^{-2}
1 atmosphere	76	14.70	2.116×10^{3}
1 centimeter mercury[a]	1	0.194 3	27.85
1 pound per square inch	5.171	1	144
1 pound per square foot	3.591×10^{-2}	6.944×10^{-3}	1

[a]   At 0°C and at a location where the free-fall acceleration has its "standard" value, 9.806 65 m/s^2.

Table A.2

Symbols, Dimensions, and Units of Physical Quantities

Quantity	Common Symbol	Unit[a]	Dimensions[b]	Unit in Terms of Base SI Units
Acceleration	**a**	m/s^2	L/T^2	m/s^2
Amount of substance	n	MOLE		mol
Angle	θ, ϕ	radian (rad)	1	
Angular acceleration	$\boldsymbol{\alpha}$	rad/s^2	T^{-2}	s^{-2}
Angular frequency	ω	rad/s	T^{-1}	s^{-1}
Angular momentum	**L**	kg·m^2/s	ML^2/T	kg·m^2/s
Angular velocity	$\boldsymbol{\omega}$	rad/s	T^{-1}	s^{-1}
Area	A	m^2	L^2	m^2
Atomic number	Z			

continued

Table A.2

Symbols, Dimensions, and Units of Physical Quantities *continued*				
Quantity	**Common Symbol**	**Unit**[a]	**Dimensions**[b]	**Unit in Terms of Base SI Units**
Capacitance	C	farad (F)	Q^2T^2/ML^2	$A^2 \cdot s^4/kg \cdot m^2$
Charge	q, Q, e	coulomb (C)	Q	$A \cdot s$
Charge density				
Line	λ	C/m	Q/L	$A \cdot s/m$
Surface	σ	C/m^2	Q/L^2	$A \cdot s/m^2$
Volume	ρ	C/m^3	Q/L^3	$A \cdot s/m^3$
Conductivity	σ	$1/\Omega \cdot m$	Q^2T/ML^3	$A^2 \cdot s^3/kg \cdot m^3$
Current	I	AMPERE	Q/T	A
Current density	$\mathbf{J}$	A/m^2	Q/T^2	A/m^2
Density	ρ	kg/m^3	M/L^3	kg/m^3
Dielectric constant	κ			
Length	ℓ, L	METER	L	m
Position	$x, y, z, \mathbf{r}$			
Displacement	$\Delta x, \Delta \mathbf{r}$			
Distance	d, h			
Electric dipole moment	$\mathbf{p}$	C · m	QL	$A \cdot s \cdot m$
Electric field	$\mathbf{E}$	V/m	ML/QT^2	$kg \cdot m/A \cdot s^3$
Electric flux	Φ_E	V · m	ML^3/QT^2	$kg \cdot m^3/A \cdot s^3$
Electromotive force	$\mathcal{E}$	volt (V)	ML^2/QT^2	$kg \cdot m^2/A \cdot s^3$
Energy	E, U, K	joule (J)	ML^2/T^2	$kg \cdot m^2/s^2$
Entropy	S	J/K	$ML^2/T^2 \cdot K$	$kg \cdot m^2/s^2 \cdot K$
Force	$\mathbf{F}$	newton (N)	ML/T^2	$kg \cdot m/s^2$
Frequency	f	hertz (Hz)	T^{-1}	s^{-1}
Heat	Q	joule (J)	ML^2/T^2	$kg \cdot m^2/s^2$
Inductance	L	henry (H)	ML^2/Q^2	$kg \cdot m^2/A^2 \cdot s^2$
Magnetic dipole moment	$\boldsymbol{\mu}$	N · m/T	QL^2/T	$A \cdot m^2$
Magnetic field	$\mathbf{B}$	tesla (T) (= Wb/m^2)	M/QT	$kg/A \cdot s^2$
Magnetic flux	Φ_B	weber (Wb)	ML^2/QT	$kg \cdot m^2/A \cdot s^2$
Mass	m, M	KILOGRAM	M	kg
Molar specific heat	C	J/mol · K		$kg \cdot m^2/s^2 \cdot mol \cdot K$
Moment of inertia	I	kg · m^2	ML^2	$kg \cdot m^2$
Momentum	$\mathbf{p}$	kg · m/s	ML/T	$kg \cdot m/s$
Period	T	s	T	s
Permeability of free space	μ_0	N/A^2 (= H/m)	ML/Q^2T	$kg \cdot m/A^2 \cdot s^2$
Permittivity of free space	ϵ_0	C^2/N · m^2 (= F/m)	Q^2T^2/ML^3	$A^2 \cdot s^4/kg \cdot m^3$
Potential	V	volt (V) (= J/C)	ML^2/QT^2	$kg \cdot m^2/A \cdot s^3$
Power	$\mathcal{P}$	watt (W) (= J/s)	ML^2/T^3	$kg \cdot m^2/s^3$
Pressure	P	pascal (Pa) (= N/m^2)	M/LT^2	$kg/m \cdot s^2$
Resistance	R	ohm (Ω) (= V/A)	ML^2/Q^2T	$kg \cdot m^2/A^2 \cdot s^3$
Specific heat	c	J/kg · K	$L^2/T^2 \cdot K$	$m^2/s^2 \cdot K$
Speed	v	m/s	L/T	m/s
Temperature	T	KELVIN	K	K
Time	t	SECOND	T	s
Torque	τ	N · m	ML^2/T^2	$kg \cdot m^2/s^2$
Velocity	$\mathbf{v}$	m/s	L/T	m/s
Volume	V	m^3	L^3	m^3
Wavelength	λ	m	L	m
Work	W	joule (J) (= N · m)	ML^2/T^2	$kg \cdot m^2/s^2$

[a] The base SI units are given in uppercase letters.

[b] The symbols M, L, T, and Q denote mass, length, time, and charge, respectively.

Table A.3

Table of Atomic Masses[a]						
Atomic Number Z	Element	Symbol	Chemical Atomic Mass (u)	Mass Number (*Indicates Radioactive) A	Atomic Mass (u)	Percent Abundance	Half-Life (If Radioactive) $T_{1/2}$
0	(Neutron)	n		1*	1.008 665		10.4 min
1	Hydrogen	H	1.007 94	1	1.007 825	99.988 5	
	Deuterium	D		2	2.014 102	0.011 5	
	Tritium	T		3*	3.016 049		12.33 yr
2	Helium	He	4.002 602	3	3.016 029	0.000 137	
				4	4.002 603	99.999 863	
				6*	6.018 888		0.81 s
3	Lithium	Li	6.941	6	6.015 122	7.5	
				7	7.016 004	92.5	
				8*	8.022 487		0.84 s
4	Beryllium	Be	9.012 182	7*	7.016 929		53.3 days
				9	9.012 182	100	
				10*	10.013 534		1.5×10^6 yr
5	Boron	B	10.811	10	10.012 937	19.9	
				11	11.009 306	80.1	
				12*	12.014 352		0.020 2 s
6	Carbon	C	12.010 7	10*	10.016 853		19.3 s
				11*	11.011 434		20.4 min
				12	12.000 000	98.93	
				13	13.003 355	1.07	
				14*	14.003 242		5 730 yr
				15*	15.010 599		2.45 s
7	Nitrogen	N	14.006 7	12*	12.018 613		0.011 0 s
				13*	13.005 739		9.96 min
				14	14.003 074	99.632	
				15	15.000 109	0.368	
				16*	16.006 101		7.13 s
				17*	17.008 450		4.17 s
8	Oxygen	O	15.999 4	14*	14.008 595		70.6 s
				15*	15.003 065		122 s
				16	15.994 915	99.757	
				17	16.999 132	0.038	
				18	17.999 160	0.205	
				19*	19.003 579		26.9 s
9	Fluorine	F	18.998 403 2	17*	17.002 095		64.5 s
				18*	18.000 938		109.8 min
				19	18.998 403	100	
				20*	19.999 981		11.0 s
				21*	20.999 949		4.2 s
10	Neon	Ne	20.179 7	18*	18.005 697		1.67 s
				19*	19.001 880		17.2 s
				20	19.992 440	90.48	
				21	20.993 847	0.27	
				22	21.991 385	9.25	
				23*	22.994 467		37.2 s
11	Sodium	Na	22.989 77	21*	20.997 655		22.5 s
				22*	21.994 437		2.61 yr

continued

Table A.3

Table of Atomic Masses [a] *continued*

Atomic Number Z	Element	Symbol	Chemical Atomic Mass (u)	Mass Number (*Indicates Radioactive) A	Atomic Mass (u)	Percent Abundance	Half-Life (If Radioactive) $T_{1/2}$
(11)	Sodium			23	22.989 770	100	
				24*	23.990 963		14.96 h
12	Magnesium	Mg	24.305 0	23*	22.994 125		11.3 s
				24	23.985 042	78.99	
				25	24.985 837	10.00	
				26	25.982 593	11.01	
				27*	26.984 341		9.46 min
13	Aluminum	Al	26.981 538	26*	25.986 892		7.4×10^5 yr
				27	26.981 539	100	
				28*	27.981 910		2.24 min
14	Silicon	Si	28.085 5	28	27.976 926	92.229 7	
				29	28.976 495	4.683 2	
				30	29.973 770	3.087 2	
				31*	30.975 363		2.62 h
				32*	31.974 148		172 yr
15	Phosphorus	P	30.973 761	30*	29.978 314		2.50 min
				31	30.973 762	100	
				32*	31.973 907		14.26 days
				33*	32.971 725		25.3 days
16	Sulfur	S	32.066	32	31.972 071	94.93	
				33	32.971 458	0.76	
				34	33.967 869	4.29	
				35*	34.969 032		87.5 days
				36	35.967 081	0.02	
17	Chlorine	Cl	35.452 7	35	34.968 853	75.78	
				36*	35.968 307		3.0×10^5 yr
				37	36.965 903	24.22	
18	Argon	Ar	39.948	36	35.967 546	0.336 5	
				37*	36.966 776		35.04 days
				38	37.962 732	0.063 2	
				39*	38.964 313		269 yr
				40	39.962 383	99.600 3	
				42*	41.963 046		33 yr
19	Potassium	K	39.098 3	39	38.963 707	93.258 1	
				40*	39.963 999	0.011 7	1.28×10^9 yr
				41	40.961 826	6.730 2	
20	Calcium	Ca	40.078	40	39.962 591	96.941	
				41*	40.962 278		1.0×10^5 yr
				42	41.958 618	0.647	
				43	42.958 767	0.135	
				44	43.955 481	2.086	
				46	45.953 693	0.004	
				48	47.952 534	0.187	
21	Scandium	Sc	44.955 910	41*	40.969 251		0.596 s
				45	44.955 910	100	
22	Titanium	Ti	47.867	44*	43.959 690		49 yr
				46	45.952 630	8.25	

continued

Table A.3

Atomic Number Z	Element	Symbol	Chemical Atomic Mass (u)	Mass Number (*Indicates Radioactive) A	Atomic Mass (u)	Percent Abundance	Half-Life (If Radioactive) $T_{1/2}$
(22)	Titanium			47	46.951 764	7.44	
				48	47.947 947	73.72	
				49	48.947 871	5.41	
				50	49.944 792	5.18	
23	Vanadium	V	50.941 5	48*	47.952 254		15.97 days
				50*	49.947 163	0.250	1.5×10^{17} yr
				51	50.943 964	99.750	
24	Chromium	Cr	51.996 1	48*	47.954 036		21.6 h
				50	49.946 050	4.345	
				52	51.940 512	83.789	
				53	52.940 654	9.501	
				54	53.938 885	2.365	
25	Manganese	Mn	54.938 049	54*	53.940 363		312.1 days
				55	54.938 050	100	
26	Iron	Fe	55.845	54	53.939 615	5.845	
				55*	54.938 298		2.7 yr
				56	55.934 942	91.754	
				57	56.935 399	2.119	
				58	57.933 280	0.282	
				60*	59.934 077		1.5×10^6 yr
27	Cobalt	Co	58.933 200	59	58.933 200	100	
				60*	59.933 822		5.27 yr
28	Nickel	Ni	58.693 4	58	57.935 348	68.076 9	
				59*	58.934 351		7.5×10^4 yr
				60	59.930 790	26.223 1	
				61	60.931 060	1.139 9	
				62	61.928 349	3.634 5	
				63*	62.929 673		100 yr
				64	63.927 970	0.925 6	
29	Copper	Cu	63.546	63	62.929 601	69.17	
				65	64.927 794	30.83	
30	Zinc	Zn	65.39	64	63.929 147	48.63	
				66	65.926 037	27.90	
				67	66.927 131	4.10	
				68	67.924 848	18.75	
				70	69.925 325	0.62	
31	Gallium	Ga	69.723	69	68.925 581	60.108	
				71	70.924 705	39.892	
32	Germanium	Ge	72.61	70	69.924 250	20.84	
				72	71.922 076	27.54	
				73	72.923 459	7.73	
				74	73.921 178	36.28	
				76	75.921 403	7.61	
33	Arsenic	As	74.921 60	75	74.921 596	100	
34	Selenium	Se	78.96	74	73.922 477	0.89	
				76	75.919 214	9.37	
				77	76.919 915	7.63	

continued

Table A.3

Table of Atomic Masses [a] *continued*

Atomic Number Z	Element	Symbol	Chemical Atomic Mass (u)	Mass Number (*Indicates Radioactive) A	Atomic Mass (u)	Percent Abundance	Half-Life (If Radioactive) $T_{1/2}$
(34)	Selenium			78	77.917 310	23.77	
				79*	78.918 500		$\leq 6.5 \times 10^4$ yr
				80	79.916 522	49.61	
				82*	81.916 700	8.73	1.4×10^{20} yr
35	Bromine	Br	79.904	79	78.918 338	50.69	
				81	80.916 291	49.31	
36	Krypton	Kr	83.80	78	77.920 386	0.35	
				80	79.916 378	2.28	
				81*	80.916 592		2.1×10^5 yr
				82	81.913 485	11.58	
				83	82.914 136	11.49	
				84	83.911 507	57.00	
				85*	84.912 527		10.76 yr
				86	85.910 610	17.30	
37	Rubidium	Rb	85.467 8	85	84.911 789	72.17	
				87*	86.909 184	27.83	4.75×10^{10} yr
38	Strontium	Sr	87.62	84	83.913 425	0.56	
				86	85.909 262	9.86	
				87	86.908 880	7.00	
				88	87.905 614	82.58	
				90*	89.907 738		29.1 yr
39	Yttrium	Y	88.905 85	89	88.905 848	100	
40	Zirconium	Zr	91.224	90	89.904 704	51.45	
				91	90.905 645	11.22	
				92	91.905 040	17.15	
				93*	92.906 476		1.5×10^6 yr
				94	93.906 316	17.38	
				96	95.908 276	2.80	
41	Niobium	Nb	92.906 38	91*	90.906 990		6.8×10^2 yr
				92*	91.907 193		3.5×10^7 yr
				93	92.906 378	100	
				94*	93.907 284		2×10^4 yr
42	Molybdenum	Mo	95.94	92	91.906 810	14.84	
				93*	92.906 812		3.5×10^3 yr
				94	93.905 088	9.25	
				95	94.905 842	15.92	
				96	95.904 679	16.68	
				97	96.906 021	9.55	
				98	97.905 408	24.13	
				100	99.907 477	9.63	
43	Technetium	Tc		97*	96.906 365		2.6×10^6 yr
				98*	97.907 216		4.2×10^6 yr
				99*	98.906 255		2.1×10^5 yr
44	Ruthenium	Ru	101.07	96	95.907 598	5.54	
				98	97.905 287	1.87	
				99	98.905 939	12.76	
				100	99.904 220	12.60	

continued

Table A.3

Atomic Number Z	Element	Symbol	Chemical Atomic Mass (u)	Mass Number (*Indicates Radioactive) A	Atomic Mass (u)	Percent Abundance	Half-Life (If Radioactive) $T_{1/2}$
(44)	Ruthenium			101	100.905 582	17.06	
				102	101.904 350	31.55	
				104	103.905 430	18.62	
45	Rhodium	Rh	102.905 50	103	102.905 504	100	
46	Palladium	Pd	106.42	102	101.905 608	1.02	
				104	103.904 035	11.14	
				105	104.905 084	22.33	
				106	105.903 483	27.33	
				107*	106.905 128		6.5×10^6 yr
				108	107.903 894	26.46	
				110	109.905 152	11.72	
47	Silver	Ag	107.868 2	107	106.905 093	51.839	
				109	108.904 756	48.161	
48	Cadmium	Cd	112.411	106	105.906 458	1.25	
				108	107.904 183	0.89	
				109*	108.904 986		462 days
				110	109.903 006	12.49	
				111	110.904 182	12.80	
				112	111.902 757	24.13	
				113*	112.904 401	12.22	9.3×10^{15} yr
				114	113.903 358	28.73	
				116	115.904 755	7.49	
49	Indium	In	114.818	113	112.904 061	4.29	
				115*	114.903 878	95.71	4.4×10^{14} yr
50	Tin	Sn	118.710	112	111.904 821	0.97	
				114	113.902 782	0.66	
				115	114.903 346	0.34	
				116	115.901 744	14.54	
				117	116.902 954	7.68	
				118	117.901 606	24.22	
				119	118.903 309	8.59	
				120	119.902 197	32.58	
				121*	120.904 237		55 yr
				122	121.903 440	4.63	
				124	123.905 275	5.79	
51	Antimony	Sb	121.760	121	120.903 818	57.21	
				123	122.904 216	42.79	
				125*	124.905 248		2.7 yr
52	Tellurium	Te	127.60	120	119.904 020	0.09	
				122	121.903 047	2.55	
				123*	122.904 273	0.89	1.3×10^{13} yr
				124	123.902 820	4.74	
				125	124.904 425	7.07	
				126	125.903 306	18.84	
				128*	127.904 461	31.74	$> 8 \times 10^{24}$ yr
				130*	129.906 223	34.08	$\leq 1.25 \times 10^{21}$ yr

continued

Table A.3

Table of Atomic Masses[a] *continued*

Atomic Number Z	Element	Symbol	Chemical Atomic Mass (u)	Mass Number (*Indicates Radioactive) A	Atomic Mass (u)	Percent Abundance	Half-Life (If Radioactive) $T_{1/2}$
53	Iodine	I	126.904 47	127	126.904 468	100	
				129*	128.904 988		1.6×10^7 yr
54	Xenon	Xe	131.29	124	123.905 896	0.09	
				126	125.904 269	0.09	
				128	127.903 530	1.92	
				129	128.904 780	26.44	
				130	129.903 508	4.08	
				131	130.905 082	21.18	
				132	131.904 145	26.89	
				134	133.905 394	10.44	
				136*	135.907 220	8.87	$\geq 2.36 \times 10^{21}$ yr
55	Cesium	Cs	132.905 45	133	132.905 447	100	
				134*	133.906 713		2.1 yr
				135*	134.905 972		2×10^6 yr
				137*	136.907 074		30 yr
56	Barium	Ba	137.327	130	129.906 310	0.106	
				132	131.905 056	0.101	
				133*	132.906 002		10.5 yr
				134	133.904 503	2.417	
				135	134.905 683	6.592	
				136	135.904 570	7.854	
				137	136.905 821	11.232	
				138	137.905 241	71.698	
57	Lanthanum	La	138.905 5	137*	136.906 466		6×10^4 yr
				138*	137.907 107	0.090	1.05×10^{11} yr
				139	138.906 349	99.910	
58	Cerium	Ce	140.116	136	135.907 144	0.185	
				138	137.905 986	0.251	
				140	139.905 434	88.450	
				142*	141.909 240	11.114	$>5 \times 10^{16}$ yr
59	Praseodymium	Pr	140.907 65	141	140.907 648	100	
60	Neodymium	Nd	144.24	142	141.907 719	27.2	
				143	142.909 810	12.2	
				144*	143.910 083	23.8	2.3×10^{15} yr
				145	144.912 569	8.3	
				146	145.913 112	17.2	
				148	147.916 888	5.7	
				150*	149.920 887	5.6	$>1 \times 10^{18}$ yr
61	Promethium	Pm		143*	142.910 928		265 days
				145*	144.912 744		17.7 yr
				146*	145.914 692		5.5 yr
				147*	146.915 134		2.623 yr
62	Samarium	Sm	150.36	144	143.911 995	3.07	
				146*	145.913 037		1.0×10^8 yr
				147*	146.914 893	14.99	1.06×10^{11} yr
				148*	147.914 818	11.24	7×10^{15} yr

continued

Table A.3

Table of Atomic Masses[a] *continued*

Atomic Number Z	Element	Symbol	Chemical Atomic Mass (u)	Mass Number (*Indicates Radioactive) A	Atomic Mass (u)	Percent Abundance	Half-Life (If Radioactive) $T_{1/2}$
(62)	Samarium			149*	148.917 180	13.82	$> 2 \times 10^{15}$ yr
				150	149.917 272	7.38	
				151*	150.919 928		90 yr
				152	151.919 728	26.75	
				154	153.922 205	22.75	
63	Europium	Eu	151.964	151	150.919 846	47.81	
				152*	151.921 740		13.5 yr
				153	152.921 226	52.19	
				154*	153.922 975		8.59 yr
				155*	154.922 889		4.7 yr
64	Gadolinium	Gd	157.25	148*	147.918 110		75 yr
				150*	149.918 656		1.8×10^{6} yr
				152*	151.919 788	0.20	1.1×10^{14} yr
				154	153.920 862	2.18	
				155	154.922 619	14.80	
				156	155.922 120	20.47	
				157	156.923 957	15.65	
				158	157.924 100	24.84	
				160	159.927 051	21.86	
65	Terbium	Tb	158.925 34	159	158.925 343	100	
66	Dysprosium	Dy	162.50	156	155.924 278	0.06	
				158	157.924 405	0.10	
				160	159.925 194	2.34	
				161	160.926 930	18.91	
				162	161.926 795	25.51	
				163	162.928 728	24.90	
				164	163.929 171	28.18	
67	Holmium	Ho	164.930 32	165	164.930 320	100	
				166*	165.932 281		1.2×10^{3} yr
68	Erbium	Er	167.6	162	161.928 775	0.14	
				164	163.929 197	1.61	
				166	165.930 290	33.61	
				167	166.932 045	22.93	
				168	167.932 368	26.78	
				170	169.935 460	14.93	
69	Thulium	Tm	168.934 21	169	168.934 211	100	
				171*	170.936 426		1.92 yr
70	Ytterbium	Yb	173.04	168	167.933 894	0.13	
				170	169.934 759	3.04	
				171	170.936 322	14.28	
				172	171.936 378	21.83	
				173	172.938 207	16.13	
				174	173.938 858	31.83	
				176	175.942 568	12.76	
71	Lutecium	Lu	174.967	173*	172.938 927		1.37 yr
				175	174.940 768	97.41	
				176*	175.942 682	2.59	3.78×10^{10} yr

continued

Table A.3

Table of Atomic Masses[a] *continued*

Atomic Number Z	Element	Symbol	Chemical Atomic Mass (u)	Mass Number (*Indicates Radioactive) A	Atomic Mass (u)	Percent Abundance	Half-Life (If Radioactive) $T_{1/2}$
72	Hafnium	Hf	178.49	174*	173.940 040	0.16	2.0×10^{15} yr
				176	175.941 402	5.26	
				177	176.943 220	18.60	
				178	177.943 698	27.28	
				179	178.945 815	13.62	
				180	179.946 549	35.08	
73	Tantalum	Ta	180.947 9	180*	179.947 466	0.012	8.152 h
				181	180.947 996	99.988	
74	Tungsten (Wolfram)	W	183.84	180	179.946 706	0.12	
				182	181.948 206	26.50	
				183	182.950 224	14.31	
				184*	183.950 933	30.64	$>3 \times 10^{17}$ yr
				186	185.954 362	28.43	
75	Rhenium	Re	186.207	185	184.952 956	37.40	
				187*	186.955 751	62.60	4.4×10^{10} yr
76	Osmium	Os	190.23	184	183.952 491	0.02	
				186*	185.953 838	1.59	2.0×10^{15} yr
				187	186.955 748	1.96	
				188	187.955 836	13.24	
				189	188.958 145	16.15	
				190	189.958 445	26.26	
				192	191.961 479	40.78	
				194*	193.965 179		6.0 yr
77	Iridium	Ir	192.217	191	190.960 591	37.3	
				193	192.962 924	62.7	
78	Platinum	Pt	195.078	190*	189.959 930	0.014	6.5×10^{11} yr
				192	191.961 035	0.782	
				194	193.962 664	32.967	
				195	194.964 774	33.832	
				196	195.964 935	25.242	
				198	197.967 876	7.163	
79	Gold	Au	196.966 55	197	196.966 552	100	
80	Mercury	Hg	200.59	196	195.965 815	0.15	
				198	197.966 752	9.97	
				199	198.968 262	16.87	
				200	199.968 309	23.10	
				201	200.970 285	13.18	
				202	201.970 626	29.86	
				204	203.973 476	6.87	
81	Thallium	Tl	204.383 3	203	202.972 329	29.524	
				204*	203.973 849		3.78 yr
				205	204.974 412	70.476	
		(Ra E″)		206*	205.976 095		4.2 min
		(Ac C″)		207*	206.977 408		4.77 min
		(Th C″)		208*	207.982 005		3.053 min
		(Ra C″)		210*	209.990 066		1.30 min

continued

Table A.3

Table of Atomic Masses[a] *continued*

Atomic Number Z	Element	Symbol	Chemical Atomic Mass (u)	Mass Number (*Indicates Radioactive) A	Atomic Mass (u)	Percent Abundance	Half-Life (If Radioactive) $T_{1/2}$
82	Lead	Pb	207.2	202*	201.972 144		5×10^4 yr
				204*	203.973 029	1.4	$\geq 1.4 \times 10^{17}$ yr
				205*	204.974 467		1.5×10^7 yr
				206	205.974 449	24.1	
				207	206.975 881	22.1	
				208	207.976 636	52.4	
		(Ra D)		210*	209.984 173		22.3 yr
		(Ac B)		211*	210.988 732		36.1 min
		(Th B)		212*	211.991 888		10.64 h
		(Ra B)		214*	213.999 798		26.8 min
83	Bismuth	Bi	208.980 38	207*	206.978 455		32.2 yr
				208*	207.979 727		3.7×10^5 yr
				209	208.980 383	100	
		(Ra E)		210*	209.984 105		5.01 days
		(Th C)		211*	210.987 258		2.14 min
				212*	211.991 272		60.6 min
		(Ra C)		214*	213.998 699		19.9 min
				215*	215.001 832		7.4 min
84	Polonium	Po		209*	208.982 416		102 yr
		(Ra F)		210*	209.982 857		138.38 days
		(Ac C′)		211*	210.986 637		0.52 s
		(Th C′)		212*	211.988 852		0.30 μs
		(Ra C′)		214*	213.995 186		164 μs
		(Ac A)		215*	214.999 415		0.001 8 s
		(Th A)		216*	216.001 905		0.145 s
		(Ra A)		218*	218.008 966		3.10 min
85	Astatine	At		215*	214.998 641		≈ 100 μs
				218*	218.008 682		1.6 s
				219*	219.011 297		0.9 min
86	Radon	Rn					
		(An)		219*	219.009 475		3.96 s
		(Tn)		220*	220.011 384		55.6 s
		(Rn)		222*	222.017 570		3.823 days
87	Francium	Fr					
		(Ac K)		223*	223.019 731		22 min
88	Radium	Ra					
		(Ac X)		223*	223.018 497		11.43 days
		(Th X)		224*	224.020 202		3.66 days
		(Ra)		226*	226.025 403		1 600 yr
		(Ms Th$_1$)		228*	228.031 064		5.75 yr
89	Actinium	Ac		227*	227.027 747		21.77 yr
		(Ms Th$_2$)		228*	228.031 015		6.15 h
90	Thorium	Th	232.038 1				
		(Rd Ac)		227*	227.027 699		18.72 days
		(Rd Th)		228*	228.028 731		1.913 yr
				229*	229.031 755		7 300 yr
		(Io)		230*	230.033 127		75.000 yr

continued

Table A.3

Table of Atomic Masses[a] *continued*

Atomic Number Z	Element	Symbol	Chemical Atomic Mass (u)	Mass Number (*Indicates Radioactive) A	Atomic Mass (u)	Percent Abundance	Half-Life (If Radioactive) $T_{1/2}$
(90)	Thorium	(UY)		231*	231.036 297		25.52 h
		(Th)		232*	232.038 050	100	1.40×10^{10} yr
		(UX$_1$)		234*	234.043 596		24.1 days
91	Protactinium	Pa	231.035 88	231*	231.035 879		32.760 yr
		(Uz)		234*	234.043 302		6.7 h
92	Uranium	U	238.028 9	232*	232.037 146		69 yr
				233*	233.039 628		1.59×10^5 yr
				234*	234.040 946	0.005 5	2.45×10^5 yr
		(Ac U)		235*	235.043 923	0.720 0	7.04×10^8 yr
				236*	236.045 562		2.34×10^7 yr
		(UI)		238*	238.050 783	99.274 5	4.47×10^9 yr
93	Neptunium	Np		235*	235.044 056		396 days
				236*	236.046 560		1.15×10^5 yr
				237*	237.048 167		2.14×10^6 yr
94	Plutonium	Pu		236*	236.046 048		2.87 yr
				238*	238.049 553		87.7 yr
				239*	239.052 156		2.412×10^4 yr
				240*	240.053 808		6 560 yr
				241*	241.056 845		14.4 yr
				242*	242.058 737		3.73×10^6 yr
				244*	244.064 198		8.1×10^7 yr

[a] Chemical atomic masses are from T. B. Coplen, "Atomic Weights of the Elements 1999," a technical report to the International Union of Pure and Applied Chemistry, and published in *Pure and Applied Chemistry*, 73(4), 667–683, 2001. Atomic masses of the isotopes are from G. Audi and A. H. Wapstra, "The 1995 Update to the Atomic Mass Evaluation," *Nuclear Physics*, A595, vol. 4, 409–480, December 25, 1995. Percent abundance values are from K. J. R. Rosman and P. D. P. Taylor, "Isotopic Compositions of the Elements 1999", a technical report to the International Union of Pure and Applied Chemistry, and published in *Pure and Applied Chemistry*, 70(1), 217–236, 1998.

Appendix B • Mathematics Review

These appendices in mathematics are intended as a brief review of operations and methods. Early in this course, you should be totally familiar with basic algebraic techniques, analytic geometry, and trigonometry. The appendices on differential and integral calculus are more detailed and are intended for those students who have difficulty applying calculus concepts to physical situations.

B.1   Scientific Notation

Many quantities that scientists deal with often have very large or very small values. For example, the speed of light is about 300 000 000 m/s, and the ink required to make the dot over an i in this textbook has a mass of about 0.000 000 001 kg. Obviously, it is very cumbersome to read, write, and keep track of numbers such as these. We avoid this problem by using a method dealing with powers of the number 10:

$$10^0 = 1$$
$$10^1 = 10$$
$$10^2 = 10 \times 10 = 100$$
$$10^3 = 10 \times 10 \times 10 = 1000$$
$$10^4 = 10 \times 10 \times 10 \times 10 = 10\ 000$$
$$10^5 = 10 \times 10 \times 10 \times 10 \times 10 = 100\ 000$$

and so on. The number of zeros corresponds to the power to which 10 is raised, called the **exponent** of 10. For example, the speed of light, 300 000 000 m/s, can be expressed as 3×10^8 m/s.

In this method, some representative numbers smaller than unity are

$$10^{-1} = \frac{1}{10} = 0.1$$

$$10^{-2} = \frac{1}{10 \times 10} = 0.01$$

$$10^{-3} = \frac{1}{10 \times 10 \times 10} = 0.001$$

$$10^{-4} = \frac{1}{10 \times 10 \times 10 \times 10} = 0.000\ 1$$

$$10^{-5} = \frac{1}{10 \times 10 \times 10 \times 10 \times 10} = 0.000\ 01$$

In these cases, the number of places the decimal point is to the left of the digit 1 equals the value of the (negative) exponent. Numbers expressed as some power of 10 multiplied by another number between 1 and 10 are said to be in **scientific notation.** For example, the scientific notation for 5 943 000 000 is 5.943×10^9 and that for 0.000 083 2 is 8.32×10^{-5}.

When numbers expressed in scientific notation are being multiplied, the following general rule is very useful:

$$10^n \times 10^m = 10^{n+m} \tag{B.1}$$

where n and m can be *any* numbers (not necessarily integers). For example, $10^2 \times 10^5 = 10^7$. The rule also applies if one of the exponents is negative: $10^3 \times 10^{-8} = 10^{-5}$.

When dividing numbers expressed in scientific notation, note that

$$\frac{10^n}{10^m} = 10^n \times 10^{-m} = 10^{n-m} \qquad\qquad (B.2)$$

Exercises

With help from the above rules, verify the answers to the following:

1. $86\ 400 = 8.64 \times 10^4$
2. $9\ 816\ 762.5 = 9.816\ 762\ 5 \times 10^6$
3. $0.000\ 000\ 039\ 8 = 3.98 \times 10^{-8}$
4. $(4 \times 10^8)\,(9 \times 10^9) = 3.6 \times 10^{18}$
5. $(3 \times 10^7)\,(6 \times 10^{-12}) = 1.8 \times 10^{-4}$
6. $\dfrac{75 \times 10^{-11}}{5 \times 10^{-3}} - 1.5 \times 10^{-7}$
7. $\dfrac{(3 \times 10^6)\,(8 \times 10^{-2})}{(2 \times 10^{17})\,(6 \times 10^5)} = 2 \times 10^{-18}$

B.2   Algebra

Some Basic Rules

When algebraic operations are performed, the laws of arithmetic apply. Symbols such as x, y, and z are usually used to represent quantities that are not specified, what are called the **unknowns.**

First, consider the equation

$$8x = 32$$

If we wish to solve for x, we can divide (or multiply) each side of the equation by the same factor without destroying the equality. In this case, if we divide both sides by 8, we have

$$\frac{8x}{8} = \frac{32}{8}$$

$$x = 4$$

Next consider the equation

$$x + 2 = 8$$

In this type of expression, we can add or subtract the same quantity from each side. If we subtract 2 from each side, we obtain

$$x + 2 - 2 = 8 - 2$$

$$x = 6$$

In general, if $x + a = b$, then $x = b - a$.

Now consider the equation

$$\frac{x}{5} = 9$$

If we multiply each side by 5, we are left with x on the left by itself and 45 on the right:

$$\left(\frac{x}{5}\right)(5) = 9 \times 5$$

$$x = 45$$

In all cases, *whatever operation is performed on the left side of the equality must also be performed on the right side.*

The following rules for multiplying, dividing, adding, and subtracting fractions should be recalled, where a, b, and c are three numbers:

	Rule	**Example**
Multiplying	$\left(\dfrac{a}{b}\right)\left(\dfrac{c}{d}\right) = \dfrac{ac}{bd}$	$\left(\dfrac{2}{3}\right)\left(\dfrac{4}{5}\right) = \dfrac{8}{15}$
Dividing	$\dfrac{(a/b)}{(c/d)} = \dfrac{ad}{bc}$	$\dfrac{2/3}{4/5} = \dfrac{(2)(5)}{(4)(3)} = \dfrac{10}{12}$
Adding	$\dfrac{a}{b} \pm \dfrac{c}{d} = \dfrac{ad \pm bc}{bd}$	$\dfrac{2}{3} - \dfrac{4}{5} = \dfrac{(2)(5) - (4)(3)}{(3)(5)} = -\dfrac{2}{15}$

Exercises

In the following exercises, solve for x:

Answers

1. $a = \dfrac{1}{1 + x}$ $x = \dfrac{1 - a}{a}$

2. $3x - 5 = 13$ $x = 6$

3. $ax - 5 = bx + 2$ $x = \dfrac{7}{a - b}$

4. $\dfrac{5}{2x + 6} = \dfrac{3}{4x + 8}$ $x = -\dfrac{11}{7}$

Powers

When powers of a given quantity x are multiplied, the following rule applies:

$$x^n x^m = x^{n+m} \tag{B.3}$$

For example, $x^2 x^4 = x^{2+4} = x^6$.

When dividing the powers of a given quantity, the rule is

$$\frac{x^n}{x^m} = x^{n-m} \tag{B.4}$$

For example, $x^8/x^2 = x^{8-2} = x^6$.

A power that is a fraction, such as $\frac{1}{3}$, corresponds to a root as follows:

$$x^{1/n} = \sqrt[n]{x} \tag{B.5}$$

For example, $4^{1/3} = \sqrt[3]{4} = 1.5874$. (A scientific calculator is useful for such calculations.)

Finally, any quantity x^n raised to the mth power is

$$(x^n)^m = x^{nm} \tag{B.6}$$

Table B.1 summarizes the rules of exponents.

Table B.1

Rules of Exponents

$$x^0 = 1$$
$$x^1 = x$$
$$x^n x^m = x^{n+m}$$
$$x^n/x^m = x^{n-m}$$
$$x^{1/n} = \sqrt[n]{x}$$
$$(x^n)^m = x^{nm}$$

Exercises

Verify the following:

1. $3^2 \times 3^3 = 243$

2. $x^5 x^{-8} = x^{-3}$

3. $x^{10}/x^{-5} = x^{15}$
4. $5^{1/3} = 1.709\,975$ (Use your calculator.)
5. $60^{1/4} = 2.783\,158$ (Use your calculator.)
6. $(x^4)^3 = x^{12}$

Factoring

Some useful formulas for factoring an equation are

$$ax + ay + az = a(x + y + z) \qquad \text{common factor}$$
$$a^2 + 2ab + b^2 = (a + b)^2 \qquad \text{perfect square}$$
$$a^2 - b^2 = (a + b)(a - b) \qquad \text{differences of squares}$$

Quadratic Equations

The general form of a quadratic equation is

$$ax^2 + bx + c = 0 \tag{B.7}$$

where x is the unknown quantity and a, b, and c are numerical factors referred to as **coefficients** of the equation. This equation has two roots, given by

$$x = \frac{-b \pm \sqrt{b^2 - 4ac}}{2a} \tag{B.8}$$

If $b^2 \geq 4ac$, the roots are real.

Example 1

The equation $x^2 + 5x + 4 = 0$ has the following roots corresponding to the two signs of the square-root term:

$$x = \frac{-5 \pm \sqrt{5^2 - (4)(1)(4)}}{2(1)} = \frac{-5 \pm \sqrt{9}}{2} = \frac{-5 \pm 3}{2}$$

$$x_+ = \frac{-5 + 3}{2} = -1 \qquad x_- = \frac{-5 - 3}{2} = -4$$

where x_+ refers to the root corresponding to the positive sign and x_- refers to the root corresponding to the negative sign.

Exercises

Solve the following quadratic equations:

Answers

1. $x^2 + 2x - 3 = 0$ $x_+ = 1$ $x_- = -3$
2. $2x^2 - 5x + 2 = 0$ $x_+ = 2$ $x_- = \frac{1}{2}$
3. $2x^2 - 4x - 9 = 0$ $x_+ = 1 + \sqrt{22}/2$ $x_- = 1 - \sqrt{22}/2$

Linear Equations

A linear equation has the general form

$$y = mx + b \tag{B.9}$$

where m and b are constants. This equation is referred to as being linear because the graph of y versus x is a straight line, as shown in Figure B.1. The constant b, called the **y-intercept,** represents the value of y at which the straight line intersects the y axis. The constant m is equal to the **slope** of the straight line. If any two points on the straight line are specified by the coordinates (x_1, y_1) and (x_2, y_2), as in Figure B.1, then

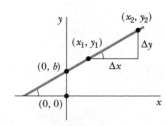

Figure B.1

the slope of the straight line can be expressed as

$$\text{Slope} = \frac{y_2 - y_1}{x_2 - x_1} = \frac{\Delta y}{\Delta x} \qquad \text{(B.10)}$$

Note that m and b can have either positive or negative values. If $m > 0$, the straight line has a *positive* slope, as in Figure B.1. If $m < 0$, the straight line has a *negative* slope. In Figure B.1, both m and b are positive. Three other possible situations are shown in Figure B.2.

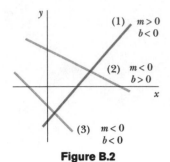

Figure B.2

Exercises

1. Draw graphs of the following straight lines:
   (a) $y = 5x + 3$ (b) $y = -2x + 4$ (c) $y = -3x - 6$

2. Find the slopes of the straight lines described in Exercise 1.

Answers   (a) 5   (b) -2 (c) -3

3. Find the slopes of the straight lines that pass through the following sets of points:
   (a) $(0, -4)$ and $(4, 2)$ (b) $(0, 0)$ and $(2, -5)$ (c) $(-5, 2)$ and $(4, -2)$

Answers   (a) $3/2$ (b) $-5/2$ (c) $-4/9$

Solving Simultaneous Linear Equations

Consider the equation $3x + 5y = 15$, which has two unknowns, x and y. Such an equation does not have a unique solution. For example, note that $(x = 0, y = 3)$, $(x = 5, y = 0)$, and $(x = 2, y = 9/5)$ are all solutions to this equation.

If a problem has two unknowns, a unique solution is possible only if we have *two* equations. In general, if a problem has n unknowns, its solution requires n equations. In order to solve two simultaneous equations involving two unknowns, x and y, we solve one of the equations for x in terms of y and substitute this expression into the other equation.

Example 2

Solve the following two simultaneous equations:

$$(1) \qquad 5x + y = -8$$
$$(2) \qquad 2x - 2y = 4$$

Solution From Equation (2), $x = y + 2$. Substitution of this into Equation (1) gives

$$5(y + 2) + y = -8$$
$$6y = -18$$
$$y = \boxed{-3}$$
$$x = y + 2 = \boxed{-1}$$

Alternate Solution Multiply each term in Equation (1) by the factor 2 and add the result to Equation (2):

$$10x + 2y = -16$$
$$\underline{2x - 2y = 4}$$
$$12x = -12$$
$$x = \boxed{-1}$$
$$y = x - 2 = \boxed{-3}$$

Two linear equations containing two unknowns can also be solved by a graphical method. If the straight lines corresponding to the two equations are plotted in a conventional coordinate system, the intersection of the two lines represents the solution. For example, consider the two equations

$$x - y = 2$$
$$x - 2y = -1$$

These are plotted in Figure B.3. The intersection of the two lines has the coordinates $x = 5$, $y = 3$. This represents the solution to the equations. You should check this solution by the analytical technique discussed above.

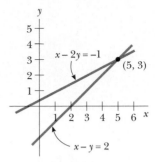

Figure B.3

Exercises

Solve the following pairs of simultaneous equations involving two unknowns:

Answers

1. $x + y = 8$ $x = 5, y = 3$
    $x - y = 2$

2. $98 - T = 10a$ $T = 65, a = 3.3$
    $T - 49 = 5a$

3. $6x + 2y = 6$ $x = 2, y = -3$
    $8x - 4y = 28$

Logarithms

Suppose that a quantity x is expressed as a power of some quantity a:

$$x = a^y \tag{B.11}$$

The number a is called the **base** number. The **logarithm** of x with respect to the base a is equal to the exponent to which the base must be raised in order to satisfy the expression $x = a^y$:

$$y = \log_a x \tag{B.12}$$

Conversely, the **antilogarithm** of y is the number x:

$$x = \text{antilog}_a y \tag{B.13}$$

In practice, the two bases most often used are base 10, called the *common* logarithm base, and base $e = 2.718\,282$, called Euler's constant or the *natural* logarithm base. When common logarithms are used,

$$y = \log_{10} x \qquad (\text{or } x = 10^y) \tag{B.14}$$

When natural logarithms are used,

$$y = \ln x \qquad (\text{or } x = e^y) \tag{B.15}$$

For example, $\log_{10} 52 = 1.716$, so that $\text{antilog}_{10} 1.716 = 10^{1.716} = 52$. Likewise, $\ln 52 = 3.951$, so $\text{antiln } 3.951 = e^{3.951} = 52$.

In general, note that you can convert between base 10 and base e with the equality

$$\ln x = (2.302\,585) \log_{10} x \tag{B.16}$$

Finally, some useful properties of logarithms are

$$\left.\begin{array}{l} \log(ab) = \log a + \log b \\ \log(a/b) = \log a - \log b \\ \log(a^n) = n \log a \\ \qquad \ln e = 1 \\ \qquad \ln e^a = a \\ \ln\left(\dfrac{1}{a}\right) = -\ln a \end{array}\right\} \text{ any base}$$

B.3 Geometry

The **distance** d between two points having coordinates (x_1, y_1) and (x_2, y_2) is

$$d = \sqrt{(x_2 - x_1)^2 + (y_2 - y_1)^2} \tag{B.17}$$

Radian measure: The arc length s of a circular arc (Fig. B.4) is proportional to the radius r for a fixed value of θ (in radians):

$$s = r\theta$$
$$\theta = \frac{s}{r} \tag{B.18}$$

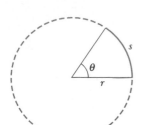

Figure B.4

Table B.2 gives the areas and volumes for several geometric shapes used throughout this text:

Table B.2

Useful Information for Geometry			
Shape	**Area or Volume**	**Shape**	**Area or Volume**

Rectangle — Area = ℓw

Sphere — Surface area = $4\pi r^2$, Volume = $\dfrac{4\pi r^3}{3}$

Circle — Area = πr^2 (Circumference = $2\pi r$)

Cylinder — Lateral surface area = $2\pi r \ell$, Volume = $\pi r^2 \ell$

Triangle — Area = $\frac{1}{2} bh$

Rectangular box — Surface area = $2(\ell h + \ell w + hw)$, Volume = ℓwh

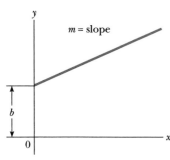

Figure B.5

The equation of a **straight line** (Fig. B.5) is

$$y = mx + b \tag{B.19}$$

where b is the y intercept and m is the slope of the line.

The equation of a **circle** of radius R centered at the origin is

$$x^2 + y^2 = R^2 \tag{B.20}$$

The equation of an **ellipse** having the origin at its center (Fig. B.6) is

$$\frac{x^2}{a^2} + \frac{y^2}{b^2} = 1 \tag{B.21}$$

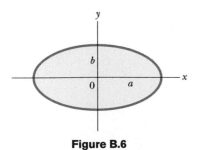

Figure B.6

where a is the length of the semimajor axis (the longer one) and b is the length of the semiminor axis (the shorter one).

The equation of a **parabola** the vertex of which is at $y = b$ (Fig. B.7) is

$$y = ax^2 + b \tag{B.22}$$

The equation of a **rectangular hyperbola** (Fig. B.8) is

$$xy = \text{constant} \tag{B.23}$$

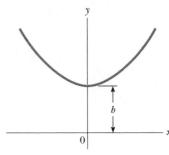

Figure B.7

B.4   Trigonometry

That portion of mathematics based on the special properties of the right triangle is called trigonometry. By definition, a right triangle is one containing a 90° angle. Consider the right triangle shown in Figure B.9, where side a is opposite the angle θ, side b is adjacent to the angle θ, and side c is the hypotenuse of the triangle. The three basic trigonometric functions defined by such a triangle are the sine (sin), cosine (cos), and tangent (tan) functions. In terms of the angle θ, these functions are defined by

$$\sin \theta \equiv \frac{\text{side opposite } \theta}{\text{hypotenuse}} = \frac{a}{c} \tag{B.24}$$

$$\cos \theta \equiv \frac{\text{side adjacent to } \theta}{\text{hypotenuse}} = \frac{b}{c} \tag{B.25}$$

$$\tan \theta \equiv \frac{\text{side opposite } \theta}{\text{side adjacent to } \theta} = \frac{a}{b} \tag{B.26}$$

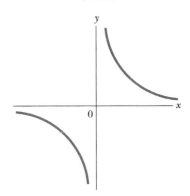

Figure B.8

The Pythagorean theorem provides the following relationship among the sides of a right triangle:

$$c^2 = a^2 + b^2 \tag{B.27}$$

From the above definitions and the Pythagorean theorem, it follows that

$$\sin^2 \theta + \cos^2 \theta = 1$$

$$\tan \theta = \frac{\sin \theta}{\cos \theta}$$

The cosecant, secant, and cotangent functions are defined by

$$\csc \theta \equiv \frac{1}{\sin \theta} \qquad \sec \theta \equiv \frac{1}{\cos \theta} \qquad \cot \theta \equiv \frac{1}{\tan \theta}$$

The relationships below follow directly from the right triangle shown in Figure B.9:

$$\sin \theta = \cos(90° - \theta)$$

$$\cos \theta = \sin(90° - \theta)$$

$$\cot \theta = \tan(90° - \theta)$$

Some properties of trigonometric functions are

$$\sin(-\theta) = -\sin \theta$$

$$\cos(-\theta) = \cos \theta$$

$$\tan(-\theta) = -\tan \theta$$

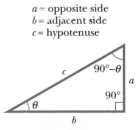

a = opposite side
b = adjacent side
c = hypotenuse

Figure B.9

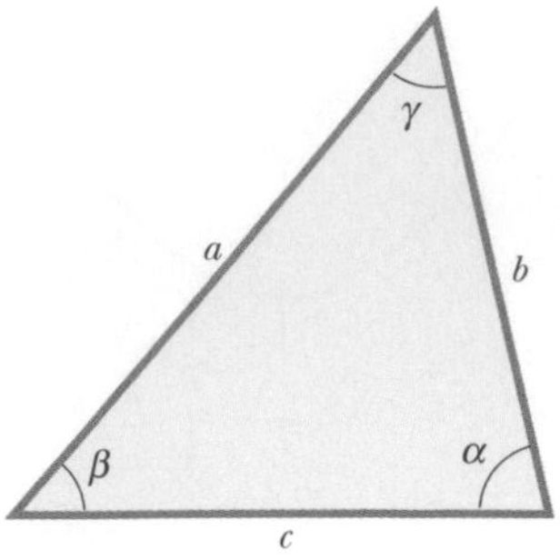

Figure B.10

The following relationships apply to *any* triangle, as shown in Figure B.10:

$$\alpha + \beta + \gamma = 180°$$

Law of cosines
$$a^2 = b^2 + c^2 - 2bc\cos\alpha$$
$$b^2 = a^2 + c^2 - 2ac\cos\beta$$
$$c^2 = a^2 + b^2 - 2ab\cos\gamma$$

Law of sines
$$\frac{a}{\sin\alpha} = \frac{b}{\sin\beta} = \frac{c}{\sin\gamma}$$

Table B.3 lists a number of useful trigonometric identities.

Table B.3

Some Trigonometric Identities

$$\sin^2\theta + \cos^2\theta = 1 \qquad\qquad \csc^2\theta = 1 + \cot^2\theta$$

$$\sec^2\theta = 1 + \tan^2\theta \qquad\qquad \sin^2\frac{\theta}{2} = \tfrac{1}{2}(1 - \cos\theta)$$

$$\sin 2\theta = 2\sin\theta\cos\theta \qquad\qquad \cos^2\frac{\theta}{2} = \tfrac{1}{2}(1 + \cos\theta)$$

$$\cos 2\theta = \cos^2\theta - \sin^2\theta \qquad\qquad 1 - \cos\theta = 2\sin^2\frac{\theta}{2}$$

$$\tan 2\theta = \frac{2\tan\theta}{1 - \tan^2\theta} \qquad\qquad \tan\frac{\theta}{2} = \sqrt{\frac{1 - \cos\theta}{1 + \cos\theta}}$$

$$\sin(A \pm B) = \sin A\cos B \pm \cos A\sin B$$
$$\cos(A \pm B) = \cos A\cos B \mp \sin A\sin B$$
$$\sin A \pm \sin B = 2\sin[\tfrac{1}{2}(A \pm B)]\cos[\tfrac{1}{2}(A \mp B)]$$
$$\cos A + \cos B = 2\cos[\tfrac{1}{2}(A + B)]\cos[\tfrac{1}{2}(A - B)]$$
$$\cos A - \cos B = 2\sin[\tfrac{1}{2}(A + B)]\sin[\tfrac{1}{2}(B - A)]$$

Example 3

Consider the right triangle in Figure B.11, in which $a = 2$, $b = 5$, and c is unknown. From the Pythagorean theorem, we have

$$c^2 = a^2 + b^2 = 2^2 + 5^2 = 4 + 25 = 29$$

$$c = \sqrt{29} = \boxed{5.39}$$

To find the angle θ, note that

$$\tan\theta = \frac{a}{b} = \frac{2}{5} = 0.400$$

From a table of functions or from a calculator, we have

$$\theta = \tan^{-1}(0.400) = \boxed{21.8°}$$

where $\tan^{-1}(0.400)$ is the notation for "angle whose tangent is 0.400," sometimes written as arctan (0.400).

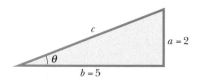

Figure B.11 (Example 3).

Exercises

1. In Figure B.12, identify (a) the side opposite θ (b) the side adjacent to ϕ. Then find (c) $\cos\theta$ (d) $\sin\phi$ (e) $\tan\phi$.

   Answers   (a) 3   (b) 3   (c) $\frac{4}{5}$ (d) $\frac{4}{5}$ (e) $\frac{4}{3}$

2. In a certain right triangle, the two sides that are perpendicular to each other are 5 m and 7 m long. What is the length of the third side?

   Answer   8.60 m

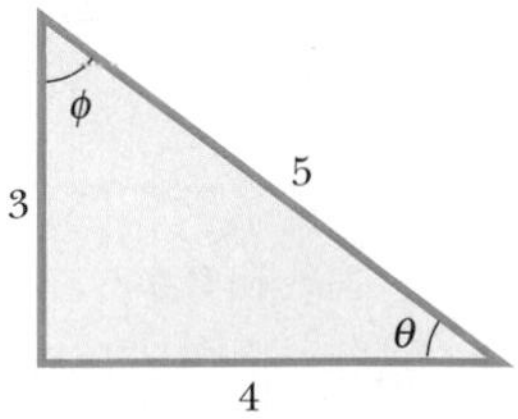

Figure B.12

3. A right triangle has a hypotenuse of length 3 m, and one of its angles is 30°. What is the length of (a) the side opposite the 30° angle (b) the side adjacent to the 30° angle?

Answers (a) 1.5 m (b) 2.60 m

B.5   Series Expansions

$$(a + b)^n = a^n + \frac{n}{1!}a^{n-1}b + \frac{n(n-1)}{2!}a^{n-2}b^2 + \cdots$$

$$(1 + x)^n = 1 + nx + \frac{n(n-1)}{2!}x^2 + \cdots$$

$$e^x = 1 + x + \frac{x^2}{2!} + \frac{x^3}{3!} + \cdots$$

$$\ln(1 \pm x) = \pm x - \tfrac{1}{2}x^2 \pm \tfrac{1}{3}x^3 - \cdots$$

$$\sin x = x - \frac{x^3}{3!} + \frac{x^5}{5!} - \cdots$$

$$\cos x = 1 - \frac{x^2}{2!} + \frac{x^4}{4!} - \cdots$$

$$\tan x = x + \frac{x^3}{3} + \frac{2x^5}{15} + \cdots \qquad |x| < \pi/2$$

x in radians

For $x \ll 1$, the following approximations can be used[1]:

$$(1 + x)^n \approx 1 + nx \qquad \sin x \approx x$$
$$e^x \approx 1 + x \qquad \cos x \approx 1$$
$$\ln(1 \pm x) \approx \pm x \qquad \tan x \approx x$$

B.6   Differential Calculus

In various branches of science, it is sometimes necessary to use the basic tools of calculus, invented by Newton, to describe physical phenomena. The use of calculus is fundamental in the treatment of various problems in Newtonian mechanics, electricity, and magnetism. In this section, we simply state some basic properties and "rules of thumb" that should be a useful review to the student.

First, a **function** must be specified that relates one variable to another (such as a coordinate as a function of time). Suppose one of the variables is called y (the dependent variable), the other x (the independent variable). We might have a function relationship such as

$$y(x) = ax^3 + bx^2 + cx + d$$

If a, b, c, and d are specified constants, then y can be calculated for any value of x. We usually deal with continuous functions, that is, those for which y varies "smoothly" with x.

The **derivative** of y with respect to x is defined as the limit, as Δx approaches zero, of the slopes of chords drawn between two points on the y versus x curve. Mathematically, we write this definition as

$$\frac{dy}{dx} = \lim_{\Delta x \to 0} \frac{\Delta y}{\Delta x} = \lim_{\Delta x \to 0} \frac{y(x + \Delta x) - y(x)}{\Delta x} \qquad \text{(B.28)}$$

where Δy and Δx are defined as $\Delta x = x_2 - x_1$ and $\Delta y = y_2 - y_1$ (Fig. B.13). It is important to note that dy/dx *does not* mean dy divided by dx, but is simply a notation of the limiting process of the derivative as defined by Equation B.28.

---

[1] The approximations for the functions $\sin x$, $\cos x$, and $\tan x$ are for $x \leq 0.1$ rad.

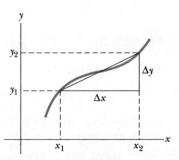

Figure B.13

A useful expression to remember when $y(x) = ax^n$, where a is a *constant* and n is *any* positive or negative number (integer or fraction), is

$$\frac{dy}{dx} = nax^{n-1} \tag{B.29}$$

If $y(x)$ is a polynomial or algebraic function of x, we apply Equation B.29 to *each* term in the polynomial and take $d[\text{constant}]/dx = 0$. In Examples 4 through 7, we evaluate the derivatives of several functions.

Special Properties of the Derivative

A. Derivative of the product of two functions   If a function $f(x)$ is given by the product of two functions, say, $g(x)$ and $h(x)$, then the derivative of $f(x)$ is defined as

$$\frac{d}{dx} f(x) = \frac{d}{dx} [g(x) h(x)] = g\frac{dh}{dx} + h\frac{dg}{dx} \tag{B.30}$$

B. Derivative of the sum of two functions   If a function $f(x)$ is equal to the sum of two functions, then the derivative of the sum is equal to the sum of the derivatives:

$$\frac{d}{dx} f(x) = \frac{d}{dx} [g(x) + h(x)] = \frac{dg}{dx} + \frac{dh}{dx} \tag{B.31}$$

C. Chain rule of differential calculus   If $y = f(x)$ and $x = g(z)$, then dy/dz can be written as the product of two derivatives:

$$\frac{dy}{dz} = \frac{dy}{dx}\frac{dx}{dz} \tag{B.32}$$

D. The second derivative   The second derivative of y with respect to x is defined as the derivative of the function dy/dx (the derivative of the derivative). It is usually written

$$\frac{d^2y}{dx^2} = \frac{d}{dx}\left(\frac{dy}{dx}\right) \tag{B.33}$$

Example 4

Suppose $y(x)$ (that is, y as a function of x) is given by

$$y(x) = ax^3 + bx + c$$

where a and b are constants. Then it follows that

$$y(x + \Delta x) = a(x + \Delta x)^3 + b(x + \Delta x) + c$$

$$y(x + \Delta x) = a(x^3 + 3x^2\Delta x + 3x\Delta x^2 + \Delta x^3) + b(x + \Delta x) + c$$

so

$$\Delta y = y(x + \Delta x) - y(x) = a(3x^2\Delta x + 3x\Delta x^2 + \Delta x^3) + b\Delta x$$

Substituting this into Equation B.28 gives

$$\frac{dy}{dx} = \lim_{\Delta x \to 0} \frac{\Delta y}{\Delta x} = \lim_{\Delta x \to 0} [3ax^2 + 3x\Delta x + \Delta x^2] + b$$

$$\frac{dy}{dx} = \boxed{3ax^2 + b}$$

Example 5

Find the derivative of

$$y(x) = 8x^5 + 4x^3 + 2x + 7$$

Solution Applying Equation B.29 to each term independently, and remembering that $d/dx\,(\text{constant}) = 0$, we have

$$\frac{dy}{dx} = 8(5)x^4 + 4(3)x^2 + 2(1)x^0 + 0$$

$$\frac{dy}{dx} = \boxed{40x^4 + 12x^2 + 2}$$

Example 6

Find the derivative of $y(x) = x^3/(x + 1)^2$ with respect to x.

Solution We can rewrite this function as $y(x) = x^3(x + 1)^{-2}$ and apply Equation B.30:

$$\frac{dy}{dx} = (x + 1)^{-2} \frac{d}{dx}(x^3) + x^3 \frac{d}{dx}(x + 1)^{-2}$$

$$= (x + 1)^{-2} 3x^2 + x^3(-2)(x + 1)^{-3}$$

$$\frac{dy}{dx} = \frac{3x^2}{(x + 1)^2} - \frac{2x^3}{(x + 1)^3}$$

Example 7

A useful formula that follows from Equation B.30 is the derivative of the quotient of two functions. Show that

$$\frac{d}{dx}\left[\frac{g(x)}{h(x)}\right] = \frac{h\dfrac{dg}{dx} - g\dfrac{dh}{dx}}{h^2}$$

Solution We can write the quotient as gh^{-1} and then apply Equations B.29 and B.30:

$$\frac{d}{dx}\left(\frac{g}{h}\right) = \frac{d}{dx}(gh^{-1}) = g\frac{d}{dx}(h^{-1}) + h^{-1}\frac{d}{dx}(g)$$

$$= -gh^{-2}\frac{dh}{dx} + h^{-1}\frac{dg}{dx}$$

$$= \frac{h\dfrac{dg}{dx} - g\dfrac{dh}{dx}}{h^2}$$

Some of the more commonly used derivatives of functions are listed in Table B.4.

B.7 Integral Calculus

We think of integration as the inverse of differentiation. As an example, consider the expression

$$f(x) - \frac{dy}{dx} = 3ax^2 + b \tag{B.34}$$

which was the result of differentiating the function

$$y(x) = ax^3 + bx + c$$

in Example 4. We can write Equation B.34 as $dy = f(x)\,dx = (3ax^2 + b)\,dx$ and obtain $y(x)$ by "summing" over all values of x. Mathematically, we write this inverse operation

$$y(x) = \int f(x)\,dx$$

For the function $f(x)$ given by Equation B.34, we have

$$y(x) = \int (3ax^2 + b)\,dx = ax^3 + bx + c$$

where c is a constant of the integration. This type of integral is called an *indefinite integral* because its value depends on the choice of c.

A general **indefinite integral** $I(x)$ is defined as

$$I(x) = \int f(x)\,dx \tag{B.35}$$

where $f(x)$ is called the *integrand* and $f(x) = dI(x)/dx$.

For a *general continuous* function $f(x)$, the integral can be described as the area under the curve bounded by $f(x)$ and the x axis, between two specified values of x, say, x_1 and x_2, as in Figure B.14.

The area of the blue element is approximately $f(x_i)\,\Delta x_i$. If we sum all these area elements from x_1 and x_2 and take the limit of this sum as $\Delta x_i \rightarrow 0$, we obtain the *true*

Table B.4

Derivative for Several Functions
$\dfrac{d}{dx}(a) = 0$
$\dfrac{d}{dx}(ax^n) = nax^{n-1}$
$\dfrac{d}{dx}(e^{ax}) = ae^{ax}$
$\dfrac{d}{dx}(\sin ax) = a\cos ax$
$\dfrac{d}{dx}(\cos ax) = -a\sin ax$
$\dfrac{d}{dx}(\tan ax) = a\sec^2 ax$
$\dfrac{d}{dx}(\cot ax) = -a\csc^2 dx$
$\dfrac{d}{dx}(\sec x) = \tan x\sec x$
$\dfrac{d}{dx}(\csc x) = -\cot x\csc x$
$\dfrac{d}{dx}(\ln ax) = \dfrac{1}{x}$

Note: The symbols a and n represent constants.

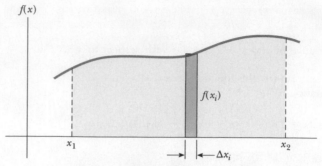

Figure B.14

area under the curve bounded by $f(x)$ and x, between the limits x_1 and x_2:

$$\text{Area} = \lim_{\Delta x \to 0} \sum_i f(x_i)\,\Delta x_i = \int_{x_1}^{x_2} f(x)\,dx \qquad (B.36)$$

Integrals of the type defined by Equation B.36 are called **definite integrals.**

One common integral that arises in practical situations has the form

$$\int x^n\,dx = \frac{x^{n+1}}{n+1} + c \qquad (n \neq -1) \qquad (B.37)$$

This result is obvious, being that differentiation of the right-hand side with respect to x gives $f(x) = x^n$ directly. If the limits of the integration are known, this integral becomes a *definite integral* and is written

$$\int_{x_1}^{x_2} x^n\,dx = \frac{x^{n+1}}{n+1}\bigg]_{x_1}^{x_2} = \frac{x_2^{\,n+1} - x_1^{\,n+1}}{n+1} \qquad (n \neq -1) \qquad (B.38)$$

Examples

1. $\displaystyle\int_0^a x^2\,dx = \frac{x^3}{3}\bigg]_0^a = \frac{a^3}{3}$

2. $\displaystyle\int_0^b x^{3/2}\,dx = \frac{x^{5/2}}{5/2}\bigg]_0^b = \tfrac{2}{5}b^{5/2}$

3. $\displaystyle\int_3^5 x\,dx = \frac{x^2}{2}\bigg]_3^5 = \frac{5^2 - 3^2}{2} = 8$

Partial Integration

Sometimes it is useful to apply the method of *partial integration* (also called "integrating by parts") to evaluate certain integrals. The method uses the property that

$$\int u\,dv = uv - \int v\,du \qquad (B.39)$$

where u and v are *carefully* chosen so as to reduce a complex integral to a simpler one. In many cases, several reductions have to be made. Consider the function

$$I(x) = \int x^2 e^x\,dx$$

This can be evaluated by integrating by parts twice. First, if we choose $u = x^2$, $v = e^x$, we obtain

$$\int x^2 e^x\,dx = \int x^2\,d(e^x) = x^2 e^x - 2\int e^x x\,dx + c_1$$

Now, in the second term, choose $u = x$, $v = e^x$, which gives

$$\int x^2 e^x \, dx = x^2 e^x - 2xe^x + 2\int e^x \, dx + c_1$$

or

$$\int x^2 e^x \, dx = x^2 e^x - 2xe^x + 2e^x + c_2$$

The Perfect Differential

Another useful method to remember is the use of the *perfect differential,* in which we look for a change of variable such that the differential of the function is the differential of the independent variable appearing in the integrand. For example, consider the integral

$$I(x) = \int \cos^2 x \sin x \, dx$$

This becomes easy to evaluate if we rewrite the differential as $d(\cos x) = -\sin x \, dx$. The integral then becomes

$$\int \cos^2 x \sin x \, dx = -\int \cos^2 x \, d(\cos x)$$

If we now change variables, letting $y = \cos x$, we obtain

$$\int \cos^2 x \sin x \, dx = -\int y^2 \, dy = -\frac{y^3}{3} + c = -\frac{\cos^3 x}{3} + c$$

Table B.5 lists some useful indefinite integrals. Table B.6 gives Gauss's probability integral and other definite integrals. A more complete list can be found in various handbooks, such as *The Handbook of Chemistry and Physics,* CRC Press.

Table B.5

Some Indefinite Integrals (An arbitrary constant should be added to each of these integrals.)	
$\int x^n \, dx = \dfrac{x^{n+1}}{n+1}$ (provided $n \neq -1$)	$\int \dfrac{dx}{\sqrt{a^2 - x^2}} = \sin^{-1}\dfrac{x}{a} = -\cos^{-1}\dfrac{x}{a}$ $(a^2 - x^2 > 0)$
$\int \dfrac{dx}{x} = \int x^{-1} \, dx = \ln x$	$\int \dfrac{dx}{\sqrt{x^2 \pm a^2}} = \ln(x + \sqrt{x^2 \pm a^2})$
$\int \dfrac{dx}{a + bx} = \dfrac{1}{b} \ln(a + bx)$	$\int \dfrac{x \, dx}{\sqrt{a^2 - x^2}} = -\sqrt{a^2 - x^2}$
$\int \dfrac{x \, dx}{a + bx} = \dfrac{x}{b} - \dfrac{a}{b^2} \ln(a + bx)$	$\int \dfrac{x \, dx}{\sqrt{x^2 \pm a^2}} = \sqrt{x^2 \pm a^2}$
$\int \dfrac{dx}{x(x + a)} = -\dfrac{1}{a} \ln \dfrac{x + a}{x}$	$\int \sqrt{a^2 - x^2} \, dx = \dfrac{1}{2}\left(x\sqrt{a^2 - x^2} + a^2 \sin^{-1}\dfrac{x}{a} \right)$
$\int \dfrac{dx}{(a + bx)^2} = -\dfrac{1}{b(a + bx)}$	$\int x\sqrt{a^2 - x^2} \, dx = -\dfrac{1}{3}(a^2 - x^2)^{3/2}$
$\int \dfrac{dx}{a^2 + x^2} = \dfrac{1}{a} \tan^{-1}\dfrac{x}{a}$	$\int \sqrt{x^2 \pm a^2} \, dx = \dfrac{1}{2}[x\sqrt{x^2 \pm a^2} \pm a^2 \ln(x + \sqrt{x^2 \pm a^2})]$
$\int \dfrac{dx}{a^2 - x^2} = \dfrac{1}{2a} \ln \dfrac{a + x}{a - x}$ $(a^2 - x^2 > 0)$	$\int x(\sqrt{x^2 \pm a^2}) \, dx = \dfrac{1}{3}(x^2 \pm a^2)^{3/2}$
$\int \dfrac{dx}{x^2 - a^2} = \dfrac{1}{2a} \ln \dfrac{x - a}{x + a}$ $(x^2 - a^2 > 0)$	$\int e^{ax} \, dx = \dfrac{1}{a} e^{ax}$
$\int \dfrac{x \, dx}{a^2 \pm x^2} = \pm\dfrac{1}{2} \ln(a^2 \pm x^2)$	$\int \ln ax \, dx = (x \ln ax) - x$

continued

Table B.5

Some Indefinite Integrals (An arbitrary constant should be added to each of these integrals.) *continued*

$$\int xe^{ax}\, dx = \frac{e^{ax}}{a^2}(ax - 1)$$

$$\int \frac{dx}{a + be^{cx}} = \frac{x}{a} - \frac{1}{ac}\ln(a + be^{cx})$$

$$\int \sin ax\, dx = -\frac{1}{a}\cos ax$$

$$\int \cos ax\, dx = \frac{1}{a}\sin ax$$

$$\int \tan ax\, dx = -\frac{1}{a}\ln(\cos ax) = \frac{1}{a}\ln(\sec ax)$$

$$\int \cot ax\, dx = \frac{1}{a}\ln(\sin ax)$$

$$\int \sec ax\, dx = \frac{1}{a}\ln(\sec ax + \tan ax) = \frac{1}{a}\ln\left[\tan\left(\frac{ax}{2} + \frac{\pi}{4}\right)\right]$$

$$\int \csc ax\, dx = \frac{1}{a}\ln(\csc ax - \cot ax) = \frac{1}{a}\ln\left(\tan\frac{ax}{2}\right)$$

$$\int \sin^2 ax\, dx = \frac{x}{2} - \frac{\sin 2ax}{4a}$$

$$\int \cos^2 ax\, dx = \frac{x}{2} + \frac{\sin 2ax}{4a}$$

$$\int \frac{dx}{\sin^2 ax} = -\frac{1}{a}\cot ax$$

$$\int \frac{dx}{\cos^2 ax} = \frac{1}{a}\tan ax$$

$$\int \tan^2 ax\, dx = \frac{1}{a}(\tan ax) - x$$

$$\int \cot^2 ax\, dx = -\frac{1}{a}(\cot ax) - x$$

$$\int \sin^{-1} ax\, dx = x(\sin^{-1} ax) + \frac{\sqrt{1 - a^2 x^2}}{a}$$

$$\int \cos^{-1} ax\, dx = x(\cos^{-1} ax) - \frac{\sqrt{1 - a^2 x^2}}{a}$$

$$\int \frac{dx}{(x^2 + a^2)^{3/2}} = \frac{x}{a^2\sqrt{x^2 + a^2}}$$

$$\int \frac{x\, dx}{(x^2 + a^2)^{3/2}} = -\frac{1}{\sqrt{x^2 + a^2}}$$

Table B.6

Gauss's Probability Integral and Other Definite Integrals

$$\int_0^\infty x^n e^{-ax}\, dx = \frac{n!}{a^{n+1}}$$

$$I_0 = \int_0^\infty e^{-ax^2}\, dx = \frac{1}{2}\sqrt{\frac{\pi}{a}} \qquad \text{(Gauss's probability integral)}$$

$$I_1 = \int_0^\infty xe^{-ax^2}\, dx = \frac{1}{2a}$$

$$I_2 = \int_0^\infty x^2 e^{-ax^2}\, dx = -\frac{dI_0}{da} = \frac{1}{4}\sqrt{\frac{\pi}{a^3}}$$

$$I_3 = \int_0^\infty x^3 e^{-ax^2}\, dx = -\frac{dI_1}{da} = \frac{1}{2a^2}$$

$$I_4 = \int_0^\infty x^4 e^{-ax^2}\, dx = \frac{d^2 I_0}{da^2} = \frac{3}{8}\sqrt{\frac{\pi}{a^5}}$$

$$I_5 = \int_0^\infty x^5 e^{-ax^2}\, dx = \frac{d^2 I_1}{da^2} = \frac{1}{a^3}$$

$$\vdots$$

$$I_{2n} = (-1)^n \frac{d^n}{da^n} I_0$$

$$I_{2n+1} = (-1)^n \frac{d^n}{da^n} I_1$$

B.8 Propagation of Uncertainty

In laboratory experiments, a common activity is to take measurements that act as raw data. These measurements are of several types—length, time interval, temperature, voltage, etc.—and are taken by a variety of instruments. Regardless of the measurement and the quality of the instrumentation, **there is always uncertainty associated with a physical measurement.** This uncertainty is a combination of that associated with the instrument and that related to the system being measured. An example of the former is the inability to exactly determine the position of a length measurement between the lines on a meter stick. An example of uncertainty related to the system being measured is the variation of temperature within a sample of water so that a single temperature for the sample is difficult to determine.

Uncertainties can be expressed in two ways. **Absolute uncertainty** refers to an uncertainty expressed in the same units as the measurement. Thus, a length might be expressed as (5.5 ± 0.1) cm, as was the length of the computer disk label in Section 1.7. The uncertainty of ± 0.1 cm by itself is not descriptive enough for some purposes, however. This is a large uncertainty if the measurement is 1.0 cm, but it is a small uncertainty if the measurement is 100 m. To give a more descriptive account of the uncertainty, **fractional uncertainty** or **percent uncertainty** is used. In this type of description, the uncertainty is divided by the actual measurement. Thus, the length of the computer disk label could be expressed as

$$\ell = 5.5 \text{ cm} \pm \frac{0.1 \text{ cm}}{5.5 \text{ cm}} = 5.5 \text{ cm} \pm 0.018 \qquad \text{(fractional uncertainty)}$$

or as

$$\ell = 5.5 \text{ cm} \pm 1.8\% \qquad \text{(percent uncertainty)}$$

When combining measurements in a calculation, the uncertainty in the final result is larger than the uncertainty in the individual measurements. This is called **propagation of uncertainty** and is one of the challenges of experimental physics. As a calculation becomes more complicated, there is increased propagation of uncertainty and the uncertainty in the value of the final result can grow to be quite large.

There are simple rules that can provide a reasonable estimate of the uncertainty in a calculated result:

Multiplication and division: When measurements with uncertainties are multiplied or divided, add the *percent uncertainties* to obtain the percent uncertainty in the result.

Example: The Area of a Rectangular Plate

$$A = \ell w = (5.5 \text{ cm} \pm 1.8\%) \times (6.4 \text{ cm} \pm 1.6\%) = 35 \text{ cm}^2 \pm 3.4\%$$
$$= (35 \pm 1) \text{ cm}^2$$

Addition and subtraction: When measurements with uncertainties are added or subtracted, add the *absolute uncertainties* to obtain the absolute uncertainty in the result.

Example: A Change in Temperature

$$\Delta T = T_2 - T_1 = (99.2 \pm 1.5)°C - (27.6 \pm 1.5)°C = (71.6 \pm 3.0)°C$$
$$= 71.6°C \pm 4.2\%$$

Powers: If a measurement is taken to a power, the percent uncertainty is multiplied by that power to obtain the percent uncertainty in the result.

Example: The Volume of a Sphere

$$V = \tfrac{4}{3}\pi r^3 = \tfrac{4}{3}\pi(6.20 \text{ cm} \pm 2.0\%)^3 = 998 \text{ cm}^3 \pm 6.0\%$$
$$= (998 \pm 60) \text{ cm}^3$$

Notice that uncertainties in a calculation always add. As a result, an experiment involving a subtraction should be avoided if possible. This is especially true if the measurements being subtracted are close together. The result of such a calculation is a small difference in the measurements and uncertainties that add together. It is possible that the uncertainty in the result could be larger than the result itself!

Appendix C • Periodic Table of the Elements

Group I	Group II	Transition elements							
H 1 1.007 9 $1s$									
Li 3 6.941 $2s^1$	**Be** 4 9.0122 $2s^2$								
Na 11 22.990 $3s^1$	**Mg** 12 24.305 $3s^2$								
K 19 39.098 $4s^1$	**Ca** 20 40.078 $4s^2$	**Sc** 21 44.956 $3d^14s^2$	**Ti** 22 47.867 $3d^24s^2$	**V** 23 50.942 $3d^34s^2$	**Cr** 24 51.996 $3d^54s^1$	**Mn** 25 54.938 $3d^54s^2$	**Fe** 26 55.845 $3d^64s^2$	**Co** 27 58.933 $3d^74s^2$	
Rb 37 85.468 $5s^1$	**Sr** 38 87.62 $5s^2$	**Y** 39 88.906 $4d^15s^2$	**Zr** 40 91.224 $4d^25s^2$	**Nb** 41 92.906 $4d^45s^1$	**Mo** 42 95.94 $4d^55s^1$	**Tc** 43 (98) $4d^55s^2$	**Ru** 44 101.07 $4d^75s^1$	**Rh** 45 102.91 $4d^85s^1$	
Cs 55 132.91 $6s^1$	**Ba** 56 137.33 $6s^2$	57-71*	**Hf** 72 178.49 $5d^26s^2$	**Ta** 73 180.95 $5d^36s^2$	**W** 74 183.84 $5d^46s^2$	**Re** 75 186.21 $5d^56s^2$	**Os** 76 190.23 $5d^66s^2$	**Ir** 77 192.2 $5d^76s^2$	
Fr 87 (223) $7s^1$	**Ra** 88 (226) $7s^2$	89-103**	**Rf** 104 (261) $6d^27s^2$	**Db** 105 (262) $6d^37s^2$	**Sg** 106 (266)	**Bh** 107 (264)	**Hs** 108 (269)	**Mt** 109 (268)	

Symbol — **Ca** 20 — Atomic number
Atomic mass † — 40.078
$4s^2$ — Electron configuration

*Lanthanide series

La 57 138.91 $5d^16s^2$	**Ce** 58 140.12 $5d^14f^16s^2$	**Pr** 59 140.91 $4f^36s^2$	**Nd** 60 144.24 $4f^46s^2$	**Pm** 61 (145) $4f^56s^2$	**Sm** 62 150.36 $4f^66s^2$

**Actinide series

Ac 89 (227) $6d^17s^2$	**Th** 90 232.04 $6d^27s^2$	**Pa** 91 231.04 $5f^26d^17s^2$	**U** 92 238.03 $5f^36d^17s^2$	**Np** 93 (237) $5f^46d^17s^2$	**Pu** 94 (244) $5f^66d^07s^2$

▫ Atomic mass values given are averaged over isotopes in the percentages in which they exist in nature.
 † For an unstable element, mass number of the most stable known isotope is given in parentheses.
 †† Elements 110, 111, 112, and 114 have not yet been named.
†††For a description of the atomic data, visit *physics.nist.gov/atomic*

			Group III	Group IV	Group V	Group VI	Group VII	Group 0
							H 1 1.007 9 $1s^1$	**He** 2 4.002 6 $1s^2$
			B 5 10.811 $2p^1$	**C** 6 12.011 $2p^2$	**N** 7 14.007 $2p^3$	**O** 8 15.999 $2p^4$	**F** 9 18.998 $2p^5$	**Ne** 10 20.180 $2p^6$
			Al 13 26.982 $3p^1$	**Si** 14 28.086 $3p^2$	**P** 15 30.974 $3p^3$	**S** 16 32.066 $3p^4$	**Cl** 17 35.453 $3p^5$	**Ar** 18 39.948 $3p^6$
Ni 28 58.693 $3d^84s^2$	**Cu** 29 63.546 $3d^{10}4s^1$	**Zn** 30 65.39 $3d^{10}4s^2$	**Ga** 31 69.723 $4p^1$	**Ge** 32 72.61 $4p^2$	**As** 33 74.922 $4p^3$	**Se** 34 78.96 $4p^4$	**Br** 35 79.904 $4p^5$	**Kr** 36 83.80 $4p^6$
Pd 46 106.42 $4d^{10}$	**Ag** 47 107.87 $4d^{10}5s^1$	**Cd** 48 112.41 $4d^{10}5s^2$	**In** 49 114.82 $5p^1$	**Sn** 50 118.71 $5p^2$	**Sb** 51 121.76 $5p^3$	**Te** 52 127.60 $5p^4$	**I** 53 126.90 $5p^5$	**Xe** 54 131.29 $5p^6$
Pt 78 195.08 $5d^96s^1$	**Au** 79 196.97 $5d^{10}6s^1$	**Hg** 80 200.59 $5d^{10}6s^2$	**Tl** 81 204.38 $6p^1$	**Pb** 82 207.2 $6p^2$	**Bi** 83 208.98 $6p^3$	**Po** 84 (209) $6p^4$	**At** 85 (210) $6p^5$	**Rn** 86 (222) $6p^6$
110†† (271)	111†† (272)	112†† (285)		114†† (289)				

Eu 63 151.96 $4f^76s^2$	**Gd** 64 157.25 $5d^14f^76s^2$	**Tb** 65 158.93 $5d^14f^86s^2$	**Dy** 66 162.50 $4f^{10}6s^2$	**Ho** 67 164.93 $4f^{11}6s^2$	**Er** 68 167.26 $4f^{12}6s^2$	**Tm** 69 168.93 $4f^{13}6s^2$	**Yb** 70 173.04 $4f^{14}6s^2$	**Lu** 71 174.97 $5d^14f^{14}6s^2$
Am 95 (243) $5f^76d^07s^2$	**Cm** 96 (247) $5f^76d^17s^2$	**Bk** 97 (247) $5f^86d^17s^2$	**Cf** 98 (251) $5f^{10}6d^07s^2$	**Es** 99 (252) $5f^{11}6d^07s^2$	**Fm** 100 (257) $5f^{12}6d^07s^2$	**Md** 101 (258) $5f^{13}6d^07s^2$	**No** 102 (259) $6d^07s^2$	**Lr** 103 (262) $6d^17s^2$

Appendix D • SI Units

Table D.1

SI Units		
	SI Base Unit	
Base Quantity	**Name**	**Symbol**
Length	Meter	m
Mass	Kilogram	kg
Time	Second	s
Electric current	Ampere	A
Temperature	Kelvin	K
Amount of substance	Mole	mol
Luminous intensity	Candela	cd

Table D.2

Some Derived SI Units				
Quantity	**Name**	**Symbol**	**Expression in Terms of Base Units**	**Expression in Terms of Other SI Units**
Plane angle	radian	rad	m/m	
Frequency	hertz	Hz	s^{-1}	
Force	newton	N	$kg \cdot m/s^2$	J/m
Pressure	pascal	Pa	$kg/m \cdot s^2$	N/m^2
Energy; work	joule	J	$kg \cdot m^2/s^2$	$N \cdot m$
Power	watt	W	$kg \cdot m^2/s^3$	J/s
Electric charge	coulomb	C	$A \cdot s$	
Electric potential	volt	V	$kg \cdot m^2/A \cdot s^3$	W/A
Capacitance	farad	F	$A^2 \cdot s^4/kg \cdot m^2$	C/V
Electric resistance	ohm	Ω	$kg \cdot m^2/A^2 \cdot s^3$	V/A
Magnetic flux	weber	Wb	$kg \cdot m^2/A \cdot s^2$	$V \cdot s$
Magnetic field	tesla	T	$kg/A \cdot s^2$	
Inductance	henry	H	$kg \cdot m^2/A^2 \cdot s^2$	$T \cdot m^2/A$

All Nobel Prizes in physics are listed (and marked with a P), as well as relevant Nobel Prizes in Chemistry (C). The key dates for some of the scientific work are supplied; they often antedate the prize considerably.

1901 (P) *Wilhelm Roentgen* for discovering x-rays (1895).

1902 (P) *Hendrik A. Lorentz* for predicting the Zeeman effect and *Pieter Zeeman* for discovering the Zeeman effect, the splitting of spectral lines in magnetic fields.

1903 (P) *Antoine-Henri Becquerel* for discovering radioactivity (1896) and *Pierre* and *Marie Curie* for studying radioactivity.

1904 (P) *Lord Rayleigh* for studying the density of gases and discovering argon.
(C) *William Ramsay* for discovering the inert gas elements helium, neon, xenon, and krypton, and placing them in the periodic table.

1905 (P) *Philipp Lenard* for studying cathode rays, electrons (1898–1899).

1906 (P) *J. J. Thomson* for studying electrical discharge through gases and discovering the electron (1897).

1907 (P) *Albert A. Michelson* for inventing optical instruments and measuring the speed of light (1880s).

1908 (P) *Gabriel Lippmann* for making the first color photographic plate, using interference methods (1891).
(C) *Ernest Rutherford* for discovering that atoms can be broken apart by alpha rays and for studying radioactivity.

1909 (P) *Guglielmo Marconi* and *Carl Ferdinand Braun* for developing wireless telegraphy.

1910 (P) *Johannes D. van der Waals* for studying the equation of state for gases and liquids (1881).

1911 (P) *Wilhelm Wien* for discovering Wien's law giving the peak of a blackbody spectrum (1893).
(C) *Marie Curie* for discovering radium and polonium (1898) and isolating radium.

1912 (P) *Nils Dalén* for inventing automatic gas regulators for lighthouses.

1913 (P) *Heike Kamerlingh Onnes* for the discovery of superconductivity and liquefying helium (1908).

1914 (P) *Max T. F. von Laue* for studying x-rays from their diffraction by crystals, showing that x-rays are electromagnetic waves (1912).
(C) *Theodore W. Richards* for determining the atomic weights of sixty elements, indicating the existence of isotopes.

1915 (P) *William Henry Bragg* and *William Lawrence Bragg*, his son, for studying the diffraction of x-rays in crystals.

1917 (P) *Charles Barkla* for studying atoms by x-ray scattering (1906).

1918 (P) *Max Planck* for discovering energy quanta (1900).

1919 (P) *Johannes Stark*, for discovering the Stark effect, the splitting of spectral lines in electric fields (1913).

1920 (P) *Charles-Édouard Guillaume* for discovering invar, a nickel–steel alloy with low coefficient of expansion.
(C) *Walther Nernst* for studying heat changes in chemical reactions and formulating the third law of thermodynamics (1918).

1921 (P) *Albert Einstein* for explaining the photoelectric effect and for his services to theoretical physics (1905).
(C) *Frederick Soddy* for studying the chemistry of radioactive substances and discovering isotopes (1912).

1922　(P) *Niels Bohr* for his model of the atom and its radiation (1913).

　　　(C) *Francis W. Aston* for using the mass spectrograph to study atomic weights, thus discovering 212 of the 287 naturally occurring isotopes.

1923　(P) *Robert A. Millikan* for measuring the charge on an electron (1911) and for studying the photoelectric effect experimentally (1914).

1924　(P) *Karl M. G. Siegbahn* for his work in x-ray spectroscopy.

1925　(P) *James Franck* and *Gustav Hertz* for discovering the Franck–Hertz effect in electron–atom collisions.

1926　(P) *Jean-Baptiste Perrin* for studying Brownian motion to validate the discontinuous structure of matter and measure the size of atoms.

1927　(P) *Arthur Holly Compton* for discovering the Compton effect on x-rays, their change in wavelength when they collide with matter (1922), and *Charles T. R. Wilson* for inventing the cloud chamber, used to study charged particles (1906).

1928　(P) *Owen W. Richardson* for studying the thermionic effect and electrons emitted by hot metals (1911).

1929　(P) *Louis Victor de Broglie* for discovering the wave nature of electrons (1923).

1930　(P) *Chandrasekhara Venkata Raman* for studying Raman scattering, the scattering of light by atoms and molecules with a change in wavelength (1928).

1932　(P) *Werner Heisenberg* for creating quantum mechanics (1925).

1933　(P) *Erwin Schrödinger* and *Paul A. M. Dirac* for developing wave mechanics (1925) and relativistic quantum mechanics (1927).

　　　(C) *Harold Urey* for discovering heavy hydrogen, deuterium (1931).

1935　(P) *James Chadwick* for discovering the neutron (1932).

　　　(C) *Irène* and *Frédéric Joliot-Curie* for synthesizing new radioactive elements.

1936　(P) *Carl D. Anderson* for discovering the positron in particular and antimatter in general (1932) and *Victor F. Hess* for discovering cosmic rays.

　　　(C) *Peter J. W. Debye* for studying dipole moments and diffraction of x-rays and electrons in gases.

1937　(P) *Clinton Davisson* and *George Thomson* for discovering the diffraction of electrons by crystals, confirming de Broglie's hypothesis (1927).

1938　(P) *Enrico Fermi* for producing the transuranic radioactive elements by neutron irradiation (1934–1937).

1939　(P) *Ernest O. Lawrence* for inventing the cyclotron.

1943　(P) *Otto Stern* for developing molecular-beam studies (1923) and using them to discover the magnetic moment of the proton (1933).

1944　(P) *Isidor I. Rabi* for discovering nuclear magnetic resonance in atomic and molecular beams.

　　　(C) *Otto Hahn* for discovering nuclear fission (1938).

1945　(P) *Wolfgang Pauli* for discovering the exclusion principle (1924).

1946　(P) *Percy W. Bridgman* for studying physics at high pressures.

1947　(P) *Edward V. Appleton* for studying the ionosphere.

1948　(P) *Patrick M. S. Blackett* for studying nuclear physics with cloud-chamber photographs of cosmic-ray interactions.

1949　(P) *Hideki Yukawa* for predicting the existence of mesons (1935).

1950　(P) *Cecil F. Powell* for developing the method of studying cosmic rays with photographic emulsions and discovering new mesons.

1951　(P) *John D. Cockcroft* and *Ernest T. S. Walton* for transmuting nuclei in an accelerator (1932).

　　　(C) *Edwin M. McMillan* for producing neptunium (1940) and *Glenn T. Seaborg* for producing plutonium (1941) and further transuranic elements.

1952　(P) *Felix Bloch* and *Edward Mills Purcell* for discovering nuclear magnetic resonance in liquids and gases (1946).

1953　(P) *Frits Zernike* for inventing the phase-contrast microscope, which uses interference to provide high contrast.

1954　(P) *Max Born* for interpreting the wave function as a probability (1926) and other quantum-mechanical discoveries and *Walther Bothe* for developing the coincidence method to study subatomic particles (1930–1931), producing, in particular, the particle interpreted by Chadwick as the neutron.

1955 (P) *Willis E. Lamb, Jr.*, for discovering the Lamb shift in the hydrogen spectrum (1947) and *Polykarp Kusch* for determining the magnetic moment of the electron (1947).

1956 (P) *John Bardeen, Walter H. Brattain*, and *William Shockley* for inventing the transistor (1956).

1957 (P) *T.-D. Lee* and *C.-N. Yang* for predicting that parity is not conserved in beta decay (1956).

1958 (P) *Pavel A. Čerenkov* for discovering Čerenkov radiation (1935) and *Ilya M. Frank* and *Igor Tamm* for interpreting it (1937).

1959 (P) *Emilio G. Segrè* and *Owen Chamberlain* for discovering the antiproton (1955).

1960 (P) *Donald A. Glaser* for inventing the bubble chamber to study elementary particles (1952).

(C) *Willard Libby* for developing radiocarbon dating (1947).

1961 (P) *Robert Hofstadter* for discovering internal structure in protons and neutrons and *Rudolf L. Mössbauer* for discovering the Mössbauer effect of recoilless gamma-ray emission (1957).

1962 (P) *Lev Davidovich Landau* for studying liquid helium and other condensed matter theoretically.

1963 (P) *Eugene P. Wigner* for applying symmetry principles to elementary-particle theory and *Maria Goeppert Mayer* and *J. Hans D. Jensen* for studying the shell model of nuclei (1947).

1964 (P) *Charles H. Townes, Nikolai G. Basov*, and *Alexandr M. Prokhorov* for developing masers (1951–1952) and lasers.

1965 (P) *Sin-itiro Tomonaga, Julian S. Schwinger*, and *Richard P. Feynman* for developing quantum electrodynamics (1948).

1966 (P) *Alfred Kastler* for his optical methods of studying atomic energy levels.

1967 (P) *Hans Albrecht Bethe* for discovering the routes of energy production in stars (1939).

1968 (P) *Luis W. Alvarez* for discovering resonance states of elementary particles.

1969 (P) *Murray Gell-Mann* for classifying elementary particles (1963).

1970 (P) *Hannes Alfvén* for developing magnetohydrodynamic theory and *Louis Eugène Félix Néel* for discovering antiferromagnetism and ferrimagnetism (1930s).

1971 (P) *Dennis Gabor* for developing holography (1947).

(C) *Gerhard Herzberg* for studying the structure of molecules spectroscopically.

1972 (P) *John Bardeen, Leon N. Cooper*, and *John Robert Schrieffer* for explaining superconductivity (1957).

1973 (P) *Leo Esaki* for discovering tunneling in semiconductors, *Ivar Giaever* for discovering tunneling in superconductors, and *Brian D. Josephson* for predicting the Josephson effect, which involves tunneling of paired electrons (1958–1962).

1974 (P) *Anthony Hewish* for discovering pulsars and *Martin Ryle* for developing radio interferometry.

1975 (P) *Aage N. Bohr, Ben R. Mottelson*, and *James Rainwater* for discovering why some nuclei take asymmetric shapes.

1976 (P) *Burton Richter* and *Samuel C. C. Ting* for discovering the J/psi particle, the first charmed particle (1974).

1977 (P) *John H. Van Vleck, Nevill F. Mott*, and *Philip W. Anderson* for studying solids quantum-mechanically.

(C) *Ilya Prigogine* for extending thermodynamics to show how life could arise in the face of the second law.

1978 (P) *Arno A. Penzias* and *Robert W. Wilson* for discovering the cosmic background radiation (1965) and *Pyotr Kapitsa* for his studies of liquid helium.

1979 (P) *Sheldon L. Glashow, Abdus Salam*, and *Steven Weinberg* for developing the theory that unified the weak and electromagnetic forces (1958–1971).

1980   (P) *Val Fitch* and *James W. Cronin* for discovering CP (charge-parity) violation (1964), which possibly explains the cosmological dominance of matter over antimatter.

1981   (P) *Nicolaas Bloembergen* and *Arthur L. Schawlow* for developing laser spectroscopy and *Kai M. Siegbahn* for developing high-resolution electron spectroscopy (1958).

1982   (P) *Kenneth G. Wilson* for developing a method of constructing theories of phase transitions to analyze critical phenomena.

1983   (P) *William A. Fowler* for theoretical studies of astrophysical nucleosynthesis and *Subramanyan Chandrasekhar* for studying physical processes of importance to stellar structure and evolution, including the prediction of white dwarf stars (1930).

1984   (P) *Carlo Rubbia* for discovering the W and Z particles, verifying the electroweak unification, and *Simon van der Meer,* for developing the method of stochastic cooling of the CERN beam that allowed the discovery (1982–1983).

1985   (P) *Klaus von Klitzing* for the quantized Hall effect, relating to conductivity in the presence of a magnetic field (1980).

1986   (P) *Ernst Ruska* for inventing the electron microscope (1931), and *Gerd Binnig* and *Heinrich Rohrer* for inventing the scanning-tunneling electron microscope (1981).

1987   (P) *J. Georg Bednorz* and *Karl Alex Müller* for the discovery of high-temperature superconductivity (1986).

1988   (P) *Leon M. Lederman, Melvin Schwartz,* and *Jack Steinberger* for a collaborative experiment that led to the development of a new tool for studying the weak nuclear force, which affects the radioactive decay of atoms.

1989   (P) *Norman Ramsay* for various techniques in atomic physics; and *Hans Dehmelt* and *Wolfgang Paul* for the development of techniques for trapping single-charge particles.

1990   (P) *Jerome Friedman, Henry Kendall* and *Richard Taylor* for experiments important to the development of the quark model.

1991   (P) *Pierre-Gilles de Gennes* for discovering that methods developed for studying order phenomena in simple systems can be generalized to more complex forms of matter, in particular to liquid crystals and polymers.

1992   (P) *George Charpak* for developing detectors that trace the paths of evanescent subatomic particles produced in particle accelerators.

1993   (P) *Russell Hulse* and *Joseph Taylor* for discovering evidence of gravitational waves.

1994   (P) *Bertram N. Brockhouse* and *Clifford G. Shull* for pioneering work in neutron scattering.

1995   (P) *Martin L. Perl* and *Frederick Reines* for discovering the tau particle and the neutrino, respectively.

1996   (P) *David M. Lee, Douglas C. Osheroff,* and *Robert C. Richardson* for developing a superfluid using helium-3.

1997   (P) *Steven Chu, Claude Cohen-Tannoudji,* and *William D. Phillips* for developing methods to cool and trap atoms with laser light.

1998   (P) *Robert B. Laughlin, Horst L. Störmer,* and *Daniel C. Tsui* for discovering a new form of quantum fluid with fractionally charged excitations.

1999   (P) *Gerardus 'T Hooft* and *Martinus J. G. Veltman* for studies in the quantum structure of electroweak interactions in physics.

2000   (P) *Zhores I. Alferov* and *Herbert Kroemer* for developing semiconductor heterostructures used in high-speed electronics and optoelectronics and *Jack St. Clair Kilby* for participating in the invention of the integrated circuit.

2001   (P) *Eric A. Cornell, Wolfgang Ketterle,* and *Carl E. Wieman* for the achievement of Bose–Einstein condensation in dilute gases of alkali atoms.

2002   (P) *Raymond Davis Jr.* and *Masatoshi Koshiba* for the detection of cosmic neutrinos and *Riccardo Giacconi* for contributions to astrophysics that led to the discovery of cosmic x-ray sources.

Answers to Odd-Numbered Problems

CHAPTER 23

1. (a) $+160$ zC, 1.01 u   (b) $+160$ zC, 23.0 u
   (c) -160 zC, 35.5 u   (d) $+320$ zC, 40.1 u
   (e) -480 zC, 14.0 u   (f) $+640$ zC, 14.0 u
   (g) $+1.12$ aC, 14.0 u   (h) -160 zC, 18.0 u

3. The force is $\sim 10^{26}$ N.

5. (a) 1.59 nN away from the other
   (b) 1.24×10^{36} times larger
   (c) 8.61×10^{-11} C/kg

7. 0.872 N at 330°

9. (a) 2.16×10^{-5} N toward the other   (b) 8.99×10^{-7} N away from the other

11. (a) 82.2 nN   (b) 2.19 Mm/s

13. (a) 55.8 pN/C down   (b) 102 nN/C up

15. 1.82 m to the left of the negative charge

17. $-9Q$ and $+27Q$

19. (a) $(-0.599\hat{\mathbf{i}} - 2.70\hat{\mathbf{j}})$ kN/C
    (b) $(-3.00\hat{\mathbf{i}} - 13.5\hat{\mathbf{j}})$ μN

21. (a) $5.91 k_e q/a^2$ at 58.8°   (b) $5.91 k_e q^2/a^2$ at 58.8°

23. (a) $[k_e Qx/(R^2 + x^2)^{3/2}]\hat{\mathbf{i}}$ (b) As long as the charge is symmetrically placed, the number of charges does not matter. A continuous ring corresponds to n becoming larger without limit.

25. 1.59×10^6 N/C toward the rod

27. (a) $6.64\hat{\mathbf{i}}$ MN/C   (b) $24.1\hat{\mathbf{i}}$ MN/C   (c) $6.40\hat{\mathbf{i}}$ MN/C
    (d) $0.664\hat{\mathbf{i}}$ MN/C, taking the axis of the ring as the x axis

31. (a) 93.6 MN/C; the near-field approximation is 104 MN/C, about 11% high   (b) 0.516 MN/C; the point-charge approximation is 0.519 MN/C, about 0.6% high

33. $-21.6\hat{\mathbf{i}}$ MN/C

37. (a) 86.4 pC for each
    (b) 324 pC, 459 pC, 459 pC, 432 pC
    (c) 57.6 pC, 106 pC, 154 pC, 96.0 pC

39.

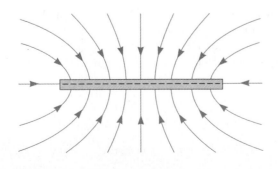

41. (a)

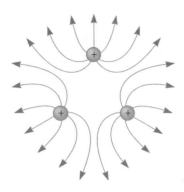

   The field is zero at the center of the triangle.
   (b) $1.73 k_e q \hat{\mathbf{j}}/a^2$

43. (a) 61.3 Gm/s^2 (b) 19.5 μs   (c) 11.7 m   (d) 1.20 fJ

45. K/ed in the direction of motion

47. (a) 111 ns   (b) 5.68 mm   (c) $(450\hat{\mathbf{i}} + 102\hat{\mathbf{j}})$ km/s

49. (a) 36.9°, 53.1°   (b) 167 ns, 221 ns

51. (a) 21.8 μm   (b) 2.43 cm

53. (a) mv^2/qR (b) $mv^2/2^{1/2}qR$ oriented at 135° to the x axis

55. (a) 10.9 nC   (b) 5.44 mN

57. 40.9 N at 263°

59. $Q = 2L \sqrt{\dfrac{k(L - L_i)}{k_e}}$

63. $-707\hat{\mathbf{j}}$ mN

65. (a) $\theta_1 = \theta_2$

67. (a) 0.307 s   (b) Yes. Ignoring gravity makes a difference of 2.28%.

69. (a) $\mathbf{F} = 1.90(k_e q^2/s^2)(\hat{\mathbf{i}} + \hat{\mathbf{j}} + \hat{\mathbf{k}})$ (b) $\mathbf{F} = 3.29(k_e q^2/s^2)$ in the direction away from the diagonally opposite vertex

CHAPTER 24

1. (a) 858 N·m^2/C   (b) 0   (c) 657 N·m^2/C

3. 4.14 MN/C

5. (a) aA (b) bA (c) 0

7. 1.87 kN·m^2/C

9. (a) -6.89 MN·m^2/C   (b) The number of lines entering exceeds the number leaving by 2.91 times or more.

11. $-Q/\epsilon_0$ for S_1; 0 for S_2; $-2Q/\epsilon_0$ for S_3; 0 for S_4

13. $E_0 \pi r^2$

15. (a) $+Q/2\epsilon_0$ (b) $-Q/2\epsilon_0$

17. -18.8 kN·m^2/C

19. 0 if $R \le d$; $\dfrac{2\lambda}{\epsilon_0}\sqrt{R^2 - d^2}$ if $R > d$

21. (a) 3.20 MN·m^2/C   (b) 19.2 MN·m^2/C   (c) The answer to (a) could change, but the answer to (b) would stay the same.

23. 2.33×10^{21} N/C

25. $-2.48\ \mu$C/m^2

27. 5.94×10^5 m/s

29. $\mathbf{E} = \rho r/2\epsilon_0$ away from the axis

31. (a) 0   (b) 7.19 MN/C away from the center

33. (a) ~1 mN   (b) ~100 nC   (c) ~10 kN/C   (d) ~10 kN·m^2/C

35. (a) 51.4 kN/C outward   (b) 646 N·m^2/C

37. 508 kN/C up

39. (a) 0   (b) $5\,400$ N/C outward   (c) 540 N/C outward

41. $\mathbf{E} = Q/2\epsilon_0 A$ vertically upward in each case if $Q > 0$

43. (a) $+708$ nC/m^2 and -708 nC/m^2 (b) $+177$ nC and -177 nC

45. 2.00 N

47. (a) $-\lambda, +3\lambda$ (b) $3\lambda/2\pi\epsilon_0 r$ radially outward

49. (a) 80.0 nC/m^2 on each face   (b) $9.04\hat{\mathbf{k}}$ kN/C   (c) $-9.04\hat{\mathbf{k}}$ kN/C

51. $\mathbf{E} = 0$ inside the sphere and within the material of the shell. $\mathbf{E} = k_e Q/r^2$ radially inward between the sphere and the shell. $\mathbf{E} = 2k_e Q/r^2$ radially outward outside the shell. Charge $-Q$ resides on the outer surface of the sphere. $+Q$ is on the inner surface of the shell. $+2Q$ is on the outer surface of the shell.

53. (b) $Q/2\epsilon_0$ (c) Q/ϵ_0

55. (a) $+2Q$ (b) radially outward   (c) $2k_e Q/r^2$ (d) 0   (e) 0   (f) $3Q$ (g) $3k_e Q/r^2$ radially outward   (h) $3Qr^3/a^3$ (i) $3k_e Qr/a^3$ radially outward   (j) $-3Q$ (k) $+2Q$ (l) See below.

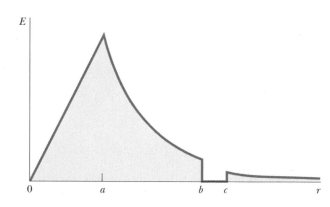

57. (a) $\rho r/3\epsilon_0$; $Q/4\pi\epsilon_0 r^2$; 0; $Q/4\pi\epsilon_0 r^2$, all radially outward   (b) $-Q/4\pi b^2$ and $+Q/4\pi c^2$

59. $\theta = \tan^{-1}[qQ/(2\pi\epsilon_0 dmv^2)]$

61. For $r < a$, $\mathbf{E} = \lambda/2\pi\epsilon_0 r$ radially outward. For $a < r < b$, $\mathbf{E} = [\lambda + \rho\pi(r^2 - a^2)]/2\pi\epsilon_0 r$ radially outward. For $r > b$, $\mathbf{E} = [\lambda + \rho\pi(b^2 - a^2)]/2\pi\epsilon_0 r$ radially outward.

63. (a) σ/ϵ_0 away from both plates   (b) 0   (c) σ/ϵ_0 away from both plates

65. $\sigma/2\epsilon_0$ radially outward

69. $\mathbf{E} = a/2\epsilon_0$ radially outward

73. (b) $\mathbf{g} = GM_E r/R_E^3$ radially inward

CHAPTER 25

1. 1.35 MJ

3. (a) 152 km/s   (b) 6.49 Mm/s

5. (a) $-600\ \mu$J   (b) -50.0 V

7. 38.9 V; the origin

9. $+260$ V

11. (a) $2QE/k$ (b) QE/k (c) $2\pi\sqrt{m/k}$ (d) $2(QE - \mu_k mg)/k$

13. (a) 0.400 m/s   (b) the same

15. (a) 1.44×10^{-7} V   (b) -7.19×10^{-8} V   (c) -1.44×10^{-7} V, $+7.19 \times 10^{-8}$ V

17. (a) 6.00 m   (b) $-2.00\ \mu$C

19. -11.0 MV

21. 8.95 J

25. (a) no point at a finite distance from the charges   (b) $2k_e q/a$

27. (a) $v_1 = \sqrt{\dfrac{2m_2 k_e q_1 q_2}{m_1(m_1 + m_2)}\left(\dfrac{1}{r_1 + r_2} - \dfrac{1}{d}\right)}$

$v_2 = \sqrt{\dfrac{2m_1 k_e q_1 q_2}{m_2(m_1 + m_2)}\left(\dfrac{1}{r_1 + r_2} - \dfrac{1}{d}\right)}$

(b) faster than calculated in (a)

29. $5k_e q^2/9d$

31. 0.720 m, 1.44 m, 2.88 m. No. The radii of the equipotentials are inversely proportional to the potential.

33. 7.26 Mm/s

35. $\left[\left(1 + \sqrt{\dfrac{1}{8}}\right)\dfrac{k_e q^2}{mL}\right]^{1/2}$

37. (a) 10.0 V, -11.0 V, -32.0 V   (b) 7.00 N/C in the $+x$ direction

39. $\mathbf{E} = (-5 + 6xy)\hat{\mathbf{i}} + (3x^2 - 2z^2)\hat{\mathbf{j}} - 4yz\hat{\mathbf{k}}$; 7.07 N/C

41. $E_y = \dfrac{k_e Q}{y\sqrt{\ell^2 + y^2}}$

43. (a) C/m^2 (b) $k_e \alpha[L - d\ln(1 + L/d)]$

45. -1.51 MV

47. $k_e \lambda(\pi + 2\ln 3)$

49. (a) 0, 1.67 MV   (b) 5.84 MN/C away, 1.17 MV   (c) 11.9 MN/C away, 1.67 MV

51. (a) 450 kV   (b) $7.51\ \mu$C

53. 253 MeV

55. (a) -27.2 eV   (b) -6.80 eV   (c) 0

59. $k_e Q^2/2R$

63. $V_2 - V_1 = (-\lambda/2\pi\epsilon_0)\ln(r_2/r_1)$

69. (b) $E_r = (2k_e p\cos\theta)/r^3$; $E_\theta = (k_e p\sin\theta)/r^3$; yes; no   (c) $V = k_e py(x^2 + y^2)^{-3/2}$; $\mathbf{E} = 3k_e pxy(x^2 + y^2)^{-5/2}\hat{\mathbf{i}} + k_e p(2y^2 - x^2)(x^2 + y^2)^{-5/2}\hat{\mathbf{j}}$

71. $V = \pi k_e C \left[R\sqrt{x^2 + R^2} + x^2 \ln\left(\dfrac{x}{R + \sqrt{x^2 + R^2}} \right) \right]$

73. (a) 8 876 V   (b) 112 V

CHAPTER 26

1. (a) 48.0 μC   (b) 6.00 μC

3. (a) 1.33 μC/m^2 (b) 13.3 pF

5. (a) 5.00 μC on the larger and 2.00 μC on the smaller sphere   (b) 89.9 kV

7. (a) 11.1 kV/m toward the negative plate.
(b) 98.3 nC/m^2 (c) 3.74 pF   (d) 74.7 pC

9. 4.42 μm

11. (a) 2.68 nF   (b) 3.02 kV

13. (a) 15.6 pF   (b) 256 kV

15. 708 μF

17. (a) 3.53 μF   (b) 6.35 V and 2.65 V   (c) 31.8 μC on each

19. 6.00 pF and 3.00 pF

21. (a) 5.96 μF   (b) 89.5 μC on 20 μF, 63.2 μC on 6 μF, 26.3 μC on 15 μF and on 3 μF

23. 120 μC; 80.0 μC and 40.0 μC

25. 10

27. 6.04 μF

29. 12.9 μF

31. (a) 216 μJ   (b) 54.0 μJ

33. (a) Circuit diagram:

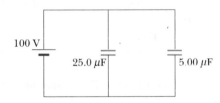

Stored energy = 0.150 J
(b) Potential difference = 268 V
Circuit diagram:

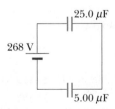

35. (a) 1.50 μC   (b) 1.83 kV

39. 9.79 kg

43. (a) 81.3 pF   (b) 2.40 kV

45. 1.04 m

47. (a) 369 pC   (b) 118 pF, 3.12 V   (c) -45.5 nJ

49. 22.5 V

51. (b) -8.78×10^6 N/C·m; $-5.53 \times 10^{-2}\, \hat{\mathbf{i}}$ N

55. (a) 11.2 pF   (b) 134 pC   (c) 16.7 pF   (d) 66.9 pC

57. (a) $-2Q/3$ on upper plate, $-Q/3$ on lower plate
(b) $2Qd/3\epsilon_0 A$

59. 0.188 m^2

61. (a) $C = \dfrac{\epsilon_0 A}{d}\left(\dfrac{\kappa_1}{2} + \dfrac{\kappa_2 \kappa_3}{\kappa_2 + \kappa_3} \right)$ (b) 1.76 pF

63. (b) $1/C$ approaches $\dfrac{1}{4\pi\epsilon_0 a} + \dfrac{1}{4\pi\epsilon_0 b}$

65. (a) $Q_0^2\, d(\ell - x)/(2\ell^3 \epsilon_0)$ (b) $Q_0^2\, d/(2\ell^3 \epsilon_0)$ to the right
(c) $Q_0^2/(2\ell^4 \epsilon_0)$ (d) $Q_0^2/(2\ell^4 \epsilon_0)$

67. 4.29 μF

69. (a) The additional energy comes from work done by the electric field in the wires as it forces more charge onto the already-charged plates.   (b) $Q/Q_0 = \kappa$

71. 750 μC on C_1 and 250 μC on C_2

73. 19.0 kV

75. $\frac{4}{3}C$

CHAPTER 27

1. 7.50×10^{15} electrons

3. (a) 0.632 $I_0\tau$ (b) 0.999 95 $I_0\tau$ (c) $I_0\tau$

5. $q\omega/2\pi$

7. 0.265 C

9. (a) 2.55 A/m^2 (b) 5.31×10^{10} m^{-3} (c) 1.20×10^{10} s

11. 0.130 mm/s

13. 500 mA

15. 6.43 A

17. (a) 1.82 m   (b) 280 μm

19. (a) $\sim 10^{18}$ Ω (b) $\sim 10^{-7}$ Ω (c) ~ 100 aA, ~ 1 GA

21. $R/9$

23. $6.00 \times 10^{-15}/\Omega \cdot$m

25. 0.181 V/m

27. 21.2 nm

29. 1.44×10^{3}°C

31. (a) 31.5 n$\Omega \cdot$m   (b) 6.35 MA/m^2 (c) 49.9 mA
(d) 659 μm/s   (e) 0.400 V

33. 0.125

35. 67.6°C

37. 7.50 W

39. 28.9 Ω

41. 36.1%

43. (a) 5.97 V/m   (b) 74.6 W   (c) 66.1 W

45. 0.833 W

47. $0.232

49. 26.9 cents/d

51. (a) 184 W   (b) 461°C

53. $\sim\$1$

55. (a) $Q/4C$ (b) $Q/4$ and $3Q/4$ (c) $Q^2/32C$ and $3Q^2/32C$ (d) $3Q^2/8C$

59. Experimental resistivity = $1.47\ \mu\Omega \cdot m \pm 4\%$, in agreement with $1.50\ \mu\Omega \cdot m$

61. (a) $(8.00\hat{i})$ V/m   (b) $0.637\ \Omega$ (c) 6.28 A
(d) $(200\hat{i})$ MA/m^2

63. $2\,020°C$

65. (a) 667 A   (b) 50.0 km

67.

Material	$\alpha' = \alpha/(1 - 20\alpha)$
Silver	$4.1 \times 10^{-3}/°C$
Copper	$4.2 \times 10^{-3}/°C$
Gold	$3.6 \times 10^{-3}/°C$
Aluminum	$4.2 \times 10^{-3}/°C$
Tungsten	$4.9 \times 10^{-3}/°C$
Iron	$5.6 \times 10^{-3}/°C$
Platinum	$4.25 \times 10^{-3}/°C$
Lead	$4.2 \times 10^{-3}/°C$
Nichrome	$0.4 \times 10^{-3}/°C$
Carbon	$-0.5 \times 10^{-3}/°C$
Germanium	$-24 \times 10^{-3}/°C$
Silicon	$-30 \times 10^{-3}/°C$

69. No. The fuses should pass no more than 3.87 A.

73. (b) 1.79 PΩ

75. (a) $\dfrac{\epsilon_0 \ell}{2d}(\ell + 2x + \kappa\ell - 2\kappa x)$

(b) $\dfrac{\epsilon_0 \ell v\,\Delta V(\kappa - 1)}{d}$ clockwise

CHAPTER 28

1. (a) $6.73\ \Omega$ (b) $1.97\ \Omega$

3. (a) $4.59\ \Omega$ (b) 8.16%

5. $12.0\ \Omega$

7. Circuit diagram:

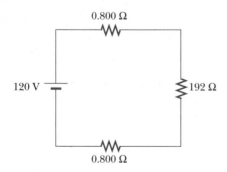

power 73.8 W

9. (a) 227 mA   (b) 5.68 V

11. (a) 75.0 V   (b) 25.0 W, 6.25 W, and 6.25 W; 37.5 W

13. 1.00 kΩ

15. 14.2 W to 2 Ω, 28.4 W to 4 Ω, 1.33 W to 3 Ω, 4.00 W to 1 Ω

17. (a) $\Delta t_p = 2\Delta t/3$ (b) $\Delta t_s = 3\Delta t$

19. (a) $\Delta V_4 > \Delta V_3 > \Delta V_1 > \Delta V_2$
(b) $\Delta V_1 = \mathcal{E}/3,\ \Delta V_2 = 2\mathcal{E}/9,\ \Delta V_3 = 4\mathcal{E}/9,\ \Delta V_4 = 2\mathcal{E}/3$
(c) $I_1 > I_4 > I_2 = I_3$ (d) $I_1 = I, I_2 = I_3 = I/3, I_4 = 2I/3$
(e) I_4 increases while $I_1, I_2,$ and I_3 decrease
(f) $I_1 = 3I/4, I_2 = I_3 = 0, I_4 = 3I/4$

21. 846 mA down in the 8-Ω resistor; 462 mA down in the middle branch; 1.31 A up in the right-hand branch

23. (a) -222 J and 1.88 kJ   (b) 687 J, 128 J, 25.6 J, 616 J, 205 J
(c) 1.66 kJ of chemical energy is transformed into internal energy

25. 50.0 mA from a to e

27. starter 171 A; battery 0.283 A

29. (a) 909 mA   (b) -1.82 V $= V_b - V_a$

31. (a) 5.00 s   (b) 150 μC   (c) 4.06 μA

33. $U_0/4$

37. (a) 6.00 V   (b) 8.29 μs

39. (a) 12.0 s   (b) $I(t) = (3.00\ \mu\text{A})e^{-t/12.0\ \text{s}}$;
$q(t) = (36.0\ \mu\text{C})\left(1 - e^{-t/12.0\ \text{s}}\right)$

41. 0.302 Ω

43. 16.6 kΩ

45.

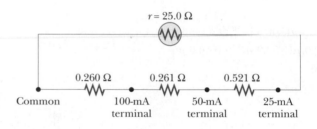

47. 145 Ω, 0.756 mA

49. (a) 12.5 A, 6.25 A, 8.33 A   (b) No; together they would require 27.1 A.

51. (a) $\sim 10^{-14}$ A   (b) $V_h/2 + (\sim 10^{-10}$ V) and $V_h/2 - (\sim 10^{-10}$ V), where V_h is the potential of the live wire, $\sim 10^2$ V

53. (a) either 3.84 Ω or 0.375 Ω (b) impossible

55. (a) $\mathcal{E}^2/3R$ (b) $3\mathcal{E}^2/R$ (c) in the parallel connection

57. (a) $R \to \infty$ (b) $R \to 0$ (c) $R = r$

59. 6.00 Ω; 3.00 Ω

61. (a) 4.40 Ω (b) 32.0 W, 9.60 W, 70.4 W   (c) 48.0 W

63. (a) $R \le 1\,050\ \Omega$ (b) $R \ge 10.0\ \Omega$

65. (a) 9.93 μC   (b) 33.7 nA   (c) 334 nW   (d) 337 nW

67. (a) 40.0 W   (b) 80.0 V, 40.0 V, 40.0 V

69. (a) 0.991   (b) 0.648   (c) Insulation should be added to the ceiling.

71. (a) 0 in 3 kΩ and 333 μA in 12 kΩ and 15 kΩ (b) 50.0 μC
(c) $(278\ \mu\text{A})\,e^{-t/180\ \text{ms}}$ (d) 290 ms

73. (a) $\ln(\mathcal{E}/\Delta V) = (0.011\,8)t + 0.088\,2$ (b) 84.7 s, 8.47 μF

75. $q_1 = (240\ \mu\text{C})(1 - e^{-1\,000t/6})$; $q_2 = (360\ \mu\text{C})(1 - e^{-1\,000t/6})$

CHAPTER 29

1. (a) up   (b) out of the plane of the paper   (c) no deflection   (d) into the plane of the paper

3. negative z direction

5. $(-20.9\hat{j})$ mT

7. 48.9° or 131°

9. 2.34 aN

11. 0.245 T east

13. (a) 4.73 N   (b) 5.46 N   (c) 4.73 N

15. 1.07 m/s

17. $2\pi r I B \sin\theta$ up

19. 2.98 μN west

21. 18.4 mA·m^2

23. 9.98 N·m clockwise as seen looking down from above

27. (a) 118 μN·m   (b) $-118\ \mu$J $\leq U \leq 118\ \mu$J

29. (a) 49.6 aN south   (b) 1.29 km

31. 115 keV

33. $r_\alpha = r_d = \sqrt{2}r_p$

35. 4.98×10^8 rad/s

37. 7.88 pT

39. $m = 2.99$ u, either $^3_1\text{H}^+$ or $^3_2\text{He}^+$

41. (a) 8.28 cm   (b) 8.23 cm; ratio is independent of both ΔV and B

43. (a) 4.31×10^7 rad/s   (b) 51.7 Mm/s

45. (a) 7.66×10^7 rad/s   (b) 26.8 Mm/s   (c) 3.76 MeV
    (d) 3.13×10^3 rev   (e) 257 μs

47. 70.1 mT

49. 1.28×10^{29} m^{-3}, 1.52

51. 43.3 μT

53. (a) The electric current experiences a magnetic force.

55. (a) -8.00×10^{-21} kg·m/s   (b) 8.90°

57. (a) $(3.52\hat{\mathbf{i}} - 1.60\hat{\mathbf{j}})$ aN   (b) 24.4°

59. $(2\pi/d)(2m_e\,\Delta V/e)^{1/2}$

61. 0.588 T

63. 0.713 A counterclockwise as seen from above

65. 438 kHz

67. 3.70×10^{-24} N·m

69. (a) 0.501 m   (b) 45.0°

71. (a) 1.33 m/s   (b) Positive ions moving toward you in magnetic field to the right feel upward magnetic force, and migrate upward in the blood vessel. Negative ions moving toward you feel downward magnetic force and accumulate at the bottom of this section of vessel. Thus both species can participate in the generation of the emf.

CHAPTER 30

1. 12.5 T

3. (a) 28.3 μT into the paper   (b) 24.7 μT into the paper

5. $\dfrac{\mu_0 I}{4\pi x}$ into the paper

7. 26.2 μT into the paper

9. (a) $2I_1$ out of the page   (b) $6I_1$ into the page

11. (a) along the line ($y = -0.420$ m, $z = 0$)
    (b) $(-34.7\hat{\mathbf{j}})$ mN   (c) $(17.3\hat{\mathbf{j}})$ kN/C

13. (a) $4.5\,\dfrac{\mu_0 I}{\pi L}$ (b) stronger

15. $(-13.0\hat{\mathbf{j}})\ \mu$T

17. $(-27.0\hat{\mathbf{i}})\ \mu$N

19. (a) 12.0 cm to the left of wire 1   (b) 2.40 A, downward

21. 20.0 μT toward the bottom of the page

23. 200 μT toward the top of the page; 133 μT toward the bottom of the page

25. (a) 6.34 mN/m inward   (b) greater

27. (a) 0   (b) $\dfrac{\mu_0 I}{2\pi R}$ tangent to the wall in a counterclockwise sense   (c) $\dfrac{\mu_0 I^2}{(2\pi R)^2}$ inward

29. (a) $\frac{1}{3}\mu_0 b r_1^2$ (b) $\dfrac{\mu_0 b R^3}{3r_2}$

31. 31.8 mA

33. 226 μN away from the center of the loop, 0

35. (a) 3.13 mWb   (b) 0

37. (a) 11.3 GV·m/s   (b) 0.100 A

39. (a) 9.27×10^{-24} A·m^2 (b) down

41. 0.191 T

43. 2.62 MA/m

45. (b) 6.45×10^4 K·A/T·m

47. (a) 8.63×10^{45} electrons   (b) 4.01×10^{20} kg

49. $\dfrac{\mu_0 I}{2\pi w} \ln\left(1 + \dfrac{w}{b}\right)\hat{\mathbf{k}}$

51. 12 layers, 120 m

53. 143 pT away along the axis

59. (a) 2.46 N up   (b) 107 m/s^2 up

61. (a) 274 μT   (b) $(-274\hat{\mathbf{j}})\ \mu$T   (c) $(1.15\hat{\mathbf{i}})$ mN
    (d) $(0.384\hat{\mathbf{i}})$ m/s^2 (e) acceleration is constant
    (f) $(0.999\hat{\mathbf{i}})$ m/s

63. 81.7 A

65. $\dfrac{\mu_0 I_1 I_2 L}{\pi R}$ to the right

69. $\dfrac{\mu_0 I}{4\pi}(1 - e^{-2\pi})$ out of the plane of the paper

71. $\frac{1}{3}\rho\mu_0\omega R^2$

73. (a) $\dfrac{\mu_0 I(2r^2 - a^2)}{\pi r(4r^2 - a^2)}$ to the left   (b) $\dfrac{\mu_0 I(2r^2 + a^2)}{\pi r(4r^2 + a^2)}$ toward the top of the page

CHAPTER 31

1. 500 mV

3. 9.82 mV

5. 160 A

7. (a) 1.60 A counterclockwise   (b) 20.1 μT   (c) up

9. (a) $(\mu_0 I L/2\pi)\ \ln(1 + w/h)$ (b) $-4.80\ \mu$V; current is counterclockwise

11. 283 μA upward

13. (68.2 mV) $e^{-1.6t}$, tending to produce counterclockwise current

15. 272 m

17. $(0.422 \text{ V}) \cos \omega t$

19. (a) eastward (b) 458 μV

21. (a) 3.00 N to the right (b) 6.00 W

23. 360 T

25. (a) 233 Hz (b) 1.98 mV

27. 2.83 mV

29. (a) $F = N^2 B^2 w^2 v / R$ to the left (b) 0 (c) $F = N^2 B^2 w^2 v / R$ to the left

31. 145 μA

33. 1.80 mN/C upward and to the left, perpendicular to r_1

35. (a) 7.54 kV (b) The plane of the coil is parallel to **B**.

37. $(28.6 \text{ mV}) \sin(4\pi t)$

39. (a) 110 V (b) 8.53 W (c) 1.22 kW

41. (a) $(8.00 \text{ mWb}) \cos(377t)$ (b) $(3.02 \text{ V}) \sin(377t)$
(c) $(3.02 \text{ A}) \sin(377t)$ (d) $(9.10 \text{ W}) \sin^2(377t)$
(e) $(24.1 \text{ mN} \cdot \text{m}) \sin^2(377t)$

43. (b) Larger R makes current smaller, so the loop must travel faster to maintain equality of magnetic force and weight. (c) The magnetic force is proportional to the product of field and current, while the current is itself proportional to field. If B becomes two times smaller, the speed must become four times larger to compensate.

45. $(-2.87\hat{\mathbf{j}} + 5.75\hat{\mathbf{k}}) \text{ Gm/s}^2$

47. (a) Doubling N doubles the amplitude. (b) Doubling ω doubles the amplitude and halves the period. (c) Doubling ω and halving N leaves the amplitude the same and cuts the period in half.

49. 62.3 mA down through 6.00 Ω, 860 mA down through 5.00 Ω, 923 mA up through 3.00 Ω

51. $\sim 10^{-4}$ V, by reversing a 20-turn coil of diameter 3 cm in 0.1 s in a field of 10^{-3} T

53. (a) 254 km/s (b) 215 V

55. 1.20 μC

57. (a) 0.900 A (b) 0.108 N (c) b (d) no

59. (a) $a\pi r^2$ (b) $-b\pi r^2$ (c) $-b\pi r^2/R$ (d) $b^2 \pi^2 r^4/R$

61. (a) 36.0 V (b) 600 mWb/s (c) 35.9 V (d) 4.32 N $\cdot$ m

65. 6.00 A

67. (a) $(1.19 \text{ V}) \cos(120\pi t)$ (b) 88.5 mW

71. $(-87.1 \text{ mV}) \cos(200\pi t + \phi)$

CHAPTER 32

1. 19.5 mV

3. 100 V

5. $(18.8 \text{ V}) \cos(377t)$

7. -0.421 A/s

9. (a) 188 μT (b) 33.3 nT $\cdot$ m^2 (c) 0.375 mH (d) B and Φ_B are proportional to current; L is independent of current

11. 0.750 m

13. $\varepsilon_0/k^2 L$

15. (a) 0.139 s (b) 0.461 s

17. (a) 2.00 ms (b) 0.176 A (c) 1.50 A (d) 3.22 ms

19. (a) 0.800 (b) 0

21. (a) 6.67 A/s (b) 0.332 A/s

23. $(500 \text{ mA})(1 - e^{-10t/s})$, $1.50 \text{ A} - (0.25 \text{ A}) e^{-10t/s}$

25. 0 for $t < 0$; $(10 \text{ A})(1 - e^{-10\,000t})$ for $0 < t < 200$ μs; $(63.9 \text{ A}) e^{-10\,000t}$ for $t > 200$ μs

27. (a) 5.66 ms (b) 1.22 A (c) 58.1 ms

29. 0.064 8 J

31. 2.44 μJ

33. 44.2 nJ/m^3 for the **E**-field and 995 μJ/m^3 for the **B**-field

35. (a) 0.500 J (b) 17.0 W (c) 11.0 W

37. 2.27 mT

39. 1.73 mH

41. 80.0 mH

43. (a) 18.0 mH (b) 34.3 mH (c) -9.00 mV

45. $(L_1 L_2 - M^2)/(L_1 + L_2 - 2M)$

47. 20.0 V

49. 608 pF

51. (a) 135 Hz (b) 119 μC (c) -114 mA

53. (a) 6.03 J (b) 0.529 J (c) 6.56 J

55. (a) 4.47 krad/s (b) 4.36 krad/s (c) 2.53%

57. $L = 199$ mH; $C = 127$ nF

59. (b) $\mu_0 J_s^2/2$ away from the other sheet (c) $\mu_0 J_s$ and zero (d) $\mu_0 J_s^2/2$

61. (a) -20.0 mV (b) $-(10.0 \text{ MV/s}^2)t^2$ (c) 63.2 μs

63. $(Q/2N)(3L/C)^{1/2}$

65. (a) $L \approx (\pi/2)N^2 \mu_0 R$ (b) ~ 100 nH (c) ~ 1 ns

71. (a) 72.0 V; b
(b)

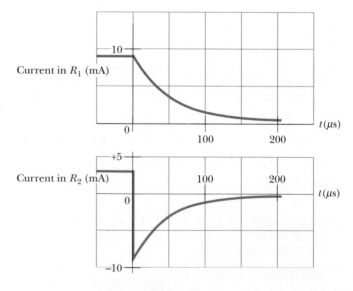

(c) 75.2 μs

73. 300 Ω

75. (a) It creates a magnetic field. (b) The long narrow rectangular area between the conductors encloses all of the magnetic flux.

77. (a) 62.5 GJ (b) 2 000 N

79. (a) 2.93 mT up (b) 3.42 Pa (c) clockwise as seen from above (d) up (e) 1.30 mN

CHAPTER 33

1. $\Delta v(t) = (283 \text{ V}) \sin(628t)$

3. 2.95 A, 70.7 V

5. 14.6 Hz

7. 3.38 W

9. (a) 42.4 mH (b) 942 rad/s

11. 5.60 A

13. 0.450 Wb

15. (a) 141 mA (b) 235 mA

17. 100 mA

19. (a) 194 V (b) current leads by 49.9°

21. (a) 78.5 Ω (b) 1.59 kΩ (c) 1.52 kΩ (d) 138 mA (e) −84.3°

23. (a) 17.4° (b) voltage leads the current

25. 1.88 V

27.

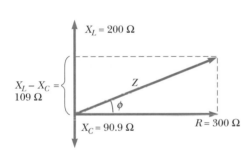

29. (a) either 123 nF or 124 nF (b) 51.5 kV

31. 8.00 W

33. (a) 16.0 Ω (b) −12.0 Ω

35. $\sqrt{\dfrac{800 \, \rho \mathscr{P} d}{\pi (\Delta V)^2}}$

37. 1.82 pF

39. (a) 633 fF (b) 8.46 mm (c) 25.1 Ω

41. 242 mJ

43. 0.591 and 0.987; the circuit in Problem 23

45. 687 V

47. 87.5 Ω

49. (a) 29.0 kW (b) 5.80 × 10⁻³ (c) If the generator were limited to 4 500 V, no more than 17.5 kW could be delivered to the load, never 5 000 kW.

51. (b) 0; 1 (c) $f_h = (10.88RC)^{-1}$

53. (a) 613 μF (b) 0.756

55. (a) 580 μH and 54.6 μF (b) 1 (c) 894 Hz (d) ΔV_{out} leads ΔV_{in} by 60.0° at 200 Hz. ΔV_{out} and ΔV_{in} are in phase at 894 Hz. ΔV_{out} lags ΔV_{in} by 60.0° at 4 000 Hz. (e) 1.56 W, 6.25 W, 1.56 W (f) 0.408

57. 56.7 W

59. 99.6 mH

61. (a) 225 mA (b) 450 mA

63. (a) 1.25 A (b) Current lags voltage by 46.7°.

65. (a) 200 mA; voltage leads by 36.8° (b) 40.0 V; $\phi = 0°$ (c) 20.0 V; $\phi = -90.0°$ (d) 50.0 V; $\phi = +90.0°$

67. (b) 31.6

71.

f (Hz)	X_L (Ω)	X_C (Ω)	Z (Ω)
300	283	12 600	12 300
600	565	6 280	5 720
800	754	4 710	3 960
1 000	942	3 770	2 830
1 500	1 410	2 510	1 100
2 000	1 880	1 880	40.0
3 000	2 830	1 260	1 570
4 000	3 770	942	2 830
6 000	5 650	628	5 020
10 000	9 420	377	9 040

(b)

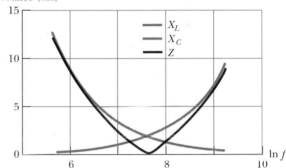

73. (a) 1.84 kHz

(b)

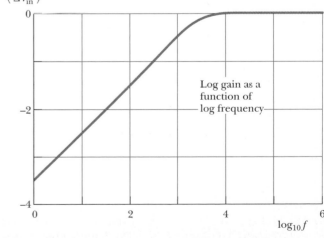

CHAPTER 34

1. (a) $(3.15\hat{\mathbf{j}})$ kN/C   (b) $(525\hat{\mathbf{k}})$ nT   (c) $(-483\hat{\mathbf{j}})$ aN

3. 2.25×10^8 m/s

5. (a) 6.00 MHz   (b) $(-73.3\hat{\mathbf{k}})$ nT
   (c) $\mathbf{B} = [(-73.3\hat{\mathbf{k}})$ nT$]\cos(0.126x - 3.77 \times 10^7 t)$

7. (a) 0.333 μT   (b) 0.628 μm   (c) 477 THz

9. 75.0 MHz

11. 3.33 μJ/m^3

13. 307 μW/m^2

15. 3.33×10^3 m^2

17. (a) 332 kW/m^2 radially inward   (b) 1.88 kV/m
    and 222 μT

19. (a) $\mathbf{E} \cdot \mathbf{B} = 0$ (b) $(11.5\hat{\mathbf{i}} - 28.6\hat{\mathbf{j}})$ W/m^2

21. 29.5 nT

23. (a) 2.33 mT   (b) 650 MW/m^2 (c) 510 W

25. (a) 540 V/m   (b) 2.58 μJ/m^3 (c) 773 W/m^2
    (d) 77.3% of the intensity in Example 34.5

27. 83.3 nPa

29. (a) 1.90 kN/C   (b) 50.0 pJ   (c) 1.67×10^{-19} kg·m/s

31. (a) 11.3 kJ   (b) 1.13×10^{-4} kg·m/s

33. (a) 134 m   (b) 46.9 m

35. (a) away along the perpendicular bisector of the line segment joining the antennas   (b) along the extensions of the line segment joining the antennas

37. (a) $\mathbf{E} = \frac{1}{2}\mu_0 c J_{max}[\cos(kx - \omega t)]\hat{\mathbf{j}}$

    (b) $\mathbf{S} = \frac{1}{4}\mu_0 c J_{max}^2[\cos^2(kx - \omega t)]\hat{\mathbf{i}}$

    (c) $I = \dfrac{\mu_0 c J_{max}^2}{8}$ (d) 3.48 A/m

39. 545 THz

41. (a) 6.00 pm   (b) 7.50 cm

43. 60.0 km

45. 1.00 Mm = 621 mi; not very practical

47. (a) 3.77×10^{26} W   (b) 1.01 kV/m and 3.35 μT

49. (a) $2\pi^2 r^2 f B_{max} \cos \theta$, where θ is the angle between the magnetic field and the normal to the loop   (b) The loop should be in the vertical plane containing the line of sight to the transmitter.

51. (a) 6.67×10^{-16} T   (b) 5.31×10^{-17} W/m^2
    (c) 1.67×10^{-14} W   (d) 5.56×10^{-23} N

53. 95.1 mV/m

55. (a) $B_{max} = 583$ nT, $k = 419$ rad/m, $\omega = 126$ Grad/s;
    $\mathbf{B}$ vibrates in xz plane   (b) $\mathbf{S}_{av} = (40.6\hat{\mathbf{i}})$ W/m^2
    (c) 271 nPa   (d) $(406\hat{\mathbf{i}})$ nm/s^2

57. (a) 22.6 h   (b) 30.6 s

59. (a) 8.32×10^7 W/m^2 (b) 1.05 kW

61. (a) 1.50 cm   (b) 25.0 μJ   (c) 7.37 mJ/m^3
    (d) 40.8 kV/m, 136 μT   (e) 83.3 μN

63. 637 nPa

65. $\epsilon_0 E^2 A/2m$

67. (a) 16.1 cm   (b) 0.163 m^2 (c) 470 W/m^2 (d) 76.8 W
    (e) 595 N/C   (f) 1.98 μT   (g) The cats are nonmagnetic

and carry no macroscopic charge or current. Oscillating charges within molecules make them emit infrared radiation.   (h) 119 W

69. 4.77 Gm

CHAPTER 35

1. 299.5 Mm/s

3. 114 rad/s

5. (c) 0.055 7 °

9. 23.3°

11. 15.4°; 2.56 m

13. 19.5° above the horizon

15. (a) 1.52   (b) 417 nm   (c) 474 THz   (d) 198 Mm/s

17. 158 Mm/s

19. 30.0° and 19.5° at entry; 19.5° and 30.0° at exit

21. 3.88 mm

23. 30.4° and 22.3°

25. $\sim 10^{-11}$ s; between 10^3 and 10^4 wavelengths

29. 0.171°

31. 86.8°

33. 27.9°

35. 4.61°

37. (a) 33.4°   (b) 53.4°   (c) There is no critical angle.

39. 1.000 08

41. 1.08 cm $< d <$ 1.17 cm

43. Skylight incident from above travels down the plastic. If the index of refraction of the plastic is greater than 1.41, the rays close in direction to the vertical are totally reflected from the side walls of the slab and from both facets at the bottom of the plastic, where it is not immersed in gasoline. This light returns up inside the plastic and makes it look bright. Where the plastic is immersed in gasoline, total internal reflection is frustrated and the downward-propagating light passes from the plastic out into the gasoline. Little light is reflected up, and the gauge looks dark.

45. Scattered light leaving the photograph in all forward horizontal directions in air is gathered by refraction into a fan in the water of half-angle 48.6°. At larger angles you see things on the other side of the globe, reflected by total internal reflection at the back surface of the cylinder.

47. 77.5°

49. 2.27 m

51. (a) 0.172 mm/s   (b) 0.345 mm/s   (c) northward at 50.0° below the horizontal   (d) northward at 50.0° below the horizontal

53. 62.2%

55. 82 reflections

57. (b) 68.5%

59. 27.5°

61. (a) It always happens.   (b) 30.3°   (c) It cannot happen.

63. 2.36 cm

67. 1.93

69. (a) $n = [1 + (4t/d)^2]^{1/2}$ (b) 2.10 cm (c) violet

71. (a) 1.20 (b) 3.40 ns

CHAPTER 36

1. $\sim 10^{-9}$ s younger

3. 35.0 in.

5. 10.0 ft, 30.0 ft, 40.0 ft

7. (a) 13.3 cm, -0.333, real and inverted (b) 20.0 cm, -1.00, real and inverted (c) no image is formed

9. (a) -12.0 cm; 0.400 (b) -15.0 cm; 0.250 (c) upright

11. (a) $q = 45.0$ cm; $M = -0.500$ (b) $q = -60.0$ cm; $M = 3.00$ (c) Image (a) is real, inverted, and diminished. Image (b) is virtual, upright, and enlarged. The ray diagrams are like Figures 36.15a and 36.15b, respectively.

13. At 0.708 cm in front of the reflecting surface. Image is virtual, upright, and diminished.

15. 7.90 mm

17. (a) a concave mirror with radius of curvature 2.08 m (b) 1.25 m from the object

19. (a) 25.6 m (b) 0.058 7 rad (c) 2.51 m (d) 0.023 9 rad (e) 62.8 m from your eyes

21. 38.2 cm below the top surface of the ice

23. 8.57 cm

25. (a) 45.0 cm (b) -90.0 cm (c) -6.00 cm

27. 1.50 cm/s

29. (a) 16.4 cm (b) 16.4 cm

31. (a) 650 cm from the lens on the opposite side from the object; real, inverted, enlarged (b) 600 cm from the lens on the same side as the object; virtual, upright, enlarged

33. 2.84 cm

37. (a) -12.3 cm, to the left of the lens (b) 0.615 (c)

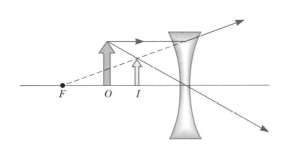

39. (a) 7.10 cm (b) 0.074 0 mm (c) 23.3 MW/m^2

41. (a) $p = \dfrac{d}{2} \pm \sqrt{\dfrac{d^2}{4} - fd}$ (b) Both images are real and inverted. One is enlarged, the other diminished.

43. 1.24 cm

45. 21.3 cm

47. -4.00 diopters, a diverging lens

49. -3.70 diopters

51. -575

53. (a) -800 (b) image is inverted

55. (a) virtual (b) infinity (c) 15.0 cm, -5.00 cm

57. -40.0 cm

59. (a) 23.1 cm (b) 0.147 cm

61. (a) 67.5 cm (b) The lenses can be displaced in two ways. The first lens can be displaced 1.28 cm farther away from the object, and the second lens 17.7 cm toward the object. Alternatively, the first lens can be displaced 0.927 cm toward the object and the second lens 4.44 cm toward the object.

63. $q = 5.71$ cm; real

65. 0.107 m to the right of the vertex of the hemispherical face

67. 8.00 cm
Ray diagram:

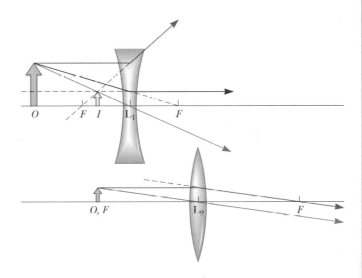

69. 1.50 m in front of the mirror; 1.40 cm (inverted)

71. (a) 30.0 cm and 120 cm (b) 24.0 cm (c) real, inverted, diminished with $M = -0.250$

73. -75.0

75. (a) 44.6 diopters (b) 3.03 diopters

77. (a) 20.0 cm to the right of the second lens, -6.00 (b) inverted (c) 6.67 cm to the right of the second lens, -2.00, inverted

CHAPTER 37

1. 1.58 cm

3. (a) 55.7 m (b) 124 m

5. 1.54 mm

7. (a) 2.62 mm (b) 2.62 mm

9. 11.3 m

11. (a) 10.0 m (b) 516 m (c) Only the runway centerline is a maximum for the interference patterns for both frequencies. If the frequencies were related by a ratio of small integers k/ℓ, the plane could by mistake fly along the kth side maximum of one signal where it coincides with the ℓth side maximum of the other.

13. (a) 13.2 rad   (b) 6.28 rad   (c) 0.012 7 degree
    (d) 0.059 7 degree

15. (a) 1.93 μm   (b) 3.00λ (c) maximum

17. 48.0 μm

19. (a) 7.95 rad   (b) 0.453

21. (a) and (b) 19.7 kN/C at 35.0°   (c) 9.36 kN/C at 169°

23. $10.0 \sin(100\pi t + 0.927)$

25. $26.2 \sin(\omega t + 36.6°)$

27. $\pi/2$

29. $360°/N$

31. (a) green   (b) violet

33. 0.500 cm

35. no reflection maxima in the visible spectrum

37. 290 nm

39. 4.35 μm

41. 39.6 μm

43. $1 + N\lambda/2L$

45. 1.25 m

47. (a) $\sim 10^{-3}$ degree   (b) $\sim 10^{11}$ Hz, microwave

49. $20.0 \times 10^{-6}\ °\text{C}^{-1}$

51. 3.58°

53. 1.62 km

55. 421 nm

59. (a) $2(4h^2 + d^2)^{1/2} - 2d$ (b) $(4h^2 + d^2)^{1/2} - d$

61. $y' = (n - 1)tL/d$

63. (a) 70.6 m   (b) 136 m

65. 1.73 cm

67. (a) 4.86 cm from the top   (b) 78.9 nm and 128 nm
    (c) 2.63×10^{-6} rad

69. 0.505 mm

CHAPTER 38

1. 4.22 mm

3. 0.230 mm

5. three maxima, at 0° and near 46° on both sides

7. 51.8 μm wide and 949 μm high

9. 0.016 2

11. 1.00 mrad

13. 3.09 m

15. violet; between 186 m and 271 m

17. 13.1 m

19. Neither. It can resolve objects no closer than several centimeters apart.

21. 0.244 rad = 14.0°

23. 7.35°

25. 5.91° in first order, 13.2° in second order, 26.5° in third order

27. (a) 478.7 nm, 647.6 nm, and 696.6 nm   (b) 20.51°, 28.30°, and 30.66°

29. (a) 12 000, 24 000, 36 000   (b) 11.1 pm

31. (a) 2 800 grooves   (b) 4.72 μm

33. (a) 5 orders   (b) 10 orders in the short-wavelength region

35. 93.4 pm

37. 14.4°

39. 5.51 m, 2.76 m, 1.84 m

41. (a) 54.7°   (b) 63.4°   (c) 71.6°

43. 1.11

45. 60.5°

47. 36.9° above the horizon

49. (a) 6   (b) 7.50°

51. 632.8 nm

53. (a) 25.6°   (b) 19.0°

55. 545 nm

57. (a) 3.53×10^3 grooves/cm   (c) Eleven maxima

59. 4.58 μm $< d <$ 5.23 μm

61. 15.4

63. (a) 41.8°   (b) 0.593   (c) 0.262 m

67. (b) 3.77 nm/cm

69. (b) 15.3 μm

71. $\phi = 1.391\ 557\ 4$ after seventeen steps or fewer

73. $a = 99.5\ \mu$m $\pm 1\%$

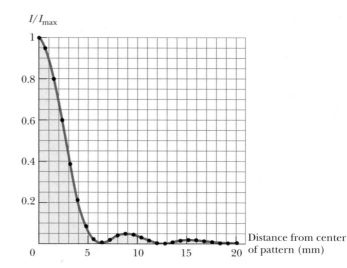

CHAPTER 39

5. $0.866c$

7. (a) 64.9/min   (b) 10.6/min

9. 1.54 ns

11. $0.800c$

13. (a) 39.2 μs   (b) accurate to one digit

15. (a) 20.0 m   (b) 19.0 m   (c) $0.312c$

17. (a) 21.0 yr   (b) 14.7 ly   (c) 10.5 ly   (d) 35.7 yr

19. (c) 2.00 kHz   (d) $\pm 0.075\ 0$ m/s ≈ 0.2 mi/h

21. $0.220c = 6.59 \times 10^7$ m/s

23. (a) 17.4 m   (b) 3.30°

25. (a) 2.50×10^8 m/s   (b) 4.97 m   (c) -1.33×10^{-8} s

27. $0.960c$

29. (a) 2.73×10^{-24} kg·m/s   (b) 1.58×10^{-22} kg·m/s
     (c) 5.64×10^{-22} kg·m/s

31. 4.50×10^{-14}

33. $0.285c$

35. (a) 5.37×10^{-11} J   (b) 1.33×10^{-9} J

37. 1.63×10^3 MeV/c

39. (a) 938 MeV   (b) 3.00 GeV   (c) 2.07 GeV

41. 8.84×10^{-28} kg and 2.51×10^{-28} kg

45. (a) 3.91×10^4 (b) $u = 0.999\,999\,999\,7c$ (c) 7.67 cm

47. 4.08 MeV and 29.6 MeV

49. $\sim 10^{-15}$

51. 0.842 kg

53. 4.19×10^9 kg/s

55. (a) 26.6 Mm   (b) 3.87 km/s   (c) -8.34×10^{-11}
     (d) 5.29×10^{-10} (e) $+4.46 \times 10^{-10}$

57. (a) a few hundred seconds   (b) $\sim 10^8$ km

59. (a) $0.800c$ (b) 7.50 ks   (c) 1.44 Tm, $-0.385c$ (d) 4.88 ks

61. 0.712%

63. (a) $0.946c$ (b) 0.160 ly   (c) 0.114 yr   (d) 7.50×10^{22} J

65. (a) 76.0 min   (b) 52.1 min

67. yes, with 18.8 m to spare

69. (b) For u small compared to c, the relativistic expression
     agrees with the classical expression. As u approaches c, the
     acceleration approaches zero, so that the object can never
     reach or surpass the speed of light.
     (c) Perform $\int (1 - u^2/c^2)^{-3/2}\,du = (qE/m)\int dt$ to obtain
     $u = qEct(m^2c^2 + q^2E^2t^2)^{-1/2}$ and then
     $\int dx = \int qEct(m^2c^2 + q^2E^2t^2)^{-1/2}\,dt$ to obtain
     $x = (c/qE)\,[(m^2c^2 + q^2E^2t^2)^{1/2} - mc]$

75. 1.82×10^{-3} eV

CHAPTER 40

1. 5.18×10^3 K

3. 1.30×10^{15}/s

5. (a) 5.75×10^3 K   (b) 504 nm

7. (a) 2.57 eV   (b) 12.8 μeV   (c) 191 neV   (d) 484 nm
     (visible); 9.68 cm and 6.52 m (radio waves)

9. 2.27×10^{30} photons/s

11. 7.73×10^3 K

13. (a) 296 nm, 1.01 PHz   (b) 2.71 V

15. (a) only lithium   (b) 0.808 eV

17. (a) 1.90 eV   (b) 0.216 V

19. 8.41 pC

21. 1.78 eV, 9.47×10^{-28} kg·m/s

23. 70.0°

25. (a) 43.0°   (b) 602 keV, 3.21×10^{-22} kg·m/s   (c) 278 keV,
     3.21×10^{-22} kg·m/s

27. (a) 0.101 nm   (b) 81.1°

29. 0.004 86 nm

31. By this definition, ionizing radiation is the ultraviolet light,
     x-rays, and γ rays with wavelength shorter than 124 nm;
     that is, with frequency higher than 2.42×10^{15} Hz.

33. 397 fm

35. (a) 0.709 nm   (b) 414 nm

37. (a) $\sim$100 MeV or more   (b) No. The kinetic energy of the
     electron is much larger than the typical binding energy, so
     the electron would immediately escape.

39. (b) No. $\lambda^{-2} + \lambda_C^{-2}$ cannot be equal to λ^{-2}.

41. (a) 14.9 keV or, neglecting relativistic correction, 15.1 keV
     (b) 124 keV

43. $v_{\text{phase}} = u/2$

45. (a) 993 nm   (b) 4.96 mm   (c) If its detection forms part
     of an interference pattern, the neutron must have passed
     through both slits. If we test to see which slit a particular
     neutron passes through, it will not form part of the
     interference pattern.

47. 2.27 pA

49. Within 1.16 mm for the electron, 5.28×10^{-32} m for the
     bullet

51. The electron energy must be $\sim 100mc^2$ or larger. The
     proton energy can be as small as $1.001mc^2$, which is within
     the range well described classically.

53. (b) 519 am

55. (a)

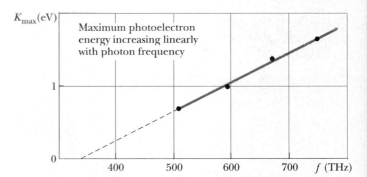

     (b) 6.4×10^{-34} J·s $\pm$ 8%   (c) 1.4 eV

57. (a) 191 MeV   (b) 9.20 MeV

67. 2.81×10^{-8}

69. (a) The light is unpolarized. It contains both horizontal
     and vertical electric field oscillations. (b) The interfer-
     ence pattern appears, but with diminished overall inten-
     sity. (c) The results are the same in each case. (d) The
     interference pattern appears and disappears as the polar-
     izer turns, with alternately increasing and decreasing con-
     trast between the bright and dark fringes. The intensity on
     the screen is precisely zero at the center of a dark fringe
     four times in each revolution, when the filter axis
     has turned by 45°, 135°, 225°, and 315° from the vertical.

(e) Looking at the overall light energy arriving at the screen, we see a low-contrast interference pattern. After we sort out the individual photon runs into those for trial 1, those for trial 2, and those for trial 3, we have the original results replicated: The runs for trials 1 and 2 form the two blue graphs in Figure 40.24, and the runs for trial 3 build up the red graph.

CHAPTER 41

1. (a) 126 pm   (b) 5.27×10^{-24} kg·m/s   (c) 95.5 eV
3. (a) 0.434 nm   (b) 6.00 eV
5.

(a)

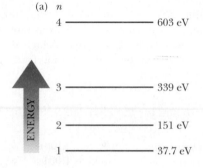

(b) 2.20 nm, 2.75 nm, 4.12 nm, 4.71 nm, 6.60 nm, 11.0 nm

7. 0.793 nm
9. 6.17 MeV, 202 fm, a gamma ray
11. 0.513 MeV, 2.05 MeV, 4.62 MeV; yes
19. at $L/4$ and at $3L/4$
21. (a) 0.196   (b) 0.609   (c) The classical probability, 1/3, is very different.
23. (a) $E = \hbar^2/mL^2$ (b) Requiring $\int_{-L}^{L} A^2(1 - x^2/L^2)^2\, dx = 1$ gives $A = (15/16L)^{1/2}$ (c) $47/81 = 0.580$

25. (a)

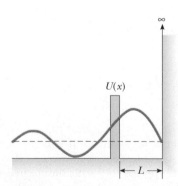

(b) $2L$

27. (a) 0.010 3   (b) 0.990
29. by 0.959 nm, to 1.91 nm
31. 1.35

35. 600 nm
37. (b) The acceleration is equal to a negative constant times the excursion from equilibrium. The frequency is $\dfrac{1}{2\pi}\sqrt{\dfrac{k}{\mu}}$.

39. $\sim 10^{-10^{30}}$

41.

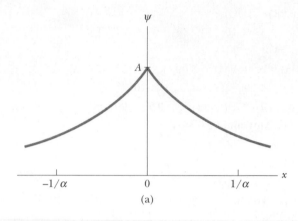

(a)

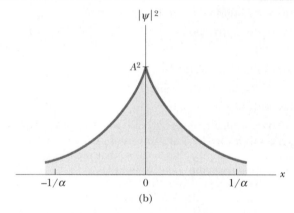

(b)

(c) The wave function is continuous. It shows localization by approaching zero as $x \rightarrow \pm\infty$. It is everywhere finite and can be normalized.   (d) $A = \sqrt{\alpha}$ (e) 0.632

43. 0.029 4
45. (a) 434 THz   (b) 691 nm   (c) 165 peV or more
49. (a)

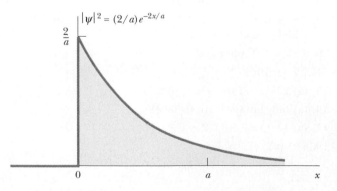

(b) 0   (c) 0.865

51. (a) $-7k_ee^2/3d$ (b) $h^2/36m_ed^2$ (c) 49.9 pm
(d) 280 pm is 5.62 times larger.

53. (a) $A = (2/17L)^{1/2}$ (b) $|A|^2 + |B|^2 = 1/a$

55. 2.25

CHAPTER 42

1. (a) 5 (b) no; no

3. (b) 0.846 ns

5. (a) 2.19 Mm/s (b) 13.6 eV (c) -27.2 eV

7. (a) 0.212 nm (b) 9.95×10^{-25} kg·m/s
(c) 2.11×10^{-34} kg·m^2/s (d) 3.40 eV (e) -6.80 eV
(f) -3.40 eV

9. (a) 3.03 eV (b) 411 nm (c) 732 THz

11. 4.42×10^4 m/s

13. (a) $E_n = -54.4$ eV$/n^2$ for $n = 1, 2, 3, \ldots$

n	$E(\text{eV})$
∞	0
4	-3.40
3	-6.05
2	-13.6
1	-54.4

ENERGY

(b) 54.4 eV

15. (a) 2.89×10^{34} kg·m^2/s (b) 2.74×10^{68}
(c) 7.30×10^{-69}

17. (a) $\Delta p \geq \hbar/2r$ (b) Choosing $p \approx \hbar/r$,
$E = K + U = \hbar^2/2m_er^2 - k_ee^2/r$ (c) $r = \hbar^2/m_e k_ee^2 = a_0$
and $E = -13.6$ eV, in agreement with the Bohr theory.

19. (b) 0.497

21. It does, with $E = -k_ee^2/2a_0$

23. (a)

n	ℓ	m_ℓ	m_s
3	2	2	$1/2$
3	2	2	$-1/2$
3	2	1	$1/2$
3	2	1	$-1/2$
3	2	0	$1/2$
3	2	0	$-1/2$
3	2	-1	$1/2$
3	2	-1	$-1/2$
3	2	-2	$1/2$
3	2	-2	$-1/2$

(b)

n	ℓ	m_ℓ	m_s
3	1	1	$1/2$
3	1	1	$-1/2$
3	1	0	$1/2$
3	1	0	$-1/2$
3	1	-1	$1/2$
3	1	-1	$-1/2$

25. $\ell = 4$

27. (a) 2 (b) 8 (c) 18 (d) 32 (e) 50

29. (a) 3.99×10^{17} kg/m^3 (b) 81.7 am (c) 1.77 Tm/s
(d) $5.91 \times 10^3 c$

31. $n = 3$; $\ell = 2$; $m_\ell = -2, -1, 0, 1,$ or 2; $s = 1$; $m_s = -1, 0,$
or 1, for a total of 15 states

33. The $4s$ subshell is filled first. We would expect $[\text{Ar}]3d^44s^2$
to have lower energy, but $[\text{Ar}]3d^54s^1$ has more unpaired
spins and lower energy according to Hund's rule. It is the
ground-state configuration of chromium.

35. Aluminum

37. (a) $1s, 2s, 2p, 3s, 3p, 4s, 3d, 4p, 5s, 4d, 5p, 6s, 4f, 5d, 6p, 7s$
(b) Element 15 should have valence $+5$ or -3, and it
does. Element 47 should have valence -1, but it has valence $+1$. Element 86 should be inert, and it is.

39. 18.4 T

41. 124 V

43. 0.072 5 nm

45. iron

47. 28.2 THz, 10.6 μm, infrared

49. 3.49×10^{16} photons

51. (a) 217 nm (b) 93.1 nm

53. (a) 609 μeV (b) 6.9 μeV (c) 147 GHz, 2.04 mm

55. $r_n = (0.106 \text{ nm})n^2$, $E_n = -6.80$ eV$/n^2$, for $n = 1, 2, 3, \ldots$

57. The classical frequency is $4\pi^2 m_e k_e^2 e^4/h^3 n^3$.

59. (a) 1.57×10^{14} m$^{-3/2}$ (b) 2.47×10^{28} m^{-3}
(c) 8.69×10^8 m^{-1}

61. (a) 4.20 mm (b) 1.05×10^{19} photons
(c) 8.82×10^{16}/mm^3

63. $3h^2/4mL^2$

67. 5.39 keV

69. 0.125

71. 9.79 GHz

73. (a) $\sim -10^6$ m/s^2 (b) ~ 1 m

CHAPTER 43

1. (a) 921 pN toward the other ion (b) -2.88 eV

3. (a) $(2A/B)^{1/6}$ (b) $B^2/4A$ (c) 74.2 pm, 4.46 eV

5. ~ 10 K

7. 5.63 Trad/s

11. 2.72×10^{-47} kg·m^2

13. 0.358 eV

15. $(18.4 \ \mu\text{eV})J(J + 1)$, where $J = 0, 1, 2, 3, \ldots$

17. 558

19. (a) 4.60×10^{-48} kg·m^2 (b) 1.32×10^{14} Hz
(c) 0.074 1 nm

21. 6.25×10^9

23. -7.84 eV

27. An average atom contributes 0.981 electron to the conduction band.

29. (a) 2.54×10^{28} electrons/m^3 (b) 3.15 eV

31. 5.28 eV

35. (a) 1.10   (b) 1.47×10^{-25}, much smaller

37. All of the Balmer lines are absorbed, except for the red line at 656 nm, which is transmitted.

39. 1.24 eV or less; yes

41. $-0.021\,9$ eV, 2.81 nm

43. 4.4 V

45. (a) 2.98 mA   (b) 67.1 Ω (c) 8.39 Ω

47. 203 A to produce a magnetic field in the direction of the original field

49. (a) 6.15×10^{13} Hz   (b) 1.59×10^{-46} kg·m^2
(c) 4.79 μm or 4.96 μm

51. 7

55. (a) 0.350 nm   (b) -7.02 eV   (c) $-1.20\hat{\mathbf{i}}$ nN

57. (a) r_0 (b) B (c) $(a/\pi)[B/2\mu]^{1/2}$
(d) $B - (ha/\pi)[B/8\mu]^{1/2}$

59. (b) No. The fourth term is larger than the sum of the first three.

CHAPTER 44

1. $\sim 10^{28}$; $\sim 10^{28}$; $\sim 10^{28}$

3. (a) 27.6 N   (b) 4.17×10^{27} m/s^2 away from the nucleus
(c) 1.73 MeV

5. (a) 455 fm   (b) 6.04 Mm/s

7. (a) 1.90 fm   (b) 7.44 fm

9. 16.0 km

11. Z magic: He, O, Ca, Ni, Sn, Pb. N magic: isotopes of H, He, N, O, Cl, K, Ca, V, Cr, Sr, Y, Zr, Xe, Ba, La, Ce, Pr, Nd, Pb, Bi, Po.

13. (a) $^{36}_{16}$S, $^{6}_{2}$He   (b) $^{6}_{2}$He is unstable. Two neutrons must be removed to make it the stable $^{4}_{2}$He.

15. They agree with the figure.

17. (a) $^{139}_{55}$Cs   (b) $^{139}_{57}$La   (c) $^{139}_{55}$Cs

19. greater for $^{15}_{7}$N by 3.54 MeV

21. 200 MeV

23. (a) 491 MeV   (b) 179%, -53.0%, -24.6%, -1.37%

25. 1.16 ks

27. (a) 1.55×10^{-5}/s, 12.4 h   (b) 2.39×10^{13} atoms
(c) 1.88 mCi

29. 9.47×10^9 nuclei

31. 2.66 d

33. 4.27 MeV

35. 9.96×10^3 yr

37. $^{3}_{1}$H atom $\rightarrow$ $^{3}_{2}$He atom $+ \bar{\nu}$; 18.6 keV

39. (a) $e^- + p \rightarrow n + \nu$ (b) $^{15}_{8}$O atom $\rightarrow$ $^{15}_{7}$N atom $+ \nu$
(c) 2.75 MeV

41.

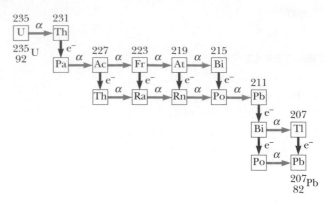

43. (a) X: 3.66 days, Y: 10.6 h   (b) ^{224}Ra, ^{212}Pb   (c) three

45. -2.64 MeV

47. (a) $^{197}_{79}$Au $+ ^{1}_{0}$n $\rightarrow$ $^{198}_{80}$Hg $+ ^{0}_{-1}$e $+ \nu$ (b) 7.89 MeV

49. 8.005 3 u; 10.013 5 u

51. 165 MeV

53. (a) 29.2 MHz   (b) 42.6 MHz   (c) 2.13 kHz

55. (a) 5.70 MeV   (b) 3.27 MeV, exothermic

57. 1.88 MeV

59. (a) cannot occur   (b) can occur   (c) K_e between 0 and 156 keV

61. (a) conservation of energy   (b) electric potential energy of the nucleus   (c) 1.20 MeV

63. (a) $^{141}_{59}$Pr   (b) 2.32 MeV   (c) 97.2%

65. (a) 61.8 Bq/L   (b) 40.3 d

67. 5.94 Gyr

69. 2.20 μeV

73. 0.400%

75. 2.64 min

CHAPTER 45

1. 0.387 g

3. $^{144}_{54}$Xe, $^{143}_{54}$Xe, $^{142}_{54}$Xe

5. $^{1}_{0}$n $+ ^{232}$Th $\rightarrow$ ^{233}Th $\rightarrow$ ^{233}Pa $+ e^- + \bar{\nu}$,
^{233}Pa $\rightarrow$ ^{233}U $+ e^- + \bar{\nu}$

7. (a) 8.68 MeV (b) The proton and the boron nucleus have positive charges. The colliding particles must have enough kinetic energy to approach very closely in spite of their electric repulsion.

9. 5.80 Mm

11. about 3 000 yr

13. 6.25×10^{19} Bq

15. (a) 2.30×10^{-14} $Z_1 Z_2$ J   (b) 0.144 MeV

17. (a) 2.22 Mm/s   (b) $\sim 10^{-7}$ s

19. (a) 1.7×10^7 J   (b) 7.3 kg

21. 1.30×10^{25} ^{6}Li; 1.61×10^{26} ^{7}Li

23. 1.66×10^3 yr

25. (a) 2.5 mrem per x-ray
(b) 5 rem/yr is 38 times 0.13 rem/yr

27. 2.09×10^6 s

29. 1.14 rad

31. (a) 3.12×10^7 (b) 3.12×10^{10} electrons

33. (a) about 8 min (b) 27.6 min (c) 30 min $\pm$ 30%

35. $\sim 10^3$ Bq

37. (b) R/λ

39. (a) 8×10^4 eV (b) 4.62 MeV and 13.9 MeV
(c) 1.03×10^7 kWh

41. (a) 177 MeV (b) $K_{Br} = 112$ MeV, $K_{La} = 65.4$ MeV
(c) $v_{Br} = 15.8$ Mm/s, $v_{La} = 9.20$ Mm/s

43. 482 Ci, less than the fission inventory by on the order of a hundred million times

45. 2.56×10^4 kg

47. (a) 2.65 GJ (b) The fusion energy is 78.0 times larger.

49. (a) 4.91×10^8 kg/h $= 4.91 \times 10^5$ m^3/h (b) 0.141 kg/h

51. (a) 15.5 cm (b) 51.7 MeV (c) The number of decays per second is the decay rate R, and the energy released in each decay is Q. Then the energy released per unit time interval is $\mathscr{P} = QR$. (d) 227 kJ/yr (e) 3.18 J/yr

53. 14.0 MeV or, ignoring relativistic correction, 14.1 MeV

55. (a) 2.24×10^7 kWh (b) 17.6 MeV (c) 2.34×10^8 kWh
(d) 9.36 kWh (e) Coal is cheap at this moment in human history. We hope that safety and waste-disposal problems can be solved so that nuclear energy can be affordable before scarcity drives up the price of fossil fuels.

57. (b) 26.7 McV

59. (a) 5×10^7 K (b) 1.94 MeV, 1.20 MeV, 1.02 MeV, 7.55 MeV, 7.30 MeV, 1.73 McV, 1.02 MeV, 4.97 MeV, 26.7 MeV (c) Most of the neutrinos leave the star directly after their creation, without interacting with any other particles.

61. (b) 1.00 μs (c) 2.83 km/s (d) 14.6 μs (e) yes, 107 kilotons of TNT

CHAPTER 46

1. 453 ZHz; 662 am

3. 118 MeV

5. $\sim 10^{-18}$ m

7. 67.5 MeV, 67.5 MeV/c, 16.3 ZHz

9. (a) 0.782 MeV (b) $v_e = 0.919$ c, $v_p = 380$ km/s (c) The electron is relativistic; the proton is not.

11. $\Omega^+ \rightarrow \overline{\Lambda}^0 + K^+$ $\quad$ $\overline{K}_S^0 \rightarrow \pi^+ + \pi^-$ $\quad$ $\overline{\Lambda}^0 \rightarrow \overline{p} + \pi^+$
$\overline{n} \rightarrow \overline{p} + e^+ + \nu_e$

13. (b) The second violates strangeness conservation.

15. (a) $\overline{\nu}_\mu$ (b) ν_μ (c) $\overline{\nu}_e$ (d) ν_e (e) ν_μ (f) $\overline{\nu}_e + \nu_\mu$

17. (a), (c), and (f) violate baryon number conservation. (b), (d), and (e) can occur. (f) violates muon–lepton number conservation.

19. (a) ν_e (b) ν_μ (c) $\overline{\nu}_\mu$ (d) $\nu_\mu + \overline{\nu}_\tau$

21. (b) and (c) conserve strangeness. (a), (d), (e), and (f) violate strangeness conservation.

23. (a) not allowed; violates conservation of baryon number (b) strong interaction (c) weak interaction (d) weak interaction (e) electromagnetic interaction

25. (a) K^+ (b) Ξ^0 (c) π^0

27. 9.26 cm

29. (a) 3.34×10^{26} e$^-$, 9.36×10^{26} u, 8.70×10^{26} d
(b) $\sim 10^{28}$ e$^-$, $\sim 10^{29}$ u, $\sim 10^{29}$ d. You have zero strangeness, charm, truth, and beauty.

31. $m_u = 312$ MeV/c^2; $\quad m_d = 314$ MeV/c^2

33. (a) The reaction $\overline{u}d + uud \rightarrow \overline{s}d + uds$ has a total of 1 u, 2 d, and 0 s quarks originally and finally. (b) The reaction $\overline{d}u + uud \rightarrow \overline{s}d + uus$ has a net of 3 u, 0 d, and 0 s quarks before and after. (c) $\overline{u}s + uud \rightarrow \overline{s}u + \overline{s}d + sss$ shows conservation at 1 u, 1 d, and 1 s quark. (d) The process $uud + uud \rightarrow \overline{s}d + uud + \overline{d}u + uds$ nets 4 u, 2 d, 0 s quarks initially and finally; the mystery particle is a Λ^0 or a Σ^0.

35. a neutron, udd

37. (a) $-e$, antiproton (b) 0, antineutron

39. (a) $0.160c$ (b) 2.82×10^9 ly

41. (a) 590.07 nm (b) 597 nm (c) 661 nm

43. 3.15 μW/m^2

45. (a) $\sim 10^{13}$ K (b) $\sim 10^{10}$ K

47. (b) 11.8 Gyr

49. (a) 1.61×10^{-35} m (b) 5.38×10^{-44} s (c) yes

51. 19 eV/c^2

53. one part in 50 000 000

55. (a) $2Nmc$ (b) $3^{1/2}Nmc$ (c) method (a)

57. 5.35 MeV and 32.3 MeV

59. 1 116 MeV/c^2

61. 70.4 MeV

63. 2.52×10^3 K

65. (a) Z^0 boson (b) gluon or photon

Credits

Photographs

Chapter 23. **704:** M. Shilin/M. Newman/Photo Researchers, Inc. **706:** Courtesy of Resonance Research Corporation **708:** Rolin Graphics, Inc. **709:** Charles D. Winters **710:** © 1968 Fundamental Photographs **711:** Courtesy of AIP Niels Bohr Library/E. Scott Barr Collection **716:** Courtesy Johnny Autery **724:** both, Courtesy of Harold M. Waage, Princeton University **725:** Courtesy of Harold M. Waage, Princeton University

Chapter 24. **739:** Getty Images **742:** public domain **751:** Courtesy of Harold M. Waage, Princeton University

Chapter 25. **762:** © Keith Kent/Photo Researchers, Inc. **779:** Courtesy of Harold M. Waage, Princeton University **783:** Rei O'Hara/Black Star/PictureQuest **783:** Greig Cranna/Stock, Boston, Inc./PictureQuest **791:** E.R. Degginer/H. Armstrong Roberts

Chapter 26. **795:** Paul Silverman/Fundamental Photographs **799:** Courtesy of Harold M. Waage, Princeton University **810:** Adam Hart-Davis/SPL/Custom Medical Stock **812:** © Loren Winters/Visuals Unlimited **813:** George Semple

Chapter 27. **831:** Telegraph Color Library/Getty Images **835:** © Bettmann/CORBIS **837:** Henry Leap and Jim Lehman **838:** SuperStock **844:** Courtesy of IBM Research Laboratory **847:** George Semple

Chapter 28. **858:** George Semple **869:** George Semple **870:** AIP ESVA/W.F. Maggers Collection **881:** George Semple **884:** Henry Leap and Jim Lehman

Chapter 29. **894:** James King-Holmes/Science Photo Library/Photo Researchers, Inc. **895:** North Wind Picture Archives **897:** Henry Leap/Jim Lehman **909:** Henry Leap/Jim Lehman **912:** Bell Telephone Labs/Courtesy of Emilio Segre Visual Archives **913:** Courtesy of Lawrence Berkeley Laboratory/University of California **917:** Courtesy of Central Scientific Company

Chapter 30. **926:** Defense Threat Reduction Agency (DTRA) **931:** Richard Megna/Fundamental Photographs **933:** Leinard de Selva/CORBIS **933:** Richard Megna/Fundamental Photographs **938:** Henry Leap/Jim Lehman **952:** left,

Leon Lewandowski **952:** right, High Field Magnet Laboratory, University of Nijmegen, The Netherlands **952:** Courtesy Argonne National Laboratory **957:** Courtesy of Central Scientific Company

Chapter 31. **967:** Michael Melford/Getty Images **968:** By kind permission of the President and Council of the Royal Society **972:** Charles D. Winters **979:** Richard Megna/Fundamental Photographs **985:** Photo by Brent Romans www.Edmunds.com **988:** John W. Jewett, Jr. **991:** Courtesy of Central Scientific Company **995:** Ross Harrison Koty/Getty Images

Chapter 32. **1003:** Jack Hollingsworth/Getty Images **1005:** North Wind Picture Archives **1015:** © by Braun GmbH, Kronberg **1022:** Courtesy of J. Rudmin

Chapter 33. **1033:** Lester Lefkowitz/Getty Images **1053:** George Semple **1053:** UPI/CORBIS **1053:** George Semple **1063:** Eddie Hironaha/Getty Images

Chapter 34. **1066:** upper left, NASA/CXCISAO; upper right, Palomar Observatory; lower right, VLA/NRAO; lower left, WM Keck Observatory **1067:** North Wind Picture Archives **1068:** Hulton-Deutsch Collection/CORBIS **1077:** Courtesy of NASA **1081:** Ron Chapple/Getty Images **1084:** George Semple **1086:** Philippe Plailly/SPL/Photo Researchers, Inc. **1088:** Amos Morgan/Getty Images **1090:** © Bill Banaszewski/Visuals Unlimited

Chapter 35. **1092:** David Muench/CORBIS **1094:** Mark D. Phillips/Photo Researchers, Inc. **1098:** both, Courtesy of Henry Leap and Jim Lehman **1101:** a) Courtesy of NASA; b, c) George Semple **1102:** top, both Courtesy Digital Light Processing, Texas Instruments, Inc.; bottom, Henry Leap/Jim Lehman **1108:** Courtesy of Rijksmuseum voor de Geschiedenis der Natuurwetenschappen. Courtesy Niels Bohr Library. **1110:** David Parker/Science Photo Library/Photo Researchers, Inc. **1113:** Courtesy of Henry Leap and Jim Lehman **1114:** left, Dennis O'Clair/Getty Images; right, Hank Morgan/Photo Researchers, Inc. **1117:** Courtesy of U.S. Air Force, Langley Air Force Base **1120:** Roy Atkeson Photo Library **1122:** Courtesy Edwin Lo

Chapter 36. **1126:** Don Hammond/CORBIS **1128:** George Semple **1130:** Courtesy of Henry Leap/Jim Lehman **1131:** Ken Kay/Fundamental Photographs **1133:** top, Henry Leap/Jim Lehman **1133:** Courtesy of Thompson Consumer Electronics **1135:** David Rogers **1136:** NASA **1138:** © 1990 Paul Silverman/Fundamental Photographs **1144:** Henry Leap/Jim Lehman **1149:** Henry Leap/Jim Lehman **1155:** From Lennart Nilsson, in collaboration with Jan Lindberg, *Behold Man: A Photographic Journey of Discovery Inside the Body*, Boston, Little, Brown & Co. 1974. **1160:** George Semple **1161:** © Tony

University of Tokyo **1523:** Courtesy Lawrence Berkeley Laboratory, University of California, Photographic Services **1527:** Courtesy of Michael R. Dressler **1528:** Courtesy of Brookhaven National Laboratory **1535:** Courtesy of CERN **1536:** Courtesy of Fermi National Accelerator Laboratory **1537:** AT&T Bell Laboratories **1548:** Maarten Schmidt/ Palomar Observatory/California Institute of Technology **1548:** Maarten Schmidt/Palomar Observatory/California Institute of Technology Photograph **1549:** Anglo-Australian Telescope Board **1549:** Anglo-Australian Telescope Board

Tables and Illustrations

This page constitutes an extension of the copyright page. We have made every effort to trace the ownership of all copyrighted material and to secure permission from copyright holders. In the event of any question arising as to the use of any material, we will be pleased to make the necessary corrections in future printings. Thanks are due to the following authors, publishers, and agents for permission to use the material indicated.

Chapter 30. **966:** Calvin and Hobbes © Watterson. Reprinted with permission of Universal Press Syndicate. All rights reserved.

Chapter 36. **1166:** © 2003 Sidney Harris **1169:** The Far Side® by Gary Larson © 1985 FarWorks, Inc. All Rights Reserved. Used with permission.

Chapter 38. **1241:** © 2003 by Sidney Harris

Chapter 39. **1283:** © 2003 by Sidney Harris

Chapter 40. **1320:** ROSE IS ROSE reprinted by permission of United Features Syndicate, Inc.

Chapter 41. **1340:** Based on a drawing from P.K. Hansma, V.B. Elings, O. Marti, and C. Bracker, *Science* 242:209, 1988. © 1988 by AAAS. **1342:** from C.W. Sherwin, *Introduction to Quantum Mechanics,* New York, Holt, Reinhart and Winston, 1959. Used with permission.

Chapter 46. **1552:** © 2003 by Sidney Harris

Index

Locator notes: **boldface** indicates a definition; *italics* indicates a figure; *t* indicates a table

Aberrations
  chromatic, *1153*, 1165
  spherical, *1132*, *1152*–1154, 1164
Absolute uncertainty, **A.29**
Absorption
  stimulated, *1384*–1385
Absorption spectroscopy, **1353**, 1408–1411
Abundances, isotopic, 1442, 1448, 1484, 1537, A.4*t*–A.13*t*
AC. *See* Alternating-current
AC–DC converters, *1053*–1054
  and general relativity, *1273*–1274
Acceleration (**a**),
  of electric charges, 988, 1079–1080
  relativistic, 1268
Accelerators, *1242*, 1465
Acceptor atoms, **1423**
Accommodation, eye, **1156**, 1158
Activation energy (E_a), 1434
Activity. *See* Decay rate
Addition
  and differentiation, A.24
  and uncertainty, A.29
AFM. *See* Atomic force microscopes
Air
  dielectric strength of, 782, 791, *812*
  reduction of pollution in, *783*, 1475, 1487
Algebra, A.15–A.19
  matrix, 870
Alkali metals, 1380
Allowed transitions, *1381*
Alloys, 1414
Alpha (α) decay, **1337**, 1452–*1453*, *1457*–1460
  and tunneling, 1337–1338
Alpha (α) particles, 1337, 1457
  radiation damage by, 1496
  scattering of, from nucleus, *1355*
Alternating-current (AC) circuits, 1033–1065
  capacitors in, 1041–1043
  inductors in, 1038–1040
  power in, 1047–1049
  resistors in, 1034–1038
  series *RLC*, 1043–1047
  sources for, 1034
Alternating-current (AC) generators, **982**–984
Ammeters, **879**
Amorphous solids, 1229
Ampère, André-Marie, *933*, 959
Ampere (A), **833**, **932**

Ampère–Maxwell law, **943**, 989
Ampère's law, 933–938, **934**
Amplifiers, 1339, 1428
Amplitude (A)
  of simple harmonic motion, 1018
  voltage (ΔV_{max}), **1034**
Analyzers, polaroid, 1227
Analyzing problems. *See also* Problem-solving strategies
Anderson, Carl, 1514, 1516, 1519
Angle of deviation (δ), *1109*
Angle of incidence (θ_1), *1099*
Angle of reflection (θ_1'), *1099*
Angle of refraction (θ_2), *1102*
Angular frequency (ω)
  in AC circuits, 1034
  of simple harmonic motion, 1018, 1021
Angular magnification (m), **1159**
  of microscopes, 1161–1162
  of telescopes, 1163–1164
Angular momentum (**L**)
  and atomic magnetic moments, 945
  and orbital quantum numbers, 1367–1368
  spin (**S**), 946
Antennas, *1079*–1080, *1133*
Antiderivatives. *See* Integration
Antilogarithms, **A.19**
Antimatter, 1452
Antineutrinos ($\bar{\nu}$), *1461*, 1516, 1519*t*
Antineutrons ($\bar{n}$), 1515, 1519*t*
Antiparticles, 1513–1515, 1519*t*
Antiprotons ($\bar{p}$), 1515, 1519*t*
Antiquarks, 1529–1530*t*
Apertures, circular, *1215*–1216
Apollo 11 [spacecraft], 1100, 1118–1119, 1387
Approximating
  and limiting techniques, 721
  with rays, **1097**–1098
  and series expansions, 1250, 1269, A.23
  and small angles, 1138–1139, 1180
  and small values, A.23
Aqueous humor, *1155*
Arago, Dominique, 1207
Area
  of geometric shapes, A.20*t*
Areas under curves. *See* Integration
Art history, 1502
Artificial radioactivity, **1465**
Astigmatism, **1158**

Astronomy and astrophysics. *See also* Cosmology; Relativity; Stars; Sun; Universe
  and age of solar system, 1465
  and lasers, 1387
  and nebulae, 1477
Atmospheres
  and polarization by scattering in, *1231*
Atomic bombs, 1507, 1508, 1510
Atomic cohesive energy (U_0), **1412**
Atomic force microscopes (AFM), *1341*
Atomic mass, A.4*t*–A.13*t*, A.30*t*–A.31*t*
Atomic nucleus. *See* Nucleus
Atomic number (Z), **1375**, **1441**, A.4*t*–A.13*t*, A.30*t*–A.31*t*
  effective (Z_{eff}), 1381–1382
Atomic physics, 1351–1397
  and Bohr's model of the atom, 1356–1361
  and early models of the atom, 1355–1356
  and the exclusion principle, 1374–1380
  and lasers, 1385–1389
  and the periodic table, 1374–1380
  and physical interpretation of quantum numbers, 1367–1374
  and quantum model of the atom, 1361–1364
  and spectra of gases, 1352–1354, 1380–1383
  and transitions, 1383–1385
  and wave functions for hydrogen, 1364–1367
Atomic spectra, 1222, 1352–1354, 1380–1383
  and line splitting, 1359, 1361, 1368–1370
Atomic spectroscopy, 1220, 1353–1354
Atoms. *See also* Hydrogen atom
  acceptor, **1423**
  constituents of, 1512
  donor, **1422**
  magnetic dipole moments of, 944–946
  Rydberg, 1361
Audio systems, 1053, 1056
Audiovisual devices. *See also* Movies; Musical instruments
  compact discs (CDs), 1203, *1221*, *1236*
  digital video discs (DVD), 1203
  and magnetic recording, 951
  pickup coils, 971–*972*
  radio, 813, 1051, 1061
  television, 728, 765, *900*, 1156–1157

Auroras, 910
Autobiography [Cellini], 1117
Automobiles
 and energy, 828
 and hybrid drive systems, *985*
 intermittent windshield wipers for, 877
 taillights of, 1100–*1101*
Average power ($\overline{\mathcal{P}}$), **1047**
Avogadro's number (N_A), 1443

Back-of-the-envelope calculations. *See
 also* Approximating
Background radiation, 1497
 cosmic microwave, 1088, 1537–*1538*
Bainbridge mass spectrometers, 911
Balances
 current, 933
 equal-arm, 987–*988*
 torsional, 895, 1077
Ball-park figures. *See also*
 Approximating
Balmer, Johann Jacob, 1354, 1359
Balmer series, *1354, 1358*
Band theory of solids, 1418–*1420*
Bardeen, John, 1427, 1431
Barrier height (U), **1336**
Barrier penetration, **1336**. *See also*
 Tunneling
Baryon number (B), **1520**–1522
Baryons, **1519***t*
 conservation of, 1520–1522
 quark composition of, 1527–1531*t*,
 1533
Base of logarithms, **A.19**
Base quantities, A.32*t*
Batteries, *858*–861
 chemical energy in, 799, 845–846
 recharging, 1014–*1015*, 1061
BCS theory of superconductivity,
 1431–1432
Beam splitters, *1194*
Beat frequency (f_b)
 and quantum particles, *1304–1306*
Becquerel, Henri, 1441, 1452
Becquerel (Bq), **1454**
Bednorz, J. Georg, 1431
Bell Laboratories, *1537*
Beryllium
 electronic configuration of, 1375–1377*t*
Beta (β) decay, 1452–*1453, 1457*,
 1460–1463
Beta (β) particles, 1460. *See also*
 Electrons; Positrons
Betatrons, 1001
Bias of *p–n* junctions, *1425*
Big Bang, 1088, 1536–1539. *See also*
 Cosmology
Binding energy
 nuclear, 1443, 1445, **1447**–*1448*, 1481
 semiempirical formula for, **1449**–1450
Binomial expansions, 1250, 1269, A.23

Biophysics. *See also* Health; Medicine;
 DNA, 1403
 eyes, 1080, 1087, *1155*–1159, 1216–*1217*
 and MRI, *1469*
 and radiation damage, 1469–1470,
 1486, 1495–1497
Biot, Jean-Baptiste, 927
Biot–Savart law, **927**–932
Birefringent materials, *1229*
Black bodies, **629**, 1076, *1286*
Black holes, **1275**, 1282
Blackbody radiation, 1285–1291, *1286*,
 1341–1343
 cosmic background, 1088, 1537–*1538*
Blue shifts, 1281
Bode plots, 1064
Bohr, Niels, *1356, 1371*
 and correspondence principle, 1360
 and liquid-drop model of nucleus, 1448
 and model of hydrogen atom, 957,
 1356, 1359
 and quantum mechanics, 1285
Bohr magneton (μ_B), **946**, 1373
Bohr radius (a_0), **1358**
Boltzmann distribution law
 and blackbody radiation, 1289
Bonaparte, Napoleon, 1502
Bond lengths, 1405–1406
Bonds, molecular, 1399–1403
Born, Max, 1285, 1323
Boron
 electronic configuration of, 1375–1377*t*
Bosons, 1431, 1513
Bottom quarks (b), **1530**–1531*t*
Bottomness, 1530*t*–1531
Bound systems, 1334
Boundary conditions, **1330**
 and particle-in-a-box, 1326–1327,
 1331–1332
Brackett series, 1354
Bragg, William Lawrence, 1225
Bragg's law, **1225**
Brattain, Walter, 1427
Bremsstrahlung, *1381*–**1382**, 1489
Brewster, David, 1228
Brewster's angle, 1119, *1228*
Brewster's law, **1228**
Brookhaven National Laboratory, *1500*,
 1530, 1532, 1535
Bubble chambers, 918, *1498, 1514*,
 1523, 1528
Buckminsterfullerene, *1414*
Bunsen, Robert, 870
Buoyant forces (**B**), 782

Calculus, A.23–A.28. *See also* Chain rule;
 Definite integrals; Derivatives;
 Differential equations;
 Differentiation; Gradient;
 Integration; Partial derivatives
 development of, A.23

Cameras, *1153*–1155
 electron bombardment CCD, 1296
 lightmeters for, 1295
 pinhole, 1239
Capacitance (C), 795–830
 calculation of, 797–802
 equivalent (C_{eq}), 803–805
Capacitive reactance (X_C), **1042**, 1056
Capacitors, *795*–**796**
 in AC circuits, *1041*–1043
 charging of, 873–876
 circuits of, *800*, 802–806, 870, 873
 construction of, 812*t*–814
 cylindrical, *801*–802
 discharging of, 876–878
 electrolytic, *813*
 parallel-plate, *767*, 798–800
 spherical, *802*
 variable, *813*
Carbon
 as covalent solid, 1413–1414
 electronic configuration of,
 1376–1377*t*, 1413
 isotopic abundance of, 1442
 radioactive, 1455, 1461–1463
Carbon dating, 1462–1463
Carbon dioxide
 and global warming, 1410–1411
Carbon dioxide lasers, 1387–1388
Carbon monoxide, 1405*t*–1406
Carlson, Chester, 784
Cars. *See* Automobiles
Cataracts, 1174
Cathode ray tubes (CRT), *727*–728
Cavendish, Henry, 711
Cavities
 electric fields in, 780
CCD. *See* Charge-coupled devices
CD. *See* Compact discs
Cellini, Benvenuto, 1117
Central maxima, **1206**
Ceramics
 superconductivity in, *844*
CERN, 1518, 1532, *1535*
Chadwick, James, 1441, 1444,
 1475–1476
Chain reactions, self-sustained,
 1483–**1484**
 natural, 1484
Chain rule, A.24
Challenger [space shuttle], 1517
Chamberlain, Owen, 1515
Characteristic x-rays, *1381*–**1382**
Charge carriers, **833**
 and Hall effect, 914–916
Charge-coupled devices (CCD), 1154,
 1296, 1319
Charge density (ρ, σ, or λ), **720**
Charge number. *See* Atomic number
Charges. *See* Electric charge
Charm (C), **1530**–1531*t*
Charon, 1217–*1218*

Chemical energy
  in batteries, 799, 845–846
Chemistry. *See* Physical chemistry
Chernobyl accident, 1477, 1486
Chips, microelectronic, 1428–1429
Chromatic aberration, *1153*, 1165
Chu, Steven, 1388–1389, 1397
Ciliary muscles, eye, 1156, 1158
Cinema. *See* Movies
Circles, **A.20**
Circuit breakers, *880*
Circuit diagrams, **802**
Circuit elements, 796. *See also* Batteries;
       Capacitors; Diodes; Inductors;
       Junctions; Resistors; Transformers;
       Transistors
Circuit symbols
  for AC sources, *1034*
  for batteries, ***802***
  for capacitors, ***802***
  for diodes, *1054*
  for ground, *710*
  for inductors, *1006*
  for resistors, *845*
  for switches, ***802***
  for transformers, *1051*
Circuits. *See also* Kirchhoff's rules; Parallel
       combinations; Series combinations
  alternating-current, 1033–1065
  of capacitors, *800*, 802–806, 870, 873
  direct-current, 858–893
  filter, ***1055–1056***
  household, 865, *880*–882
  integrated, *1398*, ***1428–1429***
  LC, ***1015***–1020
  LRC, *1020*–1022, 1043–1047, 1049–1051
  RC, **873**–878
  of resistors, *845*
  RL, 1006–1011, ***1007***
  short, 881–*882*
Cladding, in optical fibers, 1114
Classical physics. *See also* Quantum
       mechanics
Clocks
  moving, 1255
Cloud chambers, **1498**
Coaxial cables, 801, *840*, 1012–1013
Cockroft, John, 1441
Coefficients
  Hall (R_{H}), **915**
  of linear absorption (μ), 1509–1510
  of reflection (R), **1336**
  of resistivity, temperature (α), 837*t*,
       843, 855
  of transmission (T), **1336**
Cohen-Tannoudji, Claude, 1397
Coherence, **1177**
  and lasers, 1384–1385
Coils
  Helmholtz, *963*
  pickup, 971–*972*
  primary, 949, 968–*969*

Coils (*Continued*)
  Rogowski, *993*–994
  saddle, *963*
  search, 999
  secondary, 949, 968–*969*
Collective model of the nucleus, 1452
Colliders, *1500*, *1511*, 1532, ***1535***
Collins, S. C., 1030
Collisions. *See also* Scattering
  and Compton effect, 1297–1300
Color charge, **1532–1534**
Color force, **1533**. *See also* Strong force
Common logarithms, A.19
Commutators, 984
Compact discs (CDs), 1203, *1221*, *1236*
Compactification, 1543
Compasses, 895
  dip, 962
Complementarity principle, **1302**
Complex conjugates, 1323
Complex numbers, 1323
Compound microscopes, **1160**–1162
Compton, Arthur Holly, *1297–1299*
Compton effect, 1297–1300, **1298**, 1537
Compton shift equation, **1298**
Compton wavelength, **1298**
Concave mirrors, ***1131***–1133. *See also*
       Mirrors
Conceptualizing problems. *See also*
       Problem-solving strategies
Condon, Edward U., 1459
Conduction
  electrical, 710, 841–843, 1420–1424
Conduction bands, *1421–1422*
Conduction electrons, 709, 841
Conductivity (σ), 835, 842
  of solids, 1412–1414
Conductors, **709**
  electric potential due to charged,
       778–781
  in electrostatic equilibrium, 750–752
  motion in magnetic fields of, 970,
       973–977, 980–981
  perfect, 1430
  resistivity of, 837*t*
Cones, eye, 1156
Confinement
  inertial, *1493*
  magnetic, **1490**–1492
Confinement time (τ), **1490**
Conseil Européen pour la Recherche
       Nucléaire. *See* CERN
Conservation laws. *See also* Conservation
       of energy; Conservation of
       momentum
  of baryon number, **1520**–1522
  of electric charge, **708**, 870
  of lepton number, **1522**–1523
  of strangeness, 1523–**1524**
Conservation of energy
  and Kirchhoff's rules, 870
  and Lenz's law, 978–979

Conservation of energy (*Continued*)
  and motional emf, 975, 977
  relativistic, 1268–1272
  and uncertainty principle, 1310–1311
  violations of, 1311, 1334, 1336, 1517
Conservation of momentum
  relativistic, 1267–1268, 1270
Conservative fields, **768**
Constructive interference, 1177–**1180**,
       1190
  and quantum particles, *1304*
Contact lenses, 1169
Control rods, *1486*
Conversion of units, A.1*t*–A.2*t*
Converters, AC–DC, *1053*–1054
Convex mirrors, *1134*. *See also* Mirrors
Cooper, Gordon, 1278
Cooper, Leon N., 1431
Cooper pairs, 1431
Coordinate systems
  plane polar, 1330
  spherical, 1330, 1362
  transformations between, Galilean,
       1247
  transformations between, Lorentz,
       1262–1267, **1263**
Corneas, *1155*
Corner cube reflectors, 1119
Cornu, Marie, 1199
Corona discharge, **780**–782
Correspondence principle, **1360**, 1395
Cosine function, A.21. *See also*
       Trigonometric functions
  and law of cosines, A.22
  and power factor (cos ϕ), **1048**, 1051
Cosmic background radiation, 1088,
       1537–*1538*
Cosmic rays, 1082
  and background radiation, 1497
  and carbon dating, 1462
  in Earth's magnetic field, 910
  and mesons, 1516
  and time dilation, *1255*
Cosmology, 1536–1542
Coulomb, Charles, 705, ***711***
Coulomb (C), **711**, **933**
Coulomb constant (k_e), **711**
Coulomb's law, **711**–715
  and alpha decay, 1338
  and nuclei, 1444–*1446*
Covalent bonds, **1401**–*1402*
Covalent solids, 1413–1414
Crab nebula, 1477
Critical angle (θ_c), **1111**–*1112*
Critical ignition temperature
       (T_{ignit}), **1489**
Critical reactors, **1485**
Critical temperature (T_c), 844–845*t*,
       1430*t*–1432
Critically damped oscillators, 1022
CRT. *See* Cathode ray tubes
Crystal spectrometers, *1298*

Crystalline lenses, *1155*
Crystalline solids, 1229
  diffraction of x-rays by, *1224–1225*
Curie, Marie, *1452*
Curie, Pierre, 951, *1452*
Curie (Ci), **1454**
Curie temperature (T_{Curie}), 951*t*
Curie's constant (C), **952**
Curie's law, 951–**952**
Current, electric. *See* Electric current
Current, fluid. *See* Flow
Current balances, 933
Current density (J), **835**, 842
Cutoff frequency (f_c), **1293**
Cutoff wavelength (λ_c), **1294**
Cyclotron frequency (ω), **908**–909
Cyclotrons, 908, *913*–914

Damped oscillators, 1020–*1022*
Dark energy, 1541–1542
Dark matter, 1520, 1541
D'Arsonval galvanometers, *879, 907*
Daughter nucleus, **1457**
Davisson, Clinton, 1301–1302
Davisson–Germer experiment,
    1301–1303, 1307
DC. *See* Direct-current
de Broglie, Louis, 1285, *1301*
de Broglie wavelength (λ), **1301**
  of particle-in-a-box, 1327
de Maricourt, Pierre, 895
Dead Sea Scrolls, 1463
Debye, Peter, 1297
Decay constants (λ), **1453**
Decay processes, radioactive, 1456–1464*t*
Decay rate (R), **1454**
Declination, magnetic, *954*
Dees, cyclotron, *913*
Defibrillators, *810*, 825
Definite integrals, **A.26**, A.28*t*. *See also*
    Integration
Degrees of freedom, 1362, 1404
Delta (Δ) particles, 1519*t*, 1525–1526
Demagnetization, *950*
Density (ρ), 9*t*–10, 422*t*
  ion (n), **1490**
  of nuclei, 1444
  surface charge (σ), **720**, *779*, 818
  of the Universe, 1540–1541
Density-of-states function ($g(E)$),
    1416–*1418*
Deoxyribonucleic acid (DNA), *1403*
Depletion regions, *1424*–1425
Depth of field, **1154**
Derivatives, **A.23**–A.25*t*. *See also*
    Differentiation
  of exponential functions, 1366
  as limits of discrete changes, 1306
  partial, 773, 1069–1070
  second, A.24

Derived quantities, A.32*t*
Destructive interference, 1177–**1180**,
    1190
  and quantum particles, *1304*
Deuterium
  as fuel, 1488–1489, 1493–1495
Devices. *See* Audiovisual devices;
    Electronic devices; Mechanical
    devices; Optical devices
Diagrams. *See* Energy-level diagrams;
    Graphs
Diamagnetism, **947**–948, 952
  perfect, 952, 1031, 1430
Diamonds, 1112, 1413
Dielectric constant (κ), **810**, 812*t*
Dielectric strength, **812***t*
  of air, 782, 791, *812*
  and line spectra, 1353
Dielectrics, 796, **810**–814
  atomic description of, 817–820
Differential equations
  partial, 1362
  second-order, 1069
  separable, 875, 1007–1008, 1323, 1362
Differentials, perfect, A.27
Differentiation, A.23–A.25. *See also*
    Derivatives
Diffraction, 1069, 1097, *1178*. *See also*
    Interference
  of electrons, 1301–1303, 1307–*1309*
  of x-rays by crystals, *1224–1225*
Diffraction-grating spectrometers,
    1219
Diffraction gratings, 1187, **1217**–1224
  resolution of, 1221–1222
Diffraction patterns, *1188*, 1205–1241.
    See also Interference
  defined, *1206–1207*
  from slits, *1207*–1214
Diffuse reflection, *1098*–**1099**
Digital cameras, 1154
Digital micromirror devices (DMD),
    1101–*1102*, 1220
Digital video discs (DVD), 1203
Dimensions, A.2*t*–A.3*t*
Diodes, 838, 1053–**1054**
  junction, *1424*–1425
  light-emitting (LEDs), *1426*
Diopters, **1158**
Dipole antennas, 1079–*1080*
Dipole–dipole forces, 1399, 1402
Dipole–induced-dipole forces, 1402
Dipoles. *See* Electric dipoles; Magnetic
    dipoles
Dirac, Paul, 1285, 1371–1373, *1513*,
    1517
Dirac sea, 1513–*1514*
Direct-current (DC) circuits, 858–893
  and electrical meters, 879–880
  and electrical safety, 881–882
  and emf, 859–861

Direct-current (DC) circuits (*Continued*)
  and household wiring, 880–882
  and Kirchhoff's rules, 869–873
  and *RC* circuits, 873–878
  resistor combinations in, 862–869
Direct-current (DC) generators, *984*
Direction of propagation, 1069
Disintegration energy (Q), **1457**, 1460,
    1462, 1466, 1482
Dispersion, *1109*–1111, 1239
Dispersion forces, 1402–1403
Displacement current (I_d), *942–944*, **943**
Dissociation energy, *1401*
Distance (d), **A.20**
Distribution functions
  Fermi–Dirac ($f(E)$), *1415*–1416, 1438
Diverging mirrors, *1134*. *See also* Mirrors
Division
  and uncertainty, A.29
DMD. *See* Digital micromirror devices
DNA. *See* Deoxyribonucleic acid
Domain walls, **949**
Domains, *949*
Donor atoms, **1422**
Doped semiconductors, 1422–1424
Doping, **1422**
Doppler, Christian Johann, 522
Doppler effect
  and Compton effect, 1297–1298
  relativistic, for light, 1262, 1548
Double rainbows, *1094*, 1111
Double-refracting materials, *1229*
Double-slit interference patterns. *See*
    Young's double-slit experiment
Doublets, 1371, *1410*, 1451
Down quarks (d), **1529**–1531*t*
Drains, transistor, *1427*
Drift chambers, **1499**, 1525
Drift speed (v_d), **834**, 841–842
Drude, Paul, 841
Dual nature of light and matter,
    1300–1302
Ductility, 1414
DVD. *See* Digital video discs
Dynodes, 1295–1296

e (Euler's number), 875, A.19
Ear thermometers, *1290*
Ears. *See also* Biophysics
Earth
  magnetic field of, 895, *910, 953–954*
  orbital velocity of, 1248–1250
  Van Allen belts of, *910*
Eddy currents, **986**–988
  in transformers, 987, 1051
Edison, Thomas, 1053
Effective atomic number (Z_{eff}),
    1381–1382
Eigenzeit. *See* Proper time intervals
Eightfold way, **1527**–*1528*

Einstein, Albert, *1244, 1251*
  and blackbody radiation, 1290
  and development of special relativity, 988, 1243, 1245, 1249–1250
  and general relativity, 1195, 1273–1275
  and photoelectric effect, 1093, 1095–1097, 1293–1294
  and quantum mechanics, 1285
  and special relativity, 1249–1252, 1254, 1263
  and unified theory, 1542
Elastic scattering, 1466
Electric charge (q), 707–709
  distributions of, 719–723, 746–750, 774–778
  electromagnetic radiation from accelerating, 988, 1079–1080, 1356, 1360, 1381–1382
  and electrostatic equilibrium, 750–752
  fractional, 1529
  fundamental (e), 708, 712, 781–782
  induction of, 709–711
  point, **711**, 724, 768–771
  positive and negative, **707**
  source and test, 715–*716*
  of subatomic particles, 712*t*, 1442–1443*t*
Electric current (I), 831–857, **833**
  average (I_{av}), **832**
  direct, **859**
  displacement (I_d), *942*–944, **943**
  eddy, *986*–988
  induced, 968–970
  instantaneous (i), **833**, 1034
  magnetic field of, 927–932
  magnetic force on, 900–904
  model of, 833–835
  root-mean-square (I_{rms}), **1037**
  surface, 1430–*1431*
Electric current density (J), **835**, 842
Electric dipole moment (**p**), **815**, 905
  induced, *817*
Electric dipoles, *719*
  electric field lines for, *724, 772–773, 942*
  in electric fields, *815*–817
  electric potential due to, *769, 773–774*
  electromagnetic waves from oscillating, 1079–*1080*
  equipotential surfaces for, *772–773*
Electric discharge, 1353
Electric field lines. *See* Field lines
Electric fields (**E**), 706–738, 717*t*
  in capacitors, 799–802
  of charge distributions, 719–723
  defined, **715**–719
  electric dipoles in, *815*–817
  and electric potential, 772–774
  energy density of, 808, 1012
  and Gauss's law, 743–745

Electric fields (**E**) (*Continued*)
  induced, *981*–982
  motion of charged particles in, 725–728
  potential difference in uniform, 765–768
  and superposition principle, 718–719, 744
Electric flux (Φ_E), **740**–743
  and Gauss's law, 743–745
Electric forces (**F**$_e$), **711**–715
Electric generators. *See* Generators
Electric potential (V), 762–794
  defined, **764**
  due to charge distributions, 774–778
  due to charged conductors, 778–781
  due to point charges, 768–771
  obtaining, from electric fields, 772–774
  in uniform electric fields, 765–768
Electrical conduction, 710
  model of, 841–843
Electrical engineering. *See also* Circuits; Electronic devices
  and generators, 782–*783*, 982–984, 994–*995*, 1034
  and household wiring, *880*–882
  and magnetic recording, 951
  and motors, **984**–986
  and power generation, *967*, 982–986, 1034, 1272, *1479*, 1485
  and solar energy, 1078–1079, 1193, 1426
  and sound recording, 971–*972*
  and transformers, 846–847, 987, *1033*, **1051**–1054
Electrical induction, 709–711, **710**
Electrical meters, *879*–880
  loading of, 889
Electrical power, 845–849
  generation of, *967*, 982–986, 1034, 1272, *1479*, 1485
  transmission of, 1051–1054
Electrical safety, 881–882
Electrical transmission, *831*, 846–847
  and electric generators, 982
  and motors, 984
  and superconductors, 1031
  and transformers, 1051–1054
Electrolytes, 813
Electromagnetic forces, 1512–1513*t*, *1536*–1537
Electromagnetic radiation, 1079–1080
  from accelerated charges, 988, 1079–1080, 1356, 1360, 1381–1382
Electromagnetic spectrum, **1080**–1082
Electromagnetic waves, 988, *1066*–1091
  energy carried by, 1074–1076
  and photons, 1300–1301
  plane, 1069–1073
  production of, 1079–1080

Electromagnetism, 704–1091. *See also* Capacitance; Circuits; Dielectrics; Electric current; Electric fields; Electric potential; Electromagnetic waves; Faraday's law; Gauss's law; Inductance; Magnetic fields; Resistance
  defined, 705
  history of, 705, 895–896, 927, 968
Electromotive force. *See* emf
Electron affinity, **1400**
Electron beams
  bending of, with magnetic fields, *909*
Electron bombardment CCD cameras, 1296
Electron capture, **1462**
Electron configurations, A.30*t*–A.31*t*
Electron gas, 841, *1414*
Electron guns, 728
Electron microscopes, *1303*–1304
Electron spin, **1371**. *See also* Spin
Electron spin resonance, 1469
Electron volt (eV), **764**–765, 1270
Electron–hole pairs, *1422*
Electronic configurations, 1375–1379*t*
  and ionic bonding, 1400
Electronic devices. *See also* Circuits; Diodes; Electrical meters; Generators; Semiconductor devices; Sound recording; Transistors
  AC–DC converters, *1053*–1054
  burglar alarms, 1295
  charge-coupled devices (CCD), 1154, 1296, 1319
  compact discs (CDs), *1203, 1221, 1236*
  digital micromirror devices (DMD), 1101–*1102*, 1220
  digital video discs (DVD), 1203
  grating light valves (GLV), 1220
  lightmeters, 1295
  microwave ovens, 816, 1085
  photocopiers, *784*
  radiation detectors, 1497–1500
  smoke detectors, *1459*–1460
  stud-finders, *813*
Electrons
  e/m_e ratio of, 912
  free or conduction, 709, 841
  as fundamental particles, 912, 1519*t*, 1531*t*
  and magnetic dipole moments of atoms, 944–946
  photo-, **1291**–1294
  scattering of x-rays by, 1297–1300
  and spin, 945–946
Electrostatic equilibrium, **750**–752
  applications of, 782–784
Electrostatic precipitators, *783*
Electroweak theory, **1534**, *1536*–1537
Elementary-particle physics. *See* Subatomic physics

Elementary particles, 1531*t*. *See also* Electrons; Leptons; Muons; Neutrinos; Quarks; Tau particles
Elements, A.4*t*–A.13*t*
  extraterrestrial, 1353–1354
  periodic table of, 1374–1380, A.30*t*–A.31*t*
  transuranic, 1465
Ellipses, *A.20*
Emerson, Ralph Waldo, 1552
emf (ε), **859**–861
  back, 985, 1004, 1040
  induced, 968–970, 981–982, 1372
  motional, **973**–977
  self-induced, **1004**
Emission
  spontaneous, *1384*, 1386
  stimulated, **1384**–1386
Emission spectroscopy, **1353**
Emissivity (e), 1286
Endothermic reactions, **1466**
Energetic favorability, **1401**
Energy. *See also* Conservation of Energy, Kinetic Energy
  activation (E_a), 1434
  atomic cohesive (U_0), **1412**
  binding, 1447–1450
  and blackbody radiation, 1287
  chemical, 799, 845–846
  dark, 1541–1542
  disintegration (Q), **1457**
  dissociation, *1401*
  electric, 807–810, 1015–1019
  and electromagnetic waves, 1074–1076
  Fermi (E_F), **1415**–1417*t*
  ionic cohesive (U_0), **1412**
  ionization, 1358, *1380*
  magnetic, 1011–1013, 1015–1019
  mass as a form of, 1270, 1272
  of particle-in-a-box, *1328*–1330, 1332
  quantization of, 1288, 1328–1330
  reaction (Q), **1466**–1467*t*
  relativistic, 1268–1272
  rest (E_R), **1270**
  and rotational motion, 1405
  and simple harmonic motion, 1341–1343
  threshold, **1466**
  total (E), **1270**
  and uncertainty principle, 1310–1311
  units of, A.2*t*
  of wave packets, 1306–1307
Energy bands, 1418–*1420*
Energy conversion
Energy density
  in electric fields (u_E), **808**, 1012, 1074
  in electromagnetic waves (u), 1074–1075, 1287–1288
  in magnetic fields (u_B), 1011–1012, 1074
Energy gaps (E_g), *1421–1423*

Energy-level diagrams, *1288*
  for finite square wells, *1335*
  for gamma decay, *1464*
  for hydrogen atoms, *1358*, 1380–*1381*
  for infinite square wells, *1328*, *1335*
  for lasers, *1386*
  for molecules, *1404*, *1407*, *1409*
  for simple harmonic oscillators, *1342*
  for tunneling, *1336*
Energy states, 1328
  of molecules, 1403–1411
  negative, 1513–1514
Engineering. *See also* Electrical engineering; Materials science
Envelopes
  for diffraction patterns, *1188*, *1213*, *1241*
  for oscillatory motion, *1306*
Equal-arm balances, 987–*988*
Equations. *See also* Differential equations; Kinematic equations
  coefficients of, **A.17**
  Galilean transformation, **1247**
  lens makers', **1143**
  linear, 870, A.17–A.19
  Lorentz transformation, 1262–1267, **1263**
  Maxwell's, **988**–989, 1067–1069, 1245, 1247
  mirror, **1132**–1133
  quadratic, A.17
  thin lens, 1143–**1144**
  transcendental, 1318
Equilibrium separation, 1400
Equipartition of energy theorem, 1287, 1289
Equipotential surfaces, *766*
  and field lines, *772*–773, *779*
Equivalence principle, **1274**
Equivalent capacitance (C_{eq}), 803–805
Equivalent resistance (R_{eq}), **863**–868
Escher, M. C., *1167*
Eta (η) particles, 1519*t*
Ether, 1247–1250
Euler's number (e), A.19
European Laboratory for Particle Physics. *See* CERN
European Organization for Nuclear Research. *See* CERN
Events, *1247*, 1251–1252
  and Lorentz transformation equations, 1262–1264
Exchange particles, **1513**
Excited states, **1328**
  and quarks, 1525–1526
  and radioactive decay, 1458, 1464, 1466
Exclusion principle, 1352, **1374**
  exceptions to, 1431, 1532
  and free-electron theory of metals, 1416
  and molecular bonding, 1399, 1402
  and the periodic table, 1374–1380
Exothermic reactions, **1466**

Expectation values ($\langle x \rangle$), **1325**, 1333, 1366
Experiments. *See also* Instrumentation; Measurement; Thought experiments
  Davisson-Germer, 1301–1303
  Michelson–Morley, 1248–1250
  Millikan oil-drop, *781*–782
  and null results, 1250
  Phipps–Taylor, 1372
  Stern–Gerlach, 1371–*1372*, 1373–1374
  Young's double-slit, 1177–1182, 1307–*1309*
Exponents, **A.14**, A.16*t*–A.17
  and uncertainty, A.29
Exposure times, 1153
Extraordinary (E) rays, *1229*–1230
Extrinsic semiconductors, **1424**
Eyepiece lenses, 1160–*1163*
Eyes, *1155*–1159. *See also* Biophysics
  laser surgery on, 1387
  resolution of, 1216–*1217*
  sensitivity of, 1080, 1087, *1156*, 1290

f-numbers, 1154
Factoring, A.17
Far point of the eye, **1156**
Farad (F), **797**
Faraday, Michael, *968*
  and capacitance, 797
  and electric fields, 715, 723
  and electromagnetic induction, 705, 896, *968*
Faraday disks, 994–*995*
Faraday's law of induction, 967–1002, 989
  applications of, 971–973
  defined, 968–**970**
  and electromagnetic waves, 1072–1073
  and inductance, 1005, 1014
  and inductive reactance, 1040
  and superconductivity, 1430
Farsightedness, *1157*
Faust [Goethe], 1504
Femtometer (fm), 1444
Fermat, Pierre de, 1114
Fermat's principle, **1114**–*1115*
Fermi, Enrico, 1441, 1461, 1481, *1484*, 1508
Fermi (fm), **1444**
Fermi energy (E_F), **1415**–1417*t*, 1421
Fermi temperature (T_F), 1438
Fermi–Dirac distribution function ($f(E)$), *1415*–1416, 1438
Fermilab, 1520, 1531, 1535–1536
Fermions, 1374, 1431
  quarks as, 1529
Ferromagnetism, **947**–951, **949**
FET. *See* Field-effect transistors
Feynman, Richard, 730, 1313, *1517*
Feynman diagrams, *1517*, *1533*
Fiber optics, *1114*
Field-effect transistors (FET), **1427**–1428

Field forces, 715
  electrical, 715–719
  magnetic, 896–900
  superposition of, 718–719
Field lines, 723
  closed, *930, 942*
  electric, *723–725, 751, 942*
  and equipotential surfaces, *772–773, 779*
  magnetic, *896, 930, 942,* 1430–*1431*
Field particles, **1513**
Filter circuits, *1055–1056*
Finite square wells, *1334–1336*
Finnegans Wake [Joyce], 1529
First-order maxima, 1180
Fission, 757, 1272, *1447,* **1481**–1483
Fission fragments, **1481**
Fizeau, Armand H. L., 1097, 1118
Flat mirrors, *1127–1130. See also* Mirrors
Flavors of quarks, **1529**–1531*t*
Flow
  rate of, 832
Fluctuations of nucleons, 1518
Fluorine
  electronic configuration of, *1376*–1377*t*
Flux
  electric (Φ_E), **740**–743
  magnetic (Φ_B), **940**–941
Flux compression, 994
Flux linking, 1014
Focal length (f), *1133, 1143*
Focal point (F), *1133*
Food preservation, *1503*
Forbidden regions, *1336,* 1420
Forbidden transitions, 1381
Force (**F**)
  color, **1533**
  conservative, 768
  dipole–dipole, 1399, 1402
  dipole–induced-dipole, 1402
  dispersion, 1402–1403
  electric ($\mathbf{F}_e$), **711**–715
  electromagnetic, 1512–1513*t, 1536–1537*
  electroweak, **1534,** *1536–1537*
  fundamental, 707, 1512–1513*t, 1536–1537*
  Lorentz, 910, **989**
  magnetic ($\mathbf{F}_B$), 896–900
  mediation of, 1513, 1516
  relativistic, 1268
  restoring, 1341
  saturated, 1447
  strong, 1518, 1524, **1532**–1533, *1536–1537*
  units of, A.1*t*
  weak, 1512–1513*t,* 1524, *1536–1537*
Force constant. *See* Spring constant
Forward bias, *1425*
Fourier analysis, 1195
Fourier Transform Infrared
  Spectroscopy (FTIR), 1195

Fourier transforms, 1195
Fractional uncertainty, **A.29**
Franklin, Benjamin, 707
Fraunhofer diffraction patterns, *1207–1210*
Free-electron theory of metals, 709, 841, 1415–1418
Frequency (f or ω)
  cutoff (f_c), **1293**
  cyclotron (ω), **908**–909
  Larmor precessional (ω_p), *1468*–1469
  natural (ω_0), 1018
  quantization of standing-wave, 1330
  resonance (ω_0), **1049**
Fresnel, Augustin, 1207
Fresnel diffraction patterns, 1207
Fresnel lenses, *1146*
Friction
  and electric current, 834
Fringe fields, *818–819*
Fringes, **1178**–1180, 1208
  shifts in, 1249–1250
Frisch, Otto, 1481
Front–back reversals, *1128*
FTIR. *See* Fourier Transform Infrared Spectroscopy
Fuel elements, **1485**–*1486*
Full width at half maximum, 1240
Fuller, R. Buckminster, 1414
Functions, **A.23**
Fundamental constants, 1289
Fundamental forces, 707, 1512–1513*t*
  and history of Universe, *1536–1537*
Fundamental particles. *See* Elementary particles
Fundamental quantities
  electric charge (e), 708, 712, 781–782
Fuses, 880
Fusion [nuclear], 910, 960, 1272, *1447,* **1487**–*1495*
  and tunneling, 1338

Gabor, Dennis, 1222
Gabrielse, Gerald, 1395
Galilean relativity, **1246**–1248
Galilean transformation equations, **1247**
Galilei, Galileo
  and speed of light, 1096
  and telescopes, 1172
Galvanometers, *879,* **907**
Gamma (γ) decay, 1452–*1453, 1457, 1460*
Gamma (γ) rays, *1081*–1082
  gravitational red-shifts of, 1274
  radiation damage by, 1496
Gamow, George, 1459
Gaps, energy, *1421–1423*
Gases. *See also* Materials science
  atomic spectra of, 1352–1354
  inert, 1378–1379
Gasoline engines, 1482
Gates, transistor, *1427*
Gauge bosons, **1513**

Gauge theory, 1513
Gauss, Karl Friedrich, *742,* 959
Gauss (G), 899
Gaussian surfaces, 743
Gauss's law, 739–761, 988–989
  for charge distributions, 746–750, 1362
  defined, 743–**745**
  formal derivation of, 752–753
  for gravitation, 761, 1540
  for magnetism, **941**–942, 989
Gauss's probability integral, A.28*t*
Gedanken experiments. *See* Thought experiments
Geiger, Hans, 790, 1355, 1441
Geiger counters, **1498**–*1499*
Geiger tubes, 792
Gell-Mann, Murray, *1527–1529*
Gems, 1112–1113
General relativity. *See* Relativity, general
Geometric optics, 1094–1175, **1097.** *See also* Optics; Reflection; Refraction; Wave optics
Geometry, A.20–A.21
Geophysics. *See also* Earth
  and global positioning system (GPS), 1281
  and global warming, 1410–1411
  and greenhouse effect, 1411
Gerlach, Walter, 1371
Germer, L. H., 1301–1302
GFI. *See* Ground-fault interrupters
Gilbert, William, 705, 895
Glashow, Sheldon, 1534
Glasses, eye, *1157–1158*
Global positioning system (GPS)
  and general relativity, 1281
  and special relativity, 1281
Gluinos, 1543
Gluons, 1513, **1533**
GLV. *See* Grating light valves
Goeppert-Mayer, Maria, *1451*
Goethe, Johann Wolfgang von, 1504
Goudsmit, Samuel, 1371–1372
GPS. *See* Global positioning system
Gradient [mathematical operator] (∇), 773
Graphical methods, A.18–*A.19*
Graphite, 1414
  as moderator, 1484
  STM image of, *1340*
Graphs. *See also* Energy-level diagrams
  Bode plots, 1064
  space–time, *1259*–1262
Grating light valves (GLV), 1220
Gravitation. *See also* Relativity, general
  Gauss's law for, 761

Gravitational forces (**F**$_g$), 1512–1513t, *1536–1537*
Gravitational lenses, *1275*
Gravitational mass, 1273
Gravitational potential energy (U_g), 770
Gravitational radius, 1282
Gravitational waves, 1195
Gravitons, 1513, *1518*
Gray (Gy), 1497
Greenhouse effect, 1411, 1487
Ground, electrical, **710**
Ground-fault interrupters (GFI), 881, *971*
Ground states, **1328**
Group speed (v_g), **1306**–1307
Gurney, R. W., 1459

Hadrons, **1518**–1519t
Hafele, J. C., 1255
Hahn, Otto, 1356, 1441, 1481
Half-life ($T_{1/2}$), A.4t–A.13t. *See also* Time constant
  of radioactive decay, *1453*–**1454**
  of *RC* circuits, 878
  and time dilation, 1255–*1256*
Half-power points, *1050*–1051
Half-wave antennas, *1079*
Half-wave rectifiers, 1054–*1055*
Hall, Edwin, 914
Hall coefficient (R_H), **915**
Hall effect, *914*–916
Hall field, 914
Hall voltage (ΔV_H), **914**
Halogens, 1379–1380
Hand with Reflection Sphere [Escher], *1167*
Harmonics, 1330
Health. *See also* Biophysics; Medicine
  and radioactivity, 1469–1470, 1486, 1495–1497
Hearing. *See* Ears
Heat (Q)
  units of, A.2t
Heat capacity (C), 796
Heat sinks, 846
Heisenberg, Werner, 1285, *1310*, 1322, 1331
Heisenberg uncertainty principle. *See* Uncertainty principle
Helical motion, *908*
Helium. *See also* Alpha particles
  abundance of, 1537
  discovery of, 1353–1354
  electronic configuration of, 1375, 1377t
Helium–neon lasers, *1386*
Helmholtz coils, *963*
Henry, Joseph, 705, 896, 968, *1005*, 1024
Henry (H), **1005**
Hertz, Heinrich, 705, 989, *1068*–1069, 1095

Higgs bosons, **1535**
High-pass filters, *1055*–1056
High-temperature superconductors, 1431–1432
History of physics
  Ampère, André-Marie, *933*
  atomic models, 1355–1361
  Bohr, Niels, *1356*
  Compton, Arthur Holly, *1297*
  Coulomb, Charles, *711*
  Curie, Marie, *1452*
  de Broglie, Louis, *1301*
  Dirac, Paul, *1513*
  Einstein, Albert, *1251*
  electricity, 705
  Faraday, Michael, *968*
  Fermi, Enrico, *1484*
  Feynman, Richard, *1517*
  Gauss, Karl Friedrich, *742*
  Gell-Mann, Murray, *1527*
  Goeppert-Mayer, Maria, *1451*
  Heisenberg, Werner, *1310*
  Henry, Joseph, *1005*
  Hertz, Heinrich, *1068*
  Huygens, Christian, *1108*
  Kirchhoff, Gustav, *870*
  light, 1067, 1093, 1095–1096
  magnetism, 705, 895–896, 927, 968
  Maxwell, James Clerk, *1067*
  nuclear physics, 1441
  Oersted, Hans Christian, *895*
  Ohm, Georg Simon, *835*
  optics, 1093
  particle physics, 1516–1518
  Pauli, Wolfgang, *1375*
  Planck, Max, *1288*
  relativity, 1243, 1245, 1250–1251, 1273
  Schrödinger, Erwin, *1331*
  Tesla, Nikola, *1053*
  Thomson, Joseph John, *1355*
  Yukawa, Hideki, *1516*
Holes, **1421**–*1422*
  and antiparticles, 1514
Holography, **1222**–*1224*
Hooke's law, 907
Hubble, Edwin, 1262, 1539
Hubble constant (H), **1539**–1541
Hubble Space Telescope [satellite], *1205*
  resolution of, 1217
Hubble's law, *1539*
Humason, Milton, 1539
Hund's rule, **1376**
Huygens, Christian, 1093, 1095–1096, *1107*–1108
Huygens's principle, **1107**–1109, 1178, 1208
Hybrid drive systems, *985*
Hydrogen
  electronic configuration of, 1375, 1377t
  isotopic abundance of, 1442
  line spectra of, *1353*–1354

Hydrogen (*Continued*)
  and MRI, 1469
  in the Universe, *1351*
Hydrogen atom
  Bohr's model of, 731, 791, 852, 1356–1361
  quantum model of, 1361–1364
  Rutherford's model of, *1355–1356*
  Thomson's model of, 759, *1355*
  wave functions for, 1364–1367
Hydrogen bonds, *1403*
Hyperbolas, **A.21**
Hyperopia, *1157*
Hysteresis, magnetic, *950*
  in transformers, 1051

"Ice Man," *1440*, 1463
Illusions, optical, *1130*, *1174*
Image (I), **1127**
  real, **1127**–1128
  sign conventions for, 1134t, 1139t, 1144t
  virtual, **1127**–1128
Image distance (q), **1127**
Image formation, 1126–1175
  and aberration, *1132*, *1152–1153*
  by cameras, 1153–1155
  by compound microscopes, 1160–1162
  by eyes, *1155*–1159
  by flat mirrors, 1127–1130
  by lenses, 1141–1152
  by refraction, *1138*–1141
  by simple magnifiers, *1159–1160*
  by spherical mirrors, 1131–1138
  by telescopes, 1162–1165
Impedance (Z), **1045**–1046t
Impedance matching, 861, 890, 1053, 1064
Impedance triangles, *1045*
Incoherence, **1177**
Indefinite integrals, **A.25**, A.27t–A.28t. *See also* Integration
Independent-particle model. *See* Shell model of the nucleus
Index of refraction (n), **1104**t–1107, 1230t
  and chromatic aberration, *1153*
  and dispersion, *1109*
  measurement of, *1124*–1125
  and polarizing angle, *1228*
Induced dipole moments, *817*
Inductance (L), 1003–1032
  defined, **1005**
  mutual, **1013**–1015
  self-, **1004**–1006
Induction
  electrical, 709–711, **710**
  magnetic, *967*–973
Induction furnaces, 999
Induction welding, 991
Inductive reactance (X_L), **1040**

Inductors, **1006**
  in AC circuits, *1038*–1040
Inelastic scattering, 1466
Inert gases, 1378–1379
Inertial confinement, *1493*
Inertial mass, 1273
Inertial reference frames, 1246
  and twin paradox, 1258
Infinite square wells, *1326*, 1332
Infrared waves, **1080**–*1081*
Instantaneous electric current (i), **833**,
    1034
Instantaneous voltage (Δv), 1034
Instrument Landing System, 1198
Instrumentation. *See also* Experiments;
      Laboratories; Lenses; Measurement;
      Medicine; Microscopes;
      Observatories; Satellites
  accelerators, *1242*
  ammeters, **879**
  balances, 895, 933, 987–*988*, 1077
  beam splitters, *1194*
  betatrons, 1001
  bubble chambers, 918
  cathode ray tubes (CRT), 727–*728*
  clocks, 1255
  colliders, *1500*, *1511*, 1532, *1535*
  compasses, 895
  cyclotrons, 908, **913**–914
  electron guns, 728
  electrostatic precipitators, *783*
  flowmeters, 915–916, 924
  galvanometers, *879*, **907**
  Geiger tubes, 792
  Helmholtz coils, *963*
  interferometers, **1194**–1196, *1249*
  International Thermonuclear
      Experimental Reactor (ITER),
      1491–*1492*
  JET tokamak, 1491
  JT-60U tokamak, 1491–1492
  keratometers, 1169
  Large Electron–Positron (LEP)
      Collider, *1535*
  Large Hadron Collider (LHC), 1535
  lasers, *1387*–1389
  light detectors, 1295–1296
  magnetic bottles, *910*, 921
  National Spherical Torus Experiment
      (NSTX), *1491*
  Nova Laser Facility, 1493–*1494*
  Omega Laser Facility, 1493
  oscilloscopes, 728
  periscopes, 1117
  Relativistic Heavy Ion Collider (RHIC),
      1500, *1511*, 1532, 1535
  spectrometers, *911*–912, *1219*, *1298*,
      1442
  Stanford Linear Accelerator (SLAC),
      1530
  Stanford Linear Collider, 1535
  STAR detector, *1500*, *1511*

Instrumentation (*Continued*)
  Super Kamiokande, *1521*–1522, 1550
  Super Proton Synchrotron, 1535
  synchrotrons, 913, 1535
  telescopes, **1162**–1165, *1205*, 1217,
      1238, 1539–1540
  Tevatron, 1535
  thermometers, 843, *1290*
  Thomson's apparatus, *912*
  Tokamak Fusion Test Reactor (TFTR),
      1491
  tokamaks, 960, **1490**–*1491*
  Van de Graaff generators, 782–*783*
  voltmeters, *811*, **879**–880
Insulators, **709**
  and band theory of solids, *1421*
  resistivity of, 837*t*
Integrated circuits, *1398*, **1428**–*1429*
Integration, A.25–A.28*t*
  along a path, 763, 902, 931, 934, 939,
      1073
  of inverse functions, 875–876, 1007,
      1269
  over a surface, 741–742
  partial, 1333, A.26–A.27
  and separation of variables, 686,
      1038
  of vectors, 720, 902, 928
Intensity (I), 1183
  of absorption spectra, 1410
  in diffraction grating patterns, *1219*
  in double-slit diffraction patterns,
      1212–1214
  in double-slit interference patterns,
      1182–*1184*
  of electromagnetic waves, 1074–**1075**,
      1183
  and photoelectric effect, 1292–1294
  of polarized beams, 1227
  in single-slit diffraction patterns,
      1210–*1212*
Interference, 1069, 1176–1204. *See also*
      Diffraction
  conditions for, 1177
  constructive, 1177–**1180**, 1190
  destructive, 1177–**1180**, 1190
  from a double-slit, 1177–*1184*
  from multiple slits, 1188
  and quantum particles, 1304–1306
  in thin films, *1189*–1194
  from a triple-slit, 1186–1187
Interference microscopes, 1202
Interferograms, 1195
Interferometers, **1194**–1196, *1249*
  and hysteresis cycles, 951
  and resistors, 845
  rotational, 1344
  translational, 1343–1344
  vibrational, 1343–1344
Internal resistance (r), **860**
Intrinsic semiconductors, *1422*
Invariant mass (m), 1267, **1270**

Inverse-square laws
  electrical, 711
  and potential energy, 770
Io, 1096
Iodine, 1486, 1500–1502
Ion chambers, **1498**–*1499*
Ion density (n), **1490**
Ionic bonds, **1400**–*1401*
Ionic cohesive energy (U_0), **1412**
Ionic solids, 1411–1413
Ionization energy, **1358**, *1380*
Iridescence, *1176*, *1238*, 1313–1314
Irises, eye, *1155*
Irradiance, 1074, 1394
Isobars, mirror, 1473
Isotones, 1451
Isotopes, **1442**, A.4*t*–A.13*t*
  radioactive, 1455–1456

J/Ψ particles, 1530
Jackets, in optical fibers, 1114
Jacob's ladder, *851*
Jensen, Hans, 1451
Jewett, Frank Baldwin, *912*
Joliot-Curie, Irène and Frédéric,
      1441
Joule heating, 846
Joyce, James, 1529
Jumpers, 868–*869*
Junction diodes, **1424**–1425
Junction rule, **869**–873
Junction transistors, *1427*–1428
Junctions, **864**
Jupiter. *See also* Planetary motion
  magnetic field of, 954
  moon of, 1096
  and speed of light, 1096

K capture, **1462**
K shell, 1375, 1382–1383
Kamerlingh-Onnes, Heike, 844
Kaons (K), 1519*t*, *1523*
Keating, R. E., 1255
Keck Observatory, 1165
Keratometers, 1169
Kilby, Jack, 1428
Kinematic equations
  for motion of charged particles in
      electric fields, 726–727, 767
Kinetic energy (K). *See also*
      Conservation of energy; Energy;
      Potential energy
  and charges in electric fields, 766
  and photoelectric effect, 1292–1295
  and radioactive decay, 1457
  relativistic, 1268–1270, 1299
  and Schrödinger wave equation, 1331
Kirchhoff, Gustav, *870*, 1314
Kirchhoff's rules, **869**–873
  and RL circuits, 1007

Laboratories. *See also* Instrumentation; Observatories
Bell, *1537*
Brookhaven, *1500*, 1530, 1532, 1535
CERN, 1518, 1532, *1535*
Fermilab, 1520, 1531, 1535–1536
Lawrence Livermore, 1493–*1494*
Laguerre polynomials, 1418
Lambda (Λ) particles, 1519*t*, *1523*
Land, Edwin H., 1226
Landau, Lev Davidovich, 1395
Larmor precessional frequency (ω_p), *1468*–1469
Laser cooling, *1389*, 1397
Laser Interferometer Gravitational-Wave Observatory (LIGO), 1195–*1196*
Laser printers, *784*
Laser trapping, *1388–1389*
Lasers, **1385**–1386
applications of, 1386–1389, 1493
Lateral magnification (M), **1128**, 1132, 1145
of microscopes, 1161–1162
Laue patterns, *1224–1225*
Law of cosines, A.22
Law of reflection, *1099*
from Fermat's principle, 1115
from Huygens's principle, *1108*–1109
Law of sines, A.22
Lawrence, Ernest O., 913
Lawrence Livermore National Laboratory, 1493–*1494*
Laws of physics. *See also*
Ampère–Maxwell law; Ampère's law; Biot–Savart law; Boltzmann distribution law; Bragg's law; Brewster's law; Conservation laws; Conservation of energy; Conservation of momentum; Correspondence principle; Coulomb's law; Curie's law; Equipartition of energy theorem; Exclusion principle; Faraday's law; Fermat's principle; Galilean transformation equations; Gauss's law; Hooke's law; Hubble's law; Huygens's principle; Inverse–square laws; Kinematic equations; Kirchhoff's rules; Law of atmospheres; Law of reflection; Lennard–Jones law; Lens makers' equation; Lenz's law; Lorentz force law; Malus's law; Maxwell's equations; Mirror equation; Ohm's law; Principle of complementarity; Principle of equivalence; Snell's law of refraction; Standard Model; Stefan's law; String theory; Superposition principle; Thin lens equation; Uncertainty principle; Unified theory; Wave equations

Lawson, J. D., 1490
Lawson's criterion, **1490**–1491, 1493
Lawson's number, *1490*
LC circuits, *1015*–1020
LED. *See* Light-emitting diodes
Length (x, y, z, r, l, d, or h)
units of, A.1*t*
Length contraction, **1258**–*1259*
Lennard–Jones law
potential for, *1400*, 1434
Lens makers' equation, **1143**
Lenses, 1141–1152. *See also* Refraction
aberrations in, *1152–1153*
anti-reflective coatings for, *1193*
combinations of, 1149–1152
contact, 1169
converging, *1144–1145*
diverging, *1144–1146*
eyepiece, 1160–*1163*
Fresnel, *1146*
gravitational, *1275*
objective, 1160–*1163*
testing of, 1192
thick, *1142*–1143
thin, *1143*–1146
zoom, 1172
Lenz, Heinrich, 977
Lenz's law, **977**–981
and eddy currents, 986
and inductance, 1004–1005
Lepton number (L), **1522**–1523
Leptons, 1519*t*–**1520**
conservation of, 1522–1523
Levitation, magnetic, *952*, 1430
Lifetimes (Δt), mean, 1516–1517, 1525–*1526*, 1533
Light, 1092–1241. *See also* Optics; Speed of light
and diffraction patterns, 1205–1241
dual nature of, 1095–1096, 1300–1302
as electromagnetic waves, 1068, 1070, 1249
and general relativity, *1273–1275*
history of, 1067, 1093, 1095–1096
and image formation, 1126–1175
and interference, 1176–1204
and laws of geometric optics, 1094–1125
monochromatic, **1177**
polarization of, 1225–1232
spectrum of, **1080**–*1081*
white, 1109–*1110*
world-lines for, *1259*–1260
Light-emitting diodes (LEDs), *1426*
Lightbulbs, 863, 868–*869*, *1284*
Lightmeters, 1295
Lightning, *704*, *716*, *762*
Lightyear (ly), 1257
LIGO. *See* Laser Interferometer Gravitational-Wave Observatory
Line density. *See* Flux

Line of stability, *1446*–1447
and radioactive decay, 1456–*1457*, 1481
Line spectra, **1352**–*1354*
Linear absorption coefficient (μ), 1509–1510
Linear charge density (λ), **720**
Linear equations, A.17–A.19
and Kirchhoff's rules, 870
Linear momentum (**p**)
relativistic, 1267–1268, 1270
Linearly polarized light, **1226**
Linearly polarized waves, **1069**
Lines, *A.20*
Liquid crystal displays, 1232
Liquid-drop model of the nucleus, 791, 1444, **1448**–1450
and fission, 1481–*1482*
Lithium
electronic configuration of, 1375–1377*t*
as moderator, 1494–*1495*
Live wires, *880*
Livingston, M. S., 913
Lloyd, Humphrey, 1188
Lloyd's mirror, *1188*
Load resistance (R), **860**
matching of, 861, 890, 1053
Logarithms, 1454, **A.19**
Loop rule, **869**–873
Lorentz, Hendrik, A., 1263
Lorentz force law, 910, **989**
Lorentz transformation equations, 1262–1267, **1263**
Low-pass filters, *1056*
LRC circuits, *1020*–1022, 1043–1047, 1049–1051
Lyman series, 1354, *1358*

M-theory, **1543**
Macroscopic states
vs. quantum states, 1291
Madelung constant (α), **1412**
Magic numbers, **1450**–1451
Magnes, 895
Magnetic bottles, *910*, 921
Magnetic braking, 987–*988*
Magnetic confinement, **1490**–1492
Magnetic declination, *954*
Magnetic dipole moment (μ), **905**
of atoms, 944–946*t*, 1368–1370
of nucleons, 1468
Magnetic dipoles
in magnetic fields, 879, *904*–907
in nonuniform magnetic fields, *919*
Magnetic field lines. *See* Field lines
Magnetic field strength (**H**), **947**
Magnetic fields (**B**), 894–925, 899*t*
applications of charged particles moving in, 910–914
current loops in, 904–907
defined, 896–900

Magnetic fields (**B**) (*Continued*)
  due to current-carrying wires, 927–*933*, 935–936
  due to current loops, 930–932
  due to solenoids, *938–940*
  energy density of, 1011–1012
  motion of charged particles in, 907–910, *1453*
  motion of conductors through, 970, 973–977, 980–981
  sources of, 926–966
  and superconductivity, 1430–1432
Magnetic flux (Φ_B), **940**–941
Magnetic flux density. *See* Magnetic fields
Magnetic forces (**F**$_B$), 896–900
  between current-carrying conductors, *932–933*
  on current-carrying conductors, 900–904
Magnetic hysteresis, *950*
Magnetic induction, *967–973. See also* Magnetic fields
Magnetic levitation, *952*, 1430
Magnetic permeability (μ_m), **948**
Magnetic poles
  and monopoles, 895, 942, 989
  north and south, 895
Magnetic recording, 951
Magnetic resonance imaging (MRI), 845, *1469*–1470. *See also* Nuclear magnetic resonance
Magnetic susceptibility (χ), **947**–948t
Magnetism in matter, 944–953
  classification of, 947–953
Magnetite, 705, 895
Magnetization (**M**), **946**–947
  remanent, 950
  saturation, 953
Magnetization curves, *950*
Magnetons
  Bohr (μ_B), **946**, 1373
  nuclear (μ_n), **1467**–1468
Magnetrons, 1085–1086
Magnets
  bar, 895–*897*, *931*, *938*, *942*
  superconducting, 845, 1490
Magnification
  angular (*m*), **1159**, 1161–1164
  lateral (*M*), **1128**, 1132, 1145, 1161–1162
Magnifiers, simple, *1159–1160. See also* Microscopes; Telescopes
Magnifying glasses, 1159–*1160*
Malus, E. L., 1227
Malus's law, **1227**
Maricourt, Pierre de, 895
Mariner 10 [spacecraft], *1077*
Marsden, Ernest, 790, 1355, 1441
Mass (*m*).
  atomic, A.4t–A.13t
  as form of energy, 1270, 1272
  gravitational *vs.* inertial, 1273
  invariant, 1267, **1270**

Mass (*m*) (*Continued*)
  nuclear, 1442–1443t
  reduced, 1343
  "relativistic," 1267
  units of, A.1t
Mass defect, **1481**
Mass–energy equivalence, 1447
Mass number (*A*), **1442**, A.4t–A.13t
Mass spectrometers, *911*–912, 1442
Materials science. *See also* Friction; Gases; Optics; Solids
  and crystalline *vs.* amorphous materials, 1229
  and electrical properties of solids, 709, 835–845, 838, 1031
  and magnetic properties of solids, 947–952, 1031
  and neutron activation analysis, 1501–*1502*, 1507
  and optical properties of materials, 1104–1107, 1230, *1230*
Mathematics, A.14–A.29. *See also* Addition; Algebra; Approximating; Calculus; Complex numbers; Division; Equations; Geometry; Logarithms; Measurement; Multiplication; Operators; Series expansions; Subtraction; Trigonometric functions; Units
Matrix algebra, 870
Matrix mechanics, 1310, 1331
Matter
  fundamental particles of, 912
Maxima of intensity, 1180, 1187–*1188*, 1206, 1219
Maximum angular separation (θ_{max}) [apertures], 1215
Maxwell, James Clerk, *1067*
  and electromagnetic waves, 1067–1068, 1093, 1095
  and electromagnetism, 705, 896, 942, 944
  and Maxwell's equations, 988, 1067–1068
Maxwell's equations, **988–989**
  and electromagnetic waves, 1067–1069
  and special relativity, 1245, 1247
Mean free path (ℓ), 842
Measurement. *See also* Experiments; Instrumentation
  of distance, 1387
  of electric current, *879*
  of magnetic fields, 914
  of potential difference, 879–880
  of speed of light, *1096–1097*
  uncertainty in, A.28–A.29
  of wavelength of light, 1181
Measurements
  of electric current, 879
Mechanical devices
  balances, 895, 987–*988*, 1077
  photocopiers, *784*

Mechanical equivalent of heat, 606–*607*
Mechanics. *See also* Energy; Force; Momentum; Motion; Quantum mechanics; Statistical mechanics
Media for wave propagation
  and ether, 1247–1250
Mediation of forces, 1513, 1516
Medicine. *See also* Biophysics; Health
  and blood flowmeters, 915–916, 924
  and contact lenses, 1169
  and cyclotrons, 913
  and defibrillators, *810*, 825
  and eyeglasses, *1157–1158*
  and eyes, 1157–1159, 1169, 1174
  and fiber optics, 1114
  and laser surgery, 1387–1388
  and MRI, 845, *1469*–1470
  and PET scans, 1507, *1515*
  and radiation therapy, 1502
  and radioactivity, 1456, 1500–1501
  and sunglasses, *1081*, 1229
Meissner, W. Hans, 1430
Meissner effect, *952*, 1031, 1430
Meitner, Lise, 1481
Melting points, 1412–1414
Mendeleev, Dmitri, 1376
Mercury [element]
  line spectra of, *1353*–1354
  superconductivity of, 844
Mercury [planet]. *See also* Planetary motion
Mesons, **1516–1519**t
  quark composition of, 1529–1531t, *1533*
Metal detectors, 1051
Metal-oxide-semiconductor field-effect transistors (MOSFET), *1427*–1428
Metallic solids, 1414
Metals
  alkali, 1380
  and band theory of solids, *1420*–1421
  charge carriers in, 915
  free-electron theory of, 1415–1418
Metastable states, 1384, *1386*
Michelson, Albert A., 1194, 1248–1250
Michelson interferometers, *1194*–1196, *1249*
Michelson–Morley experiment, 1248–1250
Microchips, *1398*
Microprocessors, 1428–1429
Microscopes
  atomic force, *1341*
  compound, **1160**–1162
  electron, *1303*–1304
  interference, 1202
  scanning tunneling, *1321*, 1338, *1340*–1341
Microwave ovens, 816, 1085
Microwaves, **1080**–*1081*
  cosmic background, 1088, 1537–*1538*

Millikan, Robert, 708, 781–782
Minima of intensity, **1206**
Mirages, 1121
Mirror equation, **1132**–1133
Mirror isobars, 1473
Mirrors, *1127–1138. See also* Reflection
  concave, *1131*–1133
  convex, *1134*
  flat, *1127*–1130
  Lloyd's, *1188*
  parabolic, *1164*
  partially reflecting, 1386
  ray diagrams for, 1134–1136
  spherical, *1131–1138*
Missing mass, 1520, 1541
Models
  of electric current, 833–835
  of electrical conduction, 841–843,
    1415–1418
  of hydrogen atom, 759
  of the nucleus, 791, 1444, 1448–1452
  of solids, 1418–*1420*
Moderators, **1480**, *1486*
Modern physics, 1242–1552. *See also*
    Quantum mechanics; Relativity
Molar specific heats. *See also* Specific
    heat
  at constant volume (C_V), 1343–1344,
    1403
Molecules
  bonding in, 1399–1403
  energy states of, 1403–1411
  polar *vs.* nonpolar, **816**
  rotational motion of, 1404–1406
  spectra of, 1408–1411
  vibrational motion of, 1406–1408
Moment of inertia (I)
  of molecules, 1404–1405
Momentum (**p**). *See also* Angular
    momentum; Linear momentum
  in electromagnetic waves, 1076–1079,
    1301
  and nuclear reactions, 1466
  of particle-in-a-box, 1327
  and radioactive decay, 1457
  relativistic, 1267–1268, 1270
  and uncertainty principle,
    1309–1310
Monochromatic light, **1177**
  and lasers, 1385
Monopoles, magnetic, 895, 942,
    989
Moon. *See also* Planetary motion
  measurement of distance to, 1387
  planetary, 1096, 1217–*1218*
Moore, Gordon, 1429
"Moore's law," *1429*
Morley, Edward W., 1248–1250
Morse, Samuel, 959
Morse potential, 1438
Moseley, Henry G. J., 1383, 1391

MOSFET. *See* Metal-oxide-
    semiconductor field-effect
    transistors
Motion. *See* Oscillatory motion;
    Planetary motion; Precessional
    motion; Rotational motion; Waves
Motion pictures. *See* Movies
Motors, **984**–986. *See also* Generators
  homopolar, 995
Mount Palomar Observatory, 1540
Mount Wilson Observatory, 1539
Movies
  digital, 1101
  *M*A*S*H*, 1233
  sound recording for, 1295
  Star Wars Episode II: Attack of the Clones,
    1101
MRI. *See* Magnetic resonance imaging
Mu (μ) mesons. *See* Muons
Müller, K. Alex, 1431
Multiplication
  and differentiation, A.24
  scalar (dot) product of two vectors,
    741
  and uncertainty, A.29
Muons (μ), **1516**, 1519t
  gravitational red-shifts of, 1274
  length contraction of, 1259
  and time dilation, *1255–1256*
Musical instruments
  electric, 971–*972*
Mutual inductance (M), **1013**–1015
Mutual induction, 1013
Myopia, 1157–*1158*

n-type semiconductors, *1423*
Nanotechnology, **1335**–1336, 1424
Napoleon, 1502
Natural frequency (ω_0)
  of LC circuits, 1018
Natural logarithms, 875, A.19
Natural radioactivity, **1465**t
Near point of the eye, **1156**, *1159*
Nearsightedness, **1157**–*1158*
Ne'eman, Yuval, 1527
Negative charges, **707**
Neon
  electronic configuration of,
    1376–1377t
  line spectra of, *1353–1354*
Neutral wires, *880*
Neutrinos (ν), **1461**, 1519t
Neutron activation analysis, **1501**–*1502*,
    1507
Neutron capture, **1480**–1481
Neutron number (N), **1442**
Neutron stars, 1473
Neutrons
  interactions involving, 1480–1481
  radiation damage by, 1496

Neutrons (*Continued*)
  structure of, 1468
  thermal *vs.* fast, 1480
Newton, Isaac
  and development of calculus, A.23
  and gravitation, 1273
  and optics, 1093, 1095, 1191
  and telescopes, 1164
  and time, 1252
Newtonian mechanics. *See* Mechanics
Newton's laws of motion. *See* Second law
    of motion
Newton's rings, *1191–1192*
Nitrogen
  electronic configuration of, 1376–**1377**t
NMR. *See* Nuclear magnetic resonance
Nobel prizes, A.33–A.36
Nonconservative forces
  and induced electric fields, 981
Noninertial reference frames
  and general relativity, 1273–1274
  and twin paradox, 1258
Nonohmic materials, 835, 838
Nonpolar molecules, **816**
Normal modes
  on strings fixed at both ends, 1288
Normalization, **1324**, 1331, 1342
Noyce, Robert, 1428
Nuclear binding energy, 1443, 1445,
    1447–*1448*
  semiempirical formula for, **1449**–1450
Nuclear fission. *See* Fission
Nuclear forces, **1445**, 1512–1513t,
    1532–1534
  and alpha decay, 1337–1338
  and mesons, 1516
Nuclear fusion. *See* Fusion [nuclear]
Nuclear magnetic resonance (NMR),
    1467–1469, **1468**. *See also* **Magnetic
    resonance imaging**
Nuclear magneton (μ_n), **1467**–1468
Nuclear physics, 1440–1478. *See also*
    Nucleus; Radioactivity; Subatomic
    physics
  applications of, 1479–1510
  history of, 1441
  and nuclear models, 1448–1452
Nuclear reactions, **1465**–1467t
Nuclear reactors
  fission, 1272, *1479*, 1483–1487
  fusion, 1493–*1495*
Nuclear spin quantum number (I),
    1467
Nucleons, **1442**
Nucleus, **1355**. *See also* **Radioactivity**
  collective model of, 1452
  discovery of, 1355
  liquid-drop model of, 791, 1444,
    1448–1450, 1481–*1482*
  magnetic dipole moment of, 946
  parent *vs.* daughter, **1457**

Nucleus (*Continued*)
  properties of, 1441–1447
  shell model of, 1450–1452
  size and structure of, 1444–1445
  stability of, 1445–1447

Object, optical (*O*), *1127*
  virtual, **1149**
Object distance (*p*), *1127*
Objective lenses, 1160–*1163*
Observatories. *See also* Instrumentation;
    Laboratories; Satellites; Telescopes
  Keck, 1165
  Mount Palomar, 1540
  Mount Wilson, 1539
  Yerkes, 1165
Ochsenfeld, Robert, 1430
Oersted, Hans Christian, 705, *895*, 927,
    933
Ohm, Georg Simon, *835*
Ohm (Ω), **836**
Ohmic materials, 835, 838
Ohm's law, **835**, *838*
Oil-drop experiment, *781–782*
Omega (Ω) particles, 1519*t*, *1528*
Open-circuit voltage (ε), **860**
Operators, 1325
Optic axis, *1229*
Optical activity, **1232**
Optical devices, 1114. *See also* Cameras;
    Lenses; Microscopes; Mirrors;
    Telescopes
Optical fibers, *1114*, 1388
Optical illusions, *1130, 1174*
Optical molasses, 1388
Optical resolution, *1214–1217*
Optical stress analysis, *1230*
Optical traps, *1388–1389*
Optical tweezers, 1388–1389
Optics, 1092–1241. *See also* Diffraction;
    Dispersion; Electromagnetic waves;
    Image formation; Interference;
    Light; Polarization; Reflection;
    Refraction; Speed of light
  geometric, 1094–1175
  history of, 1093
  wave, 1176–1241
Orbital magnetic quantum number
    (*m_ℓ*), **1363***t*, 1368–1370
Orbital quantum number (*ℓ*), **1363***t*,
    1367–1368
Orbitals, 1375
Order number (*m*), **1180**
Order-of-magnitude (~) calculations. *See
    also* Approximating
Ordinary (O) rays, *1229*–1230
Original quark model, 1529–1530
Orion nebula, *1351*
  damped, 1020–*1022*
Oscilloscopes, 728

Overdamped oscillators, *1022*
Oxygen
  electronic configuration of, *1376–1377t*
  paramagnetism of, *952*

p–n junctions, **1424**–*1425*
p-type semiconductors, *1423*–**1424**
Pair production, *1514*
  of quarks, *1533–1534*
Paleomagnetism, 954
Parabolas, *A.21*
Parabolic mirrors, *1164*
Paradoxes
  moving clocks, 1255
  pole-in-the-barn, 1260–*1261*
  speed of light, 1248
  twin, *1257–1259*
Parallax, 1167
Parallel combinations
  of capacitors, *803*–804
  of resistors, 862–869, *864*
Paramagnetism, **947**–948, 951–952
Paraxial rays, **1132**
Parent nucleus, **1457**
Partial derivatives, 773, 1069–1070. *See
    also* Differentiation
Partial integration, 1333, 1366,
    A.26–A.27
Particle-in-a-box, 1326–1330
  and Schrödinger wave equation,
    1331–1333
Particle physics, 1511–1552
  and antiparticles, 1513–1515
  and classification of particles,
    1518–1520, 1527–1528
  and conservation laws, 1520–1523
  and the fundamental forces,
    1512–1513
  measurements in, 1524–1527
  and mesons, 1516–1518
  and quarks, 1529–1534
  and the standard model, 1534–1536
  and strange particles, 1523–1524
Particle–wave duality, 1300–1302, 1323
Paschen series, 1354, *1358*
Path difference (δ), *1180*, 1182–1183
Pauli, Wolfgang, *1371*, 1374–*1375*, 1461
Pauli exclusion principle. *See* Exclusion
    principle
Penzias, Arno, 1088, *1537–1538*
Percent uncertainty, **A.29**
Perfect diamagnetism, 952, 1031
Perfect differentials, A.27
  of uniform circular motion, 908
Periodic table of the elements,
    1374–1380, **1378**, A.30*t*–A.31*t*
Periscopes, 1117
Permeability, magnetic (*μ_m*), **948**
Permeability of free space (*μ_0*), **928**,
    943

Permittivity of free space (ε_0), **712**, 943
PET scans, 1507
Pfund, A. H., 1124
Phase (*ωt + φ*)
  in AC circuits, 1035
  and reflection, 1188–*1189*
  and wave packets, 1305
Phase constant (*φ*)
  and interference, 1182–1183
  in series *RLC* circuits, **1044**–1045
  of simple harmonic motion, 1018
Phase speed (*v_phase*), **1306**
Phases of matter. *See also* Gases; Solids
Phasor diagrams, *1035*
  for circuits, *1035*–1036, *1039,
    1041*–1042, *1044–1046*
  for diffraction patterns, *1210–1211*
  for interference patterns, *1186–1188*
Phasors, **1036**, 1184–1188
Phillips, William, 1397
Phipps, 1372
Phipps–Taylor experiment, 1372
Photoconductors, 784
Photocopiers, *784*
Photoelectric effect, 1095, **1291–1296**
Photoelectric photometry, 1296
Photoelectrons, **1291–1294**
Photographic emulsions, **1497–1498**
Photomultiplier tubes, *1295*–1296, 1500
Photons, 1096, **1270**, 1290, **1293**
  and electromagnetic waves, 1300–1301
  virtual, 1517
Phototubes, 1295
Photovoltaic cells, 1193, 1426
Physical chemistry.
  and batteries, 799, 845–846, *858*–861,
    1014–*1015*, 1061
  and chemical energy, 799, 845–846
  and periodic table, 1374–1380
  and scrubbers, *783*
  and surfactants, 816
Physical optics. *See* Wave optics
Physics. *See also* Astronomy and
    astrophysics; Biophysics;
    Engineering; Geophysics; History
    of physics; Laws of physics; Physical
    chemistry
Pi (*π*) mesons. *See* Pions
Pickup coils, *971–972*
Piezoelectric crystals, 1340
Pinch effect, 960
Pinhole cameras, 1239
Pioneer 10 [spacecraft], *1508*
Pions (*π*), **1516**–1519*t*
Planck, Max, *1288*
  and blackbody radiation, 1288, 1290,
    1343
  and quantum mechanics, 1285
Planck length, **1543**
Planck's constant (*h*), 945, 1096,
    1288–1289

Plane polar coordinates (r, θ), 1330
Plane-polarized light, **1226**
Plane waves, **1069**
Plasma, 910, 1488
  balls of, *739*
  quark–gluon, **1532**, 1536
Plates of capacitors, 796
Plug-in problems. *See also* Problem-
      solving strategies
Pluto.
  moon of, 1217–*1218*
Point charges, **711**
  electric field lines for, *724, 772*
  electric potential due to, 768–771,
      769
  equipotential surfaces for, *772*
Poisson, Simeon, 1207
Polar coordinates (r, θ), 1330
Polar molecules, **816**
Polarization, 1069, 1225–1232
  by double refraction, *1229–1230*
  induced, *816*
  linear, **1069**
  of molecules, 816
  by reflection, 1227–1229
  rotation of plane of, 1232
  by scattering, 1230–*1231*
  by selective absorption, *1226*–1227
Polarization angle (θ_p), *1228*
Polarizers, 1226
Polarizing angle, 1119
Polaroid materials, 1226
Pole-in-the-barn paradox, 1260–*1261*
Population inversions, **1385–1386**
Position (**x**)
  and uncertainty principle, 1309–1310
Positive charges, **707**
Positron-emission tomography (PET)
      scans, 1507, *1515*
Positrons, **1452**, 1513–1515
Potential difference (ΔV), **764**
  due to point charges, 768–771
  stopping, **1292**
  in uniform electric fields, 765–768
Potential energy (U). *See also*
      Conservation of energy; Electric
      potential; Energy; Kinetic energy;
      Lennard–Jones law
  and barrier penetration, 1336–1339
  in capacitors, 807–810, 876, 878
  chemical, 799, 845–846
  electric, 763–766, **764**, 1015–*1017*
  of electric dipoles in electric fields,
      815
  in ionic solids, *1412*
  of magnetic dipoles in magnetic fields,
      906, 1468
  of molecular bonds, 1399–*1400*
  of nucleons, 1445–*1446, 1488*
  and Schrödinger wave equation, 1331
  and square wells, *1326,* 1332, 1334–1336
  and tunneling, 1336–1339

Power ($\mathcal{P}$)
  in AC circuits, 1037, 1047–1051
  average ($\overline{\mathcal{P}}$), **1048**
  electrical, 845–849, 860
  radiated by black bodies, 1286
Power factor (cos ϕ), **1048**, 1051
Power of a lens (P), **1158**
Power transfer
  by electricity, *831,* 846–847
Powers. *See* Exponents
Poynting vectors (**S**), **1074**–1079
Precessional frequency (ω_p)
  Larmor, *1468*–1469
Precessional motion, *1468*–1469
Presbyopia, **1158**
Pressure (P)
  radiation, 1076–1079
  units of, A.2t
Primary coils
  of AC transformers, *1051*
  and Faraday's law of induction,
      968–*969*
  of Rowland rings, 949
Primary maxima, 1187–*1188*
Principal axes
  of spherical mirrors, **1131**
Principle of complementarity, **1302**
Principle of equivalence, **1274**
Principle quantum number (n),
      1362–1363t
Prisms, *1109–1111*
Probability
  and decay processes, 1452
  and distribution functions, 1302, 1416
  and electron diffraction, 1308
  and Gauss's probability integral, A.28t
  and quantum mechanics, 1322–1326,
      1364
Probability amplitude (Ψ, ψ), **1323**. *See
      also* Wave functions
Probability density ($|\psi|^2$), *1324*
  for finite square well, *1335*
  for hydrogen atoms, 1364–*1367*
  for particle-in-a-box, *1328*
  radial, 1365–*1367*
Problem-solving strategies
  for capacitors, 806
  for electric fields, 720–721
  for electric potential, 775
  for Kirchhoff's rules, 871
  for thin-film interference, 1192
Products. *See* Multiplication
Propagation of uncertainty, A.28–**A.29**
Proper length (L_p), **1258**–*1259*
Proper time intervals (Δt_p), 1253–1256,
      1254, 1259
Proportional counters, **1498**
Proton–proton cycle, **1487**
Protons, 1519t
  half-life of, 1520–1522
Pupils, eye, *1155*
Pythagorean theorem, 1254, A.21

Q value. *See* Disintegration energy (Q);
      Reaction energy (Q)
QCD. *See* Quantum chromodynamics
QED. *See* Quantum electrodynamics
Quadratic equations, A.17
Quality factor (Q), **1051**
Quantization
  of angular momentum, 945, 1404
  of electric charge, 708
  of energy, 1288, 1328–1330, 1358, 1451
  of space, 1368–*1369*, 1372–1374
Quantum chromodynamics (QCD),
      1533–1534
Quantum corrals, *1321*, 1335
Quantum dots, **1335**–1336
Quantum electrodynamics (QED), 1533
Quantum mechanics, 1321–1350. *See
      also* Mechanics
  and barrier penetration, 1336–1339
  and boundary conditions, 1330
  and ferromagnetism, 949
  and finite square well, 1334–1336
  and infinite square well, 1326–1333
  introduction to, 1284–1320
  and magnetic dipole moments of
      atoms, 945–946
  and particle-in-a-box, 1326–1333
  and probability, 1322–1326
  relativistic, 1372, 1513
  and Schrödinger wave equation,
      1331–1333
  and simple harmonic oscillators,
      1341–1344
  and tunneling, 1336–1339
Quantum numbers, **1288**
  baryon number (B), **1520**–1522
  bottomness, 1530t–1531
  charm (C), **1530**
  lepton number (L), **1522**–1523
  nuclear spin (I), **1467**
  orbital (ℓ), **1363**, 1367–1368
  orbital magnetic (m_ℓ), **1363**, 1368–1370
  physical interpretation of, 1367–1374
  principal (n), **1362**
  rotational (J), **1404**
  spin magnetic (m_s), **1371**–1374
  strangeness (S), 1523–**1524**
  topness, 1530t–1531
  vibrational (v), **1407**
Quantum particles, **1304**–1307
Quantum states, **1288**
  vs. macroscopic states, 1291
Quantum statistics, 1415
Quark–gluon plasma, **1532**, 1536
Quarks, 712, 1512, **1529**–1532
  color charge of, 1532–1534
Quarter-wave plates, 1240
Quasars, 1539, *1548*

Rad. *See* Radiation absorbed dose
Radar, police, 1278–*1279*

Radial probability density function ($P(r)$), 1365–*1367*

Radian (rad), **A.20**

Radiation. *See also* Electromagnetic radiation; Radioactivity
background, 1497
cosmic background, 1088
dual nature of, 1300–1302
thermal, 845–846
uses of, 1500–1503

Radiation absorbed dose (rad), **1496**

Radiation damage, 1486, 1495–1497
by MRI, 1469–1470
by x-rays, 1470, 1496

Radiation detectors, 1497–1500

Radiation dosage, 1496–1497*t*

Radiation equivalent in man (rem), **1497**

Radiation pressure (P), 1076–1079

Radiation therapy, 1502

Radio, 813, 1051, 1061
waves, discovery of, 1068–1069
waves, in electromagnetic spectrum, **1080**–*1081*

Radioactive waste, 1486

Radioactivity, **1452**–1456. *See also* Alpha decay; Beta decay; Gamma decay; Nuclear reactions; Radiation
decay processes of, 1456–1464*t*
natural *vs.* artificial, **1465***t*

Radium, 1452, 1454–1455, 1459

Radon, 1475, 1497

Rail guns, *926, 964*

Rainbow holograms, 1224

Rainbows, *1094*, 1109–1111

Ray approximation, **1097**–1098

Ray diagrams
for mirrors, *1134*–1136
for thin lenses, 1145–*1149*

Rayleigh, John William Strutt, Lord, 1233

Rayleigh–Jeans law, *1287*

Rayleigh's criterion, *1214*

Rays, **1069**
gravitational deflection of light, *1273–1275*
ordinary *vs.* extraordinary, *1229*–1230
paraxial, **1132**

RBE. *See* Relative biological effectiveness

RC circuits, **873**–878
as filters, *1055–1056*
as rectifiers, 1054–*1055*

Reactance
capacitive (X_C), **1042**, 1056
inductive (X_L), **1040**

Reaction energy (Q), **1466**–1467*t*

Reber, Grote, 1235

Rectification, **1054**

Rectifiers, **1054**–1056

Red shifts, 1262, 1274, 1279, 1539–*1540*

Reduced mass (μ), 1343, **1404**

Reference frames
inertial, 1246
noninertial or accelerating, 1258, 1273–1274

Reflecting telescopes, **1162**, *1164*

Reflection, 1069, *1098–1102. See also* Mirrors
and Huygens's principle, *1108*–1109
images formed by, 1127–1138
law of, *1099, 1108*–1109, 1115
phase changes upon, 1188–*1189*
polarization by, 1227–1229
total internal, **1111**–*1114*

Reflection coefficients (R), **1336**

Reflection gratings, 1217–1218

Refracting telescopes, **1162**–*1163*

Refraction, 1069, **1102**–1107. *See also* Lenses
and Huygens's principle, 1108–*1109*, 1178
images formed by, 1138–1141
index of, **1104***t*, *1124*–1125
polarization by double, *1229–1230*
Snell's law of, **1105**, 1108–*1109, 1115*

Reines, Frederick, 1461

Relative biological effectiveness (RBE), **1496***t*

Relativity, Galilean, **1246**–1248

Relativity, general, 1273–1275
history of, 1273

Relativity, special, 1244–1283
consequences of, 1251–1262
history of, 1243, 1245, 1250–1251
and Maxwell's equations, 988
and spin, 945
and wave equations, 1372, 1513

Rem. *See* radiation equivalent in man

Reproduction constant (K), **1484**–1486

Resistance (R), 831–857
defined, 835–840, **836**
equivalent (R_{eq}), **863**–868
internal (r), **860**
load (R), **860**
temperature dependence of, 841, 843–844

Resistivity (ρ), **836**–837*t*, 842
temperature dependence of, 843–*844*

Resistors, *837–838*
in AC circuits, 1034–1038
composition, 838
shunt (R_p), 879
wire-wound, 838

Resolution, optical, *1214*–1217, 1221–1222

Resolving power (R), 1221–**1222**

Resonance
and lasers, 1384
in *LC* circuits, 1016
in series *RLC* circuits, **1049**–1051

Resonance frequency (ω_0), **1049**

Resonance particles, **1525**–1527

Resonant tunneling devices, *1338–1339*

Resonant tunneling transistors, *1339*

Rest energy (E_R), **1270**, 1443

Restoring forces, 1341

Retardation plates, 1240

Retinas, *1155*–1156

Retroreflection, 1100–*1101*, 1387

Revere, Paul, 1235

Reverse bias, *1425*

Richter, Burton, 1530

Right-hand rule
and magnetic forces, 898, *905, 930*

Rings
Newton's, *1191–1192*
Rowland, *949*

Ripple, 1055

RL circuits, 1006–1011, *1007*

rms. *See* Root-mean-square

Rocketry. *See* Spacecraft

Rods, eye, 1156

Roemer, Ole, 1096, 1118

Roentgen, Wilhelm, 1224

Roentgen (R), **1496**

Rogowski coils, *993*–994

Romognosi, Gian Dominico, 895

Root-mean-square electric current (I_{rms}), **1037**

Root-mean-square voltage (ΔV_{rms}), **1038**

Rotational motion
in molecules, 1404–1406

Rotational quantum number (J), **1404**

Rowland rings, *949*

Rubbia, Carlo, 1518

Rutherford, Ernest, 790, 1355, 1441, 1444, 1452, 1465

Rydberg, Johannes, 1354, 1359

Rydberg atoms, 1361

Rydberg constant (R_H), **1354**

Saddle coils, *963*

Salam, Abdus, 1534

Sandage, Allan R., 1540

Satellites. *See also* Instrumentation; Observatories; Spacecraft
attitude control of, 907
Wilkinson Microwave Anisotropy Probe (WMAP), 1536, 1538

Saturated forces, 1447

Saturation, magnetic, 950

Savart, Félix, 927

Scalar product, 741

Scanning tunneling microscopes (STM), *1321*, 1338, *1340*–1341

Scattering, **1230**, **1466**
and Compton effect, 1297–1300
polarization by, 1230–*1231*

Schreiffer, John R., 1431

Schrödinger, Erwin, 1285, 1322–1323, *1331*

Schrödinger wave equation, time-independent, 1323, **1331**–1333, 1361–1362

Schwinger, Julian, 1517
Science fiction
 Verne, Jules, 1167
Scientific notation, **A.14**–A.15
Scintillation counters, **1499**
Scrubbers, *783*
Search coils, 999
Second law of motion
 and charged particles, 725
 relativistic, 1268
Secondary coils
 of AC transformers, *1051*
 and Faraday's law of induction,
  968–*969*
 of Rowland rings, 949
Secondary maxima, 1187–*1188*
Segrè, Emilio, 1515
Selection rules, **1381**, 1405, 1407
Selectrons, 1543
Self-inductance, **1004**–1006
Semiconductor devices, 1424–1429. *See
  also* Diodes; Integrated circuits;
  Light-emitting diodes; Transistors
Semiconductor-diode detectors,
  1499–1500
Semiconductors, 709, 844, **1420**
 and band theory of solids, 1421–1424
 charge carriers in, 915, 1421–1422
 doped, 1422–1424
 extrinsic, **1424**
 intrinsic, *1422*
Semiempirical binding-energy formula,
  1449–1450
Series combinations
 of capacitors, *804*–806
 of resistors, *862*–869
Series expansions, mathematical, 1250,
  1269, A.23
Series limits, **1354**
Seurat, Georges, 1235
Shadows, 1097
Shell model of the nucleus, 1450–1452,
  1451
Shells, electronic, 1363*t*, 1375–1376
Shielding, electronic charge, 1375,
  1381
Shockley, William, 1427
Short circuits, 881–*882*
Shunt resistors (R_p), 879
SI system of units, A.1*t*–A.3*t*, A.32*t*
Side maxima, **1206**
Sievert (Sv), 1497
Sight. *See* Eyes
Sigma (Σ) particles, 1519*t*, 1523
Silicon, *1413*
 and energy, *1017*
 in molecules, 1406–1408
 quantum mechanical, 1341–1344, 1349
Simultaneity, 1252–1253
Sine function, A.21. *See also*
  Trigonometric functions
 and law of sines, A.22

Skin effect, 999
Sky color, *1092, 1231*
Slipher, Vesto Melvin, 1539
Slits
 diffraction of light waves from,
  1207–1214
 interference of light waves from,
  1177–*1184*, 1186–1188, 1212–*1213*
Slopes of graphs, *A.17*
Smoke detectors, *1459*–1460
Smoot, George, 1538
Snell, Willebrord, 1105
Snell's law of refraction, **1105**
 and Brewster's law, 1228
 from Fermat's principle, *1115*
 from Huygens's principle,
  1108–*1109*
Soap, 816
Soap bubbles, *1191, 1194*
Sodium chloride
 ionic bonding in, 1400–*1401*,
  1411–1413
Solar cells, 1193, 1426
Solar energy, 1078–1079
Solar sailing, *1077*, 1086
Solenoids, *938*–940, *982*, 1006
Solid solutions, 1414
Solids, 1398–1439. *See also* Materials
  science
 band theory of, 1418–1420
 bonding in, 1411–1414
 crystalline *vs.* amorphous, 1229
 electrical conduction in, 1412–1414,
  1420–1424
 free-electron theory of, 1415–1418
 and semiconductors, 1424–1429
 and superconductivity, 1430–1432
Somatic damage by radiation, 1496
Sommerfeld, Paul, 1371
Sound recording
 and pickup coils, 971–*972*
Source charges, **715**–*716*
Sources, transistor, *1427*
Space quantization, 1368–*1369*,
  1372–1374
Space shuttles. *See also* Spacecraft
 Challenger, 1517
Space–time, curvature of, 1274–*1275*
Space–time coordinates (x, y, z, t), 1263
Space–time graphs, *1259*–1262
Spacecraft. *See also* Satellites; Space
  shuttles
 Apollo 1100, 1118–1119
 Mariner 1077
 Pioneer 1508
Spark chambers, **1499**
Special relativity. *See* Relativity, special
Specific heat (c). *See also* Heat capacity
 molar, at constant volume (C_V),
  1343–1344, 1403
Spectra
 atomic, 1222, 1352–1354, 1380–1383

Spectra (*Continued*)
 Doppler shifts of, 1539
 electromagnetic, **1080**–1082
 molecular, 1403–1411
 visible, **1109**
Spectrometers
 crystal, *1298*
 diffraction-grating, *1219*
 mass, *911*–912, 1442
Spectroscopy, 1195
 atomic, 1220, 1353–1354
Specular reflection, *1098*–**1099**
Speed (v)
 drift (v_d), **834**, 841–842
 group (v_g), **1306**–1307
 phase (v_{phase}), **1306**
 units of, A.1*t*
 of wave packets, 1306–1307
Speed of light (c)
 and electromagnetic waves, 988,
  1069–1070
 and ether, 1248–1250
 and Lorentz transformation equations,
  1265, 1268
 in materials, 1103–1105
 measurement of, *1096*–1097
 and photons, 1293
 and range of nuclear force, 1518
 and special relativity, 1245, 1247–1248
Spherical aberration, *1132, 1152*–1154,
  1164
Spherical coordinates (r, θ, ϕ), 1330,
  1362
Spherical mirrors, *1131*–1138. *See also*
  Mirrors
Spherical waves, **1069**, *1098*
Spin, **945**–946
Spin angular momentum (**S**), 946,
  1372–*1373*
 nuclear, 1467–*1468*
 unpaired, 1376
Spin magnetic quantum number (m_s),
  1371–1374
Spin–orbit coupling, 1394
 nuclear, 1451
Spontaneous decay, **1457**
Spontaneous emission, *1384*, 1386
Spring constant (k)
 effective molecular, 1406–1408*t*
Springs
 torsional, 907
Square barriers, *1336*–1339
Square waves, *1058*
Square wells, *1326*, **1332**, 1334–1336,
  1451
Squarks, 1543
Standard Model, **1534**–1536
Standing waves
 and blackbody radiation, 1286
 electromagnetic, 1069
 vs. interference patterns, 1178
 and wave packets, *1306*

Stars. *See also* Astronomy; Sun
  evolution of, 1448
  line spectra of, 1354
  neutron, 1473
  red shifts of, 1262, 1274, 1279
States of matter. *See also* Gases; Solids
Stationary states, **1356**
Statistical mechanics.
  quantum mechanical, 1415
Stefan–Boltzmann constant (σ), 1286
Stefan's law, **1286**
Steradian, **752**–753
Stern, Otto, 1371
Stern–Gerlach experiment, 1371–*1372*, 1373–1374
Stimulated absorption, *1384*–1385
Stimulated emission, **1384**–1386
STM. *See* Scanning tunneling microscopes
Stopping potential (ΔV_s), **1292**
Strange particles, 1523–1524
Strange quarks (s), **1529**–1531t
Strangeness (S), 1523–**1524**
Strassman, Fritz, 1356, 1441, 1481
Stress
  optical analysis of, *1230*
String theory, 1542–**1543**
Strong force, 1518, 1524, **1532**–1533, *1536*–1537
Strontium, 1497
Strutt, John William, Lord Rayleigh, 1233
Stud-finders, *813*
Subatomic physics, 1270, 1272, 1282–1283. *See also* Nuclear physics; Particle physics; Radioactivity
Subshells, **1363**t, 1375
Subtraction
  and uncertainty, A.29
Sun. *See also* Astronomy; Stars
  chromosphere of, 1200
  and discovery of helium, 1353
  fusion in, 1338, 1488
  magnetic field of, 1369
  radiation from, 1078
Sunday Afternoon on the Island of La Grande Jatte [Seurat], *1235*
Sunglasses, *1081*, 1229
Sunsets, *1231*
Superconductors, *844*–845, **1430**–1432
  in magnets, 1490
  and Meissner effect, *952*, 1031
  resistance of, 1030–1031
Supernovae, 1541, *1549*–1550
Superposition
  of electric potentials, 769
  of fields, 718–719, 744
Superposition principle
  for electric fields, 718–719, 744
  for electromagnetic waves, 1071–1072
  and wave packets, 1305
Supersymmetry (SUSY), **1543**
Surface charge density (σ), **720**, *779*, 818

Surface currents, 1430–*1431*
Surfactants, 816
Susceptibility, magnetic (χ), **947**–948t
SUSY. *See* Supersymmetry
Symbols for quantities, A.2t–A.3t
  in circuits, *710*
Symmetry breaking, *1535*, 1537
Synchrotrons, 913, 1535
Systems
  mechanical *vs.* electrical, 1021t

Tangent function, A.21. *See also* Trigonometric functions
Tau (τ) particles, 1519t, 1530
Taylor, 1372
Telecommunications, 1114. *See also* Movies; Radio; Television
  and antennas, *1079*–1080, *1133*
Telescopes, **1162**–1165. *See also* Hubble Space Telescope
  reflecting, **1162**, *1164*
  refracting, **1162**–*1163*
  resolution of, 1217
  Very Large Array, *1238*
Television, 728, 765, *900*, 1156–1157
Temperature (T). *See also* Heat; Internal energy
  and blackbody radiation, *1286*–*1287*
  critical (T_c), 844–845t, 1430t–1432
  critical ignition (T_{ignit}), **1489**
  Curie (T_{Curie}), **951**t
  and electric current, 834
  Fermi (T_F), 1438
  and laser cooling, *1389*, 1397
  and resistance, 841, 843–844
Temperature coefficient of resistivity (α), 837t, **843**, 855
Terminal voltage, 860–861
Tesla, Nikola, *1053*
Tesla (T), **899**
Test charges, **715**–*716*
Theory of Everything. *See* Unified theory
Thermal neutrons, **1480**
Thermal radiation, 845–846, **1285**
Thermometers
  resistance, 843
  Stefan's-law, *1290*
Thermonuclear fusion reactions, **1487**
Thin films
  interference in, *1189*–1194
Thin lens equation, 1143–**1144**
Thomson, George Paget, 1302
Thomson, Joseph John, 759, *912*, *1355*
Thomson's apparatus, *912*
Thoreau, Henry David, 1117
Thought experiments
  for relativity of time, *1252*–*1253*
  for time dilation, 1257–*1258*
Three Mile Island accident, 1486
Threshold energy, **1466**

Time (t)
  confinement (τ), **1490**
  and general relativity, 1274
  interval proper (Δt_p), 1253–1256, **1254**, 1259, 1525
  relativity of, 1247, 1252–1253
  and uncertainty principle, 1310–1311
  units of, A.1t
Time constant (τ). *See also* Half-life
  of *RC* circuits, **875**–876
  of *RL* circuits, **1008**, 1010–1011
Time dilation, 1253–1257, **1254**
Time-independent Schrödinger wave equation, 1323, **1331**–1333, 1361–1362
Time projection chambers, 1499
Ting, Samuel, 1530
Tokamaks, 960, **1490**–1492
Tomonaga, Sin Itiro, 1517
Toner, 784
Top quarks (t), **1530**–1531t
Topness, 1530t–1531
Toroids, *936*
Torque (τ)
  on electric dipoles in electric fields, 815, *817*, 905
  on magnetic dipoles in magnetic fields, 879, *904*–907
  and potential energy, 815
  and vector (cross) products, 815
Torquers, 907
Torsional balances, 895, 1077
Total energy (E), **1270**
Total internal reflection, **1111**–*1114*
Tracers, radioactive, 1500–*1501*
Track detectors, **1499**
Transcendental equations, 1318
Transformers, 846–847, *1033*, **1051**–1054
  and eddy currents, 987, 1051
  and hysteresis, 1051
  ideal, *1051*
  step-up *vs.* step-down, 1051
Transistors, 838, *1427*–1428
  resonant tunneling, 1339
Transitions
  allowed *vs.* forbidden, *1381*
  spontaneous *vs.* stimulated, 1383–*1385*
Transmission axis, 1226
Transmission coefficients (T), **1336**
Transmission gratings, 1217
Transportation. *See* Automobiles; Satellites; Space shuttles; Spacecraft
Transuranic elements, 1465
Trigonometric functions, A.21–A.23
  identities for, 1037, 1039, 1041, 1048, 1183, 1305, A.21–A.22t
Tritium, 1488–1489, 1493–1495
Tunneling, **1336**–1339
  and alpha decay, *1459*
Turbines, 983, 1485
Twin paradox, 1257–*1259*

Uhlenbeck, George, 1371–1372
Ultraviolet catastrophe, 1287
Ultraviolet waves, *1081*
Uncertainty, A.28–**A.29**
Uncertainty principle, 1309–1312, **1310**
 and space quantization, 1369
 and violation of conservation of energy, 1311, 1334, 1336, 1517
Unified theory, 1534, *1536*, 1542–1543
Units. *See also specific quantity*
 conversion of, A.1*t*–A.2*t*
 SI, A.1*t*–A.3*t*, A.32*t*
 U.S. customary, A.1*t*–A.2*t*
Universe. *See also* Big Bang; Cosmology
 expansion of, 1262, *1539–1541*
 history of, *1536–1537*
Unknowns, **A.15**
Unpolarized light, **1226**
Up quarks (u), **1529**–1531*t*
Upsilon (Υ) particles, 1531
Uranium
 isotopic abundance of, 1484
 and radioactivity, 1452, 1459, 1465*t*, 1481–1484
 supply of, 1487

Vacuum tubes, 1424
Valence bands, *1421–1422*
Van Allen belts, *910*
Van de Graaff, Robert J., 782
Van de Graaff generators, 782–*783*
Van der Meer, Simon, 1518
Van der Waals bonds, **1402**–1403
Variables. *See* Unknowns
Vector model of space quantization, **1369**
Velocity (**v**)
 Galilean transformation equation for, **1247**
 Lorentz transformation equations for, 1264–1267
Velocity selectors, *911*
Venus.
 magnetic field of, 954
Verne, Jules, 1167
Vibrational quantum number (*v*), **1407**
Virtual objects, **1149**
Virtual photons, 1517
Visible spectrum, **1109**
Vision. *See* Eyes
Volt (V), **764**
Voltage, **764**. *See also* Potential difference
 breakdown, 812
 Hall (ΔV_H), **914**
 instantaneous (Δ*v*), 1034
 open-circuit (ε), **860**

Voltage (*Continued*)
 root-mean-square (ΔV_rms), **1038**
 terminal, 860–861
Voltage amplitude (ΔV_max), **1034**
Voltage drops, 862
Voltmeters, *811*, **879**–880
Volume (*V*)
 of geometric shapes, A.20*t*
Volume charge density (ρ), **720**
Volume stress. *See also* Pressure
von Laue, Max, 1224
von Neumann, John, 1027

W particles, 1513, 1534–1535
Walden [Thoreau], 1117
Walton, Ernest, 1441
Water
 diamagnetism of, *952*
 as fuel source, 1488
 as moderator, 1484
 as polar molecule, *816*–817
Wave equations
 linear, 1070
 relativistic, 1372, 1513
 Schrödinger, 1323, **1331**–1333, 1361–1362
Wave fronts, **1069**, *1097*
Wave functions
 for covalent bonds, 1401–*1402*
 for finite square wells, *1335*
 for hydrogen atom, 1364–1367
 for metals, 1418–*1419*
 for particle-in-a-box, *1328*, 1332–1333
 for tunneling, *1336*
Wave optics, 1176–1241. *See also* Diffraction; Geometric optics; Interference; Optics; Polarization
Wave packets, *1305*–1307
Wavelength (λ)
 Compton, **1298**
 cutoff (λ_c), **1294**
 de Broglie, **1301**
Wavelets, 1107
Wave–particle duality, 1300–1302, 1323
 gravitational, 1195
 plane, **1069**
 spherical, **1069**, *1098*
 square, *1058*
Weak charge, 1534
Weak force, 1512–1513*t*, 1524, *1536*–1537
Weber, Wilhelm, 959
Weber (Wb), 940
Weinberg, Steven, 1534
Welding, induction, 991

Wells, potential, **1332**
Wheeler, John, 1274
White light, 1109–*1110*
Wien's displacement law, **1287**
Wilkinson Microwave Anisotropy Probe (WMAP) [satellite], 1536, 1538
Wilson, Charles, 1297
Wilson, Robert, 1088, *1537*–1538
Wiring, household, *880*–882
WMAP. *See* Wilkinson Microwave Anisotropy Probe
Work (*W*)
 by conductors moving in magnetic fields, 974
 by a conservative force, 763–764
 in electric fields, 726, 763–764, 807, 815–817
 and electric generators, 982
 in magnetic fields, 934
 and motors, 984–985
 units of, A.2*t*
Work function (φ), **1294***t*
 and charged particles in electric fields, 726
 and charged particles in magnetic fields, 899
 relativistic, 1268–1269
World-lines, *1259*–**1260**

X-rays, *1081*
 atomic spectra of, *1381–1383*
 characteristic, *1381*–**1382**
 diffraction of by crystals, *1224–1225*
 radiation damage by, 1470, 1496
 scattering of by electrons, *1297–1300*
Xerography, *784*
Xi (Ξ) particles, 1519*t*

y-intercepts, *A.17*
Yerkes Observatory, 1165
Young, Thomas, 1093, 1095, 1178
Young's double-slit experiment, 1177–1182
 for electrons, 1307–*1309*
Yukawa, Hideki, *1516*, 1533

Z particles, 1513, 1534–1535
Zeeman effect, 1369–*1370*
Zero-point energy, **1328**, 1342, 1407
Zero-point motion, 1407
Zeroth-order maxima, 1180
Zonules, eye, 1156
Zoom lenses, 1172
Zweig, George, 1529

Standard Abbreviations and Symbols for Units

Symbol	Unit	Symbol	Unit
A	ampere	K	kelvin
u	atomic mass unit	kg	kilogram
atm	atmosphere	kmol	kilomole
Btu	British thermal unit	L	liter
C	coulomb	lb	pound
°C	degree Celsius	ly	lightyear
cal	calorie	m	meter
d	day	min	minute
eV	electron volt	mol	mole
°F	degree Fahrenheit	N	newton
F	farad	Pa	pascal
ft	foot	rad	radian
G	gauss	rev	revolution
g	gram	s	second
H	henry	T	tesla
h	hour	V	volt
hp	horsepower	W	watt
Hz	hertz	Wb	weber
in.	inch	yr	year
J	joule	Ω	ohm

Mathematical Symbols Used in the Text and Their Meaning

Symbol	Meaning
$=$	is equal to
$\equiv$	is defined as
$\neq$	is not equal to
$\propto$	is proportional to
$\sim$	is on the order of
$>$	is greater than
$<$	is less than
$\gg(\ll)$	is much greater (less) than
$\approx$	is approximately equal to
Δx	the change in x
$\displaystyle\sum_{i=1}^{N} x_i$	the sum of all quantities x_i from $i = 1$ to $i = N$
$\lvert x \rvert$	the magnitude of x (always a nonnegative quantity)
$\Delta x \rightarrow 0$	Δx approaches zero
$\dfrac{dx}{dt}$	the derivative of x with respect to t
$\dfrac{\partial x}{\partial t}$	the partial derivative of x with respect to t
$\displaystyle\int$	integral